ROUTLEDGE LIBRARY EDITIONS:
ECOLOGY

I0462193

Volume 6

ARCTIC AND ALPINE ENVIRONMENTS

ARCTIC AND ALPINE ENVIRONMENTS

Edited by
JACK D. IVES AND ROGER G. BARRY

Routledge
Taylor & Francis Group

LONDON AND NEW YORK

First published in 1974 by Methuen & Co Ltd

This edition first published in 2020
by Routledge
2 Park Square, Milton Park, Abingdon, Oxon OX14 4RN

and by Routledge
52 Vanderbilt Avenue, New York, NY 10017

Routledge is an imprint of the Taylor & Francis Group, an informa business

© 1974 Jack D. Ives and Roger G. Barry

British Library Cataloguing in Publication Data
A catalogue record for this book is available from the British Library

ISBN: 978-0-367-36640-7 (Set)
ISBN: 978-0-429-35088-7 (Set) (ebk)
ISBN: 978-0-367-35251-6 (Volume 6) (hbk)
ISBN: 978-0-367-35254-7 (Volume 6) (pbk)
ISBN: 978-0-429-33082-7 (Volume 6) (ebk)

Publisher's Note
The publisher has gone to great lengths to ensure the quality of this reprint but points out that some imperfections in the original copies may be apparent.

Disclaimer
The publisher has made every effort to trace copyright holders and would welcome correspondence from those they have been unable to trace.

Arctic and Alpine Environments

An arctic mountain landscape showing an intensely glaciated and glacierized mountain area south of Inugsuin Fiord on the northeast coast of Baffin Island (70°N). Total relief is of the order of 1700 m. Photograph by Jack D. Ives.

Arctic and Alpine Environments

Edited by Jack D. Ives

and Roger G. Barry

Contributing authors:

Donald Alford, John T. Andrews, Roger G. Barry, W. Dwight Billings,
J. Brian Bird, Nel Caine, Hermann Flohn, Robert F. Grover,
F. Kenneth Hare, Geoffrey Hattersley-Smith, Helmut Heuberger,
Robert S. Hoffmann, Wilfred M. Husted, Jack D. Ives,
James A. Larsen, Olav H. Løken, Áskell Löve, Doris Löve,
Robert McGhee, Donald K. Mackay, Paul S. Martin,
Ludger Müller-Wille, Harvey Nichols, William S. Osburn Jr,
Gunnar Østrem, John L. Retzer, Samuel Rieger,
H. Olav Slaymaker, Claudia C. Van Wie, Peter Wardle,
Patrick J. Webber.

METHUEN

First published 1974
by Methuen & Co Ltd
11 New Fetter Lane London EC4P 4EE
© *1974 Jack D. Ives and Roger G. Barry*
Printed in Great Britain by
William Clowes and Sons Limited
London, Colchester and Beccles

ISBN 0 416 65980 2

Distributed in the USA by
Harper & Row Publishers Inc.
Barnes & Noble Import Division

Carl Troll, the 'Grand Old Man of the High Mountains' on the occasion of the meeting of the International Geographical Union Commission on High-Altitude Geoecology, Calgary, in August 1972. Photograph by Jack D. Ives.

Dedication to Professor Dr Carl Troll

Former Director of the Geographical Institute, University of
Bonn; Past Chairman of the Commission of High-Altitude
Geoecology, International Geographical Union;
Former President of the International Geographical Union

Few environments convey greater inspiration to naturalists than the arctic or alpine. Because of the sparse vegetation their geological substrates are usually more easily observed than those of temperate or tropical environments, and their climatic and edaphic conditions are more extreme. Their flora and fauna are comparatively poor in species but those occurring there, in many cases, display striking adaptations, and the zonation and differentiation of their vegetation can often be explained by climatic and edaphical factors. These environments consequently offer exceptional opportunities for ecological studies.

No serious student of high mountains or arctic areas can have missed the name of Carl Troll. Born on the slopes of the Alps he early developed an intense interest in natural history, and on entering the University of Munich he turned to botany, geology and geography. His earliest scientific publications were devoted to plant physiology, phyto-geography and glacial geomorphology. Having achieved comprehensive knowledge of geography and vegetation in central Europe, Troll then extended his work to other parts of the world. During three years he travelled widely in the mountainous parts of South America, producing among other things some excellent surveys of the zonation of vegetation and human occupation. Later he travelled through most of the high moun-tain areas of the Americas, studying glaciation, solifluction, meteorology and vegetation. Next he joined the unfortunate German expedition to Nanga Parbat, from which he was one of the two survivors. He has also made extensive studies in Asia and Africa.

Carl Troll's numerous publications deal, *inter alia*, with glacial geomorphology, glaciology, climatology, landscape ecology, photogrammetry, solifluction, vegetation zonation, and life forms in plants. Several of his contributions, such as the treatises on 'Thermoisopletendiagramme', 'Frostwechselhäufigkeit', 'Strukturboden, Solifluktion und Frostklimaten der Erde' and 'Die assymmetrische Aufbau der Vegetationszonen und Vegetationsgürteln auf der Nord- und Sudhalbkugel', have become classics. In these he eloquently demonstrated the differences between alpine and arctic conditions and the uniqueness of tropical high mountains.

Carl Troll's scientific achievements have secured him a prominent place among the pioneers of natural history, as evidenced by the numerous honorary doctorates and other awards conferred upon him. But he is worthy of recognition also from at least two other points of view. The first of these concerns his activity as teacher and inspirer. Many of the present generation of German geographers have been taught and inspired by Troll. But his influence has extended far beyond this. My own work on the vegetation of the high East African mountains, for instance, was largely inspired by a lecture given by Troll in Uppsala, and he has afterwards followed my work – like that of many others – with kind interest. His wide knowledge of and interest in the biological and geographical work of his many pupils and colleagues has also enabled Carl Troll to perform a very valuable function in the third capacity I would like to stress, namely as administrator and coordinator of geographical and biological studies in many parts of the world, and as editor for their publication. His activities within the International Geographical Union which included its presidency and in its Commission on High-Altitude Geoecology are relevant here, as well as his function as editor for *Erdkunde* and several other journals and publications.

Last but not least Carl Troll is appreciated by his colleagues in many countries as a generous and kind friend. His understanding and helpful collaboration and mediation has meant much to promote international collaboration in science, not least for those engaged in the study of high mountains or arctic areas. It is therefore singularly fitting that this book on arctic and alpine environments should be dedicated to him, the Grand Old Man of the High Mountains.

Olaf Hedberg

Universitets Institution för Systematik Botanik
Uppsala

Contents

Plates

Plates 1 to 47 will be found at the end of the book

Frontispiece
  An arctic mountain landscape showing an intensely glaciated and glacierized mountain area south of Inugsuin Fiord on the northeast coast of Baffin Island

Dedication
  Carl Troll, the 'Grand Old Man of the High Mountains', on the occasion of the meeting of the International Geographical Union Commission on High-Altitude Geoecology, Calgary, in August 1972

1. The forest–tundra ecotone showing the transition from open lichen-woodland of the Subarctic to full tundra, central Labrador–Ungava
2. Loveland Pass, Colorado, showing the alpine transition from upper montane forest to alpine tundra
3. Krummholz, or dwarfed Engelmann spruce, at the upper reaches of the forest–tundra ecotone on the east slope of the Front Range, Colorado Rocky Mountains
4. Avalanche released by artillery fire, December 1971, near Silverton in the San Juan Mountains of southwestern Colorado
5. The redistribution of snow by the wind in mountainous terrain
6. The glacier burst, or jökulhlaup, from Skeidararjökull, southeast Iceland, in August 1954, as seen before and at the climax of the event
7. Oblique air photograph of large-scale polygonal ground on the Donnelly Dome area, Alaska
8. Massive ground ice of the Mackenzie Delta area overlain by zone containing intrusive ice
9. Open-system, or East Greenland type pingo
10. East Greenland pingo in process of destruction
11. The Inuvik 'utilidor', one solution to the problems of sewerage disposal and the provision of piped water in the permafrost zone

xi

Acknowledgements

The editors gratefully acknowledge the cooperation and enthusiasm of the contributors in helping, despite their many other commitments, to produce this book. Our colleagues in INSTAAR have assisted with many helpful suggestions and criticisms apart from their direct contributions.

For various forms of technical help we are indebted to the following persons: Jill Williams for organizing and preparing much of the glossary; Dr Doris Löve and also Dr Waltraud Brinkmann in translating Chapter 6B; Carol Haverkampf, Jane Bradley, Jennifer Skiwot, Johnna McKnight, Cheryl Welch and Dorothy Dean for secretarial and clerical support; Kathleen Salzberg for assistance on editorial items; Vija Handley, Nancy Stonington and Cathy Verhulst for drafting work; Betsy Palmer for help with indexing. Our collaboration with Methuen has been greatly facilitated by the editorial expertise of Jane Hunter.

The editors and authors wish to thank the following individuals and organizations for permission to reproduce figures, plates and tables.

INDIVIDUALS

R. Armstrong for Plate 4; Dr C. E. Braun for Plate 35; Dr R. A. Bryson for fig. 7A.5; F. E. Burbidge for fig. 2A.12; Dr H. L. Crutcher for fig. 2A.6; Dr R. Daubenmire for fig. 8A.3; Dr L. A. Douglas for fig. 13A.1; Dr I. Hustich for fig. 7A.1; Dr W. Karlén for Plate 17; G. A. McKay, B. F. Findlay and H. A. Thompson for fig. 2A.7; Dr J. R. Mackay for figs. 4A.9, 4A.10, and Plates 8 and 11; Dr W. C. Mahaney for fig. 13B.2; Dr B. Michel for figs. 3A.4, 3A.5 and 3A.6; Dr F. Müller for fig. 11A.8 and Plates 9 and 10; P. Parnell for fig. 17.1b; Dr T. Péwé for Plate 7; Dr E. Reiter for figs. 2B.2, 2B.5 and 2B.6; E. Schneider for Plate 20; Dr R. G. Steadman for fig. 14A.2; Dr L. W. Swan for fig. 1.2; J. A. Teeri for Plate 25; A. P. Underhill for Plate 21; Dr L. A. Viereck for the quotation on p. 174–5 and for fig. 4A.8; and Dr R. B. Weeden for fig. 9.13.

PUBLISHERS

Academic Press, Inc. (New York) for fig. 9.20; Duke University Press for figs. 7B.4; 8A.3, 9.10, 9.19 and 11A.10; Elsevier Publishing Company (Amsterdam) for figs. 2A.2, 11A.3 and 11A.9; W. de Gruyter & Co. (Berlin) for fig. 1.1; Harper & Row, Publishers, Inc. (New York) for fig. 8B.2; Indiana University Press for fig. 1.2; Thomas Nelson & Sons Ltd (London) for figs. 9.1a and 9.2; Pergamon Press, Inc. (New York) for fig. 10B.1; Politekens Forlag (Copenhagen) for fig. 9.1b; Rhodos (Copenhagen) for figs. 4B.4 and 4B.5; W. B. Saunders Co. (Philadelphia) for figs. 9.7 and 9.18; Springer-Verlag (Heidelberg) for fig. 7B.1; Stanford University Press for fig. 15A.2 (© 1967 by the Board of Trustees of the Leland Stanford Junior University); F. Steiner Verlag (Weisbaden) for fig. 2D.2; University of Washington Press for fig. 2D.4; Van Nostrand Reinhold Co. for fig. 9.17 (©1963 by Litton Educational Publishing, Inc.); Williams & Wilkins Co. (Baltimore) for figs. 13B.3 and 13B.4 (© 1952 by The Williams & Wilkins Co., Baltimore); and Yale University Press for fig. 6A.9.

ORGANIZATIONS

American Association for the Advancement of Science for figs. 5.1 and 8B.1 (© 1969 and 1971, and 1972, by the American Association for the Advancement of Science); American Geophysical Union for figs. 4A.4 (© 1958 by the American Geophysical Union), 5.2, 5.4 and 6A.5; *American Journal of Physiology* for fig. 9.3; Arctic Institute of North America for figs. 4B.1 and 4B.2; *Biological Bulletin* for figs. 9.4 and 9.8; Canadian National Committee of the International Hydrological Decade for fig. 3A.1; Dept of Energy, Mines and Resources (Ottawa) for Plates 12, 13, 15 and 18; Florida State Museum for fig. 9.9; Geological Society of America for figs. 6A.1 and 10B.4; Geological Survey of Canada, Dept of Energy, Mines and Resources (Ottawa) for fig. 3A.8; Geologische Bundesanstalt (Vienna) for fig. 6B.2; Glaciological Society for fig. 4B.3; Information Canada for fig. 7A.2; *Journal of Applied Physiology* for fig. 9.5; Long Island Biological Association for fig. 9.15; National Museum of Canada (Ottawa) for Plates 41 and 42; National Research Council of Canada for figs. 3B.4 and 12A.3; *New Zealand Journal of Botany* for figs. 7B.2, 7B.3 and 7B.4; Optical Society of America for fig. 2C.1; Soil Science Society of America for fig. 13A.2; Svenska Geofysiska Föreningen for fig. 2A.9; Swedish Natural Science Research Council for fig. 10B.5; *Systematic Zoology* for fig. 9.11; US Army Engineering Topographic Laboratories (Fort Belvoir) for fig. 2A.11; Western Snow Conference for fig. 3B.3; Wilson Ornithological Society for table 9.9; and Zoological Society of Southern Africa for fig. 2C.3.

Preface

At a time when mankind is showing increasing concern over the impact of human technology on the quality of the environments in which we live, we find ourselves alarmingly ignorant of the interaction between atmosphere, biosphere and lithosphere. Large sections of the world have been so extensively altered by man from the 'natural' state that the already complex task of learning to understand the dynamics of the world's major ecosystems has become increasingly difficult. The arctic and alpine regions contain vast tracts of terrain as yet little disturbed; in this sense, therefore, they are of considerable interest to the ecologist, although comparatively neglected by him. Moreover, arctic and alpine ecosystems are not only regarded as being relatively simple – in the sense of possessing a small range of physiognomic types, few plant and animal species and apparently being dominated by a single major climatic parameter (temperature) – but the fauna and flora are liable to exist under environmental stress and are thus prone to irreversible disturbance under the impact of man's activities. This is especially serious in the realm of resource exploitation and environmental pollution.

The idea of preparing this book was primarily an outgrowth of the interdisciplinary cooperation in research and teaching at the Institute of Arctic and Alpine Research, University of Colorado. Particular stimulus has been provided by a graduate seminar on arctic and alpine environments given by Institute faculty since 1968. Many of our other contributors are associated with the Institute through its Scientific Advisory Committee, and we have also invited the collaboration of specialists from various other institutions. Their keen interest in the project, and valuable contributions to it, are greatly appreciated.

The major objective of this collaborative effort is to provide a comprehensive statement of our present knowledge of arctic and alpine environments and to identify the gaps in our understanding of them. We consider first the definitions of 'arctic' and 'alpine' to be found in the literature, and the problems they present, individually or in common, to the field scientist. Then, in three major sections, the environments, the physical processes operating in them, and their biota, are discussed. Environmental aspects include climate, hydrology and ice mantle, both present and past. Biota are considered under the headings of treeline, vertebrate zoology, vegetation, the flora and its origin, palaeoecology and the question of zoological extinctions. Then geomorphic processes and pedology are considered and, finally, man's response to these environments is treated in terms of early settlement, physiology and the impact of modern technology.

A review of the table of contents will show certain obvious omissions. We have purposely omitted specific treatment of the arctic marine environment because our focus is on terrestrial comparisons. Many basic aspects of geophysics and geology have been excluded also because they are strictly unique in geographical occurrence. Certain omissions, such as invertebrate ecology, result from our inability to find an author willing to undertake the difficult task of synthesis. Hopefully, this can be remedied in a future edition. While many of the problems considered are common also to the Antarctic, no specific treatment of this region has been included. It has formed the subject of several recent books and monographs whereas the Arctic, by comparison, has been neglected.

Editorial policy has not sought for consistency of treatment of the material. The chapters and sections range from broad-scale reviews to more specialist accounts which include material published, at least in part, for the first time. We can make virtue out of (some) necessity in this regard, for it provides a stimulating variety of approach to the different themes. The occurrence of some disagreements illustrates the relative immaturity of our topic and the existence of research frontiers.

The book is aimed primarily at the senior undergraduate and graduate student in ecology, in its broadest sense, as well as professional colleagues in one or other specialist branch of the environmental sciences. In addition, we hope to have provided a useful reference for the resource developer in industry and government, as well as for those interested in the preservation of these fascinating landscapes.

Jack D. Ives
Roger G. Barry

Institute of Arctic and Alpine Research
University of Colorado

CHAPTER 1

Introduction

Roger G. Barry and Jack D. Ives
Institute of Arctic and Alpine Research,
University of Colorado

> As we do not usually lump . . . non-polar, non-alpine and non-tree-covered
> vegetation of the remainder of the globe under one heading, one wonders
> why a collective term should be applied to the vegetation of arctic and
> alpine regions in the word 'tundra', unless we understand its use as an
> involuntary expression of our ignorance.   (Beschel, 1970, p. 86)

The terms 'arctic' and 'alpine' have been used repeatedly to describe physical and
biological environments and their flora and fauna, as well as geographical regions.
Attempts to define these environments and their characteristics have concentrated on
their visual appearance – the treelessness of the tundra – and their severe climate, from
the point of view of man. We shall return to the question of definitions below, but first
let us consider in very general terms some of the major features of the high latitude and
high altitude parts of the world.

Comparison of arctic and alpine areas

Extensive areas of the Arctic in North America and northern Asia are formed by the
low-lying treeless plains of the tundra which range from wet meadow to dry heath and
rocky fellfields. The proportion of wet ground is small, however, in the fiord terrain of
the eastern Canadian Arctic and Greenland and in the mountains of Alaska, Anadyr, and
Kamchatka, and very large in the other areas. The alpine areas of the world (fig. 1.1)
are of course markedly non-contiguous and are exposed to a wide variety of regional
climatic controls. Nevertheless, the broad comparability of the physiognomy of alpine
and arctic vegetation is readily apparent, although xeric and mesic alpine communities
usually form a more complex small-scale mosaic due to the nature of the terrain, and the
extensive wet tundras are largely lacking.

It is commonly argued that cool summers are the basic cause of the treeless character
of both tundra environments, and that mean annual temperature likewise provides an

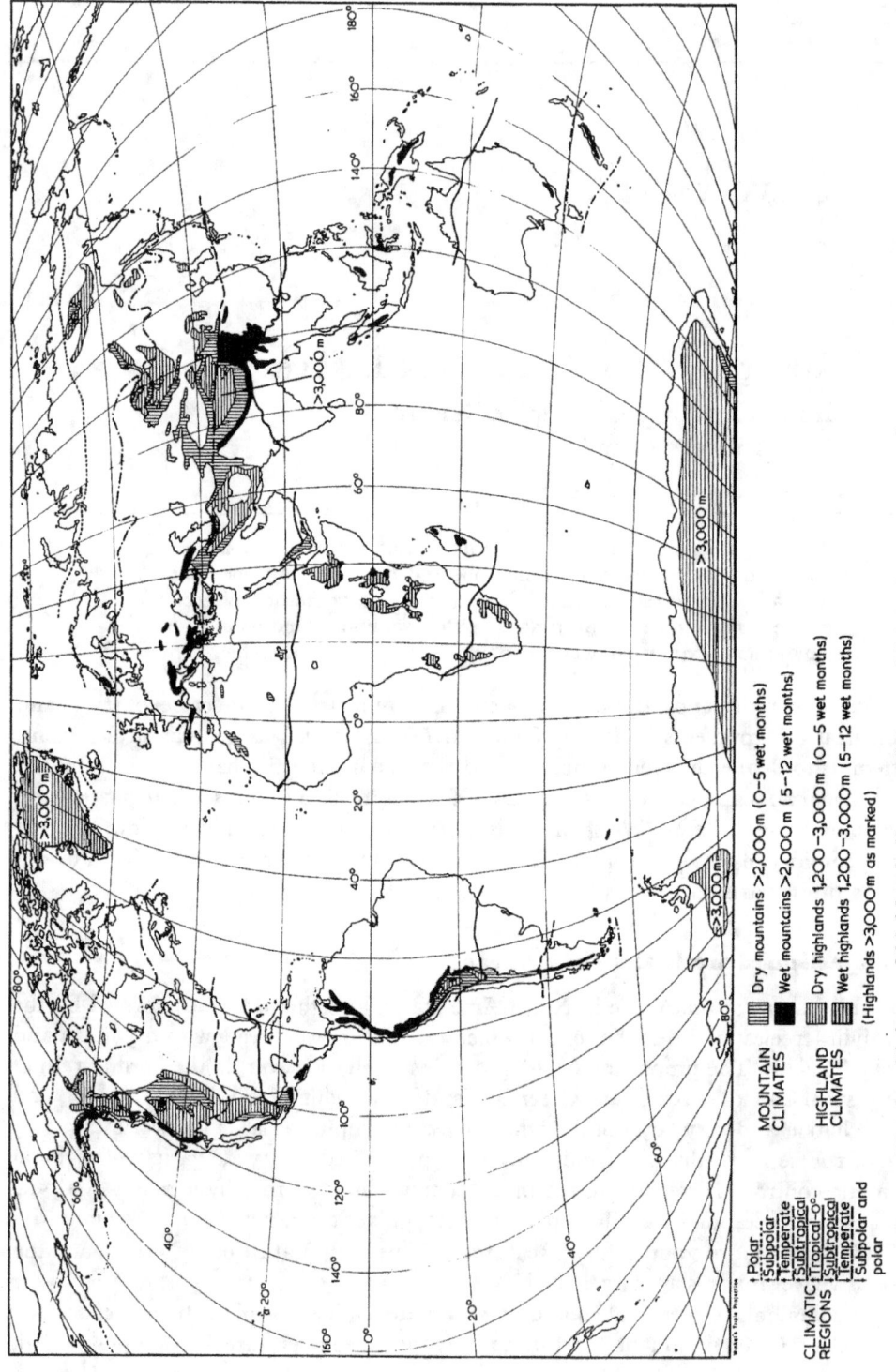

Fig. 1.1. Alpine and highland climates of the world (after N. Creutzberg, in Blüthgen, 1966).

indication of the occurrence of permafrost throughout the Arctic and, as is being increasingly recognized, in many alpine areas also. However, quantitative comparisons and evaluations are almost wholly lacking and we cannot assume that the apparent similarities derive from the same environmental controls. The climatic environments of the arctic and alpine regions are similar in terms of mean annual temperature and snow cover duration, but little else. For up to six months of the year the Arctic receives virtually

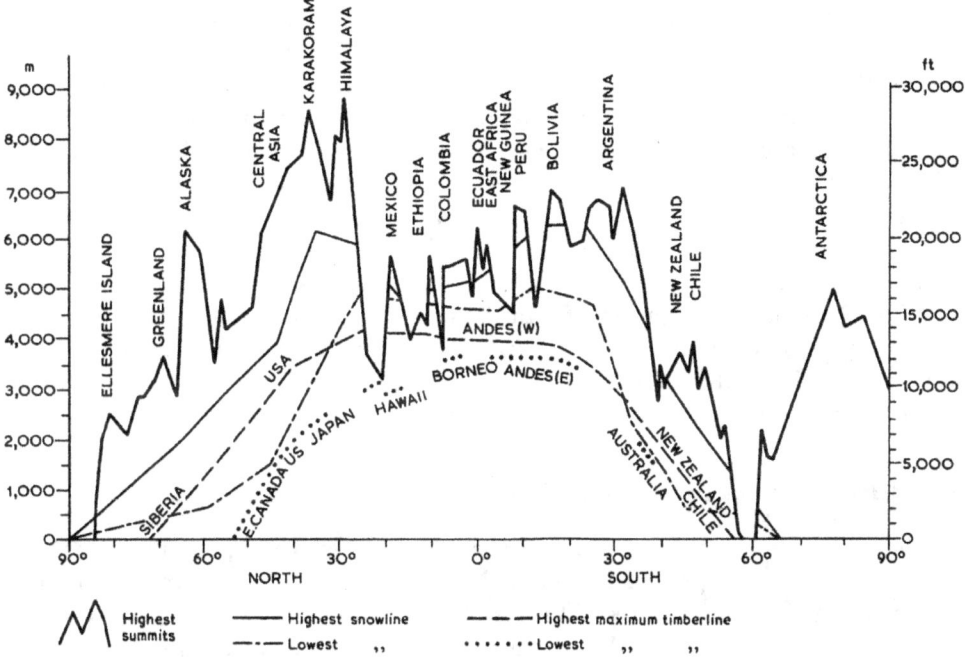

Fig. 1.2. Global cross-section of the alpine regions, showing the highest summits, snowline and timberline (partly after Swan, 1967).

no solar radiation, whereas alpine areas experience regular diurnal and seasonal regimes appropriate to their latitude. The Arctic is no more windy than most mid-latitude areas, while many alpine sites experience extreme maximum and high mean wind speeds. Much of the Arctic receives little snowfall whereas certain alpine areas have among the highest totals of the world.

In spite of the need for quantitative comparative studies of environmental characteristics and processes in these environments, a serious difficulty is posed by the problem of scale. The heterogeneity in the alpine environment resulting from slope, orientation, and superficial mantle is displayed in microclimate, geomorphic activity, hydrology, fauna and flora, so that proper sampling encounters conceptual problems which may have no real solution.

On the gross scale, arctic and alpine environments can be treated as a continuum from high latitude/low altitude to low latitude/high altitude conditions, as suggested by world timberline and snowline elevations (fig. 1.2). Eventually, it should be possible to formulate this relationship more precisely in terms of an energy exchange model. Evaluation of the energy transfers between organisms and their environment, on the one hand, and amongst the organisms themselves, on the other, is the basis of modern eco- · systems analysis.

Productivity studies at arctic and alpine tundra sites (see Webber, Chapter 8B; Hoffman, Chapter 9) indicate that temperature is the primary limiting factor and that net primary production is in the same range as that of deserts. However, the importance of below ground production necessitates care in interpreting the available data. Webber's data indicate greater primary production in the Colorado alpine ($200-250$ g m^{-2} yr^{-1}) than at Point Barrow (100 g m^{-2} yr^{-1}) but net photosynthetic efficiency is apparently about 1 per cent in each case. Although a detailed evaluation in these terms is not yet possible for tundra environments, the groundwork for this is being laid as several of the following chapters show.

Finally, we should point to certain features which are unique to one or other of the tundra environments. The principal ones result from geographical location and altitude as they affect seasonal regime and atmospheric density, respectively. Thus, the 'polar night' has important consequences for the fauna of the Arctic, while at altitudes above 4–5 km hypoxia is a severe hindrance to man's activities, at least. Nevertheless, this constraint is mitigated by the fact that man and animals can move into and out of the alpine environment much more readily than in the case of the Arctic.

The delimitation of the arctic and alpine environments

We have given some indication of the characteristics which impart meaning to the concept of arctic and alpine environments. Now the question of the delimitation of these environments can be considered. Much has been written on this topic, and since the drawing of precise boundaries is largely an intellectual exercise we will limit our discussion to an indication of the types of approach that have been made and some of the problems encountered.

First, we must briefly review the derivation and meaning of the terms arctic and subarctic, alpine and subalpine, since they are a source of much semantic disagreement (see Löve, 1970, from whom much of this discussion derives).

'*Arctic*' is derived from the Greek word *Arktos* ('bear'). In its Latin equivalent, this occurs in the names of the constellations *Ursae major* and *minor* by which the North Polar Star can be located. Subsequently *arctic* came to refer to high northern latitudes. The term *alpine* originated from the Latin name *Alpes* for the snow-covered (white) mountain summits of northern Italy. Later, the adjective was applied to inhabitants of this region and finally to any high snow-covered mountains. As Löve (1970, p. 64) points out, *subarctic* and *subalpine* imply (in an etymological sense) areas which are subdivisions of the arctic zone and of the alpine belt, respectively. Ambiguity arises because the Latin prefix *sub* can have both a prepositional and adverbial meaning. In the former context, *sub* means under or below and is easily understood in such words as submarine, subglacial, subterranean. In scientific literature, however, *sub* usually has an adverbial connotation which implies a lessening in force of the substantive. Thus there is an implied meaning of 'not quite', 'almost', 'somewhat', as in *subacute, subspecies, subcordate*. A *subzone*, therefore, is not a zone below, but a part of a zone. Logically then, the *Subarctic* is part of the *arctic zone* and the *Subalpine* is part of the *alpine belt*. Nevertheless it is sometimes difficult not to think of the Subalpine as the belt next below the Alpine, and the Subarctic as a zone next below the Arctic, since that is their position on the map, and this thinking is frequently reflected in the direct or implied definitions of many scientists (see tables 1.1 and 1.2 and UNESCO, 1970).

The Arctic Circle forms the southern boundary of the Arctic in an astronomical and cartographic context, but this criterion bears little relationship to climatic or other parameters. A meteorological approach to the problem is equally unsatisfactory in other respects. The limit of arctic airmasses, for example, fluctuates markedly from day-to-day and with major longitudinal differences, apart from seasonal and interannual variations (Bryson, 1966; Barry, 1967; Krebs and Barry, 1970).

Because of such difficulties, it has long been customary to use biological or geographical features as indicators of the extent of the Arctic. The most common descriptor of the tundra is that of treelessness (Tikhomirov, 1962). Here, 'tree' refers to a perennial woody plant with a single central stem at least 2 m tall. *Tundra*, which in Russian–Finnish usage referred to the treeless zone of northern Eurasia, especially Finland, was subsequently applied to its counterpart in northern North America.[1]

It is generally considered that the vegetational life form represents the integrated effect of primarily climatic parameters. In parts of the Subarctic, however, the activities of man have greatly modified the vegetation cover. In Eurasia, especially, centuries of timber cutting and reindeer browsing have decimated wide areas of pine and birch, respectively. Moreover, the delimitation of treeline is open to wide differences in interpretation in various sectors of the Arctic (see Larsen, Chapter 7A; Wardle, Chapter 7B). A further complication has been introduced into the literature through the use of climatic parameters which purportedly correlate with the arctic treeline. Thus, the climatic category of tundra recognized by Köppen (1936) – identified by the fact that no summer month attains a mean temperature of $10°C$ – represents at best a first approximation to the limit of tundra vegetation since the correlation was based on imperfectly known plant distributions and meagre climatic data. Adjustments to the temperature criterion, such as that of Nordenskjöld and Mecking (1928, p. 73) afford little real improvement (see Hare, 1951). A more physically based approach was indicated by Hare (1950) who showed that the southern limits of tundra and of the forest-tundra ecotone in Labrador–Ungava correspond to annual potential evapotranspiration values of 31 cm and 35 cm, respectively, as determined by Thornthwaite (1948). The Thornthwaite equation for potential evapotranspiration is an empirical one involving only mean temperatures and day length, but Budyko (1956) developed a more fundamental approach involving net radiation (R_0) for a saturated surface with an albedo of 0.18 and the energy necessary to evaporate the mean annual precipitation (Lr). The radia-

[1] A useful summary of the various definitions of tundra which have been proposed is given by Dagon (1966).

tional index of dryness (R_0/Lr) has a value less than 0.33 for tundra and less than 0.45 for forest-tundra and alpine meadow areas according to Budyko. The existence of a relationship between the atmosphere and the arctic treeline is suggested by the coincidence of this boundary with the summer position of the arctic front (Bryson, 1966; Krebs and Barry, 1970; see Barry and Hare, Chapter 2A). These types of evidence lend greater confidence to the use of a treeline as an index of environmental factors.

Another geographical indicator of arctic conditions is the occurrence of permafrost. Again it is determined by climatic factors, but it is also affected by depth of snow cover, vegetation cover, soil moisture and aspect (see Ives, Chapter 4A). Correlations with mean annual air temperature have been attempted, but the relationship is extremely crude at best. In North America, east of Hudson Bay, there is good field evidence to indicate a close correlation between treeline and southern limit of permafrost, but in northwestern Canada and Alaska the two limits diverge by several hundred kilometres, with the permafrost limit to the south. In contrast, wet maritime tundras (e.g. coastal Iceland, the Aleutian Islands, northwestern Ireland) lie well outside the permafrost zone. Aside from this general problem, the southern limit of permafrost itself is very imperfectly known at present. A further problem arises from the probable imbalance between permafrost and present climate. In some areas permafrost would probably not reform under present climatic conditions if it were destroyed, and the time lag in the response of this indicator is likely to be rather greater than that of vegetation cover.

In the case of alpine areas, the boundary has been drawn primarily with respect to treeline since climatic data are invariably sparse in mountainous terrain and their representativeness is often problematic (see Barry and Van Wie, Chapter 2C). The 10 °C isotherm for mean daily temperature of the warmest month is probably a reasonable indicator of the treeline although wind is a significant control in restricting tree growth. A difficulty arises in many parts of the world, however, from man's modification of the vegetation and also possible climatic fluctuations, such that the potential treeline for the present climate may not be known. Moreover, in the tropical mountains the life forms and ecological relations of the arboreal plants make delimitation of a treeline problematic (Troll, 1972, 1973a, b).

While the southern margin of the Arctic and the lower limit of the alpine belt are fairly readily delimited by treeline on the scale of an atlas map, the subarctic and subalpine areas of the world are less easily defined, for both practical and semantic reasons. Table 1.1 compares various schemes for the zonation of the Arctic. In the Soviet Union, botanists recognize arctic and subarctic subzones within the tundra zone as summarized in table 1.2. Tundra, which possesses a more or less continuous vegetative cover on mesic sites, comprises plant associations with a well-developed moss cover and low-growing shrubs, such as dwarf birch and willow, in the Subarctic of the USSR, whereas in the arctic zone the dwarf shrubs do not form distinct synusiae. Thus defined, the Arctic/Subarctic boundary approximately follows the 6°C July isotherm, although in continental Siberia it lies to the north of this isotherm and in the sector under Atlantic influence to the south of it. The forest-tundra ecotone (*lyesotundra* in Russian) typically forms a mosaic of boreal forest conifers, fully grown in sheltered sites, tree islands or krummholz,

Table 1.1. Schematic review of various opinions on continental zonation in North America and Eurasia (from D. Löve, 1970).

Vegetation	MERRIAM (1898) *North America*	WEAVER and CLEMENTS (1938) *N North America*	SEMENOVA-TYAN-SHANSKAIA (1964) *North America*	AHTI (1964) *Ontario, Canada*	HUSTICH (1949) *N Quebec, Canada*	ROUSSEAU (1952) *Quebec, Canada*	HARE (1968) *Quebec, Canada*	SJORS (1963) *NE North America NW Eurasia*	AHTI et al. (1968) *NW Europe*	LAVRENKO and SOCHAVA (1976) *European USSR*	MEUSEL et al. (1965) *N Hemisphere*	HOLUB and JIRASEK (1967) *N Hemisphere*
Treeless tundra	Arctic tundra	Arctic tundra	Arctic desert / Spotty tundra / Tundra	Arctic tundra	Arctic tundra	Arctic tundra	Arctic tundra	Arctic tundra	Arctic tundra	Arctic desert / Spotty tundra / Tundra	Arctic tundra	Arctic tundra
— Treeline —												
Open to closed coniferous forest	Hudsonian	Boreal forest	Forest-tundra / Northern taiga	Forest-tundra / Hemiarctic	Forest-tundra / Taiga	Hemiarctic / Subarctic taiga	Forest-tundra / Open boreal woodland	Hemiarctic and birch woodland / Northern boreal	North boreal	Forest-tundra / Northern taiga	Boreal	Supra-septentrional
Closed, mainly coniferous forest	Canadian		Middle taiga	North boreal / Middle boreal		Temperate conifer forest	Main boreal	Main boreal / Southern boreal	Middle boreal / South boreal	Middle taiga		Infra-septentrional
Closed, mainly mixed forest	Transition		Southern taiga	South boreal / Hemiboreal	Southern spruce forest	Temperate forest mixed	Great Lakes and St Lawrence mixed forest	Boreo-nemoral	Hemiboreal	Southern taiga	Temperate or boreo-meridional / Sub-meridional	Supra-meridional
Deciduous forest	Upper Austral		Nemoral forest			Deciduous forest		Nemoral	Boreo-meridional	Nemoral	Meridional	Infra-meridional
Steppe, grassland, or desert	Lower Austral		Prairie or desert								Boreo-subtropical	
Mainly evergreen forest	Tropical		Tropical forest					Tropical			Tropical	Tropical

Overall categorizations (bottom bands of the original):

- **MERRIAM:** BOREAL — AUSTRAL — TROPICS
- **WEAVER and CLEMENTS:** ARCTIC — BOREAL
- **SEMENOVA-TYAN-SHANSKAIA:** ARCTIC — Subarctic — Taiga — Nemoral — TEMPERATE REGIONS — TROPICS
- **AHTI:** ARCTIC — SUBARCTIC — BOREAL
- **HUSTICH:** ARCTIC — Subarctic — Boreal — TEMPERATE
- **ROUSSEAU:** Arctic — Subarctic — TEMPERATE
- **HARE:** ARCTIC — BOREAL or SUBARCTIC — TEMPERATE
- **SJORS:** ARCTIC — Subarctic — BOREAL — NEMORAL — TROPICS
- **AHTI et al.:** ARCTIC — BOREAL — MERIDIONAL
- **LAVRENKO and SOCHAVA:** ARCTIC — SUBARCTIC — TAIGA — NEMORAL
- **MEUSEL et al.:** ARCTIC — BOREAL — MERIDIONAL — TROPICS
- **HOLUB and JIRASEK:** ARCTIC — SEPTENTRIONAL — MERIDIONAL — TROPICS

Table 1.2. Zonal division of vegetation in the USSR (after
Alexandrova, 1970).

Zone	Subzone		
Arctic — Polar desert	Northern variant Southern variant		
Arctic / Sub arctic — Tundra	Arctic tundra	Northern variant Southern variant	
	Subarctic tundra	Typical tundra Southern tundra	
Forest-tundra	Northern forest-tundra		
	Southern woodland		
	Open boreal woodland		
Forest	Northern taiga		

and tundra on exposed ridge sites and in other xeric habitats (see Plate 1). South of
this ecotone, the northern part of the boreal forest (taiga) consists for the most part of
open lichen woodland. Sources other than those in the USSR equate Subarctic with the
forest-tundra and northern boreal forest (table 1.1) and we shall adopt this convention.
The ecotonal status of the Subarctic, so defined, poses an interesting question. In
Labrador–Ungava it forms an extensive subzone, consisting of a mosaic of tiny islands of
the tundra and taiga ecosystems, stretching southward to the limit of economic timber.
It is not clear whether it possesses full ecosystem integrity or is better regarded as a
mosaic of discrete elements from the tundra and taiga zones.

The alpine forest-tundra ecotone is very limited in vertical extent and is sometimes
virtually absent (see Plates 2 and 3). Ideally, it consists of a belt of candelabra and
flagged conifers, upright and prostrate (krummholz) tree islands and dwarf shrubs,
separated by stretches of alpine meadow and heath. A comparison between European
and North American terminology is shown in table 1.3. This also illustrates the different
usage of the terms *subalpine* and *montane* in phytogeographical literature. There appears to
be a basic weakness in the western United States usage of subalpine and montane (see
Löve, 1970; Weaver and Clements, 1938) in that the subalpine-montane ecotone, in this
sense, cuts through a vegetational structure which possesses broad physiognomic
integrity – the montane equivalent of the circumpolar coniferous forests. In this volume
subalpine is taken to mean the altitudinal forest-tundra ecotone, unless otherwise
specified. Löve's (1970) exposition could be countered on the grounds that in many
mountain areas trees are full grown to the tree limit so that no ecotone (the subalpine
belt) exists. However, spatial discontinuities are not uncommon in geological, pedologi-
cal and biological distributions, and such irregularities do not invalidate this approach
to the problem.

Table 1.3. Schematic review of various opinions on alpine zonation in North America and Eurasia (from Löve, 1970).

North American vegetation type	MERRIAM (1898) W North America	WEAVER and CLEMENTS(1938) W North America	DANSEREAU (1957) E North America	BRUSSELS CONGRESS(1910) Europe	MEUSEL et al. (1965) NW Eurasia	FAVARGER (1956) The Alps	Eurasiatic vegetation type
Treeless alpine tundra	ALPINE	ALPINE	ALPINE	ALPINE	ALPINE — Nival or upper / Middle / Lower	ALPINE	Snow, rock, polsters / Mats, polsters / Grass Heath Tundra
— Alpine treeline — Krummholz Meadow Heath — Forestline —	HUDSONIAN	FOREST-TUNDRA ECOTONE	SUBALPINE		SUBALPINE		— Alpine treeline — Krummholz Meadow Heath — Forestline —
Closed conifer forest (mainly spruce and fir)	CANADIAN	SUBALPINE	MONTANE	MONTANE Upper	MONTANE Upper	SUBALPINE	Closed conifer forest (mainly spruce, fir, larch-sylviculture)
Closed, mainly mixed coniferous and deciduous forest	TRANSITION	MONTANE		Middle / Lower	Middle / Lower	MONTANE	Closed, mixed forest (spruce, fir, pine, oak, beech)
Deciduous or pine and oak forest	UPPER AUSTRAL	FOOTHILLS	LOWLANDS and PLAINS	COLLINE	COLLINE	COLLINE	Deciduous forest (oak, walnut, beech—some agriculture)
Steppe, prairie, desert	LOWER AUSTRAL	PLAINS and LOWLANDS		PLANAR	PLANAR		Parkland or steppe, mostly cultivated

The High Arctic is typically rock desert with only lichen vegetation, although small pockets of flowering plants and bog occur in suitable sites. The distribution of polar desert soils (Charlier, 1969) compares reasonably well with simple climatic data, but physical parameters such as the dryness ratio and indices of the moisture balance (Bovis and Barry, 1973) show a poor fit. At the present, climatic information is inadequate to permit an unambiguous analysis to be carried out.

Qualification is necessary over the usage of the term High Arctic. Polunin (1960) distinguishes three broad vegetation zones: low-Arctic – continuous cover and frequent shrubs; middle-Arctic – more or less continuous herbaceous cover, with lichens, on most of the lowlands; and high-Arctic – whose vegetation is limited to favourable sites. The three zones of Porsild (1957) are closely comparable (see fig. 1.3). Porsild refers to the northernmost zone as rock desert or fellfield, which corresponds approximately with the polar desert category of Alexandrova (1970). High Arctic is used here in a broad climatic–geographical sense except where a specific botanical definition is given. It will, therefore, generally include some areas of middle-Arctic in Polunin's terminology (cf. fig. 8B.7).

The counterpart of polar desert in high alpine areas is not an exact equivalent. It is sometimes called the *nival* belt, but a preferable term is the *aeolian* belt (Swan, 1963,

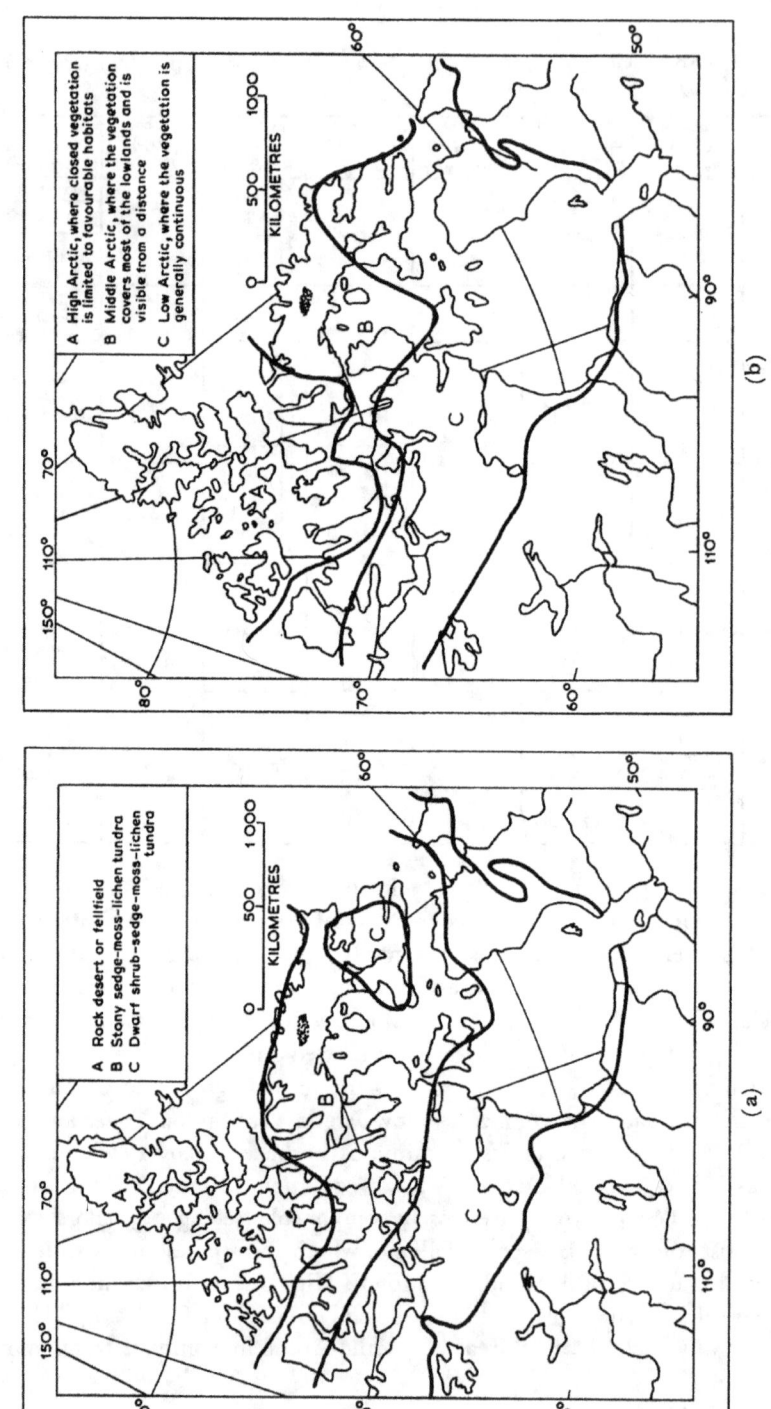

Fig. 1.3. Vegetation zones of the Arctic of North America (after P. J. Webber, based on (a) Porsild, 1957, and (b) Polunin, 1960).

1967). This lies above the limit of vascular plants and includes snowfields (a nival phase) as well as exposed rock and rock-covered ice (a terrestrial phase). The source of nutrition for the biota is wind transport of organic matter, a mechanism which is probably of relatively less importance in polar deserts. The aeolian belt is particularly extensive in the Himalaya and the Andes. In arid situations in the Himalaya even lichens are scarce. In line with this terminology, Troll (1972a) proposed that the high mountain landscapes between timberline and the snowline be designated as 'subnival'.

The variety of usages of the terms to which we have referred, and their frequent overlap, is not unusual in rapidly developing subject areas. It would be unwise to attempt any new definitive statements at this time, however, since the results being obtained through the studies of the International Biological Programme (IBP) Tundra Biome will undoubtedly greatly enhance our view of these environments and revise our thinking on them.

The basis of the IBP work is the holistic ecosystem approach (see Webber, Chapter 8B, and Ives and Barry, Chapter 18). Nevertheless, since many of the components and processes in ecosystem models must be treated as 'black-boxes' in our present state of knowledge, systematic single discipline studies remain indispensable to the total picture. Moreover, the student and professional scientist is faced with the absence of any comprehensive study of arctic and alpine environments.

The following chapters, therefore, bring together for the first time our present knowledge of these environments along systematic lines. Even so, the many areas of interrelationship and interaction will become increasingly apparent, especially in progressing from considerations in the abiotic to the biotic realm. It is in these areas of interaction that many of the most intriguing research problems lie.

References

AHTI, T. (1964) Macrolichens and their zonal distribution in boreal and arctic Ontario, Canada. *Ann. Bot. Fenn.*, **1**, 1–35.

AHTI, T., HÄMET-AHTI, L. and JALAS, J. (1968) Vegetation zones and their sections in northwestern Europe. *Ann. Bot. Fenn.*, **5**, 169–211.

ALEXANDROVA, V. (1970) Vegetation and primary productivity in the Soviet Subarctic. *In:* FULLER, W. A. and KEVAN, P. G. (eds.) *Productivity and Conservation in Northern Circumpolar Lands*, 93–114. IUCN Pub. (n.s.) 16. Morges, Switzerland: Int. Union Conserv. Nature.

BARRY, R. G. (1967) Seasonal location of the arctic front over North America. *Geogr. Bull.*, **9**, 79–95.

BESCHEL, R. E. (1970) The diversity of tundra vegetation. *In:* FULLER, W. A. and KEVAN, P. G. (eds.) *Productivity and Conservation in Northern Circumpolar Lands*, 85–92. IUCN Pub. (n.s.) 16. Morges, Switzerland: Int. Union Conserv. Nature.

BLÜTHGEN, J. (1966) *Allgemeine Klimageographie* (2nd edn). Berlin: W. de Gruyter. 720 pp.

BOVIS, M. J. and BARRY, R. G. (1973) A climatological analysis of north polar deserts. *In:* SMILEY, T. L. and ZUMBERGE, J. H. (eds.) *Polar deserts and modern man* (Sympos., Amer. Ass. Adv. Sci), 23–31. Tucson: Univ. of Arizona Press.

BRYSON, R. A. (1966) Airmasses, streamlines and the boreal forest. *Geogr. Bull.,* **8**, 228– 69.

BUDYKO, M. I. (1956) *The Heat Balance of the Earth's Surface.* Leningrad: Gidromet. Izdat. (English edn, trans. N. A. Stepanova, Washington DC, US Dept of Commerce, 1958, 259 pp.)

CHARLIER, R. E. (1969) The geographic distribution of polar desert soils in the northern hemisphere. *Bull. Geol. Soc. Amer.,* **80**, 1985–96.

DAGON, R. R. (1966) Tundra – a definition and structural description. *Polar Notes,* **6**, 22–34.

DANSEREAU, P. (1957) *Biogeography.* New York: Ronald Press. 294 pp.

FAVARGER, C. (1956) Flore et végétation des Alpes, 1. Neuchâtel and Paris: Delachaux & Niestlé. 271 pp.

HARE, F. K. (1950) Climate and zonal divisions of boreal formation in eastern Canada. *Georgr. Rev.,* **40**, 615–35.

HARE, F. K. (1951) Some climatological problems of the Arctic and the Sub-Arctic. *In:* MALONE, T. F. (ed.) *Compendium of meteorology,* 952–64. Amer. Met. Soc.

HARE, F. K. (1968) The Arctic. *Quart. J. R. Met. Soc.,* **94**, 439–59.

HOLUB, J. and JIRASEK, V. (1967) Zur Vereinheitlichung der Terminologie in der Phytogeographie. *Folia Geobot. Phytotax.,* **2**, 69–113.

HUSTICH, I. (1949) Phytogeographical regions of Labrador. *Arctic,* **2**, 36–44.

KÖPPEN, W. (1936) Das geographische System der Klimate. *In:* KÖPPEN, W. and GEIGER, R. (eds.) *Handbuch der Klimatologie,* 1. C. Berlin Bornträger. 44 pp.

KREBS, J. S. and BARRY, R. G. (1970) The arctic front and the tundra–taiga boundary in Eurasia. *Geogr. Rev.,* **60**, 548–54.

LAVRENKO, E. M. and SOCHAVA, B. (1956) *Karta rastitel'nosti evropeiskoi chasti, SSSR* (Map of the vegetation of the European parts of the USSR). Leningrad.

LÖVE, D. (1970) Subarctic and subalpine: where and what? *Arct. Alp. Res.,* **2**, 63–73.

MERRIAM, C. H. (1898) Life zones and crop zones of United States. *US Dept of Agric. Biol. Surv. Bull.,* **10**, 1–79.

MEUSEL, H., JÄGER, E. and WEINERT, E. (1965) *Vergleichende Chorologie der zentraleuropäischen Flora,* I. Jena: Gustav Fischer.

NORDENSKJÖLD, O. and MECKING, L. (1928) The geography of polar regions. *Amer. Geogr. Soc. Spec. Pub.,* **8**. 359 pp.

POLUNIN, N. (1960) *Introduction to Plant Geography.* London: Oxford Univ. Press. 640 pp.

PORSILD, E. A. (1957) Natural vegetation and flora. *In: Atlas of Canada,* Map 38. Tech. Surv. Ottawa: Dept of Mines.

ROUSSEAU, J. (1952) Les zones biologiques de la peninsule Québec–Labrador et l'hemiarctique. *Can. J. Bot.,* **30**, 436–74.

SEMENOVA-TYAN-SHANSKAIA, A. M. (1964) Map in North America. *In: Fizikogeografischeskii Atlas Mira* (Physico-geographical atlas of the world), 150–1. Moscow.

SJÖRS, H. (1963) Amphiatlantic zonation. Nemoral to Arctic. *In:* LÖVE, A. and LÖVE, D. (eds.) *North Atlantic Biota and their History,* 109–25. New York: Pergamon Press.

SWAN, L. W. (1963) The aeolian zone. *Science,* **140**, 77–8.

SWAN, L. W. (1967) Alpine and aeolian regions of the world. *In:* WRIGHT, H. E. JR and OSBURN, W. S. JR (eds.) Arctic and alpine environments, *Proc. VII Congress INQUA,* **10**, 29–54. Bloomington: Indiana Univ. Press.

THORNTHWAITE, C. W. (1948) An approach towards a rational classification of climate. *Geogr. Rev.*, **38**, 55–94.

TIKHOMIROV, B. A. (1962) The treelessness of the tundra. *Polar Rec.*, **11**, 24–30.

WEAVER, J. E. and CLEMENTS, F. E. (1938) *Plant ecology*. New York: McGraw-Hill. 601 pp.

WEBBER, P. J. (1971) *Gradient Analysis of the Vegetation Around the Lewis Valley, North-Central Baffin Island, Northwest Territories, Canada*. Unpub. Ph.D. thesis, Queen's Univ., Kingston. 366 pp.

Manuscript received May 1972.

Additional references

TROLL, C. (1972) Geoecology and the world-wide differentiation of high-mountain eco-systems. *In:* TROLL, C. (ed.) Geoecology of the High-Mountain regions of Eurasia, *Erdwiss. Forsch.*, **4**, 1–13. Wiesbaden: Franz Steiner Verlag.

TROLL, C. (1973a) The upper timberlines in different climatic zones. *Arct. Alp. Res.*, **5** (3), part 2, A3–A18.

TROLL, C. (1973b) High-mountain belts between the polar caps and the equator: their definition and lower limit. *Arct. Alp. Res.*, **5** (3), part 2, A19–A28.

UNESCO (1973) *Ecology of the subarctic regions*. Paris: UNESCO. 364 pp.

Present environments

Arctic climate

Roger G. Barry and F. Kenneth Hare

Institute of Arctic and Alpine Research, University of Colorado,
and Department of Geography, University of Toronto

The climate of the Arctic, perhaps contrary to popular opinion, is not easily characterized in simple terms. Conventional seasonal names can be applied only in some restricted sense and their duration is irregular in space and in time, both within a year and from one year to the next. This variety exists on a large spatial scale and over tens of metres. It is as well to begin, therefore, by noting the seasonal character of different types of surface. These may be summarized as follows:

	Land	Sea
Winter (6–8 months)	Snow cover. Some bare rock.	Snow cover on pack ice. A few open water patches (leads, polynyas).
Spring (1 month)	Unstable snow cover. Some melt ponds in south, bare rock.	Unstable snow cover on pack. Minimum open water.
Summer (2–3 months)	Wet and dry tundra, lakes, bare rock, snowbeds, glaciers and ice caps.	Puddles on ice, numerous leads, marginal seas more open.
Autumn (1–2 months)	Snow cover, lakes freezing, some bare rock.	Snow on pack ice, maximum open water in marginal seas.

The approximate maximum extent of sea ice in March and the August limits are shown in fig. 2A.1.

The existence of a mosaic of arctic surfaces on land and sea must be borne in mind in the subsequent discussion, although we are not able here to take detailed account of the climatic consequences. Apart from the effects of large-scale land features, which include modification of synoptic systems, the development of föhn winds on lee slopes of mountains, and cold air drainage on slopes at night or in winter generally, it is also important to recognize that low solar altitude makes even minor topographic features responsible for major differences in topoclimate and microclimate. The implications of such differences with respect to surface energy budgets are of particular interest (see, for example, Wendler, 1971).

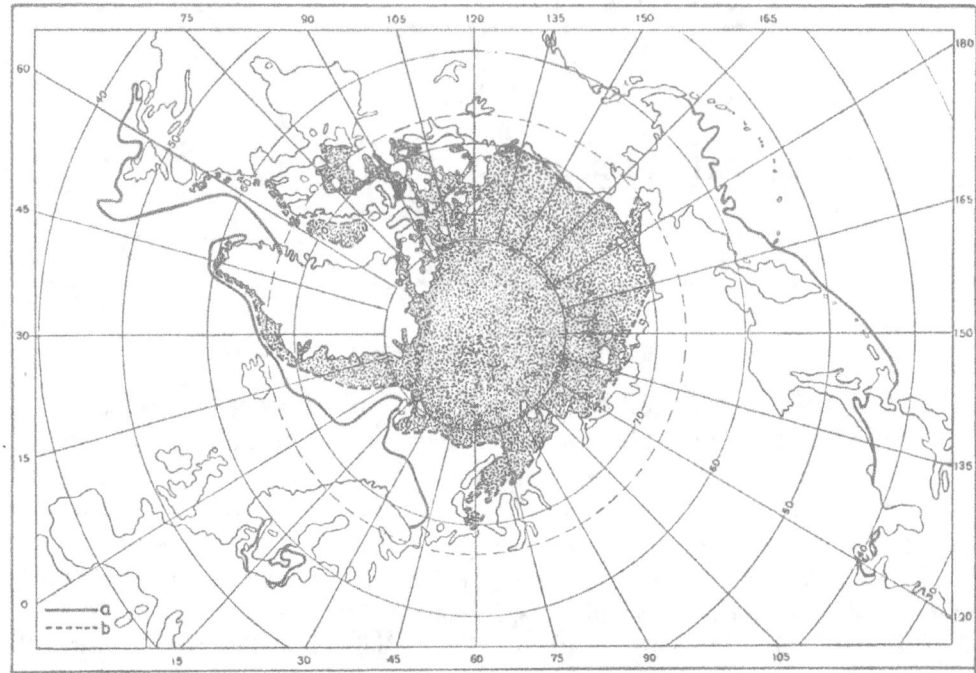

Fig. 2A.1. Sea ice limits in (a) March and (b) August (from Budyko, 1971).

1. The energy balance

There is no lack of solar energy in the Arctic. Annual values of global solar radiation along the northern coasts of the Queen Elizabeth Islands and across the pack ice to northernmost Siberia (Dolgin, 1970) are near 70 kcal cm^{-2} of horizontal surface (70 kly yr^{-1}). This is about half the receipt along the United States–Canadian border and a third of that in the subtropical deserts. However, the high reflectivity (albedo) of the snow and ice surfaces keeps absorption low even during the summer, especially in June when the sun is strongest (see table 2A.1). There is, moreover, a net long-wave cooling of the surface at all seasons. Consequently, the net radiation income is small chiefly due to the role of the snow and ice surfaces.

The energy-balance equation for unit horizontal area at the earth's surface is

$$R_n = I(1 - a) + R\downarrow - R\uparrow = H + LE + G \qquad \dots \dots \dots \quad (1)$$

where R_n is net radiation (radiation balance in Russian usage), I is global solar radiation; a is albedo, or reflectivity of surface, for all solar wavelengths, $R\downarrow$ and $R\uparrow$ are atmospheric downward and terrestrial upward long-wave radiation, H is convective atmospheric heat flux, LE is convective atmospheric latent heat flux (E = evaporation), and G is subsurface heat flux in soil or water. Energy partitioning is highly dependent on surface properties; on wet tundra meadows in summer 70 per cent of R_n goes into latent heat flux.

Table 2A.1. Estimated normal values of radiative fluxes (with albedos), North Pole area and Baker Lake, NWT (64°18'N, 96°00'W) (kcal cm^{-2}).

	Flux, etc.	Jan.	Feb.	Mar.	Apr.	May	June	July	Aug.	Sept.	Oct.	Nov.	Dec.	Year
North Pole	Global solar (I)	—	—	—	8.1	17.5	19.2	13.5	8.3	2.2	—	—	—	68.8
	Albedo (%) (a)	—	84	84	81	82	77	65	69	83	85	—	—	—
	Absorbed solar (I–aI)	—	—	—	1.5	3.1	4.4	4.7	2.6	0.4	—	—	—	16.7
	R↓–R↑	-2.0	-1.9	-1.9	-2.2	-1.5	-1.2	-1.1	-1.1	-1.2	-1.3	-1.9	-2.1	-19.4
	Net radiation (R$_n$)	-2.0	-1.9	-1.9	-0.7	+1.6	+3.2	+3.6	+1.5	-0.8	-1.3	-1.9	-2.1	-2.7
Baker Lake	Global solar (I)	0.6	2.5	7.6	13.7	17.3	15.5	14.5	9.5	4.9	2.9	0.9	0.1	89.9
	Albedo (%) (a)	79	80	79	75	65	34	21	18	21	41	59	73	—
	Absorbed solar (I–aI)	0.1	0.5	1.5	3.4	6.0	10.4	11.7	7.8	3.9	1.7	0.3	0.0	47.7
	R↓–R↑	-2.4	-2.1	-2.8	-3.4	-3.5	-4.1	-4.8	-4.1	-2.8	-2.1	-2.3	-2.2	-36.6
	Net radiation (R$_n$)	-2.2	-1.6	-1.3	+0.0	+2.5	+6.3	+7.0	+3.7	-1.2	-0.4	-2.0	-2.2	11.1

Sources: Marshunova and Chernigovskiy, 1966; Hay, 1970.

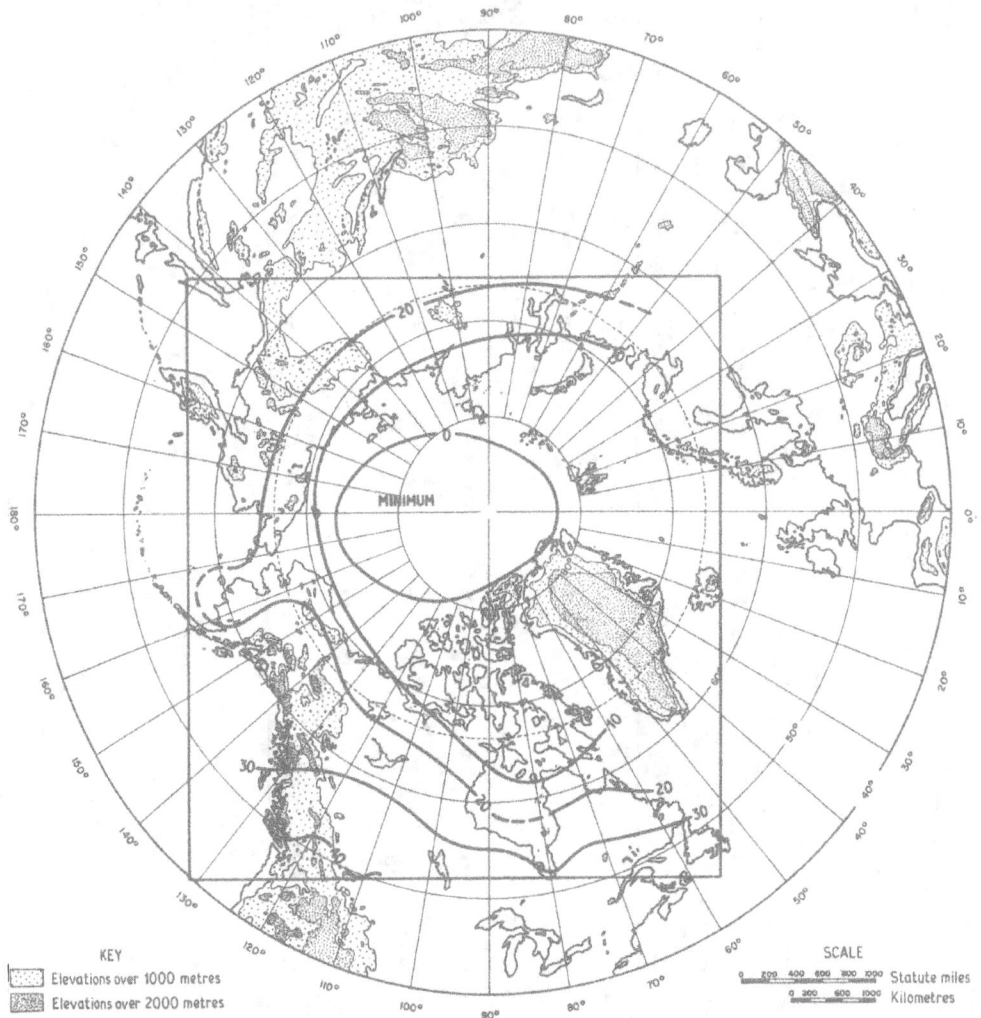

Fig. 2A.2. Mean annual net radiation (kcal cm^{-2}). Tentative isolines based on data of Marshunova and Chernigovskiy (1966), Dolgin (1970) and Hay (1970).

The true arctic landscapes – those north of the limit of tree growth – have a net annual radiation income of less than 15 kly yr^{-1}. The boreal forest landscapes have a radiation income in the range 20–35 kly yr^{-1}, the forest-tundra ecotone (between it and the tundra; see Larsen, Chapter 7A) between 15 and 20 kly yr^{-1}. These figures are applicable to those areas in which equilibrium has been reached, i.e. where postglacial boundary changes have come to an end.

At sea, where information is more scattered, much the same figures apply. The Arctic Ocean, with its permanent pack ice cover, is believed to have small negative net radiation values (down to −4 kly yr^{-1}), and the marginal seas with seasonal ice cover have values up to 15 kly yr^{-1} (Marshunova and Chernigovskiy, 1966) although Vowinckel and Orvig (1966) find small positive values even in the central Arctic Ocean. Fig. 2A.2 shows, in tentative form, the mean annual net radiation field over the Arctic. In table 2A.1 we have compiled estimates of the various radiative fluxes for the North Pole and for a representative Canadian station.

Table 2A.2. Annual mean heat exchanges over the Arctic Seas (kcal cm^{-2}).

Sea	Layer	Process 1	2	3	4	5	6	7	8
Arctic Ocean	Atmosphere	+19	+28	−126	+65	+6	+4	+3	—
	Surface	+31	−28	—	—	—	−4	−3	+5
	Net surface radiation: +2.6 Net loss to space: −76								
Norwegian and	Atmosphere	+21	+39	−138	+30	+4	+16	+27	—
Barents seas	Surface	+50	−39	—	—	—	−16	−27	+32
	Net surface radiation: +11.5 Net loss to space: −66								

Source: Vowinckel and Orvig, 1966.

Key to processes
1. I$(1 - a)$, absorbed solar radiation.
2. Net infra-red flux, surface to atmosphere.
3. Net infra-red flux, atmosphere to space.
4. Heating due to atmospheric advection of sensible heat.
5. Heating due to atmospheric advection of latent heat.
6. Convective heat flux, surface to atmosphere.
7. Convective latent heat flux, surface to atmosphere.
8. Flux of heat from sea to surface.

It comes as a surprise to most people that the arctic surface is thus not, on an annual basis, radiatively cooled, except for parts of the permanent pack ice. In effect the strong radiative cooling of winter is offset, or a little better, by heating during the long days of summer. The balance thus achieved requires that the mean annual surface temperature be low – of the order of −10 °C to −15 °C – and that there be strong year-round cooling of the middle and upper troposphere, by long-wave (infra-red) radiation to space. Much heat is transferred horizontally (i.e. advection) by wind or ocean currents. If these horizontal fluxes converge there is local heating, expressed in equation (1) by contributions to the terms H and LE (from the wind) or G (from the sea). In such a case the upward radiative flux from the ground, R↑, is correspondingly increased to achieve balance, and the net radiation is hence reduced.

To complete the picture we need to consider the role of the atmosphere. In table 2A.2 we present estimates of the various exchange processes for the Arctic Ocean, and for the Norwegian and Barents seas. The figures are spatial averages, and were computed (Vowinckel and Orvig, 1966) using lower estimates of summer albedo than those assumed in table 2A.1.

From these figures we conclude that energy exchanges and conversions at the surface of the Arctic Ocean are quite small, because the pack ice both reflects back much of the

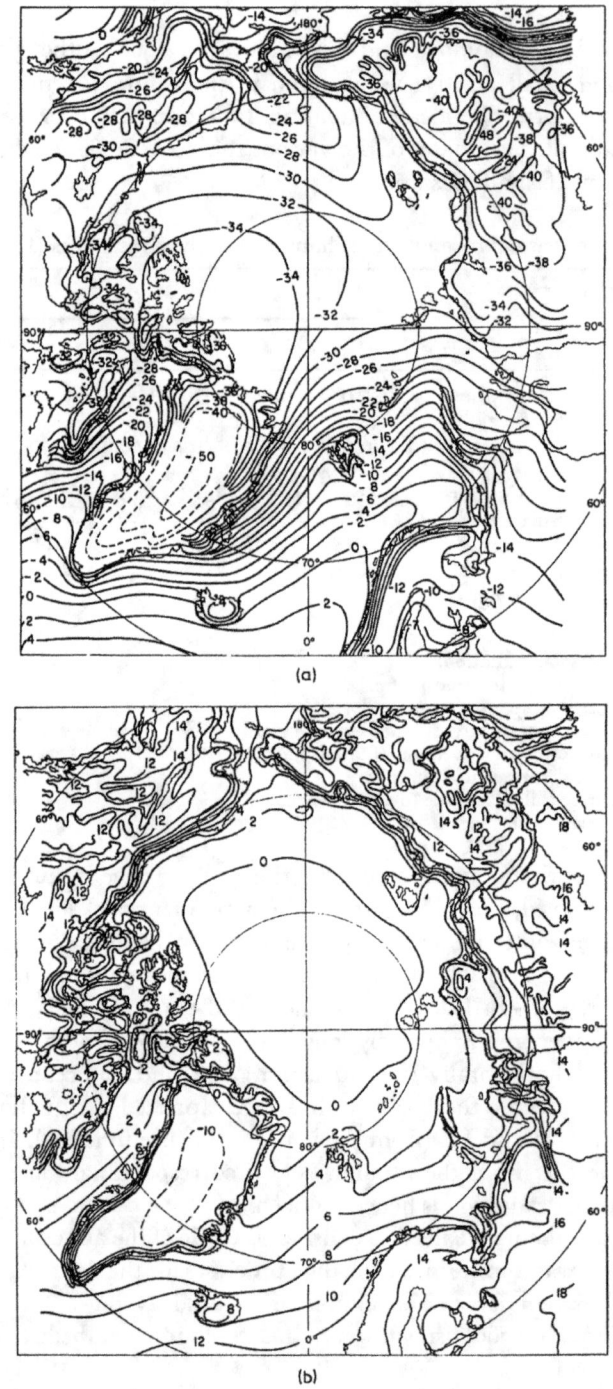

(a)

(b)

Fig. 2A.3. Mean daily temperature (°C) for (a) January, (b) July (after Prik, 1959, from Orvig, 1970).

solar radiation and inhibits the flow of heat from sea to air. Convective heat losses at the surface, and gains from the sea, are greater over the Norwegian and Barents seas, where there is much open water. Over northern lands and ice caps relations are more like those over the Arctic Ocean.

There is nevertheless a strong, year-round heat loss by radiation from the atmosphere to space over all northern areas. This cooling proceeds mainly from the level of the arctic inversion (see p. 27), especially from the tops of clouds and moist layers. The heat used in this process is derived not only by convection or radiation from below, but by advection: warm airstreams are brought into the Arctic by travelling cyclones (discussed below) and these provide the largest source – about 55 per cent – of the heat transferred to space. In effect a vigorous atmosphere overlies a relatively quiescent surface.

The measure of heat storage is, of course, temperature. Figs. 2A.3a and 2A.3b give mean daily air temperature for January and July over northern regions. The duration of the thaw season (air temperature above 0°C) is about four months along the arctic treeline and a few weeks, at the most, over the Arctic Ocean.

2. Large-scale circulation

As we have just seen, the heat balance over the north requires a vigorous atmospheric circulation, at least above the sluggish surface. The low equilibrium temperatures also require, from hydrostatic arguments, that the vertical decrease of pressure be unusually rapid, at least up to 12 km (above which in summer the Arctic is not cold). Hence in the middle and upper troposphere and lower stratosphere, the north is permanently covered by a giant centre of low pressure. This is the core of the circumpolar westerly ('Ferrel') vortex, whose winds permanently cover middle-latitudes. The horizontal heat transports made necessary by the radiative cooling of the arctic atmosphere are brought about by the travelling disturbances of the westerlies, which frequently penetrate the Arctic. These disturbances also bring the daily weather changes, which are often dramatic.

MEAN TROPOSPHERIC FLOW

Figs. 2A.4a and 2A.4b show the mean height in geopotential metres of the 300 millibar (mb) constant pressure surface for January and July. The resultant flow of the winds at this level is almost parallel to the contours, lower heights being to the left of the wind vector. Speeds are proportional to the gradient (i.e. the closer the contours the stronger the wind) and inversely proportional to the sine of the latitude, according to the relations

$$\bar{u} = -\frac{g}{f}\frac{\partial \bar{Z}}{\partial y}; \qquad \bar{v} = \frac{g}{f}\frac{\partial \bar{Z}}{\partial x}$$

where $\bar{u}$ and $\bar{v}$ are the components of the resultant flow from west and south respectively, g is the acceleration of gravity, $\bar{Z}$ is the mean height of the pressure surface, x and y are distance towards east and north respectively, and f is the Coriolis parameter ($2\omega \sin\phi$, where ϕ is latitude and ω is the angular velocity of the earth's rotation).

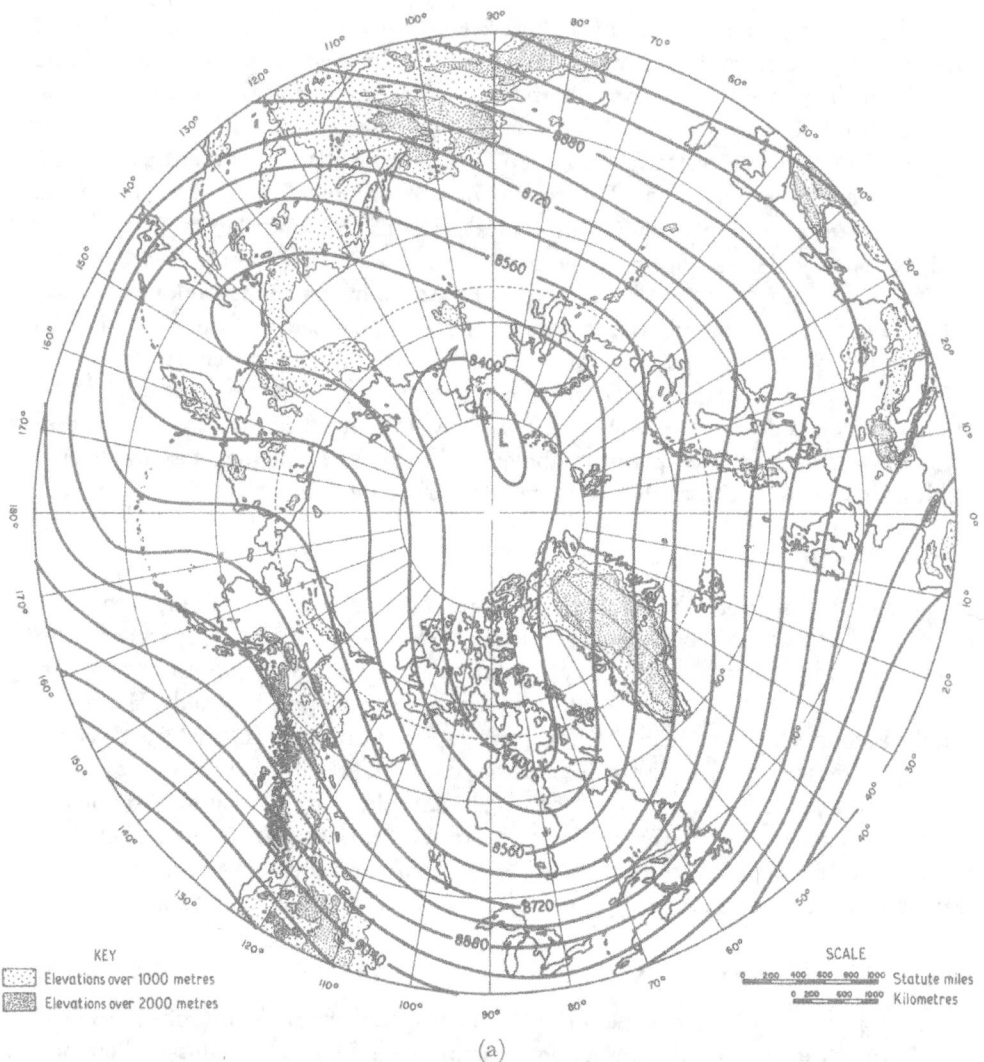

(a)

Fig. 2A.4. Mean height (gpm) of the 300 mb surface for (a) January, (b) July (after Crutcher and Meserve, 1970). The resultant wind blows almost parallel to the contours at a speed proportional to the gradient with lower heights to the left of the wind vector.

The 300 mb surface is roughly the level of strongest wind, and the entire layer between 3 and 15 km tends to resemble this surface in flow pattern. Clearly the tropospheric circulation consists of a vast circumpolar westerly flow on both charts. In July the apparent centroid of the vortex is north of Ellesmere Island, but in January the central area is complex, with three apparent centroids – over Lancaster Sound, north of Novaya Zemlya, and northwest of the Sea of Okhotsk. The January circulation is far more intense, kinetic energy at this level being more than three times greater than that of July.

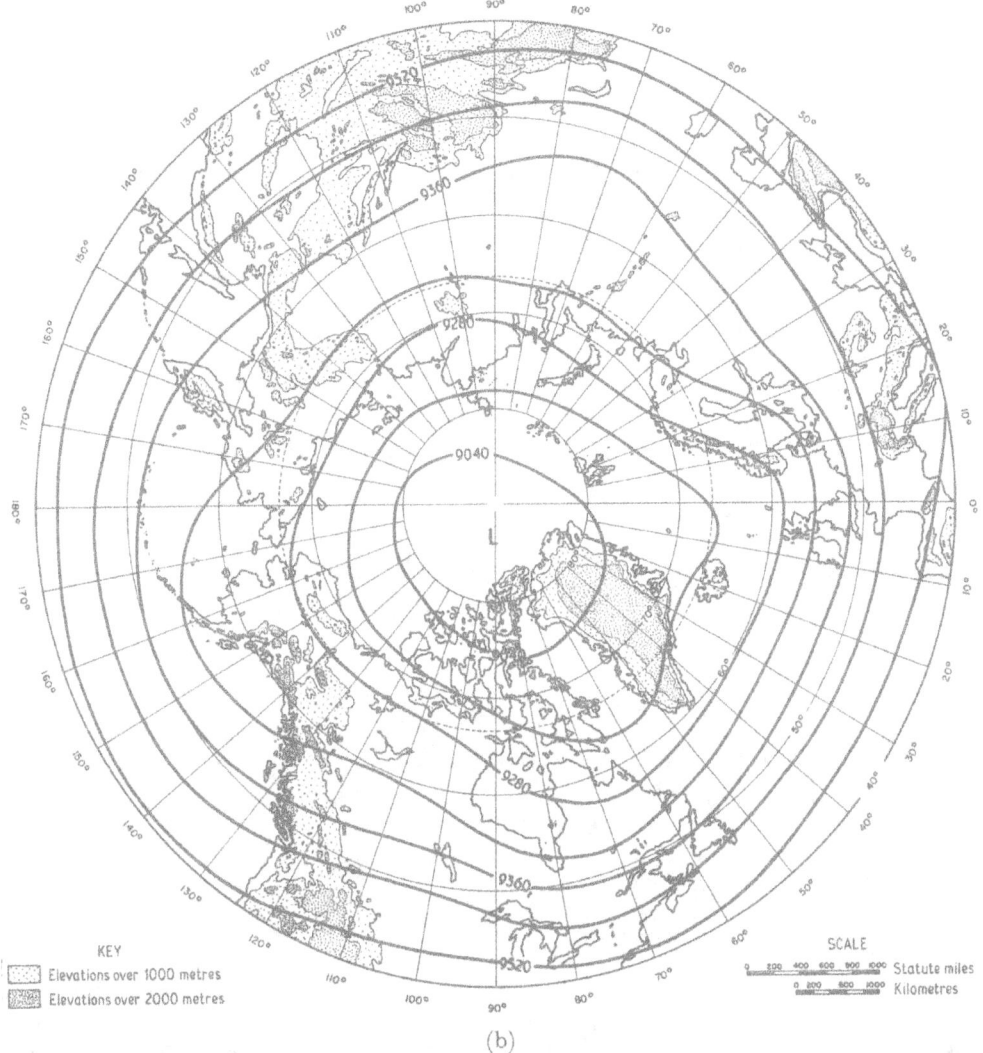

(b)

Fig. 2A.4. (*cont.*)

Strong standing disturbances (long westerly waves) are apparent on both charts. In January there are deep troughs of low pressure extending southward towards Japan and along the 80°W meridian, and a weaker trough extends towards the Black Sea. Of the intervening ridges, by far the most powerful is the tilted system over western Alaska. In July the standing waves are much less vigorous, but that over eastern Canada is still clearly visible. Fourier analysis by Van Mieghem (1961), at 500 mb, shows that these waves are dominated in high latitudes by circumpolar wave number 2, which shifts eastwards in summer, westwards in winter, with summer amplitude being less than half that of winter.

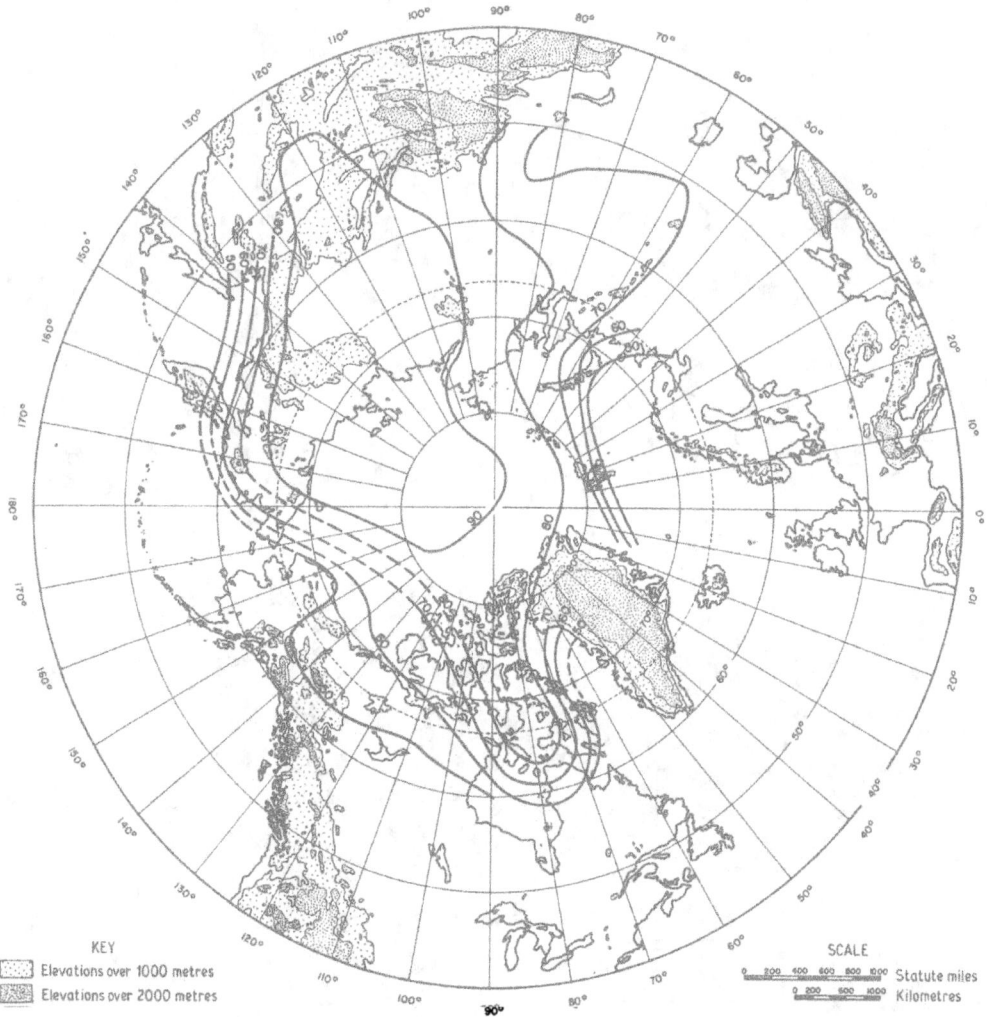

Fig. 2A.5. The frequency of low-level temperature inversions in January. Tentative isolines based on analyses by Putnins and Stepanova (1956), Vowinckel and Orvig (1967) and Bilello (1966).

In practice the upper tropospheric flow is usually more complex than these mean fields suggest. In particular one can normally distinguish a distinct maximum of westerly component not far from the coast of the Arctic Ocean. This arctic jet stream corresponds to the tendency for a well-marked arctic front to separate the true arctic airstreams from the main westerly current. The arctic front and jet stream play a critical role in the climates of the northern world, and are equally critical in the mid-latitude areas of central North America and eastern Asia. They vary greatly in position from season to season, and of course from day to day.

The average location of the arctic front in July has been shown to coincide closely with the boreal forest/tundra boundary in North America and in Eurasia (Bryson, 1966; Krebs and Barry, 1970) although in eastern North America this relationship seems to be less certain (Barry, 1967a). In part the difference of opinion here relates to the frontal definition. However, net radiation at the surface in July shows a broad minimum roughly parallel to the locus of the front and some 600–1000 km north of it (Hare, 1968) which appears to reflect the cloud distribution associated with the frontal zone. This minimum accords better with the mean front identified by Barry over southern Baffin Island than the position along the treeline in Labrador–Ungava proposed by Bryson.

The characteristic thermal structure of the arctic troposphere in January is dominated by the presence of an intense and virtually permanent inversion in the lowest levels. Vowinckel (1965) shows that this structure occurs with a frequency of 80–90 per cent in the Polar Basin during the winter (fig. 2A.5). There is commonly an intense nival (ground) inversion 300–500 m in depth, capped by a more nearly isothermal layer up to about 1.5 km (850 mb). Temperatures at the top of the inverted layer average −20°C to −25°C even near the Pole, whereas at the surface temperatures tend to be below −30°C over pack ice and continental interior alike. In northeast Siberia the mean temperature difference between the surface and the warmest level is of the order of 20°C. Over the open water of the Bering, Norwegian and Barents seas, however, surface air temperatures are in the range from 0°C to −10°C, and the inversion is absent much of the time. A sharp change takes place over the Polar Basin in May, and from then until August the lower troposphere lapse rates are usually normal, although shallow inversions still occur at times over cold water or ice surfaces.

THE STRATOSPHERE

The arctic tropopause separates the troposphere, whose behaviour is governed largely by radiative imbalances involving the earth's surface, clouds, water vapour and carbon dioxide, from the exceedingly dry stratosphere. In this layer, extending from about 10 km to 40 or 50 km, ozone and carbon dioxide are the chief absorbers and emitters of radiation, and convective exchanges with the lower levels are trivial. Hence the polar stratosphere becomes warm in midsummer, being above −40°C between 25 and 35 km because of ozone absorption of the continuous insolation, and cold in winter (below −70°C between 25 and 30 km). Hydrostatic adjustments to this great temperature contrast produce annually reversing mean wind systems.

In January at the 30 mb level (approximately 23–24 km) there is a strong, cold vortex centred north of Novaya Zemlya, with troughs across eastern Siberia and eastern North America. The associated *polar-night westerlies* are very unstable, and strong temperature variations due to vertical motion occur from day to day. Finally the vortex collapses in February, March or April in sudden warming events that may destroy the central cold in a few days. During the summer the pattern is one of quiet easterly flow round a warm, stable polar anticyclone, the undisturbed flow lasting from late May until late August, when westerlies begin to appear around the rapidly cooling polar area. In September and

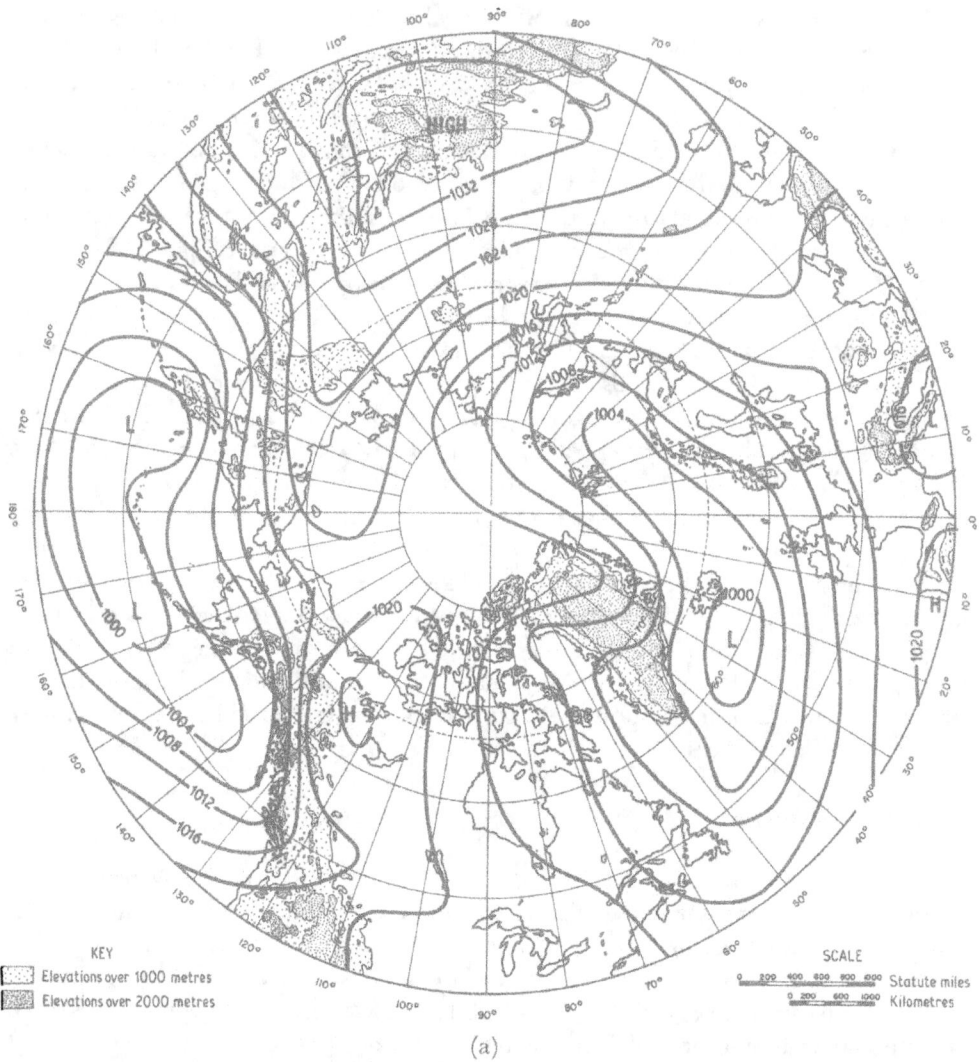

(a)

Fig. 2A.6. Mean msl pressure (mb) for (a) January, (b) July (from Crutcher, unpublished).

early October these are fairly slack and undisturbed, but in late October a strong ridge develops over Alaska, and thereafter the disturbed westerly polar-night circulation is resumed.

MEAN FLOW AT THE SURFACE

Below 2 km, and especially at sea level, the mean flow becomes far more complex than in the mid- and upper troposphere. This arises from the great differences in radiative

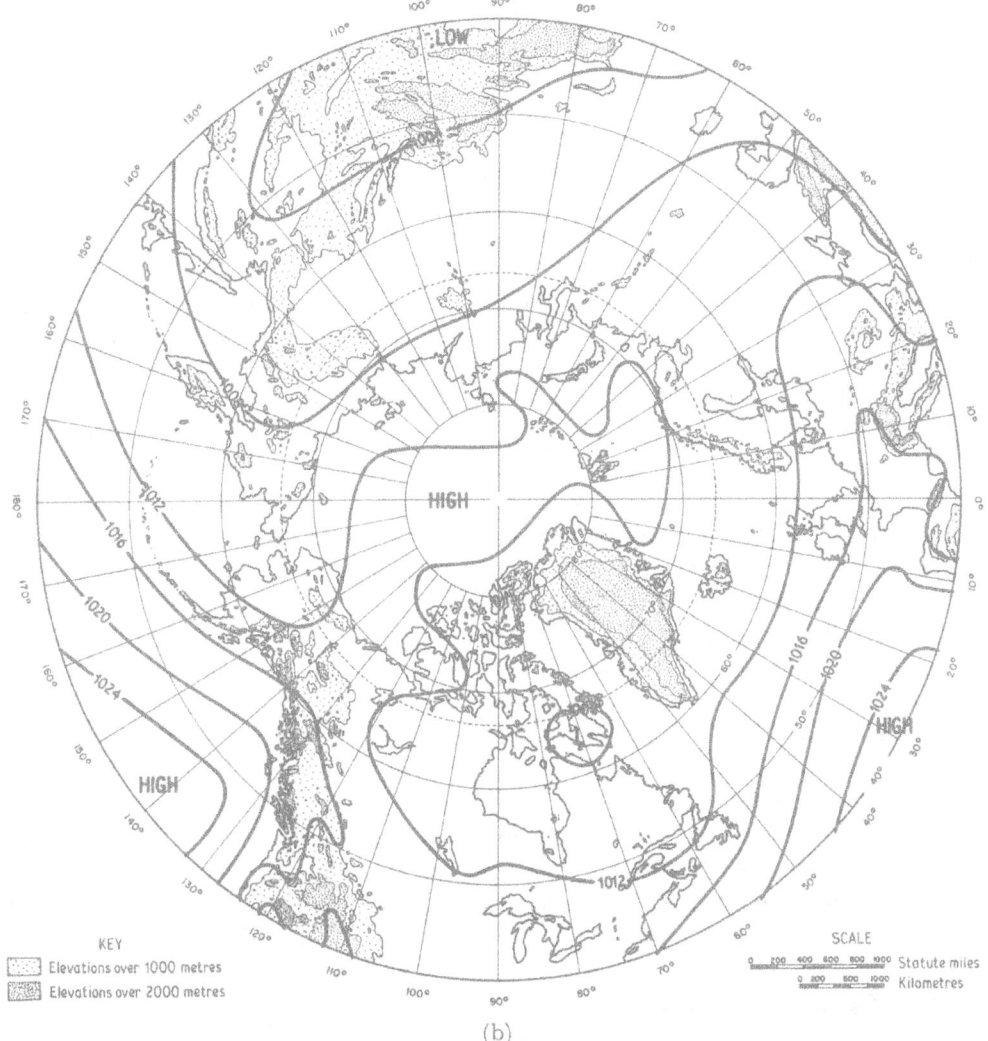

(b)

Fig. 2A.6. (*cont.*)

regime and thermal storage between the continental landmasses and the northern oceans (though in winter the pack ice effectively adds the central Arctic Ocean to the landmasses as far as thermodynamic influences are concerned).

Figs. 2A.6a and 2A.6b show mean sea level pressure (in millibars) for January and July. Low level resultant flow is parallel to the isobars except for a component down-gradient near the surface. The simplicity of the 300 mb charts is entirely lacking.

In January (and for the whole winter period) the chart is dominated by the great oceanic low-pressure systems, called the Aleutian and Icelandic lows. The Icelandic low is elongated so as to stretch effectively from south of Greenland to Novaya Zemlya, and the Aleutian system similarly extends from Kamchatka to the Gulf of Alaska. Both lows represent the 'graveyards' of countless Atlantic and Pacific cyclones. It is to be noted that the Aleutian low is effectively cut off from the Arctic Ocean by the high ground of Alaska and northeast Siberia, whereas the Icelandic low extends its influence to the Pole itself. Along the poleward sides of these lows easterly winds are habitual, but these polar easterlies are not in fact very extensive. More significant are the two vast monsoonal outpourings of arctic air on the western flanks, over eastern Asia and Canada. On the Atlantic side, warm southwesterly currents penetrate the arctic coasts and seas north of Scandinavia and Russia.

High pressure cells appear over the continental landmasses in winter. By far the larger is the Siberian high, centred over the inner plateaus and basins of Siberia and Mongolia. The location of many of the weather stations in this area in high valleys, which retain intensely cold air, has led to uncertainty as to the validity of some of the reported extreme pressures reduced to MSL (Walker, 1967). While this problem certainly exists, there is no doubt as to the reality of high pressure. The smaller cell lies over the Mackenzie–Yukon area. The two are linked by a ridge across the Alaskan–east Siberian sector of the Arctic Ocean. A ridge extends southward over Greenland. The continental highs are associated with very cold surface conditions, and intensified inversions.

During summer the Aleutian low disappears, and the Icelandic low drifts westward as a rather formless area of low pressure centred over southern Baffin, with troughs towards Keewatin and Iceland. Pressure is low over central Siberia, and a weak ridge across the Arctic Ocean separates this belt from the Canadian low. For about two months in spring, this cross-Arctic ridge is strong enough to constitute a true arctic anticyclone; only in April and May, in fact, does the map look like the traditional picture of northern pressure distribution.

It is apparent from figs. 2A.6a and 2A.6b that the surface wind regime over the Arctic is very complex. In two areas alone are there highly persistent currents. These are the northerly and northwesterly flows west of the Icelandic and Aleutian lows. In other areas, and at all seasons, the mean flow is weak, and the daily variations large.

3. Synoptic-scale circulation

As in the westerly belt of mid-latitudes, synoptic-scale disturbances control the daily weather events of northern areas. Both the intensity and the frequency of such disturbances tend to be lower in the north, but there are conspicuous exceptions. We recognize the following classes of disturbance:

(1) mobile wave cyclones of the arctic frontal belt;
(2) cold lows;
(3) anticyclones.

The spatial distributions of cyclone centres in February and August, taken as typical of winter and summer regimes, are shown in figs. 2A.7a and 2A.7b. The unit on these charts is percentage frequency of occurrence within areas of 650,000 km². Representative

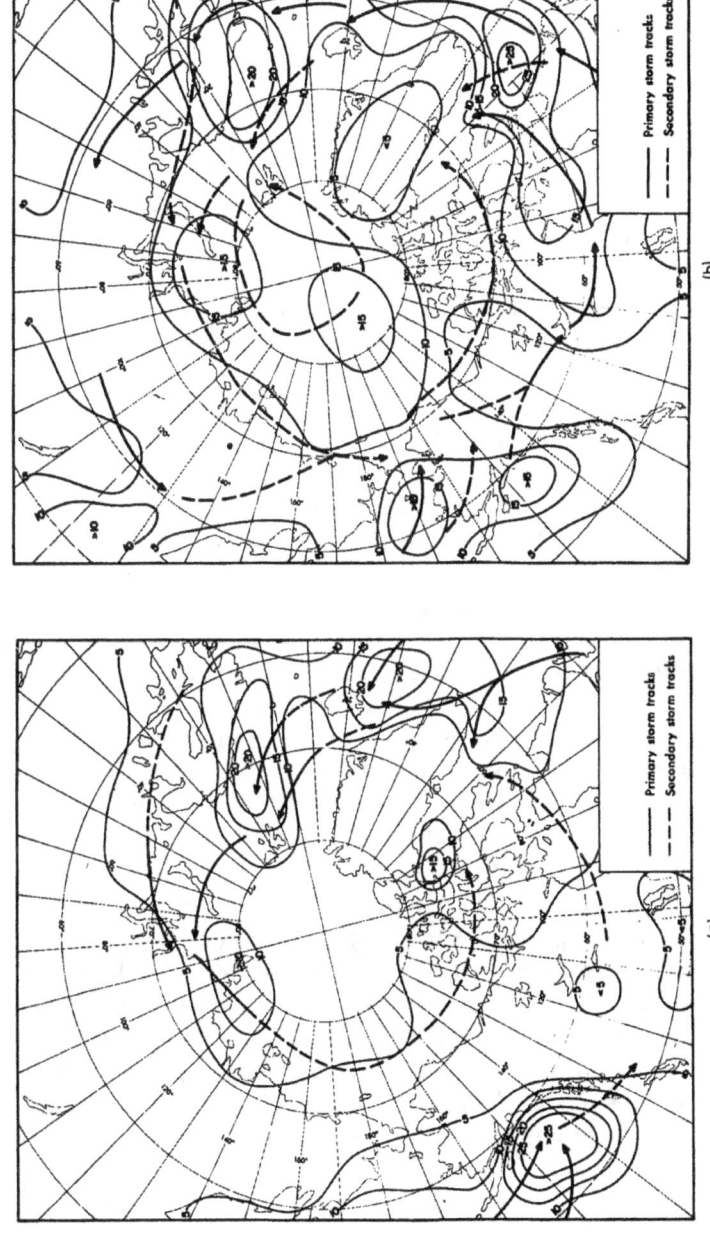

Fig. 2A.7. Percentage frequency of cyclone centres in squares of 650,000 km² and representative tracks in (a) February, (b) August (from McKay *et al.*, 1970).

tracks for the travelling systems are indicated, but it should be remembered that the cold lows typically move slowly and erratically, and that in central areas of the Arctic no characteristic direction of motion is observed for much of the year.

In February very high frequencies occur over the Gulf of Alaska, and in three sub-centres of the Icelandic low – south of Greenland, north of Norway and north of Cape Chelyuskin. These represent zones to which deep and intense Pacific (Gulf of Alaska) and Atlantic cyclones, respectively, tend to move after occlusion. Such systems are often slow-moving, and may persist for some days. As they drift eastward along the arctic coast of Scandinavia, Russia and western Siberia, they carry mild Atlantic air over these northern land areas. They also bring to the Atlantic flank of the Arctic Ocean and its bounding seas periods of high wind, cloudiness and snowfall quite unlike the winter weather across the Pole. Minimum winter cyclone frequencies occur over Siberia across the Chukchi Sea and northern Alaska into the Mackenzie–Yukon area.

The centre of high cyclone frequency over northern Baffin Bay near the North Water is in part an effect of the Greenland barrier: small lee-depressions occur in the lee of the ice sheet, against the eastern flank of which easterly flow is maintained for much of the winter. Lows also stagnate in this area after moving northward from the Labrador Sea (Putnins, 1970).

In August, as throughout summer, cyclone frequencies are highest in a broad belt extending along the 60th parallel from Keewatin to mid-Atlantic, and thence across the Norwegian Sea to the Novaya Zemlya area, and finally to the central pack ice of the Arctic Ocean. Another belt of high frequency lies in the vicinity of southwestern Alaska.

At all seasons it is along the margins of the Arctic, i.e. in the Arctic frontal belt, that mobile wave cyclones of the familiar westerly kind are commonest. Where such systems can involve moist air in their warm sectors, they may bring extensive cloud and light to moderate snow or (in summer only) rain, even to the heart of the pack ice area. The more characteristic cyclones of high latitudes, however, are the cold lows, which are cold-cored, rather small, roughly circular vortices that drift slowly about the central Arctic for up to a week or so, sometimes ending their days by plunging southward into the westerly belt. Fig. 2A.8 illustrates their distribution in terms of 1000–500 mb thickness centres (Flohn, 1952). These systems are most evident above 700 mb as they may be associated with low or high pressure at the surface. Some develop from old occluded mid-latitude cyclones while others originate within the Arctic. They bring only thin, high cloud and light snow in high latitudes, and in winter may even be cloud-free.

Anticyclones also occur in arctic latitudes. Mention has already been made of the persistent continental anticyclone over Siberia, and of the less persistent high pressure common in the northwest of North America. Anticyclones are less common over the Arctic Ocean, but between March and early June slow-moving highs typically lie over the Ocean, with a tendency for a ridge to extend south across Canada, west of Hudson Bay. Blocking highs are common over Scandinavia and occur less frequently over Alaska and the Greenland area.

Before leaving this topic it should be noted that many characteristics of arctic weather cannot be altogether accounted for by synoptic-scale events. Comparisons of the

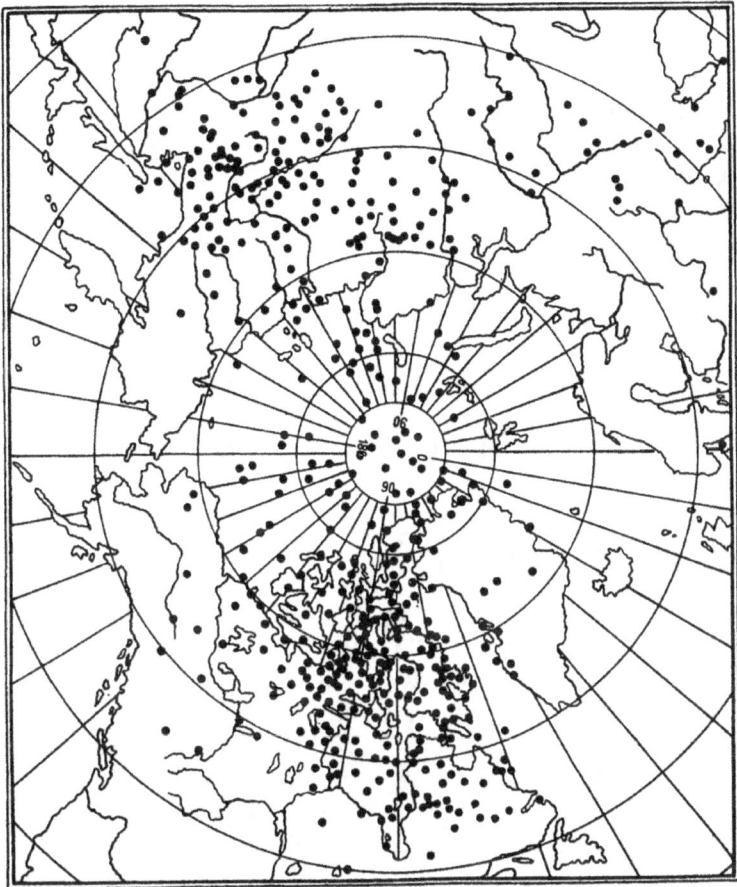

Fig. 2A.8. The distribution of cold centres for the 1000–500 mb layer (with a temperature difference of at least 8°C (=160 gpm) from the surroundings) for winters 1948–9 to 1951–2 (from Flohn, 1952).

occurrence of rain days in summer at 'neighbouring' stations in Ellesmere Island and eastern Baffin Island (Barry and Jackson, 1969; Andrews *et al.*, 1970) suggest that the complexity of land/sea distribution and orography together probably with subsynoptic-scale meteorological factors, often make it hard to discern the synoptic control. The available station network is not sufficient, however, to study this question further at present.

4. Regional characteristics

The energy exchanges and circulation characteristics in the Arctic produce a wide variety of climatic conditions and these are now examined in an albeit rather arbitrary regional setting.

GREENLAND

The vast extent of Greenland – over 2 million km² – is more than 80 per cent covered by the ice sheet and peripheral glaciers. The plateau ice sheet generally exceeds 1200 m elevation, and in its highest parts rises to over 3000 m. Coastal mountains also reach this elevation. This mass is large enough to exert a significant influence on the circulation over a considerable sector of the hemisphere.

It was formerly believed that the circulation regime over the ice sheet was dominated by a quasi-permanent 'glacial anticyclone' (Hobbs, 1945) arising from the radiative cooling effects of the ice and snow surface. Field experience and synoptic data led to the refutation of this theory (Loewe, 1936; Matthes, 1946; Matthes and Belmont, 1950) and modern work indicates a much more varied regime, especially in the south and along the west coast. Anticyclonic curvature of MSL isobars occurred, however, on 50–65 per cent of days in January 1947–58 over northeastern Greenland (Meteorological Office, 1964, fig. 44).

The principal effect of the ice sheet is to steer depressions from the west and southwest northward along the west coast, and in some situations to cause a low approaching Cape Farewell to split into two with one system moving northward into Baffin Bay and the other moving off towards Iceland (Walden, 1959; Putnins, 1970). This development also promotes outbreaks of arctic air along the east coast. In spite of this barrier effect lows cross Greenland at the 3 km (700 mb) level and above, causing disruption of the temperature inversion over the ice sheet and giving snowfall. The work of Matthes and Belmont suggested that depressions mainly cross the ice sheet in the southern part of Greenland. However, the subsequent investigations of Hamilton (1958) showed that even at 'Northice' (78°N, 38.5°W) there is frequent synoptic activity in winter, although the frontal structure is weak and significant snowfall occurs only on two or three days a month. Here the annual accumulation is only about 12 cm w.e. Maximum values of over 50 cm w.e. occur on the western slope at about 2500 m, decreasing eastward to about 25 cm at the highest elevations and less near the coast (Diamond, 1958).

The ice sheet climate is dominated in winter by a surface temperature inversion averaging 400 m in depth (Putnins, 1970). Based on data for 'Station Centrale' (71°N, 40.5°W; 2993 m) during the 1949–51 expedition of P. E. Victor, the mean inversion intensity for November–March was almost 10°C. In summer the inversion frequency is much reduced and the average intensity is only about 2.5°C. However, Putnins notes that the influence of the ice sheet on the vertical temperature distribution is evident, even to a certain degree in summer, up to the 500 mb level. Brockamp (1966) shows that free-air temperatures over Egedesminde and Kap Tobin in June average 8°C warmer than at 'Station Centrale'. This cold is not, of course, transferred by air moving off the Greenland ice sheet into the North Atlantic due to the effect of adiabatic warming.

The winter storms which lead to advection of maritime air into the interior are reflected in large interdiurnal temperature changes. The mean annual value (for 0001 GMT) was 6.3°C at 'Station Centrale' with a maximum average value of 8.4°C in November compared with respectively, 4.8°C and 7.7°C (in January) at 'Northice'. Miller

(1956) demonstrates that the heat conserved in the air and snow during storm periods supplies 60 per cent of the energy for long-wave radiation loss during interstorm conditions. In winter the low conductivity of the snow is its most significant property, but in summer its high albedo is also a critical factor. The albedo averages 85 per cent for fresh snow, declining to about 75 per cent during interstorm periods. Highest temperatures on the ice sheet above 2500 m are about $-15°$ to $-20°C$ in winter and $-3°C$ in summer. Mean maxima are $-30°$ to $-35°C$ in winter months and $-5°$ to $-10°C$ in summer. Mean minima are $-40°$ to $-45°C$ in winter with extremes below $-60°C$. These conditions are made more severe by the wind regime. The inversion conditions during interstorm periods give rise to katabatic flow from the higher parts of the ice sheet toward the coast, veered some $30-40°$ from the direction of maximum slope (Hamilton, 1958). The mean speed was 4 m s^{-1} during interstorm periods and 5 m s^{-1} during storm periods at 'Eismitte' (71°N, 40.5°W; 3000 m) in 1930–1, giving windchill values of 2200 kcal m^{-2} hr^{-1}, and 2000 kcal m^{-2} hr^{-1}, respectively (Miller, 1956).

The climates of the fiords and ice-free coastal areas exhibit significant differences from the interior. In winter stagnant pools of cold air collect in sheltered localities and are only disrupted by strong cyclonic situations combined with gravity flow from the ice sheet. The frequency of calms is generally 20–30 per cent rising to 50 per cent at sheltered stations, although föhn gales also affect the heads of many fiords.

The exposed coast of southern Greenland, especially in the southwest, has a winter climate akin to that of Iceland. The January mean temperature at Ivigtut, for example, is $-5°C$. Although reputedly stormy, this area is no more so than the coasts of northwest Europe. In spring and summer the southwest coast of Greenland is affected by the 'storis' drift from the East Greenland current. This ice reaches 65°N along the west coast in June–July. The cold water and ice keep mean temperatures in the summer months to between about 7°C and 10°C.

The most distinctive area is the polar desert of northern Greenland. This is a region of high continentality (Fristrup, 1961) with a monthly mean temperature (1952–6) of $-32.5°C$ in March and 4.2°C in July at Nord (81.5°N, 16.5°W) although interdiurnal variability reflects the coastal situation. The mean values for 0001 GMT are 3–4°C for winter months at Brønlunds Fiord (82°N, 30.5°W) (Putnins, 1970). There are about 70 frost-free days at Brønlunds Fiord, which is remarkable for its latitude, although Barry and Jackson (1969) reported an average of 65 frost-free days at Tanquary Fiord, Ellesmere Island. Annual precipitation is less than 20 cm, with a maximum in November–December. Large areas of Peary Land are blown free of snow even in winter.

LAND AREAS IN THE NORTH ATLANTIC – BARENTS SEA SECTOR

The most striking feature of this sector is its extreme (relative) winter warmth related to the interaction of the predominantly southwest to westerly airflow and branches of the North Atlantic Drift. The air temperature anomaly with respect to the latitudinal mean is $+25°C$ in January over the eastern Norwegian Sea at 65–70°N. However, this influence decreases sharply beyond the North Cape of Norway. The Barents Sea is shallow and its waters cold so that the northern and eastern portions become ice-covered in late winter and spring. Ice conditions in the Norwegian Sea in spring and summer are highly variable from year to year as a result of variable synoptic patterns and their associated heat budget regimes. Details of such influences can be found in the work of Gagnon (1964) and the broader survey of Vowinckel and Orvig (1970). Weather and climatic conditions in the sea areas are detailed in Meteorological Office (1964). Here we are concerned primarily with the islands and adjacent landmasses.

The area is stormy at all times of year (see fig. 2A.7), but particularly in the cold season when the depressions are intense and the airmass contrasts more marked. In winter about half of the depressions move northeastward via the Denmark Strait, whereas in summer their track is more commonly eastward towards the Faeroe Islands and Scandinavia. The southeast Barents Sea is also affected by one or two depressions per month during April–October moving from the Baltic Sea.

The cyclonic activity gives heavy precipitation in the south of Iceland, 209 cm at Vik and probably two to four times as much on the southern slopes of the Myrdasjökull and Vatnajökull ice caps. At low levels much of the precipitation occurs as rain. Even in November–March only 40–50 per cent of the days with precipitation have snowfall. The amounts decrease northward with an average of only 47 cm annually at Akureyri (65°N).

The temperature regime in Iceland and Jan Mayen is of the 'pure oceanic' type. The mean monthly temperature on Jan Mayen (71°N, 8°W) ranges between 5.5°C and −5.2°C with absolute extremes of only 16°C and −18°C. At Akureyri on the north coast of Iceland, the January and July means are −1.6°C and 10.9°C respectively, while on the northern margin of this sector at Isfjord, Spitsbergen (78°N, 14°E), mean temperatures range from −11.9°C in March to 5.0°C in July. On Spitsbergen the proximity of the arctic pack and the attenuation of Atlantic influences are evident. The south and west coasts of Iceland are kept ice-free all year by the Irminger Current, but the northern coast may be affected by pack ice in late winter and spring. Indeed, the last few years have witnessed a persistence of this ice into summer not matched since the 1890s (Marshall, 1968) with severe consequences for the fisheries.[1]

The climate of the region is very much affected by the depression tracks. Lows passing south of Iceland give rise to mainly cold, dry weather in the northerly airflow, whereas those moving through the Denmark Strait are associated with mild rainy conditions. Blocking patterns over Scandinavia, which are most frequent in spring, exert a strong influence on weather in this sector (Rex, 1950). Mean temperature anomalies in winter months are +6° to +8°C over Iceland, Jan Mayen and Spitsbergen when a blocking anticyclone is situated over northwest Europe (fig. 2A.9) causing strong southerly flow components in the North Atlantic. The same pattern occurs in summer although the mean anomalies are then only +2°C over Iceland.

Temperatures in northern Scandinavia are similar to those in Jan Mayen in winter and Iceland in summer. The range in monthly means at Vardø (70°N, 31°E) is between −5.2°C in February and 9.7°C in August. There is an autumn maximum and late

[1] On occasion heavy ice has been present off the south coast.

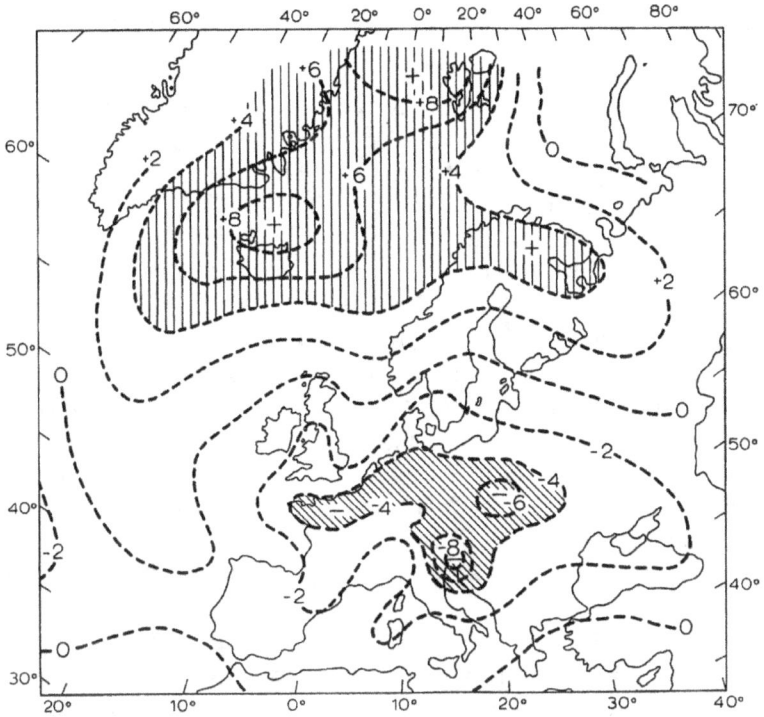

Fig. 2A.9. Mean air temperature anomaly (°C) during anticyclonic blocking over Scandinavia in winter (after Rex, 1950).

winter–spring minimum of precipitation as in much of the Canadian Arctic. The annual total averages only 43 cm in contrast to 74 cm at Jan Mayen, which is closer to the main depression tracks and is in a less sheltered location than Vardø. At Murmansk, 150 km to the southeast, there is a slight increase in continentality – mean temperature ranging between −11.4°C in February and 13.4°C in July. Annual precipitation is 39 cm, but here there is already a winter–spring minimum and late summer maximum. In spite of the low total, precipitation falls on 190 days in the year (similar to southern Iceland) and, except in spring, skies are cloudy. In autumn there is a 90 per cent or more probability of overcast and in summer advection fog is common.

For our purposes, the areas of the northern USSR eastward to the Urals can be considered transitional between this sector and northern Siberia, though some features are distinctive. Winter thaws are much less frequent than in the Kola Peninsula. The February mean at Naryan-Mar (68°N, 53°E) is −17.2°C. Summer is rather shorter and cooler than to the west with a tendency for cold spells to occur, but there is still a three-month frost-free period (Dolgin, 1970). The precipitation regime here is similar to that at Murmansk, with almost identical amounts, although totals increase southward. Of special note is the snow cover which has a mean depth of 70 cm or more in the northern pre-Urals.

NORTHWEST SIBERIA AND THE ARCTIC ISLANDS (NOVAYA ZEMLYA –
TAIMYR PENINSULA)

This sector obviously extends over a variety of climate zones from Severnaya Zemlya in the Arctic Ocean to the continental Subarctic of the lower Ob and Yenisei rivers. Alisov (in Borisov, 1965) shows the Taimyr Peninsula as 'Atlantic Arctic' whereas the montane area of Krasnoyarsk is designated 'Siberian Subarctic'. Atlantic influences diminish sharply east of Novaya Zemlya. Mean winter isotherms run more or less north–south through the islands from 78°N to 60°N with a mean isotherm of -15°C in this location in February. Sea ice is permanent in the northern part of the Kara Sea, although the southern part is open from July into October (see fig. 2A.1).

The continental character of the climate of interior northwest Siberia is indicated by annual airmass frequencies of 33 per cent for continental arctic air and 52 per cent for continental temperate air. In winter the general surface airflow is southerly or south-westerly around the Siberian anticyclone, whereas in summer it is predominantly westerly. Depressions from the general direction of the Black Sea affect the area especially in winter. In summer, when the arctic frontal zone lies east–west along the tundra/boreal forest boundary (Krebs and Barry, 1970), depressions on a North Atlantic–Barents Sea track are more important.

Precipitation ranges from 15–20 cm in the tundra[1] to about 25–50 cm in the forest-tundra and northern taiga, of which half falls in summer associated with depressions on the arctic front. The Urals are by no means a negligible barrier, especially with respect to moisture supply, in spite of the fact that they rise only to between 1200–1900 m. Borisov (1965) notes that the precipitation maximum occurs at rather low altitudes on their windward side. Only some 20 per cent of the total precipitation falls as snow. Nevertheless, snow depth averages 40–50 cm on the high arctic tundra, increasing to 80 cm in the forest-tundra, including the Taimyr Peninsula. In the northern taiga average depths east of the Yenisei River reach 125–150 cm in the month with greatest thickness. The snow cover persists from mid-October to the beginning of June (about 250 days) around the lower Ob, and for 280 days or more on Severnaya Zemlya.

The mean temperature of the winter months, December–March, is between -21° and -25°C at Dikson Island (73.5°N, 80°E), only 3° warmer than at Salekhard (66.5°N, 67°E). Stronger winds create more severe conditions at the coast, however. There is a 25–30 per cent frequency of winds exceeding 11 m s^{-1} along the northwest coast of the Taimyr Peninsula, compared with 10 per cent or less in the interior. The probable annual wind maximum at Dikson Island is 39 m s^{-1}. The effect of storms together with local bora effects on Novaya Zemlya produce even higher annual extremes of 50 m s^{-1} (Anapolskaia and Zavialova, 1970).

The spring snowmelt is slow and temperatures do not rise above freezing until June. Although there are about 160 days between attaining 0°C in spring and the autumn cooling to 0°C, the frost-free period is only 30–40 days in the tundra compared with

---

[1] Bochkov and Struzer (1970) report that precipitation averages have been adjusted upwards by 40–50 per cent along the arctic coast to fit moisture balance data.

60–90 days in the forest-tundra. The contrast with the region west of the Urals is pronounced (Dolgin, 1970). The soil thaws only 1.4–1.6 m and thick permafrost is widespread in this sector of the tundra.

NORTHEAST SIBERIA – THE TAIMYR PENINSULA TO THE KOLYMA RIVER

Northeast Siberia is characterized by one of the most extreme continental climates on earth. The greatest seasonal range of conditions is experienced in the subarctic taiga. Here the winter climate is dominated by the Siberian anticyclone which undoubtedly determines the climatic regime regardless of problems as to its real intensity (see p. 30). With the approach of winter this establishes itself first near Lake Baikal and then extends westward and northeastward (Dmitriev, 1970). Independent cells are frequently formed near 67°N, 100°E and 67°N, 140°E as a result of such ridging (Keegan, 1958). Even so, the main centre may weaken and be displaced southeastward for about 12 per cent of the time in November and December according to Dmitriev's study for 1948–65 (see fig. 2A.10d). This is rare in January–February, but in March the frequency of the pattern rises sharply to 22 per cent heralding the return of spring (table 2A.3).

Table 2A.3. Frequency of various states of the Siberian anticyclone, 1948–65 (from Dmitriev, 1970).

	Strong anticyclone and extension (%)	Meridional block (%)	Col across the Arctic coast (%)	Major break (%)	Total cases
November	37	30.5	21.5	11	176
December	25.5	30	31	13	188
January	27	33	33	7	190
February	30	24	38	7	162
March	21	22.5	34.5	22	177
Total	28	28	32	12	893

The winter radiation inversion has a frequency of at least 90 per cent over most of the area (fig. 2A.5). Its mean depth is 1500–1600 m in the subarctic zone where the average intensity is 15°C. The extreme cold at the surface (see fig. 2A.11) is intensified in the valleys and basins of Yakutia – a January mean temperature (1943–60) of −47.2°C is reported at Oimyakon (64°N, 143°E) and a similar severity persists during November–March. The mean annual absolute minimum temperature, which is −65°C at Oimyakon, is generally 15–17°C lower than on hill summits in central and eastern Yakutia where the basin–summit elevation difference exceeds 300 m (Gol'tsberg, 1970). The inversion is rarely disturbed, although when it is the temperature change is dramatic. Borisov (1965, p. 156) cites twenty-one years of January records at Yakutsk showing this:

Wind speed (m s^{-1})	0–1	6–7	10
Temperatures (°C)	−43.1	−36.3	−17.6

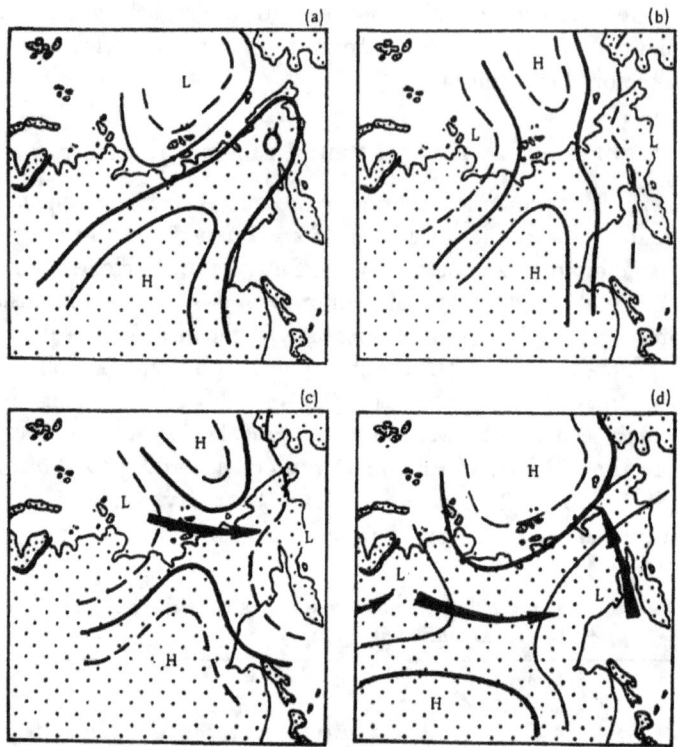

Fig. 2A.10. Synoptic patterns of the Siberian winter anticyclone, 1948–65 (from Dmitriev, 1970). (a) Strong anticyclone and extension. (b) Meridional block. (c) Col across the arctic coast. (d) Major block.

In spite of the extreme dryness associated with the anticyclone and the absence of moisture sources, fog occurs at Yakutsk on about one-third of days in December and January. Apparently the inversion persistence, together with the combustion products and moisture given off by the city, combine to create ideal conditions for ice fog occurrence at below −40°C temperatures. This phenomenon has been intensively studied at Fairbanks, Alaska.

Along the arctic coast and over the islands the synoptic conditions are more varied. Meridionally elongated high cells and a high over the East Siberian Sea (figs. 2A.10b,d) account for about 40 per cent of the circulation regimes in winter (see table 2A.3) but the other two patterns give strong west to southwesterly flow along the coast with travelling disturbances (figs. 2A.10a,c). Little precipitation derives from these systems, but the wind makes the temperatures less bearable even though they are higher than in the interior. The February mean is −29.9°C at Ostrov Kotelnii (75°N), for example. The mean wind velocity is much less than to the west of Cape Chelyuskin but averages 5–6 m s^{-1} in this coastal region. Mean windchill in the coldest month of the year exceeds 1800 kgcal m^{-2} hr^{-1} in the coastal tundra from 85°E to 175°E (Falkowski and Hastings, 1958).

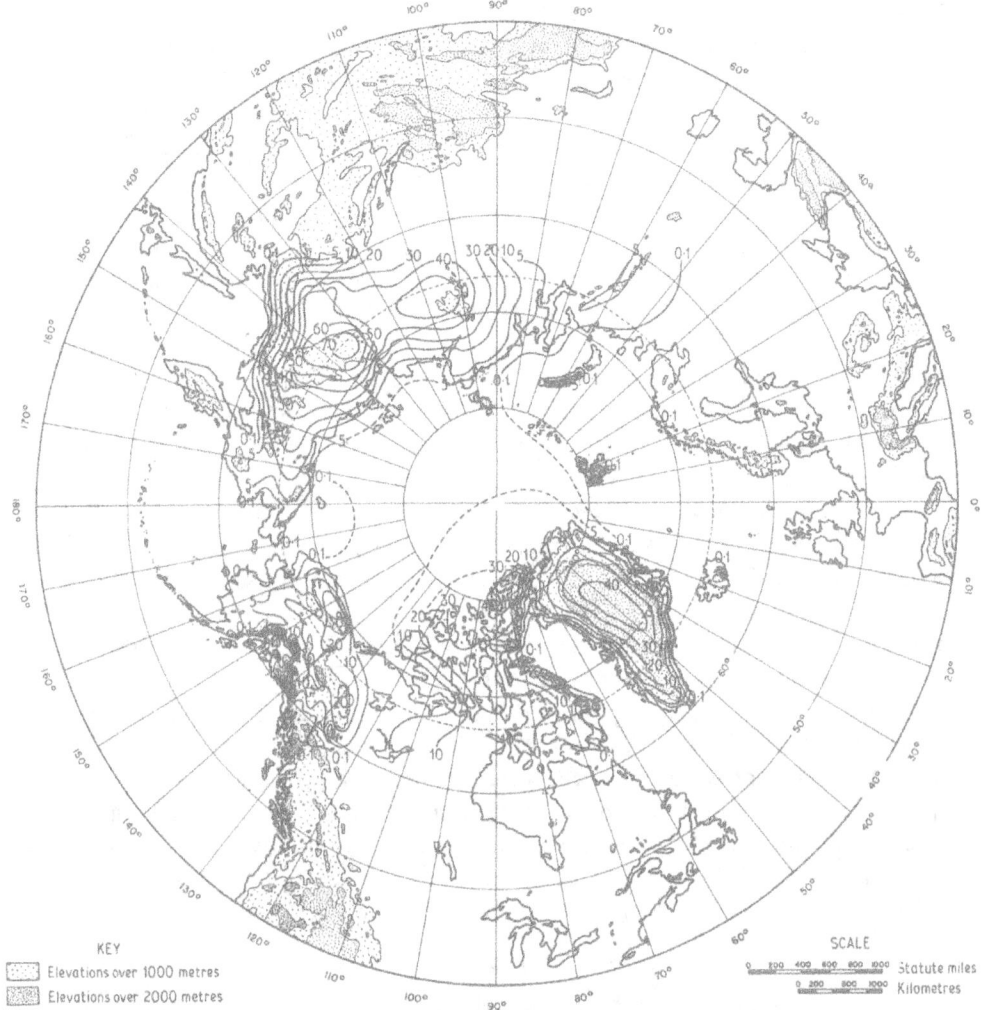

Fig. 2A.11. The percentage frequency of January temperatures below −40°C
(from Hastings, 1961).

Winter precipitation is light throughout this whole area and especially east of the
Lena. Khatanga (72°N, 102°E) receives under 10 cm in the period November–April,
about one-third of the annual average total, while Verkhoyansk gets only 4 cm – one-
quarter of the annual precipitation.[1] The snow cover is correspondingly thin, decreasing
from 50–60 cm in the Subarctic of the central Siberian uplands to 30 cm or less near the
coast and east of the Lena River.

---

[1] Adjustments of +10 to +20 per cent to measured precipitation are necessary in Yakutia (Bochkov and
Struzer, 1970). Cf. pp. 45 and 48.

A short transition in late April–May ushers in summer, except in the coastal area and the islands. The Siberian anticyclone weakens from the east where the heating occurs earliest. The snow cover disappears in early May at the latitude of Yakutsk and before the end of the month at 70°N (contrast Canada!) except in the mountain ranges east of the Lena. In summer cyclones cross the area from the west and southwest. The arctic front in July has a median position in the centre of the forest-tundra ecotone zone (Krebs and Barry, 1970). Maxima of cyclone occurrence are located about 60–66°N, 130°E and 63–70°N, 196°E (Reed and Kunkel, 1960). Maxima of anticyclone frequency in July–September are over the Taimyr Peninsula and East Siberian Sea. Cells move southward from both areas, but especially the latter, causing cold outbreaks of arctic air. In this area, as in western Siberia, the main contrast in summer conditions is latitudinal. On the New Siberian Islands and in the immediate coastal tundra, temperatures are most of the time in the range 0–4°C and relative humidity 95–100 per cent in July and August (Zavialova, 1970). The mean daily maximum temperatures are only 8°C in July and August at Sagastyr (75°N, 126°E). The coastal strip in July is on average some 2–4°C colder by day, 1–2°C by 'night', than locations 6–10 km inland (Gol'tsberg, 1970). At Khatanga mean daily maxima in June, July and August are 8°C, 15°C and 8°C respectively, while at Yakutsk the July mean is 19°C and the mean daily maximum 23°C.

Summer precipitation over the New Siberian Islands is only 4 cm in July–August. Over the mainland, however, larger amounts fall in both the arctic tundra and the Subarctic. Khatanga receives 15 cm in the four months (June–September) with mean temperatures above 0°C, and Yakutsk 13 cm in the same period. Dry spells associated with the movement of anticyclone cells from the north (or less commonly the south) are frequent. These conditions with low relative humidity and relatively high temperatures are frequently referred to as *sukhovei*, implying a dry wind, although the wind is usually light and the cause of the dryness is subsidence on the margin of the high pressure cell (Lydloph, 1964).

NORTHEAST SIBERIA, ALASKA AND THE YUKON

This highly accidented part of the northland takes the form of a giant isthmus linking the Eurasian and North American continents, broken only by the narrow Bering Strait. On both Asian and American sides of this exiguous divide the land is rugged or mountainous, with deep intervening valleys – the Anadyr, Yukon, Kobuk and Kuskokwim. Isolated coastal lowlands confront the Arctic Ocean, especially in Alaska, but the Pacific shore is everywhere hilly or mountainous. The overall relief is sufficient to make the area an effective divide between arctic and Pacific maritime climates. Within the area the inner valley systems have a continental regime less extreme than that of inner Siberia, but the Pacific coast climates are far milder, especially in Alaska. The arctic lowlands, however, have a full arctic climate (Conover, 1960).

Winter is a season of great contrasts. The dominating circulation regimes (Putnins, 1966–9) between November and March are those where deep low pressure lies over the Bering Sea, the Gulf of Alaska and the Sea of Okhotsk. Intense Pacific cyclones invade these seas from the south every few days, occasionally penetrating interior regions, but rarely pushing as far north as the arctic coasts. Pressure remains high over the continental interiors, with relatively high pressure across Alaska and northeast Siberia. Hence central and northern regions tend to be covered by easterly arctic airstreams. Occasionally these reach the Pacific coasts, especially over or west of the Bering Strait, but for much of the season the Pacific coast and the mountainous regions behind are blanketed by moist, cool, cloudy maritime airstreams.

Along the arctic coast, and over the coastal plains, the winter regime is dominated by rarely broken intense cold, with air temperatures below $-30°C$ and often moderate winds, usually from the east. Cloudiness is low, and snow light and infrequent, but blowing snow is frequent. High windchill makes overland movement uncomfortable. Interruptions of the regime are rarely long continued, but occasional breakdowns of the normal circulation patterns bring much warmer Pacific air to the north Alaskan coast in some winters – for a brief respite.

The interior valley systems – Anadyr, Yukon, Kobuk and Kuskokwim, with tributaries – have a remarkable winter regime. The relief favours the trapping of air below the summit level of the surrounding hills, and the valleys become filled with trapped, still air that often almost ignores the overlying inversion and the synoptically controlled circulation. Fairbanks (65°N, 148°W), for example, has 55 per cent of all observations with calm in January, and mean wind speeds are below 2 m s^{-1} nearly everywhere. Very low temperatures occur in the trapped pools. Mean January temperatures are below $-20°C$ in almost all areas, Fairbanks and Markovo (64.5°N, 170.5°E) having $-24°C$. Great cold develops, however, chiefly when the Siberian high expands across the region and the inversion intensifies. Temperatures may fall below $-40°C$ for one or two weeks.

Ice fog is a major hazard under such conditions (Benson, 1969). Combustion products from furnaces (domestic and industrial) and car exhausts overload the frigid air with water vapour, and below $-35°C$ freezing nuclei are abundant. Ice fog becomes dense and persistent, accumulating in addition the various pollutants emitted by the towns. The Fairbanks region, and to a lesser extent Whitehorse, YT, have an additional problem due to aircraft exhausts, now that the volume of winter air traffic is high.

Only light snow occurs in the inland winter, falling from overrunning Pacific airstreams during cyclonic passages. The snow cover, however, is usually soft, fairly deep and lacks wind crusts (Benson, 1967).

The Pacific coastal climates are much stormier, and snowfall is frequent. Excessive falls occur on the southeast- and south-facing slopes of the Alaskan Range, Kodiak Island, the Kenai Peninsula and the Alaskan Peninsula. Much lower falls occur on the Bering Sea coasts, and in the partly protected southern basins, such as Cook Inlet and Penzhinskaya Guba. Anchorage, for example, has a total winter snowfall of only 152 cm, equivalent to about 15 cm of rain. The exposed coasts are mild for their latitude, temperatures rising from about $-15°C$ on the Kamchatkan shore to near 0°C in the Alaskan panhandle. Again the deep inlets are less fortunate. Anchorage has a January mean of $-10.9°C$.

Between April and June the controlling circulation gradually changes, and warmer air begins to penetrate more deeply. The chief feature of spring throughout the area, however, is the brilliant sunshine, cloudiness being at its annual minimum. Fairbanks, for example, has over 300 hours of sun in each of April, May and June. Whitehorse and Anchorage do almost as well. Spring air temperatures climb rapidly, and are far above those of corresponding longitudes east of the Mackenzie. In April and May the differential along a latitude between inland Alaska and the eastern Canadian Arctic is of the order of 15°C. Snowmelt is rapid, interior Yukon and Alaska clearing in mid-May, the Arctic coast plain and the Anadyr basin two to four weeks later.

In the summer and autumn months (July–September) the area is traversed by many cyclones, the main arctic frontal belt lying across it in these months (Krebs and Barry, 1970; Reed and Kunkel, 1960). Cloudiness over the entire area rises rapidly after mid-June, and precipitation from travelling cyclones becomes progressively more frequent, reaching its maximum in August. This rainfall is mostly of the moderate, warm-frontal type, but heavy thunderstorms occur inland on about 10 days in July and August. Air temperatures at the peak of the summer are near 15°C in interior Alaska, Yukon and the upper Anadyr Valley, and there is typically a frost-free period of 70–110 days. Along the Pacific coast of Alaska it is a little cooler, and the tundra coasts ringing the cloudy, cold surface of the Bering Sea are much cooler. Along the arctic coast mean daily temperatures fail to attain 10°C in any month.

MACKENZIE–KEEWATIN

The land east of the Mackenzie River contains within it Great Bear Lake and Great Slave Lake as well as numerous smaller water bodies. Most weather stations are located at the coast or beside lakes and care is needed in assessing the representativeness of their records. The coastal lowlands give way to uplands, which exceed 500 m in places.

The tundra 'Barren Grounds' (delimited in fig. 7A.2) have long severe winters with mainly clear skies, when the circulation is dominated by the Mackenzie high pressure cell, and short cool summers when arctic air continues to dominate the region. The mean daily temperature is below 0°C from late September to late May, and the frost-free period is only about 60 days (although at Yellowknife it is 112 days). Mean daily minima in midwinter are below −30°C (−36.4°C in January at Baker Lake, 64°N, 96°W, and −35.4°C in February at Coppermine on the arctic coast) and the percentage frequency of minima below −45°C at Baker Lake exceeds that at Alert (82.5°N)! (See table 2A.4.) The predominant north to northwest winds make conditions more severe. They result in a mean windchill of 1900 kg cal m^{-2} hr^{-1} in January in Keewatın and blowing snow conditions on about 90 days in the year. Temperatures are almost as severe in the interior of Mackenzie District. The brief spring is greatly delayed by the melting of snow and ice and then by the cold seas and lakes. In July and August daily means exceed 10°C away from the arctic coast and daily maxima average about 15°C. Summer is much warmer however, in the interior. In the Mackenzie Valley, for example, there are on average at least 20 days with daily maxima exceeding 25°C (Kendrew and Currie, 1955). This same area also receives over 300 hours of sunshine in June.

Precipitation amounts are small; about 20 cm in the north, increasing to 30 cm in the south, although Hare and Hay (1971) demonstrate from moisture balance considerations that the measured totals are probably 20 per cent less than the true amount (cf. results for Siberia, pp. 38 and 41). Most of the precipitation falls during July–September in association with depressions on the arctic front. Snowfall, which reaches its maximum in October–November, contributes some 30–40 per cent of the annual total. In winter, the predominantly anticyclonic circulation and the low vapour content of the cold air preclude significant amounts of precipitation. When snowy spells do occur in Keewatin the 700 mb mean maps indicate anomalous southeasterly flow components (Brinkmann and Barry, 1972). The synoptic pattern shows a tropospheric trough over western North America and another off the east coast. Depressions can then cross Keewatin. The snow cover generally reaches a maximum depth of 50–65 cm although it is subject to much blowing and drifting. In northern Keewatin snow cover persists into the second half of June.

THE CANADIAN ARCTIC ARCHIPELAGO

The Canadian Arctic Archipelago extends over 15° of latitude but the climatic characteristics are relatively homogeneous. Apart from the eastern margin the relief is small and the whole area is interlaced with channels. The eastern margin from Ellesmere Island and Axel Heiberg Island, through Devon Island and eastern Baffin Island, is mountainous with summits and ice caps rising to 2500 m in Ellesmere Island and generally 1800 m further south. Devon and Baffin islands are treated in the next section, however, since their climates are considerably different from the area to the west and north.

From December–April virtually all of the sea is frozen with the exception of Lancaster Sound, the North Water of northern Baffin Bay and, in some years, a few small coastal leads. Maximum extent of sea ice is usually reached in March and thicknesses are generally of the order of 1.7–2 m except where rafting and pressure ridging has occurred. Mean daily temperatures reach their minimum in February or March – (Resolute −33.7°C in February, Eureka −37.6°C in March)[1] – and although these values are similar to averages in Keewatin, the mean daily minima, which are below −40°C in Ellesmere Island, are lower. The area is in fact marked by persistent, rather than extreme, cold. Table 2A.4 compares frequencies of minima below −45°C at several Canadian Arctic stations. The difference between coastal Alert and 'inland' Eureka is striking, but observations at Lake Hazen (81°49′N, 71°18′W) during the IGY year show that minima are even lower at sheltered locations in the interior of the larger islands (Jackson, 1959). In December 1957 minima of −45°C or below were recorded on all but three days. Also, while conditions appear from table 2A.4 to be comparable at Eureka and Snag, the frequency of temperatures below −40°C is more than twice as great at the former station (see fig. 2A.11).

These temperature conditions reflect the light winds and low cloudiness of the area in winter. Typically, calm is reported for 30 per cent of all observations, and less than

[1] 1951–60.

Table 2A.4. Percentage frequency of minima below −45°C (−50°F) and extreme records (°C), 1950–60 (from Hagglund and Thompson, 1964).

	Alert 82.5°N, 62°W	Eureka 80°N, 85.5°W	Snag 62°N, 140°W	Baker Lake 64°N, 96°W
Nov.			2.0 (−49)	
Dec.	0.3 (−46)	10.6 (−49)	7.7 (−52)	0.3 (−46)
Jan.	1.9 (−48)	18.7 (−51)	17.7 (−61)	5.0 (−48)
Feb.	1.8 (−47)	21.2 (−52)	6.4 (−52)	3.9 (−50)
Mar.	2.3 (−48)	16.8 (−53)	2.6 (−51)	1.0 (−47)
Apr.	0.7 (−46)	0.3 (−46)		

10 per cent of winds are above 13 m s^{-1} (30 mph), in response to the infrequent occurrence of cyclonic activity. The year-to-year variability, as measured by the standard deviation of monthly mean temperatures (Kendall and Anderson, 1966) is only 4–6°C over the area from January–March. In December and April it is 6–7°C, reflecting a greater incidence of breakdown of the surface inversion due to changes in circulation intensity and cloud cover. Warm air advection is a less common cause of variability in this area.

Cloudiness averages 40 per cent or less during the winter months and much of this is middle and high cloud associated with upper cold lows. Obviously, precipitation is very light in view of the near absence of active cyclones and the very low vapour content. The mean mixing ratio at the surface is about 0.25 g kg^{-1} and the total precipitable water content is only 0.15 cm (Barry and Fogarasi, 1968). Snowfall is very light – between 2 and 5 cm in depth in each month. The average depth of maximum snow cover is less than 30 cm in the northwest of the Archipelago increasing to 50 cm in southeastern Ellesmere (Potter, 1965). Higher values occur generally around Baffin Bay due to greater cyclonic activity. An important weather feature is blowing snow which occurs on about 90 per cent of occasions with wind speeds over 13 m s^{-1} and on 50 per cent of occasions with a wind of 9–13 m s^{-1}. Between October and March there are, on average, 68 days with blowing snow (and visibility below 9.6 km) at Resolute (75°N, 85°W) (Thompson, 1965).

The returning sun, which is above the horizon ten hours per day by 1 March at 70°N, has only a gradual effect on temperatures due to the high surface albedo. 'Spring' (May–June) is a time of continuing cold, dry, sunny weather. The area is now dominated by a surface high pressure cell (see p. 30) and average wind speeds are consequently much reduced. The snow cover persists on average to 15–20 June in the southern area and after 30 June in northern Ellesmere (Potter, 1965). In some years it is not removed in the highest latitudes until the end of July. This gives a mean duration of 260 days in the south and over 300 days in the extreme north.

Summer (July–August) sees a considerable change. Frontal depressions move along the arctic coastline from Siberia–Alaska and either cross to northern Baffin Bay or continue along to northern Ellesmere Island. Moisture provided by the open water areas and melting snow helps to maintain local fog/stratus areas even in the absence of cyclonic convergence since little uplift is necessary to saturate the air. Fog (with visibility below 1 km) occurs on about half the days in July and August at Isachsen (79°N, 104°W) and Resolute. Temperatures, after rising above freezing, remain mainly around 5°C although in sheltered inland locations extremes of 15–18°C are recorded on a few days in most summers, especially where there are föhn effects (Barry and Jackson, 1969). The standard deviation of monthly mean temperatures in summer is only 2–3°C indicating low interannual variability. Precipitation falls mainly as rain except on the mountain areas; indeed, 25–33 per cent of the annual precipitation occurs in this form. Occasional daily falls of 25 mm or more have been recorded at most stations, but there is great year-to-year variability in monthly totals related to the frequency of depressions. However, while most precipitation is related to cyclone passages, orographic effects are important (Jackson, 1961) and there is great spatial variability even as to the occurrence or non-occurrence of days with measurable precipitation (Barry and Jackson, 1969).

The first snow cover may occur before 15 August in the Queen Elizabeth Islands, and the average date is 31 August in central Ellesmere. Nevertheless, it is September–November when stormy conditions with moderate snowfall occur. Average precipitation for this three-month period is 3–5 cm throughout the Archipelago. Temperatures fall steadily from the end of August and by November most of the sea is frozen over, cutting off the local moisture supply to the atmosphere.

THE EASTERN CANADIAN ARCTIC AND SUBARCTIC

There are steep climatic gradients within the eastern Arctic and Subarctic related to circulation characteristics as well as to topography and the effects of the various water surfaces. Nevertheless, the area is distinctive in terms of the degree of maritime influence which gives rise to relatively mild winter temperatures and high snowfall. It is in this sector that the arctic treeline is furthest south (see Larsen, Chapter 7A) and that the forest-tundra ecotone of the Labrador–Ungava lake plateau is most extensively developed. The former characteristic is in essence related to the deep, quasi-permanent trough in the tropospheric flow pattern (see fig. 2A.4) and thickness pattern over eastern North America, whereas the latter is, in part at least, due to the gradual increase in elevation southward to the Laurentide Scarp.

In winter depressions cross Labrador–Ungava and the Gulf of St Lawrence after moving either eastward from the Great Lakes and Prairies or northward from the Atlantic Coast. Many continue northeastward towards Iceland but others swing north-ward over the Labrador Sea towards Baffin Bay (Forsdyke, 1955). In exceptional cases they may move north to affect Ellesmere Island, as in January 1958 (Thomas and Titus, 1958). Barry (1960, 1967b, 1968) showed that such occurrences have a major impact on the temperature and moisture flux conditions in winter months over central Labrador–Ungava giving temperatures near or above freezing and significant snowfall (or freezing rain). The winter isotherms reflect the overall effect of these circulation characteristics. They run approximately north–south parallel to the coasts of Labrador and Baffin Island. The mean January temperature at Frobisher (−26.1°C) is little lower than at Nitchequon (−22.2°C) in central Labrador–Ungava.

In general, maximum snowfall occurs in October–December, and this is especially marked along the east shore of Hudson Bay where open water continues to provide a heat and moisture source until the end of December or early January (Hare and Montgomery, 1949; Burbidge, 1951) as shown in fig. 2A.12. Snowfall is also heavy in southeastern Baffin Island as a result of moist easterly airflow ahead of the frequent northward-moving depressions and augmentation due to topographic effects. This limited area receives 20–30 per cent of its annual precipitation in December–March. Cape Dyer (67°N, 62°W) records over 400 cm of snowfall (1960–4) between September and February. Snow accumulation on the Devon Island ice cap (Koerner, 1966) shows a southwest to north-east gradient demonstrating the effect of cyclonic activity in Baffin Bay and possibly also some intensification of the totals due to open water areas in winter in northern Baffin Bay – the North Water especially (Dunbar, 1970).

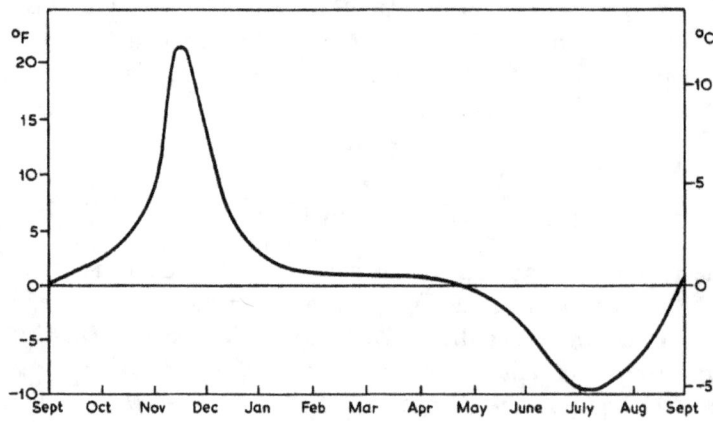

Fig. 2A.12. The average warming of arctic air crossing Hudson Bay (from Burbidge, 1949).

It is only since 1960 that shielded Nipher gauges have been officially used to measure snowfall. Prior to that a standard conversion of o.1 times measured snowdepth was used to determine precipitation. Undoubtedly the totals have been underestimated. Findlay (1969) shows that on the basis of water yield estimates in central Labrador–Ungava the 'true' average precipitation is probably 15 cm (21 per cent) greater than that recorded due to various inaccuracies in the measurements. A further problem is created by the great year-to-year variability. For example, at Port Harrison (Inoucdjouac) the annual snowfall has varied from 70 to 464 cm in seventeen years of measurement.

Spring in the eastern Arctic is not unlike that in the Archipelago, although fog and stratus are common earlier around the Davis Strait and Hudson Strait in response to the open water. Further south, where the season is more advanced, temperatures generally rise above freezing by 10 May in central Labrador–Ungava and the snow cover has largely gone by 10 June (except in northwest Ungava and northern Labrador).

In Labrador–Ungava summers are cloudy (averaging 70–75 per cent) though mean July temperature is about 11–13°C. Along the Labrador Coast, however, the low sea temperature and ice in the Labrador Current keep temperatures, particularly daily maxima, depressed. The effect of Hudson Bay is similarly pronounced along its eastern shore. In Baffin Bay ice lingers late in the summer, especially in the Home Bay area where it may not clear until late August or early September (see Dunbar, 1972).

The maximum precipitation falls throughout this area in the period June–September with generally 50–60 per cent of the annual total. About 9 cm is received in each month in central Labrador–Ungava with amounts diminishing northwestward. In Baffin Island, Frobisher receives a total of 21 cm, almost all as rain, in these four months while further north only 11 cm falls at Clyde (70°N). Generally, August totals are the highest as cyclonic activity is then at its maximum. Twenty-four hour totals may exceed 3 cm. On 19 August 1960, record falls occurred over Baffin Island as a cyclone with warm, moist maritime tropical air aloft moved northward into Foxe Basin. Falls were recorded of 2.4 cm at Frobisher Bay and 3.6 cm at Clyde (Thomas and Thompson, 1962).

5. Outstanding problems

The large-scale characteristics of the atmospheric circulation in the Arctic are now well established, but many regional and local problems remain to be solved, such as the role of the North Water in local weather and climate. The importance of the pack ice for world climate is well known but the question of its long-term stability is still debated, and observational data on trends in thickness are not reliable. Even such basic climatic parameters as precipitation amounts are uncertain.

As development of the Arctic progresses, there will be a major need for reliable information on local wind conditions and temperature inversions, especially, with respect to town sites and mining activities, while all arctic activities may have to take account of climatic changes, both naturally-occurring and man-influenced, and associated changes in snow cover and sea ice (see Larsen and Barry, Chapter 5, p. 267; Nichols, Chapter 11A). The historical and more remote past provide numerous illustrations of the amplitude of changes which can occur and the time scales over which they may persist.

ACKNOWLEDGEMENTS

R.G.B. wishes to acknowledge grants from Army Research Office, Durham, and the National Science Foundation (Office of Polar Programs) for support of INSTAAR climatological research in the eastern Canadian Arctic, 1969–74.

References

ANAPOLSKAIA, L. E. and ZAVIALOVA, I. N. (1970) Some wind characteristics in the Arctic. *In:* TRESHNIKOV, A. F. (ed.) *Problems of the Arctic and Antarctic*, 440–7. Jerusalem: Israel Progr. Sci. Transl. (Original Russian edn, Leningrad, Gidromet. Izdat., 1968–9.)

ANDREWS, J. T., BARRY, R. G. and DRAPIER, L. (1970) An inventory of the present and past glacierization of Home Bay and Okoa Bay, east Baffin Island, NWT, Canada, and some climatic and palaeoclimatic considerations. *J. Glaciol.*, **9** (57), 337–62.

BARRY, R. G. (1960) A note on the synoptic climatology of Labrador–Ungava. *Quart. J. R. Met. Soc.*, **86**, 557–65.

BARRY, R. G. (1967a) Seasonal location of the arctic front over North America. *Geogr. Bull.*, **9**, 79–95.

BARRY, R. G. (1967b) Variations in the content and flux of water vapour over north-eastern North America during two winter seasons. *Quart. J. R. Met. Soc.*, **93**, 535–43.

BARRY, R. G. (1968) Vapour flux divergence and moisture budget calculations for Labrador-Ungava. *Cah. Géogr. Quebec*, **12** (25), 91–102.

BARRY, R. G. and FOGARASI, S. (1968) Climatology studies of Baffin Island, Northwest Territories. *Inland Waters Branch Tech. Bull.*, 13. Ottawa. 106 pp.

BARRY, R. G. and JACKSON, C. I. (1969) Summer weather conditions at Tanquary Fiord, NWT, 1963–1967. *Arct. Alp. Res.*, **1**, 169–80.

BENSON, C. (1967) Polar regions snow cover. *Physics of Snow and Ice*, **1** (2), 1039–63. Inst. Low Temp. Sci., Univ. of Hokkaido.

BENSON, C. (1969) The role of air pollution in arctic planning and development. *Polar Rec.*, **14** (93), 783–90.

BILELLO, M. A. (1966) Survey of Arctic and Subarctic temperature inversions. *US Army Tech. Rep.*, 161. Hanover: US Army CRREL. 35 pp.

BOCHKOV, A. P. and STRUZER, L. R. (1970) Estimation of precipitation as water balance element. *In:* Symposium on World Water Balance, *Int. Ass. Sci. Hydrol. Pub.* 92, Vol. 1, 186–93. UNESCO.

BORISOV, A. A. (1965) *Climates of the USSR*. Edinburgh: Oliver & Boyd. 255 pp. (Original Russian edn, Moscow, 1959.)

BRINKMANN, W. A. R. and BARRY, R. G. (1972) Palaeoclimatological aspects of the synoptic climatology of Keewatin, Northwest Territories. *Palaeogeogr., Palaeoclim., Palaeoecol.*, **11**, 77–91.

BROCKAMP, B. (1966) Das Grönländische Inlandeis. *Erdkunde*, **20**, 208–11.

BRYSON, R. A. (1966) Air masses, streamlines and the boreal forest. *Geogr. Bull.*, **8**, 228–69.

BUDYKO, M. I. (1971) *Klimat i Zhizn*. Leningrad: Gidromet. Izdat. 472 pp.

BURBIDGE, F. E. (1949) *The Modification of Continental Polar Air over Hudson Bay and Eastern Canada*. Unpub. M.Sc. thesis, McGill Univ. 233 pp.

BURBIDGE, F. E. (1951) The modification of continental polar air over Hudson Bay. *Quart. J. R. Met. Soc.*, **77**, 365–74.

CONOVER, J. H. (1960) Macro- and microclimatology of the arctic slope of Alaska. *US Army Tech. Rep.*, EP-139. Natick, Mass.: HQ Quartermaster Res. Eng. Command. 65 pp.

CRUTCHER, H. L. and MESERVE, J. M. (1970) *Selected Level Heights, Temperatures and Dew Points for the Northern Hemisphere*. NAVAIR 50-1C-52 (rev.). Washington, DC: Naval Weather Service Command.

DIAMOND, M. (1958) Air temperature and precipitation on the Greenland ice cap. *SIPRE Res. Rep.*, 43. US Army Snow, Ice and Permafrost Res. Estab. 9 pp.

DMITRIEV, A. A. (1970) Situation characterizing various states of the Siberian anticyclone and its northeastern extension in the cold period of the year. *In:* TRESHNIKOV, A. F. (ed.) *Problems of the Arctic and Antarctic* 29–32, 448–55. Jerusalem: Israel Progr. Sci. Transl. (Original Russian edn, Leningrad, Gidromet. Izdat., 1968–9.)

DOLGIN, I. M. (1970) Subarctic meteorology. *In: Ecology of the Subarctic regions*, 41–61. Paris: UNESCO.

DUNBAR, M. (1970) The geographical position of the North Water. *Arctic*, **22**, 438–41.

FALKOWSKI, S. J. and HASTINGS, A. D. JR (1958) Windchill in the northern hemisphere. *US Army Tech. Rep.*, EP-82. Natick, Mass.: Env. Prot. Res. Div., HQ Quartermaster Res. Eng. Command. 9 pp.

FINDLAY, B. F. (1969) Precipitation in northern Quebec and Labrador: an evaluation of measurement techniques. *Arctic*, **22**, 140–50.

FLOHN, H. (1952) Zur Aerologie der Polargebiete (The aerology of polar regions). *Met. Rund.*, **5**, 81–7 and 121–8. (English transl., Appendix 1 to *Status Report*, 6, Contract AF 19(604)–1141, Arctic Met. Res. Group, McGill Univ., 1955.)

FORSDYKE, A. G. (1955) Depressions crossing Labrador and the St Lawrence Basin. *Prof. Notes*, **7**(113). London: Met Office. 44 pp.

FRISTRUP, B. (1961) Climatological studies of some high arctic stations in north Greenland. *Folia Geogr. Danica*, **9**, 67–78.

GAGNON, R. M. (1964) Types of winter energy budgets over the Norwegian Sea. *Pub. in Met.*, 64. Arctic Met. Res. Group, McGill Univ. 75 pp.

GOL'TSBERG, I. A. (1970) *Microclimate of the USSR*. Jerusalem: Israel Progr. Sci. Transl. 236 pp. (Original Russian edn, Leningrad, Gidromet. Izdat., 1967.)

HAGGLUND, M. G. and THOMPSON, H. A. (1964) A study of sub-zero Canadian temperatures. *Can. Met. Mem.*, 16. Toronto: Met. Branch. 77 pp.

HAMILTON, R. A. (1958) The meteorology of north Greenland during the midwinter period. *Quart. J. R. Met. Soc.*, **84**, 355–74.

HARE, F. K. (1968) The Arctic. *Quart. J. R. Met. Soc.*, **94**, 439–59.

HARE, F. K. and HAY, J. E. (1971) Anomalies in large-scale annual water balance over northern North America. *Can. Geogr.*, **15**, 79–94.

HARE, F. K. and MONTGOMERY, M. R. (1949) Ice, open water and winter climate in the eastern Arctic of North America. *Arctic*, **2**, 78–89, 149–64.

HASTINGS, A. D. JR (1961) Atlas of the Arctic Environment. *US Army Rep.*, RER-33. Natick, Mass.: Env. Prot. Res. Div., HQ Quartermaster Res. Eng. Command. 22 pp.

HAY, J. E. (1970) *Aspects of the Heat and Moisture Balance of Canada*. Unpub. Ph.D. thesis, Univ. of London. 212 pp.

HOBBS, W. M. (1945) The Greenland glacial anticyclone. *J. Met.*, **2**, 143–53.

JACKSON, C. I. J. (1959) Coastal and inland weather contrasts in the Canadian Arctic. *J. Geophys. Res.*, **64**, 1451–5.

JACKSON, C. I. J. (1961) Summer precipitation in the Queen Elizabeth Islands. *Folia Geogr. Danica*, **9**, 140–53.

KEEGAN, T. J. (1958) Arctic synoptic activity in winter. *J. Met.*, **15**, 513–21.

KENDALL, G. R. and ANDERSON, S. R. (1966) Standard deviation of monthly and annual mean temperatures. *Climatol. Stud.*, 4. Toronto: Met. Branch. 18 pp.

KENDREW, W. G. and CURRIE, B. W. (1955) *The Climate of central Canada*. Ottawa: Queen's Printer. 194 pp.

KOERNER, R. M. (1966) Accumulation on the Devon Island ice cap, Northwest Territories, Canada. *J. Glaciol.*, **6**(45), 383–92.

KREBS, J. S. and BARRY, R. G. (1970) The arctic front and the tundra–taiga boundary in Eurasia. *Geogr. Rev.*, **60**, 548–54.

LOEWE, F. (1936) The Greenland ice cap as seen by a meteorologist. *Quart. J. R. Met. Soc.*, **62**, 359–77.

LYDOLPH, P. E. (1964) The Russian sukhovey. *Ann. Ass. Amer. Geogr.*, **54**, 291–309.

MCKAY, G. A., FINDLAY, B. F. and THOMPSON, H. A. (1970) A climatic perspective of tundra areas. *In:* FULLER, W. A. and KEVAN, P. G. (eds.) *Productivity and Conservation in Northern Circumpolar Lands*, 10–33. IUCN Publ. (n.s.) no. 16. Morges, Switzerland: Int. Union Conserv. Nature.

MARSHALL, N. (1968) The icefields around Iceland in spring 1968. *Weather*, **23**, 368–76.

MARSHUNOVA, M. S. and CHERNIGOVSKIY, N. T. (1966) Numerical characteristics of the radiation regime in the Soviet Arctic. *In:* FLETCHER, J. O. (ed.) *Proceedings of the Symposium on the Arctic Heat Budget and Atmospheric Circulation*, 281–97. Mem. RM-5233-NSF. Santa Monica, Calif.: RAND Corp.

MATTHES, F. E. (1946) The glacial anticyclone theory examined in the light of recent meteorological data from Greenland, Part I. *Trans. Amer. Geophys. Union*, **27**, 329–41.

MATTHES, F. E. and BELMONT, A. D. (1950) The glacial anticyclone theory examined in the light of recent meteorological data from Greenland, Part II. *Trans. Amer. Geophys. Union*, **31**, 174–82.

METEOROLOGICAL BRANCH (1966) *Temperature and Precipitation Data from DEW Line Stations*. CDS 5/66. Toronto: Met. Branch.

METEOROLOGICAL OFFICE (1964) *Weather in Home Fleet Waters, Vol. 1: Northern Seas*, Part 1. MO 732a. London: HMSO. 265 pp.

MILLER, D. H. (1956) The influence of snow cover on local climate in Greenland. *J. Met.*, **13**, 112–20.

ORVIG, S. (1970) *Climates of the Polar Regions*. Vol. 14 of LANDSBERG, H. E. (ed.) *World Survey of Climatology*. Amsterdam: Elsevier. 370 pp.

POTTER, J. G. (1965) Snow cover. *Climatol. Stud.*, 3. Toronto: Met. Branch. 69 pp.

PRIK, Z. M. (1959) Mean position of surface pressure and temperature distribution in the Arctic. *Trudy Arkt. Nauch.-Issled. Inst.*, **217**, 5–34 (in Russian).

PUTNINS, P. (1966) The sequences of baric weather patterns over Alaska. *Studies on the Meteorology of Alaska*, 1st Interim Rep. Environmental Data Service, ESSA. 81 pp.

PUTNINS, P. (1967) Extremely cold weather spells in Alaska. *Studies on the Meteorology of Alaska*, 2nd Interim Rep. Environmental Data Service, ESSA. 53 pp.

PUTNINS, P. (1968) Some aspects of the atmospheric circulation over the Alaskan area. *Studies on the Meteorology of Alaska*, 3rd Interim Rep. Environmental Data Service, ESSA. 57 pp.

PUTNINS, P. (1969) Weather situations in Alaska during the occurrences of specific baric weather patterns. *Studies on the Meteorology of Alaska*, Final Rep. Environmental Data Service, ESSA. 267 pp.

PUTNINS, P. (1970) The climate of Greenland. *In:* ORVIG, S. (ed.) *Climates of the Polar Regions*, 3–128. Vol. 14 of LANDSBERG, H. E. (ed.) *World Survey of Climatology*. Amsterdam: Elsevier.

PUTNINS, P. and STEPANOVA, N. A. (1956) The climate of the Eurasian northlands. *In:* *The Dynamic North*, I, Part IV. OP-03A3. US Navy. 104 pp.

REED, R. J. and KUNKEL, B. A. (1960) The Arctic circulation in summer. *J. Met.*, **17**, 489–506.

REX, D. F. (1950) The effect of Atlantic blocking action upon European climate. *Tellus*, **2**, 196–211, 275–301.

THOMAS, M. K. and THOMPSON, H. A. (1962) Heavy rainfall in the Canadian Arctic during August 1960. *Weatherwise*, **15**, 153–8.

THOMAS, M. K. and TITUS, R. L. (1958) Abnormally mild temperatures in the Canadian Arctic during January 1958. *Mon. Wea. Rev.*, **86**, 19–22.

THOMPSON, H. A. (1965) *The climate of the Canadian Arctic*. Toronto: Met. Branch. 32 pp.

VAN MIEGHEM, J. (1961) Zonal harmonic analysis of the northern hemisphere geostrophic wind field. *Int. Union. Geod. Geophys. Monogr.*, 8. Paris. 57 pp.

VOWINCKEL, E. (1965) The inversion over the Polar Ocean. *Pub. in Met.*, 72. Arctic Met. Res. Group, McGill Univ. 30 pp.

VOWINCKEL, E. and ORVIG, S. (1966) The heat budget over the Arctic Ocean. *Arch. Met. Geophys. Biokl.*, B, 14, 303–25.

VOWINCKEL, E. and ORVIG, S. (1967) The inversion over the Polar Ocean. *WMO Tech. Note*, 87, 39–59. Geneva.

VOWINCKEL, E. and ORVIG, S. (1970) The climate of the north polar basin. *In:* ORVIG, S. (ed.) *Climates of the Polar Regions*, 129–252. Vol. 14 of LANDSBERG, H. E. (ed.) *World Survey of Climatology*. Amsterdam: Elsevier.

WALDEN, H. (1959) Statistisch – synoptische Untersuchung über das Verhalten von Tiefdruckgebieten im Bereich von Grönland. *Dt. Wetterd. Seewetteramt*, 20. 69 pp. Hamburg.

WALKER, J. M. (1967) Subterranean isobars. *Weather*, **22**, 296–7.

WENDLER, G. (1971) An estimate of the heat balance of a valley and hill station in central Alaska. *J. Appl. Met.*, **10**, 684–93.

WILSON, C. V. (1967) Climatology. Introduction, Northern Hemisphere I. *Cold Regions Science and Engineering Monogr.*, I–A3. Hanover: US Army CRREL. 141 pp.

WILSON, C. V. (1969) Climatology of the Cold Regions, Northern Hemisphere II. *Cold Regions Science and Engineering Monogr.*, I–A3b. Hanover: US Army CRREL. 158 pp.

ZAVIALOVA, I. N. (1970) The temperature–humidity regime of the Arctic in the summer period. *In:* TRESHNIKOV, A. F. (ed.) *Problems of the Arctic and the Antarctic* 29–32, 186–95. Jerusalem: Israel Progr. Sci. Transl. (Original Russian edn, Leningrad, Gidromet. Izdat., 1968–9.)

Manuscript received November 1971.

Additional references

BARRY, R. G., BRADLEY, R. S. and JACOBS, J. D. (1974) Synoptic climatological studies in the Baffin Island area. *Proc. 24th Alaskan Sci. Conf. (Climate of the Arctic)*. Fairbanks: Univ. of Alaska Press (in press).

BRADLEY, R. S. (1973) Seasonal climatic fluctuations on Baffin Island during the period of instrumental records. *Arctic.* **26**, 230–43.

BURNS, B. M. (1972) *The climate of the Mackenzie Valley – Beaufort Sea, Vol. 1, Climatol. Stud.*, 24. Toronto: Dept of Environment. 227 pp.

FOGARASI, S. (1972) Weather systems and precipitation characteristics over the Arctic Archipelago in the summer of 1968. *Water Resources Branch, Sci. Ser.*, 16. Ottawa: Dept of Environment. 116 pp.

OUTCALT, S. I. (1973) The simulation of diurnal surface thermal contrast in sea ice and tundra terrain. *Arch. Met. Geophys. Biokl.*, B, **21**, 147–56.

WELLER, G. *et al.* (1972) The tundra microclimate during snow-melt at Barrow, Alaska. *Arctic*, **25**, 291–300.

Contribution to a comparative meteorology of mountain areas

Hermann Flohn

Meteorological Institute, University of Bonn

1. Introduction

Meteorology is growing at a faster pace than ever before in its history. This is evident in every meteorological magazine and at every international conference. New methods enable us to measure both point values and two-dimensional patterns by microwaves in radar meteorology as well as by cloud photographs and radiation data from satellites. Although the efficiency of computers is ten times greater than it was a few years ago, meteorology still places the highest demands on calculation speed, capacity, and accuracy. Since, with the help of the computer, prognoses have been put on a physical–mathematical basis, they have been improving, according to objective criteria, in range and accuracy.

Because of these developments, however, most emphasis has been placed on large-scale processes in the atmosphere. In the cloud photographs from the geo-stationary ATS-satellites, the astronauts' marvellous slides, and the automatically evaluated pictures from the sun-synchronous meteorological satellites in the visible as well as in the infra-red part of the spectrum, we see the atmosphere as a whole. We see the interaction between tropical and extratropical processes and the similar features of tropical and extratropical cyclones, in spite of basic differences in structure.

Mountain meteorology has been somewhat overlooked in the rush of these new developments. Investigations on the local scale (areas of 1–10 km diameter) and in the mesoscale (10–100 km diameter) have been overshadowed by spectacular research, whose fundamental importance cannot be questioned, on large-scale problems as well as microscale turbulence studies and by work in precipitation and aerosol physics. The importance of the *scale problem* (Flohn, 1954) has emerged with this recent boom in theoretical work on large-scale patterns and processes. What role do micro- and mesoscale processes play in the hierarchy of motion systems? Is there an energy cascade that includes all scales from the laminar, millimetre-thick boundary layer up to the earth's atmosphere? (Proportional to the earth's radius the atmosphere is also very thin; 99 per cent is concentrated in a layer 30 km thick and 40,000 km long.)

Mountain areas offer us one of the best opportunities to study the interaction between the different stages in this hierarchy of processes. Here we can measure the processes of the heat balance down to the lowest 1–10 cm of the surface layer. Horizontal differences in this layer produce the shallow (few metres thick) slope breezes which can cover many hundreds of metres. These winds acting together make up the valley and mountain breezes which have been studied for more than one hundred years. In turn, the sum of the valley and mountain breezes produces weak, but for weather very effective, thermally driven, diurnal, large-scale circulations covering large mountain ranges (see p. 66). These winds were first quantitatively studied by Burger and Ekhart (1937) in the Alps, where they were found to have a horizontal length of approximately 500–1000 km. All these processes occur, in forms varied by the earth's topography and radiative and thermal properties of the snow cover, in all mountain areas.

2. Vertical distribution of precipitation

In contrast to many erroneous theories of the nineteenth century, properly installed and reliable stations in the Alps show an increase in the amount of precipitation with height up to 3000–3500 m, above which few representative stations exist. This is actually a surprising result seemingly inconsistent with the fact that the absolute water-vapour content of the atmosphere at 3000 m is only about one-third of that at sea level. Similar, but not so reliable, evidence from other mountain areas in temperate and subtropical zones yields the same results as in the Alps. In the Hindu Kush (Flohn, 1969a), the Pamirs, and the northwestern Himalaya (Flohn, 1970a), ridge stations at 3200–4200 m show a real increase in precipitation with height.

In the tropics, however, especially in the equatorial region within 10° latitude of the equator, the pattern is different. In these areas, according to Weischet (1969), there is a zone of maximum precipitation at 1000–1400 m perhaps with the exception of continental mountains. This zone has been found in Java, Sumatra, Ceylon, west Africa, central America, and the tropical Andes, and is especially striking on Kilimanjaro (fig. 2B.1). Weischet differentiates a tropical convective type of rain from an extratropical advective type. Under conditions of purely convective vertical mass exchange, the precipitation maximum should lie at the level where the drops from the clouds begin to evaporate, i.e. the cloud base, which near the equator lies at 400–700 m at coasts and 600–1000 m inland. Because in orographic rain, updraughts of 1–3 m s^{-1} reduce the fall speed of the drops – in cumulonimbus (Cb) cells updraughts of $\geqslant$10 m s^{-1} can even transport large drops upwards – it is understandable that the zone of maximum rain lies a few hundred metres above the cloud base. In the region of the trade wind inversion, at Kilimanjaro for example, areas above 2000 m are arid. Even in the area of the Intertropical Convergence (ITC) with its cumulonimbus clusters, the amount of precipitation in areas above 3000 m decreases to about 100 cm yr^{-1}, which amounts to 10–30 per cent of the amount in the maximum zone (Mt Kenya and Mt Cameroon). Why does the amount of precipitation in mid-latitude mountains increase with height? This is primarily because the effects of the orographically caused lifting get stronger as the wind velocity (**V**) increases with height. The amount of precipitation along a mountain profile

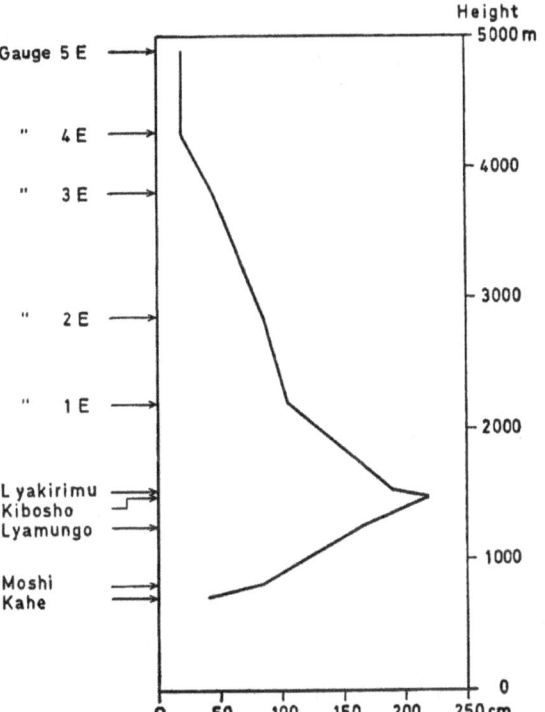

Fig. 2B.1. Average yearly amount of rain, 1959–67, on the southeast slope of Kilimanjaro (3°S).

is dependent on the total vapour amount which condenses per unit time during transport. It is then proportional to the vertical velocity component, w, and to

$$\iint_{p\,t} \mathbf{V} q \, dp \, dt$$

where q is specific humidity, p is pressure, t is time. Havlik (1969) has tested this hypothesis and found that in the Alps, at least during extended storms, the water vapour transport increases up to 700 m. Thus the decrease in specific humidity, q, is more than compensated by the increase in wind with height. In Japan this is the case even in the mean. In contrast, the winds in the tropics decrease with height above the Ekman layer (about 1000 m). Thus the decrease in water content is more effective. Sarker (1966, 1967) has based a dynamic model of the orographic precipitation on the escarpment of the Western Ghats, India, on similar hypotheses. An application to mountains in different climates has not yet been undertaken.

3. Energy balance and climate

The physical processes which, together with advection, produce climate can be quantitatively described by the terms of the radiation and heat budget. These expressions have changed classical descriptive climatology into a physical science which, with the help of computer models, can predict (Flohn, 1965) the climatic effect of technical projects and, therefore, predict nonexistent climates. The problem of climate modification has become very important in the last few years, and we must meet its challenge.

Table 2B.1. Radiation and heat budget at mountain stations (ly day^{-1}).

	Elevation (m)	Albedo	Net radiation Q	Sensible heat U_L	Latent heat $L \cdot E$	Ablation U_B	Month
Sary Tasch, 40°N, 73°E	3150	.17	321	178	89	54	Sept.
Kara-Kul Lake, Pamir, 39°N, 73.5°E	3990	.22	324	243	29	52	Aug.
Koshagyl, Aksu Valley, Pamir	3710	.21	294	244	0	50	Aug.
Fedtschenko Glacier, catchment area	4900	.66	142	−37	116	7	July–Aug.
Fedtschenko Glacier, ice	3880	.36	496	−397	−41	(934)*	July–Aug.
Fedtschenko Glacier, near tongue	2900	.16	374	202	95	65	July–Aug.
Turkestan Mountains, Kum-Bel-Pass	3150	.14	348	248	50	50	Sept.
Turkestan Mountains, northern slope 33°	3150	.20	145	85	19	41	Sept.
southern slope 31°	3150	.15	426	304	47	75	Sept.
Chogo Lungma Glacier (Karakoram)	4300	.23	391	−41	10	434*	July
Paramo de Cotopaxi (0.5°S, 78.5°W) (Ecuador)	3570	.22	201	47	154	0	July
Chukhung (28°N, 87°E) (Nepal–Himalaya)	4750	.16	332	199	121	12	Apr.
Tujuksu Glacier (Tien Shan)	3475	—	137	−38	25	150*	July–Sept.

*Ablation as residual.

Much valuable information about the changes in radiation terms with the height of a mountain has been given by Sauberer and Dirmhirn (1958). Direct solar radiation was at one time considered the most important term in describing these changes, then global (or total) radiation and today net radiation. The changes of net radiation with height are rather complex. Naturally the increase of snow cover (which has an albedo of 0.6–0.9 in contrast to the 0.1–0.2 albedo of vegetation) with height plays an important role in this relationship. The snow cover reduces the net radiation with height significantly. We urgently need, therefore, reliable measurements of the net radiation in mountain areas, especially in mountain areas within different climatic zones.

Accurate measurements are even more important for the heat budget, especially for the flux of sensible and of latent heat (evaporation) and their relationship to each other (the Bowen ratio). This relationship fluctuates very greatly, often with box fluxes changing signs. Table 2B.1 gives several interesting evaluations of such measurements. Data have been computed for the mountains of Soviet Central Asia (Aisenshtat, 1966), the Karakoram Mountains (Untersteiner, 1958a,b), the Tien Shan (Skeib, 1962), the Himalaya (Haeckel *et al.*, 1970; Kraus, 1971), and for the Paramo in equatorial Ecuador (Korff, 1971). Schichlinski (1966) has determined the fluxes of latent and sensible heat in the Caucasus, and found strong variations related to the duration of snow cover (table 2B.2).

Table 2B.2. Heat fluxes and elevation in the Caucasus; yearly averages (ly day^{-1}) (after Schichlinski, 1966).

		Elevation *(m)* 200	600	1000	1500	2000	2500	3000	3500	4000
Greater Caucasus	L·E	55	73	84	79	74	64	55	45	33
	U_L	74	62	50	39	29	21	14	7	2
Lenkoran Mountains	L·E	112	99	79	58	38	22			
(38.5°N, 48.5°E)	U_L	48	56	66	78	90	99			

The sensible heat flux, with its horizontal variation, produces primarily thermally driven circulations which make up the entire hierarchy of diurnal wind systems in our mountains. However, as soon as these diurnal circulations produce rain in addition to the clouds – and only this irreversible process is determinant – a heat source is produced which is strong enough to maintain thermal circulations even when the primary heat sources are long gone (Flohn, 1970b; Flohn and Fraedrich, 1966; Fraedrich, 1968).

We must then consider the heat budget of the atmosphere rather than that of the earth's surface. Measurements from satellites are most useful for this study. The radiation budget (Q_{atm}) of the atmosphere is composed of different radiation fluxes just as the heat budget ΔW of an air column is influenced by different horizontal and vertical heat fluxes:

$$Q_{atm} = S_0(1 - a_p) - (S + H)_s(1 - a_s) + (E\uparrow - G\downarrow)_s - E_0$$

The first two terms on the right are short-wave contributions, the last two are long-wave.

$$\Delta W = Q_{atm} + U_L + L \cdot P + div\left(\int_0^\infty c_p \rho T \mathbf{V}\, dz\right)$$

Where Q_{atm} is the radiation budget of a vertical air column, ΔW is the heat budget of an air column, S_0 is solar radiation at the upper boundary of the atmosphere, $(S + H)_s$ is global radiation at the earth's surface, $(E\uparrow - G\downarrow)_s$ is effective terrestrial radiation, a_p is planetary albedo, a_s is terrestrial (surface) albedo, E_0 is the long-wave emission at the upper boundary of the atmosphere (E_0 and a_p are measured by meteorological satellites), $G\downarrow$ is downward long-wave radiation from the atmosphere, U_L is sensible heat flux at ground level, P is precipitation, L is latent heat of condensation, c_p is specific heat of constant pressure, ρ is density, T is temperature, **V** is horizontal wind vector, and z is the vertical coordinate.

The heat of condensation liberated by precipitation $(L \cdot P)$ is, even with moderate amounts of precipitation (10 mm $\simeq 600$ Ly), larger than the flux of sensible heat U_L. However, this energy is first used with heat storage ΔW to maintain the moist-adiabatic lapse rate. Only what is then left over is available for heat advection (divergence term $c_p \, \rho \mathbf{T V}$). We now know that this heat source in the humid, tropical mountains, along with the orographically triggered rains, plays a very important role in the tropical circulation. In addition to the classical Hadley circulation in a meridional plane, we have the concept of a zonal (Walker) circulation (Bjerknes, 1969) whose interaction is only partly explained (Flohn, 1971).

The radiation and heat balance terms which occur in these fundamental equations of theoretical climatology (or 'climatonomy': Lettau, 1967) can be formulated as non-dimensional basic parameters.

Radiation:	surface albedo	a_s	$= \dfrac{R\uparrow}{S + H}$	(short-wave)
	Ångström ratio	Å	$= \dfrac{E\uparrow - G\downarrow}{E\uparrow}$	(long-wave)
	planetary albedo	a_p	$= \dfrac{R_0\uparrow}{S_0\downarrow}$	(short-wave)
Heat budget:	Bowen ratio	Bo	$= \dfrac{U_L}{U_V}$	(U = turbulent heat fluxes)
	Budyko ratio	Bu	$= \dfrac{Q}{L \cdot P}$	(aridity index)
Water budget:	runoff ratio	RR	$= \dfrac{N}{P}$	(N = runoff)
Dynamics:	surface Rossby no.	Ro_s	$= \dfrac{V}{Z_0 f}$	(z_0 = roughness parameter)
	thermal Rossby no.	Ro_T	$= \dfrac{V_{xT}}{r\Omega}$	(V_{xT} = thermal zonal wind)

Thus on the basis of measurements of the radiation and heat balances at surface in the Pamirs (Aisenshtat, 1966) and the radiation flux at the upper boundary of the atmosphere, as measured by satellites (Raschke and Bandeen, 1970), we can estimate the heat budget of the atmosphere (Q_{atm}) for summer in an air column over the Tibetan highlands. In contrast to the estimates made by Gutman and Schwerdtfeger (1965) in the Altiplano (Peru–Bolivia) in which the advection was calculated from data from only one aerological station, the total advection is here evaluated under stationary conditions $(\Delta W = 0)$ since it is estimated as the residual of the area-averaged heat balance.

Radiation budget
surface (July):

Q_s = 320 ly day^{-1} (Aisenshtat)

extraterrestrial S_0 = 990 ly day^{-1} $a_p = 0.32$ (Raschke)

$S_0(1 - a_p)$ = 675 ly day^{-1}

E_0 = 545 ly day^{-1} (Raschke)

Q_0 = 675 − 545 = 130 ly day^{-1}

atmosphere Q_{atm} = $Q_0 - Q_s$ = 130 − 320 = −190 ly day^{-1}

Heat budget (July): U_L = 150 ly day^{-1} (arid 250 ly day^{-1}, humid 50 ly day^{-1}) (Aisenshtat)

obs. precipitation   P   = 120 mm month^{-1} = 4 mm day^{-1}

L·P   = 240 ly day^{-1}

heat storage ΔW   = 0

export of heat $\text{div}(c_p T)\mathbf{V}$ = 150 + 240 − 190 = 200 ly day^{-1}

When we combine the important terms for the troposphere in Tibet (considering a layer between 580 and 100 mb = 480 mb or g cm^{-2}) and calculate an equivalent temperature change, we find a cooling due to a negative net radiation of 190 ly day^{-1} equivalent to −1.65 °C day^{-1} and a heating due to the flux of sensible heat from the ground and the latent heat released by precipitation of 390 ly day^{-1} or +3.4 °C day^{-1}. Thus the advective heat export results in a net cooling of 200 ly day^{-1} or −1.74 °C day^{-1}. For this calculation a minimum value was used for precipitation, namely the average for the

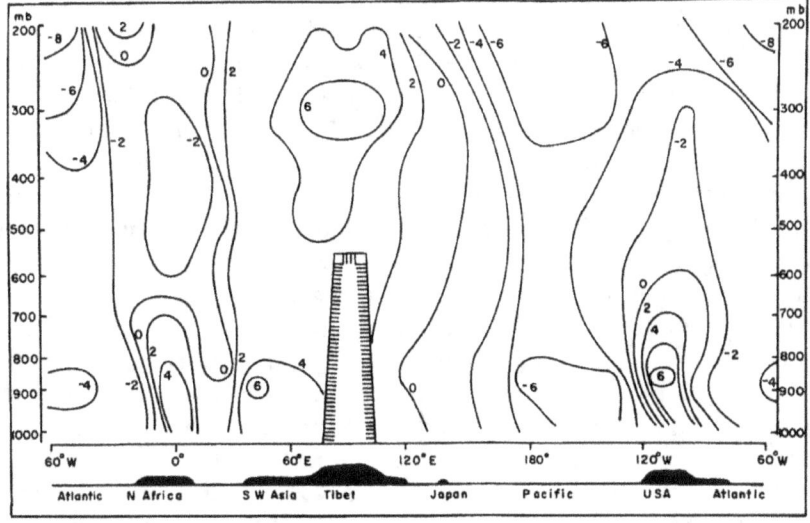

Fig. 2B.2. Zonal cross-section of temperature (°C; deviations from latitudinal average) along 32°N in July.

relatively dry valley stations of La-Sa,[1] Chiang-tsu (Gyangtse), Ch'ang-tue, Pa-t'ang, K'ang-ting (Tatsienlu), and Hsi-ning, varying between 80 and 160 mm per month (see Flohn, 1968a, table 20). The heat balance is certainly larger in areas of heavier orographical rain, and more heat is available for export. The plains of Assam, for example, have an area average daily precipitation of over 15 mm, at the edge of the Himalaya even near 25 mm, which release about 900 and 1500 ly day^{-1} respectively. At the same time the values of U_L remain very small. In arid western Tibet, if Leh can be considered representative for this area, $L \cdot P$ is probably as low as 50–100 ly day^{-1}. On this evidence, it is clear that the term $L \cdot P$ is, as assumed earlier (Flohn, 1955), of significant importance for the energy budget and circulation of the atmosphere on a large scale. Fig. 2B.2 shows the thermal anomaly (Flohn, 1968a) produced in Tibet in a latitudinal cross-section at 32°N in July. Similar estimates are now possible in many mountains as an atmospheric heat source depends primarily on the term $L \cdot P$.

4. Diurnal circulations

In spite of a recent survey (Flohn, 1969b), there is already something new to say about the thermal circulation on the local- and mesoscales. The very detailed observations in Dischma Valley near Davos, Switzerland (Urfer-Henneberger, 1970) show that in a small valley the change from night-time to daytime circulation takes place very quickly. That these circulations can take on very different forms according to their horizontal and vertical extension, azimuth, and slope grade, geometry of their radiation flux (Heyne, 1969), and vegetation, is immediately clear when we consider the complicated three-dimensional pattern of global radiation. Examples have been presented from Dischma Valley, Switzerland (Turner, 1968) and the Baye of Montreux (Kasser et al., 1970). However, the horizontal differences in the effective terrestrial radiation and the net radiation Q are unfortunately less well known. Since the flux of sensible heat (U_L) depends primarily on Q, we must expect very large local differences for the hierarchical system of slope and mountain and valley winds. The steep, narrow, and thickly forested valleys near Mt Rainier, Washington (Buettner and Thyer, 1966; Thyer, 1966), develop systems different from the wide, mostly open Inn Valley near Innsbruck, Austria, where the classical studies of A. Wagner and his collaborators were carried out. The well-conceived dynamical model by Buettner and Thyer, although it was discontinued in its initial stages, has confirmed our empirical knowledge.

Troll (1951) has often described the effect of these thermal circulations on the precipitation and vegetation in mountain valleys (fig. 2B.3). Examples of this relationship can be found in many tropical and subtropical mountain areas (e.g. the Andes from Venezuela to Bolivia, in Ethiopia, the Asiatic mountains from Hindu Kush and Karakoram to the Himalaya and the meridional gorges of upper Burma). Similar, if not quite so distinct, examples of this relationship exist in the Alps, especially in the large, longitudinal valleys and their branches. It is at least qualitatively understandable that the daytime

---

[1] Place names transliterated according to *The Times Atlas of the World*. Old names are shown in parentheses where the name has changed.

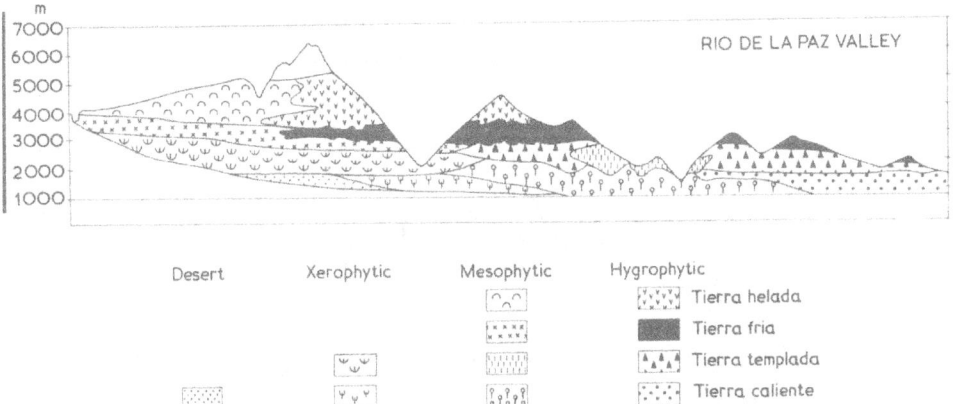

Fig. 2B.3. Profile of vegetation along Rio de la Paz (Bolivia, 16°S) (after Troll, 1951).

upslope circulation (rain on slopes, aridity in the valleys) so fully overpowers the night-time downslope circulation when one considers the differences in the horizontal distribution of Q and U_L between day and night, as well as the diurnal changes in the vertical stability. A numerical model of this phenomenon would be very welcome.

The only known example of the climatic effect of a night circulation comes from the Cauca Rift in Colombia. In the large rift valleys of the Colombian Andes, diurnal wind systems develop leading to a concentration of the night rain at the rift floor (relative heights under 500 m), and the afternoon rain on the upper slopes (especially over 1500 m). Fig. 2B.4 shows the relative distribution of precipitation according to time of day (Trojer, 1959; Flohn, 1968b). Evidently two conditions are here present – a constant moist-unstable stratification of the equatorial atmosphere which releases latent energy

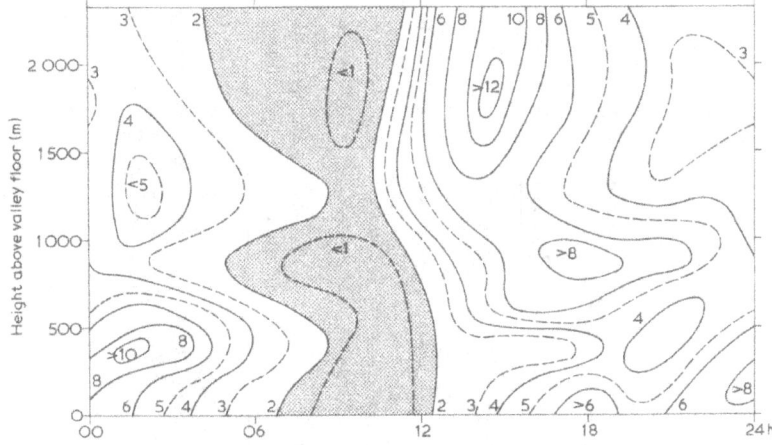

Fig. 2B.4. Daily cycle of rain amount (in per cent of average), east side of the Cauca rift (Colombia, 5°N) (data from Trojer, 1959).

Table 2B.3. Annual water budget estimates from runoff data (after Flohn, 1970c).

Mountains	Catchment	Station	Lat. (°N)	Long. (°E)	Area (km²)	R (cm)	RF (%)	P_meas (cm)	P_est (cm)	E_est (cm)
Hindu Kush	Panjshir	Gulkahar	35.2	69.4	3,400	54	60	?	99	45
Hindu Kush	Kunar	Chitral	35.8	71.8	10,300	64	56	35	100	37
Hindu Kush	Swat	Kalam	35.5	72.6	2,020	139	70	100	187	48
Hindu Kush	Swat	Chakdarra	34.6	72.0	5,770	92	65	88	139	47
Koh-i-Baba	Helmand	Kajakai Dam	32.3	65.1	32,500	19	35	(15)	58	39
Karakoram	Gilgit	Gilgit	35.9	74.3	12,100	75	65	12	120	45
Karakoram	Hunza	Near Gilgit	35.9	74.3	13,100	92	65	10	140	48
Karakoram	Indus	Darband	34.4	72.8	166,000	45	55	13	91	46
Pamir	Amu-Dar'ya	Nizhniy Pyandzh	37.2	68.6	107,000	34	55	(35)	69	35
Pamir	Gunt	Khorog	37.5	71.5	13,700	23	50	20	55	32

R = runoff (cm yr^{-1}).
RF = runoff factor (estimated).
P = precipitation (cm yr^{-1}) measured from existing stations or estimated (as an average from RF and three different empirical formulae).
E = evaporation, estimated from (P_{est} − R).

even at night, and a rift floor wide enough to permit the development of cumulonimbus systems (10–20 km) which develop as a result of night-time surface wind convergence. Precipitation maxima, resulting from the convergence of night-time land breezes, have been verified at Lake Victoria (Flohn and Fraedrich, 1966; Fraedrich, 1968) and Lake Titicaca (Kessler and Monheim, 1968).

The effect of the diurnal circulations – including the large-scale mountain circulations (see section 5 of this chapter) – on the horizontal rainfall pattern is not to be underestimated. Convergence and divergence of these circulations cause many local and regional anomalies. Since, in mountain areas with little population, observation stations are often situated only in valleys, the area average of rainfall is considerably underestimated (the so-called 'Troll effect'). The most striking example of this lack of representativity is in the Karakoram mountains with their giant glaciers up to 75 km in length. The five existing climatic stations all lie in desertlike large valleys which have yearly rainfall totals of 8–16 cm yr^{-1}. The glaciers, however, require extremely large amounts of snow, estimated by glaciologists at 3.6 m, or even 8 m, water equivalent per year (Schneider, 1969). Calculations based on the runoff from rivers (Flohn, 1969a) yield an area average (runoff) of 90–150 cm yr^{-1} and thus, in high areas, a precipitation of at least 2–3 m. More abundant runoff data measured under the aegis of the International Hydrological Decade are now available. These data show that in many mountain areas precipitation measurements, as a result of the Troll effect (Flohn, 1970c), are utterly unrepresentative (table 2B.3). Not infrequently precipitation values have been smaller than actual runoff measurements in the same catchment.

5. Large-scale mountain-induced circulations

Although superimposed large-scale mountain-induced circulations are generally weak (horizontal magnitude 0.1–1 m s^{-1}, vertical 1 cm s^{-1}), they can have a great effect on the horizontal pattern of cloud cover and precipitation. In the Alps such an effect appears to be exceptional. The study by Burger and Ekhart (1937) deserves a repetition using the improved data from RAWINSONDE stations. Bleeker and Andre (1951) have demonstrated the correlation of night-time divergence over the Rocky Mountains with night-time thunderstorms in the prairie states. In the daytime in summer the convection zone over the mountains corresponds well with the cloud-free divergence zone on both sides of the mountains (especially evident on the eastern side), even though the cirrus anvils of the towering cumulonimbus cells are transported by the jet stream far to the east. In November 1968, I observed a cloud-free divergence zone over a distance of more than 400 km along the east of the Andes, within the region of 8 km deep east winds. At 1000 local time a chain of well-developed cumulonimbus cells hovered over the 4000 to 5000 m high mountain ridge. Series of satellite pictures of the Himalaya confirm a cloud-free border area even during the summer monsoons. At least on the average, about 60 per cent of the border zone is cloud-free, while the other 40 per cent is covered by the cloud fields of the monsoon storms (Flohn, 1968a).

An especially well-developed large-scale mountain-induced circulation exists in the highlands of Tibet, an area of approximately 2×10^6 km² encircled by the Himalaya, Karakoram, Hindu Kush, and Kunlun Mountains (4500 m average height). Both the lower convergence and the upper divergence of this diurnal system were substantiated by a calculation of difference vectors between 0600 and 1800 local time (fig. 2B.5). This yielded an average vertical movement of 1.4 cm s^{-1} at the level assumed to be non-divergent (500 mb) (Flohn, 1968a, 1970b). Here we have a thermal circulation as described by the circulation theorem of V. Bjerknes, superimposed on the planetary wind systems. In connection with this, the existence of large-scale slope-parallel wind systems which develop from the geotriptic winds in the surface friction layer (Lettau, 1967) must be mentioned. Such very persistent systems develop at the western slope of the Andes in Peru and in the Great Plains in the United States. Here a balance between

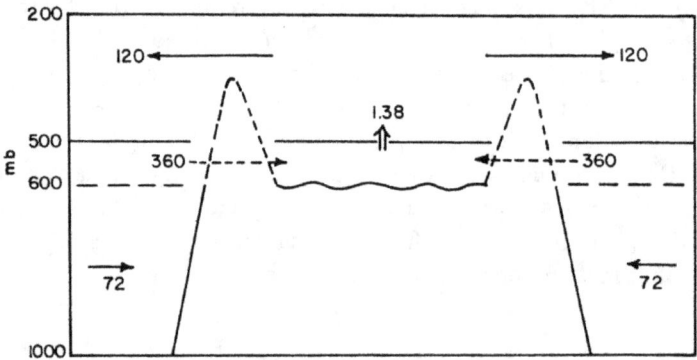

Fig. 2B.5. Model of the daytime circulation above Tibet, cm s^{-1}. Average inflow across the mountains concentrated between 600 and 500 mb.

the thermally induced pressure gradient perpendicular to the slope, the surface friction, and the Coriolis forces develops. Lettau's 'thermo-tidal theory' deserves a verification in other suitable areas. He himself applied the theory to a continental ice sheet of the Greenland type where it indicates a cyclonic geotriptic wind system around a cold core.

A detailed study of the Tibetan Highlands showed that even in the daily average (calculated from data from 0600 and 1800 local time) an all-day thermal circulation exists (Flohn, 1968a, table 23; Kasahara, 1967) whose lower branch at 700 mb has an inflow component of 120 cm s^{-1}. In agreement with this, the local night-time mountain winds at many stations in the north and south are either fully suppressed (Flohn, 1968a, fig. 5) or, at lower levels, dominated by a steady, upward flow (Flohn, 1970a, fig. 2) at 0600 local time. This is a very unusual finding – a seasonal, all day, thermally induced circulation with a horizontal length of 2000 km and a width of 1000 km. What is the origin of the energy to maintain this circulation in the night-time, when the mountains are supposed to cool off in relation to the free atmosphere? The main source of energy is evidently the latent energy released by the rain from the towering cumulonimbus cells (L·P). This mountain-triggered convection is maintained until late in the evening in all tropical mountain areas. The afternoon rain maximum is delayed until 1800–2200 local time.

In southeastern Tibet many stations show shower activity throughout the entire night. Only in the morning do these stations show a minimum (Flohn, 1968a, tables 3, 4) as in the Andes (fig. 2B.4). The liberated latent heat in the upper troposphere is maintained as a heat source the entire night, so that the local night circulations caused

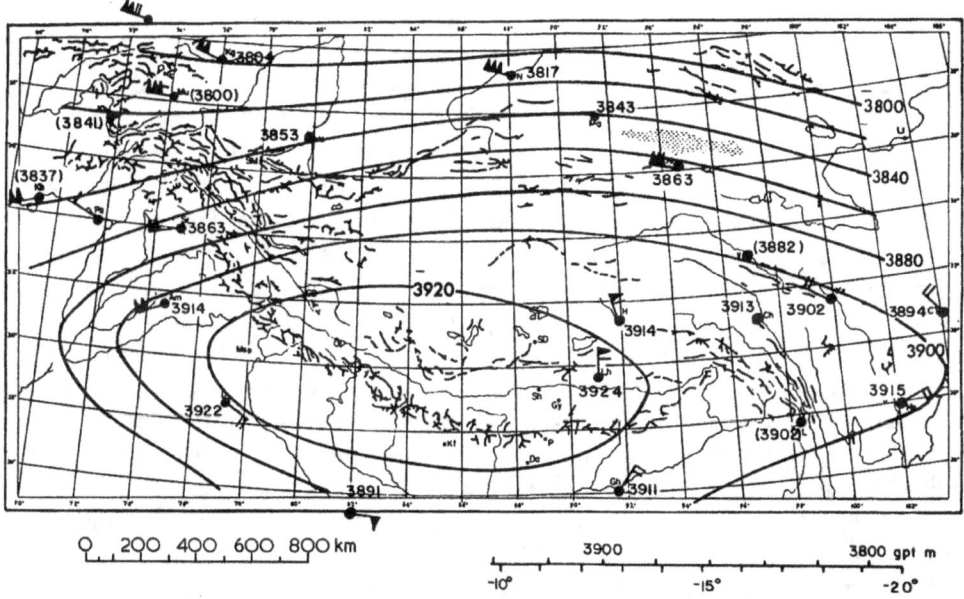

Fig. 2B.6. Relative topography of the 300/500 mb layer with thermal wind, Tibet, July–August. Scale lower right: thickness in geopotential metres and average temperature (tropical standard atmosphere −19°C).

by the flux of sensible heat are either totally suppressed or reduced to a shallow surface layer. That a heat island in the upper troposphere over southern Tibet really exists is supported by a carefully corrected map of the relative topography of the 300/500 mb layer drawn with direct observations of the thermal wind (fig. 2B.6). The direction of these shear winds remains the same at both the 0600 and 1800 observations and thus verifies the constancy of the heat centre between the night-time and the daytime circulations. The few infra-red satellite pictures to which I have had access confirm the persistence of the cumulonimbus convection until midnight.

Mountain-triggered convection has apparently everywhere a minimum in the morning. Unfortunately the regular photographs of the sun-synchronous ESSA weather satellites are taken at approximately 0900 local time – right during this minimum – so that orographical convection derived from these pictures is often underestimated. It is entirely possible that 30–40 per cent of the hot towers of the large cumulonimbus clusters which cause the vertical heat transport in tropical circulations are more or less orographically 'fixed' – always occurring in a place set by the orographically-triggered diurnal circulations (Flohn, 1970b).

6. Concluding remarks

All comparative studies show that in the tropical and subtropical mountains the local thermal circulations are, in principle, similar to but stronger and more dominant than those in the Alps. At these latitudes, the superimposed system of mountain winds influences the weather and climate greatly. In the case of Tibet, the large-scale circulations completely dominate the smaller ones. The cloud pattern and rain distribution is more or less correlated to this scalar hierarchy of diurnal wind systems. V. Bjerknes's classical circulation theorem is clearly demonstrated. The determinant role of the relief of mountains – even for the thermal circulations – causes difficulty in the formulation of mathematical circulation models on local and regional scales. Kasahara (1967), in his large-scale NCAR model, concluded that the purely dynamic effect of mountains as, for example, Bolin (1950) formulated according to potential vorticity, is at most secondary. If this effect is eliminated, then the thermal effect of land and water distributions is left as determinant (Kasahara, 1967). Our empirical observations show that the *effect of mountains* on the atmospheric circulation is, in reality, primarily *thermal*. It is possible, either by introducing a theoretical heat source over the mountains in a dry atmosphere or by including the hydrological cycle from an orographic viewpoint, to simulate this effect with sophisticated, realistic models.

With regard to the heat budget, which with the help of satellite radiation measurements (Raschke and Bandeen, 1970) can be reliably estimated, the determinant role is apparently played by latent heat liberated by orographically induced rain. Because of the relatively slow operation of radiation processes, even area rainfalls of 1–2 cm day^{-1} act as effective heat sources in the free atmosphere. The heat so liberated is not only locally important (anticyclonic deviation of the upper winds), but also plays a role in advective heat export. This is also true in winter in subarctic latitudes where the mountains of Norway, Alaska and Greenland produce large amounts of orographic rain.

Our observations have been purposely limited to large and mesoscale processes. Global comparison of microclimatic processes and their variations with latitude caused by radiation geometry is not possible at the present time. Even though it is, in principle, possible to derive the direct irradiation of slopes as a function of latitude and azimuth (Heyne, 1969), it is much more difficult to derive a similar relationship for diffuse radiation, and really complicated for long-wave radiation fluxes. Without question, the study of local variations in the net radiation in mountains (Aisenshtat, 1966; Schichlinski, 1966) is one of the most important challenges of physical climatology.

References

AISENSHTAT, B. (1966) Investigations of the heat budget of central Asia. *In:* BUDYKO, M. I. (ed.) *Current Problems of Climatology*, 87–129 (in Russian). Leningrad.

BJERKNES, J. (1969) Atmospheric teleconnections from the equatorial Pacific. *Mon. Wea. Rev.*, **97**, 163–72.

BLEEKER, W. and ANDRE, M. J. (1951) On the diurnal variation of precipitation, particularly over central USA, and its relation to large-scale orographic circulation systems. *Quart. J. R. Met. Soc.*, **77**, 260–71.

BOLIN, B. (1950) On the influence of the earth's orography on the general character of the westerlies. *Tellus*, **2**, 184–95.

BUETTNER, K. J. K. and THYER, N. (1966) Valley winds in the Mount Rainier area. *Arch. Met. Geophys. Biokl.*, B, **14**, 125–47.

BURGER, A. and EKHART, E. (1937) Über die tägliche Zirkulation im Bereich der Alpen. *Gerlands Beitr. Geophys.*, **49**, 341–67.

FLOHN, H. (1954) Grundsätzliche Probleme der Wettervorhersage. *Met. Abhandl.*, **2** (3), 189–98. Berlin.

FLOHN, H. (1955) Zur vergleichenden Meteorologie der Hochgebirge. *Arch. Met. Geophys. Biokl.*, B, **6**, 193–206.

FLOHN, H. (1965) Probleme der theoretischen Klimatologie. *Naturwiss. Rund.*, **18**, 385–92.

FLOHN, H. (1968a) Contributions to a meteorology of the Tibetan Highlands. *Atmosph. Sci. Pap.*, 130. Colorado State Univ.

FLOHN, H. (1968b) Ein Klimaprofil durch die Sierra Nevada de Merida (Venezuela). *Wetter und Leben*, **20**, 181–91.

FLOHN, H. (1969a) Zum Klima und Wasserhaushalt des Hindukushs und der benachbarten Gebirge. *Erdkunde*, **23**, 205–15.

FLOHN, H. (1969b) Local wind systems. *In:* FLOHN, H. (ed.) *General Climatology*, 139–71. Vol. 2, of LANDSBERG, H. E. (ed.) *World Survey of Climatology*. Amsterdam: Elsevier.

FLOHN, H. (1970a) Beiträge zur Meteorologie des Himalaya. *In:* HELLMICH, W. (ed.) *Khumbu Himal*, **7**, 25–45.

FLOHN, H. (1970b) Climatic effects of local circulation in tropical and subtropical latitudes. *Proceedings of the Symposium on Tropical Meteorology, Honolulu*, J VI.

FLOHN, H. (1970c) Comments on water budget investigations, especially in tropical and subtropical mountain regions. *In:* Symposium on World Water Balance, *Int. Ass. Sci. Hydrol. Pub.* 93, Vol. 2, 251–62. UNESCO.

FLOHN, H. (1971) Tropical circulation patterns. *Bonner Met. Abhandl.*, **15**. 55 pp.

FLOHN, H. and FRAEDRICH, K. (1966) Tagesperiodische Zirkulation und Niederschlagsverteilung am Victoria-See (Ostafrika). *Met. Rund.*, **19**, 157–65.

FRAEDRICH, K. (1968) Das Land- und Seewindsystem des Victoria-Sees nach aerologischen Daten. *Arch. Met. Geophys. Biokl.*, A, **17**, 186–206.

GUTMAN, G. J. and SCHWERDTFEGER, W. (1965) The role of latent and sensible heat for the development of a high pressure system over the subtropical Andes, in the summer. *Met. Rund.*, **18**, 69–75.

HAECKEL, H., HAECKL, K. and KRAUS, H. (1970) Tagesgänge des Energiehaushaltes der Erdoberfläche auf der Alp Chukhung im Gebiet des Mt Everest. *In:* HELLMICH, W. (ed.) *Khumbu Himal*, **7** (2), 61–133.

HAVLIK, D. (1969) Die Höhenstufe maximaler Niederschlagssummen in den Westalpen. *Freiburger Geogr. Hefte*, **7**. 76 pp.

HEYNE, H. (1969) Diagramme zur Bestimmung der extraterrestrischen Hangbestrahlung. *Mitteilungen aus dem Institut für Geophysik und Meteorologie der Universität zu Köln*, **10**. 61 pp.

KASAHARA, A. (1967) The influence of orography on the global circulation patterns of the atmosphere. *In:* REITER, E. R. and RASMUSSEN, J. L. (eds.) Proceedings of the Symposium on Mountain Meteorology, *Atmosph. Sci. Pap.*, **122**, 193–221. Colorado State Univ.

KASSER, P., SCHRAM, P. K. and THAMS, J. C. (1970) Die Strahlungsverhältnisse im Gebiet der Baye de Montreux. *Veröff. Schweiz. Met. Zentral.*, **17**. 46 pp.

KESSLER, A. and MONHEIM, F. (1968) Der Wasserhaushalt des Titicacasees nach neueren Messergebnissen. *Erdkunde*, **22**, 275–83.

KORFF, H. C. (1971) Messungen zum Wärmehaushalt in den äquatorialen Anden. *Ann. Met.* (n.s.), 5, 99–102.

KRAUS, H. (1971) Der Tagesgang des Energiehaushaltes in einem Hochgebirgstal. *Ann. Met.* (n.s.), 5, 103–6.

LETTAU, H. H. (1967) Small to large-scale features of boundary layer structures over mountain slopes. *In:* REITER, E. R. and RASMUSSEN, J. L. (eds.) Proceedings of the Symposium on Mountain Meteorology, *Atmosph. Sci. Pap.*, 122, 1–74. Colorado State Univ.

RASCHKE, E. and BANDEEN, W. R. (1970) The radiation balance of the planet Earth from radiation measurements of the satellite Nimbus II. *J. Appl. Met.*, **9**, 219–38.

SARKER, R. P. (1966) A dynamical model of orographic rainfall. *Mon. Wea. Rev.*, **94**, 555–72.

SARKER, R. P. (1967) Some modifications in a dynamical model of orographic rainfall. *Mon. Wea. Rev.*, **95**, 673–84.

SAUBERER, F. and DIRMHIRN, I. (1958) Das Strahlungsklima. *In:* Klimatographie von Österreich, *Österreichische Akademie der Wissenschaften, Denkschrift*, **3** (1), 13–102. Vienna.

SCHICHLINSKI, E. M. (1966) On the energy balance in the Caucasus. *In:* BUDYKO, M. I. (ed.) *Current Problems in Climatology*, 130–46 (in Russian). Leningrad.

SCHNEIDER, H. J. (1969) Minapin – Gletscher und Mensch im NW-Karakorum. *Die Erde*, **100**, 266–86.

SKEIB, G. (1962) Zum Strahlungs – und Wärmehaushalt des zentralen Tujuksu-Gletschers im Tienschan-Gebirge. *Zeit. Met.*, **16**, 1–9.

THYER, N. H. (1966) A theoretical explanation of mountain and valley winds by a numerical method. *Arch. Met. Geophys. Biokl.*, A, **15**, 318–47.

TROJER, H. (1959) Fundamentos para una zonificación meteorólogica y climatológica del trópico y especialmente de Colombia. *Cenicafé*, **10** (8), 289–373.

TROLL, C. (1951) Die Lokalwinde der Tropengebirge und ihr Einfluss auf Niederschlag und Vegetation. *Bonner Geogr. Abhandl.*, **9**, 124–82.

TURNER, H. (1968) Die globale Hangbestrahlung als Standortsfaktor bei Aufforstungen in der subalpinen Stufe. *Schweiz. Zeit. Forstwesen*, **4–5**, 335–52.

UNTERSTEINER, N. (1958a) Glazial-meteorologische Untersuchungen im Karakorum, I. Strahlung. *Arch. Met. Geophys. Biokl.*, B, **8**, 1–30.

UNTERSTEINER, N. (1958b) Glazial-meteorologische Untersuchungen im Karakorum, II. Wärmehaushalt, *Arch. Met. Geophys. Biokl.*, B, **8**, 137–71.

URFER-HENNEBERGER, CH. (1970) Neuere Beobachtungen uber die Entwicklung des Schönwettersystems in einem V-förmigen Alpental (Dischmatal bei Davos). *Arch. Met. Geophys. Biokl.*, B, **18**, 21–42.

WEISCHET, W. (1969) Klimatologische Regeln zur Vertikalverteilung der Niederschläge in Tropengebirgen. *Die Erde*, **100**, 287–306. Berlin.

Manuscript received March 1971.

Additional references

CRAMER, P. (1972) Potential temperature analysis for mountainous terrain. *J. Appl. Met.*, **11**, 44–50.

FRAEDRICH, K. (1972) A simple climatological model of the dynamics and energetics of the nocturnal circulation at Lake Victoria. *Quart. J. R. Met. Soc.*, **98**, 322–35.

HERRMANN, R. (1971) Zur regional hydrologischen Analyse und Gliederung der nordwestlichen Sierra Nevada de Santa Marta (Kolumbien). *Giessener Geogr. Schrift*, **23** (1). 88 pp.

KRAUS, H. (1971) A contribution to the heat and radiation budget in the Himalayas. *Arch. Met. Geophys. Biokl.*, A, **20**, 175–82.

KRISHNAMURTI, T. N. *et al.* (1973) Tibetan high and upper tropospheric tropical circulations during northern summer. *Bull. Amer. Met. Soc.*, **54**, 1234–50.

LETTAU, H. (1969) Evapotranspiration climatonomy. *Mon. Wea. Rev.*, **97**, 691–9.

THOMPSON, B. W. (1966) The mean annual rainfall of Mt Kenya. *Weather*, **21**, 48–9.

WORLD METEOROLOGICAL ORGANIZATION (1973) *Distribution of precipitation in mountainous areas:* Vol. 1, 228 pp.; Vol. 2, 587 pp. WMO No. 326. Geneva.

Topo- and microclimatology in alpine areas

Roger G. Barry and Claudia C. Van Wie

Institute of Arctic and Alpine Research,
University of Colorado

1. Introduction

As indicated in the preceding section, the climate of mountainous areas cannot be described solely in terms of macro- and mesoscale processes. The effect of rugged topography is to create countless *topoclimates* (Thornthwaite, 1954; Geiger, 1965, p. 455), which differ widely from one another in response to slope and aspect effects. Thus, windiness is highly variable over short distances and the amount of solar radiation or moisture received in a particular locality may bear little relation to measured or expected values for that altitude. Mountain areas usually present such a mosaic of heterogeneous facets that a conventional climatological analysis is likely to be misleading. Indeed, although there are obvious gross differences between the Coast Ranges and the Rockies, or between the Pyrenees and the Alps, for example, it is questionable whether the concept of a 'mountain climate' has much validity or value. This is not a problem which climatologists have adequately treated, primarily because of the remoteness of most alpine areas and the difficulties of collecting pertinent data.

Climatic conditions at or near ground surface are especially important in the alpine environment since they largely determine the distribution of plants as well as that of animals and insects that feed upon or shelter within the vegetation. Study of these climates is the concern of microclimatology, or perhaps preferably in this case of *ecoclimatology* (Geiger, 1965, p. 2). Such small-scale climates result from the interaction of local surface features with the prevailing regional meteorological controls.

In this section we shall discuss the main factors which influence the topo- and microclimates of alpine areas and the processes by which these conditions come to differ, sometimes markedly, from the regional climatic characteristics.

2. Effects of altitude

The basic conditions of alpine climate arise from the decrease in air density and in the number of dust and haze particles with height and the consequent increase in solar radiation (fig. 2C.1). In the Alps, Sauberer and Dirmhirn (1958) indicate that daily mean values of solar radiation for clear skies are 21 per cent higher at 3000 m than at 200 m in June and 33 per cent higher in December. The altitudinal increase in radiative transmission is proportionally greater in the short wavelength (ultra-violet) region of

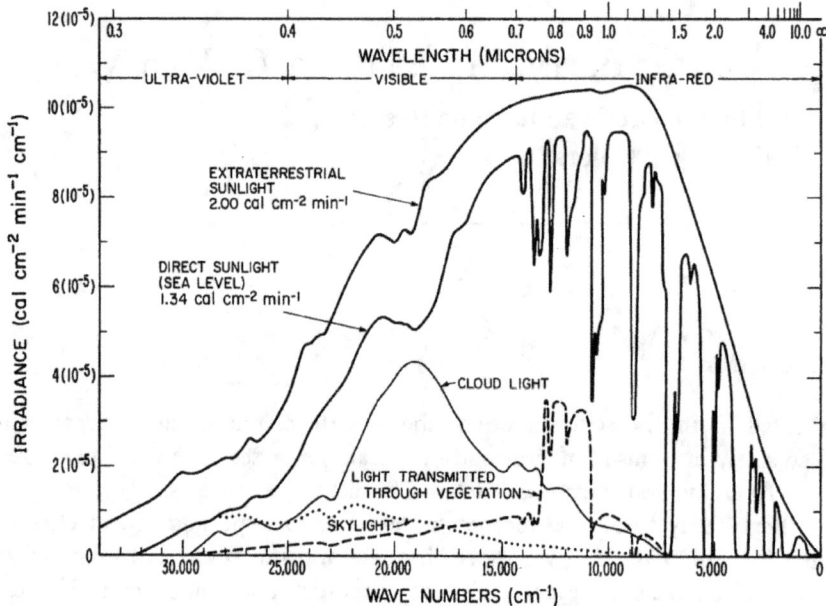

Fig. 2C.1. The intensity of direct solar radiation on a surface perpendicular to the beam at sea level compared with extraterrestrial radiation (from Gates *et al.*, 1965). Cloud and sky light at sea level are also shown. The spectrum refers to mean solar distance and a solar constant of 2.0 cal cm^{-2} min^{-1}.

the spectrum (see Kondratiev, 1969; Robinson, 1966, for further discussion), but the quantitative differences in ultra-violet energy received at different altitudes are still controversial (see Caldwell, 1968; Gates and Janke, 1966; Sauberer, 1955; and Schulze, 1960). Gates and Janke (1966) compute a 50 per cent increase in clear-sky ultra-violet radiation (<0.32 μm) at 3650 m, compared with sea level, for an optical airmass of 1.05 and a 120 per cent increase for an optical airmass of 2.0. (Optical airmass is a measure of the path length of radiation through the atmosphere. It is given approximately by the secant of the angle of the path from the zenith multiplied by the ratio of the actual pressure to sea level pressure.) However, Sauberer (1955) reports only a 34 per cent increase in global ultra-violet radiation between 0.28–0.315 μm from sea level to 3000 m, while for the *biologically effective* irradiance in these wavelengths (which is weighted to the shortest wavelength, decreasing exponentially with increasing wavelength), Caldwell (1968) demonstrated increases of only 4 per cent from sea level to 3650 m for an optical airmass of 1.05 and of 50 per cent for an airmass of 2.0. Caldwell also found an absolute decrease in sky ultra-violet radiation between 0.28–0.315 μm with increasing elevation above 1500 m due to reduced scattering in the thinner atmosphere.

Cloud cover drastically modifies the picture. Thams (1961) shows that in the Alps diffuse radiation increases up to a particular threshold of cloudiness – about 50 per cent cloudiness at a surface altitude of 400 m, 90 per cent cloudiness at 2000 m – beyond which there is a decrease. Under overcast skies the daily mean solar radiation at 3000 m in the Alps is about 1.5 times that at 200 m in both June and December (Sauberer and Dirmhirn, 1958) due to the effect of clouds in increasing the diffuse radiation. Commonly, the cloudiness over the mountains is greater than that over the plains so that the *total* solar radiation is thereby reduced. Caldwell (1968) suggested that the total solar radiation at 3050 m in the Colorado Front Range is about the same as that on the adjacent high plains at 1500 m as a result of this difference in cloudiness; and for annual totals, at least, there is little or no change between 2600 and 3750 m, although on a seasonal basis altitudinal variations seem to occur (Barry, 1973).

Net long-wave (infra-red) radiation probably changes little with altitude since the decreasing air density reduces the downward component from the atmosphere and the lower surface temperatures reduce the upward component. Again, however, the actual picture is greatly complicated by cloud cover, and the virtual absence of data on long-wave or net radiation rules out significant discussion. On an annual basis for the Tirol, Fliri (1971) shows values of the two long-wave components as follows:

	Upward component	Downward component	Net long-wave
2000 m	260	200	−60 kcal cm^{-2}
500 m	290	220	−70

Increase in altitude in the free air is, on average, associated with a temperature decrease (lapse rate) of approximately 6.5°C km^{-1} but there are significant seasonal differences in the major climatic regimes (Hastenrath, 1968). The lapse rate along a mountain slope is similar to that in the free air above about 2000 m (Steinhauser, 1967), although the two are not identical, and at lower levels the divergence may be considerable. The results of a number of studies show that mountain summits are on average 1–2°C warmer than the free air by day in summer, and 1–2°C colder than the free air by night in summer and throughout the day in winter (von Ficker, 1913; Hänsel, 1962; Samson, 1965). The diurnal temperature range decreases steadily with increasing altitude. In the free air it is typically about 1°C at 3000 m. On the east slope of the Front Range, Colorado, the diurnal range of screen temperature decreases from 11°C in January and 12°C in July at 2200 m to 6°C and 8°C, respectively, at 3750 m (Barry, 1973). This decrease is associated with the general increase of wind speed with height in middle latitudes, which accelerates turbulent mixing of air over the mountain slopes.

While information on free air conditions may be a useful indication of mountain winds and temperature (see, for example, Coulter, 1967, on New Zealand), it must be noted

that comparisons in the earlier literature may be misleading. Thus, for instance, observations analysed by E. Wahl (cited in Lettau, 1967) demonstrate that mountain summit
wind speeds are *less* than corresponding free air values except for winds of less than 2–3
ms^{-1}, contradicting earlier findings of the converse. For winds of 20 ms^{-1} the median
speed on a summit is 13 ms^{-1}. This study, based on data from six observatories and
four RAWINSONDE stations, confirms that mountain peaks have only a limited effect
on wind velocity.

3. Effects of slope and aspect

Slope angle and aspect are key determinants of topo- and microclimates. Differences
in radiation receipts resulting from topographic factors are reflected in soil and air
temperatures, snow cover duration, and soil moisture and consequently in the distribution of vegetation. In the Alps settlements and cultivation are commonly located on
sunny (*adret*) south-facing slopes while the shaded (*ubac*) north-facing slopes remain
under forest (Garnett, 1937). Marr (1961) shows that in the Colorado Front Range plant
species generally occur at lower elevations on north-facing than on south-facing slopes
in response to radiation and soil moisture differences (see also Wardle, Chapter 7B).

Direct solar beam radiation for clear skies has been computed as a function of slope
and aspect for 50°N by Frank and Lee (1966). As shown in table 2C.1, there is a four-fold
difference between north- and south-facing slopes of 45° (cf. table 2B.1). Differences
between slopes are reduced when, due to low solar altitude or cloud cover, diffuse
radiation accounts for a large proportion of the total radiation.

The radiation receipt of slopes can be determined from a combination of astronomical,
meteorological and topographical information. The details are discussed by Geiger
(1969). The computations involve (1) determination of the extraterrestrial radiation for

Table 2C.1. Radiation index as a function of slope and aspect at
50°N (after Frank and Lee, 1966).

| | Radiation index for given slope expressed as percentage of radiation index for horizontal surface | | |
| | Slope | | |
Aspect	10% (5°43')	50% (26°43')	100% (45°)
N	90	56	33
NNE	91	59	38
NE	93	69	54
ENE	96	84	76
E	100	100	98
ESE	104	114	117
SE	106	125	130
SSE	108	131	139
S	109	133	142

the appropriate date, time and latitude; (2) evaluation of the direct beam radiation for clear skies assuming an appropriate transmission coefficient (0.6–0.9) depending on the atmospheric turbidity; (3) computation of the direct radiation received on a slope of given azimuth and inclination; (4) determination of the transmitted diffuse or sky radiation for a solar altitude (for slopes up to 30° this can be considered independently of azimuth); and (5) calculation of the effect of neighbouring mountains screening a portion of the horizon.

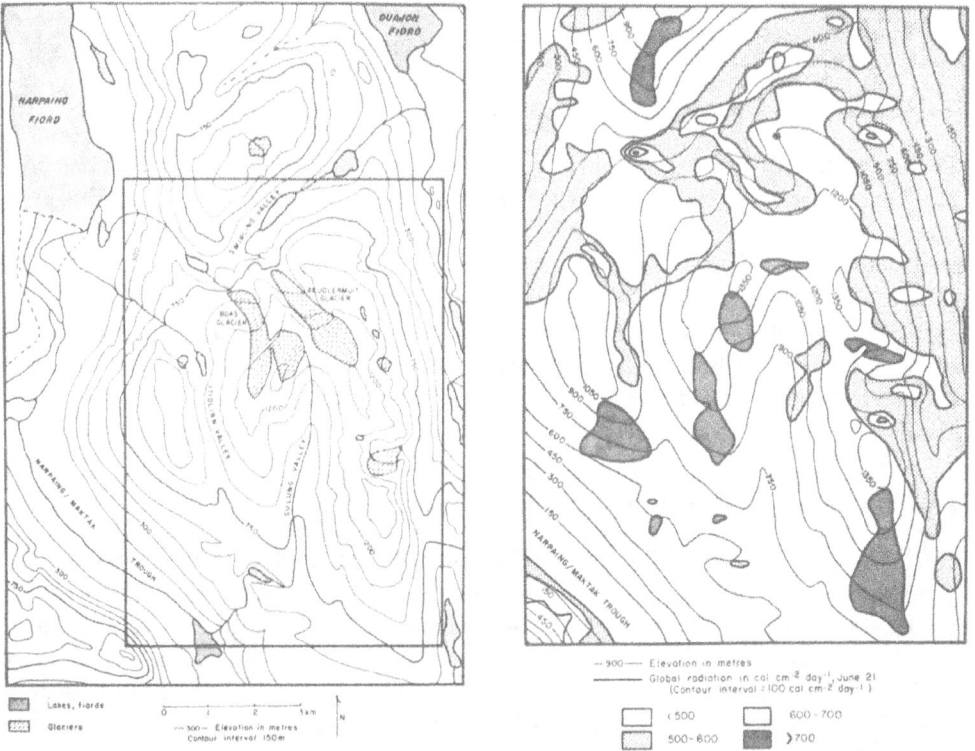

Fig. 2C.2. Computed global radiation for 21 June under clear skies for a mountainous area in eastern Baffin Island, NWT (from Williams *et al.*, 1972).

Garnier and Ohmura (1968, 1970) have detailed computational means of determining items 1 to 4, and a computer programme, based on their procedures and also incorporating screening effects, has been developed by Williams, Barry and Andrews (1972). Fig. 2C.2 illustrates a computer-produced map of slope radiation for an area in eastern Baffin Island. A further modification of this map would be to include the effect of radiation reflected from facing slopes, above or below. This is important for steep slopes, especially when there is a snow cover, and it may, to some extent, compensate for screening effects in valley situations. Such maps may in future provide an objective basis for the optimum siting of meteorological stations in mountainous areas.

Radiation differences on slopes are clearly reflected by soil temperatures. In one extreme situation at an elevation of 2070 m in the Öztal, Austria, Turner (1958) reported an absolute surface temperature of 80°C in black humus on a southwest slope of 35° compared with one of only 23°C on a nearby northeast slope. Shreve (1924) found that soil maximum temperatures showed a greater response to aspect than minima in the Santa Catalina Mountains near Tucson, Arizona. Averages of eighteen weekly readings in summer at 7.6 cm were as follows:

	Mean maximum (°C)	Mean minimum (°C)
2750 m north-facing	15	12
2750 m south-facing	25	16
2150 m north-facing	26	11
2150 m south-facing	33	16

Data reported by Marr (1961) in the Front Range of Colorado (table 2C.2) suggest that slope differences in soil temperature at 15 cm decrease with altitude. However, these observations relate to gentle slopes on a ridge. Ground temperature data collected in the same area by Ives (unpublished) indicated that at 3750 m elevation in February 1972 there were differences of 5–6°C at 4 m depth between a level site and a 40° north slope. In the Northwest Himalaya, Mani (1962, p. 71) notes that soil temperatures in summer are almost the same on north and south slopes above 4500 m. The changes in relative roles of turbulent heat fluxes and of radiative transfers with increasing altitude clearly need further study.

4. Effects of ground cover and surface properties

As discussed in Chapter 2B the net radiation available at the surface is partitioned into sensible heat transferred to the ground and the air and the latent heat of vaporization used in evaporation. The relative proportions depend on the surface cover, the season and the time of day, apart from the meteorological and hydrological factors. Specific studies of these variables in alpine areas are so few that worthwhile generalization is impossible at present. The selectivity of the following is determined, therefore, by the availability of data.

A striking feature of the alpine tundra is the relationship between snow cover and plant distribution. In winter bare sites are not uncommon on exposed ridges while shallow drifts accumulate around krummholz and in the lee of boulders, elongated parallel to the prevailing wind direction and often in an echelon pattern. In snow accumulation sites the underlying vegetation and ground are protected from the extreme cold and the drying winds, while in early summer a moisture source is available. The role of microenvironments related to snow cover and moisture conditions is evident from the results of a vegetation ordination on Niwot Ridge in the Colorado Front Range by Webber (see Chapter 8B, fig. 8B.8). In addition to soil moisture effects, the occurrence of snowbanks also shortens the growing season and modifies the soil temperature regime. Table 2C.3 shows the temperature gradient near the edge of a snowbank in June at 5000 m on the south slope of the Himalayas. As a snowbank recedes during the melt period, the

Table 2C.2. Temperature data for the Colorado Front Range, 1953 (after Marr, 1961).

Altitude (m)	Aspect[1]	Vegetation[2]	Mean daily max. and min. air temp. (°C) Jan.		July		Mean weekly max. and min. soil temp. at 15 cm (°C) Jan.		July	
2195	S	Ponderosa pine	11	−2	28	14	5	0	28	18
2195	N	Douglas fir–Ponderosa pine	6	−3	28	13	−1	−3	24	16
2590	S	Ponderosa pine–Douglas fir	6	−5	24	12	3	−2	26	14
2590	N	Douglas fir–Ponderosa pine–Aspen	3	−6	24	11	−2	−4	23	13
3050	S	Lodgepole pine	−1	−8	12	7	−2	−6	22	12
3050	N	Lodgepole pine	−1	−8	20	7	−4	−6	20	12
3750	S	Buttercup meadow	−9	−14	13	6	−7	−12	13	6
3750	N	Kobresia meadow	−8	−14	13	4	−7	−11	13	7

[1] Slopes are between 7° and 10°.
[2] Sites are in clearings in the forest belts.

Table 2C.3. Soil temperature (°C) in relation to a snowbank at 5000 m in the Himalaya in June (from Mani, 1962).

Distance from snow margin (m)	Air Max. temp.	Air Min. temp.	5 cm depth Max. temp.	5 cm depth Min. temp.	20 cm depth Max. temp.	20 cm depth Min. temp.
10	5	−10	8	−8	3	2
100	7	−1.5	11	4	7	6
200	12	7.5	16	5	10	8

newly exposed vegetation has a progressively shorter time in which to reach maturity and this has a pronounced effect on the vegetational composition of communities and on the phenology of the individual species (Billings and Bliss, 1959). Holway and Ward (1963) found that the basal plant cover in the centre of a snowbank depression in Rocky Mountain National Park, Colorado, was only 14 per cent, compared with 20 per cent in the outer area, and that the number of dominants was reduced from three species in the outer area to a single one in the centre.

The presence of a vegetation cover considerably modulates diurnal and annual ground temperature fluctuations. Moreover, this effect is evident within the plant canopy. In the equatorial mountains of east Africa, Coe (1967) measured a diurnal range of 13.3°C in the outer leaves of a *Festuca* tussock compared with 2.1°C in the basal disc (see fig. 2C.3). Insects have evolved that are well adapted to utilize this microclimatic contrast. Larvae of two species of Lepidoptera (*Gorgopis* sp. and *Metarctia* sp.) build silken tubes on grass from the basal part of the tussock to the outer portion of a leaf. Small spines on the side of the larvae allow them to move up and down the tube as the air temperature changes.

In the early morning and late evening, the larvae can be found at the outer edge of the leaves; during the hot part of the day and at night the larvae move to the region of less extreme temperature at the leaf base.

In areas with no vegetation, the temperature changes of the soil surface are more extreme. In the high African mountains, where the diurnal temperature variations are

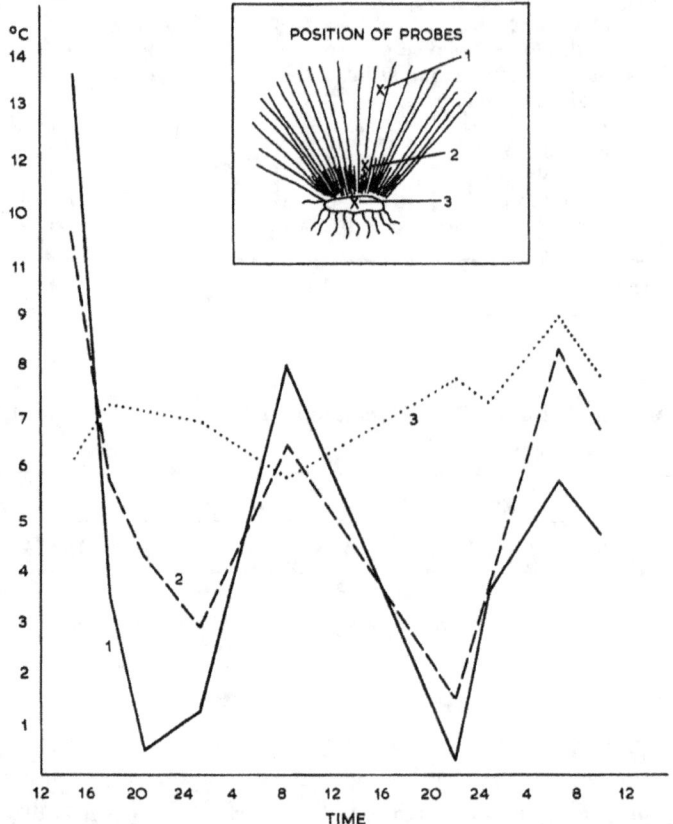

Fig. 2C.3. Diurnal temperature variations in a tussock of *Festuca pilgeri* in the alpine zone of Mt Kenya, March 1966 (from Coe, 1967).

large, needle ice regularly forms at night. Seedlings succeed better in vegetated areas, in spite of root competition, than in areas of open soil where frost heaving occurs daily (Hedberg, 1964). For further discussion of the effects of soil cover on vegetation, see Billings, Chapter 8A, section 2.

Rock surfaces respond even more rapidly to solar heating. Swan (1952) has observed that lizards, which appear on rock surfaces above the treeline (4100 m) on Mt Orizaba, Mexico, shortly after sunrise, will scurry for cover during cloudy intervals[1] until the

[1] The effect of cloud shadow on the surface energy budget has recently been analysed by Lambert (1970).

rocks warm up. Underneath rocks temperature and humidity conditions remain relatively constant and such niches are of great importance to alpine insects, even well above the permanent snowline in the Himalaya. Mani (1962, p. 63) cites diurnal air temperature ranges of 24°C in June at 4000–4500 m on the south slopes of the Himalaya compared with only 5–7°C in insect niches beneath rocks. The relative humidity is rarely below 90 per cent in such niches due to the absence of insolation and air movement.

Boulders and vegetation clumps also have a significant microscale effect on airflow by setting up small eddies on the lee side. As we have already noted, the distribution of snow drifts is a result of this. Studies by Gloyne (1955) indicate that in the lee of an obstacle eddy (rotor) flow is set up for a distance of ten to fifteen times the height of the obstacle downwind. A vegetation barrier of 50 per cent density will reduce the wind speed by 80 per cent for distances of three to five times the height of the barrier. This effect is demonstrated by measurements made on Jan Mayen on a tussock of *Juncus trifidus* by Warren-Wilson (1959). With a wind of 50 cm s^{-1} at the base on the windward side, equivalent speeds at the base in the plant centre and on the lee side were only 10 cm s^{-1} and 3 cm s^{-1}, respectively.

5. Conclusions

The typical range of topo- and microclimatic variations encountered in alpine areas is now reasonably well known, in terms of conventional climatic parameters. More work is undoubtedly necessary, however, on energy budget characteristics. Apart from this, the outstanding need is for model formulations to provide close approximations to the climatic conditions at individual sites on the basis of a limited number of topographic parameters and others describing surface properties. Until this is available, the biologist will have to rely on short-term site measurements or more general empirical findings, interpreted with respect to the types of relationship we have described.

References

BARRY, R. G. (1973) A climatological transect on the east slope of the Front Range, Colorado. *Arct. Alp. Res.*, 5, 89–110.

BILLINGS, D. W. and BLISS, L. C. (1959) An alpine snowbank environment and its effect on vegetation, plant development and productivity. *Ecology*, 40, 388–97.

CALDWELL, M. M. (1968) Solar ultraviolet radiation as a ecological factor for alpine plants. *Ecol. Monogr.*, 38, 243–68.

COE, M. J. (1967) Microclimate and animal life in the equatorial mountains. *Zoolog. Africana*, 4, 101–28.

COULTER, J. D. (1967) Mountain climate. *Proc. New Zealand Ecol. Soc.*, 14, 40–67.

FLIRI, F. (1971) Neue klimatologische Querprofile der Alpen – ein Energiehaushalt. *Ann. Met.*, (n.s.) 5, 93–7.

FRANK, E. C. and LEE, R. (1966) Potential solar beam irradiation on slopes: tables for 30–50° latitude. *US Dept of Agric., Forest Serv. Res. Pap.*, RM 18. 116 pp.

GARNETT, A. (1937) Insolation and relief. *Trans. Inst. Brit. Geogr.*, **5**. 71 pp.

GARNIER, B. J. and OHMURA, A. (1968) A method of calculating the direct short-wave radiation income of slopes. *J. Appl. Met.*, **7**, 796–800.

GARNIER, B. J. and OHMURA, A. (1970) The evaluation of surface variations in solar radiation income. *Solar energy*, **13**, 21–34.

GATES, D. M. and JANKE, R. (1966) The energy environment of the alpine tundra. *Oecol. Planta.*, **1**, 39–62.

GATES, D. M., KEEGAN, H. J., SCHLETER, J. C. and WEIDER, V. R. (1965) Spectral properties of plants. *Appl. Optics*, **4**, 11–20.

GEIGER, R. (1965) *The Climate Near the Ground* (4th edn). Cambridge, Mass.: Harvard Univ. Press. 611 pp.

GEIGER, R. (1969) Topoclimates. *In:* FLOHN, H. (ed.) *General Climatology*, 105–17. Vol. 2 of LANDSBERG, H. E. (ed.) *World Survey of Climatology*. Amsterdam: Elsevier.

GLOYNE, R. W. (1955) Some effects of shelter-belts and windbreaks. *Met. Mag.*, **84**, 272–81.

HÄNSEL, C. (1962) Die Unterschiede von Temperatur und Relative Feuchtigkeit zwischen Brocken und umgebender freier Atmosphäre. *Zeit. Met.*, **16**, 248–52.

HASTENRATH, S. L. (1968) Der regionale und jahrzeitliche Wandel des vertikalen Temperaturgradienten und sein Behandlung als Wärmhaushaltproblem. *Met. Rund.*, **21**, 46–51.

HEDBERG, O. (1964) Features of Afro-Alpine plant ecology. *Acta Phytogeogr. Suecia.* 144 pp. Uppsala.

HOLWAY, J. G. and WARD, R. T. (1963) Snow and meltwater effects on an area of Colorado alpine. *Amer. Midl. Nat.*, **69**, 189–97.

KONDRATIEV, K. YA. (1969) *Radiation in the Atmosphere*, 485–502. New York: Academic Press.

LAMBERT, J. L. (1970) The thermal response of a plant canopy to drifting cloud shadows. *Ecology*, **51**, 143–9.

LETTAU, H. H. (1967) Small to large-scale features of boundary layer structures over mountain slopes. *In:* REITER, E. R. and RASMUSSEN, J. L. (eds.) Proceedings of the Symposium on Mountain Meteorology, *Atmosph. Sci. Pap.*, 122, 1–74. Colorado State Univ.

MANI, M. S. (1962) *Introduction to High Altitude Entomology*, 1–73. London: Methuen.

MARR, J. W. (1961) Ecosystems of the east slope of the Front Range in Colorado. *Univ. of Colorado Stud., Ser. in Biol.*, 8. 134 pp.

ROBINSON, N. (1966) *Solar Radiation*. Amsterdam: Elsevier. 347 pp.

SAMSON, C. A. (1965) A comparison of mountain slope and radiosonde observations. *Mon. Wea. Rev.*, **95**, 327–30.

SAUBERER, F. (1955) Über die natürliche Ultraviolettstrahlung. *Wetter und Leben*, **7**, 80–5.

SAUBERER, F. and DIRMHIRN, I. (1958) Das Strahlungsklima. *In:* Klimatographie von Österreich, *Österreichische Akademie der Wissenschaften, Denkschrift*, **3** (1), 13–102. Vienna.

SCHULZE, R. (1960) Zur biologischen Wirkung der Langwelligen Ultraviolett-Strahlung in den Alpen. *Strahlentherapie*, **III**, 392–8.

SHREVE, F. (1924) Soil temperature as influenced by altitude and slope exposure. *Ecology*, **5**, 128–36.

STEINHAUSER, F. (1967) Methods of evaluation and drawing of climatic maps in mountainous countries. *Arch. Met. Geophys. Biokl.*, B, **15**, 329–58.

SWAN, L. W. (1952) Some environmental conditions influencing life at high altitude. *Ecology*, **33**, 109–11.

THAMS, J. C. (1961) The influence of the Alps on the radiation climate. *In:* Recent progress in photobiology, *Proc. 3rd Int. Conf.*, 76–91. Amsterdam: Elsevier.

THORNTHWAITE, C. W. (1954) Topoclimatology. *In: Proceedings of the Toronto Meteorological Conference 1953*, 227–32. London: Roy. Met. Soc.

TURNER, H. (1958) Maximaltemperaturen oberflächennäher Bodenschichten an der alpinen Waldgrenze. *Wetter und Leben*, **10**, 1–12.

VON FLICKER, H. (1913) Temperature Differenz zwischen freier Atmosphäre und Berggipfeln. *Met. Zeit.*, **30**, 278–89.

WARREN-WILSON, J. (1959) Notes on wind and its effect in arctic-alpine vegetation. *J. Ecol.*, **47**, 415–27.

WILLIAMS, L. D., BARRY, R. G. and ANDREWS, J. T. (1972) Application of computed global radiation for areas of high relief. *J. Appl. Met.*, **11**, 526–33.

Manuscript received February 1971.

Additional reference

KARRASCH, H. (1973) Microclimatic conditions in the Alps. *Arct. Alp. Res.*, **5** (3), part 2, A55–A64.

Snow

Donald Alford

Institute of Arctic and Alpine Research,
University of Colorado

1. Introduction

Snow is a crystalline form of solid atmospheric precipitation formed by the sublimation
and condensation of water vapour onto minute nucleii. These nucleii result from the
freezing of supercooled water droplets or occur as terrestrial dust, normally clay minerals.
While conditions conducive to snow formation in the atmosphere exist everywhere over
the surface of the earth, snow crystals can reach the ground only in those areas where
temperatures near or below freezing exist at the surface during all or part of each year.
In the high latitudes this condition is met everywhere, but at low latitudes only areas
above some critical elevation receive snow. In general, two types of snow cover can be
distinguished: these are (1) the 'seasonal' cover, which is a more-or-less cyclical feature,
being deposited and removed on an annual basis (see, for example, the maps of Dickson
and Posey, 1967), and (2) the 'permanent' cover, whose duration may exceed tens of
thousands of years.

At any given instant in time, snow-covered areas are separated from those which are
snow-free by the snowline. In terms of the seasonal snow cover, the snowline is transient
and fluctuates in response to variations in snow accumulation and surface energy
exchange. Annually, the snowline reaches some maximum position with regard to both
latitude and elevation; the long-term mean of this position is the boundary separating
seasonal from permanent snow deposits. In the most general terms, the snowline slopes
upward away from the Poles where it is, at least theoretically, at its lowest elevation on
the earth's surface (since the South Pole is at an elevation of some 3000 m, this condition
is approximated only in the northern hemisphere). The snowline also commonly slopes
upward inland from windward coasts of continents, where a plentiful moisture supply
acts as a depressing factor. Locally, this general picture is complicated by surface features
such as mountain ranges or large bodies of inland water such as the Great Lakes in the

central United States, both of which act to modify accumulation patterns and/or the intensity of surface energy exchange processes. Figs. 2D.1 and 2D.2 show two aspects of the snowline on a global basis. Fig. 2D.1 illustrates the annual frequency and duration of snow cover, while fig. 2D.2 shows the dependence of the snowline on latitude and elevation above sea level and, to a lesser degree, continentality. Both maps demonstrate the strong control exercised over the existence and persistence of the snow cover by polar and mountain regions.

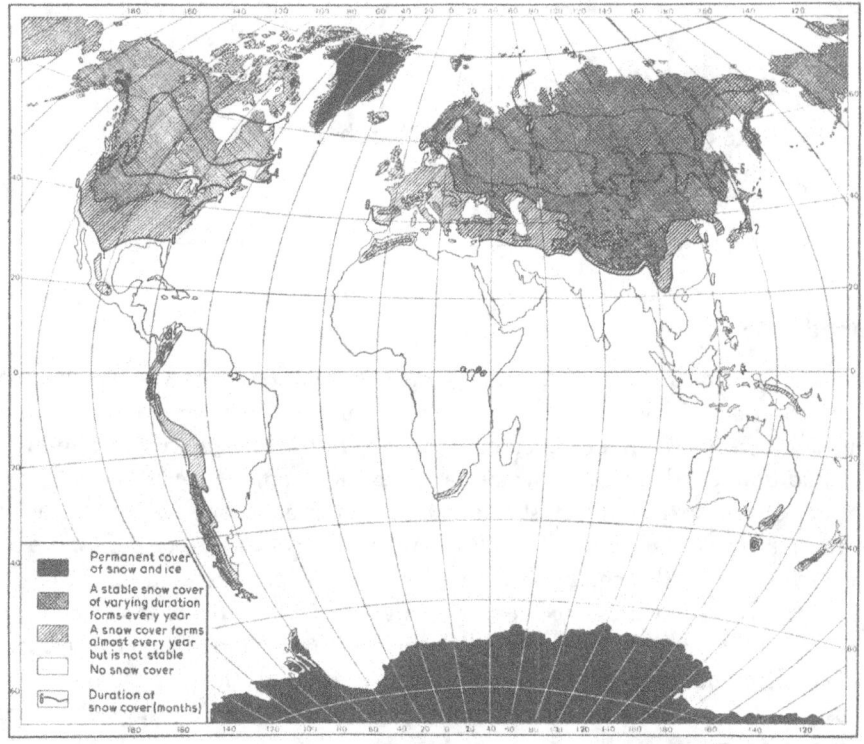

Fig. 2D.1. World distribution of snow cover (after Mellor, 1964b).

Following deposition on the earth's surface, snow consists of a porous, permeable aggregate of ice grains. Individually, these grains can be predominantly single crystals or they can consist of several crystals bonded together. The intergranular pores are filled with air and water vapour and, if the snow is wet, the grains are coated with liquid water. Perhaps the most characteristic attribute of snow is the inherent thermodynamic instability of the material. At normal ambient environmental temperatures, snow remains in a virtually continuous state of flux. The physical properties inevitably change with time. The rate at which this change takes place and the direction in which it proceeds are governed by the texture and structure of the deposit, the rate at which energy is transferred into the snowpack, the slope of the energy gradient and the proximity to the melting point at which the changes are taking place. Ultimately, however, one of two endpoints is reached. Either the energy input becomes sufficiently high to convert the snow to liquid water, or a combination of energy and overburden pressure produce an elimination of void space, increasing the density to that approaching ice. The change from snow to ice has been arbitrarily defined as the point where connections between intergranular pores no longer exist. This normally occurs at densities between $0.80\,\mathrm{g\,cm^{-3}}$ and $0.85\,\mathrm{g\,cm^{-3}}$.

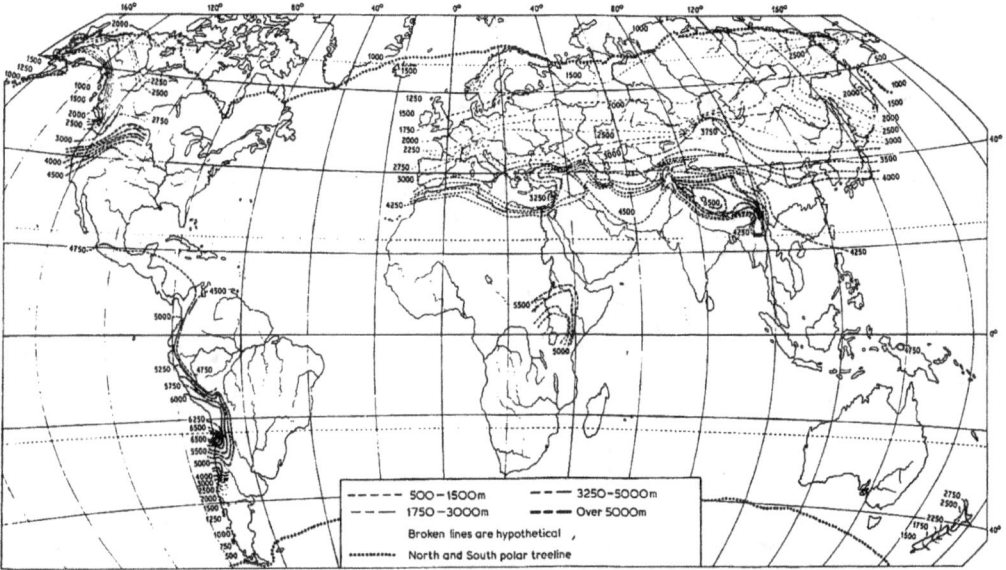

Fig. 2D.2. World distribution of snowline elevations (from Hermes, 1965). Isolines show the approximate elevation of the permanent snow cover. In most areas, this is a hypothetical line since mountain summits do not reach the elevations indicated by the lines.

The concepts 'arctic' and 'alpine' are not readily applicable to a discussion of snow except in so far as they describe areas of the earth's surface where snow accumulation and persistence are normally higher than the average. Snow is one of the most characteristic elements of both polar and mountain environments and, as such, no discussion of either is complete without some understanding of the way in which the occurrence of snow relates to the other aspects of these environments. However, it is not possible, at least on the basis of our present understanding, to differentiate 'arctic' snow from 'alpine' snow in terms of any quantitative criteria. Snow responds to variations in its energy level, to mechanical stresses such as overburden pressure, and to processes such as wind transport which produce both geometric and thermodynamic alteration. In order to discuss the bulk properties of a snow deposit in terms of a specific environment, it is necessary that the environment be well defined in terms of the processes of energy, mass and momentum transfer. Sufficient variation exists in terms of these parameters in both alpine and arctic regions to suggest that no 'type' of arctic or alpine snow exists. For this reason snow will be discussed primarily in terms of properties and processes. This general approach will be more useful than an attempt to describe the nature of snow deposits at specific geographical locations. Based upon the general considerations outlined below and a knowledge of the intensity of the relevant environmental parameters at any specific geographic location, the reader should be able to assess at least approximately the state of any property of interest at that location. For those who are interested in pursuing a particular aspect of snow in greater detail than is presented here, references to some of the more important papers on the subject are included.

2. The origin and deposition of snow crystals

Snow crystals originate in clouds at temperatures below the freezing point of water. The nucleation temperature may vary from within a few degrees of the melting point of ice to as low as −40°C. This range in temperature results from variations in airmass properties and the efficiency of nucleii present.

Clouds are normally composed of water droplets with radii between 10 and 50 μm and terminal velocities up to 30 cm s^{-1}. It has been shown (Dorsey, 1948) that bulk water almost always supercools 5–10°C before freezing while micrometre-sized droplets may be supercooled to −40°C before freezing. While temperatures of −40°C and lower are

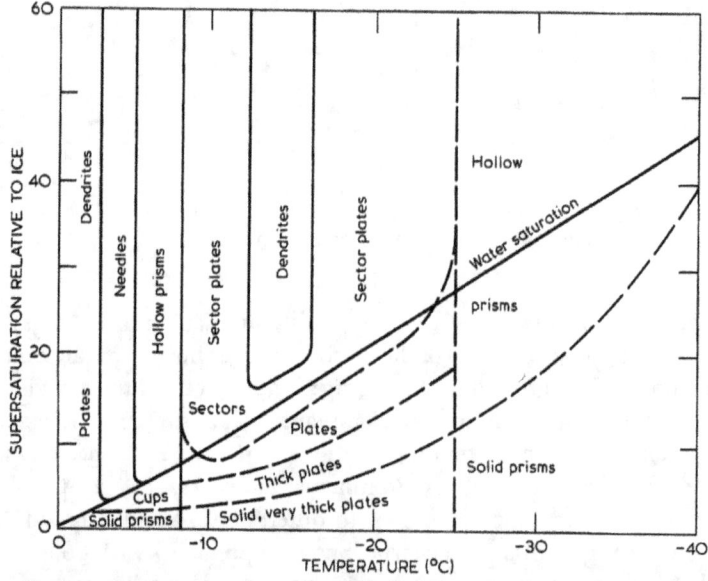

Fig. 2D.3. Mason's scheme for the variation of crystal form with temperature and supersaturation (per cent) relative to a flat ice surface (from Mellor, 1964a). The sensitivity of initial crystal form to formation temperature may result from the temperature dependence of molecular surface diffusion to the growth faces.

not uncommon at the higher cloud levels (Nakaya, 1954; Gold and Power, 1952) snow crystals may be found in clouds much warmer than this. Mason (1962) suggests that this may be ascribed to the presence of nucleating agents composed of terrestrial dusts, primarily silicate minerals of the clay and mica variety. These dust particles act to decrease the amount of supercooling required to freeze a cloud droplet. Although the way in which these nucleating agents act to decrease the amount of supercooling

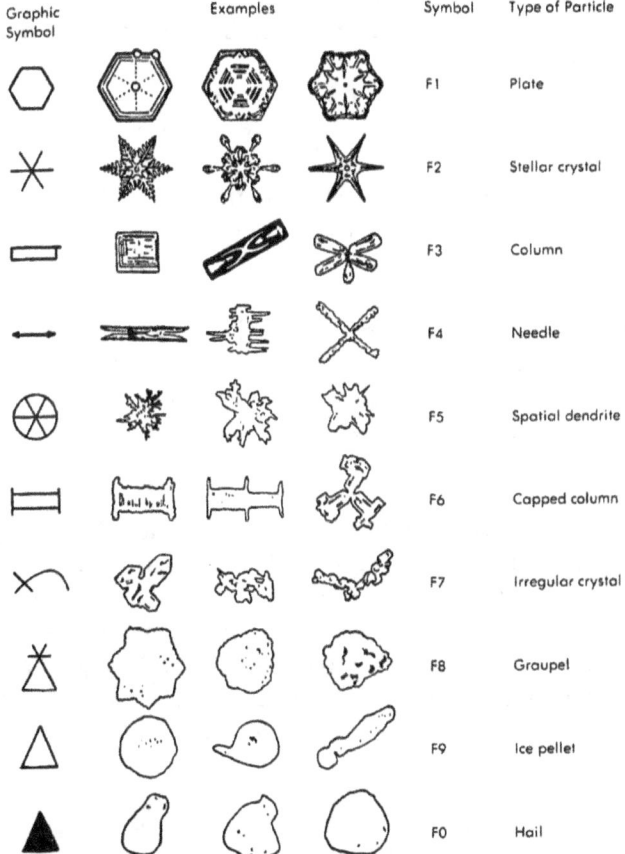

Graphic Symbol	Examples			Symbol	Type of Particle
				F1	Plate
				F2	Stellar crystal
				F3	Column
				F4	Needle
				F5	Spatial dendrite
				F6	Capped column
				F7	Irregular crystal
				F8	Graupel
				F9	Ice pellet
				F0	Hail

Fig. 2D.4. The international classification scheme for all forms of solid precipitation (from LaChapelle, 1969). Hail and ice pellets are amorphous forms of ice; graupel is a transition form between crystalline snow and hail.

required to freeze a cloud droplet has not been completely determined, it is believed that this property is in some way related to the similarities which exist between the structure of the crystal lattice of the nucleii and of ice. An ice crystal has hexagonal symmetry, as do the most efficient nucleating agents. Kumai (1951), using an electron microscope to study the nucleii of natural snowflakes, found that nearly 90 per cent were soil particles, 70 per cent of which were identifiable as one of the clay minerals (either kaolinite, montmorillonite or closely related forms).

Another mechanism has been proposed to explain the initiation of crystal growth in clouds whose temperature is close to the freezing point, since none of the artificial nucleii studied are effective at temperatures above about $-5°C$. During the growth stages of a snow crystal, small pieces are often torn off by collision and turbulent air motion. These crystal fragments slowly fall and may in turn serve as nucleii at temperatures above those required for the activation of mineral nucleii. Alternatively, high-level cirrus may 'seed' lower clouds. In theory, ice crystal fragments should serve as the most efficient nucleii and propagate crystal formation into portions of a cloud in which the mineral nucleii would not be effective.

Once germinated upon a nucleus, the snow crystal continues to grow by diffusion of water vapour to, and condensation upon, its surface. Snow crystals settling to the ground surface display a wide variety of shape and form. This variety has been ascribed by Mason (1962) to the cloud temperature at the time of crystal formation and to the subsequent temperature and saturation environment(s) through which the crystal passes during its fall. Mason's scheme for the dependence of crystal form on temperature and degree of supersaturation is shown in fig. 2D.3. The growth of an ice crystal by sublimation and condensation at the expense of atmospheric water vapour is made possible by the fact that, at all temperatures at which it can exist, ice has a lower vapour pressure than liquid water. Air which is saturated with respect to water is supersaturated with respect to ice and, under these conditions, an ice crystal will be surrounded by an excess of water molecules. More of these fall onto the surface of the crystal than are ejected by it and growth therefore takes place.

During its fall to earth, a snow crystal may undergo considerable change. Variations in temperature and humidity with altitude produce changes in growth rate and form; evaporation or melting may occur. The flake may be 'recycled' by turbulence and, if it is delicate, fragmented. In spite of the apparent myriad of forms produced by these processes, however, it has been possible to devise a classification scheme for snowflakes based on a limited number of forms. Fig. 2D.4 shows the basic groups into which snow crystals have been divided, together with the symbols used to refer to them in the International Snow Classification.

3. Post-depositional changes

Snow crystals falling to the surface of the earth under calm conditions construct a delicate, cellular mass. Because the environment within the deposit is seldom, if ever, similar to that in which the constituent crystals were formed, both the texture and structure of the mass undergo almost continuous alteration. This alteration continues until the snow is either removed by melt and evaporation or converted into relatively dense ice. This change in texture and structure is referred to as 'metamorphism' by most workers in snow physics today. (The term 'diagenesis' has been proposed, e.g. Benson, 1962, to describe the changes which take place in snow deposits where the liquid phase is not a factor. This term has not gained wide acceptance.) Several attempts have been made to classify snow metamorphism. Two of these are reviewed here.

The fact that metamorphism has occurred in a snow deposit is recognized by morphological changes which have taken place and by the presence of crystal forms which are distinctive in terms of one or more metamorphic processes. A classification of snow metamorphism based on the morphology of the crystals present was developed primarily by Swiss glaciologists during the 1930s (de Quervain, 1945; Eugster, 1952; Bader and Kuroiwa, 1962). In terms of this classification, four endpoints may be reached; each is associated with a dominant process. The four types of metamorphism are referred to as (1) destructive, (2) constructive, (3) melt and (4) pressure.

1. *Destructive metamorphism*

This occurs most often in new dry snow deposits. The name derives from the fact that commonly, crystallographic elements (i.e. faces, vertices, edges) are destroyed. This type of metamorphism involves a rearrangement of ice in the immediate vicinity of each crystal due to surface-free energy gradients produced by the initial crystal geometry. Within a few days after deposition the original crystal shape is almost completely lost. Grains become rounded and weakly bonded, most grains consist of only one crystal. The predominant grain size is commonly quite close to 1 mm. One result of destructive metamorphism is an increase in density because void space is eliminated as the crystal geometry becomes simpler. Initially, bulk strength may be lost as the interlocking arms of the original crystals are destroyed, but a prolongation of the process will result in the growth of very strong bonds, particularly if the grains are forced closer together by the plastic densification caused by the load of subsequent snowfalls.

2. *Constructive metamorphism*

This occurs as a result of the extensive transfer of mass in the vapour phase, initiated by differences in temperature and vapour concentration. It is characterized by the growth of grains to significantly more than 1 mm with the development of crystallographic elements. The rate at which constructive metamorphism takes place is partially dependent upon the density of the deposit, being slow if the density is high. If the density is below about 0.3 g cm^{-3}, however, it can become very rapid when the vapour transfer is accelerated by convective mass transfer under the influence of a negative temperature gradient (colder at the top than at the bottom). The number of intergranular bonds per unit volume in certain layers of the deposit may increase and, from the irregular sub-rounded grains produced by destructive metamorphism, crystals of partly idiomorphic shape grow. In an advanced stage, hollow cuplike crystals are formed with the c-axis parallel to the axis of the cups, each cup often exhibiting a steplike growth pattern. The end product of this type of metamorphism is often referred to as *depth hoar*, which is a snow type associated with a marked decrease of bulk mechanical strength.

3. *Melt metamorphism*

In the presence of liquid water, temperature gradients and their effects all but vanish. Morphological changes in the snow are produced by the transfer of latent heat by percolating meltwater and by the constrictive effect of the interstitial water film coating each grain. All crystallographic elements quickly disappear and grains may grow to sizes ranging from 5 to 15 mm. Melt metamorphism is usually characterized by a succession of freeze and thaw events which alternate in response to the direction of the surface energy flux. During the freezing portion of this cycle, intergranular bonds are formed as the water, which is retained preferentially at points of grain contact, is frozen. The bulk mechanical strength is greatly increased in those layers in which this freezing occurs. This strength, however, lasts only until such time as liquid water is again produced at, or near, the surface of the deposit. A marked decrease in mechanical strength will then take place as the intergranular bonds are again destroyed by percolating meltwater. This cyclic increase and decrease of strength is particularly apparent to those travelling over a snow cover undergoing melt metamorphism. A crust which will support a large vehicle at certain times may deteriorate in the interval of a few hours to the point where a man on skis may have difficulty in making progress.

Percolating meltwater is not equally distributed in all layers of a snow deposit mainly due to variations in the mean grain size. Fine grained layers are less permeable than coarse grained ones, thereby impeding the passage of the meltwater and leading to the formation of ice lenses upon re-freezing.

In the accumulation zones of temperate glaciers, melt metamorphism leads to the formation of firn (Fr. *névé*), which is commonly defined as snow which has survived one melt season. Firn may be ultimately converted to glacier ice by additional melt metamorphism during succeeding ablation seasons or by plastic deformation under the load of additional accumulation.

4. *Pressure metamorphism*

Pressure metamorphism is the dominant process where thermodynamic processes, such as grain and bond growth by vapour diffusion and surface migration, are inhibited by very low ambient temperatures or very low permeability. In this case, shifting and rotation of individual grains, presumably caused by the weight of overlying deposits, produce a more closely packed structure. Such metamorphism proceeds slowly and it may take decades (or even centuries) to convert the initial low density snow to ice.

The classification of snow metamorphism outlined above has not been universally accepted. An alternative system has been proposed by Sommerfeld and LaChapelle (1970) to correct what they feel to be weaknesses in terminology and to describe the process in more generic terms. The classification they propose is:

I. *Unmetamorphosed snow*
A. No wind action: many fragile snow crystal forms easily distinguishable; little difference from snow in air.
B. Windblown: shards and splinters of original snow crystals; parts of original forms may be recognizable but whole forms very uncommon.
C. Surface hoar.

II. *Equi-temperature metamorphism*

A. Decreasing grain size.
 1. Beginning: original snow crystal shapes recognizable, but corners show rounding and fine structure has disappeared.
 2. Advanced: very few indistinct plates and fragments recognizable; grains show distinct rounding.
B. Increasing grain size.
 1. Beginning: no original snow crystal shapes recognizable; grains show a distinct equi-dimensional tendency; a few indistinct facets may be visible.
 2. Advanced: larger equi-dimensional grains present; a strong tendency toward uniform grain size; faceting generally absent.

III. *Temperature-gradient metamorphism*

A. Early: the result of a strong thermal gradient on new-fallen snow; associated with the first snowfalls of the season.
 1. Beginning: angular or faceted grains common; stepped surfaces not visible.
 2. Partial: medium-sized angular grains predominate; poorly formed steps visible.
 3. Advanced: medium to large angular grains predominate; well-developed facets and steps visible; a few filled or hollow cups may be found.
B. Late: the result of a strong thermal gradient acting on snow in the later stages of equi-temperature metamorphism.
 1. Beginning: medium to large angular or faceted grains predominate; some stepped surfaces visible.
 2. Advanced: large grains predominate; many very fragile hollow cups or lattice grains; very deep steps.

IV. *Firnification*

A. Melt-freeze metamorphism.
 1. Limited: single thaw–freeze cycle and limited gain in ice density.
 2. Advanced: repeated thaw–freeze cycles and appreciable gain in density and mechanical strength; density range 600–700 kg m^{-3}.
B. Pressure metamorphism.
 1. Beginning: grains deformed and rearranged by pressure; density range 700–800 kg m^{-3}.
 2. Advanced: pore spaces become non-communicating; permeability zero; density range 800–830 kg m^{-3}.

The two classification schemes described above have been criticized on both technical and semantic grounds, and it is quite likely that additional classifications of snow metamorphism will be forthcoming as our knowledge of this complicated process increases. Whatever scheme is finally accepted, however, it will have to incorporate all of the complex interactions of the process by means of clearly defined terminology.

4. The causes of metamorphism

Because an acceptable classification of snow metamorphism can come only from an understanding of the basic aspects of the process, it is useful to discuss these briefly at this point.

The characteristic property of snow is its inherent thermodynamic instability. From the lithologic point of view, snow is simply the sedimentary and/or metamorphic form of the mineral ice. In common with all crystalline minerals, ice possesses a definite melting point. Unlike most minerals on the earth's surface, ice exists relatively close to its melting point even in the coldest natural environments. The surface-free energy will always be high and surface molecules should be highly mobile. These molecules migrate under the influence of the surface-free energy gradients, which in part are established by crystal morphology; both vapour pressure and free energy gradients are inversely related to the radius of surface curvature. A sphere will be the most stable form that a single ice crystal can assume.

For illustration, let us assume that a single snow crystal is enclosed in a small, dry, air space, and, further, that the air temperature is relatively close to the melting point ($\geqslant -50°C$ or $\geqslant 0.8\,T_m$ where T_m is the melting point in degrees kelvin). At this temperature, the molecules will have a high rate of thermal oscillation around their mean lattice position, and the lattice will be close to breakdown. The molecules are constrained from moving from fixed lattice positions by the lattice force field, which is weakest near the surface of the crystal. This leads to the development of a layer of highly mobile molecules on the surface of the crystal which has been referred to as a 'quasi-liquid' layer (Nakaya and Matsumoto, 1953). Within this layer, individual molecules and groups of molecules will have a relatively large freedom of movement and will tend to migrate in the direction of the steepest energy gradient. In terms of any individual crystal, this migration will take place by surface diffusion from regions of high surface-free energy, which will commonly be convex areas with a high radius of curvature, to regions of low surface-free energy (concave areas with a low radius of curvature). In addition, random thermal agitation will impart sufficient velocity to individual molecules to kick them out of the surface layer into the surrounding air space, where they become a component of the vapour phase. With an increase in the vapour concentration of the surrounding air space, molecules begin to impinge back on the surface of the crystal. Some of these strike the surface with a low velocity at points where the thermal agitation is also low at that moment and these are re-incorporated into the boundary layer. Both surface diffusion and vapour diffusion work toward the establishment of equilibrium conditions. However, since temperatures may fluctuate rapidly, particularly in the near-surface layers of the snow deposit, a true equilibrium may not necessarily result and the ice crystals will oscillate through many states of quasi-equilibrium.

The metamorphic processes in the bulk snow cover proceed in essentially the same fashion as the individual crystals discussed above. The natural snow deposit is both porous and permeable, and energy and vapour pressure gradients exist as a consequence of temperature differences. While the size range within any given layer is small, a range does exist. Because larger crystals may be expected to have a larger radius of curvature than smaller crystals, they may also be expected to have a lower surface-free energy. Large crystals will therefore be expected to grow at the expense of smaller ones as molecules are lost from the surface of the smaller crystals and captured by the larger. Similarly, when two crystals of unequal diameter are in contact, the larger may grow at the expense of the smaller by surface diffusion. Because of the relative instability of the smaller crystals, a gradual decrease in the range of crystal sizes present occurs. This homogenization of grain size increases conditions favourable for constructive metamorphism by increasing the porosity of the deposit.

While there is still disagreement among workers in the field of snow physics concerning certain aspects of the metamorphic process, a certain amount of experimental evidence has been obtained which accords with the general picture presented above. De Quervain (1945) enclosed individual snow crystals in small, sealed air spaces and observed the changes which took place with time and variations in the ambient air temperature. In this experiment, the crystals gradually approached a spherical shape during a fifty-four-day period of observation. Variations in the ambient air temperature had the effect of accelerating or decelerating the rate at which the crystal morphology changed. One primary point of disagreement concerns the relative importance of vapour and surface diffusion in metamorphism. Ramseier and Keeler (1966) have concluded that surface diffusion is relatively unimportant as a result of experiments involving ice spheres placed in an isothermal silica oil bath. The silica oil bath had the effect of suppressing vapour diffusion. They found no perceptible mass transfer between spheres, suggesting that surface diffusion does not contribute significantly to the metamorphic process. Certainly, much more work is needed on this problem before a final answer will be possible.

5. Thermal processes in snow deposits

Physical characteristics of a snow deposit undergo continual change in direct response to changing thermodynamic conditions. These thermodynamic changes are primarily the result of variations in the intensity of meteorological parameters and the nature of the surface energy exchange processes. In combination with energy transferred at the base of the snow deposit, the meteorological influences help to determine the rate and type of metamorphism occurring in the deposit.

Except for precipitation, all meteorological processes may be regarded as effects of the transfer of momentum, heat or water vapour between the atmosphere and the snow deposit. It is necessary that the nature of these transfer mechanisms be at least qualitatively understood if one is to understand how they promote metamorphic changes within a snow deposit. The continuously changing thermal and physical properties of a snow deposit, and variations in the intensity of the transfer processes at its boundaries, make the theory of heat flow in snow much more complicated than that for more homogenous materials.

Snow is a mixture of ice, air, water vapour and, sometimes, liquid water. The relative

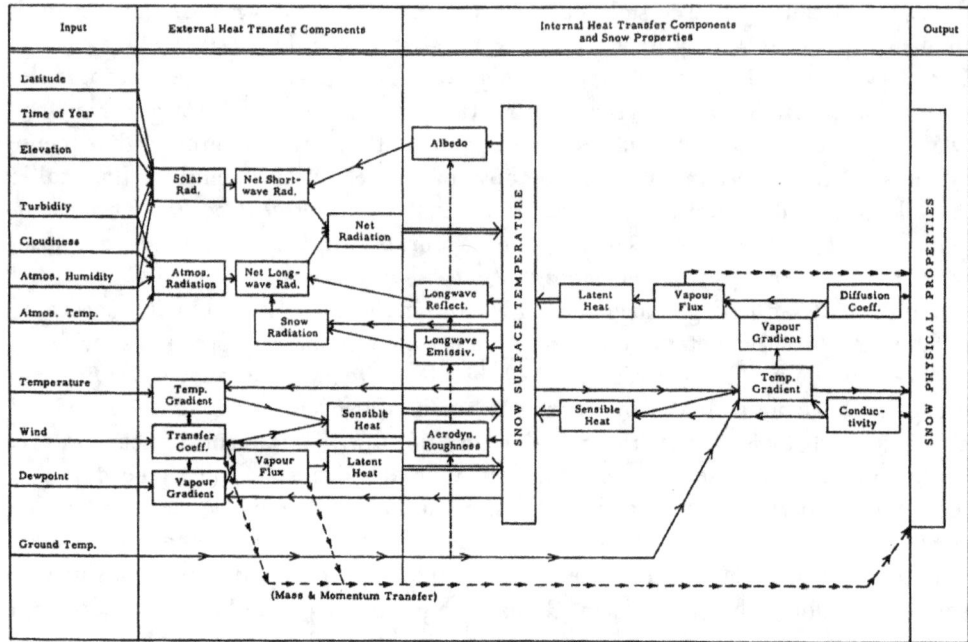

Fig. 2D.5. Heat transfer components and relationships in a snow deposit (from Portman and Ryznar, 1961).

percentage composition is not fixed but rather varies with time and location. The thermal properties of each of these constituents are different and the relative contribution of each to energy transfer processes depends upon the structure and texture of the deposit and the proximity of the material to the melting point. Fig. 2D.5 illustrates the processes influencing energy exchange in the earth-snow-atmosphere system and their inter-relationships. While this basic model is sufficiently complex, it is further complicated by the fact that the contribution of each of the parameters is a continuous variable, and a change in one modifies the effect of the others to some extent. In this discussion, we will consider the mechanisms of convective, conductive and radiative energy transfer separately in spite of the fact that such a distinction does not normally exist in nature.

RADIATIVE TRANSFERS

The radiation flux at the surface of a snow deposit depends on the position of the sun (time of day and season of year), snow properties, surface characteristics and on meteorological conditions (atmospheric temperature, water vapour content, turbidity and cloudiness). Together, these factors determine the radiation exchange which plays the major role in the thermal regime of a snow deposit.

Of the several components which together constitute the radiation balance of a snow deposit, radiant energy from the sun (direct plus diffuse) is the largest. Ninety-nine per cent of solar radiation is between 0.15 and 4.0 μm in the electromagnetic spectrum with maximum intensity in the visible bands at about 0.5 μm (Sellers, 1965). Of the total radiation incident on the surface of a snow deposit, some is reflected at the surface and the rest penetrates into the deposit. The ratio of reflected to incident radiation, or *albedo*, is at least partially dependent upon properties of the snow deposit (Eckel and Thams, 1939; Liljequist, 1965). Newly fallen snow may reflect 80–90 per cent of the incident solar radiation, while old granular snow may reflect as little as 40 per cent. In general, albedo decreases with increasing density, increasing crystal size, an increase in the incident angle of the radiation and with increasing surface temperature. The presence of liquid water or some contaminant, such as windblown dust particles, also serve to decrease the albedo.

A second source of radiant energy is that emitted by the atmosphere as long-wave (infra-red) radiation. This is included in the electromagnetic spectrum between about 3 and 80 μm (Sellers, 1965) and results primarily from absorption of terrestrial infra-red radiation by water vapour and carbon dioxide in the atmosphere. Snow is very nearly a perfect black body for long-wave radiation, absorbing nearly all such radiation and emitting the maximum possible radiation for its surface temperature in accordance with Stefan's Law (Falkenberg, *in* Eckel and Thams, 1939). The maximum amount of energy which can be lost by the snow through the mechanism of long-wave radiation can be easily calculated since the surface temperature of snow can never rise above 0°C. Variations in the amount of long-wave radiation incident upon the snow surface depend almost entirely on the amount of cloudiness and the temperature of the clouds. For a cloudless atmosphere, variations in the amount of this radiation depend upon the temperature and water vapour content of the total thickness of the atmosphere. A number of techniques have been devised to calculate the net long-wave radiation flux. These are summarized by Sellers (1965).

The net radiation at the snow–air interface is expressed as the difference between the total all-wave radiation flux downward to the snow surface (incoming solar and atmospheric long-wave) and the total all-wave flux upward from the snow surface (reflected solar and long-wave emission).

If all radiation fluxes directed toward the snow surface are considered positive and those directed away are considered negative, the radiation balance may be written:

$$R_n = S (1 - \alpha) - (L_o - L_i)$$

where R_n is the net all-wave radiation gain (+) or loss (−) by the snow (usually positive during the day and negative at night); α is the albedo of the snow surface; S is the incident short-wave radiation; L_o is the outgoing long-wave radiation; and L_i is the incident long-wave atmospheric radiation.

The net radiation (R_n) must balance the sum of:

1. the energy involved in the change of state of snow at and near its surface (M);
2. the conduction of heat in the snow (G); and
3. the turbulent transfer of sensible (H) and latent heat (LE) between the atmosphere and the snow surface.

Within a snow deposit, radiation is a rather poor form of energy transfer. The penetration of short-wave radiation into snow is some exponential function of depth, such that:

$$I_z = I^* e^{-kz} \quad \text{(Hoeck, 1952)}$$

where I_z is the radiation intensity at depth z; I^* is the short-wave radiation incident at the surface; e is the base of the natural logarithms; and k is the absorption coefficient of snow for short-wave radiation.

There is little long-wave energy transfer within the snow deposit. For this form of energy transfer to be effective, large and persistent surface temperature differences must exist. Normally, this condition is not met in a snow deposit where only small temperature differences are present.

CONVECTIVE HEAT TRANSFERS

Convective energy transfers in a snow deposit involve the movement of air and water vapour into and through the deposit, carrying both sensible and latent heat. Turbulence in the surface layer of the atmosphere is largely responsible for the convective transfer of energy across the air–snow interface. Within the deposit, convective transfer also occurs as a result of temperature or pressure gradients which may develop in response to influences within the deposit or external to it. As might be expected, convective transfer is highly dependent upon the porosity of the deposit and is most effective at lower densities.

Near-surface meteorological influences promoting convective energy transfer are the wind speed profile, and the air temperature and vapour pressure gradients, all measured within a few metres of the surface. Within the deposit, the most critical characteristic is a temperature gradient in which temperatures decrease toward the upper surface.

For convective transfers to occur between the atmosphere and the snow deposit, near-surface air temperatures ideally must decrease with height above the surface. However, the air temperature lapse rate over a snow surface is normally characterized by an inversion, in which temperatures increase with height. The thickness of this inversion is largely determined by the intensity and duration of the long-wave back radiation. This condition will inhibit turbulence at the surface and, in general, decrease the amount of energy available for transfer into the deposit. Under calm conditions, heat transfer must proceed by diffusion along temperature gradients. Since diffusion can only proceed from zones of higher temperature toward those of lower temperature (a 'negative' gradient) the amount of convective energy transfer which takes place is determined by the type and magnitude of the temperature gradient which exists between the base and upper surface of the snow deposit. If the snow temperatures are higher near the surface of the deposit than at the base, convection can occur only if it is 'forced' by near-surface wind shear.

CONDUCTIVE HEAT TRANSFERS

Conduction is usually the primary heat transfer process when energy passes through a large thickness of dense, unaspirated snow. It takes place through the crystal matrix and, to a lesser degree, through the interstitial air. The coefficient of thermal conductivity may be expected to vary with the structure and texture of the deposit; it should increase as intergranular bond growth increases the continuity of the ice matrix. Conduction can only take place when a temperature gradient exists and cannot therefore be a factor in isothermal snow deposits. A number of experimental determinations of thermal conductivity in snow have been made, usually as some function of density. A wide range of values for this parameter have been obtained which Mellor (1964a) suggests may be at least in part explained by the difficulty of eliminating the effects of heat transfer by vapour diffusion.

The most detailed studies of the energy flux into natural snow deposits have been conducted in the dry snow cover of the polar ice caps. There it has been found that subsurface temperatures fluctuate periodically in response to the cyclic variation of surface air temperature. Temperature changes with depth in cold dry snow have been interpreted as a problem in conduction through a solid medium without phase changes. This assumption is justifiable in terms of the low temperatures and ice crusts, which inhibit air movement through the relatively high-density snow. While these conditions do not necessarily apply to the seasonal snowpack, the empirical calculations discussed should apply during those periods when free water is not present and basal energy exchange is not a significant factor.

For a cold, dry snow deposit in which no heat is introduced geothermally, one may assume that the temperature at any given time and depth does not vary appreciably and that radiation, which is effective only near the surface, does not contribute appreciably to the energy balance. In this case, the problem reduces to one of linear conduction in a semi-infinite solid according to the equation:

$$\frac{\partial^2 \theta}{\partial z^2} - \frac{1}{a}\frac{\partial \theta}{\partial t} = 0$$

where the diffusivity a is given by a $= k/\gamma c$ (k = conductivity, γ = snow density, c = specific heat).

It is necessary to develop a solution for the boundary conditions:

$$z = 0 \qquad \theta = f(t)$$
$$z \to \infty \qquad \theta = \theta_m$$

θ_m is the mean annual surface temperature for the location in question and f(t) is some harmonic function of the annual temperature cycle.

For a snow deposit in which temperature changes take place only at one surface then:

$$\theta_0, t = \theta_m + A_0 \sin(2\pi nt)$$

where θ_0, t is the surface temperature at time t (t is measured from the moment in spring when surface temperature rises above the annual mean temperature, i.e. t = 0 when $\theta_{z=0} = \theta_m$); θ_m is the mean annual surface temperature; A_0 is the amplitude of the surface temperature wave; and n is the wave frequency (one cycle/year); and the solution of the third equation is

$$\theta_z, t = \theta_m + A_0\, e^{-z} \left(\frac{\pi n}{a}\right)^{1/2} \sin\left(2\pi n t - z\left(\frac{\pi n}{a}\right)^{1/2}\right)$$

In this case, therefore, the temperature at any depth, z, in the snow deposit will fluctuate sinusoidally around the annual mean air temperature, reduced below that of the surface wave by a factor:

$$e^{-z}\left(\frac{\pi n}{a}\right)^{1/2}$$

and lagging behind the surface wave by an angle of $z(\pi n/a)^{1/2}$ radians (fig. 2D.6). At any depth in the snow deposit, the wave amplitude and time lag are given by

$$A_z = A_0\, e^{-z} \left(\frac{\pi n}{a}\right)^{1/2}$$

$$t = \frac{z}{2(\pi n a)^{1/2}}$$

and the temperature wave travels into the deposit with an apparent velocity:

$$V = 2(\pi n a)^{1/2}$$

This approach is not directly applicable to the seasonal snow deposit for a variety of reasons. In terms of the applicable boundary conditions, z does not approach ∞ and θ can no longer be defined in terms of θ_m because of the transient annual nature of the deposit. Also, energy is introduced into the deposit at both its upper and lower surfaces which will lead to more than a single value of θ and A_0. In spite of this, however, the model discussed above does provide some insight into the manner in which the temperature at any given level in a snow deposit fluctuates, so long as the deposit in question contains no liquid water. It should be possible to define θ, and therefore A_0, in terms of any desired time period.

While it is possible to approximate the relationship between air temperature and snow temperature for any given period of time in terms of a model in solid conduction, this should not be interpreted as indicating that this transfer mechanism is either the sole or dominant means whereby energy is transferred into and through a snow deposit. As we have discussed above, radiation, convection and conduction all play a part in the energy exchange between a snow deposit and its environment. The dominant process will vary, depending upon the nature of the relevant environmental influences and the portion of the deposit with which one is concerned, but in terms of our present intepretation of the available empirical data it seems that valid extrapolations can be made in terms of conduction alone. This fact has been used by Benson (1962) and Mock and Weeks (1965) on the Greenland and Antarctic ice sheets to determine the mean annual air temperature over extensive areas for which no pre-existing climatological records were available. In both cases, temperatures were obtained from the dry snowpack at a depth of 10 m, which

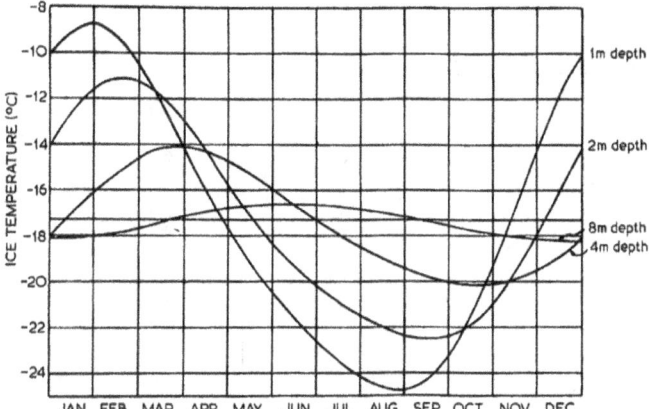

Fig. 2D.6. Annual temperature waves (smoothed) at various depths in the upper layers of an Antarctic ice shelf, illustrating the decreasing effect of the surface temperature fluctuation with depth (from Mellor, 1964a; after Schytt, 1958).

is assumed to be the depth of zero temperature amplitude, and these values were then used to construct maps of areal mean annual temperature variations as some function of latitude, longitude or elevation. In terms of surface energy exchange, it is possible to suggest that we have more climatological information for the accumulation areas of Greenland and Antarctica than for any other comparably-sized area of the earth's surface.

6. Mechanical properties

The mechanical properties of snow have received considerable attention from both theoretical and empirical workers in the field of snow physics. The results are difficult to summarize, both because they are extensive and because they are ambiguous.

The study of the mechanical properties of snow is, in spite of the many research papers published on the subject, a relatively new area of scientific investigation. The first workers in the field were interested in improving avalanche defence construction techniques and devising more efficient methods for removing snow from roads. More recent studies have involved the construction of airfields on the surface of the Greenland and Antarctic ice sheets, and the design of structures on and beneath the surface of these ice sheets. Because an understanding of the stability of a snow deposit lying on a mountain slope relates directly to our ability to forecast avalanche potential, many of the published studies of snow mechanics have originated with workers concerned with the problem of avalanching.

While snow mechanics was originally approached as an analogue of soil mechanics, it was quickly discovered that there were some fundamental differences between soil and snow and that generalizations were not necessarily possible or desirable. These differences stem mainly from basic thermodynamical and structural differences which will not be outlined here. Because of the wide range of quantitative relationships which

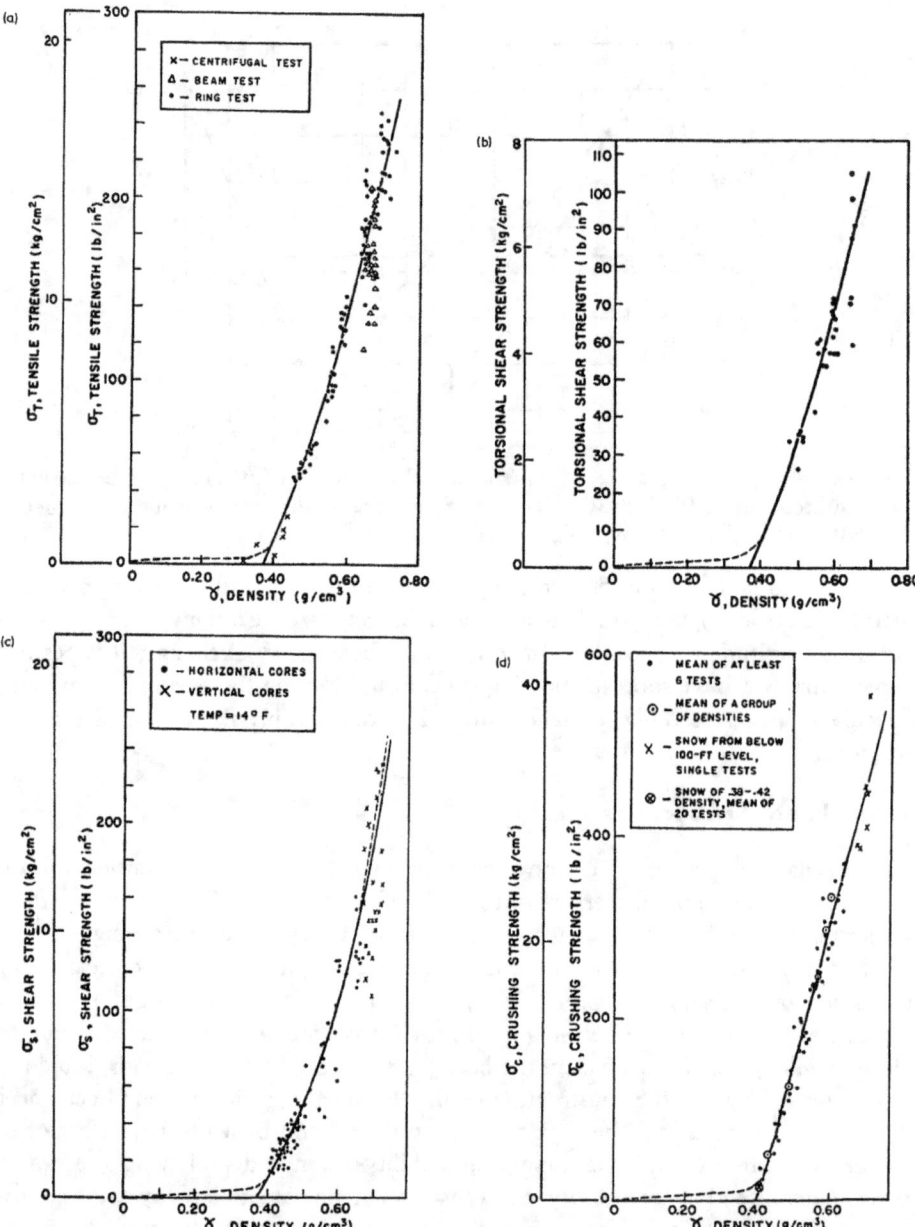

Fig. 2D.7. Some mechanical properties of snow (from Butkovitch, 1956). (a) Tensile strength of snow measured as some function of density. (b) Torsional shear strength of snow measured as some function of density. (c) Shear strength of snow measured as some function of density. These measurements were obtained by shearing out the central section of a cylinder (double shear). (d) Crushing strength of snow measured as some function of density.

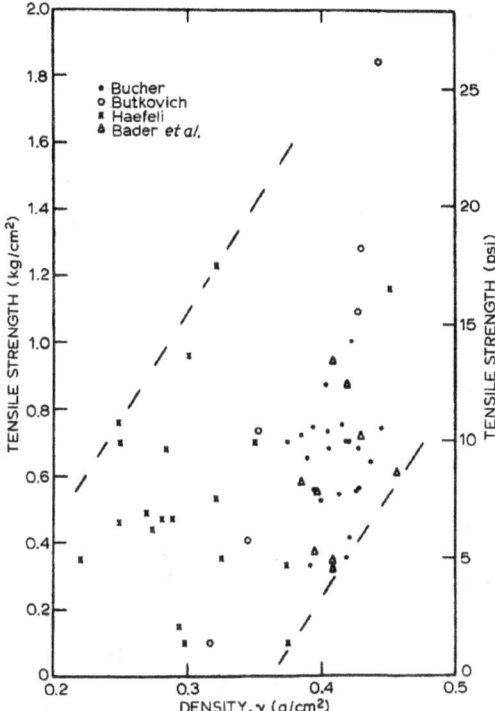

Fig. 2D.8. Tensile strength of snow measured as some function of density for low-density snow (from Mellor, 1964a). Measured by the centrifugal method. Note the much greater scatter of these data than those for the higher density snow shown in fig. 2D.7.

have been presented in the literature, it is more useful to present most of the information in a qualitative fashion.

The strength characteristics of snow deposits are derived primarily from the number of intergranular bonds present and the temperature of those bonds. As the number of intergranular bonds per unit volume decreases or their temperature increases, strength will commonly decrease. It has been found that a useful index to the number of intergranular bonds per unit volume is the bulk density of the snow. Most workers have successfully correlated bulk density with the various measured strength properties at the higher densities (fig. 2D.7). Density does not appear to be a useful index parameter at lower densities, however (fig. 2D.8). It may be possible to explain this in one of several ways. All strength properties are temperature-dependent and, therefore, the degree of thermodynamic stability which has been achieved by the snow deposit can be expected to influence the strength characteristics of that deposit. Since low-density snows are commonly composed of more geometrically complex grains than are the higher density snows, it should also prove less stable in terms of small temperature fluctuations and gradients. Secondly, the packing geometry of the snow grains varies with density, such that as density increases, a 'closest packing' relationship is approximated. When an external mechanical stress is applied to a snow deposit in which closest packing does not exist, the individual grains are free to shift in relation to one another. This readjustment of the geometry of the matrix is sufficient to modify many of the mechanical properties of the mass.

There are other aspects of a natural snow deposit which make it extremely difficult to obtain valid measurements of mechanical properties, irrespective of the density. In general the inhomogeneity of snow deposits makes it difficult to obtain a sample which reproduces the bulk properties of the entire deposit. Snow is anisotropic (i.e. the value of any property often changes directionally) and, in testing, the rate at which a stress is applied greatly influences the nature of the resulting strain.

In spite of the difficulties which have been experienced in developing a general quantitative model which will accurately describe snow mechanics, four types of process can be distinguished qualitatively. These are *collapse mechanics*, in which some form of total failure of the intergranular bonds occurs. Since ultimate strength in tension, compression and shear is given by the resistance of the deposit to structural collapse, the characteristic feature of collapse mechanics is a sudden loss of cohesion and the breaking of intergranular bonds with or without the development of cracks. In tension and pure shear, the snow mass often breaks in one plane; in compression it sometimes crumbles but, when unconfined, often shows the conical failure indicating shear surfaces. A second type of deformation under stress is *creep mechanics*, which occurs at stresses smaller than those leading to structural collapse. Here, intergranular bonds are not broken but are rather plastically deformed, so that there is no sudden change in the cohesion of the mass. Perhaps the most common example of this type of deformation is found in the 'slow flow' of snow from a roof, forming a curl of snow around the eaves at the roof edge. If a sudden structural collapse occurs in a snow deposit lying on a mountain slope, a continuing rapid deformation may occur. This constitutes *flow mechanics*. While in collapse mechanics the entire mass constitutes the rigid unit, in flow mechanics the rigid units are individual grains or groups of grains. Depending upon the liquid water content of the mass, these may either become more completely disaggregated or, if sufficient moisture is present, re-aggregate into larger coherent masses. Flow mechanics is involved in rapid sliding along shear planes within a snow mass and is characteristic of all types of avalanches. It is also a factor in snow-ploughing, ski-ing and the movement of oversnow vehicles. We have already mentioned that the nature of the deformation is, in part, related to the rate at which the stress is applied. In this respect, there is a fourth type of mechanics called *elasticity mechanics*, which is essentially creep under the influence of an instantaneous stress. Snow is not normally elastic but it will deform in this way if the stress is applied very rapidly, as in an explosion. This type of deformation is relevant to any undertaking involving the propagation of high energy into and through a snow deposit, such as seismological depth sounding on the Antarctic and Greenland ice sheets. Elastic deformation causes a very rapid attenuation of the original energy source.

The relationship between snow mechanics and snow properties still requires much study. Mechanical testing procedures need to be refined and standardized in terms of the existing models so that the time-, stress-, and temperature-dependent nature of snow properties are incorporated in the results. With the possible exception of mass, all snow properties can be expected to change during testing and to be influenced by the experimental design. Because of this, unless extreme care is exercised during testing, unique results will be obtained in place of what is actually a series of continuous variables. It seems quite probable that it is this fact which has contributed largely to the wide variations in the experimental results obtained by workers in the field of snow mechanics.

Two excellent summaries of the present state of our understanding of snow mechanics are given by Mellor (1964) and Bader *et al.* (1939).

7. The importance of snow to man

Our approach in this discussion of the material snow has been that solutions to specific problems can come only in terms of the application of general models. Because of this, we have largely avoided consideration of the material in terms of any specific geographic location. It may be useful, however, to conclude by indicating a few of the specific impacts that snow has on the activities of man in order to show where the application of the principles discussed above may produce worthwhile results in the future. Snow influences the activities of man primarily from the standpoint of the potential hazards which may be associated with the material, from the economic standpoint and in terms of water supply.

Avalanches are the most spectacular example of the potential hazards caused by snow accumulation (Plate 4). They are likely to occur wherever there are deep accumulations of snow on steep mountain slopes. No reliable estimates of the number of avalanches which occur annually in all the earth's mountain ranges are available, but a rough estimate, based upon the known density and frequency of avalanche occurrence in Switzerland – a country in which they have been intensively studied – suggests that between 10^5 and 10^6 avalanches may occur each year over the globe. Obviously, the majority of these do not occur in areas occupied by man. Their primary impact is geomorphic and hydrologic. Inevitably, however, the activities of man are conducted in areas in which the potential for avalanching exists, and it is in those areas that property damage, injuries and fatalities occur.

Understanding the causes of avalanches involves a knowledge of virtually all the processes which affect a snow deposit. The physical environment in which the snow crystals constituting the deposit are formed is of great significance, since the crystal geometry at the time of deposition determines the degree to which the grains interlock. The initial mechanical strength of the deposit is largely determined by this interlocking. Metamorphic processes act to destroy the initial crystal geometry. This may lead to either an increase or decrease in the intergranular bonding, which in turn may either increase or decrease the strength of the snow mass. The snow's own weight introduces internal stresses to the material and, since snow creeps readily under sustained loading, compressive bulk stresses cause it to compact gradually to higher bulk density, thus tending to increase its strength. Hence the mechanical properties of snow tend to change with time, the rate being determined by the original snow type and by prevailing weather conditions, which control the thermal state of the deposit.

Both the bulk and shear deformation rates depend on the structure and temperature of the snow; in general, deformation increases as snow density decreases and temperature increases. Stresses, which for any point in a snow mass are largely determined by the inclination of the slope and the weight of the overlying snow, largely control the deformation rate. When stresses become critically high, the snow may rupture spontaneously. Such a failure may occur either by an increase of stress following the deposition of succeeding snowfalls or by a decrease in strength following a temperature rise or a period of adverse metamorphism of the structure of the deposit. Once such a failure has occurred, it may propagate spontaneously allowing a mass of snow to slide downhill, dislodging the delicately balanced snow which lies in its path. The conversion of potential to kinetic energy, as large masses of snow fall through long vertical distances, creates one of the more impressive catastrophic forces of nature.

On a worldwide basis, avalanche activity probably fluctuates primarily in response to climatic changes, but avalanche hazard is directly related to the population of mountain regions and to the extent of human activity in those regions. While this hazard is still minimal if one considers only the number of people who live permanently in such regions, it is greatly increased if all those who are transient in the mountains are taken into consideration. These transients are mostly those driving on roads which traverse high mountain ranges and those who go to the mountains for recreational purposes, primarily ski-ing and snowmobiling. Because winter recreation in the mountains is a growing sport and industry, it is likely that property damage and loss of life from avalanche activity will show a steady upward trend in the future.

There are few useful estimates of the economic impact of snow (Maunder, 1970). It is obvious, however, that snow must be considered a parameter in the economic equation. On the positive side, water released by the melting snowpack used to generate hydroelectric power and agricultural products grown on lands irrigated by snow meltwater are two immediate examples. In the latter context, the Imperial Valley of California is completely dependent upon irrigation water produced by melting snow in the nearby Sierra Nevada Mountains. Virtually all agriculture in the semi-arid western United States is dependent to some extent on water supplied by the melting mountain snowpack.

Many regions of the United States, Canada and large sections of Europe derive a large annual income from snow-related recreation, such as ski-ing. In the United States alone, the number of skiers has increased from approximately 50,000 twenty years ago to over 4 million today. In addition to those areas which derive income directly from skiers in pursuit of their sport, a large number of peripheral industries, which manufacture and distribute equipment used in winter sports activities, provide jobs for many individuals.

On the negative side of the economic coin, the disruption of transportion and communication by heavy snowfalls has both regional and national consequences (Rooney, 1969). The economic gains resulting from the constructive use of meltwater runoff are sometimes offset by the floods resulting from the too-rapid melting of the mountain snowpack in the spring, which may cause extensive damage and loss of life.

We have mentioned the importance of snow as a water supply source in terms of its economic implications. From any standpoint involving man's dependence on water, it is difficult to overemphasize the importance of snow in the hydrologic cycle. However, there appear to be no reliable estimates of the amount of water which is deposited annually in the form of snow and removed by melt on a continental or worldwide scale.

In part this stems from the basic problem of measuring snowfall in arctic and alpine areas. Surveys and water balance considerations indicate that measured precipitation may be at least 35 per cent too low in much of the Arctic and Subarctic of North America (Findlay, 1969; Findlay and McKay, 1972; Hare and Hay, 1971). Soviet comparisons of shielded gauges in open and sheltered sites show that when the mean temperature is below $-20°C$ the gauges in open sites underestimate the totals by 80 per cent for a monthly mean wind speed of 2.5 m s^{-1} (at 2 m) and by 200 per cent for a mean speed of 5 m s^{-1} (Struzer et al., 1965). As far as accumulation is concerned, snow course measurements provide useful and extensive data for mountain areas in the United States (for example, Brown and Peck, 1962; Grant and Schleusner, 1961). Even in the absence of such quantitative data, however, certain aspects of the role of snow in the hydrologic cycle are worth a limited discussion in qualitative terms.

Quite obviously, the importance of snow as a source of meltwater in any given region is largely dependent upon the proximity of that region to the actual or theoretical snowline. Thus, while a river such as the Yukon in Alaska may be almost wholly dependent upon melting snow to sustain its flow, tropical rivers derive only a very small percentage of water from snowmelt. About 50 per cent of the total annual volume of flow of the Missouri River, which originates in the mountains of southwestern Montana and northern Wyoming, is supplied by melting snow, and similar estimates have been made for rivers draining the Sierra Nevada Mountains in California. The importance of the melting of mountain snowpacks to streamflow has recently been emphasized by attempts by the United States Government to increase the total volume of streamflow by 'seeding' clouds in winter frontal systems moving across selected mountain ranges, thus increasing the depth of the annual snow accumulation.

From the management standpoint, the hydrologic potential of the mountain snowpack presents a number of largely unexplored potential advantages. Perhaps as much as 90 per cent of all water derived from this source is deposited on less than 10 per cent of the surface area of a catchment area of major rivers such as the Columbia or Missouri. Annual variations in the timing and volume of water produced by the mountain snowpack are much less than in those rivers which are more dependent upon rainfall. This increases the reliability of streamflow forecasting for those rivers. Perhaps the primary research need in this area at the present time is a more thorough study of the contribution of meltwater from the alpine belt of major mountain ranges. The limited data we have at the present time are derived largely from the montane belt, and it is on these data that much of the forecasting is presently based. It is possible at least to suggest that the alpine belt may ultimately prove to be more important as a snow storage and meltwater production zone.

The study of snow is only one part of the science of glaciology. Serious investigations of this material are relatively recent and its importance to man is only now being realized.

From the scientific standpoint, there is still much to be learned and, in terms of effective management of snow as a natural resource, we have not even begun to scratch the surface of its potential.

References

BADER, H. and KUROIWA, D. (1962) The physics and mechanics of snow as a material. *Cold Regions Science and Engineering Monogr.*, II-B. Hanover: US Army CRREL. 79 pp.

BADER, H., HAEFELI, R., BUCHER,,E., NEHER, J., ECKEL, O. and THAMS, C. (1939) Der Schnee und seine Metamorphase (Snow and its metamorphism). *Beitr. zur. Geologie der Schweiz*, Geotech. Serie, Hydrol., Lieferung 3. Berne. 313 pp. (English edn, Transl. 14, US Army Snow, Ice and Permafrost Res. Estab., 1954.)

BAGNOLD, R. (1941) *The Physics of Blown Sand and Desert Dunes*. London: Methuen. 265 pp.

BENSON, C. (1962) Stratigraphic studies in the snow and firn of the Greenland ice sheet. *SIPRE Res. Rep.*, 70. US Army Snow, Ice and Permafrost Res. Estab. 93 pp.

BRADLEY, C. and ALFORD, D. (1967) Yellowstone field research expedition report. *In: Final Rep., 7th Yellowstone Field Res. Expedition*, Pub. 45, 29–43. Atmosph. Sci. Res. Center, State Univ. of New York.

BROWN, M. J. and PECK, E. L. (1962) Reliability of precipitation measurements as related to exposure. *J. Appl. Met.*, **1**, 203–7.

BUCHER, E. and ROCH, A. (1946) *Reibungs-und Packungswiderstande bei raschen Schnee-bewegungen*. Mitteilungen des Eidgenoss. Davos-Weissfluhjoch: Inst. für Schnee und Lawinenforschung. 9 pp.

BUTKOVITCH, T. (1956) Strength studies of high-density snow. *SIPRE Res. Rep.*, 18. US Army Snow, Ice and Permafrost Res. Estab. 19 pp.

DE QUERVAIN, M. (1945) Schnee als kristallines Aggregat. *Experimentia*, 1. 7 pp.

DICKSON, R. R. and POSEY, J. (1967) Maps of snow-cover probability for the northern hemisphere. *Mon. Wea. Rev.*, **95**, 347–53.

DORSEY, N. (1948) The freezing of supercooled water. *Trans. Amer. Phil. Soc.*, **38** (3), 247–328.

ECKEL, O. and THAMS, C. (1939) Untersuchungen über Dichte-Temperatur- und Strahlungsverhältnisse der Schneedecke in Davos (Investigations on conditions of density, temperature and radiation of the Davos snowcover). *In:* BADER, H. *et al.* (eds.) Der Schnee und seine Metamorphose (Snow and its metamorphism). *Beitr. zur Geologie der Schweiz*, Geotech. Serie, Hydrol., Lieferung 3. Berne. (English edn, Transl. 14, US Army Snow, Ice and Permafrost Res. Estab., 1954, 247–301.)

EUGSTER, H. (1952) Beitrag zu einer Gefugeanalyse des Schnees. *Beitr. zur Geologie der Schweiz*, Geotech. Serie, Hydrol. Berne. 64 pp.

FINDLAY, B. F. (1969) Precipitation in northern Quebec and Labrador: an evaluation of measurement techniques. *Arctic*, **22**, 140–50.

FINDLAY, B. F. and MCKAY, G. A. (1972) Climatological estimation of Canadian snow resources. *In:* ADAMS, W. P. and HELLEINER, F. M. (eds.) *International Geography 1972*, Vol. 1, 138–41. Toronto: Univ. of Toronto Press.

GOLD, J. and POWER, B. (1952) Correlation of snow-crystal type with estimated temperature of fusion. *J. Met.*, **9** (6), 447.

GRANT, L. O. and SCHLEUSNER, R. A. (1961) Snowfall and snowfall accumulation near Climax, Colorado. *Proc. 29th Western Snow Conf.*, 53–65.

HARE, F. K. and HAY, J. E. (1971) Anomalies in large-scale annual water balance over northern North America. *Can. Geogr.*, **16**, 79–94.

HERMES, K. (1965) Der Verlauf der Schneegrenze. *Geogr. Taschenbuch* 1964–5, 58–71.

HOECK, E. (1952) Der Einfluss der Strahlung und der Temperatur auf den Schmelzprozess der Schneedecke (Influence of radiation and temperature on the melting process of the snow cover). *Beitr. zur Geologie der Schweiz*, Geotech. Serie, Hydrol., Lieferung 8. (English edn, Transl. 49, US Army Snow, Ice and Permafrost Res. Estab., 1958, 1–36.)

KONDRATIEVA, A. (1945) Teploprovodnost snegovogo pokrova i fizicheskie protsessy proiskhodiaschie v nem pod vlianiem tempeeraturnogo gradienta (Thermal conductivity of the snow cover and physical processes caused by the temperature gradient). *In: Physical and Mechanical Properties of Snow and Their Utilization in Airfield and Road Construction*. Moscow–Leningrad: Akad. Nauk SSR. 13 pp. (English edn, Transl. 22, US Army Snow, Ice and Permafrost Res. Estab., 1954.)

KUMAI, M. (1951) Electron-microscope study of snow-crystal nuclei. *J. Met.*, **8**, 151–6.

LACHAPELLE, E. (1969) *Field Guide to Snow Crystals*. Seattle: Univ. of Washington Press. 101 pp.

LILJEQUIST, G. (1956) Short-wave radiation. *In: Scientific Results of the Norwegian–British–Swedish Antarctic Expedition*, Vol. II, 1–109. Oslo: Norsk Polarinstitut.

MASON, B. (1962) *Clouds, Rain and Rainmaking*. Cambridge: Cambridge Univ. Press. 145 pp.

MAUNDER, W. J. (1970) *The Value of the Weather*. London: Methuen. 388 pp.

MELLOR, M. (1964a) Properties of snow. *Cold Regions Science and Engineering Monogr.*, III-A. Hanover: US Army CRREL. 105 pp.

MELLOR, M. (1964b) Snow and ice on the earth's surface. *Cold Regions Science and Engineering Monogr.*, II-C-1. Hanover: US Army CRREL. 163 pp.

MOCK, S. and WEEKS, W. (1965) The distribution of ten-meter temperatures on the Greenland ice sheet. *Cold Regions Science and Engineering Res. Rep.*, 170. Hanover: US Army CRREL. 21 pp.

NAKAYA, U. (1954) Formation of snow crystals. *SIPRE Res. Pap.*, 3. US Army Snow, Ice and Permafrost Res. Estab. 12 pp.

NAKAYA, U. and MATSUMOTO, A. (1953) Evidence of the existence of a liquidlike film on ice surfaces. *SIPRE Res. Pap.*, 4. US Army Snow, Ice and Permafrost Res. Estab. 6 pp.

PORTMAN, D. and RYZNAR, E. (1961) Thermodynamic studies of a snow cover in northern Michigan. *SIPRE Res. Rep.*, 74. US Army Snow, Ice and Permafrost Res. Estab. 73 pp.

RAMSEIER, R. and KEELER, C. (1966) The sintering process in snow. *J. Glaciol.*, **6** (45), 421–4.

ROONEY, J. F. JR (1969) The economic and social implications of snow and ice. *In:* CHORLEY, R. J. (ed.) *Water, Earth and Man*, 389–401. London: Methuen.

SCHYTT, V. (1958) Snow and ice temperatures in Dronning Maud Land. *In: Glaciology II, Scientific Results of the Norwegian–British–Swedish Antarctic Expedition*, Vol. IV. Oslo: Norsk Polarinstitut. (Cited in MELLOR, 1964a, p. 67.)

SELLERS, W. (1965) *Physical Climatology*. Chicago and London: Univ. of Chicago Press. 272 pp.

SOMMERFELD, R. and LACHAPELLE, E. (1970) The classification of snow metamorphism. *J. Glaciol.*, **9** (55), 3–17.

STRUZER, L. R., NECHAYER, I. N. and BOGDANOVA, E. G. (1965) Systematic errors of measurements of atmospheric precipitation. *Soviet Hydrol.*, **5**, 500–4.

Manuscript received June 1971.

Additional references

IVES, J. D., HARRISON, J. C. and ALFORD, D. L. (1972) *Development of methodology for evaluation and prediction of avalanche hazard in the San Juan mountain area of southwestern Colorado.* INSTAAR-14-06-7155, US Bureau of Reclamation. Denver: Dept of Interior. 88 pp.

IVES, J. D., HARRISON, J. C. and ARMSTRONG, R. A. (1973) *Development of methodology for evaluation and prediction of avalanche hazard in the San Juan mountain area of southwestern Colorado.* INSTAAR-14-06-7155-2, US Bureau of Reclamation. Denver: Dept of Interior. 122 pp.

JUISTO, J. E. and WEICKMANN, H. K. (1973) Types of snowfall. *Bull. Amer. Met. Soc.*, **54**, 1148–62.

MARTINELLI, M. JR (1973) Snow-fence experiments in alpine areas. *J. Glaciol.*, **12** (65), 275–90.

MIKHEL, V. M., RUDNEVA, A. V. and LIPOVSKAYA, V. I. (1971) *Snowfall and Snow Transport during Snowstorms over the USSR.* Jerusalem: Israel Progr. Sci. Transl. 174 pp. (Original Russian edn, Leningrad, Glav. Uprav. Gidromet. Sluzhby, 1969.)

TRABANT, D. and BENSON, C. S. (1972) Field experiments on the development of depth hoar. *Geol. Soc. Amer. Mem.*, 135, 309–22.

Arctic hydrology

Donald K. Mackay and Olav H. Løken

Inland Waters Branch,
Department of Environment, Ottawa

1. Introduction

Arctic hydrology is simply a regional expression of the science of water. In the arctic region, the various elements in the hydrologic cycle differ in intensity, magnitude, and significance when compared with those in mid-latitudinal or equatorial regions. These differences are largely determined by variations in the regional heat balances, by the atmospheric circulation (see Barry and Hare, Chapter 2A), and by the presence of permafrost.

Arctic hydrology is rather difficult to isolate and discuss without reference to subarctic and even temperate regions because the majority of large rivers flowing into the Arctic Ocean have their headwaters in mid-latitudes far south of the treeline and because of the southerly extent of discontinuous permafrost. Major arctic drainage systems, therefore, display the integrated effects of flow through a variety of climatic and vegetative regions that are non-arctic in character. Although accepted boundaries of the Arctic are treated herewith in a somewhat loose manner, the scope of this section is limited in other ways, notably by the treatment of various parts of hydrology in other chapters. For example, moisture balance and snowmelt hydrology are also discussed in Chapter 3B; glaciology, a specialized branch of hydrology, is examined in Chapter 4 and in sections on permafrost (Ives, 4A) and on geomorphic processes in the Arctic (Bird, 12A) deal with hydrology pertaining to ground ice, soil moisture, and some aspects of groundwater. The emphasis here thus lies in a review of present knowledge and research in the fields of surface water hydrology, groundwater, and water management.

The literature on arctic hydrology is surprisingly extensive although somewhat disparate due to economic, geographic and local factors. A large amount of work has been done in arctic Alaska and in the Soviet northlands, yet vast areas of the Canadian Arctic have still to be investigated by hydrologists. A recent spur to research in these areas has come from oil and gas discoveries in the Mackenzie Delta region and the Arctic Islands; the extraction of oil and gas and its transportation to markets is indirectly but significantly related to the hydrology of the Arctic. A comprehensive listing of publications dealing with arctic water resources has been published by the Institute of Water Resources, University of Alaska (Hartman and Carlson, 1970); it contains almost 3000 items and the compilers admit it is not complete. A review of the limnological problems of arctic North America is provided by Livingstone (1966).

2. Surface water hydrology

Estimation of runoff, discharge, and other surface water parameters is normally based on long-term point measurements; in the Arctic, unfortunately, most available records are of short duration and often intermittent. Moreover, in northern Canada and Alaska there are large gaps in the hydrologic data networks which preclude any major discussion of rivers and lakes in these areas. The longest arctic discharge records available are those for major USSR rivers where observations began in the 1930s or earlier (CNC/IHD, 1969).

Mean annual surface runoff in the Arctic is shown in fig. 3A.1. The map is necessarily incomplete owing to a lack of suitable data principally for Greenland and some of the northern islands. In areas where estimates have been made, the isolines should be regarded as first approximations only.

A comparison of runoff in arctic North America and in the Soviet northland suggests some basic differences in patterns and intensities. Runoff in the continuous permafrost zone of the USSR is considerably greater than in the North American zone, the former averaging roughly 10 cm/year higher than the latter. The western mountain chains of North America appear to have less influence on runoff in the Arctic than have the Urals, where significant increases are apparent. The limited runoff in the vast permafrost area of eastern Siberia reflects lower precipitation and the intensified continentality of the landmass, in contradistinction to eastern Canda where the presence of Hudson Bay influences precipitation runoff conditions.

Various components of the water budget in arctic areas have been studied by Dingman (1966), Skakaliskiy (1966), Walker and McCloy (1969), Findlay (1969), and others. The results suggest that water budget components are quite variable with changes in latitude continentality, relief, geology, integration of surface water systems, and other factors. The summary characteristics can only be loosely described in general terms. In tundra regions, the total runoff may average 60–70 per cent of the precipitation; evaporation would thus be in the order of 30–40 per cent assuming precipitation measurements are correct. The groundwater contribution is probably less than 10 per cent of the total runoff but this can increase in the discontinuous permafrost zone to as much as 20–40 per cent of the total, with consequent changes in other components.

In the tundra, runoff in streams rises sharply in late spring due to rapid melting of the winter snowpack. Flow is almost entirely confined to a five-month period, May to September, with appropriate time period adjustments for latitude and continentality.

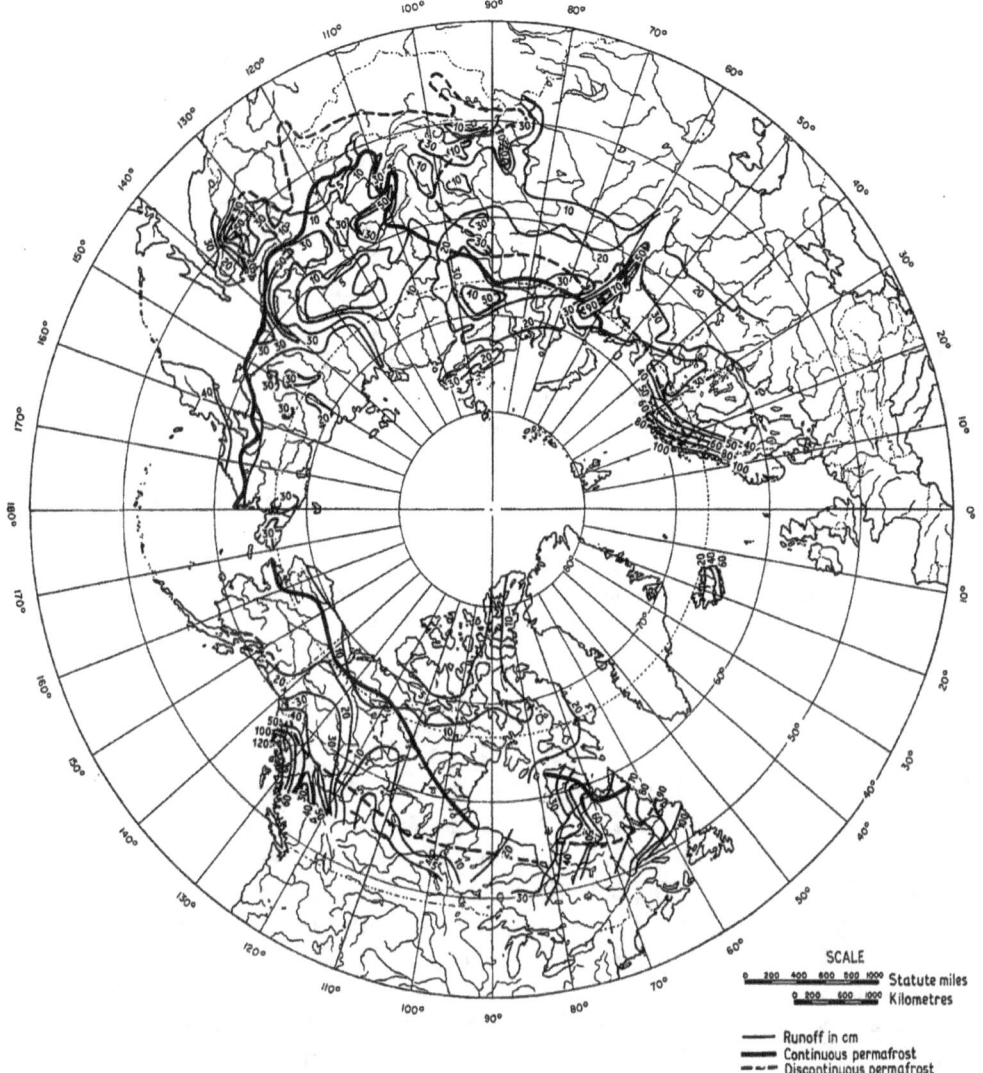

Fig. 3A.1. Mean annual surface runoff (cm) in the Arctic (based on Dreyer, 1968, for the USSR; CNC/IHD, 1969, for Canada; and Chernogaeva, 1970, for Europe). The limits of continuous and discontinuous permafrost are also shown (cf. Ives, Chapter 4A).

Although snowmelt is the main runoff source, intense summer rainstorms have produced flood conditions in some arctic basins exceeding those caused by spring snowmelt (Sokolov, 1966). Major rivers rising in southerly latitudes and flowing through the tundra zone are less susceptible to rainfall flooding because of the ratio of rainstorm area to drainage basin area. The runoff of major rivers is also more equably distributed throughout the year than that of streams originating and contained within the tundra zone.

MAJOR RIVERS

Some of the world's largest rivers flow into the Arctic Ocean (table 3A.1). The Lena, for example, drains about 2.6 million km² along its 4260 km route to the Laptev Sea; moreover, it is navigable over most of this length during the summer high-water season. Another major Siberian river system, the Ob-Irtysh, is 3400 km long and, in its lower reaches, expands to widths approaching 20 km. The third major Siberian river, the Yenisei, is roughly the same length as the Ob-Irtysh but flows at an annual rate that exceeds the latter by more than 40 per cent. These three rivers drain an area roughly comparable to that of the conterminous United States.

The two largest rivers of arctic North America, the Mackenzie and the Yukon, have a combined drainage area equivalent to that of the Lena River. The Mackenzie catchment alone covers nearly one-fifth of the total land area of Canada.

The distribution of flow in major arctic rivers is shown in the same table. Peak discharge occurs earliest in rivers of the European USSR and latest in Canadian and Eastern Siberian rivers. Flow in the peak month averages 32 per cent of the yearly total with both extremes, the highest (46 per cent, Back) and the lowest (14 per cent, Mackenzie), being in Canada. The relatively stable annual regime of the Mackenzie River is due to the regulating effect of Great Slave Lake. Open season flow of the rivers in table 3A.1 is approximately 80 per cent of their annual total.

LAKES

Lakes are common features of the Arctic although their distribution is extremely irregular. For example, in the Canadian Arctic the Shield country west of Hudson Bay is 'littered' with lakes, while further north on the Arctic Islands they are rare or even absent over large areas. The following section discusses some of their characteristic features and properties; the discussion, however, is limited to lakes in the Arctic and the hydrologically very interesting lakes in the Antarctic are therefore excluded.

Arctic lakes are generally small and the largest are all on the mainland. In Canada there are only ten lakes north of 60°N which exceed 1000 km²; Great Bear (31,326 km²) and Great Slave (28,568 km²) lakes are by far the largest of these (Gilliland, in press). There are, however, 198 lakes with a surface area of more than 100 km². The biggest lakes on the Arctic Islands are Netilling (5525 km²) and Amadjuak (3105 km²) on Baffin Island, and Lake Hazen (518 km²) on Ellesmere Island. The number of lakes is greatest on the flat coastal plain of Alaska, adjacent parts of Canada and, in general, in flat drift-covered areas such as Keewatin, central Labrador–Ungava and southwestern Baffin Island.

Formation

With one important exception, the origin of arctic lakes is similar to the origin of lakes in other parts of the world, although some regional variations in the mode of formation can be seen.

Table 3A.1. Distribution of flow (monthly percentage) in arctic rivers.

	Drainage area (10 km^2)	Jan.	Feb.	Mar.	Apr.	May	June	July	Aug.	Sept.	Oct.	Nov.	Dec.	Mean annual flow ($m^3\ s^{-1}$)
Alaska														
Yukon	7,670	1.8	1.5	1.2	1.2	10.8	_24.6_	17.6	15.9	13.2	13.2	3.2	2.0	6,210
Canada														
George	386	1	1	1	1	4	_37_	19	11	11	8	4	2	793
Back	858	1	1	1	1	1	11	_46_	18	10	6	3	1	524
Mackenzie	18,425	6	6	5	4	7	13	_14_	11	10	9	8	7	10,800
Porcupine	541	1	1	0	0	22	_30_	16	16	10	2	1	1	447
USSR														
Indigirka	3,050	0.22	0.11	0.06	0.04	1.6	30.0	_30.5_	26.4	11.5	2.6	0.70	0.43	1,550
Kolyma	3,610	0.42	0.28	0.23	0.20	7.9	_38.2_	19.6	15.3	12.2	3.5	1.1	0.74	2,240
Lena	24,300	1.3	0.97	0.70	0.57	2.3	_37.6_	20.0	13.8	12.5	7.4	1.7	1.4	16,300
North Dvina	3,480	2.5	2.0	1.8	5.7	_34.3_	17.8	7.5	5.6	6.0	7.5	6.0	3.4	3,380
Ob-Irtysh	24,300	3.0	2.5	2.1	2.2	9.9	21.9	_19.9_	15.0	9.3	7.0	4.1	3.4	12,200
Onega	557	2.8	2.4	2.1	7.6	_34.2_	14.4	7.5	5.3	6.2	7.8	6.2	3.6	483
Pechora	2,480	1.6	1.2	1.1	2.0	22.6	_34.0_	10.7	5.5	6.8	8.1	3.9	2.4	3,360
Yenisei	24,400	2.3	2.1	2.0	1.9	14.1	_35.6_	13.1	8.8	8.4	6.9	2.9	2.3	17,800

Note: Months of maximum flow are underlined.

The truly unique arctic lakes are the thermokarst ones, shallow water bodies formed by the melting of the permafrost. These occur in great numbers on the flat coastal plains of the northern coast of Alaska, an area which has become a classic environment for the study of such lakes. An intensive literature has developed about these lakes and their unique yet controversial mode of formation, and Black (1969) provides a valuable review of the current state of knowledge.

Most other lakes are results of the Pleistocene ice sheets which covered almost all of the Arctic. In the central part of regions occupied by the former ice sheets, active ice-scouring formed depressions now filled with lakes. The area west of Hudson Bay is a typical example. In areas marginally situated relative to the former ice sheets are moraine lakes and kettle lakes; both types were created by an ice marginal deposition. Such lakes are common but not restricted to the peripheral zone, as dead ice deposits are common near the centre of the former ice sheets, e.g. Labrador–Ungava, and numerous lakes are usually found in these areas.

Ice-dammed lakes are more common in arctic regions than elsewhere and in Greenland this type accounts for all large lakes. Ice-dammed lakes have been particularly well studied on Axel Heiberg Island where Maag (1969) found 125 in a 500 km² area. He also studied the altitude distribution of these lakes, with the greatest number lying below the equilibrium line on the local glaciers. In Alaska and the adjacent parts of Canada, Post and Mayo (1971) identified a total of 750 ice-dammed lakes. Greenland and Spitzbergen also have numerous examples with Britannia Sø in east Greenland being one of the best known (Lister and Wyllie, 1957). Most ice-dammed lakes are subjected to periodic emptying and are thus ephemeral features. Many lakes existed during the final phases of the Wisconsin glaciation and although the water has disappeared, evidence of these is still conspicuous in terms of sediments and shore features.

Most of the Arctic has been glaciated and has risen isostatically after deglaciation. As a result of this many low-lying depressions have been filled with water and now appear as lakes. Häggblom (1963) and MacLaren (1967) describe such lakes located on Spitsbergen and Baffin Island.

The last type of lake is closely related to the lakes on Ellesmere Island, described by Hattersley-Smith and Serson (1964), where a glacier from a tributary valley has advanced and blocked the inner part of the fiord creating Lake Tuborg. Such lakes have a very distinct aquatic environment with extremely high vertical stability of the water column.

Physical limnology

The long periods with continuous ice cover and the strong seasonal variation in incoming radiation are fundamental features of arctic lakes. Few lakes retain their ice cover for twelve months, but the majority are ice-covered for more than nine months (fig. 3A.7 on p. 123 shows the mean ice thickness formed during the winter). In general the ice is thicker in the High Arctic than further south. This is partly a reflection of lower temperatures, but the higher snowfall in the south which forms an insulating snow layer on top of the lake ice also contributes to the lesser ice thickness.

The formation and dissipation of the ice cover has a fundamental influence on the heat budget of arctic lakes because a very large part of the energy flux that reaches the lake surface is used to melt the ice cover, thus reducing the summer temperature. An important exception to this appears in areas where small shallow lakes are not well integrated into a larger drainage system and therefore, together with a small adjacent land area, can be considered as a closed system with little throughflow. In such cases the water temperature can reach 10–15°C during the summer months even at 70°N latitude.

While the depth of the lakes is important for the summer temperature it also greatly influences the winter conditions as the whole water column will freeze if the lake is too shallow. Freezing to the bottom affects the survival of many life forms during the winter, and it also means that there may be no continuous thawed zone under the lake, hence the lake may not be a recharge or discharge area for groundwater.

Chemical limnology
Little systematic study has been made of arctic lakes, but in general there appear to be two major factors influencing the chemical composition of the water: (1) precipitation, and (2) the local bedrock. The dissolved sediments are highest in the area of low precipitation and in areas of intensive evaporation. Many arctic areas are underlain by Archean rock formation and the amount of dissolved materials is low. However, other lakes, e.g. in the Arctic Islands in Canada, contain large amounts of salts derived from the easily dissolved young sediments.

Biological limnology
The productivity of arctic lakes is low due to the cold climate, the nutrient content is generally low and they must be regarded as oligotrophic. As most of the area was glaciated the fauna has become re-established in a short period of time. The poorly integrated drainage systems have further retarded this process, and there is no major fishery or other harvesting except where migration takes place from the sea along major rivers.

RIVER AND LAKE ICE CONDITIONS

Each year, for more than six months, arctic water bodies are either ice-covered or subject to the transitory phases of ice formation and decay. The significance of ice conditions in northern hydrology is well expressed by the mass of Russian, North American, and Scandinavian literature devoted to it. The literature is too extensive to list; however, a useful review and discussion of the winter regime of rivers and lakes is provided by Michel (1971). Lake freeze-up, ice growth and break-up has been described, and discussed with reference to climatic conditions by Williams (1968). Michel and Ramseier (1971) have classified river and lake ice according to the history of formation, its physical properties and texture. Pounder (1965) has surveyed the basic structure and properties of both freshwater and sea ice and his book provides considerable information on their occurrence and movement.

In the following sections on processes, ice formation, growth and decay on arctic rivers

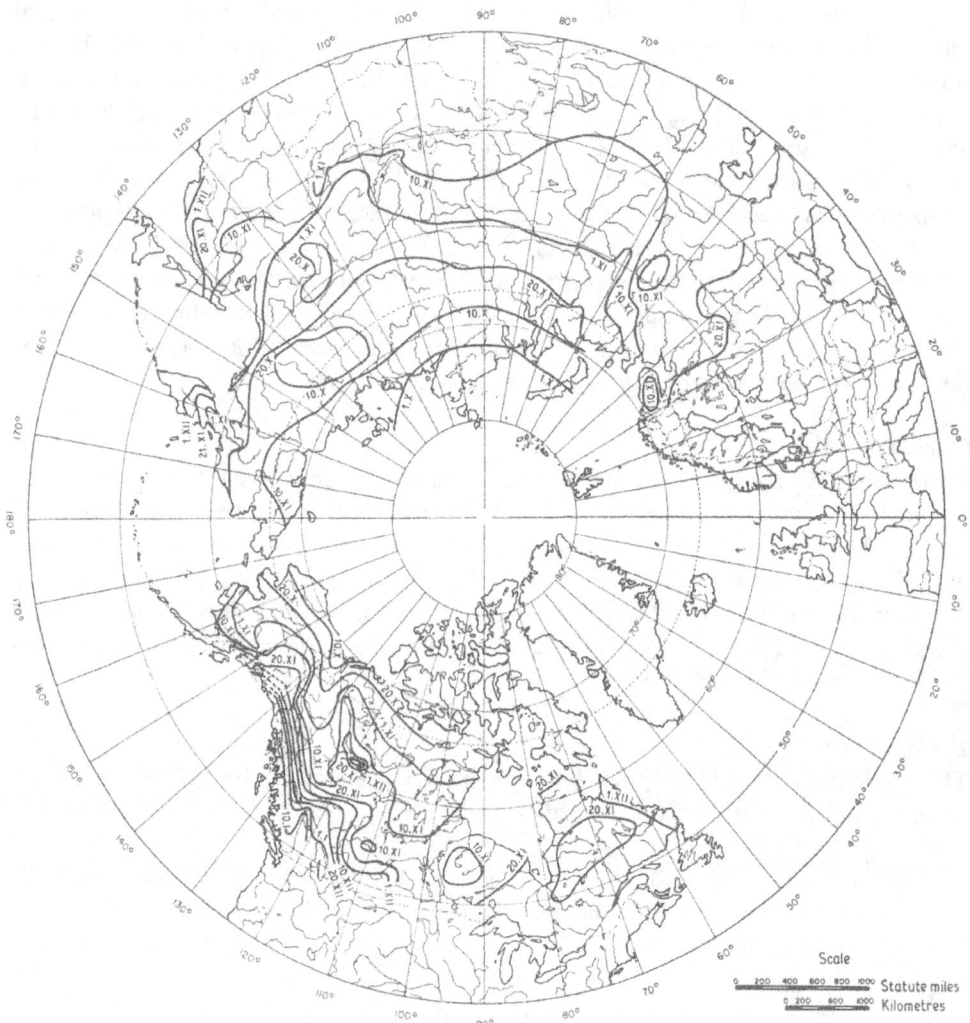

Fig. 3A.2. Mean dates of freeze-up in arctic rivers (based on Allen, 1964; Gerasimov, 1964; and Johnson and Hartman, 1969).

and lakes are described in general terms. The progress of river freeze-up and break-up is shown in figs. 3A.2 and 3A.3.

Freeze-up

Formation of an ice cover depends on such factors as the heat energy stored in a water body, the heat exchange with the atmosphere and the change in heat storage due to inflow and outflow. The process begins in autumn when heat losses to the atmosphere exceed gains. As the upper water layer is cooled, convectional mixing is induced until the water body becomes isothermal at 4°C (the temperature of maximum density).

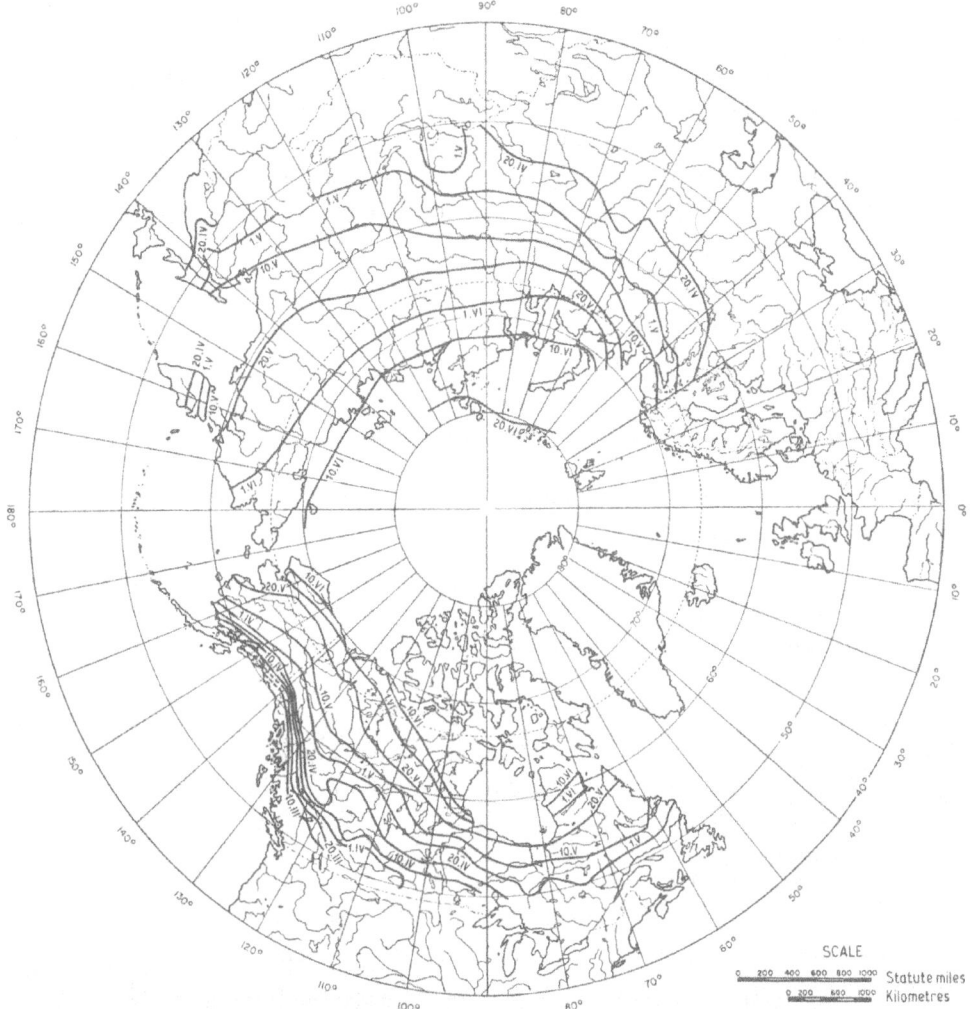

Fig. 3A.3. Mean dates of break-up in arctic rivers (based on Allen, 1964, for initial breaking of river ice; Gerasimov, 1964, for opening of rivers; and Johnson and Hartman, 1969).

Further cooling lowers densities, due to the physical characteristics of water, thus allowing the cooled water to remain at the surface. When the surface water temperature reaches 0°C, ice cover formation ensues (see Brewer, 1958).

The foregoing explanation of the mixing–cooling process is an oversimplification; under real conditions many other factors (fig. 3A.4) are involved. In rivers, for example, turbulent flow may cause near-isothermal conditions at temperatures below 4°C. Mixing, and a consequent increase in the thickness of the cooling layer, is also caused by wind and wave action, particularly on bigger lakes and rivers. Heat losses to the atmosphere are very complex involving radiation, convection, advection, evaporation and conduction, while the temperature of the water body may also be affected by precipitation, inflow–outflow and conduction of heat through the bottom.

Initial formation of an ice cover begins adjacent to river and stream banks and around the perimeters of lakes. The relative warmth of the water body and the resulting convection causes a flow of cold air from the land surfaces cooling the water body most appreciably in the region adjacent to the land. In calm near-shore areas lateral mixing is also limited, and heat losses are greater per unit volume in the shallows than elsewhere.

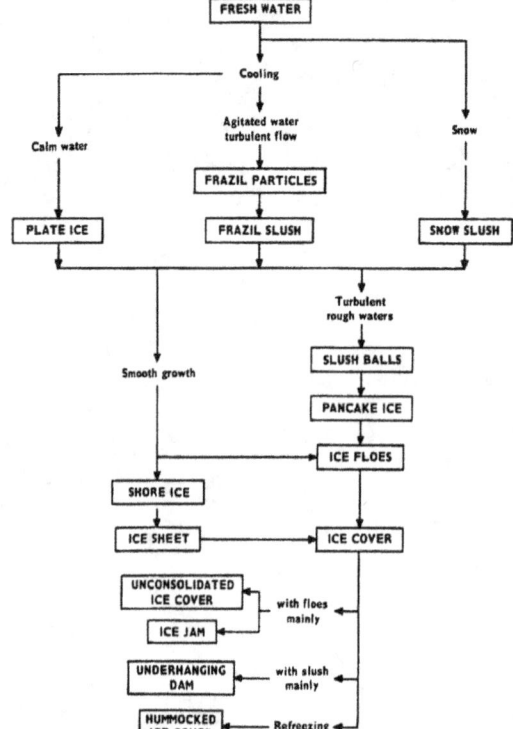

Fig. 3A.4. Processes of ice formation in rivers and lakes (from Michel, 1971).

Gradually, ice extends farther out from the shore until an ice cover is formed completely. This process is also subject to many variations, particularly on rivers where turbulent flow can alter the situation.

In turbulent sections of rivers, great quantities of *frazil ice* are produced. This ice, drifting into calmer stretches, agglomerates rapidly into patches of slush that can eventually cover the surface. Persistence of low temperatures will lead to solidification of the slush and the development of a continuous ice cover. Michel (1963) and Williams (1959) examined various aspects of frazil ice formation and discussed some of the problems associated with it. An unusual river ice condition caused by deposition of frazil has been described by Gold and Williams (1963).

Adding to the complexity of ice development in rivers is the formation of *anchor ice* in supercooled water on the bottoms of some reaches (fig. 3A.5). Formed often at night in clear shallow rivers with rocky beds, anchor ice is occasionally released by solar radiation and rises to the surface to be incorporated at some later date in the formation of a solid ice cover. Anchor ice can also be important in flow reduction on rivers as described by Foulds (1967) and Devik (1964). The growth of anchor ice on hydraulic structures and its adherence to specific materials has been investigated by Zagirov (1966). There seems to be no clear distinction between frazil and anchor ice types and few observations have been made on the latter in the literature. The process of anchor ice formation remains unclear and requires further investigation.

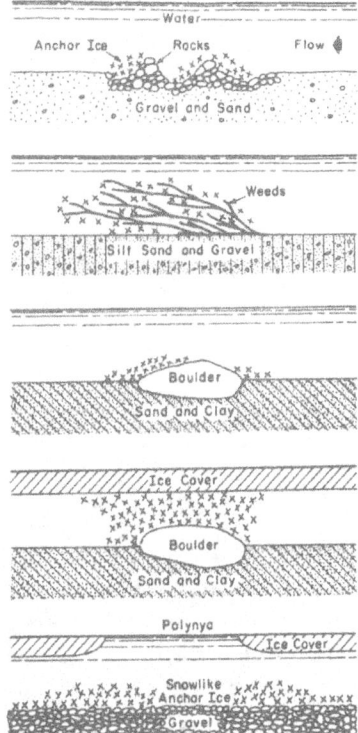

Fig. 3A.5. Types of anchor-ice deposits (from Michel, 1971).

The winter ice regime

The onset of winter ice conditions is assumed to begin when rivers and lakes are fully covered by ice. This generally occurs in late October in mid-Arctic regions, slightly later in more southerly areas and earlier in the High Arctic. It should be mentioned that some high arctic lakes are perennially frozen while others will lose only part of their ice cover during the summer season.

Northward-flowing rivers that extend through wide latitudinal ranges may exhibit a drop in ice cover levels as runoff is progressively reduced by the onset of freezing temperatures in the southern extremities of their basins. The change in ice levels may be exposed by shearing along cut banks and other sections where the river bottom falls off steeply, or by flexure and dishing of the ice cover. Near the termini of some rivers whose flow is widely distributed through delta channels, ice cover heaving may occur in the early part of winter. This is in consequence of ice growth in shallow channel mouths, constricting the flow and creating pressure on the undersurface of the ice cover.

Water bleeding from thermal contraction and strain-induced cracks can add appreciably to river and lake ice thickness as the winter progresses. Fifteen per cent or more of ice-cover thickness may be due to this phenomenon although the percentage is highly variable from year to year. The thickness of 'white' ice depends on both the distribution of snow and the amount of water reaching the surface. The types of ice associated with river ice covers are shown in fig. 3A.6.

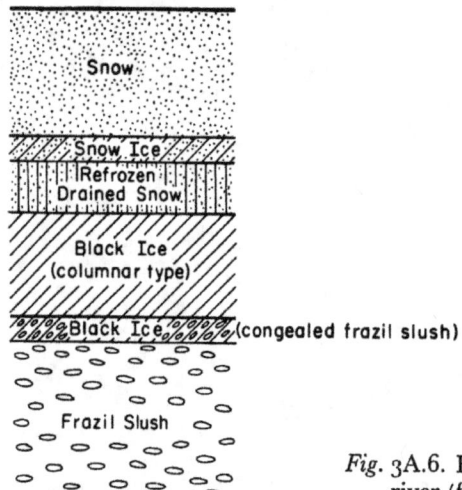

Fig. 3A.6. Forms of ice in the ice cover of a river (from Michel, 1971).

Ice covers normally reach maximum thickness during March–April in arctic regions. Thickness can vary from more than 2 m in higher latitudes to a metre or less in the discontinuous permafrost zone. Again, the variation from year to year is considerable as it is a function of the integrated effects of winter weather conditions. Williams (1966) has examined various approaches to the prediction of ice thickness and to the rate of ice growth on freshwater lakes and, in an earlier paper (Williams 1963), has developed probability charts for their estimation. As an example of ice thicknesses in arctic and subarctic regions, fig. 3A.7 shows observations made in March 1966 for northern Canada.

Break-up

On major north-flowing arctic rivers, the progress of break-up follows a basic pattern. The initial stage is generated by radiational melting of snow lying on riverbanks and adjacent slopes. Meltwater from these areas moves down onto the ice cover concentrating along its edges. As melt intensities and discharge increase, shore leads develop along channel edges, enlarging at tributary junctions where inflow is greatest.

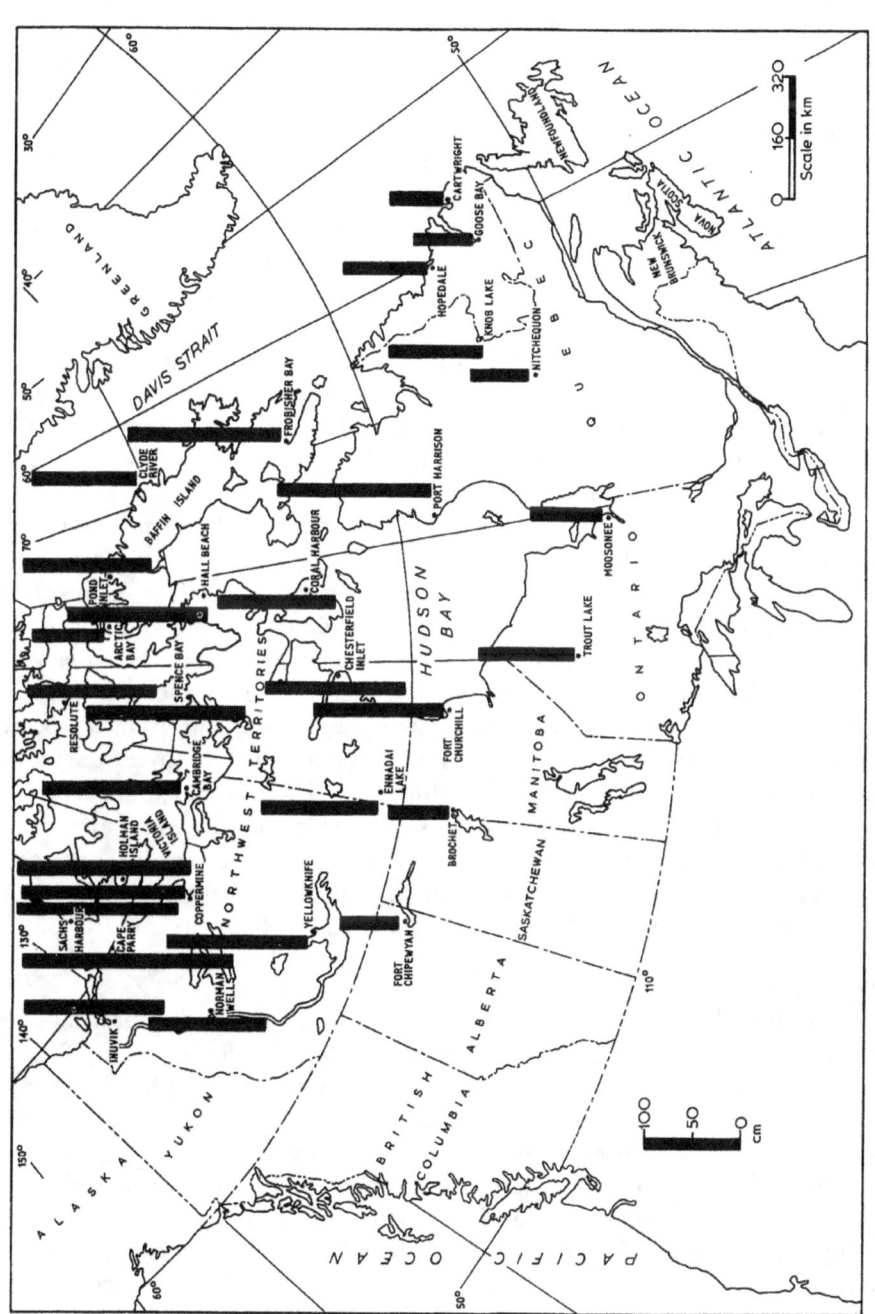

Fig. 3A.7. Ice thickness (cm) in Canada, March 1966 (based on data from Met. Branch, Dept of Transport, Toronto).

As the river stage rises, the decaying ice cover is fractured and broken into large rafts and pans. Transverse leads open where the flow of tributaries enters the main channel. Initial ice movement begins when a rising stage frees some ice sections from the banks. Movement downstream is intermittent and slow due to blockages of pans caused by ice remaining shorefast or anchored in shallow reaches of the river.

The rate of downstream movement increases as abrasion and melting reduce the ice cover to a mass of smaller pans and ice wreckage. The ice moves down-channel until natural constrictions in the width or bed cause jamming. Masses of ice wreckage queue up at the jam sites with clear reaches developing upstream. During this phase, surging runoff from tributary streams is an effective ice-clearing agent since masses of ice are grounded or deposited high along channel banks in stretches above the jams. As the pressure of impounded water combines with the processes of ice abrasion, melting and candling, the jams give way and the downstream surge and recession that follows strands much of the remnant ice cover on the river's banks. Ice jamming is thus a significant ice clearance mechanism and can, under certain conditions, contribute to the acceleration of the break-up process. Bolsenga (1968) has reviewed and summarized published information on river ice jam characteristics and the known methodology of jam prevention and removal.

Lake ice disintegration has been described in detail by Williams (1966). In general, the process begins as meltwater from the surrounding terrain flows onto the ice surface concentrating along thermal contraction cracks and opening holes in the cover where these cracks conjoin. The ice cover develops a patch-like appearance and begins to darken as meltwater gathers in pools at various locations on its surface. As melt proceeds, shore leads form around the lake perimeter, eventually freeing the ice cover from the lake bottom and shore. When the ice cover floats free, water drains through the cracks and joints caused by incipient melt, and the ice surface reassumes its white appearance. Progressive melting and candling increases the water content of the ice, once again darkening the cover. When this stage is reached any major weather disturbance accompanying melting temperatures will cause the disintegration of the remnant ice cover.

Prediction of freeze-up and break-up

A number of methods for the estimation of freeze-up based on air temperatures have been developed, one of the most successful being that formulated by Rodhe (1952). Mackay (1960) has used Rodhe's method in conjunction with flow-travel times to estimate freeze-up in the Mackenzie Delta. Bilello (1964) expanded on the technique to derive prediction curves for a number of sites in northern Canada.

In the Soviet Union, where operation of waterways is of great economic significance, forecasting of ice conditions has received a considerable amount of attention. An ice-forecasting manual prepared by Shulyakovskii (1963) examines methods of freeze-up, ice growth and break-up prediction on rivers and inland lakes. Stress is placed on heat-energy budgets, techniques of introducing information on atmospheric circulation into long-range forecasting, and on ice jam prediction. An example of the use of atmospheric circulation indices in forecasting freeze-up one to two months in advance is given by Yefremova (1963). Rymsha and Donchenko (1965) have applied energy budget methods to compute heat losses, turbulent mixing factors and ice formation intensities on a number of the regulated rivers and reservoirs in the Soviet Union. Standard statistical procedures have been used by Ginzburg (1969) to determine and map probabilities of river freeze-up and break-up occurring on given dates throughout the country.

In North America, the energy budget approach has been applied with some success by Wardlaw (1953), Lauzier and Graham (1958), Newbury (1968), and others, but the links between large-scale atmospheric circulation processes and ice condition phenomena have received little consideration from investigators.

Prediction of break-up is a much more difficult task than prediction of freeze-up. Freeze-up depends almost entirely on heat losses to the atmosphere from a relatively uniform water surface. Break-up, however, is complicated by variations in the albedo of snow and ice surfaces, irregularities in ice thicknesses and rapid increases in river stage. These added factors increase the difficulties for successful estimation of break-up dates.

Poliakova (1969) investigated heat fluxes for the atmosphere–ice and ice–water interfaces as an aid to forecasting break-up of the Lena River. The success of this method depends not only on accurate weather forecasting, but on intensive data collection at appropriate river sites as well. Williams (1963) has developed a simplified heat exchange formula combining degree-day and solar radiation terms for estimation of break-up on small sheltered lakes. Later Williams (1971) compared the accuracy of forecasting based on mean dates, melting degree-days and regression of break-up on air temperature records; the results are only applicable to continental lakes where wind and river inflow effects are relatively unimportant. Mackay (1963) has correlated break-up at location in the Mackenzie Delta with break-up at Peel and Mackenzie river sites.

In the final analysis, estimation of freeze-up and break-up is tied inextricably to weather forecasting. As improvements in the range and accuracy of weather forecasts are made, the problems in estimation of freeze-up and break-up will diminish concomitantly.

3. Groundwater

The occurrence and distribution of groundwater in the Arctic is profoundly affected by permafrost (see Ives, Chapter 4A). The presence of a thick impermeable frozen layer under much of the arctic landmass imposes formidable constraints on the volume of materials suitable for groundwater containment and on the movement of groundwater within drainage systems. Nevertheless, groundwater is widely distributed throughout the arctic region and could prove of great economic significance in the future development of the North.

Groundwater may be characterized by its mode of occurrence as suprapermafrost, intrapermafrost or subpermafrost water (fig. 3A.8). Suprapermafrost water is present seasonally in the active layer and perennially under lakes and rivers that do not freeze to the bottom. Intrapermafrost water occurs in thaw pockets and pipes called 'taliks', which may be present under alluvial flood plains, drained or shallow lakes and swamps,

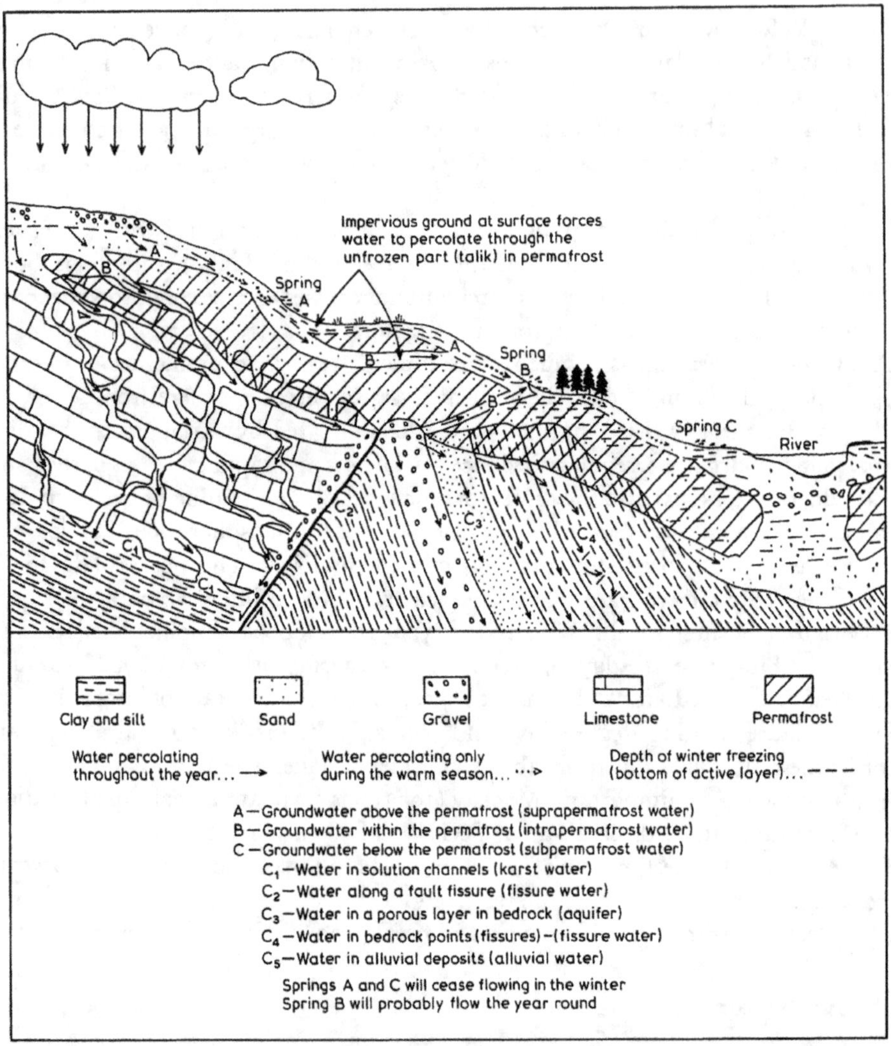

Fig. 3A.8. The occurrence of groundwater in permafrost areas (after Owen, 1967).

and in other locations where thermally anomalous conditions persist. Subpermafrost water is found at variable levels depending on the thickness of permafrost, which can reach depths of more than 600 m near the shores of the Arctic Ocean but which thins southwards towards the perimeter of the arctic region.

Permafrost constrains infiltration and groundwater movement and this is reflected to a degree in the numbers of lakes and swamps that dot the ill-drained surfaces of some arctic terrain. Groundwater recharge begins during spring melt and reaches the subpermafrost zone through unfrozen zones in the permafrost beneath deep lakes and rivers. Recharge of the active layer ensues after moisture deficiencies arise due to evapotranspiration and drainage, and is thus mainly a function of the summer precipitation. The active layer under tundra cover reaches maximum thickness and maximum soil moisture capacity as early as the latter half of June in the extreme north of European Russia (Skakaliskiy, 1966) and as late as the last two weeks of August in the Mackenzie Delta region.

Groundwater discharge in summer is by the processes of evaporation and transpiration and through seeps and springs. In winter, surface effusions of groundwater in the form of *aufeis* (or *icings*) and frost mounds are prevalent in some river flood plains and low-lying swampy areas. Surface icings are most common on beds of shallow streams and rivers with steep gradients, but may also occur in other areas where the pressure head in the active layer increases with freeze-back in autumn and winter. Subpermafrost water is discharged through deep wells penetrating permafrost, and naturally through springs that may be perennial in nature. It should be noted that the zones of groundwater occurrence in permafrost are interconnected and subject to interaction with changes in surface water conditions, thus creating a loosely integrated system in the arctic region.

Williams (1970), in a comprehensive survey and discussion of groundwater in Alaska's permafrost regions, emphasizes the inadequacy of present knowledge and suggests that an intensive effort should be made to collect geophysical and hydrological data associated with subsurface conditions in order to understand the principles of groundwater occurrence in permafrost regions more fully. Alaskan supplies are dependent on local environmental and geologic factors which may be further complicated by the presence of relic permafrost. Measurements of total permafrost thickness have been recorded for fewer than 1500 of the estimated 15,000 wells, drillholes and shafts in Alaska's permafrost regions. Williams also suggests that further research should be carried out on the chemical and thermal properties of groundwater, on ground ice and on the surface manifestations of groundwater movement such as aufeis, pingos, springs and the baseflow of rivers.

An assessment of groundwater in Canada's arctic regions could only be speculative at the present time owing to the lack of data and the absence of pertinent research. Owen (1967) describes the character of the northern hydrogeological region as delineated by the southern limit of discontinuous permafrost. Detailed information is provided on the volumes of water pumped from mines and the chemical analyses of water from springs and wells, all of which are located in discontinuous permafrost. According to Owen, no Canadian settlements in the continuous permafrost zone are supplied permanently by groundwater.

The present spate of drilling activity by the oil industry in the Arctic Islands and the continuous permafrost zone of mainland Canada should yield much new information about groundwater in the Arctic, particularly in the subpermafrost zones of the sedimentary basins. Studies of the Great Slave Lake, Mackenzie River and Yukon River basins indicate that a negligible amount of groundwater is effluent in the rivers of the Interior Plains and Cordillera (Brandon, 1965).

In the Soviet Union, increasing population and industrial development in the far north has amplified the need for more basic information and research into groundwater supplies in permafrost areas. Ponomarev and Tolstikhin (1959) review the state of groundwater hydrology in the Soviet northland and discuss the occurrence and distribution of groundwater in fifty-five hydrogeological regions affected by permafrost. The authors maintain that large local reserves of fresh groundwater are located in suitable areas under thick permafrost such as those under the towns of Yakutsk and Vilyuisk, and in the Chulyma Basin. It is suggested that if geothermal gradients are the same, the freshwater zone is theoretically thicker under permafrost than elsewhere, due to thickness being dependent on ground temperature differences at the level of zero annual amplitude.

Three zones of water exchange are distinguished in the study: (1) the zone of intensive water exchange, characterized by folded mountain regions, (2) the weak water exchange of lowland areas, and (3) the intermediate intensities of the highlands. In the western lowlands of the permafrost region, highly mineralized waters of marine origin are present due to flooding by the sea during the Boreal transgression. Where dry land sections along the sea coast have been extensively uplifted following the marine transgression, seawater has been replaced by freshwater. Ponomarev and Tolstikhin also include a useful list of some 900 references related to groundwater hydrology in the Soviet Union.

4. Water management

In the world's populated regions, mounting water quality and supply problems have created an awareness and public acceptance of the need for proper management of water resources. In some areas with seemingly inexhaustible water resources, rapid industrial and population growth have outstripped supplies, a situation that has become critical through indiscriminate dumping of waste materials into the water systems. The degradation of the physical and biological environments caused by the pollution of streams, rivers and lakes has led to the partial destruction of some ecosystems and the significant alteration of others. Such ecological deterioration has profound implications for the social and economic activities of man.

In many regions, man is faced with curtailing economic growth on the one hand and with the adoption of remedial measures on the other in order to raise water quality standards to tolerable levels. The costs of pollution abatement are, in many cases, staggering, and these costs emphasize the desirability of prevention rather than cure. One region where prevention is still largely possible and where wise management of water resources will yield beneficial results, both in the economic sphere and in the maintenance of its ecological character, is the Arctic.

The arctic environment imposes stringent constraints on the development and utilization of water resources and on the disposal of waste materials. For example, the potability of water in shallow tundra lakes can be affected in winter by the increased concentration of salts due to the freezing process; permafrost underlying the arctic terrain creates difficulties in the utilization of groundwater; and waste materials are not readily assimilated by biologic processes under low temperature conditions. Fortunately, arctic waters are still relatively clean and the minor pollution problems that presently exist are amenable to control at an acceptable cost. If the problems of pollution and misuse common to other areas are to be avoided, strenuous efforts in the research and inventory fields must precede large-scale industrial development.

Lotspeich (1969), in a discussion of Alaskan water resources and the effects of future industrial development, has stressed the need for a new approach based on a fundamental understanding of ecosystem dynamics. The fragility of arctic tundra and taiga eco-systems enhances the desirability of gaining a complete understanding of the ecological response to new economic activities. If the ecological response can be determined then, Lotspeich suggests, technological innovation and imaginative development concepts can control pollution and avoid the ultimate destruction of the environment.

In the Soviet Arctic where resource development is more advanced than in other sectors of the region, matters of pollution control and abatement are more acute. The utilization and conservation of water resources has been widely discussed in the Soviet Union.

L'vovich (1969) deplores the haphazard character of past water resource development, and articulates the need for a closer examination of the interrelationships between water and other components of the physical environment involved in the hydrologic cycle. As L'vovich points out, soils, vegetation, geological structure, landforms, climate and the biosphere all may suffer as a consequence of poor water management. In his view, it is vital for the Soviet Union and others to adopt a complete, all-inclusive approach to water use and conservation based on the theoretical principle of the unity of water re-sources as expressed by the hydrologic cycle. Grin and Koronkevich (1969) have analysed the total available water supply in terms of its two major components, groundwater and surface water. Methods of increasing this total through long-term stream regulation and through the recirculation of water used for municipal and industrial purposes are dis-cussed. A plea is made for the complete elimination of industrial wastes and sewage from water systems, but this seems hardly compatible with large-scale economic develop-ment in northern regions. Kritskiy and Menkol (1968) have examined the multiple use aspects of water management and its effects on engineering hydrology. In Siberia, for example, the rationalization of power grids and the utilization of waterways must be reconciled with the differing hydrologic regimes of such rivers as the Angara and Yenisei. These and other multiple-use problems, together with the studies necessary for the optimization of water management objectives, are detailed in the paper.

In Canada's Arctic, the problems of water management and pollution control are localized and relatively insignificant at the present time by virtue of the small number of inhabitants, the wide scatter of settlements and the negligible development that has taken place. Like Alaska and the Soviet northland, however, the arctic region of Canada is faced with the immediate prospect of large-scale exploration and development of its mineral resources. The main thrust lies in the search for oil and gas. The discovery of economic deposits and their exploitation will create great water management problems because of possible leakages or spills from conveyance systems or at drill sites.

The need to prevent, control and abate pollution, to strike a rational balance between conservation and water use, to determine the role of water in ecosystem dynamics and to examine the effects of water diversion can only be satisfied by an intensive research programme. Development of hydroelectric power sites, extensions to navigation on waterways, and protection of fisheries and recreational environments are other issues requiring hydrological input. The scientific investigation of many of these problems is dependent on an expanded information base and on the establishment of a more comprehensive hydrometeorological data network. It is apparent that the present state of arctic hydrology is inadequate to deal with many water resource questions arising out of accelerated resource development in the far north.

References

ALLEN, W. T. R. (1964) Break-up and freeze-up dates in Canada. *Dept of Transport Met. Branch Circular*, 4116, Ice 17. 201 pp.

BILELLO, M. A. (1964) Ice prediction curves for lake and river locations in Canada. *Cold Regions Science and Engineering Res. Rep.*, 129. Hanover: US Army CRREL. 12 pp.

BLACK, R. F. (1969) Thaw depressions and thaw lakes. A review. *Biul. Peryglacjal.*, **3** (3), 131–50.

BOLSENGA, S. J. (1968) River ice jams, a literature review. *US Lake Survey Res. Rep.*, 5–5. Great Lakes Res. Center. 568 pp.

BRANDON, L. V. (1965) Groundwater hydrology and water supply in the district of Mackenzie, Yukon Territory, and adjoining parts of British Columbia. *Geol. Surv. Can. Pap.*, 64–39. 102 pp.

BREWER, M. C. (1958) The thermal regime of an arctic lake. *Trans. Amer. Geophys. Union*, **39**, 278–84.

CANADIAN NATIONAL COMMITTEE/INTERNATIONAL HYDROLOGICAL DECADE (1969) *Hydrological Atlas of Canada: Preliminary Maps*. Ottawa: CNC/IHD.

CHERNOGAEVA, G. M. (1970) Water resources of Europe. *Bull. Int. Ass. Sci. Hydrol.*, **4**, 67–76.

DEVIK, O. (1964) Present experiences in ice problems connected with the utilization of water power in Norway. *J. Hydraul. Res.*, **2** (1), 25–40.

DINGMAN, S. L. (1966) Characteristics of summer runoff from a small watershed in central Alaska. *Water Resources Res.*, **2** (4), 751–4.

DREYER, N. N. (1968) Water resources of major economic regions of the RSFSR and the other Union Republics. *Voprosy Geografii*, **73**, 63–9. (English transl., *Soviet Geogr.*, 1969, 137.)

FINDLAY, B. F. (1969) Precipitation in northern Quebec and Labrador: an evaluation of measurement techniques. *Arctic*, **22** (2), 140–50.

FOULDS, D. M. (1967) Niagara River ice control. *Proc. Eastern Snow Conf.* (Niagara Falls), 35–45.

GERASIMOV, J. P. (1964) *Fiziko-Geograficheskiy Atlas Mira* (Physical Geographic Atlas of the World). Moscow: Acad. Sci. USSR and State Geol. Cte. 298 pp. (Contents listed in *Soviet Geogr*, 1965, **6**, 5–6, 403 pp.)

GILLILAND, J. A. (in press) *Atlas of Canadian fresh-water lakes*. Ottawa: Inland Waters Branch, Dept of Environment.

GINZBURG, B. M. (1969) Probability characteristics of river freezeup and ice breakup dates. *Gidromet. Nauch.-Issled. Tsentr. SSR*, **40**, 21–39 (in Russian). Leningrad.

GOLD, L. W. and WILLIAMS, G. P. (1963) An unusual ice formation on the Ottawa River. *J. Glaciol.*, **4** (35), 569–73.

GRIN, A. M. and KORONKEVICH, N. I. (1969) Principles of construction of long-term water-management balances. *Soviet Geogr.*, **10** (3), 119–37.

HÄGGBLOM, A. (1963) Sjöar på Spitsbergen Nordosrland. *Medd. Geografiska Institutionen vid Stockholms Universitet*, **149**, 76–105.

HARTMAN, C. W. and CARLSON, R. F. (1970) *A Bibliography of Arctic Water Resources.* Inst. Water Res., Univ. of Alaska.

HATTERSLEY-SMITH, G. and SERSON, H. (1964) Stratified water of a glacial lake in northern Ellesmere Island. *Arctic*, **17** (2), 108–11.

JOHNSON, P. R. and HARTMAN, C. W. (1969) *Environmental Atlas of Alaska.* Inst. Water Res., Univ. of Alaska, 111 pp.

KRITSKIY, S. N and MENKEL, M F. (1968) Water management in the USSR and problems in engineering hydrology. *Soviet Hydrol.*, **1**, 1–35.

LAUZIER, L. M. and GRAHAM, R. D. (1958) Computation of ice potentials and heat budget in the Gulf of St Lawrence *Atlantic Oceanographic Group, Manuscript Ser.*, 11. 7 pp.

LISTER, H. and WYLLIE, P. J. (1957) The geomorphology of Dronning Louise Land. *Medd. om Grønland*, **158** (1). 73 pp.

LIVINGSTONE, D. A. (1966) Alaska, Yukon, Northwest Territories and Greenland. *In:* FREY, D. G. (ed.) *Limnology in North America*, 559–75. Madison: Univ. of Wisconsin Press.

LOTSPEICH, F. B. (1969) Water pollution in Alaska: present and future. *Science*, **166**, 1239–44.

L'VOVICH, M. I. (1969) Scientific principles of the complex utilization and conservation of water resources. *Soviet Geogr.*, **10** (3), 95–118.

MAAG, H. (1969) Ice dammed lakes and marginal glacial drainage on Axel Heiberg Island (Canadian Arctic Archipelago). *Axel Heiberg Island Res. Rep.* McGill Univ. 147 pp.

MACKAY, J. R. (1960) Freeze-up and break-up prediction of the Mackenzie River, NWT, Canada. *In: NAS–NRC Symposium on Quantitative Methods in Geography.* Chicago: ONR. 43 pp.

MACKAY, J. R. (1963) The Mackenzie Delta area. *Geogr. Branch Mem.*, 8. Ottawa. 202 pp.

MACLAREN, I. A. (1967) Physical and chemical characteristics of Ogac Lake, a land-locked fiord on Baffin Island. *J. Fish. Res. Board Can.*, **24** (5), 981–1015.

MICHEL, B. (1963) Theory of formation and deposition of frazil ice. *Proc. Eastern Snow Conf.*, 27–35.

MICHEL, B. (1971) Winter regime of rivers and lakes. *Cold Regions Science and Engineering Monogr.*, III–B1a. Hanover: US Army CRREL. 131 pp.

MICHEL, B. and RAMSEIER, R. O. (1971) Classification of river and lake ice. *Can. Geotech. J.*, **8** (1), 36–45.

NEWBURY, R. (1968) *The Nelson River: A Study of Subarctic River Processes.* Unpub. Ph.D. thesis, Dept of Civil Engineering, John Hopkins Univ. 318 pp.

OWEN, E. B. (1967) Northern hydrogeological region. *In:* BROWN, I. C. (ed.) Groundwater in Canada, *Geol. Surv. Can. Econ. Geol. Rep.* 24. Ottawa. 228 pp.

POLIAKOVA, K. N. (1969) Short-range forecasting of ice breakup in the middle course of Lena River. *Gidromet. Nauch.-Issled. Tsentr. SSR*, **40**, 89–98 (in Russian).

PONOMAREV, V. M. and TOLSTIKHIN, N. I. (1959) Groundwater in permafrost areas. *In: Principles of Geocryology*, Chap. 8, Part 1. Moscow: Akad. Nauk SSR. (English transl. by C. de Leuchtenberg, NRC TT-1138, Canada, 1964, 98 pp.)

POST, A. and MAYO, L. R. (1971) Glacier dammed lakes and outburst floods in Alaska. *In: Hydrologic Investigation Atlas HA-455*, US Geol. Surv. Washington, DC. 10 pp.

POUNDER, E. R. (1965) *The Physics of Ice*. Oxford: Pergamon Press. 151 pp.

RODHE, B. (1952) On the relation between air temperature and ice formation in the Baltic. *Geogr. Ann.*, **34**, 175–202.

RYMSHA, V. A. and DONCHENKO, R. V. (1965) Investigations and calculations of freezing of rivers and reservoirs. *Soviet Hydrol.*, **4**, 372–84.

SHULYAKOVSKII, L. G. *et al.* (1963) *Manual of Forecasting Ice-Formation for Rivers and Inland Lakes.* Leningrad: GIMIZ Gidromet. Izdat. (English transl., Jerusalem, Israel Progr. Sci. Transl., 1966, 245 pp.)

SKAKALISKIY, B. G. (1966) Basic geographical and hydrochemical characteristics of the local runoff of natural zones in the European territory of the USSR. *Soviet Hydrol.*, **4**, 389–434.

SOKOLOV, A. A. (1966) On the excess of maximum discharges of summer-fall rain floods over discharges of spring high water. *Soviet Hydrol.*, **5**, 476–82.

WALKER, H. J. and MCCLOY, J. M. (1969) Morphologic change in two arctic deltas. *Arctic Inst. N. Amer., Res. Pap.*, 49. 91 pp.

WARDLAW, R. L. (1953) A study of heat losses from a water surface as related to winter navigation. *Proc. Eastern Snow Conf.* (Albany, NY), **2**, 8–15.

WILLIAMS, G. P. (1959) Frazil ice. A review of its properties with a selected bibliography. *Engin. J.*, **42** (11), 55–60.

WILLIAMS, G. P. (1963) Probability charts for predicting ice thickness. *Engin. J.*, **46**, 31–5.

WILLIAMS, G. P. (1968) Freeze-up and break-up of fresh water lakes. *In: Proceedings of Conference on Ice Pressures Against Structures* (Laval University, Que., November 1966). NRC Tech. Mem. 92, NRC 9851. 247 pp.

WILLIAMS, G. P. (1971) Predicting the date of lake ice breakup. *Water Resources Res.*, **7** (1), 323–33.

WILLIAMS, J. R. (1970) Groundwater in the permafrost regions of Alaska. *US Geol. Surv. Prof. Pap.*, 696. Washington, DC. 83 pp.

YEFREMOVA, N. D. (1963) Freezing of the Votkinskiy reservoir and methods of its forecasting. *Soviet Hydrol.*, **5**, 450–8.

ZAGIROV, F. G. (1966) Formation of anchor ice on bodies of various structure. *Soviet Hydrol.*, **1**, 99–101.

Manuscript received August 1971.

Alpine hydrology

H. Olav Slaymaker

Department of Geography,
University of British Columbia

1. Introduction

Those components which seem to be essential to the concept of alpine hydrology are high radiant energy fluxes, high moisture fluxes in the form of rain and snow, discharge hydrographs greatly influenced by snowmelt and/or glacier melt processes, high available relief, poorly developed regolith and a low percentage ground cover by trees. Each of these six factors may occur outside the alpine zone, but it is their distinctive association which goes to form alpine hydrology. The discussion that follows will attempt to describe these components and, in the concluding paragraph, to outline the major ways in which alpine hydrology differs from arctic hydrology.

While no attempt is made to formulate a rigorous definition of alpine hydrology, the use of the term implies a particular spatial scale of investigation. The classical studies of French geographers in delimiting regional subdivisions within the Alps of Savoie and Dauphiné emphasized the contrasts between 'ubac' (shaded) and 'adret' (sunny) slopes. These are mesoscale environmental systems which give the distinctive alpine flavour to the high alpine region of southeastern France. The understanding of hydrologic processes can be approached at many different spatial scales (Meier, 1970; Slaymaker, 1969), but it is probable that alpine hydrology can best be studied at the mesoscale. The hydrologic process at a site within the alpine region is commonly not different in kind from site processes in other environments and fails to give a distinctive alpine signal. At the same time, larger-scale studies tend to include non-alpine problems within them and these distort the nature of the alpine hydrologic response. It is the discrimination of mesoscale spatial entities, such as the ubac and the adret, with their characteristic energy, mass and moisture balances, and the understanding of ways in which each of these environmental systems receives, transforms and releases energy, moisture and mass which are the two most difficult challenges of alpine hydrology.

For the purpose of this chapter, fig. 3B.1, which shows an area of the Pacific Ranges of

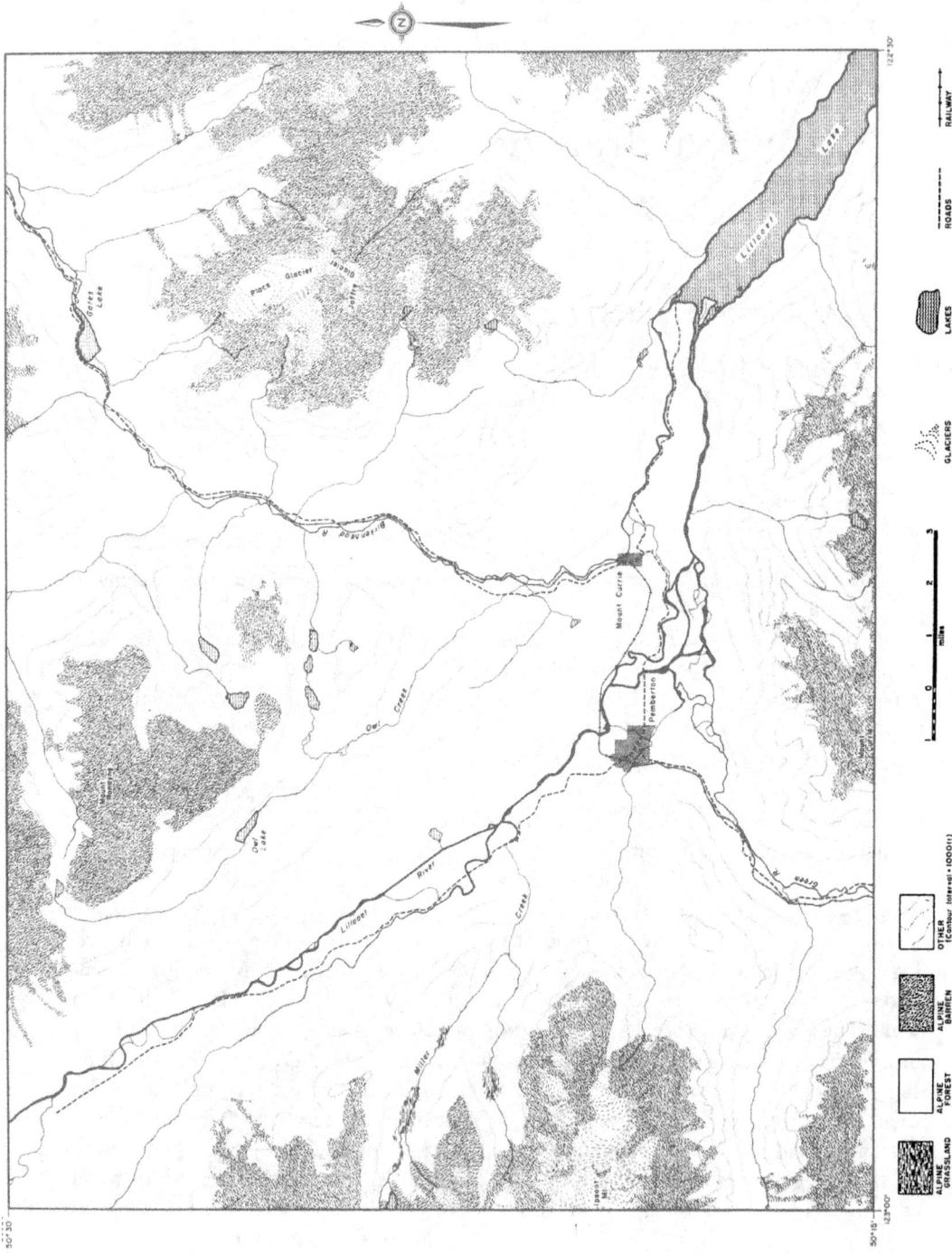

Fig. 3B.1. Distribution of alpine and subalpine regions in the vicinity of Pemberton, British Columbia.

the Coast Mountains in British Columbia, is helpful in illustrating the context of alpine hydrology. Glaciers, alpine lakes, barren rock surfaces, alpine and subalpine forest,[1] and alpine and subalpine grassland form the categories which have been differentiated in the field as environmental systems which give an alpine hydrologic response. The area of the map which has been left white and the lower-lying lakes display a non-alpine hydrologic response. These areas are covered with forest stands of western hemlock, red cedar and Douglas fir or, as along the flood plain of the Lillooet River, are cleared and cultivated lands; such areas are excluded from the present discussion.

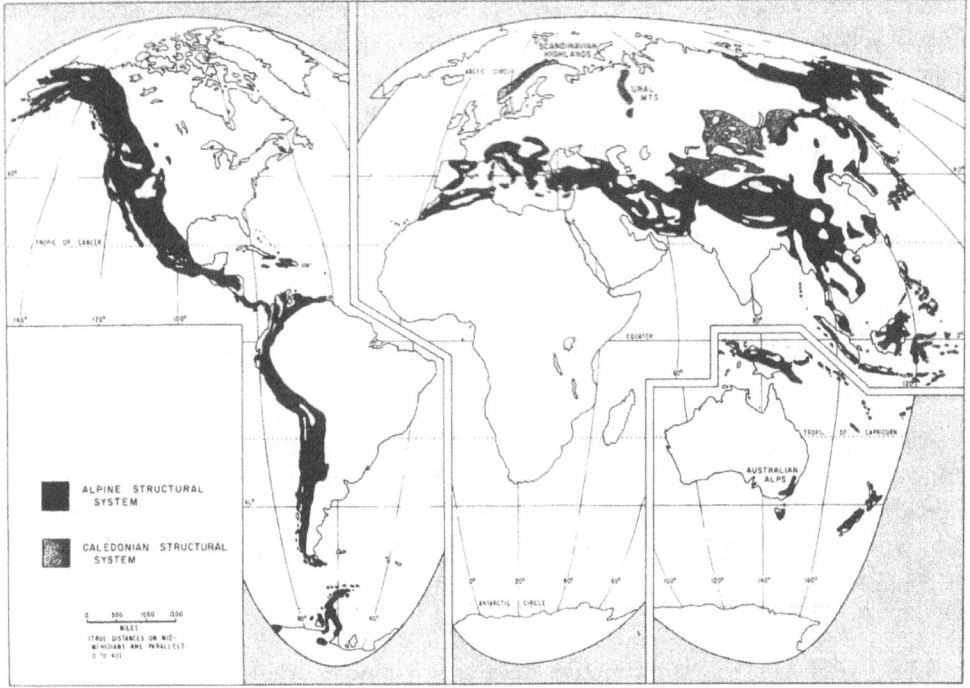

Fig. 3B.2. Global distribution of major alpine regions.

It is apparent that, in the restricted sense of the previous paragraphs, alpine hydrologic regions have a global distribution pattern which is closely correlated with the alpine orogenic (or young fold mountain) belts. The only major exceptions to this generalization are the Scandinavian Highlands, the Ural Mountains, Mongolia and the Australian Alps (fig. 3B.2). In this chapter, the western Cordillera of North America, the European Alps and the Scandinavian Highlands will receive disproportionate attention, as it is with these regions that the writer is familiar.

2. Alpine surfaces

Using the categories provided in fig. 3B.1, some generalizations about the albedo and transmissivity of the various surfaces can be made.

[1] *Edit. note.* This usage differs from that proposed in Chapter 1.

Glaciers. The albedo of a glacier may vary from 20 per cent (dirty glacier ice) to 46 per cent (clean glacier ice) through to 95 per cent (fresh snow cover in the accumulation zone). This means that, especially in the autumn period when a glacier receives fresh snow at its upper end and has extensively exposed medial moraine near its snout, there is an extremely variable energy reflecting surface. In addition, the transmissivity of glacier ice is greater than that of snow, so that while 10 per cent of the incident radiation which is not reflected will reach 40 cm in glacier ice, it will reach only 12 cm in dry snow and 4 cm in wet snow (Geiger, 1965).

Alpine lakes. Those alpine lakes which are influenced by the presence of glaciers have a higher albedo (15–20 per cent) than those in unglacierized basins (5–10 per cent). The high sediment content of the lakes which are fed by glaciers makes the contrast, and this sediment content also influences the transmissivity of the lake. Lakes in glacierized basins may allow 10 per cent of the incident radiation to penetrate up to 5 m while clear lakes will allow penetration of a similar percentage of incident radiation to 20 m. The seasonal variation, a function of ice and snow cover, is even greater than that due to sediment concentration in the water.

Barren rock surfaces. Depending on their colour, these have an albedo of 5–15 per cent.

Alpine and subalpine grassland. These have an albedo of 12–25 per cent, and form the high altitude meadows known as alps in Switzerland.

Alpine and subalpine forest. Included under this category are the widely scattered alpine fir and mountain hemlock, which cover less than 50 per cent of the ground surface, and also the extensive alder growth along the tracks of snow avalanches, the presence of which is suggested by the long thin extensions reaching comparatively low altitudes in fig. 3B.1. The albedo of this surface varies from 5–20 per cent depending on the percentage of conifer cover.

Snow surface. One of the distinctive features of alpine hydrology is the pervasive influence of snow, its accumulation, storage and melt. An understanding of the thermal and radiative properties of snow is therefore of some importance in the following discussion (see Alford, Chapter 2D). According to Mellor (1964), there are four main ways in which the transfer of heat through snow takes place:

(i) conduction through ice grains;
(ii) convection through the pores between ice grains;
(iii) vapour diffusion;
(iv) radiation.

Research has traditionally centred on conduction and radiation to the comparative neglect of the other two processes. Indeed 'the determination of the heat flow within and from a snow cover is dependent to a great extent upon a reasonable description of the solar energy absorbed, transmitted and reflected' (Dunkle and Bevans, 1956), and this reasonable description is still difficult to make in quantitative terms. More recently, Yen

(1962) and Vinje (1967) have attempted the assessment of the role of the three non-conductive forms of heat transfer in snow. It is clear that the most important controls of heat transfer are the average grain size, the free water content and the development of laminar layers within the snow, and that significant transmission of radiation is limited to less than one metre.

3. Energy balance

One half of the total weight of the atmosphere is below 5000 m; 22 per cent is below 2000 m. As the atmosphere acts as a filter to the solar radiation passing through it, radiation intensity and composition vary with changes in elevation (see Flohn, Chapter 2B; Barry and Van Wie, Chapter 2C). In alpine regions, the influence of aspect and slope angle are of primary significance. Frank and Lee (1966) have shown that over the period of a year, the radiation index (a theoretical measure of radiation intensity) at 50°N differs by a factor of four between north-facing and south-facing slopes of 45° angle (see table 2C.1).

SUMMER ENERGY BALANCE

The summer energy regime in alpine environments has been discussed by Gates and Janke (1966) and Terjung *et al.* (1969). The basic energy balance equation is of the form

$$R_N = G + H + LE$$

where R_N is the net radiation heat transfer; G is the soil heat transfer; H is the sensible heat transfer; and LE is the gain or loss of heat by evaporation or condensation.

The work of Terjung *et al.* is of particular interest because they make explicit consideration of the scale problem and the need to define a homogeneous mesoscale environment.

In their study, an area of alpine tundra in the White Mountains of California at about 3580 m was isolated. On 17 July 1968, direct and diffuse solar radiation on a horizontal surface and on slopes of 45° and 90° for the cardinal directions, net radiation, radiant temperatures and actual surface temperatures were recorded. The data indicated that this alpine tundra experienced some extremely high energy fluxes. The thermal fluctuations were far greater than the data on air temperature would suggest, as little of the heat is stored in the air. It is quickly radiated away and so does not influence the air temperatures. For 17 July 1968, R_N was 540 cal cm^{-2}, G was 60 cal cm^{-2}, H was 160 cal cm^{-2}, and LE was 330 cal cm^{-2}. The discussion by Miller (1955) of the snow climate on clear summer days in the Sierra Nevada provides interesting comparative data.

WINTER ENERGY BALANCE

The winter energy regime for alpine environments has been discussed by Miller (1955), Williams (1965) and Anderson (1968). Two distinctive problems will be mentioned here, namely the energy regimes of a snowpack and an ice- and snow-covered lake.

The basic energy balance equation for a snowpack is

$$R_N = G + H + LE + M + WE$$

where M is the energy used in melting snow or in freezing the surface layer of the snow-pack, and WE is the heat content of the gain or loss in water equivalent of the surface layer. The usefulness of this equation has been well established for conditions at a site (US Army Corps of Engineers, 1955, 1956), but the extension of site data over a variable alpine basin is a less certain procedure. The three most important problems are the adjustment of radiation data for the effect of subalpine and alpine tree cover, the adjustment of site wind data for the effect of physiography and variable basin wind strengths, and the variable relationship between liquid water retention capacity and the physical characteristics of snow (Alford, 1967). If the purpose of the investigation is to simulate the snowmelt hydrograph, then adjustments can be made by changing constants in the wind function and in the net radiation function. Such adjustments do not, however, aid in the understanding of the spatial variability of energy fluxes over an alpine snowpack. Hutchinson (1966) and Williams (1965) have compared evaporation rates from snow and soil surfaces.

Alpine lakes commonly have a cover of ice and snow in winter. Under these circumstances, three separate energy budgets are relevant:

for the lake waters

$$R_{N_w} = G_w + H_w + \Delta F$$

for the ice

$$R_{N_i} = G_i + H_i + LE_i$$

and for the snow

$$R_{N_s} = G_s + H_s + LE_s + P$$

where R_{N_w} is the net radiation within the water body, R_{N_i} is the net radiation within the ice mass, R_{N_s} is the net radiation within the snowpack, G_w is the sensible heat transfer in water, G_i is the sensible heat transfer in ice, G_s is the sensible heat transfer in snow, H_w is the sensible heat exchange through water to ice and between water and bottom sediments, H_i is the sensible heat exchange through water to ice and sensible heat conducted from ice to snow, H_s is the sensible heat conducted from ice to snow and sensible heat exchange with the air, ΔF is the advection of heat in and out of the lake by tributary streams entering and leaving lake, LE_i is the change in amount of latent heat of freezing stored in ice, LE_s is the change in amount of latent heat of freezing stored in snow and latent heat exchange with the air, and P is the sensible or latent heat added in rain or snow.

Few comprehensive studies of winter energy budgets of alpine lakes are available. Reference is therefore made to the work of Scott and Ragotzkie (1961) on Lake Mendota, Wisconsin, on the assumption that their description of the sequence of events involved in a winter energy budget of a snow- and ice-covered lake will not be greatly different from that of an alpine lake. While the quantitative value of the energy terms, and the duration of each period, is different, the relative significance of the energy terms in each period is likely to be similar. There are seven periods recognized as follows:

(1) The cooling period. This is a period of rapid heat loss through sensible and latent heat exchange with the air.

(2) The ice build-up period. The dominant energy term during this period is sensible heat exchange through water to the ice in the form of melting of the floating ice. This is a very short period in alpine lakes.

(3) The ice growth period; the rapid growth of ice thickness and the slow warming of the lake waters. During this period the conduction of sensible heat from the ice to the overlying snow is the most important energy term.

(4) The equilibrium period. Deeper snow accumulation and higher air temperatures inhibit the increase in ice thickness. Sensible heat exchange with the air and conduction of sensible heat from the ice to snow dominate the budget. During rapid snowmelt towards the end of this period, sensible heat exchange and condensation become progressively more important.

(5) The wastage period. The lowered albedo and higher net radiation lead to rapid ice melt. Change in the amount of latent heat of freezing stored in the snow and ice, sensible heat exchange through water to the ice and sensible heat exchange with the air are the key processes. The importance of the latent heat of melting of ice has been illustrated in the case of Chandler Lake, Alaska (quoted in Hutchinson, 1957).

(6) The break-up period. Similar processes continue until the ice thickness is 20 cm or less. Most of the break-up occurs in one day.

(7) The initial warm-up period. A sudden change in the energy budget occurs when the ice disappears. Latent and sensible heat exchange with the air become the two dominant processes.

SEASONAL ENERGY ACCUMULATION

Sharpe (1970) has directed attention to the need to consider definition of 'effective climate' relative to the alpine timberline of the Colorado Rockies in terms of both energy and moisture availability. The approach is not a new one as it makes use of a fairly conventional Thornthwaite water balance model. The intrinsic interest lies in its application to an alpine and subalpine environment, via six pairs of climate stations chosen

Table 3B.1. Thornthwaite water balance terms as a function of elevation, near Colorado timberline (after Sharpe, 1970).

Elevation (m)	Potential evapotranspiration (mm)	Actual evapotranspiration (mm)		Water deficit (mm)	
		Seedlings	Trees	Seedlings	Trees
	PE	AE_0	AE	D_0	D
2850	420	250	340	175	90
3400	355	265	320	90	40
3750	300	200	250	100	50

because of their closeness to the timberline in Colorado. The values of potential evapo-transpiration calculated from these station records (table 3B.1) drop off very rapidly as the timberline is crossed. The interpretation is that potential evapotranspiration as a cumulative seasonal energy index is the key control of alpine timberline.

Other energy budget studies can be found on alpine lakes (Blaney, 1958), alpine shrub communities (Bliss, 1960), alpine terrain parameters (Peck and Pfankuch, 1963), and ubac and adret slopes (Loup, 1969).

4. Mass balance

Complete energy balance measurements are difficult to obtain over long time periods in the alpine zone. Equipment is sensitive and costly, and while valuable information, such as that of Terjung *et al.* (1969), can be gained from one day's intensive study, it is probable that greater reliance will have to be placed on less sophisticated instrument-ation operated over longer time periods (Hagen, 1965; Jakhelln, 1965; Church and Kellerhals, 1970). In this respect, mass balance studies, especially of glaciers, but also of perennial snow cover, are good examples. They provide longer-term indicators of the net effect of variable energy transfers over time and space. Data on mass balance are usually collected in such a way that they provide spatial integration of a large number of site processes.

Mass balance studies of snow and ice are also of great practical significance in that the differences between accumulation and ablation on an annual basis represent a change in the major freshwater reservoir at the earth's surface. Potential hydroelectricity, irrigation and domestic water supplies in semi-arid regions depend on the delicate balance of perennial ice and snow volumes.

Because temperate glaciers are normally out of equilibrium with their energy environ-ment, their annual budget is rarely balanced. Standard techniques of mass budget measurement on glaciers are discussed by Meier (1962) and the usual form of the mass budget equation is:

$$A = A_c - A_b = \int_{t_1}^{t_2} a.dt = \int_{t_1}^{t_2} (a_c - a_b)\, dt$$

where the budget year begins at t_1 and ends at t_2, A_c is accumulation for the budget year, A_b is ablation for the budget year, a is the mass budget, a_c is accumulation, and a_b is ablation over a short period of time within the budget year.

Many examples of glacier mass budgets are available in the literature (e.g. Meier and Post, 1962). In Canada, an intensive programme of mass budget studies has been initiated under the direction of the Glaciology Subdivision of the Inland Waters Branch in Ottawa (Østrem, 1966; Stanley, 1970). An east to west transect of five glacier basins in British Columbia and Alberta has been established from Ram River in Alberta to Sentinel Glacier in British Columbia. All five glaciers showed a negative budget during 1965–9, with the greatest deficit in 1967. By contrast, Berendon Glacier in northern British Columbia has shown a positive budget since 1967. On the Hintereisferner in Austria, a

10 km² glacier with mean height at about 3000 m, a dominantly negative mass budget was recorded from 1952 to 1961 (Hoinkes and Rudolph, 1962), but more recently the budget has balanced (Hoinkes *et al.*, 1968). A similar trend has been noted in Swedish Lapland (Schytt, 1962) and in southern Norway (Østrem, 1962, and Østrem and Pytte, 1968; see also Østrem, Chapter 4C).

One of the more interesting questions related to the mass budget method is to find out what is the lag time between changes in the energy environment and the response of the glacier mass budget on an annual basis (LaChapelle, 1961; Marcus, 1964). The behaviour of surging glaciers (Robin, 1969) complicates such a calculation, but it seems likely that the lag time is as much as five to fifteen years, based on northern hemisphere mean temperature trends.

5. Water balance

GLACIERIZED BASINS

Water discharge is the most useful spatial integrator over an entire alpine basin, both glacierized and unglacierized. Tangborn (1966) and Lang (1968) have pioneered the use of hydrologic parameters in the assessment of the reliability of glacier mass balances. In Tangborn's (1968) study of discharge from the glaciers of the Thunder Creek drainage basin (at about 2500 m) and the South Cascade Glacier (at about 2000 m) in the State of Washington, a negative mass budget was inferred for both areas for 1920–44. For the period 1944–65, however, the higher glaciers of the Thunder Creek basin showed a positive mass budget while the South Cascade Glacier showed only a reduced negative budget.

The method adopted by Tangborn makes use of only a part of the conventional water balance equation

$$Q = P \pm E \pm S$$

where Q is discharge, P is precipitation, E is evaporation (or condensation), and S is a storage term. This he is able to do by first of all defining a watertight drainage basin, so that major groundwater complications are avoided. Soil moisture storage can be ignored on an annual basis, and at all events is small in alpine basins. The most drastic assumption made is that evaporation and condensation terms are not large enough to alter the budget appreciably. The water balance equation then reduces to the simple form

$$Q = P \pm S_g$$

where S_g is the mass budget of the glacier in water equivalent terms. The method, considering the major assumptions within it, gives surprisingly useful results, and has been adopted with minor modifications in Canada (Stanley, 1970) and in Norway (Pytte, 1967).

Each of the terms in the water balance equation, when evaluated for an alpine environment, is associated with a number of distinctive alpine problems, and, instead of discussing the whole water balance equation, it would seem more useful to look at each term in turn.

PRECIPITATION IN ALPINE REGIONS

Mountains strongly affect the motion of airmasses and the mode of distribution of the precipitation (Hövind, 1965). This is discussed in detail in Chapter 2B by Flohn. In general the spatial variability of precipitation is greater in small, mountainous basins than over extensive plains areas (Apollov et al., 1964). The implication is that in the typical alpine basin, more gauges are required to give a standard error of precipitation assessment similar to that in lowland areas (Jeffrey, 1964, 1965; Rainbird, 1965). It would seem necessary to suggest that requirements for accuracy of areal precipitation depth measurement in an alpine basin should be relaxed by comparison with lowland standards (cf. Hutchinson, 1969). Hoover and Leaf (1967) have discussed the interception process in subalpine forest, but in alpine basins as a whole the process is not of major significance.

The spatial variability of precipitation due to orography has been classically investigated by Spreen (1947). He successfully evaluated seasonal and annual precipitation amounts in mountainous terrain in terms of a small number of orographic controls. Studies such as those of Schermerhorn (1967) in western Washington and Oregon follow closely the pattern of analysis used by Spreen. Subsequent to Spreen's (1947) study, it has been repeatedly demonstrated (e.g. Walker, 1961; Williams and Peck, 1962; and Keeler, 1969) that in some synoptic situations, particularly those associated with 'cold lows' at higher levels, the distribution of individual storm precipitation is effectively independent of terrain. Keeler (1969) showed that on Mt Logan, Yukon, precipitation is effectively the same between about 2500 m and 5400 m. He interprets this as being due to the importance of frontal conditions aloft with no orographic processes associated. It is clear that large errors could occur if, for example, Spreen's results were used to estimate spatial variability of precipitation in individual storms in the Colorado Rockies or to estimate seasonal and annual precipitation totals in entirely different alpine regions.

Keeler's (1969) discussion of precipitation gradients inland from Yakutat, Alaska, to Mt Logan is of interest not only for the data from high levels. He shows the rapid decrease of precipitation from 341 cm at sea level to 65 cm at 3350 m on the windward side; there is a decrease of 65 cm per 1000 m of elevation between 1000 m and 2500 m on the windward side and an increase at the same rate between 1000 m and 2500 m on the leeward side of Mt Logan (cf. Marcus and Ragle, 1970). Solomon et al. (1968) and Sporns (1964) discuss alternative approaches to spatial precipitation estimation in mountainous terrain. Walkotten and Patric (1967) show the elevation effects on rainfall near Hollis, Alaska.

SNOW ACCUMULATION IN ALPINE REGIONS

Meiman (1970) has commented that snow accumulation is probably the weakest link in our knowledge of the snow hydrologic cycle. Studies such as those of Hansen and Ffolliott (1968) in the Beaver Creek Watershed, Arizona, on the effects of strip cutting on snow accumulation, provide little in the way of physical understanding of snow accumulation processes, even though they are of unquestioned practical significance. The results 'are necessarily limited to the type of weather conditions encountered' (Hansen and Ffolliott, 1968, p. 11). A more promising approach to the problem is presently being investigated by Fitzharris (1970). In a study of snow accumulation on Mt Seymour, north of Vancouver, British Columbia, he focuses attention on the processes operating during snowfall, as follows:

(1) Orographic effects. A net orographic effect is clear in that snowpack ranges from 0 at sea level to a maximum on Grouse Mountain at 1150 m of 7.2 m, recorded in 1946. More significant is the fact that this orographic effect varies from storm to storm.

(2) Variations in freezing level. The snow-rain boundary on the mountain is very sensitive to temperature changes as winter temperatures on Mt Seymour are close to 2°C.

(3) Variations in level of melting. Snow may fall up to 300 m below the freezing level before it melts.

(4) Turbulence generated by rough surfaces. The presence of forest margins is usually thought to induce local turbulence which influences the pattern of snow accumulation. In alpine regions, terrain roughness is probably more important in influencing spatially variable accumulation patterns.

(5) Snow interception by trees. In the alpine zone, trees are scattered rather widely and this is not as important a process as in those areas where there is a continuous crown cover.

(6) Formation of rime. Berndt and Fowler (1969) have reported riming that has added 7–10 cm to winter precipitation.

There are many practical implications of the high spatial variability of precipitation and snow accumulation in alpine regions. The work of Schaerer (1970) on maximum ground snow loads in alpine regions of British Columbia and Alberta from Mt Seymour to Lake Louise is illustrated in fig. 3B.3. The data plotted are for 1969, and it is interesting to note two major contrasts: the rate of change of snow load with elevation and the absolute amounts of snow involved.

The density of the snowpack in a west coast alpine location such as that of Mt Seymour means that large snow creep pressures develop. Mackay and Mathews (1966) have discussed the implications of such pressures on deformation of structures and trees.

Drifting of snow may cause redistribution (Plate 5) and concentration of snow into basins which may form avalanches or snowdrift glaciers. Kotlyakov and Plam (1966) have carried out investigations in the Elburz Mountains, Iran, between 3680 and 4060 m. Out of a winter total of 103.6 cm of snowfall recorded on a level firn field, 27.6 cm drifted, and while distances drifted were rarely greater than 100 m, this was sufficient to increase the avalanche hazard. Martinelli (1959) has studied the same phenomenon in the Colorado Front Range between 3500 and 3800 m, and noted 15 m deep snow accumulation pockets separated by shallow snow areas of only 1 to 2 m depth. The hydrological

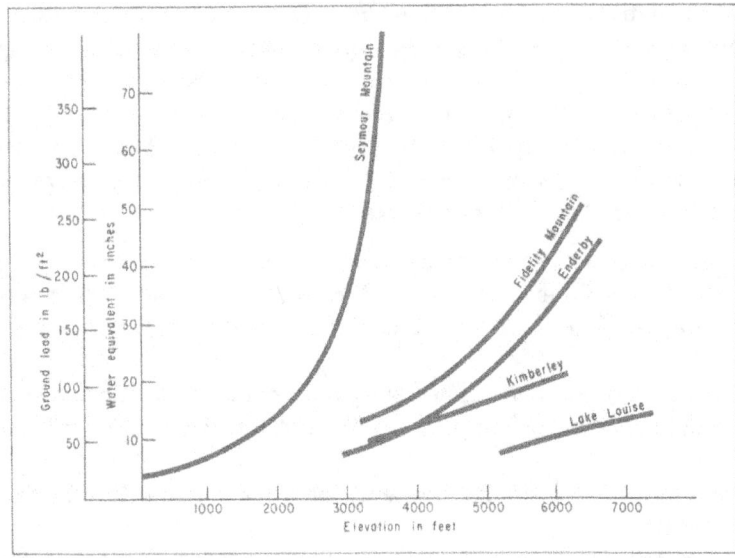

Fig. 3B.3. Variation of ground snow loads in British Columbia (from Schaerer, 1970).

significance of avalanches has been examined by Iveronova (1966), Judson (1966) and Sosedov and Seversky (1966).

Avalanche hazard may be reduced by preventing the accumulation of this drifting snow above natural depressions leading towards well-developed gullies (e.g. LaChapelle, 1966). At the same time snowpack management may require that additional snow be

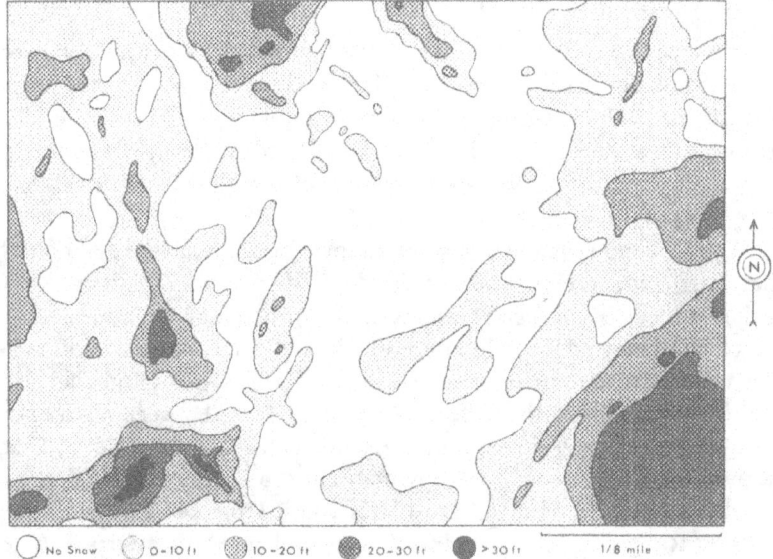

Fig. 3B.4. Snowpack depth in Marmot Creek, Alberta (from McKay, 1970).

stored at high altitude to allow greater water resource potential during late summer (Martinelli, 1967).

The spatially variable pattern of snow accumulation and the processes which redistribute the snow lead to a snowpack pattern which may look like fig. 3B.4. This is a photogrammetric survey of the Marmot Creek watershed in Alberta made in April 1964, and the snow depths plotted are thought to be accurate to the order of ± 1.5 m (± 5 feet). Barnes and Bowley (1968) have discussed the use of satellite photography in snow distribution inventory and Meier (1973) has applied it to snowmelt estimates.

SNOWMELT

Studies of snowmelt are characterized by at least three broad approaches. These are the energy budget approach, the multiple regression approach and the temperature index approach. The Snow Simulation and Reservoir Regulation model (Rockwood, 1958) employs all three approaches in one. The basic equation for the snowpack energy budget has already been introduced and it has been commented that a broad enough data base is rarely available in the alpine zone. Riley et al. (1969) have modified the required meteorological data base, but their most versatile model remains demanding for alpine work. The sources of heat involved in the melting of snow are net radiation, sensible heat transfer from air to snow, latent heat of water vaporization by condensation from the air, conductive heat transfer, and heat content of rainwater. Rantz (1964) attempted to evaluate the snowmelt process on the basis of data from the nearest weather stations and produced the very simple relationship

$$M_s = \frac{\sum H}{203B}$$

where M_s is snowmelt in inches, $\sum H$ is the algebraic sum of all the heat components in cal cm^{-2}, and B is thermal quality (0.95 to 0.97).

The multiple regression approach is well illustrated by Garstka et al. (1959) in which total snowmelt is correlated with whatever relevant meteorologic data are available. The approach may give good results for a given region, but it is not necessarily (or even usually) transferable. Spurious correlations, such as those indicated by the US Army Corps of Engineers (1956, 1960), may arise.

Thirdly, degree-days, mean temperature and maximum temperature are frequently used as linear predictors of snowmelt (Anderson and Rockwood, 1970). This is the simplest and often, in practice, the most effective method. According to the US Army Corps of Engineers (1960), the daily springtime melt (in cm) as a function of mean and maximum daily temperature for open sites is approximately

$$M_s = 0.27 \, (T_{mean} + 4.4)$$

$$M_s = 0.18 \, (T_{max} + 2.8)$$

applicable in the range of $T_{mean} = 1.1°$ to $15.5°C$ and $T_{max} = 6.7°$ to $24.4°C$. Simmons (1961) provides an illustration of the effectiveness of this method for the Fraser River basin, but it seems doubtful that it would be very useful in an alpine basin as the scatter of points in the original derivation of these relationships is extreme. Observations in Miller Creek, British Columbia, in summer 1971 (table 3B.2) indicate the extreme rapidity with which the snow melted during the period 9–20 July, when an average of at least 5 cm per day was removed from the meadows and the snowline rose by 550 m on the adret slopes and by 350 m on the ubac slopes. Hendrick *et al.* (1971) provide useful comments on the reliability of watershed snowmelt data and illustrate procedures for incorporating factors of basin topography and vegetation cover into computational models.

Table 3B.2. Snowmelt in Miller Creek, British Columbia, during July and early August 1971.

	Subalpine meadows	Lower limit of snow (m)	
		Ubac slopes	Adret slopes
1 July	Snowpack depth mean of 0.8 m	750	1100
9 July	Snowpack depth mean of 0.6 m	850	1250
20 July	No snow	1200	1800
5 August	No snow	1600	2300

SOIL MOISTURE, INFILTRATION AND OVERLAND FLOW

The high available relief, steep slopes, thin soils (see Retzer, Chapter 13B) and low tree cover of alpine regions inhibit the storage of moisture in the soil, reduce the infiltration process to comparative unimportance and encourage the process of overland flow (Neal, 1938). Haupt (1967) has made some critical experimental observations on twelve small plots at about 2000 m near Reno, Nevada. He made winter study of infiltration, overland flow and soil movement on six frozen and six snow-covered plots. In particular, he showed the following features:

(1) a rapidly melting snowpack over soil containing dense frost may accelerate runoff and produce significant overland flow;
(2) the presence of stalactite soil frost (or needle ice) increases infiltration of snowmelt;
(3) the presence of porous concrete frost reduces infiltration rate and increases overland flow;
(4) exposed rock encourages overland flow.

In recent years, the significance of overland flow in forested (Whipkey, 1965) and temperate vegetated landscapes has been shown to have been exaggerated (Kirkby and Chorley, 1967; but cf. Pierce, 1967). It is apparent, however, that in alpine regions overland flow must be highly significant although little published material exists. Dickinson and Whiteley's (1970) extension of Hewlett's (1961) concept of the contributing basin area is particularly appropriate to alpine basins. As the snowpack melts at progressively higher elevations during the spring and summer as a response to net radiation conditions, so the contributing basin area expands proportionately. Sudden decreases in net radiation at critical elevations can rapidly decrease the effective basin area.

During periods of particularly rapid snowmelt such as that which occurred during the second and third weeks of July 1971 at Miller Creek (table 3B.2), the overland flow phenomenon is observable within a restricted elevation band of about 30–40 m below the snowline. This zone of overland flow retreats upslope following the snowline and is particularly well developed on the ubac slopes above about 750 m, where rocky outcrops and thin soils are dominant.

The process of interflow, whereby the water moves through the soil and vegetation layers rather rapidly and emerges as surface runoff at a percoline downslope or from the banks of a stream, is also of importance. The thickness and nature of the soil and the vegetation cover are critical in determining the actual response time. The rapid response time of interflow and even more rapid response time of overland flow are important factors in the interpretation of alpine stream hydrographs.

6. Alpine lakes

According to Hutchinson (1957), alpine lakes are of three kinds, if classified on the basis of temperature stratification. There are temperate lakes with surface temperature above 4°C in summer and below 4°C in winter; subpolar lakes, with surface temperatures above 4°C only for a short time during the summer; and polar lakes where the surface temperature is always below 4°C.

The summer heat income, or the amount of heat needed to raise a lake from the isothermal condition at 4°C to the highest observed summer heat content, has been considered for a number of alpine lakes. Table 3B.3, which is derived from sensible temperature measurements, should not be confused with an energy budget because the latent heat of melting of ice is not included in the calculation.

In shallow lakes, which are ice-covered for a part of the year, Bilello (1968) has shown the importance of the heat budget of lake bottom sediments. During a twenty-five day period in December 1960 at Seneca Lake in Upper Michigan, the water temperature

Table 3B.3. Summer heat income in alpine lakes (from Hutchinson, 1957).

	Latitude	Altitude (m)	Area (km²)	Summer heat income (cal cm⁻²)
Scandinavian Highlands				
Flakevatn	60°49′	1,448	3	6,000
Canadian Rockies				
Bow	51°45′	1,990	3.6	7,500
Minnewanka	51°16′	1,454	13	25,900
European Alps				
Oeschinen	46°30′	1,581	1.2	10,400
Western Cordillera, USA				
Tahoe	39°09′	1,890	499	34,800
Tibet				
Manasarovar	30°40′	4,602	558	26,000

near the bottom of the lake increased from 0.3° to 3.0°C, and the bulk of this heat gain came from the sediments underlying the lake.

Johnston (1922) and Mathews (1956) provide case studies of alpine lake hydrology in the Canadian Cordillera.

7. Discharge regime in alpine regions

Alpine rivers respond rapidly to changes in the energy balance of the basin and to precipitation (Kellerhals, 1970). Infiltration rates are commonly such as to maximize the rapid response processes of overland flow and interflow, the gradients are steep, and deep groundwater seepage is quantitatively less important. In areas of extensive karst development, for example as in the Alberta Rockies, such a generalization does not hold and local lithologic peculiarities need to be investigated in order to confirm this point.

Because of the rapid response of the alpine river, it can be said that the hydrograph is typically extremely irregular during spring through autumn (Dingman *et al.*, 1971) and then apparently becomes more regular during winter (fig. 3B.5). Three major factors in alpine discharge regimes can be differentiated: precipitation, snowmelt and glacier melt. In the more coastal alpine streams of British Columbia and in western

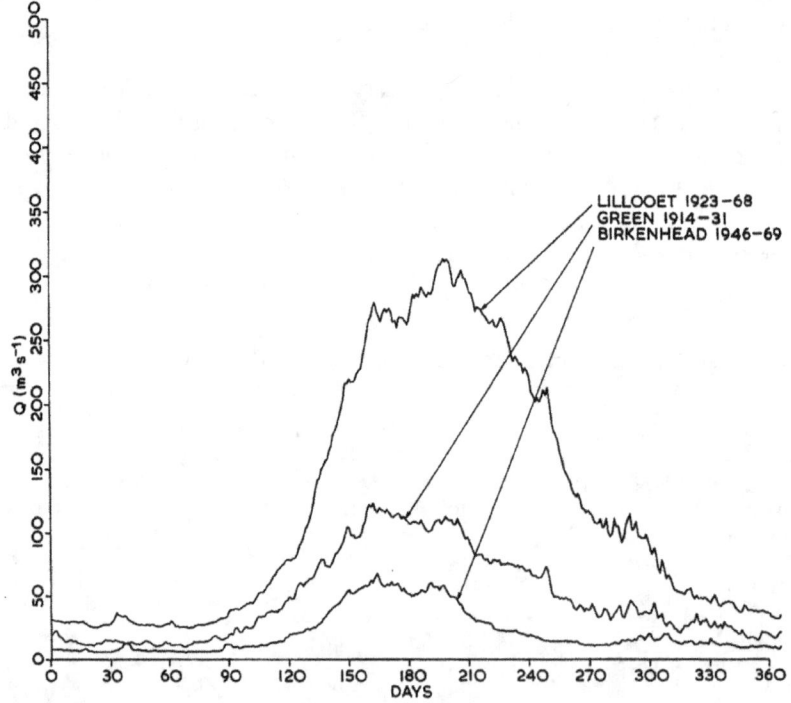

Fig. 3B.5. Mean daily discharge for the Lillooet, Green and Birkenhead rivers, British Columbia (days numbered from 1 January to 31 December).

Washington State (Fahnestock, 1963) all three influences on discharge hydrographs are clearly shown. In the more continental Alberta Rockies, by contrast, only the snowmelt and glacier melt peaks are prominent.

Fig. 3B.6 shows the discharge hydrograph for Miller Creek, British Columbia, for a period of high surplus energy in July 1971. The peak discharge of 10.45 m³ s⁻¹ at 1900 hours on 1 August represents a discharge intensity of 0.45 m³ s⁻¹ km⁻². Mathews (1964a) reported a peak discharge of 14 m³ s⁻¹ on 16 July 1949 in the Sunwapta River basin, Alberta. This represented a discharge intensity of 0.5 m³ s⁻¹ km⁻² for a slightly larger basin than that of Miller Creek. Fahnestock recorded a maximum discharge intensity

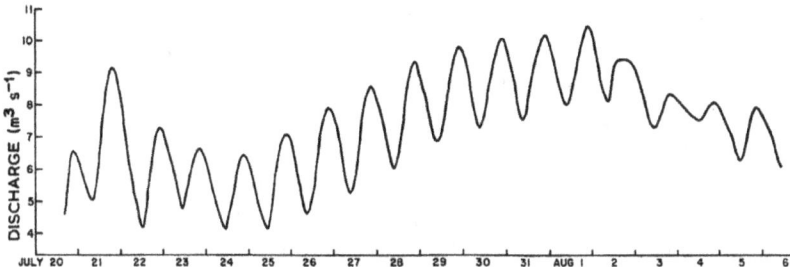

Fig. 3B.6. Discharge hydrograph for the Miller Creek, British Columbia, during July and August 1971.

for the White River basin, Washington, of 0.64 m³ s⁻¹ km⁻² in July 1959. The other noteworthy feature of fig. 3B.6 is the large daily fluctuation in discharge, with the minimum recorded around 0900–1000 and the maximum around 1900–2000 hours. In general there is no direct relationship between precipitation and runoff in any season except the late summer. This observation led Mathews (1964a) to introduce his simple but effective predictive equation for the Sunwapta River mean daily discharge (Q) which has the form

$$Q = b(\textstyle\sum T_n k^n + a'/b') \qquad (0 \leqslant n \leqslant 11)$$

where T_n is the mean daily temperature at Jasper, Alberta, on the day n days before the day of flow measurement, a' and b' are coefficients determined by regression analysis, and k is a recession coefficient with value less than 1.

Mathews also reported an anomalously high discharge event during the period 24 September to 5 October 1956. As this was a period of near-freezing weather, he concluded that the event must be due to the release of meltwater previously ponded by the glacier. In this sense, the event could be categorized as a *jökulhlaup* (Thorarinsson, 1939; Richardson, 1968). Plate 6 illustrates the dramatic impact of such an event.

Even mean daily flows can differentiate basins in terms of glacier and snowmelt contribution. Figure 3B.5 shows the snowmelt peak of June well developed in each of three alpine basins. In the case of the Lillooet River the July glacier melt peak is the major peak, while for the Green and Birkenhead, with progressively less glacierization, the glacier melt peak is subdued. The autumn rains peaks are present in all three hydrographs. Problems of alpine discharge estimation are discussed by Riggs and Moore (1965) and Eiselstein (1967), and Julian *et al.* (1967) have attempted to predict water yield on the basis of physiography.

8. Alpine water quality and sediment yield

The study of Rainwater and Guy (1961) for the Chamberlin Glacier area of northern Alaska is a classic of its kind in its careful delineation of uniform source areas. The quality of water in the form of precipitation, surface glacier ice, subsurface glacier ice, glacial meltwater (high and low water), non-glacial river water and lake water in an alpine arctic environment was separately determined. They showed that dissolved solids ranged from 7 parts per million in rainwater to 54 parts per million in the non-glacial tributary to the alpine lake. By further analysis of the saturation indices and the hardness of the waters, it was inferred that very rapid dissolution of minerals occurs downstream from glaciers in the alpine zone. Hembree and Rainwater (1961) and Kunkle and Meiman (1967) provide valuable data on this topic.

Rates of solution as reported from alpine regions vary from 26 tons yr^{-1} km^{-2} (Rapp, 1960) to 135 tons yr^{-1} km^{-2} (Jackli, 1957, quoted in Leopold *et al.*, 1964). It is clear that the lithology influences very greatly the variation in this value. As far as suspended sediment is concerned, and conservatively assuming a two-month period of active suspended sediment movement, published data for glacierized basins indicate a range of 18 tons yr^{-1} km^{-2} (Mathews, 1964b) to 912 tons yr^{-1} km^{-2} (Østrem and Pytte, 1968) for eleven basins ranging from 3.7 to 56 km^2 in area. These are in general underestimates, but indicate the order of magnitude of suspended sediment movement involved. The variability in the data reflects differences in subglacial lithology and in the nature of the morainic material available for transport. The bedload transport rates are very difficult to evaluate but they range up to c. 6000 tons yr^{-1} km^{-2} for the White River basin of 20 km^2.

It is clear that most of the above rates are highly influenced by the fact that the basins are glacierized (Lanser, 1958; Mathews, 1956). Data from unglacierized alpine basins, for example in the Alberta Rockies (McPherson, 1971) and the Colorado Front Range (Caine, Chapter 12B), suggest rates of sediment movement one to three orders of magnitude lower than those of the glacierized basins. Any attempt to evaluate mean denudation rates for glacierized alpine basins from sediment movement data is completely spurious as the figures are inflated by high sediment availability. For purposes of comparison it is however interesting to make computations of total sediment yield in terms of millimetres of surface lowering per year equivalent. If this is done, we find a range of reported data from alpine sediment movement ranging from 0.001 to 4.0 mm yr^{-1} (Caine, Chapter 12B; Fahnestock, 1963).

9. A model for alpine hydrology

A model of the hydrology of alpine regions must consider ways in which the primary inputs of precipitation and solar radiation interact with a number of different surfaces.

The number of surfaces of different thermal and roughness characteristics is infinite; the task is therefore to isolate mesoscale environmental systems within which there is some greater homogeneity of response to the primary inputs than is the case for each surface separately. The following readily suggest themselves as being reasonably homogeneous; they are also highly characteristic of alpine regions:

(1) Glaciers.
(2) Snowpacks.
(3) Alpine and subalpine meadows; adret slopes.
(4) Alpine and subalpine meadows; valley bottoms.
(5) Alpine and subalpine forest; ridge tops.
(6) Alpine and subalpine forest; adret slopes.
(7) Alpine and subalpine forest; ubac slopes.
(8) Alpine lakes.
(9) River channels.
(10) Morainic landforms, partially vegetated.
(11) Bedrock ridges.
(12) Bedrock adret slopes.
(13) Bedrock ubac slopes.

In order that it may be possible adequately to understand and manage the alpine regions from the point of view of their hydrology (cf. Jeffrey, 1968), there is need for intensive study of these distinctive alpine environmental systems. Table 3B.4 summarizes some of the hydrologically significant features about these thirteen systems.

10. Major contrasts between alpine and arctic hydrology

Of the six factors regarded as essential to the recognition of alpine hydrologic regions (see p. 136), only those associated with high moisture fluxes and high available relief are irrelevant to the definition of arctic hydrology. The relative importance of the other four factors is quite different and so is the magnitude and frequency with which they occur. It could also be said that the presence of continuous permafrost, the importance of the active layer and the role of groundwater in the production of significant quantities of base flow are all important additional factors in arctic hydrology.

Whereas alpine hydrology in lower latitudes is influenced by high energy fluxes throughout the year on account of the low turbidity of the atmosphere, arctic hydrologic processes are controlled by the extraordinarily contrasted energy regimes of summer and winter. The surface runoff season, during which recognizable fluvial processes are highly active, is extremely short by comparison with alpine runoff regimes.

In summary, it would seem that high radiant energy fluxes for a short summer season, discharge hydrographs greatly influenced by snowmelt and/or glacier melt processes during that short summer, poorly developed soils but with regolith occasionally very deep (e.g. Mackenzie Delta), trees almost absent, continuous permafrost, the importance of the active layer and major groundwater contributions to runoff are the key factors in

Table 3B.4. Summary of major hydrologic processes operating in mesoscale alpine environmental systems.

	$>10^3$ years	$10-10^3$ years	$10^{-1}-10$ years	$<10^{-1}$ (c. 1 month)
(1) Glaciers	Water storage term in hydrologic cycle	Glacier retreat and advance	Ablation, accumulation	Physics of glacier motion, surging
(2) Snowpacks	—	—	Snow metamorphism firn ice (see 1)	Melt, storage, accumulation
(3) Alpine lakes	Glacial scour and lake formation, sedimentation and meadow formation (see 6)	Sedimentation	Temperature stratification, draining events, ice formation and break-up, sedimentation regime	Seiches, daily level changes, sediment movement into and through lake, evaporation
(4) Mountain streams	Glacial erosion	Downcutting, supply (creep)	Discharge regime, sediment supply (talus)	Runoff concentration, 'tumbling' flow, rapid response to precipitation and evaporation
(5) Morainic mounds	Glacial deposition	Degradation of morainic slopes	Chemical weathering, frost action, vegetational change	Infiltration, interflow
(6) Alpine and sub-alpine meadows; valley bottoms	Sedimentation in glacial lakes (see 3)	Dissection by streams, flood plain development, soil formation	Periodic inundation, jökulhlaups (see 1), vegetational change	Precipitation, infiltration, interflow, baseflow
(7) Alpine and subalpine meadows; adret slope	Postglacial soil development	Slope degradation, channel dissection	Vegetational change, mass wasting	Precipitation, infiltration, interflow
(8) Alpine and subalpine treed slopes, adret	Postglacial soil development, soil creep, gullying	Slope degradation, channel dissection	Frost action, mass wasting	Interception, precipitation, infiltration, interflow
(9) Alpine and subalpine treed slopes, ubac and ridge top	Postglacial soil development, slides and flows of earth and mud	Slope degradation	Frost action, mass wasting	Overland flow, precipitation, interception, infiltration, interflow
(10) Alpine barren	Glacial erosion postglacial rockfall and talus, slope formation	Slope degradation	Frost action, mass wasting	Overland flow, precipitation

any assessment of arctic hydrology. Implications are that the study of arctic hydrology is not so influenced by the spatial scale problem, in that the spatial variability of energy, mass and water fluxes is not so great as in the alpine regions. More significant is the time scale of the study and the recognition of the variability of these same fluxes over time.

References

ALFORD, D. (1967) Density variations in alpine snow. *J. Glaciol.*, **6**, 495–6.

ANDERSON, E. A. (1968) Development and testing of snow pack energy balance equation. *Water Resources Res.*, **4**, 19–37.

ANDERSON, J. A. and ROCKWOOD, D. (1970) Runoff synthesis for rain on snow basin. *Proc. Western Snow Conf.*, **38**, 82–90.

APOLLOV, B. A., KALININ, G. P. and KOMAROV, V. D. (1964) *Hydrological Forecasting.* Jerusalem: Israel Progr. Sci. Transl. 338 pp.

BARNES, J. C. and BOWLEY, C. J. (1968) Snow cover distribution as mapped from satellite photography. *Water Resources Res.*, **4**, 252–72.

BERNDT, H. W. and FOWLER, W. B. (1969) Rime and hoar frost in upper slope forests of eastern Washington. *J. Forest.*, **67**, 92–5.

BILELLO, M. A. (1968) Water temperatures in a shallow lake during ice formation, growth and decay. *Water Resources Res.*, **4**, 749–60.

BLANEY, H. F. (1958) Evaporation from free water surfaces at high altitudes. *Trans. Amer. Soc. Civ. Engr.*, **123**, Pap. 2925.

BLISS, L. C. (1960) Transpiration rates of arctic and alpine shrubs. *Ecology*, **41**, 386–9.

CHURCH, M. A. and KELLERHALS, R. (1970) Stream gauging techniques for remote areas using portable equipment. *Inland Waters Branch Tech. Bull.*, **25**, Ottawa: Dept of Energy, Mines and Resources. 89 pp.

DICKINSON, W. T. and WHITELEY, H. (1970) Watershed areas contributing to runoff. *Int. Ass. Sci. Hydrol. Pub.* (General Assembly of Wellington), 96, 12–26.

DINGMAN, S. L., SAMIDE, H. R., SABOE, D. L., LYNCH, M. J. and SLAUGHTER, C. W. (1971) Hydrologic reconnaissance of the Delta River and its drainage basin. *Cold Regions Science and Engineering Res. Rep.*, 262. Hanover: US Army CRREL. 83 pp.

DUNKLE, R. V. and BEVANS, J. T. (1956) An approximate analysis of the solar reflectance and transmittance of a snow cover. *J. Met.*, **13**, 212–16.

EISELSTEIN, L. M. (1967) A principal component analysis of surface runoff data from a New Zealand alpine watershed. *Proc. Int. Hydrol. Symp.* (Fort Collins), **1**, 479–89.

FAHNESTOCK, R. K. (1963) Morphology and hydrology of a glacial stream, White River, Mount Rainier, Washington. *US Geol. Surv. Prof. Pap.*, 422-A. Washington, DC. 70 pp.

FITZHARRIS, B. B. (1970) *Implications of Spatial and Temporal Variations in Snowfall.* Unpub. MS, Univ. of British Columbia.

FRANK, R. C. and LEE, R. (1966) Potential solar beam irradiation on slopes. *US Dept of Agric., Forest Serv. Res. Pap.*, RM 18. 116 pp.

GARSTKA, W. U., LOVE, L., GOODELL, B. C. and BERTLE, F. A. (1959) *Factors Affecting Snowmelt and Streamflow.* US Dept of Interior Bur. Reclamation, and US Dept of Agric. Forest Serv. 187 pp.

GATES, D. M. and JANKE, R. (1966) The energy environment of the alpine tundra. *Oecol. Planta.*, **1**, 39–61.

GEIGER, R. (1965) *The Climate Near the Ground.* Cambridge, Mass: Harvard Univ. Press. 611 pp.

GILBERT, R. E. (1969) *Some Aspects of the Hydrology of Ice-Dammed Lakes.* Unpub. M.A. thesis, Univ. of British Columbia.

HAGEN, I. (1965) Planning of hydrological observations in catchment areas with partially glacier covered tracts. *Int. Ass. Sci. Hydrol. Pub.* (General Assembly of Quebec), 68, 760–71.

HANSEN, E. A. and FFOLLIOTT, P. F. (1968) Observations of snow accumulation and melt in demonstration cuttings of Ponderosa pine in central Arizona. *US Dept of Agric., Forest Serv. Res. Pap.*, RM 111. 12 pp.

HAUPT, H. F. (1967) Infiltration, overland flow and soil movement on frozen and snow covered plots. *Water Resources Res.*, **3**, 145–61.

HEMBREE, C. H. and RAINWATER, F. H. (1961) Chemical degradation on opposite flanks of the Wind River Range, Wyoming. *US Geol. Surv. Water Supply Pap.*, 1535E. 9 pp.

HENDRICK, R. L., FILGATE, B. D. and ADAMS, W. M. (1971) Application of environmental analysis to watershed snowmelt. *J. Appl. Met.*, **10**, 418–29.

HEWLETT, J. D. (1961) Some ideas about storm runoff and baseflow. *US Dept of Agric., Forest Serv., Southeast Forest Experimental Station Ann. Rep.*, 61–6.

HOINKES, H. C. and RUDOLPH, R. (1962) Mass balance studies on the Hintereisferner, Ötztal Alps, 1952–1961. *J. Glaciol.*, **4**, 266–80.

HOINKES, H. C., HOWORKA, F. and SCHNEIDER, W. (1968) Glacier mass budget and mesoscale weather in the Austrian Alps 1964–1966. *Int. Ass. Sci. Hydrol. Pub.* (General Assembly of Berne), 79, 241–54.

HOOVER, M. D. and LEAF, C. F. (1967) Process and significance of interception in Colorado sub-alpine forest. *In:* SOPPER, W. E. and LULL, H. W. (eds.) *Proceedings of a Symposium on Forest Hydrology*, 213–24. Oxford: Pergamon Press.

HOVIND, E. L. (1965) Precipitation distribution around a windy mountain peak. *J. Geophys. Res.*, **70**, 3271–8.

HUTCHINSON, B. A. (1966) A comparison of evaporation from snow and soil surfaces. *Bull. Int. Ass. Sci. Hydrol.*, **11**, 34–42.

HUTCHINSON, G. E. (1957) *A Treatise on Limnology: Geography, Physics and Chemistry*, Vol. 1. New York: John Wiley. 1015 pp.

HUTCHINSON, P. (1969) Estimation of rainfall in sparsely gauged areas. *Bull. Int. Ass. Sci. Hydrol.*, **14**, 101–19.

IVERONOVA, M. I. (1966) Le rôle hydrologique des avalanches. *Int. Ass. Sci. Hydrol. Pub.* (General Assembly of Davos), 69, 73–7.

JAKHELLN, A. (1965) Design of hydrometeorological network for estimation of snow accumulation in Norway. *Int. Ass. Sci. Hydrol. Pub.* (General Assembly of Quebec), 68, 736–47.

JEFFREY, W. W. (1964) Watershed research in the Saskatchewan River headwaters. *Proc. Hydrol. Symp.*, **4**, 79–132. Ottawa: Nat. Res. Council.

JEFFREY, W. W. (1965) Experimental watersheds in the Rocky Mountains, Alberta, Canada. *Int. Ass. Sci. Hydrol. Pub.* (General Assembly of Berkeley), 63, 502–21.

JEFFREY, W. W. (1968) Watershed management problems in British Columbia. *Water Resources Bull.*, **4**, 58–70.

JOHNSTON, W. A. (1922) Sedimentation in Lake Louise, Alberta, Canada. *Amer. J. Sci.*, **4**, 376–86.

JUDSON, A. (1966) Snow cover and avalanches in the high alpine zone of western United States. *Proc. Int. Conf. Low Temp. Sci.* (Conference on Physics of Snow and Ice), 1151–68.

JULIAN, R. W., YEVJEVICH, V. and MOREL-SEYTOUX, H. J. (1967) Prediction of water yield in high mountain watersheds based on physiography. *Colorado State Univ.*, *Hydrol. Pap.*, 22. 20 pp.

KEELER, C. M. (1969) Snow accumulation on Mount Logan, Yukon Territory, Canada. *Water Resources Res.*, **5**, 719–23.

KELLERHALS, R. (1970) *Runoff Concentration in Steep Channel Networks.* Unpub. Ph.D. thesis, Univ. of British Columbia. 262 pp.

KIRKBY, M. J. and CHORLEY, R. J. (1967) Throughflow, overland flow and erosion. *Bull. Int. Ass. Sci. Hydrol.*, **12**, 5–21.

KOTLYAKOV, V. M. and PLAM, M. Y. (1966) The influence of drifting on snow distribution in the mountains and its role in the formation of avalanches. *Int. Ass. Sci. Hydrol. Pub.* (General Assembly of Davos), 69, 53–60.

KUNKLE, S. H. and MEIMAN, J. R. (1967) Water quality of mountain watersheds. *Colorado State Univ., Hydrol. Pap.*, 21. 53 pp.

LACHAPELLE, E. (1961) Energy exchange measurements on the Blue Glacier, Washington. *Int. Ass. Sci. Hydrol. Pub.* (General Assembly of Helsinki), 54, 302–10.

LACHAPELLE, E. (1966) The control of snow avalanches. *Sci. Amer.*, **214**, 92–101.

LANG, H. (1968) Relations between glacier runoff and meteorological factors observed on the outside of the glacier. *Int. Ass. Sci. Hydrol. Pub.* (General Assembly of Berne), 79, 429–39.

LANSER, O. (1958) Réflexions sur les débits solides en suspension des cours d'eau glaciaires. *Bull. Int. Ass. Sci. Hydrol.*, **4**, 37–43.

LEOPOLD, L. B., WOLMAN, M. G. and MILLER, J. P. (1964) *Fluvial Processes in Geomorphology.* San Francisco: Freeman. 522 pp.

LOUP, J. (1969) Durée de la couverture nivale sur un adret et sur un ubac dans la région d'Embrun (Hautes Alpes). *Rev. Géogr. Alp.*, **57**, 593–608.

MCKAY, G. A. (1970) Precipitation. *In:* GRAY, D. M. (ed.) *Handbook on Principles of Hydrology*, Chap. 2. 111 pp.

MACKAY, J. R. and MATHEWS, W. H. (1966) Observations on pressures exerted by creeping snow, Mount Seymour, British Columbia, Canada. *Proc. Int. Conf. Low Temp. Sci.* (Conference on Physics of Snow and Ice), 1185–98.

MCPHERSON, H. J. (1971) Downstream changes in sediment character in a high energy mountain stream channel. *Arct. Alp. Res.*, **3**, 65–79.

MARCUS, M. G. (1964) Climate–glacier studies in the Juneau Icefield region, Alaska. *Univ. of Chicago, Dept of Geog. Res. Pap.*, 88. 128 pp.

MARCUS, M. G. and RAGLE, R. H. (1970) Snow accumulation in the Icefield Ranges, St Elias Mountains, Yukon. *Arct. Alp. Res.*, **2**, 277–92.

MARTINELLI, M. (1959) Alpine snowfields – their characteristics and management possibilities. *Int. Ass. Sci. Hydrol. Pub.* (Symposium of Hannoversch-Munden), 48, 120–7.

MARTINELLI, M. (1967) Possibilities of snowpack management in alpine areas. *In:* SOPPER, W. E. and LULL, H. W. (eds.) *Proceedings of a Symposium on Forest Hydrology*, 225–31. Oxford: Pergamon Press.

MARTONNE, E. DE (1933) *The Geographical Regions of France* (transl. H. C. Brentnall). London: Heinemann. 224 pp.

MATHEWS, W. H. (1956) Physical limnology and sedimentation in a glacial lake. *Bull. Geol. Soc. Amer.*, **67**, 537–52.

MATHEWS, W. H. (1964a) Discharge of a glacial stream. *Int. Ass. Sci. Hydrol. Pub.* (General Assembly of Berkeley), 63, 502–21.

MATHEWS, W. H. (1964b) Sediment transport from Athabasca Glacier, Alberta. *Int. Ass. Sci. Hydrol. Pub.* (General Assembly of Berkeley), 65, 155–65.

MEIER, M. F. (1962) Proposed definitions for glacier mass budget terms. *J. Glaciol.*, **4** (33), 252–63.

MEIER, M. F. (1964) Ice and glaciers. *In:* CHOW, V. T. (ed.) *Handbook of Applied Hydrology.* New York: McGraw-Hill.

MEIER, M. F. (1970) Glaciers and snowpacks: some common problems in non-polar glaciology. *Proc. 1st Int. Seminar for Hydrol. Professors*, 729–36.

MEIER, M. F. and POST, A. S. (1962) Recent variations in mass net budgets of glaciers in western North America. *Int. Ass. Sci. Hydrol. Pub.* (Symposium of Obergurgl), 58, 63–77.

MEIMAN, J. R. (1970) Snow accumulation related to elevation, aspect and forest canopy. *Proc. of a Workshop Seminar on Snow Hydrol.*, 35–47. Can. Nat. Comm./Int. Hydrol. Decade.

MELLOR, M. (1964) Thermal properties and radiation characteristics. *In:* Properties of Snow, Chap. VI. *Cold Regions Science and Engineering Monogr.*, III-A1. Hanover: US Army CRREL. 105 pp.

MILLER, D. H. (1955) Snow cover and climate in the Sierra Nevada, California. *Univ. of California, Pub. in Geography*, 11. 218 pp.

NEAL, J. H. (1938) The effects of degree of slope and rainfall characteristics on runoff and erosion. *Missouri Agric. Exp. Station Res. Bull.*, 280.

ØSTREM, G. (1961–2) Nigardsbreen hydrologi. *Norsk Geogr. Tidssk.*, **18**, 156–202.

ØSTREM, G. (1966) Mass balance studies on glaciers in Western Canada, 1965. *Geogr. Bull.*, **8**, 81–107.

ØSTREM, G. (1969) Correlation studies and regression analysis. *In:* PYTTE, G. (ed.) Glasiologiske undersøkelser i Norge 1968, *Norges vassdrags og elektrisitetsvesen Rapport*, 5/69. 149 pp.

ØSTREM, G. and PYTTE, R. (1968) Glasiologiske undersøkelser i Norge, 1967. *Norges vassdrags og elektrisitetsvesen Rapport*, 4/68. 131 pp.

PECK, E. L. and PFANKUCH, D. J. (1963) Evaporation rates in mountainous terrain. *Int. Ass. Sci. Hydrol. Pub.* (General Assembly of Berkeley), 62, 267–78.

PIERCE, R. S. (1967) Evidence of overland flow on forested watersheds. *In:* SOPPER, W. E. and LULL, H. W. (eds.) *Proceedings of a Symposium on Forest Hydrology*, 247–54. Oxford: Pergamon Press.

PYTTE, R. (1967) Glasiohydrologiske undersøkelser i Norge 1966. *Norges vassdrags og elektrisitetsvesen Rapport*, 2/67. 83 pp.

RAINBIRD, A. F. (1965) Precipitation basic principles of network design. *Int. Ass. Sci. Hydrol. Pub.* (General Assembly of Quebec), 67, 19–30.

RAINWATER, F. H. and GUY, H. P. (1961) Some observations on the hydrochemistry and sedimentation of the Chamberlin Glacier area, Alaska. *US Geol. Surv. Prof. Pap.*, 414–C. 14 pp.

RANTZ, S. E. (1964) Snowmelt hydrology of a Sierra Nevada stream. *US Geol. Surv. Water Supply Pap.*, 1779-R. 36 pp.

RAPP, A. (1960) Recent development of mountain slopes in Karkevagge and surroundings, northern Scandinavia. *Geogr. Ann.*, **42A**, 71–200.

RICHARDSON, D. (1968) Glacier outburst floods in the Pacific Northwest. *US Geol. Surv. Prof. Pap.*, 600-D, D79–86.

RIGGS, H. C. and MOORE, D. O. (1965) A method of estimating mean runoff from ungauged basins in mountainous regions. *US Geol. Surv. Prof. Pap.*, 525-D, 199–202.

RILEY, J. P., CHADWICK, D. G. and EGGLESTON, K. O. (1969) *Snow melt simulation.* Logan, Utah: Utah Water Res. Lab. 33 pp.

ROBIN, G. DE Q. (1969) Initiations of glacier surges. *Can. J. Earth Sci.*, **6**, 919–28.

ROCKWOOD, D. M. (1958) Columbia Basin stream flow routing by computer. *J. Waterways and Harbors Div.*, **84**, Pap. 1874. Amer. Soc. Civ. Engr.

SCHAERER, P. A. (1970) Variation of ground snow loads in British Columbia. *Proc. Western Snow Conf.*, **38**, 44–8.

SCHERMERHORN, V. P. (1967) Relations between topography and annual precipitation in western Oregon and Washington. *Water Resources Res.*, **3**, 707–11.

SCHYTT, V. (1962) Mass balance studies in Kebnekajse. *J. Glaciol.*, **4**, 281–6.

SCOTT, J. T. and RAGOTZKIE, R. A. (1961) Heat budget of an ice covered inland lake. *Univ. of Wisconsin, Dept of Met. Tech. Rep.*, 6. 41 pp.

SHARPE, D. M. (1970) The effective climate in the dynamics of alpine timberline ecosystems in Colorado. *Publ. in Climatol.*, 23. 82 pp. Centerton, NJ.

SIMMONS, G. E. (1961) Snowmelt runoff, Fraser River Basin. *Proc. Hydrol. Symp.*, **1**, 227–59. Ottawa: Nat. Res. Council.

SLAYMAKER, H. O. (1969) The scale problem in hydrology. *In:* BOWEN, E. G., CARTER, H. and TAYLOR, J. (eds.) *Geography at Aberystwyth*, 68–86. Cardiff: Univ. of Wales Press.

SOLOMON, S. F., DENOUILLIEZ, J. P., CHART, E. J., WOOLEY, J. A. and CADON, C. (1968) The use of square grid system for computer estimation of precipitation, temperature and runoff. *Water Resources Res.*, **4**, 919–29.

SOSEDOV, I. S. and SEVERSKY, I. V. (1966) On the hydrological role of snow avalanches in the northern slope of the Zailiysky Alatau (Tien Shan). *Int. Ass. Sci. Hydrol. Pub.* (General Assembly of Davos), 69, 78–85.

SPORNS, V. (1964) On the transposition of short duration rainfall intensity data on mountainous regions. *Arch. Met. Geophys. Biokl.*, B, 13, 438–42.

SPREEN, W. C. (1947) A determination of the effect of topography on precipitation. *Trans. Amer. Geophys. Union*, **28**, 286–90.

STANLEY, A. D. (1970) Combined balance studies at selected glacier basins in Canada. *Proc. of a Workshop Seminar on Glaciers*, 5–9. Can. Nat. Comm./Int. Hydrol. Decade.

TANGBORN, W. V. (1966) Glacier mass budget measurements by hydrologic means. *Water Resources Res.*, **2**, 105–10.

TANGBORN, W. V. (1968) Mass balance of some North Cascade glaciers as determined by hydrologic parameters, 1920–1965. *Int. Ass. Sci. Hydrol. Pub.* (General Assembly of Berne), 79, 267–74.

TERJUNG, W. H., KICKERT, R. N., POTTER, G. L. and SWARTS, S. W. (1969) Energy and moisture balances of an alpine tundra in mid-July. *Arct. Alp. Res.*, **1**, 247–66.

THORARINSSON, S. (1939) Hoffellsjökull, its movements and drainage. *Geogr. Ann.*, **21**, 189–215.

US ARMY CORPS OF ENGINEERS (1955) *Lysimeter Studies of Snowmelt*. Res. Note, 25. Portland, Oreg.: North Pacific Div.

US ARMY CORPS OF ENGINEERS (1956) *Snow Hydrology*. Portland, Oreg.: North Pacific Div. 437 pp.

US ARMY CORPS OF ENGINEERS (1960) *Runoff from Snowmelt*. EM 1110-2-1406. Washington, DC. 75 pp.

VINJE, T. E. (1967) Some results of micrometeorological measurements in Antarctica. *Arch. Met. Geophys. Biokl.*, A, 16, 31–43.

WALKER, E. R. (1961) A synoptic climatology for parts of the Western Cordillera. *Arctic Met. Res. Group, Sci. Rep.*, 8. McGill Univ. 218 pp.

WALKOTTEN, W. J. and PATRIC, J. H. (1967) Elevation effects on rainfall near Hollis, Alaska. *US Dept of Agric., Forest Serv. Res. Pap.*, PNW 53, 1–7.

WHIPKEY, R. Z. (1965) Subsurface storm flow from forested slopes. *Bull. Int. Ass. Sci. Hydrol.*, **10**, 74–85.

WILLIAMS, G. P. (1965) Evaporation from water, snow and ice. *Proc. Hydrol. Symp.*, **2**, 31–47. Ottawa: Nat. Res. Council.

WILLIAMS, P. JR and PECK, E. L. (1962) Terrain influences on precipitation in the intermountain west as related to synoptic situations. *J. Appl. Met.*, **1**, 343–7.

YEN, Y. C. (1962) Effective thermal conductivity of ventilated snow. *J. Geophys. Res.*, **67**, 1091–8.

Manuscript received November 1971.

Additional references

BEHRENS, H. *et al.* (1971) Study of the discharge of alpine glaciers by means of environmental isotopes and dye tracers. *Zeits. Gletscherkunde und Glazialgeol.*, **7**, 79–102.

CLAGUE, J. J. and MATHEWS, W. H. (1973) The magnitude of jökulhlaups. *J. Glaciol.*, **12**, 501–4.

FÖHN, P. M. B. (1973) Short-term snow melt and ablation derived from heat and mass balance measurements. *J. Glaciol.*, 275–89.

KEELER, C. M. (1971) Snow and ice. *Eos*, **52**, 295–302.

SLATT, R. M. (1972) Geochemistry of meltwater streams from nine Alaskan glaciers. *Bull. Geol. Soc. Amer.*, **83**, 1125–31.

SLAYMAKER, H. O. and MCPHERSON, H. J. (1972) *Mountain Geomorphology*. Vancouver: Tantalus Research Ltd. 274 pp

UNESCO (1973) *International Symposia on the Role of Snow and Ice in Hydrology, September, 1972, Banff, Canada*. UNESCO/WMO. 2 vols.

Permafrost

Jack D. Ives

Institute of Arctic and Alpine Research,
University of Colorado

1. Introduction

During the early part of the twentieth century a climatic warming trend produced a marked impact upon the environments of the arctic regions, especially in areas bordering the North Atlantic: glaciers thinned and receded, the level of the world's oceans showed a minute rise, the southern boundary of perennially frozen ground in the Soviet Union moved northward, the Icelandic and Greenlandic cod fisheries were displaced toward the Pole. Much speculation was presented to the effect that a similar trend could be detected in other parts of the world, although the trend was either less marked, or the quantitative data less profuse, or a combination of the two, to inhibit the progress from speculation to scientific fact.

Since mid-century evidence has been accumulating from widely scattered areas in the northern hemisphere to indicate that a reversal of the earlier trend is underway. While it may be too soon to devote a great deal of energy to proving that the southern limit of perennially frozen ground is pushing equatorward, we are faced with the startling evidence that Iceland has experienced a series of extremely bad sea ice years which, if continued over even a very few more years, will have serious economic consequences.

These remarks serve to emphasize that climatic changes do occur within the life span of man and that they do have significant consequences in terms of man's total environment. They also state an assumption, perhaps overly self-evident, upon which this chapter is based: the presence of perennially frozen ground, which incidentally underlies more than one-fifth of the world's land area, is dependent upon world and regional climate. Perhaps less self-evident, but also implied, is the assumption that environmental responses to one of its inherent driving factors, climate, will frequently show a significant time lag, and that some responses will be more sluggish than others. Thus the permafrost of our planet is the summation of present world climate and the climatic history of the last several million years operating upon widely variegated terrain.

Permafrost, as defined by Muller (1947), is ground that remains frozen through at least two succeeding winters and the intervening summer, regardless of whether the frozen material is wet peat, dry gravel or bedrock. He coined the term from the more cumbersome expression *permanently frozen ground*. It is a synonym for *perennially frozen ground* and has won wide acceptance in the western world. There is some current opposition (Corte, 1969) and *perennially frozen ground* is occasionally receiving greater favour. The term *permafrost* is retained here, however, because of its brevity and literary 'bite'.

In a recent review on 'Geocryology and engineering' Corte (1969) indicates that literature surveys refer to the existence of several thousand published articles in the general field of geocryology, which includes permafrost as part of the 'subscience dealing with processes and phenomena of, or relating to, freezing temperatures occurring in nature' (Corte, 1969, p. 122). Despite this it must be admitted that great gaps exist in our knowledge of the detailed geographic distribution of permafrost and of the precise relationships between the various climatic parameters, climatic history, controlling terrestrial factors and the subsurface thermal gradient.

2. Factors affecting permafrost distribution

If a secular change in surface conditions, whether resulting from changing climate, natural or man-induced vegetation and soil modifications, or variations in surface or groundwater distribution, causes the mean annual air temperature to fall below 0°C the depth of winter frost penetration will tend to exceed the depth of summer thaw. As a consequence an initial thin layer of permafrost will form below the level of seasonal frost. It will thicken progressively year by year until a balance is struck between heat flow from the earth's interior and heat loss into the atmosphere through the ground/ air interface. The geothermal gradient is of the order of 1°C for every 30–60 m depth[1] so that, given a stable surface temperature, the lower limit of permafrost development 'ultimately approaches an equilibrium depth at which the temperature increase due to internal earth heat just offsets the amount by which the freezing point exceeds the mean surface temperature' (Lachenbruch, 1967, p. 833). It must be emphasized, however, that heat conduction in earth materials is extremely slow and, depending upon the magnitude of the surface temperature change, it may take thousands of years before a new thermal quasi-equilibrium is attained.

In contrast, the upper limit of the permafrost layer is controlled primarily by annual seasonal fluctuations about the mean. Thus while there exists a natural simplifying tendency to describe conditions, including depth to permafrost, thickness of the permafrost layer and temperatures within the permafrost as static, it must be borne in mind that actual conditions show constant change since the ground thermal regime is per-

---

[1] Gradients as steep as 1°C/15 m depth have been recorded in the USSR (R. J. E. Brown, personal communication, 1972).

sistently seeking to approximate impulses derived from the diurnal and seasonal march of air and ground surface temperature interactions.

From the foregoing it will be seen that permafrost occurrence will be an imperfect reflection of the mean annual air temperature of a particular site, the two depending primarily upon the energy balance at the ground/air interface. Climate in the broadest sense, therefore, is the predominant control of permafrost distribution; thus in the simplest possible terms the lower the mean annual air temperature of a given locality the thicker will be the permafrost layer, assuming that the mean annual air temperature has remained within a relatively narrow range for a significant period of time, say, thousands of years. Thus the world's greatest permafrost thicknesses (in excess of 500 m)

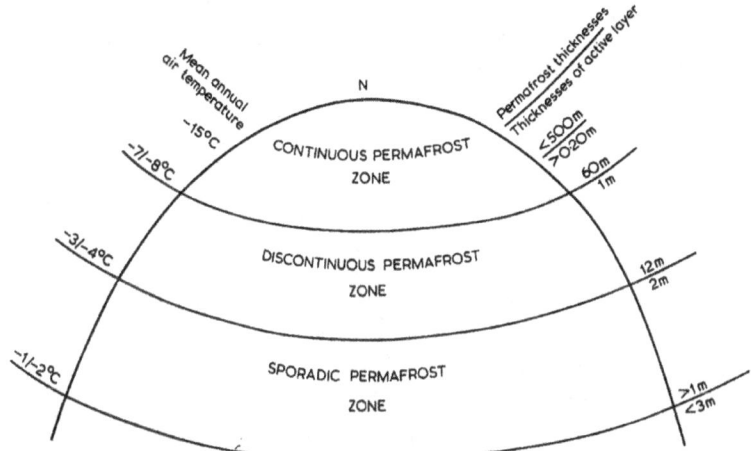

Fig. 4A.1. Theoretical distribution of permafrost in a perfect terrestrial northern hemisphere. Limiting data on mean annual air temperature, thickness of permafrost and thickness of active layer are shown.

are to be found in high latitudes, in cold continental interiors or at extremely high altitudes. Since very little work has been directed toward the study of permafrost in high mountains, or, for that matter, anywhere in the southern hemisphere, the first and major part or this chapter is devoted to a review of arctic and subarctic permafrost conditions. Fig. 4A.1 illustrates schematically the distribution of permafrost in a perfect, terrestrial northern hemisphere; temperature and thickness data are approximations of known fact. When the schematic distribution is compared with the map showing generalized permafrost distribution in a real world (fig. 4A.2) several qualifying factors become apparent. These are best illustrated by stating that while mean annual air temperature can be cited as a broad indicator of permafrost distribution, a number of important and interrelated secondary factors must be evaluated. Thus the character of the ground surface, or cover type, will have an important impact upon the energy exchange across the ground/air interface. In an extreme case deep water, either ocean or a lake with mean depth in excess of about 3 m, which is the maximum thickness of winter

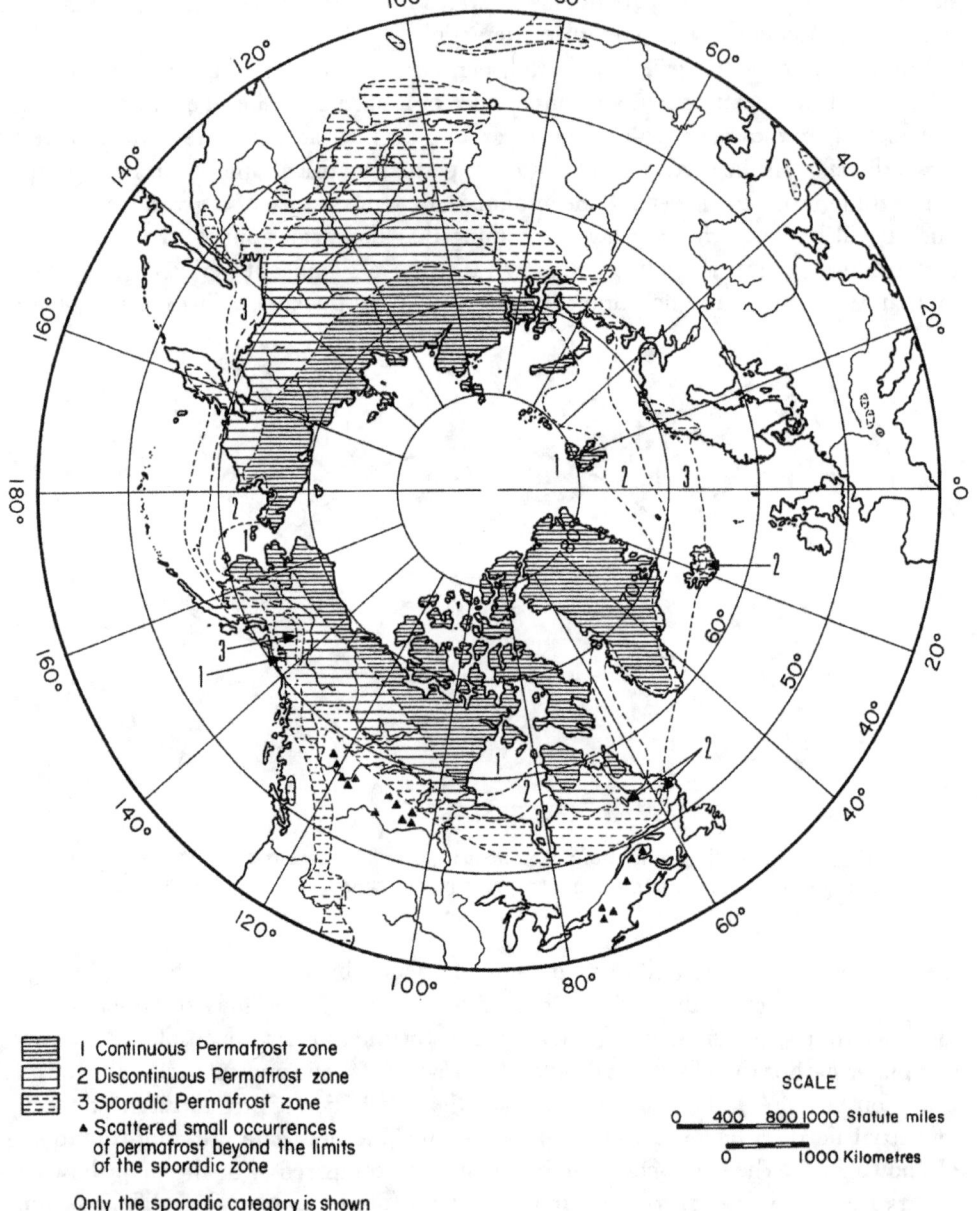

I Continuous Permafrost zone
2 Discontinuous Permafrost zone
3 Sporadic Permafrost zone
▲ Scattered small occurrences
  of permafrost beyond the limits
  of the sporadic zone

Only the sporadic category is shown
in high mountain areas south of
the main arctic zone

SCALE

0   400   800 1000 Statute miles

0   1000 Kilometres

Fig. 4A.2. Generalized map of the distribution of permafrost in the northern hemisphere
(compiled from various sources).

ice development, will totally inhibit permafrost formation even under the most severe climatic regime. A partial exception to this is the known existence of thin permafrost beneath shallow seas (e.g. Laptev Sea and northern Canada) where saltwater temperatures at −2°C occur. Similarly glacier ice and extensive winter snow accumulation,

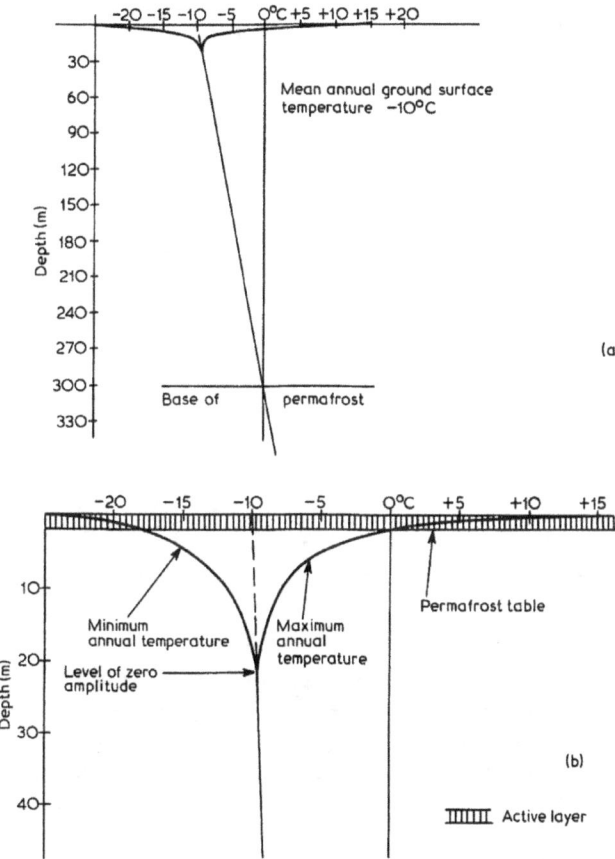

Fig. 4A.3. Schematic ground temperature profiles illustrating the relationships between (a) thermal gradient and the base of permafrost, and (b) the active layer and the level of zero amplitude.

through blanketing the ground surface with an effective layer of insulation, will tend to produce a considerable degree of inhibition.[1] Vegetation type, soil colour, soil moisture and site aspect are all important secondary factors since they affect the thermal characteristics of the surface which may be described as the energy filter through which the subsurface thermal gradient is established. It is also self-evident that both primary and secondary factors are interdependent variables; soil type and moisture content affect

[1] In a strict sense, glacier ice with temperature below the pressure-melting point should be considered permafrost. This interpretation has been used in preparation of the map, fig. 4A.2.

vegetation development, which in turn affects the pattern of winter snow accumulation, which influences soil moisture and vegetation. These interrelationships are so complex in terms of the energy exchange at the surface that we have little definitive knowledge of the absolute importance of the individual parameters, and in many instances even their relative importance is hard to determine.

The relationships between ground temperature gradient and permafrost thickness are illustrated in fig. 4A.3; this assumes a thermal gradient of 1°C per 30 m and a mean

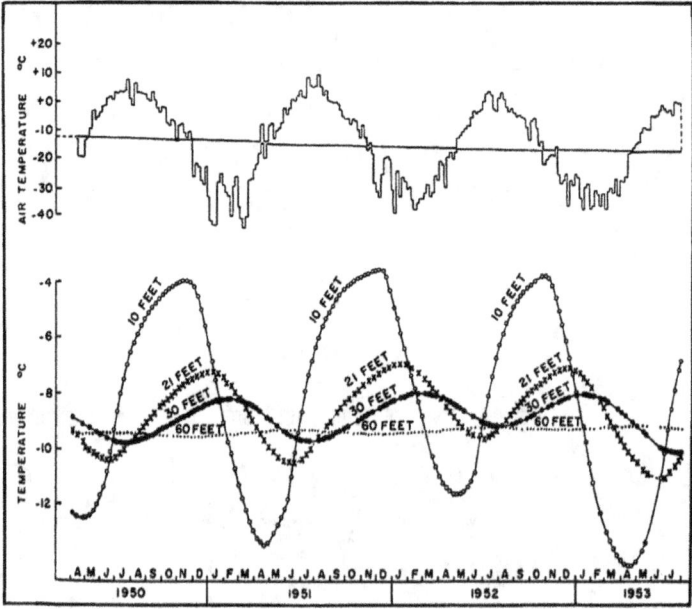

Fig. 4A.4. A group of time-temperature curves for various depths in permafrost near Barrow, Alaska. The corresponding air temperatures are given for the same period using a five-day average and a datum line of −12.2°C (mean annual air temperature at Barrow) (from Brewer, 1958).

annual surface temperature of −10°C. In this instance the assumptions, drawn from conditions typical of much of the Alaska north slope, indicate a permafrost thickness of about 300 m. The temperature range shown in the upper part of fig. 4A.3b indicates the manner in which the seasonal march of air and surface temperature affects the temperature of the upper 20 m or so of the ground. Seasonal temperature fluctuations attenuate rapidly with depth, and below 20–30 m little change is detectable; this level is known as the *level of zero amplitude*. It has been suggested that the annual[1] temperature at this depth approximates the mean annual surface temperature, which in turn will give a rough indication of total permafrost thickness. However, there is much disagreement on this point and the relationships are crude at best.

[1] The word 'annual' is inserted here since slight changes of temperature at this depth may occur over long periods of time.

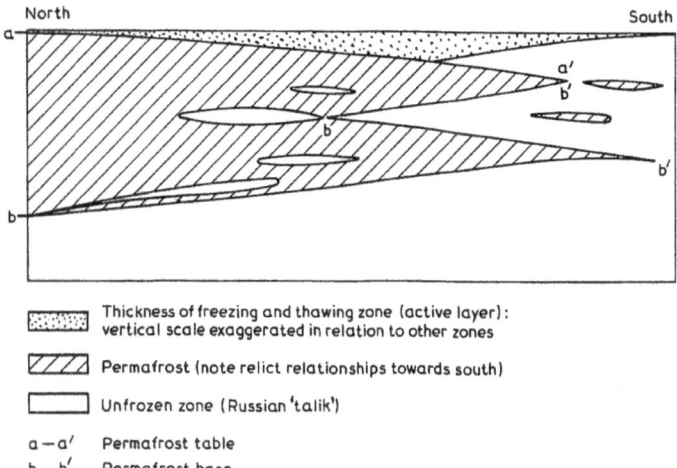

Thickness of freezing and thawing zone (active layer):
vertical scale exaggerated in relation to other zones

Permafrost (note relict relationships towards south)

Unfrozen zone (Russian 'talik')

a — a'   Permafrost table
b — b'   Permafrost base

Fig. 4A.5. Schematic profile through high-latitude, continental interior illustrating the manner in which permafrost thins towards the south, the unfrozen zone, and deep, relict wedges of permafrost. Permafrost table $a - a'$; permafrost base $b - b'$.

As mentioned above, temperature changes at the surface attenuate rapidly with depth; also, penetration of this change is relatively slow so that with increasing depth the occurrence of the lowest (or highest) ground temperature of a given year lags progressively behind the coldest (or warmest) month as recorded above the surface in a meteorological

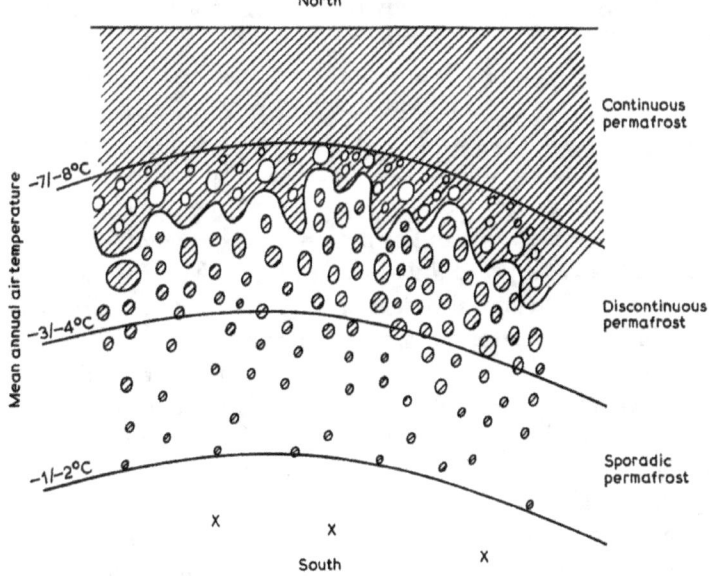

Fig. 4A.6. Schematic plan view of fig. 4A.5 introducing the general concepts of zonal division of the permafrost.

screen. This lag may exceed five or six months and accounts for the fact that below-surface water pipes in cold regions experience the greatest risk of freezing sometimes as late as the following summer season. Fig. 4A.4 shows a group of time/temperature curves with corresponding air temperature obtained from Point Barrow, Alaska (Brewer, 1958).

Fig. 4A.3 also shows that the shallow, uppermost layer alternately freezes and thaws from winter to summer. This is known as the active layer. Its thickness depends upon the local climate and the same variety of secondary factors that influence total permafrost thickness. In the High Arctic the active layer may be only a few centimetres thick; near the southern limits of permafrost, and in dry situations, thicknesses may attain 5 m in rare instances, although generally a maximum thickness close to 3 m should be anticipated. Peaty soil and fine grained clays with a high moisture content will invariably show a thin active layer relative to drier, coarse-grained soils in the same area.

Fig. 4A.5 shows a schematic cross-section through high latitude continental terrain indicating the manner in which permafrost thins with decreasing latitude; fig. 4A.6 indicates the plan view of the same terrain and introduces the general concepts of zonal division of the permafrost region into continuous, discontinuous and sporadic zones.

3. Problems of mapping permafrost

There has been some disagreement in North America over the justification of dividing the permafrost region into three zones. Black's map (1954) attempts this although Brown (1968, 1969) has argued that, since any distinction between *discontinuous* and *sporadic* would be highly arbitrary and impossible to relate to actual field conditions, it is more rational to include Black's sporadic zone within the southern limit of the discontinuous zone. This stance is most readily justifiable for central Canada, for instance, where terrain conditions are relatively uniform. East of Hudson Bay, however, the landscape is extremely rugged and elevations of hilltops increase toward the south. Here tiny patches of permafrost may be expected to occur several hundred kilometres south of areas where permafrost underlies any significant proportion of the total area (the Shickshock Mountains of southeast Quebec and the occasional high New England summits provide the extreme case). Any attempt, therefore, to provide a small-scale map of the southern limit of discontinuous permafrost according to Brown's criteria, loses value. With this view in mind, a sporadic zone has been retained for the present chapter; its cartographic delineation may be arbitrary, but its inclusion provides for greater flexibility when the objective is to depict the broad scale world pattern. Even then it would seem necessary to show individual spot occurrences that are found in extreme southerly locations.

Fig. 4A.5 indicates a southerly transition between a condition where the entire soil thickness above the permafrost table freezes and thaws annually to one where the active layer fails to penetrate to the permafrost table. This condition implies that the southernmost part of the permafrost wedge is a relic of a former colder climate or that the permafrost is degrading. M. I. Sumgin (1937) developed the theory of *degradation of permafrost* whereby he regarded permafrost as continually changing and not established once and for all. He considered that permafrost is in disharmony with present-day climatic conditions, a relic from the ice ages. It is now known, however, that aggradation is occurring in some areas whilst degradation has been recorded in others. Observations of aggradation range as far afield as the Lena Delta, central Alaska and the Mackenzie Delta. Sumgin's theory, therefore, while providing some useful insights, is not wholly acceptable.

Accumulating data, from numerous deep drill holes from the Soviet Union, indicate the existence of two or more major permafrost wedges partially separated from the main, upper wedge by an unfrozen zone. This condition may be a consequence of climatic change on a scale of hundreds of thousands of years as exemplified by the Neogene glacial–interglacial fluctuations, or it may result from a change in sediments. Lack of similar evidence from North America may be a reflection of the dearth of temperature data rather than the absence of that particular phenomenon.

This brings us to the question of how the presence of permafrost is determined. Obviously, in areas of high groundwater content the existence of ice lenses in the ground, either exposed in mining or construction cuts, or in riverbanks and sea coasts, gives a definitive indication of permafrost presence. Natural and artificial cuts, however, are relatively rare in the vast arctic and subarctic regions of the globe. Also in areas of bedrock outcrop or very dry soil conditions the telltale presence of ice may be difficult to detect. In addition, such superficial manifestations tell us little about the thickness of the permafrost layer. The only certain way to determine permafrost presence and thickness is through the measurement of ground temperatures at various depths and the construction of an accurate thermal profile. This usually involves extremely expensive drilling and the installation of thermocouples, or thermistors, and their maintaining, either manually or on automatic recorders, over several years duration. When extensive drilling is undertaken, for instance, the very process of the drilling operation can significantly raise temperatures in the vicinity of the drill hole so that several years may elapse before pre-drilling temperature equilibrium is regained. Ground temperature data, therefore, are very expensive. Add to this the enormous logistical problems of drilling in undeveloped arctic and subarctic terrain and we see the basic reason for the sparsity of North American data. In the Soviet Union, colonization of the permafrost zone began earlier and has continued on a much larger scale, so that many more data have been accumulated in large part out of sheer economic necessity. Even so, Soviet maps of permafrost distribution depend to a relatively large extent on two types of extrapolation: 1) direct extrapolation from one drill site to the next, and 2) indirect extrapolation based upon studies of the relationships between ground temperature and the primary and secondary factors influencing permafrost distribution. In practice many scientists have been tempted to use an assumed, but only partially substantiated, relationship between mean annual ground temperature and mean annual air temperature as a basis for mapping permafrost limits on a continental scale, since air temperature data derived from the international network of weather observing stations are readily available.

Let us consider this relationship. Mean annual air temperature is a synthesis of numerous thermometer readings in a meteorological screen. The great advantage of this type of data is that standardization of instrumentation and observing practices render it strictly comparable around the world and that it is readily available and thus inexpensive. Even then, accuracy of maps showing mean annual isotherms depends upon density of weather observing stations; and mean annual air temperature is essentially derived data and the mean at any one site will differ from the mean annual ground temperature by several degrees, depending upon the thermal properties of the ground and surface characteristics, and especially snowfall amount and duration. The mean annual ground temperature is generally higher than the mean annual air temperature at the same location; this is further complicated by a pronounced geographical variation in this difference, dependent upon the factors mentioned above, so that we have to deal with a general range of difference from 8°C down to zero (the absolute maximum difference recorded is 11°C in Kolyma which experiences a heavy snow cover for a major part of the year and a large annual air temperature range of 55–60°C). Differences of 4–8°C are very considerable in terms of regional climate and, if not taken into account, could result in cartographic misrepresentation of permafrost limits by several hundred kilometres. This is best illustrated by consideration of the Soviet experience. Mean annual air temperature varies with latitude; other factors being equal, a northerly decrease of 1°C per 150–200 km has been observed. This has been referred to as the *latitude temperature stage* by Soviet scientists. Toward the southern limits of permafrost it is approximately 180 km: in northern European USSR it is 120 km, while in eastern Siberia it may be as much as 200 km. If a short-period climatic fluctuation occurs with, for instance, a 2°C air temperature amplitude, the amount of displacement of the southern boundary of permafrost in eastern Siberia will be 350 km. The rate of displacement will be approximately 17 km per year with a 40-year period, or 5–8 km per year with a 100-year period. Thus the problem is not only complicated by the accuracy of the mean annual air isotherms and the point to point correlation difficulties between mean annual air and mean annual ground temperature, but also by the fact of short-period climatic change. An additional area of confusion lies in the realization that mean annual air temperatures, although generally lower than mean annual ground temperatures, are not invariably so. Again, snowfall amount, timing and duration are the complicating factors. Consider a situation whereby autumn snowfall is delayed so that an insulating layer of snow is laid down after an area has experienced severely low air temperatures. If the same locality receives abnormally late spring snowfalls, then heating of the colder-than-average ground is delayed. Hypothetical variations on this theme will indicate the magnitude of the uncertainty when faced with the task of extrapolating permafrost boundary conditions on a continental scale. A further example is afforded by permafrost conditions on the Aldan Plateau (Yakutskaya, USSR). Here the water divides are frequently underlain by highly weathered rock mantled by coarse grained residual material. Summer precipitation exceeds half the annual total; infiltration of relatively warm rainwater occurs with resultant warming of the coarse material. As a result, permafrost is frequently absent from such areas; ground temperature means vary in the range +1° to +2°C when from the regional climatic data figures of −1° to −4°C would be anticipated. In the same general area this latter range of mean annual ground temperature has been recorded where the surface is underlain by fine grained soils. Nevertheless, as with many scientific problems of this type, generalizations are made, extrapolations, if clearly enunciated, are justifiable, since there is no alternative. But with increase in ground temperature data through time, the generalizations are made less and less coarse and the small-scale maps become closer and closer approximations to reality.

Within the Soviet experience, therefore, maps showing permafrost distribution are composites depending upon a relatively large amount of ground temperature observations, analysis of relationships between ground temperatures and the primary and secondary factors affecting permafrost distribution, and extrapolations from maps of mean annual isotherms, snow depth and duration and so on. In North America we depend heavily upon the Soviet experiences, especially through much of northern Canada since Alaskan ground temperature conditions are better known. I intend to use the Canadian example, therefore, as a means of illustrating several of the problems involved in the mapping of permafrost on a continental scale that have been discussed in more general terms above.

The current official Canadian map of permafrost distribution produced by Brown (1968) represents a considerable achievement and is in large part a cartographic synthesis of many years of permafrost research carried out by Brown and his co-workers (Brown, 1960, 1964, 1965, 1966, 1967, 1968; Pihlainen, 1962; Johnston et al., 1963; Legget et al., 1966). It is based both upon extensive field reconnaissance, detailed permafrost surveys in small areas, notably the lower Mackenzie and northern Manitoba, and upon the extrapolation of Soviet experience to northern and central Canada. For much of the delineation of the permafrost zone boundaries, reliance is placed upon the basic assumption that the absolute southernmost limit of permafrost shows a strong relationship to the mean annual air isotherm of $(-1$ to $-2°C)$ $30°F$ and that the southern limit of continuous permafrost relates to the mean annual air isotherm of $(-8°C)$ $17°F$. Adjustments of these boundaries are made to fit with the ground temperature data and other definitive field observations where available.

Since ground temperature data are particularly scarce east of James Bay, the permafrost zonal limits across Labrador–Ungava and Baffin Island are made to follow smoothly the relevant mean air isotherms. Even north and west from James Bay, where isotherms and permafrost zonal lines diverge slightly, the latter are shown as smooth lines. A further anomaly is the indication of permafrost patches in peat bogs up to 300 km south of the 'southern limit of permafrost'. A closer examination of the southeastern sector of Brown's map is most instructive in the present context. The examination will take place in two steps: 1) analysis of the justification of mean annual isotherm data for the eastern Canadian Arctic and Subarctic; and 2) comparison of permafrost zonal limits, as shown on the map, with knowledge of local terrain conditions and of relationships between permafrost occurrence and the secondary environmental parameters.

(1)   Consideration of the form of the mean annual air isotherm for $-1°$ to $-2°C$ shows that over a distance of nearly 1400 km it is dependent upon only fourteen weather

observing stations, the majority of which are located at, or close to, sea level. The great massif of the Mealy Mountains, for instance, that rises to over 1200 m, is effectively eliminated since no climatic data are available. The basic isotherm, therefore, becomes a gross generalization with little detailed factual underpinning.

(2) Ground temperature studies in central Labrador–Ungava (Ives, 1960a, 1962; Annersten, 1963; Thom, 1969) indicate that there is a strong relationship between vegetation, topography and permafrost occurrence in this area. Specifically, where windswept ridges occur above local treeline, permafrost, up to 60–100 m thick, occurs. Below treeline the microclimate changes drastically since the taller vegetation acts as a trap for blowing snow in winter; snow and vegetation together so redress the potentially negative energy balance that permafrost is either non-existent or scattered in tiny patches, and usually relic. Yet throughout the general study area the difference in mean annual air temperature between forested valley bottom and windswept tundra ridge is probably less than 2°C. Using this knowledge and Hare's (1959) map of the cover types of Labrador–Ungava, a still schematic map of potential permafrost distribution in Labrador–Ungava was constructed (Ives, 1962) (fig. 4A.7). It is contended that, while highly speculative, it is probably one approximation nearer reality than the comparable portion of the official map.

Since we believe that the current patterns of climate, including snow fall and duration, and vegetation and permafrost distribution are in reasonable balance, an interesting question remains: how to account for relic permafrost well below treeline in areas where it could not possibly have formed within the last few centuries? As a working hypothesis it is assumed that since the climate of the region has not changed significantly over the last 6000 years, that is, the period since Wisconsin ice finally melted, then the distribution of vegetation must have changed, which in turn implies that the winter snow accumulation pattern has also changed. The residual (Labradorean) Laurentian ice sheet probably wasted away rapidly from the vast tract of central Labrador–Ungava about 6000 years ago (Ives, 1960b; Morrison, 1970; Bryson *et al.*, 1969; Prest, 1970; Andrews, Chapter 6A). It is suggested that this thin residual ice sheet melted sufficiently rapidly that some decades or centuries elapsed before the present boreal forest and forest–tundra ecotone could become established. During this period, scattered permafrost developed, later to be covered by forest with its attendant winter snowpack, which in turn resulted in progressive, but very slow, degradation of the permafrost so that the remaining scattered patches are historical and ecological relics.

Conditions in Labrador–Ungava, where permafrost and treeline are causally related, contrast markedly with conditions in northwest Canada and Alaska. In the northwest, it is widely known that the southernmost extent of permafrost reaches several hundred kilometres south of the arctic treeline. The contrasts between the northwest and Labrador–Ungava probably relates to the considerably heavier winter snowfall of the latter area; or a combination of lower winter temperatures and relatively light snowfall of the former.

In Chapters 2A and 7A, Barry and Hare and Larsen have referred to relationships

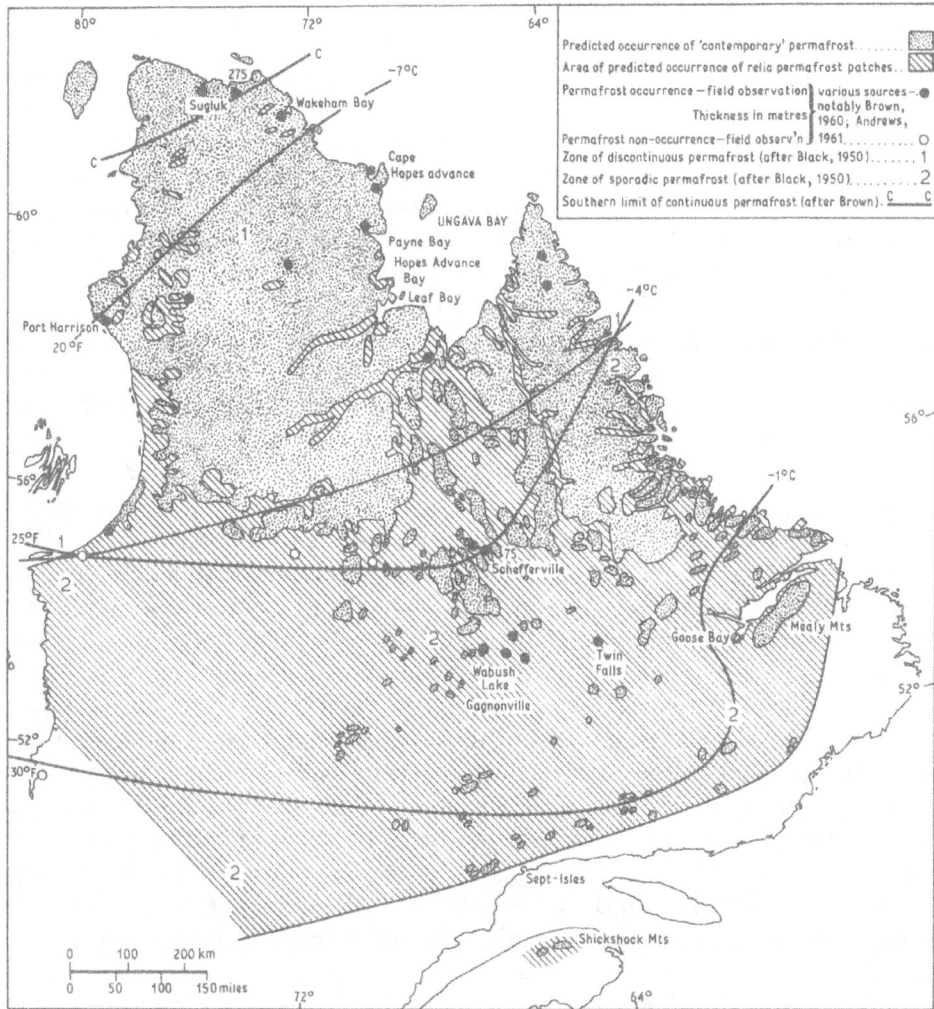

Fig. 4A.7. Map of potential distribution of permafrost in Labrador–Ungava (from Ives, 1962.) Superimposed is Brown's (1968) southern limit of continuous permafrost and the 25°F mean annual air isotherm that he uses to divide his discontinuous zone into a northern area with widespread permafrost, and a southern subzone classified as the southern fringe of the permafrost region. Note that the high Mealy Mountain massif lies beyond Brown's southern limit (= −1°C/30° Fair isotherm).

between the mean position of airmass boundaries and treeline. Bryson (1966) and Barry (1967) have discussed these relationships extensively, and Krebs and Barry (1970) have applied their working assumptions to the Soviet Union. In the present context, the lack of fit between airmass boundaries and treeline in Labrador–Ungava and in eastern Siberia is worthy of consideration when discussing contrasts in treeline location and permafrost boundary conditions between various sectors of the Subarctic.

EXAMPLES OF DATA ON PERMAFROST OCCURRENCE

Many Soviet publications (Kudryavtsev, 1959) indicate that thickness of permafrost varies within a range from 0.1–0.2 m to 600–800 m, based upon actual temperature observations, or reliable extrapolations to permafrost base from observations on thermo-electric sensing cables installed to several hundred metres depth. It is generally believed that the 600–800 m values are quite probably not the maximum, and that in high mountainous regions, where the mean annual ground temperature may fall well below $-15\,°C$, maximum thicknesses in excess of 1000 to 1500 m may occur.[1]

In Canada recorded maximum thicknesses are generally much less than those of the Soviet Union. Almost 400 m thickness has been estimated for Resolute on Cornwallis Island. This estimate was based upon extrapolation from direct thermocouple measurements to a depth of 190 m (Misener, 1965; Cook, 1958). The site, however, is close to sea level and to a shallow lake, and also experienced over 80 m of marine submergence during the past 11,000 years so that greater thicknesses may be anticipated for central Cornwallis Island (Lachenbruch, 1957). G. Jacobsen (personal communication) has reported a provisional estimate of 500 m thickness from a deep installation at Winter Harbour on Melville Island. Greater thicknesses may be anticipated for the northern Queen Elizabeth Islands. Judging from mean annual air temperature data available from the Lake Hazen area, Ellesmere Island, it appears that the plateau above the lake may experience a mean of $-23\,°C$; this could imply a permafrost thickness up to 1000 m.

The Canadian Arctic mainland, however, has not provided any comparable figures except in the Yukon where Brown (1969) reports that 'permafrost is probably nearly 1000 ft (305 m) thick' northeast of Dawson. Mackay (1968) also reports thicknesses in excess of 300 m in the sediments of the Pleistocene Mackenzie Delta. Péwé (personal communication, 1972) reports thicknesses in excess of 600 m in northern Alaska. An exception is the recording of a thickness greater than 275 m near Sugluk in northwest Labrador–Ungava (Brown, 1968). This is all the more interesting since this site lies on the dividing line between continuous and discontinuous permafrost as shown on Brown's map. Thus with future deep-seated thermal exploration in the Yukon and Canadian Arctic Archipelago, North American maximum thicknesses may be shown to approach more closely the Soviet maxima than hitherto suspected.

This comparison and contrast between the permafrost zones of North America and the Soviet Union has been part of a debate concerning the history of permafrost development over several decades. Initially it was strongly argued that areas underlain by the greatest thicknesses of permafrost coincided with areas that remained ice-free throughout the Neogene glaciations. There was a general tendency to assume that, since large tracts of Siberia had never experienced glacier ice cover while it was held that the whole of the Canadian Arctic, excluding much of the Yukon, had been glacially inundated, then greater permafrost thicknesses would occur in Siberia than in northern Canada. This argument was widely accepted despite the lack of ground temperature data from

---

[1] A permafrost thickness of 1500 m has been reported from northwestern Yakutia – apparently an unusual situation where supercooled brine penetrated to this depth in limestone (R. J. E. Brown, personal communication, 1972).

northern Canada and despite the very scanty knowledge of glacial history. Recent glacial geomorphological research in Canada (Ives and Andrews, 1963; Falconer *et al.*, 1965; Ives and Buckley, 1969; Løken, 1966; Andrews, 1970; Blake, 1966; Prest, 1970; see also Andrews, Chapter 6A, and Ives, Chapter 10B) would indicate that extent of glacial ice, both vertically and horizontally, was much less than hitherto had been assumed. Large parts of the arctic archipelago carried only thin local ice caps and significant areas may have remained unglaciated, especially during the classical Wisconsin stade. In addition, temperatures measured to the base of the Greenland Inlandis at Camp Century indicate ice basal temperatures of $-13°C$ (Dansgaard *et al.*, 1971), which in turn would imply the existence of a considerable thickness of permafrost beneath at least the 'cold' northern section of the ice sheet. In contrast, recent work in Siberia indicates more extensive former ice sheets than had been assumed. In addition the extensive marine submergence of vast areas of northern Siberia during glacial stades (Kind, 1967; Zubakov, 1969) and widespread marine and lacustrine drowning of northern Canada (Andrews, 1970; Prest, 1970) would further complicate this type of speculation. Nevertheless, the general principle would appear to hold true: great thicknesses of permafrost require tens of thousands, if not hundreds of thousands, of years to develop; consequently maximum thicknesses will only be found where former ice sheets did not provide a long continued and thick insulating layer. To this must be added absence of submergence beneath sea and lake water. Thus parts of Alaska and the Yukon, Queen Elizabeth Islands, northern Greenland and the eastern mountain fringe of Baffin Island may be expected to provide greatest permafrost thicknesses in North America. The problem of permafrost and extent of glacial ice in Siberia is discussed in a most illuminating way by Gerasimov and Markov. This work in translation (Hope, 1968) is accompanied by Gravé's paper on the earth's permafrost beds.

4. Permafrost and vegetation

Because of the difficulty of determining permafrost occurrence over wide areas by direct observations, many attempts have been made to establish the interrelationships between vegetation and permafrost. One objective has been to facilitate permafrost mapping through air photograph interpretation of the vegetation type, or in a slightly broader sense, of the cover type (Benninghoff, 1952).

It is generally recognized that since the vegetation cover forms the boundary between lithosphere and atmosphere, it will have an important effect upon the energy exchange between them and hence upon the soil temperature. Vegetation influences the amount of radiant energy absorbed or reflected by the surface; it protects the soil surface from wind, influences the accumulation of snow and finally, to an important degree, it determines the moisture exchange between air and soil. The absorption of water by vegetation, evapotranspiration, the precipitation of dew upon vegetation, and the decrease in evaporation by vegetation-covered surfaces, most definitely have a critical effect on the thermal regime of the underlying soil and bedrock, although each factor does not necessarily work in the same direction. The processes of moisture exchange through a vegetative cover are very complex and do not readily yield to quantitative estimation. The problem is further complicated by the multiplicity in vegetation forms and variation through time. Forest, shrub, meadow, marsh and tundra forms all have noticeably different influences. Kudryavtsev (1959) in discussing this general problem points out that all of this indicates that vegetation cover, in contrast to snow cover, cannot be regarded simply as some additional thermal resistance. The following remarks are largely synthesized from his work.

Vegetation cover protects the soil from winter cooling and summer heating, thus serving to reduce the amplitude of the annual temperature range. In the more southern permafrost regions the decrease in summer temperatures will be greater than the winter effect, while the reverse is true for the more northern regions. Vegetation in the south, therefore, has a predominantly cooling effect on the soil, and in the north a warming effect. Variations occur, however, under different conditions: for instance, in the Soviet Far East it has been shown that with a snow cover of up to 20 cm removal of the forest will result in a cooling of the soil, but with a snow cover much in excess of 20 cm the reverse will occur once deforestation is effected. The increase in annual temperature amplitude due to removal of the forest may reach 10 to 13°C.

In northern regions the insulating effect of grass and shrub vegetation increases the mean annual temperature of the soil by not more than 1 to 2°C. The role of moss cover in the establishment of the soil temperature regime is determined by its low thermal conductivity, high moisture capacity and hygroscopic properties. The moisture capacity of hypnum moss may exceed 350 per cent and for the sphagnums 1300 to 5000 per cent in relation to dry weight. It has been estimated that to evaporate the moisture required to produce 1 gm of dry moss requires 8000 calories. Under a moss cover, therefore, there is a greatly reduced temperature amplitude. During the winter the conductivity of frozen moss increases greatly. This results in a 1 to 3°C decrease in the mean annual soil temperature. In central Labrador–Ungava differences in temperature at 25 cm depth under an undisturbed Cladonia lichen mat and a neighbouring site where the soil was stripped bare amounted to 11°C within a three-week period (E. Mortenson, personal communication, 1958).

Viereck (1970) has provided valuable analysis of the effect of vegetation succession on ground temperature regime along the Chena River in the Fairbanks area of subarctic Alaska. Since this study interrelates vegetation type, soil type and moisture content and climate with permafrost development in the discontinuous zone, the summary and conclusions of his paper are quoted in full:

> Forest succession on the Chena River begins with a shrub stage of willow on a coarse gravel substrate. The water-holding capacity of this soil is low, and during periods of prolonged drought, the wilting point may be reached to some depth in the soil. Freezing and thawing in this soil are rapid, and the soil becomes cold in winter and warm during summer. Mineral nutrient supply is low in the coarse gravel.
>
> Frequent flooding adds silt and sand to the surface, creating a silty loam soil. Balsam poplars, which enter into the succession at the same time as the willows, grow

rapidly and form pure stands of large trees. A completely new vegetation understory develops as the willows mature or are shaded out by the poplars. Soil temperature fluctuations are less than in the willow stage, and freezing and thawing are slower as a result of the insulating effect of the leaf litter and the slower rate of heat conductivity in the silty loam soil. The percentage of moisture retained in the soil is greater and moisture does not appear to be limiting plant growth. Mineral nutrients are more abundant in the silt deposits. Tree growth is rapid and a well-developed shrub and herb layer occurs in the stand.

At some point in the development of the balsam poplar stands, seedlings of white spruce become established. These grow slowly at first, but within 75 to 100 years of their establishment they over-top the balsam poplar which by then have reached maturity and are rapidly disappearing from the stands. As the balsam poplar leaf fall becomes less and as the spruce litter accumulates, a continuous and thick moss mat develops in the stand. This begins to bring about a significant change in the soil temperature regime. The moss and organic layer acts as a more efficient insulator during summer than winter, and soil temperatures remain cooler throughout the summer. If the spruce stands are far enough from the river, silt deposition stops or becomes very small and the organic layer continues to increase. Soil nutrients are still abundant at this stage, and growth of the white spruce is rapid for the first 100 years. As thawing becomes progressively slower in the soil, a permafrost layer develops. This prevents drainage and creates a wetter soil on which sphagnum mosses may develop. Moss growth is rapid, and tree growth becomes slower either as a result of the colder soil temperatures or the waterlogged nature of the soil.

These conditions favor black spruce over white spruce, and the former becomes established in the stand. As succession continues, the permafrost layer rises closer to the surface. If the organic layer has developed sufficiently, the mineral layer may be entirely frozen and all tree roots become located in the organic layer. Nutrients in the organic layer are depleted, and some elements such as phosphorus, nitrogen, and manganese may become limiting. Soils are cold throughout the summer and water-logged when not frozen. Black spruce grow slowly with an understory of ericaceous shrubs. Sphagnum mosses find optimum conditions for growth, thus creating an even thicker moss layer in a waterlogged condition.

It can be seen that succession proceeds from the near xeric conditions of the gravel bar to the mesic balsam poplar and white spruce stands. Autogenic changes, brought about primarily by the development of thick organic layer and interaction with the cold subarctic climate, result in the development of a permafrost layer. Succession proceeds from the mesic conditions of the balsam poplar and white spruce stands to a hydric condition of a slow-growing black spruce stand with a thick saturated sphagnum mat on a permafrost table only 20 to 30 cm below the moss surface (Viereck, 1970, p. 25).

Further south in Alaska, in Mt McKinley National Park, Viereck (1965) provides another illustrative detail of the vegetation/ground temperature interrelationship. In this case the growth of permafrost mounds under white spruce, *Picea glauca*, growing in silty clay, is believed to relate to the summer insulation effect of a thickening moss mat and from soil cooling in winter as the result of a thin snow layer under the trees in question. Subsequent disturbance of the moss mat will cause melting of the ice lens, collapse of the mound, and often result in death of the tree. Fig. 4A.8 gives a schematic outline of the cycle of events described by Viereck (1965, p. 266, fig. 3).

Enough has been said to indicate the complexities of the relationship between vegetation and ground temperature regime. On a regional scale, therefore, vegetation patterns cannot be used to predict permafrost conditions. This conclusion has already been

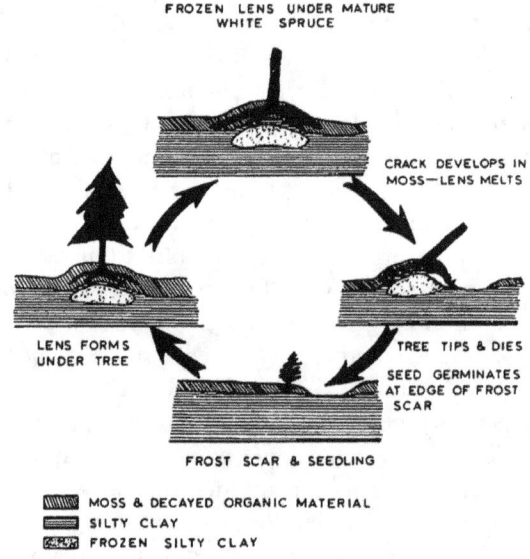

Fig. 4A.8. Cycle of development and collapse of the mounds (from Viereck, 1965). Seedlings germinate on frost scar or bare surface created by the melting of the frozen core in the mound.

anticipated by the earlier contrast between the position of treeline and permafrost limits in southeastern and northwestern Canada. On a local scale, however, the situation is very different. Two contrasting cases are used to illustrate this.

In the Fairbanks area of interior Alaska (discontinuous zone) Péwé (1954, 1966b) explains that vegetation on alluvial fans and colluvial slopes is distinctive, and some trees are stunted because of shallow rock zone, low soil temperatures and high soil moisture. A scrubby wood of stunted black spruce (*Picea mariana*), with knee-high shrubs of dwarf birch (*Betula glandulosa*), Labrador tea (*Ledum decumbens*), blueberry (*Vaccinium uliginosum*) and a thick moist carpet of cranberry (*Vaccinium vitis idaea*), cottongrass (*Eriophorum* spp.), much peat moss (*Sphagnum* spp.) and caribou moss (*Cladonia* spp.) grows on gentle slopes where drainage is poor and permafrost generally is at a depth of 0.5 to 1.2 m. Willows (*Salix* spp.) and alders (*Alnus* spp.) are found in poorly developed water courses. In contrast, slightly steeper slopes support white spruce

(*Picea glauca*) and quaking aspen (*Populus tremuloides*) where drainage is better and permafrost is lacking or is at a depth of more than 1.2 m. With the accumulation of detailed knowledge of this type, extrapolations of probable permafrost occurrence can be made from air photograph interpretation of cover types for the local area (Plate 7).

Brown (1969), amongst many others, has shown that at its southernmost limits permafrost tends to be restricted to peatlands, especially in areas of relatively slight or moderate relief. Microrelief in peatlands ranges from hummocks 0.5 m high and peat plateaus and ridges 1 m high to palsas over 6 m high. In the peatlands the extent of permafrost patches varies from a few square metres to several hectares. Depth to permafrost varies from a minimum of about 46 cm to a general maximum of about 90 cm, with occasional depths as much as 170 cm. The palsas, distinct frozen peat mounds, are conspicuous from low-flying aircraft and on air photographs, and have been used to map the general outline of the southern limit of permafrost although permafrost patches without palsas have been found along the southernmost fringe (Brown, 1969; Rapp and Annersten, 1969). During surface and helicopter-borne reconnaissance, Brown was able to determine that along the southern fringe small islands of permafrost could be expected beneath the drier sphagnum areas but not under the sedge or wet sphagnum areas. At the edge of the dry sphagnum areas the permafrost table either dipped down very steeply or formed a nearly vertical face (Brown, 1969, p. 40). On a somewhat smaller scale the present writer, as indicated above, was tempted to use Hare's (1959) map of cover types in Labrador–Ungava to construct his map of potential permafrost distribution (fig. 4A.7).

The possibility of using surface expressions other than vegetation for air photograph interpretation of permafrost conditions would seem even more limited. Geomorphic features, such as pingos, polygons, solifluction lobes, thermokarst, either only occur in areas where the occurrence of permafrost can be assumed from the prevailing very low mean annual air temperatures, or could be fossil forms left over from a colder climate of the more distant past. Generally, there is no surface expression for permafrost existence along the southern fringe, apart from the exceptions already noted. This implies that especial care should be taken in the discontinuous and sporadic zones before initiation of construction projects.

5. Ice in the ground

The ice content of perennially frozen ground is probably the single most important feature affecting twentieth century man's development of the northlands. Ice exists in various forms and has definite distribution characteristics and the question of its origin is frequently highly controversial. There is no universally accepted ground ice classification. Some scientists traditionally exclude buried glacier ice, although in many instances it is impossible to differentiate between various forms of buried ice, especially at the reconnaissance level or where metamorphism has occurred. Péwé (1966a) groups the various types of ground ice into five categories: a) pore ice, b) segregated ice, c) foliated or ice-wedge ice, d) pingo ice and e) buried ice. Mackay (1966) has developed a more comprehensive classification based upon earlier work by Shumskii (1959). While Mackay's classification has been developed specifically for application to the Mackenzie Delta (fig. 4A.9) its comprehensive nature renders it highly appropriate for inclusion in the present chapter (Mackay, 1966, p. 61).

Mackay's paper (1966) deals primarily with segregated ice, and since this is one of the more interesting and problematic categories of ground ice some of his findings will be

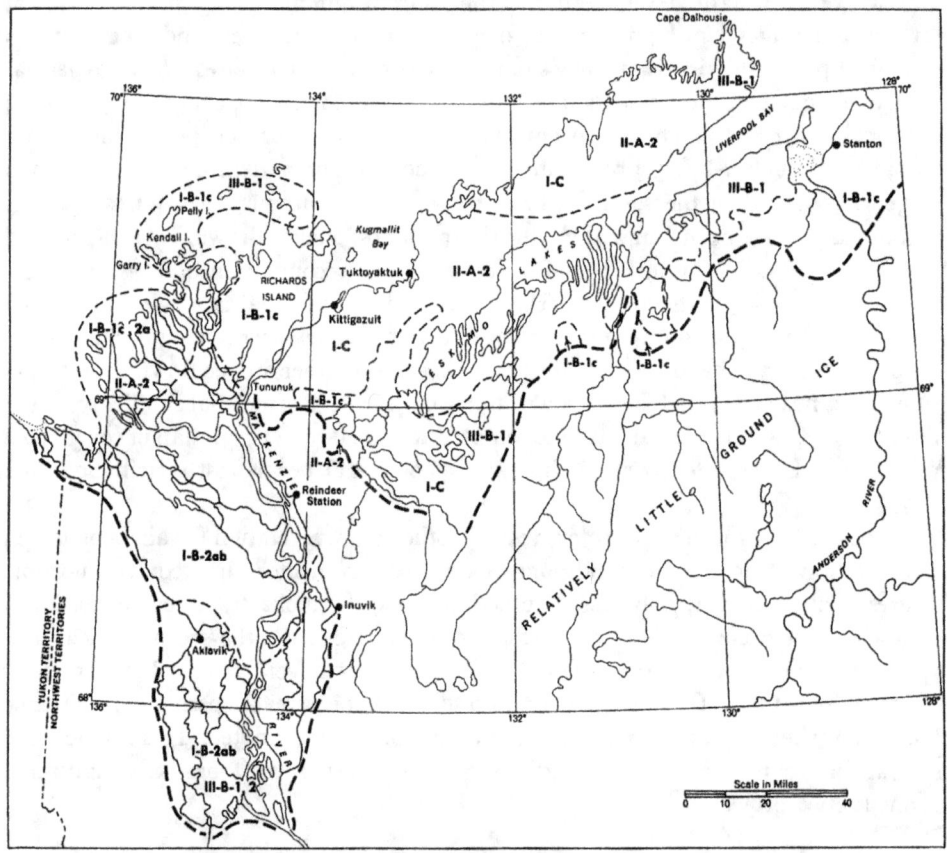

Fig. 4A.9. Types of ground ice in the Mackenzie Delta area (from Mackay, 1966). The legend corresponds to the classification given in table 4A.1. Only broad categories were mapped and the symbols within a boundary refer to the entire area.

enlarged upon. Segregated ice is a broad term used to describe soil with a high ice content; it is an inclusive term embracing both the 'soil' and the ice, in a similar manner to the engineering use of the term 'soil' rather than the pedologic use. Ice lenses within segregated ice are usually visible to the unaided eye, but the total amount of ice present can be easily underestimated. In soils where lensing is minimal the distinction from pore ice can be made by thawing a soil sample and determining whether or not excess water is present since by definition pore ice occurs in saturated or undersaturated soil

Table 4A.1. Classification of ground ice.

I. **Soil ice** (formed by *in situ* freezing of local concentration of soil moisture)
   - A. NEEDLE ICE (pipkrake, mushfrost, etc.): on surface of ground, or immediately beneath vegetation mat, upper few cm of soil, or individual stones, which are forced upwards on freezing and collapse under gravity upon thawing; a short-term or diurnal phenomenon.
   - B. SEGREGATED (Taber)* ICE: supersaturated frozen soil.
     1. Epigenetic ice (formed later than enclosing material)
        - (a) in active layer
        - (b) at base of active layer
        - (c) in permafrost
     2. Penecontemporaneous (formed concurrently with surface aggradation)
        - (a) aggradation from progressive sedimentation
        - (b) aggradation from organic accumulation
        - (c) aggradation from burial by soil movement
   - C. PORE (cement, bonding) ICE: usually formed in saturated or under-saturated frozen soil.

II. **Intrusive ice** (formed by freezing of 'injected' water)
   - A. PINGO ICE.
     1. Open system (freezing of groundwater under a hydraulic gradient)
     2. Closed system (freezing of expelled soil moisture)
   - B. SHEET ICE (freezing of intruded water in a sill-like sheet).

III. **Vein (foliated) ice** (freezing of water and vapour in thermal contraction cracks)
   - A. VEIN ICE (usually representing a single episode of freezing).
   - B. REPEATED VEIN ICE (ice wedges).
     1. Epigenetic (formed later than enclosing material)
     2. Penecontemporaneous (formed concurrently with surface aggradation)

IV. **Extrusive ice** (formed subaerially)
   - A. River icings (overflow from a river channel constricted during autumn and winter freeze-up). Also termed *aufeis* or *naled*.
   - B. Ice fans (freezing of water issuing from spring into cold air).

V. **Sublimation ice**
   - A. Crystals in open cavities, caverns.
   - B. Crystals in closed cavities.

VI. **Buried ice**
   - A. Buried snowbank (by slumping, rockfall, deposition of glacial moraine, etc.).
   - B. Sea, lake, river ice (by fluvial aggradation, ice-push ridging, slumping, rockfall, etc.).
   - C. Glacier ice.

---

* Péwé (1966a) suggested that the name 'Taber ice' be given to segregated ice in honour of Taber's pioneer studies (Taber, 1916, 1917, 1918, 1930).

(Black, 1954). Also, segregated ice is more typical of fine grained and organic soils, while pore ice tends to occur in coarser sands and gravels.

There is usually little difficulty in differentiating between segregated ice and intrusive ice because the latter is relatively pure and occurs under rather distinctive geomorphic conditions. The widely distributed vein ice (wedge ice) of tundra polygons is also quite distinct from segregated ice because of its vertical foliation, structure and inclusions, although ambiguous cases may occur, especially if there has been replacement of foliated ice where it intersects segregated ice (Mackay, 1966, p. 68) (Plate 8).

Vein ice or wedge ice describes the large masses of ice that grow in thermal contraction cracks. Black (1964) has shown that ice-wedges grow by small annual increments over several thousands of years, and are frequently associated with the large-scale polygonal surface patterns so conspicuous over much of the area underlain by continuous permafrost; in fact, the latter are the surface expression of the former. Upon thawing, because of climatic warming, these forms will sometimes be filled with blown sand, gravel or other slumped material to occur as the fossil ice-wedges found in the sand and gravel pits of more southerly latitudes that experienced periglacial conditions during the various glacial maxima of the Neogene glaciations. Recent studies of ice-wedges (Lachenbruch, 1962, 1967) indicate that their active growth requires a mean annual air temperature of −6 to −8°C.

Extrusive ice, such as river icings and ice fans built through winter flow of perennial springs, or in the vicinity of hot springs, is both morphologically and structurally distinct from segregated ice. The same is true of sublimation ice.

'Pingo' is an Eskimo word meaning conical hill and was originally proposed by Porsild (1938) as a technical term for ice-cored hills. Mackay (1962) showed that pingos, perhaps the most dramatic surface expression of permafrost, are stable, intrapermafrost ice-cored hills. The majority of the Mackenzie Delta pingos (fig. 4A.10) were formed more than one thousand years ago as an indirect result of the shoaling of lakes by geomorphic or climatic processes. As the lake level gradually lowered and lake ice froze to the bottom in winter an impermeable permafrost layer was formed on top of unfrozen saturated sediments beneath. Downward development of permafrost must be in predominantly fine to medium grained sands not susceptible to extensive ice segregation. Thus the pressure of expelled pore water, trapped in a closed system, was relieved by upward doming of the thinnest part of the overlying and developing permafrost layer. This would usually coincide with the deepest part of the original lake. Finally water in the dome froze to form the pingo ice-core; later rupturing of the up-domed sedimentary cover may expose the ice core to insolation and initiate progressive degradation of the feature and its ultimate collapse. Actual growth rates probably involved tens of years. Mackay's (1962) theory of closed-system pingo formation is shown diagramatically as fig. 4A.10.

As many as 1400 pingos, perhaps the largest group in the world, occur in the Pleistocene Mackenzie Delta, although at least 70 have been observed in the modern delta, usually within a few metres of present sea level. Theoretically, artificial draining of a present-day lake in the Mackenzie Delta area, provided it was underlain by fine to medium grained sands, would initiate a pingo (Mackay, personal communication, 1967).

Pingos also occur under an open-system regime (Müller, 1959) where up-welling groundwater freezes on approaching the surface to cause up-doming of near-surface

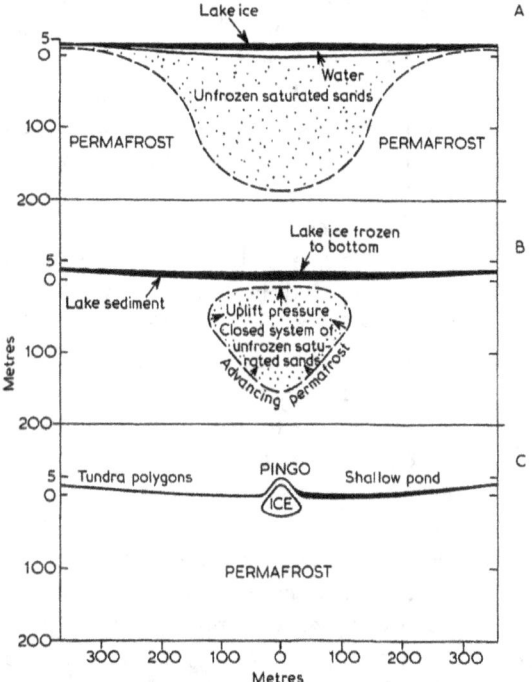

Fig. 4A.10. Three stages in the formation of 'closed-system' or Mackenzie-type pingos (modified after Mackay, 1962). A vertical exaggeration of 5× has been used for the height above zero in order to show the lake ice and the open pool of water. (A) A broad shallow lake has an open pool of water in winter with a frozen annulus around it. No permafrost lies beneath the centre of the lake. (B) Prolonged shoaling has caused the lake ice to freeze to the bottom in winter and induced downward aggradation of permafrost. Infilling has raised the lake bottom a small distance. The deepest part of the lake, which has the thinnest permafrost, is gradually domed up to relieve the hydrostatic pressure. (C) The pingo ice core, being within permafrost, is a stable feature. The old lake bottom is occupied by tundra polygons and shallow ponds. Because of scale changes in the diagram, the volume of the ice core should not be construed as showing a direct relationship to the initial volume of unfrozen material.

sediments. Flood plains are common situations (Plates 9 and 10), but occasional pingos have been found in bedrock; such forms were initially studied by Müller in northeast Greenland. More recent work in the Queen Elizabeth Islands (Pissart, 1967) has revealed the existence of many linear and irregular forms which also are probably pingos. There has also been much controversy over the recognition of fossil pingos (Pissart, 1965; Bik, 1968).

According to Mackay (1966) epigenetic segregated ice occurs in two main forms in the Mackenzie Delta sediments: in areas of Pleistocene age sediments in which permafrost has developed during a prolonged period of refrigeration, and in areas of recent sediments where permafrost is developing under the present climatic regime. When fine grained sediments freeze, usually from the surface down in a cooling climatic trend, ice lenses usually grow parallel to the freezing plane, i.e. normal to the direction of heat flow. Under conditions of extremely rapid freezing or impermeable sediments, ice lensing may be inhibited. Analysis of field samples of soils containing segregated ice (Mackay, 1966, p. 63) using the following grain size limits – sand greater than 0.06 mm, silt 0.06 to 0.004 mm, and clay less than 0.004 mm – suggests that soils with less than 30 per cent clay have tended to freeze in bands rather than to have soil dispersed as inclusions in clear or 'dirty' ice. Mackay goes on to explain that a major area of epigenetic ice formation is that part of the Pleistocene Mackenzie Delta where the sediments were deposited offshore in depths that probably exceeded the contemporary thickness of winter sea ice formation. Mackay and Stager (1964) have shown that such sediments of Garry and Kendall islands were deformed by glacial ice thrusting after they had been frozen. It seems likely, therefore, that since the last ice advance is dated at approximately 25,000 BP (Müller, 1962) much of the epigenetic ice in the Pleistocene sediments formed more than 25,000 years ago, and possibly much before this.

The most conspicuous indication of the presence of abundant segregated ice is the widespread slump features seen along over-steepened coasts or parallel to undercut riverbanks (see Bird, Chapter 12A). These slumps resemble earth flows of more temperate climates, although the pattern of slumping is quite distinct especially since large voids occur but very little slump debris. If the ice content is exceptionally high, a scarp face will retreat with little actual transport of thawed slump debris. Rate of annual retreat of an active scarp face ranges from 2 to 5 m; multiple slumps are numerous; ice content ranges from several hundred to over one thousand per cent of volume of dry soil.

So far emphasis has been placed upon the major or more dramatic forms of ice associated with permafrost, such as pingos and large tabular ground ice sheets. Numerous other forms occur over vast areas of the northern world and only a few examples can be mentioned briefly here.

Palsas are small peat mounds with frozen cores that are extremely widespread along the discontinuous and sporadic permafrost zones, frequently forming the southernmost surface expression of permafrost occurrence (Salmi, 1970). Maximum dimensions are of the order of 20 m diameter and 4 to 6 m in height, although average size is much less. Their occurrence is largely restricted to peat accumulation areas and wet marshlands in general.

In addition to ice masses in the ground, with or without surface expression, there are many surficial ice forms, seasonal or perennial, that are characteristic of the world's permafrost regions. Especially widespread are the various forms of icing (Russian, *naled;* German, *aufeis*) that accumulate through freezing of spring or river water that continues running through the winter period. In some instances ice mounds up to 20 m in thickness may accumulate. Largest ice accumulations are to be found in the Verkhoyansk region of Siberia where entire valley floors may be covered. Another form results from relatively mild mountain winter weather, whereby snow melts during the day, the meltwater drains into a depression – a road cut for instance – and freezes during the late afternoon and following night, the process repeating itself on numerous occasions. This form is observed in the Colorado Rocky Mountains. Icings can produce serious transportation problems. Depending upon rate of water flow and winter and summer temperature regimes, they may survive the following summer, and have frequently resulted in false reports of glaciers.

THERMOKARST

The introduction of this chapter stressed the essential short-term instability of ground temperature characteristics resulting from climatic change. When a climatic warming trend causes a northerly shift of the southern limit of permafrost in a relatively dry area, the results may be scarcely noticeable. In wet areas, however, and especially in areas characterized by large quantities of variegated ground ice, the effect may be dramatic. Melting of ice in the ground (thermokarst) will cause significant surface disruptions, especially where ice distribution is irregular. Closed pits, thaw lakes, slumps, buckling of vegetation cover, and on slopes, gullying and extensive mass movement, may completely disrupt the pre-thermokarst surface.

Thermokarst resulting from climatic warming is usually regional in extent. Local areas of thermokarst may result from human disturbances, such as poorly planned construction, deforestation, widespread use of tracked vehicles on tundra surfaces, or by natural events such as forest fires. Thermokarst occurrence should serve as a natural warning that unstable ground temperature regimes in areas of high water content are especially susceptible to human interference. It is from this extreme condition that the term 'fragile tundra' has been developed by the conservationist opponents of arctic development projects, such as the Alaska pipeline.

6. Alpine permafrost

It has been emphasized earlier that topography, aspect, snow distribution and vegetation become determining factors for the existence or absence of permafrost in the discontinuous and sporadic zones. This was discussed particularly with reference to conditions in southern Labrador–Ungava where patches of permafrost are likely to occur beneath high, treeless, windswept hilltops as far south as the crest of the Laurentian Scarp overlooking the Gulf of St Lawrence. South of the St Lawrence permafrost probably occurs beneath the summit of Mount Jacques Cartier in the Shickshock Mountains (L. E. Hamelin, personal communication, 1960) and certainly beneath the summit of Mt Washington (a well log shows that permafrost was encountered between 50 and 150 ft/15 and 45 m: Parker H. Vincent, personal communication, 1972). By extension it seems likely that Mt Katahdin and many other high windswept New England summits that rise above the treeline are underlain by permafrost. Thus a north–south transect from Schefferville in latitude 54°50′ N to Mt Washington in latitude 44°16′ N should provide a transition across the subarctic sporadic permafrost zone into what is perhaps better described as an alpine-sporadic zone. In the latter the term sporadic is obviously related to altitude and topography. Similarly it should be possible to recognize a southern depression of the boundaries of all three permafrost zones, continuous, discontinuous and sporadic, or at least the last two, along the crestline of the Rocky Mountains and the Coast Range through British Columbia and Alberta deep into the United States, at least as far south as southern Colorado. This concept has been recognized for the Western Cordillera on most maps showing generalized permafrost distributions, although this has not been the case for the New England and eastern Canadian mountains, possibly because of the minute relative area of the summits in question and the lack of definite continuity that is typical of the Cordillera. Recently the occurrence of small patches of permafrost have been reported from the summit areas of Mt Fuji, Japan (Higuchi and Fujii, 1971) and Mauna Kea, Hawaii (Woodcock, 1974). It would seem highly likely that continued exploration will reveal many similar occurrences on high mountains throughout the world.

The paucity of ground temperature data from high mountain regions is understandable in terms of difficulty of access, especially for drilling equipment, and general absence of economic pressure. Despite this, prospectors, miners and engineers in mineral-rich alpine areas, such as western Colorado, learned over one hundred years ago that their mining shafts penetrated perennially frozen ground, although such data have been qualitative and have not reached the scientific literature in any systematic manner. Nevertheless, the high mountain regions of the world, by their very altitude, penetrate climatic belts that are the partial equivalents of the arctic and subarctic latitudinal permafrost zones. It would also seem appropriate to determine whether or not the broad relationships between mean annual air and ground temperature, used for preliminary extrapolative mapping of permafrost latitudinal zones, apply in the case of altitude.

In their work on the geographical distribution of seasonally frozen ground and permafrost, Baranov (1952, 1959) and Kudryavtsev (1950, 1959) extend this approach and predict that permafrost will underlie many of the world's high mountains. Baranov (1959) makes an interesting comparison between latitudinal variations in snowline height and lower altitudinal limit of permafrost. He points out that the mean annual air temperature at the snowline is +1.5 °C at the equator, +0.5 °C on the southern slopes of the Himalayas, −2.8 °C on the northern slopes, −3.9 °C in the Karakoram range, −4 °C in the Alps, −6 °C in the Turkestan range and on the Pamir Plateau, and −11 °C on the hills of Novaya Zemlya. He goes on to assume that the permafrost boundary in the tropical-equatorial zone will be located above the snowline, but that the fall in elevation of the former with increasing latitude will be more rapid than that of the snowline, so that the two lines will cross somewhere in mid-latitudes. It is emphasized that Baranov may be criticized for gross oversimplification, but since so little ground temperature data, or even simple descriptive information, is available, these generalizations make a reasonable first approach. It would be instructive to map the location of the −2 °C and −8 °C mean annual air isotherms on a global scale and systematically check the two lines for value as indicators of limits for potential permafrost boundaries. Attention is drawn, however, to the earlier discussion (pp. 166–71) of relationships between mean annual air isotherms and permafrost limits in eastern Canada.

Reconnaissance permafrost studies, based upon INSTAAR's Mountain Research Station on the east slope of the Front Range, Colorado Rocky Mountains, will now be described and used as an illustration of permafrost conditions in an alpine environment.

As mentioned above, the existence of perennial ice in the Colorado alpine has been common knowledge for over one hundred years. The earliest known reference in the scientific literature is a paper by Weiser (1875) in which he described permanent ice in a mine on Mt Elbert (maximum elevation 4396 m, lower permafrost limit according to Baranov, 1959, 4000 m), although Retzer deserves credit for being the first to collect systematically this type of descriptive information and, using his own field observations in addition, to hazard a map of permafrost distribution in the central Rocky Mountains (Retzer, 1956). Johnson and Billings (1962), from research based upon shallow pits dug in middle and late August during three summers, describe permafrost conditions in alpine bog soils and beneath unsorted stripes at depths of less than 1 m on the Beartooth Plateau on the Montana–Wyoming border (altitude 10,500 ft; 3230 m, 45°00′N, 109°30′W). However, Ives and Fahey (1971) were the first to provide actual ground temperatures over several years in Colorado alpine permafrost. Most of the following discussion is taken from that paper.

The INSTAAR Mountain Research Station is situated in the subalpine forest belt (Marr, 1961; for discussion of ecosystem terminology see Löve, 1970) at 2930 m. The Continental Divide along which individual summits exceed 4000 m lies 12 km due west. This section of the Divide, known as the Indian Peaks, forms a relatively straight north–south ridge extending 20 km northward from Arapaho Peak (4086 m) to Sawtooth Mountain (3785 m). Its eastern slope is dissected by a series of well-developed cirques that open out into deep U-shaped valleys. Between the glacial troughs broad, gently sloping interfluves extend approximately east–west; Niwot Ridge, which lies immediately above the Mountain Research Station, is the most extensive having an above-treeline-area of about 20 km². Treeline lies at about 3400 m, above which the forest-tundra ecotone extends for a further 100 to 150 m finally merging into alpine tundra meadow, fellfield, semi-permanent snowbank and rock outcrop above about 3500 m. Benedict (1970) has provided much information on the periglacial landforms of the alpine area, and the ecosystems of the Front Range have been described by Marr (1961) who has also tabulated climatic data for the period 1951–64 (Marr, 1967; Marr et al., 1968a, b.) The basic climatic parameters for Niwot Ridge Station, altitude 3750 m, based upon eighteen years of continuous record, are provided in table 4A.2 (Barry, 1973).

From the viewpoint of permafrost occurrence, the most significant climatological data are the relatively low mean annual air temperature and the very high mean wind speed for the winter months. The high winds ensure that the broad ridge crests extending above treeline are largely snow-free for much of the snow accumulation period. April–May experiences the greatest snow cover, but by then the period of most intensive heat loss is past and the heavy wet spring snow probably serves to reduce the warming process

Table 4A.2. Climatic data for Niwot Ridge Station (3750 m), 1952–70.

Mean annual air temperature	−3.8°C
Mean July temperature	+8.5°C
Mean January temperature	−13.2°C
Mean monthly temperature range	21.7°C
Mean annual precipitation	1021 mm
Mean annual wind speed	10.3 ms^{-1}
Mean wind speed for Nov., Dec., Jan., Feb., Mar.	13.4, 13.6, 13.9, 11.8, 12.1 ms^{-1}
Average frost-free season	47 days [1]

of the early summer by presenting a high albedo surface to the May–June-early July high-angle sun and by consuming much available energy during this period in its melting process.

My first acquaintance with the Niwot Ridge alpine together with familiarity with the available climatic data (table 4A.2) prompted an immediate comparison with conditions in central Labrador–Ungava (Ives, 1960a, 1961, 1962), although altitudinal and latitudinal differences were noted together with their possible impact upon available solar radiative energy and annual temperature regime. Nevertheless, the situation was sufficiently tempting to prompt the installation of thermistors in a variety of sites (fig. 4A.11 provides an example of data derived from one site). The full programme of ther-

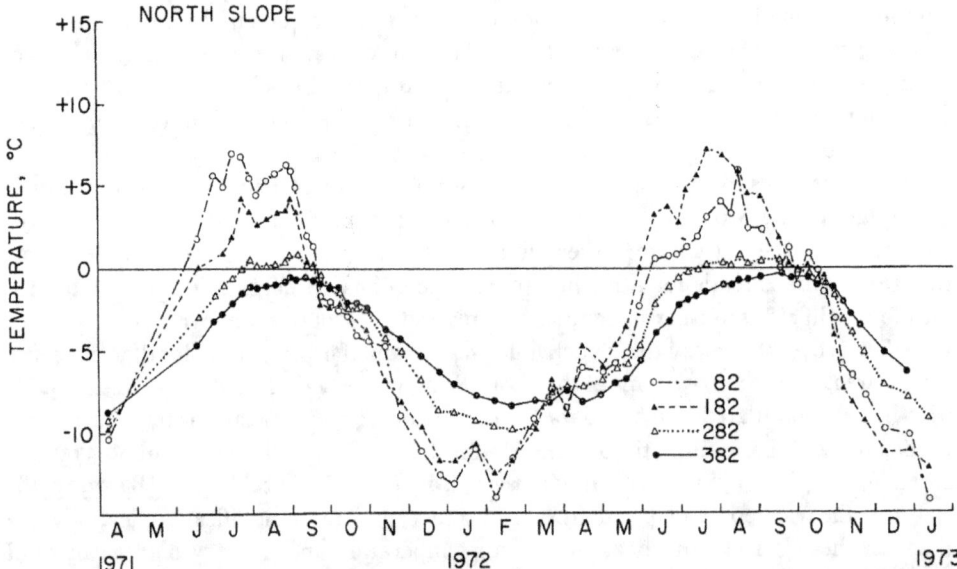

Fig. 4A.11. Ground temperature record from April 1971 to January 1973 for a north-facing site at 3750 m. Following June 1971, thermistors at 82, 182, 282 and 382 cm were read weekly during the summer and approximately once every 2 weeks during the winter. The 382-cm thermistor ceased to function in December 1972.

mistor installation is still incomplete and so far drilling has been limited to depths of 5 m through a range of altitude from 3500 m to 4300 m, the highest installation being on the summit of nearby Mt Evans. In addition Fahey operated a climatological station in conjunction with detailed frost heave studies at 3500 m on Niwot Ridge (Ives and Fahey, 1971; Fahey, 1971). At this particular site, on the tread of a large turf-banked lobe (Galloway, 1961; Benedict, 1970) with a high water table, depth to permafrost each September/October for three successive years has been less than 2 m. On nearby dry sites thermistors installed to depths of 4.5 m recorded above freezing temperatures at the same period. The 1970–1 winter afforded the first opportunity of temperature readings at other higher sites, the extreme being −8.7°C in March 1972 at a depth of 382 cm (altitude 3750 m) on a steep north-facing slope. While many more data are required the following initial conclusions seem warranted:

(1) At the 3500 m level under wet sites, permafrost occurs beneath an active layer less than 2 m thick. Freeze-back to the permafrost table was completed by early March in 1968 and by mid-February in 1969. Under nearby dry sites no below-freezing temperatures were recorded in October 1969 down to depths of 5 m, although temperatures at this depth had fallen below the freezing point by mid-December 1969. Mean annual air temperature in a standard screen for this elevation, based upon a single twelve-month period of observation, is −0.8°C. Thus, at the 3500 m level permafrost is probably restricted to wet sites, although it may occur under dry sites with an active layer in excess of 5 m.
(2) At the 3750 to 3800 m level permafrost occurs extensively under wet sites with an active layer less than 2 m thick.
(3) Under the highest summits in the Front Range (4100 to 4400 m) with an extrapolated minimum mean annual air temperature of −9°C, permafrost probably occurs to considerable thicknesses and can be classified as the alpine equivalent of the continuous arctic permafrost zone.

Various reports of ice in the ground in late summer have been collected for a number of sites in western Colorado, and most have been acquired through the courtesy of Dr John Retzer (personal communication, 1971). Buildings installed on the summits of Pikes Peak and Mt Evans, both above 4000 m, have experienced appreciable disturbance, particularly the former, and one building was completely destroyed. On Niwot Ridge itself many surface characteristics give indirect evidence of permafrost occurrence:

(1) active patterned ground;
(2) persistence of small ponds throughout summer;
(3) persistence of small springs high on the ridge flanks; here it is inferred that the only possible water source is melting ground ice.

Finally, mechanical excavations at 3140 m in the valley bottom south of Niwot Ridge indicate the presence of permafrost. The site is well below treeline and may indicate the persistence of relic permafrost that has been preserved from a cooler period or, as in the case of Labrador–Ungava, may have resulted from a lower treeline following withdrawal of the Pinedale Glacier about 9000 years ago.

From the foregoing it would appear that permafrost, at least sporadic patches of perennially frozen ground, should be anticipated down to levels as low as 3500 m in windswept areas of the Colorado Rockies, and that relic permafrost patches may exist several hundred metres lower still. This contrasts with the 4000 m figure given by Baranov (1959). In exceptional circumstances perennial ice may be found at even lower elevations, as for instance in the case of interstitial ice associated with active rock glaciers that penetrate to elevations as low as 2800 m. If the general limit of 3500 m is accepted then this coincides with a mean annual air temperature of about $-1\,^{\circ}\mathrm{C}$.

In a recent paper Furrer and Fitze (1970) have collected numerous scattered observations on reports of permafrost occurrence in the Alps as well as presenting their own observations. As in Colorado, the absence of ground temperature data below a few metres depth is remarkable. Their general conclusion is that for a large section of the Alps discontinuous permafrost can be expected above 2300 m, coinciding approximately with the mean annual air isotherm of $-1\,^{\circ}\mathrm{C}$; above 2500 m continuous permafrost can be anticipated, this elevation coinciding with the mean annual air isotherm of $-2\,^{\circ}\mathrm{C}$. Their conclusions agree in general with the Colorado observations, although Furrer and Fitze's equating of the limit of continuous permafrost with the $-2\,^{\circ}\mathrm{C}$ isotherm appears ambiguous since the arctic equivalent is usually taken as -8 to $-9\,^{\circ}\mathrm{C}$. Presumably also the great variation in aspect and snow cover resulting from the extremely mountainous topography will cause considerable local departures from the general figures given.

It is apparent that systematic study of permafrost in the alpine regions of the world has barely begun. This area of research is very worthy of greater attention, especially with the rapid growth in human impact on the alpine areas of the world as the demand for recreational space accelerates. It should also provide an additional avenue for comparative arctic–alpine research.

7. Problems of human development in permafrost areas

Passing reference has been made to the collapse of the surface in permafrost regions due to natural causes. Of particular note are the slumps and depressions of segregated ice areas, pingo degradation due to rupture of the crest, undercutting of riverbanks, collapse of ice mounds beneath white spruce trees (Viereck, 1965). Very spectacular landscape changes may develop, especially where ice content is high. These changes will occur within a stable permafrost climate; a warming trend will naturally accelerate the process of thermokarst formation. The degree of change can be appreciated best when it is understood that the ratio of ice to soil in the uppermost 10 to 30 m of the ground may be in excess of 10 : 1. Much alarm has been expressed in recent years by those concerned over man's impact on the 'fragile' tundra. The word 'fragile' has been introduced largely because of the reaction to surface disturbance that will upset the energy balance at the ground/air interface. This is particularly critical in areas of high ice content although almost irrelevant in coarse grained sediments and areas underlain predominantly by bedrock.[1] Even during the last century the Russian colonization of Siberia met with serious technical difficulties. The log cabin of the early settler, with its fireplace, thawed the permafrost beneath the structure and, in extreme cases, groundwater was tapped which entered the house. The ultimate condition was a wrecked house filled with a mass of ice.

Today, of course, much larger buildings are being constructed in the permafrost regions of the world (Brown, 1970). The biggest problem to be overcome is the prospect of differential thawing of permafrost beneath the building with its attendant partial collapse of the foundations. This is equally true of road and airport surfaces, dams storage tanks and pipelines for oil, water or sewerage (Plate 11). Two basic engineering approaches are employed: an active approach, usually reserved for areas of thin permafrost, whereby the underlying ice is deliberately melted; and a passive approach, where an attempt is made to insulate the permafrost to preserve it, and so conserve its engineering *strength*. The passive approach may be achieved in two ways, depending upon local conditions: 1) by elevating buildings on concrete, wooden or steel piles, the bases of which are sunk deep into the permafrost, so that an air space remains between the building and the ground surface; and 2) by constructing a thick insulating pad, usually of gravel, although artificial materials may be used in conjunction with gravel, so that once again the permafrost is protected from the heat of the building. A northern town, such as Inuvik in the Mackenzie Delta, is conspicuous for its elevated and electrically heated water and sewerage lines. The Soviet Union has made spectacular progress in city and utility construction in permafrost regions; but the price is high and costly mistakes are made. Perhaps the most dramatic prospect of costly human impact is the proposed trans-Alaska pipeline, although we may finally have reached a level of environmental concern in North America such that pipeline construction will be delayed until governments and people are satisfied that a sufficiently high level of control will be developed to ensure an acceptable minimal level of disturbance. This aspect of permafrost in the tundra regions will be developed more fully in Chapter 17B.

ACKNOWLEDGEMENTS

Several colleagues have read and criticized the original manuscript of this chapter and provided ideas and factual information hitherto unpublished. In particular I would like to thank Drs Roger J. E. Brown, J. Ross Mackay, Troy Péwé and Leslie A. Viereck. The incompleteness of the coverage, of course, rests with the writer. Alpine permafrost research in Colorado was supported by a National Science Foundation grant through the Tundra Biome, IBP.

[1] In exceptional circumstances, however, massive ice layers some tens of metres thick have been observed in glaciofluvial deposits in the Canadian Arctic Islands, in jointed limestone and in the weathered zone of other rock types: all of these represent potential engineering problems.

References

ANDREWS, J. T. (1961) Permafrost in southern Labrador–Ungava. *Can. Geogr.*, **3**, 219–31.

ANDREWS, J. T. (1970) A geomorphological study of post-glacial uplift with particular reference to Arctic Canada. *Inst. Brit. Geogr. Spec. Pub.*, **2**. 156 pp.

ANNERSTEN, L. J. (1963) Investigations of permafrost in the vicinity of Knob Lake, 1961–62. *McGill Sub-Arctic Res. Pap.*, **16**, 51–143.

BARANOV, I. YA. (1952) The transition zone between seasonal freezing and permafrost within the USSR and neighbouring countries (the southern boundary of permafrost). Moscow: Rukopis',. fondy In-ta merzlotoved (in Russian).

BARANOV, I. YA. (1959) Geographical distribution of seasonally frozen ground and permafrost (transl. A. Nurklik). *In: Principles of Geocryology*, Part I, Chap. VII, 193–219. NRC Tech. Transl., 1121. Ottawa: Nat. Res. Council.

BARRY, R. G. (1967) Seasonal location of the Arctic Front over North America. *Geogr. Bull.*, **9** (2), 79–95.

BENEDICT, J. B. (1970) Downslope soil movement in a Colorado Alpine Region: rates, processes and climatic significance. *Arct. Alp. Res.*, **2** (3), 165–226.

BENNINGHOFF, W. S. (1952) Interaction of vegetation and soil frost phenomena. *Arctic*, **5**, 34–44.

BIK, M. J. J. (1968) Morphoclimatic observations on prairie mounds. *Zeit. Geomorphol.*, **12** (4), 409–69.

BLACK, R. F. (1950) Permafrost. *Annual Rep., Board of Regents, Smithsonian Inst.*, 273–301. Washington, DC: GPO.

BLACK, R. F. (1954) Permafrost – a review. *Bull. Geol. Soc. Amer.*, **65**, 839–56.

BLACK, R. F. (1964) Periglacial studies in the United States, 1959–1963. *Biul. Peryglacjal.*, **14**, 5–29.

BLAKE, W. JR. (1966) End moraines and deglaciation chronology in northern Canada, with special reference to southern Baffin Island. *Geol. Surv. Can. Pap.*, 66–26.

BREWER, M. C. (1958) Some results of geothermal investigations of permafrost in northern Alaska. *Trans. Amer. Geophys. Union*, **36**. 503 pp.

BROWN, R. J. E. (1960) The distribution of permafrost and its relation to air temperatures in Canada and the USSR. *Arctic*, **13**, 163–77.

BROWN, R. J. E. (1964) *Permafrost Investigations on the Mackenzie Highway in Alberta and Mackenzie District.* NRC 7885. Ottawa: Div. Building Res., Nat Res. Council. 27 pp.

BROWN, R. J. E. (1965) *Permafrost Investigations in Saskatchewan and Manitoba.* NRC 8375. Ottawa: Div. Building Res., Nat. Res. Council. 36 pp.

BROWN, R. J. E. (1966) The relation between mean annual air and ground temperatures in the permafrost region of Canada. *Proc. Int. Permafrost Conf.*, 241–7. NAS/NRC Pub. 1287. Ottawa: Nat. Acad. Sci./Nat. Res. Council.

BROWN, R. J. E. (1967) Comparison of permafrost conditions in Canada and the USSR. *Polar Rec.*, **13** (87), 741–51.

BROWN, R. J. E. (1968) Permafrost Map of Canada. *Can. Geogr. J.*, **2**, 56–63.

BROWN, R. J. E. (1969) Factors influencing discontinuous permafrost in Canada. *In:* PÉWÉ, T. L. (ed.) *The Periglacial Environment: Past and Present*, 11–53. Montreal: McGill–Queens Univ. Press.

BROWN, R. J. E. (1970) *Permafrost in Canada.* Toronto: Univ. of Toronto Press. 234 pp.

BRYSON, R. A. (1966) Air masses, streamlines and the boreal forest. *Geogr. Bull.*, **8** (3), 228–69.

BRYSON, R. A., WENDLAND, W. M., IVES, J. D. and ANDREWS, J. T. (1969) Radiocarbon isochrones on the disintegration of the Laurentide Ice Sheet. *Arct. Alp. Res.*, **1** (1), 1–14.

BYLINSKY, E. N. (1968) Transgression in the Quaternary period over the Russian Lowland and its correlation with continental glaciations. *In:* BELOV, N. A. *et al.* (eds.) *Cenozoic History of the Polar Basin and its Influence on the Development of the Landscapes of the Northern Territories* (transl. D. Löve). Div. Biogeogr. Abstract Collection, 25–6. Leningrad: NIL Frontier Geol., Geog. Soc. USSR.

COOK, F. A. (1958) Temperatures in permafrost at Resolute, NWT. *Geogr. Bull.*, **12**, 5–18.

CORTE, A. E. (1969) Geocryology and engineering. *Rev. Eng. Geol.*, II, 119–85. Geol. Soc. Amer.

DANSGAARD, W., JOHNSON, S. J., CLAUSEN, H. B. and LANGWAY, C. C. JR. (1971) Climatic record revealed by the Camp Century ice core. *In:* TUREKIAN, K. K. (ed.) *Late Cenozoic Ice Ages*, 37–56. New Haven: Yale Univ. Press.

FAHEY, B. D. (1971) *A Quantitative Analysis of Freeze-Thaw Cycles, Frost Heave Cycles and Frost Penetration in the Front Range of the Rocky Mountains, Boulder County, Colorado.* Unpub. Ph.D. dissertation, Univ. of Colorado. 305 pp.

FALCONER, G., IVES, J. D., LØKEN, O. H. and ANDREWS, J. T. (1965) Major end moraines in eastern and central Arctic Canada. *Geogr. Bull.*, **7** (2), 137–53.

FURRER, G. and FITZE, P. (1970) Beitrag zum permafrost problem in den Alpen. *Vierteljahrsschrift der Naturforschenden Gesellschaft in Zurich*, **115** (3), 353–68.

GALLOWAY, E. W. (1961) Solifluction in Scotland. *Scot. Geogr. Mag.*, **77** (2), 75–87.

HARE, F. K. (1959) A photo-reconnaissance survey of Labrador–Ungava. *Geogr. Branch Mem.*, 6. Ottawa. 83 pp.

HIGUCHI, K. and FUJII, Y. (1971) Permafrost at the summit of Mt Fuji, Japan. *Nature*, **230**, 521.

HOPE, E. R. (1968) The earth's permafrost beds (transl. N. A. Gravé). *In:* GERASIMOV, I. P. and MARKOV, K. K. (eds.) *Permafrost and ancient glaciation.* Transl. T499R. Ottawa: Def. Res. Inf. Serv., Def. Res. Board.

IVES, J. D. (1960a) Permafrost in central Labrador–Ungava. *J. Glaciol.*, **3** (28), 798–90.

IVES, J. D. (1960b) The deglaciation of Labrador–Ungava. *Ann. Geogr. Quebec*, **4** (8), 324–43.

IVES, J. D. (1961) A pilot project for permafrost investigations in central Labrador–Ungava. *Geogr. Branch Pap.*, 28. Ottawa. 26 pp.

IVES, J. D. (1962) Iron mining in permafrost, central Labrador–Ungava. *Geogr. Bull.*, **17**, 66–77.

IVES, J. D. and ANDREWS, J. T. (1963) Studies in the physical geography of north-central Baffin Island. *Geogr. Bull.*, **19**, 5–48.

IVES, J. D. and BUCKLEY, J. T. (1969) Glacial geomorphology of Remote Peninsula Baffin Island, NWT, Canada. *Arct. Alp. Res.*, **1** (2), 83–96.

IVES, J. D. and FAHEY, B. D. (1971) Permafrost occurrence in the Front Range, Colorado Rocky Mountains, USA. *J. Glaciol.*, **10** (58), 105–11.

JOHNSON, P. L. and BILLINGS, W. D. (1962) The alpine vegetation of the Beartooth Plateau in relation to cryopedogenic processes and patterns. *Ecol. Monogr.*, **32** (2), 105–35.

JOHNSTON, G. H., BROWN, R. J. E. and PICKERSGILL, D. N. (1963) *Permafrost Investigations at Thompson, Manitoba.* NCC 7568. Ottawa: Div. Building Res., Nat. Res. Council. 51 pp.

KIND, N. V. (1967) Radiocarbon chronology in Siberia. *In:* HOPKINS, D. M. (ed.) *The Bering Land Bridge*, 172–92. Stanford: Stanford Univ. Press.

KREBS, J. S. and BARRY, R. G. (1970) The arctic front and the tundra–taiga boundary in Eurasia. *Geogr. Rev.*, **60**, 548–54.

KUDRYAVTSEV, V. A. (1950) *Temperature Regions in the Permafrost Zones of the USSR.* Moscow: Rukopis', fondy In-ta merzlotoved.

KUDRYAVTSEV, V. A. (1959) Temperature thickness and discontinuity of permafrost (transl. G. Belkov). *In: Principles of Geocryology*, Part I, Chap. VIII, 219–73. NRC Tech. Transl. 1187. Ottawa: Nat. Res. Council.

LACHENBRUCH, A. H. (1957) Thermal effects of the ocean on permafrost. *Geol. Soc. Amer. Bull.*, **68**, 1515–30.

LACHENBRUCH, A. H. (1962) Mechanics of thermal contraction cracks and ice-wedge polygons in permafrost. *Geol. Soc. Amer. Spec. Pap.*, 70. 65 pp.

LACHENBRUCH, A. H. (1967) Permafrost. *In: Encyclopedia of Geomorphology*, 833–9.

LACHENBRUCH, A. H., GREENE, G. W. and MARSHALL, B. V. (1966) Permafrost and geothermal regimes. *In:* WILIMOVSKY, N. J. and WOLFE, J. N. (eds.) *Environment of the Cape Thompson Region, Alaska*, 149–63. Oak Ridge, Tenn.: US Atomic Energy Commission.

LEGGET, R. F., BROWN, R. J. E. and JOHNSTON, G. H. (1966) Alluvial fan formation near Aklavik, Northwest Territories, Canada. *Bull. Geol. Soc. Amer.*, **77**, 15–30.

LØKEN, O. H. (1966) Baffin Island refugia older than 54,000 years. *Science*, **153** (3742), 1378–80.

LÖVE, D. (1970) Subarctic and Subalpine: where and what? *Arct. Alp. Res.*, **2** (1), 63–72.

MACKAY, J. R. (1962) Pingos of the Pleistocene Mackenzie Delta area. *Geogr. Bull.*, **18**, 21–63.

MACKAY, J. R. (1966) Segregated epigenetic ice and slumps in permafrost, Mackenzie Delta area, NWT *Geogr. Bull.*, **8** (1), 59–80.

MACKAY, J. R. (1968) Discussion on the theory of pingo formation by water expulsion in a region affected by subsidence. *J. Glaciol.*, **7** (50), 346–50.

MACKAY, J. R. and STAGER, J. K. (1964) Thick tilted beds of segregated ice, Mackenzie Delta area. Abstracts of Papers, *Int. Geogr. Congress* (London), 131.

MARR, J. W. (1961) Ecosystems of the east slope of the Front Range in Colorado. *Univ. of Colorado Stud.*, Ser. in Biol., 8. 134 pp.

MARR, J. W. (1967) Data on mountain environments, I: Front Range, Colorado. Sixteen sites, 1952–3. *Univ. of Colorado Stud.*, Ser. in Biol., 27. 110 pp.

MARR, J. W., CLARK, J. M., OSBURN, W. S. and PADDOCK, M. W. (1968a) Data on mountain environments, II: Front Range, Colorado. Four climax regions, 1953–58. *Univ. of Colorado Stud.*, Ser. in Biol., 28. 169 pp.

MARR, J. W., JOHNSON, A. W., OSBURN, W. S. and KNORR, O. A. (1968b) Data on mountain environments, III: Front Range, Colorado. Four climax regions, 1959–1964. *Univ. of Colorado Stud.*, Ser. in Biol., 29. 181 pp.

MISENER, A. D. (1955) Heat flow and depth of permafrost at Resolute Bay, Cornwallis Island, NWT, Canada. *Trans. Amer. Geophys. Union*, **36**, 1055–60.

MORRISON, A. (1970) Pollen diagrams from interior Labrador. *Can. J. Bot.*, **48**, 157–75.

MÜLLER, F. (1959) Beobachtungen über pingos. *Medd. om Grønland*, **153** (3), 1–127.

MÜLLER, F. (1962) Analysis of some stratigraphic observations and radio-carbon dates from two pingos in the Mackenzie Delta area, NWT. *Arctic*, **15** (4), 278–88.

MULLER, S. W. (1947) *Permafrost or Permanently Frozen Ground and Related Engineering Problems*. Ann Arbor, Mich.: Edwards Bros. 231 pp.

PÉWÉ, T. L. (1954) Effect of permafrost on cultivated fields. *US Geol. Surv. Bull.*, **989**, 315–51.

PÉWÉ, T. L. (1966a) *Ice Wedges in Alaska*. Paper presented at Int. Conf. on Permafrost, Purdue Univ.

PÉWÉ, T. L. (1966b) *Permafrost and its Effects on Life in the North*. Corvallis: Oregon State Univ. Press. 40 pp.

PIHLAINEN, J. A. (1962) *Inuvik, NWT, Engineering Site Information*. NRC 6557. Ottawa: Div. Building Res., Nat. Res. Council. 18 pp.

PIHLAINEN, J. A. and JOHNSTON, G. H. (1963) *Guide to a Field Description of Permafrost*. Tech. Mem. 79, NRC 7576. Ottawa: Can. Ass. Comm. on Soil and Snow Mechanics, Nat. Res. Council. 23 pp.

PISSART, A. (1965) Les pingos des hautes Fagnes: les problems de leur genèse. *Ann. Soc. Geol. Belg.*, **88** (5–6), 277–89.

PISSART, A. (1967) Les pingos de l'île Prince Patrick (76°N–120°W). *Geogr. Bull.*, **9** (3), 189–217.

PORSILD, A. E. (1938) Earth mounds in unglaciated Arctic northwestern America. *Geogr. Rev.*, **28**, 46–58.

PREST, V. K. (1970) Quaternary geology of Canada. *In:* Geology and Economic Minerals of Canada, *Econ. Geol. Rep.*, 1 (5th edn), 676–764. Ottawa: Dept of Energy, Mines and Resources.

RAPP, A. and ANNERSTEN, L. (1969) Permafrost and tundra polygons in northern Sweden. *In:* PÉWÉ, T. L. (ed.) *The Periglacial Environment: Past and Present*, 65–91. Montreal: McGill–Queens Univ. Press.

RETZER, J. L. (1956) Alpine soils of the Rocky Mountains. *J. Soil. Sci.*, **7** (1), 22–32.

SALMI, M. (1970) Investigations on palsas in Finnish Lapland. *In: Ecology of the subarctic regions*, 143–50. Ecol. and Conserv. Ser. Paris: UNESCO.

SHUMSKII, P. A. (1969) Ground (subsurface) ice (transl. C. de Leuchtenberg). *In: Principles of Geocryology*, Part I. Chap. 9. NRC Tech. Transl. 1130. Ottawa: Nat. Res. Council. 118 pp.

SUMGIN, M. I. (1937) *Perennially Frozen Soil Within the USSR* (2nd edn). Moscow: Izd. Akad. Nauk. 379 pp.

TABER, S. (1916) The growth of crystals under external pressure. *Amer. J. Sci.*, 4th ser., **41**, 532–56.

TABER, S. (1917) Pressure phenomena accompanying the growth of crystals. *Proc. Nat. Acad. Sci.*, **3**, 279–302.

TABER, S. (1918) Surface heaving caused by segregation of water forming ice crystals. *Eng. News-Record*, **81**, 683–6.

TABER, S. (1930) The mechanics of frost heaving. *J. Geol.*, **38**, 303–17.

THOM, B. G. (1969) New permafrost investigations near Schefferville, PQ. *Rev. Géogr. Montreal*, **23** (3), 317–27.

VIERECK, L. A. (1965) Relationship of white spruce to lenses of perennially frozen ground, Mount McKinley National Park, Alaska. *Arctic*, **18**, 262–7.

VIERECK, L. A. (1970) Forest succession and soil development adjacent to the Chena River in interior Alaska. *Arct. Alp. Res.*, **2** (1), 1–26.

WEISER, S. (1875) Permanent ice in a mine in the Rocky Mountains. *Phil. Mag.*, 4th ser., **49** (322).

WOODCOCK, A. H. (1974) Permafrost and climatology of a Hawaii volcano crater. *Arct. Alp. Res.*, **6**, 49–62.

ZUBAKOV, V. A. (1969) La chronologie des variations climatiques au cours du Pleistocène en Sibéria occidentale. *Rev. Géogr. Phys. Géol. Dyn.*, **11**, 315–24.

Manuscript received March 1971.

Additional references

CAREY, K. L. (1973) Icings developed from surface water and ground water. *Cold Regions Science and Engineering Monogr.*, III-D-3. Hanover, NH: US Army CRREL. 74 pp.

FAHEY, B. D. (1974) Seasonal frost heave and frost penetration measurement in the Indian Peaks region of the Colorado Front Range. *Arct. Alp. Res.*, **6**, 63–76.

IVES, J. D. (1973) Permafrost and its relationship to other environmental parameters in a midlatitude, high altitude setting, Front Range, Colorado Rocky Mountains. *In:* NATIONAL ACADEMY OF SCIENCE, *Permafrost* (North American Contribution, 2nd Int. Conference), 121–5. Washington, DC.

NATIONAL ACADEMY OF SCIENCE (1973) *Permafrost* (North American Contribution, 2nd Int. Conference). Washington, DC. 783 pp.

WASHBURN, A. L. (1973) *Periglacial Processes and Environments*. New York: St Martin's Press. 320 pp.

Present arctic ice cover

Geoffrey Hattersley-Smith[1]

Defence Research Board,
Ottawa

1. Introduction

Since the end of the Second World War air photographs have become available of all ice cover in the Arctic, except perhaps in the central parts of the Greenland ice sheet (*Inlandsis*). There has also been a great expansion of glaciological work in the Arctic that can be dated from the International Geophysical Year, 1957–8. Thus it is now possible to make a reasonably accurate assessment of the areal extent of the glaciers, even if the standard of geodetic control is inadequate in some regions. A considerable amount of data on the classification, regime, depth and movement of arctic glaciers has also become available in the last two decades. All these aspects were very ably reviewed by Sharp (1956), but the new information now at hand justifies an updated account of present arctic glaciation.

2. Distribution of glaciers

As Sharp (1956) pointed out, 'the present development of glaciers in the Arctic is clearly related to sources of moisture, storm paths and topography, more or less in that order of importance'. Figs. 4B.1 and 4B.2 show the geographical distribution of arctic glaciers, and table 4B.1 shows their areal extent according to the best available information: for Greenland (Bauer, 1967), arctic Canada (Ommaney, 1971), Iceland, Jan Mayen and Svalbard (Sharp, 1956), USSR (Grosval'd and Kotlyakov, 1969), Alaska and Yukon Territory (Sharp, 1956). We have followed Sharp in including some glaciers that should more properly be regarded as subarctic; these are the glaciers of Iceland and Jan Mayen and of southern Alaska as far south as latitude 56°30′N.

The vast size of the Inlandsis and associated glaciers of Greenland greatly exceeds the total area and total volume of all other arctic ices since the mean ice depth over

[1] Present affiliation: British Antarctic Survey, London.

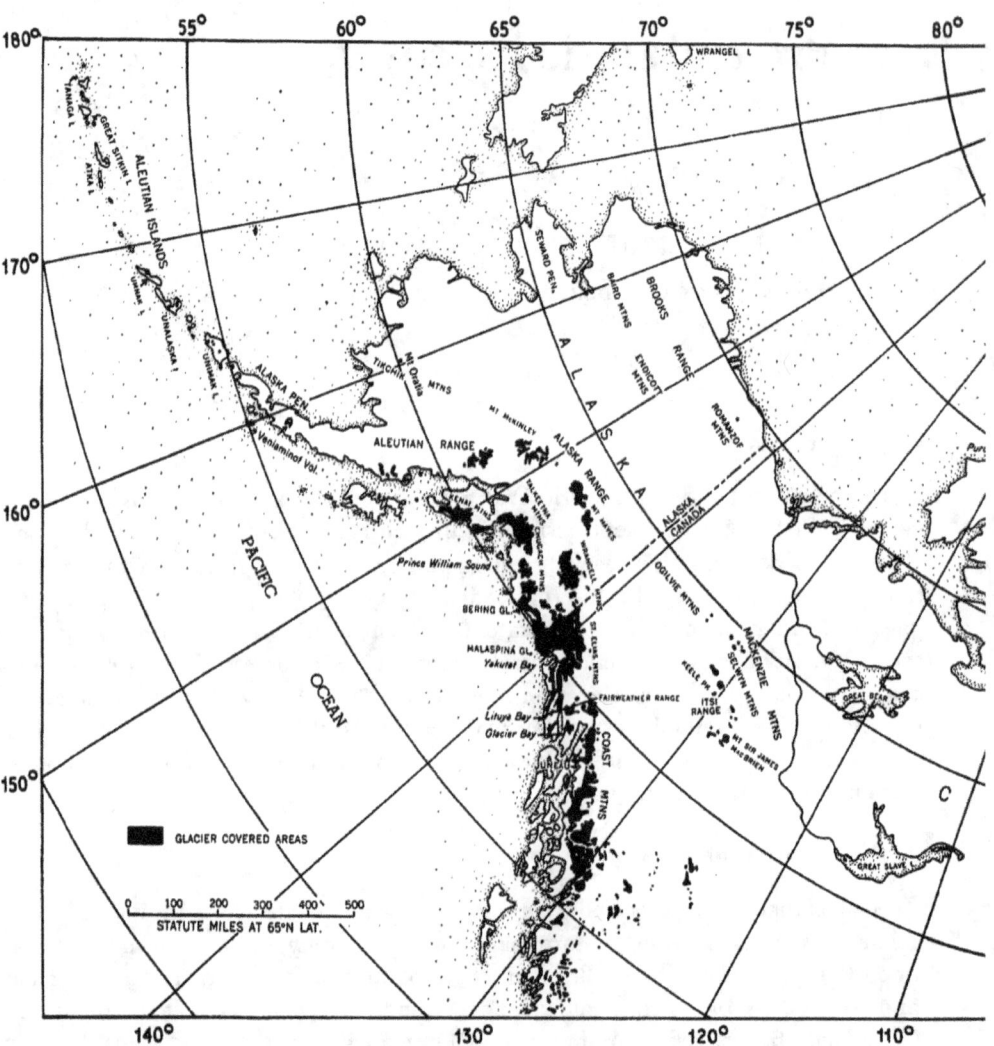

Fig. 4B.1. Location map of arctic glaciers in the western hemisphere (from Sharp, 1956)

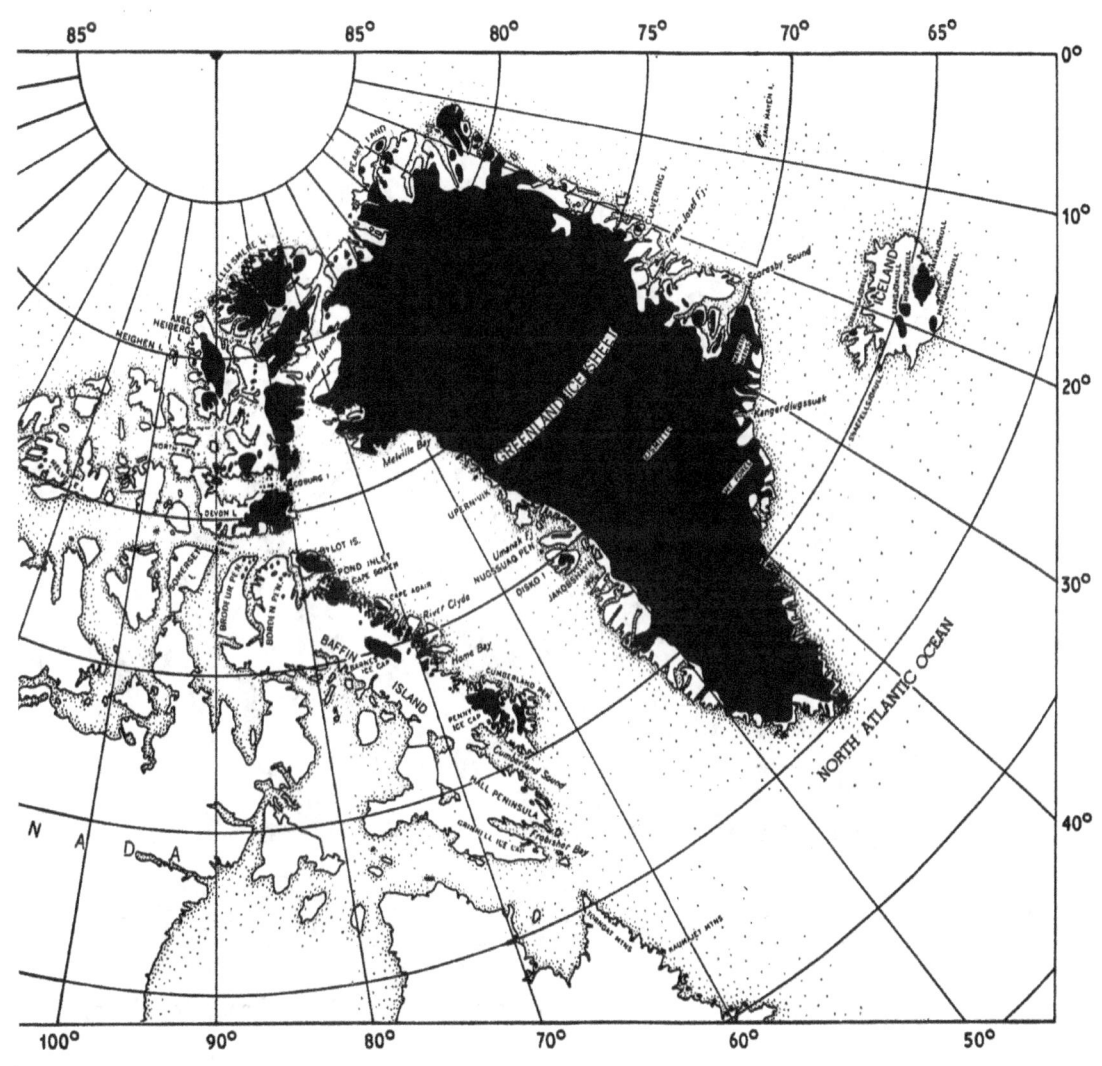

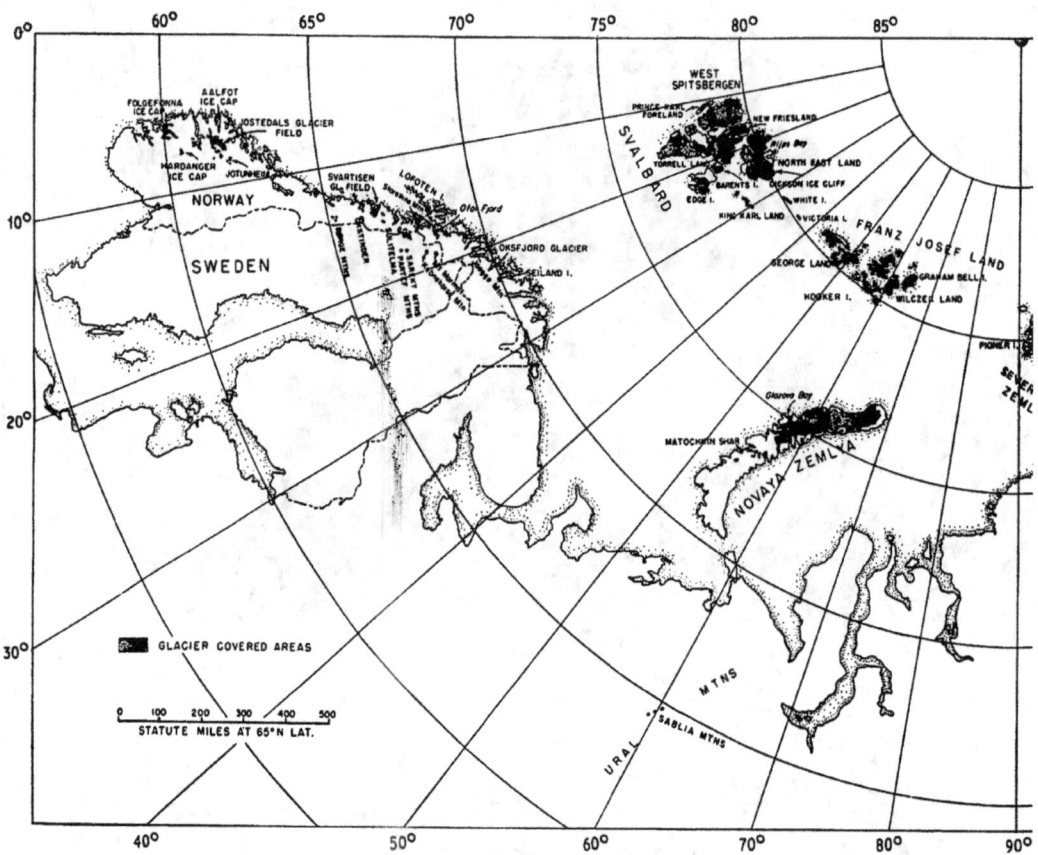

Fig. 4B.2. Location map of arctic glaciers in the eastern hemisphere (from Sharp, 1956)

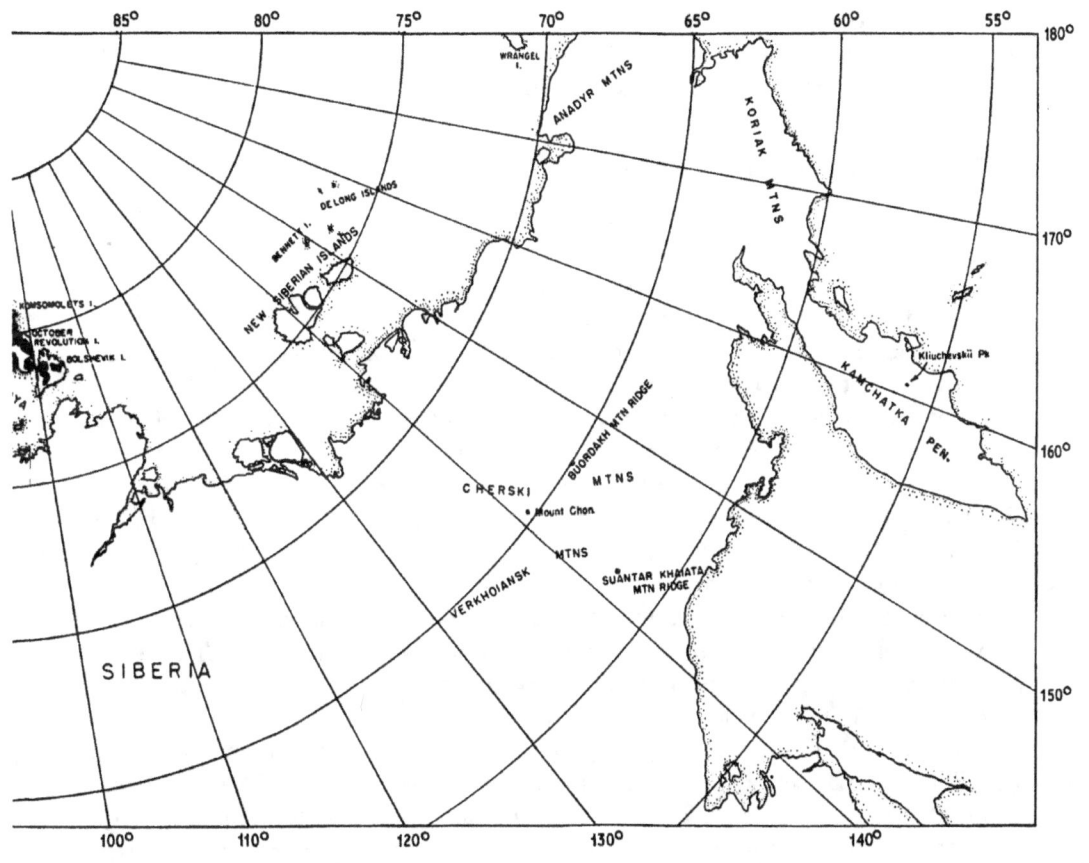

Greenland is the order of 2000 m, whereas the greatest ice depths recorded in very limited areas elsewhere probably do not exceed about 900 m, as for example in northern Elles-mere Island (Evans *et al.*, 1969). Although Helk (1966) rightly cautioned against too great reliance on figures for area and volume of the Inlandsis because of inadequate surveys, the figure of 2.6 million km³ for the total volume of ice in Greenland gives an order of magnitude equivalent to 8 per cent of the total volume of ice in the world, and more than five times the total volume of fresh water in all lakes and rivers (Bauer, 1967). The figure for Greenland may be compared with the figure of 13,140 km³ for the volume of ice in the glaciers of the Russian Arctic (Grosval'd and Kotlyakov, 1969), the only other arctic area for which we could find an estimate.

3. Classification of arctic glaciers

MORPHOLOGICAL CLASSIFICATION

Various morphological classifications have been proposed (for example, Ahlmann, 1940), and in the Arctic almost every morphological type is represented. Although such classification serves a broad and useful purpose for regional studies and map-making, the detailed glaciological studies of the last few decades have emphasized the require-ment for a classification that helps to explain the behaviour of glaciers rather than one that merely describes their surface form. For the present purpose the list of commonly accepted glacier types given by Armstrong *et al.* (1973) provides a basis for description that is adequate in most respects. But a few comments on, and additions to, the terms are in order.

The term *ice sheet* is used to include *ice caps* that are grounded on rock and *ice shelves* that are floating. Because of its vast size the Greenland ice sheet merits the special term of *inland ice sheet* or, in Danish, *Inlandsis*, which ranks as a geographic name. The term *ice cap* is widely used in the Arctic, although it does not give a true picture of most land ice sheets which are more or less broken by *nunataks* and mountain ranges. The term *ice field*, widely used in Canada, now appears to have gained international acceptance (UNESCO/IASH, 1970a); it is perhaps less misleading in a descriptive sense than ice cap, but it is nevertheless unfortunate that the same term should be used in sea-ice nomenclature. Until comparatively recently *ice shelves* were usually regarded as features peculiar to the Antarctic, but a few authorities, notably Koch (1928) and Wordie (1950), seem to have realized that ice shelves may also occur in the Arctic. In fact, the first description of floating *ice islands* established that they originate from ice shelves, especially those off Ellesmere Island (Koenig *et al.*, 1952). Of the other typically polar forms *piedmont glaciers* and *ice streams* deserve some mention. The former are the lobe-shaped extensions of valley glaciers where they debouch over broad lowlands at the base of mountains. With their vertical or very steep *ice walls* they are characteristic features of the landscape of Greenland near the margins of the Inlandsis and of the intermontane areas of Ellesmere and Axel Heiberg islands. *Ice streams*, defined as part of an ice sheet in which the ice flows more rapidly than and not necessarily in the same direction as the surrounding ice, occur only where the bedrock is completely inundated by ice, and are best seen near the margins of the Inlandsis. The common glacier form of alpine regions, the *cirque glacier*, is well represented in the Arctic (Plate 13), although the extent to which active cirque erosion is proceeding under present conditions is open to dispute (Rudberg, 1963). In many of the main fiords of the Arctic, particularly in north Greenland, *glacier tongues* constitute the floating extensions of valley glaciers. They normally end at *ice fronts* from which *icebergs* are calved, or in rare cases as in northern Ellesmere Island they may merge with ice shelves. Of local importance are glacier forms classified as *snow patches* and *taryn ice*. *Snow patches* are isolated bodies of snow and *firn* lying above or below the regional *snowline* that persist through one or more summers.

Table 4B.1. Areal distribution of arctic and some subarctic glaciers.

Region	Area (km²)	
Greenland		
Inlandsis	1,726,400	
Independent glaciers	76,200	
Total	———	1,802,600
Canadian Arctic Islands		
Ellesmere, land glaciers	80,000	
ice shelves	500	
Axel Heiberg	11,735	
Devon	16,200	
Coburg	225	
Meighen	85	
Melville	160	
North Kent	152	
Baffin	37,000	
Bylot	5,000	
Total	———	151,100
North American Mainland		
Alaska	67,000	
Yukon and Mackenzie	12,060	
Total	———	79,100
Svalbard		
North East Land	46,200	
West Spitsbergen and other islands	11,100	
Iceland	11,900	
Jan Mayen	115	
Total	———	69,300
Arctic USSR		
Franz Josef Land	13,735	
Novaya Zemlya	24,300	
Severnaya Zemlya	17,500	
East Siberia	406	
Koryak and Kamchatka Mountains	1,516	
Other areas	476	
Total	———	57,900
Grand Total		2,160,000

If they persist long enough, they may pass through the firn stage to become *perennial ice masses*. Snow patches are especially prone to persist as fillings in creek bottoms and narrow gullies where they may develop into *taryn* ice. These forms are of course very sensitive to climatic change, even from one summer to the next. In periods of increasing cold, ice sheets may be generated by the gradual coalescence of snow patches and taryn ice.

GEOPHYSICAL CLASSIFICATION

A geophysical classification of glaciers was proposed by Ahlmann (1948) on the basis of the temperature regime, the meltwater behaviour and the texture of the firn. Under this classification *temperate glaciers* are considered to be at the pressure melting point throughout their mass, except for a thin layer near the surface which is chilled below the freezing point in the winter. Movement of free water with subsequent recrystallization is therefore possible throughout the firn, with corresponding effect on the firn texture. In *polar glaciers* the temperature is below freezing to a considerable depth even in summer, at least in the accumulation area. Polar glaciers are subdivided into *high-polar glaciers* where no meltwater is formed even in summer, and *subpolar glaciers* where some surface melting and meltwater percolation do occur. Ahlmann's distinction between temperate and polar glaciers is valid and useful, but as a basis for classification it does not take into account the fact that in a large glacier system, extending over thousands of metres in elevation, accumulation processes may vary from temperate to high-polar according to altitude.

GEOPHYSICAL AND GEOLOGICAL CLASSIFICATION

A classification put forward by Benson (1962) recognized that glaciers originate as sedimentary rocks subject to various diagenetic processes due to temperature and sedimentary load. The classification is based on the *facies* concept, and Benson summarized the characteristics of the *glacier facies* as follows:

(i)   The *ablation facies* extends from the snout or margin of the glacier to the firn line, or the highest elevation to which the snow cover recedes during the melt season.

(ii) The *soaked facies* becomes wet throughout during the melt season and extends from the firn line to the uppermost limit of complete wetting, the *saturation line*.

(iii) The *percolation facies* is subjected to localized percolation of meltwater from the surface without becoming wet throughout. Percolation can occur in snow and firn of subfreezing temperatures with only the pipe-like percolation channels being at the melting point. A network of ice pipes, lenses and layers forms when refreezing occurs. This facies extends from the saturation line to the *dry snow line*.

(iv) The *dry snow facies* includes all the glacier lying above the dry snow line; negligible melting occurs in this facies.

The facies express the response of glacier to climate, and long-term changes in size, elevation and net accumulation (or ablation) of the various zones can be used as climatic indicators. Glaciers are typified by the number (from one to four) of facies present. At one extreme, on certain low-lying ice sheets during the present climatic cycle, only the *ablation facies* may be represented, and at the other extreme, as for example on the Inlandsis, all four facies may be represented.

Müller (1962a) refined the facies concept by the recognition of a *percolation zone A* in which the meltwater normally refreezes within the snowpack of the reference year; a

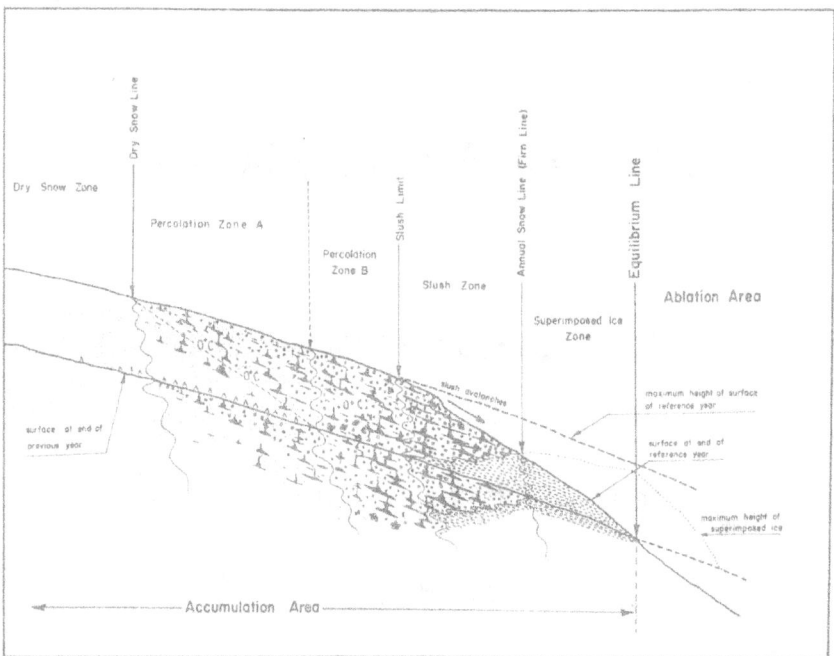

Fig. 4B.3. Schematic zonation of the accumulation area of an arctic glacier (from Müller, 1962a).

percolation zone B in which a proportion of the meltwater is normally lost to the firn of previous years; a *slush zone* in which slush avalanches are an important cause of loss of material to form accumulation outliers within the ablation area; and a *superimposed ice zone*, situated between the firn line and the *equilibrium line*, in which accumulation occurs solely by the refreezing of meltwater (fig. 4B.3).

4. Regime of arctic glaciers

GREENLAND

Just as the Greenland ice sheet dominates the glacier scene in the Arctic, so has the knowledge gained from it profoundly influenced the course of glaciological research in the Arctic (and elsewhere). In fact the present section is largely about glacier problems in

Greenland, and space allows only a brief review of the regime of the ice sheet. Figs. 4B.4 and 4B.5, taken from Fristrup (1966), summarize essential data on ice thickness and snow accumulation for the ice sheet.

There is a considerable difference between the topography of the bedrock beneath the southern and northern domes of the ice sheet. The southern part of the ice sheet covers buried mountains, rising to about 1000 m above sea level, to a maximum ice thickness of 850 m. On the other hand, in the central and northern parts of the ice sheet

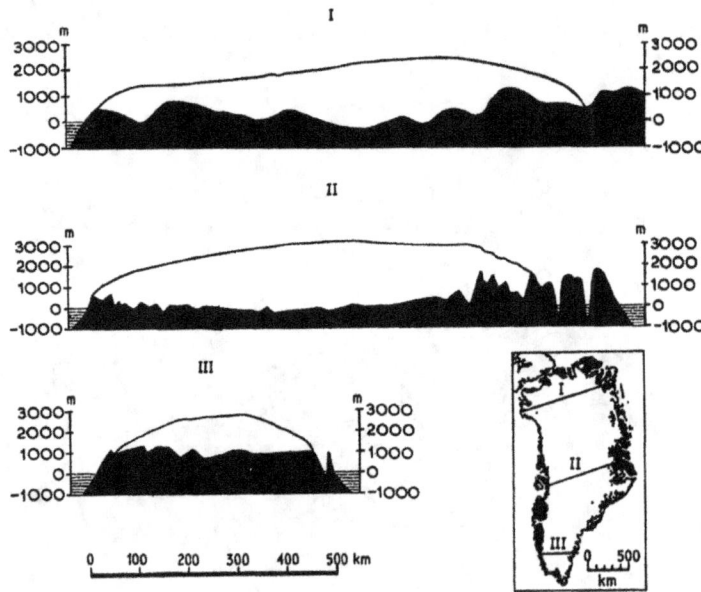

Fig. 4B.4. Three sections through the Greenland ice sheet (from Fristrup, 1966). (I) Thule air base to Kap Georg Cohn. (II) Disko Bugt to Cecilia Nunatak. (III) JAD Jensens Nunatakkar to Skjoldungen. Vertical exaggeration approximately 1:40.

the ice reaches thicknesses up to 3200 m, and the bedrock is at sea level or down to 400 m below sea level over wide areas.

In order to calculate the mass balance of the ice sheet, Bauer (1955) assumed an average altitude of the snowline of 1390 m, a mean ablation of 1100 mm H_2O, and a mean accumulation of 310 mm H_2O. He also allowed for wastage due to iceberg production by classifying the numerous tidewater glaciers draining the ice sheet according to rates of movement taken as 3–10 m day^{-1}, 3 m day^{-1} and 30 m year^{-1} according to their activity. He reached the conclusion that the Greenland ice sheet has an average annual deficit of 84 km^3 H_2O. Subsequently Benson (1962), from his work in Greenland, revised upward to 340 mm H_2O Bauer's figure for mean annual accumulation, and he also extended the accumulation area to take into account the fact that superimposed ice forms well below the snowline. Benson concluded, as had Loewe (1936), that the mass balance of the ice sheet is slightly on the positive side. However, with the uncertainties inherent in the calculations, the ice sheet could well be essentially in equilibrium.

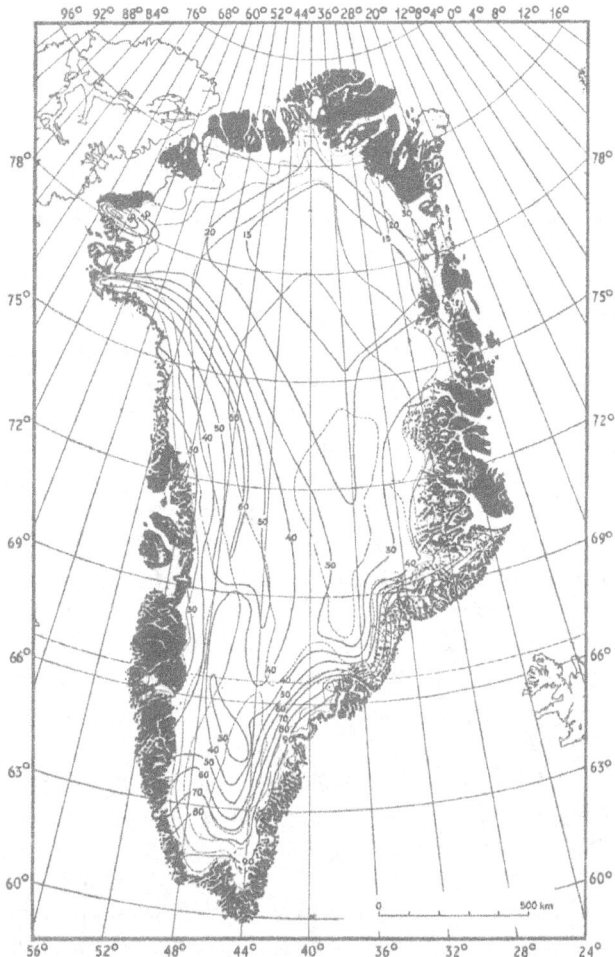

Fig. 4B.5. Map showing annual accumulation (cm H_2O) on the Greenland ice sheet (from Fristrup, 1966, based on work by C. S. Benson, R. W. Gerdel, H. Bader, C. Langway, and others).

Of the numerous separate ice bodies and glaciers in Greenland, mention will only be made of those in Peary Land, where the mean annual precipitation is only 125 mm H_2O (Fristrup, 1952). Koch (1928) concluded on glacial geological evidence that the Inlandsis had never extended to cover this region. The ice cover of northern Peary Land, with its own dynamics and regime, originated locally; there has been little recent change in the positions of glacier termini and ice margins in this most northern land (Davies and Krinsley, 1962).

CANADIAN ARCTIC ISLANDS

Knowledge of the glaciers of the Canadian Arctic Islands up until 1954 was well reviewed by Baird (1955). The first detailed glaciological studies in this whole region were undertaken on the Barnes ice cap, Baffin Island, in 1950. Baird and Ward (1952) deduced a slightly negative mass balance for the ice cap, and concluded that it was nourished entirely by increments of superimposed ice. Subsequent investigations showed that on the highest parts of the ice cap, at an altitude of about 1130 m, firn often remains at the end of a season, but nevertheless demonstrated predominantly negative mass balances for the years 1961-6, leading to decrease in mass of the ice cap with very little marginal recession, since the margin is protected from melting by drift snow (Sagar, 1966; Løken and Sagar, 1968). However, it seems certain from geomorphological and lichenological studies that ice and snow was much more extensive north of the ice cap (Plate 12) within the last 300 to 400 years (Ives, 1962). The orographic increase of precipitation induced by the ice cap itself is, according to Løken and Andrews (1966), a major component of the winter budget at the present time. They also state that the Barnes ice cap is a remnant of the Wisconsin ice sheet which did not disappear during the Hypsithermal (cf. also, Ives and Andrews, 1963). On the Penny ice cap in southeastern Baffin Island, where elevations reach 2100 m, Ward and Baird (1954) recorded the firn line at an altitude of about 1550 m and the equilibrium line at 1380 m. Radar soundings in 1965 showed a very irregular bedrock topography with maximum ice depths to 500 m (Weber and Andrieux, 1970). Of particular interest in this general area of eastern Baffin Island are paradoxical relationships between ice cover, topography and climate which cannot readily be explained (Andrews et al., 1970). The Grinnell and Terra Nova ice caps, also in southern Baffin Island, reach a maximum altitude of 870 m, and in 1952-3 were reported to be in slight recession; their mode of nourishment is similar to that of the Barnes ice cap (Mercer, 1956).

In the Queen Elizabeth Islands, glaciological studies have been conducted in northern Ellesmere Island since 1953 when investigations were started on the Ward Hunt ice shelf, one of the main sources of floating ice islands (Hattersley-Smith et al., 1955). The ice shelf, which has a thickness of from 20 to 70 m (Evans et al., 1969), was shown to originate from accumulation of superimposed ice on thick pads of landfast sea ice within the last 3000 years (Crary, 1960). The lower part of the ice shelf is largely composed of brackish ice, which interdigitates with sea ice (Lyons et al., 1971). The mass balance at the surface of the ice shelf is very close to equilibrium (Hattersley-Smith and Serson, 1970), and the ice shelf is spreading slightly under its own weight (Dorrer, 1971). Massive calving in recent years is thought to be mainly due to unusual tidal movements (Holdsworth, 1971). In the interior of northern Ellesmere Island, where the ice sheets reach an altitude of about 1900 m, mean annual accumulation is estimated at 160 mm H_2O and the mean altitude of the equilibrium line is at 1200 m (Hattersley-Smith, 1963a, 1967). Negative mass balances were measured on the Gilman Glacier, one of the main outlet glaciers on the south side of the central ice cap, in three out of four years between 1956 and 1961 (Sagar, 1964), probably as a result of generally warmer summers since about 1925 (Hattersley-Smith, 1963a). In the same period the smaller glaciers of northern Ellesmere Island receded, while the major glaciers were advancing slightly (Hattersley-Smith, 1969).

On Axel Heiberg Island a major programme of glaciological research and glacier mapping (p. 214) has been conducted since 1959 on the Akaioa ice cap (formerly 'McGill ice cap') and glaciers on its southwest side. About 30 per cent of the island is covered by ice, mainly included in the Akaioa ice cap and another large ice cap to the south, but the maximum ice depth measured is only 470 m (Redpath, 1961). The Akaioa ice cap reaches a maximum altitude of about 1900 m, and the mean annual accumulation is as high as 370 mm, or more than twice that in northern Ellesmere Island (Müller, 1963b). Mass balance figures for the White Glacier, which has a well-defined accumulation basin clearly separated from the ice cap, show years of highly negative values and years of small positive or zero values (Müller, 1962b, and unpublished data, 1971). Although the smaller glaciers of the area are receding at the snouts, the larger glaciers are advancing. Thus the Thompson Glacier, one of the largest outlet glaciers of the Akaioa ice cap with a length of 38 km measured from the ice divide at an elevation of 1600 m, is advancing at the rate of about 20 m year^{-1} (Müller, 1962b).

On the Devon Island ice cap, which reaches a maximum altitude of 1885 m, there is a great contrast between the glacier regimes of the northwestern and southeastern parts, although the mass balances may be similar (Koerner, 1966). During the period 1961–6, the northwestern part which was studied in more detail had a slightly negative mass balance and it is thought likely that the whole ice cap had a mean negative mass balance during the preceding thirty years (Koerner, 1970).

The small ice cap in Meighen Island and the four small ice caps on Melville Island, the most westerly of the ice caps in the Queen Elizabeth Islands, nowhere exceed 150 m in thickness and are essentially stagnant. The mass balances have been negative in most years since observations were started in 1959. It is evident that the existence of the ice caps is sensitively related to recent climatic change (Arnold, 1965; Paterson, 1969, and unpublished data, 1971).

ICELAND, JAN MAYEN AND SVALBARD

The glaciers of Iceland are characteristic of a moist maritime environment and are subject to a high rate of metabolism, as shown by the large values of measured accumulation and ablation (Ahlmann, 1939; Bauer, 1956). Ice caps, of which Vatnajökull with an area of about 7800 km^2 is the largest, are the dominant glacier form. Elevations on Vatnajökull range from 1750 m (2100 m in the limited segment of Öraefajökull in the south) almost down to sea level, but elevations are lower on other ice caps which terminate at some height above sea level (Sharp, 1956). Some outlet glaciers are extremely active, although receding under the present negative regime. Sigbjarnarson (1970) showed that runoff in Icelandic rivers exceeds precipitation, and concluded that the additional runoff is due to glacier recession. The regime of some of the Icelandic ice caps and glaciers is intimately linked with the vulcanism of the region (p. 211).

In Jan Mayen glaciers are limited to the slopes of the high volcanic peak of Beerenberg (2280 m), from whose mantle of ice and snow about fifteen valley or wall-sided glaciers extend to the lower levels, even to tidewater in the case of the Weyprechtbreen and several other glaciers (Jennings, 1948). The annual precipitation on Jan Mayen is of the order of 750 mm H_2O, and rime formation may play a significant part in the accumulation processes at high altitude around the rim of the Beerenberg (Kinsman and Sheard, 1963). A general advance of the glaciers since about 1954 has been attributed almost entirely to a continuing increase in precipitation since about 1920, with a time lag in the response of the glaciers (Lamb *et al.*, 1962).

In West Spitsbergen, the largest island of Svalbard, there are three main icefield areas with elevations of 600 to 900 m; the icefields reflect the rolling relief of the underlying dissected plateaus. There are also many small ice caps, valley and cirque glaciers typical of alpine glaciation, with peaks reaching elevations of 1200 to 1600 m, especially on the west coast. Outlet glaciers up to 40 km long reach tidewater in many places or coalesce to form piedmont glaciers (Sharp, 1956). However, it is the ice cover of North East Land that has been best studied, probably of any area in the European Arctic. Icefields and ice caps, with thicknesses up to at least 350 m, and a number of outlet glaciers cover about 65 per cent of the island (Schytt, 1964). The ice cover of North East Land was classified as subpolar by Ahlmann (1935), although the large amount of meltwater and the isothermal condition of the firn on West Ice suggest that this ice cap may be near temperate in condition (Glen, 1941). In West Spitsbergen the Fourteenth of July Glacier is indeed considered to be temperate throughout (Ahlmann, 1935). Up until 1956, the larger ice masses of Svalbard were considered to be relatively inactive or even stagnant, but recent work has shown that there is in fact a considerable turnover of mass on the ice caps of North East Land, where the equilibrium line is at an elevation of from 300 to 450 m above sea level (Schytt, 1964).

USSR

A useful survey of present-day glaciers in the USSR (Grosval'd and Kotlyakov, 1969) divides the arctic and subarctic glaciers between three regions. The first glacier region includes Franz Josef Land, Novaya Zemlya, Severnaya Zemlya, the Subpolar and Polar Urals, the Byrranga Mountains (Taimyr Peninsula) and the islands of the De-Longa group; this region is under the influence of cyclones coming from the North Atlantic Ocean (see p. 32). The second region includes the outlying ranges of the East Siberian Mountains, which are also under the influence of the North Atlantic Ocean and, in some seasons of the year, of the Pacific Ocean as well. The third region includes Kamchatka Peninsula and the Koryakskiy Mountains on the coast of the Pacific Ocean, which is the main source of nourishment of the glaciers.

Ice caps are the typical glacier forms in the islands of Franz Josef Land, Severnaya Zemlya, and the islands of the De-Longa group; the ice probably does not exceed 200 m in thickness and on the smaller islands it is much thinner. A number of islands of Franz Josef Land are completely ice-covered, and on other islands a system of ice domes and small outlet glaciers are associated with the main ice caps (cf. Plate 13). The equilibrium line is reported to range from 200 to 600 m above sea level. However, at some places off Severnaya Zemlya glacier ice merges with the pack ice with apparent formation of small ice shelves (Dibner, 1955), implying that locally and in certain periods the equilibrium line may be at sea level. The elevation of none of these islands exceeds 800 m above sea level, which is well below the dry snow line in most years. Superimposed ice therefore plays an important role in accumulation processes.

On Novaya Zemlya the ice cover is almost confined to the more rugged northern half of the island, where elevations reach 1000 m. The glaciers are of the mountain rather than ice cap type; nunataks rise above the ice fields and valley glaciers reach tidewater on both coasts. The equilibrium line is at about the 700 m level, and in the accumulation area the soaked facies predominates over the percolation facies. The glaciers of the Byrranga Mountains immediately to the south are of very limited extent.

In the Subpolar and Polar Urals and in the East Siberian Mountains the ice cover consists predominantly of cirque glaciers. Elevations in the Urals reach 1800 m with the equilibrium line ranging from the 600 m level in the north and west to the 1400 m level in the east and south; the accumulation areas lie almost entirely below the saturation line. In the East Siberian Mountains, where the climate is continental and precipitation is light, complete melting of snow normally occurs and accumulation is in the form of superimposed ice. On the Kamchatka Peninsula, where volcanic mountain cones rise to 5000 m, star-shaped glaciers cover the upper slopes in this region of maritime climate and high precipitation.

It appears that the arctic glaciers of the USSR have in general been subject to negative mass balance in this century, and that the observed improvement in glacier nourishment over the past decade will not halt the widespread recession that has been well documented.

ALASKA AND YUKON TERRITORY

Icefields, valley and piedmont glaciers are exceptionally well displayed in Alaska and the southwestern corner of the Yukon Territory from tidewater to the highest elevation in the region, namely 6200 m near the summit of Mt McKinley. The ice cover was well described by Sharp (1956). Additions to knowledge since Sharp's review was published, and recent trends in research on Alaskan and Yukon glaciers, can perhaps best be exemplified by a brief summary of fieldwork in three selected areas.

The Romanzof Mountains of the Brooks Range in northeastern Alaska, situated only 70 km from the arctic coast, constitute the most northerly ice-covered area on the North American mainland. Partly for this reason and partly because of its very well defined drainage basin, the McCall Glacier in these mountains was chosen for detailed investigation in the International Geophysical Year. This glacier is 8 km long and stems from three cirques at an altitude of about 2700 m above sea level; it terminates at an altitude of 1250 m. The accumulation area of the glacier is well within the saturation facies, for the temperature of the glacier at depth is only −1 °C (Orvig and Mason, 1963). It appears that the glacier is roughly in a state of balance at the present time with a relatively low mass turnover (Keeler, 1959).

In southwestern Alaska the ice cover of Mt Wrangell is of very different type. After detailed investigations, Benson (1968) concluded that the facies parameters in the summit area at 4000 to 4300 m are similar to those near the dry snow line on the Greenland ice sheet. The mean annual temperature at this altitude is −20°C, and the annual accumulation more than 1000 mm H_2O.

Of wide scope have been the investigations, continuous since 1961, in the St Elias Mountains which form part of the North American Cordillera between latitudes 59 and 62° N. These mountains rise as a high barrier between the Pacific Ocean and the continental interior, and comprise a number of parallel ranges over a distance of nearly 200 km between the Gulf of Alaska and the central Yukon Territory. The topography is that of a high alpine region with the major valleys filled by the most extensive ice cover in continental North America (Wood, 1963). The results of meteorological work along a transect in the St Elias Mountains showed great diversity in summer temperatures, as might be expected in an area of high relief and varied surface characteristics (Marcus, 1965). Glaciological work has been concentrated in the Icefield Ranges on the east side of the St Elias Mountains, particularly in the area of the ice divide and on the Kaskawulsh Glacier, where maximum ice depths up to nearly 1000 m have been reported (Dewart, 1970). All glacier facies are represented in the area, including the dry snow facies on the highest snowfields above the level of about 4500 m (Grew and Mellor, 1969). On the north and central arms of the Kaskawulsh Glacier, however, only the ablation facies to an elevation of about 2000 m, the saturation facies to an elevation of about 2400 m, and the percolation facies are represented between the glacier terminus at an elevation of 810 m and the highest part of the accumulation area at 2760 m (Wagner, 1969). Classen and Clarke (1971) showed that the temperature at depth may reach the melting point at least locally in the glaciers of the area, a factor that may be important in initiating glacier surges for which several glaciers in the Icefield Ranges are noted (pp. 216–17). Apart from these surging glaciers and apart from some of the largest trunk glaciers, most glaciers in southeastern Alaska have been receding over the last fifty years (Miller, 1970).

5. The role of arctic glaciers

GLACIERS AS A GEOMORPHOLOGICAL AGENT

The answer to the classical question of the extent to which glaciers erode or alternatively protect the bedrock depends on the amount of sliding over the bed that is possible. As far as polar glaciers are concerned, an investigation of the physics of the basal one metre layer of an ice mass is perhaps the most important single requirement in glaciology at the present time (Crary, 1970). Direct measurements in the deep corehole at Camp Century, Greenland, gave −13°C as the temperature at the bottom of the ice at a depth of 1400 m (Weertmann, 1968). Whether cold glaciers, such as the Inlandsis, slip on their beds, and if so by what mechanism, is still not certain. However, evidence is mounting that the ice may be at the pressure melting point throughout the main parts of many polar and subpolar ice caps and in the peripheral parts of polar ice sheets, depending on the permeability of the firn to meltwater (Schytt, 1969), and Müller (1963a) showed that for the tongue area of a polar glacier the temperature may be at the pressure melting point at least for the lower half of the glacier's depth. This means that in the Arctic 'temperate' glacier behaviour can occur, with the ice slipping over its bed by a combination of flow past larger obstacles and regelation through smaller ones (Weertmann, 1957).

It is certainly difficult, without the conventional explanation of glacier scour and plucking, to account for such features as, for example, the bottom profiles of some fiords in northern Ellesmere Island where overdeepening extends for many kilometres behind present ice fronts (Evans *et al.*, 1969), or the closed deep basins of the Nansen Sound fiord system that occur where trunk glaciers formerly coalesced to cause thickening of ice masses and presumably intensified erosion (Hattersley-Smith, 1969). The submarine geomorphology of eastern Baffin Island (see Plate 12) cannot be explained without invoking glacial erosion (Løken and Hodgson, 1971). A mechanism of fiord erosion by floating ice and a geophysical explanation of the fact that fiord basins deepen inland were presented by Crary (1966). But the mechanism by which floating ice erodes bottom material is irrelevant to the action of grounded ice, and Crary brought forward no evidence on the importance of ice erosion in shaping the original valley occupied by a fiord. In fact the mechanism and extent of bottom erosion by grounded ice are still matters for hypothesis. There can be no doubt, however, of the importance of lateral widening of valleys by marginal glacial streams (Rudberg, 1963, for example).

In many arctic areas it is well recognized that the effect of an ice cover is almost wholly protective. These are areas where the ice cover is thin, the bottom relief is subdued, and the ice is not channelled in valleys; examples occur in the vicinity of the Barnes ice cap, northern Baffin Island, where recent glacial recession exposed patterned ground well-preserved beneath the ice (Falconer, 1962), and on the periphery of the Meighen ice cap, Northwest Territories, where Arnold (1965) found well preserved plant material uncovered by the ice.

In Iceland glacier floods, or *jökulhlaups*, in the Grimsvötn and Katla areas are geomorphologically important (Thorarinsson and Sigurdsson, 1947; Thorarinsson, 1954). They are caused by the eruption of subglacial volcanoes and/or by melting induced by an abnormal heat flow from a non-erupting volcanic area with formation initially of a large subglacial lake. When the lake exceeds a certain level or extends to the edge of the ice cap, the flood bursts forth carrying large quantities of debris and ice over the outwash areas peripheral to the ice. In recent times outbursts of floodwaters have occurred in the Grimsvötn area at intervals of six to ten years, and the Katla area has erupted twice a century since 1580. Glacier floods with a similar periodicity to those in the Grimsvötn area have been reported from the Sukkertoppen area of southwestern Greenland, but these are caused by the emptying of a glacier-dammed lake (Helk, 1966). Similar examples can be cited from Alaska and arctic Canada, as for example near Tanquary Fiord, northern Ellesmere Island (Hattersley-Smith, 1969). Formation of *aufeis* downvalley frequently accompanies glacier floods in high arctic areas (see Plate 6).

GLACIERS AS CLIMATOLOGICAL INDICATORS

Since the pioneer work in Greenland by Sorge (1935), great strides have been made in the interpretation and areal correlation of near-surface glacier stratigraphy (Benson,

1962). The new techniques of drilling, particularly with the cable-suspended electro-mechanical drill (Langway, 1970a), and of oxygen isotope analysis applied to ice cores (Dansgaard, 1954) have made it possible to probe and interpret the stratigraphy of the Greenland ice sheet down to bedrock. The work performed on the 411 m ice core from Site 2 (Langway, 1970b) and on the 1440 m ice core from Camp Century (Dansgaard et al., 1969; Johnsen et al., 1970), which constitutes a landmark in glaciology, relies on the fact that from the concentration of the ^{18}O content, annual layers within the ice can be identified at least for the last few thousand years; in ice that is much older seasonal differences tend to have been eliminated by molecular diffusion and plastic thinning of the layers. For the deeper ice in the 1440 m core it was possible to calculate the age of the various increments by assuming a mean annual accumulation, deduced from studies of the upper part of the core, and by allowing for attenuation of layers from consideration of a simple ice flow model, which important theoretical studies have made possible (Nye, 1959, 1963). A climatic record was then produced by plotting the $\delta^{18}O$ data against the calculated ages, and the record was not only in agreement with known climatic events dated on other evidence, but also provided an unbroken sequence of climatic change over a period of about 100,000 years. At the same time, from very detailed studies of the upper part of the ice core representing a time span of about 800 years, certain periodicities in temperature fluctuation were recognized, including one cycle of 120 years, and these periodicities can be extrapolated over the next decades to provide long-term climatic predictions for the first time. It is predicted that the present cooling trend will continue for the next one or two decades and will be followed by a tendency to warmer conditions culminating in the first half of the twenty-first century.

The climatic record from the Greenland ice sheet provides a baseline against which the results of studies of the variations of glaciers across the Arctic (p. 214) can be assessed.[1]

GLACIERS AS PART OF THE ECOSYSTEM

Under this heading a few general remarks are offered on the effects of ice cover as a barrier in limiting the movement of animals and on the effect of glacial meltwater on plant habitats (see also Löve and Löve, Chapter 10A, and Ives, Chapter 10B).

In Greenland the Inlandsis severely limits the migration of musk-oxen and caribou by confining their movements to the peripheral ice-free land. In the Canadian Arctic Archipelago the ice cover is not so extensive as to pose a serious problem to the movement of these animals, and certainly not to the movement of foxes and wolves. The writer has found musk-ox remains on glaciers and nunataks in northern Ellesmere Island, and on Ward Hunt Island. In the case of the arctic fox, Vibe (1967) has documented its move-ment far out on the Inlandsis, and has even suggested that it may migrate between north-east and northwest Greenland via the Inlandsis. The present author has observed fox tracks at all localities on ice caps and glaciers in northern Ellesmere Island.

With regard to plant habitats two main points may be made. As shown by Porsild (1955), the richness of the flora in the Canadian High Arctic (and doubtless elsewhere in the Arctic) is controlled rather by the availability of water from melting glaciers and

---

[1] See Andrews, Chapter 6A, p. 300.

snow patches than by altitude, at least up to the level of about 700 m. In the main valleys, however, glacial rivers can cause considerable erosion of well-vegetated areas, especially below glacier-dammed lakes, such as the one near Tanquary Fiord cited above. The sudden drainage of this lake in 1965 caused extensive damage to peat bogs, the vegetation of which provides prime grazing for musk-oxen (Hattersley-Smith, 1969).

GLACIERS FROM AN ECONOMIC VIEWPOINT

Ice caps and glaciers could theoretically provide unlimited water supply, and seasonally at least water power, if the need should ever arise. In many areas of the Arctic that have an ice cover, as for example in the Canadian Arctic Archipelago, precipitation is very light and natural or artificial lakes fed by glacial rivers could become of key importance in the siting of arctic stations that require a large and reliable water supply. At the present time the freshwater needs of most stations near ice-covered arctic areas are provided by lakes of quite limited extent that are supplied by the melting of local snow cover. Water consumption is extremely low, and investigations are needed into improved water engineering and treatment before the water potential of the ice caps and glaciers of the Arctic can be exploited (Sater, 1969, p. 114).

Discharges from ice caps and glaciers at present merely constitute a hazard rather than serve any useful purpose. Fortunately glacier surges and ice floods in the Arctic have so far been confined to areas that are well removed from human habitations or works. But, although the area is somewhat south of the arctic region, it should be noted that an ice avalanche from the Salmon Glacier in northern British Columbia claimed the lives of a number of workers at a mining camp near the terminus of the glacier in 1965. In Alaska and the Yukon Territory, where a number of surging glaciers have been identified (Horvath and Field, 1969), a threat could be posed to oil pipeline construction, and discrimination will be necessary in the selection of routes.

However, it is the tidewater glaciers calving icebergs that pose the main threat for the immediate future. In the North Atlantic Ocean the icebergs originate mainly from west Greenland, particularly from the glaciers of Disko and Melville bays and from the Humboldt Glacier. In the Arctic Ocean the occasional calving of ice islands provides a real, if uncommon, threat in the channels of the Canadian Arctic Archipelago and the Beaufort Sea, and also off the east and west coasts of Greenland on the rare occasions when ice islands escape from the Arctic Ocean. Exploration and development of oil and gas reserves in Alaska, the Canadian Arctic Archipelago and the islands of the European Arctic will cause a continuous increase in shipping and offshore activity in seas far north of the traditional area of iceberg hazard in the North Atlantic Ocean. The main danger here is not so much impact with ships as impact with offshore drilling rigs and damage to wellheads and pipelines, for it is necessary to take account of the bottom scouring action of icebergs and ice islands. It should prove possible, by means of a coordinated system of surveillance by aircraft and probably also by satellite, to provide warning of massive calvings of ice such as occurred, for example, from the Franz Josef Land glaciers in the 1870s and 1920s (Sandford, 1955) and from the Ward Hunt ice shelf in 1961-2 (Hattersley-Smith, 1963b; Nutt, 1966). An experiment in tracking icebergs off the coast of Labrador from aircraft gave measured rates of drift of 40 km day^{-1} (Jones and Diehl, 1971), and the feasibility of tracking massive icebergs from satellite photography has already been demonstrated in the Antarctic (Swithinbank, 1969).

6. Observational and research requirements

GLACIER INVENTORY AND GLACIER VARIATIONS

The International Hydrological Decade (IHD) 1965–74 has provided a stimulus to obtaining accurate data on the area and volume of arctic glaciers, and of their response to seasonal and climatic change. A worldwide inventory of perennial ice and snow masses is a major project of the IHD, and of arctic countries of the USSR (Avsyuk and Krenke, 1968) and Canada (Ommaney, 1969, 1971) have been particularly active in the project. The basic requirements for glacier inventory (UNESCO/IASH, 1970a) are shown in table 4B.2. The main objective is to obtain an accurate assessment of the world's freshwater resources. At the same time the inventory provides an ideal framework for measurement of glacier variations, and for all glaciers a datum to which new information can be added as it becomes available. In complementary programmes to measure glacier variations and mass balance, a number of glaciers across the Arctic (and elsewhere) have been selected for detailed studies, and special guides have been prepared to establish the parameters needed for assessment of regime and mass balance, and to standardize measuring techniques (UNESCO/IASH, 1969, 1970b). It is to be hoped that records can be maintained on these glaciers beyond the end of the IHD, and that critical gaps in the network can be filled, as for example in the ice-covered areas of central and southern Ellesmere Island.

GLACIER MAPPING

The standard of mapping glaciers for most areas of the Arctic is still very unsatisfactory. Thus contour lines are usually inaccurate and form lines may be misleading, while surface features such as glacial streams, crevasses and moraines are seldom shown. There is an urgent need for standard conventions in depicting glacial features, and colour symbolization should be much more widely exploited to show relief, contour lines and glacier margins (Ewing and Marcus, 1966; Henoch, 1969). Two series of maps of glaciers in the Canadian Arctic Archipelago show what can be done with modern techniques of air photogrammetry and automatic plotting, supplemented by accurate ground control. These are the very detailed maps on scales of 1:5000 and 1:10,000 of glaciers on western Axel Heiberg Island, which employ microrelief shading and a variety of symbols and hachures to depict glacial features (Blachut and Müller, 1966), and maps on the scale of 1:25,000 of the small ice cap on Meighen Island on which subglacial bedrock contours were included (Arnold, 1966). The need for similar large-scale maps of other representative glaciers, including their complete drainage basins, was emphasized at a Symposium on Glacier Mapping in 1965; such maps are particularly necessary to support studies of the glacier–climate relationship (Hattersley-Smith, 1966).

Table 4B.2. Data sheet for Canadian glacier inventory.

Province or territory: _____

Mountain area: _____

Hydrological basin:

 1st order : _____
 2nd order: _____
 3rd order: _____
 4th order: _____

Sources: Maps _____

 Map title and number: _____
 Compiled by: _____
 Date: _____
 Scale: _____
 Contour interval: _____
 Reliability:
  Vertical: _____
  Horizontal: _____

Sources: Photographs

 Type: _____
 Serial number: _____
 Date: _____
 Flying height: _____
 Focal length: _____
 Remarks: _____

Work done on glacier:	References
O	
1	
2	
3	
4	
5	
6	
7	
8	
9	

	Other photos	Special	Moraine	Lake
O				
1				
2				
3				
4				
5				
6				
7				
8				
9				

Remarks _____
_____
_____
_____

Region and basin identification
Glacier number
Longitude
Latitude
UTM
Orientation: Accumulation area
 Ablation area
Highest glacier elevation (m)
Lowest glacier elevation (m): Exposed
 Total
Elevation of snowline (m)

Region and basin identification
Glacier number
Date of snowline
Mean accumulation area elevation (m)
 Accuracy rating
Mean ablation area elevation (m)
 Accuracy rating
Maximum length (km): Ablation area
 Exposed
 Total
Mean width of main stream (km)

Region and basin identification
Glacier number
Surface area (km²):
 Exposed
 Total
 Accuracy rating
Area of ablation (km²)
 Accuracy rating
Accumulation area ratio (%)
Mean depth (m)
Volume (km³) in ice
 Accuracy rating
Classification and description

Region and basin identification
Glacier number
Comments: Special
 General
Glacier name

Data compiled by: _____
 Date: _____
 Supervisor: _____
 Centre: _____

DEEP DRILLING

The main requirement in deep drilling is to extend to other areas of the Arctic the operational methods and core analysis techniques developed at Camp Century, Greenland (Johnsen et al., 1970). The Inlandsis by its very size must act as a damper on climatic change, so that the effect of minor climatic fluctuations on ice core stratigraphy and chemistry should be less marked for the Inlandsis than for the other very much smaller arctic ice caps, although the much shorter ice cores obtainable from the latter would represent a comparatively limited time span. Nevertheless, on ice caps in northern Ellesmere Island, for example, where the mean annual accumulation is about 160 mm H_2O (Hattersley-Smith, 1963a) and the maximum ice depth 900 m (Evans et al., 1969), the deepest ice core might represent a time span of more than 5000 years, allowing for attenuation of annual ice layers at depth. This would provide a climatic record extending back into the Hypsithermal that is more detailed and relevant to areas bordering the Arctic Ocean than any previous record.

RADIO ECHO SOUNDING

The radio echo sounding technique for the survey of arctic glaciers and ice caps was fully proven during experiments in Greenland in 1963-4 (Waite, 1966; Robin et al., 1969). In these experiments the radar was operated from a vehicle moving over the ice, but in 1966 the equipment was installed in a light aircraft for successful airborne sounding of glaciers and ice caps in northern Ellesmere Island (Evans and Robin, 1966; Robin et al., 1969). Unlike the seismic and gravity sounding techniques, radio echo sounding can provide a continuous profile of the bedrock surface. A further great advantage of the system is its capacity to detect echoes from small irregularities within the ice mass, particularly within the top few hundred metres (Robin et al., 1969). The system has now reached a stage of development that makes it technically possible to obtain ice thicknesses down to depths of at least 2200 m for arctic glaciers, provided that, in the case of valley glaciers, the valley is not so narrow as to introduce interference by oblique echoes from valley walls (Evans et al., 1969). Improvements in the system may well close the gaps in the information obtainable at present. A much more accurate assessment of the volume of glacier ice in the Arctic should eventually come from the new technique. At the same time, reflecting horizons traced over long distances within the ice will provide valuable information on how ice deforms and flows in the Greenland ice sheet and elsewhere. The internal layers may also have climatic significance if they are in fact formed of slightly denser ice caused by unusually warm summer temperatures at the time of deposition, as suggested by Robin et al. (1969).

GLACIER SURGES

Most of the glaciers that are known to surge are of the valley type, although the greatest surge occurred from the ice cap of North East Land to form a new glacier Brasvellbreen

(Glen, 1941). It has been suggested that continental ice sheets can surge, and in so doing trigger an ice age (Wilson, 1964; Flohn, 1969). The far-reaching implications of the Wilson hypothesis have focused attention on the cause of glacier surges which has become perhaps the most important glaciological problem awaiting solution (Müller, 1969). It appears that glacier surges are periodically triggered by some mechanism that is not directly attributable to net mass balance, as for example changes in the meltwater distribution or of bedrock temperature, but this mechanism is still unknown (Glaciology Panel, Committee on Polar Research, 1967; Meier and Post, 1969). Apart from Bråsvell-breen in recent years, the following arctic and subarctic glaciers are among those known to have surged: Steele Glacier (Stanley, 1969) and other glaciers in Alaska and the Yukon Territory (Horvath and Field, 1969); Otto Glacier, Ellesmere Island (Hattersley-Smith, 1964; Konecny, 1966); the Bruärjokull, Iceland (Thorarinsson, 1969); and the Negri Glacier, West Spitsbergen (Liestøl, 1969). A detailed review of surging glaciers in the Arctic (and elsewhere) is needed, and precise measurements should be made on several glaciers that are likely to surge in the near future. Special studies are required of the longitudinal profile, ice thickness, temperature of the ice and bedrock, quantity of water in and below the glacier, and discharge of meltwater (Müller, 1969). In addition, the geological effects of surges should be studied, particularly in order to provide criteria for recognizing the glacial deposits of past surges (Rutter, 1969).

7. Conclusion

In the last twenty years the outstanding advances in glaciology have been: the application of physical theory to the dynamics of glaciers and ice sheets (Nye, 1959) and their response to seasonal and climatic change (Nye, 1960); a refined scheme of glacier classification (Benson, 1962); the application of radio isotope analysis to the interpretation of glacier stratigraphy and chronology (Dansgaard, 1954); the use of radio waves for airborne glacier sounding (Evans and Robin, 1966); and the development of deep-drilling equipment (Langway, 1970a) and thermal probes (Aamot, 1970). Arctic glaciers, and particularly the Inlandsis of Greenland, provided much of the data on which new concepts were based and served as a proving ground for new techniques. The logistic support provided by aircraft and oversnow vehicles and the availability of air photographs for reconnaissance and mapping have allowed planning and execution of fieldwork on a scale and of a quality inconceivable thirty years ago. A recent review of glaciology in the Arctic took all these aspects into account (Glaciology Panel, Committee on Polar Research, 1967).

For the future the most fruitful lines of field investigation are seen as follows: the extension of deep drilling to other areas of the Arctic besides Greenland, in order to provide long-term climatic records and to investigate physical processes in the basal layers of glacial ice; wide-ranging radio echo sounding of arctic glaciers and ice caps with the objective of producing maps that show bedrock contours as well as surface features; intensive studies of glacier surges when opportunities offer, and application of glacier flow theory to explain their mechanism and to define the characteristics of periglacial deposits caused by surges; and the use of satellite photography to monitor glacier calving and also seasonal changes in snow cover.

References

AAMOT, H. W. C. (1970) Self-contained thermal probes for remote measurements within an ice sheet. *Int. Ass. Sci. Hydrol. Pub.* (Commission of Snow and Ice), 86, 63–8.

AHLMANN, H. W. (1935) The Fourteenth of July Glacier. *Geogr. Ann.*, **17**, 167–218.

AHLMANN, H. W. (1939) The regime of Hoffellsjökull. *Geogr. Ann.*, **21**, 171–88.

AHLMANN, H. W. (1940) The relative influence of precipitation and temperature on glacier regime. *Geogr. Ann.*, **22**, 188–205.

AHLMANN, H. W. (1948) Glaciological research on the North Atlantic coasts. *Roy. Geogr. Soc. Res. Ser.*, **1**. 83 pp.

ANDREWS, J. T., BARRY, R. G. and DRAPIER, L. (1970) An inventory of the present and past glacierization of Home Bay and Okoa Bay, east Baffin Island, NWT, Canada, and some climatic and palaeoclimatic considerations. *J. Glaciol.*, **9**, 337–62.

ARMSTRONG, T. E., ROBERTS, B. B. and SWITHINBANK, C. W. M. (1973) *Illustrated Glossary of Snow and Ice* (2nd edn). Spec. Pub. 4. Cambridge: Scott Polar Res. Inst. 60 pp.

ARNOLD, K. C. (1965) Aspects of the glaciology of Meighen Island, Northwest Territories, Canada. *J. Glaciol.*, **5**, 399–410.

ARNOLD, K. C. (1966) The glaciological maps of Meighen Island, NWT. *Can. J. Earth Sci.*, **3**, 903–8.

AVSYUK, G. A. and KRENKE, A. N. (1968) The beginning of the Soviet glaciological investigations in the IHD programme. *IUGG (Int. Ass. Sci. Hydrol.) Reports and Discussions*, 292–9.

BAIRD, P. D. (1955) Glaciological research in the Canadian Arctic. *Arctic*, **8**, 96–108.

BAIRD, P. D. and WARD, W. H. (1952) The glaciological studies of the Baffin Island Expedition, 1950. *J. Glaciol.*, **2**, 2–23.

BAUER, A. (1955) The balance of the Greenland ice sheet. *J. Glaciol.*, **2**, 456–62.

BAUER, A. (1956) Contribution à la connaissance du Vatnajökull, Islande. *Jökull*, **5–6**, 3–19.

BAUER, A. (1967) Nouvelle estimation du bilan de masse de l'Indlandsis du Groenland. *Deep-Sea Res.*, **14**, 13–17.

BENSON, C. S. (1962) Stratigraphic studies in the snow and firn of the Greenland ice sheet. *SIPRE Res. Rep.*, 70. US Army Snow, Ice and Permafrost Res. Estab. 93 pp.

BENSON, C. S. (1968) Glaciological studies on Mount Wrangell, Alaska, 1961. *Arctic*, **21**, 127–52.

BLACHUT, T. J. and MÜLLER, F. (1966) Some fundamental considerations on glacier mapping. *Can. J. Earth Sci.*, **3**, 747–60.

CLASSEN, D. F. and CLARKE, G. K. C. (1971) Basal hot spot on a surge type glacier. *Nature*, **229**, 481–3.

CRARY, A. P. (1960) Arctic ice island and ice shelf studies, Part II. *Arctic*, **13**, 32–50.

CRARY, A. P. (1966) Mechanism for fiord formation indicated by studies of an ice-covered inlet. *Bull. Amer. Geol. Soc.*, **77**, 911–29.

CRARY, A. P. (1970) Presidential address. *Int. Ass. Sci. Hydrol. Pub.* (Commission of Snow and Ice), 86, x-xvi.

DANSGAARD, W. (1954) The O^{18} abundance in atmospheric water and water vapour. *Geochim. Cosmochim. Acta*, **6**, 241–60.

DANSGAARD, W., JOHNSEN, S. J., MØLLER, J. and LANGWAY, C. C. JR. (1969) One thousand centuries of climatic record from Camp Century on the Greenland Ice Sheet. *Science*, **166**, 377–81.

DAVIES, W. E. and KRINSLEY, D. B. (1962) The recent regimen of the ice cap margin in north Greenland. *Int. Ass. Sci. Hydrol. Pub.* (Commission of Snow and Ice), 58, 119–30.

DEWART, G. (1970) Confluence of the north and central arms of the Kaskawulsh Glacier. *Icefield Ranges Research Project, Scientific Results*, **2**, 77–102. Amer. Geogr. Soc.

DIBNER, V. D. (1955) The origin of the floating ice islands. *Priroda*, **3**, 89–92. (English transl. by E. R. Hope, T176R, Defence Res. Board, Canada.)

DORRER, E. (1971) Movement of the Ward Hunt Ice Shelf, Ellesmere Island, NWT, Canada. *J. Glaciol.*, **10**, 211–26.

EVANS, S. and ROBIN, G. DE Q. (1966) Glacier depth-sounding from the air. *Nature*, **210**, 883–5.

EVANS, S., GUDMANDSEN, P., SWITHINBANK, C., HATTERSLEY-SMITH, G. and ROBIN, G. DE Q. (1969) Glacier sounding in the polar regions: a symposium. *Geogr. J.*, **135**, 547–63.

EWING, K. J. and MARCUS, M. G. (1966) Cartographic representation and symbolization in glacier mapping. *Can. J. Earth Sci.*, **3**, 761–70.

FALCONER, G. (1962) Patterned ground under ice fields. *J. Glaciol.*, **4**, 238–9.

FLOHN, H. (1969) Ein geophysikalisches Eiszeit-Modell. *Eiszeitalter und Gegenwart*, **20**, 204–31.

FRISTRUP, B. (1952) Climate and glaciology of Peary Land, north Greenland. *Int. Ass. Sci. Hydrol. Pub.* (Commission of Snow and Ice), 1, 185–93.

FRISTRUP, B. (1966) *The Greenland Ice Cap*. Copenhagen: Rhodos. 312 pp.

GLACIOLOGY PANEL, COMMITTEE ON POLAR RESEARCH (1967) Glaciology in the Arctic. *Trans. Amer. Geophys. Union*, **48**, 759–67.

GLEN, A. R. (1941) A sub-arctic glacier cap: the West Ice of North East Land. *Geogr. J.*, **98**, 65–76, 135–46.

GREW, E. and MELLOR, M. (1969) High snowfields of the St Elias Mountains. *Icefield Ranges Research Project, Scientific Results*, **1**, 75–88. Amer. Geogr. Soc.

GROSVAL'D, M. G. and KOTLYAKOV, V. M. (1969) Present glaciers in the USSR and some data on their mass balance. *J. Glaciol.*, **8**, 9–22.

HATTERSLEY-SMITH, G. (1963a) Climatic inferences from firn studies in northern Ellesmere Island. *Geogr. Ann.*, **45**, 139–51.

HATTERSLEY-SMITH, G. (1963b) The Ward Hunt Ice Shelf: recent changes of the ice front. *J. Glaciol.*, **4**, 415–24.

HATTERSLEY-SMITH, G. (1964) Rapid advance of glacier in northern Ellesmere Island. *Nature*, **201**, 176.

HATTERSLEY-SMITH, G. (1966) The Symposium on Glacier Mapping. *Can. J. Earth Sci.*, **3**, 737–42.

HATTERSLEY-SMITH, G. (1967) Canada. *Ann. Int. Geophys. Year*, **41**, 71–4.

HATTERSLEY-SMITH, G. (1969) Glacial features of Tanquary Fiord and adjoining areas of northern Ellesmere Island, NWT. *J. Glaciol.*, **8**, 23–50.

220   GEOFFREY HATTERSLEY-SMITH

HATTERSLEY-SMITH, G. and SERSON, H. (1970) Mass balance of the Ward Hunt Ice Shelf: a 10 year record. *J. Glaciol.*, **9**, 247–52.

HATTERSLEY-SMITH, G., CRARY, A. P. and CHRISTIE, R. L. (1955) Northern Ellesmere Island, 1953 and 1954. *Arctic*, **8**, 2–36.

HELK, J. V. (1966) Glacier mapping in Greenland. *Can. J. Earth Sci.*, **3**, 771–4.

HENOCH, W. E. S. (1969) Topographic maps of Canada in glaciological research. *Can. Cartogr.*, **6**, 118–30.

HOLDSWORTH, G. (1971) Calving from Ward Hunt Ice Shelf, 1961–62. *Can. J. Earth Sci.*, **8**, 299–305.

HORVATH, E. V. and FIELD, W. O. (1969) References to glacier surges in North America. *Can. J. Earth Sci.*, **6**, 831–40.

IVES, J. D. (1962) Indications of recent extensive glacierization in north-central Baffin Island, NWT. *J. Glaciol.*, **4**, 197–205.

IVES, J. D. and ANDREWS, J. T. (1963) Studies in the physical geography of north-central Baffin Island, NWT. *Geogr. Bull.*, **19**, 5–48.

JENNINGS, J. N. (1948) Glacier retreat in Jan Mayen. *J. Glaciol.*, **1**, 167–81.

JOHNSEN, S. J., DANSGAARD, W., CLAUSEN, H. B. and LANGWAY, C. C. (1970) Climatic oscillations 1200–2000 AD. *Nature*, **227**, 482–3.

JONES, G. H. S. and DIEHL, C. H. H. (1971) Iceberg tracking off Labrador. *Nature*, **229**, 189–90.

KEELER, C. M. (1959) Notes on the geology of the McCall valley area. *Arctic*, **12**, 87–97.

KINSMAN, D. J. J. and SHEARD, J. W. (1963) The glaciers of Jan Mayen. *J. Glaciol.*, **4**, 439–48.

KOCH, L. (1928) Contributions to the glaciology of north Greenland. *Medd. om Grønland*, **46**, 1–77.

KOENIG, L. S., GREENAWAY, K. R., DUNBAR, M. and HATTERSLEY-SMITH, G. (1952) Arctic ice islands. *Arctic*, **5**, 67–103.

KOERNER, R. M. (1966) Accumulation on the Devon Island ice cap, Northwest Territories, Canada. *J. Glaciol.*, **6**, 383–92.

KOERNER, R. M. (1970) The mass balance of the Devon Island ice cap, Northwest Territories, Canada, 1961–66. *J. Glaciol.*, **9**, 325–36.

KONECNY, G. (1966) Applications of photogrammetry to surveys of glaciers in Canada and Alaska. *Can. J. Earth Sci.*, **3**, 783–98.

LAMB, H. H., PROBERT-JONES, J. R. and SHEARD, J. W. (1962) A new advance of the Jan Mayen glaciers and a remarkable increase of precipitation. *J. Glaciol.*, **4**, 355–66.

LANGWAY, C. C. (1970a) Features and measurements of deep polar ice cores obtained by various drilling methods. *Int. Ass. Sci. Hydrol. Pub.* (Commission of Snow and Ice), 86, 77.

LANGWAY, C. C. (1970b) Stratigraphic analysis of a deep ice core from Greenland. *Geol. Soc. Amer. Spec. Pap.*, 125. 186 pp.

LIESTØL, O. (1969) Glacier surges in West Spitsbergen. *Can. J. Earth Sci.*, **6**, 895–8.

LOEWE, F. (1936) Höhenverhältnisse und massenhaushalt des Grönländischen Inlandeises. *Beitr. Geophys.*, **46**, 317–30.

LØKEN, O. H. and ANDREWS, J. T. (1966) Glaciology and chronology of fluctuations of the ice margin at the south end of the Barnes Ice Cap, Baffin Island, NWT. *Geogr. Bull.*, **8**, 341–59.

LØKEN, O. H. and HODGSON, D. A. (1971) On the submarine geomorphology along the east coast of Baffin Island. *Can. J. Earth Sci.*, **8**, 185–95.

LØKEN, O. H. and SAGAR, R. B. (1968) Mass balance observations on the Barnes Ice Cap, Baffin Island, Canada. *Int. Ass. Sci. Hydrol. Pub.* (Commission of Snow and Ice), 79, 282–91.

LYONS, J. B., SAVIN, S. M. and TAMBURI, A. J. (1971) Basement ice, Ward Hunt Ice Shelf, Ellesmere Island, Canada. *J. Glaciol.*, **10**, 93–100.

MARCUS, M. G. (1965) Summer temperature relationships along a transect in the St Elias Mountains. *Univ. of Colorado Stud., Ser. in Earth Sci.*, 3, 15–30.

MEIER, M. F. and POST, A. (1969) What are glacier surges? *Can. J. Earth Sci.*, **6**, 807–18.

MERCER, J. H. (1956) The Grinnell and Terra Nivea ice caps, Baffin Island. *J. Glaciol.*, **2**, 653–6.

MILLER, M. M. (1970) 1946–1962 survey of the regional pattern of Alaskan glacier variations. *Nat. Geogr. Soc. Res. Rep., 1961–1962 Projects*, 167–89.

MÜLLER, F. (1962a) Zonation in the accumulation area of the glaciers of Axel Heiberg Island, NWT, Canada. *J. Glaciol.*, **4**, 302–10.

MÜLLER, F. (1962b) Glacier mass-budget studies on Axel Heiberg Island, Canadian Arctic Archipelago. *Int. Ass. Sci. Hydrol. Pub.* (Commission of Snow and Ice) 58, 131–42.

MÜLLER, F. (1963a) Englacial temperature measurements on Axel Heiberg Island, Canadian Arctic Archipelago. *Int. Ass. Sci. Hydrol. Pub.* (Commission of Snow and Ice), 61, 168–80.

MÜLLER, F. (1963b) Investigations in an ice shaft in the accumulation area of the McGill Ice Cap. *Axel Heiberg Island Res. Rep. Prelim. Rep., 1961–1962*, 27–36. McGill Univ.

MÜLLER, F. (1969) Introduction: seminar on the causes and mechanics of glacier surges. *Can. J. Earth Sci.*, **6**, iii–iv.

NUTT, D. C. (1966) The Drift of ice island WH-5. *Arctic*, **19**, 244–62.

NYE, J. F. (1959) The motion of ice sheets and glaciers. *J. Glaciol.*, **3**, 493–507.

NYE, J. F. (1960) The response of glaciers and ice sheets to seasonal and climatic changes. *Proc. Roy. Soc. London*, A, 256, 559–84.

NYE, J. F. (1963) Correction factor for accumulation measured by the thickness of the annual layers in an ice sheet. *J. Glaciol.*, **4**, 785–8.

OMMANEY, C. S. L. (1969) A study in glacier inventory. The ice masses of Axel Heiberg Island, Canadian Arctic Archipelago. *Axel Heiberg Res. Rep.* McGill Univ. 105 pp.

OMMANEY, C. S. L. (1971) The Canadian glacier inventory. *In: Glaciers, Proc. of Workshop Seminar, 1970*, 23–30. Can. Nat. Comm./Int. Hydrol. Decade.

ORVIG, S. and MASON, R. W. (1963) Ice temperatures and heat flux, McCall Glacier, Alaska. *Int. Ass. Sci. Hydrol. Pub.* (Commission of Snow and Ice), 61, 181–8.

PATERSON, W. S. B. (1965) The Meighen Ice Cap, Arctic Canada: accumulation, ablation and flow. *J. Glaciol.*, **8**, 341–52.

PORSILD, A. E. (1955) The vascular plants of the western Canadian Arctic Archipelago. *Nat. Mus. Can. Bull.*, 135. 226 pp.

REDPATH, B. B. (1961) Seismic operations. *In: Jacobsen-McGill Arctic Research Expedition to Axel Heiberg Island, Prelim. Rep. 1959–1960*, 101–7. Montreal.

ROBIN, G. DE Q., EVANS, S. and BAILEY, J. T. (1969) Interpretation of radio echo sounding in polar ice sheets. *Phil. Trans. Roy. Soc., London*, **265**, 437–505.

RUDBERG, S. (1963) Morphological processes and slope development in Axel Heiberg Island, Northwest Territories, Canada. *Nachrichten der Akademie der Wissenschaften in Göttingen, 2. Matematische-physikalische Klasse*, **14**, 211–18.

RUTTER, N. W. (1969) Comparison of moraines formed by surging and normal glaciers. *Can. J. Earth Sci.*, **6**, 991–9.

SAGAR, R. B. (1964) Meteorological and glaciological observations on the Gilman Glacier, northern Ellesmere Island, 1961. *Geogr. Bull.*, **6**, 13–56.

SAGAR, R. B. (1966) Glaciological and climatological studies on the Barnes Ice Cap, 1962–64. *Geogr. Bull.*, **8**, 3–47.

SANDFORD, K. S. (1955) Tabular icebergs between Spitsbergen and Franz Josef Land. *Geogr. J.*, **121**, 164–70.

SATER, J. E. (1969) *The Arctic Basin* (rev. edn). Washington, DC: Arctic Inst. N. Amer. 337 pp.

SCHYTT, V. (1964) Scientific results of the Swedish Glaciological Expedition to Nordaustlandet, Spitsbergen. 1957–1958, Part II. Glaciology: previous knowledge, accumulation and ablation. *Geogr. Ann.*, **46**, 245–81.

SCHYTT, V. (1969) Some comments on glacier surges in eastern Svalbard. *Can. J. Earth Sci.*, **6**, 867–74.

SHARP, R. P. (1956) Glaciers in the Arctic. *Arctic*, **9**, 78–117.

SIGBJARNARSON, G. (1970) On the recession of Vatnajökull. Paper presented at meeting of Glaciological Society and Icelandic Glaciological Society, Iceland, June 1970. *Ice*, **34**, 13.

SORGE, E. (1935) Glaziologische Untersuchungen in Eismitte. *Wissenschaftliche ergebnisse der Deutschen Grönland-Expedition Alfred Wegener 1929 und 1930–31*, **3**, 68–270.

STANLEY, A. (1969) Observations of the surge of Steele Glacier, Yukon Territory, Canada. *Can. J. Earth Sci.*, **6**, 819–30.

SWITHINBANK, C. W. M. (1969) Giant icebergs in the Weddell Sea. *Polar Rec.*, **14**, 477–8.

THORARINSSON, S. (1954) The Jökulhlaup (Glacier burst) from Grimsvötn in July, 1954. *Jökull*, **4**, 37.

THORARINSSON, S. (1969) Glacier surges in Iceland, with special references to the surges of Bruarjökull. *Can. J. Earth Sci.*, **6**, 875–82.

THORARINSSON, S. and SIGUROSSON, S. (1947) Volcano-glaciological investigations in Iceland during the last decade. *Polar Rec.*, **5**, 60–6.

UNESCO/IASH (1969) Variations of existing glaciers. *Tech. Pap. in Hydrol.*, 3. 19 pp.

UNESCO/IASH (1970a) Perennial ice and snow masses: a guide for compilation and assemblage of data for a world inventory. *Tech. Pap. in Hydrol.*, 1. 59 pp.

UNESCO/IASH (1970b) Combined heat. ice and water balances at selected glacier basins. *Tech. Pap. in Hydrol.*, 5. 20 pp.

VIBE, C. (1967) Arctic animals in relation to climatic fluctuations. *Medd. om Grønland*, **170**, 1–227.

WAGNER, W. P. (1969) Snow facies and stratigraphy on the Kaskawulsh Glacier. *Icefield Ranges Research Project, Scientific Results*, 1, 55–62. Amer. Geogr. Soc.

WAITE, A. H. (1966) International experiments in glacier sounding, 1963 and 1964. *Can. J. Earth Sci.*, **3**, 887–92.

WARD, W. H. and BAIRD, P. D. (1954) Studies in glacier physics on the Penny Ice Cap, Baffin Island, 1953. Part I: A description of the Penny Ice Cap, its accumulation and ablation. *J. Glaciol.*, **2**, 342–55.

WEBER, J. R. and ANDRIEUX, P. (1970) Radar soundings on the Penny Ice Cap, Baffin Island. *J. Glaciol.*, **9**, 49–54.

WEERTMAN, J. (1957) On the sliding of glaciers. *J. Glaciol.*, **5**, 33–8.

WEERTMAN, J. (1968) Comparison between measured and theoretical temperature profiles of the Camp Century, Greenland, borehole. *J. Geophys. Res.*, **73**, 2691–700.

WILSON, A. T. (1964) Origin of ice ages: an ice shelf theory for Pleistocene glaciation. *Nature*, **201**, 147–9.

WOOD, W. A. (1963) The Icefield Ranges Research Project. *Geogr. Rev.*, **53**, 163–84.

WORDIE, J. M. (1950) Barrier *versus* shelf. *J. Glaciol.*, **8**, 416–20.

Manuscript received January 1972.

Present alpine ice cover

Gunnar Østrem

Norwegian Water Resources
and Electricity Board, Oslo

1. Introduction

Large amounts of fresh water are stored as solid ice in the Arctic and in the alpine regions of the world. The areal distribution of the ice is extremely irregular, and this is especially true of the comparatively small ice masses of the alpine regions, most of which cannot even be shown on maps of the scale permissible in this book. With this limitation in mind, figs. 4B.1 and 2 show glacier distribution in the northern hemisphere. The overwhelming dominance of the Greenland Inlandsis, with far greater mass than all other extra-Antarctic ice added together, will be readily apparent (see Hattersley-Smith, Chapter 4B). Alpine glaciers occur in practically all the world's principal mountain ranges, from the equatorial massifs of East Africa to the tiny drift glaciers of the Colorado Rocky Mountains, so that a rational description of their distribution and condition faces the same problems as an account of alpine permafrost (see Ives, Chapter 4A). Because of this difficulty, this section will concentrate on a conceptual discussion of alpine glaciers together with consideration of past and present conditions and vitally interrelated aspects such as regional glaciation levels and regional snowlines.

During the period of approximately 25,000 to 8000 years ago ice masses similar to the present Greenland Inlandsis covered vast areas of the European and North American continents. At the same time the strictly alpine ice masses were both more extensive and more numerous. These ice masses produced striations in bedrock and drumlins in softer materials as they moved across the country, and they left moraines, eskers and other deposits behind them when they melted. Many of these features have proved to be of value to human activity – eskers and outwash deposits are good aquifers, moraines contain building materials, etc. For this reason and for purely scientific ones, geographers and geologists are interested in studies of glacial landforms both in areas where no ice occurs at present and in areas of contemporary glacierization.

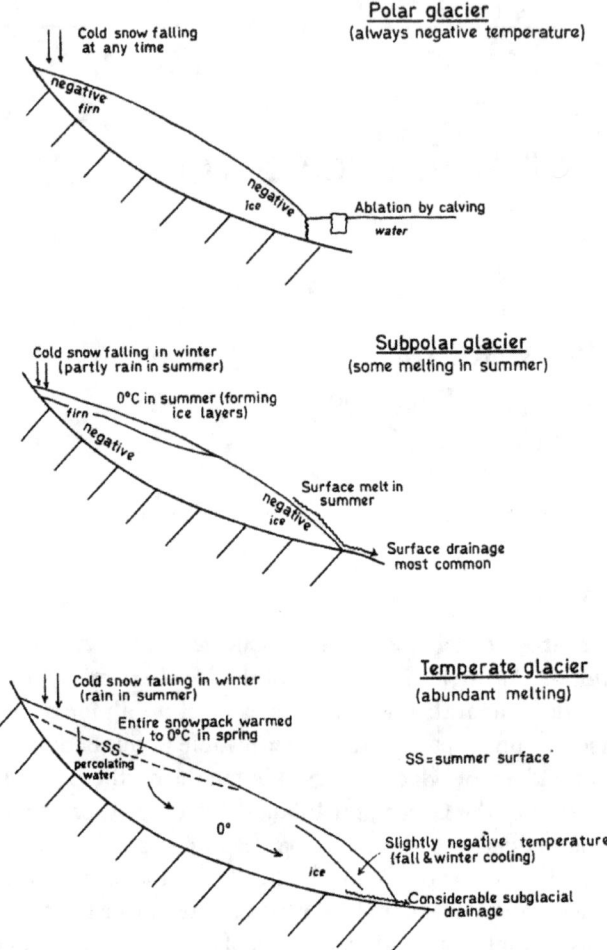

Fig. 4C.1. Three glacier types, according to the thermal classification system (a modified Ahlmann system). Polar glaciers are in regions of no positive temperatures. Ablation takes place mainly by calving, partly by evaporation. Subpolar glaciers experience limited melting during a fairly short summer season. This causes settling and refreezing in the upper parts, some surface melt and runoff from the lower parts. The main glacier body is always at sub-zero temperatures. Temperate glaciers are in general at melting point except at the tongue where ice temperatures are slightly below zero during certain times of the year.

The concept *alpine glacier* probably has its origin in the type of glacier that is most common in the Alps, where numerous valley glaciers drain high mountain accumulation basins. To some people the term 'alpine glacier' may also be associated with a glacier situated in an area of less severe climate than is the case for high arctic glaciers. Thus the term contains both a definition of topography (valley glacier) as well as a possible indication of climatic conditions. Consequently, the term may not be strictly adequate to define a particular type of glacier, although it has been used widely and is still used by geographers and geomorphologists.

2. Glacier classification

Various classification systems have been proposed to divide glaciers into separate groups (see also Hattersley-Smith, Chapter 4B). Some of these systems are based upon thermal conditions within the glacier, others deal with the outline configuration and/or the morphology of the ice body. From a glaciological point of view a thermal classification may be the most satisfactory, because it is based upon measurable physical conditions and not on more or less subjective judgements, which is the case for morphological classification systems. Which of the two systems is used, however, may be of less importance – in fact they may be used simultaneously because there is very little relation between thermal conditions and glacier shape.

The first classification system based upon thermal conditions was proposed by Ahlmann (1935). According to this system glaciers are divided into three main groups:

(1) *Polar glaciers*, which are present in areas where no melt takes place, the temperature being always below freezing point. Consequently, in the extreme form of these glaciers snow transforms into ice without passing through the liquid phase. This is the opposite of what occurs on other glaciers where liquid water often refreezes, forming ice layers in the firn basin or superimposed ice on the glacier tongue (see below). The ice masses that are produced in a polar glacier would create a continuously growing ice body, unless material was removed by other means than melt. Positive temperatures never occur, and consequently no melt can ever take place.

Removal of 'excess ice' from a polar glacier occurs through the formation of icebergs; where ice streams meet the sea, large pieces break off. Apart from evaporation, this mechanism is the only ablation process possible for a polar glacier. The best known example is the Antarctic Inland Ice where ice breaks off into the ocean and large pieces are carried away by sea currents as icebergs. In the northern hemisphere several glaciers on arctic islands are of the polar type, in general draining their ice into the Arctic Ocean. Such glaciers are found on northern Ellesmere Island, NWT, on northern Greenland, Svalbard and the Russian Arctic Islands.

(2) *Subpolar glaciers* are also characterized by low temperatures, although some melting takes place during a short summer period and certain quantities of meltwater are produced. However, the amount of energy that is transferred to the glacier during the melt season is too small to compensate for the great 'cold storage' in the glacier, originating from low temperatures during a long, cold winter period.

Positive air temperatures may be encountered on the entire glacier surface during certain periods in the summer, and this causes some melt, even in the upper firn basin. The meltwater so produced will percolate down into the cold snow for a relatively short distance and refreeze at shallow depths. When the water freezes – thus forming ice layers in the snow – energy is released (80 cal g^{-1} of freezing water) which will warm the snow, but be insufficient to increase the temperature of the entire snowpack to 0°C. Thus the remainder of the previous winter's snow will retain negative temperatures at the end of the summer. Consequently, the main part of the firn basin will remain at sub-zero temperatures all year round. In lower areas of the glacier, percolating meltwater will freeze onto the ice surface – i.e. to the cold glacier surface buried under the previous winter's snow. Thus a layer of different ice is formed on top of the old glacier ice; this ice is termed *superimposed ice*. It has smaller crystals than glacier ice and air bubbles are arranged differently. The energy released by formation of superimposed ice is insufficient to increase the ice temperature by any considerable amount: it remains below zero.

On the lowest parts of these glaciers, all winter snow may disappear and meltwater channels form on the ice surface. The underlying ice is so cold that running meltwater can hardly melt its way down to the bottom of the glacier. In other words, *if* meltwater could penetrate into the ice, part of it would freeze there and hinder further down-cutting. Therefore, abundance of meltwater channels on the glacier surface is a fairly typical feature on subpolar glaciers. Such channels can be seen on temperate glaciers also, but in general meltwater will more easily cut its way down in temperate ice, thus forming intraglacial or subglacial drainage systems. Such systems are less pronounced for subpolar glaciers.

(3) *Temperate glaciers* are generally situated in areas of fairly long or warm summers, so that they receive considerable amounts of energy during the melt season, sufficient to produce abundant meltwater all over the glacier. This water, in turn, will be able to warm up the whole snowpack from the previous winter's accumulation by percolating, refreezing and subsequent release of energy at various depths. The refreezing process is in fact almost the only means of vertical heat transport into a glacier. For most temperate glaciers the quantity of meltwater produced at the surface is more than sufficient to raise the entire snowpack temperature to 0°C. The excess water – i.e. the water that does not freeze within the snow – will penetrate further down and eventually drain from the glacier.

The considerable amounts of meltwater that are produced by ice melt on the glacier tongue have very little thermal impact on the solid glacier ice. However, this ice was formed from old snow (firn) at the freezing point (as explained above); thus it will hold a temperature at or very close to the pressure melting point. Theoretically, the entire glacier would therefore have a temperature at or very close to 0°C.

At the beginning of the winter some energy will be lost from the exposed ice by out-going radiation, which will lower its temperature. Consequently, if temperature measurements are made in the solid ice on the tongue of a temperate glacier, slightly negative temperatures will be encountered in the upper layers. This is the explanation for the fact that stakes drilled into the ice may freeze solid, even in the middle of summer.

A sketch showing the three glacier types and their temperature conditions is given as fig. 4C.1. This thermal classification system is very simplified and several attempts have been made to make it more comprehensive by adding more groups between the three main ones. Compare for example Avsiuk (1955), Benson (1962), Müller (1962) and Stenborg (1965). It should also be borne in mind that at extreme altitudes, such as occur in the Himalaya or Andes, temperature conditions will prevail such that alpine glaciers may possess the high-mountain equivalent to the subpolar, and even polar, type.

Classification systems based upon glacier morphology have developed gradually during several years; some of them were more or less designed to fit glaciers in a given region, and served this purpose very well. In Europe, for example, Swiss types, Norwegian types, etc., were enumerated. Nevertheless, the demands for worldwide systematic glacier inventories in recent years have made it highly desirable to develop a standard morphologic system that can be applied to glaciers in all regions.

Several attempts were made to standardize the fairly comprehensive vocabulary used in glacier classification. This was not a simple task, because the same term could have a different meaning when used by different authors. During the International Hydrological Decade, this standardization made a great step forward in connection with glacier inventories. Firm borders were established between selected terms, and a complete system of definitions was published by UNESCO (1970). According to this system all kinds of glaciers should be divided in the following main groups:

(1) *Continental ice sheet* which inundates areas of continental size.
(2) *Icefield* which is defined as an ice mass of sheet or blanket type of a thickness not sufficient to obscure the subsurface topography.
(3) *Ice cap* defined as a dome-shaped ice mass with radial flow.
(4) *Outlet glacier* which drains an ice sheet or ice cap, usually of valley glacier form; the catchment area may not be clearly delineated.
(5) *Valley glacier* is a mass of ice that flows down a valley; the catchment area is well defined.
(6) *Mountain glacier* is a term that would include cirque glaciers or niche glaciers and even the crater type glacier. This type also includes groups of small units as well as ice aprons.

In addition to these six types of 'real' glaciers there is room provided for *glacierets* (small ice masses developed from snow drifting, avalanches, etc., and that may exist for only a few years), *ice shelves*, which are floating ice masses mainly nourished by glaciers, and, finally, *rock glaciers*.

No space has been allocated for 'alpine' glaciers in this system, but it seems that this type would probably best fit in among valley glaciers and mountain glaciers (groups 5 and 6). Drift glaciers (Ural-type) as described by Dolgushin (1961) and Outcault and MacPhail (1965) may fit into group 6 or the category of glacieret, according to their permanence and activity (Plate 14).

In the classification system there are also various terms introduced to describe in detail the outline or morphology of the basins. For example, a *simple basin* is defined as a single accumulation basin that is feeding one glacier, whereas a *compound basin* is defined as two or more individual accumulation basins feeding one glacier, etc. For further details see UNESCO (1970, pp. 16–18).

3. The glaciation level

The glaciation level is defined as the critical summit elevation which is necessary to produce a glacier on a mountain.

If solid precipitation in the form of snow, hail or rime exceeds the total summer melt, the snowpack will grow thicker and thicker and, provided topographic conditions are favourable, a glacier will form. There are obviously two major factors that govern the formation of glaciers: the amount of winter accumulation, and the extent of summer melt.

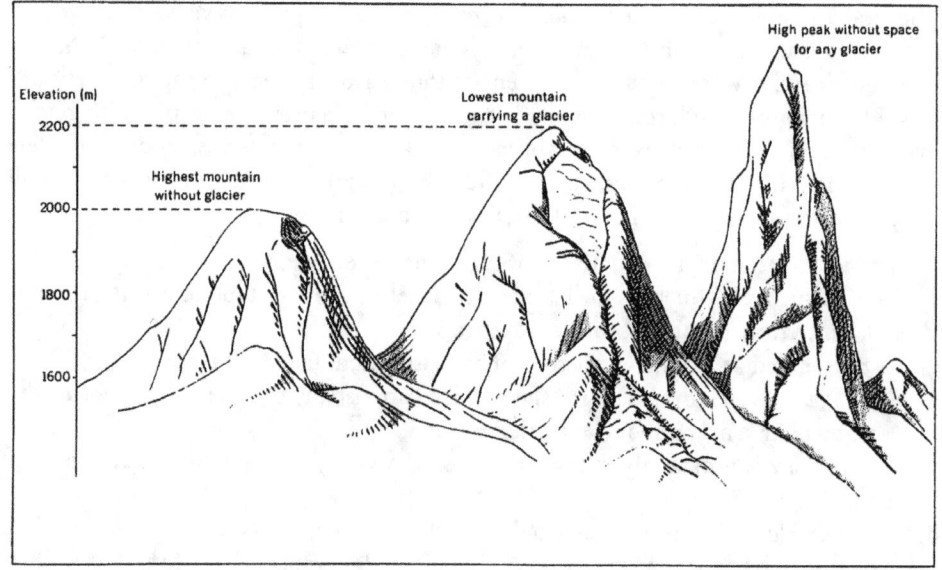

Fig. 4C.2. Method of determining the critical height of 'glaciation level'. Only mountains of favourable topographic conditions are considered when inventories are made from information on topographic maps (see text).

Precipitation increases generally towards higher altitudes and towards maritime areas, while the proportion of solid to liquid precipitation increases with increasing altitude and/or increasing latitude. It is most likely, therefore, that heavy snow precipitation will occur at high altitudes and particularly in coastal areas. The other main factor, the summer melt, will normally decrease towards higher altitudes. Consequently, by considering the two factors simultaneously, glaciers should be most common in high mountain areas under maritime climatic conditions. However, there will exist a lower level under which the precipitation (or more correctly, the snow accumulation) no longer exceeds the summer melt, and no glacier can form. Conversely, above this critical height glaciers may form, provided topographical conditions are favourable. This critical height has been named differently by various authors. Brückner (1887) was the first to indicate that such a critical level must exist and he termed it 'die Schnee-linie'. Enquist (1916) used the term 'Vergletscherungsgrenze' – *limit of glacierization* – for the same concept. When Ahlmann (1924) determined critical summit heights for Scandinavia he used the term 'le niveau de glaciation'; later he used the English term *glaciation limit*. This term has unfortunately been used also to indicate the outer limit of former

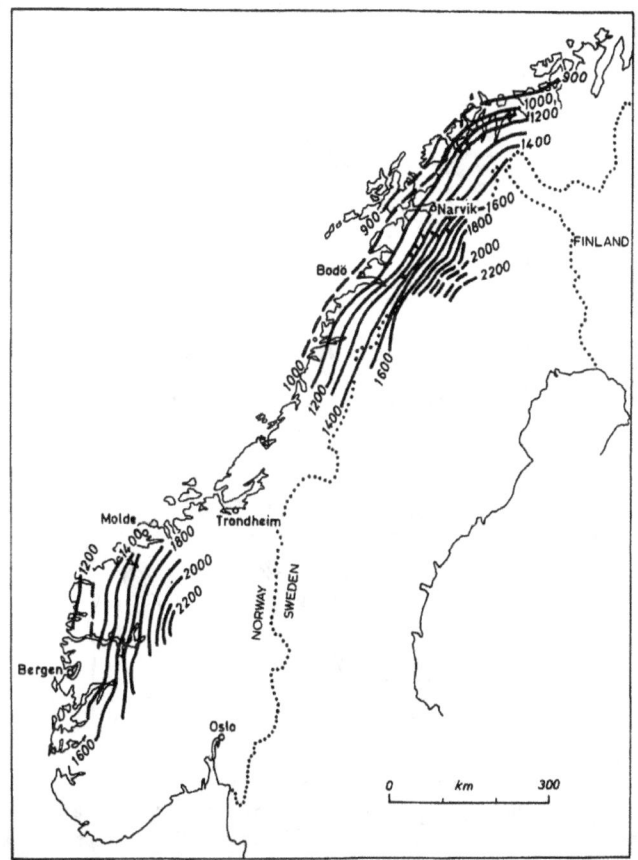

Fig. 4C.3. Height (m) of the glaciation level in Scandinavia. Continuous lines are based upon several determinations taken from recent topographic maps. Broken lines are based upon single determinations (in areas of few glaciers) or indicate interpolated heights. For details see Østrem, 1964.

glaciation (as shown on maps) of Pleistocene glaciers, for example. To avoid confusion, it is now proposed that the concept should be termed *glaciation level* (see fig. 4C.2).

It may be difficult to determine in detail the critical mountain elevation that is necessary to produce a glacier (it is not the same as the height of the glacier); on north-facing slopes it may be higher than on south-facing slopes, for instance. For practical reasons, it has been agreed that the height of the glaciation level shall be determined from the *summit* elevation of mountains where glaciers are formed. For example, if one mountain 2000 m high carries no glacier, although topographic conditions are favourable (i.e. sufficient space on the mountain slopes), and in the same area another mountain with a summit elevation of 2200 m does carry a glacier, it is assumed that the critical height in this case would be in the order of 2100 m (compare fig. 4C.2). Note that the glacier itself will be situated at a lower elevation – the only height used in the determinations is the summit elevation, thus the method has been called 'the summit method'.

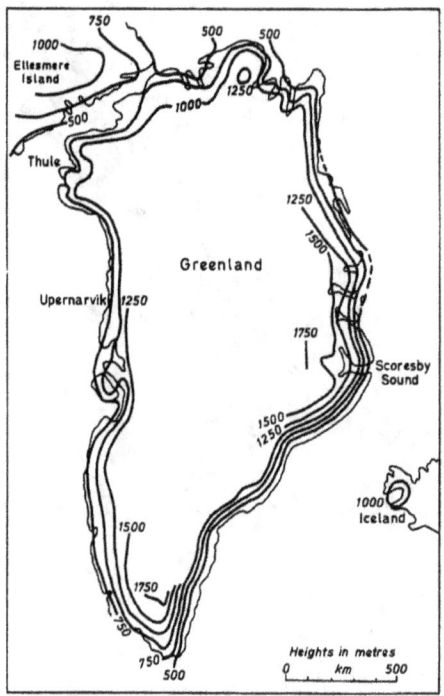

Fig. 4C.4. Height of the glaciation level in Greenland and adjacent areas (according to A. Weidick, personal communication, 1971). There are three areas of maximum height, indicating a high degree of continentality: the highland in the south, inland areas west of Scoresby Sound and on Peary Land in the north.

To determine the height of the glaciation level it is necessary to examine good topographic maps, find the highest mountain in a given area without glaciers (but with favourable topography) and locate the lowest mountain that carries a glacier in the same area. It is then assumed that the critical height lies somewhere between the two mountain summit elevations.

Several authors have investigated the height of the glaciation level in various parts of the world. This is a time-consuming work and, due to lack of good topographic maps in many areas, this kind of study has been limited to certain areas only. Examples of

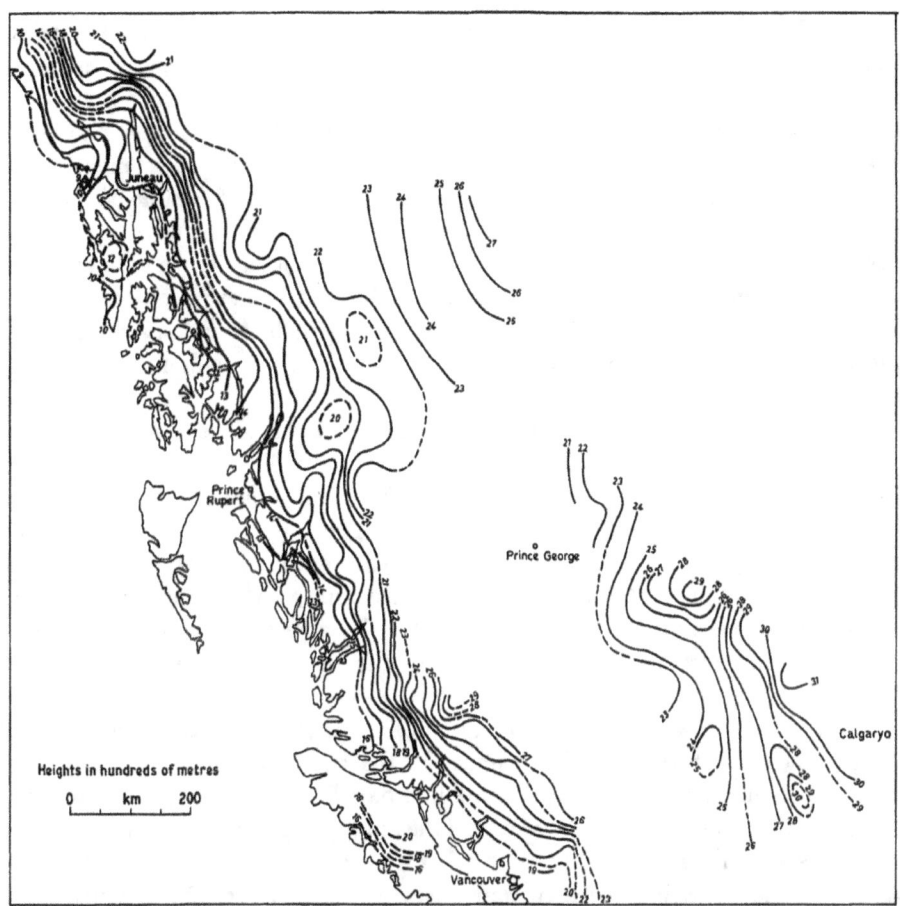

Fig. 4C.5. Height of the glaciation level in western Canada and part of Alaska. The general pattern of increased levels from the coast inland is disturbed in several places. Most prominent is a decrease in glaciation level east of Prince Rupert, where moist airmasses seem to have easier access through a comparatively lower section of the coast mountains. Note also that the glaciation level decreases when passing from west to east across the interior plateau south of Prince George.

results from such studies are shown in figs. 4C.3, 4C.4 and 4C.5. From these results it can be seen clearly that the height of the glaciation level, although generally decreasing towards higher latitudes, depends more on variations in continentality than on latitude. The concept 'glaciation level', therefore, may also be used as a convenient indicator to determine the degree of continentality for a given region.

4. Snowlines

In glaciological literature and elsewhere the term *snowline* is often used, but various authors have used the same term to define different concepts (see also Heuberger,

Chapter 6B). This has naturally caused much confusion, and in recent years the term has been given prefixes to define more clearly what the author means when he uses the term. Expressions like 'present snowline', 'climatic snowline', 'orographic snowline', 'temporary snowline', 'regional snowline', have been introduced, but they may still not be understood easily unless they are defined in more detail. Some of the main concepts will now be explained, to indicate those expressions that should if possible be avoided.

The *transient snowline* is defined as the border at any given time between a snow-free area and a snow-covered area, either on a glacier or on adjacent ground. This means that the transient snowline may be close to sea level in wintertime, when the whole country is covered by snow, and it may be situated far up in the mountains in the summertime when the lowlands are snow-free and only the highest mountains are still snow-covered.

Although this definition is very simple, direct observation of the position of the transient snowline and its variations with time would be a difficult task on glacier-free ground, because more or less extensive snowbanks would complicate the picture. They would remain a long time after the transient snowline proper had passed through the area. Further, variations in topography turn the theoretical, almost horizontal line into an extremely complicated 'line' or surface cutting through the landscape. In most cases it has proved impossible to make direct observations of transient snowline elevations, except in autumn when new snow covers the mountains down to a well defined altitude.

On the glacier, however, it is in general much easier to observe the slowly migrating transient snowline throughout the summer. In the spring the whole glacier surface is covered by white, clean snow. During the first part of the summer a small portion of glacier ice will be visible on the lower part of the tongue and, in most cases, the transition between exposed ice and the previous winter's snow will be clearly visible on air photographs as a line across the glacier. In many cases this line almost coincides with a contour line, thus making it possible to determine the height of the transient snowline as of the date of photography. Throughout the summer more and more ice is being exposed, thus pushing the transient snowline to higher altitudes on the glacier surface. The transition between exposed ice and the previous winter's snow is always easily determined, but when the transient snowline is situated at higher elevations, e.g. in the firn area, the contrast between firn and the previous winter's snow is less pronounced. However, the border can still be seen on most good air photographs and its height can be determined by comparison with a topographic map (Plate 15).

At the end of the summer, the transient snowline will be situated at a certain, relatively high, elevation on the glacier. This highest position may be termed 'the annual snowline' for that particular season – particularly if its height could be determined for a glacier-free area. Due to the cooling effect of ice and snow, it will be lower on the glacier than on bare ground. The vertical distance between them is supposed to be in the order of 100 m (von Klebelsberg, 1948, p. 31). The first snowfall at the end of the summer will immediately extinguish this highest position of that season's transient snowline and bring it down to a lower altitude; eventually down to sea level in cold regions.

The final height of the transient snowline will be different from year to year according to the relative amount of winter snow and summer melt in various years. If one could establish the *average* elevation of the highest (final) transient snowline on bare ground throughout several years (ideally for a thirty-year period) this elevation would be termed the *climatic snowline* for that area in the said period. Such determinations would require extensive end-of-summer-season air photography during several years, or good height determinations in a year when 'normal' conditions have prevailed for both winter snow cover and summer melt – conditions that will almost never occur. Consequently, the establishment of the height of a climatic snowline on glacier-free ground is indeed a difficult task. For glaciers it is less difficult because determination of the transient snowline is so much easier there. Assuming that a certain constant vertical relation exists between corresponding snowlines on glaciers and on adjacent glacier-free ground, one can use observations on glaciers to determine the climatic snowline for a given area.

There may also be a similar relation between the glaciation level and the climatic snowline. Charlesworth (1957, p. 11) states that the former, determined by the summit method (see above), is situated some 100 m above the climatic snowline for a given area. If both methods are used – *deducting* 100 m from the height of the glaciation level for a given area, and *adding* 100 m to the average highest annual position of the transient snowline on glaciers in the area (averaged over thirty years) – one should be able to obtain at least a fairly good upper and lower limit for the height of the climatic snowline.

As indicated, the term 'snowline' has been given several slightly different definitions in glaciological literature. This is very unfortunate because misinterpretations may easily arise. Therefore, it should be stressed that if the term 'snowline' (without prefix) is used it has to be clearly defined in the context. The lower limit of the previous winter's snow on a glacier, as deduced from visual inspection or from good air photographs, should be termed *transient snowline* – a term which should also be easy to understand and to use in a consistent manner. The risk of misinterpretation of the term is small and it is recommended that this be used whenever possible and particularly in cases where the migrating border between snow cover and uncovered ground is discussed (Plate 15).

The *equilibrium line* is by definition a line (or a zone) on the glacier where the previous winter's snow just disappeared during the subsequent summer. At the equilibrium line the glacier mass balance equals zero – i.e. the mass gain equals exactly the mass loss during the glaciological year in question. Above the equilibrium line there is an excess of winter snow, consequently the glacier increases its mass there. Below the equilibrium line the mass loss in the summer season exceeds the total winter accumulation, so that glacier is losing mass in that area. When the glacier is in a steady state condition, the mass gain above the snowline equals exactly the mass loss below it.

The equilibrium line is not a visible line on the glacier, so to establish its position it is necessary to perform detailed field studies of the mass exchange at numerous points on the glacier surface. Such studies involve measurements of *winter accumulation* (snow depth measurements and snow density observations) and *summer ablation* (the amount of mass removed from the glacier, generally by melt and evaporation). Based upon field studies and subsequent calculations it is then possible to determine the height of the equilibrium line for a given glacier; it is not always directly related to the position of the transient snowline.

In the literature, the term *firn line* is frequently used. This term also suffers from the fact that various authors have given a slightly different definition for it (Ahlmann, 1948, has also used the term 'firn limit'). In most cases this term coincides with the above-mentioned equilibrium line. However, because the expression may be misinterpreted as the upper edge of firn on a glacier, i.e. the border between firn (or ice) and the last winter's snow (which is identical to the transient snowline described above), it is recommended that the term 'firn line' also be clearly defined in the context. (To differentiate between 'firn line' in the sense of equilibrium line, and 'firn line' defined as the lower limit of last winter's snow, some authors have introduced the term *firn edge* as being valid for the latter concept.)

For glaciological mass balance studies and in connection with glacier hydrology, the term *equilibrium line* has a particular importance, and it can hardly be misunderstood. Therefore it is recommended to use this term wherever applicable.

5. Glacier mass balance and glacier variations

If all the snow falling on a glacier during one winter season completely melts during the subsequent summer, the glacier is said to be in a *steady state condition* because its total mass is kept constant for that year. This does not happen very often. For most years either the winter snow or the summer melt will be larger, thus increasing or decreasing the total mass of the glacier. If the glacier has increased its total mass during one winter and the subsequent summer, we say that the mass balance has been positive during that *glaciological year*. Similarly, a mass loss is termed negative mass balance.

Such annual variations, relatively small compared to the total mass of the whole glacier, are hard to detect without detailed glaciological studies. However, if the glacier is exposed to mass loss during several years, i.e. the mass balance happens to be negative for a long period, this will inevitably result in such a considerable decrease in glacier volume that this can be directly observed. The glacier will gradually become thinner and eventually disappear completely from areas that were earlier covered by permanent ice and snow. The most obvious and easily measurable feature connected with glacier decrease is a retreat of the glacier tongue, and, hydrologically, an increased water discharge in the glacier stream compared to normal flow conditions.

In other periods it may happen that glaciers gradually grow, due to several years of positive mass balance. Then their volumes increase and particularly the tongues begin to move forward. Glacier retreat has been one of the most widespread glaciological phenomena during the last few decades over most of the world, but only fifty to sixty years ago most North European glaciers showed a very marked advance (see Hoinkes, 1968, for example). The most impressive glacier advances, however, occurred in the 1750s. Large amounts of damage were then reported from several places in the world where the 'abominable ice' advanced down into previously unglacierized valleys and destroyed arable land before the frightened inhabitants. Farmers in western Norway lost so much of their land that they wrote to the King to obtain tax exemption due to extensive destruction caused by advancing glacier tongues (Grove, 1972).

GLACIER FLUCTUATIONS

Glacier fluctuations have taken place at all times because they are closely connected with fluctuations in climate. The recent tendency for glacier retreat is apparently connected with a rise in summer temperature in the northern hemisphere. This tendency, however, seems to have been somewhat weaker during the last decades, and some glaciologists are of the opinion that we reached or even passed the optimal point during the 1950s. This would mean that glaciers should gradually stop their present retreat, possibly pass through some years of steady state conditions, and finally turn into a period of common glacier advance. It is almost a fiction, however, to predict anything in this respect, but from glacier terminus observations on more than one hundred glaciers in the Alps, it seems that an increasing number of glaciers are now stagnant than was the case a few years ago. Also the number of advancing glaciers shows a definite increasing tendency, and this is probably a strong confirmation of the above assumption of a deterioration in present climate.

Although such terminus observations are restricted to the annual position of the glacier tongue, which is not year-by-year directly connected to the mass balance, this kind of statistic is valuable for long-term studies of glacier variations. To facilitate data storage and retrieval on a worldwide basis, an international organization to record glacier fluctuations has been established to collect and publish all available data in this respect (Kasser, 1967). This organization is supported by UNESCO.

There is always a considerable time lag between variations in mass balance and subsequent reactions at the glacier tongue – advance or retreat. Therefore, to study the annual mass variations for glaciers in a given region it is necessary to carry out detailed mass balance measurements on selected glaciers, representative of the region. By such studies it is possible to obtain data of annual glacier variations and to discover trends in glacier fluctuations. However, detailed mass balance measurements are fairly expensive and it is necessary, therefore, to select glaciers very carefully and be sure that continuous studies will be carried out for a long unbroken series of years.

STUDY METHODS

There are in principle two different methods to obtain mass balance data for a glacier. The simplest is to make a survey of the glacier surface each autumn; a total study should then be made of the remaining snow from last winter's accumulation. This equals the amount of water in solid form that is added to the glacier body. Similarly, one must assess the net mass loss, i.e. determine the total ice melt that has taken place during the summer. This is mainly by stake observations in the ablation area. Thus, by simple subtraction, the glacier's mass balance can be obtained for that year. If the two masses are equal the glacier has been in equilibrium during the year in question. If the amount of remaining snow exceeds the amount of ice melt, the mass balance has been positive, and vice versa.

Such a method to obtain the glacier mass balance has been used for several years in the Alps and elsewhere. However, by making only this single general survey per year it is almost impossible to detect whether a positive balance is a result of unusually high winter accumulation or of slight summer melt. These two important factors cannot be separated by this method, only their *net* influence on the glacier is determined.

For scientific reasons, particularly for the study of relationships between climate and glacier behaviour, it is important to perform *separate* measurements of the glacier's 'income' (winter accumulation) and its 'expenses' (summer ablation). It is then possible to consider the reason for a 'positive' year – whether it is a result of unusually heavy winter snow or a relatively cool summer.

It is obvious that this more complete mass balance study gives a far better picture of reasons and effects; winter and summer conditions are treated separately and their individual influence on glacier behaviour can be studied in more detail. However, this method involves at least two general glacier surveys: one in spring, and one in autumn. At the first occasion the total winter snow accumulation must be measured; at the latter, all the summer melt.[1] In the following pages some details will be shown related to the fieldwork and methods of reporting results.

Accumulation measurements

In a complete mass balance study, a detailed survey must be made of the total mass of snow that has fallen on the entire glacier surface during each winter season. Due to wind action the snow will form an uneven cover on the glacier, and the local snow distribution pattern may change from year to year. It is, therefore, necessary, each spring, to perform snow depth measurements at a large number of single points distributed over the glacier surface to map all snow thickness variations. In practice, this is carried out by using a snow probe (ramsonde). This is pushed vertically into the snow and thicknesses are easily measured, provided the previous summer surface can be distinguished. This is normally the case because the firn (defined as snow that has survived at least one melt season) has a much higher density than the previous winter's snow and it is generally covered by an icy crust formed during late summer. This so-called *summer surface* forms a distinctive datum above which fresh snow rests as a white, less dense blanket.

In the lower part of the glacier it is especially easy to determine the thickness of the previous winter's snow because the solid ice is easily detected by the probe and almost no possibilities exist for misinterpretation. The snow survey is normally carried out along traverses or probing profiles that are traced on a map.

The density of the snow is not always the same in the upper and lower part of a glacier, so during probing it is necessary also to determine the mean snow density in various

---

[1] A serious problem may still remain if correlations are required between mass balance data collected on this stratigraphic basis and hydrological data collected on an annual (fixed date) system (Mayo *et al.*, 1972).

areas. For this reason two or three pits are dug from the surface down to the ice or the previous year's summer surface. Continuous snow sampling is performed in the pit wall and the mean snow density obtained by weighing known volumes of snow. Density results from a single pit are then regarded as representative of a certain part of the glacier. Based upon this information and the probe results a map of snow distribution on the glacier can be constructed. If this snow survey is made at the very end of the winter, the map will show the total winter accumulation on the glacier, i.e. the glacier's total 'income'. This amount, previously called winter accumulation, is now termed the *winter balance*.

Ablation measurements

During the warm season large amounts of the winter accumulation will melt away again so that glacier ice will be exposed on the lower part, and probably some firn on the middle and upper part, whereas certain quantities of the previous winter's snow will remain on the highest parts of the glacier at the end of the summer. This remaining snow forms the nourishment of the glacier in its highest areas, and it is equal to the amount of remaining snow that is measured to obtain the glacier mass balance according to the above-mentioned 'simple method' (see p. 237).

On the lower part of the glacier all winter snow has normally disappeared and, in addition, some of the old glacier ice has also melted away. The amount of ice melt is normally measured against vertical poles drilled into the glacier. Such ablation stakes are made of aluminium, steel, bamboo or plastic. However, it must be ascertained that the stakes are fixed securely in the glacier, otherwise false readings may be obtained due to vertical stake movement relative to the glacier (stakes may start to sink, and readings become unreliable). In most cases, however, stakes will freeze solidly into the glacier due to slightly negative temperatures there (compare p. 228 and fig. 4C.1).

Based upon readings of a sufficient number of stakes distributed over the glacier surface, reliable figures can be obtained of the ice melt. This, together with previously obtained data on the water equivalent of all snow that has now disappeared, forms a base to calculate the total ablation in the ablation area.

In the accumulation area, ablation calculations must be based upon spring data on snow cover, and a further measurement. The difference equals the snow that has disappeared, and the amount is expressed as water equivalent. A map similar to the above-mentioned map of accumulation distribution can be constructed for the ablation distribution (see fig. 4C.6). By surface measurements of corresponding areas on these maps, tables are constructed to show variations in accumulation and ablation versus altitude. Compare table 4C.1. When adding up corresponding figures in this table the glacier mass balance can be expressed either as a total volume of water (m³), or as a 'specific water equivalent' (m). This is a figure that indicates the thickness of a water layer that would be formed if total volumes of water were distributed evenly over the entire glacier surface.

When mass balance results are reported, a table like the above-mentioned is very useful. To visualize the results, the content of the table may as well be plotted in graph

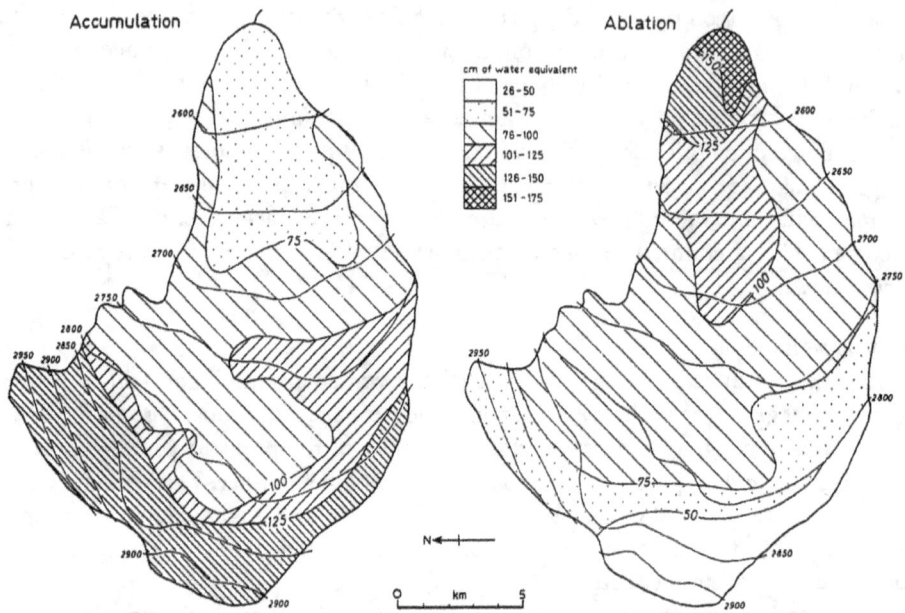

Fig. 4C.6. Example of accumulation and ablation maps. Highest winter accumulation is found at the highest elevations, whereas maximum melt takes place at the tongue. Results from volume calculation based on these maps are shown in table 4C.1.

form (see fig. 4C.7). From this graph the *activity index* (or 'energy of glacierization' as proposed by Shumskiy, 1964, p. 443) can be easily determined. It is defined as the slope of the balance curve (curve b_n in fig. 4C.7). This slope proves to be steeper for glaciers in arctic or continental regions than for glaciers in temperate, maritime regions.

The most important results from mass balance investigations are the final sum figures shown in table 4C.1. There it can be seen whether the glacier has grown or decreased during the glaciological year in question. If the glacier selected as representative within a given region is enlarging, it is presumed that most (or all) other glaciers in the same

Table 4C.1. Mass balance for Ram River Glacier, Alberta, 1966.

Height interval	Area (km²)	Accumulation (10⁶ m³)	Ablation (10⁶ m³)	Mass balance Total (10⁶ m³)	Specific (m)
>2900	0.11	0.153	0.041	0.112	1.01
2800–2900	0.46	0.516	0.291	0.225	0.63
2700–2800	0.75	0.684	0.621	0.063	0.13
2600–2700	0.42	0.295	0.431	−0.136	−0.30
<2600	0.11	0.059	0.159	−0.100	−0.89
Sum	1.85	1.707	1.543	+0.164	+0.19

Elevation of equilibrium line = 2730 m.

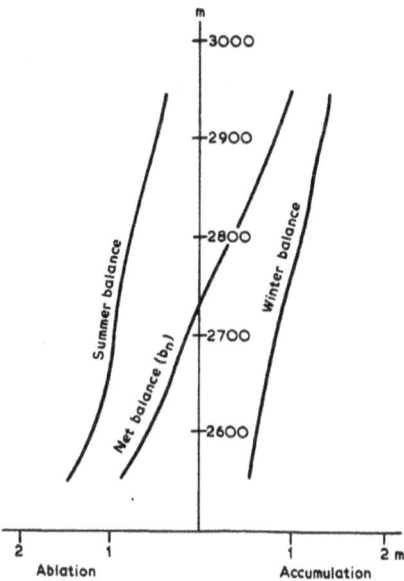

Fig. 4C.7. Graphic representation of mass balance variations with height, Ram River Glacier, Alberta, 1966. Compare the maps shown in fig. 4C.6. The diagram is based upon figures in table 4C.1.

region are also growing – thus catching a part of the annual precipitation which is added to the glacier mass (in solid form) and consequently withheld from the river discharge. This time-honoured concept is being challenged, however, following recent and more intensive glacier investigations (Dugdale, 1972; Alford, 1972). Both Alford and Dugdale have demonstrated that great response variations occur among groups of glaciers in the same general area, where previously one glacier would have been considered 'representative' of the entire group.

MASS BALANCE–IMPACT ON RIVER HYDROLOGY

Rivers draining glacier-free basins reflect variations in annual precipitation by corresponding variations in total annual discharge. Although a certain part of the precipitation disappears by evaporation, or adds to the groundwater storage, a wet year will undoubtedly produce a higher discharge than a dry year, and discrepancies between variations in total precipitation and total runoff are generally fairly small.

For glacier-covered basins, however, this general rule does not apply. A dry year could in fact produce a very high discharge because summers in dry years are often hot, and this will cause a large melt on glaciers so that the deficit in precipitation may well be compensated by meltwater. Consequently, a glacier-fed stream has a different hydrology from streams in glacier-free basins (see also Slaymaker, Chapter 3B).

Long-term discharge measurements are made in a great number of streams all over the world. The data are used for various purposes; in many cases they form a base for water power planning or irrigation calculations. If data were obtained in a glacier-fed stream for a series of years when glaciers were continuously shrinking, this would mean that the observed annual water volumes also included 'extra' water originating from negative glacier mass balance. Consequently, the average annual water discharge would be higher than if the glacier had been stationary. It is quite clear that if plans for utilization of water were based only upon the observed values, the data would be too 'optimistic'.

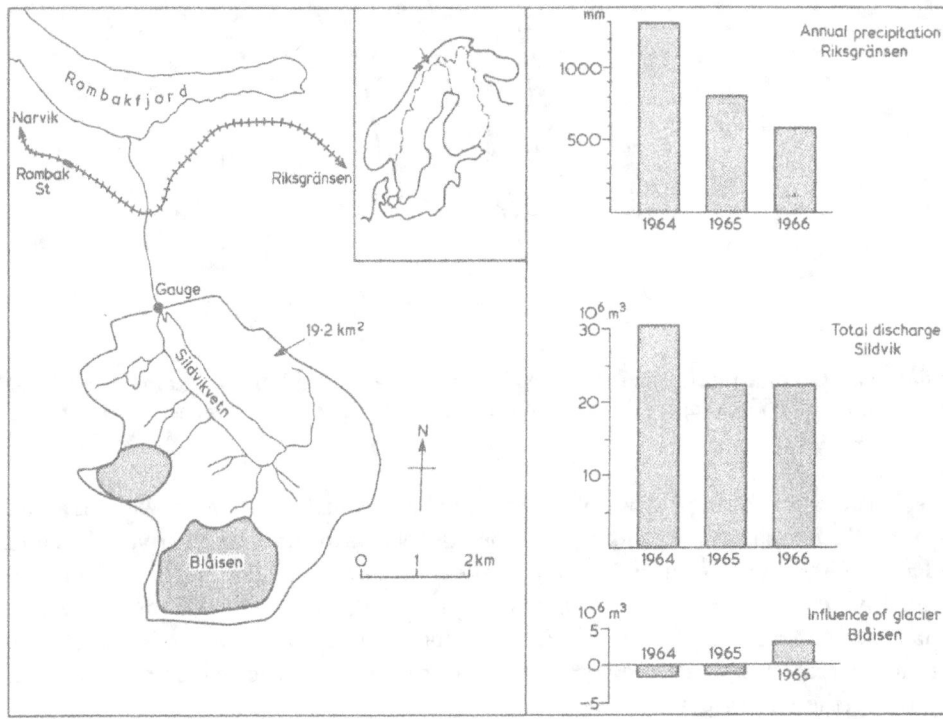

Fig. 4C.8. Impact of glacier mass balance on river hydrology. Annual precipitation shown in upper histogram indicates a successive decrease in precipitation throughout the years 1964–6. Middle histogram, giving annual water discharge totals for the same years, demonstrate an unexpected high discharge in 1966. Lower histogram shows that this was due to glacial melt, yielding an extra contribution of water that is not connected to the precipitation. A negative mass balance for the glacier is shown as a positive amount of water in the stream.

One cannot believe that glaciers would continue to retreat indefinitely, supplying rivers with water from their ice storage. On the contrary, glaciers may start to grow again, thus transforming a part of the annual precipitation into ice which will then be stored for a a long time within the glacier. The river will then lose a considerable part of its annual discharge, and all plans based solely on earlier river observations must be revised drastically.

It is obvious that it is highly desirable to conduct glacier mass balance studies for glacierized basins that are important potential areas for future water power production or irrigation. Data from such studies make it possible to correct observed discharge values for annual glacier variations. For example, if the observed annual discharge in the river equals Q and the glacier has shown a negative mass balance (M), this latter amount must be deducted from Q to obtain the theoretical discharge, D, that would have been observed if the glacier had been in equilibrium:

$$D = Q - M$$

Similarly, in years of positive mass balance, when the glacier has withdrawn water from the basin by adding snow and ice to its own mass, a positive correction must be made:

$$D = Q + M$$

An example of such calculations is shown in table 4C.2, where the size of glacier influence on river hydrology can also be seen. In one year (1969) extra meltwater from the glacier accounted for 25 per cent of total annual discharge. This amount of water would not have been supplied if the glacier was in a steady state condition.

Table 4C.2. Glacier influence on annual river discharge (sample figures from Nigardsbreen, W. Norway).

Year	Observed runoff (10^6 m^3)	Glacier mass balance (10^6 m^3)	'Corrected' runoff
1964	166.9	+40.2	207.1
1965	154.2	+37.1	191.3
1969	249.3	−61.7	187.6
1970	210.9	−21.1	189.8

Total basin area = 64 km^2
Glacier-covered part = 47 km^2 (or 73%)

Water discharge measurements observed at a gauging station in Jostedalen, Norway, are greatly influenced by glacier mass balance. In the years 1964–5 the glacier withheld approximately 25 per cent of total annual discharge from the river, whereas in the years 1969–70 large 'extra' water quantities originated from negative glacier mass balance. Observed figures from these years would give a too 'optimistic' base for water power calculations – one cannot expect a glacier to retreat indefinitely.

'Corrected' runoff figures are calculated from the assumption that the glacier is in a steady state condition, thus its hydrologic influence is eliminated. These figures are used as a base for planning water power installations.

6. Glacier movement

The normal life of a glacier includes downslope movement; the ice produced in the upper parts is slowly moving down to lower areas where it melts (or for polar glaciers, breaks off as icebergs). The movement is in fact a definite condition; if there is no movement, the ice mass is no longer a glacier.[1]

For obvious reasons it is very difficult to observe movement mechanisms within the glacier. Surface movement studies, however, are performed relatively easily by placing

---

[1] There is a risk of oversimplification in this definition since small ice and snow patches on steep slopes will show some degree of movement under the combined influence of gravity and melt/recrystallization processes.

markers on the surface and observing their displacement at certain time intervals. The results of such measurements show that for valley glaciers the highest surface velocity is found in the middle of the glacier; velocity decreases towards the valley sides and also towards the terminus. Velocities range from a few millimetres per day to several metres per day.

SURFACE MOVEMENT AND BOTTOM SLIP

Several theories have been developed concerning ice velocity distribution with depth. Some glaciologists are inclined to the view that the glacier is moving as a rigid block, i.e. the speed is the same all the way down to the bottom. If this is the case the glacier must slide on the bedrock at a velocity corresponding to that observed at the surface. Other scientists believe that the velocity distribution is similar to conditions in a river, i.e. that velocity is slowly decreasing towards the bottom, where it should approach zero. Field studies have shown, however, that intraglacial movement, *shearing*, in general takes place in the bottom layers of polar and subpolar glaciers. This means that most of the glacier body is gliding on relatively thin ice layers along internal shear planes near the bottom. Such internal movements destroy individual ice crystals and, although not directly observed, it can be proved by crystallographic studies of ice samples taken from the basal layers. However, it is quite clear that temperate glaciers are also sliding on their beds. The ratio between shearing and sliding is not exactly known, but this topic is under discussion among glaciologists. Polar glaciers may be frozen solid to the bed so *all* movement must then take place along internal shear planes.

On the tongue of some glaciers part of the ice becomes stagnant and shear planes reach the surface, resulting in at least partial overriding of the stagnant ice. If the till cover becomes thick enough, and the active glacier retreats, an ice-cored shear-plane moraine is formed (see further below).

Surface velocity differences between the middle of the glacier and the sides indicate that shear planes may also be present along the sides of the glacier. However, the mechanical stress–tension between ice in the middle and ice along the sides often creates *crevasses*, a fairly common feature on glaciers. Also, where a glacier flows across steps or abrupt changes of slope in the underlying bedrock, crevasses are formed on the surface. A special crevasse is formed along the back wall (in the firn area) of most glaciers. There parts of the snow and firn are hanging on the bedrock, whereas the rest of the glacier is moving from it. Consequently a fairly deep crevasse opens parallel to the back wall. This feature is termed *bergschrund*, based upon a German expression. For slow-moving glaciers in the Arctic many crevasses as well as the bergschrund may be filled with drifting snow, so that they are invisible most of the year.

Quantities of bedrock material are continuously falling onto the glacier surface either from steep rock walls in the firn area, or from nunataks, or from the valley sides. Part of the material falls into the bergschrund or in crevasses further down. This material is then carried on the glacier surface or within the ice. Such material is often seen as long parallel stripes or bands on the ice surface on the lower part of the glacier. If a valley glacier receives tributaries *medial moraines* are often formed. Long valley glaciers may have several parallel bands of medial moraines.

All rock material so transported by the glacier will eventually be deposited at its tongue or, in the case of very fine material, it is carried further down by meltwater streams. Moraine formation and moraine classification are dealt with in a following section.

GLACIER SURGES

A special kind of glacier movement is the so-called glacier surge. A surge is defined as a more or less sudden and unexpected rapid movement of a glacier. Its front inundates areas that have not been glacier-covered for a long time, and simultaneously the firn basins are drained so that the surface in these areas drops considerably. The reasons for glacier surges are not known, but it seems that one and the same glacier is inclined to surge at certain intervals, each interval being several decades, whereas other glaciers in the same area do not surge. After surging the glacier seems to become more or less stagnant for a long time, the tongue melts away and the snow accumulates again in the firn basins until a state of supposed instability is reached. For some unknown reason the surging mechanism is suddenly triggered again and the glacier repeats its rapid flow down the valley. The total time involved in one complete surge may be a few months, a year or more. Data on surging glaciers have been relatively scarce in the past, but interest has recently increased. Much of our present knowledge on glacier surges is collected in the *Canadian Journal of Earth Science*, no. 6 (1969).

7. Erosion and sediment production

It is obvious that glacier movement causes wear on the underlying bedrock. It can be shown that pieces of rock are picked up from the bedrock by the glacier and frozen into basal ice layers. Other rocks are delivered by frost action and rockfall into the bergschrund. If many rock particles are moved by the glacier, particularly when they are frozen in the ice at its very bottom, or when they are rolling between the ice and the bedrock, this material will act as a very efficient sandpaper. Materials thus moved by the glacier will, of course, eventually be deposited at the tongue, but during the transport process many of the particles are moved in direct contact with the bedrock so that more material is loosened and brought forward. Due to the enormous weight of the overlying ice much of the material is crushed and ground into finer and finer fragments. These particles are so small that they can be moved easily by meltwater. Consequently, most of the meltwater leaving the glacier at the tongue is more or less silty. The existence of suspended sediment in the meltwater stream in fact can be taken as a proof of glacier movement.

Fairly large amounts of small rock particles can be moved by the water. Recent studies have shown that several hundred tons of suspended sediment can be produced per km² of glacier during a single summer season.

Suspended sediment concentrations show great variations through time. During days of rapidly increasing discharge the total amount of transported sediments may be several hundred times as large as corresponding figures in days of low water discharge. In order to observe total sediment transport it is therefore necessary to perform sediment sampling several times a day throughout the entire melt season.

Recent studies have shown that the total annual sediment transport from one and the same glacier does not show particularly great variations from year to year. On the other hand, variations from glacier to glacier may be considerable, even between glaciers situated in the same geological formation and under similar climatological conditions. Further, there seems to be no simple correlation between water discharge and suspended sediment concentration so that sediment transport rating curves cannot be constructed for glacier streams (Østrem, 1971).

MORAINES

The coarse material that cannot be moved by running water is deposited on the ground in front of the ice. If the glacier is stationary for a long series of years, each year's deposition accumulates on top of that from previous years so that a *moraine ridge* is slowly formed just at the outer edge of the ice. Such moraine ridges are characteristic of most glaciers. Ridges may also form when an advancing glacier pushes forward, so that loose deposits are pushed away almost as by a bulldozer and when the ice melts a ridge is left on the ground. This feature is generally called a *push moraine*, and it may be distinguished from the above-mentioned depositional moraine by till fabric studies. If a retreating glacier makes small advances during a long period of retreat, a series of conformable moraine arcs may be found in front of the tongue (Plate 16). Under certain conditions, apparently by subaquatic ice action, special moraine forms can be developed, such as cross-valley moraines (Andrews 1963), Rögen-type moraines (Hoppe, 1952), etc.

Ice-cored moraine ridges are fairly large ridges that contain buried ice. Such ice cores are thought to be formed in two ways:

(a) Till is brought to the surface along shear planes and is deposited on stagnant ice (former glacier ice). If the thickness of the till increases sufficiently it will prevent the stagnant ice from melting. After glacier retreat, this stagnant ice is hidden under a thick till layer and the deposit forms an ice-cored moraine which may persist for hundreds or thousands of years.

(b) Morainic material may be deposited on a snowbank, adjacent to the ice front. If the deposit becomes thick enough it will also in this case prevent the buried snowbank (that easily transforms into ice) from melting. After glacier retreat the feature becomes an ice-cored moraine.

Ice-cored moraines are frequently formed in arctic areas and in areas of continental climate in the Subarctic. Whether they contain glacier ice (often reported from arctic areas – cf. Goldthwait, 1951) or snowbank ice (this type is abundant in Scandinavia –

cf. Østrem, 1964) can be decided by crystallographic studies on samples of the buried ice. An example of a typical ice-cored moraine is shown in Plate 17. Some ice-cored moraines may have the superficial appearance of rock glaciers (see below) and can be misinterpreted as such (Barsch, 1971). Where the ice core is derived from burial of snow, it is sometimes possible to obtain an absolute age through collecting organic matter that has accumulated in the snowbank and subjecting it to radiometric dating (Østrem, 1964, 1965).

ROCK GLACIERS

A feature typical of many arctic and alpine environments is the rock glacier (Plates 18 and 19). It is defined as a glacier-like mass of rocks that is slowly moving downslope (UNESCO, 1970, p. 17; Flint, 1971, p. 273). Movement is explained by the presence of interstitial ice, or in certain cases the deposit is resting upon remnants of a glacier that has stagnated and been buried by talus accumulation. Presumably both forms of ice may occur within a single feature. When a rock glacier for any reason is no longer moving, it is said to be inactive, or fossil. It is frequently difficult to determine whether a rock glacier is active or inactive because the rate of movement is only fractions of a metre per year and active field surveys would be required in each particular case. The snout, or front of the fossil rock glacier will sometimes show evidence of reactivation through steepening and consequent rolling of rocks under gravity. In this case fresh material in the steepened front will contrast with weathered and lichen-covered boulders further back.

Rock glaciers are usually found in areas where ice glaciers are approaching their limits, either as the equatorward extensions of glacierized areas or along their lowest altitudinal extent. They are clearly related to areas with cold climates and may represent sporadic, localized occurrences of permafrost (see Ives, Chapter 4A); they are probably restricted to areas of high continentality. Several scientists have observed rock glacier behaviour in various areas (Wahrhaftig and Cox, 1959; White, 1971), but so far no systematic attempt has been made to map their distribution. According to proposals from the IHD all rock glaciers should be included in future glacier inventories, as is done in several countries. Interior Alaska, European Alps, Canadian Rocky Mountains, southeastern Baffin Island and the San Juan Mountains and Front Range of the Colorado Rocky Mountains are some areas where rock glaciers have been observed and studied recently.

7. Concluding remarks

Glaciers may be regarded as one of man's natural resources. Meltwater from glaciers consists mainly of 'old' water originating from snow that fell several centuries ago. Consequently this snow was not influenced by the present day's industrial air contamination and glacier meltwater is thus, in principal, very clean and may in the future be an important source of uncontaminated drinking water.

As a curiosity, it should be mentioned that it has been seriously proposed to move icebergs from Antarctica to cities in Australia or on the American continent, where clean water is scarce. Glacier ice would then be used directly as a drinking water supply. Studies have been made of the economy of this project – it was thought to tow big icebergs from Antarctica to Los Angeles – calculations were made of the loss by melting and of the price per gallon of drinking water at the destination (Weeks and Campbell, 1973).

We have already seen that glacier meltwater may contribute to the river discharge in glacierized areas. Experience has shown that such glacier meltwater has a favourable influence on the discharge pattern: in dry summers many rivers will carry little water, resulting in decreased water power production; but dry summers are often also hot summers, so extra meltwater will then originate from the glaciers, thus compensating for the lack of water in the other streams. Conversely, cool summers are generally also wet summers, in which case most streams carry relatively large quantities of water and, at the same time, glaciers may release less water than normal and store it in solid form for a future year.

Meltwater from glaciers is today used in certain areas for irrigation. In dry areas, for example in some deep valleys in central Norway, most creeks will dry out completely during the summer, whereas those originating at small glaciers will continue to carry water all summer, and these are then utilized for irrigation.

Glaciers may also serve as a scientific archive, making studies of past climatic oscillations possible. It has been mentioned that an annual net accumulation takes place in the upper parts of glaciers and, provided the summer crusts are well developed, a stratigraphic study will yield data to determine variations in the net accumulation which in turn are a result of variations in climate. Glacier stratigraphy could therefore be a useful tool to study climatic oscillations. An even better method is based upon the stable oxygen isotope ^{18}O which has proved to be invaluable to determine temperature conditions in the atmosphere at the time when precipitation was formed. If, for example, snow is forming at low temperatures, the ratio $^{18}O/^{16}O$ will be less than if the precipitation forms at higher temperatures (see Chapters 5 and 6A).

From detailed analysis of an ice core taken from the Greenland inland ice at Camp Century, the Danish glaciologist Dansgaard has been able to detect both annual variations in temperature (i.e. summer snowfalls and winter snowfalls) and long-term climatic variations that coincide well with earlier known climatic oscillations, but the core reveals climatic data as far back as 100,000 years before present (Dansgaard et al., 1971).

Finally, it should be mentioned that glaciers in many areas are regarded as picturesque additions to the landscape, and viewpoints are arranged on highways or tourist routes to allow urban man to contemplate the beauty of large ice masses. In some places, glaciers have in fact been commercialized: sightseeing trips are arranged with oversnow vehicles, and tourists are shown crevasses, moulines, etc. In other places, the permanent snow in the accumulation areas is utilized for summer ski-ing, a new, popular use of old glaciers!

References

AHLMANN, H. W. (1924) Le niveau de glaciation comme fonction de l'accumulation d'humidité sous forme solide. *Geogr. Ann.*, **6**, 223–72.

AHLMANN, H. W. (1935) Contribution to the physics of glaciers. *Geogr. J.*, **86** (2), 97–107.

AHLMANN, H. W. (1948) Glaciological research on the North Atlantic Coast. *Roy. Geogr. Soc. Res. Ser.*, 1. 83 pp.

ALFORD, D. (1972) *Cirque Glaciers of the Colorado Front Range: Aspects of a Glacier Environment.* Unpub. Ph.D. dissertation, Univ. of Colorado. 157 pp.

ANDREWS, J. T. (1963) Cross-valley moraines of the Rimrock and Isortoq River valleys, Baffin Island, NWT. A descriptive analysis. *Geogr. Bull.*, **19**, 49–77.

AVSIUK. G. A. (1955) Temperaturnoje sostojanie lednikov. *Izvestia, Ser. Geogr.*, 1, 14–31 (in Russian). Moscow: Akad. Nauk SSSR.

BARSCH, D. (1971) Rock glaciers and ice-cored moraines. *Geogr. Ann.*, **53A**, 203–6.

BENSON, C. S. (1962) Stratigraphic studies in the snow and firn of the Greenland ice sheet. *SIPRE Res. Rep.*, 70. US Army Snow, Ice and Permafrost Res. Estab. 92 pp.

BRÜCKNER, E. (1887) Die Höhe der Schneelinie und ihre Bestimmung. *Met. Zeits.*, **4**, 31–2.

CHARLESWORTH, J. K. (1957) *The Quaternary Era*, Vol. 1. London: Edward Arnold. 591 pp.

DANSGAARD, W., JOHNSEN, S. J., CLAUSEN, H. B. and LANGWAY, C. C. (1971) Climatic record revealed by the Camp Century ice core. *In:* TUREKIAN, K. K. (ed.) *Late Cenozoic Glacial Ages*, 37–54. New Haven: Yale Univ. Press.

DOLGUSHIN, L. D. (1961) The main features of modern glaciation of the Urals. *Int. Ass. Sci. Hydrol. Pub.* (Symposium of Helsinki), 54, 335–48.

DUGDALE, R. E. (1972) A statistical analysis of some measures of the state of a glacier's 'health'. *J. Glaciol.*, **11** (61), 73–80.

ENQUIST, F. (1916) Der Einfluss des Windes auf der Verteilung der Gletscher. *Bull. Geol. Inst. Uppsala*, **14**, 1–108.

FLINT, R. F. (1971) *Glacial and Quaternary Geology.* New York: John Wiley. 892 pp.

GOLDTHWAIT, R. P. (1951) Development of end moraines in east-central Baffin Island. *J. Geol.*, **59** (6), 567–77.

GROVE, J. M. (1972) The incidence of landslides, avalanches, and floods in western Norway during the Little Ice Age. *Arct. Alp. Res.*, **4** (2), 131–8.

HOINKES, H. C. (1968) Glacier variation and weather. *J. Glaciol.*, **7** (49), 3–19.

HOPPE, G. (1952) Hummocky moraine regions with special reference to the interior of Norrbotten. *Geogr. Ann.*, **34**, 1–72.

KASSER, P. (1967) *Fluctuations of Glaciers 1959–1965. A Contribution to the International Hydrological Decade.* IASH and UNESCO. 52 pp.

KLEBELSBERG, R. VON (1948) *Handbuch der Gletscherkunde und Glazialgeologie*, Vol. 1. Vienna: Springer-Verlag. 403 pp.

MAYO, L. R., MEIER, M. F. and TANGBORN, W. V. (1972) A system to combine stratigraphic and annual mass-balance systems: a contribution to the International Hydrological Decade. *J. Glaciol.*, **11** (61), 3–14.

MÜLLER, F. (1962) Zonation in the accumulation area of the glaciers of Axel Heiberg Island, NWT, Canada. *J. Glaciol.*, **4**, 302–10.

ØSTREM, G. (1964) Ice-cored moraines in Scandinavia. *Geogr. Ann.*, **64A**, 228–337.

ØSTREM, G. (1965) Problems of dating ice-cored moraines. *Geogr. Ann.*, **47A**, 1–38.

ØSTREM, G. (1966) The height of the glaciation limit in southern British Columbia and Alberta. *Geogr. Ann.*, **48A**, 126–38.

ØSTREM, G. (1971) Sambandet mellom vassföring og slamtransport i bre-elver (Correlation between water discharge and sediment transport in glacier streams). *In:* ZIEGLER, T. (ed.) *Slamtransport undersøkelser i norske bre-elver 1970* (Sediment transport studies in Norwegian glacier streams 1970), 72–82. Rep. 1/72. Hydrol. Div., Norwegian Water Res. and Electr. Board.

OUTCAULT, S. I. and MACPHAIL, D. D. (1965) A survey of neoglaciation in the Front Range of Colorado. *Univ. of Colorado Stud., Ser. in Earth Sci.*, 4. 124 pp.

SHUMSKIY, P. A. (1964) *Principles of Structural Glaciology*. New York: Dover Publications. 497 pp. (Original Russian edn., 1955.)

STENBORG, T. (1965) Terminology of the thermal characteristics of glaciers. *Geogr. Ann.*, **47A**, 182–4.

UNESCO (1970) Perennial ice and snow masses, A guide for compilation and assemblage of data for a world inventory. *Tech. Pap. in Hydrol.*, **1**. 59 pp.

WAHRHAFTIG, C. and COX, A. (1959) Rock glaciers in the Alaska Range. *Bull. Amer. Geol. Soc.*, **70**, 383–436.

WEEKS, W. F. and CAMPBELL, W. J. (1973) Icebergs as a fresh-water source: an appraisal. *J. Glaciol.*, **12** (65), 207–33.

WHITE, S. E. (1971) Rock glacier studies in the Colorado Front Range, 1961 to 1968. *Arct. Alp. Res.*, **3** (1), 43–64.

Manuscript received April 1972.

Additional references

KOTLYAKOV, V. M. (1973) Snow accumulation on mountain glaciers. *In: International Symposia on the Role of Snow and Ice in Hydrology, Banff, 1972.* UNESCO-4. UNESCO/WMO. 6 pp.

WENDLER, G. and ISHIKAWA, N. (1973) Experimental study of the amount of ice melt using three different methods. *J. Glaciol.*, **12** (66), 399–410.

Past environments

CHAPTER 5

Palaeoclimatology

James A. Larsen and Roger G. Barry

Center for Climatic Research, University of Wisconsin, and
Institute of Arctic and Alpine Research, University of Colorado

1. Introduction

Climatic change is recognized to have had a significant influence on the evolution and geographical distribution of the earth's species and it may similarly have affected the history of civilizations. However, the amplitude of climatic changes, the manner in which they occur – abruptly or gradually – and their causes are still controversial matters. The discussion here will be concerned mainly with the evidence on which presently accepted chronologies of climatic change in high northern latitudes are based and with the atmospheric processes that may be involved.

The evidence that large circumpolar continental areas have been covered by ice sheets during several periods of the earth's history is voluminous and need not be considered here. In full-glacial episodes, of the Quaternary particularly, much of the Arctic and especially the Subarctic was deep under glacial ice. Thus, much of the stratigraphic record of these episodes comes from areas beyond the ice, necessitating that our treatment be concerned with more than just high latitudes. Also, it is from areas near the former ice margins that much critical evidence of climatic fluctuations is obtained. The present arctic fringes are important too in view of their sensitivity to historically recent fluctuations in climate, as is evident in the history of Iceland and Greenland since the period of the Viking settlements.

Before considering the evidence for climatic fluctuations in recent geological time, some qualitative indication of the magnitude of the changes involved seems appropriate. Our knowledge of the conditions that prevailed in the most distant past is naturally the least complete. Nevertheless, it seems that ice ages recur at intervals of 2 to 2.5×10^8 years, persisting for only 5–10 per cent of geological time (Fairbridge, 1972). For at least the last 500 million years conditions generally (at least over the 18 per cent of the earth's surface lying north of 40°N) were warmer than they are today: by approximately 6°C according to Brooks (1951). Periodically, within the major ice age epochs, great ice sheets form in high latitudes and spread out to cover as much as 40 per cent of the earth's land surface, with accompanying temperature drops approaching 14°C in middle latitudes according to some estimates (Sellers, 1965). The glacial epochs are quasi-periodic and short-lived, with each major ice advance ('pleniglacial') lasting less than 10,000 years (Fairbridge, 1972). During a glacial period, the ice sheets seem to be in the process either of advancing or disintegrating. Similarly, the interglacial peaks persist for only about 10,000 years in middle latitudes, though for an even shorter span and with a phase lag in polar regions according to Fairbridge's model. The polar regions likewise appear to show the earliest evidence of glacial onset (Andrews *et al.*, 1972).

The Neogene ice age began at least 1–2 million and possibly more than 6 million years ago. According to recent evidence (see Andrews, Chapter 6A), it was characterized by eight glacial advances within the last 700,000 years (Kukla, 1970), the last of which ended about 8000 years ago in central North America and 9000 years ago in Sweden. As discussed below, the ice persisted until much later in northern Canada. In view of the great ice masses still remaining in both polar regions the question remains to be answered whether the Quaternary ice age has ended or whether the current period (Holocene) is another interglacial. Most specialists in the field support the latter viewpoint (Kukla *et al.*, 1972).

2. Quaternary changes

There is a large and rapidly growing body of evidence from many sources that the climate characteristics of northern regions have not been constant even since the climatic change that brought dissolution of the last of the Pleistocene ice over the northern part of the hemisphere (with, of course, the exception of the Greenland ice sheet and the Barnes ice cap on Baffin Island). Much of the evidence exists in the form of isotopic ratios in deep cores from the Greenland ice sheet, from marine sediments on the floor of the Arctic Ocean and elsewhere, from fossil soils, tree ring measurements, archaeological sites, and a variety of other sources. Some of these recent findings are now considered.

EVIDENCE FROM ISOTOPIC RATIOS

Urey (1947) first proposed use of the $^{18}O/^{16}O$ ratio to determine the temperature of formation of fossil shell carbonates, thus indicating ocean temperatures at the time the animal built the shell. The method has been employed rather extensively to obtain indications of ocean temperatures in the Pleistocene, Tertiary, and even the Mesozoic. Emiliani (1966) has employed the evidence available from shells in shallow ocean sediments to construct a generalized temperature curve for the surface water of the central Caribbean and the equatorial Atlantic. The temperature curves reveal that the dates of the oceanic temperature fluctuations, as determined by ^{14}C, $^{230}Th/^{238}U$, and $^{41}K/^{40}Ar$ isotope methods, correlate well with other estimated dates of the late, middle, and early stages of the last readvance of continental glaciers known as the Wisconsin glaciation (in North America) and the Würm (in Europe). Emiliani further demonstrated that temperature minima in the shell data correlate highly with summer solar insolation minima calculated on the basis of secular changes in the earth's orbital parameters (obliquity, eccentricity, longitude of perihelion) in its relationship to the sun. The variation in the amount of solar radiation received by a point on the earth's surface due to these changes was calculated long ago by Milankovitch (1920, 1930, 1969) and forms the basis of an astronomical theory of climatic change that will be discussed in greater detail below.

Dansgaard (1954, 1961) later proposed that the $^{18}O/^{16}O$ ratio could also be employed with glacier ice for indicating past climatic conditions. The concentration of ^{18}O in precipitation, particularly at high latitudes, is determined largely by temperature of formation, with decreasing temperatures of formation leading to decreasing content of ^{18}O in rain or snow. In 1966 the US Army Cold Regions Research and Engineering Laboratory succeeded in obtaining a core 12 cm in diameter through the entire Greenland ice sheet at a point where the sheet is 1390 m thick. Some 1600 ice samples from 218 different increments along the upper 300 m of the core were analysed by mass spectrometry to obtain $^{18}O/^{16}O$ ratios. Approximate ages of the core increments were obtained by considering their depth in relation to generally accepted flow patterns of ice in glaciers and certain other assumptions. Using the data obtained, a chronology for the past 70–100,000 years was then constructed by Dansgaard et al. (1969). The general trend of the curve (fig. 5.1) is in agreement with other reported climatic changes in other parts of the world during the past 75,000 years. The conclusion reached was that the method is capable of furnishing continuous climatic records for at least the past several hundred millennia. There are, however, various problems in attempting to infer absolute temperatures from the record (Budd et al., 1971).

Taking into consideration the large volume of ice that builds up in continental glaciers during a major glacial period (47×10^6 km^3 water equivalent), Dansgaard and Tauber (1969) calculated that the fluctuations in $^{18}O/^{16}O$ ratios in the world's oceans (of a volume 1370×10^6 km^3), resulting from release of the excess ^{18}O bound up in glacier ice, would account for about 70 per cent of the fluctuations in $^{18}O/^{16}O$ ratios observed by Emiliani in the shells of marine animals. Thus, they argue that the fluctuations in the curve obtained by Emiliani are not so much the result of temperature changes (30 per cent) as of the amount of ^{18}O in water (as ice) resting on the continents during glaciations and released during interglacials (70 per cent). They conclude, therefore, that the data from marine sediments are predominantly an indicator of palaeoglaciation and not of palaeotemperatures, although their value as such is at least equally great. 'If this interpretation is correct the palaeoglaciation curve shows that within the last 425,000 years we have had 7 or 8 major, independent glaciations of the continents, each of a magnitude similar to the maximum of the last glaciation, and each with a time spacing of the order of 40–50,000 years. With a total length of the Pleistocene of 2 to 3 million years, we should expect perhaps 40 major and independent glaciations during this period, rather than the 4 or 6 subdivided glaciations usually recognized (Dansgaard and Tauber, 1969, p. 502). Subsequently, Emiliani (1971b) has demonstrated that Dansgaard and Tauber

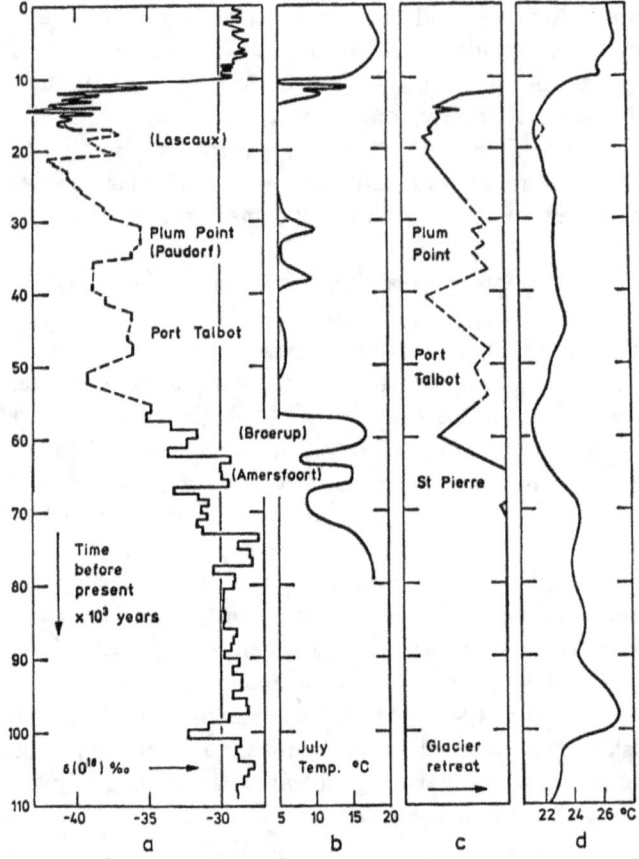

Fig. 5.1. Climatic variations during the last 70,000 or 100,000 years estimated by various methods. (a) The $\delta^{18}O/\delta^{16}O$ variations in the Greenland ice core. (b) A ^{14}C dated pollen study in Holland (Van der Hammen *et al.*, 1967). (c) A ^{14}C dated study of deposits in the Ontario–Erie basins showing the edge of the Laurentide ice sheet (Goldthwaite *et al.*, 1965). (d) An oxygen-isotope study of deep sea cores showing generalized temperature curve for surface water of the central Caribbean (Emiliani, 1966). European names of interstadials are given in parentheses (from Dansgaard *et al.*, 1969).

probably assumed too light an isotopic value for the ice sheets. Palaeontological data indicate a glacial/interglacial temperature range of 5–6° in the equatorial Atlantic, 3–4° in the equatorial Pacific.

Broecker and Van Donk (1970) indicate that there exists a rather pronounced periodicity in the $^{18}O/^{16}O$ ratios presented by Emiliani (see fig. 5.2). The ^{18}O content of the foraminifera oscillates between rather uniform limits with the increase in ^{18}O content taking place gradually over a long period of time. 'If the ^{18}O record does indeed reflect the extent of glacial ice, it suggests that glacial periods are characterized by relatively long periods of more or less continuous ice growth followed by relatively short periods during which the ice is destroyed' (Broecker and Van Donk, 1970, p. 170).

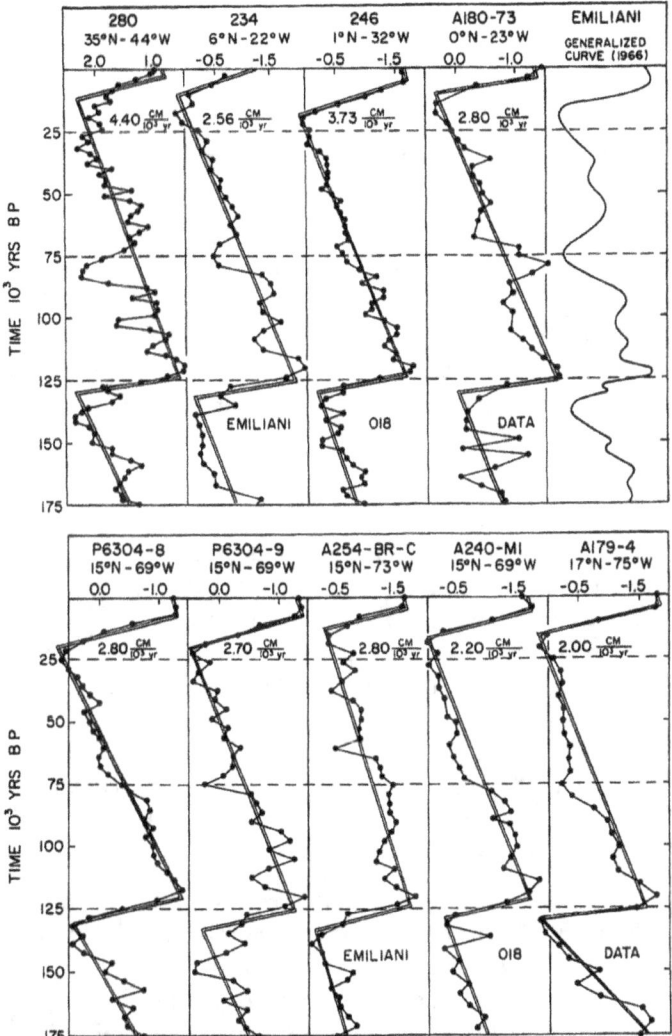

Fig. 5.2. Plots of $\delta^{18}O/\delta^{16}O$ versus time: the upper part of nine of the eleven North Atlantic and Caribbean cores analysed by Emiliani (1955, 1964, 1966). The time scale was obtained by assigning an age of 127,000 years to the midpoint of termination of glaciation and assuming the sedimentation rate to be constant. Since sedimentation rates during cold periods are probably somewhat different from those during warm periods, the age of the point intersected by the dotted line for 75,000 years could be in error by ±10,000 years. The extent of this distortion would vary from core to core. The idealized curve of Emiliani (1966) is given for comparison (from Broecker and Van Donk, 1970).

This general scheme of glacial formation and dissolution is in accord with the supposition that although the rate of deposition of snow might change abruptly, long periods would be required to accumulate the depth of ice characteristic of major glaciations. The data presented by Broecker and Van Donk suggest a growth phase of the glacial ice of about 20,000 years or more, a period in close agreement with that postulated by Barry (1966), who estimates that at least this interval would be required to build an accumulation of ice 2000 m thick over Labrador–Ungava.

THE QUESTION OF AN ICE-FREE ARCTIC OCEAN: THE RECORD FROM FORAMINIFERA

The suggestion that the Arctic Ocean periodically becomes free of ice, thus providing an evaporative source for the moisture necessary for at least the initiation of glacial growth to some critical size, was made many years ago by Ewing and Donn (1956). Later evidence from abundance of marine foraminifera in sediments on the ocean floor indicates that the Arctic Ocean was not ice-free during initiation of at least the Pleistocene glaciations (Ericson *et al.*, 1964; Donn and Ewing, 1966; Ericson and Wollin, 1968, 1970; Morin *et al.*, 1970; Herman, 1970). Studies of extensive core collections have led Clark (1971, p. 3313) to the same conclusion:

> The Arctic Ocean probably froze during the northward migration of the continents to their present position, and this shift accompanied continental glaciation . . . While glaciers have waxed and waned on the continents, the Arctic Ocean has remained frozen. The principal change in the Arctic Ocean ice during the Late Cenozoic has been attainment of greater and lesser thickness in response to little understood atmospheric conditions. During the last 700,000 to 1 million years the Arctic Ocean has had thinner ice than during any comparable time since at least the Middle Pliocene. The faunal evidence further suggests that the Arctic Ocean has never been warmer than it is today for at least 3.5 million years. This means that the effects of an ice-free Arctic Ocean are unknown because there may never have been an ice-free Arctic Ocean as it is now, in its present geographical position.

Clark's evidence is taken from 531 sediment cores obtained from the Arctic Ocean floor using the ice island T-3 as the coring platform. Foraminifera abundance increases during times of thin ice, declines to zero when the ice is thick. The data show high foraminifera abundance for brief intervals about 1.2 and 2.1 million years ago. Since about 700,000 years ago abundance has remained high. Essentially equilibrium conditions with a fairly thin ice cover persisted through the Yarmouth, Illinoian, Sangamon, and Wisconsin phases of the Pleistocene. There has been no obvious correlation with the classical continental cycles of the northern continents. The work does indeed suggest that the Arctic Ocean has not been a factor in the growth or melting of Pleistocene continental glaciers.

THE CLOSE OF THE PLEISTOCENE

It is apparent from a number of sources of evidence that a major portion of the temperature rise that ended the Pleistocene about 11,000 years ago could have taken place within a time interval as short (in geological time) as 1000 years in duration (Broecker *et al.*, 1960; Emiliani, 1971a). There is, however, the possibility that the rapid tem-

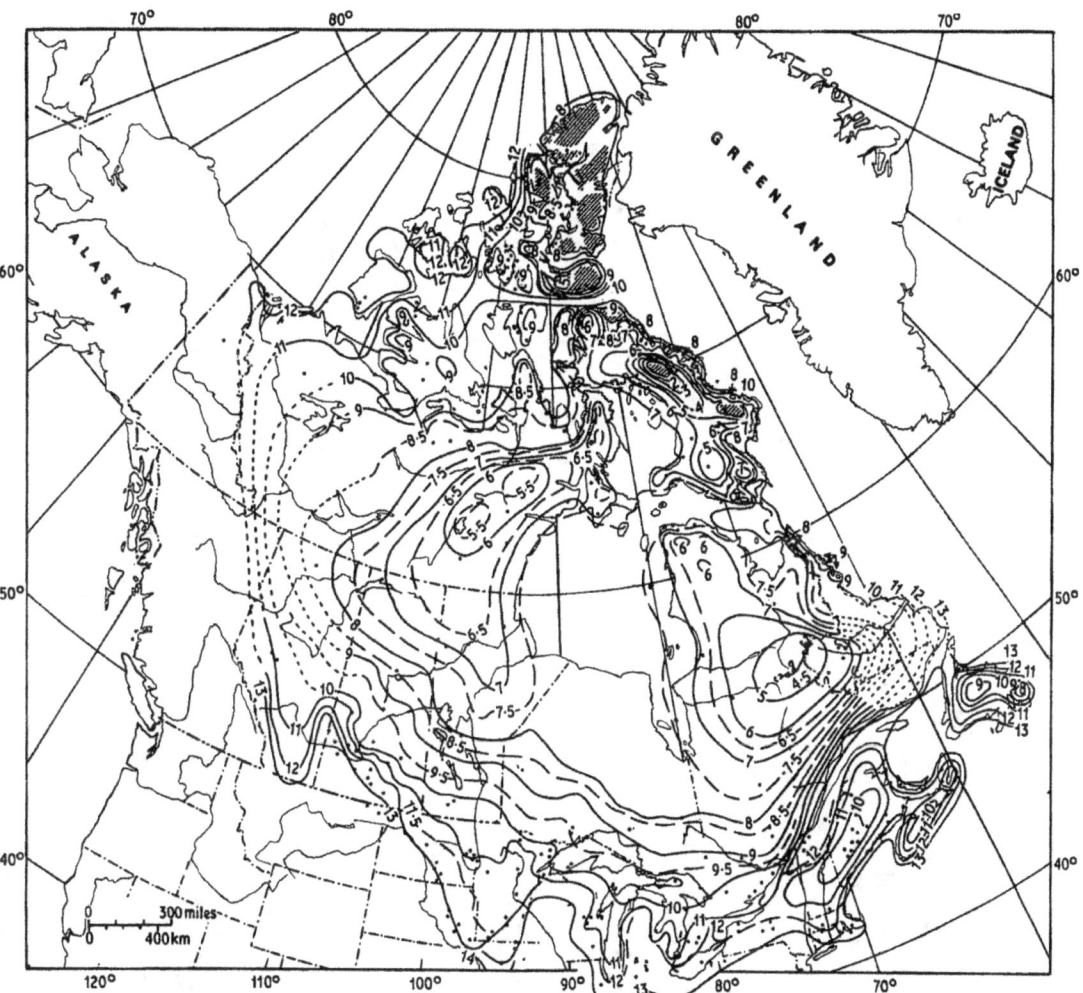

Fig. 5.3. Radiocarbon isochrones of the retreat of the Laurentide ice. Isochrone location based on ^{14}C dates, coastline location, moraine location and other field evidence (from Bryson *et al.*, 1969).

perature change at that time was associated in part with a rapid decrease in albedo brought about by thawing of the northern North Atlantic ice when northern summer came to coincide with terrestrial perihelion (Emiliani, 1966). Other evidence, indeed, suggests that the major warming change took place about 20,000 years ago. Final retreat of the ice sheet from a brief, late advance during the Wisconsin could reasonably have coincided with the trends demonstrated for the Mediterranean by Emiliani (1971a). There a final warming occurred about 11,000 to 12,000 years ago, leading to the so-called postglacial optimum. This latter phase then persisted until cooling began 3500–4000 years ago. This coincides with the trend deduced from analysis of ^{14}C dates from a

Table 5.1. Significant dates of recent (12,000 BP* – present) climatic episodes from the analysis of 620 ¹⁴C dates by Bryson et al. (1970, see discussion in text) and correlation with supposed dates of other known climatic events during the periods indicated.

Episode	Date of significant change (Bryson et al., 1970)	Period of max. of major episodes (Bryson et al., 1970)	Approx. bracketing dates by previous authors†	Description of episode derived from previous studies‡
Pre-Boreal				Generally warm and dry with a 600-year very cold period (10,800–10,200 BP) known from European evidence
Boreal	9650 BP (±240 yrs)§	9140 BP at max (±280 yrs)	10,000	Cool and dry Cool and moist (Europe)
Atlantic	8450 BP (±320 yrs)	7730–5980 BP (±260) (±530)	9000	Cool and dry (North America, Europe) Thermal maximum ('Climatic Optimum') both hemispheres. Warm and dry
	4680 BP (±490 yrs)		5500	Final deglaciation (Canada); max. glacial retreat (Alaska)
Sub-Boreal	2890 BP (±510 yrs)	3970–3480	3000	Intense droughts (Europe); cool and dry (North America) 900–500 BC. Cold, wet period (Europe); northern forest retreat (2470 BP: Bryson et al., 1965)
Sub-Atlantic				Drought (southwestern US)
Recent	1690 BP (±410 yrs) (AD 280)			AD 1000–1250. Vikings colonize Greenland during warm period AD 1550–1700. Greenland colony abandoned. 'Little Ice Age'

* BP refers to 'Before the Present'; dates are from Blytt–Sernander sequence (Nilsson, 1964): see fig. 1 in Godwin (1966).
† Derived from fig. VIII.4 in Barry and Chorley (1968).
‡ Many references available including Nairn (1961), Schwarzbach (1963), Lamb (1966), Lamb et al. (1966).
§ Standard deviation of the observations.

wide variety of sources gathered by Bryson *et al.* (1969). Radiocarbon isochrones (fig. 5.3) show that the last great event of the Wisconsin glacial stage in North America was the disintegration of the Laurentide ice sheet which occurred between 13,000 and 6000 years ago, although the apparent retreat rate and mass loss presents something of a problem in terms of energy requirements according to Andrews (1973). The ice sheet retained its identity as a distinct unit until about 8400 years ago, then disintegrating rapidly by about 8000 years ago. Three remaining remnant ice sheets centred over Keewatin, Labrador–Ungava, and Foxe Basin–Baffin Island persisted until about 5500 years ago. The Barnes ice cap in central Baffin Island is a survivor from that time.

POST-GLACIAL CHANGES

The various curves depicting climatic events of the past 100,000 years shown in fig. 5.1 uniformly agree on the general rise of temperatures beginning about 11,000 years ago. It is of interest, however, that each presents a rather individualistic interpretation of events since that time. In an attempt to discern a consensus regarding the dates of significant recent climatic change, Bryson *et al.* (1970) reviewed all ^{14}C dates in ten volumes of the journal *Radiocarbon* (1959–68). These workers collected the 620 dates marking recurrence surfaces, breaks in stratigraphy, sea level maxima and minima, species maxima, and other events indicative of discontinuities of one kind or another that could result from significant environmental change. A frequency histogram of the data proved multi-modal, tending to aggregate in certain 'preferred' time periods. These were, by inference, periods of significant climatic change. It is of interest that the data correspond rather well with the episodes of what is termed the Blytt–Sernander sequence in Europe (Nilsson, 1964). The terminology of post-glacial climatic chronology proposed by Blytt and Sernander has therefore been adopted in table 5.1 summarizing the major postglacial changes.

3. Mechanisms of climatic change

Mitchell (1966) points out that all potential causes of climatic change can be classified into three basic categories: 1) autovariation in which instabilities lead to fortuitous fluctuations of indefinitely long duration; 2) autovariation in the ocean-atmosphere system with thermodynamic feedback causing sustained anomalies; 3) changes in the rate of absorption of solar energy due to atmospheric conditions (volcanic dust, CO_2) or changes in the intensity and distribution of the radiation itself (as a result of earth-orbital variations). Theoretical evidence exists in support of all three possibilities; the last two seem, however, to be supported by most convincing evidence and they will be discussed here.

RADIATION CHANGES

Milankovitch (1930, 1969) computed changes in radiation reaching the earth at various latitudes as a result of changes in the obliquity of the ecliptic, eccentricity of the orbit, and changes in time of year at which the earth is nearest the sun in its elliptical orbit. He found, for example, that for the last 10,000 years the radiation received in the northern hemisphere in summer has been decreasing and the radiation received in winter increasing. The range from winter to summer was some 10 to 15 per cent more in middle

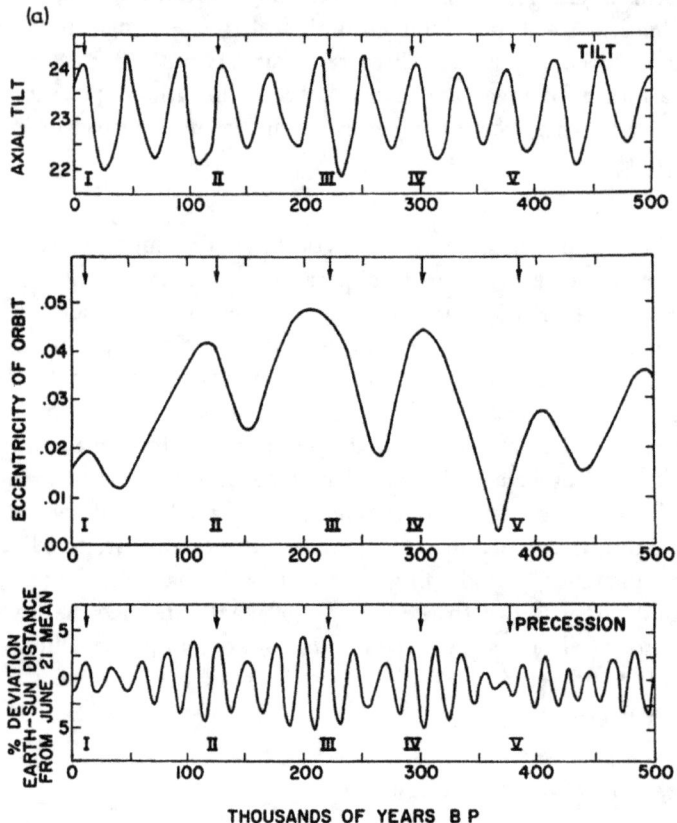

Fig. 5.4a. Variation of the earth's axial tilt (top), orbital eccentricity (middle), and 21 June sun–earth distance (bottom) over the last half-million years (based on the calculations of Vernekar, 1968). The arrows and Roman numerals indicate the timing of the sudden terminations seen in the $\delta^{18}O$ record from deep sea cores (from Broecker and Van Donk, 1970).

latitudes 10,000 years ago than today. In high latitudes excess summer radiation was not fully compensated by reduced winter radiation. As a result, latitudes north of 60°N received about 1.6 per cent more radiation over the year than at the present time; or 0.9 per cent more if averaged over the area north of 50°N (Sawyer, 1966).

Broecker and Van Donk (1970) illustrate the changes in distribution and intensity of solar radiation received at the surface of the earth during the past 500 millennia and show the relationships of these variations with the deglaciation of the far northern continental regions as determined by the marine $^{18}O/^{16}O$ ratios in foraminifera shells (fig. 5.4). While noting the dangers inherent in techniques of association based on 'proof by curve matching', they nevertheless conclude that the persistent coincidence of the radiational variations and the glaciations are more than fortuitous, and that changes in the earth's orbital parameters lead to climatic change. The sawtooth shape of the climatic cycle may be the consequence of modulating influences due to other causes, perhaps the lag effect introduced by the slow cooling of the oceans, for example. Broecker

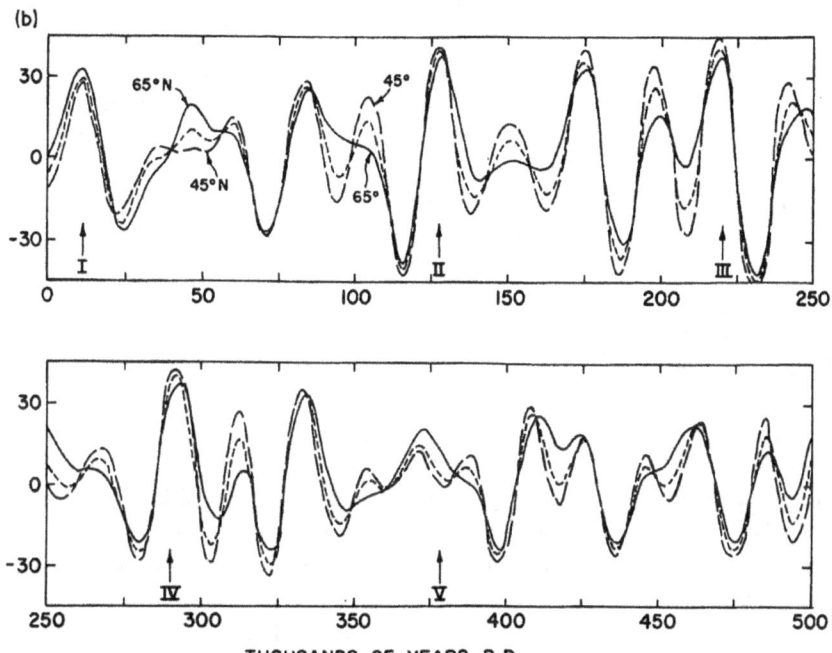

THOUSANDS OF YEARS B P

Fig. 5.4b. Summer insolation curves for the past 500,000 years for latitudes 45°N (long dash), 55°N (dash), and 65°N (solid) based on calculations made by Vernekar (1968). The changes are expressed in units of langleys per day difference from the mean caloric summer radiation. The arrows and Roman numerals indicate the times at which sudden terminations occurred in the $\delta^{18}O$ record from deep sea cores (from Broecker and Van Donk, 1970).

and Van Donk emphasize the need for more precise data to test the matching between insolation and climatic variation more fully.

A number of investigators have come to conclusions somewhat at odds with those discussed above. Kutzbach *et al.* (1968) estimated the effect of astronomical variations in incoming solar radiation on the thermal Rossby number, an expression that takes into account both the total effect of rotation and of differential heating of the atmosphere. They employed Milankovitch's climatic model (which does not include north–south heat transport) to convert secular changes in the incoming radiation gradients into secular changes in meridional surface temperature gradients. Thermal Rossby numbers were estimated and, during several periods in the Pleistocene, the summer thermal

Rossby numbers rose to wintertime values. Because the Milankovitch climatic model probably overestimates the direct effect of radiational changes, they also derived an empirical calibration of the relationship between annual radiational variations and the average meridional tropospheric temperature gradient (annual variation in the north–south 1000–300 mb thickness gradient between 20° and 70°N). It was found that for the past 500,000 years (at 1000-year intervals) the variations in latitudinal distribution of radiation (as calculated with the Milankovitch astronomical model) could cause significant fluctuations in the thermal Rossby numbers. The model indicated that changes of about ±5 per cent from present-day average summer thermal winds and ±2 per cent from present-day winter thermal winds were possible. Kutzbach *et al.* argue that changes in differential heating brought about by variations in orbital parameters might have been sufficient to cause shifts in planetary wave numbers directly. Even if the effects were small, the changes and subsequent feedback should have played a significant role in the Pleistocene. They add: 'One of the more frequent criticisms of the Milankovitch-type glaciation chronologies is that there are many more fluctuations in the radiational curves than glacial and interglacial periods: the concept of an atmosphere that undergoes transitions at certain critical thermal Rossby numbers may answer these criticisms' (Kutzbach *et al.*, 1968, p. 137).

Similarly, using a stochastic model of an ocean–atmosphere system incorporating a radiation factor, Mitchell (1966) showed that, when this factor was varied according to the Croll–Milankovitch sequence of computed values, the model results gave a sequence suggestive of the Pleistocene chronology of Ericson, Ewing and Wollin (1964) and others. This type of analysis points to the need to take feedback effects into account.

Acceptance of the view that climatic change is due to Milankovitch–type variations in solar energy received at the earth's surface is not universal. There has long existed an alternate opinion, based more recently on numerical simulation of the earth–atmosphere system, that temperature changes resulting from radiational variations would be insufficient to bring about major climatic shifts required for glaciation.

Shaw and Donn (1968), for example, employed a thermodynamic model to obtain quantitative determinations of changes in surface temperatures caused by variations in insolation as calculated by Milankovitch. Their results indicate that, at latitude 65°N, a mean cooling of 1.4°C below the present-day temperature would occur at times of radiation minima, in their opinion a rather small change to have triggered glacial climates. They conclude that, rather than being the cause of the Pleistocene glaciation, the temperature changes recorded in marine sediments, for example, were the result of variations in the amount of glacial ice over the northern portions of the northern hemisphere.

An alternative conclusion has been reached by Sellers on the basis of a global climatic model (1973) of the thermodynamic energy equation for the earth–atmosphere system, a model allowing for seasonal coupling and interaction between continents and oceans. Sellers found that a decrease in the solar constant of about 0.9 per cent would be sufficient to decrease the global average temperature by 5°C. Increasing the solar constant by 1 per cent gives a temperature rise of 0.9°C. It is of some significance that the final equilibrium state appeared to depend on the initial state prevailing when the calculations were begun. This implies that there are a multitude of possible steady-state equilibrium conditions for the system. Furthermore, the equilibrium conditions that derive from a given set of initial values do not seem to be completely reversible.

Sellers's conclusion is that there are at least two steady states possible with the present solar constant: one corresponds to present conditions or slightly warmer and the other to a glacial or ice age type climate, with temperatures at high northern latitudes 5–10°C lower than they are today. It appears likely that most of the climate variability during the Pleistocene can be reproduced by the model without changing the solar constant, the amount of carbon dioxide or particulate matter in the atmosphere, the cloud cover, the orbital relationship between the earth and sun, or any other variable. What apparently is needed, however, is an impulse or perturbation, perhaps volcanic activity or an extended regional climatic anomaly, to shift the climate from one steady-state condition to another. One of the observations to come out of Sellers's model is the extreme sensitivity of high latitudes of the northern hemisphere to factors which would produce climatic change. A decrease in the solar constant of 0.6 per cent in the model results in a temperature drop at 65°N of 3.5°C in February and 4.1°C in August. The idea of climatic sensitivity in higher latitudes is also seen in the work of Fairbridge (1961, 1972), Budyko (1971), and others. Although changes of such magnitude in the solar constant would result in climatic change, Sellers points out that most of the natural climatic variability which has occurred on the earth, at least during the Pleistocene epoch, lies within the range of climates possible with the existing solar constant, atmospheric composition, cloud cover, and orbital relationships. The question to be asked, says Sellers, is whether or not the earth system always rests precariously on the brink of an ice age. Dwyer and Petersen (1973) in a model similar to that of Sellers, but incorporating time-dependent influences, particularly oceanic heat capacity and transport, show that a change of only 2 per cent in the solar constant can produce a large change (10–15°C) in sea level temperature. It will be interesting to see whether subsequent numerical models bear out these latest conclusions.

OCEAN-ATMOSPHERE INTERACTIONS

Fletcher (1966) and Budyko (1966) both consider that the present arctic ice cover is in a delicate balance with the prevailing climate-forming factors. Budyko has calculated that small positive temperature anomalies in summer would cause the arctic pack ice to disappear in a few years, after which the Arctic Ocean would remain ice-free. He shows, for example, that a positive anomaly of the summer temperature of 4°C would melt ice 4 m thick in four years under the present mean climatic conditions. A smaller initial temperature anomaly, however, might effect the same end. A decrease in area of ice at the free surface of the ocean would result in increased amounts of absorbed radiation which would in turn intensify the melting.

Vowinckel and Orvig (1970) also point out that the polar atmosphere is at present adjusted in the best possible way for conservation of energy in winter. Thermal stratification and clouds make the winter inversion most effective. They conclude, 'the Polar Ocean is at present in a delicate radiational balance, and relatively minor variations in any term can instigate a gradually intensifying process, ending either in complete freeze-over or in complete melting' (Vowinckel and Orvig, 1970, p. 221). According to Maykut and Untersteiner's (1969, 1971) model, a 10 per cent increase in either incoming or long-wave radiation in the Arctic would result in a 50 per cent decrease in present ice thickness.

Applying his same methods to the pre-Pleistocene, Budyko postulates that the central Arctic was free of ice the year around. For the ocean to freeze, however, it was sufficient for the air temperature of the Arctic to decrease for only a relatively brief period. When the development of ice exceeded some critical value, the freezing became self-perpetuating sustaining its existence and increasing in dimensions on land and ocean independently of external factors and merely as a result of climatic conditions effected by the ice cover itself. While the geological evidence discussed above does not support the hypothesis of an open Polar Ocean during the glacial maxima, this pattern is theoretically possible according to the semi-empirical work of Fletcher (1966) on energy exchanges.

The consequences of an ice-free Arctic have recently been evaluated in three numerical models. Using a zonally symmetric model, MacCracken (1970) found that the results contradicted the Ewing–Donn hypothesis. An ice-free Arctic Ocean is not stable and would rapidly develop an ice cover even in the absence of continental ice sheets. The circulation pattern showed a central low pressure surrounded by a continental anti-cyclone with increased precipitation only over the ocean. The other two studies are more general. Warshaw and Rapp (1973) use the Mintz–Arakawa two-level model to simulate winter conditions with areas of arctic sea ice replaced by water at −1°C. The results indicate decreased vertical stability and lower pressure over the Arctic Basin due to a removal of mass. The westerlies are slightly stronger in higher latitudes than for the ice-in case, but are weaker in middle latitudes. In a similar experiment with the British Meteorological Office model, Newson (1973) found a similar weaking of the westerlies and an unexpected decrease in temperature over northern hemisphere continents. These results lay the foundations for a first realistic assessment of Late Tertiary climate (although this was not the specific intention of the investigators).

4. Recent circulation changes and their implications

The palaeoclimatic record is beginning to provide guidelines as to the climatic changes likely to result from changes in either the total energy input to the earth–atmosphere or in its latitudinal distribution. On a shorter time scale, historical climatic data afford a basis for assessing such interactions.

Lamb (1965) points out that, during periods of increased westerly flow, global increases in energy may be generally available, heating the tropics as well as northern latitudes. Recent evidence substantiates this. Beginning in the early 1800s and culminating in the 1930s, the mean North Atlantic circulation became stronger, the mean position of the Icelandic low moved northward, sea ice common around Iceland in previous times became rare, and ports were ice-free many months of the year (Krist-jannson, 1969). After about 1940, the trend reversed. Circulation has weakened, the Icelandic low has moved southward, and Icelandic shipping is harassed by sea ice. Ocean currents tend to follow the patterns of the winds, with about a twenty-year lag. Lamb cites evidence that the Gulf Stream crossed the North Atlantic at a lower latitude before 1850 than since, a displacement parallel to the winds. He concludes that these shifts in ocean and atmospheric circulation patterns are the result of a deficit in the energy available of about 1 per cent before 1800 compared to later. Mitchell (1966) determined a net global warming trend of the order of 0.5 °C beginning in the 1880s and culminating about 1940 with the strongest warming around 1920. Thereafter, the warming has given way to a cooling trend which is still continuing. This reversal corres-ponds to the increase and subsequent decrease in the strength of the westerlies over the North Atlantic (Lamb and Johnson, 1966). Kirch (1970) suggests that the post-1940 cooling was slow in high latitudes until the 1960s due to the limited extent of the pack ice. Subsequently, the cooling seems to have been more rapid, especially in the Canadian Arctic with more persistent summer ice in Baffin Bay (Bradley and Miller, 1972; Dun-bar, 1972; Bradley, 1973).

Longitudinal shifts in the position of the troughs in the upper westerlies are also apparent during periods of climatic change. Lamb (1966) cites evidence for a westward movement of the troughs over Europe during the climatic deterioration from 1300 to 1600, and an eastward movement during the general improvement of climate from 1700. The consequence of shifts in the mean position of troughs and ridges in terms of modifi-cation of climatic patterns and weather conditions has been illustrated by Brinkmann and Barry (1972). Employing data from Keewatin and Labrador–Ungava because of the palaeoecological significance of these areas (see Nichols, Chapter 11A), they determined the mean monthly composite patterns of surface weather anomalies and 700 mb height anomalies, then related these to different positions of the mean 700 mb trough over eastern Canada (the mean tropospheric trough normally lies over Baffin Island and Labrador–Ungava). The results show both distinct similarities and differences between the two regions associated with a westward displacement of the trough. On a monthly time scale, high snowfall in winter occurs simultaneously in the two areas although circulation patterns involved are distinct. Spells of cool summer weather can affect both areas simultaneously, but such weather patterns are usually dry in Keewatin and wet in Labrador–Ungava. This suggests that vegetation changes which are determined pri-marily by summer precipitation should not be expected to be in phase in Keewatin and northern Ungava. However, temperature differences between the areas with respect to trough position in summer are probably too slight to be of consequence. Since the advance and retreat of the boreal forest in postglacial time (Bryson et al., 1965, 1970; Nichols, 1970; Nichols, Chapter 11A) is probably a response to temperature changes, the shifts in forest border in both areas probably occurred simultaneously. Brinkmann and Barry also indicate that, on synoptic climatological grounds, it seems evident that deep snow accumulations presaging development of a continental ice sheet would probably develop first over Keewatin because of the different circulation patterns associated with winter snowfall in the two areas. Later, the low summer temperatures required for the snow cover to persist through the warm season would probably be associated with similar climatic patterns in the two areas.

5. The glacial climate

The hypothetical characteristics of the glacial and nonglacial circulation patterns and the nature of the shift between them has been discussed in some detail by Lamb et al. (1966) and Lamb and Woodroffe (1970). Changes in circulation patterns occurring in initial phases of a glacial period could take place abruptly due to a shift from one quasi-stable state to another. For example, Dansgaard et al. (1972) point to a pronounced decrease in $^{18}O/^{16}O$ ratio in the Camp Century core reflecting a shift of climate to one of glacial severity within perhaps 100 years. The cause of this event is unknown. Recovery from this decrease was equally rapid. The onset of the Wisconsin is apparent at 73,000 BP when there was a less dramatic, but sustained, cooling coincident with less summertime solar radiation. Lamb and Woodroffe (1970) proposed that rapid glacierization and ice build-up could occur in 5000 years or less, and perhaps only 1000 years, although such rapidity of ice growth has been questioned in view of the snow accumulation that would be required (Barry et al., 1971). Many such detailed questions remain to be resolved. Nevertheless, it is clear that once an extensive cover of snow and ice persisted, the increased albedos and cooling effect of the snow cover would modify the thickness pattern (see Barry and Hare, Chapter 2A) with consequent changes in the hemispheric circulation (Lamb, 1955). The magnitude of the temperature change required for the initiation of glaciation, and of the climatic patterns associated with the accumulation of an ice sheet over the northern regions of the globe, still remains uncertain. Loewe (1971) points out some of the factors involved in full understanding of these events. It is generally agreed that the Quaternary ice sheets were brought about by a cooling of the atmosphere. But, since a general temperature decrease would result in a lowered content of water vapour in the air, there must be an explanation of the mechanism by which both ample moisture and low temperatures could be achieved simultaneously. Loewe concludes that for growth of the Keewatin and Labrador–Ungava ice sheets, with present-day rates of precipitation, a summer cooling of even 6°C would be insufficient to initiate the events required. According to his view, an initial cooling exceeding 6°C, or a decrease of 6°C with increased precipitation would be needed for inception of the Laurentide ice sheet. For Baffin Island also, Andrews et al. (1972) have shown the importance of increased winter snowfall for an extension of cirque glacierization.

It is apparent that the atmospheric conditions associated with the onset of glaciation remain to be determined. However, a clearer picture of conditions at the last glacial maximum is now available. A reconstruction based on thermodynamic principles by Lamb and Woodroffe (1970) suggests that the westerlies were displaced southward and that there was enhanced meridional flow, with moisture for the North American and European ice sheets coming from the Gulf of Mexico and the subtropical North Atlantic on the one hand, and from the Mediterranean and the subtropical Atlantic on the other.

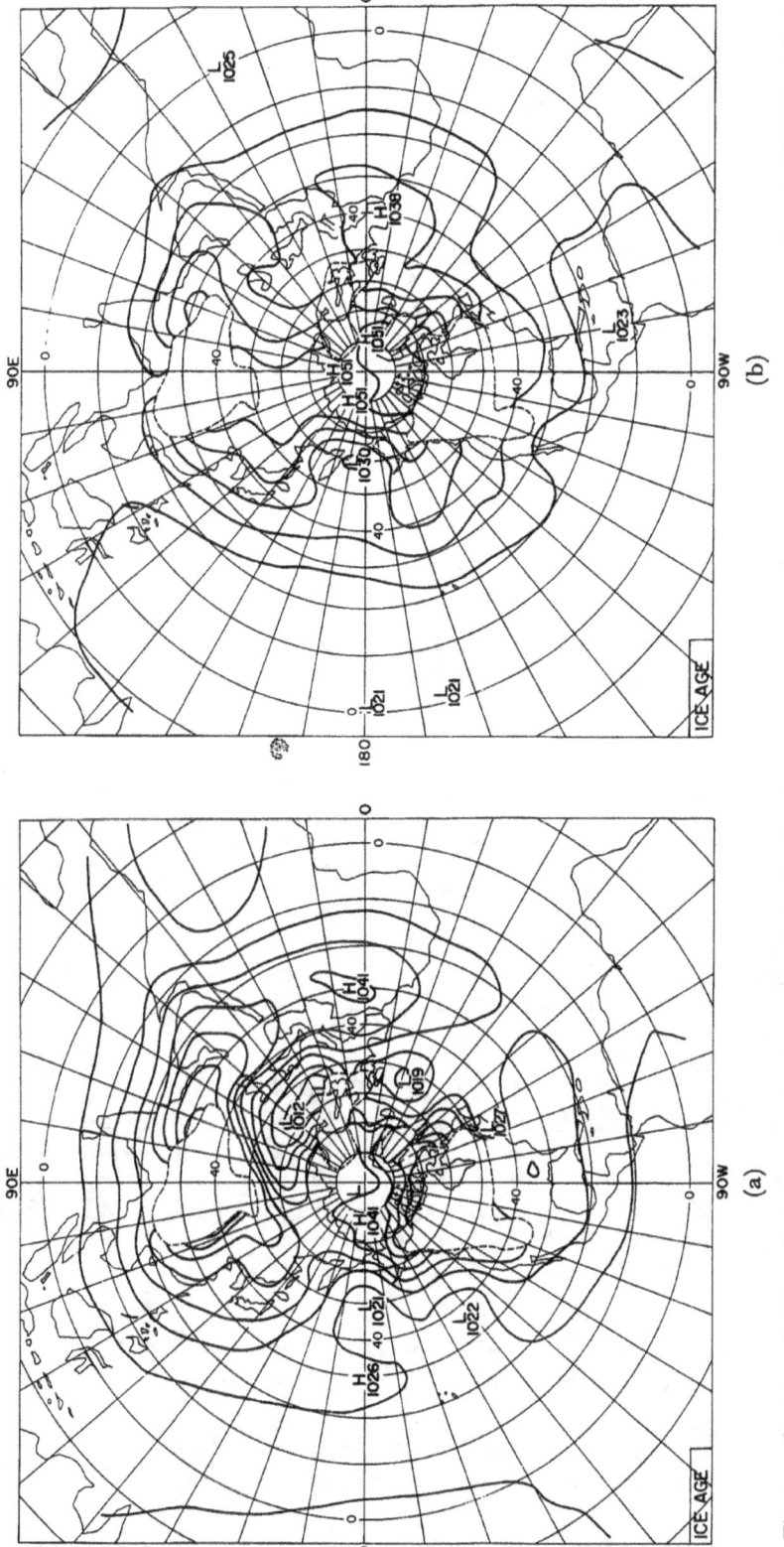

Fig. 5.5. Computed thirty-day mean-sea-level pressure (mb) for (a) January, (b) July, at the last glacial maximum, using the NCAR global circulation model with modified boundary conditions (from Williams *et al.*, 1973). Areas above 1.5 km outlined by dashed line. Actual pressure values are essentially arbitrary.

A simulation using a numerical global circulation model gives somewhat different results (Barry, 1973; Williams *et al.*, 1973). The NCAR six-layer model was run for control cases in January and July and for the last glacial maximum (20,000 BP) in these two months with appropriate boundary conditions for ice extent and topography, snow

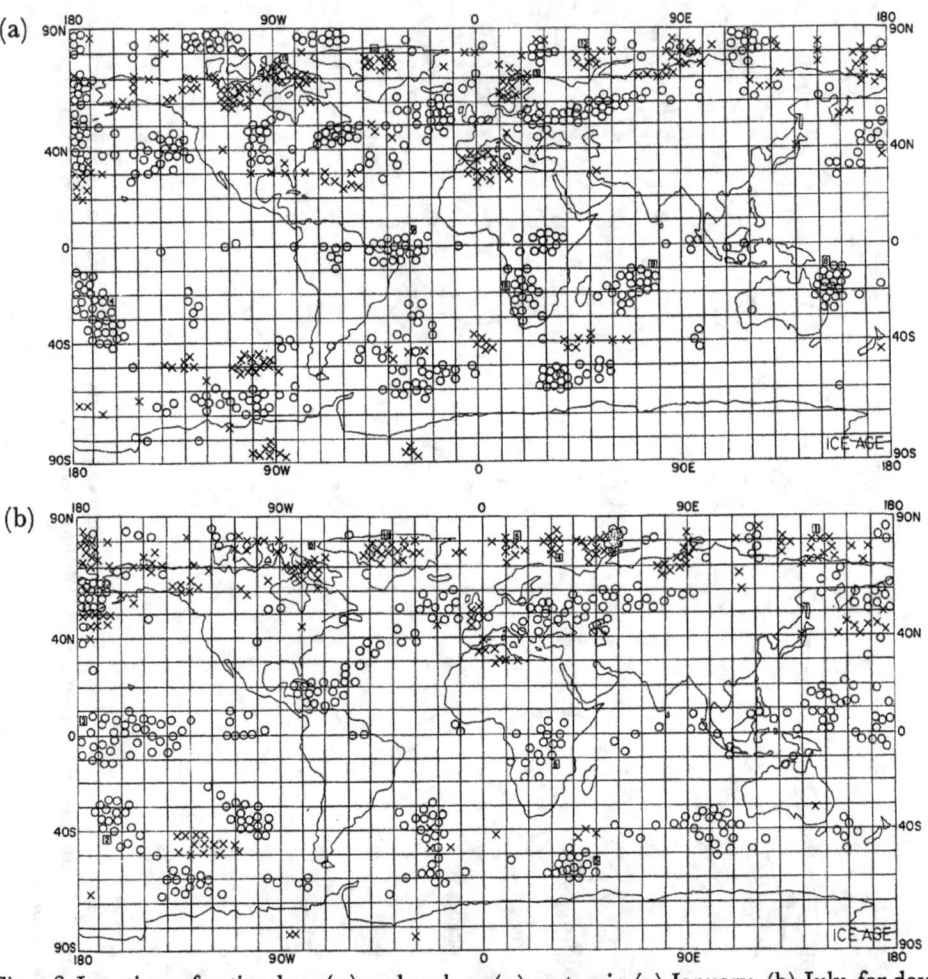

Fig. 5.6. Locations of anticyclone (×) and cyclone (○) centres in (a) January, (b) July, for days 51–80 of a simulation of conditions at the last glacial maximum (from Williams *et al.*, 1973). Numbers denote additional centres within a cluster.

cover, land surface albedos and sea surface temperature. The sea temperatures remain fixed in the individual experiments while temperatures on the land are computed from a surface energy balance. Cloudiness at 3 and 9 km is model-derived.

The simulations for the glacial maximum indicate that the mean zonal wind in July in mid-latitudes of the northern hemisphere was comparable with present winter conditions. The upper westerlies are shown not to have been displaced south of the

Laurentide ice sheet although the Icelandic and Aleutian lows are 10° south of their present locations and the ice sheets and sea ice displaced the frontal zones southward (fig. 5.5). In July, high pressure developed over northern Siberia and the Arctic Basin. There were storm tracks across the North Atlantic and from eastern Europe into Asia, but with virtually no cyclonic activity near the Laurentide ice sheet (fig. 5.6).

Analysis of precipitation shows that the amounts indicated in the control cases are much greater than observed. However, the magnitudes in the July glacial maximum case are dramatically reduced in the northern hemisphere. There is also a slight reduction in January, most pronounced between 0–10°N and 55–70°N. These preliminary investigations indicate that palaeoclimatic modelling will provide a wealth of fascinating material for comparison with observational evidence.

6. Conclusions

There are still major gaps in our understanding of the dynamics of the global circulation and the relative significance of the various climatogenic factors. Influences as yet unrecognized may play highly significant or crucial roles in triggering climatic change. The theory of continental drift was, for example, long held in disrepute, but recent evidence has brought it forcefully back into position as a leading contender among causal hypotheses for climatic change on the geological time scale (Crowell and Frakes, 1969; Crawford and Daily, 1971). Moreover, the evidence of pre-Pleistocene glaciation in high latitudes (see Andrews, Chapter 6A) suggests that continental drift may be more important than was formerly believed for Cainozoic and Quaternary climatic events: Flohn (1969) derives a synthetic geophysical ice age model based on the theory of continental drift that 'needs neither a change of the solar constant, nor, necessarily, an influence of a changing orbit of the earth; neither a change in gravitation nor, with the exception of the temperature-dependent constituent of water vapour, a change in the composition of the earth's atmosphere'. Much new evidence has recently been obtained on climatic change for arctic areas deriving from glacial geomorphology, glaciology and palynology (Andrews et al., 1972; Blake, 1970; Ritchie and Hare, 1971; Schytt et al., 1968; Vasari et al., 1972). Continual reassessment of existing syntheses is therefore essential, along the lines presented by Kukla et al. (1972). This is all the more important in order to provide input for and a check on numerical modelling results. Simulation experiments may hold the key to many otherwise unresolvable problems. Certainly, the modelling results so far obtained show several unexpected effects, demonstrating that the interactions of the earth–atmosphere system are too complex to be treated by simple inferential reasoning. Even so, new problems will arise. Fletcher (1969) showed that warming in much of northern high latitudes appears to coincide with increasing Antarctic Ocean ice, whereas MacCracken's (1970) model demonstrates the opposite. Probably, the former result is merely coincidental on a limited time scale. Close liaison between field scientists, meteorologists and modellers will be necessary to identify the most probable answers to the many critical questions of palaeoclimatology.

ACKNOWLEDGEMENT

Dr John Kutzbach contributed careful reading of early versions of the manuscript and offered many valuable suggestions.

References

ANDREWS. J. T. (1973) The Wisconsin Laurentide ice sheet: dispersal centers, problems of rates of retreat, and climatic implications, *Arct. Alp. Res.*, **5** (3), 185–99.

ANDREWS, J. T., BARRY, R. G., BRADLEY, R. S., MILLER, G. H. and WILLIAMS, L. D. (1972) Past and present glaciological responses to climate in eastern Baffin Island. *Quat. Res.*, **2**, 303–14.

BARRY, R. G. (1966) Meteorological aspects of the glacial history of Labrador–Ungava with special reference to atmospheric vapour transport. *Geogr. Bull.*, **8**, 319–40.

BARRY, R. G. (1973) The conditions favouring glacierization and deglacierization of North America from a climatological viewpoint. *Arct. Alp. Res.*, **5** (3), 171–84.

BARRY, R. G. and CHORLEY, R. J. (1968) *Atmosphere, Weather and Climate.* London: Methuen. 319 pp.

BARRY, R. G., IVES, J. D. and ANDREWS, J. T. (1971) A discussion of atmospheric circulation during the last ice age. *Quat. Res.*, **1** (13), 415–18.

BLAKE, W. JR (1970) Studies of glacial history in Arctic Canada: I. Pumice, radiocarbon dates, and different postglacial uplift in the eastern Queen Elizabeth Islands. *Can. J. Earth Sci.*, **7** (2), 634–64.

BRADLEY, R. S. and MILLER, G. H. (1972) Recent climatic change and increased glacierization in the eastern Canadian Arctic. *Nature*, **237**, 385–7.

BRINKMANN, W. A. R. and BARRY, R. G. (1972) Palaeoclimatological aspects of the synoptic climatology of Keewatin, Northwest Territories. *Palaeogeogr., Palaeoclimatol., Palaeoecol.*, **11**, 77–91.

BROECKER, J., EWING, M. and HEEZEN, B. C. (1960) Evidence for an abrupt change in climate close to 11,000 years ago. *Amer. J. Sci.*, **258**, 429–48.

BROECKER, W. S. and VAN DONK, J. (1970) Insolation changes, ice volumes, and the ^{18}O record in deep sea cores. *Rev. Geophys. Space Phys.*, **8** (1), 169–98.

BROOKS, C. E. P. (1951) Geological and historical aspects of climatic change. *In:* MALONE, T. F. (ed.) *Compendium of Meteorology*, 1004–18. Amer. Met. Soc.

BRYSON, R. A., BAERREIS, D. A. and WENDLAND, W. M. (1970) The character of late-glacial and post-glacial climatic changes. *In:* DORT, W. JR and JONES, J. K. JR (eds.) *Pleistocene and Recent Environments of the Central Great Plains*, 53–74. Univ. of Kansas, Dept of Geol., Spec. Pub. 3. Lawrence: Univ. of Kansas Press.

BRYSON, R. A., IRVING, W. N. and LARSEN, J. A. (1965) Radiocarbon and soil evidence of former forest in the southern Canadian tundra. *Science*, **147** (3653), 46–8.

BRYSON, R. A., WENDLAND, W. M., IVES, J. D. and ANDREWS, J. T. (1969) Radiocarbon isochrones on the disintegration of the Laurentide Ice Sheet. *Arct. Alp. Res.*, **1** (1), 1–14.

BUDD, W. F., JENSSEN, D. and RADOK, U. (1971) Reinterpretation of deep ice temperature. *Nature*, **232**, 84–5.

BUDYKO, M. I. (1966) Polar ice and climate. *In:* FLETCHER, J. O. (ed.) *Proceedings of the Symposium of the Arctic Heat Budget and Atmospheric Circulation*, 3–21. Mem. RM-5233-NSF. Santa Monica: RAND Corp.

BUDYKO, M. I. (1971) *Klimat i Zhizn.* Leningrad: Gidromet. Izdat. 472 pp.

CLARK, D. L. (1971) The Arctic Ocean ice cover and its late Cenozoic history. *Bull. Amer. Geol. Soc.*, **82**, 3313–23.

CRAWFORD, A. R. and DAILY, B. (1971) Probable non-synchroneity of late Precambrian glaciations. *Nature*, **230**, 111–12.

CROWELL, J. C. and FRAKES, L. A. (1969) Phanerozoic glaciation and the causes of ice ages. *Amer. J. Sci.*, **268**, 193–224.

DANSGAARD, W. (1954) The O^{18} abundance in fresh water. *Geochim. & Cosmochim. Acta*, **6**, 241–60.

DANSGAARD, W. (1961) The isotopic composition of natural waters. *Medd. om Grønland*, **165** (2), 1–120.

DANSGAARD, W. and TAUBER, H. (1969) Glacier oxygen-18 content and Pleistocene ocean temperatures. *Science*, **166**, 499–502.

DANSGAARD, W., JOHNSON, S. J., MOLLER, J. and LANGWAY, C. C. JR (1969) One thousand centuries of climatic record from Camp Century on the Greenland Ice Sheet. *Science*, **166**, 377–81.

DANSGAARD, W., JOHNSEN, S. J., CLAUSEN, H. B. and LANGWAY, C. C. JR (1972) Speculation about the next glaciation. *Quat. Res.*, **2**, 396–8.

DONN, W. L. and EWING, M. (1966) A theory of ice ages, III. *Science*, **152**, 1706–11.

DUNBAR, M. (1972) Increasing severity of ice conditions in Baffin Bay and Davis Strait and its effect on the extreme limits of ice. *In:* KARLSSON, T. (ed.) *Sea Ice*, 87–93. Reykjavik: Nat. Res. Council.

DWYER, H. A. and PETERSEN, T. (1973) Time-dependent global energy modeling. *J. Appl. Met.*, **12**, 36–42.

EMILIANI, C. (1955) Pleistocene temperatures. *J. Geol.*, **63**, 538–78.

EMILIANI, C. (1964) Palaeotemperature analysis of the Caribbean cores A 254-BR-C and CP-28. *Bull. Amer. Geol. Soc.*, **75**, 129–44.

EMILIANI, C. (1966) Isotopic paleotemperatures. *Science*, **154** (3751), 851–7.

EMILIANI, C. (1969) Interglacial high sea levels and the control of Greenland ice by the precession of the equinoxes. *Science*, **166**, 1503–4.

EMILIANI, C. (1971a) The last interglacial: paleotemperatures and chronology. *Science*, **171**, 571–3.

EMILIANI, C. (1971b) The amplitude of Pleistocene cycles at low latitudes and the isotopic composition of glacial ice. *In:* TUREKIAN, K. K. (ed.) *The Late Cenozoic Glacial Ages*, 183–97. New Haven: Yale Univ. Press.

ERICSON, D. B. and WOLLIN, G. (1968) Pleistocene climates and chronology in deep-sea sediments. *Science*, **162**, 1227–34.

ERICSON, D. B. and WOLLIN, G. (1970) Pleistocene climates in Atlantic and Pacific oceans: a comparison based on deep-sea sediments. *Science*, **167**, 1483–4.

ERICSON, D. B., EWING, M. and WOLLIN, G. (1964) Sediment cores from the Arctic and subarctic seas. *Science*, **144** (3623), 1183–92.

EWING, M. and DONN, W. L. (1956) A theory of ice ages. *Science*, **123**, 1061–6.

FAIRBRIDGE, R. W. (1961) Convergence of evidence on climatic change and ice ages. *Ann. New York Acad. Sci.*, **95**, 542–79.

FAIRBRIDGE, R. W. (1972) Climatology of a glacial cycle. *Quat. Res.*, **2**, 283–302.

FLETCHER, J. O. (1966) The Arctic heat budget and atmospheric circulation. *In:* FLETCHER, J. O. (ed.) *Proceedings of the Symposium on the Arctic Heat Budget and Atmospheric Circulation.* Mem. RM-5233-NSF. Santa Monica: RAND Corp.

FLETCHER, J. O. (1969) *Ice Extent on the Southern Ocean and its Relation to World Climate.* Mem. RM-5793-NSF. Santa Monica: RAND Corp. 108 pp.

FLOHN, H. (1969) Ein geophysikalisches Eiszeit-Modell. *Eiszeitalter u. Gegenwart*, **20**, 204–31.

GODWIN, H. (1966) Introductory address. *In: World Climate from 8000 to o BC, Proceedings of the International Symposium*, 3–14. London: Roy. Met. Soc.

GOLDTHWAIT, R. P., DREIMANNIS, A., FORSYTH, J. L., KARROW, P. F. and WHITE, G. W. (1965) Pleistocene deposits of the Erie lobe. *In:* WRIGHT, H. E. JR and FREY, D. G. (eds.) *The Quaternary of the United States*, 85–112. Princeton: Princeton Univ. Press.

HERMAN, Y. (1970) Arctic paleo-oceanography in late Cenozoic time. *Science*, **169**, 474–7.

KIRCH, H.-D. (1970) Klimaschwankungen in der Arktis. *Met. Abh.*, **110** (4). 86 pp.

KRISTJANNSON, L. (1969) The ice drifts back to Iceland. *New Sci.*, **41** (639), 508–9.

KUKLA, J. (1970) Correlation between loesses and deep-sea sediments. *Geol. Fören, Stockholm Förh.*, **92**, 148–80.

KUKLA, J., MATTHEWS, R. K. and MITCHELL, J. M. JR (1972) The end of the present interglacial. *Quat. Res.*, **2**, 261–9.

KUTZBACH, J. E., BRYSON, R. A. and SHEN, W. C. (1968) An evaluation of the thermal Rossby number in the Pleistocene. *Met. Monogr.*, **8** (30), 134–8.

LAMB, H. H. (1955) Two-way relationships between the snow or ice limit and 1000–500 mb thicknesses in the overlying atmosphere. *Quart. J. R. Met. Soc.*, **81**, 172–89.

LAMB, H. H. (1965) Climatic changes and variations in the atmospheric and ocean circulations. *Geol. Rund.*, **54** (1), 486–503.

LAMB, H. H. (1966) *The Changing Climate.* London: Methuen. 236 pp.

LAMB, H. H. and JOHNSON, A. I. (1966) *Secular Variations of the Atmospheric Circulation Since 1750.* Geophys. Mem., 110. London: HMSO. 125 pp.

LAMB, H. H. and WOODROFFE, A. (1970) Atmospheric circulation during the last ice age. *Quat. Res.*, **1**, 29–58.

LAMB, H. H., LEWIS, R. P. W. and WOODROFFE, A. (1966) Atmospheric circulation and the main climatic variables between 8000 and o BC: meteorological evidence. *In: World Climate from 8000 to o BC, Proceedings of the International Symposium*, 174–217. London: Roy. Met. Soc.

LOEWE, F. (1971) Considerations regarding the origin of the Quaternary ice sheet of North America. *Arct. Alp. Res.*, **3** (4), 331–44.

MACCRACKEN, M. C. (1970) *Tests of Ice Age Theories using a Zonal Atmospheric Model.* UCRL-72803. Livermore: Lawrence Radiation Lab., Univ. of Calif. 58 pp.

MAYKUT, G. A. and UNTERSTEINER, N. (1969) *Numerical Prediction of the Thermodynamic Response of Arctic Sea Ice to Environmental Changes.* Mem. RM-6093-PR. Santa Monica: RAND Corp. 173 pp.

MAYKUT, G. A. and UNTERSTEINER, N. (1971) Some results of a time-dependent thermodynamic model of sea ice. *J. Geophys. Res.*, **76** (6), 1550–75.

MILANKOVITCH, M. (1920) *Théorie Mathématique des Phénomenes Thermique Produits par la Radiation Solaire.* Paris: Gauthier Villars.

MILANKOVITCH, M. (1930) Mathematische Klimalehre und Astronomische Theorie der Klimaschwankungen. *Handbuch der Klimatologie*, Bd. 1, Teil A., A1-A176. Berlin: Springer-Verlag.

MILANKOVITCH, M. (1969) *Canon of Insolation and the Ice Age Theory.* Jerusalem: Israel Progr. Sci. Transl. 484 pp.

MITCHELL, J. M. JR (1966) Stochastic models of air-sea interaction and climatic fluctuation. *In:* FLETCHER, J. O. (ed.) *Proceedings of the Symposium on the Arctic Heat Budget and Atmospheric Circulation*, 45–74. Mem. RM-5233-NSF. Santa Monica: RAND Corp.

MORIN, R. W., THEYER, F. and VINCENT, E. (1970) Pleistocene climates in the Atlantic and Pacific Oceans: a reevaluated comparison based on deep-sea sediments. *Science*, **169**, 365–6.

NAIRN, A. E. M. (1961) *Descriptive Palaeoclimatology.* New York: Interscience Publishers, Inc. 381 pp.

NEWSON, R. L. (1973) Response of a general circulation model of the atmosphere to removal of the Arctic ice-cap. *Nature*, **241**, 39–40.

NICHOLS, H. (1970) Late Quaternary pollen diagrams from the Canadian Arctic barren grounds at Pelly Lake, northern Keewatin, NWT. *Arct. Alp. Res.*, **2**, 43–61.

NILSSON, T. (1964) Standartpollendiagramme und C-14 datierungen aud dem Agerods Mosse in Mittleren Schonen. *Pub. Inst. Mineral., Palaeontol.*, 124–52.

RITCHIE, J. C. and HARE, F. K. (1971) Late-Quaternary vegetation and climate near the arctic tree line of northwestern North America. *Quat. Res.*, **1**, 331–42.

SAWYER, J. S. (1966) Possible variations in the general circulation of the atmosphere. *In: World Climate from 8000 to 0 BC*, 218–29. London: Roy. Met. Soc.

SCHWARZBACH, M. (1963) *Climates of the Past.* London: Van Nostrand. 328 pp.

SCHYTT, V., HOPPE, G., BLAKE, W. JR and GROSSWALD, M. G. (1968) The extent of the Würm Glaciation in the European Arctic. *Int. Ass. Sci. Hydrol. Pub.* (IUGG Berne), 79, 207–16.

SELLERS, W. D. (1965) *Physical climatology*, 197–228. Chicago: Univ. of Chicago Press.

SELLERS, W. D. (1973) A new global climatic model. *J. Appl. Met.*, **12**, 241–54.

SHACKLETON, N. (1967) Oxygen isotope analyses and Pleistocene temperatures reassessed. *Nature*, **215**, 15–17.

SHAW, D. M. and DONN, W. M. (1968) Milankovitch radiation variations; a quantitative evaluation. *Science*, **162**, 1270–2.

UREY, H. C. (1947) The thermodynamic properties of isotopic substances. *J. Chem. Soc.*, 562–81.

VAN DER HAMMEN, J., MAARLEVELD, A. D., VOGEL, J. C. and ZAGWIN, W. H. (1967) Stratigraphy, climatic succession and radiocarbon dating of the last glacial in the Netherlands. *Geol. en Mijnbouw*, **46**, 79–95.

VASARI, Y., HYVÄRINEN, H. and HICKS, S. (1972) Climatic changes in arctic areas during the last ten-thousand years. *Acta Univ. Oulu*, Ser. A, Sci. Rerum Natural. 3, Geol. 1. 511 pp.

VERNEKAR, A. D. (1968) Long-period global variations of incoming solar radiation. *In: Research on the Theory of Climate*, Vol 2. Hartford: Travelers Research Center, Inc. 289 pp.

VOWINCKEL, E. and ORVIG, S. (1970) The climate of the North Polar Basin. *In:* ORVIG, S. (ed.) *Climates of the Polar Regions*, 129–252. Vol. 14 of LANDSBERG, G. E. (ed.) *World Survey of Climatology*. Amsterdam: Elsevier.

WARSHAW, M. and RAPP, R. R. (1973) An experiment on the sensitivity of a global circulation model. *J. Appl. Met.*, **12**, 43–9.

WILLIAMS, J., BARRY, R. G. and WASHINGTON, W. M. (1973) Simulation of the climate at the Last Glacial Maximum using the NCAR global circulation model. *Inst. Arct. Alp. Res. Occas. Pap.*, 5. Univ. of Colorado. 39 pp.

Manuscript received May 1971; revised April 1973.

Cainozoic glaciations and crustal movements of the Arctic

John T. Andrews

Institute of Arctic and Alpine Research,
University of Colorado

1. Introduction

The subject matter included within Cainozoic glaciations and crustal movements of the Arctic is immense both in terms of the area covered and the period of time under review. These practical difficulties are heightened and aggravated by an overall sparsity of data and widely divergent theories, not to mention the fact that approximately half the land area lies in the USSR (fig. 6A.1) and much Russian literature is not readily accessible to Western scientists. With these provisos the scope of this chapter can be briefly outlined.

The story does not open at the Pliocene/Pleistocene boundary because it is becoming increasingly evident that glaciation in many areas preceded this time line. Consequently the first concern is with the tectonic events of the Tertiary and the arrangement of the landmasses relative to each other and to the position of the rotational pole. Miocene glaciations are then discussed together with a consideration of the causes of the glacial/interglacial periodicities that characterize the Quaternary proper. The actual number and timing of these events is then taken up largely through the detailed stratigraphic records that have been deciphered in Alaska and Siberia and from ocean cores. A more detailed treatment of events is possible for the last glacial stage – the Wisconsin or Würm. Glacials and interglacials involve a mass transfer of ice/water in response to the build-up and removal of loads. The response of the arctic regions to glacio-isostatic and eustatic crustal movements are elaborated and used to locate the main centres of glaciation in the region.

BROAD GEOGRAPHICAL SETTING

The arctic regions (fig. 6A.1) consist of a complex interfingering of land and ocean. The Lomonosov Ridge divides the Arctic Ocean into two major basins, the Canada Basin

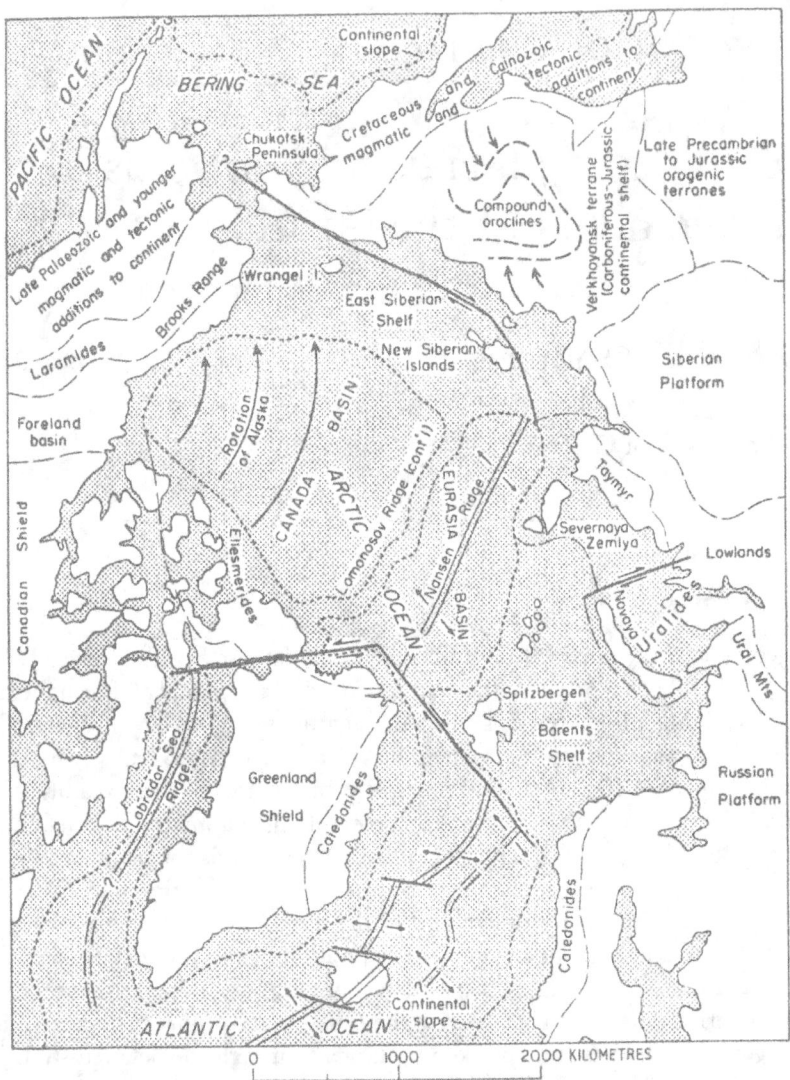

Fig. 6A.1. Tectonic elements of the Arctic related to the post-Triassic opening of the Arctic Ocean (from Hamilton, 1970).

and the Eurasia Basin. Rather narrow seaways connect the Arctic Ocean with the Atlantic and Pacific oceans. High plateau rims of the Laurentian Shield front the Labrador Sea and Baffin Bay in the eastern Canadian Arctic where elevations occasionally reach 1800 to 2000 m above sea level. Similar uplifted rims of the other shields occur in Greenland and in west Norway. However, over large areas of the Arctic relief and elevation are low. This applies to sections of the Canadian Arctic Archipelago, the Canadian Arctic mainland coast and large sections of the USSR. Most major topographic barriers

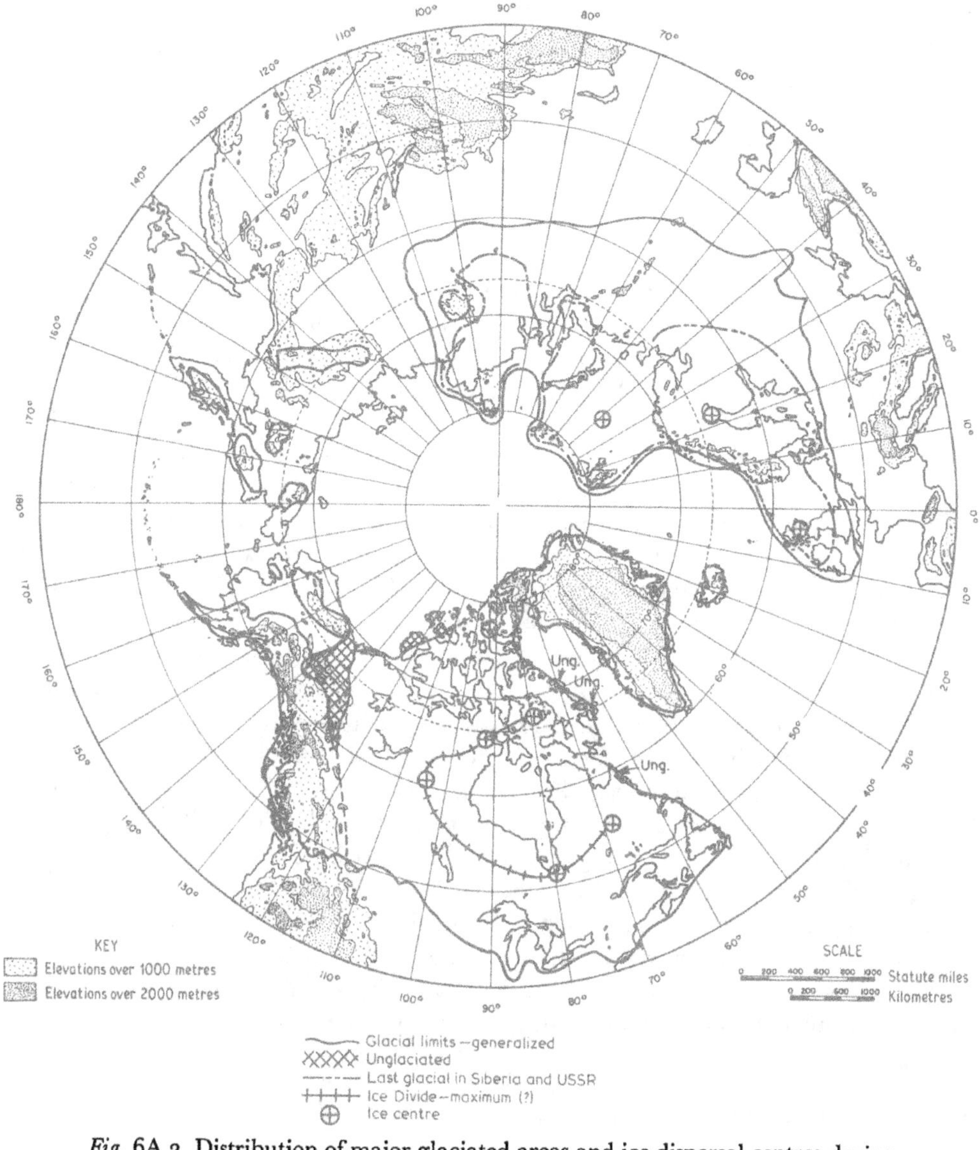

Fig. 6A.2. Distribution of major glaciated areas and ice dispersal centres during the last glaciation.

trend north–south in the form of the Ural Mountains, the Coast Ranges and Rocky Mountains and the uplifted rims of the eastern Canadian Arctic seaboard, but in Alaska significant topographic elements such as the Brooks Range run east–west.

The current state of glacierization is discussed in Chapter 4B by Hattersley-Smith. Suffice it to note that the Arctic is dominated by the mass of the Greenland ice sheet that has probably been quasi-stable for the whole Quaternary, if not since the Miocene.

This contrasts with the oscillations of the major northern hemisphere Pleistocene ice sheets such as the Laurentide, Fennoscandian, Barents Sea and Siberian as well as the smaller ice caps and mountain glaciers in Alaska, the Yukon, Iceland and the northern USSR (fig. 6A.2). The nature of this contrast in equilibrium is one of the major unanswered questions of arctic Quaternary glaciology and climatology.

THE PRE-QUATERNARY SETTING

One of the most dramatic events of the last decade has been the realization that during the Mesozoic and Cainozoic significant changes occurred in the basic configuration of land and sea, and that these changes also affected the latitudinal position of landmasses relative to the rotational pole. The terms continental drift and polar wandering have been applied to these movements. The first indicates a movement of the continents relative to each other due to the break-up and impaction of large lithospheric plates; the second term refers to the effects of large-scale slippage of segments of the earth's crust resulting in apparent motion of the rotational pole. In addition to these changes, but probably in part related to them, there has been significant block uplift and mountain building in parts of the Arctic during the Early and Mid-Cainozoic. The broad tectonic elements in the Mesozoic and Cainozoic have recently been discussed in terms of the Arctic by Hopkins (1967a), Hamilton (1970), Harland (1969) and Hallam (1971) and Thorsteinsson and Tozer (1970). Fig. 6A.1 is one interpretation of the tectonic elements of the Arctic Basin related to the post-Triassic opening of the Arctic Ocean.

There are two main facets of any attempt to 'explain' Cainozoic glaciations. The first is the need for the establishment, through the mechanism of continental drift, of a broad framework of oceans and continents at a sufficiently high latitude to produce potential snow/ice generating regions. These movements resulted in unidirectional climatic changes. In this sense Greenland has been very efficient in realizing this potential. Superimposed on the geological framework have been secondary climatic fluctuations related to changes in insolation, extent of arctic sea ice, random processes, etc., as discussed in Chapter 5 by Larsen and Barry. This section is really concerned with the establishment of the necessary platforms for continental glacierization.

CONTINENTAL DRIFT AND ATTENDANT TECTONIC MOVEMENTS

Hamilton (1970) notes that the Arctic Ocean probably opened in post-Palaeozoic time but the seaways into the Atlantic and Pacific date from Late Cretaceous time and subsequently. The development of the arctic basins and the formation of the north–south connecting seaways are associated with oceanic ridges, seafloor spreading and related transform faults (fig. 6A.1). The Eurasia Basin is spreading out from the Nansen Ridge at a rate of c. 0.85 cm yr^{-1}. The Nansen Ridge has been offset from the major Mid-Atlantic Ridge by a fault which Hamilton (1970) indicates as running in the channel between Ellesmere Island and northwest Greenland imparting a counterclockwise motion of Greenland relative to Canada. Another oceanic ridge separates Labrador and Baffin Island from Greenland.

Hallam (1971) suggests that the first major opening of the North Atlantic occurred in Late Jurassic time with a deep ocean trench extending north–south in the Atlantic and curving northwest to separate Labrador and Greenland. By Late-Cretaceous time the deep ocean trench reached north to the latitude of southern Baffin Island and an arm extended north–east between Greenland and Great Britain. A shallow, epicontinental sea extended northwards in the present site of Baffin Bay and between Norway and Greenland. The present extent of the deep ocean trenches is shown in fig. 6A.1 as the northernmost extent of their respective ridges. Extensive areas of shallow shelf occur in the vicinity of Spitsbergen and the Barents Sea, off East Siberia and northward from the Canadian Queen Elizabeth Islands.

The extension of the rift into Baffin Bay is in part shown by the presence of Tertiary volcanic units along the east coast of Baffin Island in the vicinity of Cape Dyer ($c.$ 67°N). Clarke and Upton (1971) assign the units to the lower Palaeocene based on a radiometric date and on the flora of associated shale and coal beds. The basalts are considered correlative to the basalts in West Greenland. Hallam's model (1971, p. 152) for the movements related to the rifting are attractive in that they explain the present topography of the eastern Canadian Arctic, which (as noted earlier) consists of an uplifted plateau rim fronting the east coast and a gradual decline of elevation toward the interior (Hudson Bay and Foxe Basin). In this interpretation a major isostatic uplift occurs initially because of increased heat flow in the vicinity of the rift. The uplift is probably associated with normal faulting parallel to the rift. Subsequently, as the margins move away from each other, subsidence occurs because of the need for isostatic adjustments. Clarke and Upton (1971) indicate that uplift of 400 m occurred during the Early to Mid-Tertiary. Geomorphic evidence indicates that this figure might be nearer 600 m. Evidence for subsidence is not so clear but the presence of major trenches, cutting across the continental shelf of eastern Baffin Island for 50 km or so in locations where the maximum vertical extent of the ice on the outer coast is 300 to 600 m, might be related to this aspect of Hallam's hypothesis as these trenches could not have been cut by glacial action.

In the Queen Elizabeth Islands the Mid-Cainozoic (Tozer, in Thorsteinsson and Tozer, 1970) was marked by the Eurekan Orogeny, during which time folding and thrusting took place in the region of the Sverdrup Basin and Ellesmere Island whereas normal faulting (Thorsteinsson and Tozer, 1970, p. 587) affected areas such as Baffin, Devon, Somerset, Cornwallis and Bathurst islands.

The tectonic events of the Cainozoic and their effects on the Beringia question have been treated by Hopkins (1967b). Southern Alaska, the Aleutian Islands and the Kamchatka Peninsula were active areas during the entire Cainozoic but '. . . most of Beringia, to the north, seems to have been tectonically quiet during Early and Middle Tertiary time and to have undergone little topographic change except prolonged, slow subaerial erosion, a process that had reduced much of the region to a peneplain by Late Miocene time' (Hopkins, 1967b, p. 453). At this time a narrow seaway was opened between Alaska and Siberia, and Pacific molluscs migrated across the Arctic Ocean into the

Atlantic Ocean. During the Late Miocene, Siberia and Canada extended much further north than they do today (Hopkins, 1967b, p. 457); furthermore, Greenland and Canada were still joined with Baffin Bay, an enclosed gulf, projecting northward to 74°N. The Beaufort Formation of the Queen Elizabeth Islands is interpreted as Miocene/Pliocene in age (Hopkins, 1967b) but Tozer, (in Thorsteinsson and Tozer, 1970) refers to it as Pliocene/Early Pleistocene. The Bering seaway closed after a short interval and then reopened 3.5 million years ago.

POLAR WANDERING

The relationship between polar wandering and glacierization of the northern hemisphere was discussed by Cox (1968). Palaeomagnetic evidence shows that there have been northward movements of northern hemisphere continents. During the Triassic the Pole was located at c. 50°N, 150°W (Hamilton, 1970). It moved into the Arctic Basin in the Early Cainozoic and has remained within the confines of the Basin ever since. The early movement of the Pole into the Arctic Basin indicates that there is no direct relationship between polar wandering and Cainozoic glaciations of the arctic regions, although it is suggested that the movement of the South Pole to the vicinity of the antarctic continent was more critical. This aspect is discussed later. Cox (1968) concluded that the precise position of the Pole is not critical for glacierization of antarctic and arctic areas and that shifts of 5° to 15° could probably be accommodated without major glaciological responses.

CLIMATIC EVIDENCE AND MID-CAINOZOIC GLACIATIONS

The flora and fauna of Early Tertiary units throughout the Arctic indicate the region had a warm temperate climate. On Baffin Island a recently discovered outcrop of Palaeogene age (Andrews et al., 1972) at 730 m above sea level, lying unconformably on the Precambrian bedrock, contained pollen of tulip trees and other flora indicative of a warm temperate environment. This was followed by increasingly severe conditions and Hamilton (1968) indicates that the Atlantic Arctic had a cold temperate climate by the middle of the Miocene.

The Alaskan flora indicates that '. . . the major climatic deterioration in the Neogene of Alaska took place in the latter half of the Miocene when a drop in the July average of about 7°C took place within a time span of about 4 million years' (Wolfe and Leopold, 1967, p. 203). More recently, Wolfe (1971, p. 27) has concluded from vegetation studies that '. . . a major and rapid climatic deterioration occurred in the Oligocene', and another was defined during Late Eocene time. These sudden shifts in climate are becoming increasingly apparent in the geological record and are most dramatically evidenced by significant glaciations during the Mid-Tertiary. Bandy et al. (1969) have documented glacial marine units and cold water foraminifera off the south coast of Alaska dated at 13 to 14 million years ago. This result substantially agrees with the data of Denton and Armstrong (1969) who obtained radiometric dates of 9 to 10 million years on volcanic rocks interbedded with glacial deposits. Other tills are dated in the period 8.8 to 2.7

million years ago. Recently, Herman (1970) has discussed a sequence of cores from the Arctic Basin which are dated by means of their palaeomagnetic stratigraphy. Ice-rafted debris is present in bottom sections of the cores, and Herman states '. . . presence of ice-rafted detritus in the core bottoms indicates that high latitude glaciation was underway prior to 6 million years ago' (Herman, 1970, p. 476). However, there is discussion (Hunkins *et al.*, 1971) on the dating of this core, and glaciation may not have begun until *c.* 3 million years ago.

Elsewhere around the North Atlantic there is also accruing evidence for glacierization prior to the Pleistocene. In Iceland (Einarsson *et al.*, 1967) the section at Tjörnes indicates at least *ten glaciations* within the last 3 to 1.9 million years ago and the *Glomar Challenger's* (Geotimes, 1970) drilling programme off Newfoundland recovered cores which indicate that continental glaciation occurred 3 million years ago.

DISCUSSION

The events of the Tertiary period discussed above clearly indicate that much of the present topographic and climatic framework for the Arctic was already established in Mid- to Late Tertiary times. The combined effects of polar wandering and continental drift led to a pronounced 'continentality index' for both polar regions. In the case of the southern hemisphere a very high proportion of the land in high latitudes is centred around the Pole, whereas in the northern hemisphere the continentality index peaks at about 60°N and declines poleward (Cox, 1968, fig. 1).

The northward movement of the continents, the development of connections to the Arctic Ocean from the Pacific and Atlantic oceans, and the uplift of the shields in Greenland and the eastern Canadian seaboard must all have had a pronounced effect on the climate. However, these changes were relatively gradual and it is necessary to explain the abrupt deterioration in the Miocene that produced not only major floral changes but initiated glaciation in sections of the Arctic (see Chapter 10A).

It has been suggested that these changes were largely the result of worldwide cooling caused by the development of the antarctic ice sheet during the Early Eocene (Margolis and Kennett, 1970; Le Masurier, 1972). The evidence is in part from the presence of ice-rafted debris in cores from the oceans and in part related to radiometric dates on hyaloclastites from West Antarctica. The latter work (Le Masurier, 1972) indicates the presence of a thick ice sheet in West Antarctica since Eocene time. Mercer (1968a) has, however, argued for at least one major deglaciation of West Antarctica during the Cainozoic based on terrestrial stratigraphy. The deep-sea cores were interpreted by Margolis and Kennett (1970) as indicating glacial conditions in the Antarctic during the Lower Eocene to the Middle Miocene. This was followed by a warm interval before another deterioration set in strongly during the Pliocene and Pleistocene.

The glaciation of Antarctica must have had a profound effect on the climate of the Arctic, primarily by lowering overall world temperature and thus providing conditions under which glaciation of the northern hemisphere was possible. Another significant factor in the history of the Arctic must have been the beginning of glaciation on Greenland. To date no radiometric or deep-sea core data have provided specific evidence on this event. However, certain facts can be deduced from the geological record. The first is that the continued development of Baffin Bay and Davis Strait in the Early Tertiary must have had a major effect on the climate of Greenland – specifically the climate changed from continental to maritime. The opening of the ocean was probably associated with a change in the cyclonic storm tracks which would have a greater tendency to be directed northward from eastern Canada into Baffin Bay. The opening of the rift was associated with isostatic uplift and tilting of the rim of the rift (Kerr, 1970; Hallam, 1971) and in these areas the rim was uplifted 400 to 600 m with attendant normal and block faulting. A similar effect must also have been introduced along the uplifted rim edge of Labrador and Baffin Island. These topographic and climatic changes were superimposed on a general world cooling as noted earlier and it is suggested that the Greenland ice sheet developed as a result of these factors, probably in Miocene time. Mercer (1968b) has also suggested a pre-Pleistocene age for the growth of the Greenland ice sheet. Ice caps and glaciers probably developed along the eastern Canadian Arctic at the same time, but were of more limited extent.

2. Quaternary glaciations: pre-last glaciation

With increasing data the Tertiary is no longer viewed as a period of gradual temperature decline leading to the generation of continental ice sheets once a critical threshold was passed – rather the Tertiary exhibited marked climatic fluctuations. At the moment, at least, the wavelength and amplitude of these changes appear to have been rather broad and low compared to the unstable character of Quaternary time. In particular, the Quaternary has been marked by severe glacial/interglacial oscillations operating on a time scale of 100,000 years or so. Increasingly sophisticated methods of analysis and dating are showing that the traditional mid-continental record of the United States, which embodies four major glaciations, is probably a poor world standard. Analysis of ocean cores and the extremely detailed loess stratigraphy in Czechoslovakia suggest that as many as sixteen glaciations may have affected the northern hemisphere during the last 1.8 million years (Kukla, 1969).

Attempts to explain the oscillatory nature of Quaternary climate are numerous. Recently, the good agreement between uranium series dates of coral reefs related to high sea level stands and insolation variations determined from Milankovitch's (1941) theory has been stressed (Broecker, 1968; Veeh and Chappell, 1970; Kukla, 1969). On the other hand, Shaw and Donn (1968) have argued that the computed changes are too small to provide the necessary trigger action. It is possible that the apparent regularity of the glacial/interglacial sequence is based on a random frequency model in which a sequence of years of particular type (either glacial or interglacial) occur together to give persistence to a trend which then exceeds a critical threshold so that the climate switches from one mode to the other rather abruptly (Curry, 1962; Bryson and Wendland, 1967).

The Tjörnes section in Iceland and the ocean core record of Herman (1970) are alike in indicating that northern regions have probably experienced more than four major glaciations. Core T3-67-11 shows eight major oscillations in the last 700,000 years and four or five significant events prior to that date. The Icelandic record indicates three glaciations in the last 1 million years. As the Kansan glaciation is dated about 600,000 BP by association with one of the Pearlette-like ash beds, a correlation between the Icelandic and USA record is suggested, at least for this part of the record.

The terrestrial and marine stratigraphies of arctic areas are very varied in both detail and site. In view of the many areas where detailed work has not been attempted, because of the factors of distance and logistics, no definitive statement on the Quaternary geology of the Arctic can yet be formulated. Nevertheless, there are some interesting comparisons to be made between different areas. Basic data will first be outlined and then discussed.

LABRADOR AND EASTERN BAFFIN ISLAND, NWT, CANADA

Research during the 1950s and 1960s in the fiord country of northernmost Labrador (Ives, 1958; Løken, 1962; Andrews, 1963) and along the coast of eastern Baffin Island (Mercer, 1956; Ives, 1963; Løken, 1966; Ives and Borns, 1972; Ives and Buckley, 1969; England and Andrews, 1972; and Pheasant, 1971) indicates that there are at least two, and possibly three major surficial units in the area. These units have been mapped on the basis of lithology and weathering criteria. The boundaries between the zones slope seaward (see Løken, 1962, fig. 4) and analysis indicates that these slopes are compatible with those of large outlet glaciers with basal shear stresses between 0.5 and 1.0 bar (Pheasant, 1971). From youngest to oldest they have been named the Saglek glaciation, the Koroksoak glaciation and the Torngat glaciation. Correlations between Labrador and Baffin Island are based on similar degrees of weathering and vertical separation. Until recently, no firm date could be attached to these units but work along the northern Cumberland Peninsula coast (Pheasant, 1971) and near Clyde Fiord (Løken, 1966) suggests that the Saglek glaciation and the youngest Zone (III) of Pheasant embraces the entire Wisconsin. Løken (1966) noted that weathering of a bedrock knoll, that had been a nunatak during a glacial phase which was dated by association with a glacio-marine delta at >54,000 BP (Y-1703), had similar weathering characteristics to the weathered till and bedrock that comprise the deposits of the Koroksoak glaciation (fig. 6A.3). More recently, marine shells from surficial cliffs along the outer foreland of Narpaing Fiord have been dated by uranium series at 137,000 ± 10,000 BP. This deposit is associated with the outermost moraines that form the boundary between the Saglek and Koroksoak glaciations. Studies of various weathering parameters indicate that most differ in the ratio of 1:5 between the two surficial units. There is, therefore, a very real possibility that a major lacuna exists in the record. The age of the Koroksoak glaciation, based on the weathering data, is tentatively estimated at between 200,000 and 500,000 years old. Chemical and X-ray analyses on the three units confirm the considerable difference between the Saglek and Koroksoak deposits. The latter show a development of clay minerals, principally illite and hydrobiotite and an increase in Fe_2O_3, whereas the Saglek deposits show little sign of clay mineral formation. There is less distinction

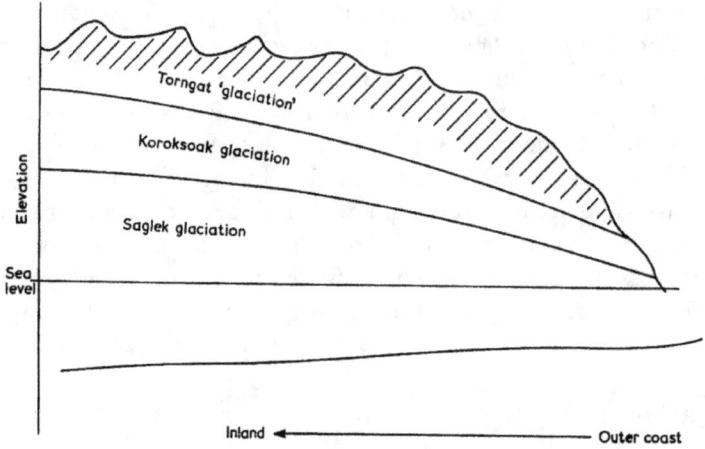

Fig. 6A.3. Schematic cross-section of the weathering zones in coastal Labrador and eastern Baffin Island, NWT, Canada.

between the Koroksoak deposits and the upper zone (the Torngat 'glaciation' of fig. 6A.3). Clay minerals are more abundant and the percent of ferric oxide increases – the differences amount to ratios of between 1 : 1.2 and 1 : 1.6. However, in view of the possible age of the Koroksoak glaciation, these ratios indicate a considerable antiquity (*c.* 400,000 to 1 million years) for the uppermost zone. Determinations of the amount of Fe_2O_3 in Mid-Wisconsin and Late Wisconsin tills indicate that the percent of ferric iron as a function of time follows a log–log function. Fig. 6A.4 presents existing data and extrapolation of the function dates the Koroksoak at 450,000 and the Torngat at 680,000 BP. Error limits on these dates must be rather large but the graph confirms the weathering data of Pheasant (1971) discussed above.

Whereas there is no doubt as to the glacial character of the deposits in the intermediate zone of fig. 6A.3, there is considerable debate as to the interpretation of the uppermost zone which Ives (1963) has argued was at one time glaciated. Løken (1962) considers that there is no unequivocal evidence to support the existence of the 'Torngat glaciation'. In certain areas the evidence supports one or other of these interpretations. Along the northern rim of Cumberland Peninsula, Baffin Island, Clark and Upton (1971) noted no evidence of erratics on the Tertiary basalts that fringe the outermost coast for 150 km. Ives and Borns (1972) noted Precambrian erratics up to elevations of 600 m above sea level in the vicinity of Cape Dyer, but no erratics were found on the basalt above that elevation. In Labrador, Tomlinson (1963) noted indisputable granite-gneiss erratics on the volcanic bedrock that forms the summits of the Kaumajet Mountains between 900 and 1200 m. Further to the south near the Kiglipait Mountains, Johnson (1969) has observed garnetiferous gneiss resting on anorthosites at 1033 m. The upper limit of the Koroksoak glaciation is identified at 975 m on the same mountain.

No precise figures can be given of the altitude of the various zones as they all slope

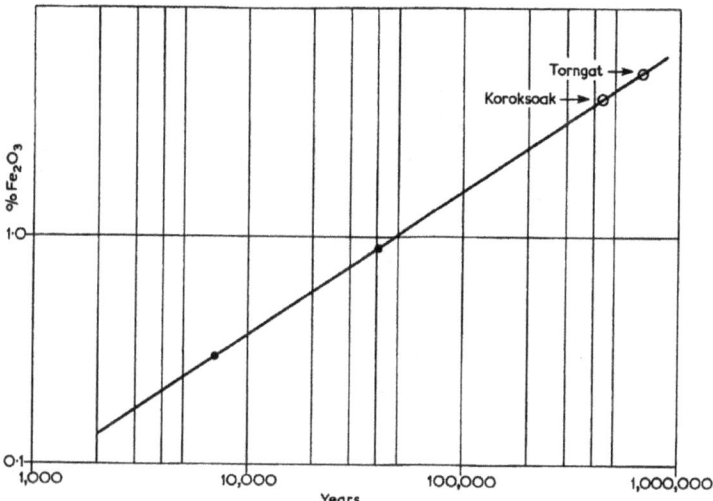

Fig. 6A.4. Tentative model of the relationship between Fe_2O_3 and age of the deposit.

seaward. The vertical extent of the Koroksoak weathered unit is commonly between 100 to 250 m. In eastern Baffin Island, in the vicinity of Narpaing Fiord, the Saglek trimline falls from 825 to 150 m over 60 km with a convex profile that is paralleled by the upper boundary of weathered till that descends from 1080 to 600 m.

These results suggest that there may be two lithologic units above the Koroksoak till – the first and youngest consists of scattered erratics lying on a surface that is intensely weathered; the uppermost zone is unglaciated and is also marked by extensive mountain-top detritus and the presence of tors. In many areas it may prove very difficult to distinguish between these two units and hence detailed studies are required in areas where erratics can be positively identified (for further discussion of weathering zones, see Ives, Chapter 10B, pp. 627–30).

QUEEN ELIZABETH ISLANDS, ARCTIC CANADA

Information on the extent of glaciations older than the Wisconsin are limited although there are indications that there was a more extensive earlier glaciation. Smith (1961) and Christie (1967) have discussed the distribution of erratics on the northeastern section of Ellesmere Island and the relationship of the Ellesmere and Greenland ice sheets. On the basis of erratics Christie (1967) delimits a western margin to a once more extensive Greenland ice sheet that penetrated 20 km onto Ellesmere Island. He also reports an absence of erratics on Miller Island at the head of Lady Franklin Bay above 460 m. As the elevation of land at the westernmost limit of the Greenland ice sheet rises up to 900 m, the absence of erratics *east* of this border is probably not significant. Smith (1961) cites erratics as evidence that northeast Ellesmere Island was totally inundated at least once. Likewise, Hattersley-Smith (1969) states that: 'These observations indicate considerable inundation by glacial ice.' Neither Smith (1961) nor Christie (1967) mention any vertical differences in weathering.

Further west, Prest (1970) regards deposits and erratics on Banks Island, Prince Patrick, Eglinton and Melville islands as evidence for an older glaciation. Savile (1961) and Mather (1968) have discussed floral and faunal evidence for unglaciated enclaves in this area during Wisconsin time and earlier (see also Ives, Chapter 10B).

WESTERN CANADIAN ARCTIC AND THE YUKON TERRITORY

Extensive unglaciated regions existed during the Quaternary in the western Yukon (fig. 6A.2). A review of the evidence is contained in Prest (1970, p. 702). In much of the Yukon there is evidence for one or more glaciations that were more extensive than the Wisconsin glaciation. In the central Yukon the Nansen Drift (Bostock, 1966) is extensively weathered. This unit extends 144 km beyond the limits of the Late Wisconsin (McConnell) ice. Vernon and Hughes (1966) observed scattered erratics above a clearly defined glaciation limit which in turn pre-dated the last major glacial episode. Older glaciated terrain has been mapped by Fyles (1966) to the west of the Mackenzie Delta as far as the Alaskan border.

ALASKA

In their review of pre-Wisconsin glaciations of Alaska, Péwé, Hopkins and Giddings (1965) delimit at least one and sometimes two glaciations. The youngest is considered correlative with the Illinoian glaciation of the mid-continent on the basis of uranium series dates. In their interpretation the equivalent of the Sangamon interglacial ended approximately 70,000 years ago. In the area of northern Alaska the Illinoian glaciation extended 50 to 100 km beyond the limits of the Wisconsin glaciers. Lying beyond this older drift, fronting the arctic coastal plain and extending toward the Mackenzie Delta is a pre-Illinoian glaciation. This is presumably correlated (?) with the older glaciated terrain described by Fyles (1966). In south-central Alaska the Illinoian limit (Péwé et al., 1965, p. 359) was closer to the edge of the Wisconsin glaciers. The marine stratigraphy of coastal Alaska contains three units considered to represent pre-Sangamon eustatic sea level maxima, thus implying as many interglacials (Hopkins, 1967a).

SIBERIA AND USSR

General statements on the Quaternary stratigraphy of Siberia are contained in Kind (1967) and Zubakov (1969). Zubakov (1969) distinguishes four major glacial pulses prior to the Kazantseva (Sangamon) interglacial. The four advances are assigned to two major glaciations. The Siberian ice sheet was considerably more extensive during Dnepr (Illinoian) time (fig. 6A.2) when '. . . a large ice sheet covered the northern part of the Russian Plain, the islands of the Barents Sea and perhaps the shelf of this sea, the northern half of the West Siberian lowland, the shelf of the Kara Sea, the northern part of the middle Siberian highland, the Taimyr peninsula, and northern Novaya Zemlya' (Artyushkov, 1971, p. 1381). The isostatic depression associated with this ice sheet reached a maximum recorded deflection of 400–500 m.

GREENLAND

There is considerable evidence in Greenland for the existence of nunataks during the maximum of the last glaciation but few statements on the existence of glaciated terrain older than this event. The northern part of Peary Land (c. 83°N) was not glaciated north of a belt of morainic material (Koch, 1928) apart from local ice caps which covered the high ground. More recent investigators (Davies, in Smith, 1961, p. 22) have not challenged this concept. Washburn (1965) believed that nunataks projected above the ice surface in the vicinity of Mesters Vig, northeast Greenland. Kelly (1969) describes nunatak moraines from the vicinity of Nordre Stromfjord (c. 67°38′N) which he says (p. 27) '... are presumably of Late Wisconsin/Würm age. Summit autochthonous boulder fields are not considered to be significantly older than this.' This statement may be questioned in view of the results from the eastern Canadian Arctic (see Ives, Chapter 10B). In western Greenland Weidick (1968) indicates that the ice cover was virtually complete over the region and the nunatak moraines are clearly judged to be Late Wisconsin in age. In the recent 'Quaternary map of Greenland' Weidick (1971) shows the distribution of the uppermost and/or outermost erratic trains. This evidence does suggest the presence of nunatak areas in Greenland during the Quaternary.

Most regions of the Arctic contain adequate stratigraphic evidence for at least one glaciation that was more extensive than the regional correlatives of the Wisconsin glaciation. In Alaska, the Yukon and the USSR at least two older Quaternary glaciations are recognized. This number is considerably less than that observed from fluctuations of sediments and fauna in ocean cores (Herman, 1970) and from the loess record (Kukla, 1969), and Iceland. No summary chart is attempted because the older deposits are poorly dated, only their relative stratigraphic positions are known and this record probably contains lacunae.

3. Wisconsin glaciation

There is a dramatic increase in the amount of data pertaining to the last interglacial and glacial. This is in part related to preservation of the deposits and more specifically to the availability of materials for various radiometric dating techniques, notably uranium series dates on carbonates and ^{14}C dates on an extensive list of organic materials. Correlations of sequences throughout the Arctic are possible and can be based on 'absolute' dating methods. However, it is unwise to believe that the increase in the number of dates has done away with all problems, rather the problems are ones of detail involving the placing of glacial events within limits imposed by the errors inherent in radiometric dating methods, and of gaining insight into the extent of the time transgressive nature of most boundaries based on geologic–climatic data. Climatic problems associated with initiating glacierization of the Laurentide region have been recently discussed by Lamb and Woodroffe (1970), Loewe (1971) and Brinkmann and Barry (1972).

A basic problem is the actual timing of the Wisconsin glaciation. Many authors (Péwé *et al.*, 1965; Goldthwait *et al.*, 1965; Dansgaard *et al.*, 1971) suggest that the maximum Early Wisconsin advance occurred about 70,000 BP. In this interpretation the Sangamon interglacial ended shortly before 70,000 BP and this led Lamb and Woodroffe (1970) to postulate the build-up of the Laurentide ice sheet in the matter of 1000 to 5000 years – a conclusion that was disputed by Barry *et al.* (1971). Considerable insight into the chronology of the Wisconsin glaciation is being provided from areas well outside

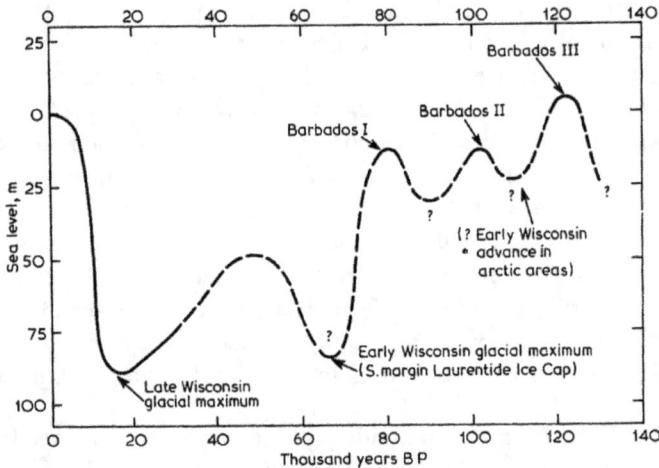

Fig. 6A.5. Eustatic sea level during the last 130,000 years (after Broecker and Van Donk, 1970).

the limits of continental glaciation. The records of major ice fluctuations are being interpreted from the sympathetic eustatic sea level oscillations in response to large-scale mass transfers of H_2O to and from the oceans. A recent interpretation of these data is given by Broecker and Van Donk (1970) and shown in modified form as fig. 6A.5. Mörner (1971) gives a similar overall treatment but shows sea level returning to its present elevation at 100,000 and 80,000 BP, whereas Broecker and Van Donk (1970) indicate sea level was *c*. −12 m. A broadly similar interpretation is adopted by Veeh and Chappell (1970) on the basis of dated coral sequence in New Guinea. The dated eustatic sea level fluctuations (fig. 6A.5) indicate four glacial stades within the Wisconsin. The sequence can be divided into two moderate glaciations between 120,000 and 90,000 BP with an intervening interstade when sea level was between −12 and −30 m; a very significant and major interstade at 80,000 BP when sea level rose between −12 and −8 m; and then two major glaciations peaking at 65,000 and 18,000 BP with a long, cool interstade between them with sea level −18 to −40 m, finally followed by the Holocene and the rapid rise of sea level to its present position.

The length of the Sangamon interglacial is shown as 50,000 yr by Veeh and Chappell

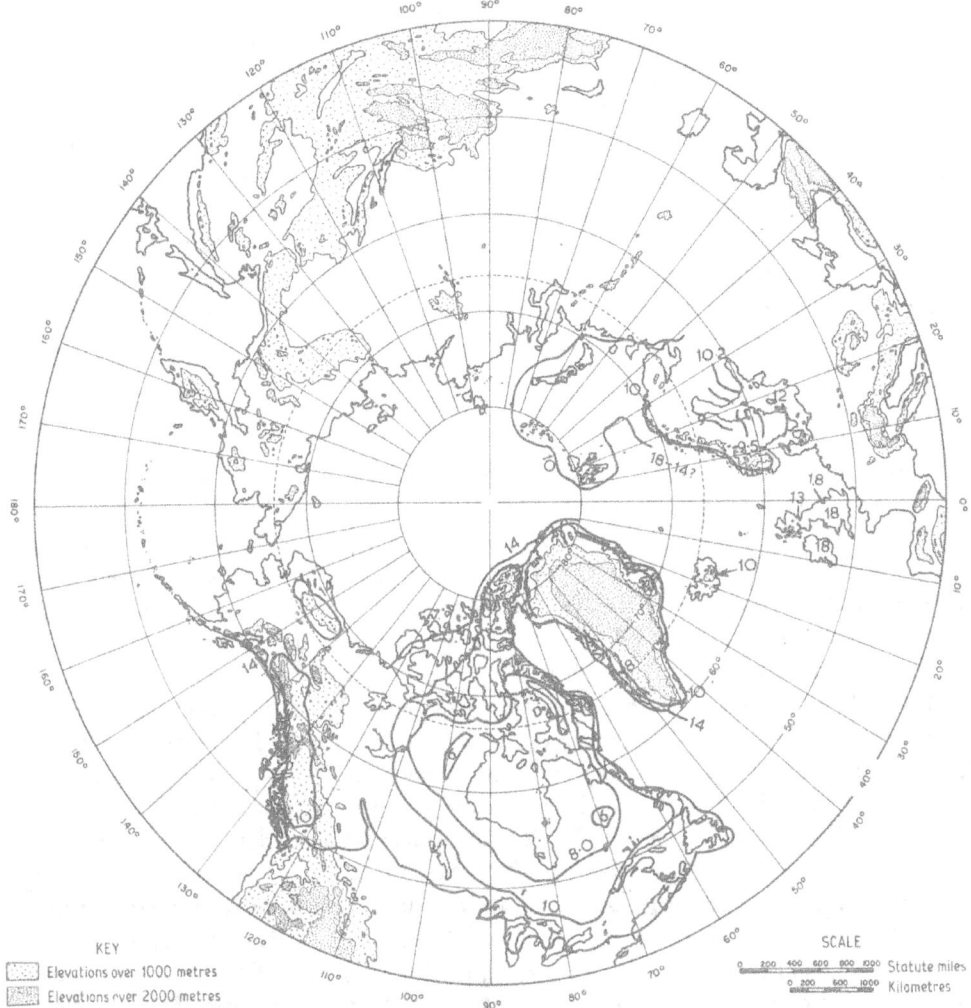

Fig. 6A.6. Isochrones of deglaciation (contours × 10³ years BP).

(1970, fig. 3C) and of much shorter duration (10,000 years?) by Broecker and Van Donk (1970, fig. 7). In view of the uncertainties in the chronology and the fact that the interstade at 80,000 BP may have been quite warm, there is a distinct possibility that in some areas deposits mapped as Illinoian are in reality Early Wisconsin. The problem is to a certain extent one of definition (see fig. 6A.5). However, in terms of arctic regions the critical implication of fig. 6A.5 is the suggestion of considerable Early Wisconsin glacierization of the Arctic and Antarctica. A world sea level at −30 m is equivalent to the state of world glacierization at 9000 BP and −12 m is compatible with the extent of ice at 7500 BP (figs. 6A.6 and 6A.7).

This lengthy preamble is necessary because the interpretation of the Wisconsin glaciation and the Sangamon interglacial are undergoing major revisions. In the next sections the evidence for the last interglacial and glacial are reviewed and interpreted. Recent syntheses of the material are included in Péwé et al. (1965), Flint (1971), Prest (1970) and Zubakov (1969). These data and others from the eastern Canadian Arctic and western Siberia are summarized in table 6A.1. Note that we would not expect a time parallelism between events in the High Arctic and those along the southern margin of the Laurentide ice sheet. The Wisconsin glaciation will obviously be time-transgressive along the 2000 km between the initial gathering grounds in Canada and final terminal position in Wisconsin and Ohio. Thus the standard starting point for the Wisconsin at 70,000 BP is not applicable to arctic areas.

LAST INTERGLACIAL

Few radiometric dates have been obtained on interglacial materials throughout the arctic region, and their age is usually inferred from the fact that they are frequently overlain by glacial deposits assigned to the last glaciation. ^{14}C dates are >59,000 BP for the Kazantseva interval (Zubakov, 1969) and >53,000 BP for deposits in the Hudson Bay lowland (McDonald, 1969). In Alaska the marine equivalent of the terrestrial interglacial sequences is the Pelukian transgression (Hopkins, 1967a) which has been radiometrically dated at 100,000 ± 8000 BP (^{230}Th/^{238}U) and 78,000 ± 5000 BP (^{226}Ra/^{238}U) (in Hopkins, 1967b). The Pelukian warm interval was interrupted by a period when sea level was below present and permafrost developed. This suggests a partial correlation with the Barbados record of sea level fluctuations between 120,000 and 80,000 BP. In Baffin Island, NWT, deposits of the Flitaway interglacial have been studied intensively by Terasmae et al. (1966). The deposits consist of detrital plant materials probably deposited in river backwaters. Preservation of the material was excellent and macrofossils of Betula, Ledum groenlandicum, Ledum decumbens and Sphagnum were identified amongst others. The palaeoclimatic interpretation of the macrofossils and pollen content is for July temperatures 1° to 4°C higher than present (i.e. 7° to 10°C), a 20–25 days longer growing season and a total annual precipitation of 0.38 m, an increase of c. 0.13 m over today's values. The disposition of the interglacial materials around the Barnes ice cap, the occurrence of interglacial pollen in the shear moraines, and the higher summer temperatures inferred from the botanical evidence, indicate that the Barnes ice cap did not survive this interglacial, although to date it has survived the Holocene.

In Hudson Bay the work of McDonald (1969, 1971) has greatly expanded our knowledge of the interglacial deposits; subtill marine clays are overlain by peat beds which in turn are overlain by two tills.

EARLY AND MID-WISCONSIN GLACIATION

Various concepts and chronologies pertaining to Wisconsin time are outlined in table 6A.1 and figs. 6A.7 and 6A.8. As noted earlier, the Wisconsin glaciation, being a geologic/

climatic unit, is not time-parallel everywhere and covered a greater length of time in arctic areas than in those regions peripheral to the ice borders at their maximum extent.

Arctic glacial chronologies (table 6A.1) are marked in many instances by an early advance, an interstadial period and then a readvance. A feature of these chronologies is that the Early Wisconsin advance was considerably more extensive than the Late Wisconsin advance (the latter is correlative to the 'classical' Wisconsin). A reverse pattern is recorded along the southern margin of the Laurentide ice sheet. The Reid glaciation in the eastern Yukon (Bostock, 1966) extended 64 km further than the Mc-Connell glaciation (table 6.1); in Alaska the Knik glaciation was 5 to 10 km more extensive than the succeeding Naptowne glaciation; in western Siberia the Norilka glaciation was considerably less extensive than the earlier Ermakovian glaciation; and in eastern Baffin Island, Canada, the earliest Wisconsin glaciers extended downfiord 20 km or so further than the Mid-Wisconsin glaciers, which in turn were larger than the Late Wisconsin glaciers (table 6A.1 and fig. 6A.8). This pattern seems to trend in the opposite direction to the sawtooth model of Broecker and Van Donk (1970), based on ocean core data, which shows a continuing deterioration and cooling until the Late Wisconsin glacial maximum. The difference in pattern probably represents a continued southward expansion of the major ice sheets and glaciers and a decline in size of arctic glaciers. This pattern is interpreted as reflecting an increasing aridity in the Arctic over the course of a full glacial sequence.

One of the most instructive areas for Wisconsin chronology is the fiords and outer coast of northern Cumberland Peninsula, eastern Baffin Island. There a large fiord glacier overrode a fossiliferous marine silt. Aragonite shells, recently released from the permafrost, were dated by the ionium-deficiency method at $137,000 \pm 10,000$ BP (B. Szabo, personal communication, 1971). Glacio-marine deltas associated with the Alikdjuak stade (table 6A.1) are now between 70 and 85 m above sea level. Further north along the eastern Baffin Island coast, Løken (1966) has described a similar situation at Cape Aston where marine shells were ^{14}C dated at >54,000 BP. High glacio-marine deltas at about 70 m above sea level along the outermost east coast have also been noted by Ives and Buckley (1969) and England and Andrews (1973). These features lie along a synchronous postglacial isobase indicating that they are correlative and date from the Early Wisconsin. Stable isotope analysis of $\delta^{18}O$ and $\delta^{13}C$ of marine shells indicates positive values of both suggestive of cold seas with high salinities.

A warm marine phase is recorded in the fauna of the region and the deposit has been dated at $68,000 \pm 5000$ BP (B. Szabo, personal communication, 1970). A delta in the mid-section of the fiord contains shell fragments dredged up during a subsequent readvance. The fauna included *Chlamys islandicus*, considered a postglacial climatic optimum indicator (Andrews, 1972) and the concentration of $\delta^{18}O$ and $\delta^{13}C$ is similar to those occurring at 8000 BP. Nothing is known of the events between about 60,000 and 40,000 BP at which date the fiord glaciers readvanced (fig. 6A.8), although they were considerably less extensive than their earlier counterparts. Retreat was underway by 24,000 BP and between this latter date and 10,500 BP there are no ^{14}C dates from eastern Baffin Island. Andrews and Ives (1972) have recently argued that the Cockburn re-

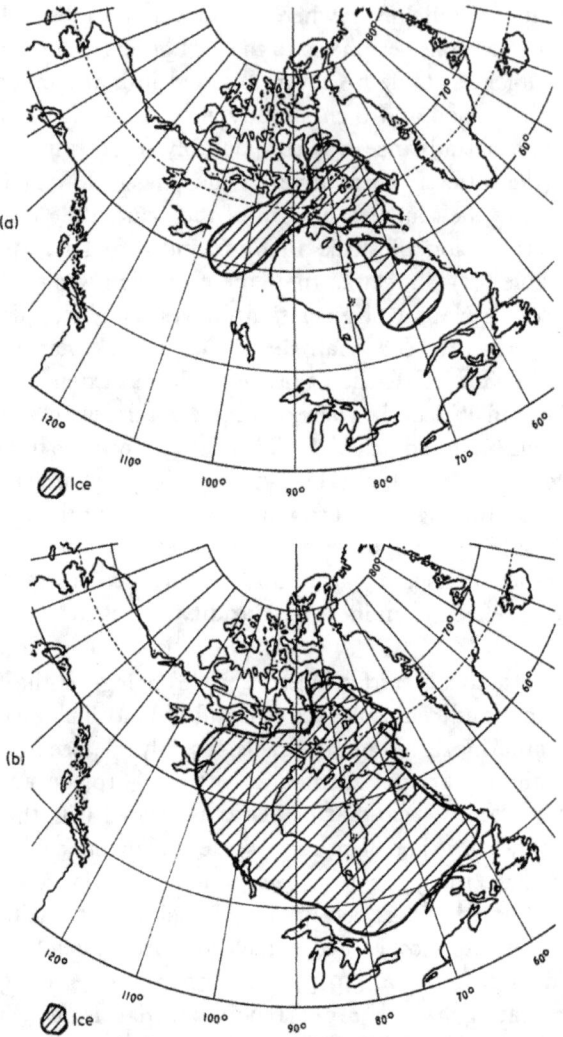

Fig. 6A.7. Suggested four-stage development of the Laurentide ice sheet. (a) 110,000 BP, Early Wisconsin, sea level −20 m, (b) 90,000 BP, sea level − 30 m, (c) 80,000 BP, St Pierre, sea level −12 m, (d) *c.* 40,000 BP, sea level −45 m.

advance of 8000 BP may in fact represent the maximum readvance of Late Wisconsin ice and that eastern Baffin Island may not have responded in phase with other margins of the Laurentide ice sheet. The considerable extent of the Alikdjuak stade glaciers and the late date of the Cockburn Moraines may both be related to the need for precipitation as the major control on glacierization in this region.

In Hudson Bay the time/distance diagrams of fig. 6A.8 show different interpretations (McDonald, 1969, 1971; Prest, 1970) as to the possible age of sediments which in places

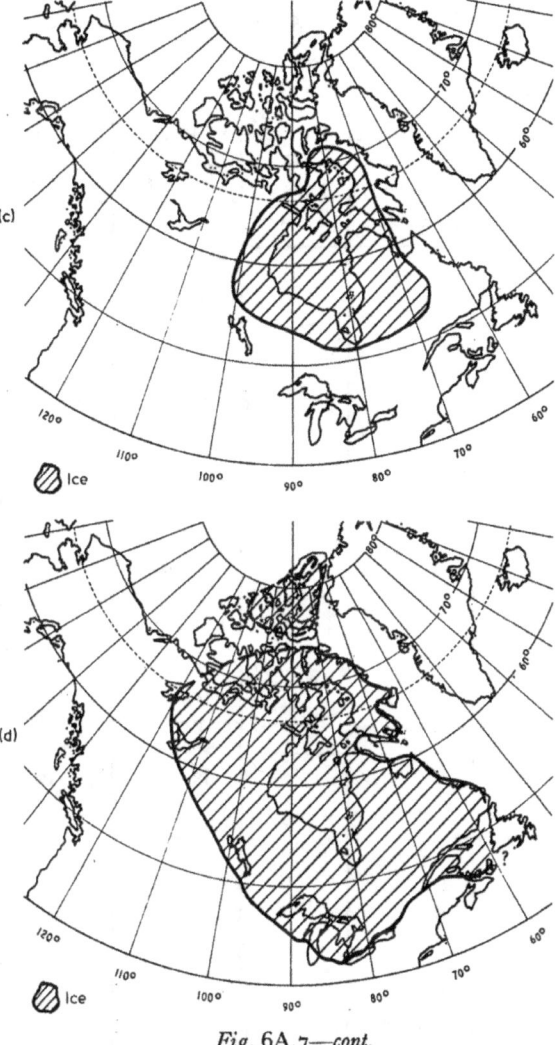

Fig. 6A.7—*cont.*

separate two tills within the Hudson Bay Lowlands (McDonald, 1969). The suggestion
that the interval might be equivalent to the Port Talbot interstade (McDonald, 1969) is
not in accord with Prest's interpretation, and the beds might be better correlated with the
St Pierre interval. A feature of the interstade at about 80,000 BP is its warm nature
(Mörner, 1971). Considerable ice recession is thus suggested – in agreement with the
return of sea level to −12 m to form the Barbados I terrace (Broecker and Van Donk,
1970). The postulated extent of the Laurentide ice sheet for various periods within the
Wisconsin is illustrated as fig. 6A.7. Note that no evidence exists in the record of the ice
sheet for it to have shrunk very much around 30,000 BP; in size it approximated the
situation at 12,000 BP. On this ground the suggested high sea level stands at 30,000 BP

Table 6A.1. Correlation chart, arctic region (from Flint, 1971, and other sources).

Date (BP)	Alaska marine sequence	Brooks Range	Cook Inlet	S. W. Yukon	E. Yukon	Western Siberia
		Fan Mt glaciation	Alaskan glaciation	Neoglacial	Neoglacial	
10,000	Krusensternian			Slims non-glacial		Postglacial
20,000		Irkillik glaciation	Naptowne glaciation	Kluane glaciation	McConnell glaciation	Norilka glaciation (Sartan)
40,000				Boutellier nonglacial		Kargy interstade
					?	
60,000				Icefield glaciation		Ermakovian glaciation (Zyrianka)
			Knik glaciation	Silver nonglacial		
80,000				Shakwak glaciation ↓ ?	Reid glaciation	
100,000	— — ? — —					
	— — ? — —					
120,000	Pelukian - 2 high stands					

* Dates given as quoted, others inferred.

W. Baffin Island	E. Baffin Island	Hudson Bay Lowlands	USSR 66½°N and 50°E
Postglacial	Neoglacial	Postglacial	
	Climatic optimum		
	Cockburn stade	Cochrane readvance	Holocene
	Interstade		
	Napiat stade	Till	Valdai II glaciation
	?		
		Stratified sediments	Karukula interglaciation
Foxe glaciation Saglek glaciation	Quajon interstade		
	?	Till	Valdai I glaciation
		Missinaibi interval	Mikulin interglacial
	Alikdjuak stade		
Flitaway interglacial			Moscow glaciation

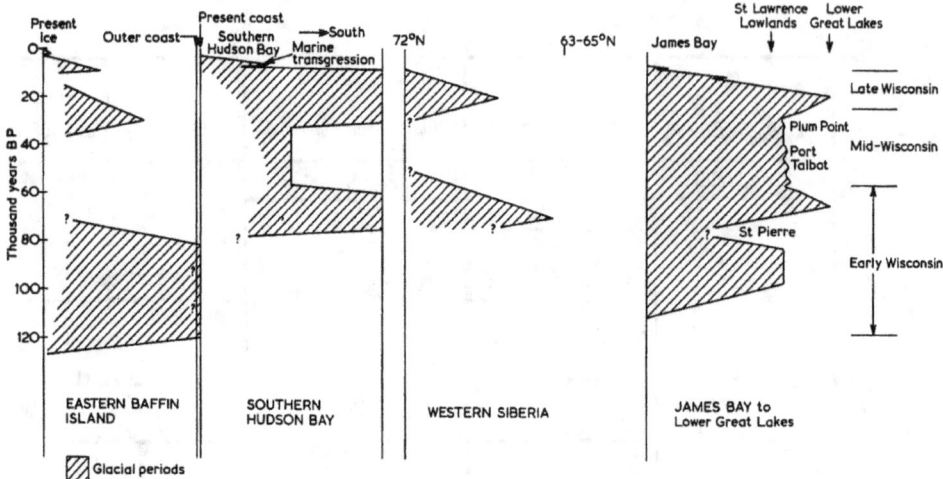

Fig. 6A.8. Time/distance diagrams for various sites showing timing and extent of major glacial advances and retreats.

are discounted and are probably caused by dating problems (see Curray, 1965; Milliman and Emery, 1968; Mörner, 1971; Andrews, 1970). There may well have been a shrinkage of the Laurentide ice sheet to a size indicated on fig. 6A.7 – the subsequent readvance during the Late Wisconsin could well explain the large number of marine shells incorporated into Late Wisconsin till. ^{14}C determinations on this shell material result in dates that vary from >41,000 to 30,000 BP. While these might be considered minimum dates because of the problem of contamination, at least some of them no doubt do actually document the long cool interstade of Mid-Wisconsin age (fig. 6A.8, table 6A.1). As noted by Bird (1967) 'old' ^{14}C dates have a distinct, peripheral distribution.

DEVELOPMENT OF THE LAURENTIDE ICE SHEET AND LOCATION OF GLACIAL CENTRES

The distribution of glacial ice inferred on fig. 6A.7 requires commentary in terms of conflicting theories regarding the initial, major source areas for the Laurentide ice sheet. Flint (1971) repeats his original concept and suggests that the Laurentide ice sheet developed by the growth of ice sheets in the eastern 'highlands' that form the eastern seaboard of Canada (Flint, 1971, fig. 18.8). This hypothesis had earlier been challenged by Ives (1957) and Ives and Andrews (1963) who believe that the major gathering grounds for the Laurentide ice sheet were not the eastern uplifted rims of the Canadian Shield but were rather the immense broad, high plateaus of central Labrador–Ungava, Baffin Island, and to a lesser extent Keewatin. These plateaus consist of broadly rolling topography at elevations between 500 and 700 m. On Baffin Island the present regional snowline lies only 100 to 200 m above this surface (Andrews and Miller, 1972) and during the last phase of the Neoglacial (300 ± BP) large areas of these uplands were mantled by thin, permanent snow patches. Ives (1962) estimated that 70 per cent of interior Baffin Island was affected. Andrews (1961) placed the regional snowline at about 1800 m above sea level today over central Labrador–Ungava, or about 1200 m above the land surface. Glacierization of these interior plateaus is considered to have occurred very rapidly as the result of a lowering of the regional snowline, largely by the coalescing of snowbanks and icefields. Lamb and Woodroffe (1970) have adopted a similar argument.

The amount of climatic change required to initiate widespread glacierization is not great. In Baffin Island a lowering of the regional snowline by 100 to 200 m is reasonable. Barry (1966) has suggested that for Labrador–Ungava a lowering of summer temperature by 3.3°C would probably be sufficient to initiate glacierization with present-day average summer cloudiness. Moisture transport similar to present winter conditions would then be adequate to build up an ice sheet 2000 m thick in 20,000 years. However, on the basis of estimates of freezing level from upper-air soundings, Loewe (1971) argues that a cooling of 6°C with present precipitation would have been insufficient for glacierization. The effect of glacio-isostatic recovery cannot be ignored in analyses of snowline changes, and Tanner (1965) has stressed its potential importance in accounting for the periodicity of glaciations/interglacials.

With a lowering of the regional snowline it is considered that the first major ice centres would develop over Baffin Island and the eastern Queen Elizabeth Islands. These in turn would increase the tendency for lower temperatures because of the change in albedo, and they would be followed by glacierization of central Labrador–Ungava, and lastly by an ice cover over Keewatin. If the sea level data of Broecker and Van Donk (1970) are approximately correct, they imply that by 110,000 BP the three major northern centres of outflow of the Laurentide ice sheet had already developed (fig. 6A.7a). If the maximum temperatures occurred at 125,000 BP then it appears that an ice sheet the size of fig. 6A.7a developed within 5000 to 10,000 years. By 90,000 BP eustatic sea level indicates the ocean was −30 m and the Laurentide ice sheet is pictured to have grown to the size indicated on fig. 6A.7b. Although the ice sheet shrank during the warm St Pierre interval (fig. 6A.7c) Hudson Bay remained ice-covered until about 8000 BP.

The location of centres of maximum ice thickness should correspond to glacio-isostatic uplift centres, although there is some evidence in Fennoscandia that these centres migrate in response to changing ice conditions during deglaciation (Mörner, 1969). Andrews and Barnett (1972) studied the areas inside the perimeter of the Laurentide ice sheet where the projection of isostatically deformed marine and lake strandlines intersected. It has long been tacitly understood that the area of Hudson Bay was a major, if not the major, dispersal centre for the Laurentide ice sheet (Flint, 1957) but our research indicated the presence of five uplift foci (fig. 6A.2) that are arranged in a broad U around the margin of Hudson Bay. In general, the foci lie on the height of land and well into the interior of the mainland areas. The exceptions are a major focus over southern Hudson Bay and James Bay and a smaller one over Foxe Basin (fig. 6A.2). In this interpretation Hudson Bay and Hudson Strait are significant centres of ice inflow. The south Hudson Bay and Foxe Basin centres must have come into operation after the glacierization of

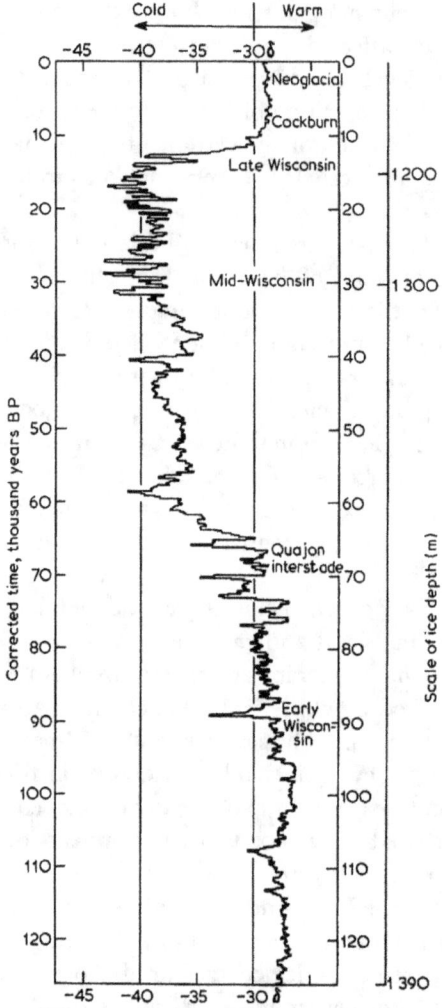

Fig. 6A.9. δ^{18}O variations from the Camp Century ice core (after Dansgaard *et al.*, 1971).

the surrounding high plateaus. McDonald (1971) has presented evidence for a westward (northwestward?) shift of the main glacial centres during the Wisconsin. His interpretation of the Port Talbot ice margin is different from that adopted here (fig. 6A.7). Andrews (1969) and Blake (1970) have also shown the existence of a separate ice sheet over the Queen Elizabeth Island (fig. 6A.2).

   The scheme of development and the corresponding climatic implications suggested in this chapter require testing from other independent sources. One such source is the δ^{18}O record from the Camp Century ice core, Greenland (Dansgaard *et al.*, 1971) shown as fig. 6A.9. The terminology has been modified to take into account the ideas of Prest (1970) and the information from eastern Baffin Island. There is some doubt as to the

Table 6A.2. Late Wisconsin and Holocene chronologies.

Date BP × 10³	E. Baffin Island	W. Baffin Island	West Greenland	Brooks Range	Cook Inlet	N.E. St Elias Mts, Yukon	N. Norway	General marine indicators
	Lewis (G)	Lewis (G)	Several phases	Fan Mt (G)	Tunnel II (G) Tunnel I (G)	Neoglaciation	(G)	Cool
	King	King	Several stages	Alapah Mr (G)	Several phases Iustumena		(G)	Warmer
	Flint	Flint					(G)	Warm
5	Optimum							
	Isortoq	Isortoq	Mt Keglen (G)		Tanya III (G) Tanya II (G)			
	Cockburn (G) Strandline I	(deglaciation of Foxe Basin)	Fiord (G) Avatdleq (G) Taserqat (G)	Anivik Lake (G)	Tanya I (G)	Slims nonglacial interval	As-Ski Moraines Tromso-Lyngen Moraines	Cool
10	?			Antler Valley (G)			Allerod	
	?					←Maximum expansion	(G)	
	Independent cirque glacier moraines			Anayaknaurak (G)	Skilak stade Killey stade	Kluane glaciation	Lista substage (G)	
15				Banded Mountain (G)	Mooschorn stade			
20	Glacial retreat					→	Egga Moraines	

G = Glacial phase
Mr = Moraine

reliability of the time scale on this figure, and it has been suggested that it may be incorrect in places by as much as 50 per cent. Compression of the time scale may have resulted from changes in accumulation on the ice sheet which are not taken into account in the flow model (Dansgaard *et al.*, 1971). It must be noted in passing that their attempted correlation of a climatic sequence from the Greenland ice cores should have included an analysis of arctic chronologies with less reliance placed on correlation with 'standard' chronologies from the southern margins of the last major European and North American ice sheets. This point will be returned to later.

LATE WISCONSIN CHRONOLOGIES

In the areas fronting the southern margin of the Laurentide ice sheet the maximum advance of the ice occurred at approximately 18,000 BP (Willman and Frye, 1970). A similar date is recorded for the Sartan glaciation of Siberia (Kind, 1967), and the Egga Moraines of northern Norway (Andersen, 1968), which lie out on the continental shelf, are assigned a similar age. Table 6A.2 lists selected glacial chronologies across the Arctic and is a more detailed treatment of material in table 6A.1 for the last 20,000 years. Denton (1970) has proposed that a major climatic amelioration occurred shortly after 14,000 BP and that this terminated the Late Wisconsin glaciation in areas not under the influence of the large ice sheets. However, he also notes that these glaciers reached maximum size at about 14,000 BP, which is a significant lag behind the Late Wisconsin maximum in southern areas of the ice sheet. In northern areas of Canada the history of the Late Wisconsin is known in considerable detail for the retreat phase, but there are some major questions that still require investigation. On the northwestern margin of the Laurentide ice sheet ^{14}C dates indicate retreat had started about 13,000 BP (Fyles, 1963). Along the east coast of Baffin Island, Andrews and Ives (1972) have commented on the general absence of ^{14}C dates older than 8500 BP even in outer coastal areas which were unglaciated during this readvance. In some locations, such as Inugsuin Fiord, dates on marine limits in the outer fiord are of the same age as those at the fiord head. The paucity of dates older than the Cockburn Moraines (*c.* 8000 BP, see table 6A.2) may indicate that these moraines mark the maximum Late Glacial advance of the eastern Baffin Island fiord glaciers (Andrews and Ives, 1972). Thus, although there is a temporal correlation between the Cockburn Moraines and the Cochrane readvance of southern James Bay the two responses are of very different character – the latter being the terminal advance of the southern Laurentide ice margin, and the former being the Late Glacial maximum.

Table 6A.2 indicates the rather general nature of the glacial advance at the Boreal/Atlantic transition. In the Arctic it has been noted from Alaska around to west Greenland where Weidick's Fiord Stage is directly correlative. It apparently did not occur east of the Rocky Mountains in the St Elias Range (Denton and Stuiver, 1967) and it is absent in northern Norway. A readvance at about 7000 BP is also recorded in west Greenland, eastern Baffin Island and in the Cook Inlet region.

These fluctuations of the margins of major ice sheets, outlet glaciers and valley glaciers appear to be roughly synchronous within the limits of the ^{14}C method. The Cockburn

Phase and its equivalents appear decisively time-parallel. In eastern Baffin Island this was an extremely significant event that produced moraines of quite distinct size. Surprisingly the Greenland ice core record (Dansgaard *et al.*, 1971) shows very little evidence for climatic reversals during the last 10,000 years. The period between 10,000 and 9000 BP was one of lower $\delta^{18}O$ concentrations and the Cockburn Phase could be a lag response to this condition. Andrews and Ives (1972) have also suggested that the readvance of the eastern Baffin Island outlet glaciers might have been in response to the opening of Davis Strait and Baffin Bay and increased precipitation along the eastern edge of the Laurentide ice sheet. The Greenland record is of course a temperature record, *not* one of precipitation amounts.

Isochrones of deglaciation from the last major glaciation are illustrated as fig. 6A.6. More detailed treatment and discussion of the North American data appear in Bryson *et al.* (1969) and in Prest (1970). The distribution of ice is shown in fig. 6A.2. One of the more interesting suggestions of recent years has been that a large ice sheet once covered the shallow shelf of the Barents Sea and was contiguous with the Fennoscandia ice sheet in northern Norway and extended onto the western Russian Plain (Schytt *et al.*, 1967; Hoppe, 1970). This hypothesis had been proposed by Corbel in 1960 and partly documented by Grosval'd in 1963. The concept of an ice sheet based on the shelf explains the widespread glacio-marine sediments that are found over wide areas of the northern European Plain (Grosval'd, 1970). At the same time it is worth noting that large parts of the Laurentide and Queen Elizabeth Islands ice sheets were located over former sea basins and channels. Indeed, Mercer (1970) has suggested an ice sheet may have existed on the North American arctic continental shelf. No evidence in support of this hypothesis has been discovered to date. The Barents Sea ice sheet disappeared quickly and by 10,000 BP moraines were deposited only 3 to 4 km from the margin of existing glaciers. The ice sheet probably disintegrated because of a combination of rising eustatic sea level and a rise in atmospheric and oceanic temperatures (see Ives, Chapter 10B, pp. 622–5).

Removal of the larger Laurentide and Scandinavian ice sheets was not so sudden. In Greenland, Weidick (1972) has estimated the rate of retreat of the fiord glaciers was 25 to 50 m yr^{-1}. In eastern Baffin Island the retreat was slower and averaged about 12 m yr^{-1} but on the Canadian mainland the isochrone maps indicate that retreat rates were considerably higher and varied between 80 and 250 m yr^{-1}. Major moraine systems that fall within the 10,000 BP isochrone are mapped on the new Glacial Map of Canada (1968) indicating that the retreat of the Laurentide ice sheet was not continuous over the period from 10,000 BP to the final phase of disintegration between 8000 and 7000 BP. Various interpretations have been placed on the timing of these events that included the deposition of the Cockburn Moraines and Cochrane readvance, the deglaciation of Hudson Bay and the deglaciation of Foxe Basin. In some ways the evidence is contradictory, but Andrews and Ives (1972) argue that this was in part caused by the very rapidity of the sequence and uncertainties in ^{14}C ages within this period. The deglaciaton of Hudson Bay was hastened by the rapid penetration of a tongue of the sea due south along the −200 m trench that runs north–south in Hudson Bay (Andrews and Falconer, 1969). Although it has been argued that the Cochrane readvance was a local response to deglaciation of Hudson Bay (Prest, 1970; McDonald, 1971; Andrews and Ives, 1972) the widespread Cockburn Moraines of the eastern and north-central Arctic and Greenland were clearly in response to a significant mass balance change. Indeed these moraine systems may represent one of the most continuous ice margin deposits yet delimited. The Cockburn Phase appears indistinguishable in time from evidence of an Early Holocene warm period, reported from Alaska (McCulloch and Hopkins, 1966) and the western Canadian Arctic (Mackay, 1971). By about 6700 BP only residual ice masses were left, and these no longer occupied marine positions but were based on interior, continental locations over Keewatin, central Labrador–Ungava, Baffin Island, and the high Queen Elizabeth Islands. The ice was active on Baffin Island and on the Queen Elizabeth Islands, but it retreated and wasted *in situ* over Keewatin and central Labrador–Ungava, and probably had disappeared by about 6000 BP. Glacial advances between 8000 and 5000 BP have been recorded from certain areas in the Arctic (table 6A.2), the most common being a readvance about 6700 to 7000 BP that has been noted in northern Greenland (Davis, 1961), Queen Elizabeth Islands and Baffin Island (Andrews *et al.*, 1970), west Greenland (Weidick, 1972) and Alaska (Karlstrom, 1964).

In many areas detailed interpretation of Late Glacial and postglacial events is made possible by detailed appraisals of the relationship between glacial fluctuations and the isostatic rebound of the formerly depressed crust. Andrews (1970), Schytt *et al.* (1967) and Weidick (1971), among others, have interpreted the Late Glacial history of Baffin Island, Barents Sea and West Greenland very largely through such studies. The relationship between Eskimo archaeological sites and relative sea levels has been explored by Andrews, McGhee and McKenzie-Pollack (1971); the results indicate that a knowledge of postglacial isostatic recovery can be used as a reconnaissance chronological tool in arctic areas. This facet of arctic glacial research is dealt with in greater detail in the next major section of this chapter.

Most arctic Holocene chronologies (i.e. the period covered by the last 10,000 years) indicate a 'non-glacial' interval of varying length. In certain parts of the Arctic, notably Greenland, this interval is represented by a situation when the inland ice had retreated to a position less extensive than today. In eastern Baffin Island deposits of Late Glacial age have low, rounded forms and contrast strongly with the ice-cored moraines that date from the Neoglacial. The intervening interval between Late and Neoglaciation lasted approximately 1000 to 2500 years at most sites, although in the Yukon it may have been considerably longer (table 6A.2). Fossil evidence (Terasmae and Craig, 1958) indicates that temperatures were higher about 5500 BP than today (see also Nichols, Chapter 11A). The Neoglaciation commenced about 5000 BP at least at some sites. The rapid retreat of the western margin of the Barnes ice cap was terminated by a significant readvance dated at 4700 ± BP (Andrews, 1970) and in the mountains of eastern Baffin Island small icefields and valley glaciers established and advanced across the lateral moraines of the fiord glacier stage (Ives and Andrews, 1963; Andrews and Ives, 1972). Weidick (1972) cites evidence for a significant readvance of the Greenland ice sheet which is dated 'somewhat after 6000 BP' (Weidick, 1972). Marine bivalves from raised marine deposits throughout the North Atlantic Arctic (Andrews, 1972) do not show a climatic deterioration at this time. Indeed, the marine optimum in terms of bivalve growth rates, biomass, faunal composition, and isotopic composition occurred at 3500 BP (Andrews, 1972; Davies and Andrews, n.d.). A considerable component of the 'optimum' appears to be related to an increase in salinity. Nichols (Chapter 11A) has suggested that there was a short, very sharp drop in temperature around 5000 BP. In the Greenland core (Dansgaard *et al.*, 1971, fig. 6) the oxygen isotopic variations above and below a smooth curve become much more noticeable after 5000 BP and there are considerable periods when low (cool) isotopic precipitation accumulated.

The Neoglaciation in arctic areas is marked by a number of distinct phases (table 6A.2). At most localities an early and late phase can be separated, and in others as many as four distinct readvances can be recognized on the basis of lichenometry and weathering criteria (e.g. Carrara and Andrews, 1972). The last 5000 years have been marked by a variable climate, and it is within this period that the various arctic people spread from Alaska across arctic Canada to Greenland in several distinct folk migrations. The timing of the migrations appear to have been synchronous with major glacial/deglacial events (McGhee, Chapter 15A). These people inhabited fairly distinct beaches that are now raised some distance above present sea level (Andrews *et al.*, 1971). It is also in this period that the pollen and glacial evidence can be compared (see Nichols, Chapter 11A).

In the last 3000 years or so there have been marked changes in the fauna of arctic coastal sites. Certain species have been forced to move south due to a decrease in sea temperature and a decrease in salinity caused by the influx of summer meltwater into coastal waters. The present distribution of glaciers in the Arctic (Hattersley-Smith, Chapter 4B) is largely a response to climatic changes of the last 5000 BP. The pattern of glacierization has changed as well, and apart from Greenland, the majority of ice bodies exist at high elevations where they form small icefields or individual glaciers.

4. Postglacial isostatic recovery of arctic areas

One effect of the large Quaternary ice sheets was to impose considerable stresses on the earth's crust. As the loads were imposed for a considerable period of time an isostatic adjustment to the load occurred, mainly by the outflow of plastic material at depths of 100 km or so. There are no observational data on the form of the response curve to loading, but there is a growing body of information on the form of postglacial isostatic recovery. Exploration in arctic areas has largely occurred in the last two decades and coincided with the advent of ^{14}C dating; these factors, together with the preservation of organic materials due to the slowness of arctic weathering processes, have led to a rapid increase in knowledge of postglacial recovery in the Canadian Arctic (Andrews, 1970; Bird, 1967; Blake, 1970), the region of the Barents Shelf (Schytt *et al.*, 1967; Grosval'd, 1970) and Greenland (Washburn and Stuiver, 1962; Fredskild, 1969; Lasca, 1969; Weidick, 1971). At the same time additional information has become available from regions that have been studied since the nineteenth century, namely northern Norway and Scotland. Einarsson (1966) has published data on Iceland.

Fig. 6A.10 shows the pattern of postglacial emergence throughout the Arctic for

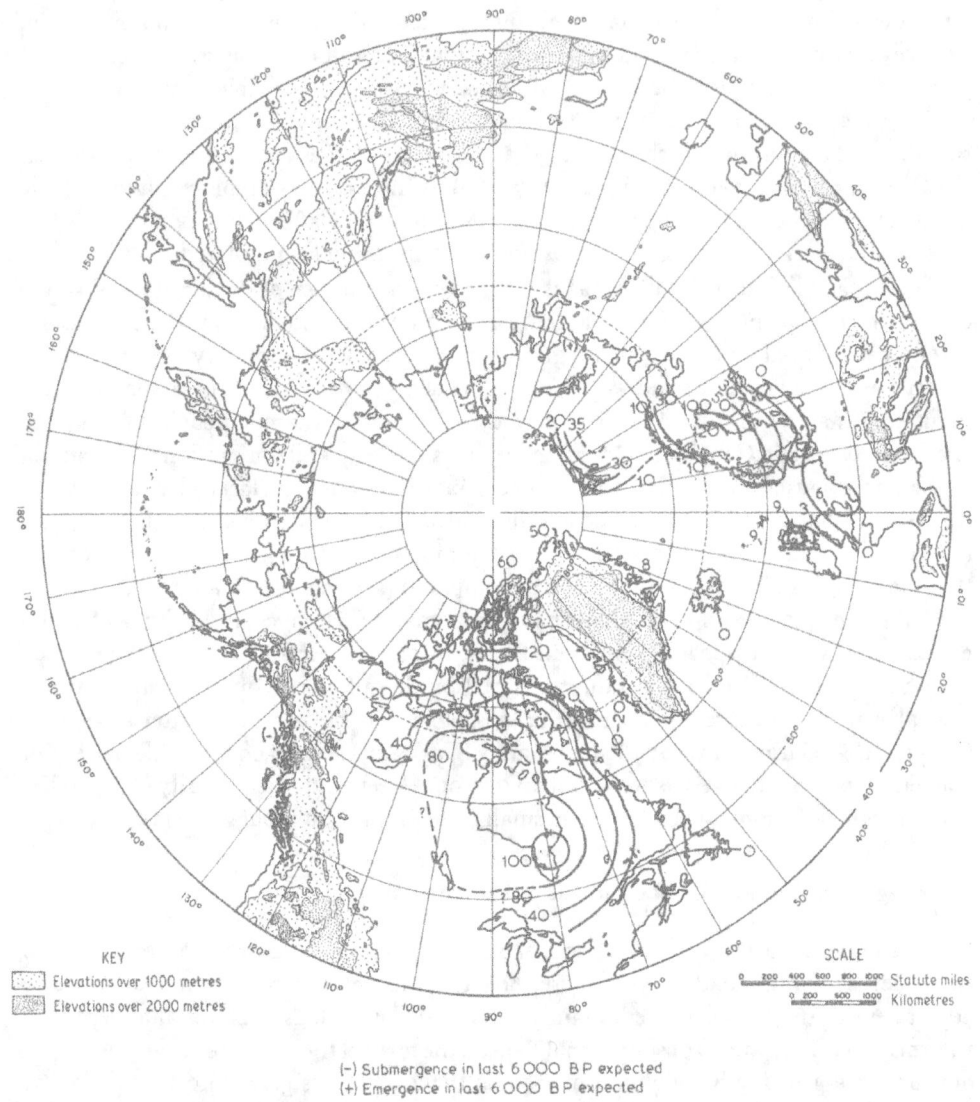

KEY
Elevations over 1000 metres
Elevations over 2000 metres

SCALE
0   200 400 600 800 1000 Statute miles
0   200   600   1000 Kilometres

(-) Submergence in last 6 000 B P expected
(+) Emergence in last 6 000 B P expected

Fig. 6A.10. Isobases on emergence for the last 6000 to 7500 years.

the period 6000 to 7500 BP, based on many sources, but principally Schytt *et al.* (1967), Donner (1970) and Andrews (1970). No overall statement on postglacial emergence is yet possible for Greenland although Weidick (1971) has analysed existing data. The map indicates the existence of three uplift centres in arctic Canada – one over south Hudson Bay, one on the northwest coast of Hudson Bay and the other located over the Queen Elizabeth Islands. These centres should be compared to those determined from the intersection of strandline dips on fig. 6A.2. The maximum emergence in the last 6000 years has amounted to 100 m. Over the Barents Shelf a strandline can be distinguished by the presence of pumice; it dates from *c.* 6000 BP and indicates a maximum emergence of 35 m. The centre of uplift for the Fennoscandia ice sheet is located over the Gulf of Bothnia (fig. 6A.10) where the Littorina I shoreline reaches a maximum altitude of 120 m (Donner, 1969). This shoreline is dated at several localities at 7000 to 7500 BP and hence is older than the Canadian dates on fig. 6A.10. The maximum emergence in Canada in the last 7000 to 7500 years has been 160 to 200 m, nearly double the Gulf of Bothnia value. The difference in the amount of postglacial recovery over the last 7000

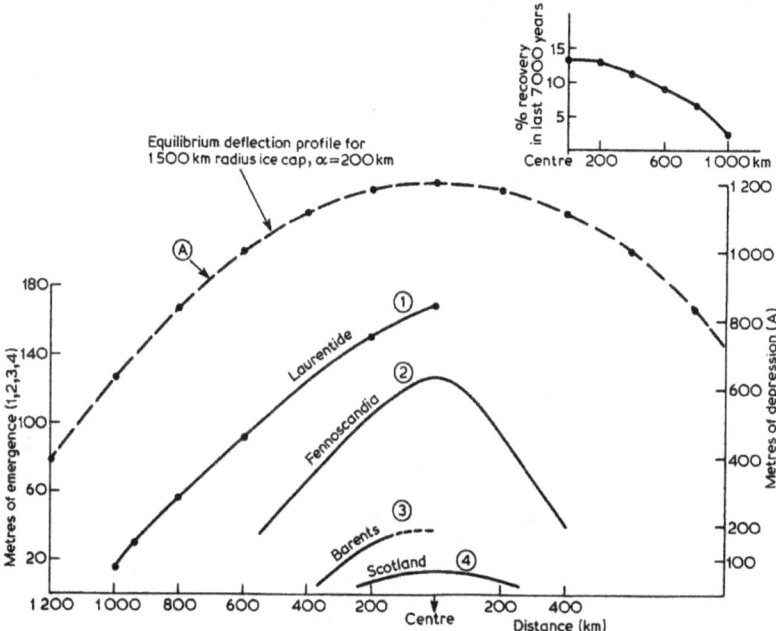

Fig. 6A.11. Cross-sections across the four major ice sheets showing the amount of emergence in the last 7000 years and the amount of equilibrium (A) under a 1500 km ice sheet. Inset, upper right, shows for the Laurentide ice sheet the ratio of rebound in the last 7000 years/ total crustal deflection.

years is shown on fig. 6A.11 as a series of cross-sections. Also included on the diagram is the equilibrium deflection for the Laurentide ice sheet based on a radius of 1500 km, a maximum thickness of 4500 m and a crustal flexural parameter α of 200 km (Walcott, 1970). The dimensions of the flexural parameter were determined from field evidence on eastern Baffin Island, which indicated that the crustal depression at the margin of the ice sheet was of the order of 110 m. The depression extends out beyond the limit of the load by $\pi\alpha/2 = 300$ km. Thus postglacial uplift is not sufficient by itself to indicate a region was glaciated. Late Wisconsin glaciation of the outermost Queen Elizabeth Islands is not necessary to explain the presence of raised beaches on the islands.

The data on fig. 6A.11 are nested in order of the size of the former ice sheet. In the last 6000 years the centre of the 500 km radius highland Britain ice sheet has risen 12 m or so,

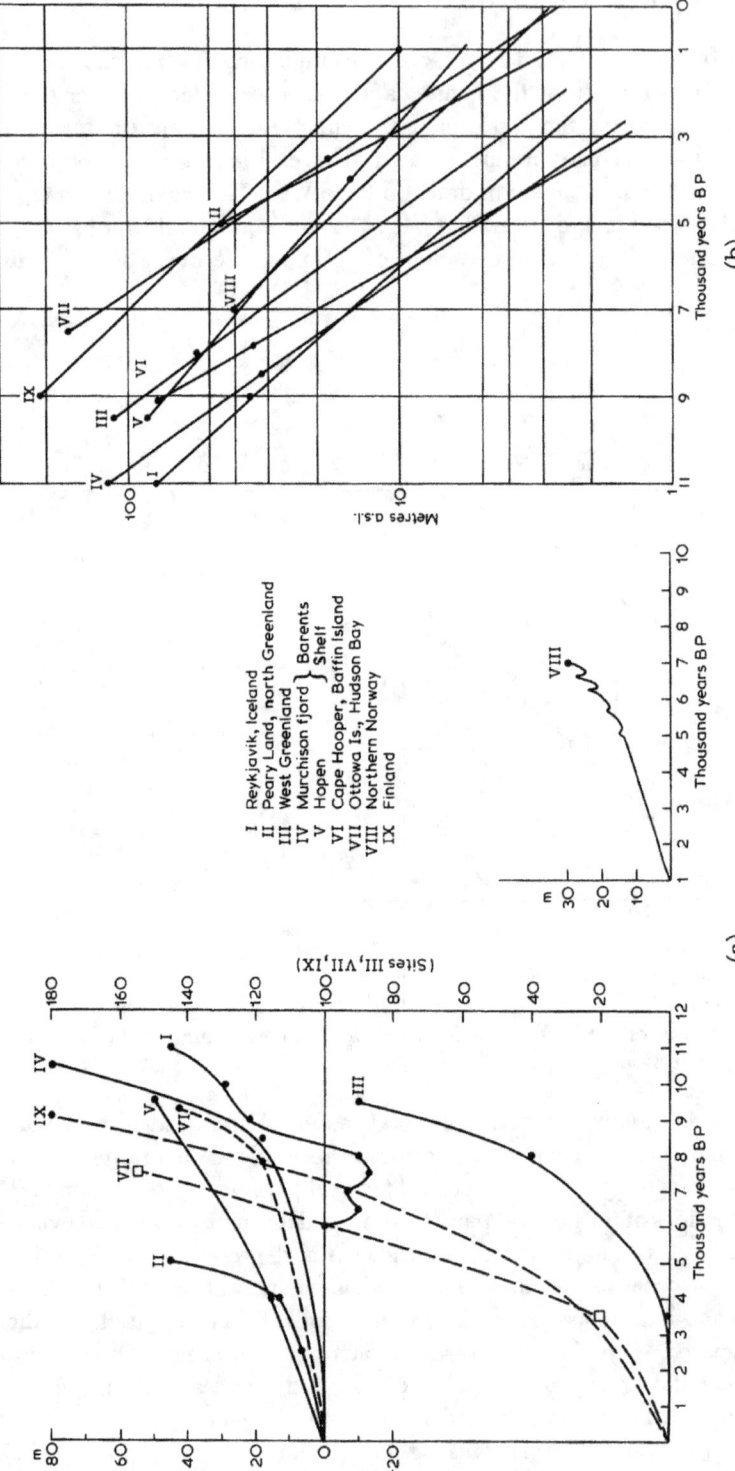

Fig. 6A.12. (a) Emergence curves for nine sites in the Arctic; the right-hand scale refers to sites III, VII and IX. (b) Postglacial uplift curves for the same sites plotted on semi-log paper showing the general exponential decay response of the crust to glacial unloading.

compared to 170 m for the area in southern Hudson Bay. The inset of fig. 6A.11 indicates how the central parts of the Laurentide ice sheet have risen proportionally more during the last 6000 years than the outer margins. A more useful comparison would be to ascertain if there are differences in the amount of postglacial uplift at some specified distance from the former margins of these ice sheets, because at such sites the actual ice load should not have differed greatly. At a distance 500 km in from the former maximum limit the amount of postglacial recovery is very similar, ranging between 12 to 20 m of emergence over the last 7000 years.

The advent of ^{14}C dating led to studies on the form of postglacial emergence and uplift and the construction of time/elevation graphs such as illustrated in fig. 6A.12. Because world sea level was lower than present during Late Glacial and postglacial time, the observed elevation of a dated raised marine deposit provides only a minimum estimate of the amount of postglacial isostatic uplift. Sites ice-free since 10,000 BP have actually risen about 30 m more than recorded. In arctic areas, postglacial emergence and/or uplift curves have been produced for the Barents Shelf, Iceland, northern Norway, northern and West Greenland, and arctic Canada. On fig. 6A.12a nine postglacial emergence curves are illustrated, and on fig. 6A.12b the eustatic sea level correction has been added and the uplift curves plotted on semi-log paper. Several authors have commented on the mathematical form of glacio-isostatic recovery (Washburn and Stuiver, 1962; Andrews, 1968). Many recovery curves can be closely approximated by an expression: $U = C(1 - e^{-kt})$, where U is the uplift at time t, C is the maximum observed postglacial uplift, and k is the decay constant. In arctic Canada k is roughly 0.44×10^{-4} years. Certain of the emergence curves, such as curves I and VII, show one or more sequences of marine regressions and transgressions. This is caused by changes in the *relative* rates of eustatic sea level change and glacio-isostatic recovery. The detailed fluctuations recorded in the curve from northern Norway are not apparent in curves V and VI, although they should be because both have slower rates of uplift than the northern Norway site over the last 7000 years. The absence of evidence for a fluctuating relative sea level is the rule at most sites so far recorded in the Canadian Arctic. Fluctuations have been inferred because of an unusually well developed strandline (e.g. Blake, 1970) but stratigraphic evidence has not been reported. A recent marine transgression is stratigraphically observable along the northern Cumberland Peninsula coast where soils and *in situ* moss have been covered by shingle ridges and beach sands.

The majority of uplift curves on fig. 6A.12b are approximately parallel to each other, indicating that the characteristic response times of the curves are quite similar. Apparent half-lives for the curves vary between 1000 and 1700 years (curves II and V respectively). Variations within the group are related to the nature of the glacio-isostatic load, the structure of the crust and mantle, location of the site relative to the load, the deglacial history, and the effect of hydro-isostatic forces caused by the Late Glacial and postglacial eustatic sea level rise.

Current crustal motions in the Arctic related to glacio-isostatic recovery have been studied with a variety of evidence, including folk legends, destruction of former Eskimo sites, ^{14}C dates, stratigraphy, tide gauge and precise levelling. The highest rate of recovery has been estimated at 1.2 m (100 yr)$^{-1}$ for southern Hudson Bay (Webber *et al.*, 1970). In the Gulf of Bothnia current uplift is about 1.0 m (100 yr)$^{-1}$. The current rate of uplift decreases away from these former ice centres (see Andrews, 1970) and submergence is occurring at sites along their outermost periphery. Evidence from eastern Baffin Island was referred to above, and Bird (1967) notes a similar phenomenon along the coast of Banks Island in the northwestern Canadian Arctic. Sections of the Greenland coast are apparently sinking (e.g. Nielsen, 1952). This might be related to renewed loading of the crust by the expansion of the ice sheet during the Neoglacial.

No mention has been made in this section of the variation in the height of the marine limit throughout the region. The marine limit is by definition the highest deposit or feature related to a former, and higher, relative sea level. In arctic Canada the marine limit increases from 0 m along sections of the outermost coasts (i.e. Cape Dyer, Baffin Island) to 280 m along the southeast side of Hudson Bay (Archer, 1968). However, in at least certain sections there is indisputable evidence that at any site there are one or more 'marine limits' of different ages. An excellent example of this occurs between outer Narpaing and Quajon fiords, eastern Baffin Island (*c.* 68°N) (Pheasant, 1971). Major glacio-marine deposits and erosion forms are found at elevations of 210, 80, 25 and 18 m. The 80 m feature is dated at 137,000 ± 10,000 BP; the 210 m feature as been dated at >100,000 BP by the amino acid technique, and may be 400,000 years old as it falls within the Zone II weathering limits (see earlier section and fig. 6A.4); whereas the 18 m wave-cut terrace immediately south is dated at 24,000 ± 800 BP. The postglacial marine limit is, in fact, at 5 m asl. The occurrence of high raised marine deposits along outer coasts has also been reported from further north along the eastern Baffin Island coast where Løken (1966) identified a glacio-marine delta at 80 m which dated >54,000 BP. On the outer coast of West Greenland at the same latitude as the Narpaing/Quajon deposits, Kelly (1969, p. 27) noted: 'Net sea level displacement at the outer coast has been 150 ± 5 m . . . In the outer coastal regions there are extensive outcrops of marine sediments. Their stratigraphy is complex due to facies variation and recycling during sea level recession.' Kelly makes no inferences about the likely age of the deposits but as he states that the nunatak moraines are '. . . presumably of Late Wisconsin/Würm age' it would appear that the 150 m deposits are considered to be Late Glacial in age.

5. Conclusions

The Cainozoic history of the Arctic is critical to a proper understanding of the mechanisms and timing of the major glaciations. The critical threshold for initiation of glaciation in arctic areas is naturally much lower than in areas far south of the initial accumulation centres. Nevertheless, much of our current concept of the Cainozoic, and of the Quaternary in particular, has been developed without adequate knowledge of the history of the initial accumulation areas. In particular the obvious non-parallelism of 'glaciations' between arctic and southern areas must be noted. Data from arctic and antarctic cores (Dansgaard *et al.*, 1971, and Epstein *et al.*, 1970) both err, in my judgement, by not allowing any time for the gradual build-up of the Laurentide ice sheet. In this chapter it has been proposed that the last or Wisconsin glaciation began in the Arctic about <130,000 ± BP. The evidence for such a conclusion is based on field evidence from the eastern Canadian Arctic and an analysis of eustatic sea level. The Arctic is very sensitive to climatic change and the threshold between glacial and non-glacial modes is low. For example, the elevation difference between glacier-filled cirques and empty south-facing cirques in the Okoa Bay area, eastern Baffin Island, is only 200 m. An understanding of arctic environments in the broad sense is essential to an understanding of the dramatic climatic changes of the last few million years.

References

ANDERSEN, B. G. (1968) Glacial geology of Western Troms, North Norway. *Norg. Geol. Unders.*, 256. 160 pp.

ANDREWS, J. T. (1961) The development of scree slopes in the English Lake District and central Quebec–Labrador. *Cah. Géogr. Qué.*, 5, 219–31.

ANDREWS, J. T. (1963) End moraines and late-glacial chronology of the northern Nain–Okak section of Labrador. *Geogr. Ann.*, 45, 158–71.

ANDREWS, J. T. (1968) Postglacial rebound in Arctic Canada: similarity and prediction of uplift curves. *Can. J. Earth Sci.*, 5, 39–47.

ANDREWS, J. T. (1969) The pattern and interpretation of restrained, postglacial and residual rebound in the area of Hudson Bay. *In:* HOOD, P. J. (ed.) Earth Science Symposium on Hudson Bay, *Geol. Surv. Can. Pap.*, 68–53, 49–62.

ANDREWS, J. T. (1970) A geomorphological study of post-glacial uplift with particular reference to Arctic Canada. *Inst. Brit. Geogr. Spec. Pub.* 2. 157 pp.

ANDREWS, J. T. (1972) Recent and fossil growth rates of marine bivalves, Canadian Arctic, and late-Quaternary Arctic marine environments. *Palaeogeogr., Palaeoclimat., Palaeoecol.*, 11, 157–76.

ANDREWS, J. T. and BARNETT, D. M. (1972) Analysis of strandline tilt directions in relation to ice centers and postglacial crustal deformation, Laurentide ice sheet. *Geogr. Ann.*, A, 54, 1–11.

ANDREWS, J. T. and FALCONER, G. (1969) Late-glacial and postglacial history and emergence of the Ottawa Islands, Hudson Bay, NWT: evidence on the deglaciation of Hudson Bay. *Can. J. Earth Sci.*, 6, 1263–76.

ANDREWS, J. T. and IVES, J. D. (1972) Late- and postglacial events (>10,000 BP) in the eastern Canadian Arctic with particular reference to the Cockburn Moraines and the break-up of the Laurentide Ice Sheet. *In:* VASARI, Y. *et al.* (eds.) Climatic Changes in Arctic Areas During the Last 10,000 Years, *Acta Univ. Oulu*, Ser. A, Sci. Rerum Natural. 3, Geol. 1, 149–71.

ANDREWS, J. T. and MILLER, G. H. (1972) Quaternary history of northern Cumberland Peninsula, Baffin Island, NWT, Part IV: Maps of the present glaciation limit and lowest equilibrium line altitude for north and south Baffin Island. *Arct. Alp. Res.*, 4, 45–59.

ANDREWS, J. T., BUCKLEY, J. T. and ENGLAND, J. H. (1970) Late-glacial chronology and glacio-isostatic recovery, Home Bay, East Baffin Island, Canada. *Bull. Geol. Soc. Amer.*, **81**, 1123–48.

ANDREWS, J. T., MCGHEE, R. and MCKENZIE-POLLACK, L. (1971) Comparison of elevations of archaeological sites and calculated sea levels in Arctic Canada. *Arctic*, **24**, 210–28.

ANDREWS, J. T., GUENNEL, G. K., WRAY, J. L. and IVES, J. D. (1972) An early Tertiary outcrop in north-central Baffin Island, Northwest Territories, Canada: environment and significance. *Can. J. Earth Sci.*, **9**, 233–8.

ARCHER, D. R. (1968) The upper marine limit in the Little Whale River, New Quebec. *Arctic*, **21**, 153–60.

ARTYUSHKOV, E. V. (1971) Rheological properties of the crust and upper mantle according to data on isostatic movements. *J. Geophys. Res.*, **76**, 1376–90.

BANDY, O. L., BUTLER, E. A. and WRIGHT, R. C. (1969) Alaskan Upper Miocene marine glacial deposits and the *Turborotalia pachyderma* datum plane. *Science*, **166**, 607–9.

BARRY, R. G. (1966) Meteorological aspects of the glacial history of Labrador–Ungava with special reference to atmospheric vapour transport. *Geogr. Bull.*, **8**, 319–40.

BARRY, R. G., IVES, J. D. and ANDREWS, J. T. (1971) A discussion of atmospheric circulation during the Last Ice Age. *Quat. Res.*, **1**, 415–18.

BIRD, J. B. (1967) *The Physiography of Arctic Canada.* Baltimore: Johns Hopkins Univ. Press. 336 pp.

BLAKE, W. JR (1970) Studies on the glacial history in Arctic Canada, I. Pumice, radiocarbon dates, and differential postglacial uplift in the eastern Queen Elizabeth Islands. *Can. J. Earth Sci.*, **7**, 634–64.

BOSTOCK, H. S. (1966) Notes on glaciation in central Yukon Territory. *Geol. Surv. Can. Pap.*, 65–36.

BRINKMANN, W. A. R. and BARRY, R. G. (1972) Palaeoclimatological aspects of the synoptic climatology of Keewatin, Northwest Territories, Canada. *Palaeogeogr., Palaeoclimat., Palaeoecol.*, **11**, 77–91.

BROECKER, W. S. (1968) In defense of the astronomical theory of glaciation. *In:* MITCHELL, J. M. JR (ed.) Causes of Climatic Change, *Met. Monogr.*, **8** (30), 139–41. Amer. Met. Soc.

BROECKER, W. S. and VAN DONK, J. (1970) Insolation changes, ice volumes and the O^{18} record in deep-sea cores. *Rev. Geophys. Space Phys.*, **8**, 169–98.

BRYSON, R. A. and WENDLAND, W. M. (1967) Tentative climatic patterns for some late glacial and postglacial episodes in central North America. *In:* MAYER-OAKES, W. (ed.) *Life, Land and Water*, 271–98. Winnipeg: Manitoba Univ. Press.

BRYSON, R. A., WENDLAND, W. M., IVES, J. D. and ANDREWS, J. T. (1969) Radio-carbon isochrones on the disintegration of the Laurentide ice sheet. *Arct. Alp. Res.*, **1**, 1–14.

CARRARA, P. E. and ANDREWS, J. T. (1972) The Quaternary history of northern Cumberland Peninsula, Baffin Island, NWT, I: The late and Neoglacial deposits of the Akudlermiut and Boas glaciers. *Can. J. Earth Sci.*, **9**, 403–14.

CHRISTIE, R. L. (1967) Reconnaissance of the surficial geology of northeastern Ellesmere Island, Arctic Archipelago. *Geol. Surv. Can. Bull.*, 138. 50 pp.

CLARKE, D. B. and UPTON, B. G. J. (1971) Tertiary basalts of Baffin Island: field relations and tectonic setting. *Can. J. Earth Sci.*, **8**, 248–58.

COX, A. (1968) Polar wandering, continental drift, and the onset of Quaternary glaciation. *In:* MITCHELL, J. M. JR (ed.) Causes of Climatic Change, *Met. Monogr.*, **8** (30), 112–25. Amer. Met. Soc.

CURRAY, J. R. (1965) Late Quaternary history, continental shelves of the United States. *In:* WRIGHT, H. E. JR and FREY, D. G. (eds.) *The Quaternary of the United States*, 723–36. Princeton: Princeton Univ. Press.

CURRY, L. (1962) Climatic change as a random series. *Ann. Ass. Amer. Geogr.*, **52**, 21–31.

DANSGAARD, W., JOHNSEN, S. J., CLAUSEN, H. B. and LANGWAY, C. C. JR (1971) Climatic record revealed by the Camp Century ice core. *In:* TUREKIAN, K. K. (ed.) *Late Cenozoic Ice Ages*, 37–56. New Haven: Yale Univ. Press

DAVIES, T. T. and ANDREWS, J. T. (n.d.) *Ultrastructure, Growth Rates and δO^{18} Composition of Certain Postglacial Arctic Bivalves* (in preparation).

DAVIS, W. E. (1961) Glacial geology of northern Greenland. *Polarforschung*, **5**, 94–103.

DENTON, G. H. (1970) Late Wisconsin glaciation in northwestern North America: ice recession and origin of paleo-Indian Clovis complex. *1st Meeting, Amer. Quat. Ass., Bozeman* (abstract), 34–5.

DENTON, G. H. and ARMSTRONG, R. L. (1969) Miocene-Pliocene glaciations in southern Alaska. *Amer. J. Sci.*, **267**, 1121–42.

DENTON, G. H. and STUIVER, M. (1967) Late Pleistocene glacial stratigraphy and chronology, northeastern St Elias Mountains, Yukon Territory, Canada. *Bull. Geol. Soc. Amer.*, **78**, 485–510.

DONNER, J. J. (1969) A profile across Fennoscandia of Late Weichselian and Flandrian shorelines. *Soc. Sci. Fennica* (Comm. Physico-Math), **36**, 1–23.

DONNER, J. J. (1970) Land/sea level changes in Scotland. *In:* WALKER, D. and WEST, R. G. (eds.) *Studies in the Vegetational History of the British Isles*, 23–39. Cambridge: Cambridge Univ. Press.

EINARSSON, T. (1966) Late- and post-glacial rise in Iceland and subcrustal viscosity. *Jökull*, **III** (16A), 157–66.

EINARSSON, T., HOPKINS, D. M. and DOELL, R. D. (1967) The stratigraphy of Tjörnes, northern Iceland and the history of the Bering Land Bridge. *In:* HOPKINS, D. M. (ed.) *The Bering Land Bridge*, 312–25. Stanford: Stanford Univ. Press.

ENGLAND, J. H. and ANDREWS, J. T. (1973) *The Quaternary History of Northern Cumberland Peninsula, Part II: A Preliminary Wisconsin Chronology for Broughton Island* (submitted).

EPSTEIN, S., SHARP, R. P. and GOW, A. J. (1970) Antarctic Ice Sheet: stable isotope analyses of Byrd Station Cores and interhemispheric climatic implications. *Science*, **168**, 1570–2.

FLINT, R. F. (1957) *Glacial and Pleistocene Geology*. New York: John Wiley. 553 pp.

FLINT, R. F. (1971) *Glacial and Quaternary Geology*. New York: John Wiley. 892 pp.

FREDSKILD, B. (1969) A postglacial standard pollen diagram from Peary Land, North Greenland (1). *Pollen et Spores*, **11**, 573–83.

FYLES, J. G. (1963) Surficial geology of Victoria and Stefansson Islands, District of Franklin. *Geol. Surv. Can. Bull.*, 101. 38 pp.

FYLES, J. G. (1966) Quaternary stratigraphy McKenzie Delta and Arctic Coastal Plain. *In:* Rep. Activities, *Geol. Surv. Can. Pap.*, 66–1. 30 pp.

GEOTIMES (1970) Deep Sea Drilling Project Leg 12. *Geotimes*, **15**, 10–14 (by the scientific staff).

GOLDTHWAIT, R. P., DREIMANIS, A., FORSYTH, J. L., KARROW, P. F. and WHITE, G. W. (1965) Pleistocene deposits of the Erie Lobe. *In:* WRIGHT, H. E. JR and FREY, D. G. (eds) *The Quaternary of the United States*, 85–98. Princeton: Princeton Univ. Press.

GROSVAL'D, M. G. (1970) Glaciological and geomorphological research in the Eurasian Arctic with special reference to the Barents Sea. *Musk-Ox*, **17**, 1–9. Univ. of Saskatchewan.

HALLAM, A. (1971) Mesozoic geology and the opening of the North Atlantic. *J. Geol.*, **79**, 129–57.

HAMILTON, W. (1968) Cenozoic climatic change and its cause. *In:* MITCHELL, J. M. (ed.) Causes of Climatic Change, *Met. Monogr.*, **8** (30), 128–33. Amer. Met. Soc.

HAMILTON, W. (1970) The Uralides and the motion of the Russian and Siberian Platforms. *Bull. Geol. Soc. Amer.*, **81**, 2553–76.

HARLAND, W. B. (1969) Contribution of Spitsbergen to understanding of tectonic evolution of North Atlantic Region. *In:* KAY, M. (ed.) *North Atlantic – Geology and Continental Drift. A Symposium*, 817–51. Tulsa: Amer. Ass. Petrol. Geol.

HATTERSLEY-SMITH, G. (1969) Glacial features of Tanquary Fiord and adjoining areas of northern Ellesmere Island, NWT. *J. Glaciol.*, **8**, 23–50.

HERMAN, Y. (1970) Arctic paleo-oceanography in late Cenozoic time. *Science*, **169**, 474–7.

HOPKINS, D. M. (1967a) Quaternary marine transgressions in Alaska. *In:* HOPKINS, D. M. (ed.) *The Bering Land Bridge*, 47–90. Stanford: Stanford Univ. Press.

HOPKINS, D. M. (1967b) The Cenozoic history of Beringia–a synthesis. *In:* HOPKINS, D. M. (ed.) *The Bering Land Bridge*, 451–840. Stanford: Stanford Univ. Press.

HOPPE, G. (1970) The Würm ice sheets of northern and Arctic Europe. *Acta Geogr. Lodz.*, **24**, 205–15.

HUNKINS, K., BÉ, A. W. H., OPDYKE, N. and SAITO, T. (1971) Arctic paleooceanography in late Cenozoic times. *Science*, **174**, 962–3.

IVES, J. D. (1957) Glaciation of the Torngat Mountains, northern Labrador. *Arctic*, **10**, 67–87.

IVES, J. D. (1958) Glacial geomorphology of the Torngat Mountains, northern Labrador. *Geogr. Bull.*, **12**, 47–75.

IVES, J. D. (1962) On the vertical extent of glaciation in northeastern Labrador-Ungava: discussion. *Can. Geogr.*, **6**, 115–18.

IVES, J. D. (1963) Field problems in determining the maximum extent of Pleistocene glaciation along the eastern Canadian seaboard – a geographer's point of view. *In:* LÖVE, A. and LÖVE, D. (eds.) *North Atlantic Biota and their History*, 337–54. Oxford: Pergamon Press.

IVES, J. D. and ANDREWS, J. T. (1963) Studies in the physical geography of north-central Baffin Island, Northwest Territories. *Geogr. Bull.*, **19**, 5–48.

IVES, J. D. and BORNS, H. W. JR (1972) Thickness of the Wisconsin Ice Sheet in southeast Baffin Island, Arctic Canada. *Zeits. Gletscherk., Glaziol.*, **7**, 17–20.

IVES, J. D. and BUCKLEY, J. T. (1969) Glacial geomorphology of Remote Peninsula, Baffin Island, NWT, Canada. *Arct. Alp. Res.*, **1**, 83–96.

JOHNSON, P. J. JR (1969) Deglaciation of the central Nain–Okak Bay section of Labrador. *Arctic*, **22**, 373–94.

KARLSTROM, T. V. N. (1964) Quaternary geology of the Kenai lowland and glacial history of the Cook Inlet region, Alaska. *US Geol. Surv. Prof. Pap.*, 443. Washington, DC. 69 pp.

KELLY, M. (1969) Quaternary geology of Nordre Strømfjord and its environs. *In:* Rep. Activities 1968, *Grønlands Geol. Unders.*, Rep. 19. 27 pp.

KERR, J. W. (1970) Today's topography and tectonics in northeastern Canada, Abstract. *Can. J. Earth Sci.*, **7**, 570.

KIND, N. V. (1967) Radiocarbon chronology in Siberia. *In:* HOPKINS, D. M. (ed.) *The Bering Land Bridge*, 172–92. Stanford: Stanford Univ. Press.

KOCH, L. (1928) Physiography of Northern Greenland. *In: Greenland*, **1**, 491–519. Copenhagen and London.

KUKLA, J. (1969) The cause of the Holocene climate change. *Geol. en Mijnbouw*, **48**, 307–34.

LAMB, H. H. and WOODROFFE, A. (1970) Atmospheric circulation during the last Ice Age. *Quat. Res.*, **1**, 29–58.

LASCA, N. P. (1969) The surficial geology of Skeldal, Mesters Vig, Northeast Greenland. *Medd. om Grønland*, **176** (3). 56 pp.

LE MASURIER, W. E. (1972) Volcanic record of Antarctic glacial history: implications with regard to Cenozoic sea levels. *In:* PRICE, R. J. and SUGDEN, D. E. (eds.) Polar geomorphology, *Inst. Brit. Geogr. Spec. Pub.*, 4, 59–74. London.

LOEWE, F. (1971) Considerations on the origin of the Quaternary ice sheets of North America. *Arct. Alp. Res.*, **3** (4), 331–44.

LØKEN, O. H. (1962) On the vertical extent of glaciation in northeastern Labrador-Ungava. *Can. Geogr.*, **6**, 106–15, 118–19.

LØKEN, O. H. (1966) Baffin Island refugia older than 54,000 years. *Science*, **153**, 1378–80.

MCCULLOCH, D. and HOPKINS, D. M. (1966) Evidence for an early recent warm interval in northwestern Alaska. *Bull. Geol. Soc. Amer.*, **77**, 1089–108.

MCDONALD, B. C. (1969) Glacial and interglacial stratigraphy, Hudson Bay Lowlands. *In:* HOOD, P. J. (ed.) Earth Science Symp. (Hudson Bay), *Geol. Surv. Can. Pap.*, 68–53, 78–99.

MCDONALD, B. C. (1971) Late Quaternary stratigraphy and deglaciation in eastern Canada. *In:* TUREKIAN, K. K. (ed.) *Late Cenozoic Glacial Ages*, 331–54. New Haven: Yale Univ. Press.

MACKAY, J. E. (1971) The origin of massive icy beds in permafrost, western Arctic Coast, Canada. *Can. J. Earth Sci.*, **8**, 397–422.

MARGOLIS, S. V. and KENNETT, J. P. (1970) Antarctic glaciation during the Tertiary recorded in Sub-Antarctic deep-sea cores. *Science*, **170**, 1085–7.

MATHER, W. J. (1968) Muskox bone of possible Wisconsin age from Banks Island, Northwest Territories. *Arctic*, **21**, 260–6.

MERCER, J. H. (1956) Geomorphology and glacial history of southernmost Baffin Island. *Bull. Geol. Soc. Amer.*, **67**, 553–70.

MERCER, J. H. (1968a) Glacial geology of the Reedy glacier area, Antarctica. *Bull. Geol. Soc. Amer.*, **79**, 471–86.

MERCER, J. H. (1968b) The discontinuous glacio-eustatic fall in Tertiary sea level. *Palaeogeogr., Palaeoclimat., Palaeoecol.*, **5**, 77–86.

MERCER, J. H. (1970) A former ice sheet in the Arctic Ocean. *Palaeogeogr., Palaeoclimat., Palaeoecol.*, **8**, 19–27.

MILANKOVITCH, M. (1941) *Canon of Insolation and the Ice-Age Problem*. Jerusalem: Israel Progr. Sci. Transl. (transl. 1969). 484 pp.

MILLIMAN, J. D. and EMERY, K. O. (1968) Sea levels during the past 35,000 years. *Science*, **162**, 1121–3.

MÖRNER, N.-A. (1969) The late Quaternary history of the Kattegatt Sea and the Swedish west coast. *Sver. Geol. Unders.*, C, 640. 487 pp.

MÖRNER, N.-A. (1971) The position of the ocean level during the interstadial at about 30,000 BP – a discussion from a climatic-glaciologic point of view. *Can. J. Earth Sci.*, **8**, 132–43.

NIELSEN, E. (1952) A determination of the subsidence of the land at Angmagssalik. *Medd. om Grønland*, **136** (2). 10 pp.

PÉWÉ, T. L., HOPKINS, D. M. and GIDDINGS, J. L. (1965) The Quaternary geology and archaeology of Alaska. *In:* WRIGHT, H. E. JR and FREY, D. G. (eds.) *The Quaternary of the United States*, 355–76. Princeton: Princeton Univ. Press.

PHEASANT, D. A. (1971) *The Glacial Chronology and Glacio-Isostasy of the Nårpaing–Quajon Fiord Area, Cumberland Peninsula, Baffin Island*. Unpub. Ph.D. thesis, Univ. of Colorado. 232 pp.

PREST, V. K. (1970) Quaternary geology of Canada. *In:* Geology and Economic Minerals of Canada. *Geol. Surv. Can., Econ. Geol. Rep.*, 1 (5th edn), 675–764. Ottawa.

SAVILE, D. B. O. (1961) The botany of the north-western Queen Elizabeth Islands. *Can. J. Bot.*, **39**, 909–42.

SCHYTT, V., HOPPE, G., BLAKE, W. JR and GROSSWALD, M. G. (1967) The extent of the Würm glaciation in the European Arctic. *Int. Ass. Sci. Hydrol. Pub.* (General Assembly of Berne), 79, 207–16.

SHAW, D. M. and DONN, W. L. (1968) Milankovitch radiation variations: a quantitative evaluation. *Science*, **162**, 1270–2.

SMITH, D. I. (1961) The glaciation of northern Ellesmere Island. *Folia Geogr. Danica*, **9**, 224–32.

TANNER, W. F. (1965) Cause and development of an ice age. *J. Geol.*, **73**, 413–30.

TERASMAE, J. and CRAIG, B. G. (1958) Discovery of fossil *Ceratophyllum demersum* L. in Northwest Territories, Canada. *Can. J. Bot.*, **36**, 567–9.

TERASMAE, J., WEBBER, P. J. and ANDREWS, J. T. (1966) A study of late Quaternary plant-bearing beds in north-central Baffin Island, Canada. *Arctic*, **19**, 296–318.

THORSTEINSSON, R. and TOZER, E. T. (1970) Geology of the Arctic Archipelago. *In:* Geology and Economic Minerals of Canada, *Geol. Surv. Can., Econ. Geol. Rep.*, 1, 547–90.

TOMLINSON, R. F. (1963) Pleistocene evidence related to glacial theory in northeastern Labrador. *Can. Geogr.*, **7**, 83–90.

VEEH, H. H. and CHAPPELL, J. (1970) Astronomical theory of climatic change: support from New Guinea. *Science*, **167**, 862–5.

VERNON, P. and HUGHES, O. L. (1966) Surficial geology, Dawson, Larsen Creek and Nash Creek Map-Areas, Yukon Territory. *Geol. Surv. Can. Bull.*, 136. 25 pp.

WALCOTT, R. I. (1970) Flexural rigidity, thickness and viscosity of the lithosphere. *J. Geophys. Res.*, **75**, 3941–54.

WASHBURN, A. L. (1965) Geomorphic and vegetational studies in Mesters Vig district, Northeast Greenland – General Introduction. *Medd. om Grønland*, **166** (1). 60 pp.

WASHBURN, A. L. and STUIVER, M. (1962) Radiocarbon dated post-glacial delevelling in northern Greenland and its implications. *Arctic*, **15**, 66–73.

WEBBER, P. J., RICHARDSON, J. W. and ANDREWS, J. T. (1970) Postglacial uplift and substrate age at Cape Henrietta Maria, southeastern Hudson Bay, Canada. *Can. J. Earth Sci.*, **7**, 317–25.

WEIDICK, A. (1968) Observations on some Holocene glacier fluctuations in West Greenland. *Medd. om Grønland*, **165** (6). 194 pp.

WEIDICK, A. (1971) Holocene shorelines and glacial stages in Greenland – an attempt at correlation. *Repp. Grønlands Geol. Unders.*, 41. 39 pp.

WEIDICK, A. (1972) Notes on Holocene glacial events in Greenland. *In:* VASARI, Y. *et al.* (eds.) Climatic Changes in Arctic Areas During the Last 10,000 Years, *Acta Univ. Oulu*, Ser. A, Sci. Rerum Natural. 3, Geol. 1, 177–202.

WILLMAN, H. B. and FRYE, J. C. (1970) Pleistocene stratigraphy of Illinois. *Illinois State Geol. Surv. Bull.*, 94. 204 pp.

WOLFE, J. A. (1971) Tertiary climatic fluctuations and methods of analysis of Tertiary floras. *Palaeogeogr., Palaeoclimat., Palaeoecol.*, **9**, 27–57.

WOLFE, J. A. and LEOPOLD, E. B. (1967) Neogene and early Quaternary vegetation of northwestern North America and Northeastern Asia. *In:* HOPKINS, D. M. (ed.) *The Bering Land Bridge*, 193–206. Stanford: Stanford Univ. Press.

ZUBAKOV, V. A. (1969) La chronologie des variations climatiques au cours du Pleistocène en Sibéria occidentale. *Rev. Géogr., Phys. Géol. Dyn.*, **11**, 315–24.

Manuscript received July 1971.

Additional references

LUNDQUIST, J. (1971) The Interglacial deposit at the Leveåniemi mine, Svappavaara, Swedish Lapland. *Sver. Geol. Unders.*, C, 658. 163 pp.

MILLER, G. H. and DYKE, A. S. (1974) Proposed extent of Late Wisconsin Laurentide ice on eastern Baffin Island. *Geology*, **2**, 125–30.

PHEASANT, D. R. and ANDREWS, J. T. (1973) Wisconsin glacial chronology and relative sea-level movements, Narpaing Fiord Broughton Island area, eastern Baffin Island, NWT. *Can. J. Earth Sci.*, **10**, 1621–41.

RAMPTON, V. (1971) Late Quaternary vegetational and climatic history of the Snag-Klutlan area, southwestern Yukon Territory, Canada. *Bull. Geol. Soc. Amer.*, **82**, 959–78.

SKINNER, R. G. (1973) Quaternary stratigraphy of the Moose River Basin, Ontario. *Geol. Surv. Can. Bull.*, 225. 77 pp.

TEN BRINK, N. (1974) Glacio-isostasy: new data from West Greenland and geophysical implications. *Bull. Geol. Soc. Amer.*, **85**, 219–28.

Alpine Quaternary glaciation

Helmut Heuberger

Geographical Institute,
University of Munich

1. Fundamentals of the distribution of Ice Age mountain glaciations

The term 'alpine', which is derived from the Alps, is identical with the German word 'Hochgebirge' or 'high mountains'. On the basis of a morphological definition, the lower boundary of the alpine zone is the limit of the occurrence of active surface solifluction (Furrer and Fitze, 1970) and the Pleistocene snowline (Troll, 1955). The term 'alpine' ('Hochgebirge') is therefore closely connected with the distribution of the Pleistocene glaciation outside the Arctic.

SNOWLINE PROBLEMS

Pleistocene alpine glaciation occurred only in areas where mountains rose above the Pleistocene snowline (based on the orographic snowline). Ideally, as in the case of glaciers, the orographic snowline separates the snow accumulation zone from the zone with net ablation. Its long-term position thus follows the firn line.

The methods used to determine the orographic snowline even on recent glaciers are accurate only to within ±50 m. They are estimates rather than accurate calculations. Reliable results are only obtained with small glaciers. In North American literature the climatic (or regional) snowline (Flint, 1957, pp. 47–9; 1971, pp. 63–70) is therefore determined from the *glaciation limit*, rather than from the orographic snowline (Enquist, 1916; Østrem, 1964), or from similar methods (Wahrhaftig and Birman, 1965). However, in central European literature the derivation of the climatic snowline from the orographic snowline is preferred (von Klebelsberg, 1948–9, pp. 27–35; Louis, 1954–5; Kaiser, 1966, pp. 264–6; see however, Østrem, Chapter 4C, pp. 230–6).

Since glaciers develop in the most favourable orographic locations, their firn limit (as a long-term average) and consequently the orographic snowline run below the glaciation limit. The glaciation limit is determined by the lowest glacierized summits and thus by locations which, as a result of wind effects, are extremely unfavourable to glacierization. This method is more appropriate for plateau ice caps (Scandinavia) than for typical alpine glaciers. For this reason Wahrhaftig and Birman (1965) determined the climatic snowline from the lowest peaks with glaciers still existing on their south-facing (sunny) slopes. The orographic snowline runs across *all* glaciers, even the smallest and lowest. The glaciation limit, on the other hand, does not touch weakly glacierized areas at all if none of the peaks are glacierized on their sunny slopes. Most important of all, the orographic snowline for a former glaciation can be determined for a much larger number of points and with more confidence than the glaciation limit because mountain-top ice does not always leave unequivocal traces (see Ives, Chapter 10B). For a comparative study of the Ice Age snowline, the determination of the orographic snowline is therefore the best basis (see also, Østrem, Chapter 4C).

The course of the Ice Age and present-day snowlines right across the climatic zones is best seen in the Cordillera system of the Americas. As opposed to the profile by Klute (1928, p. 75; modified in Flint, 1957, p. 47), the surveys by Wilhelmy (1957) and Kaiser (1966) are especially helpful in the compilation of a new snowline profile along the Cordilleras to the points where the Pleistocene mountain glaciation reached the Pacific Ocean and where the line merges into inland ice sheets.

The course of the present-day snowline (see fig. 2D.1) and its dependence on climate will not be discussed here. The northern culminations of the curves are missing since over long stretches in the northern tropics and subtropics the mountain ranges of the Cordilleras reach neither the present-day snowline nor the Pleistocene snowline. However, similar to the South American Puna de Atacama, there is a snowline culmination in the northern hemisphere in southwest Tibet above 6000 m; between the north side of the eastern Himalaya (28°N) and the south side of the western Kuen Lun (35°N). North of Nepal between 30° and 35°N it crosses even the 6400 m contour (map in von Wissmann, 1960).

The reconstruction of Pleistocene snowlines is based on morphologically visible Ice Age evidence. Most reconstructions are therefore restricted to the last glaciation. For the older glaciations the basis for comparison is much more hypothetical. The snowline of the last glaciation runs more or less parallel to the present-day snowline. This suggests that the temperature was generally lower than today and that during the last glaciation the disposition of the climatic zones was, in principle, similar to the present.

However, looking at the fine details, there are large fluctuations in the snowline depression of the last glaciation as compared to the present-day snowline. According to fig. 6B.1 the snowline depression reaches a maximum of approximately 1400 m on Nevada de Aconquija (on the east side of the Andes in Argentina) and on Mt Shasta (Cascade Range, 41.5°N). These maxima are situated in humid regions. The minimum of 400 to 500 m is reached in the Cordillera Huayhuash of Peru. In continental montane Asia the maxima of the snowline depression (1000 to 1150 m) are concentrated in the humid meridional river valleys east of the Himalaya, the minima (less than 400 m) in especially dry, continental mountain parts: central Pamir, Kuen Lun, Nan Shan, east of the

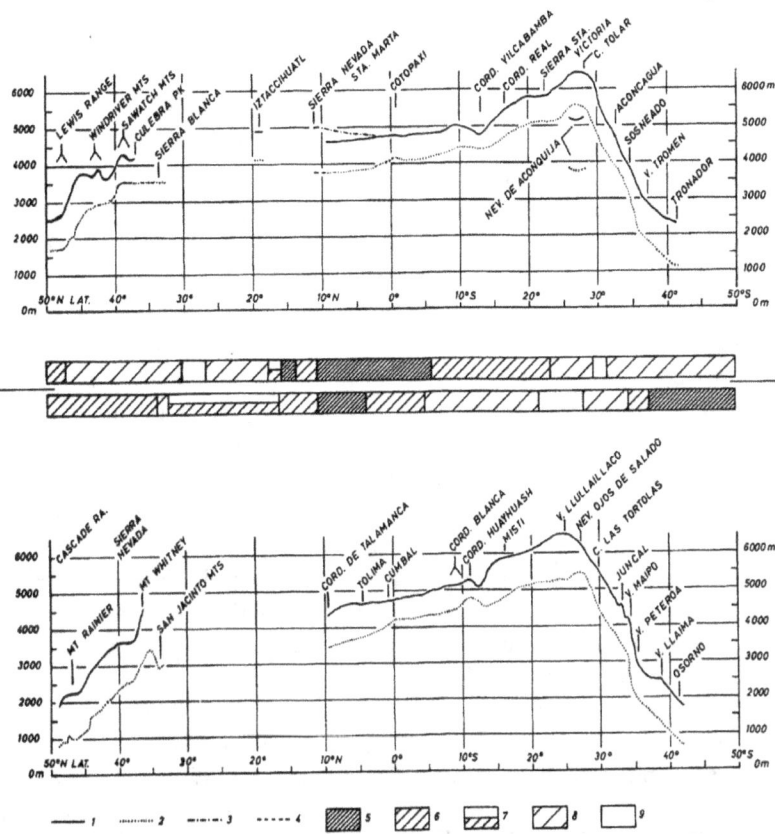

Fig. 6B.1. Present snowline (1) and that for the last glaciation (2) compared with precipitation for sections from the eastern Andes to the Rocky Mountains (above) and from the western Andes to the Coast Ranges (below). Snowlines based on the orographic snowlines: after Kaiser, 1966; Richmond, 1965a, b; Louis, 1927 (United States); White, 1962 (Mexico); Wilhelmy, 1957; Hastenrath, 1971; Weischet, 1970; Lauer, 1968; Flint and Fidalgo, 1964 (South America). Precipitation: after Walter and Lieth, 1960–7; Troll, 1964 (map of Troll and Paffen); Lauer, 1952. (3) Central Cordillera (present snowline). (4) Central Cordillera (snowline for the last glaciation). (5) Moist climate all year. (6) Seasonally moist (7–10 months). (7) Seasonally moist (4–6 months). (8) Few months dry (most months in the Rocky Mountains), weakly humid. (9) Dry climate all year.

Karakoram, Transhimalaya (von Wissmann, 1960, pp. 1328–30). In the humid tropics the snowline depression is smaller than in the humid maritime temperate latitudes. For comparison with fig. 1.2, the snowline depression in the Alps (44–48°N) is about 1200 m, decreasing to about 800 m towards the continental southeastern end (von Klebelsberg, 1948–9, pp. 682–3); the snowline depression on Mt Kenya (equator) is about 700 m (Baker, 1967, fig. 12 and p. 69, in opposition to p. 70).

Although the snowline of the last glaciation may be hypothetical for many areas in fig. 6B.1 (and even for some of the above-mentioned figures), the general trend is clear and indeed it was recognized by Machatschek (1913). The snowline depression generally decreases:

1. from the temperate latitudes towards the subtropics and tropics;
2. from the humid and maritime towards the dry and continental regions.

In the North American Cordilleras the maximum snowline depression of the last glaciation is related to a more uniform depression of the Ice Age snowline compared with that of the present day. This agrees well with the smaller annual movements of the high and low pressure systems off the North American west coast during the last glaciation (Lamb and Woodroffe, 1970, pp. 44–53; Lamb, 1971, pp. 134–40). This characteristic caused the year-round humid climatic provinces in the north and the arid belts further south to be more pronounced and therefore in more contrast to one another than today. In the Andes the snowline of the last glaciation between 13° and 29°S runs lower than the present-day snowline. The snowlines are therefore not converging towards the culmination, as one would expect; instead they diverge. Wilhelmy (1957, pp. 302–7) believes that this anomaly is due to a weakening of the dry climates of the Andean Mountains during the last glaciation, due to reduced incoming radiation and increased precipitation. Suzuki (1971, p. 41) concludes from this a northward displacement of the westerlies to about 20°S during the last glaciation. Against this is the fact that the culmination of the Ice Age snowline, as opposed to the present-day snowline, is not displaced northward; on the western side it is even shifted to the south. In the Sierra Nevada de Santa Marta (central Andes, 11°N) present-day and Ice Age snowlines diverge markedly because the mountains reach into the northern arid belt today, but were completely within the inner-tropical rainbelt during the last glaciation (Wilhelmy, 1957, p. 307).

Thus, anomalies in the snowline depression appear first where the boundary between the humid and arid climate zones has shifted since the last Ice Age. Such shifts are possible not only horizontally, but also vertically. According to Wilhelmy (1957, pp. 303–4) the highest zone, the *tierra fria*, expanded downward at the expense of the next lower zone, the *tierra templada*, during the last glaciation when there was pronounced snowline depression in the dry areas of the Andes. The lowest zone, the *tierra caliente*, hardly shifted at all. *Tierra fria* and *tierra caliente* thus moved closer together at the expense of the intermediate *tierra templada* (see fig. 2B.3).

It is possible that, even outside of the region of glacio-isostasy, uplift since the last glaciation may have influenced the separation between Ice Age and present-day snowline (Machatschek, 1944). But such uplift is usually within the error of snowline calculations (approximately ±100 m). For the Himalaya, however, von Wissmann (1960, pp. 1318, 1319, 1326, 1327) found a negative anomaly of snowline depression of *c.* 300 to 400 m. From this he concluded that the Himalaya rose by about that amount since the last Ice Age. For the Mt Everest region and south of it (Solu Valley, Rauje) I determined in 1966 (unpublished) a snowline depression for the last glaciation of 900 to 1000 m where von Wissmann (1960, pp. 1330, 1343) had assumed a depression of only 700 m based on my earlier estimates. However, this refutes von Wissmann's hypothesis for only one area.

MAJOR TYPES OF PLEISTOCENE MOUNTAIN GLACIATION

The altitude of the present snowline in relation to the height of the mountains permits large piedmont glaciers only in arctic and subarctic areas (in Alaska down to about 60°N). The most southerly transection glacier systems[1] and glaciers extending to sea level are found in Canada southward to 55°N and in Patagonia northward to 47°S. Between these two limits transection glaciers at present occur only on the western edge of the Pamirs (the Fedtschenko Glacier) and in the Karakoram (the Siachen and Hispar Glaciers).

The Pleistocene snowline depression facilitated the more frequent development of such glacier types during the last Ice Age and allowed them to form far into the present temperate latitudes. Piedmont and transection glaciers reached southward to 36°N in the American Cordilleras: Sierra Nevada, the San Juan Mountains (Wahrhaftig and Birman, 1965, pp. 305–6; Flint, 1971, pp. 472–5). In Chile, the piedmont glaciers formed the lake region northward to 39°S (Lauer, 1968; Weischet, 1970, pp. 268–74). The large valley glaciers of the Cordillera Blanca (Peru, around 9°S) reached the western foothills (Santa Valley), where the glacier snout from the Cedro Valley (8½°S), at 1800 m, penetrated deepest of all into the Tropics (Kinzl, 1935, p. 50).

In the New Zealand Alps, there were transection glaciers and piedmont glaciers northward to 42°S (Flint, 1971, pp. 687–8). In Europe, the Alps produced a glacier system with large piedmont glaciers down to 45°N. The Rhône Glacier was then 360 km long, and the Inn Glacier 340 km (von Klebelsberg, 1948–9, p. 685). There were also transection glaciers, in part with piedmont glaciers, in the mountainous areas of Asia. The longest Pleistocene glacier of interior Asia was found in the Shaksgam Glacier (36–37°N, 76–77°E), possibly extending 220 km in length, and the southernmost piedmont glaciers were to be found on the northern side of the eastern Himalaya (in the Mt Everest area) at 29°N (von Wissmann, 1960, pp. 1350, 1364).

THE SURFACE AREAS AND LENGTHS OF PLEISTOCENE AND PRESENT GLACIERS: MORPHOLOGICAL FACTORS

Information on the surface area of Pleistocene mountain glaciations is still quite vague. According to Flint (1971, pp. 76–9), however, the real relationship between the Pleistocene continental ice sheets and the separate mountain glaciations was similar to that of today: the glaciated area outside the large ice sheets, Alaska and the mountains of northern Siberia amounted to around $10^6 km^2$ and therefore to about 2.3 per cent of the then glaciated surface of the world. Of the presently glaciated area of the world, c. 1.6 per cent is outside the ice sheets of polar and subpolar areas.

The difference between present and Pleistocene glacier areas in the various mountains varies considerably. In Europe, excluding Scandinavia and the USSR, the presently glaciated area (3633 km^2) occupies about 9.3 per cent of the corresponding Pleistocene glaciation. In contrast, the glaciers of the Asiatic mountains excluding the USSR

---

[1] The German term here is *Eisstromnetz* – a system of connecting valley glaciers with ice overflowing passes between the valleys. Ahlmann (1948) refers to this as transection glacier.

($95,850$ km^2), cover just over 50 per cent of the area glaciated in the Pleistocene (Flint, 1971, pp. 76–9). However uncertain the Asiatic figures may be, they nevertheless illustrate the marked difference in the type of glaciation between the Alps and the inner Asiatic mountains. In part, this difference between the mountains of Europe and those of interior Asia is due to the considerably smaller snowline depression in many of the dry areas of inner Asia (von Wissmann, 1960, pp. 1348–65). Nevertheless, in some of these mountains morphological characteristics have a decisive effect: depth and steepness of the valleys and the lack of elevated level land are a basis for a large accumulation area. Here valley glaciers are fed first and foremost by snow and ice avalanches. The considerable amount of debris broken out of the steep walls collects in the form of a continuous cover on the glacier surfaces, protects the ice from ablation and acts as a stabilizing factor on the fluctuations of such glaciers (compare the debris-covered Pleistocene continental glacier-type on the east side of the Andes north of 40°S, according to Polanski, 1965, pp. 463–5).

Wiche (1960, pp. 194–6, map 2) demonstrated such an influence due to morphological conditions in comparing the Gilgit Range in the Karakoram (36°N, 74°E) with the west side of the adjacent Haramosh further east. From Haramosh (7406 m) and Malubiting (7452 m), the west side drops off from the summits by enormously steep faces into deep valleys. Here, the Baskei and Mani glaciers were not even 50 per cent longer during Pleistocene than at present. The snowline depression of about 1000 m did not significantly increase the accumulation. In contrast, the much lower Gilgit Range (its highest peak barely over 5000 m) has expansive areas of flat land between 3000 and 5000 m. Here, the snowline depression of 1200–1300 m caused a glaciation many times larger during the Pleistocene than at present.

Due to the same causes, the presently 58 km long Batura Glacier in the Hunza valley of the northwestern Karakoram, with many peaks above 7000 m, was about 30 km longer during the Pleistocene (snowline depression $c.$ 1000 m), but the 18 km long Minapin Glacier was only 8 km longer than today. There are no indications of any observable mountain uplift since the last Ice Age (Schneider, 1959, pp. 212–16).

Plate 20 demonstrates such a comparison: in front of Numbur (6959 m) and Karyolung (6511 m), with their steep south-facing walls, are foothills with summits barely over 5000 m and broad, shallow cirques above 4200 m. The presently 7 km long, debris-covered glacier of Numbur was only twice as long during the last Ice Age as at present. The Pleistocene glacier of the neighbouring valley to the west, the highest peak of which (5313 m) today carries a tiny glacier (the only one in the valley), reached about 7 km in length. In both valleys it is easy to recognize the Pleistocene end moraines. The Pleistocene glaciation of the flat landforms above 4200 m had a different result, due to a snowline depression of about 1000 m, from that influenced by the elevated but very steep Numbur, which today is heavily glaciated.

2. Classification of Quaternary mountain glaciation

In all well known mountains which were glaciated during the Quaternary, traces of several ice ages have been found. It is possible to distinguish:

(1) the distinct, relatively well preserved deposits from the last glaciation (in the Alps, the hummocky 'young moraines' with small lakes and peatbogs, kettles, and so on);

(2) the less distinct, less well preserved deposits of older glaciations (in the Alps, the 'old moraines', and so on), which usually extend further than those of the last glaciation.

There is still no unanimous opinion as to the number of the Quaternary ice ages. The correlation, especially of the evidence for the older glacials, is still very incomplete.

THE ALPS AND THE ROCKY MOUNTAINS: A GENERAL COMPARISON

Table 6B.1 gives a hypothetical correlation for the classification of the ice ages of the two best known mountain complexes: the Rocky Mountains and the Alps. Only glacial and fluvio-glacial deposits, and interglacial deposits and dated horizons closely associated with them, have been taken into consideration, but not the equally important loess stratigraphy, the decisive profiles of which are found outside the formerly glaciated areas. As an extension of the tabulation from the Rocky Mountains, some definite data from the Sierra Nevada, California, have been included.

The earliest acknowledged subdivision of the Glacial Age emanated from the Alps (Penck, 1882; Penck and Brückner, 1909). From there, the concepts of Günz, Mindel, Riss and Würm glacials (the names, in alphabetical order, of small rivers in the northern forelands of the Alps between the larger rivers Rhine and Isar) have also been widely adopted for other mountain areas. However, thanks to radiometric data on volcanic rocks, ashes, and so on, from the northern parts of the Cordilleran system, the radiometric chronology of the older Rocky Mountain glaciations are much better known. The hypothetical equivalence of the soils of the last major interglacial as a characteristic horizon was of particular importance for the correlation of the Rocky Mountains and the Alps. In the Alps, these major interglacial soils appear for the last time in the Mindel/Riss (M/R) interglacial, in the Rocky Mountains during the Sacagawea Ridge/Bull Lake interglacial (Richmond, 1970, pp. 6, 10).

During the Quaternary, especially during its early stages, both these mountain areas were subject to uplift, as was also the Sierra Nevada. This is a general event in the young orogenic belts. Between the individual glaciations, deep erosion took place. Because of this, the deposits from the last glaciation are often sunk into the dissected deposits from the preceding glaciation, and these in turn are found between the dissected deposits of the next older one, and so on, in such a manner that the oldest deposits are often found highest up. An impressive example of this is afforded by the oldest-dated (Quaternary) till of a mountain glaciation at temperate latitudes, the Deadman Pass Till of the Sierra Nevada (table 6B.1), which at present is separated from its source area, the Ritter Range, by a 760 m deep canyon (Curry, 1966).

In the Alps, this development is best seen in the northern forelands. During each glaciation the piedmont glaciers of the Alps pushed up immense gravel sheets along the Danube, each one entrenched into the next older one of dissected outwash. The classifi-

Table 6B.1. Quaternary glaciation of the Alps and the Rocky Mountains up to the beginning of the postglacial (partly after Richmond, 1965, 1970; and Birkeland et al., 1971).

Radiometric age (years)	ALPS	ROCKY MOUNTAINS	Radiometric age (years)	Important dates from Sierra Nevada, California.
10,200	Egesen			
	Daun	Pinedale Upper Till		
	Gschnitz — Soil	Peat (Yellowstone)	13,140	
	Steinach — Weak soil			
	Chur, Schönberg — Bühl	Pinedale Middle Till		
	2) Local valley glaciers	Soil		
	Late inner alpine stades 1) glacier net (Inn Valley): late stades, Alpenvorland, to 'Ammersee'	Pinedale Lower Till		
	Würm moraines + outwash gravel sheets (Niederterrassenschotter)			
	(Main Würm) Till ‖ Gravels			
26,800–31,000	Baumkirchen, wood in lake beds	Soil, wood in lake beds (Yellowstone)	>38,000–>45,000	
29,350	Val Caltea, wood in lake beds	Bull Lake Upper Till		
33–34,000	Steingaden, wood in lake beds	Rhyolite flow (Yellowstone)	70,000	
57–65,000	Zell (Wasserburg), peat wood — Grossweil, Peat	Soil		
		Rhyolite flow, Yellowstone (against stagnant ice?)	80,000	
69,000	Soil, peats (Riss/Würm interglacial)	Bull Lake Lower Till		
	Early Würm	Rhyolite flow (Yellowstone)	120–130,000	
	Riss II (Type) Till, moraines + outwash gravel sheets (Hochterrassenschotter)	Soil (last great interglacial soil)		
	Soil	Pumice (Yellowstone)	290,000	Basalt — 280,000
	Riss I (Paar) Till	Sacagawea Ridge Till		Casa Diablo Till — >370,000
	Soil, Ferretto (last great interglacial soil)	Soil		Basalt — 440,000
	Mindel Till, moraines + outwash sheets (Jüngerer Deckenschotter)	(Upper) Pearlette-like volcanic ash	600,000	
		Bishop ash (La Sal Mts)	710,000	Bishop Tuff — 710,000
	Soil, Ceppo Conglomerate	Cedar Ridge Till		Soil
		Soil		Sherwin Till
	Günz Till, moraine + outwash gravel sheets (Älterer Deckenschotter)	Ash (Lower Pearlette-like volcanic ash?)	1,180,000?	
	Soil	Washakie Point Till		
	Donau (?Till), outwash gravel sheets	Nussbaum Alluvium		Two Teats latite — 2,700,000
	Biber: (outwash?) – gravels			Deadman Pass Till
				Andesite — 3,100,000

══ Direct stratigraphic contact with till (soils only exceptionally emphasized).

╱ Well founded, almost direct stratigraphical contact.

cation of the ice ages in the Alps is based first and foremost on the correlation of these colossal gravel sheets of the northern forelands. Of decisive importance are:

(1) the correlation according to altitude:
(2) the alteration of the deposits due to induration, particularly of the older deposits (Mindel and older);
(3) the depth and the degree of leaching of the soils;
(4) the morphological deformation of the deposits due to features of solifluction, cryoplantation, and other periglacial action during following glaciations (compare Weinberger, 1954, fig. 10).
(5) the connection of the outwash gravels with the moraines.

With respect to point (5), there are some problems in the northern forelands of the Alps. Altitudinal correlations and connections between outwash and morainic deposits are undisturbed only where deposits of older glaciations were not overridden by younger ones. On this basis no moraines of the Danube glaciation have been found, even though Danube gravel sheets, doubtlessly, were outwash gravel from alpine piedmont glaciers, as can be proved by the transport of crystalline material from the central Alps over the northern limestone Alps (Schaefer, 1951, p. 298). Also, the connection of Günz gravels with Günz moraines has so far been demonstrated to be undisturbed only in a single locality, in the eastern part of the deposits of the Pleistocene Salzach Glacier (fig. 6B.2). Here – as elsewhere in the Austrian foothills of the Alps – the Günz glaciation reached furthest out, at least in places, in contrast to its extent in other parts of the Alps. The series, illustrated in fig. 6B.2, has become classical: the Günz, Mindel, Riss and Würm deposits lie undisturbed one behind the other, and furthermore, the older are the more elevated. Still higher than the Günz gravels are the Eichwald gravels (= Donau?), in which syngenetic cryoturbation indicates that periglacial climatic conditions existed during their accumulation (Weinberger, 1955, p. 12). Here, the above-mentioned Weinberger reasoning is successful in all five points. The simple primary classification – into Penck's four glaciations plus one still older Ice Age – has also been successful for the more complicated division of the adjacent areas to the west (Graul, 1962a, 1968).

At the southern edge of the Alps, situated more than 2° latitude further south and 300–400 m lower, the correlation of the gravel terraces in relation to altitude is problematic over larger distances, because the Po plain has been sinking during the Quaternary and up to the present day (Perconig, 1956). Increasingly southward, therefore, the erosion between the individual glaciations is missing. Depth and degree of leaching of the soils is also much stronger here (as represented by ferretto soils) than on the northern edge of the Alps; however, the old soils have been extensively washed down into depressions (Venzo, 1968, p. 95, fig. 1). The morphological deformation of the deposits by periglacial action is less pronounced here than on the northern edge of the Alps. But, because of this, the moraines are also much better preserved, much more steeply graded and higher than in the north. This leads, for instance, to continued debate as to the separation of young and old moraines at the Pleistocene Lago di Garda Glacier (Fränzle, 1965; Habbe, 1968, 1969; Venzo, 1965, 1968). There, the glaciations older than Riss can in

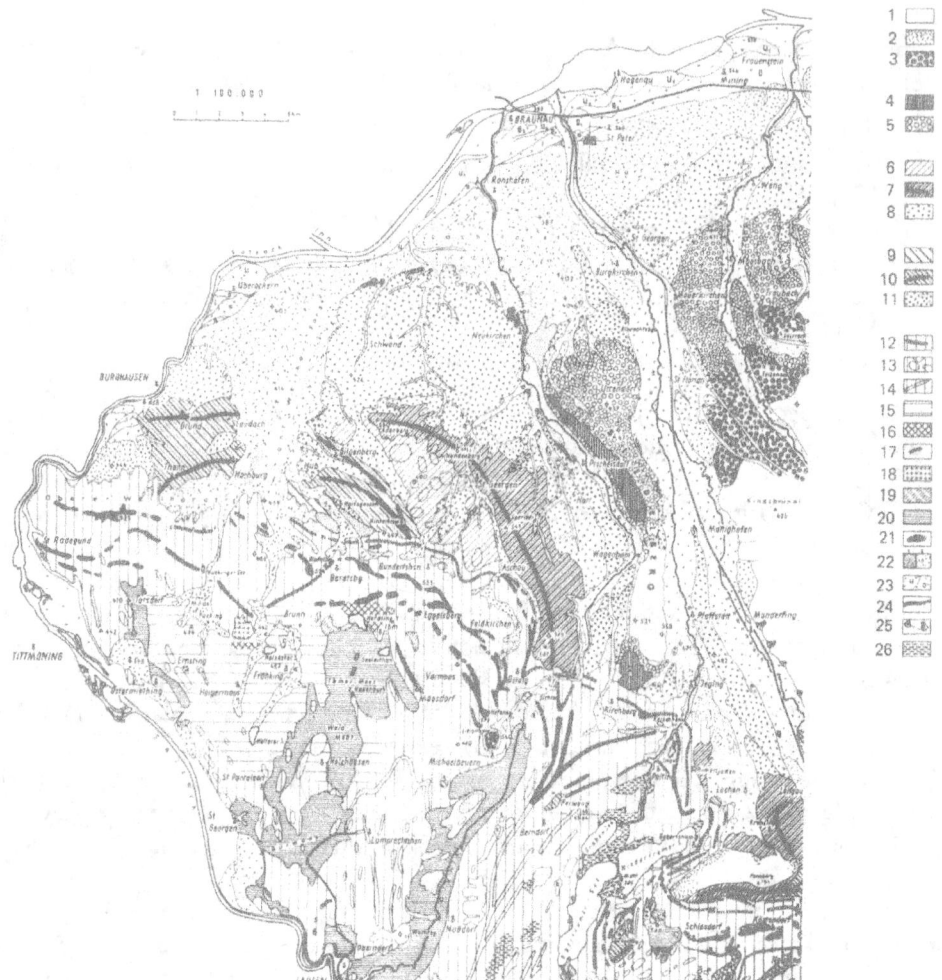

Fig. 6B.2. Deposits of the Pleistocene Salzach Glacier, eastern part, and the western part of the
Traun Glacier. Complete series of Eichwald gravel (=Donau?) – Günz – Mindel – Riss
– Würm (from Weinberger, 1955, Pl. II).

1, *base rock (Tertiary)*; 2, *glacial debris (erratics) on the base rock*; 3, *Eichwald gravels.*
GLACIAL (1): 4, *Günz moraine*; 5, *Günz outwash gravels (Älterer Deckenschotter).*
GLACIAL (2): 6, *Mindel till (ground moraine landscape, drumlinized)*; 7, *Mindel end moraine
with ridge*; 8, *Mindel outwash gravels (Jüngerer Deckenschotter).*
GLACIAL (3): 9, *Riss till (ground moraine landscape; old moraine)*; 10, *Riss end moraine with
ridge*; 11, *Riss outwash gravels (Hochterrasse).*
GLACIAL (4), *including the late and postglacials:* 12, *Würm till with end moraine*; 13,
drumlin; 14, *ground moraine (till) ridges*; 15, *old surface (pre-Würm, covered with Würm till)*; 16,
kames; 17, *osar*; 18, *kame terraces*; 19, *edge terrace and delta gravels*; 20, *laminated silts and clays,
often peats*; 21, *interstitial terrace (between 12 and 22)*; 22, *Würm outwash gravels (Niederterrasse)* –
(1) *upper 'Niederterrasse'*; (2) *lower 'Niederterrasse'*; 23, *late and postglacial river terraces*; 24,
Laufen gravel (older Würm); 25, *fan*; 26, *peats.*

practice only be identified stratigraphically, not by morphology. The south rim of the Alps thus plays a considerably lesser role for the classification of the Ice Ages than does the northern one.

In the more continental and drier Rocky Mountains, the fluvio-glacial outwash sheets are in comparison with those of the north edge of the Alps much thinner, the induration much weaker, the soil formation less pronounced, and the morphological deformation of the deposits due to periglacial action much more slight. Because of this, the moraines are more distinct and better preserved, higher and more steeply graded, and the erosion edges are sharper (Richmond, 1970, p. 5).

Without doubt, the history of the older Quaternary glaciations is more complicated in detail, as shown in the table. Richmond (1970, pp. 9, 11) drew up some important subdivisions. All of them, however, are dependent on relatively isolated observations, which cannot serve for correlations along the Alps, even if they are, at least in part, of great general importance.

CONTRIBUTIONS TO THE CORRELATION OF THE RISS – WÜRM GLACIATIONS WITH THE BULL LAKE – PINEDALE GLACIATIONS

The subdivision of the Riss into two different glacials separated by an interglacial, as shown in table 6B.1, has won general acclaim (Richmond, 1970, pp. 9–12). However, there are still complications for correlation between the Alps and the Rocky Mountains particularly as a result of the undecided standpoint of the research in the Alps, and perhaps also because of individual problematic correlations in the Rocky Mountains (Kaiser, 1966, pp. 289–90). Table 6B.1 does not follow Richmond (1970) here, but comes closer to the opinions expressed by Richmond in 1965a (pp. 226–8).

For the younger parts of the Quaternary, the correlation between Main Würm and Pinedale is best founded in respect to time. To the new data from the margins of the Alps at Steingaden, northeast of Füssen (Höfle, 1969), and at the Val Caltea in the Venezian foothills (Fuchs, 1969) are now added the much more important data from the interior of the Alps at Baumkirchen (13 km west of Innsbruck). It has been shown on the basis, so far, of six dated finds at an elevation of 655–681 m above sea level (Fliri et al., 1970, 1971), which are overlain by a further 70 m of silt and 60 m of sands and gravel, that, after a very pronounced interstadial (Fliri et al., 1970, p. 31; Schroeder-Lanz, 1971), the Inn Glacier of the Main Würm only reached the lower alpine Inn Valley considerably after 26,800 BP. Wood remnants on the top of the series indicate that Baumkirchen was ice-free again by 10,900–11,370 BP (Fliri et al., 1970, 1971). This had already been shown for considerably higher altitudes of the interior Swiss Alps for 12,000–11,370 BP (Suossa at 1700 m, south of the San Bernardino Pass, Graubünden: 13,010 BP, Zoller and Kleiber, 1971, p. 119; and Lake Hopschen at 2017 m, Simplon Pass: 12,580 BP, Welten, 1966).[1]

The chronology of the Rocky Mountain equivalent of the Würm is as yet not well known in terms of 'absolute' time. At least three stades of Pinedale glaciation are widely

---

[1] *Edit. Note:* the following four paragraphs were contributed by R. F. Madole.

recognized. However, because radiocarbon dates are few and soils and weathering criteria have failed thus far to provide correlation at the stadial level, subdivision of the Pinedale remains interpretive and subject to revision. Richmond (1965a, pp. 224–5) summarizes the Pinedale glaciation of the Rocky Mountains and later (1970, pp. 18–20) compares the Main Würm and Pinedale glaciations. Birkeland, Crandell and Richmond (1971, pp. 215–20) list existing radiometric dates for the Late Quaternary of the western United States.

Although there is regional variation in the extent of Pinedale till, in most areas Pinedale glaciers were nearly as large as their predecessors of late Bull Lake time. Ice caps and piedmont glaciers formed in parts of the northern Rockies, but valley glaciers 5 to 35 km long emanating from cirques were more typical, particularly in Utah and the southern Rocky Mountains. Moraines marking the outer limit of Pinedale glaciation are generally prominent. Those of Middle Pinedale time usually occur within 0 to 5 km of the Early Pinedale terminals and are somewhat smaller. In the larger valley systems of the southern Rocky Mountains, Late Pinedale termini tend to lie far upvalley, nearer to Neoglacial deposits than to the older Pinedale deposits, and by comparison are ill-defined in many areas.

The last major rise of pluvial Lake Bonneville suggests that Pinedale glaciation began between 25,000 and 26,000 BP (Richmond, 1965a, 1970). However, neither the time of beginning nor duration of the early stage of Pinedale glaciation is known directly from glacial deposits. Furthermore only a few widely scattered dates from early postglacial deposits overlying till or outwash exist to delimit the time of maximum extent of Middle Pinedale ice. The oldest of these comes from the Lake Yellowstone region where Richmond (1970) reports carbonaceous material in recessional kame terrace deposits to be 13,140 ± 700 BP and Baker (1969) found peaty material that postdates withdrawal of Middle Pinedale ice to be as old as 11,500 ± 350 BP. Another date for this interval comes from travertine in a deposit on outwash along the east flank of the Wind River Mountains dated 11,100 ± 300 BP (J. F. Murphy in Birkeland et al., 1971).

The end of Pinedale glaciation is also poorly defined. Madole (personal communication, 1972) obtained a date of 7690 ± 115 BP from peat on a Late Pinedale terminal moraine 3.5 km downvalley from cirques in the Colorado Front Range. A few tens of kilometres to the north, Richmond (1965b) found wood in a kettle on kame gravel upvalley from the youngest Pinedale moraine which dated 6170 ± 240 BP. Minimal dates of similar magnitude (average 6200 BP) obtained from bone associated with a probable Late Pinedale moraine were reported by Stalker (1969) from Castle River, Alberta.

In the forelands of the Alps the exact correlation between moraines from the Main Würm Maximum and those from various glaciers of the late stages of the Alpine Foreland have become problematic, since Graul (1957, 1962b) showed that the individual Würm glaciers in the forelands did not reach their maxima simultaneously.

We know little of the glaciers from the older Würm. Richmond (1970, p. 16) found no evidence that these glaciers reached the forelands of the Alps after 50,000 BP. But such indications are missing also for the time between 70,000 and 50,000 BP. The oldest organic finds from this time, coming from the edge of the Alps in Bavaria (Grossweil at

Kochel) and from the foreland of the Alps (Zell at Wasserburg), have already been referred to the older Würm on the basis of their pollen contents (*Radiocarbon*, 1967, pp. 88–90) and not to the R/W interglacial (Richmond, 1970). For the correlation Bull Lake/Upper Till – Riss II (Richmond, 1970) there are no proven arguments so far from the Alps. In particular, there are no proofs for a proper interglacial between 60,000 and 45,000 BP.

PROBLEMS CONCERNING THE INTERIOR ALPS

The advances and the retreats of the glaciers were more complicated in the Alps than in the Rocky Mountains, because in the large longitudinal valleys of the Alps the ice masses were dammed up before they could run out over the forelands. These ice masses persisted for a long time after each glaciation. Therefore, the investigation of glaciation of the inner Alps has been more thorough than for other mountains. Three of the problems will be discussed briefly below.

Ice-free periods during the interglacials

Almost all stratigraphic proofs for interglacials are found in the forelands or at the margins of the Alps. The most important evidence of freedom from ice during an interglacial in the interior Alps is the Hötting Breccia above Innsbruck. This immense, indurated colluvium occupies the following stratigraphical position (Penck, 1921; von Klebelsberg, 1948–9; Paschinger, 1950):

Upper till	*Hangendmoräne*
Inn Valley terrace sediments; silts, sands, gravels	*Inntalterrassensedimente*
Intermediate till	*Sockelmoräne*
Fossil-bearing colluvium	*Hötting Breccia*
Lower till	*Liegendmoräne*

The breccia, at 1180 m altitude (600 m above Innsbruck), contains fossils of a vegetation, which indicate a climate at least 2–3°C warmer than the present (von Klebelsberg, 1948–9, p. 437). Such a climate has not subsequently been encountered in the Alps, even during the Hypsithermal (Welten, 1958, p. 273; in opposition to Flint, 1971, p. 646). The breccia reaches to above 1900 m above sea level. By this alone it is demonstrated that the mountains around Innsbruck were largely free of ice at that time.

At least a part of the breccia was thus with certainty laid down during the interglacial. According to prevailing opinion, it was common to correlate the series in the following manner: the lower till = Mindel, the breccia = M/R interglacial, the intermediate till = Riss, the terrace sediments = R/W interglacial, the upper till= Würm (according to Paschinger, 1950, pp. 11–21). According to the dating of the base layers of the Inn Valley terrace sediments at Baumkirchen (table 6B.1), however, these were laid down only during the Würm. Mayr (1968a) has shown on the basis of proofs from the glacial and fluvio-glacial deposits in the Inn Valley terrace sediments, that the till of the interior

Alps (for instance, the intermediate till) cannot be equated with a complete glaciation of the Alps reaching out over the forelands. The age of the breccia is, therefore, not yet entirely settled.

Vertical dimensions of the Alp glaciers

The horizontal expansion of the glaciers in the Alps is particularly well known where they reach the foreland. Less well known are the vertical dimensions, the thickness of the glacier network. Some erratics are always found higher up than the limit for striae and other glacial impressions in the rocks. We do not know, however, to which glaciation the highest erratics belong. Jäckli (1965, p. 156; 1962) drew a map of the Würm glaciers in Switzerland, according to which the ice was more than 1500 m thick in places (the Rhône Valley around Visp; the Rhine Valley around Chur; in the Tessin Valley above Bellingzona; and in the Adda Valley above Lago di Como). The glacial map of the Tirol by von Klebelsberg (1935, p. 540) indicates the highest reach of the ice in the interior Alps, and thus gives even higher measurements than those of Jäckli. On this basis, the Etsch Glacier at Meran was approximately 1800 m thick. According to more recent erratic finds, the Inn Glacier as well reached 1800 m thickness at Innsbruck (Mutschlechner, 1957; Heuberger, 1952).

Late stades of Würm in the interior of the Alps

The retreat of the Würm glaciers in the Alps is not yet exactly known. Penck and Brückner's (1909) series of the Bühl, Gschnitz, and Daun stades are most frequently used today and have been extended (Mayr and Heuberger, 1968; Heuberger, 1968). Bühl (at Woergl-Kufstein) and Schönberg (at Innsbruck) are only oscillations of the active transection glacier network of the the Inn Glacier; the Rhine Glacier moved forward – probably after the Bühl stade – at least 30 km across the landslide at Flims to around Chur (Abele, 1970, pp. 351–3; Hantke, 1970, p. 14). The following stades were advances of local valley glaciers after the main glacial system had broken up. Of these, the Steinach stade and the Gschnitz stade (Gschnitz Valley south of Innsbruck) certainly still belong to the Late Glacial (before 10,200 BP), possibly so also do the Daun and Egesen stades (Patzelt, 1971), the last ones during which the glaciers of the Alps were still essentially bigger than in the Little Ice Age (see below). The glaciers of the Alps had reached their Neoglacial maximal expansion long before 9000 BP (Patzelt, 1971), but at least individual glaciers of Patagonia had already reached theirs by 11,000 BP (Mercer, 1970, pp. 13–14).

3. Postglacial glacier advances since 9000 BP: 'Neoglaciation'

The alpine glaciers were never *much* larger since the Late Preboreal than during the maxima of the Little Ice Age. Yet, in a remarkably consistent manner, they repeatedly reached that size and more frequently so than had been recognized until recently. The associated climatic conditions would have been quite dissimilar (Heuberger, 1966, 1968) although Bortenschlager and Patzelt (1969) in particular were able to obtain a good correlation between the fluctuations of both glaciers and climate through radiocarbon results of pollen analysis *above* present-day timberline, directly connected with geomorphological and stratigraphic results.

The following times of glacier advances to an extent equal to or somewhat exceeding that of the maxima of the Little Ice Age are so far known (on the basis of uncorrected radiocarbon dates):

(1) 8700–8000 BP: 'Venediger' (Bortenschlager and Patzelt, 1969; Patzelt, 1971), Venediger Mountains (Tirol/Salzburg).

(2) 6400–6000 BP: 'Frosnitz' (Bortenschlager and Patzelt, 1969; Patzelt, 1971), Venediger Mountains.
Probably 'Larstig' (Heuberger, 1966), Stubai Alps (Tirol).
Stubai Alps (Mayr, 1968b), Mt Blanc (Mayr, 1969).
Sierra Nevada (Curry, 1969).
Solifluction activity (Bachmann and Furrer, 1971), Swiss National Park.

(3) 5200–4600 BP: Northern and southern hemispheres (Mercer, 1967).

(4) 3500–3300 BP: 'Löbben' (Bortenschlager and Patzelt, 1969; Patzelt, 1971), Venediger Mountains.
Stubai Alps (Mayr, 1968b).

(5) 2850–1250 BP: 'Simming series', older part (Mayr, 1968b), Stubai Alps.
North America – Temple Lake (Rocky Mountains) etc. (Birkeland *et al.*, 1971).

(6) 1900–1200 BP (not accurately dated): 'Simming series', younger part (Mayr, 1968b), Stubai Alps.
'Arikaree' (Benedict, 1968), now renamed the 'Audubon', Colorado Front Range (Mahaney, 1972).

(7) 400–30 BP: 'Modern' (Beschel, 1961; Heuberger and Beschel, 1958).
'Little Ice Age' (Matthes, 1942: 'Lesser Ice Age').

There is already considerable worldwide evidence for glacier advances at 2850–1250 BP and 400–30 BP in particular.

Not included are glacier advances which apparently remained smaller and moraines of which were therefore obliterated by the larger advances mentioned (Kinzl, 1958), as for example the 650 BP advance of the Aletsch Glacier (Switzerland) (Oeschger and Röthlisberger, 1964; see the generalized curve of Neoglacial fluctuations of Porter and Denton, 1967, showing a glacier advance about 700 BP).

This series of glacier maxima shows that the alpine glaciers at least never disappeared completely after the Late Glacial and that the Hypsithermal was too short to leave a decisive mark on the history of alpine glaciation. Thus the term 'Neoglaciation' and its delimitation at about 5000 BP (Porter and Denton, 1967, p. 204) has become doubtful.

The *general trend* (of advance or retreat) was probably the same in all mountain areas during these different advances and intervening retreats. However, the *extent* of the individual glacier advances often varied from area to area, probably also from glacier to glacier. Hence there are no moraines post-9000 years in front of many glaciers, for example outside the moraine of the last (nineteenth, in some places twentieth century) maximum, whereas in front of others there may be several. Even within the best known Little Ice Age there are marked differences.

References

ABELE, G. (1970) Bergstürze and Flutablagerungen im Rheintal westlich Chur. *Aufschluss*, **21**, 345–59.

AHLMANN, H. W. (1948) Glaciological research on the North Atlantic coasts. *Roy. Geogr. Soc. Res. Ser.*, 1. 83 pp.

BACHMANN, F. and FURRER, G. (1971) Solifluktionsdecken im Schweizerischen National- park und ihre Beziehungen zur postglazialen Landschftsentwicklung. *Geogr. Helvetica*, **26**, 122–8.

BAKER, B. H. (1967) Geology of the Mount Kenya Area. *In: Geological Survey of Kenya, Rep.*, 79. Republic of Kenya.

BAKER, R. G. (1969) *Late Quaternary Pollen and Plant Macrofossils from Abandoned Lagoonal Sediments near Yellowstone Lake, Wyoming*. Unpub. Ph.D. dissertation, Univ. of Colorado.

BENEDICT, J. B. (1968) Recent glacial history of an alpine area in the Colorado Front Range, USA, II: Dating the glacial deposits. *J. Glaciol.*, **7** (49), 77–87.

BESCHEL, R. E. (1961) Dating rock surfaces by lichen growth and its application to glaci- ology and physiography (lichenometry). *In:* RAASCH, G. O. (ed.) *Geology of the Arctic*, Vol. 2, 1044–62. Toronto: Univ. of Toronto Press.

BIRKELAND, P. W., CRANDELL, D. R. and RICHMOND, G. M. (1971) Status of correlation of Quaternary stratigraphic units in the western conterminous United States. *Quat. Res.*, **1**, 208–27.

BORTENSCHLAGER, S. and PATZELT, G. (1969) Wärmezeitliche Klima- und Gletscher- schwankungen im Pollenprofil eines hochgelegenen Moores (2270 m) der Venediger- gruppe. *Eiszeitalter und Gegenwart*, **20**, 116–22.

CURRY, R. R. (1966) Glaciation about 3,000,000 years ago in the Sierra Nevada. *Science*, **154** (3750), 770–1.

CURRY, R. R. (1969) Holocene climatic and glacial history of the central Sierra Nevada, California. *Geol Soc. Amer. Spec. Pap.*, 123 (INQUA vol.), 1–47.

ENQUIST, A. (1916) Der Einfluss des Windes auf die Verteilung der Gletscher. *Bull. Inst. Univ. Upsala*, 14. 108 pp.

FLINT, R. F. (1957) *Glacial and Pleistocene Geology*. New York: John Wiley. 553 pp.

FLINT, R. F. (1971) *Glacial and Quaternary Geology*. New York: John Wiley. 892 pp.

FLINT, R. F. and FIDALGO, F. (1964). Glacial geology of the east flank of the Argentine Andes between latitude 39°10′ and latitude 41°20′ S. *Bull. Geol. Soc. Amer.*, **75**, 335–52.

FLIRI, F., HILSCHER, H. and MARKGRAF, V. (1971) Weitere Untersuchungen zur Chronolo- gie der alpinen Vereisung (Bänderton von Baumkirchen, Inntal, Nordtirol). *Zeit. Gletscherk. Glazialgeol.*, **6**, 5–24.

FLIRI, F., BORTENSCHLAGER, S., FELBER, H., HEISSEL, W., HILSCHER, H. and RESCH, W. (1970) Der Bänderton von Baumkirchen (Inntal, Tirol). *Zeit. Gletschkerk. Glazialgeol.*, **6**, 5–35.

FRÄNZLE, O. (1965) Die pleistozäne Klima- und Landschaftsentwicklung der nördlichen Po-Ebene im Lichte bodengeographischer Untersuchungen. *Abhand. Akad. Wiss. Lit. Mainz, Math. Naturwiss. Klasse*, **8**, 331–474.

FUCHS, F. (1969) Eine erste ^{14}C–Datierung für das Paudorf-Interstadial am Alpensüdrand. *Eiszeitalter und Gegenwart*, **20**, 68–71.

FURRER, G. and FITZE, P. (1970) Die Hochgebirgsstufe – ihre Abgrenzung mit Hilfe der Solifluktionsgrenze. *Geogr. Helvetica*, **25**, 156–61.

GRAUL, H. (1957) Sind die Jugendmoränen im nördlichen Alpenvorland gleichaltrig? *Geomorphologische Studien* (Machatschek-Festschrift), *Petermanns Geogr. Mitt.*, Suppl. 261, 209–12.

GRAUL, H. (1962a) Eine Revision der pleistozänan Stratigraphie des Schwäbischen Alpenvorlandes. *Petermanns Geogr. Mitt.*, **103**, 253–71.

GRAUL, H. (1962b) Geomorphologische Studien zum Jungquartär des nördlichen Alpenvorlandes, I. *Heidelberger Geogr. Arb.*, **9**.

GRAUL, H. (1968) The present state of Quaternary stratigraphy in the western foreland of the German Alps. *In:* RICHMOND, G. M. (ed.) Glaciation of the Alps, *Univ. of Colorado Stud, Ser. in Earth Sci.*, **7**, 3–7.

HABBE, K. A. (1968) The Riss–Würm boundary in the northwestern part of the Lake Garda terminal moraine basin of northern Italy. *In:* RICHMOND, G. M. (ed.) Glaciation of the Alps, *Univ. of Colorado Stud., Ser. in Earth Sci.*, **7**, 79–84.

HABBE, K. A. (1969) Die würmzeitliche Vergletscherung des Garda-See-Gebietes. *Freiburger Geogr. Arb.*, **3**.

HANTKE, R. (1970) Aufbau und Zerfall des würmeiszeitlichen Eisstromnetzes in der zentralen und östlichen Schweiz. *Ber. Naturforsch. Ges. Freiburg i.Br.*, **60**, 5–33.

HASTENRATH, S. L. (1971) On the Pleistocene snow-line depression in the arid regions of the South American Andes. *J. Glaciol.*, **10** (59), 255–67.

HERMES, K. (1955) Die Lage der oberen Waldgrenze in den Gebirgen der Erde und ihr Abstand zur Schneegrenze. *Kölner Geogr. Arb.*, **5**.

HERMES, K. (1964) Der Verlauf der Schneegrenze. *In:* Geogr. Taschenbuch, 58–71. Wiesbaden: F. Steiner.

HEUBERGER, H. (1952) Hochgelegene Erratika an der Südseite des Inntales westlich Innsbruck. *Zeit. Gletscherkunde Glazialgeol.*, **2**, 118–19.

HEUBERGER, H. (1966) Gletschergeschichtliche Unterschungen in den Zentralalpen zwischen Sellrain- und Ötzal. *Wiss. Alpenvereinshefte*, **20**.

HEUBERGER, H. (1968) Die Alpengletscher im Spät- und Postglazial. *Eiszeitalter und Gegenwart*, **19**, 270–5.

HEUBERGER, H. and BESCHEL, R. (1958) Beiträge zur Datierung alter Gletscherstände im Hochstubai (Tirol). *Schlern-Schriften*, **90** (Festschrift H. Kinzl), 73–100.

HÖFLE, H.-CH. (1969) Ein neues Interstadialvorkommen im Ammergebirgsvorland (Obb.). *Eiszeitalter und Gegenwart*, **20**, 111–15.

JÄCKLI, H. (1962) Die Vergletscherung der Schweiz im Würm-maximum. *Eclogae Geol. Helvet.*, **55** (2), 285–94.

JÄCKLI, H. (1965) Pleistocene glaciation of the Swiss Alps and signs of postglacial differential uplift. *Geol. Soc. Amer. Spec. Pap.*, **84**, 153–7.

KAISER, K. (1966) Probleme und Ergebnisse der Quartärforschung in den Rocky Mountains (i.w.S) und angrenzenden Gebieten. *Zeit. Geomorphol.*, **10**, 264–302.

KINZL, H. (1935) Gegenwärtige und eiszeitliche Vergletscherung in der Cordillera Blanca (Peru). *Verhandl. dtsch. Geographentages*, **25**, 41–56.

KINZL, H. (1958) Die Gletscher als Klimazeugen. *Verhandl. dtsch. Geographentages*, **31**, 222–31.

KLEBELSBERG, R. VON (1935) *Geologie von Tirol*. Berlin: Borntraeger.

KLEBELSBERG, R. VON (1948–9) *Handbuch der Gletscherkunde und Glazialgeologie*. Vienna: Springer. 2 vols.

KLUTE, F. (1928) Die Bedeutung der Depression der Schneegrenze für eiszeitliche Probleme. *Zeit. Gletscherk.*, **16**, 70–93.

LAMB, H. H. (1971) Climates and circulation regimes developed over the northern hemisphere during and since the last ice age. *Palaeogr., Palaeoclimatol., Palaeoecol.*, **10**, 125–62.

LAMB, H. H. and WOODROFFE, A. (1970) Atmospheric circulation during the last ice age. *Quat. Res.*, **1**, 29–58.

LAUER, W. (1952) Humide und aride Jahreszeiten in Afrika und Südamerika und ihre Beziehung zu den Vegetationsgürteln. *Bonner Geogr. Abhand.*, **9**, 15–98.

LAUER, W. (1968) Die Glaziallandschaft des südlichenischen Seengebietes. *Acta Geogr.*, **20** (16), 215–36. Helsinki.

LOUIS, H. (1927) Die Verbreitung von Glazialformen im Westen der Vereinigten Staaten. *Zeit. Geomorphol.*, **2**, 221–35.

LOUIS, H. (1954–5) Schneegrenze und Schneegrenzbestimmung. *Geogr. Taschenbuch*, 414–18. Wiesbaden: F. Steiner.

MACHATSCHEK, F. (1913) Die Depression der eiszeitlichen Schneegrenze. *Zeit. Gletscherk.*, **8**, 104–28.

MACHATSCHEK, F. (1944) Diluviale Hebung und eiszeitliche Schneegrenzdepression. *Geol. Rund.*, **34**, 327–41.

MAHANEY, W. C. (1972) Audubon: new name for Colorado Front Range neoglacial deposits formerly called 'Arikaree'. *Arct. Alp. Res.*, **4** (4), 355–7.

MATTHES, F. E. (1942) Lesser ice age. *In:* MEINZER, O. E. (ed.) *Physics of the Earth*, Vol. 9, 207. New York.

MAYR, F. (1968a) Über den Beginn der Würmeiszeit im Inntal bei Innsbruck. *Zeit. Geomorphol.*, **12**, 256–95.

MAYR, F. (1968b) Postglacial glacier fluctuations and correlative phenomena in the Stubai Mountains, Eastern Alps, Tyrol. *In:* RICHMOND, G. M. (ed.) Glaciation of the Alps, *Univ. of Colorado Stud., Ser. in Earth Sci.*, 7, 167–77.

MAYR, F. (1969) Die postglazialen Gletscherschwankungen des Mont Blanc-Gebietes. *Zeit. Geomorphol.*, Suppl. 8, 31–57.

MAYR, F. and HEUBERGER, H. (1968) Type areas of late glacial and post-glacial deposits in Tyrol, Eastern Alps. *In:* RICHMOND, G. M. (ed.) Glaciation of the Alps, *Univ. of Colorado Stud., Ser. in Earth Sci.*, 7, 143–65.

MERCER, J. H. (1967) Glacier resurgence at the Atlantic/sub-Boreal boundary. *Quart. J. R. Met. Soc.*, **93**, 528–34.

MERCER, J. H. (1970) Variations of some Patagonian Glaciers since the Late-Glacial: II. *Amer. J. Sci.*, **269**, 1–25.

MUTSCHLECHNER, G. (1957) Spuren der Eiszeit an der Saile bei Innsbruck. *Veröff. Museum Ferdinandeum, Innsbruck*, **37**, 83–7.

OESCHGER, H. and RÖTHLISBERGER, H. (1964) Datierung eines ehemaligen Standes des Aletschgletschers durch Radioaktivitätsmessung an Holzproben und Bemerkungen zu Holzfunden an weiteren Gletschern. *Zeit. Gletscherkunde Glazialgeol.*, **4**, 191–205.

ØSTREM, G. (1964) Glacio-hydrological investigations in Norway. *J. Hydrol.*, **2**, 101–15.

PASCHINGER, H. (1950) Morphologische Ergebnisse einer Analyse der Höttinger Breccie bei Innsbruck. *Schlern-Schriften*, **75**.

PATZELT, G. (1971) Bericht über eine glazialmorphologische Exkursionstagung in den Ostalpen vom 1. bix 6. September 1970. *Zeit. Geomorphol.*, **15**, 115–20.

PENCK, A. (1882) *Die Vergletscherung der Deutschen Alpen.* Leipzig: Barth.

PENCK, A. (1921) Die Höttinger Breccie und die Inntalterrasse nördlich Innsbruck. *Abhandl. Preuss. Akad. Wiss. 1920, Phys. math. Klasse*, 2.

PENCK, A. and BRÜCKNER, E. (1909) *Die Alpen im Eiszeitalter.* Bern: Tauchnitz. 3 vols.

PERCONIG, E. (1956) Il quaternario nella pianura Padana. *Actes INQUA*, **IV**, Rome (2), 481–524.

POLANSKI, J. (1965). The maximum glaciation in the Argentine Cordillera. *Geol. Soc. Amer. Spec. Pap.*, 84, 453–72.

PORTER, S. C. (1971) Fluctuations of Late Pleistocene alpine glaciers in western North America. *In:* TUREKIAN, K. (ed.) *The Late Cenozoic Glacial Ages*, 307–29. New Haven and London: Yale Univ. Press.

PORTER, S. C. and DENTON, G. H. (1967) Chronology of neoglaciation in the North American Cordillera. *Amer. J. Sci.*, **265**, 177–210.

RICHMOND, G. M. (1965a) Glaciation of the Rocky Mountains. *In:* WRIGHT, H. E. JR and FREY, D. G. (ed.) *The Quaternary of the United States*, 217–30. Princeton: Princeton Univ. Press.

RICHMOND, G. M. (1965b) Quaternary stratigraphy of the La Sal Mountains, Utah. *US Geol. Surv. Prof. Pap.*, 324. Washington, DC.

RICHMOND, G. M. (1970) Comparison of the Quaternary stratigraphy of the Alps and Rocky Mountains. *Quat. Res.*, **1**, 3–28.

SCHAEFER, I. (1951) Über methodische Fragen der Eiszeitforschung im Alpenvorland. *Zeit. dtsch. Geol. Ges.*, **102**, 287–310.

SCHAEFER, I. (1968) The succession of fluvioglacial deposits in the northern Alpine Foreland. *In:* RICHMOND, G. M. (ed.) Glaciation of the Alps, *Univ. of Colorado Stud., Ser. in Earth Sci.*, 7, 9–14.

SCHNEIDER, H.-J. (1959) Zur diluvialen Geschichte des NW-Karakorum. *Mitt. Geogr. Ges. München*, **44**, 201–16.

SCHROEDER-LANZ, H. (1971) War das Frühwürm (W I) eine selbständige Kaltzeit? *Mitt. Geogr. Ges. München*, **56**, 173–84.

STALKER, A. MACS. (1969) A probable late Pinedale terminal moraine in Castle River Valley, Alberta. *Geol. Soc. Amer. Bull.*, **80**, 2115–22.

SUZUKI, H. (1971) Climatic zones of the Würm glacial stage. *Bull. Dept of Geogr., Univ. of Tokyo*, 3, 35–46.

TROLL, C. (1955) Über das Wesen der Hochgebirgsnatur. *Jahrbuch des Österreichischen (Deutschen) Alpenvereins*, **80**, 142–57.

TROLL, C. (1964) Karte der Jahreszeiten-Klimate der Erde. *Erdkunde*, **18**, 5–28.

VENZO, S. (1965) Rilevamento geologico dell'anfiteatro Morenico frontale del Garda dal Chiese all'Adige. *Memorie della Società di Scienze Naturali e del Museo Civico di Storia Naturale di Milano*, **14** (1).

VENZO, S. (1968) The frontal end moraines of the Lake Garda basin and the origin of the terraces of the Po valley, northern Italy. *In:* RICHMOND, G. M. (ed.) Glaciation of the Alps, *Univ. of Colorado Stud., Ser. in Earth Sci.*, 7, 93–9.

WAHRHAFTIG, C. and BIRMAN, J. H. (1965) The Quaternary of the Pacific Mountain System in California. *In:* WRIGHT, H. E. JR and FREY, D. G. (eds.) *The Quaternary of the United States*, 299–340. Princeton: Princeton Univ. Press.

WALTER, H. and LIETH, H. (1960–7) *Klimadiagramm-Weltatlas.* Jena: G. Fischer.

WEINBERGER, L. (1954) Die Periglazial-Erscheinungen im Österreichischen Teil des eiszeitlichen Salzach-Vorlandgletschers. *Göttinger Geogr. Abhandl.*, **15** (2), 17–90.

WEINBERGER, L. (1955) Exkursion durch das österreichische Salzachgletschergebiet und die Moränengürtel der Irrsee- und Attersee-Zweige des Traungletschers. *Verhandl. Geol. Bundesanstalt*, D, 7–34. Vienna.

WEISCHET, W. (1970) Chile. *Wiss. Länderkunden*, **2–3**. Darmstadt: Wissenschaftliche Buchgemeinschaft.

WELTEN, M. (1958) Pollenanalytische Untersuchung alpiner Bodenprofile: historische Entwicklung des Bodens und säkulare Sukzession der örtlichen Pflanzengesellschaften. *Veröff. Geobot. Inst. Rübel*, **33**, 253–74. Zurich.

WELTEN, M. (1966) Simplon-Hopschensee. *Radiocarbon*, **8**, 25.

WHITE, S. E. (1962) Late Pleistocene glacial sequence for the west side of Iztaccihuatl, Mexico. *Bull. Geol. Soc. Amer.*, **73**, 935–58.

WICHE, K. (1960) Klimamorphologische Untersuchungen im westlichen Karakorum. *Verhandl. dtsch. Geographentages*, **32**, 190–203.

WILHELMY, H. (1957) Eiszeit und Eiszeitklima im den feuchttropischen Anden. *Petermanns Geogr. Mitt.*, Suppl. 262, 281–310.

WISSMANN, H. VON (1960) Die heutige Vergletscherung und Schneegrenze in Hochasien mit Hinweisen auf die Vergletscherung der letzten Eiszeit. *Abhandl. Akad. Wiss. Mainz, math. naturwiss. Klasse, 1959*, 14, 1101–407.

ZOLLER, H. and KLEIBER, H. (1971) Vegetationsgeschichtliche Untersuchungen in der montanen und subalpinen Stufe der Tessintäler. *Verhandl. Naturforsch. Ges. Basel*, **81** (1), 90–154.

Manuscript received April 1972.

Additional references

PATZELT, G. and BORTENSCHLAGER, S. (1973) Die postglazialen Gletscher- und Klimaschwankungen in der Venediegruppe (Hohe Tavern, Ostalpen). *Zeits. Geomorphol.*, Suppl. 16, 25–72.

WEBBER, P. J. and ANDREWS, J. T. (1973) Lichenometry: dedicated to the memory of the late Roland E. Beschel. *Arct. Alp. Res.*, **5** (4), 293–424.

Present biota

Ecology of the northern continental forest border

James A. Larsen

Center for Climatic Research,
University of Wisconsin

1. The forest border

POSITION

Reasonably accurate information on the position of the continental forest border from eastern to western Canada has been available in generalized form at least since the time of Louis Agassiz. His map of the natural provinces of North America (Agassiz, 1845) depicts with surprising accuracy the subarctic zone of boreal forest or taiga with its 'Canadian Fauna' and the arctic zone with its 'Hyperboreal Fauna'. The line between the two coincides with the position of the forest border as we know it today, diverging somewhat south of the line only in the region around Great Bear Lake and, eastward, in central Labrador–Ungava (Kendeigh, 1954).

In a somewhat more detailed presentation of the natural 'provinces' and 'regions' of North America, Cooper (1859) achieved a closer coincidence to the forest border with his line between 'Esquimaux' and the 'Athabascan' and 'Algonquin' regions. Later maps became increasingly accurate and treated vegetational zonation on a smaller scale, but they also began to show the consequences of differences in the interpretation of 'treeline' and 'timberline' and 'forest border'.

Hustich (1966) recognizes this problem in his study of the forest–tundra regions of the northern hemisphere: 'The concepts forest-line, or forest-limit, timberline or tree-line are all somewhat vague and difficult to define . . . In the sense employed, for instance, by North American writers, the frequently used expression "timber-line" may, thus, mean the northern limit of commercially profitable forest utilization, the polar or altitudinal limit of forest or the tree-line in general.'

In this excellent review, Hustich discusses ecological and climatic relationships, range limits, and other characteristics of species making up the arboreal stratum of the northern circumpolar forest border in North America and Eurasia. He points out that regional

Fig. 7A.1. The northern range of species of *Picea, Abies, Pinus* and *Larix,* showing the differences in tree species composition of the vegetation in the four 'corners' of the northern circumpolar forest–tundra (from Hustich, 1966). Species comprising the regional forests are these: North America – *Picea sitchensis* (Alaska), *P. mariana* and *P. glauca, Abies lasiocarpa* (Alaska), *A. balsamea, Pinus contorta* (Alaska), *P. banksiana,* and *Larix laricina.* Europe – *Picea excelsa* and *Pinus silvestris.* Eurasia – *Picea obovata, Abies sibirica, Pinus silvestris, P. sibirica, P. pumila* (Asia), *Larix sibirica* s.l., *L. dahurica* (Asia).

differences are the consequence of differences in topography, bedrock, and Quaternary history. White spruce, Siberian larch, cedar and balsam fir, for example, seem to prefer or at least not to avoid basic sedimentary bedrock. Treeline species which seem to prefer acid bedrock are jack pine, black spruce, and aspen. These preferences are revealed to some extent in the map presented by Hustich (fig. 7A.1) showing in generalized fashion the northern limits of species comprising the four genera, *Picea, Abies, Pinus* and *Larix,* in the northern circumpolar forest–tundra zone. Over much of the forest–tundra ecotone in northern Canada, surficial geology and topography are relatively homogeneous. In this region particularly, the influence of climate can be most clearly discerned since in effect these edaphic factors have been held constant by natural events.

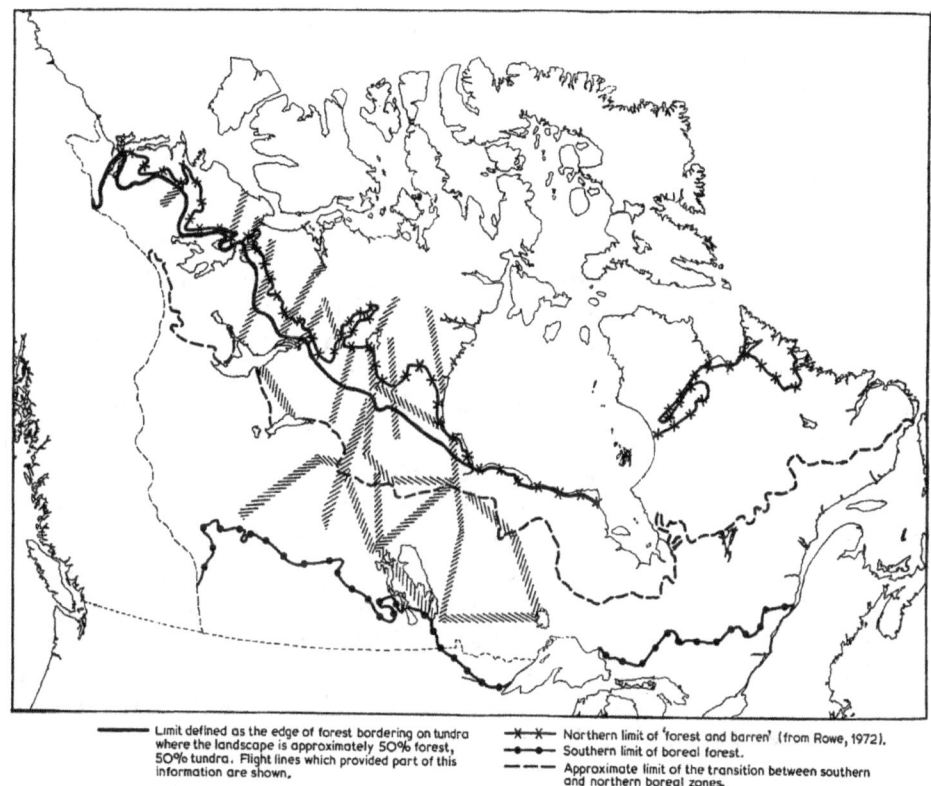

Limit defined as the edge of forest bordering on tundra where the landscape is approximately 50% forest, 50% tundra. Flight lines which provided part of this information are shown.

—X—X— Northern limit of 'forest and barren' (from Rowe, 1972).
—●—●— Southern limit of boreal forest.
— — — Approximate limit of the transition between southern and northern boreal zones.

Fig. 7A.2. Position of the northern forest border in North America.

While it is relatively easy to map the edge of the forest with an accuracy of ±25–30 km, the edge of the range of 'trees' is another matter and depends not only upon the definition of 'tree' but also on whether arboreal morphology or species is intended when the 'limit of trees' is indicated, even though one would be hard pressed to find a normally developed tree for many miles southward. On early editions of the Canadian Topographic series, certain difficulties of interpretation are apparent and various treelines have been identified in the literature using apparently different interpretations.

Fig. 7A.2 shows the approximate position of the forest border, taking as the definition of 'forest' the occupancy of at least 50 per cent of the land surface by trees. North of the forest border, by this definition, at least half of the land surface is occupied by tundra, southward at least half of the landscape is covered with plant communities in which trees are dominants. The observations upon which this line is based were made by the author during the course of canoe travel at many points along the forest border from the west shore of Hudson Bay to the western Cordillera. Additionally, photographic records of flight lines, obtained with time-lapse cameras mounted on low-level reconnaissance aircraft (lines of coverage shown by shaded areas in the figure), have provided data in a number of remote areas where accurate data would otherwise have been difficult to obtain. Fig. 7A.2 also shows the northern limit of 'forest and barren' as delineated by Rowe (1972).

CLIMATOLOGY OF THE FOREST–TUNDRA ECOTONE

The position of the forest border in northern Canada has long been regarded as a most convenient, and perhaps physically the most significant, boundary for conceptually separating the faunal, vegetational, climatic and ethnic characteristics of Arctic and Subarctic. The fairly definite correspondence between the 10°C (50°F) isotherm for the warmest month and the northern limit of trees, for example, was used by Köppen (1936) in his classification. This scheme was modified by Nordenskjöld to consider the mean temperature of the coldest month as well as the warmest (w = 9.0 − 0.1 k°C, where w is the mean temperature of the warmest month and k that of the coldest) with some improvement in fit. Thornthwaite's classification of climates by means of a relationship computed between water supply and estimated evapotranspiration was employed by Sanderson (1948) to delineate the climatic regions of Canada. Hare (1950) noted that precipitation in eastern Canada is adequate to supply the needs of growth under cool conditions and stressed the view that the northern forest growth is governed primarily by temperature. There is thus a more apparent correlation between the zonal forest divisions and thermal efficiency (potential evapotranspiration: a direct function of temperature). Boundaries between forest zonal divisions tend to follow isopleths of potential evapotranspiration, Hare noted, but are independent of moisture gradients. He reported a correspondence between forest divisions and potential evapotranspiration in Labrador–Ungava:

Division	Typical value of PE (cm) along boundaries	Dominant vegetation
Tundra		Tundra
	30–32	
Forest–tundra ecotone		Tundra and lichen woodland intermingled
	36–37	
Open boreal woodland		Lichen woodland
	42–43	
Main boreal forest		Close-forest with spruce-fir associations
	47–48	
Boreal–mixed forest ecotone		Close-forest with white and red pine, yellow birch, and other non-arboreal invaders
	50–51	
Great Lakes and St Lawrence mixed forest		Mixed forest

Hare also noted that wide variations in moisture occur within vegetational divisions without any readily apparent effect upon vegetational characteristics, and he cited other evidence (Hare, 1954) to show that thermal efficiency rather than moisture governs the distribution of northern forest types in the regions considered. The isopleth of 10°C accumulated month-degrees above the 6°C threshold, for example, shows a close correspondence to the northern forest border. More recently, Hare (1968) further analyses the relationships between northern climate and forest, asking if the forest border is the consequence of climatic conditions defined by airmass concepts or whether climatic differences across the forest–tundra ecotone are the result of differences in albedo, evaporating potential and aerodynamic roughness. He declines a categorical answer but indicates that biotic and climatic limits positioned at the forest border may be the consequence of atmospheric circulation patterns determined by overriding radiative controls, and hence both may be effects rather than one the cause and the other an effect (Barry and Hare, Chapter 2A; Larsen and Barry, Chapter 5).

In Alaska, at the western continental extreme, Hopkins (1959) found a correspondence between vegetation types with the number of degree–days above 10°C and also with the mean temperature of the coldest month. It is of interest additionally that the approximate southern limit of continuous permafrost in Canada (Brown, 1960, 1969, 1970b) coincides well with the northern limit of the forest–tundra ecotone in northwestern Canada, but in Eurasia the zone of continuous permafrost extends southward into the forested region (Brown, 1970a; Tikhomirov, 1970; see Ives, Chapter 4A).

From the point of view of airmass climatology, Stupart (1928) was at least one of the first to see the coincidence between the position of the forest–tundra ecotone and the path of the summer cyclonic storms, indicating that the latter lies just to the south of the northern limit of the trees. Reed (1959, 1960) established the mean summer position of the arctic frontal zone with improved accuracy. It is apparent that from July through October the forest border coincides with the average position of the climatic boundary between arctic air on the north and airmasses of Pacific and continental origin on the south. Bryson (1966), employing airmass frequency analysis and maps showing the position of resultant winds (average wind directions), demonstrated a close relationship between the position of the northern forest border with this mean position or modal position of the arctic front in summer, concluding that the boreal forest formation in North America corresponds to the zone over which the mean position of the arctic front oscillates during the course of the average year. Barry (1967), noting that this correspondence is most apparent in Canada west of Hudson Bay, where the forest–tundra boundary is relatively sharp, suggests that the correlation may not be as applicable in Labrador–Ungava where the location of the climatological front is, according to the result of his analysis, not as clearly a definitive demarcation zone. He notes that perhaps this has an important bearing on the fact that there exists a somewhat more extensive forest–tundra ecotone in the Labrador–Ungava region. Krebs and Barry (1970) report also that the result of frontal analysis of climate in northern Eurasia indicates a strikingly close coincidence between the forest–tundra boundary and the median location of the arctic front, especially in the case of the montane areas of the eastern Soviet Union.

Although the evidence thus strongly suggests that some important biological relation-ships must exist between such closely coincident climatic and vegetational distributions, there remains little but speculation on the mechanisms involved. Climatic inferences from the structure of the plant communities in the forest border areas have been impos-sible because of the almost total absence of definitive quantitative information on the communities involved (Larsen, 1965). The result has been general agreement that climate is, indeed, related to the position of the northern treeline, but acceptance of this line as a significant climatic boundary admittedly has involved an element of faith (Britton, 1966). Krebs and Barry (1970) state the case succinctly when they point out that the question now meriting attention concerns the precise nature of the postulated interaction between the vegetation cover and the various properties of the atmosphere.

A major step in this direction has been accomplished by Hare and Ritchie (1972) who have shown that global solar radiation per annum in Canada increases southwards from values near 90 kly at the arctic treeline to about 110 kly at the boundary between open woodland and closed forest zones (i.e. the northern forest line). More significant is absorbed solar radiation, which in the same span ranges from 50 to 55 kly at the forest–tundra treeline to about 80 kly in the northern forest proper. Zonal divisions of the vegetation appear to correlate closely with mean net radiation. With the forest strongly heated, and the tundra much less so, an intense air temperature gradient develops. The low albedo of dense forest permits more rapid warming of the forest in spring than of tundra, and hence the above-freezing season is about fifty days longer in the forest than on the tundra, even though these zones are separated by less than 400 km. The structure of the vegetation in this case markedly influences the physical climate, just as the latter provides the energy resource on which plant growth depends. Across Canada the zonal divisions of the boreal forest seem to be in close relation with the distribution of annual, and still more of warm season, net radiative heating. The arctic treeline in Canada occurs near an annual net radiation of 18–19 kly. The isoline of 16 kly is close in Alaska, but the scattered patches of spruce woodland across the neck of the Seward Peninsula, and in the valleys draining to Kotzebue Sound, lie well north of this line, and represent climatically the most poleward outliers of the boreal forest in North America together with similar groves in the Mackenzie Delta and Coppermine River Valley.

The treelessness of the Aleutians, the Alaskan Peninsula, much of Kodiak Island and parts of the southeastern slopes of the Kenai Peninsula cannot be explained on energy grounds. Very low summer temperatures, high winds and remarkably high cloudiness can be invoked qualitatively as causes. Treeless maritime tundra is common in other, similar areas of the northern hemisphere, notably in Iceland.

Hare and Ritchie also point out that standing phytomass increases from less than 5 tons ha^{-1} in arctic tundra to as high as 25 tons just north of the forest border. It then rises rapidly southward to values as high as 300–400 tons ha^{-1}. In relation to the annual net radiation, much more phytomass exists per unit of energy absorbed in the southern boreal forest than in the southern tundra.

ENVIRONMENT, SPECIES AND THE FOREST BORDER

The northern forest border obviously is a significant zone in both climatic and ecological aspects. It remains at the present time one of the least understood regions of the world, climatically and ecologically. But the problems of arctic and subarctic ecology are persistently fascinating and undoubtedly they will be studied now with increasing intensity. This will be true especially as the difficulties imposed by the cost of transport and the discomforts of prolonged fieldwork become more manageable, and the economic pressures for exploitation of the potentials of the northern regions more insistent.

Botanical exploration of Canada has been relatively intensive, with few regions that have not been visited by at least one or two botanical collectors. As a consequence, regional studies of the flora based on existing plant collections are numerous. Manuals for plant identification are sufficiently comprehensive so that conduct of field sampling is not impossibly complicated by taxonomic difficulties, which field ecologists are notoriously ill-equipped emotionally and intellectually to handle. Raup long ago (1941) pointed out that as the taxonomy of Canadian species was sufficiently well known, the next major task for botanists was to begin study of plant communities found over the Canadian landscape: 'Although a great deal of work on the structure of arctic and subarctic plant communities has been done in the Old World, in America it has hardly more than begun.' Even now, three decades later, statistically quantitative vegetational studies of the boreal forest are still limited largely to southern portions of the region: very few have been published on the forest–tundra ecotone of the Northwest Territories (Larsen, 1965, 1971a, b, 1972a, b; Maini, 1966), of northeastern Manitoba (Ritchie, 1960a, b, c, 1962), or of Labrador–Ungava (Fraser, 1956; Hustich, 1962; Maycock and Matthews, 1966).

The importance of continuing ecological studies in these areas is, however, now clearly apparent. Ecologists interested in fundamental relationships of plant communities can still find here native communities undisturbed by human land-use practices. Moreover, knowledge of baseline conditions existing in undisturbed habitats are now required so that the effects of man-induced environmental changes can be assessed and, hopefully, minimized by appropriate protective or management techniques.

The geographical range of most of the species of Canadian boreal forest and tundra is now known with some degree of certainty, but understanding of the ecological relationships is still largely intuitive. Only the ecological relationships of spruce have been outlined to a degree sufficient to afford some recognition of the physiological mechanisms involved.

It is initially of significance that little change has occurred in the position of the forest border in at least central Canada since the earliest intensive explorations which date from the latter part of the 1800s (Larsen, 1965, 1971a, 1972a, 1973). The maximum range of spruce as a species extends many miles north of the present day forest border, and it is apparent that the relict population (see Larsen and Barry, Chapter 5) making up the northernmost extension of the spruce range would furnish an ample source of seed for recolonization of the area were environmental conditions to ameliorate. The short duration of the growing season would appear to make production of mature viable seed unlikely in most years. Even though an occasional productive year may occur, conditions are likely to be unfavourable for seedling survival. Layering is without doubt the most frequent method by which black spruce is maintained in the small clumps found far beyond the forest border. Indeed, in places where clumps of dwarfed spruce have been cut by Indians or Eskimos for wood, no regeneration has taken place (Larsen, 1965).

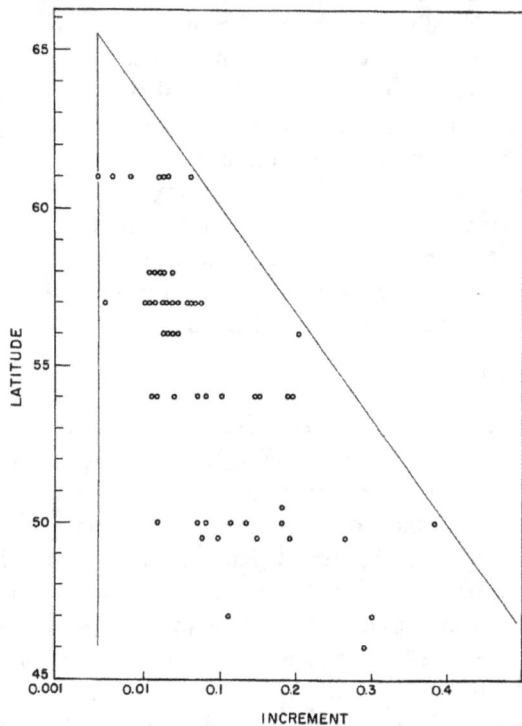

Fig. 7A.3. Relative growth rates for ten to twenty randomly selected trees in black spruce stands at the indicated latitudes in central Canada. Each dot represents the average growth rates of the trees sampled in a stand. Widest variations in growth rates are found in the southern latitudes.

The general deterioration of conditions for spruce growth northward can be seen in a comparison of growth rates of spruce in study sites at different latitudes (see fig. 7A.3). In this comparison (Larsen, 1965), rates of growth of individuals in forty-eight black spruce stands at nine lines of latitude in Ontario, Manitoba and Wisconsin were shown to vary markedly from stand to stand with a general narrowing of the range of growth rates northward. At Ennadai Lake (the northernmost study area), the rate of growth on optimum sites does not much exceed the growth rate on the poorest sites, indicating that only the most narrow range of habitable environment is available at the northern edge of the forest. Intersection of a line drawn through maximum growth rates with the minimum growth rate line occurs at approximately 65°N latitude, some 4° beyond the range of spruce in this longitude, but 2° or less beyond the known range of spruce as a species (with the exception of the outlier of spruce along the Thelon River, which may well have special microclimatic and edaphic conditions and/or a special history). Later, Mitchell (1967) analysed spruce growth rates in greater detail, using the same core material and coming to the similar conclusion that the location of the theoretical black spruce treeline coincides closely with the northern limit of spruce.

Efforts to explain variation in growth rates of trees along an environmental gradient must be multifactorial in approach; inferred single-factor causal relationships are simplistic and unconvincing. The same is true of geographical distribution of non-arboreal species. At present, in both boreal and arctic regions, the paucity of environmental data makes even the simplest correlation between environmental parameters and growth rates in trees (or of abundance in communities in the case of the smaller plants) difficult or impossible to obtain. There is, thus, no possibility as yet for decisive identification of the factors which, at one time or another, become limiting to plant growth, either in terms of arboreal form or geographical distribution. Recognizing this difficulty, however, it is nevertheless possible to discern some differences existing in environments of arctic and subarctic regions and to recognize that there are, indeed, some rather abrupt transitions in some of these factors across the forest border ecotone. Several easily discerned and possibly pertinent environmental factors should be considered in analyses of vegetational/environmental relationships within and across this forest–tundra transition zone. The next section considers a few of these factors.

2. The environment: vegetation and climate

The fundamental vegetational difference between Subarctic and Arctic, a difference that accounts for profoundly different influences upon soils and microclimate, is the presence of nearly continuous forest in the Subarctic and its total absence in the Arctic.

The arctic environment in general can be characterized as one receiving relatively low annual amounts of solar radiation and relatively low amounts of precipitation. Evaporation rates are low due to low temperatures. Runoff through streams and rivers accounts for a large proportion of water loss in the hydrological cycle. There are relatively high wind velocities at ground level due to absence of protective shielding and surface friction.

In the Subarctic there are characteristically somewhat larger amounts of radiation received in comparison to the Arctic. Precipitation is higher. More water is lost by evaporation from the ground surface and by evapotranspiration through vegetation. Total evaporation, however, is still low and runoff in streams and rivers is high.

In both regions low soil temperatures are characteristic. Permafrost is continuously present below the soil active layer in the Arctic; in most of the Subarctic of North America it is discontinuous and found with the upper surface at greater depths during summer than is the case in the Arctic. A discussion of frost phenomena is presented in a number of publications (Péwé, 1969; Brown, 1960, 1965, 1970a, b) dealing with many aspects of the physical action of frost upon soils.

The effect of soil frost phenomena upon vegetation has often been noted and Benninghoff gives the following description (1952): 'In the high latitudes roots are not only encased in frozen material for a great part of the year, but by repeated freezing and thawing, especially during the autumn freeze-up, they are heaved, torn, split by forces of great strength.'

Moreover, in the Arctic and northern Subarctic an increased plant cover does not signify an increase in mesophytism since thick vegetation usually decreases the depth of the active layer. Benninghoff states: 'Plant succession in temperate regions tends to establish more mesophytic conditions in which drainage relations are less extreme. But in regions of severe frost climate, plants commonly generate conditions of extreme lack of drainage and greatly intensified soil frost; in short, the plants frequently destroy the very environmental conditions that favor their growth.'

Interaction of soil and vegetation apparently is at least partly responsible for many of the cyclic phenomena reported in areas possessing permafrost (Billings and Mooney, 1959; Hopkins and Sigafoos, 1951; Britton, 1966), but the time factor appears to be long. Black (1954) points out that frost splitting and churning are minor processes in the far northern areas where there is lack of repeated diurnal freezing and thawing, where the active layer is thin, where sediments are predominantly monomineralic fragments of quartz and chert, and slopes are of low angle. The author (Larsen, 1965) recorded an episode of severe upheaval of the surface vegetation overlying glacial till in a low latitude arctic tussock muskeg, with evidence in the general area (southern Keewatin) of frequent activity of this type. Northward, frost scars of this variety become less evident; perhaps as a consequence of the fact that the tussock muskeg community itself becomes increasingly rare. In northern Keewatin, very little evidence of exposed bare peat is to be found.

The influence of wind may be more significant in arctic plant communities, and adaptations to persistently high wind velocities are apparent in many arctic species. Vegetation decreases the velocity of air currents within the stratum occupied by stems and leaves, tending to reduce heat loss by conduction and convection and often permitting an increase in temperature of above-ground plant parts well above that of the surrounding atmosphere during periods of high insolation. Thus, while conditions favour low root temperatures, above-ground plant organs often possess temperatures appreciably higher than the air temperature. Shading by the canopy results in a persisting ice lens beneath spruce trees at the northern edge of the forest. Roots of trees and other plants in the shade are, thus, very often colder than roots of tundra plants receiving the same amount of insolation but beneath which the permafrost surface has retreated further. This may be a significant factor in limiting tree growth at the northern forest border (see fig. 4A.8, p. 176).

Low summer temperatures and the shortness of the growing season evidently are primarily factors in the increasingly reduced number of species present northward. Savile (1961) made note of such an effect during studies of the vegetation of the Queen Elizabeth Islands: 'When the July mean temperature is about 38°F [3.3°C] a change of only a degree in either direction greatly affects the flora. The plants of Isachsen are fewer and more stunted than those of Alert and Mould Bay, where the temperature difference is scarcely measurable. On the central plateau of Somerset I., where the July mean temperature is probably ca. 37°F [2.8°C], only 11 vascular plants could be found.'

Downes (1964), in a discussion of the Arctic, points out that the low temperatures and shortness of the growing season greatly restrict the number of degree–days above 0°C, and that even this is a generous measure of the time available for growth since often the 'year is well advanced towards the solstice before the snow melts and growth can start'. He adds that 'the very small size of the heat budget is most impressive, and it seems that it must be an important and controlling feature of arctic life.'

A strong coincidence between the ranges of alpine plants and various isotherms has been demonstrated for more than one hundred species of plants in Scandinavia; outside certain critical isotherms, certain species are absent or very rare, whereas inside they are very common (Dahl, 1951). Similar coincidences between species ranges and temperature parameters have been described for European plants (Jeffree, 1960), eastern North American tree species (Leopold, 1958; Wolfe and Leopold, 1967), and for high arctic–alpine species along an altitudinal gradient (Powell, 1961). It is apparent that rather distinct features characterize the environment of the northern regions. These features have resulted in evolutionary development of species with characteristic physiological and morphological adaptations; in short, in a distinct flora directly related to environmental conditions (Bliss, 1962; Billings and Mooney, 1968). In general, it can be said that temperature is probably the most important factor limiting growth and development of vegetation in northern regions. Three temperature-dependent physiological processes must be considered most important in these relationships (Warren Wilson, 1957): 1) translocation, which is not greatly affected by temperature under normal temperate zone conditions, but which is markedly checked when plants are chilled to 0–5°C; 2) water absorption is affected similarly; 3) rate of photosynthesis increases roughly 50 per cent with an increase in temperature from 0°C to 1°C, while an equal increase in photosynthetic rate is achieved only by a 10°C increase in temperature from 20°C. The consequences of temperature dependence are probably critical to an understanding of the geographical distribution and growth rates of boreal and arctic plant species. Very fundamental biochemical processes, as well as genetic ones, are involved as shown by work demonstrating that oxidative rates of mitochondria and other physiological processes differ in populations of plant species along climatic gradients (Klikoff, 1966; Billings and Mooney, 1968). In regard to the growth of trees, particularly capacity to attain typical long-stemmed arboreal form, the relationship between assimilation rates and respiration may be significant: in trees, according to Warren Wilson (1957), respiratory loss is proportionately greater than in herbaceous plants, apparently because of 'maintenance costs of a trunk-and-branches system: some two-thirds of their total matter is devoted to this supporting structure. Probably over half their budget is spent on defense expenditure related solely to competition.' At a point where assimilation becomes insufficient to maintain an arboreal 'defense system', trees will give way to species smaller in relation to leaf area and which therefore have lower fixed respiration costs.

There can perhaps be no more suitable summary than an early statement of Went (1950), who wrote that distribution of plants is not just a question of frost damage but is correlated with many specific temperature requirements, met for each species only within a given climate:

> An analysis of the genetical basis of climatic response may provide some interesting insights into the problem of evolution and migration of species, because it will indicate how many genes have to participate to allow invasion into a new climatic territory.

A study by Jeffree (1960) resulted in a similar conclusion:

> For each species there is a fairly definite range of possible temperatures, and above or below these the plant will perish. Somewhere between are the optimal temperatures in which the plant will thrive best . . . It is unlikely that plants will be able to expand into temperature ranges as extreme as the ultimate possible in protected experiments.

3. Species and environmental parameters

Studies cited in previous pages (Larsen, 1965, 1971b, 1972c, 1973) furnish statistical data on plant communities in two areas of the forest–tundra transition zone, the Ennadai Lake area in southwestern Keewatin and the area between Fort Reliance and Artillery Lake north of the east arm of Great Slave Lake. Analyses of these data, along with data from other study areas in the boreal forest and tundra of northern Canada, have furnished information on relationships of species to airmass regimes, as well as on the dominance and diversity characteristics of plant communities studied throughout the area considered (see fig. 7A.4). Before elaborating upon the result of these analyses, brief descriptions of the three major forest-border ecotone study areas are to be presented.

VEGETATION OF THE FOREST–TUNDRA ECOTONE

Surveys of the vegetation of the forest–tundra ecotone in three areas have been conducted during the course of the study; the Ennadai Lake area in southeastern Keewatin, the area extending from Fort Reliance in the east arm of Great Slave Lake northward to Artillery and Aylmer Lakes, and from the Inuvik area northward to Reindeer Station near the mouth of the Mackenzie River.

The south end of Ennadai Lake is covered with virtually continuous black spruce (*Picea mariana*) forest. Dense stands occupy lowlands. Somewhat less dense lichen woodlands are found over well-drained upland areas, with intermediate types between these two topographic extremes. Open muskeg is frequent on low poorly drained basins, grading into sedge meadow on areas where water to depths of several inches is found throughout most of the summer, covering accumulations of plant detritus and sedge peat.

A distinct transition is apparent a few miles north of the southern extremity of Ennadai Lake. Although topography and surficial geology remain similar, summits of the hills are here devoid of trees and a treeline can be seen on the higher slopes (cf. Plate 1). Ennadai Lake is a long, narrow body of water extending from its south end some 70 km towards the northeast. Transition from forest to tundra takes place in roughly a third of

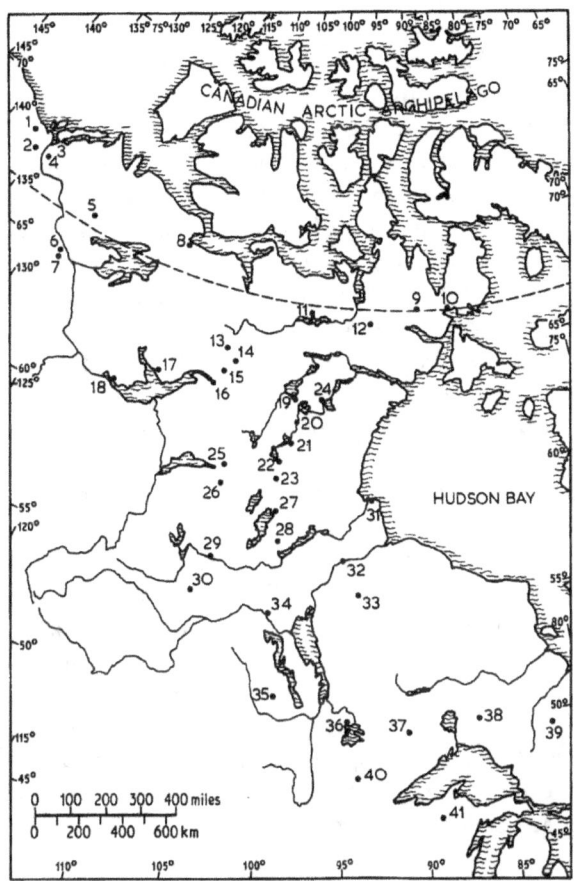

Fig. 7A.4. Study areas in which sampling of vegetational communities was conducted. 1, *Trout Lake (NWT)*; 2, *Canoe Lake*; 3, *Reindeer Station*; 4, *Inuvik*; 5, *Colville Lake*; 6, *Florence Lake*: 7, *Carcajou Lake*; 8, *Coppermine*; 9, *Curtis Lake*; 10, *Repulse Bay*; 11, *Pelly Lake*; 12, *Snow Bunting Lake*; 13, *Aylmer Lake*; 14, *Clinton-Colden Lake*; 15, *Artillery Lake*; 16, *Fort Reliance*; 17, *Yellowknife*; 18, *Fort Providence*; 19, *Dubawnt Lake*; 20, *Dimma Lake (Kazan River)*; 21, *Ennadai Lake*; 22, *Kasba Lake*; 23, *Kasmere Lake*; 24, *Yathkyed Lake*; 25, *Black Lake*; 26, *Wapata Lake*; 27, *Brochet*; 28, *Lynn (Zed) Lake*; 29, *Otter Lake*; 30, *Waskesiu Lake*; 31, *Churchill*; 32, *Ilford*; 33, *God's Lake*; 34, *Rocky Lake*; 35, *Clear Lake*; 36, *West Hawk Lake*; 37, *Raven Lake*; 38, *Klotz Lake*; 39, *Remi Lake*; 40, *Bagley (Pike Bay)*; 41, *Trout Lake*.

this distance from the south end. Only occasional small groves of spruce are visible at the far north end of the lake. Spruce groves are then found northward from the lake along the Kazan River, becoming increasingly small and more and more restricted to exceptionally favourable sites affording marginal habitat for a few dwarfed or decumbent specimens. The last groves are found near Yathkyed Lake along the Kazan and, to the west, at the south end of Dubawnt Lake (Larsen, 1965).

Virtually these same characteristics of forest–tundra ecotone exist between the northeast end of Great Slave Lake and Artillery Lake (Larsen, 1971a). Here, however, a moderate altitudinal gradient also exists; the country rises gently toward the northeast, possibly contributing to the abrupt transition to tundra. Across the 30 km or so of land separating Great Slave from Artillery Lake, along the series of trails and small lakes known as Pike's Portage, forest communities undergo a transition to tundra. Noteworthy is the observation that here both boreal and arctic species are found growing often on adjacent or nearby sites. There are strong indications that the climatic gradient is steeper here than in the Ennadai Lake region (McFadden, 1965) where southern boreal and tundra species are separated by a much wider ecotonal corridor (occupied largely by ubiquitous wide-ranging species) even though the forest itself ends abruptly (Larsen, 1973). In the Ennadai Lake area, a floristically depauperate zone exists in the southern belt of tundra, with arctic species beginning to appear in plant communities only many miles to the northward (Larsen, 1967, 1973). This belt appears to be narrower in the Artillery Lake area (Larsen, 1971).

In the continental interior west of Great Bear Lake, forest extends northward along the flat alluvial valley and delta nearly to the mouth of the Mackenzie River, but on the plateau extending away from the river the forest does not extend northward beyond a transition zone roughly between Inuvik and Reindeer Station. In this western zone, the forest border is much more irregular than in the Artillery Lake and Ennadai Lake areas, the consequence of somewhat greater differences in elevation from place to place than is the case to the eastward and also perhaps of maritime climatic influences originating in the Arctic Ocean. In the mountains west of Inuvik, timberline is found at about 300 m above sea level. Southward, approximately west of Norman Wells, treeline in the mountains lies between 3000 and 4000 feet. In both areas, the ecotone is irregular and patchy due to the effects of slope, topography, and montane climatic influences.

ENVIRONMENT: AIRMASSES AND SPECIES ABUNDANCE

Large regions are often dominated by relatively homogeneous atmospheric conditions. These stretches of atmosphere possessing roughly identical characteristics are known as airmasses, and belts of converging air from two or more 'source' regions are known as frontal zones or storm tracts. Vegetation of one type may be replaced by another across such a frontal zone.

It is possible to trace the movement of airmasses for some distance by means of distinguishing characteristics; in practice, airmass analysis consists simply of determining frequency of occurrence of airmasses of different origin over a given point of the earth's surface. Upper air data are often employed. When these are not available, approximation can be achieved with surface air temperature data. Using frequency distribution of daily maximum temperatures, Bryson (1966) fitted normal curves to North American airmass distributions in such a way as to determine the proportional influence upon climate exerted by each of four airmass types. Airmass frequencies for the upper United States and Canada (in July) are presented in the four figures below (figs. 7A.5 a–d), each of which depicts the frequency of a single airmass type. The existence of frontal zones between the major airmass types was confirmed by average wind direction, with the occurrence of convergence delimiting frontal zones.

Airmass frequency values can be correlated with the frequency of occurrence of plant species in various communities. Principal component analyses select those species showing highest correlations between some measure of abundance (frequency, relative frequency, density or dominance tabulations) and airmass frequencies.

AIRMASSES AND SPECIES

Species occurring in communities possessing high correlations with the frequency of airmass types throughout north-central regions of Canada have been analysed (Larsen, 1971b). Prior to a discussion of relationships so revealed, however, a description of the sampling method is appropriate.

Sampling at the study sites in the region was conducted most intensively in four forest communities and three tundra communities. The forest communities were those dominated by *Picea mariana* (black spruce), *P. glauca* (white spruce) and (or) *Abies balsamea* (balsam fir), *Pinus banksiana* (jack pine), and *Populus tremuloides* and (or) *Populus balsamifera*. The tundra communities can best be characterized as found commonly on three general topographic locations, rock fields (fellfields) on the summits of hills, tussock muskegs on topographically intermediate sites, and low meadows found where water accumulates to depths of 3 cm or more at least during spring and early summer.

In the case of the forest communities, as an initial requirement the arboreal stratum of each stand was dominated by black spruce, white spruce and/or balsam fir, aspen, or jack pine. In selection of stands, visual inspection was sufficient to indicate the dominant species, and sampling by the point quarter method confirmed this observation statistically.

In tundra, selection of sampling sites was primarily visual. Rock field communities are found frequently covering a large proportion of the upper parts of low, rolling hills characteristic of much of the central Canadian tundra region. Tussock muskeg is similarly easy to identify by observation; this community is a dense aggregation of plants growing in an accumulation of peat and plant detritus found usually on lower slopes. The low meadows appear as flat areas of poor drainage, and often are shallow basins more or less filled with accumulations of plant detritus.

Data obtained in all of these areas consisted of frequency tabulations (in 1 metre square quadrats) of understory species in forests and of all species in the case of tundra. Replication of sampling was practised by running as many transects through as many different examples of a given community as was possible at each study area.

COMMUNITY ANALYSIS

Individual species and groups of species possessing high correlations with airmass types have been identified in a previous paper (Larsen, 1971) and it is sufficient here to summarize by stating that the frequency values of many species attain maxima in communities

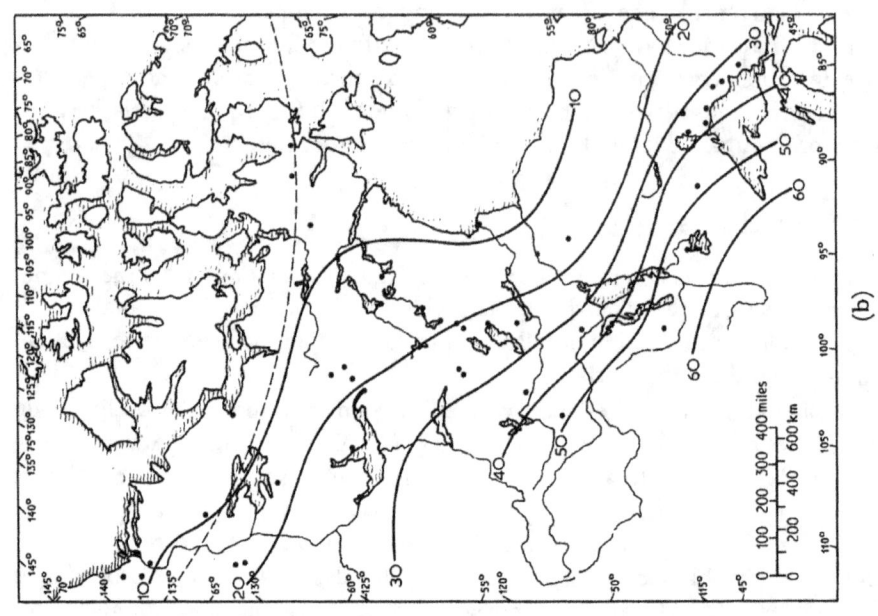

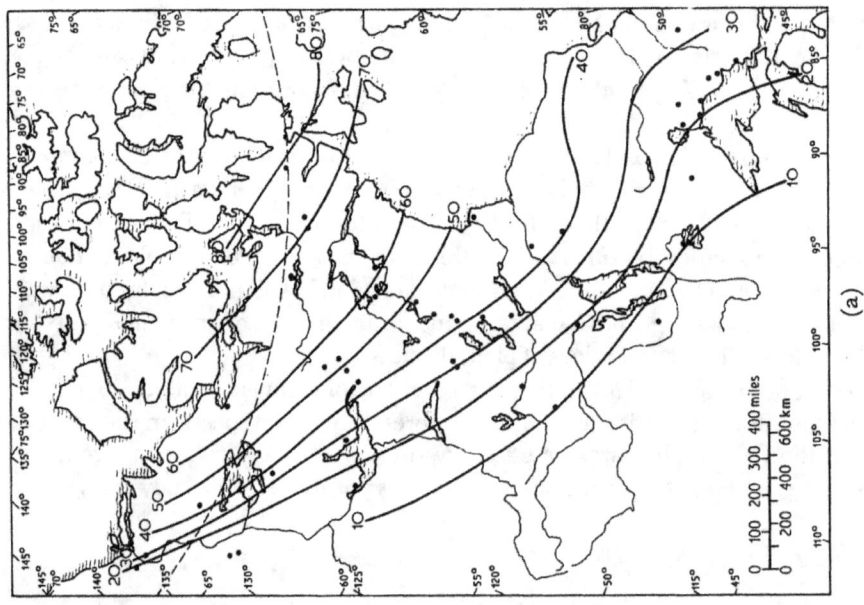

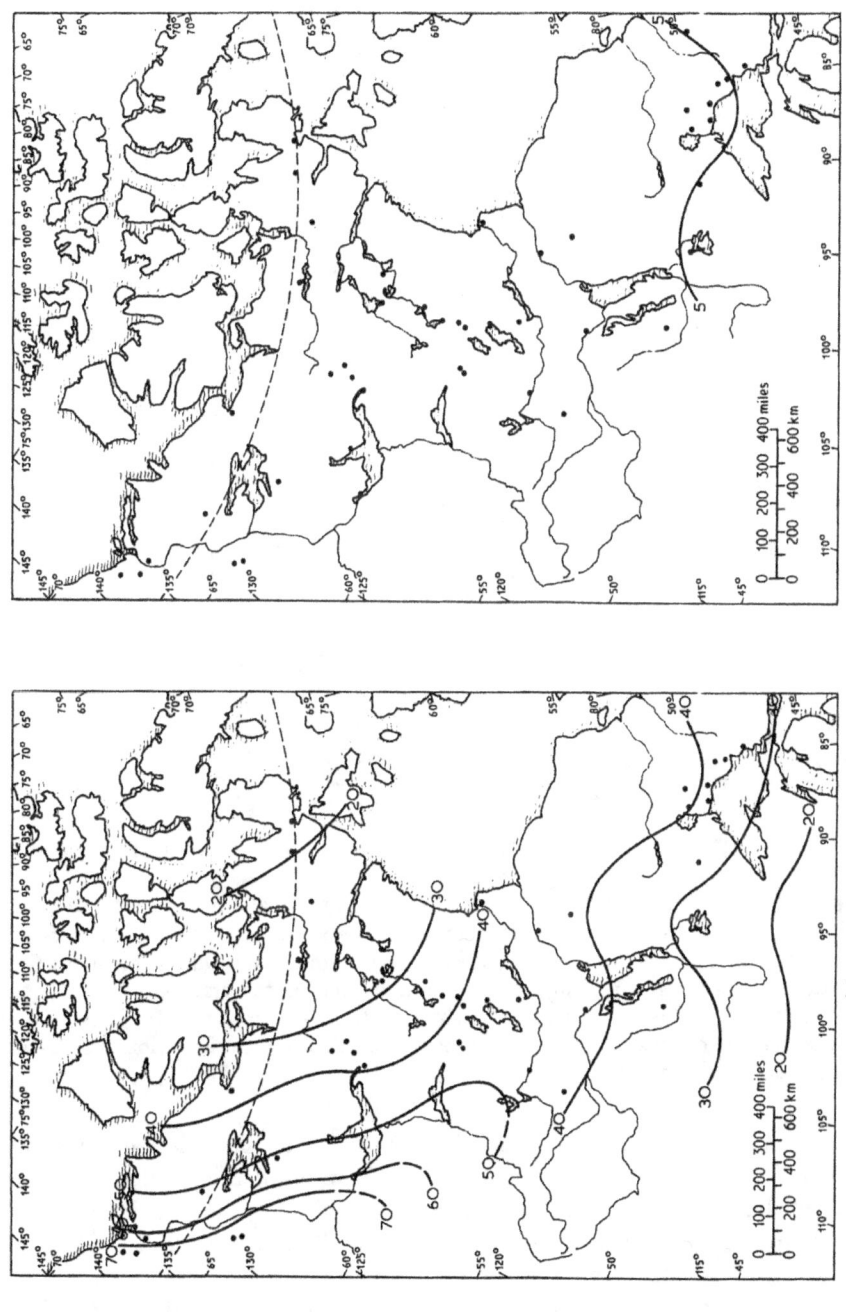

Fig. 7A.5. Isopleths of airmass frequencies (per cent) for July in central Canada and adjacent regions. Frequencies of (a) arctic, (b) Pacific, (c) Alaska–Yukon, and (d) southern airmasses are shown (after Bryson, 1966).

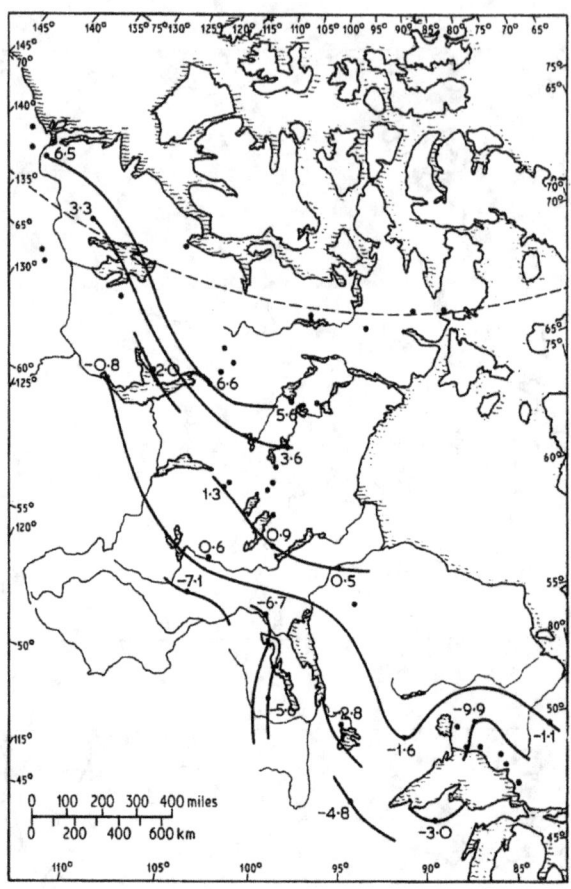

Fig. 7A.6. Isopleths of eigenvector coefficients obtained in principal component analysis of tabulations of species frequency in understory communities of black spruce stands in the region shown. Eigenvector coefficients evaluate the importance in the communities at each study site of the group of species with high loadings in the first component of the analysis, i.e. high values indicate the increasing importance of this group of species northward. Note that the orientation of the isopleths follows closely the orientation of climatic lines in the region (see fig. 7A.5).

at regionally identifiable climatic optima. The data indicate that relationships with airmass frequencies are sufficiently high and consistent from community to community to warrant the conclusion that distribution and frequency of occurrence of these species in the plant communities are a response to the regional distribution of climatic conditions.

Because the forest border represents a major climatic boundary, it would be expected that species would be related to this zone in some rather specific ways. That this is the case is shown by correlations between eigenvector coefficients (obtained in a principal component analysis) and distance from the forest border. The eigenvector coefficients are a measure of importance; thus, if species making up the eigenvector become of greater importance (if abundance values increase) as the forest border (and hence as the frontal zone between Pacific and arctic air) is approached, then a high negative correlation would be expected between eigenvector coefficients and distance from treeline. These coefficients show a distinct relationship of this kind in black spruce, jack pine, aspen, rock field, and tussock muskmeg communities as shown by means of isopleths drawn between study areas with similar eigenvector coefficient values (fig. 7A.6).

It can be seen that species correlated positively with arctic air are often positively correlated with Alaskan-Yukon air, and species positively correlated with Pacific air are often also positively correlated with southern air. Thus one suspects that species are not directly responsive to specific airmass types but rather to a northern and southern climatic regime, the northern composed of some combination of arctic and Alaska–Yukon airmasses and the southern of Pacific and southern airmasses. It should be indicated that the airmass frequencies are of value as an index to conditions because they furnish a simplified graphic delineation of the complex of environmental conditions to which the plants are exposed.

Seven species correlate with opposing airmass types in different communities. They increase in frequency northward throughout the forest and decrease northward in tundra. The species following this pattern include *Empetrum nigrum*, *Ledum decumbens*, *Rubus chamaemorus*, *Vaccinium uliginosum*, and *Vaccinium vitis-idaea*. It is evident from range maps showing geographical distribution of these species that here we have the interesting spectacle of species correlating positively with the frequency of arctic air in forested regions (decreasing in frequency southward) and with Pacific air in the tundra (decreasing in frequency northward). These species attain their highest frequencies in the communities of the forest–tundra ecotone. Whether uniquely adapted or simply widely tolerant of harsh conditions, they provide an interesting group for further study.

DOMINANCE RELATIONSHIPS OF THE VEGETATION

One of the most frequent observations made concerning characteristics of plant communities of temperate and arctic regions is that in each community some few species are more abundant than others. It is usually possible to array species of these communities in such a way that they form a series in descending order of importance. Such a series often closely follows a mathematically characterizable curve. Various niche-space relationships have been postulated for species in communities represented by various types of curves, based on assumptions regarding the manner in which species pre-empt the niche-space they occupy; these assumptions are derived from the mathematical relationships expressed in the curve.

The result of an investigation of these relationships in seven communities is presented in figs. 7A.7a,b.

It becomes apparent that tussock muskeg is distinct in that frequency values throughout all but the tail of the distribution are consistently higher for a given frequency class than is the case in either the rock field or the low meadow community. Of the forest communities, the curves in the jack pine community are lowest, aspen highest, and black and white spruce intermediate and closely coincident.

Whittaker (1970) summarizes the hypotheses so far put forth to describe niche-space relationships on the basis of different importance-value curves. Since none of the seven community curves follow any one of the various hypothetical importance-value curves closely, one can conclude (as Whittaker points out is usually the case) that while some similarity to one or another hypothesis can be discerned in each of the seven community

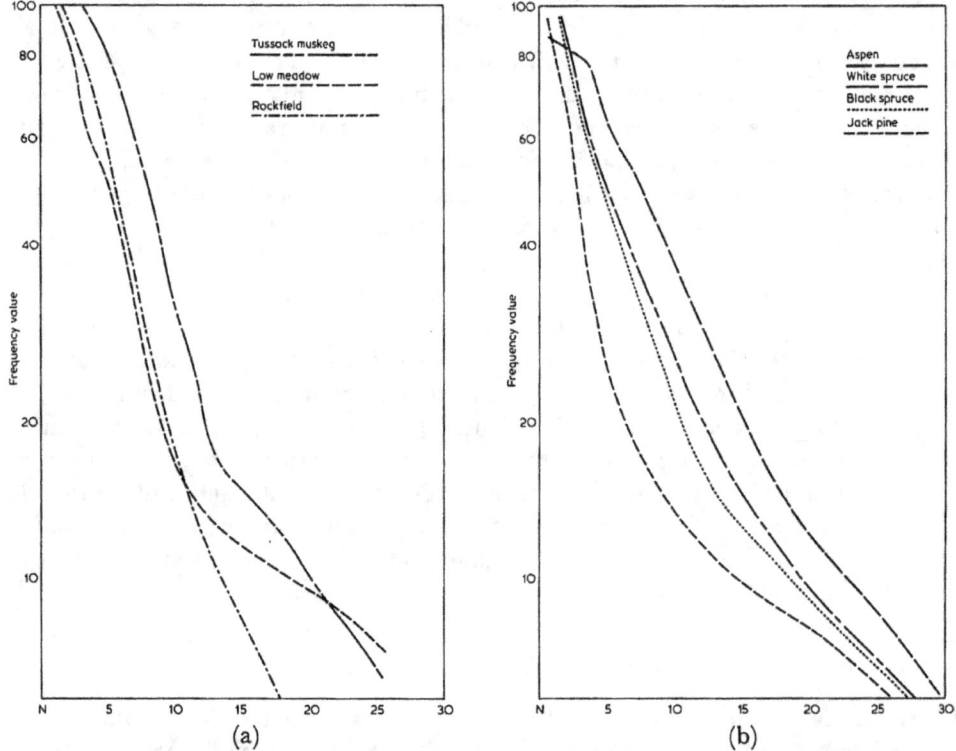

Figs. 7A.7a, b. Average frequency values (scale on left) of species in communities when the species are arranged in descending order of importance. The bottom scale indicates the position of the species in this array, from the species with highest frequency on the left to the species with lowest frequency on the right. For discussion see text.

curves, differences are sufficient to preclude any easy conclusions regarding the niche pre-emption pattern of any one. Different competitive and niche division relationships must exist in each. The closest approach to the random niche boundary hypothesis is provided by the rock field community; other tundra communities as well as the forest communities approach the lognormal distribution to some greater or lesser degree.

It can readily be discerned from the frequency distributions that the seven communities are each characterized by a few dominants, a larger number of species of intermediate importance, and an even larger number of rare species occurring often only once or twice in the data of a sampled stand. Grouping the data into logarithmically based classes makes it possible to discern the relative importance of dominant, intermediate, and rare species. Frequency class 1 includes all of the species occurring only once in a transect, class 2 includes all species with an average frequency of 10 per cent and half of those with a frequency of 15 per cent, class 3 includes all species with an average frequency of 20 or 25 per cent plus half of those with a frequency of 15 and 30 per cent, class 4 includes all species with frequencies of 35 to 60 per cent plus half those with frequencies of 30 and 65 per cent, class 5 includes all species with frequency values greater than 65 per cent, plus half of those with a frequency of 60 per cent, and these are thereby classed as dominants.

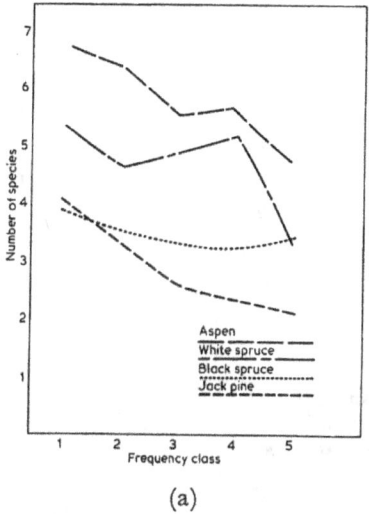
(a)

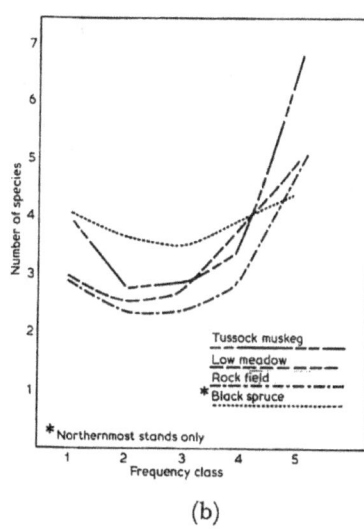
(b)

Figs. 7A.8a, b. Data for average frequency values re-ordered into logarithmic frequency classes; the frequency class 1 contains rare species with low frequency values, class 5 contains the abundant species. Noteworthy is the comparison between forest and tundra communities; in the latter there are relatively fewer rare species and more abundant species per average quadrat. The designation on the scale at left refers to the number of different species found per average quadrat.

It is now possible to discern marked differences between forest and tundra communities as well as between communities within each category (figs. 7A.8a,b). In forest communities, the number of species occurring in each quadrat is markedly highest in aspen and lowest in jack pine. All forest communities demonstrate a general decline from large numbers of infrequent species to a few frequent species. Aspen, however, has nearly equal numbers in the third and fourth class, in the latter nearly equalling the numbers in the first class. Black spruce communities have nearly equal numbers of species in each class.

The curves of tundra communities are more closely coincident and markedly distinct from forest communities in the relationship of rare to dominant species.

The distribution of species in the black spruce community is of particular interest, since of all forest communities it most closely approaches the distribution characteristic of the tundra communities. Indeed, if the distribution of the stands from Ennadai and the Artillery Lake–Fort Reliance areas are considered alone (fig. 7A.8b), the distribution even more markedly approaches the distribution curve typical of the tundra communities. It is apparent that environmental influences characteristic of the more northern areas tend to result in a community structure in which a larger number of species attain dominant positions and relatively fewer occur rarely. The ratio of dominant to rare species clearly will demonstrate a correlation with a north–south climatic gradient across the forest–tundra ecotone.

DOMINANTS AND AIRMASS FREQUENCIES

Species dominant in their respective communities often also display the highest degree of relationships to airmass frequencies. In table 7A.1 are listed (for several communities) species showing both dominance and high correlations with airmass frequencies. The relationship of species in each community to the crude environmental index afforded by airmass frequencies can be summarized by stating that the species (in table 7A.1) are

Table 7A.1. Species in various communities showing a high incidence of dominance and also high correlation with airmass frequency.

Community	Number of stands in which species occurs	Number of stands in which species ranks 1st, 2nd, 3rd, 4th, in frequency value				Correlation with airmass type
		1	3	3	4	
Black spruce						
Vaccinium vitis-idaea	104	63	22	10		0.65 (Arctic)
Empetrum nigrum	56	2	8	4	8	0.69 (Arctic)
Vaccinium uliginosum	48	9	8	10	7	0.56 (Arctic)
White spruce						
Vaccinium vitis-idaea	23	6	4	3	2	0.65 (Arctic)
Empetrum nigrum	18		4	3	3	0.76 (Arctic)
Betula glandulosa	12	4	1	1	1	0.67 (Arctic)
Ledum groenlandicum	25	1	3	1	1	0.51 (Arctic)
Cornus canadensis	23	5	1	3	1	0.48 (Pacific)
Rubus pubescens	15	3	1	2	3	0.67 (Southern)
Jack pine						
Vaccinium vitis-idaea	20	11	4	5		0.68 (Alaska–Yukon)
Maianthemum canadense	9	3			3	0.84 (Pacific)
Rock field						
Vaccinium vitis-idaea	65	32	13	6	2	0.86 (Pacific)
Ledum decumbens	45	12	8	9	9	0.68 (Pacific)
Vaccinium uliginosum	50	9	4	9	7	0.58 (Pacific)
Arctous alpina	44	6	7	8	2	0.56 (Pacific)
Empetrum nigrum	48	4	7	7	10	0.83 (Pacific)
Loiseleuria procumbens	32		1	6	2	0.60 (Pacific)

geographically wide-ranging species that become dominant in areas where the airmass to which they are positively correlated attains high frequency.

The role played by the dominant species in the black spruce community is illustrated by the data in table 7A.2. It can be seen that as the northern edge of the forest is approached (or passed), the species in the black spruce community listed in table 7A.1 (along with four other species) become of greatly increased importance. An interesting comparison can be made with the rock field species correlated with Pacific air. Here are many of the same species that are found correlated with arctic air in the black spruce community.

Table 7A.2. Average frequencies in stands of each study area of species highly correlated positively with arctic airmass frequencies:* black spruce understory communities.

	Betula glandulosa	*Empetrum nigrum*	*Ledum decumbens*	*Ledum groenlandicum*	*Salix glauca*	*Vaccinium uliginosum*	*Vaccinium vitis-idaea*	*Average*	*Number samples (stands)*
Clinton–Colden	90	100	100	60	30	90	100	81	1
Kazan River	85	65	65	95		65	95	67	1
Artillery Lake	53	71	39	73	8	73	98	59	8
Colville Lake	47	32	40	67	18	68	93	52	3
Ennadai Lake	50	49	24	60	5	75	85	50	18
Inuvik	53	35	15	73	3	90	76	49	2
Ft Reliance	7	41	8	71	19	60	77	41	7
Dubawnt Lake	65	43	25	40	10	40	43	39	2
Florence–Carcajou	40	28	17	32	3	66	37	33	5
Churchill		50		40		40	100	33	1
Yellowknife	2	8	11	83	8	7	96	31	5
Wapata Lake		2		86			96	26	16
Otter Lake	10			81			88	26	5
Black Lake		8		90			71	24	5
Ilford	3	9		80			75	24	9
Lynn Lake	1	1	2	74		1	79	22	16
Ft Providence	1	25	17	77	3	25		21	5
God's Lake				75			40	16	1
West Hawk Lake				88			45	16	2
Rocky Lake	12			41			33	12	9
Waskesiu Lake				47			37	12	3
Remi Lake				85				12	1
Trout Lake				80				11	3
Raven Lake				62			2	9	5
Klotz Lake				40			3	6	4
Clear Lake				37			7	6	3
Bagley-Pike Bay				43				6	2

* The exception is *Ledum groenlandicum* included here to illustrate significance as a widespread and generally abundant species.

Consideration of these relationships will show that *Vaccinium vitis-idaea*, *V. uliginosum* and *Empetrum nigrum* become of increased importance northward in the black spruce community, in positive correlation with arctic air, and then decline northward in the tundra rock field community, in positive correlation with Pacific air (and hence negative correlation with arctic air).

It seems apparent that this group of species is a distinctly dominant component in both the black spruce forest understory and the rock field tundra found in the forest–tundra ecotone. They evidently are particularly adapted to the ecotonal environment and constitute a significant portion of the vegetation of the ecotonal region. *Ledum groenlandicum* is of interest principally because of its continual dominance virtually throughout the black spruce forest. That this particular group of northern forest and southern tundra species lends distinctive characteristics to their respective communities in relation to diversity and genecological dynamics will be shown in the next section.

DIVERSITY, DOMINANCE, GENECOLOGY

The Shannon–Weiner function has been widely employed as a measure of species diversity in biological communities. Other statistical methods afford nearly the same result with simpler computations (Pielou, 1966, 1969; Margalef, 1968). The index employed in the discussion undertaken here, D_L, was devised to measure the number of potential interrelationships between individuals of the same and of different species rather than, as in the case of these other diversity indices, to measure information content (as an index of diversity) in the species aggregation of a community (Larsen, 1971).

The D_L index is based upon the possible number of one-to-one interactions between any given individual of one species and all other individuals of the same species (designated as P) and between any given individual and all individuals of different species (Q). Thus $P + Q$ = the total number of interrelationships existing in the average sample quadrat of a community. The value Q/P is the ratio of inter- to intraspecific relationships. To exaggerate the distribution of the Q/P ratio, $Q + P$ is multiplied by Q/P to give D_L.

Comparison of index values for forest and tundra communities is presented in fig.7A.9, including the index values obtained with the Shannon–Weiner function ($D = \sum p_i \log_2 p_i$), the number of species in the average stand (S), and the number of individual plants in the average quadrat (N). The magnitude of the value of D_L is preponderantly a function of N and of Q/P.

As McNaughton and Wolf (1970) affirm, there are two competitive interfaces in communities, the interface between individuals of the same species and the interface between individuals of different species. These interfaces influence the genetic characteristics of evolving species in different ways. Greater differences between individuals of a species (expressed as genotypic variability and ecotypic variation) will be apparent when competition has been largely intraspecific; greater uniformity in a species develops when competition during evolution has been predominantly interspecific.

Populations capable of survival in an environment characterized by widely fluctuating conditions will as a rule possess a wide range of genetic variation. These species pay a price for their ability to survive the vicissitudes of this kind of environment, however; they possess reduced competitive abilities against species adapted to occupy a narrow niche in a stable community. Species adapted to the highly variable environment of the rock fields, for example, will be poorly equipped to invade tussock muskeg communities occupied by the narrowly adapted species characteristically found there.

Of tundra communities, it is apparent that interspecific competition (as shown by the high D_L value) is most intense in tussock muskeg. It seems probable, also, that the

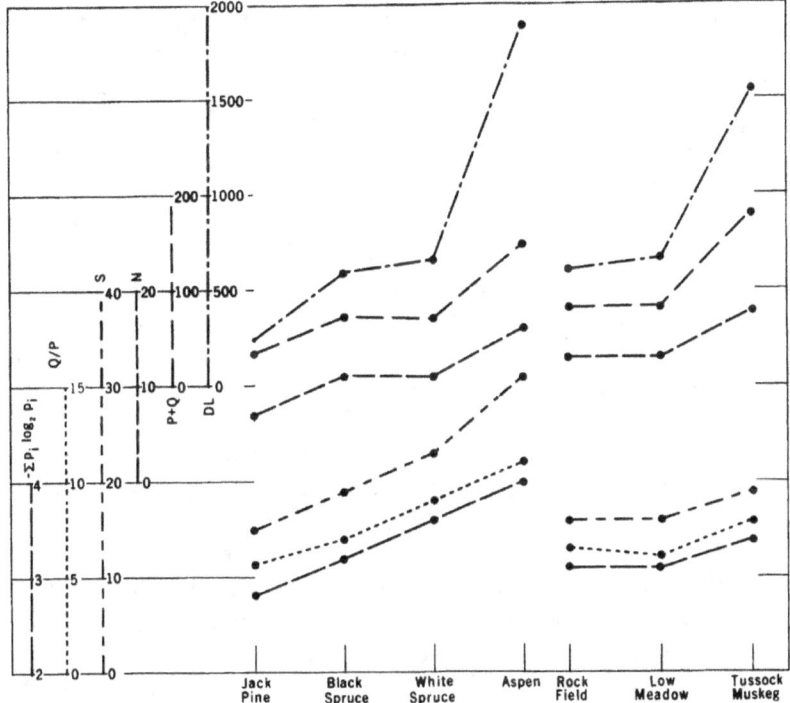

Fig. 7A.9. Average diversity values for the various communities obtained by six methods. For complete discussion see text. The scale farthest on left refers to the diversity index derived from the Shannon-Weiner function; for the values Q/P, $P+Q$, and D_L see text. S refers to the number of species present in the average stand. N refers to the number of individual plants in the average quadrat.

environment of the tussock muskeg is less variable than that of the exposed rock fields, the latter being subject to extreme soil temperatures because of the gravel substrate, and to desiccation because of topographic position and exposure to winds. It thus seems probable the species of the rock fields will be those characterized by greater variability and a wider range of genotypic expression; they will likely be predominantly genotypically variable diploid cross-pollinating species. Mosquin (1966) points out that many Canadian species are conservative and relatively static in evolution; they utilize self-pollinating mechanisms and apomixis to preserve highly adapted physiological systems once these have proved successful. It is thus conceivable that such species will be found to predominate in the tussock muskeg communities. In a study of polyploidy, Johnson and Packer (1967) interpret their data (from Alaskan tundra) as indicating that polyploid species found there most often invaded colder, wetter, highly disturbed lowland habitats of recent (postglacial) origin. It will be interesting to learn the genetic characteristics of the species in the central Canadian regions that now occupy the low meadow communities.

The distinctive character of the tussock muskeg community is of considerable interest. In geographical range, the community is found with rather sharply decreasing frequency over the landscape northward and southward of the forest–tundra ecotone, an observation first made by Tyrrell (1896) in his trips down the Dubawnt and Kazan in 1893 and 1894, and apparent in more recent studies of the northern Canadian forest and tundra (Larsen, 1965, 1967, 1971). It is, in the region studied, a community in which, in comparison to rock field and low meadow communities, there exists a greater stability in prevailing soil moisture and temperature conditions. Although frost action may be a disturbance factor, recolonization of disturbed sites by species of the original tussock muskeg community is usually rapid. By inference, the tussock muskeg possesses a wider range of available niche-space than other communities of either tundra or northern boreal forest. Competitive relationships are proportionately more interspecific in nature. It appears that the tussock muskeg community is more apt to possess species with a narrower range of genetic variation than rock field or low meadow species or species occupying spruce forest understory communities. Incidence of polyploidy should be lower if the interpretation of polyploid habitat preference by Johnson and Packer holds true.

The tussock muskeg of low latitude arctic tundra and the forest–tundra ecotone thus emerges as a distinctive community, interesting and as yet understood only in meagre outline in terms of ecology, population dynamics, and genetic character of component species. The community may constitute a diagnostic feature of significance for locating by palynological techniques the position of the forest–tundra ecotone and hence the arctic frontal climatic zone in past millenia (see Nichols, Chapter 11A).

References

AGASSIZ, L. (1845) Essai sur la géographie des animaux. *Rev. Suisse et Chron. Litt.*, **8**, 441–52, 538–55.

BARRY, R. G. (1967) Seasonal location of the arctic front over North America. *Geogr. Bull.*, **9** (2), 79–95.

BENNINGHOFF, W. S. (1952) Interaction of vegetation and soil frost phenomena. *Arctic*, **5** (1), 34–44.

BILLINGS, W. D. and MOONEY, H. A. (1959) An apparent frost hummock-sorted polygon cycle in the alpine tundra of Wyoming. *Ecology*, **40**, 16–20.

BILLINGS, W. D. and MOONEY, H. A. (1968) The ecology of arctic and alpine plants. *Biol. Rev.*, **43** (4), 481–529.

BLACK, R. F. (1954) Precipitation at Barrow, Alaska, greater than recorded. *Trans. Amer. Geophys. Union*, **35**, 203–6.

BLISS, L. C. (1956) A comparison of plant development in microenvironments of arctic and alpine tundras. *Ecol. Monogr.*, **26**, 303–37.

BLISS, L. C. (1962) Adaptations of arctic and alpine plants to environmental conditions. *Arctic*, **15** (2), 117–44.

BLISS, L. C. (1971) Arctic and alpine plant life cycles. *Ann. Rev. Ecol. System.*, **2**, 405–38.

BRITTON, M. E. (1966) Vegetation of the arctic tundra. *In:* HANSEN, H. P. (ed.) *Arctic Biology*, 67–130. Corvallis: Oregon State Univ.

BRITTON, M. E. (1970) Vegetation of the arctic tundra. *In:* HANSEN, H. P. (ed.) *Arctic Biology* (2nd edn), 26–61. Corvallis: Oregon State Univ. Press.

BROWN, R. J. E. (1960) The distribution of permafrost and its relation to air temperature in Canada and the USSR. *Arctic*, **13**, 163–77.

BROWN, R. J. E. (1965) Some observations on the influence of climate and terrain on permafrost at Norman Wells, NWT. *Can. J. Earth Sci.*, **2**, 15–31.

BROWN, R. J. E. (1969) Factors influencing discontinuous permafrost in Canada. *In:* PÉWÉ, T. L. (ed.) *The Periglacial Environment*, 11–53. Montreal: McGill–Queen's Univ. Press.

BROWN, R. J. E. (1970a) Permafrost as an ecological factor in the subarctic. *In: Ecology of the Subarctic Regions*, 129–40. Paris: UNESCO.

BROWN, R. J. E. (1970b) *Permafrost in Canada*. Toronto: Univ. of Toronto Press. 234 pp.

BRYSON, R. A. (1966) Airmasses, streamlines, and the boreal forest. *Geogr. Bull.*, **8** (3), 228–69.

COOPER, J. G. (1859) On the distribution of the forests and trees of North America, with notes on its physical geography. *Ann. Rep. Bd. Reg. Smithsonian Inst.*, 246–80.

DAHL, E. (1951) On the relation between summer temperature and the distribution of alpine vascular plants in the lowlands of Fennoscandia. *Oikos*, **3** (1), 22–52.

DOWNES, J. A. (1964) Arctic insects and their environment. *Can. Entomol.*, **96** (1–2), 279–307.

FRASER, E. (1956) The lichen woodlands of the Knob Lake area of Quebec–Labrador. *McGill Sub-Arctic Res. Pap.*, 1. Montreal: McGill Univ. 28 pp.

HARE, F. K. (1950) Climate and zonal divisions of the boreal forest formations in Eastern Canada. *Geogr. Rev.*, **40**, 615–35.

HARE, F. K. (1954) The boreal conifer zone. *Geogr. Stud.*, **1**, 4–18.

HARE, F. K. (1968) The Arctic. *Quart. J. R. Met. Soc.*, **94** (402), 439–59.

HARE, F. K. and RITCHIE, J. C. (1972) The boreal bioclimates. *Geogr. Rev.*, **62**, 333–65.

HOPKINS, D. M. (1959) Some characteristics of the climate in forest and tundra regions in Alaska. *Arctic*, **12**, 215–20.

HOPKINS, D. M. and SIGAFOOS, R. S. (1951) Frost action and vegetation patterns on Seward Peninsula, Alaska. *US Geol. Surv. Bull.*, 974–C.

HUSTICH, I. (1962) A comparison of the floras on subarctic mountains in Labrador and in Finnish Lapland. *Acta Geogr.*, **17** (2), 1–24.

HUSTICH, I. (1966) On the forest–tundra and the northern tree-lines. *Ann. Univ. Turku.*, A, II, 36 (*Rep. Kevo Subarctic Sta. 3*), 7–47.

JEFFREE, E. P. (1960) A climatic pattern between latitudes 40° and 70°N and its probable influence on biological distributions. *Proc. Linn. Soc. London*, **171**, 89–121.

JOHNSON, A. W. and PACKER, J. G. (1967) Distribution, ecology, and cytology of the Ogotoruk Creek flora and the history of Beringia. In: HOPKINS, D. M. (ed.) *The Bering Land Bridge*, 245–65. Stanford: Stanford Univ. Press.

KENDEIGH, S. C. (1954) History and evaluation of various concepts of plant and animal communities in North America. *Ecology*, **35**, 152–71.

KLIKOFF, B. G. (1966) Temperature dependence on the oxidative rates of mitochondria in *Danthonia intermedia, Pentstemon davidsonii*, and *Sitanion hystrix*. *Nature*, **212** (5061), 529–30.

KÖPPEN, W. (1936) Das geographische System der Klimate. *In:* KÖPPEN, W. and GEIGER, R. (eds.) *Handbuch der Klimatologie*, Vol. I, C. Berlin: Bornträger. 44 pp.

KREBS, J. S. and BARRY, R. G. (1970) The arctic front and the tundra–taiga boundary in Eurasia. *Geogr. Rev.*, **60** (4), 548–54.

LARSEN, J. A. (1965) The vegetation of the Ennadai Lake Area, NWT: studies in arctic and subarctic bioclimatology. *Ecol. Monogr.*, **35** (1), 37–59.

LARSEN, J. A. (1967) Ecotonal Plant Communities of the Forest Border, Keewatin, NWT, Central Canada. *Univ. of Wisconsin, Dept of Met. Tech. Rep.*, 32. ONR Contract 1202 (07).

LARSEN, J. A. (1971a) Vegetation of the Fort Reliance and Artillery Lake areas, NWT. *Can. Field Nat.*, **85** (2), 147–67.

LARSEN, J. A. (1971b) Vegetation and airmasses: boreal forest and tundra. *Arctic*, **24** (3), 177–94.

LARSEN, J. A. (1972a) Growth of spruce at Dubawnt Lake, Northwest Territories. *Arctic*, **25** (1), 59.

LARSEN, J. A. (1972b) Observations of well-developed podzols on tundra and of patterned ground within forested boreal regions. *Arctic*, **25** (2), 153–4.

LARSEN, J. A. (1972c) The vegetation of northern Keewatin. *Can. Field. Nat.*, **86** (1), 45–72.

LARSEN, J. A. (1973) Plant communities north of the forest border, Keewatin, Northwest Territories. *Can. Field Nat.*, **87**, 241–8.

LEOPOLD, E. B. (1958) Some aspects of Late-Glacial climate in eastern United States. *Zurich Geobot. Inst. Veröff.*, N34, 80–5.

MCNAUGHTON, S. J. and WOLF, L. L. (1970) Dominance and niche in ecological systems. *Science*, **167** (3915), 131–9.

MAINI, J. S. (1966) Phytoecological study of sylvotundra at Small Tree Lake, NWT. *Arctic*, **19** (3), 220–44.

MARGALEF, R. (1968) *Perspectives in Ecological Theory*. Chicago: Univ. of Chicago Press. 111 pp.

MAYCOCK, P. F. and MATTHEWS, B. (1966) An arctic forest in the tundra of northern Quebec. *Arctic*, **19** (2), 114–44.

MITCHELL, V. L. (1967) An investigation of certain aspects of tree growth rates in relation to climate in the central Canadian boreal forest. *Univ. of Wisconsin, Dept of Met. Tech. Rep.*, 33. ONR Contract 1202(07).

MOSQUIN, T. (1966) Reproductive specialization as a factor in the evolution of the Canadian flora. *In:* TAYLOR, R. L. and LUDWIG, R. A. (eds.) *The Evolution of Canada's Flora*, 43–65. Toronto: Univ. of Toronto Press.

NORDENSKJÖLD, O. and MECKING, L. (1928) The geography of the polar regions. *Amer. Geogr. Soc. Spec. Pub.*, 8. 359 pp.

PÉWÉ, T. L. (1969) *The Periglacial Environment*. Montreal: McGill–Queen's Univ. Press. 487 pp.

PIELOU, E. C. (1966) Species-diversity and pattern-diversity in the study of ecological succession. *J. Theoret. Biol.*, **10**, 370–83.

PIELOU, E. C. (1969) *An Introduction to Mathematical Ecology*. New York: John Wiley. 286 pp.

POWELL, J. M. (1961) The vegetation and micro-climate of the Lake Hazen area, northern Ellesmere Island, NWT. *Operation Hazen Rep.*, 14. Ottawa: Def. Res. Board. 112 pp.

RAUP, H. M. (1941) Botanical problems in boreal America. *Bot. Rev.*, **7**, 147–248.

REED, R. J. (1959) Arctic weather analysis and forecasting. *Univ. of Wisconsin, Dept of Met. Sci. Rep.*, 2, Occasional Rep. 11, AF Contract 19 (604)–3063. 119 pp.

REED, R. J. (1960) Principal frontal zones of the Northern Hemisphere in winter and summer. *Bull. Amer. Met. Soc.*, **41** (11), 591–8.

RITCHIE, J. C. (1960a) The vegetation of northern Manitoba, IV. The Caribou Lake Region. *Can. J. Bot.*, **38** (2), 185–97.

RITCHIE, J. C. (1960b) The vegetation of northern Manitoba, V. Establishing the major zonation. *Arctic*, **13**, 211–29.

RITCHIE, J. C. (1960c) The vegetation of northern Manitoba, VII. The lower Hayes River region. *Can. J. Bot.*, **38**, 769–88.

RITCHIE, J. C. (1962) A geobotanical survey of northern Manitoba. *Arct. Inst. N. Amer., Tech. Pap.* 9, 1–48.

ROWE. J. S. (1972) *Forest regions of Canada*. Can. For. Serv. Pub., No. 1300. Ottawa. 172 pp.

SANDERSON, M. (1948) The climates of Canada according to the new Thornwaite classification. *Sci. Agric.*, **28**, 501–17.

SAVILE, D. B. O. (1961) The botany of the northwestern Queen Elizabeth Islands. *Can. J. Bot.*, **30**, 909–42.

STUPART, R. F. (1928) The influence of Arctic meteorology on the climate of Canada especially. *In:* Problems of Polar Research, *Amer. Geogr. Soc. Spec. Pub.* 7, 38–50.

TIKHOMIROV, B. A. (1970) Forest limits as the most important biogeographical boundary in the north. *In: Ecology of the Subarctic Regions*, 35–40. Paris: UNESCO.

TYRRELL, J. B. (1896, 1897–8) Report on the Doobaunt, Kazan and Ferguson Rivers and the Northwest Coast of Hudson Bay. *Geol. Surv. Can. Ann. Rep.*, F, 1–218. (N.s. vol IX for 1896 printed 1897.)

WARREN WILSON, J. (1957) Observations on the temperatures of arctic plants and their environment. *J. Ecol.*, **45**, 499–531.

WENT, F. W. (1950) The response of plants to climate. *Science*, **112**, 489–94.

WHITTAKER, R. H. (1970) *Communities and Ecosystems*. New York: Macmillan. 162 pp.

WOLFE, J. A. and LEOPOLD, E. B. (1967) Neogene and early Quaternary vegetation of northwestern North America and northeastern Asia. *In:* HOPKINS, D. M. (ed.) *The Bering Land Bridge*, 193–206. Stanford: Stanford Univ. Press.

Manuscript received April 1971.

Alpine timberlines

Peter Wardle

Botany Division, Department of Scientific and Industrial Research,
Christchurch, New Zealand

1. Introduction

One need not be a botanist for the word timberline to evoke images of coniferous forests opening out to alpine meadows dotted with brightly coloured flowers, of dense, dark New Zealand beech forests abruptly giving way to tussock grassland, or of the more gradual transition at the upper limit of snow gums in Australia. Nor does one need to be a scientist to appreciate that timberline represents the altitude at which the climate becomes too cold to support growth of trees. Nevertheless, a closer look shows that the phenomenon of timberline is quite complex, in both manifestation and cause.

In this section, I shall briefly survey the timberlines of the world, examine possible physiological causes, and discuss temporal changes, particularly those caused directly or indirectly by man. Most timberlines are in the north temperate zone, and set the upper limits to forests of boreal affinities. Since these will be familiar to most readers, I have attempted to put the subject in a different light, by discussing the relatively fragmentary but very diverse timberlines of the tropics and south temperate zone in detail which may seem disproportionate to their extent. Even the autecology of Rocky Mountain species is considered largely with reference to experiments in the Craigieburn Range, at 43 °S in New Zealand.

To aid discussion, some working definitions are necessary. For present purposes, I shall use *alpine timberline* in a broad sense for the upper limit of trees and forest on mountains. *Tree limit* is regarded as the extreme upper limit of trees and tall shrubs. Trees growing closely together constitute a *forest*. Often, as in the New Zealand *Nothofagus* forests, the tree limit and *forest limit* coincide; or there can be parkland between these limits; or the trees in the timberline region can be stunted and deformed by climatic severity, so that a zone of *krummholz* lies above the forest limit, in which case tree limit can be taken as the level at which krummholz with tall flagged stems is replaced by low krummholz of tree species reduced to prostrate form. As a fourth possibility, woody vegetation above the forest limit can consist of plants inherently of shrub form, or *scrub*. I have found it useful to define tree limit as the upper limit of trees, krummholz, and shrubs more than about 2 m tall, and it is this component of timberline which seems to be most closely related to the altitudinal decrease or *lapse* of temperature.

Timberline is the ecotone between two entirely different habitats and types of community. Many authors from the nineteenth century onwards, including myself, have referred to the vegetation below tree limit, including both forest and ecotonal krummholz, as *subalpine*, and the low-growing vegetation above as *alpine*. However Löve (1970) favours a terminology wherein 'subalpine' is restricted to the ecotone with krummholz. The forest below then becomes included in the montane belt. This has the disadvantage that 'subalpine' would no longer indicate a regular altitudinal belt, but instead would refer only to krummholz vegetation, the altitudinal extent of which varies from zero to several hundred metres.

In many places, local circumstances prevent the potential tree limit from being attained. At most timberlines, trees reach their maximum limits only on convex topography. On concave topography and especially on flat valley floors, vegetation markedly alpine in character may descend several hundred metres below tree limit. By analogy with the nocturnal temperature inversions which occur on such sites, the boundaries between this herbaceous vegetation and the forest on the valley slopes can be referred to as *inverted timberlines*.

Variation in snow cover and exposure to wind also affect the course of timberline, as do catastrophic events such as avalanche and volcanic eruption. Finally, man has depressed timberline over wide areas, both directly through burning and cutting forests, and indirectly through his grazing animals.

2. Floristics of timberline

By far the greatest extent of mountainous country high enough to support alpine timberlines is in the north temperate zone. These northern timberlines are all floristically related, because they have been linked by continuous routes for plant migration, at least during the cold phases of the Quaternary era. The Pinaceae are the predominant trees, and *Pinus*, *Picea* and *Abies* the most widespread genera, forming timberline over thousands of kilometres (see Plate 1 and also fig. 7A.1). *Tsuga* is represented only by *T. mertensiana* (Bong.) Carr. in the western mountains of North America. *Larix*, the only deciduous genus of conifer at timberline, is represented by species in various parts of Eurasia and western North America. There are also two genera of the Cupressaceae; *Juniperus* which is very widespread although represented at timberline mainly by shrubby species, and *Chamaecyparis* which is represented only by *C. nootkatensis* Spach of western North America.

Dicotyledonous trees are also widespread at timberline in the north temperate zone, and are nearly all deciduous. *Betula pubescens* forms arctic and alpine timberlines over wide stretches of Scandinavia, *Alnus viridis* Chaix and related species occupy moist slopes subject to avalanching in the Alps and British Columbia, and *Fagus sylvatica* L. occurs at timberline in central and southern Europe and the Caucasus. *Acer*, *Sorbus*, *Populus* and *Quercus* can occur at the upper limit of tree growth, although it is sometimes uncertain whether timberlines with these genera are 'climatic' or lowered through human activity. According to W. R. Sykes (personal communication, 1971), however, the evergreen oak *Quercus semecarpifolia* Sm. forms timberlines at 4000 m in central Nepal which give every appearance of being natural. Shrubby rhododendrons, such as the European *Rhododendron ferrugineum* L., occur at some timberlines, but in the Himalayas there are also tree-forming species. In the southeastern part of the Himalayan system at approximately latitude 27°N, where there is heavy and prolonged monsoonal precipitation, these large rhododendrons dominate to form low, dense, evergreen forest (Schweinfurth, 1957; Sykes, personal communication, 1971).

Timberlines in the tropics and south temperate zone are geographically isolated and floristically diverse. Although the Eurasian mountainous belt continues to southeast Asia and beyond, it includes no land high enough to support a natural timberline south of about latitude 25°N in western China. In Formosa boreal forest of *Picea* and *Abies* is replaced above 3000 m by dense scrub of *Juniperus*, *Rhododendron* and *Berberis* which ascends to the tops of the mountains (Li, 1963). Kinabalu (4175 m) is an isolated mountain at latitude 6°N in Borneo which provides floristic links between the high altitude vegetation of Eurasia and Australasia (Van Steenis, 1964), but the bareness of its summit results from lack of soil rather than an alpine climate (see photos in Meijer, 1965).

In the Americas, boreal vegetation extends further towards the equator, and timberline on the Mexican volcanoes at latitude 19°N consists of *Pinus hartwegii* Lindl., which forms pure stands above a belt of forest of mixed conifers and hardwoods. *Juniperus monticola* Martinez forms patches of scrub above the timberline. *Alnus jorullensis* H.B.K. extends from Mexico to the province of Tucumén in northern Argentina, where it forms timberlines in the absence of *Polylepis*, which extends nearly 1000 m higher in the same province (Hueck, 1966). *Polylepis* is a rosaceous genus which is characteristic of very high altitudes in tropical Andean South America, mainly in shrubby form, but in places trees and groves occur on steep bouldery slopes to as high as 4200 m, and perhaps as high as 4900 m. More widespread at these altitudes, however, is the *páramo* vegetation which includes peculiar arborescent herbs of the genus *Espeletia* (Compositae), with imbricate, large, densely tomentose leaves which are persistent on the thick, unbranched stems (Troll, 1959). Plants with this growth form occur on high mountains in several parts of the tropics, and have been called *megaphytes* (Cotton, 1944). The páramos and *Polylepis* groves supersede, via a belt of shrubs, a 'timberline' formed by cloud forest at 3000–3250 m (Hueck, 1966; Walter and Medina, 1969). On the Andes south of latitude 36°S, *Nothofagus* forest prevails under a temperate climate and generous rainfall. At the lower altitudes, there are both deciduous and evergreen species, but the upper forests consist of pure stands of the deciduous *N. pumilio* Krasser and, locally, *N. antarctica* (Forst.f.) Oerst. which form both the alpine timberline and, with *N. betuloides* (Mirb.) Blume, the southern limit of tree growth at latitude 56°S (Hooker, 1847)[1].

Two evergreen species of *Nothofagus* form most of the timberlines in New Zealand. On the drier mountains, *N. solandri* var. *cliffortioides* (Hook.f.) Poole grows as pure stands, there being few accompanying vascular plants other than the mistletoe *Peraxilla*

---

[1] *Araucaria araucana* is important at some Patagonian timberlines (D. R. McQueen, personal communication, 1973).

(Plate 21). On the wetter mountains, both *N. solandri* and *N. menziesii* (Hook.f.) Oerst. occur, together with a much richer accompanying flora. However, *Nothofagus* is widely absent in areas deforested during Pleistocene glaciation or volcanism, and instead there are floristically complex timberlines of small trees and large shrubs, belonging to genera such as *Podocarpus* and *Dacrydium* (Podocarpaceae), *Hoheria* (Malvaceae), *Senecio* and *Olearia* (Compositae), *Dracophyllum* and *Archeria* (Epacridaceae), and *Hebe* (Scrophulariaceae). They form very dense, interlocking vegetation known locally as *subalpine scrub*, which often extends for 100–300 m above the upper limit of tall forest.

Timberline on the mountains of southeastern Australia is formed solely of *Eucalyptus niphophila* Maiden & Blakeley. Tasmanian timberlines, with at least one species of *Eucalyptus* (*E. coccifera* Hook.f.) and a deciduous *Nothofagus* (*N. gunnii* Hook.f. which is similar to *N. antarctica*), provide a fascinating link between Australia, New Zealand and Patagonia. There are also species belonging to genera endemic in Tasmania, including *Athrotaxis*, the only representative of the Taxodiaceae in the southern hemisphere.

In New Guinea even the uppermost forest is floristically rich; at the upper limits of tree growth at 4000 m on Mt Wilhelm, I listed thirteen species of trees and tall shrubs drawn from the genera *Podocarpus*, *Quintinia* (Escalloniaceae), *Rhododendron*, *Vaccinium*, *Rapanea* (Myrsinaceae) and *Olearia* (Compositae), among others. The subalpine region of New Guinea is characterized also by wide valleys and basins, supporting herbaceous swamps and tussock grassland. The latter are dotted with *Cyathea* tree ferns, with trunks which are unusually thick because of an outer fibrous layer consisting largely of adventitious roots. The stem apex and developing fronds are hidden among densely packed, long scales.

On the high mountains of equatorial Africa, ericaceous woodland and shrubland of *Erica arborea* L. and species of *Philippia* peter out at 3500–4000 m, to be replaced by tussock grassland with megaphytes such as the arborescent *Senecio keniodendron* R.E. Fr. and Th. Fr. Jr. On Mt Kenya, this species can form virtual forests between 4000 and 4300 m, and specimens occur to nearly 4700 m (Coe, 1967).

Although the Drakensburg Mountains in South Africa are high enough to support a climatic timberline, there is little forest, tree limits are low for the subtropical latitude, and the highest trees belong to genera such as *Protea*, which one would not expect at a timberline determined by low temperatures. Grassland and sclerophyllous scrub prevail, reflecting both the dry climate and a long history of burning. On the Atlas Mountains in North Africa, the forests are floristic extensions of those of southern Europe, but they have been fragmented by millenia of human interference in a dry Mediterranean climate. The forest limits, which are formed by *Quercus ilex* L., *Cedrus atlantica* Manetti, *Juniperus oxycedrus* L. and *J. thurifera* L. (Rikli, 1946), are evidently much depressed.

Several oceanic islands are high enough to support tree limits (table 7B.1), but as comparison with table 7B.2 will show, these limits are low for the latitudes. One reason for this lies in the extreme oceanic climates (see p. 379), although according to Schimper (1903) aridity sets the upper limit of pine in the Canary Islands. Another consideration is that hardier species, capable of growing at higher altitudes, have failed to reach these remote islands, most of which are recent volcanoes.

Table 7B.1. Tree limits on oceanic islands.

Island or mountain	Latitude	Tree limit	Altitude (m)	Vegetation above tree limit	Source
Pico, Azores	38°N	Woodland of *Erica azorica* Hochst. and *Juniperus brevifolia* Ant.	1500	*Calluna* heath	Tutin, 1953
Pico de Teide, Canary Islands	28°N	*Pinus canariensis* C. Sm. forest	2000	*Cytisus* and other xeromorphic legumes	Schimper, 1903, p. 771, and others
Tristan da Cunha	37°S	*Phylica arborea* Thouars (Rhamnaceae) woodland	450		Wace and Holdgate, 1958
		Blechnum palmiforme (Thouars) C.Chr., a 1m tall 'tree' fern	800	*Empetrum* heath	
Mauna Kea, Hawaii	22°N	*Sophora chrysophylla* Seem. (Papilionaceae) woodland	3000	Browse-degraded heath (*Styphelia*)	My notes
Masafuera, Juan Fernandez	34°S	*Dicksonia* tree ferns, with a few other small trees	1200	Grassland and herb-field	Skottsberg, 1953, p. 928
Campbell Island, south of New Zealand	53°S	Tall scrub of *Dracophyllum* (Epacridaceae)	200	Tussock grassland	Zotov, 1965, p. 104

Table 7B.2. Altitudes of some timberlines.

Locality	Latitude	Main species	Altitude (m)*	Source
Near Haines, Alaska	60°N	*Picea sitchensis* Carr.	900	Daubenmire, 1953
Garibaldi Park, British Columbia	50°N	*Abies lasiocarpa* Nutt.	1850–1900 (krummholz to 2250)	Wardle, 1965a
Rocky Mountains of Alberta	50°N	*Picea engelmannii* (Parry) Engelm. and *Abies lasiocarpa*	2150–2300	Wardle, 1965a
Front Range, Colorado	39°N	*Picea engelmannii*	3350–3600 (krummholz to 3750)	Wardle, 1968
Sierra Nevada, California	38°N	*Pinus albicaulis* Engelm.	3300 (krummholz to 3650)	Clausen, 1965
Iztaccihuatl, Mexico	19°N	*Pinus hartwegii* Lindl.	Mean 3950 Maximum 4100	Beaman, 1962; Wardle, 1965a
Andes of Venezuela	9°N	*Polylepis sericea* Wedd.	To over 4200	Walter and Medina, 1969
Northern Chile	19°S	*Polylepis tomentella* Wedd.	Up to 4900	Troll, 1959
Andina de Neuquén, Argentina	40°S	*Nothofagus pumilio* (Poepp. & Endl.) Krasser	1650	Dimitri, 1964
Hermite Island, Chile	56°S	*Nothofagus pumilio*	450	Hueck, 1966
Mt Hakkôda, Japan	41°N	*Abies mariesii* Mast. (*Pinus pumila* Regel scrub above timberline)	1400–1500	Yoshioka and Kaneko, 1963
S. E. Sinkiang, China	38°N(?)	*Picea likiangensis* (Franchet) Pritzel and *Abies georgei* Orr.	4500	Wang, 1961

Table 7B.2.—cont.

Locality	Latitude	Main species	Altitude (m)*	Source
Härjedalen, Sweden	63°N	Betula pubescens Ehrh.	1000	Kilander, 1965
Cairngorm Mts, Scotland	57°N	Pinus silvestris L.	600–700 (potential) 500 (actual)	Pears, 1968a
Glarus, Switzerland	47°N	Pinus cembra L. and Picea abies (L.) Karst	Mean 1900 Maximum 2000	Schimper, 1903
Poschiavo, Switzerland	46°N	Larix decidua Mill, Pinus cembra and P. abies	Mean 2250 Maximum 2300	Schimper, 1903
USSR Caucasus	41–44°N	Betula verrucosa Ehrh.	Up to 2500	Radde, 1899
Western Himalaya	29–35°N	Betula utilis Don, etc.	3500–3800 Maximum 4200	Schweinfurth, 1957
Eastern Himalaya, inner valleys	28°N	Larix griffithii Hook.f.	3800	Schweinfurth, 1957
Eastern Himalaya, outer valleys	27°N	Rhododendron spp.	3600–3800	Schweinfurth, 1957
Mt Wilhelm, New Guinea	6°S	Podocarpus compactus Wassch., etc.	3900–4100	My notes
Mt Albert Edward, New Guinea	8°S	Podocarpus compactus Wassch., etc.	3800–3850	E. Löffler, personal communication, 1970
Snowy Mountains, New South Wales	36°S	Eucalyptus niphophila Maiden & Blakely	1850–2000	My notes; A. B. Costin, personal communication, 1971
Mt Field, Tasmania	43°S	Eucalyptus coccifera Hook.f., Athrotaxis cupressoides Don, etc.	1200–1300	My notes
West Coast of South Island, New Zealand	42°S	Nothofagus menziesii (Hook.f.) Oerst.	1200–1300	My notes
Inland South Island, New Zealand	42°S	Nothofagus solandri var. cliffortioides (Hook.f.) Poole	1500–1550	My notes

* As far as possible, values represent the highest tree limits to the nearest 50 m.

The foregoing sketch indicates the great taxonomic diversity among timberline trees. Many families of dicotyledons and five families of conifers are represented, and two genera of tree ferns. On the other hand, arborescent monocotyledons are notably absent, although the wax palm *Ceroxylon utile* is said to attain 4100 m in the northern Andes (Corner, 1966). Table 7B.2 shows some representative altitudes for timberlines. Altitudes exceeding 4200 m probably need verifying with more recent survey data; this applies especially to scattered reports of timberlines exceeding 4500 m in central Asia.

The dominant trees are, of course, only one component of timberline vegetation, but it is impossible to do justice to the subordinate vascular vegetation within the confines of this chapter, let alone the cryptogams and fauna. Where timberline is abrupt, there is a correspondingly abrupt change from a forest florula to an alpine or tundra florula, and where there is a broad ecotone, there is a mixed situation, with species of the forest understory ascending within the increasingly isolated clumps of trees and krummholz, and alpine plants occupying the more exposed ground between. Alpine species can descend into the subalpine belt in habitats with reduced competition and cooler-than-normal microclimates, such as avalanche tracks, moist, shaded gullies and floors of cirques (e.g. Jeník, 1959), eroding riverbanks, bogs, and hollows where cold air accumulates and gives rise to inverted timberlines. There are fewer sites where environmental 'compensating factors' allow subalpine species to ascend into the alpine belt; perhaps most typical are sheltered crevices among coarse, blocky talus and on steep cliffs facing the midday sun. Few species are confined to the proximity of timberline (Plate 22).

There are broad differences between the subordinate vascular plants of north temperate timberlines and those of the tropics and the south temperate zone. In the former, shrubs are mostly deciduous (e.g. *Salix, Vaccinium myrtillus* L.) and most of the evergreen ones are cupressoid or ericoid (e.g. *Juniperus communis, Empetrum*). Forbs and grasses are nearly all summergreen. The deciduous timberline beeches of the southern Andes are associated with herbs and shrubs which mostly recall the florulas of north temperate timberlines in both physiognomy and taxonomic relationships, distinctively South American or Fuegian plants being unimportant (Oberdorfer, 1960, pp. 143–6). This contrasts with the 'magellanic moorland', characterized by tussocks and cushion plants belonging to subantarctic genera such as *Marsippospermum* (Juncaceae) and *Donatia* (Donatiaceae), which prevails on the hills along the fiord coast of Chile south of latitude 48°S. Here the windy oceanic climate and inhospitable substratum of granite and gneiss restrict *Nothofagus* forest to sheltered sites, mainly at low altitudes, with better soil (Godley, 1960). The situation has an obvious parallel along the western coasts of the British Isles and Norway, where moorland also dominates the landscape.

At other tropical and southern hemisphere timberlines, subordinate vegetation is as floristically diverse as the dominant trees, but physiognomically there is considerable uniformity. Large tussock grasses are characteristic, and these do not die back annually, although the green leaves are usually obscured by dead leaves and culms. Shrubs and forbs are also typically evergreen, although summergreen species are by no means absent. The prevailing evergreenness reflects a winter less severe than those experienced on the continents of the northern hemisphere, while the exceptionally mild climate of the New Guinea mountains permits epiphytic orchids to occur at timberline.

3. A preferred explanation of alpine tree limits

On a global scale, altitude of timberline depends not only on latitude, but also on continentality so that, as table 7B.2 shows, timberlines in the temperate zone interiors of North America and Asia are nearly as high as those of the tropics. This is due in part to the 'Massenerhebung effect', whereby there is a decreased temperature lapse rate and therefore higher climatic and biotic zones on larger and higher landmasses. More important is the increased annual temperature range in continental regions, for altitude of timberline is correlated with summer warmth, and scarcely influenced by winter cold (table 7B.3). Some authors have claimed that climatic tree limit in the northern hemisphere corresponds to a mean isotherm of 10°C for the warmest month of the year, and Zotov (1938) has extended this correlation to New Zealand. Even if the correspondence is less than complete, it suggests 'that a major autecological principle is involved that may be analogous to the wilting coefficient of the soil, in which some environmental complex abruptly exceeds the tolerance of all trees regardless of variation among them' (Daubenmire, 1954, p. 132). Recently (Wardle, 1971) I have attempted to identify this principle in terms of a synthesis of two early but somewhat neglected hypotheses.

The first step recapitulates ideas of Michaelis (1934a, b). Success of woody plants at high altitudes is seen to depend on the ripening of shoots, so that they can withstand unfavourable periods (which in temperate regions correspond to winter). The ripening process can be defined morphologically as the completion of growth, and the loss of the 'soft' appearance imparted by high water content, incompletely lignified cell walls, and thin cuticles. Physiologically, it is the acquisition of an ability to withstand low temperatures and desiccation, which is associated with increased osmotic concentration of cell sap, decrease of free water in the protoplasm, permeable protoplasts tolerant of considerable dehydration, and tissues undamaged when ice forms between the cells (Levitt, 1956).

An apparent anomaly, that the highest timberlines at a given latitude occur in continental regions which experience the most severe winters, can be resolved if it is realized that two consecutive processes are involved in ripening: a) completion of growth, including that of individual cells and their walls, this being dependent on summer conditions; and b) development of hardiness. Different species (or races) each develop a characteristic degree of cold tolerance, and the slight frosts of early winter suffice to induce the full development of this specific hardiness, provided that process a) has been completed (e.g. Larcher and Mair, 1968; Sakai, 1970a).

Damage through the direct effects of low temperature, however, is less important than *winter desiccation*. Whereas low temperature damage usually shows as blackening or shrivelling of actively growing shoots and leaves after a cold snap in early summer, winter desiccation shows as gradual drying and browning of foliage during prolonged spells of cold winter weather, and is mainly seen on shoots exposed to bright sunshine or strong wind. Although at the cellular level freezing and drought both involve water stress,

Table 7B.3. Monthly and annual temperate means, to illustrate 'Massenerhebung' and continentality (from US Weather Bureau publications and Garnier, 1958).

Locality	Latitude	Altitude (m)	Alt.dist.* (m)	Temperature (°C) Warmest month	Coldest month	Annual
Steffanville, Missouri	39°58'N	175		25.6	−2.0	12.0
Fort Collins, Colorado	40°35'N	1525		21.6	−3.3	8.8
Leadville, Colorado	39°15'N	3110	−460	13.8	−8.0	2.3
Rainier Paradise, Washington	46°47'N	1690	−300	11.8	−3.2	3.6
Mt Ruapehu, New Zealand	39°12'S	1120	−370	11.7	+2.2	7.0
Craigieburn Range, New Zealand†	43°08'S	1400	+30	11.0	−1.0	6.0

* *Alt.dist.* means the approximate altitudinal distance of the station from timberline ('−' indicates below timberline).
† The data for the Craigieburn Range (Morris, 1965) are not strictly comparable.

freezing by itself merely withdraws water from the protoplasts without removing it from the tissues, whereas in winter desiccation water is evaporated from the leaves during extended periods when it cannot be replaced because the conducting channels in the roots and stem bases are frozen (Sakai, 1970b). While many species of tree can become hardened to withstand very low temperatures (Sakai, 1970a), relatively few have been able to evolve the tremendous degree of winter hardiness necessary to survive winter desiccation at continental timberlines. Thus, whereas there are some thirteen species at tree limit on Mt Wilhelm (New Guinea) alone, there are only four or five species in the whole of the Colorado Rockies.

The second step in my explanation depends on the relationship between the temperature of foliage and its distance above the ground, following an idea suggested in less specific form by Daubenmire (1954, p. 131). In general, foliage temperature exceeds that of the air during daytime, often by many degrees, and tends to be slightly cooler than the air at night (e.g. Gates, 1965). Canopy foliage of trees, however, lies closer to air temperature than foliage near the ground, and therefore meets limiting conditions at lower altitudes.

Thermistor measurements in canopy shoots of a *Nothofagus solandri* tree at timberline in the Craigieburn Range, New Zealand, and in shrubs and seedlings beside it show that shaded shoots remain close to air temperature, while insolated shoots 2 m above the ground are warmer, usually by 0.5–2°C, although during exceptional moments of bright sunshine and absolute calm they become as much as 6°C warmer. In contrast, insolated shoots of the sprawling shrub *Podocarpus nivalis* Hook. within 3 cm of the ground usually exceed air temperature by 5–11°C, occasionally by more. It is therefore understandable that low woody plants ascend hundreds of metres above tree limit, especially on sunny aspects. Nevertheless, plants growing under these conditions must also be able to withstand other, less favourable aspects of the microenvironment, especially the risks of overheating, and heaving out by frost. Gates and Janke (1966), among others, have drawn attention to the high temperatures experienced by alpine plants. Soil surfaces unprotected by vegetation are especially likely to rise to high temperatures which spell tremendous heat and drought stress to seedlings and small plants. Aulitzky (1961) has measured surface temperatures up to 84°C in the subalpine zone at Obergurgl, Austria. Destruction of seedlings by frost heaving results from the wide amplitude of temperature at the soil surface, especially during the spring when the soil is at its wettest. Winter snow, on the other hand, ameliorates the environment close to the ground, provided it does not lie so late as to drastically curtail the growing season.

Tree seedlings will become established above tree limit to form krummholz, therefore, only in so far as they are able to withstand or avoid these adverse conditions. Section 5 in this chapter will show that the seedlings of some tree species such as *Nothofagus solandri* cannot do this, and towards timberline they become established only in shaded places, where the temperatures are close to ambient air temperatures. With such species, timberline usually consists of erect trees without a krummholz zone, for the environmental tolerances of mature trees are greater than those of small seedlings (Larcher, 1969), and temperatures at canopy height are not much cooler than those at seedling height if the seedlings are shaded. In other words, if the microclimate permits shade-demanding seedlings to become established in the first place, it will usually also permit them to develop into mature trees. Krummholz occurs below tree limit where factors such as strong, persistent winds steepen the temperature gradient above the ground.

4. Other physiological explanations of tree limit

In the preceding section, I have emphasized an explanation of tree limit which hinges on the ripening of shoot tissues, if only to draw attention to an avenue of investigation which has been somewhat neglected in recent decades. Currently, most emphasis in timberline investigations, and perhaps in alpine ecology generally, is on energy relationships. Carbon dioxide assimilation has received greatest attention, and several authors have suggested that tree limit represents the level where trees can no longer achieve a positive carbon balance from year to year. Boysen-Jensen (1948), for example, notes the very narrow annual rings of *Betula* at timberline, and concludes that trees are at a disadvantage compared with smaller plants at high altitudes because a larger proportion of their photosynthates becomes locked up in the unproductive tissues of the trunk. The data of Schulze *et al.* (1967) also suggest a very delicate CO_2 balance for *Pinus longaeva* Bailey at timberline. They calculated that it would take about 117 hours of photosynthesis at the peak summer rate (1.2 mg of CO_2 per gram dry weight of needle per hour at 15°C), equivalent to half of the growing season, to redress the total winter negative balance of approximately 140 mg CO_2 per gram. However, in an annual balance sheet for carbon in *Pinus cembra* L. at timberline in Austria, only a third of the carbon assimilated (as estimated from measurements of gas exchange) could be satisfactorily accounted for by respiration and increase of dry matter. Tranquillini (1959) suggested that the difference might have been transferred to the mycorrhizal fungi, and certainly, Schweers and Meyer (1970) have shown that mycorrhizae use a large proportion of the carbon assimilated by *Pinus sylvestris* L. seedlings.

At increasing altitudes there is, in addition to decreasing year-round temperatures, a decreasing period during which temperatures permit growth and development. Tranquillini (1957, 1959) found that young plants of *Pinus cembra* achieve some net assimilation from the time that they emerge from winter snow until either the snow returns or the ground freezes in the rooting zone in late autumn. However, rapid assimilation commences only when needle temperatures at midday exceed 10°C, and begins to decrease when the new shoots have formed in July. Further sharp decreases occur during September and October, because of diminishing light and the onset of hard frosts which inactivate the photosynthetic mechanism (fig. 7B.1). Seedlings do not become established where snow persists until July. *Nothofagus solandri* shows a delay in onset of growth with increasing altitude, amounting to some six weeks over a range of 500 m, and a decrease in total growth (fig. 7B.2).

Nevertheless, I doubt whether growth and CO_2 assimilation reach limiting values at tree limit – and so, apparently, does Tranquillini (1967, p. 469). There are several pointers to the contrary, one being that timberline species vary widely in their growth rates. The limited assimilation and very slow growth of species such as *Pinus longaeva* can be inter-preted as a positive adaptation to the short growing season, when viewed in the light of the hypothesis that success of woody plants at high altitudes depends on the ability to ripen their shoot tissues. The reduced potential for assimilation in the timberline material of *Picea abies* studied by Pisek and Winkler (1958) may be similarly adaptive. They found that branchlets from a tree at timberline (1840 m) had a net assimilation rate only about three-quarters of that of branchlets from a tree at 600 m, when they were tested under the same conditions. Conversely, in *Nothofagus solandri* and other species in the Craigieburn

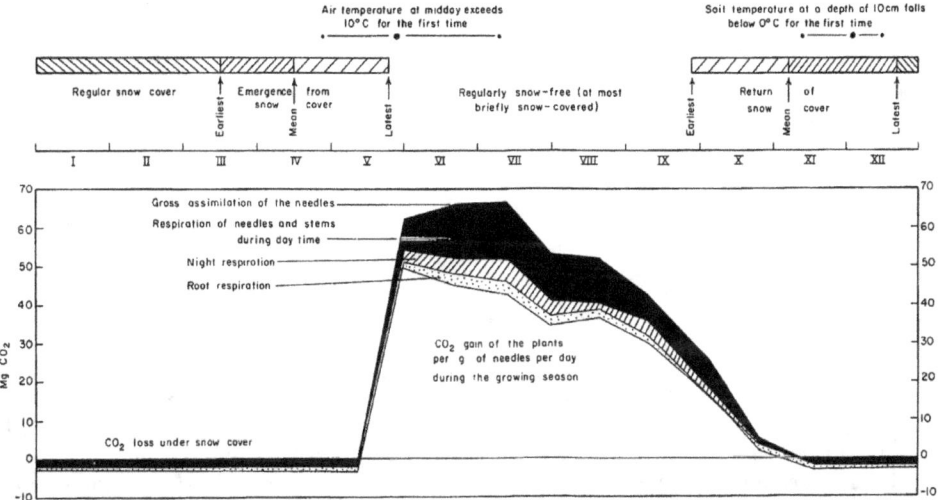

Fig. 7B.1. The upper part of the figure shows snow cover and growing seasons, measured at 1900 m in the Ötztal, Austria, from 1938–58. The lower part shows the yearly cycle of CO_2-balance in young plants of *Pinus cembra* growing near timberline between 1970 m and 2000 m in the same district. (Redrawn from Tranquillini, 1959.)

experiments, the amount of growth even 300 m above timberline can be well above the minimum necessary for survival, yet the plants eventually die.

Summergreen alpine herbs, which do not need to withstand winter conditions above ground, can produce very substantial amounts of dry matter (e.g. Scott and Billings, 1964). Also, the amount of non-photosynthetic, respiring tissue in proportion to the amount of photosynthetic tissue might well prove to be as high in many herbaceous and shrubby species as in trees. This is not to suggest that growth and assimilation are with-out significance for the timberline phenomenon. Vigour of growth is undoubtedly impor-tant in interspecific competition, and in the ability of seedlings to recover from damage of various kinds, such as frosting, and browsing by animals.

Spomer and Salisbury (1968) have suggested that trees cease where soil temperatures during the growing season are too low for growth and functioning of roots. This possibility should be tested, although I know of no obvious reason why tree roots should be absolutely limited in this way, since herbaceous alpine plants develop considerable root systems in soils much colder than those at tree limit.

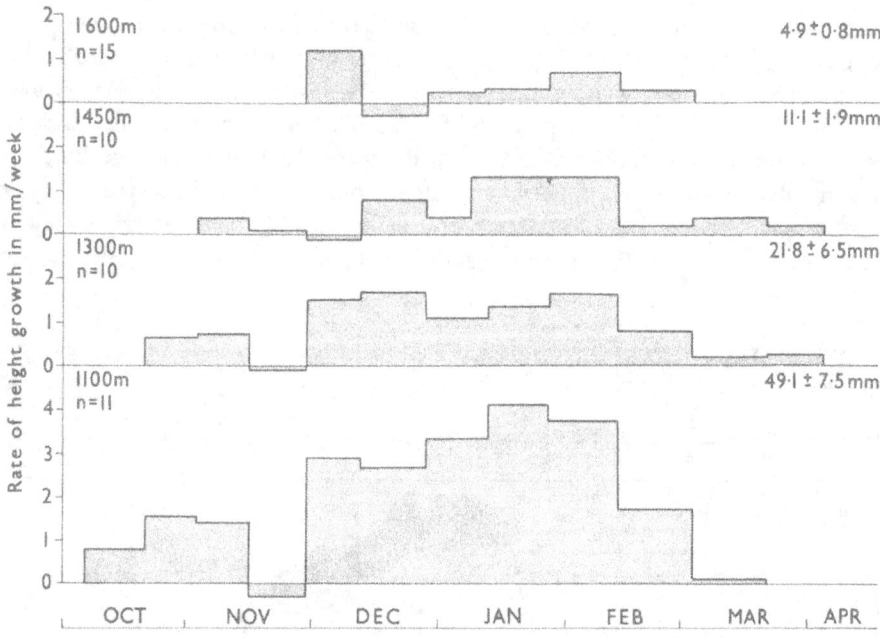

Fig. 7B.2. Rate of height growth of third year *Nothofagus* seedlings under 27 per cent daylight during summer 1965–6 (from Wardle, 1971). These germinated in the forest at 1100 m, and were transplanted to the four gardens at the beginning of their second summer. The initial peak, followed by a period with negative rate, largely results because the terminal buds decrease in length as pre-formed leaves separate off during the first, early summer flush. The mean values for total height growth during the summer (at right of figure) are significantly different between altitudes.

In discussing roots, it is appropriate to mention the observation by Moser (1966) that most timberlines are composed of trees with ectotrophic mycorrhizae, and that tree species without ectotrophic mycorrhizae do not ascend as high. The important timberline families Pinaceae, Cupressaceae, Betulaceae, Fagaceae, and Myrtaceae (*Eucalyptus*) all have ectotrophic mycorrhizae, but none were found on species forming timberline in New Guinea (A. S. Edmonds, personal communication, 1970). Among timberline species in New Zealand, only two beeches form ectotrophic mycorrhizae. However, as a whole, the species with ectotrophic mycorrhizae develop into more massive trees, show faster growth, and probably are competitively superior to those without.

The other main form of symbiotic mineral nutrition among higher plants, i.e. nitrogen fixation by microbe-containing root nodules, is on the whole poorly represented at timberline. Arborescent legumes are absent, except for *Sophora chrysophylla* which forms exceptionally low timberlines in Hawaii, while in New Zealand and New Guinea even herbaceous and shrubby legumes are virtually absent. Non-rhizobial nodules are possessed by some plants, especially in the genus *Alnus* where certain species, such as *A. viridis*, have both nitrogen fixing root nodules and ectotrophic mycorrhizae (Benecke, 1968).

It has been suggested that strong light, and in particular the proportion of ultra-violet radiation, influences the upward distribution of trees (Collaer, 1934), but this is clearly insignificant in comparison with the bearing that latitude and continentality have on the altitude of timberlines. The winter-acting factors of snow and drying winds have also been suggested as primary causes of timberline (Shaw, 1909; Griggs, 1938), but these really only modify the limits set by summer climate.

5. Discussion of specific timberlines, with reference to experiments in the Craigieburn Range

While the two principles described in section 3 of this chapter, i.e. the ability to ripen shoot tissues, and the upper limit of an aerial environment conducive to growth and ripening, seem adequate to account for the altitudinal position of nearly all natural tree limits, the detailed features of timberline in a given locality are to be explained in terms of the interaction between regional climate and microclimates on one hand, and the autecology of the timberline species on the other. These interactions are discussed in relation to various timberlines, and especially in reference to experiments in the Craigieburn Range, New Zealand (Wardle, 1971). In these experiments, I have compared the performance of the local timberline species *Nothofagus solandri* var. *cliffortioides*, with those of various overseas species.

THE *NOTHOFAGUS SOLANDRI* TIMBERLINE OF NEW ZEALAND

Seedlings sown below natural timberline (which is at 1300 m) survived and grew, although initial survival was poor unless the seedlings were quite heavily shaded (i.e. about 80 per cent of daylight being excluded). Above timberline, seedlings germinated some weeks later than below timberline, but under strong sunlight they shrivelled within a few weeks. Under shade, the seedlings grew quite vigorously during their first summer, even 300 m above the natural timberline, but at the end of the first summer their shoots were inadequately ripened, and they mostly succumbed during the winter (fig. 7B.3). A few are living after four years, but this probably reflects the somewhat ameliorated environment of the experiment and the weather during the last two growing seasons, which have been exceptionally prolonged. Seedlings germinated below timberline and transplanted to gardens above timberline, at various ages, showed much better survival; at 300 m above timberline some are still heathy after six years. This is apparently because older seedlings commence growth in the spring before seeds germinate at the same altitude, and they therefore have longer to ripen their shoots than first-year seedlings (fig. 7B.2).

Nothofagus solandri timberlines are exceptionally sharp, and usually without ecotonal krummholz. This fits the results of the experiment, which indicated that the seedlings require shade, and are therefore unlikely to become established on sites above timberline which are warmed by strong insolation. However, successful transplanting of older seedlings to above timberline, as opposed to the failure of seedlings germinated there, shows how environmental tolerance increases with age, and agrees with the observation that wherever seedlings become naturally established they can usually develop to full-sized trees. With *Nothofagus solandri*, therefore, the timberline can be related to critical events during the first summer after germination.

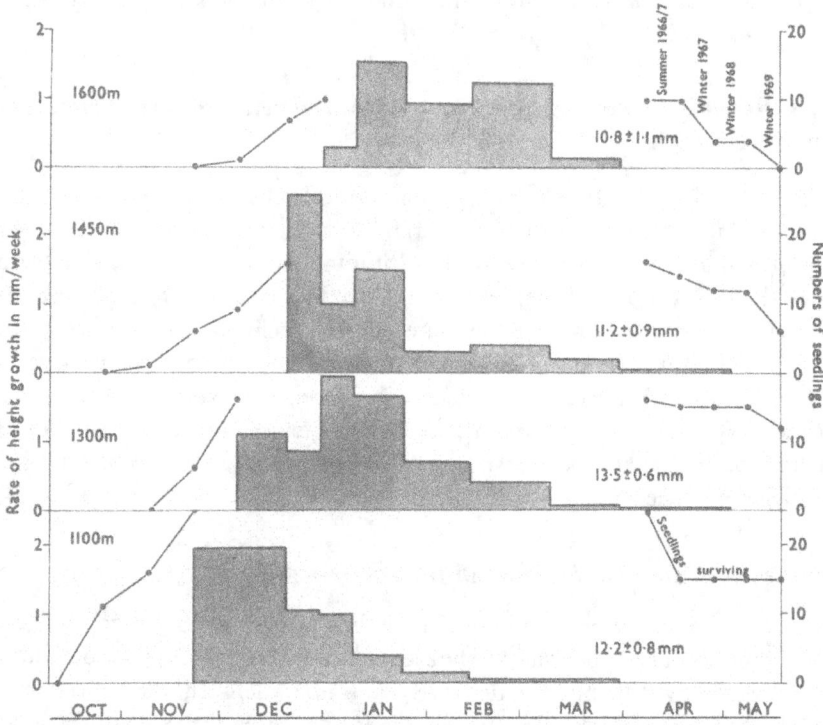

Fig. 7B.3. Rate of height growth between measurements of first year *Nothofagus solandri* seedlings under 27 per cent daylight, germinated *in situ* during summer 1966–7. Mean values for total height growth (shown with standard errors at right of histograms) are not significantly different between altitudes. Graphs for germination and for survival during the next 2.5 years are shown at left and right respectively (from Wardle, 1971).

It is also instructive to look at this timberline in the context of decreasing probability of seedlings becoming established with increasing altitude. *Nothofagus*, like the northern *Fagus*, exhibits well marked seed-years, and the frequency of good seed-years decreases towards timberline. Even within a seed-year, the quantity and proportion of sound seeds decreases towards timberline (J. Wardle, 1970), and even among sound seeds the embryos are smaller at timberline. When seed does become dispersed above the forest margin and manages to germinate there, strong insolation leads to rapid death of the resulting seedlings, through wilting and predation by *Brachaspis* grasshoppers, which feed in bright sunshine. Later, early summer frosting, winter desiccation, and failure to achieve satisfactory mycorrhizal unions take their toll. However, in treatments where most of these contingencies were avoided, failure to harden against winter desiccation still set an upper limit to the establishment of seedlings, although because of the ameliorated conditions this was higher than in nature.

Nothofagus solandri krummholz does form under two conditions, one being when prevailing winds train growing shoots without causing any winter desiccation. Presumably, the temperature of the exposed shoots is lowered, in the manner observed by Gates and Janke (1966) for krummholz of *Picea engelmannii* in the Colorado Rockies; in that case, leaf temperatures were 7.2°C higher on the sheltered side of a plant than on the top during a light wind. Krummholz also forms at the forest edge in places where snow accumulates to a depth of a metre or so, in contrast to the few centimetres which usually prevail near the forest edge under New Zealand conditions (Wardle, 1965a, p. 124). Above the deeper snowpacks, saplings suffer severe winter desiccation and dying back, but protected foliage remains green. Inadequate ripening of shoots, strong insolation and exposure to wind, and limited water content of the slender stems are probably all factors in this dying back. Neighbouring mature trees are usually unaffected.

AUSTRALIAN *EUCALYPTUS* SPECIES AT TIMBERLINE

Seedlings of timberline eucalypts are relatively fast growing, like those of *Nothofagus solandri*, and above tree limit in the Craigieburn experiments they likewise prove to be inadequately hardened at the onset of winter. The eucalypt forests in the Australian Alps open out more gradually towards timberline than do New Zealand beech forests, and this probably reflects the greater tolerance by seedlings of strong insolation shown in comparative experiments.

TIMBERLINE IN THE ROCKY MOUNTAINS OF COLORADO

Timberlines in the Colorado Rockies are very different from the *Nothofagus* timberline in New Zealand, being characterized by broad belts of krummholz, especially in the most exposed situations. One of the main species is *Picea engelmannii*, and in the Craigieburn experiments its seedlings, like those of *Nothofagus solandri*, are adversely affected by strong light at high altitudes (as Ronco, 1970, has also found in Colorado). Under shade, however, they are still surviving after six years 300 m above timberline, and after four years 450 m above. The seedlings differ from those of *Nothofagus solandri* in their exceedingly slow growth (*cf.* maximum values of 34 mm *v.* 160 mm in fourth year seedlings at 1100 m), but the shoot extension is completed quickly and relatively early in the summer, so that the seedlings have longer to complete their ripening. I suggest that this results in a great degree of hardiness which in their native habitat allows them to become established 150 m or more above tree limit. As they grow taller, however, they encounter the severe winter conditions above the snowpack; and at the same time they grow beyond the favourable summer climate close to the ground. This species, therefore, can become established and survive as krummholz above the limit of tall trees.

As well as *Picea engelmannii*, there are also *Abies lasiocarpa* Nutt., *Pinus flexilis* James and, in the south, *P. aristata* at timberline in Colorado. The ecological behaviour of *Abies lasiocarpa* is similar to that of *Picea engelmannii*, except that it seems to be less xerophytic. The pines, however, prefer southerly aspects, and also show a marked ability to grow as erect trees where accompanying *Picea* and *Abies* grow only as krummholz. I raised very few seedlings of *Pinus flexilis* in the Craigieburn experiment, but among them the better insolated ones have shown best growth. Seeds from high altitude stands of *Pinus contorta* Dougl., however, have germinated freely, even 450 m above the natural *Nothofagus* timberline. Survival has been good in all treatments so far, and although seedlings show a marked altitudinal decrease in growth rates, those receiving nearly full sunlight are growing quite vigorously, even at the highest site. *P. contorta* in Colorado reaches tree limit only as occasional cripples, and I anticipate that the Craigieburn plants will similarly cease upward growth when they are tall enough to be exposed to winter desiccation. Nevertheless, they illustrate the superior adaptation of pines to unshaded sites. It is significant that pines often have a pioneering role at timberline, as at Logan Park, Montana, where trees of *Pinus albicaulis* become established in the meadows, and serve as nuclei around which *Abies lasiocarpa* seedlings develop, leading eventually to 'islands' of *Abies* krummholz surrounding old pine trees (Habeck, 1969).

It seems to be generally accepted that members of the Pinaceae are hardier than trees of other families which form timberlines. Certainly in New Zealand their seedlings can become established at higher altitudes than those of the native tree species, although it is doubtful whether they would form an actual timberline more than 100 m or so above the native one.

THE *PINUS HARTWEGII* TIMBERLINE OF MEXICO

On the mountains near Mexico City at 19°N, *Pinus hartwegii* forms a single-species timberline of erect trees, with little or no development of krummholz. The upper forest is far more open than that of *Nothofagus solandri*, and in comparative cultivation on the Craigieburn Range, the seedlings benefit from strong insolation, as with other pines. Planted above timberline, the seedlings have relatively fast growth and the same tendency as *Nothofagus solandri* to grow for a relatively prolonged period, thus entering the winter with inadequately ripened shoots.

NEW GUINEA TIMBERLINES

The climate at high altitudes in New Guinea is uniquely moist, equable as to temperature and windless, although cloud cover and precipitation decrease somewhat during the southern hemisphere winter. The nature of timberline in this almost seasonless environment is of particular interest. On 4510 m Mt Wilhelm, the highest mountain in eastern New Guinea, dense forest with trees of considerable stature extends to about 3000 m, but above this there is gradual reduction in height, and also an increasing tendency for forest to be replaced by grassland, scrub, scattered trees and groves. As noted on p. 374, the tall shrubs and trees cease by 4000 m; at this altitude, the canopy is still generally at 3–6 m, but occasional trees are much taller. In the upper forest, trees tend to show the 'umbrella crowns' mentioned by Troll (1959) as being characteristic of humid, tropical mountains. Near tree limit there is a peculiar tendency, particularly evident in *Podocarpus compactus*, for leading shoots of saplings and young trees to lose vigour and die, and to be supplanted by lower shoots of epicormic origin. The resulting growth forms recall the krummholz of temperate zone mountains, but suggest stresses which are irregular and mild rather than seasonal and intense. Possibly the dying back is to be explained in terms of unripened shoots (reflecting the limiting aerial environment) and water stresses arising because of cold soils. In cultivation in New Zealand, *Podocarpus compactus* seedlings are unscathed by isolated hard frosts, but they dried out in the Craigieburn experiments after the winter period with successive frosts commenced.

Above tree limit on Mt Wilhelm there is grassland and stony herbfield, with a considerable shrub component, including *Eurya brassii* Kobuski (Theaceae), *Coprosma divergens* Oliver, and *Detzneria tubata* Diels, which forms a monotypic genus of the Scrophulariaceae. At tree limit, the first two species attain a height of 2 m, but they become gradually stunted with increasing altitude. No taller woody plants were found to transcend tree limit, even as seedlings, with the exception of *Drimys brassii* Smith. This plant, which reaches a height of 3 m at its upper limit at 4200 m, shows the prevailing tendency for stems to lose vigour and die back, in this case to a basal lignotuber. Young shoots arising from the surface of the lignotuber are chlorotic and unlignified, with leaves reduced to scales. In the parlance of Raunkiaer (1934), *Drimys brasii* could be said to combine the roles of hemicryptophyte and microphanerophyte.

SOPHORA CHRYSOPHYLLA FROM HAWAII

One other tropical timberline species, *Sophora chrysophylla* collected from Mauna Kea in Hawaii, was tried in the Craigieburn experiments. While it grew reasonably well in the summer, seedlings were killed by the first frost to exceed 3–4 °C, except for one plant which became partly acclimatized and lived for several years 150 m below timberline, until it was killed by unusually severe cold.

6. Inverted timberlines

The drainage of cold air down slopes at night, and its ponding in valleys, is well described by Geiger (1957), who instances differences of up to 30 °C between the floor of the Gstettneralm sink hole and the surrounding warmer slopes. In regions with marked differentiation of seasons, such temperature inversions may be unimportant during the growing season, because nights are short and temperatures are usually well above danger level. Even during the winter, dormancy and hardiness usually protect plants from frost damage. In the north temperate zone, therefore, treelessness of valley floors is generally attributed to unfavourable soil–water regimes rather than to temperature, although Billings (1954) has shown that the lower limits of woodland of *Pinus monophylla* T. & F. and *Juniperus osteosperma* (Torr.) Little on a mountain range in Nevada seem to coincide with a pronounced temperature inversion.

On tropical mountains the opposite situation prevails. In New Guinea there is no well defined growing season, and although many trees and shrubs form resting buds, there is a lack of synchronism of activity, sometimes even within a plant. Also, the resting periods are evidently short, at least in *Drimys brassii*, where unopened flower buds, flowers, young fruit and mature fruit can be seen at the same time, representing successive flushes of a single shoot. At high altitudes, therefore, phanerophytes are always liable to be damaged by the nocturnal freezing which can take place at any time of year, although the small amount of extension growth formed in each flush (e.g. about 1–2 cm in canopy shoots of *Drimys brassii* and 3–5 cm in those of *Podocarpus brassii*) may constitute a form of protection, in that the vulnerable periods are likely to be shorter than in more vigorously growing species. Below tree limit, the frosts are usually only slight, to judge from McVean's (1968) data from Mt Wilhelm, but there are reports of occasional severe frosts which damage both crops and the foliage of native trees (R. Pullen, personal communication, 1971). The effect is accentuated on valley floors and depressions subject to ponding of cold air. 'Frost grasslands' dominated by tussock grasses with meristems protected by mature and dead leaves are logically predictable under these conditions, although it must be pointed out that much of the so-called grassland is in fact swamp and bog dominated by sedges or *Gleichenia* fern, and that all of it is subject to fire. Gillison (1969, 1970) has demonstrated the pyric origin of some areas of grassland and the tendency for forest to reinvade them. Nevertheless, I see no reason to doubt that the pattern of forested slopes and herb-dominated valley floors which prevails at high altitudes in New Guinea is basically determined by incidence of frost, at least within 1000 m of tree limit. For example, on the slopes surrounding the swampy Neon basin near Mt Albert Edward, there is a difference between the induced grassland on the upper and mid slopes which tend to revert to forest, and the grassland on the lower slopes where associated woody plants are species characteristic of vegetation above tree limit.

A unique feature of the subalpine grassland is the presence, and at times abundance, of species of *Cyathea* tree ferns which are restricted to this type of vegetation. These tree ferns are undoubtedly fire resistant, but their morphological peculiarities (see section 2 of this chapter and Plate 23) may also help them to tolerate the diurnal temperature fluctuations. Susceptibility to frost also appears to underlie the failure of *Nothofagus* to exceed an altitude of 3000 m in New Guinea. To the best of my knowledge, the trees produce their new shoots in a single flush during the 'dry' season, thereby betraying their temperate origins. The young shoots are long (e.g. 10–20 cm), soft, and presumably frost tender like those of the temperate zone beeches; but whereas frosting of young shoots of temperate species is an unseasonable occurrence, frosts on the New Guinea mountains, being frequent and non-seasonal, suffice to set an upper limit to the genus.

On the mountains of equatorial East Africa, the daily temperature fluctuations are particularly severe, and although the vegetation above the ericaceous woodland does not show the topographic relationships of typical inverted timberlines, it is influenced by a similar climatic regime. Both arborescent and short stemmed megaphytes are protected by their pachycauly and large tomentose leaves, and through nyctinastic movements whereby the buds are enclosed by infolding leaves during the nightly freezing

(Hedberg, 1964). Probably the same considerations apply to the espeletias of the northern páramos, although Walter and Medina (1969) have suggested that the absence of normal trees here is due to cold soil in which diurnal warming is confined to a superficial layer.

In Australasia, treeless valley floors are a feature of mountainous areas, and seem basically related to temperature inversions (Moore, 1959). In the experiments in the Craigieburn Range of New Zealand, unshaded *Nothofagus solandri* seedlings fail to survive when transplanted to a treeless valley floor at 900 m, and it seems that death is due to desiccation of the foliage during winter. Survival of shaded seedlings at the same site suggests that this occurs when tops are heated by the sun while the roots are still frozen. Eucalypt seedlings seem somewhat hardier at this garden than those of the native *Nothofagus*, and seedlings of North American conifers, including the Mexican *Pinus hartwegii*, are fully hardy.

7. Some life forms at timberline

KRUMMHOLZ AND SCRUB

Coniferous krummholz is a very conspicuous feature of north temperate mountains, and its growth and development merit discussion in some detail. In the case of *Picea engelmannii* a healthy tree below timberline is characteristically spire-shaped, with main branches sweeping downwards from an erect central trunk. This growth form may be regarded as basic, and other growth forms as modifications imposed by severe habitat (Plate 3). In closed forest, the lowest branches usually become suppressed and die, but on spruce trees growing in open meadows they often persist and produce adventitious roots where they make contact with the ground. Once rooting occurs, such a branch is freed from domination by the original central stem and develops its own central axis. The proximal part of the branch ceases to grow and eventually dies. In this way, groups or islands of trees develop. In the most severe sites towards the extreme upper limit of *P. engelmannii*, islands are reduced to cushions of contorted stems and needles, shorn off level with the surface of the winter snowpack. I refer to these as 'cushion krummholz'. In less severe sites, in the lower part of the ecotone, erect stems with branches only on their leeward sides rise above the cushion, to produce 'flagged krummholz' (Wardle, 1968, and fig. 7B.4).

Wind-deformed krummholz is very extensive on the eastern slopes of the Rocky Mountains in Colorado and Wyoming, because of the strong, dry, westerly winds which blow persistently into the forest margin during the winter. On the Front Range in Colorado, islands of cushion krummholz die in their exposed, upwind portions, but the leeward shoots continue to grow downwind, surviving the winter in the protection of the snow which they accumulate, and rooting adventitiously as they grow. Billings (1969) has described a very interesting interaction between snow and forest just below timberline on the Medicine Bow Mountains in Wyoming. Here, accumulation of snow into deep drifts, lasting far into the summer, has given rise to a self-perpetuating pattern of alternating strips of 'ribbon forest' and 'snow-glades'. On more sheltered mountains, particularly in the Pacific Northwest of America, where winter is characterized by phenomenal depths of snow rather than strong winds, krummholz stems usually show little wind-training, although they tend to die back from the apex.

In discussing the origin of krummholz growth forms in *Picea engelmannii*, I have emphasized the importance of winter desiccation of shoots exposed above the snowpack, and like other authors (e.g. Michaelis, 1934b) think that plants are predisposed to winter

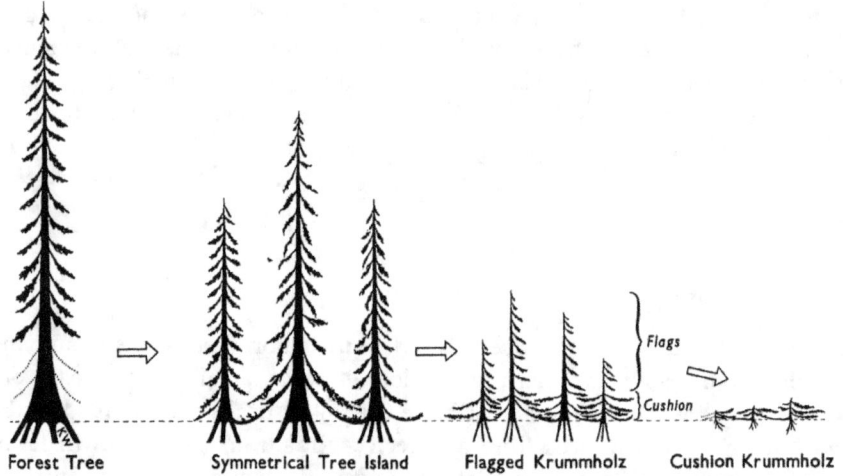

Fig. 7B.4. Derivation of common growth forms of *Picea engelmannii* increasingly deformed from left to right (from Wardle, 1968).

desiccation by being inadequately hardened, as a result of unfavourable conditions during the growing season. Other external factors can also be involved, including weight of snow, snow moulds such as *Herpotrichia*, and lesions of unknown origin on the needles. Abrasion by ice particles has also long been regarded as a factor in forming krummholz, but as Turner (1968) points out, this is not well substantiated.

Internal factors have also been proposed as inducing krummholz. Clausen (1965) suggested, *apropos* of *Pinus* and *Salix* in the Sierra Nevada of California, that krummholz growth forms are genetic variants.[1] A better known and proven example is the dwarf mountain pine (*Pinus mugo* Turra), which forms extensive scrub near timberline in the European Alps. This retains its semi-prostrate form in cultivation at low altitudes, but it is very closely related to, and sometimes regarded as conspecific with, the erect *P. uncinata* Ramond. *Pinus pumila* forms similar scrub communities at and above timberline in the far east of the Soviet Union and in Japan, and is closely related to the arborescent *P. cembra*. *Juniperus* also includes both shrub and tree species. In other species, such as *Picea engelmannii*, where dwarfing involves dying back of exposed shoots, it is clearly

[1] *Edit. note:* La Marche and Mooney (1972) have recently provided evidence that suggests the 'genetic' explanation does not apply to *P. aristata*.

imposed by the external environment. Yet even in this species there seem to be occasional genetically dwarfed individuals at timberline (Wardle, 1968). Recently Löve et al. (1970) have found that krummholz *Picea engelmannii* contains more scopulin than normal trees, and they suggest that this substance is produced in response to environmental stress (particularly by a high level of ultra-violet radiation) and that it causes stunting through its influence on the growth hormonal system.

I find it useful to distinguish between timberline vegetation of environmentally dwarfed trees and genetically determined shrubs, which I call 'krummholz' and 'scrub' respectively, not that the distinction is always clear-cut. For instance, according to E. J. Godley (personal communication, 1971) erect and prostrate forms of *Nothofagus antarctica*, which seem to be genetically different, grow side by side at timberline in the southern Andes, and certainly I have obtained procumbent seedlings of this species from seed gathered at timberline by Professor M. J. Dmitri. It may be noted that whereas krummholz growing above tree limit is non-reproductive and has no genetic future, shrubs can carry out their whole life cycle above timberline, and some genera have shown considerable speciation: e.g. *Salix*, *Vaccinium*, and in New Zealand, *Dracophyllum* and *Hebe*.

DECIDUOUS TREES AT TIMBERLINE

Most timberline species are evergreen, and this has been cited in favour of tree limit corresponding to the ability to achieve a positive assimilation of carbon dioxide, since it would be expected that deciduous species would require higher levels of energy (or CO_2) balance to compensate for the long period when they are leafless (e.g. Bliss, 1966). However, deciduousness at timberline can be seen as an evolutionary response to seasonal climates, just as it is at lower altitudes. In the tropical zones, dicotyledonous trees at high altitudes are all evergreen, with the exception of the American trans-tropical species *Alnus jorullensis*. This is also broadly true of the south temperate zone except for South America south of 36°S, where deciduous beeches form the timberline. In the South Island of New Zealand, the small deciduous tree *Hoheria glabrata* Sprague & Summerhayes locally dominates areas of timberline forest, and in Tasmania the deciduous *Nothofagus gunnii* occurs. The last has a remarkably plastic habit of growth in that it can grow erect to about 6 m, or form a tangled understory to other trees, or even occur in espalier form on rocks, with branches directed both up and down slope. In the north temperate and arctic zones, all dicotyledonous trees at timberline are deciduous, other than the Himalayan rhododendrons and *Quercus semecarpifolia*.

Few species of conifer are deciduous, but one of them, *Larix gmelini* (Rupr.) Kuz., is possibly the hardiest tree in the world, since it forms the arctic tree limit at 72°30′ N in Siberia (Hustich, 1966). *Larix lyallii*, a tree of rather restricted distribution in Montana, Alberta, British Columbia and Washington, also demonstrates its hardiness by remaining an erect tree at timberline among krummholz of *Picea* and *Abies*. According to Arno and Habeck (1972) morphological advantages possessed by *Larix lyallii* include woody buds, bare twigs in winter which trap less snow than those of evergreen species and which are therefore less likely to be broken, and flexible boles in young trees which spring back after being buried.

THE 'OLEARIA COLENSOI GROWTH FORM' AT TIMBERLINE

Olearia colensoi Hook.f. is a tall shrub of the New Zealand subalpine belt, with leaves 4–10 cm long, which are rigid, glossy and glabrous above, and covered with felted white tomentum below. They are crowded into rosettes on the ends of otherwise naked branchlets, and arranged into a close, umbrella-like canopy (Plate 24). The outer layers of the very thin bark are continually shed. On slopes, the stems on the downhill side of the plant tend to be prostrate and adventitiously rooting. I have suggested (Wardle, 1965b) that these features can be interpreted as a response to cool, cloudy climates, and in particular, as an adaptation towards maximum utilization of radiant energy.

These attributes of *O. colensoi* are characteristic, in greater or lesser degree, of the various species which make up the subalpine scrub in the parts of New Zealand which lack *Nothofagus*, and are repeated in many species of the uppermost forests in the humid tropics and subtropics. The resemblance is particularly striking in the *Rhododendron*-dominated forests of the eastern Himalayas.

8. Temporal variations in alpine timberline

NATURAL CATASTROPHES

Snow avalanche is a familiar hazard in subalpine forests. Places where winter and spring avalanches recur so frequently as to be a regular component of the environment are occupied permanently by vegetation adapted to the conditions. Communities of shrubs with tough, springy branches which bend with the impact of snow are especially characteristic below tree limit, an example being the alder shrubberies in central Europe and western North America. In New Zealand, *Nothofagus* forest is replaced by subalpine scrub of *Podocarpus nivalis*, *Phyllocladus alpinus* Hook.f. and various other shrubs on regular avalanche tracks. Most alpine regions also experience occasional winters of exceptionally heavy snow, and avalanches of abnormal severity which destroy established forest. Regeneration can be prompt, provided there is a good stocking of seedlings at the time of impact.

In New Zealand there are several localities where gales of unusual severity have overturned large areas of *Nothofagus* forest. As with avalanche damage, there has been good recovery where there were adequate numbers of seedlings, but often there is a dearth of seedlings in the uppermost hundred metres or so of forest, and quasi-permanent depression of timberline can follow catastrophic windthrow and destruction of the forest microclimate.

Catastrophic damage to vegetation can be caused by animals, especially phytophagous insects which periodically achieve plague proportions. In the United States, a bark beetle (*Dendroctonus engelmannii*) is a major pest of subalpine *Picea engelmannii* forests. On the White River Plateau in Colorado, for example, a plague in the 1940s and, apparently, another in the late eighteenth century removed most of the spruce and fir trees larger than 10 cm in diameter (Miller, 1970). In New Zealand, death of subalpine forests of *Nothofagus solandri* has been attributed to the cambium-girdling beetle *Nascioides enysii*, but there is some evidence that this occurs after a stand has been weakened by such events as damage by wind or snow. This may be in line with a suggestion, based on a study of psyllids in Australia, that outbreaks of phytophagous insects coincide with environmental stresses affecting the host plants (White, 1969).

Most or all plant-eating animals, both vertebrate and invertebrate, affect only one phase of the life cycle of dominant trees, and the effect of population explosions is therefore normally only temporary. Also, it is reasonable to suppose that between outbreaks the animals are kept in check by parasites and predators. These natural controls fail when predators are deliberately removed, as has happened with the predators of the large mammalian herbivores over most of the northern hemisphere, and where herbivores have been introduced into new areas without the appropriate predators, as in New Zealand. The implications of control of outbreaks of phytophagous insects by widespread application of nonspecific pesticides also deserves careful attention in this context.

Pathogenic fungi can also cause ecological disruptions, as in the case of the blister rust (*Cronartium ribicola*) which threatens to delay successions at some timberlines in Montana, by exterminating the 'nurse' species *Pinus albicaulis* (Habeck, 1969). I suspect that fungal root pathogens are a major factor in the death of stands of *Olearia colensoi*, the tall shrub which is widely dominant at New Zealand timberlines.

Fire is the most important of all catastrophic events affecting timberline. It is usually begun by man, but lightning fires are a natural hazard, especially in regions with dry summers. The effects last longer than damage by avalanche, windthrow or insects, because fire destroys the seedlings, and even the fallen seed if it consumes litter and humus. The literature on fire-induced successions in subalpine forests is too vast to summarize here, but attention will be drawn to a few points of particular relevance to timberline.

Regeneration is much slower than at lower altitudes, because climatic conditions are approaching the limits for establishment of trees. This is accentuated in the mountains of western North America because the two most aggressive species in fire-induced successions, *Pinus contorta* and *Populus tremuloides* Mich., ascend only to within 100–200 m of tree limit, except as rare cripples. As noted on p. 388, *Pinus albicaulis* and its relatives behave as pioneers, but usually only scattered individuals become established over a long period, and their growth is very slow.

On the Australian Alps, *Eucalyptus niphophila* can survive destruction of the crown by regenerating from its lignotuberous base, provided the coppice shoots are not destroyed by persistent browsing (Costin, 1959). In Tasmania, there is interplay between species intolerant of fire, typified by *Athrotaxis cupressoides* Don, and more tolerant species such as *Eucalyptus coccifera*.

In the New Zealand vegetation there is a pronounced absence of fire-adapted tree species and, towards timberline, recovery of both *Nothofagus* forest and subalpine scrub can require centuries, especially where grassland becomes entrenched and resists all but slow marginal encroachment by trees. In New Guinea, the fire-induced subalpine grasslands include 'islands' of forest, which are surrounded by narrow bands of very distinctive thicket vegetation. The semi-lianoid shrub *Coprosma papuensis* Oliver is often dominant, although accompanied by other shrubs, lianes and tall herbs. These thickets act both as nurse to young trees, and as fire breaks protecting the forest remnants (Wade and McVean, 1969; R. Pullen, personal communication, 1971; see also Gillison, 1970, for a careful study of the succession in similar marginal vegetation at somewhat lower altitudes).

CLIMATIC CHANGE

Timberline is the sharpest temperature-dependent boundary in nature, and might be expected to be a useful guide to climatic variations of the past, especially in the three contexts set out below.

1. *Cold climates of the Pleistocene*
These should be revealed in remains of timberline plants at low altitudes. Such evidence would seldom be unambiguous, because few species are strictly confined to timberline, being likely to occur at much lower altitudes where local conditions permit. Evidence for lower timberlines is mostly indirect, from pollen analyses indicating changes in distribution and extent of vegetation types. West (1968, p. 207) states that it is generally considered that temperatures fluctuated about 5–8°C between the temperate and cold phases of the Pleistocene. This would suggest that altitudinal zones were 900–1400 m lower than at present during the coldest stages, although it may be that the lowering was less in tropical and south temperate regions remote from major centres of glaciation.

2. *Postglacial climatic optimum or Hypsithermal*
Remains of trees above the present climatic timberline would be evidence of warmer climates in the postglacial, but there are few convincing reports of such material. In California and Nevada, dead trees of *Pinus longaeva* indicate that 2000–4000 years ago timberlines were at least 100 m higher than at present (LaMarche and Mooney, 1972). In the Cairngorm Mountains in Scotland, the natural limit of *Pinus sylvestris* L. lies between 600 m and 700 m, although it has generally been depressed through the activities of man to about 500 m (Pears, 1968a). Fossil stumps, dated by pollen analyses and radiocarbon assay, show that pine trees ascended to about 800 m during Boreal times and to 700 m during Sub-boreal times. Pears (1968b, 1969) attributes the decrease from these altitudes to increased precipitation and windiness during the Atlantic and Subatlantic periods, which caused growth of peat, impeded drainage, and exposed conditions at timberline. In this part of the British Isles, man apparently became a destructive force only relatively late in historical times.

3. *Fluctuations in climate within the life span of existing trees*
Evidence for this might be expected from dendrochronology, and from anomalous population structure. While there is good evidence for secular changes in precipitation

and temperature during the last few centuries, at least in the northern hemisphere, most of the evidence concerning timberline is ambiguous. However, the spread of tree seedlings onto timberline meadows as a result of earlier thawing of winter snow during the past century has been well established for British Columbia (Brink, 1959) and central Europe (Gams, 1954). Firbas and Losert (1949) have claimed that in the Riesengebirge timberlines fell 100–200 m from the fourteenth to the seventeenth centuries to reach their present levels, presumably because of the climatic cooling which caused glaciers to extend during the same period, culminating in the 'Little Ice Age'. In the parts of the Southern Alps of New Zealand where *Nothofagus* is absent, *Libocedrus bidwillii* Hook.f. is usually prominent at the upper limit of tall trees. The numbers of dead and moribund trees nearly always exceed the healthy, and counts of growth rings show that there has been little effective regeneration since AD 1600. Whatever the cause, it cannot be a direct result of decreased temperatures, for *Podocarpus hallii* Kirk and *Metrosideros umbellata* Cav. (Myrtaceae) in the same stands show no obvious anomalies in the structure of their populations.

MAN-INDUCED LOWERING OF TIMBERLINE

In general, any effects of climatic change at timberline during postglacial time have been obscured by the direct and indirect results of human activity. Deforestation has been intense in semi-arid regions which have supported large human populations for thousands of years. This is notably true of the vast region extending from the western Mediterranean to central Asia and in the Andes of Peru and Bolivia, where natural timberlines can scarcely be recognized. In more humid regions which have been densely inhabited for long periods, there has been extensive lowering of timberlines by fire (as in the New Guinea highlands), and by overgrazing and deliberate clearing for pasture (as in central Europe and the Sinohimalyan region). In the British Isles and Iceland, these factors have resulted in the loss of any semblance of a natural timberline.

The largest stretches of more or less undisturbed timberlines are in regions which until very recently supported only sparse populations in a primitive stage of cultural development, namely the Americas north and south of the Peruvian and Mexican centres of civilization, and Australasia. Yet, even in Australasia, fires lit by aborigines are proving to have left their mark on the structure and composition of forest up to timberline. Thus, some of the drier parts of New Zealand were completely deforested following the arrival of the first people about 1000 years ago (Molloy *et al.*, 1963), and the history of burning in Australia and Tasmania extends back for thousands of years. Recent papers are indicating that the Amerindian had no less profound effects on the forest cover of North America.

Today there is a strong interest in central Europe in restoring forest cover up to its natural limits, primarily because of the important part which forest can play in reducing the avalanche menace. The interdisciplinary research being carried out by the Austrian Federal Forestry Research Institute, at the Innsbruck Field Station for Subalpine Forest Research, is especially noteworthy. Topics covered by published papers, which are mostly to be found in the 'Mitteilungen der forstlichen Bundes-Versuchanstalt', Volumes 59 and 75, include tree physiology with special emphasis on phenology, environmental tolerances, assimilation and growth; broad and detailed mapping as an aid to assessing site potential; soil studies including tree nutrition, mycorrhizae and soil fauna; climate and microclimate including plant temperatures; and the use of natural vegetation as site indicators for afforestation. The need for this work became apparent largely because it was realized that disastrous avalanches, such as those experienced in the Tirol in 1951 and 1954, are attributable not only to exceptionally heavy snowfalls, but to a continual retreat of the timberline. In the Tirol, the timberline retreated an overall 100 vertical metres between 1774 and 1880, and the area of forest has decreased to a half (Hampel, 1961).

In North America and Australasia, emphasis so far has been on preserving existing timberlines as an important component of protective vegetation in mountain catchments, where control of flooding and erosion is the aim. Work towards restoring timberlines in these areas is still at the exploratory stage. In New Zealand an additional problem has been caused by the introduction of browsing mammals into vegetation which has evolved without them, so that in many places the uppermost forest has ceased to regenerate; lowering of the timberline by more than 100 m may well occur when existing canopy trees die.

A major thesis of this chapter is that timberline occurs at an altitude where the environmental tolerances of vascular plants, and in particular their ability to ripen their shoots so as to be able to withstand seasonally adverse conditions, are quite abruptly reached. At higher altitudes plants thrive only within the microclimate close to the ground. Timberline is therefore one of the most significant boundaries in biological nature, separating two fundamentally different ecosystems. In one, living processes take place within a depth of tens of metres, and in the other, they are usually restricted to a few centimetres. Since trees are at their climatic limits at timberline, they form a sensitive ecotone, and once this is destroyed the problems of restoring it are acute. The widespread disintegration of timberlines which has occurred, and that which threatens to occur in the future, represent a serious failure by man to attain harmony with his environment. It is ecologically comparable to the disintegration which took place at the arid limits of tall, continuous vegetation, when deserts advanced as ancient civilizations contracted and decayed.

References

ARNO, S. and HABECK, J. R. (1972) Ecology of Alpine Larch (*Larix lyallii* Parl.) in the Pacific Northwest. *Ecol. Monogr.*, **42**, 417–50.

AULITZKY, H. (1961) Standortsuntersuchungen in der subalpinen Stufe, 4. Die Bodentemperaturen in der Kampfzone oberhalb der Waldgrenze und im subalpinen Zirben – Lärchenwald. *Mitteilungen der forstlichen Bundes–Versuchsanstalt Mariabrunn*, **59**, 153–208.

BEAMAN, J. H. (1962) The timberlines of Iztaccihuatl and Popocatepetl, Mexico. *Ecology*, **43**, 377–85.

BENECKE, V. (1968) *The Root Symbiosis of Alnus viridis (Chaix) DC.* Unpub. thesis, Univ. of Canterbury, New Zealand.

BILLINGS, W. D. (1954) Temperature inversions in the pinyon-juniper zone of a Nevada mountain range. *Butler Univ. Bot. Stud.*, **11**, 112–18.

BILLINGS, W. D. (1969) Vegetational patterns near alpine timberline as affected by fire – snowdrift interactions. *Vegetatio*, **19**, 192–207.

BLISS, L. C. (1966) Plant productivity in alpine microenvironments on Mt Washington, New Hampshire. *Ecol. Monogr.*, **36**, 125–55.

BOYSEN-JENSEN, P. (1949) Causal plant-geography. *Det. kgl. Danske videnskabernes Selskab biologiske Meddelelser*, **21** (3), 1–19.

BRINK, V. C. (1959) A directional change in the subalpine forest-heath ecotone in Garibaldi Park, British Columbia. *Ecology*, **40**, 10–15.

CLAUSEN, J. (1965) Population studies of alpine and subalpine races of conifers and willows in the Californian High Sierra Nevada. *Evolution*, **19**, 56–68.

COE, M. J. (1967) *The Ecology of the Alpine Zone of Mt Kenya.* The Hague: W. Junk. 136 pp.

COLLAER, P. (1934) La rôle de la lumiere dans l'establishment de la limite superieure des forêts. *Ber. Schweiz. botan. Gesell.*, **43**, 90–125.

CORNER, E. J. H. (1966) *The Natural History of Palms.* Berkeley and Los Angeles: Univ. of California Press. 393 pp.

COSTIN, A. B. (1959) Vegetation of high mountains in Australia in relation to land use. *In:* Biogeography and ecology in Australia, *Monogr. Biol.*, **8**, 426–51.

COTTON, A. D. (1944) The megaphytic habit in the tree senecios and other genera. *Proc. Linn. Soc. London*, Session 156, 158–68.

DAUBENMIRE, R. (1953) Notes on the vegetation of forested regions of the far northern Rockies and Alaska. *Northwest Sci.*, **27**, 125–38.

DAUBENMIRE, R. (1954) Alpine timberlines in the Americas and their interpretation. *Butler Univ. Bot. Stud.*, **11**, 119–36.

DIMITRI, M. J. (1964) Los procesos de sucesión y zonatión vegetal en los bosques de montana. *Anales de Parques Nacionales, Buenos Aires*, **10**, 3–60.

FIRBAS, F. and LOSERT, H. (1949) Untersuchungen über die Entstehung der heutigen Waldstufen in den Sudeten. *Planta*, **36**, 478–506.

GAMS, H. (1954) La subdivision de l'étage alpin et ses variations séculaires et récentes dans les Alpes orientales. Extract from *Étude botanique de l'étage alpin particulierement en France.* Scientific Committee of the French Alpine Club and Executive Committee of the 8th International Botanical Congress. 6 pp. Paris.

GARNIER, B. J. (1958) *The Climate of New Zealand.* London: Edward Arnold. 191 pp.

GATES, D. M. (1965) Energy, plants and ecology. *Ecology*, **46**, 1–13.

GATES, D. M. and JANKE, R. (1966) The energy environment of the alpine tundra. *Oecol. Planta.*, **1**, 39–62.

GEIGER, R. (1957) *The Climate Near the Ground* (transl. of 2nd German edn). Cambridge, Mass.: Harvard Univ. Press. 494 pp.

GILLISON, A. N. (1969) Plant succession in an irregularly fired grassland area – Doma Peaks region, Papua. *J. Ecol.*, **57**, 415–28.

GILLISON, A. N. (1970) Structure and floristics of a montane grassland/forest transition, Doma Peaks region, Papua. *Blumea*, **18** (1), 71–86.

GODLEY, E. J. (1960) The botany of southern Chile in relation to New Zealand and the Subantarctic. *Proc. Roy. Soc. London*, B, **152**, 457–75.

GRIGGS, R. F. (1938) Timberlines in the northern Rocky Mountains. *Ecology*, **19**, 548–64.

HABECK, J. R. (1969) A gradient analysis of a timberline zone at Logan Pass, Glacier Park, Montana. *Northwest Sci.*, **43**, 65–73.

HAMPEL, R. (1961) Ökologische Untersuchungen in der subalpinen Stufe zum Zwecke der Hochlagenaufforstung. *Mitteilungen der forstlichen Bundes-Versuchsanstalt Mariabrunn*, **59**, Foreword.

HEDBERG, O. (1964) Features of afroalpine plant ecology. *Acta Phytogeogr. Suecica*, **49**, 1–144.

HOOKER, J. D. (1847) *The Botany of the Antarctic Voyage of HM Discovery Ships Erebus and Terror in the Years 1839–1843, I. Flora Antarctica*. London: Lords Commissioners of the Admiralty. 574 pp.

HUECK, K. (1966) *Dei Wälder Südamerikas*. Stuttgart: Gustav Fischer Verlag. 422 pp.

HUSTICH, I. (1966) On the forest-tundra and the northern tree-lines. *Ann. Univ. Turku.*, A, **II**, 36 (*Rep. Kevo Subarctic Sta.*, 3), 7–47.

JENÍK, J. (1959) Kurzgefasste Übersicht der Theorie der anemo-orographischen systeme. *Preslia*, **31**, 337–57.

KILANDER, S. (1965) Alpine zonation in the southern part of the Swedish Scandes. *Acta Phytogeogr. Suecica*, **50**, 78–84.

LAMARCHE, V. C. and MOONEY, H. A. (1972) Recent climatic change and development of the bristlecone pine (*P. longaeva* Bailey) krummholz zone, Mt Washington, Nevada. *Arct. Alp. Res.*, **4** (1), 61–72.

LARCHER, W. (1969) Zunahme des Frostabhärtungsmögens von *Quercus ilex* im Laufe der Individualentwicklung. *Planta*, **88**, 130–5.

LARCHER, W. and MAIR, B. (1968) Die Kalteresistenzverhalten von *Quercus pubescens*, *Ostrya carpinifolia* und *Fraxinus ornus* auf drei thermisch unterschiedlichen Standorten. *Oecol. Planta.*, **3**, 255–70.

LEVITT, J. (1956) *The Hardiness of Plants*. New York: Academic Press. 278 pp.

LI, HUI-LIN (1963) *Woody Flora of Taiwan*. Narbeth, Penn.: Livingston Publishing Co. 974 pp.

LÖVE, D. (1970) Subarctic and subalpine: where and what? *Arct. Alp. Res.*, **2**, 63–73.

LÖVE, D., MCLELLAN, C. and GAMOW, I. (1970) Coumarin and coumarin derivatives in various growth-types of Engelmann Spruce. *Svensk Bot. Tidskr.*, **64** (3), 284–96.

MCVEAN, D. N. (1968) A year of weather records at 3480 m on Mt Wilhelm, New Guinea. *Weather*, **23** (9), 377–81.

MARR, J. W. (1961) Ecosystems of the east slope of the Front Range in Colorado. *Univ. of Colorado Stud., Ser. in Biol.*, 8. 134 pp.

MEIJER, W. (1965) A botanical guide to the flora of Mt Kinabalu. *In: Symposium on Ecological Research in Humid Tropics Vegetation*, 325–64. Govt of Sarawak and UNESCO.

MICHAELIS, P. (1934a) Ökologische Studien an der alpinen Baumgrenze, IV. Zur Kenntnis des winterlichen Wasserhaushaltes. *Jahrbuch für wissenschaftliche Botanik*, **80**, 169–247.

MICHAELIS, P. (1934b) Ökologische Studien an der alpinen Baumgrenze, V. Osmotischer Wert und Wassergehalt während des Winters in den verschiedenen Höhenlagen. *Jahrbuch für wissenschaftliche Botanik*, **80**, 337–62.

MILLER, P. C. (1970) Age distributions of spruce and fir in beetle-killed forests on the White River Plateau, Colorado. *Amer. Midl. Nat.*, **83**, 206–12.

MOLLOY, B. P. J., BURROWS, C. J., COX, J. E., JOHNSTON, J. A. and WARDLE, P. (1963) Distribution of subfossil forest remains, eastern South Island, New Zealand. *New Zealand J. Bot.*, **1**, 68–77.

MOORE, R. M. (1959) Subalpine treeless grasslands in south-east Australia. Abstract *in: Proc. 9th Int. Bot. Congress* (Montreal), **2**, 268.

MORRIS, J. Y. (1965) Climate investigations in the Craigieburn Range, New Zealand. *New Zealand J. Sci.*, **8**, 556–82.

MOSER, M. (1966) Die ektotrophe Ernahrungsweise an der Waldgrenze. *Allgemeine Forstzeitung*, **77a**, 120–7.

OBERDORFER, E. (1960) *Pflanzensoziologische Studien in Chile.* Weinheim: J. Cramer. 208 pp.

PEARS, N. V. (1968a) The natural altitudinal limit of forest in the Scottish Grampians. *Oikos*, **19**, 71–80.

PEARS, N. V. (1968b) Post-glacial tree-lines of the Cairngorm Mountains, Scotland. *Trans. Bot. Soc. Edinburgh*, **40**, 361–94.

PEARS, N. V. (1969) Post-glacial tree-lines of the Cairngorm Mountains, Scotland: some modifications based on radiocarbon dating. *Trans. Bot. Soc. Edinburgh*, **40**, 536–44.

PISEK, A. and WINKLER, E. (1958) Assimilationsvermogen und Respiration der Fichte (*Picea excelsa* Link.) in verschiedener Höhenlage und der Zirbe (*Pinus cembra* L.) an der alpinen Waldgrenze. *Planta*, **51**, 518–43.

RADDE, G. (1899) Grundzüge der Pflanzenverbreitung in der Kaukasusländern. *In:* ENGLER, A. and DRUDE, O. (eds.) *Vegetation der Erde*, Part 3. Leipzig: W. Engelmann.

RAUNKIAER, C. (1934) *The Life Forms of Plants and Statistical Plant Geography* (collected papers, transl. H. Gilbert-Carter and others). London: Oxford Univ. Press. 632 pp.

RIKLI, M. (1946) *Das Pflanzenkleid der Mittelmeerländer*, Vol. II. Berne: Hans Huber. 1093 pp.

RONCO, F. (1970) Chlorosis of planted Engelmann spruce seedlings unrelated to nitrogen content. *Can. J. Bot.*, **48**, 851–3.

SAKAI, A. (1970a) Freezing resistance in willows from different climates. *Ecology*, **51**, 485–91.

SAKAI, A. (1970b) Mechanism of desiccation damage of conifers wintering in soil-frozen areas. *Ecology*, **51**, 657–64.

SCHIMPER, A. F. W. (1903) *Plant-Geography upon a Physiological Basis* (transl. W. R. Fisher). Oxford: Clarendon Press. 840 pp.

SCHROETER, C. (1926) *Das Pflanzenleben der Alpen.* Zurich: Albert Raustein. 1288 pp.

SCHULZE, E. D., MOONEY, H. A. and DUNN, E. L. (1967) Wintertime photosynthesis of bristlecone pine (*Pinus aristata*) in the White Mountains of California. *Ecology*, **48**, 1044–7.

SCHWEERS, W. and MEYER, F. H. (1970) Einfluss der Mykorrhiza auf den Transport von Assimilaten in die Wurzel. *Ber. dtsch. bot. Gesell.*, **83**, 109–19.

SCHWEINFURTH, U. (1957) Die horizontale und vertikale Verbreitung der Vegetation im Himalaya. *Bonn. geogr. Abhandl.*, **20**, 1–373.

SCOTT, D. and BILLINGS, W. D. (1964) Effects of environmental factors on standing crop and productivity of an alpine tundra. *Ecol. Monogr.*, **34**, 243–70.

SHAW, C. H. (1909) The causes of timberline on mountains; the role of snow. *Plant World*, **12**, 169–81.

SKOTTSBERG, C. (1953) The vegetation of the Juan Fernandez Islands. *In: The Natural History of Juan Fernandez and Easter Island*, Vol. II, *Botany*. Uppsala: Almquist & Wiksell. 960 pp.

SPOMER, G. C. and SALISBURY, F. B. (1968) Eco-physiology of *Geum turbinatum* and implications concerning alpine environments. *Bot. Gaz.*, **129**, 33–49.

TRANQUILLINI, W. (1957) Standortsklima, Wasserbilanz und CO_2 – Gaswechsel junger Zirben (*Pinus cembra* L.) an der alpinen Waldgrenze. *Planta*, **49**, 612–61.

TRANQUILLINI, W. (1959) Die Stoffproduktion der Zirbe (*Pinus cembra* L.) an der Waldgrenze während eines Jahr: I. Standortsklima und CO_2 – Assimilation; II. Zuwachs und CO_2 – Bilanz. *Planta*, **54**, 107–51.

TRANQUILLINI, W. (1967) Über die physiologischen Ursachen der Wald – und Baumgrenze. *Mitteilungen der forstlichen Bundes-Versuchsanstalt Wien*, **75**, 457–87.

TROLL, C. (1959) Die tropischen Gebirge. *Bonn. geogr. Abhandl.*, **25**, 1–93.

TURNER, H. (1968) Über 'Schneeschiff' in den Alpen. *Wetter und Leben*, **20**, 192–200.

TUTIN, T. G. (1953) The vegetation of the Azores. *J. Ecol.*, **41**, 53–61.

UNITED STATES WEATHER BUREAU (1931–52) *Climatological Summary of the United States.*

VAN STEENIS, C. G. G. J. (1964) Plant geography of the mountain flora of Kinabalu. *Proc. Roy. Soc. London*, B, 161, 7–38.

WACE, N. M. and HOLDGATE, N. W. (1958) The vegetation of Tristan da Cunha. *J. Ecol.*, **46**, 593–620.

WADE, L. K. and MCVEAN, D. N. (1969) Mt Wilhelm Studies, I. The alpine and subalpine vegetation. *Res. School of Pacific Stud., Dept of Biogeogr. and Geomorphol. Pub.*, BA/1. Canberra: Australian National Univ. 225 pp.

WALTER, H. and MEDINA, E. (1969) Die Bodentemperatur als ausschlaggebender Faktor für die Gliederung der subalpinen und alpinen Stufe in den Anden Venezuelas. *Ber. dtsch. bot. Gesell.*, **82**, 275–81.

WANG, CHI-WU. (1961) The forests of China with a survey of grassland and desert vegetation. *Maria Moors Cabot Foundation Pub.*, 5. Cambridge, Mass.: Harvard Univ. Press. 313 pp.

WARDLE, J. (1970) The ecology of *Nothofagus solandri*, 4. Regeneration. *New Zealand J. Bot.*, **8**, 571–608.

WARDLE, P. (1965a) A comparison of timberlines in New Zealand and North America. *New Zealand J. Bot.*, **3**, 113–35.

WARDLE, P. (1965b) Significance of xeromorphic features in humid subalpine environments in New Zealand. *New Zealand J. Bot.*, **3**, 342–3.

WARDLE, P. (1968) Engelmann spruce (*Picea engelmannii* Engel.) at its upper limits on the Front Range, Colorado. *Ecology*, **49**, 483–95.

WARDLE, P. (1971) An explanation for alpine timberline. *New Zealand J. Bot.*, **9**, 371–402.

WEST, R. G. (1968) *Pleistocene Geology and Biology.* London: Longmans. 377 pp.

WHITE, T. C. R. (1969) An index to measure weather-induced stress of trees associated with outbreaks of psyllids in Australia. *Ecology*, **50**, 905–9.

YOSHIOKA, K. and KANEKO, T. (1963) Distribution of plant communities on Mt Hakkôda in relation to topography. *Ecol. Rev.*, **16**, 71–82.

ZOTOV, V. D. (1938) Some correlations between vegetation and climate in New Zealand. *New Zealand J. Sci. Tech.*, **19**, 474–87.

ZOTOV, V. D. (1965) Grasses of the subantarctic islands of the New Zealand region. *Records of the Dominion Museum*, **5**, 101–46.

Manuscript received April 1971.

Arctic and alpine vegetation: plant adaptations to cold summer climates

W. Dwight Billings
Department of Botany,
Duke University

The word *tundra* is Russian and means 'marshy plain' or 'Siberian swamp', but the origin of the word may be in the Finn–Ugric languages. For example, 'tunturi' in Finnish is an 'arctic hill' and the Lapp 'tundar' means 'hill'. Hills in Lapland usually extend above timberline and are covered with herbaceous and low shrubby vegetation similar to that of the flat arctic plains further north. Whatever the origin, 'tundra' is used for any ecosystem in which the plant cover consists of low herbaceous, dwarf shrub, or lichen vegetation in places which have summers too cold to allow tree growth. In view of the mixed origin of the word, it is permissible to use it anywhere where such conditions exist. Billings and Mark (1961) used the word 'tundra' for southern hemisphere alpine vegetation in New Zealand – and there were some objections because of its northern hemisphere connotations. However, after a more intensive study of New Zealand alpine vegetation in central Otago, Mark and Bliss (1970) imply that 'alpine tundra' is an acceptable generic term that describes southern hemisphere alpine ecosystems dominated by lichens, herbs, and dwarf shrubs which are similar in function and structure of those north of the equator.

1. Geographical distribution of tundra vegetation

It is not possible to be precise in delimiting the boundaries of tundra vegetation because of the often broad ecotones between forest vegetation and tundra in the north (see Larsen, Chapter 7A) and the unevenness of some alpine timberlines (see Wardle, Chapter 7B), or even lack of alpine timberlines in some parts of the world. For these reasons, a good map of tundra vegetation is not possible with our present information. Map boundaries tend to be sharp and misleading. It will suffice to say that, in general, most areas beyond timberline in the Arctic or in high mountains, and which are not covered with permanent ice or snow, support mosaics of tundra vegetational types. Thus, 'timberline' is only a rough boundary at best between tundra and forest. This is particularly true in the Subarctic where large patches of tundra vegetation are embedded in the taiga and forest patches occur as 'islands' in the tundra. As the climate warms or cools, the relative amounts of land covered with forest or tundra in this broad ecotone change with the centuries. Microtopography and microclimates play an important role in determining the existence and positions of these tundra or taiga 'islands' in this broad and flexible belt between forest and 'barrens'.

While timberline is an approximate guide to tundra boundaries in the northern part of the world, it is much less reliable in the southern hemisphere. In the alpine regions of New Zealand, the *Nothofagus solandri* timberline is at too low an elevation to indicate the real boundary between alpine and subalpine conditions; the latter vegetation consists of tall grasses or other herbs, or shrubs. In places such as the western slopes of the Andes in central Chile, there is no montane forest and, thus, no timberline. Here, the lower edge of the alpine zone must be arbitrarily decided on the basis of ecological judgement. In the last analysis, what constitutes 'arctic' or 'alpine' vegetation or 'tundra' must always be a matter of judgement. In the past it has been customary to assume that the timberline-tundra ecotone is near the 10°C isotherm for the warmest summer month. This line does have value. However, it is arbitrary, its position is not always known, microclimate affects it locally, and some plants do not obey the rule. While winter temperatures and snowblast also play an important role in determining the tundra–forest boundary, the uniqueness of tundra and alpine plants lies in the fact that they are the only plants which can metabolize and complete a life cycle at low summer temperatures as well as withstand the winter winds. Thus, the presence of reproducing populations of plants such as *Silene acaulis*, *Oxyria digyna*, *Trisetum spicatum*, *Saxifraga oppositifolia*, and certain other species are quite reliable indicators of arctic or alpine environments. Fig. 8A.1, compiled from our data and other sources, shows the known locations of *Oxyria* and thus approximates the distribution of arctic and alpine vegetation in the northern hemisphere. The species distribution maps in Hultén (1962, 1968) and Porsild (1957) are very helpful in indicating the presence of such vegetation by floristic means.

Because of the moderating influence of the Arctic Ocean on temperature and the relative dryness of the High Arctic, vascular plants occur as far north as Kap Morris Jesup in Peary Land (83°39′N) at the northern tip of Greenland (Holmen, 1957). Because of aridity, many of these high arctic terrestrial ecosystems could be called arctic desert or *polar desert* (Stocker, 1963; Aleksandrova, 1971) rather than tundra in the more conventional sense. Much of the Canadian Arctic Archipelago is essentially arctic desert (Savile, 1964). South of these extremes, tundra extends around the seaward edges of the Greenland ice sheet and in a broad band across the northern edges of North America and Eurasia. Wherever mountain ranges enter the Arctic from the south, as in Alaska, Scandinavia, and the USSR, arctic tundras merge almost imperceptibly with alpine tundras. Further south in the northern hemisphere, alpine vegetation occurs in many mountain ranges and almost reaches the tropics in Nepal and southwestern China.

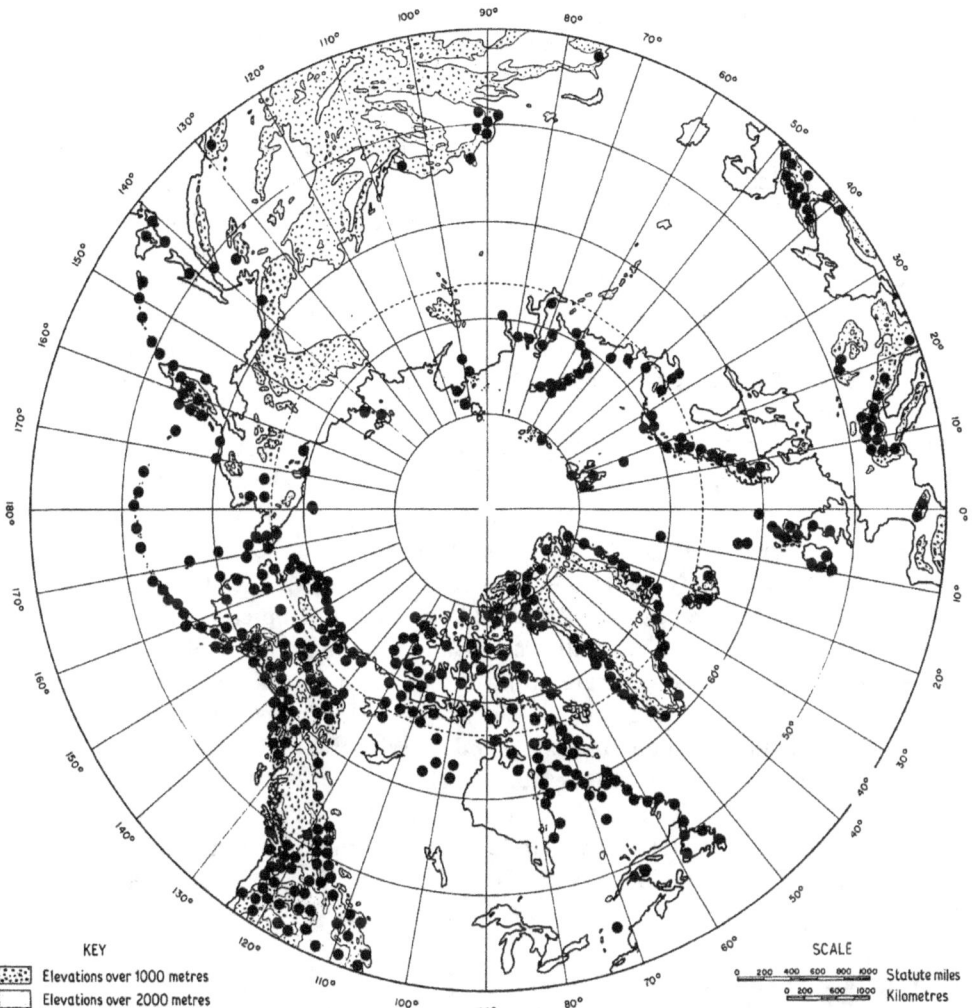

Fig. 8A.1. The distribution of a typical arctic–alpine plant species, *Oxyria digyna* (L.) Hill, based on collections in herbaria. Each dot may represent several collections. The map extends only southward to 40°N but *Oxyria* occurs as far south as 34°N in western North America and to about 28°N in the Himalaya (see Plate 27).

On the high mountains of equatorial Africa and South America are alpine regions quite different in environments, floras and vegetation from those further north. They are so different that it seems questionable as to whether the páramo of the northern Andes, for example, is really 'alpine'. Much of it is probably better considered as 'subalpine'. And yet, at its upper edges, as on Pico Espejo and Pico Bolívar, Venezuela (5002 m), depressed and scattered plants of northern genera such as *Poa*, *Draba* and *Arenaria* represent truly 'alpine' conditions. The relatively small alpine areas of the southern hemisphere in Chile, Argentina, New Zealand and Australia superficially resemble those of the north in environments and in the life forms of the plants but, except for a few bipolar species, have a flora which has evolved quite independently from different ancestral groups. Subantarctic tussock, bog, heath, or fjaeldmark vegetation predominates on most islands between 46°S and 60°S: South Georgia, Prince Edward, Marion, Îles Crozet, Îles Kerguelen, Heard, and Macquarie – but *not* on the Falklands, Campbell, or Aucklands (Greene and Greene, 1963; Greene, 1964; Huntley, 1968). Most of this vegetation can be called tundra. All land south of 60°S (but including the South Sandwich Islands and Bouvetøya) is classified by Greene (1964) as 'antarctic'. This zone extends along the shores of Palmer Peninsula (Graham Land) in Antarctica and, except for the two Angiosperms, *Colobanthus quitensis* and *Deschampsia antarctica*, which extend as far south as 68°12' (Moore, 1970), is mainly characterized by cryptogams (Gimingham and Smith, 1970).

Good (1964, p. 24) estimates the approximate land area covered by arctic or alpine vegetation. By far the largest proportion is in the northern hemisphere: almost 23.6 million km² (9.1 million miles²) compared to less than 1.3 million km² (0.5 million miles²) in the southern hemisphere.[1] In the northern hemisphere about 60 per cent of this vegetation is north of 60°N and is mostly arctic or subarctic. While alpine vegetation in the northern hemisphere covers much more land than its counterpart in the southern hemisphere, the alpine areas of North America and Eurasia are relatively small, isolated, and yet floristically rich, as compared to the great circumboreal belt of arctic tundra. Only about 0.1 per cent of the southern hemisphere tundra vegetation is south of 60°S, and much of it is alpine. In the Arctic the land is peripheral to the Arctic Ocean, which allows thawing in summer and temperatures high enough for plant growth. The reverse holds for the Antarctic with its high, icy central continent. This great ice-covered land allows little or no chance for plant establishment, and tundra is restricted to the small subantarctic islands and the mountains of Patagonia and New Zealand at relatively low latitudes.

2. Vegetation–environmental patterns

It is difficult to generalize about the intricate vegetational–environmental patterns in arctic and alpine regions. The severity of the environments for most people, animals, and plants is well known. But there is a small portion of the earth's plant species, certainly less than 3 per cent, which are adapted to the cold winters and chill summers of arctic and alpine regions. For these species, the environment is *not* severe either physically or biologically. These species, moreover, are unable to grow in those parts of the earth which have warmer and moister summers. This is because of biological competition provided by trees and tall herbaceous plants, and because of the abnormal metabolism of tundra plants caused by temperatures above about 25°C. The general relationships between arctic and alpine plants and their environments are discussed briefly in the

[1] cf. Webber, Chapter 8B.

following pages. For more detailed information it would be well to consult Bliss (1962), Tikhomirov (1963), Walter (1968), Billings and Mooney (1968), and Bliss (1971).

In arctic and alpine environments the plants are small, close to the ground, and often widely separated by bare soil or rock. Unlike the situation within a forest, the modification of microclimate by vegetation is minimal and the physical environment dominates the vegetation. This is not to say that biological influences do not exist; grazing by caribou, musk-oxen, and rodents (especially lemming outbreaks) leave their marks. Radiation and wind, however, have relatively easy access to soil and rock surfaces and to plant leaves and stems also. In such open and windy places, the effects of microenvironment are pronounced. Even a few centimetres difference in microtopography makes a marked difference in soil temperature, depth of thaw, wind effects, snow drifting, and resultant protection to leaves, buds and stems. The microtopographic effect may be caused by a beach ridge, a rock, a peat hummock, a solifluction terrace, a soil polygon rim, or by another plant itself. Whatever the microtopographic pattern, the result is an uneven distribution of snow and thus differential distribution of soil moisture, heat exchange, and protection from windblast. The frequent freeze–thaw cycles at the soil surface coupled with the common presence of permafrost not far beneath the surface results in physical frost-thrusting of soil, rocks and plants wherever water is abundant. Such soil frost activity produces various kinds of polygons, stone nets, and solifluction terraces, which tend to make the edaphic environment unstable. Such widespread soil frost activity, combined with the frequent presence of permafrost not far below the soil surface, are characteristic of arctic and alpine terrestrial environments except where soil water is in short supply. Vegetational patterns in arctic and alpine regions are strongly shaped by such soil frost activity (Raup, 1947; Drury, 1962; Johnson and Billings, 1962).

In an open vegetation of isolated cushion plants such as in a fellfield or polar desert, solar radiation reaches the soil or broken rock surface in relatively large amounts even in the Arctic. In sedge or grass turf, the vegetation insulates against heat flow thus slowing soil thawing and freezing. As a result of these vegetational differences, soil and lower level air temperatures are quite different between an open fellfield and a moist meadow or bog.

Local temperature gradients are governed by the distribution of soil moisture which is dependent largely on snowdrift pattern as determined by the interaction of wind and topography. These gradients usually are steeper in alpine locations (fig. 8A.2), but the effects are also prominent in the Arctic, particularly on the tops and sides of polar desert plateaus.

Since so much of the alpine or arctic landscape is characterized by scattered plants, there is only local modification of the microenvironment except where graminoid sods dominate. The plants themselves are at the mercy of the physical environment. Even though such scattered plants have little effect on the local environment, the microclimate within the crown of the plant itself may be considerably different from that in the open spaces between the plants. This 'phytomicroclimate' is influenced by leaf shape, size, colour pubescence, arrangement, and density. The result is an increase in plant temperature which in sunlight is often significantly higher than that of the ambient air.

Krog (1955), Warren Wilson (1957) and Tikhomirov *et al.* (1960) found this effect in arctic plants, while Salisbury and Spomer (1964) have shown it for leaves of alpine plants. Longton (1970) states that temperatures near the surface of colonies of the moss *Polytrichum alpestre* in Antarctica are often as much as 25°C higher than that of the surrounding air in sunshine. Billings and Godfrey (1967) found that temperatures within the

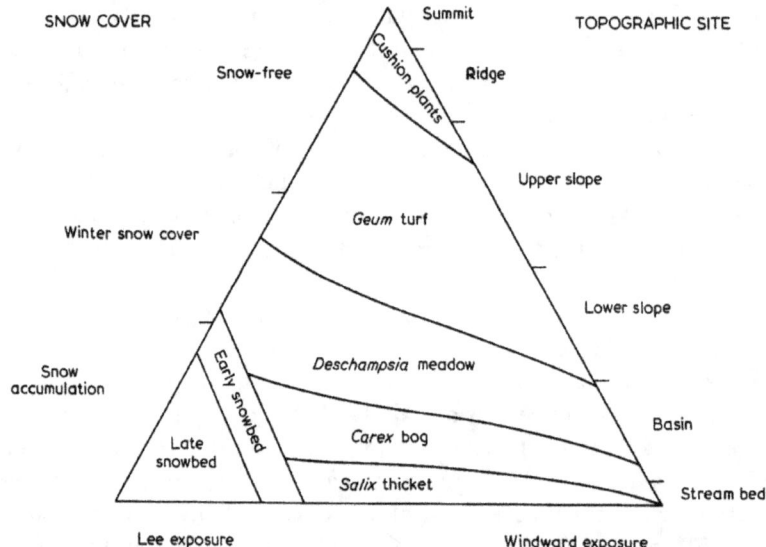

Fig. 8A.2. Diagrammatic representation of alpine vegetational pattern along topographic, wind and moisture gradients in the Beartooth Mountains, Wyoming (redrawn from Johnson and Billings, 1962).

hollow stems of some subalpine plants far exceed those of the ambient air. These heating effects in tundra plants are most pronounced on sunny, cool days and are almost negligible in cloudy weather. Some of the heat energy from the leaves and stems is convected into the air within the plant canopy and this air, protected from the chill wind a few centimetres above, becomes heated. The plant is immersed in its own warm microenvironment and its metabolic processes operate in response to it.

3. Vegetational characteristics

In the Arctic and in the mountains of the northern hemisphere, vegetation beyond timberline is usually a complex mosaic of communities arrangeable into types or continua along gentle or steep environmental gradients. A major environmental factor is topographic exposure, which influences wind. Wind and solar radiation, in turn, control temperature and the distribution of snow and its meltwater. All of this is superimposed upon geologic substratum patterns, soil frost features, and permafrost. The tolerance ranges of plant species when combined with such a variety of microenvironments result in strong patterning into vegetational mosaics on level ground and pronounced continua and/or zonation on steeper slopes. Plate 25 shows arctic tundra vegetation while Plate 26 illustrates the variety of situations to be seen on steep alpine slopes.

One of the common gradients in alpine regions is diagrammed in fig. 8A.2. The wind-swept ridges and steep upper slopes are fellfields characterized by cushion plants and dwarf subshrubs such as *Silene acaulis* and *Saxifraga oppositifolia*. Below these rocky, exposed places are dry meadows, moist meadows, and heaths at midslope. In lower protected places and on lee slopes, snowbanks and their meltwater provide enough protection and water for wet meadows, bogs, and even thickets of small willows. Snow-banks which melt very late in the summer prevent almost all plant growth, and when they do melt they reveal barren rocks and soil with possibly a few lichens and mosses. This is a common type of vegetational gradient but there are innumerable variations on the theme.

Southward from the Arctic into the middle-latitude mountain ranges of the northern hemisphere there is an increase in floristic richness, and the vegetational mosaics and gradients become more complex. However, many species extend from the Arctic down these mountain chains, so that most alpine vegetation resembles that of the Arctic both floristically and in structure. There are exceptions such as the Sierra Nevada of California (Chabot and Billings, 1972), but in Eurasia and North America most alpine mountains show a richer but similar vegetation to that of the Arctic. On entering the tropical mountains, however, alpine vegetation becomes much more complex and quite different floristically except at its upper elevational limits. The tropical alpine areas are often marked by the presence of columnar or subarborescent life forms with large woolly leaves (Hedberg, 1964; Vareschi, 1970). The tropical alpine environment is under diurnal control rather than annual, except for the presence of wet and dry seasons in places such as the northern Andes. There are great differences between day and night temperatures which require adaptations somewhat different from those occurring in the higher latitudes.

Southern hemisphere alpine vegetation shows gradients from ridge to bog somewhat similar to those in the northern hemisphere (Mark and Bliss, 1970). Floristically, though, it has little resemblance to the vegetation of the north except for a few bipolar grasses, sedges, lichens and mosses. Very dwarf shrubs are characteristic of several of the vegetational types.

It is not possible to mention all of the many papers which have described tundra vegetation. However, excellent descriptions of vegetational types, patterns, composition and structure may be located in the following publications.

Arctic: Britton (1957), Böcher (1963), Johnson *et al.* (1966), Polunin (1948).

Eurasian alpine: Dahl (1956), Gjaerevoll (1956), Braun-Blanquet and Jenny (1926), Braun-Blanquet (1948), Sukachev (1965), and many papers in the series *Beiträge zur geobotanischen Landesaufnahme der Schweiz* published by the Schweizerischen Naturfors-chenden Gesellschaft in Bern.

North American alpine: Bliss (1963), Marr (1961), Johnson and Billings (1962), Chabot and Billings (1971).

Tropical alpine: Hedberg (1964), Weberbauer (1911).

Southern hemisphere alpine: Billings and Mark (1961), Costin (1967), Mark and Bliss (1970).

Subantarctic and Antarctic: Taylor (1955), Longton (1967), Gimingham and Smith (1970).

4. Plant adaptations to arctic and alpine environments

LIFE FORMS AND GENERAL MORPHOLOGY

Outside the tropics, vegetation beyond timberline is characterized by herbaceous plants and low or prostrate shrubs. The more severe the environment, the more likely the shrubs are to be prostrate or restricted to snow accumulation areas. The amount of new wood needed by such a shrub each year is very small even compared to that in timberline trees or krummholz, because support is not needed. Prostrate shrubs are almost as well-adapted to severe tundra conditions as are herbs, and some species, such as *Salix arctica*, extend north of latitude 83°, almost to the limit of plant growth in the Arctic. A similar type of dwarf shrub adaptation has evolved independently in the subantarctic alpine flora of New Zealand, and is related there also to increasing severity of the environment (A. F. Mark, personal communication, 1970).

One of the advantages of a prostrate shrub in a severe environment is that of permanence. Its yearly maintenance costs in carbohydrates are low and reproduction need occur only at long intervals. The long-lived low shrub provides some flexibility in meeting short-term changes in the environment. In this respect, it is the equal of the herbaceous perennial which depends upon its long-lived roots and rhizomes. If the shrub is evergreen there are additional advantages, the same as are possessed by evergreen timberline trees: it is not necessary to spend food reserves on a wholly new photosynthetic apparatus each year. Hadley and Bliss (1964) also suggest that older evergreen leaves act as winter food storage organs, since lipids and proteins are mobilized and their stored energy translocated from old to new leaves during the growing season.

As well adapted as some shrub species are, the greater number of vascular plant species in the tundra are herbaceous. Almost all of these herbs are perennials with large underground root or stem storage systems; few are biennial, and fewer still are annual. Of the 96 species of vascular plants in Peary Land (Holmen, 1957) only one, *Koenigia islandica*, is an annual, and it is very small (2 to 3 cm high). Even in a moist alpine region, the Beartooth Plateau of Wyoming, only 3 species are annual in a vascular flora of 191 species – and 1 of these annuals is the same dwarf *Koenigia islandica* (Johnson and Billings, 1962). In most alpine or arctic places, annual species make up only 1 to 2 per cent of the flora. The disadvantages of the annual life form in a cold short growing season lie principally in the necessity to complete the whole life cycle from germination to seed production. Even some species which normally are annuals near timberline and are listed as annuals in floras may be represented by perennial ecotypes when they occur in tundra. For example, Williams (1970) found that in the Medicine Bow Mountains, Wyoming, all plants of *Androsace septentrionalis* at an elevation of 2600 m were annual, but at 3640 m at least 80 per cent of the plants of the same species were perennial. This species is listed by most floristic manuals as an annual. Alpine floras in relatively dry regions do have slightly higher percentages of annuals. In Howell's (1951) alpine flora of 108 taxa in the southern Sierra Nevada of California, 6 species are annual. Similarly, Chabot and Billings (1971) found only 3 species of annuals in a flora of 72 taxa in the Piute Pass areas

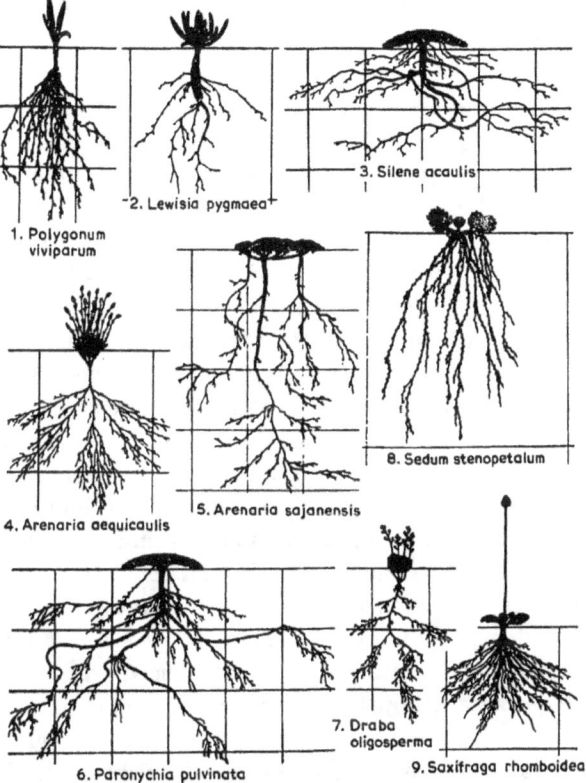

Fig. 8A.3. Silhouettes of characteristic alpine plants from the Medicine Bow Mountains, Wyoming, at 41°30′N. Squares are 1 dm on a side (from Daubenmire, 1941).

of the central Sierra Nevada: all 3 are endemic to the mountain ranges near the Pacific Coast south of Canada, and all have the dwarf size of *Koenigia islandica*.

The typical plant life form of arctic and alpine environments, then, is an herbaceous perennial with a relatively large root and/or rhizome system. These perennials are of three main types: graminoid, leafy dicot, and cushion dicot. A few small ferns and small monocot plants with bulbs also occur here and there. In a typical alpine situation, dicots generally have a deep primary root system with shoots proliferating near the soil surface (see fig. 8A.3). A few species have rhizomes which are utilized in food storage, and some species reproduce by horizontal rhizomes. Most alpine dicots in western North America reproduce by seeds rather than by rhizomes. In the High Arctic, however,

seed set may not occur except in unusually warm years and, therefore, vegetative reproduction by rhizomes or bulbils is the rule. Even within a single species, arctic ecotypes
may have rhizomes while western American alpine ecotypes do not. *Oxyria digyna* is a
good example of this. The photograph of *Oxyria* (Plate 27) shows a typical non-rhi-
zomatous plant in the Beartooth Mountains, Wyoming; this population stores carbohydrates in roots and depends upon seeds for reproduction. Any population of *Oxyria*
from northern Alaska, however, reproduces mainly by rhizomes (Mooney and Billings,
1961).

Whether characterized by rhizomes or not, the underground parts of most arctic or
alpine perennial dicots have much greater dry weight than the stems and leaves (two
to six times greater) and act as carbohydrate storage organs (Mooney and Billings, 1960;
Scott and Billings, 1964). This also holds true for plants in alpine herbfields in New
Zealand, but less so for alpine cushion plants there (Bliss and Mark, 1965). The New
Zealand alpine cushion plants are essentially dwarf evergreen shrubs, and like other
alpine evergreen shrubs store most of their food as lipids in the leaves. For example, the
leaves of *Celmisia viscosa* in New Zealand contain from 15 to 20 per cent lipids.

The herbaceous perennial, whether dicot or monocot, has another adaption which
makes the most of a short, cold growing season: the pre-formed shoot and flower bud.
This has been illustrated for almost all of the plants of northeast Greenland by Sørensen
(1941). This bud appears to be almost universal in alpine plants as well. The flower-bud
primordia are often initiated early in the growing season of the year before flowering,
and are usually well developed by the time the perennating bud is formed late in the
summer. In some species, the first indication of floral primordia is two seasons ahead of
flowering (Sørensen). The advantages are obvious: shoot elongation and flowering can
and do occur very quickly after snow-release and temperature increase. As a corollary,
flowering is thus also dependent upon environmental conditions during floral initiation
in the year preceding actual flowering. Most of these flower buds overwinter quite
close to the soil surface, and in snow-free areas are subjected to intense cold. Pre-formed
flower buds are very common in New Zealand alpine plants though most of the species
and genera are unrelated to those of the northern hemisphere tundras. Mark (1970)
found pre-formed flower buds in 81 out of the 100 species he examined in the New
Zealand mountains.

The disadvantages of the annual life form in a cold short growing season lie principally
in the necessity to complete the whole life cycle from germination to seed production.
This demands an ability to germinate and carry on metabolism at high rates at low
temperatures. In fact, although data do not seem to be available, it might be postulated
that annual plants in an arctic or alpine environment should have higher photosynthesis
and respiration rates than perennial plants from the same locations at the same temperatures.

The most severe alpine or arctic environments are of two general types: (1) late-
lying snowbank areas where the growing season is extremely short, and (2) windswept,
dry ridges or plateaus. Both of these microenvironments become more severe toward
the High Arctic and the 'high alpine'. Plants are absent in the most severe situations in

either late-lying snowbanks or exposed ridges. However, at the limits of plant growth in both cases, plants tend to be cryptogams – for the most part mosses in and below the snowbanks, and crustose and other kinds of lichens and sometimes mosses on the wind-swept, rocky barrens of the hilltops. In the late-lying snowbanks of the Scandinavian mountains, for example, Gjaerevoll (1956) reports that vascular plants are almost lacking and the sites are dominated by various moss communities. Similar situations exist throughout the Arctic and also in alpine regions, although mosses are relatively unimportant in alpine snowbanks in the western American mountains. The High Arctic north of 77° is so dry and cold (Savile, 1961, 1964; Holmen, 1957) that snowbank vegetation is considerably restricted in size and also in flora. In places snowbank vegetation may even be absent, in the sense that it occurs further south in moister climates. Often it consists of only some mosses and dwarf graminoids along miniature drainage channels below the melting snow.

Windswept ridges and plateaus provide the most severe environments in terms of temperature, drought stress, soil frost action, and wind abrasion. Beyond the limits of any vascular plants, crustose, thallose, and sometimes fruticose lichens will be found. In the High Arctic, hilltops and ridges are virtually unvegetated except for such lichens and the grey moss *Rhacomitrium lanuginosum*, with an occasional plant of *Saxifraga oppositifolia* (Savile, 1961). Polunin (1951) says that such barrens are 'desolate in the extreme' and that 'most areas have scarcely a plant to be seen'. Similar but smaller areas occur on high alpine ridges. Comparable domination of moist areas by moss carpets and dry, rock areas by crustose and fruticose lichens characterizes the maritime Antarctic (Longton, 1967).

At the limits of vascular plants in snowbanks there is apt to be a variety of life forms: graminoid, small herb, cushion plants, prostrate shrubs such as *Salix arctica* and *S. herbacea*. On windswept ridges, however, small-leaved dwarf rosette, mosslike or cushion plants with occasional grass tussocks represent vascular plants at their limit of tolerance. In the Arctic, such plants are *Saxifraga oppositifolia*, *Papaver radicatum*, *Dryas integrifolia*, *Erysimum pallasii*, *Saxifraga tricuspidata*, *S. caespitosa*, *Draba bellii*, *Cerastium arcticum*, *Luzula confusa*, *Poa abbreviata*, *Hierochloe alpina*, and others (Savile, 1961, 1964; Polunin, 1951; Holmen, 1957). Plants of these species not only can withstand the winter cold and wind without a snow cover, but can tolerate the drought of summer in the most severe environment the Arctic has to offer. Even in such extreme places, nutrients around bird perch rocks or lemming burrows allow a relatively luxuriant growth of some of these same species such as *Papaver radicatum*.

Similar rosette, mosslike or cushion plants and dwarf graminoids characterize the windswept rocks of high alpine locations. However, high on the rocky peaks or gipfels there are also miniature 'snowbeds' in the crevices of the rocks; these shelter and supply water to plants of a softer nature such as *Ranunculus* which depend upon meltwater to a greater extent than do the cushion plants. In many cases, these softer plants reach higher elevations than do the mosslike or cushion plants on the more exposed sites. Such 'snow-cranny' plants extend well up into the zone of permanent snow where their only plant associates are crustose lichens and a few moss tufts. Reisigl and Pitschmann (1958) state that *Ranunculus glacialis*, one of these 'snow-cranny' species, reaches the highest elevation of any vascular plant in the Alps at 4270 m on the Finsteraarhorn. A parallel situation exists at the other end of the world in New Zealand, where *Ranunculus grahami* exists in snow-crannies well above the permanent snowline as high as 2900 m on Malte Brun (Fisher, 1965); this is about the altitudinal equal of any vascular plant in New Zealand.

Webster (1961) lists the highest elevations known for vascular plants. The highest ones listed are *Lagotis glauca*, *Potentilla saundersiana*, *Arenaria* sp., and *Pedicularis* sp. photographed by Oleg Polunin at 5945 m on Mohala Bhanjyang, Nepal. Apparently unknown to Webster (and not noticed by Billings and Mooney, 1968, until called to our attention by Paul W. Richards) is a collection of *Arenaria musciformis* from 6222 m made by Wollaston in 1921 on the First Everest Expedition (see p. 833 in Schroeter, 1926). Since Webster's report, Swan (1967) found *Stellaria decumbens* as high as 6136 m on Makalu, also in the Himalayas. The plants on Makalu were confined to what Swan calls the 'rock-base niche', where small snowdrifts are captured and provide meltwater as a result of thermal re-radiation from the rocks. The 'rock-base niche' is the ecological equivalent of our 'snow-cranny'. This protected and water-supplying microenvironment allows vascular plants to reach their altitudinal limits in the cold, windy and dry environments of the high peaks where most snow sublimes or blows away; thus, lack of liquid water is the principal limiting factor.

As in the polar lands, lichens occur beyond the limits of vascular plants on most high mountains. This is true particularly of crustose lichens on the rocks in the mountains of Europe and North America. Mattick (1950) states that lichens exist as high as 6200 m in the Himalayas. According to Swan (1967), however, lichens are scarce or lacking at these high elevations where vascular plants occur above 6100 m. He ascribes this to the lack of surface water from snowmelt. Snowblast could also be a factor. At elevations in the Himalayas below 5750 m, subsurface liquid water from snowmelt allows the development of little 'oases' of vascular plants in the scree. Here, there are lichens tangled in with the small herbaceous plants. The relative aridity due to low temperatures and snow sublimation, thus confining plants to portions of scree slopes with underground meltwater sources or to rock-base niches, is strongly reminiscent of Savile's (1961) observations in the High Arctic on Ellef Ringnes Island.

Apparently, some lichens and mosses can withstand environments with lower 'growing' season temperatures better than vascular plants, if there is a supply of surface liquid water at some time during the year. If there is not such a surface supply, rooted vascular plants may actually do as well if not better than the cryptogams. The former situation prevails in places such as the relatively moist west coast of the Antarctic Peninsula, where the richness of the lichen and moss vegetation is remarkable (Rudolph, *in* Greene *et al.*, 1967; Gimingham and Smith, 1970). Of the c. 350 species of lichens known from Antarctica, c. 100 have been found only on the west coast of the Antarctic Peninsula. The east coast of the peninsula, being drier and colder, has only about 20 per cent as many lichen species as the whole peninsula. The lichen flora and vegetation of the rest of the continent is relatively depauperate, a fact related to cold and aridity. The same decrease in luxuriance in lichen vegetation may be seen in a transect from the relatively moist Low Arctic to the arid High Arctic.

Provided there is an occasional supply of water, lichens and mosses seem more efficient and better fitted to extremely cold summer situations than most vascular plants. There appear to be several advantages to the lichen life form in a cold environment. First, the highest daytime temperatures are close to the rock– or soil–air interface where the lichens exist. This can produce meltwater and allows photosynthesis in the lichens to operate efficiently. The very low night temperatures at the same interface do not damage lichens. Furthermore, Lange (1965) has found optimum photosynthetic temperatures for several species of alpine and antarctic lichens to be around 5°C. All of these lichens had positive net photosynthesis in light even at several degrees below 0°C (as low as −10°C in *Parmelia coreyi* from Victoria Land, Antarctica, and *Stereocaulon alpinum* from the Ötztaler Alps). Some CO_2 uptake occurred at temperatures as low as −24°C in lichens from a lowland dry site in Europe, so there appears to be little correlation of site with the ability of a lichen to take up CO_2 in the light at low temperatures. The lowland European lichens may well be more active in the winter than they are in the dry summer, while the reverse is probably true of the polar and alpine lichens.

Lastly, lichens are long-lived perennials – up to 1300 years for *Rhizocarpon geographicum* in the Alps and 4500 years in West Greenland (Beschel, 1961). As such, there is no need to produce new biomass each year. Near the limits of plant growth this is a real advantage, since favourable years for growth may be far fewer than unfavourable ones. Lichen growth may be thought of as being 'opportunistic', thus taking advantage of temporary environmental events to an extent not possible in most vascular plants. The reverse is also true: lichens can remain dormant for long periods of time at low temperatures, and when thawed resume normal physiological activity. Lange (1966) measured the photosynthesis rates of *Cladonia alcicornis* at various temperatures then froze the thalli at −15°C for periods up to 110 weeks; photosynthesis rates were normal after thawing. In the antarctic moss *Polytrichum alpestre*, Longton (Greene and Longton, 1970) found that young sporophytes frozen at −10°C for 14 months developed normally at 10°C after being slowly thawed.

PHYSIOLOGICAL ADAPTATIONS OF ARCTIC AND ALPINE VASCULAR PLANTS

Combined with morphological characteristics and the local distribution of a population of any arctic or alpine plant species in relation to microsite, a large part of the ability of a vascular plant to survive and reproduce at low temperatures depends upon the physiological characteristics of the plant as a whole. These characteristics are best seen and understood in the context of the life cycle of a typical arctic or alpine plant from the time it comes into existence as a seed until it, in time, produces viable seed or vegetative propagules.

Seed dormancy and germination

Because of the shortness and severity of the growing season, seeds may not be produced at all. If the weather does allow successful seed production, ripening usually occurs not long before the return of winter conditions. Indeed, ripening may not occur during the year of flowering but, as in *Brayia humilis* in northeast Greenland, the developing fruits may go into winter dormancy and ripen the following summer (Sørensen, 1941).

One might expect to find almost universal seed dormancy as a protective mechanism against premature germination in arctic and alpine seeds, but such is not the case. Given a suitable environment in the laboratory, fresh seeds of most tundra plants germinate easily, if sometimes slowly (Söyrinki, 1938, 1939; Chabot and Billings, 1972). Seeds of some species, however, do show dormancy mechanisms of one kind or another (Amen, 1966). Of the sixty-two alpine species listed by Amen, only about 40 per cent showed dormancy at time of harvest. *Carex*, *Trifolium*, and *Salix* are the principal genera having dormant seeds. Most dormancy is caused by seed-coat inhibition (*Carex*, *Luzula*, *Thlaspi*, *Trifolium*) and can be overcome by scarification. A few species, such as *Erythronium grandiflorum*, require a chilling period at c. 4 °C, while *Cerastium beeringianum* and *Saxifraga rhomboidea* go through after-ripening phases at higher temperatures. Amen's review indicates that seed dormancy is more common among dominant and abundant species; he suggests that such dormancy may contribute to their success.

We can conclude that intrinsic or constitutive seed dormancy is relatively rare among arctic or alpine species. Even such dormancy as does exist is easily overcome by scarification, or light. Dormancy in nature is mainly under the control of the tundra environment and its low winter temperatures, which come early and stay late. Seed usually matures too late in the season to meet the right combination of temperature and moisture over sufficient time for germination to occur.

Optimum germination temperatures, as measured by speed and completeness of germination, are surprisingly high (20–30 °C) for most arctic and alpine species. Working with *Oxyria*, Mooney and Billings (1961) found that optimum germination temperature was 20 °C. There was some germination as low as 10° and as high as 30 °C, but no germination at 3 °C. These data have been confirmed by Chabot and Billings (1972) with several Sierran populations of *Oxyria* from source elevations ranging from 3000 m to 4200 m. All achieved maximum germination at cycling temperatures averaging 20 °C or higher. In one Sierran alpine species, *Calyptridium umbellatum*, Chabot and Billings found optimum germination at temperatures of 30 °C or higher. While germination of Sierran alpine seeds of a number of species was nil or very low between 10° and 15 °C, several species from the warmer desert at the base of the Sierra germinated optimally at temperatures well below 20 °C. Seeds of some desert species germinated at almost 100 per cent at 5 °C. Desert seeds also germinated more rapidly than alpine seeds even at higher temperatures.

Mooney and Billings (1961) tested seeds of four populations of *Oxyria digyna* at temperatures which cycled diurnally between 13° and 2 °C (mean 7.5°), thus approximating summer temperatures on the Alaskan arctic coastal plain. Germination (as high as 65 per cent) took place in all populations. Germination at constant temperatures was low even at 10 °C and did not occur at lower constant temperatures. The only evidence of germination in a tundra species at a constant temperature of 5 °C seems to be that of Holtom and Greene (1967) for *Deschampsia antarctica*, one of the two antarctic angiosperm species. The other, *Colobanthus quitensis*, did not germinate at constant 5°C. However, seeds of both species germinated well if exposed to temperatures which cycled between 5° and 18°C. These results with *Oxyria*, *Deschampsia*, and *Colobanthus* confirm what apparently is a typical reaction to temperature in the germination of tundra seeds.

Field temperatures in arctic and alpine sites are not constant. The germination that occurs is thermally opportunistic, mostly at temperatures lower than the laboratory optimum, and over a much longer period of time. For example, Sørensen (1941) found that germination in several species at Eskimonaes, in northeastern Greenland, could take place in early June even though the surface soil and seeds were frozen for many hours each day. Apparently, arctic or alpine seeds germinate better at low alternating temperatures than they do at low constant temperatures; but more research is needed.

When does germination occur in the field? The best answers to this question may be found in the observations of Söyrinki (1938, 1939) and Sørensen (1941). With their observations and the available laboratory temperature data as guides, germination appears to take place a week or so after snowmelt in the early summer, after soil surface temperatures reach 10° or 15°C in the daytime, and before the soil dries out. Most germination in late summer is impeded by lack of soil moisture and low temperatures in the upland tundras, and most (but not all) of the species there produce non-dormant seed. Late soil moisture is available in most wet meadows; many (but not all) of the principal species in such places have seed dormancy mechanisms (Amen, 1966): *Carex*, *Luzula*, *Erythronium*, *Saxifraga rhomboidea*, *Deschampsia caespitosa*, *Polygonum bistortoides*. However, some plants in such wet places do produce seed that germinates the same year. Bliss (1958) found that seeds of *Salix planifolia* var. *monica* in the Rocky Mountains ripen in July and germinate immediately in the wet habitat. On the other hand, seeds of *Salix brachycarpa* do not ripen until August, when its site has become rather dry; these seeds overwinter in viable form and germinate early the next summer.

We can conclude that almost all seed germination, in alpine locations at least, takes place in early summer after snowmelt during the year following seed production. Most species lack a seed dormancy mechanism, and the elapsed year (or more) between seed production and germination is environmentally imposed. There are a few species, however, which do have an intrinsic seed dormancy mechanism of one kind or another; the commonest type is caused by a hard and impermeable seed coat. Many but not all of these species with seed dormancy are dominant and abundant in moist or wet alpine meadows.

Since tundra environments may not be suitable for germination every year, we would like to know how long such seeds may remain viable. Even under good germination conditions, there is often considerable intrapopulational variation in how long it takes a seed to germinate. In one sample of snow tussock (*Chionochloa rigida*) seed from the New Zealand mountains, Mark (1965) found (at 21°C in the dark) that after 50 days only 31.5 per cent germinated; 40 per cent had germinated at 109 days; and the last germination did not take place until 1450 days, almost 4 years after the start, at which time 94 per cent of the seed had germinated and the experiment was terminated. The delayed and prolonged germination within many Sierran alpine seed populations found by

Chabot and Billings (1972) provides additional evidence of the presence of germination polyphenism (Steiner, 1966) in alpine species. Such intermittent intrapopulational germination over a long period of time could have survival value for a tundra population near the limits of growth.

Seedling establishment

Even less is known of seedling establishment in tundras than is known about germination requirements, although some information is provided by Söyrinki (1938, 1939). Since most seed germination in the tundra occurs in early summer, the seedling has only a few weeks to develop a root system and to produce enough carbohydrates to allow survival through the following winter. There are several reasons why years may pass between episodes of successful seedling establishment of a species in a given tundra location. First, the temperature must be warm enough for germination to take place. Secondly, this must occur early enough in the summer to allow time for good growth before the return of temperatures constantly below freezing, and in some years these come unusually early. Bliss (1958) notes that seedling reproduction is more common in those species growing on deeply thawed soils in the Arctic. Thirdly, the seedlings must not be exposed to drought in the latter half of the summer before the root system has penetrated to a reliable water supply. The more severe the tundra (High Arctic, high alpine), the fewer are the years in which all of these conditions are met and in which, therefore, there is a good chance for seedling survival.

Perhaps it should not be surprising that germination temperature requirements are fairly high. If temperature thresholds were lower, germination could be triggered in a season too cold for adequate photosynthetic and respiratory rates in the seedling – or too late in a moist season in which the seeds were produced.

Seedling growth is very slow under natural conditions, and at the end of the first season the seedling is very small in most species. Wager (1938), at 68°30′N in East Greenland, found that *Oxyria* seedlings at the end of their first summer had two cotyledons and two leaves of 2.5 mm diameter. In *Saxifraga oppositifolia*, growth was so slow that true leaves did not appear until the second year. We have found that the first year of an *Oxyria* seedling in the tundra is devoted primarily to root-system establishment. Growth continues to be slow and it is usually several years before the plant is large enough and has enough carbohydrate reserves to produce flowers. Only in a miniature annual plant such as *Koenigia islandica* are all metabolites used very quickly in flowering and seed production.

Mineral nutrition

A considerable amount of information on mineral cycling is available for the Arctic. This is particularly true for aquatic systems more so than for terrestrial parts of the tundra ecosystem. However, Pieper (1964) and Schultz (1969) have shown how the cycling of phosphorus, potassium, and nitrogen in the tundra vegetation at Barrow, Alaska, are strongly correlated with the three- to four-year lemming population cycles in this same ecosystem.

It is known also that radionuclides such as Cesium-137 and Strontium-90 are accumulated readily by arctic vegetation, particularly lichens. These nuclides are then passed up the food chain to man through grazing by caribou in Alaska, and in Lapland by grazing of reindeer (Watson *et al.*, 1964; Hanson *et al.*, 1966; Potter and Barr, 1969; Miettinen, 1969) (see Osburn, Chapter 16, pp. 886–9).

However, while considerable information is available on concentrations, flow, and cycling of certain minerals, as Johnson (1969) has indicated, there really is not much knowledge as yet on the mineral requirements of individual plants and the rates and seasonality of uptake of these minerals. In regard to nitrogen, we have the classical work of Russell (1940) on the soils and plants of Jan Mayen Island. He demonstrated clearly the importance of bird and animal excreta in increasing the nitrogen content of the soil. In the polar desert region particularly, the presence and abundance of certain vascular plant species and lichens around bird perch rocks and old lemming burrows or skeletal remains of larger animals is particularly noticeable. This may not be a simple nitrogen problem but may involve calcium, phosphorus, and other elements in a complex cycle or even steady state situation. Russell suggested also that low soil temperatures are probably partly responsible for the low rates of soil bacterial activity, and thus are concerned with the relative lack of available nitrogen in many of these arctic soils. There is a noticeable absence of Leguminosae in much of the High Arctic and thus a lack of nitrogen fixation by bacterial nodules on the roots of such plants. It may be that, in the polar desert region, *Dryas integrifolia* clumps could be effective in adding nitrogen to the soil. Crocker and Major (1955), for example, in southern Alaska, found that *Dryas drummondii* mats accumulated considerable nitrogen in the soil as did *Alnus* in the early stages of succession after deglaciation. One cannot overlook, also, the possible contributions of nitrogen by blue–green algal crusts and some lichens.

Pigments
There is some evidence that arctic plants have more chlorophyll per unit leaf area than alpine ones in the same species (Mooney and Billings, 1961, for *Oxyria*, and Mooney and Johnson, 1965, for *Thalictrum alpinum*). Some of this appears to be due to ecotypic differences, but if the ecotypes are grown reciprocally in simulated arctic and alpine environments, there appears to be an interaction between light and temperature which modifies chlorophyll content phenotypically. In *Oxyria*, at least, there seems to be some survival value in lower chlorophyll content of alpine populations. But even in this species, careful experiments show that this adaptation is not entirely independent of temperature (Billings *et al.*, 1971). Intensity of light must also be involved in the interaction between environment and genotype which governs chlorophyll content. Since solar radiation intensities are much higher in alpine environments than in lowland or arctic situations, one would expect more photo-oxidation or chlorophyll at high elevations. But photo-oxidation susceptibility must have a genetic base, since we can detect Sierran Oxyrias in the phytotron by their pale leaves as compared to the darker green arctic forms growing with them under the same conditions.

The alpine ultra-violet environment and its effects on plant pigments has been studied by Caldwell (1968). Almost all uv damage to higher plants in nature is the result of uv-B (2800–3150 Å). While direct beam solar uv-B increases with altitude, sky uv-B decreases; the result is only a modest rate of increase in global uv-B with elevation. On a cloudless summer day in Colorado, Caldwell found that integrated uv-B irradiance at 4350 m was only c. 26 per cent greater than at 1670 m. Leaves of alpine plants (including those of *Oxyria digyna*) in a simulated alpine environment in the laboratory were subjected to precisely defined uv-B irradiation (1.4×10^{15} quanta cm^{-2} s^{-1} at 2967 Å for 40 minutes) without the presence of longer wavelengths. Chlorophyll breakdown and tissue destruction occurred in some of these plants. Photo-reactivation of such damage took place when leaves were exposed to longer wavelengths (2.0×10^{16} cm^{-2} s^{-1} at 4358 Å for 4.5 hours) immediately after doses of uv-B irradiation. In the field, natural filtration of uv-B by epidermal pigments (mainly flavonoid compounds) in alpine plants combined with photo-reactivation are the most likely adaptations to the uv-B climate Recently, Caldwell (1972) has extended his measurements of biologically effective uv-B irradiation to the Arctic. At Barrow, Alaska, 71°20′N, he found substantially lower intensities than would be expected at low elevations in the middle latitudes for the same solar angles. He attributes this decrease in uv-B in the Arctic to uv attenuation by the high atmospheric ozone concentrations in northerly latitudes. The effects of this lower uv intensity on arctic tundra plants and their pigments is unknown, but it can be surmised that there should be less chlorophyll destruction than in alpine situations.

The bright light and low temperatures of alpine environments which result in low chlorophyll content also allow a build-up of anthocyanins in the leaf and young stem. This can be seen in the high mountains, particularly early in the season just after snow-melt and again near the end of the season just before the leaves die back. The same effect may be observed in growth chambers by lowering the temperature while light intensity remains the same.

What are the functions of anthocyanins in the leaves of alpine plants? First, it is possible that anthocyanins act as a 'sink' for excess sugars during cool, bright weather or under conditions of poor nitrogen or phosphorus nutrition. Secondly, they may act as absorbers of ultra-violet radiation in the epidermis or mesophyll. Caldwell (1968) provides some evidence with *Sedum rosea*, an alpine species, where green epidermis transmitted almost three times as much uv as did epidermis with much anthocyanin. He also found that red leaves of *Oxyria digyna* and *Geum rossii* from the Rocky Mountains were consistently less susceptible to tissue destruction by irradiation of 2967 Å than were green leaves of the same species. Red leaves of both species had reduced uv transmission. Colourless flavonoids in the epidermis also reduced uv transmission in some species. From these results, it seems clear that an epidermis rich in anthocyanin makes a good uv filter. A third possible function of anthocyanin in alpine leaves and young stems may be the absorption of energy at other wavelengths with resultant increase in leaf temperatures; we have no evidence for this but merely suggest it as a possibility.

Photosynthesis and respiration
The key to successful adaptation to arctic or alpine environments is the development and operation of a metabolic system which can capture, store, and utilize energy at low

temperatures and in a short period of time. The short, cold tundra 'growing season' is at the opposite end of an environmental gradient from the moist, warm, yearlong environment of a tropical rainforest. In between lie the seasonal warm summer–cold winter conditions of temperate forests, meadows, and most crop land. An understanding of the workings of the photosynthetic-respiratory systems of tundra plants will be aided if we ask how such systems compare with their temperate and tropical counterparts in regard to seasonality, temperature effects, light effects, and the availability of metabolic gases and water. A start toward an answer may be found in the excellent review by Pisek (1960). Much new information on photosynthesis and respiration has been made available since Pisek's review; some of this will be presented here.

Seasonality of carbon metabolism. The photosynthetic seasons of arctic and alpine tundras are fully as short as those of desert vegetation but somewhat more regular. The principal limiting factor ʾis low temperature, and of course, in all but the windy ridge habitat, the duration of snc w cover. The snowfree period decreases with elevation (see Pisek, 1960, fig. 3; Winkler and Moser, 1967) and towards the centres of snowbeds so that, at last, any snow-free period is so short that there is no photosynthetic season. For most alpine plants, the season ranges from six to ten weeks (Bliss, 1956; Billings and Bliss, 1959) with low temperatures and snowmelt governing the start of the season; relative soil drought, return of low temperatures, shortened photoperiod, and perhaps carbohydrate accumulation govern the end of the period. In years of unusually great snow accumulation, the plants may not be uncovered at all or for only a few days of activity before the new snow and low temperatures of September return (Billings and Bliss, 1959; Moser, 1969).

Photosynthetic activity in alpine plants occurs throughout most of the snow-free period when temperatures during the daylight hours are near or above 0°C, but photosynthesis is not uniform throughout the season. Net photosynthesis of shoots is relatively low early in the season because of high respiratory rates associated with rapid growth immediately after snowmelt (Hadley and Bliss, 1964; Bliss, 1966). However, some of the respiratory carbon dioxide may be recycled by photosynthesis inside the hollow young stems of rapidly growing plants around snowbanks (Billings and Godfrey, 1967), thus making up, in part, the photosynthetic deficit during early growth.

In the Arctic, perhaps because of the longer photoperiod, plants may start growing somewhat earlier than in most alpine situations. Initiation of growth can be as early as late April or early May in *Saxifraga oppositifolia* (Plate 28) on dry sites in northern Norway, and as early as late May even at 73°30′N at Myggbukta in northeastern Greenland (Sørensen, 1941). At the latter location, flowering of *Saxifraga* occurs before the mean daily temperature reaches 0°C. Zubkof (1935) also noted that *Saxifraga oppositifolia* was the first plant to show signs of life in the spring on Novaya Zemlya at 76°14′N. Growth started before daily mean temperatures became positive, so that the metabolic period of *Saxifraga* at this high arctic location was 107 days while the duration of positive mean daily temperatures was only 78 days. Other dry site plants, notably many Cruciferae, often start growth early in the Arctic according to Sørensen. Soil drought usually puts an end to the vegetative season of these plants before low temperatures become effective. Shvetsova and Voznesenskii (1971) on the Taimyr Peninsula at 73°21′N found, of course, that photosynthesis of arctic plants goes on through the whole twenty-four hour period. Towards the end of the summer, when light intensity during the night hours decreases to 500 lux and below, the rate of photosynthesis decreases so as to be scarcely detectable. At the Taimyr station, the daily duration of photosynthesis decreased from the beginning (middle June) to the end (late August) of the growing period due to the decrease in amount of solar radiation received in each twenty-four hour period.

A common phenomenon both in arctic and alpine locations is the speeding up or telescoping of phenological events in plants released from snow cover relatively late in the summer. Plants released from snow by early or mid-June take much longer to reach maturity and have a longer photosynthetic season than those of the same species released in mid-July. This has been commented upon by Sørensen (1941) for the Arctic and by Billings and Bliss (1959) and Rochow (1969) for alpine situations. While these later plants grow faster in shorter time than plants released earlier, they are smaller and produce less dry matter.

Effects of temperature on photosynthesis and respiration. Temperatures of air and soil decrease, in a general way, with increasing latitude and altitude, so that arctic and alpine plants not only must survive low temperatures but spend much of their growing seasons in such temperatures. What are the relationships between temperature and metabolic rates in such plants?

In evaluating such temperature effects, one must attempt to separate those inherent in the genotype and those brought about by environment, since both sources of variation are present in the phenotype. There appears to be some degree of genotypic control over rates of net photosynthesis. Mooney, Wright and Strain (1964) found that plants of alpine and subalpine species in the White Mountains of California reached maximum photosynthesis at lower temperatures than did plants from lower elevations when all were grown under uniform greenhouse conditions. The same kind of situation occurs between arctic and alpine populations of both *Oxyria digyna* and *Thalictrum alpinum* (Mooney and Billings, 1961; Mooney and Johnson, 1965) with arctic populations having optima at lower temperatures than do alpine populations grown in the same environment.

Even though the genotypic effect on photosynthesis is marked, the effect of temperature acclimation on the photosynthetic rates of the phenotype is often greater. Billings *et al.* (1971), working with seventeen widely distributed populations of whole plants of *Oxyria digyna* from both arctic and alpine regions of North America and Europe, found a number of interactions between temperature of growth and genotype in the resultant gas exchange rates. For example, maximum net photosynthesis rates were higher in arctic populations than in alpine ones; this difference was increased by cold acclimation. The upper temperature compensation point in both arctic and alpine ecotypes was increased by warm acclimation. The temperature at which maximum net photosynthesis occurred was increased by warm acclimation, while maximum net photosynthesis rates were lowered by warm acclimation in all arctic populations but in only a few of the alpine ones. Alpine populations showed ideal homeostasis in net photosynthesis while arctic ecotypes showed only a low degree of partial homeostatic adjustment (see fig. 8A.4). Billings *et al.* also measured the Hill reaction rate on isolated chloroplasts from several populations. Cold acclimation increased this rate (over the warm acclimation rate) with increasing light intensity if the growth photoperiod approximated that of the field source of the population. But if arctic populations were grown at short photoperiods, cold acclimation depressed the Hill rate. Tieszen and Helgager (1968) measured Hill reaction rates in an arctic and an alpine population of the grass *Deschampsia caespitosa* grown under three different temperature regimes (photoperiod not stated). They, too, found both ecotypic and acclimation effects governing this rate and concluded that the alpine population showed greater phenotypic flexibility in the chloroplast part of its photosynthetic system. Pearcy (1969), also working with alpine ecotypes of *Deschampsia caespitosa* acclimated to warm and cold conditions, could find no consistent differences in photosynthesis that could be related to the temperatures of the source region. He found, however, that plants acclimated to low temperatures had more 'efficient' photosynthesis at low temperatures than warm-acclimated plants. The reverse was true at high temperatures. Ribulose diphosphate carboxylase activity in *Deschampsia* was higher in high elevation populations grown at low temperatures and lower at higher temperatures than those from low elevations. Pearcy concluded that acclimation effects on photosynthesis in *Deschampsia* were not due to the production of different isozymes of ribulose diphosphate carboxylase.

Alpine plants have the genetic capacity to carry on photosynthesis at lower temperatures than do plants of lowlands or warmer climates. Moser (1970) and Pisek *et al.* (1967) found that in summer *Ranunculus glacialis* and *Oxyria digyna*, nival plants in the Tirolean Alps, can carry on positive net photosynthesis at a temperature as low as −6°C, while *Citrus* and *Laurus* leaves cease photosynthesis at about −1°C. Net photosynthesis ceases when the leaves freeze; this damage is irreversible. *Ranunculus* and *Oxyria* are capable of greater supercooling than non-nival plants, thus their photosynthetic temperature minimum is at about 3°C below their freezing point.

Dark respiration rate seems to be less under the control of ecotype and more subject to temperature acclimation than is net photosynthesis. However, Mooney and Billings (1961) did find some evidence that an arctic population of *Oxyria* had higher respiration rates than an alpine one when grown under the same conditions. Björkman and Holmgren (1961) obtained similar results with leaves of an alpine population of *Solidago virgaurea* having higher dark respiration rates than those from coastal populations when grown under the same conditions. However, Billings *et al.* (1971) found relatively small genetic differences in dark respiration rates among whole plants of seventeen latitudinal populations of *Oxyria* grown side by side. But there were great increases in all of these rates after the plants had been subject to acclimation by being grown at low temperatures; this was true for all populations. Moreover, there was no ecotypic difference in dark respiration homeostasis; it was ideal in all populations (see fig. 8A.4). Ecotypic differences in total respiration rates within this species thus appear to be less marked than the effects of environmental temperature. Furthermore, Chabot and Billings (1972)

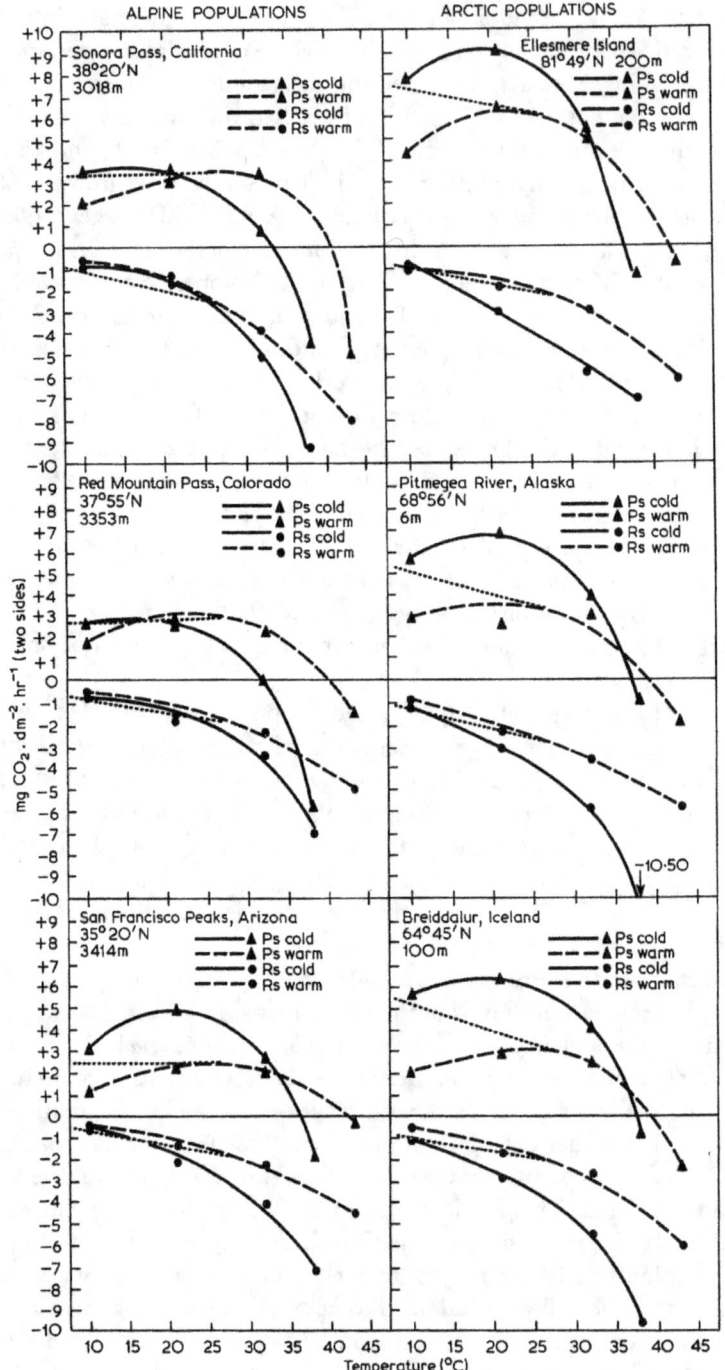

ALPINE POPULATIONS

ARCTIC POPULATIONS

Sonora Pass, California
38°20'N
3018m
△ Ps cold
△ Ps warm
● Rs cold
● Rs warm

Ellesmere Island
81°49'N 200m
△ Ps cold
△ Ps warm
● Rs cold
● Rs warm

Red Mountain Pass, Colorado
37°55'N
3353m
△ Ps cold
△ Ps warm
● Rs cold
● Rs warm

Pitmegea River, Alaska
68°56'N
6m
△ Ps cold
△ Ps warm
● Rs cold
● Rs warm

−10·50

San Francisco Peaks, Arizona
35°20'N
3414m
△ Ps cold
△ Ps warm
● Rs cold
● Rs warm

Breiddalur, Iceland
64°45'N
100m
△ Ps cold
△ Ps warm
● Rs cold
● Rs warm

mg $CO_2 \cdot dm^{-2} \cdot hr^{-1}$ (two sides)

Temperature (°C)

found that the time course of dark respiration acclimation in *Oxyria* and high elevation ecotypes of some other species was no longer than fourteen hours in adjusting to high temperatures. Warm desert plants were much slower (one to two days) in changing their respiration rates at high temperatures. The higher the source elevation of a species in the Sierra, the quicker its dark respiration acclimated to high temperatures. On the other hand, acclimation of dark respiration rates to low temperatures was slow and took seven to fourteen days in all species tested; and in every case, the low growth temperatures increased the dark respiration rate. It appears that increased respiration rates are a characteristic of plants growing in arctic or alpine areas whether or not this rate is seemingly a product of acclimation or is intrinsic. In the last analysis, the ability to keep the rate high whether by acclimation or by some other means must be genetically based. However, it has yet to be demonstrated why the increased respiration rate is of value to the tundra plant – but the rapid ability to change the rate downward with increase in temperature must have survival value for a plant with limited food reserves and only a short period in which to manufacture more. In the limited energy environment of tundra regions, it may be advantageous for a plant to be flexible enough in metabolism to follow short-term temperature changes closely; such changes are most frequent in the cold night/warm day environments of alpine regions.

One step toward an explanation of the high respiration rates in alpine plants would be to examine parts of the complex process. Klikoff (1966), for example, measured the mitochondrial oxidative rates in four populations of the grass *Sitanion hystrix* from different elevations in the Sierra Nevada. When he grew these clones under uniform conditions in a growth chamber and measured mitochondrial activity at 20°C, there was a direct relationship with source elevation: the rate increased with source elevation indicating that in this species, at least, the mitochondrial oxidative phase of respiration was genetically controlled, and apparently higher rates had survival value at 20°C for plants from higher elevations. At first these results appear to contradict the apparent lack of ecotypic difference in dark respiration rates of whole plants as reported for *Oxyria*. But perhaps they are not really contradictory but show on a finer scale just a part of the acclimation-ecotype interaction. Billings *et al.* (1971) also measured mitochondrial activity in *Oxyria* but on a latitudinal rather than altitudinal source basis (two arctic and two alpine populations), and also inserted temperature acclimation into the equation with two growth regimes: cold at 13°/7°C and warm at 30°/24°C. The results agree with Klikoff's: there is a genetic basis for increased mitochondrial oxidation rates in arctic

---

Fig. 8A.4. Effects of growth temperature on net photosynthesis (triangles) and dark respiration (dots) of three alpine and three arctic populations of *Oxyria digyna* (from Billings *et al.*, 1971). The dotted lines connect the mean growth temperatures (cold and warm) with the gas exchange rates at those same temperatures. Level connecting lines indicate compensation is 'ideal' in net photosynthesis of alpine populations; the steep angle of the dotted lines in net photosynthesis of the arctic populations indicate only 'partial' compensation. Dark respiration compensates about the same in all populations; it is 'partial' but approaches 'ideal'.

Oxyrias over alpine ones. This difference, however, was not expressed if the plants were grown in the warm regime but only if they were grown in the cold regime, although both arctic and alpine populations showed increased rates after cold acclimation (see table 8A.1). It is apparent that temperature acclimation itself is genetically based and that 'acclimation ecotypes' exist in *Oxyria* based upon certain metabolic systems, and that these ecotypes must be subject to natural selection (Billings *et al.*, 1971).

Table 8A.1. *Oxyria* root mitochondrial oxidation at 25°C in mμ atoms O_2 min^{-1} (mg protein)$^{-1}$.

	Cold regime (7/13°C)	Warm regime (24/30°C)
Red Mountain Pass, Colorado	7.16	3.12
Sonora Pass, California	7.2	1.92
Cape Thompson, Alaska	16.8	3.41
Pitmegea River, Alaska	10.4	1.43

Effects of light on photosynthesis. Being restricted to open places, it would be expected that arctic and alpine plants require higher light intensities in photosynthesis than herbaceous plants of forests and woodlands. There is evidence that they do; Godfrey (1969) found that the lower limits of alpine plants on mountainsides are set in large degree by their photosynthetic light requirement, and where this is not met because of increasing shade of taller herbaceous plants or trees, the alpine plants are excluded. Scott *et al.* (1970), using multiple regression analysis, found that CO_2 exchange of five species of subalpine meadow plants as measured in the field at an elevation of 3050 m was highly correlated (significant at 0.1 per cent level) with light intensity in every species. No other environmental variable measured was significantly correlated at that level for more than one of the species.

The photosynthetic light requirement of tundra plants is also both genetically and environmentally set. Mooney and Billings (1961) showed experimentally that arctic populations of *Oxyria* reached photosynthetic light saturation at lower light intensities than did alpine populations, thus indicating some genetic basis for light amounts necessary in photosynthesis.

Temperature–light interactions in gas exchange. In nature, temperature and light do not act independently but with each other and in concert with many other environmental factors (and environmental history) in the interaction of gene system and environment which governs photosynthetic and respiration rates. Meaningful metabolic rate measurements, whether in the field or the laboratory, must measure both of these environmental factors carefully, but should also measure water stress and wind. Unfortunately, all these factors are seldom measured simultaneously.

It is easier to measure the interaction effect in the laboratory because of better controls. In several species of alpine plants more light is needed for compensation with increasing temperature, as shown in fig. 8A.5. In other words, alpine plants appear to produce

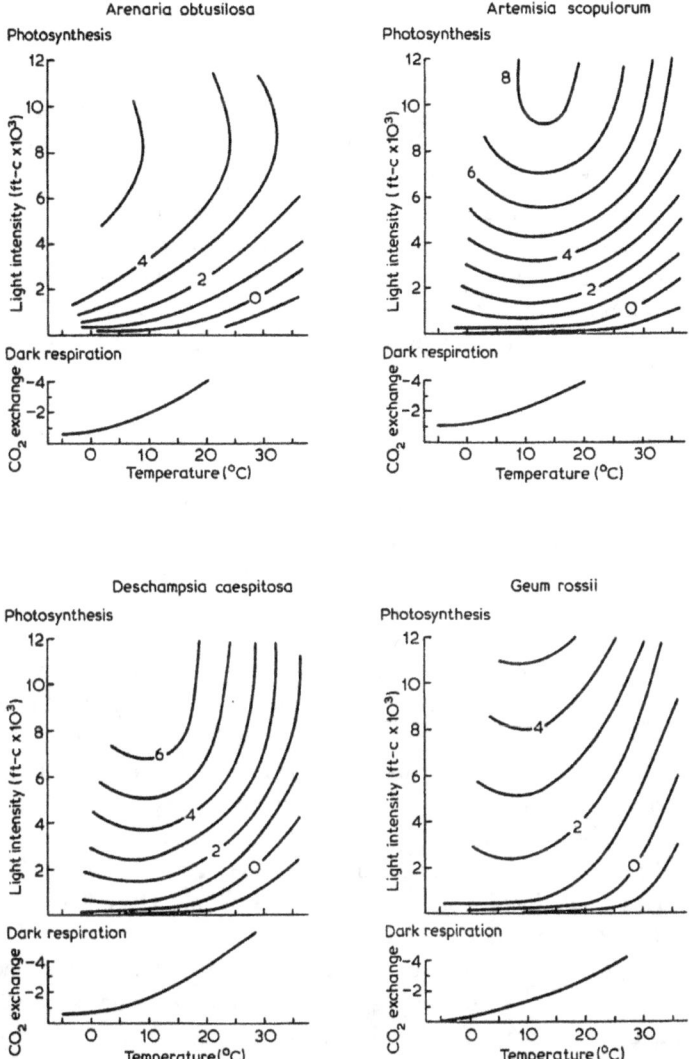

Fig. 8A.5. Laboratory measurements of net photosynthesis and dark respiration of plants of four alpine species from the Medicine Bow Mountains, Wyoming (from Scott and Billings, 1964). Measurements made as CO_2 exchange in an infra-red gas analyser system as $mgCO_2 \cdot g$ dry wt^{-1} hr^{-1}. Net photosynthesis plotted as isopleths in relation to light intensity and temperature within the plant chamber of the system.

more photosynthate at relatively low temperatures and high light intensities. Moser (1970) found essentially the same thing in *Ranunculus glacialis* in the Stubai Alps. Rastorfer (1970) reports this also for the antarctic mosses, *Bryum antarcticum* and *B. argenteum*, collected at Marble Point (72°19′S) and tested in the laboratory at McMurdo Station. In both species, light compensation points were about 1.6 mW cm^{-2} when net photosynthesis was measured at 15°C, but light compensation shifted to about one-half this value at 5°C. Light saturation in these mosses was also much lower at 5° than at 15°C.

Warren Wilson (1966) suggests that the lower light saturations in the photosynthesis of arctic plants at low temperatures may be due to accumulation of assimilates. This may be true; Chabot and Billings (1972), for example, found that chloroplasts in alpine Oxyrias from the Sierra Nevada lost most of their starch during the dark period both at temperature regimes of 11°/8°C and 23°/20°C. On the other hand, chloroplasts of *Encelia virginensis*, a species from the nearby warm desert, could transform their starch into translocatable compounds during the dark at 23°/20°C but not at 11°C/8°C (see Plate 29). Such an accumulation could lead to chloroplast breakdown and lower photosynthesis rates, and could contribute to the inability of *Encelia* to succeed in alpine temperatures.

Water stress and metabolism. Since so much of the patterning in arctic and alpine vegetation is due to local moisture gradients, it seems obvious that interactions between metabolism and water requirements must be basic to such patterning. There is no doubt that such is the case, but relatively little is known of this aspect of tundra ecology. Some plants such as *Oxyria digyna*, *Carex stans* var. *aquatilis*, and *Saxifraga cernua* require large amounts of water or can withstand flooding. At the other end of the gradient, various *Kobresia* species, *Silene acaulis*, and many ridge-top cushion plants can take considerable drought in winter and summer, particularly in alpine areas.

A distinction should be made between drought stress during the summer growing season and drought stress produced by frozen soils and dry winds of winter. Tranquillini (1964) indicates that plants in European alpine communities are seldom subjected to drought during the summer. However, during the winter in exposed, snowless sites at timberline and above, evergreen shrubs and small pioneer trees are subjected to rigorous drought stress. On sunny days such plants lose a great deal of water by cuticular transpiration in spite of closed stomates. This water cannot be replaced because of the frozen soil, and water stress becomes more severe as the winter goes on. The leaves lose much of their water and the osmotic value rises to over forty atmospheres. Shrubs covered with snow, however, actually gain water by absorption, so that the osmotic value drops during the winter. But if such plants (for example, *Rhododendron ferrugineum*) are denuded of snow in late winter, they lose water rapidly and may be killed since their drought resistance is low. On the other hand, certain dwarf or prostrate shrubs such as *Loiseleuria procumbens*, characteristic of open, snowless, windswept places, have low transpiration rates all year and tolerate the winter drought very well. According to Larcher (1963), *Loiseleuria* can also take up surface meltwater late in the winter by shallow, adventitious roots when most of the soil is frozen.

Caldwell (1970), working with *Pinus cembra* and *Rhododendron ferrugineum* on the Patscherköfel near Innsbruck, studied the effect of wind on transpiration and photosynthesis in the two species. *Rhododendron* grows almost exclusively in ravines and depressions where snow protects the shrubs from desiccation during winter and spring. *Pinus cembra* grows on the cold, frozen soils of windswept ridges. Using a porometer to measure stomatal aperture, Caldwell found that wind had very little effect on the stomates of the pine, and that transpiration increased only slightly and photosynthesis was unaffected at wind speeds up to 3.5 m s^{-1}. Even at wind speeds of 8 m s^{-1}, pine photosynthesis and transpiration were only mildly suppressed. *Rhododendron* reacted sharply to even low wind speeds by a rapid closure of all stomates and substantial reductions in photosynthesis and transpiration. While *Rhododendron* is protected by snow during the winter, during the summer wind velocities are usually sufficient at the surface of the *Rhododendron* vegetation to induce significant stomatal closure and large restrictions in gas exchange rates.

Winter drought stress also prevails in the mountains of western North America, with the same general patterning into snow-protected and snow-free environments. Lindsay (1971) measured leaf-water potential in *Picea engelmannii* at timberline (3300 m) in the Medicine Bow Mountains of Wyoming, and also in the closed forest 250 m lower in elevation. In the windswept, relatively snow-free timberline location, leaf-water potential stayed near or below −30 bars from September to early June, and only during midsummer did it rise to near −15 bars. With the drought of late summer, leaf-water potential decreased to its low winter values. In the forest location, leaf-water potential was higher throughout the year. Because of the deep snow cover in the forest, soil moisture was relatively more available throughout the winter; here, leaf-water potential did not go much below −20 bars, and this at the same time that water potential at timberline was −34 bars. *Abies lasiocarpa* showed similar results.

The dry ridge/late snowbank environmental gradient is much sharper and extreme in the mountains of western North America than it is in most of Europe. This is due, partially, to the more regular incidence of late summer drought, and the gradient is particularly pronounced in the Sierra Nevada of California. Here, the sandy granitic slopes and ridges are xeric environments in the long summer drought, but the meadows below snowbanks are usually green and moist throughout the summer (Klikoff, 1965a). Mooney *et al.* (1965) found that in the Sierra Nevada specific transpiration rates of alpine plants are closely related to the water supply of the habitat. Plants from moist sites have higher transpiration rates and little control over transpiration as compared with the more efficient water use of dry-site plants. Thus, water availability in the dry Sierran summer has much to do with the sharp vegetational patterning characteristic of the high Sierra.

Water stress which develops in these plants along the dry meadow/wet meadow gradient during the summer greatly affects photosynthesis rates, and thus productivity. Klikoff (1965b) found that photosynthesis in *Calamagrostis breweri* (in the uniform environment of a growth chamber) decreased to almost zero at water potentials below −10 bars, but photosynthesis of *Carex exserta*, characteristic of dry meadows, was operating at almost maximum values at −10 bars and at 25 per cent of maximum at water potentials as low as −20 bars.

Field measurements of metabolism. Much of our knowledge of metabolism of arctic and alpine plants has been obtained under controlled laboratory conditions. There are advantages to this but often the results have limited value in interpreting real field situations. Logistically, the measurement of gas exchange and other metabolic parameters under field conditions in the arctic tundra or high mountains is very difficult, but if we are to learn what actually goes on in these plants in their natural environments, field measurements must be made throughout both diurnal and annual cycles and compared with laboratory results. A beginning has been made on such field measurements by taking infra-red gas analysers into the mountains or tundra on a temporary or permanent basis. In 1958 Billings *et al.* (1966) made gas analyser measurements on several species

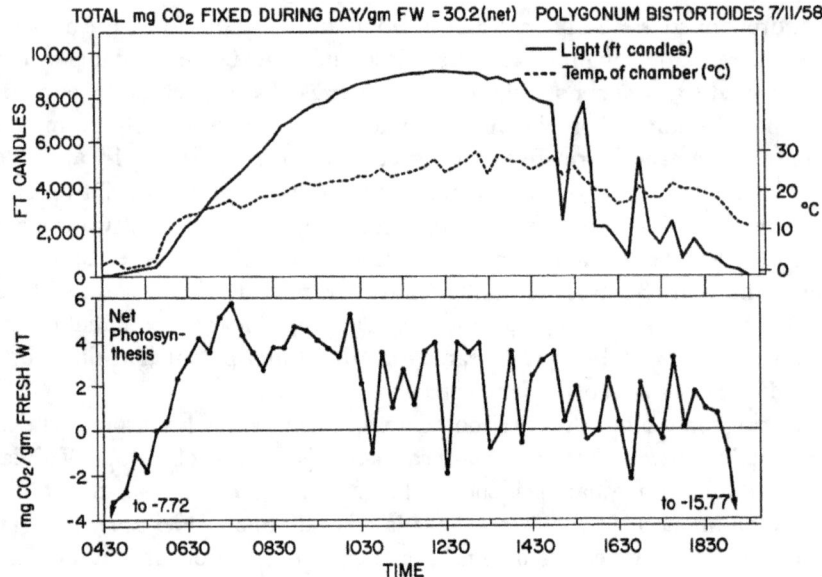

Fig. 8A.6. Net photosynthesis of a plant of *Polygonum bistortoides* from before sunrise to after sunset on a typical alpine day at 3300 m in the Medicine Bow Mountains, Wyoming, as measured with an infra-red gas analyser system (from Billings *et al.*, 1966). (1 ft.-c. = 10.76 lx.)

of alpine plants in the Medicine Bow Mountains, Wyoming, with fairly good results, even though by today's standards the equipment was somewhat primitive (Plate 30). However, the all-day runs, exemplified by that for *Polygonum bistortoides* in fig. 8A.6, showed that, under cuvette conditions at least, net photosynthesis runs an uneven course probably due to carbohydrate accumulation or stomatal closure. Since that time we have continued to make some field gas exchange measurements in the mountains (Scott and Billings, 1964; Scott *et al.*, 1970). But compared to our field measurements, those of Moser (1968, 1969; also, see Cernusca and Moser, 1969) using a permanent, automatic recording gas analyser system high in the Tirolean Alps, are elegant indeed. Moser's station is situated on a ridge on the Hoher Nebelkogel, and he has made measurements of gas exchange in *Oxyria digyna*, *Ranunculus glacialis*, and *Geum reptans in situ*. Both the American Arctic Tundra IBP Metabolic Group under Tieszen's direction at Barrow,

Alaska, and the Canadian group under Bliss at Devon Island are making gas exchange measurements in the field under permanent conditions; good results are anticipated. These will help materially in understanding tundra productivity (see Webber, Chapter 8B).

Carbohydrate and lipid storage and use
The general course of the carbohydrate cycle has been described by Russell (1940b) for arctic plants and by Mooney and Billings (1960) for alpine plants. Carbohydrates are stored in large amounts in roots and rhizomes at the end of the growing season. During the winter, because of low temperatures and dormancy, there is little use of these underground reserves. With the breaking of dormancy in early summer, the stored carbohydrates are utilized in the rapid early growth of the shoot and are depleted to a large degree. During this time, growth is so rapid that respiration far exceeds photosynthesis and food is used up faster than it can be made (Hadley and Bliss, 1964). This situation is maintained until shoot growth is 75–90 per cent completed. Then the respiration rate drops, shoot growth slows down, and replenishment of carbohydrate reserves commences. By the onset of dormancy, underground carbohydrate reserves are back to the highest level of the year.

The carbohydrate cycle is modified and changed during periods (or whole growing seasons) of cold, cloudy weather. Fonda and Bliss (1966) on Mt Washington found that an early August cold period during 1963 reduced shoot carbohydrates in *Carex bigelowii* from 22 per cent to 13 per cent and rhizome carbohydrates from 22 per cent to 17.5 per cent. However, the warmer, sunny days later in the month allowed replacement of the utilized reserves. If there had been no more sunny weather it is obvious that the plants would have entered winter with less than optimum supplies. Local differences in microsite also modify the carbohydrate cycle. Rochow (1969) found that root and shoot sugar contents in *Caltha leptosepala* were highest in those plants released from snow cover late in June. Plants released in the middle of July had lower sugar levels and were unable to increase these levels because of the short growing season remaining. Again, there appears to be some tendency for carbohydrate content to decrease with increasing elevation, as we (Mooney and Billings, 1965) found in roots of *Calyptridium umbellatum* in California.

The principal reserve foods in alpine and arctic plants are starches, sugars and lipids. In herbaceous plants these are mostly stored in roots, rhizomes, corms and bulbs, since few leaves remain during the winter. However, in low evergreen shrubs such as *Diapensia lapponica* and *Ledum groenlandicum*, Hadley and Bliss (1964) found that old stems and old leaves provided the principal storage reservoir for lipids and also for some carbohydrates. A similar situation occurs in the small evergreen alpine shrubs of New Zealand.

Flowering
Since almost all alpine and arctic perennials have pre-formed flower buds (Sørensen, 1941, and others), the flowering in a particular season depends to a great extent on environmental conditions during the previous one or two summers. The production of flowers from these buds depends upon the completion of the initial stage of leaf and stem

growth after snowmelt, upon temperature, and in some cases (particularly in arctic species or ecotypes) upon the interaction between temperature and photoperiod.

In *Oxyria digyna*, Mooney and Billings (1961) found that no populations flowered under a twelve-hour photoperiod, populations from around 40° latitude flowered at fifteen hours, but that arctic populations would flower only in very long photoperiods or in continuous light. Alpine populations flowered also under continuous light. Billings *et al.* (1965), working with seventeen *Oxyria* populations under four photoperiodic regimes in the laboratory, found the same thing with more precision. The higher the latitude of origin, the greater the number of hours of continuous light or the longer the photoperiod needed for flowering. This is true not only for the elongation and flowering of the pre-formed bud, but for the production of the pre-formed flowering primordia themselves. Both phenomena are tied to carbohydrate availability, and neither flowering nor flower primordia initiation take place until the carbohydrate level is sufficient for normal survival. At short photoperiods, arctic populations of *Oxyria* seem to have difficulty in meeting this carbohydrate requirement; this may be partially responsible for the lack of initiation or aborting of inflorescences in such ecotypes when grown under relatively short photoperiods (Billings *et al.*, 1965).

The photoperiodic effect in *Oxyria* is also tied to temperature: flowering occurs more rapidly at higher temperatures than at low temperatures under the same photoperiod. Each latitudinal population of *Oxyria* has its own combination of temperature and photoperiod; alpine populations are better adapted to shorter photoperiods and higher temperatures, and arctic populations better fitted to long photoperiods at lower temperature. Mark (1969) found that flowering in snow tussock (*Chionochloa rigida*) in New Zealand depends upon the right combination of long photoperiod and mild temperature minima. Since these relatively high minima do not occur every year, flowering in this grass tends to be irregular and dependent to a large extent on a warm summer the year before actual flowering.

Pollination

Both arctic and alpine plants face pollination problems caused by low temperatures which confine insect activities to a few weeks, and even then mainly to the sunny daylight hours (Mani, 1962). The principal agents, of course, are insects, birds and wind. Schroeter (1926) cites some data of Raddes from the high mountain flora of the Caucasus which show that 89 per cent of the species are insect pollinated and 11 per cent are wind pollinated. Schroeter's comparable figures from the Arctic (after Aurivillius) show that 32–38 per cent are wind pollinated. A few species such as *Pentstemon newberryi* and *P. davidsonii* are pollinated by humming birds in the Sierran alpine region (Grant and Grant, 1968), but these birds are lacking from the Arctic.

The principal kinds of pollinating insects in alpine locations are short-tongued bees, bumblebees, flies, butterflies and moths. Bumblebees are more common above timberline in the northern hemisphere mountains than any other kinds of bees. In many places they are the only bees present; they fly only during the daylight hours when air temperature is above about 10°C (Mani, 1962; Macior, 1968, 1970). Flies can and do work at lower temperatures and in dimmer light. As one ascends above 4000 m in the Himalaya, bee-pollinated flowers disappear completely from the biota and one finds in their place a dominance of brightly coloured flowers pollinated by Lepidoptera and Diptera (Mani, 1962). Müller (1881) found a similar situation in the Alps.

Long-tubed flowers in northern hemisphere alpine areas are usually bumblebee-pollinated, while small, flat flowers are mostly 'fly flowers' (L. W. Macior, personal communication, 1966). While some species of bumblebees (e.g. *Bombus frigidus*) are common in parts of the Arctic, Mosquin and Martin (1967) in short-term observations on Melville Island (75°N) observed bumblebees visiting only the scented flowers of *Astragalus alpinus*, *Petasites frigidus*, and perhaps *Parrya arctica*. The bees were not observed visiting *Pedicularis arctica*, *P. sudetica*, or *Oxytropis arctica*, which are unscented, or any of thirty-six other species with apparently scentless flowers, including the bright purple flowers of *Saxifraga oppositifolia* (Plate 28). This is surprising, since Hocking (1968), Kevan (1970), and J. A. Teeri (personal communication, 1971) have found flowers of the purple saxifrage to produce abundant nectar and to be fairly fragrant. Kevan found this species to be mainly dependent for pollination on Chironomids and other Diptera and, unlike most other arctic saxifrages, to be highly entomophilous and, moreover, visited by bumblebees. Mosquin and Martin (1967) and Kevan (1970) agree that Diptera are more important flower visitors in the Arctic than bees. Mosquin and Martin go further in suggesting a trend toward the evolution of scentless flowers in arctic plants. Downes (1965) has reviewed the ecology of arctic insects, including pollination relationships.

One of the striking differences between the alpine flowers of New Zealand and those of the northern hemisphere high mountains is the absence of bright-coloured flowers in alpine New Zealand and the prevalence of small, flat, white or yellow flowers. Heine (1937) explains this, at least in part, by pointing out that there were no long-tongued bees native to New Zealand, so that most insect pollination in the alpine zone there is by small insects, flies, and short-tongued bees.

Heterostyly and dioeciousness, while present in alpine locations of the middle latitudes, are less important in the Arctic, with heterostyly being rare or not known there (Mosquin, 1966). In the most severe environments, self pollination, apomixis or vivipary are often the rule and cross pollination is relatively rare.

Seed production

Seed production may or may not follow pollination depending upon whether the weather remains favourable or not. At higher latitudes and altitudes, years when any seed is produced become fewer, and even in relatively good years the amount of viable seed may be low. Holway and Ward (1965), for example, noted that several species of plants (*Trifolium parryi*, *Artemisia spithamea*, *Campanula rotundifolia*) bloomed abundantly in the alpine regions of Colorado but failed to set seed. They examined several hundred pods of the *Trifolium* and found only twenty-eight seeds. Bliss (1956) noted that seed production decreases with an increase in severity of microenvironments in the same general region, so that, for example, late snowbank populations of a given species may have poor seed-set, while seed may be produced in abundance by plants of the same species in more favourable sites nearby.

Many arctic and alpine plants, however, do produce seeds rather regularly and in apparent abundance. Such seed, depending upon the species, ripens from mid-July to September. It is usually soon dispersed, often by wind or by falling from the follicle or capsule. In general, seed ripening is too late in most species for germination in that growing season. In fact, Porsild (1951) points out that fruits of some species do not mature and dehisce until after snow arrives in the autumn, and are dispersed with the drifting snow of winter.

Vegetative reproduction

As sexual seed production decreases and becomes unreliable in the most severe environments, vegetative reproduction seems to increase. Apomixis is particularly prevalent in tundra environments (Löve, 1959). So, also, is the production of bulbils, as in *Polygonum viviparum, Saxifraga cernua*, and a number of grasses. Grass bulbils frequently develop into young plants in the inflorescence ('vivipary'); these young plants may even flower (*Trisetum spicatum* growing in controlled conditions: Clebsch, 1960). In nature, the little plants fall to the ground while vegetative, and may become established. Such vivipary has often been deemed a taxonomic character (e.g. in *Poa alpigena* var. *colpodea*), but Clebsch found in several populations of normally non-viviparous *Trisetum spicatum* that vivipary could be induced by manipulating the experimental environment.

The principal means of vegetative reproduction in both arctic and alpine plants is by rhizomes. Layering is also important in some species, particularly in cushion plants and prostrate shrubs. Intraspecific variation in rhizome production is sometimes coupled to environmental conditions. For example, Mooney and Billings (1961) found that in North America arctic forms of *Oxyria digyna* have rhizomes and often reproduce by them into clonal colonies. Western American Oxyrias south of about 60° latitude do not have rhizomes and reproduce by seed. Rhizome production in *Oxyria* is thus genetically controlled.

Onset of dormancy

Under natural conditions, tundra plants beyond the seedling stage are seldom killed by the onset of winter weather. Some plants, such as *Braya humilis* in northeastern Greenland, may be caught by such weather in flowering or fruiting condition without any damage, and resume growth and fruiting the following year (Sørensen, 1941). Normally, however, there is a period of shortening daylength, lowering temperatures, and often increasing drought which tend to 'harden' the plant and induce a dormancy which prepares it for the oncoming winter. Mooney and Billings (1961) found in all the American populations of *Oxyria digyna* with which they worked that a photoperiod of twelve hours even at high temperatures induced the formation of a perennating bud which carried the shoot primordia through experimentally maintained winter (freezing) conditions. In further work, Billings et al. (1965) found that there was some tendency for arctic and subarctic Oxyrias to initiate perennating buds at photoperiods as long as fifteen hours and to have well-developed buds by the time the photoperiod had been dropped to thirteen hours. Alpine populations from both America and Europe, however, did not form perennating buds until the photoperiod was down to twelve hours, at which daylength the bud was formed rather quickly.

A somewhat similar photoperiodic effect on hardening itself was observed by Biebl (1967) during some experiments in which he artificially shortened the summer photoperiod (early July) of some plants on Disko Island (West Greenland) to eight hours. After ten days of this treatment, plants of *Salix glauca* ssp. *callicarpaea*, *Betula nana*, *Vaccinium uliginosum* and *Empetrum hermaphroditum* had increased resistance to heat. Additionally, the leaves of *Betula nana* showed an autumnal red colouring and were significantly more resistant to a twenty-four hour freezing treatment.

Shortened photoperiod and low temperatures are not the only factors involved in the onset of dormancy in alpine plants, and perhaps in some arctic plants also. Both Bliss (1956) and Billings and Bliss (1959) noted that cessation of growth and onset of dormancy in August in the Rocky Mountain alpine zone was determined by factors other than low temperature. In the 1959 paper it was hypothesized that soil drought limited growth and hastened dormancy. Holway and Ward (1965) found just this kind of response in *Geum turbinatum* in noting that the time of dormancy up until mid-September was controlled by the year-to-year variation in soil moisture. It seems obvious that in a wet year some other factors must eventually control dormancy in this species after mid-September; most likely shortened photoperiod and low temperatures provide the stimulus for such a late move into dormancy.

Summary

How are plants adapted to the low temperatures and other stresses of arctic and alpine environments? At present it is not possible to answer this question completely. However, in brief, we can say that plants are adapted to these severe environments by employing combinations of the following general characteristics:

(1) *Life form.* Perennial herb, prostrate shrub, or lichen. Perennial herbs have greater part of biomass underground except in small rosette plants.
(2) *Seed dormancy.* Generally controlled by environment; seeds can remain dormant for long periods of time at low temperatures since they require temperatures well above freezing for germination.
(3) *Seedling establishment.* Rare and very slow; it is often several years before a seedling is safely established.
(4) *Mineral nutrition.* Not well known for individual plants, but the effects of birds and mammals on local increases and cycling in nitrogen, phosphorus, and other elements is marked. Cold soils may inhibit the fixing of nitrogen by bacteria.
(5) *Photosynthesis and respiration.*
    (a) These are at high rates for only a few weeks when temperatures and light are favourable.

(b) Optimum photosynthesis rates are at lower temperatures than for ordinary plants; rates are both genetically and environmentally controlled with phenotypic plasticity very marked. Acclimation is very important and 'acclimation ecotypes' appear to exist.

(c) Dark respiration is higher at all temperatures than for ordinary plants; rate is both genetically and environmentally controlled, with phenotypic plasticity very pronounced, i.e. low-temperature environment increases the rate at all temperatures.

(d) Alpine plants have higher light-saturation values in photosynthesis than do arctic or lowland plants; light saturation closely tied to temperature.

(6) *Drought resistance.* Most drought stress in winter in exposed sites is due to frozen soils and dry winds. It is met by decreased water potentials, higher concentrations of soluble carbohydrates, and closed stomates. Little drought resistance in snow-bank plants. Alpine plants adapted to summer drought stress can carry on photosynthesis at low water potentials; alpine or arctic plants of moist sites cannot do this.

(7) *Breaking of dormancy.* Controlled by mean temperatures near or above 0°C, and in some cases by photoperiod also.

(8) *Growth.* Very rapid even at low positive temperatures. Respiration greatly exceeds photosynthesis in early regrowth of perennials. Nitrogen and phosphorus often limiting in cold soil.

(9) *Food storage.* Characteristic of all alpine and arctic plants except annuals. Carbohydrates mostly stored underground in herbaceous perennials. Lipids in old leaves and stems of prostrate evergreen shrubs. Depleted in early growth, and usually restored after flowering.

(10) *Winter survival.* Survival and frost resistance are excellent after hardening. Cold resistance closely tied to content of soluble carbohydrates, particularly raffinose.

(11) *Flowering.* Flower buds are pre-formed the year before. Complete development and anthesis dependent upon temperature of the flowering year and also, in some cases, upon photoperiod.

(12) *Pollination.* Mostly insect pollinated in alpine regions and even in the Arctic, but to a lesser extent. Wind pollination increasingly more important with increasing latitude.

(13) *Seed production.* Opportunistic, and dependent upon temperature during flowering period and latter half of growing season.

(14) *Vegetative reproduction.* By rhizomes, bulbils, or layering. More common and important in Arctic than in alpine areas.

(15) *Onset of dormancy.* Triggered by photoperiod, low temperatures, and drought. Dormant plant extremely resistant to low temperatures.

ACKNOWLEDGEMENT

The author is grateful to Dr H. A. Mooney and Dr Brian F. Chabot for their appreciated help and suggestions; also for the support provided by the National Science Foundation, Washington, DC, under grants GB-8404 and GB-79634.

References

ALEKSANDROVA, V. D. (1971) On the principles of zonal subdivision of arctic vegetation. *Botanicheskii Zhurnal*, **56**, 3–21 (in Russian).

AMEN, R. D. (1966) The extent and role of seed dormancy in alpine plants. *Quart. Rev. Biol.*, **41**, 271–81.

BESCHEL, R. E. (1961) Dating rock surfaces by lichen growth and its application to glaciology and physiography (lichenometry). *In:* RAASCH, G. O. (ed.) *Geology of the Arctic*, **2**, 1044–62. Toronto: Univ. of Toronto Press.

BIEBL, R. (1967) Kurztag-Einflüsse auf arktische Pflanzen während der arktischen Langtage. *Planta*, **75**, 77–84.

BILLINGS, W. D. and BLISS, L. C. (1959) An alpine snowbank environment and its effect on vegetation, plant development, and productivity. *Ecology*, **40**, 388–97.

BILLINGS, W. D. and GODFREY, P. J. (1967) Photosynthetic utilization of internal carbon dioxide by hollow-stemmed plants. *Science*, **158**, 121–3.

BILLINGS, W. D. and MARK, A. F. (1961) Interactions between alpine tundra vegetation and patterned ground in the mountains of southern New Zealand. *Ecology*, **42**, 18–31.

BILLINGS, W. D. and MOONEY, H. A. (1968) The ecology of arctic and alpine plants. *Biol. Rev.*, **43**, 481–529.

BILLINGS, W. D., CLEBSCH, E. E. C. and MOONEY, H. A. (1966) Photosynthesis and respiration rates of Rocky Mountain alpine plants under field conditions. *Amer. Midl. Nat.*, **75**, 34–44.

BILLINGS, W. D., GODFREY, P. J. and HILLIER, R. D. (1965) Photoperiodic and temperature effects on growth, flowering, and dormancy of widely distributed populations of *Oxyria*. *Bull. Ecol. Soc. Amer.*, **46**, 189 (abstract).

BILLINGS, W. D., GODFREY, P. J., CHABOT, B. F. and BOURQUE, D. P. (1971) Metabolic acclimation to temperature in arctic and alpine ecotypes of *Oxyria digyna*. *Arct. Alp. Res.*, **3** (4), 277–90.

BJÖRKMAN, O. and HOLMGREN, P. (1961) Studies of climatic ecotypes of higher plants. Leaf respiration in different populations of *Solidago virgaurea*. *Kgl. LantbrHögsk. Ann.*, **27**, 297–304.

BLISS, L. C. (1956) A comparison of plant development in microenvironments of arctic and alpine tundras. *Ecol. Monogr.*, **26**, 303–37.

BLISS, L. C. (1958) Seed germination in arctic and alpine species. *Arctic*, **11**, 180–8.

BLISS, L. C. (1962) Adaptations of arctic and alpine plants to environmental conditions. *Arctic*, **15**, 117–44.

BLISS, L. C. (1963) Alpine plant communities of the Presidential Range, New Hampshire. *Ecology*, **44**, 678–97.

BLISS, L. C. (1966) Plant productivity in alpine microenvironments on Mt Washington, New Hampshire. *Ecol. Monogr.*, **36**, 125–55.

BLISS, L. C. (1971) Arctic and alpine plant life cycles. *Ann. Rev. Ecol. System.*, **2**, 405–38.

BLISS, L. C. and MARK, A. F. (1965) Alpine microenvironments and plant productivity on Rock and Pillar Range, New Zealand. *Bull. Ecol. Soc. Amer.*, **46**, 111 (abstract).

BÖCHER, T. W. (1963) Phytogeography of Middle West Greenland. *Medd. om Grønland*, **18** (3), 1–289.

BRAUN-BLANQUET, J. (1948) *La Vegetation Alpine des Pyrénées Orientales*. Barcelona: Estacion de Estudios Pirenaicos. 306 pp.

BRAUN-BLANQUET, J. and JENNY, H. (1926) Vegetationsentwicklung und Bodenbildung in der alpinen Stufe der Zentralalpen. *Neue Denkschr. schweiz. naturf. Gesell.*, **63**, 181–349.

BRITTON, M. E. (1957) Vegetation of the arctic tundra. *In:* HANSEN, H. P. (ed.) *Arctic Biology*, 26–61. Corvallis: Oregon State Univ. Press.

CALDWELL, M. M. (1968) Solar ultraviolet radiation as an ecological factor for alpine plants. *Ecol. Monogr.*, **38**, 243–68.

CALDWELL, M. M. (1970) The effect of wind on stomatal aperture, photosynthesis, and transpiration of *Rhododendron ferrugineum* L. and *Pinus cembra* L. *Zentralblatt für das gesamte Forstwesen*, **87**, 193–201.

CALDWELL, M. M. (1972) Biologically effective solar ultraviolet irradiation in the Arctic. *Arct. Alp. Res.*, **4**, 39–43.

CERNUSCA, A. and MOSER, W. (1969) Die automatische Registrierung produktions-analytischer Messdaten bei Frielandversuchen auf Lockstreifen. *Photosynthetica*, **3**, 21–7.

CHABOT, B. F. and BILLINGS, W. D. (1972) Origins and ecology of the Sierran alpine flora and vegetation. *Ecol. Monogr.*, **43**, 163–99.

CLEBSCH, E. E. C. (1960) *Comparative Morphological and Physiological Variation in Arctic and Alpine Populations of* Trisetum spicatum. Unpub. Ph.D. dissertation, Duke Univ. 241 pp.

COSTIN, A. B. (1967) Alpine ecosystems of the Australasian region. *In:* WRIGHT, H. E. JR and OSBURN, W. H. (eds.) *Arctic and alpine environments*, 55–87. Bloomington: Indiana Univ. Press.

CROCKER, R. L. and MAJOR, J. (1955) Soil development in relation to vegetation and surface age at Glacier Bay, Alaska. *J. Ecol.*, **43**, 427–48.

DAHL, E. (1956) Rondane: mountain vegetation in south Norway and its relation to the environment. *Skrifter utgitt av Det Norske Videnskaps-Akademi i Oslo*, 1, Mat.-Naturv. *Klasse*, 3, 1–374.

DAUBENMIRE, R. (1941) Some ecological features of the subterranean organs of alpine plants. *Ecology*, **22**, 370–8.

DOWNES, J. A. (1965) Adaptations of insects in the Arctic. *Ann. Rev. Entomol.*, **10**, 257–74.

DRURY, W. H. JR (1962) Patterned ground and vegetation on southern Bylot Island, North-west Territories, Canada. *Contrib. Gray Herbarium Harvard Univ.*, **190**, 1–111.

FISHER, F. J. F. (1965) The alpine *Ranunculi* of New Zealand. *New Zealand Dept of Sci. Industr. Res. Bull.*, **165**, 1–192.

FONDA, R. W. and BLISS, L. C. (1966) Annual carbohydrate cycle of alpine plants on Mt Washington, New Hampshire. *Bull. Torrey Bot. Club*, **93**, 268–77.

GIMINGHAM, C. H. and SMITH, R. I. L. (1970) Bryophyte and lichen communities in the maritime Antarctic. *In:* HOLDGATE, M. W. (ed.) *Antarctic Ecology*, **2**, 752–85. London: Academic Press.

GJAEREVOLL, O. (1956) The plant communities of the Scandinavian alpine snow-beds. *Det Kgl. Norske Videnskabers Selskabs Skrifter*, **1**, 1–405.

GODFREY, P. J. (1969) *Factors Influencing the Lower Limits of Alpine Plants in the Medicine Bow Mountains of Southeastern Wyoming*. Unpub. Ph.D. dissertation, Duke Univ. 525 pp.

GOOD, R. (1964) *The Geography of the Flowering Plants* (3rd edn). New York: John Wiley. 518 pp.

GRANT, K. A. and GRANT, V. (1968) *Hummingbirds and their Flowers*. New York: Columbia Univ. Press. 115 pp.

GREENE, S. W. (1964) The vascular flora of South Georgia. *Brit. Antarct. Surv. Rep.*, 45, 1–58.

GREENE, S. W. and GREENE, D. M. (1963) Check list of the Sub-Antarctic and Antarctic vascular flora. *Polar Rec.*, **11**, 411–18.

GREENE, S. W. and LONGTON, R. E. (1970) The effects of climate on antarctic plants. *In:* HOLDGATE, M. W. (ed.) *Antarctic Ecology*, **2**, 786–800. London: Academic Press.

GREENE, S. W. *et al.* (1967) Terrestrial life of Antarctica. *Antarctic Map Folio Series*, 5. Amer. Geogr. Soc. 24 pp.

HADLEY, E. B. and BLISS, L. C. (1964) Energy relationships of alpine plants on Mt Washington, New Hampshire. *Ecol. Monogr.*, **34**, 331–57.

HANSON, W. C., WATSON, D. G. and PERKINS, R. W. (1966) Concentration and retention of fallout radionuclides in Alaskan arctic ecosystems. *In:* ALBERG, B. and HUNGATE, F. P. (eds.) *Radioecological Concentration Processes*, 233–45. New York: Pergamon Press.

HEDBERG, O. (1964) Features of afroalpine plant ecology. *Acta Phytogeogr. Suecica*, **49**, 1–144.

HEINE, E. M. (1937) Observations on the pollination of New Zealand flowering plants. *Trans. Roy. Soc. New Zealand*, **67**, 133–48.

HOCKING, B. (1968) Insect-flower associations in the high Arctic with special reference to nectar. *Oikos*, **19**, 359–88.

HOLMEN, K. (1957) The vascular plants of Peary Land, North Greenland. *Medd. om Grønland*, **24** (9), 1–149.

HOLTOM, A. and GREENE, S. W. (1967) The growth and reproduction of antarctic flowering plants. *Phil. Trans. Roy. Soc. London*, B, 252, 323–37.

HOLWAY, J. G. and WARD, R. T. (1965) Phenology of alpine plants in northern Colorado. *Ecology*, **46**, 73–83.

HOWELL, J. T. (1951) The arctic-alpine flora of three peaks in the Sierra Nevada. *Leaflets West. Bot.*, **6**, 141–53.

HULTÉN, E. (1962) The circumpolar plants, I: Vascular cryptogams, conifers, monocotyledons. *Kungl. Svenska Vetenskapsakademiens Handlingar, Fjärde Serien*, **8** (5), 1–275.

HULTÉN, E. (1968) *Flora of Alaska and Neighbouring Territories. A Manual of the Vascular Plants*. Stanford, Calif.: Stanford Univ. Press. 1008 pp.

HUNTLEY, B. J. (1968) *A Floristic and Ecological Account of the Vegetation of Marion and Prince Edward Islands, South Indian Ocean*. Unpub. M.Sc. thesis, Univ. of Pretoria. 145 pp.

JOHNSON, A. W., VIERECK, L. A., JOHNSON, R. E. and MELCHIOR, H. (1966) Vegetation and flora. *In:* WILIMOVSKY, N. J. and WOLFE, J. N. (eds.) *Environment of the Cape Thompson Region, Alaska*, 277–354. Washington, DC: US Atomic Energy Commission.

JOHNSON, P. L. (1969) Arctic plants, ecosystems and strategies. *Arctic*, **22**, 341–55.

JOHNSON, P. L. and BILLINGS, W. D. (1962) The alpine vegetation of the Beartooth Plateau in relation to cryopedogenic processes and patterns. *Ecol. Monogr.*, **32**, 105–35.

KEVAN, P. G. (1970) *High Arctic Insect-Flower Relationships of Arthropods and Flowers at Lake Hazen, Ellesmere Island, NWT, Canada*. Unpub. Ph.D. dissertation, Univ. of Alberta. 422 pp.

KLIKOFF, L. G. (1965a) Microenvironmental influence on vegetational pattern near timberline in the central Sierra Nevada. *Ecol. Monogr.*, **35**, 187–211.

KLIKOFF, L. G. (1965b) Photosynthetic response to temperature and moisture stress of three timberline meadow species. *Ecology*, **46**, 516–17.

KLIKOFF, L. G. (1966) Temperature dependence of the oxidative rates of mitochondria in *Danthonia intermedia*, *Pentstemon davidsonii*, and *Sitanion hystrix*. *Nature*, **212**, 529–30.

KROG, J. (1955) Notes on temperature measurements indicative of special organization in arctic and subarctic plants for utilization of radiated heat from the sun. *Physiol. Planta.*, **8**, 836–9.

LANGE, O. L. (1965) Der CO_2-Gaswechsel von Flechten bei tiefen Temperaturen. *Planta*, **64**, 1–19.

LANGE, O. L. (1966) CO_2-Gaswechsel der Flechte *Cladonia alcicornis* nach langfristigem Aufenthalt bei tiefen Temperaturen. *Flora*, **156**, 500–2.

LARCHER, W. (1963) Zur spätwinterlichen Erschwerung der Wasserbilanz von Holzpflanzen au der Waldgrenze. *Berichte naturwissenschaft.-med. Vereins Innsbruck*, **53**, 125–37.

LINDSAY, J. H. (1971) Annual cycle of leaf water potential in *Picea engelmannii* and *Abies lasiocarpa* at timberline in Wyoming. *Arct. Alp. Res.*, **3**, 131–8.

LONGTON, R. E. (1967) Vegetation in the maritime Antarctic. *Phil. Trans. Roy. Soc. London*, B, **252**, 213–35.

LONGTON, R. E. (1970) Growth and productivity of the moss *Polytrichum alpestre* Hoppe in antarctic regions. *In:* HOLDGATE, M. W. (ed.) *Antarctic Ecology*, **2**, 818–37. London: Academic Press.

LÖVE, A. (1959) Origin of the arctic flora. *In:* LOWTHER, G. R. (ed.) *Problems of the Pleistocene Epoch and Arctic Area*. Montreal: McGill Univ. Museum.

MACIOR, L. W. (1968) Pollination adaptation in *Pedicularis groenlandica*. *Amer. J. Bot.*, **55**, 927–32.

MACIOR, L. W. (1970) The pollination ecology of *Pedicularis* in Colorado. *Amer. J. Bot.*, **57**, 716–28.

MANI, M. S. (1962) *Introduction to High Altitude Entomology. Insect Life Above the Timberline in the Northwest Himalaya*. London: Methuen. 302 pp.

MARK, A. F. (1965) Flowering, seeding, and seedling establishment of narrow-leaved snow tussock, *Chionochloa rigida*. *New Zealand J. Bot.*, **3**, 180–93.

MARK, A. F. (1969) Ecology of snow tussocks in the mountain grasslands of New Zealand. *Vegetatio*, **18**, 289–306.

MARK, A. F. (1970) Floral initiation and development in New Zealand alpine plants. *New Zealand J. Bot.*, **8**, 67–75.

MARK, A. F. and BLISS, L. C. (1970) The high-alpine vegetation of Central Otago, New Zealand. *New Zealand J. Bot.*, **8**, 381–451.

MARR, J. W. (1961) Ecosystems of the east slope of the Front Range in Colorado. *Univ. of Colorado Stud., Ser. in Biol.*, **8**, 1–134.

MATTICK, F. (1950) Die Flechten als ausserste Vorposten des Lebens im Gebirge. *Montagne e Uomini*, **2**, 494–6.

MIETTINEN, J. K. (1969) Enrichment of radioactivity by arctic ecosystems in Finnish Lapland. *In:* NELSON, D. J. and EVANS, F. C. (eds.) *Symposium on radioecology*, 23–31. Washington, DC: US Atomic Energy Commission.

MOONEY, H. A. and BILLINGS, W. D. (1960) The annual carbohydrate cycle of alpine plants as related to growth. *Amer. J. Bot.*, **47**, 594–8.

MOONEY, H. A. and BILLINGS, W. D. (1961) Comparative physiological ecology of arctic and alpine populations of *Oxyria digyna*. *Ecol. Monogr.*, **31**, 1–29.

MOONEY, H. A. and BILLINGS, W. D. (1965) Effects of altitude on carbohydrate content of mountain plants. *Ecology*, **46**, 750–1.

MOONEY, H. A. and JOHNSON, A. W. (1965) Comparative physiological ecology of an arctic and an alpine population of *Thalictrum alpinum* L. *Ecology*, **46**, 721–7.

MOONEY, H. A., BILLINGS, W. D. and HILLIER, R. D. (1965) Transpiration rates of alpine plants in the Sierra Nevada of California. *Amer. Midl. Nat.*, **74**, 374–86.

MOONEY, H. A., WRIGHT, R. D. and STRAIN, B. R. (1964) The gas exchange capacity of plants in relation to vegetation zonation in the White Mountains of California. *Amer. Midl. Nat.*, **72**, 281–97.

MOORE, D. M. (1970) Studies in *Colobanthus quitensis* (Kunth) Bartl. and *Deschampsia antarctica* Desv.: II. Taxonomy, distribution, and relationships. *Brit. Antarct. Surv. Bull.*, **23**, 63–80.

MOSER, W. (1968) Neues von der botanischen Forschungsstation 'Hoher Nebelkogel'/Tirol. *Vereins zum Schutze der Alpenpflanzen und-Tiere Jahrbuch*, **33**, 1–9.

MOSER, W. (1969) Die Photosyntheseleistung von Nivalpflanzen *Ber. dtsch. bot. Gesell.*, **82**, 63–4.

MOSER, W. (1970) Okophysiologische Untersuchungen an Nivalpflanzen. *Mitt. Ostalpinesche-din. Gesell. Vegetationskunde*, **11**, 121–34.

MOSQUIN, T. (1966) Reproductive specialization as a factor in the evolution of the Canadian flora. *In:* TAYLOR, R. L. and LUDWIG, T. A. (eds.) *The Evolution of Canada's Flora*, 43–65. Toronto: Univ. of Toronto Press.

MOSQUIN, T. and MARTIN, J. E. H. (1967) Observations on the pollination biology of plants on Melville Island, NWT, Canada. *Can. Field Nat.*, **81**, 201–5.

MÜLLER, H. (1881) *Die Alpenblumen, ihre Befruchtung durch Insekten und ihre Anpassung an dieselbe.* Leipzig.

PEARCY, R. W. (1969) *Physiological and Varied Environment Studies of Ecotypes of* Deschampsia caespitosa *(L.) Beauv.* Unpub. Ph.D. dissertation, Colorado State Univ. 184 pp.

PIEPER, R. D. (1964) *Production and Chemical Composition of Arctic Tundra Vegetation and their Relation to the Lemming Cycle.* Unpub. Ph.D. dissertation, Univ. of California, Berkeley.

PISEK, A. (1960) Die photosynthetischen Leistungen von Pflanzen besonderer Standorte. (a) Pflanzen der Arktis und des Hochgebirges. *Handbuch der Pflanzenphysiologie*, **5** (2), 375–414.

PISEK, A., LARCHER, W. and UNTERHOLZNER, R. (1967) Kardinale Temperaturbereiche der Photosynthese und Grenztemperaturen des Lebens der Blätter verschiedener Spermatophyten, I: Temperaturminimum der Nettoassimilation, Gefrier-und Frostschadensbereiche der Blätter. *Flora*, **157**, 239–64.

POLUNIN, N. (1948) Botany of the Canadian Eastern Arctic, Part III. Vegetation and ecology. *Nat. Mus. Can. Bull.*, **104**, *Biol. Ser.*, 32, 1–304.

POLUNIN, N. (1951) The real Arctic: suggestions for its delineation subdivisions, and characterization. *J. Ecol.*, **39**, 308–15.

PORSILD, A. E. (1951) Plant life in the Arctic. *Can. Geogr. J.*, **42**, 120–45.

PORSILD, A. E. (1957) Illustrated flora of the Canadian Arctic Archipelago. *Nat. Mus. Can. Bull.*, **146**, 1–209.

POTTER, L. D. and BARR, M. (1969) Cesium-137 concentrations in Alaskan arctic tundra vegetation, 1967. *Arct. Alp. Res.*, **1**, 147–54.

RASTORFER, J. R. (1970) Effects of light intensity and temperature on photosynthesis and respiration of two east antarctic mosses, *Bryum argenteum* and *Bryum antarcticum*. *Bryologist*, **73**, 544–56.

RAUP, H. M. (1947) The botany of southwestern Mackenzie. *Sargentia*, **6**, 1–262.

REISIGL, H. and PITSCHMANN, H. (1958) Obere Grenzen von Flora und Vegetation in der Nivalstufe der zentralen Ötztaler Alpen (Tirol). *Vegetatio*, **8**, 93–129.

ROCHOW, T. F. (1969) Growth, caloric content, and sugars in *Caltha leptosepala* in relation to alpine snowmelt. *Bull. Torrey Bot. Club*, **96**, 689–98.

RUSSELL, R. S. (1940a) Physiological and ecological studies on an arctic vegetation, II: The development of vegetation in relation to nitrogen and soil micro-organisms on Jan Mayen Island. *J. Ecol.*, **28**, 269–88.

RUSSELL, R. S. (1940b) Physiological and ecological studies on an arctic vegetation, III: Observations of carbon assimilation, carbohydrate storage and stomatal movement in relation to the growth of plants on Jan Mayen Island. *J. Ecol.*, **28**, 289–309.

SALISBURY, F. B. and SPOMER, G. G. (1964) Leaf temperatures of alpine plants in the field. *Planta*, **60**, 497–505.

SAVILE, D. B. O. (1961) The botany of the northwestern Queen Elizabeth Islands. *Can. J. Bot.*, **39**, 909–42.

SAVILE, D. B. O. (1964) General ecology and vascular plants of the Hazen Camp area. *Arctic*, **17**, 237–58.

SCHOLANDER, P. F., FLAGG, W., HOCK, R. J. and IRVING, L. (1953) Studies on the physiology of frozen plants and animals in the Arctic. *J. Cell. Compar. Physiol.*, **42**, Suppl. 1, 1–56.

SCHROETER, C. (1926) *Das Pflanzenleben der Alpen*. Zurich: Verlag von Albert Raustein. 1288 pp.

SCHULTZ, A. M. (1969) A study of an ecosystem: the arctic tundra. *In:* VAN DYNE, G. M. (ed.) *The Ecosystem Concept in Natural Resource Management*, 77–93. New York: Academic Press.

SCOTT, D. and BILLINGS, W. D. (1964) Effects of environmental factors on standing crop and productivity of an alpine tundra. *Ecol. Monogr.*, **34**, 243–70.

SCOTT, D., HILLIER, R. D. and BILLINGS, W. D. (1970) Correlation of CO_2 exchange with moisture regime and light in some Wyoming subalpine meadow species. *Ecology*, **51**, 701–2.

SHVETSOVA, V. M. and VOZNESENSKII, V. L. (1971) Diurnal and seasonal variations in the rate of photosynthesis in some plants of western Taimyr. *Int. Tundra Biome Transl.*, **2**, 1–11. (Original Russian edn, 1970, *Botanicheskii Zhurnal*, **55**, 66–76.)

SMITH, A. P. (1971) *Altitudinal Seed Ecotypes in the Venezuelan Andes*. MS, Duke Univ.

SØRENSEN, T. (1941) Temperature relations and phenology of the Northeast Greenland flowering plants. *Medd. om Grønland*, **125** (9), 1–305.

SÖYRINKI, N. (1938) Studien über die generative und vegetative Vermehrung der Samen-pflanzen in der alpinen Vegetation Petsamo-Lappland, I. *Seur. van elain. Julk.*, **11**, 1–323.

SÖYRINKI, N. (1939) Studien über die generative und vegetative Vermehrung der Samen-pflanzen in der alpinen Vegetation Petsamo-Lappland, II. *Seur. van. elain. Julk.*, **14**, 1–405.

STEINER, E. (1966) Dormant seed environment in relation to natural selection in *Oenothera*. *Bull. Torrey Bot. Club*, **95**, 140–55.

STOCKER, O. (1963) Das dreidimensionale Schema der Vegetationsverteilung auf der Erde. *Ber. dtsch. bot. Gesell.*, **76**, 168–78.

SUKACHEV, V. N. (1965) Studies on the flora and vegetation of high-mountain areas. Jerusalem: Israel Progr. Sci. Transl. (Original Russian edn, *Problemy Botaniki*, **5**, 1–293.)

SWAN, L. W. (1967) Alpine and aeolian regions of the world. *In:* WRIGHT, H. E. JR and OSBURN, W. H. (eds.) *Arctic and Alpine Environments*, 29–54. Bloomington: Indiana Univ. Press.

TAYLOR, B. W. (1955) The flora, vegetation and soils of Macquarie Island. *Australian Nat. Ant. Res. Expeditions Rep.*, B, 2, 1–192.

TIESZEN, L. L. and HELGAGER, J. A. (1968) Genetic and physiological adaptation in the Hill reaction of *Deschampsia caespitosa*. *Nature*, **219**, 1066–7.

TIKHOMIROV, B. A. (1963) *Contributions to the Biology of Arctic Plants* (in Russian). Moscow: Akad. Nauk SSSR. 154 pp.

TIKHOMIROV, B. A., SHAMURIN, V. F. and SHTEPA, V. S. (1960) The temperature of arctic plants. *Izv. Akad. Nauk SSSR, Biol. Ser.*, 3, 429–42 (in Russian).

TRANQUILLINI, W. (1964) The physiology of plants at high altitudes. *Ann. Rev. Plant Physiol.*, **15**, 345–62.

VARESCHI, V. (1970) *Flora de los Páramos de Venezuela*. Mérida, Venezuela: Universidad de los Andes, Ediciones del Rectorado. 425 pp.

WAGER, H. G. (1938) Growth and survival of plants in the Arctic. *J. Ecol.*, **26**, 390–410.

WALTER, H. (1968) *Die Vegetation der Erde in öko-physiologischer Betrachtung*, Band II. *Die gemässigten und arktischen Zonen*. Stuttgart: Gustav Fischer Verlag. 100 pp.

WARREN WILSON, J. (1957) Observations on the temperatures of arctic plants and their environment. *J. Ecol.*, **45**, 499–531.

WARREN WILSON, J. (1966) An analysis of plant growth and its control in arctic environments. *Ann. Bot.*, **30**, 383–402.

WATSON, D. G., HANSON, W. C., DAVIS, J. J. and RICHARD, W. H. (1964) Strontium-90 in plants and animals of arctic Alaska. *Science*, **144**, 1005–8.

WEBERBAUER, A. (1911) *Die Pflanzenwelt der peruanischen Anden*. Vol. 12 of ENGLER, A. and DRUDE, O. (ed.) *Die Vegetation der Erde*. Leipzig. 352 pp.

WEBSTER, G. L. (1961) The altitudinal limits of vascular plants. *Ecology*, **42**, 587–90.

WILLIAMS, L. D. (1970) *Life-form characteristics of Androsace septentrionalis, an Arctic-Alpine 'Annual', in Southeastern Wyoming*. Unpub. M.A. thesis, Duke Univ. 61 pp.

WINKLER, E. and MOSER, W. (1967) Die Vegetationzeit in zentralalpinen Lagen Tirols in Abhängigkeit von den Temperatur- und Niederschlagsverhältnissen. *Veröff. Museum Ferdinandinum, Innsbruck*, **47**, 121–47.

ZUBKOF, A. I. (1935) Duration of the vegetation period on the northern island of Novaya Zemlya. *Arctica*, **3** (in Russian with English summary). Cited in SØRENSEN (1941).

Manuscript received June 1971.

Additional references

BILLINGS, W. D. (1973) Arctic and alpine vegetations: similarities, differences, and susceptibility to disturbance. *BioScience*, **23**, 697–704.

BILLINGS, W. D. (1974) Adaptations and origins of alpine plants. *Arct. Alp. Res.*, **6** (in press).

SAVILE, D. B. O. (1972) Arctic adaptations in plants. *Res. Branch Canada Dept of Agric. Monogr.*, 6, 1–81.

Tundra primary productivity

Patrick J. Webber

Institute of Arctic and Alpine Research,
University of Colorado

1. Introduction

All animal life is dependent on the growth and production of green plants. Green plants use the energy of sunlight to manufacture organic material which is the driving force for all other life forms. The magnitude and importance of this *primary production* in the tundra environment will be described.

Here, tundra is defined as the treeless regions beyond climatic timberlines in the north (arctic tundra) and on high mountains (alpine tundra) (Bliss, 1962a; Löve, 1970).

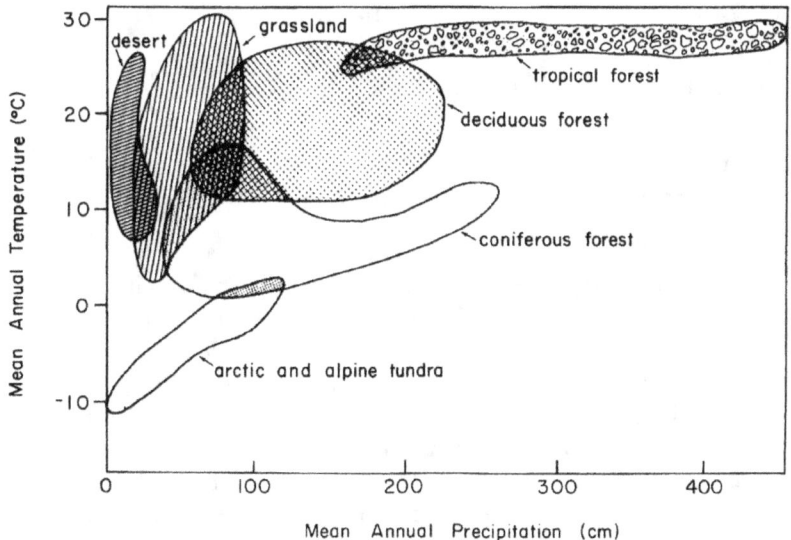

Fig. 8B.1. Six major biomes characterized in terms of their mean annual temperatures and total precipitation (after Hammond, 1972).

This is the usual definition which is more than the original meaning (Lapland treeless plains) but less than the expanded meaning of the current Tundra Biome section of the International Biological Programme (IBP) (Wielgolaski, 1972) which also includes oceanic moorland areas in cool temperate climates. Polar desert, which to some authors is not part of the tundra zone (Alexandrova, 1970), is included in this discussion.

Tundra, excluding the extremes of rock and ice, occupies about 8 million square kilometres of land, which is about 1/64th of the total earth's surface or about 1/20th of

Table 8B.1. Net primary production of major ecosystems and the earth's surface (from Whittaker, 1970).

	Area (10^6 km^2)	Net primary productivity, per unit area (dry g m^{-1} yr^{-1}) Normal range	Mean	World net primary productivity (10^9 dry tons* yr^{-1})
Lake and stream	2	100–1500	500	1.0
Swamp and marsh	2	800–4500	2000	4.0
Tropical forest	20	1000–5000	2000	40.0
Temperate forest	18	600–2500	1300	24.4
Boreal forest	12	400–2000	800	9.6
Woodland and shrubland	7	200–1200	600	4.2
Savannah	15	200–2000	700	10.5
Temperate grassland	9	150–1500	500	4.5
Tundra and alpine	8	10–400	140	1.1
Desert scrub	18	10–250	70	1.3
Extreme desert, rock and ice	24	0–10	3	0.07
Agricultural land	14	100–4000	650	9.1
Total land	149		730	109.0
Open ocean	332	2–400	125	41.5
Continental shelf	27	200–600	350	9.5
Attached algae and estuaries	2	500–4000	2000	4.0
Total ocean	361		155	55.0
Total for earth	510		320	164.0

* Metric tons (1000 kg).

the total land surface (Whittaker, 1970; Lieth, in press). The Tundra Biome is located in dry, cold areas; all other biomes are located either in warmer or wetter areas (fig. 8B.1).

The small areal extent of tundra, its very limited population, and its generally low level of biological productivity, the lowest in the world except for deserts and open oceans (table 8B.1), could be cited as reason for making the study of tundra of low priority. This is a shortsighted view which overlooks the holocoenotic and finite nature of the world, the current high human population, and the shortage of renewable resources. Further, the presence of vast, unexploited, non-renewable resources in the Arctic also necessitates an understanding of tundra in order to prevent its disruption when these resources are developed (Bliss, 1970a, b; Ives, 1970).

Much work on tundra productivity is being carried out under the auspices of the IBP (Heal, 1971; Hammond, 1972) and the best data have yet to be published. The available data used in this account are of variable quality; their limitations will be discussed and an effort made to point out areas for future study. This account is limited to terrestrial productivity.

2. Primary production

DEFINITIONS AND MEASUREMENTS

When photosynthesis is considered at the vegetation level in an ecosystem it is designated primary production. A simplified expression for photosynthesis or primary production can be written thus:

$$\text{carbon dioxide} + \text{water} + \text{light energy} \xrightarrow{\text{chlorophyll}} \text{simple sugars} + \text{oxygen}$$

Most of the sugars are changed into more complex plant components consisting of proteins, structural materials and storage compounds such as other carbohydrates and lipids. Production of these secondary compounds, which are used in plant growth and maintenance, requires energy and thus not all the energy fixed during photosynthesis is available to the animals or, in the broad sense, to the consumer organisms of the ecosystem. The unavailable energy is lost as heat of respiration during the vital activities of the plant. The total amount of energy fixed is called *gross primary production* (GP), the amount used in growth, metabolism and maintenance is due to *respiration* (R), and the difference equals the yield or *net primary production* (NP). That is:

$$GP - R = NP$$

Primary production may be expressed as the amount of energy fixed per unit area, for example calories per square metre (cal m^{-2}). When this production is rate specified, say as cal m^{-2} day^{-1}, it becomes productivity. Direct measurement of energy fixation is impossible and can only be obtained indirectly from the rates of changes of the raw materials or products in the photosynthesis equations. Odum (1971) has reviewed the various methods of measuring productivity. Measurements made at the vegetation level, in contrast to those made on a small or individual plant scale, are usually carried out over large time intervals, for example a growing season; and these measurements are almost entirely restricted to estimating net productivity because respiratory losses are difficult to assess over long periods. The most common method, especially for terrestrial vegetation, is a *harvest method* in which weight gain is measured by clipping the aerial vegetation from equivalent plots at intervals during the growing season. Thus, this method is further restricted to net *shoot* production. It is assumed that there is no grazing by herbivores and that the vegetation is not in a steady state during the sampling interval. The procedure works well when herbivores have been excluded, or their effects can be considered negligible, for annual vegetation and for perennial vegetation with distinct seasonal growth, especially if the latter is of small stature. When the plant material is dried, productivity can be expressed as dry weight gain of an area for a specific time period. Conversion to energy gain can be made through calorimetry of subsamples of the harvested, dry, plant material or from appropriate standard tables (for example, Cummings and Waycheck, 1971); these conversions are not always made and the most commonly used units of productivity remain those based on dry weight (see table 8B.2). It is difficult to measure below-ground productivity for all but annual vegetation

Table 8B.2. Common units used in the study of ecological production and energy relations.

A   Basic units of energy

1 gram-calorie (gcal or cal) = the amount of energy required to raise the temperature of one gram of water by 1° at 15°C.

Langley (ly) = 1 gcal cm^{-2}. This is the unit for incoming solar radiation and energy budget descriptions.

Kilogram-calorie (kcal or Cal) = 1000 gcal. This is the calorie referred to in human diet charts.

Watt (W) = 14.3 gcal min^{-1}; a unit of power.
= 1 Joule s^{-1}.

B   Approximate energy content per dry gram of organic material

	Kilocalories
Average terrestrial plant	4.5
Carbohydrate (sucrose)	4.0
Carbohydrate (starch and cellulose)	4.2
Crude protein	5.7
Fat (ether extract)	9.5
Crude fibre (15–30% of lignin)	4.6

C   Common units of primary productivity based on dry weight

1 g m^{-2} yr^{-1} (grams per square metre per year)
= 10 kg ha^{-1} yr^{-1} (kilograms per hectare per year)
= 0.1 m centn ha^{-1} yr^{-1} (metric centners per hectare per year)
= 1 m ton km^{-2} yr^{-1} (metric tons per square kilometre per year)
= 8.9 lb acre^{-1} yr^{-1} (pounds per acre per year)

because of the problem of removing roots and other structures from the soil and establishing which of these are living and which are of the current year. This is especially true for tundra where below-ground plant material is frequently ten times that of above ground. Consequently, very few studies do more than estimate below-ground biomass and/or standing crop at one point during the growing season; productivity discussion is then restricted to the net shoot productivities. A reasonably satisfactory way of obtaining annual net shoot productivity for tundra is to harvest at the peak of above-ground biomass. This is possible because for all but the shrubs, and a few evergreen plants, including mosses and lichens, aerial structures die each year and there is little growth past peak biomass because of the sudden truncation of the tundra growing season. The same has been suggested for grassland (Wiegert and Evans, 1964). Even when woody plants are encountered in tundra they are usually deciduous and limited in stature so that when leaves and new shoot production are harvested a fair estimate of above-ground production can be obtained; this method ignores the annual increments to previous stems which are, however, limited (Beschel and Webb, 1963; Bliss, 1966). Mosses and lichens are less easy to deal with, both from taxonomic and harvesting standpoints, and have therefore seldom been sampled. A few studies have estimated both shoot and root net productivity for tundra (Scott and Billings, 1964; Johnson and Kelley, 1970; and Dennis, 1968). Annual below-ground gain seems to be about equal to the net shoot estimates, thus doubling overall net production.

ENERGY FLOW AND CONTROLLING FACTORS

The amount of energy contained in incoming solar radiation changes with passage through the atmosphere, attenuating from a solar constant of about 2 gcal cm^{-2} min^{-1} ($=2$ ly min^{-1}) to a sea level value of about 1.34. On high mountains, say at 4500 m above sea level, insolation values may be as high as 1.8 on clear days at noon (Gates, 1965).

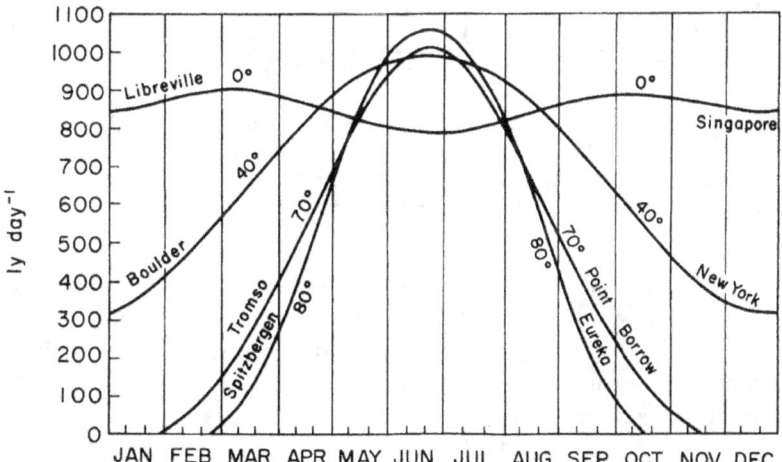

Fig. 8B.2. Daily totals of undepleted radiation on a horizontal surface for geographical latitudes as a function of time of year (adapted from Gates, 1962). Based on a solar constant of 1.94 cal cm^{-2} min^{-1}.

Solar elevation, sky conditions and local topography will further change the amount of insolation falling on vegetation (see Barry and Van Wie, Chapter 2C). Latitudinal and seasonal variation is shown in fig. 8B.2 where arctic situations, illustrated for latitudes 70° and 80°, which have twenty-four hours of sunshine in summer and none in winter, can be compared with the temperate latitudes of Boulder and Lisbon and the equatorial situation of Libreville or Singapore. Only about 45 per cent or the visible part of the solar radiation between wavelengths of 0.4 and 0.7 μm (Reifsnyder and Lull, 1965; Anderson, 1967) can be absorbed by green plants, and of this only a few percent at best can be changed into chemical energy. The fate of solar energy incident upon vegetation is nicely demonstrated in fig. 8B.3. Photosynthetic efficiency is the ratio of fixed energy to incident usable or visible energy. It is usually expressed as a percentage on the basis of the growing season, and, because most productivity is known only in net terms, *net photosynthetic efficiency* is the commonest expression. It is important always to state the basis of photosynthetic efficiency, that is whether it be for the whole year or for just the growing season,

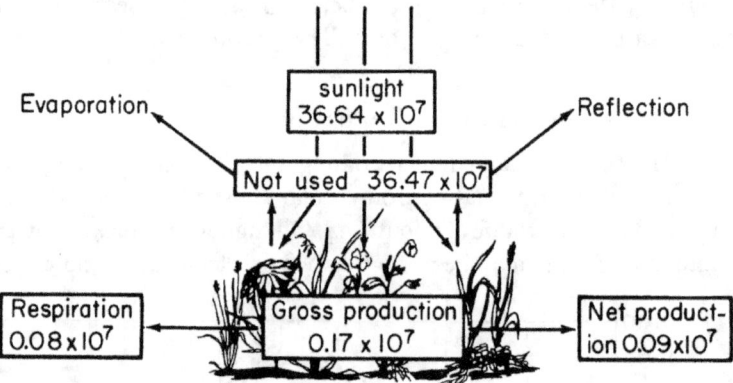

Fig. 8B.3. The fate of solar energy incident upon Alaskan north-slope tundra during an eighty-day growing season. Data, although somewhat hypothetical, are based on observations of Johnson and Kelley (1970) and a net photosynthetic efficiency for visible radiation of 1 per cent is assumed (adapted from Phillipson, 1966). Values are in cal m^{-2}.

whether it be for net or gross production and whether or not it be for total or visible radiation. Plants may actually be more efficient than published values indicate because most net productivity estimates may be minimal due to losses during sampling, especially when sampling is restricted to peak season (Ovington and Lawrence, 1967). But even then efficiency would seldom be increased by more than a fraction; a more complete consideration of efficiency can be found in Warren Wilson (1967). An energy budget for the non-plant or heterotrophic part of the ecosystem, in which the available plant energy is utilized by the consumers which are composed of the herbivores, carnivores feeding on herbivores, and the decomposer organisms, can also be constructed. The production of consumer organisms is called *secondary production*. The conversion of the available net primary production into consumer flesh is accompanied by loss of respiratory energy at each successive step in the food web of the ecosystem. Consumers appear more efficient than plants (up to 15.7 per cent in table 8B.3), but because they must be mobile to catch their food and avoid their enemies, their total respiratory

Table 8B.3. The net growth efficiencies (assimilated energy contained in growth or net secondary production) and respiration losses of some consumer (animal) populations. Values are percentages of assimilated (consumed minus undigested) energy.

Species	Consumer status	Percentage of assimilated energy		Source
		Lost in respiration	Net growth efficiency	
Various grasshoppers	herbivorous poikilotherm	84.3	15.7	Odum et al., 1962
Meadow mice	herbivorous poikilotherm	96.9	3.1	Golley, 1960
Least weasel	carnivorous homeotherm	97.6	2.4	Golley, 1960
Savannah sparrow	herbivorous homeotherm	98.6	1.4	Odum et al., 1962
Beef cattle	herbivorous homeotherm	89.1	10.9	Phillipson, 1966

losses are much greater than those of plants (table 8B.4). It may be noted from table 8B.3 that small rapidly-moving homeotherms (warm-blooded animals), like the sparrow, require relatively more energy than do poikilotherms (cold-blooded animals) or large protected and domesticated cattle. These comparisons of efficiency between plants and animals tend, however, to be misleading because the efficiencies of different processes are being compared. If exact comparisons are to be made, say for the efficiency of conversion of food into growth and storage with only constructive respiratory loss, plants and animals are about equally efficient (Warren Wilson, 1967).

Ultimately, in a balanced ecosystem all the energy contained in plants is dissipated by the consumer components of that ecosystem. It is dissipated as heat energy from respiration and this cannot be reclaimed by any organism. For this reason it is said that energy *flows* through an ecosystem and must be continually supplied from the sun through the agency of green plants. While botanists like to stress the importance of plants as energy fixers the remainder of the ecosystem is equally important to life. The consumers, especially the decomposers, function to return to the soil valuable chemicals which are in finite supply and which would remain locked in undecayed plant and animal bodies;

Table 8B.4. A comparison of respiration (R) energy losses of the producers (vegetation) of some selected ecosystems.

Ecosystem	GP $(g\ m^{-2}\ yr^{-1})$	R as % GP	Source
Wet, sedge tundra, Alaska	344	53	Johnson and Kelley, 1970
Oak-pine forest, New York	2555	57	Woodwell and Whittaker, 1968
Tropical rainforest, Puerto Rico	11,970	71	H. T. Odum and Pigeon, 1970
World mean	640	50	E. P. Odum, 1971, and Whittaker, 1970

the decomposers are nature's recyclers. Thus, in contrast to energy, material *cycles* within an ecosystem, it must be continually reused. Energy flow and material cycling are fundamental functions of ecosystems. The interactions between the different parts of an ecosystem are complex and usually in a state of balance. The ecosystem functions as a whole, and when any part of an ecosystem changes or is disturbed all other parts of

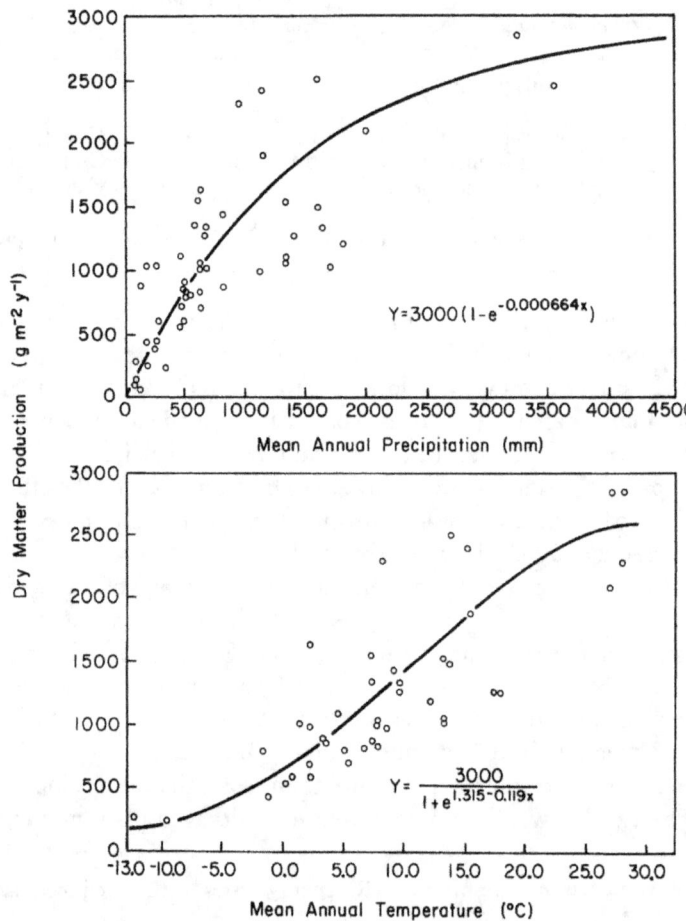

Fig. 8B.4. Graphs showing the relationships between annual precipitation and temperature and annual net primary productivity (from Lieth, in press).

the system may be affected to some degree. The interrelatedness of ecosystem components is called *holocoenosis* and is a characteristic of all ecosystems from small pond microcosms to the biosphere. It is a characteristic which we would all do well to remember (Commoner, 1971). Secondary production and the importance of energy flow and material cycling are also discussed by R. S. Hoffmann in Chapter 9.

A catalogue of the factors controlling the rate of primary production might well include all the known biotic and abiotic environmental factors. However, at the fundamental photosynthesis level, the list of limiting factors is manageable. Initially, components or resources of the photosynthesis equation are eligible; these are the amounts of light energy, carbon dioxide, water and active chlorophyll. The following factors are also considered frequently limiting to the rate of photosynthesis: leaf quantity and distribution, temperature, nutrient availability and turnover time, length of growing season, competition between and within species for available resources, grazing and trampling by animals, and the genetic capability of the plants to operate under a variety of conditions. The previous section by W. D. Billings deals with many of these factors. At the regional scale the positive correlation of water and temperature with production is illustrated in fig. 8B.4. The scattering of points suggests, however, that although one or other of these factors may frequently control production, other factors will also be controlling in some situations. The curves in fig. 8B.4 also indicate that, with high and plentiful supplies of precipitation and temperature, production levels out as other factors become more frequently limiting. This demonstrates holocoenosis in the ecosystem and Liebig's Principle of Limiting Factors, which implies that only one factor can be limiting at a time, but if that factor is supplied then some other factor becomes limiting. These observations relating temperature and water to production can be further illustrated by examining the most frequently limiting factors of the four biomes which are compared in the next subsection. In tundra, temperature is the principal growth limiting factor; in temperate grasslands, water is most frequently limiting to growth; in the temperate forests of North America temperature is limiting to the north near the boreal forest and precipitation is limiting to the west near the grasslands; in tropical rainforest temperature and precipitation regimes are adequate all year and thus some other factor will replace them. In tropical rainforest light often becomes the limiting factor as it is unable to penetrate the dense foliage in sufficient quantities to sustain photosynthesis. However, in tropical rainforests nutrients may be limiting to growth because so many are locked out of circulation in the standing vegetable matter.

3. Magnitude of production

COMPARISON WITH OTHER BIOMES

Bazilevich and Rodin (1971) and Lieth (in press) have made estimates of net productivity on a world scale. Both studies provide maps which illustrate the variation within the Arctic and by which tundra production can be set in perspective with that of other biomes. Lieth's map (fig. 8B.5) was derived from a model of world production which uses precipitation and temperature as predictors. This model used the relationships derived in fig. 8B.4. The map is an improvement over his earlier map (Lieth, 1964) which like Bazilevich and Rodin's was based on actual but very scattered production measurements of the various vegetation types. There are strong parallels between Bazilevich and Rodin's and Lieth's production distributions and there is good agreement

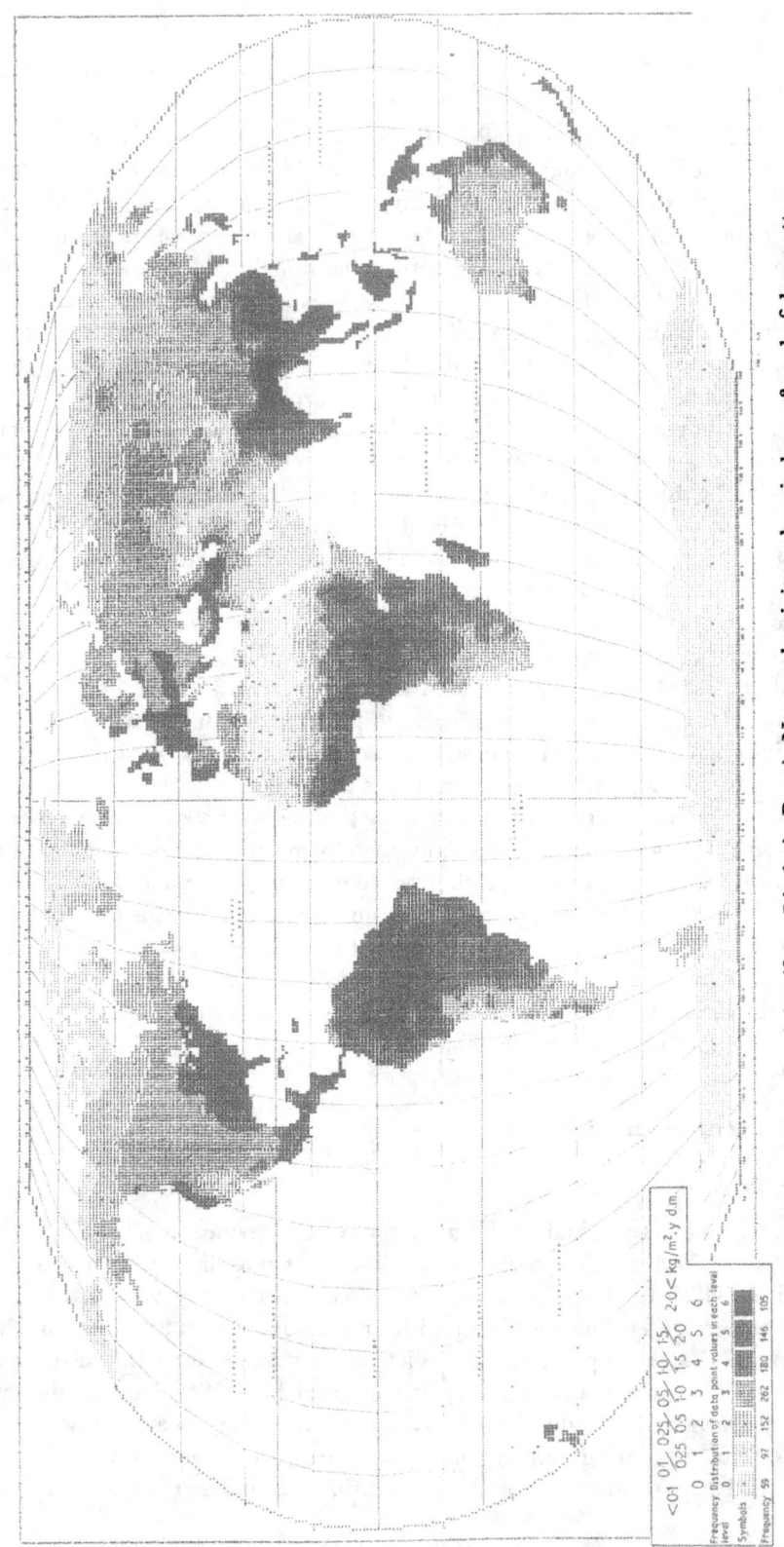

Fig. 8B.5. World terrestrial productivity (from Lieth, in Press). Net productivity values in kg m⁻² yr⁻¹ of dry matter.

between Lieth's predictions and published values (see table 8B.5). These parallels result partly from cause and effect relationships and partly from the inevitable circularity of argument involved in mapping vegetation and climates. From the maps,

Table 8B.5. A comparison of above- to below-ground vascular plant biomass ratios and annual net vascular plant production for forest, tundra and grassland ecosystems.

Ecosystem	Net production $g\ m^{-2}\ yr^{-1}$	Above: below biomass	Source
Forest			
Young, oak-pine forest, New York	1060	1 : 0.52	Woodwell and Whittaker, 1968
Mature, Scots pine plantation, England	1240	1 : 0.34	Ovington, 1957
Mature, tropical rainforest, Puerto Rico	3460	1 : 0.40	H. T. Odum and Pigeon, 1970
Tundra			
Arctic			
Polar desert, Franz Joseph Land	12*	1 : 4.83	Alexandrova, 1970
High arctic tundra, New Siberian Islands	142*	1 : 7.19	Alexandrova, 1970
Low arctic tundra, Eastern Europe	228*	1 : 4.55	Alexandrova, 1970
Alpine			
Cushion-plant fellfield, Wyoming	42	1 : 0.87†	Scott and Billings, 1964
Celmisia-Poa herbfield, New Zealand	476‡	1 : 8.90‡	Bliss and Mark, unpublished ms. *in* Bliss, 1966
Carex bigelowii meadow, New Hampshire	352*	1 : 25.59	Bliss, 1966
Grassland			
Desert grassland, New Mexico	276	1 : 2.94	Sims and Singh, 1971
Shortgrass prairie, Colorado	604	1 : 29.55	Sims and Singh, 1971
Tallgrass prairie, Colorado	1120	1 : 2.33	Moir, 1969

* Estimated as twice net shoot production.
† Standing crop not biomass.
‡ Assumed from reasonable but incomplete data.

published values and table 8B.1 the average net productivity of tundra areas in relation to temperate grassland, temperate forest and tropical forest approximate these ratios 1:4:9:14, with tundra having an average value of about 140 g m^{-2} yr^{-1}. With such low production tundra is on a par with arid ecosystems such as deserts.

Comparison of tundra with the other biomes in terms of productivity relations is restricted because data have often been collected in different ways and also because the comparative data matrices are incomplete. Tables 8B.4, 5 and 6 provide some comparisons. Table 8B.4 suggests that tundra vegetation has smaller respiratory energy losses than other natural systems. Cultivated crops (alfalfa loses 38 per cent: Thomas and Hill, 1949; 'theoretical' soybean loses 25 per cent: Gorden, 1969) do better because they have been selected and are managed. The low respiratory losses of tundra reported by

Table 8B.6. Daily net primary production and net photosynthetic efficiency from visible radiation for the growing season of some selected ecosystems.*

Ecosystem and location	Length of growing season (days)	Net production (g m⁻² day⁻¹)	Mean, insolation (cal cm⁻² day⁻¹)	Net photosynthetic efficiency	Source
Theoretical maximum crop yield	100	77	500	12.0	Loomis and Williams, 1963
Sugar cane, Hawaii	365	42	400	10.6	Monteith, 1965
Scots pine plantation, England	360	3	210	2.2	Ovington, 1961
Tropical rain-forest, Puerto Rico	365	9	383	2.1	H. T. Odum and Pigeon, 1970
Tallgrass prairie, Colorado	140	8	590	1.2	Moir, 1969
Alpine tundra, New Hampshire	69	5	463	0.84†	Bliss, 1966
World mean	—	—	—	0.25	Whittaker, 1970

* Each authority has a different method of estimating productivity and uses different assumptions concerning usable radiation. In general an attempt has been made here to keep methodology uniform, thus the above values may differ slightly from the original source; any errors are this author's.
† Estimated as twice the shoot values.

Johnson and Kelley (1970) will require further research especially in the light of the ability for individual tundra plants to have high rates of respiration at low temperatures (see previous section by Billings) although these results need not be contradictory.

Table 8B.6 illustrates the variation of net photosynthetic efficiency for different vegetation. Crops are again most efficient but, as H. T. Odum (1967) points out, man supplies energy to the crop through pest control, irrigation, and fertilization, etc.; thus these high efficiencies are artificially created. The data in table 8B.6 suggest that tundra, in spite of the large annual production difference, is only a little less efficient on a daily basis than other natural systems; this seems, on the basis of my own unpublished data, to relate to the greater structural complexity of the other vegetation types because those tundra types with greater leaf areas and densities per unit area of ground have net shoot photosynthetic efficiencies of up to 2 per cent. In effect, then, the individual efficiencies of tundra plants or the vegetation which they form are not significantly different from other plants or vegetation. Bliss (1966) first made this generalization.

Another generalization for tundra is its high caloric content when compared with other vegetation types (Golley, 1961; Bliss, 1962a, b, 1971; Jordan, 1971). Bliss (1962b) reported that evergreen alpine shrubs in New England have a mean caloric content of 5.1 kcal (g ash-free dry weight)$^{-1}$, that deciduous shrubs have a mean of 4.9 and herbs a mean of 4.6. Average values from other biomes are 3.9 for tropical rainforest (Golley, 1961), 4.1 for temperate grassland (Kucera et al., 1967), and 4.8 for temperate pine forest (Ovington, 1961). Forests with conifers, willows and alders also have high energy content (Jordan, 1971). The high caloric content in some tundra plants, especially evergreen shrubs, is a result of high lipid content (Bliss, 1962b; Hadley and Bliss, 1964). Most food storage in alpine plants is in the form of carbohydrates and is found in below-ground structures. Evergreen shrubs which have smaller root systems store the food reserves in stems and old leaves (Bliss, 1962b). An adaptive value of a high energy concentration in tundra plants may be that less energy is spent processing a concentrated food source (Jordan, 1971). In tundra this seems to fit with the rapid onset of metabolism and growth which occurs immediately in spring. Hadley and Bliss (1964) and Bliss (1971) point out that evergreens which do not exhibit this same rapid burst of spring metabolism as deciduous or herbaceous species, and yet which are common in extreme tundra, may be adapted to conserve energy by not needing to make an entirely new photosynthetic apparatus each year and by using old leaves as storage sites. Golley (1961) pointed out the increase in caloric concentration with increasing latitude which Jordan (1971) correlates with decreasing insolation during the growing season. Jordan also reports a trend of increasing caloric concentration with increasing precipitation which his data do not seem to warrant; the adaptive significance of these correlations remains highly speculative and these patterns need further examination as data become available.

The most striking characteristic of tundra is the very high proportion of the biomass which is below ground (Aleksandrova, 1958; Alexandrova, 1970; Bliss, 1962a, 1966, 1970b; Mooney and Billings, 1960; Billings and Mooney, 1968; Scott and Billings, 1964; Dennis and Johnson, 1970; Wielgolaski, 1972). Table 8B.5 compares the above:below-ground ratios of biomass, at peak above-ground biomass, for a number of ecosystems. Forests have the majority of their biomass above ground while grasslands and tundra show the opposite. The similarity between tundra and grasslands relates to the similarity of plant life form, that is, both have an absence of trees and a preponderance of grass-like hemicryptophytes. The life form spectra (sensu Raunkiaer, 1934) of tundra and grassland are, however, quite distinct from one another. Tundra has a greater proportion of low shrubs and cushion plants (chamaephytes) than grassland. A further difference between grassland and tundra is the almost complete absence of annuals in tundra. The adaptation of cushion plants to tundra seems to lie in their ability to live in the warmer, less windy environment at ground level (Bliss, 1962a). Annuals which must complete their life cycle in one year are presumably at a distinct disadvantage compared with tundra perennials with large below-ground storage, which can tide them over short growing seasons when seed production might be limited. In general there is an increase in the proportion of chamaephytes in vegetation with both latitude and altitude (Raunkiaer, 1934). The large below-ground biomass is considered by Scott and Billings (1964) to represent an adaptation to the environment, because alpine species when grown in a non-windy, non-humid and nutrient-limiting greenhouse showed higher shoot/root ratios. Dennis and Johnson (1970) interpret the large below-ground biomass in the Arctic to be an adaptation to the absence of suitable above-ground sinks (such as woody tissues) for photosynthetic products.

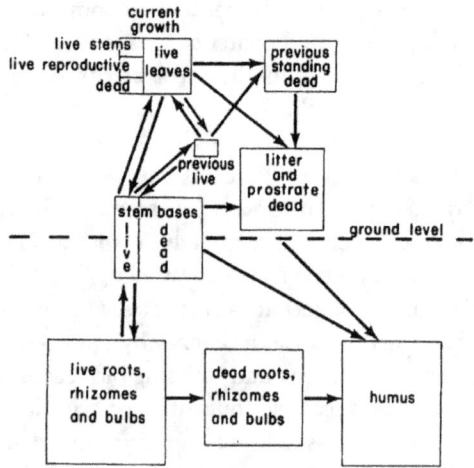

Fig. 8B.6. Vascular vegetation fractions at peak above-ground vascular plant biomass for herbaceous alpine tundra. Each fraction is represented by a box whose size is more or less proportional on a geometric scale to the dry weight of the fraction. The arrows represent carbon flow; input and outputs of carbon dioxide are not indicated.

It is likely that a careful study of the spatial and seasonal distribution of dry matter within vegetation would reveal further differences between tundra and other vegetation. These studies are being carried out within the IBP, especially in tundra areas (Wielgolaski, 1972), but they are still incomplete. Such data are also required to understand better the dynamics of both the vegetation and the utilization of the vegetation by the rest of the ecosystem. Fig. 8B.6 is a beginning in this direction, but as yet it is only a static picture based on peak-season fractions. It illustrates the importance of the belowground nutrient pools. In this example for herbaceous alpine tundra, the large 'stem base' fraction, most of which is dead sheathy material from cushion plants and tussocky hemicryptophytes, is characteristic. The picture for tundra with woody chamaephytes is similar, although the below-ground portion is reduced and the 'previous live' fraction is much larger, especially if shrubby evergreens are present.

VARIATION WITHIN THE TUNDRA BIOME

The Arctic may be divided into three vegetation zones on the basis of the plant cover on mesic or well-drained upland surfaces (fig. 8B.7). These zones are: *polar desert* with sparse, open vegetation (Plate 31); *high arctic tundra* with a more or less continuous cover of herbs

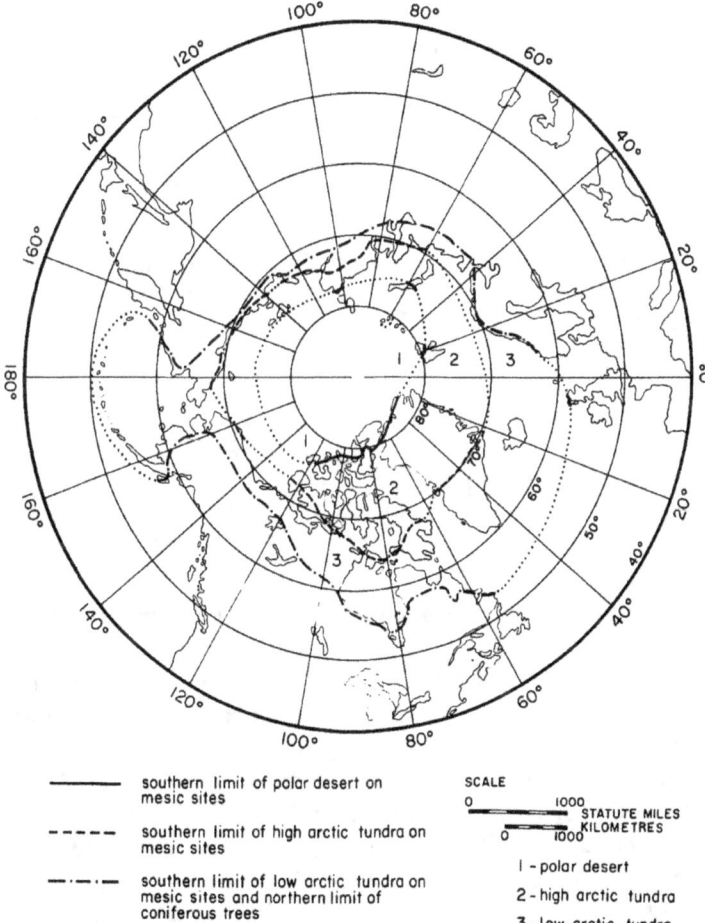

southern limit of polar desert on mesic sites

- - - - southern limit of high arctic tundra on mesic sites

— · — · — southern limit of low arctic tundra on mesic sites and northern limit of coniferous trees

SCALE

0   1000
                           STATUTE MILES
0   1000
                  KILOMETRES

1 - polar desert

2 - high arctic tundra

3 - low arctic tundra

Fig. 8B.7. The distribution of the three major types of arctic vegetation (adapted from Alexandrova, 1970).

and occasional prostrate woody species (Plate 32); and *low arctic tundra* with a good cover of low-growing shrubs such as birches and willows (Plate 33), and with polygonal bogs developed in wet sites (Plate 32). Lieth's productivity map (fig. 8B.5) predicts a net productivity of less than 100 g m^{-2} yr^{-1} for polar desert and 100–250 g m^{-2} yr^{-1} and 250–500 g m^{-2} yr^{-1} for high arctic and low arctic tundra, respectively. His map also suggests that the maritime regions of Greenland and Iceland may yield between 500 and 1000 g m^{-2} yr^{-1}. Macquarie Island in the maritime Subantarctic has yields similar to the latter (Jenkin and Ashton, 1970).

Alexandrova (1970) points out the comparison between the three arctic zones in terms of live aerial biomass is 1 : 10 : 100. In terms of net productivity the comparison is more like 1 : 5 : 15 reflecting the increase in shrub standing crop from the polar desert southwards.

There is a need for better estimates of total productivity of large areas of tundra; too often only the more productive communities have been measured rather than the impoverished communities which make up a large portion of the total arctic landscape (Webber, 1971a). There are some measurements of unproductive sites; Warren Wilson (1957) found only 3 g m^{-2} yr^{-1} in an arctic willow barren on Cornwallis Island, and

Table 8B.7. The shift in percentage by weight of the contribution to the above-ground biomass of different plant groups along the north–south gradient in the Arctic.

Plant groups	Polar desert	High arctic tundra	Low arctic tundra
Vascular plants	4	38	38
herbaceous	4	22	4
woody	0	16	34
Bryophytes (mosses and liverworts)	31	44	60
Lichens	65	18	2

Beschel (1970) estimated that the annual production of lichens on rocks, which cover so much of the Arctic, especially the polar desert, may amount to only a few milligrams per square metre. Estimates of total production of large areas can be made from good detailed vegetation maps and production estimates of each map unit; unfortunately few such maps are available.

It has already been pointed out that most estimates of tundra production have omitted the below-ground increment. A further serious omission is the production of lichens and mosses; the production of the latter may occasionally be considerable (Longton, 1970; Clarke et al., 1971). They are particularly difficult to measure because they are perennial evergreens and do not always show clear annual growth increments which would allow total ageing or harvesting of current growth. Mosses also contribute considerably to freshwater production in some tundra situations (Kalff, 1970). Even when moss and lichen production only amounts to a few grams it may be vital; although fruticose lichens only grow a few millimetres a year (Scotter, 1963) the standing crop is sufficient to support reindeer and caribou during winter. The proportion of mosses as a component of arctic vegetation increases from polar desert to low arctic tundra while that of lichens decreases along the same gradient (Alexandrova, 1970, and table 8B.7).

The latitudinal trend from forests to tundra of increasing contribution of below-ground biomass is continued from low arctic tundra to high arctic tundra. There is, however, a drop in below-ground contribution on passing into polar desert where the vegetation is open and cryoturbation increases (table 8B.5).

The present latitudinal generalizations will need strengthening as the data base grows. Frequently data do not fit the generalizations but, perhaps, rather than faulty generalization the reason for this lies in the wide range of variation present in each zone. For example, caloric content, productivity and below-ground biomass can all vary widely; it is the averages which should be considered. Variations in vegetation characteristics seen along latitudinal macrogradients are mirrored in the local microgradients; the trend of productivity and above:below-ground biomass ratios from xeric fellfield to mesic herb-tundra to shrub-tundra is one of increasing production and rising and then falling below-ground biomass balance.

Fig. 8B.8 illustrates the trends of primary production characteristics along microgradients in mid-latitude alpine tundra from Niwot Ridge, Colorado. Similar patterns have been observed for arctic sites but are less well documented. Billings and Bliss (1959) and Scott and Billings (1964) for adjacent sites in Wyoming and Bliss (1966) in New Hampshire report similar alpine patterns. The base diagram, that is, the upper three frames in fig. 8B.8 was derived by indirect compositional ordination (Whittaker, 1967) of thirty vegetation stands using the Bray and Curtis (1957) method. Full details and the rationale of ordination are not necessary here for a satisfactory interpretation. A more complete description of field and laboratory methodology is given in Webber (1971a, 1972). It is sufficient here to say that a set of three frames represent the three principal elevations of abstract vegetational space whose axes may be interpreted as complex gradients (*sensu* Whittaker, 1951) of soil moisture, winter snow cover and soil disturbance. These three gradients are considered to represent, in that sequence, the controlling microenvironmental complexes. Soil moisture is based on site water supply and soil water-holding capacity; winter snow cover is based on number of snow-free days and depth of snow cover; and disturbance is based on the burrowing activities of pocket gophers (*Thomomys talpoides*) and subsequent wind erosion. Thus the base diagram can be used to describe the distribution of a species or vegetation type within this framework of three gradients. Similarly the distribution of some vegetation parameter, such as productivity, can be described.

Annual net shoot production is highest in sites which are moist, have moderate snow cover and which have little or no disturbance; it is low in dry, snow-free sites, late snowbeds and wet sites. Along the individual gradients, production has an asymmetrical bell-shaped distribution, for example productivity goes from around 150 g m^{-2} yr^{-1} in dry sites, to 300 g m^{-2} yr^{-1} in moist sites, to around 100 in wet sites. Above-ground standing crop reaches a maximum in moist, moderately snow-covered, stable sites which support willow shrub; wet sites where dead material disappears more rapidly than in dry sites have the smallest standing crops. Maximum root biomass coincides with maximum above-ground standing crop, but wet sites have a higher root biomass than drier sites in contrast to the above-ground standing crop; snowbeds have the smallest below-ground biomass. Above:below-ground biomass ratios give still different patterns with wet, moderately snow-covered sites, with only a little disturbance, having the greatest

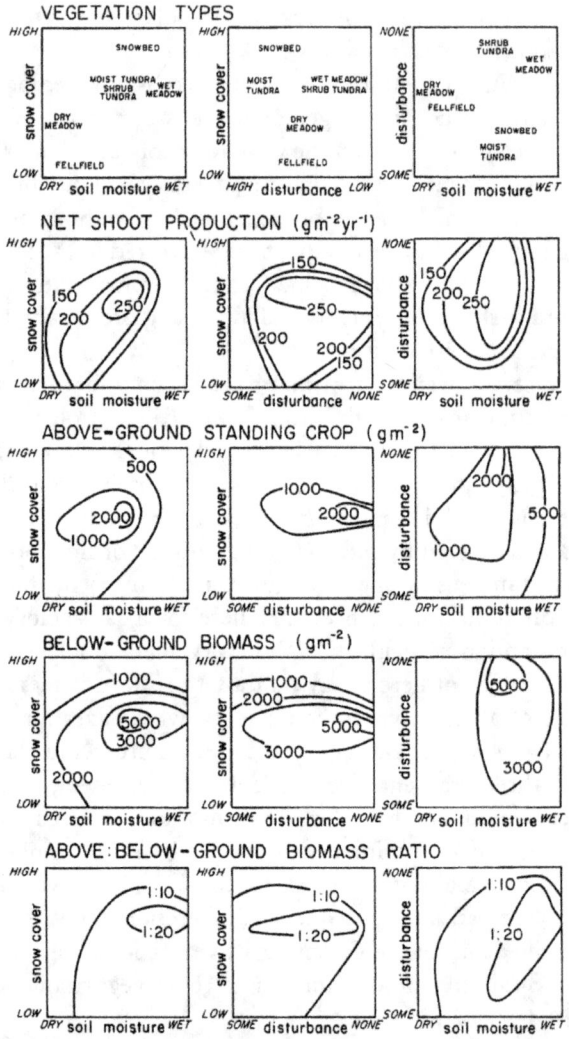

Fig. 8B.8. The distribution of six major vegetation types, net shoot production, above-ground standing crop, root biomass, and above : below-ground biomass ratios within an indirect ordination of vegetation from the Colorado alpine (from Webber, 1972). The factors correlating with the principal ones of the ordination are indicated. All data were gathered at peak above-ground vascular plant biomass.

proportion of biomass below ground; snowbeds and dry, winter snow-free sites have small (in this proportional sense), below-ground biomass. This summary of primary productivity patterns demonstrates that on Niwot Ridge the principal controlling site factors are, in order of influence: water availability, length of growing season (snow cover), exposure (snow protection), and soil stability. If the ordination is taken to a fourth dimension soil nutrients appear as a significant controlling complex, especially soil nitrogen which correlates positively with primary production. The importance of nitrogen in alpine soils is recognized by Retzer (1956) (see also Chapter 13B). It must be understood that although certain gradients, for example, snow and moisture, will be important in most tundra situations, each tundra location is unique and any ranking of microgradients will be different in each location.

The importance of nutrients as factors limiting tundra production has long been realized since the simple observations of luxuriant growth around bird perches and human habitations in the north (Polunin, 1948). In most situations the limiting nutrient is nitrogen as other minerals seem sufficient (Warren Wilson, 1957). Scott and Billings (1964) and Bliss (1966) have increased production by alpine plants by adding nitrogen to the soil, and Bliss (1971) reports that IBP studies under way on Devon Island have also increased production with the addition of both nitrogen and phosphorus. In some situations it may not be deficiencies of nitrogen in the soil but an inability of the plant to utilize it at low temperatures (Dadykin, 1954). Further work on tundra nutrient requirements will be profitable especially if production can be beneficially increased in needed situations, for example on damaged sites (Bliss, 1971). Also with regard to nutrient supply in tundra, more information is needed on rates of release from decomposition and cycling time; the IBP effort is gathering such data.

In this account of productivity variation within the tundra biome little has been said about the southern hemisphere tundra. Some information can be found (Costin, 1967; Bliss, 1969; Jenkin and Ashton, 1970; Lewis and Greene, 1970) and research is currently being expanded. Studies of southern hemisphere tundra will support some of the present generalities but they will also challenge some and provide others. Several interesting possibilities are being realized for southern hemisphere research: Bliss (1969) hypothesized that photosynthesis and respiration patterns in the New Zealand alpine are different from the northern hemisphere, and that instead of showing net carbon losses during early spring as growth gets under way, New Zealand plants seem to make growth in short bursts. Lewis and Greene (1970) have been directing a bipolar research programme to answer such questions as: are antarctic terrestrial systems poorer than their arctic counterparts because of lack of a suitable vascular flora (Lamb, 1970) or because of poorer growth conditions?

COMPARISON OF ARCTIC AND ALPINE

Arctic and alpine ecosystems appear to some workers to have more dissimilarities than similarities (Wright and Osburn, 1967, pp. v–vi) and the validity of treating them as parts of the same biome has been questioned (Hoffman and Taber, 1967). However, in this author's view arctic and alpine tundras of the northern hemisphere form a continuum (*sensu* Gleason, 1926) with no breaks, such that latitudinally separate and biologically different sites are linked by intermediate sites. The two sites, Point Barrow, Alaska, and Niwot Ridge, Colorado, in table 8B.8 are in many ways very different and yet have enough characteristics in common – such as a cold-dominated environment, floristic elements and life forms, and productivity patterns – that they can be considered, in a classification sense, parts of the same natural biome. The tundra biome also extends into the southern hemisphere where its segments (for example: central Africa – Hedberg, 1964; South America – Walter and Medina, 1969; Australasia – Costin, 1967, Mark and Bliss, 1970; Antarctica – Longton, 1967) often appear very different from, and yet still permit comparison with, their northern counterparts (Bliss, 1969; Billings, 1970).

Bliss (1971) has commented that the great geographical range of tundra precludes generalizations on its environment. However, there are sufficient biological similarities

Table 8B.8. Comparison of arctic and alpine environments and vegetation. Data adapted from Billings and Mooney (1968) with additional botanical information. The locations being compared are Point Barrow, Alaska, and Niwot Ridge, Colorado.

Component	Arctic tundra	Alpine tundra
Latitude	71°N	40°N
Altitude	7 m	3749 m
Solar radiation		
(average July intensity)	0.30 cal cm^{-2} min^{-1}	0.56 cal cm^{-2} min^{-1}
(0.4–0.7 μm for growing period)	10 × 10^7 cal m^{-2}*	9 × 10^7 cal m^{-2}
Photoperiod (maximum)	84 days	15 hr
Air temperature (July mean)	3.9°C	8.5°C
Soil temperature (maximum)	2.5°C	13.3°C
Precipitation (annual mean)	107 mm	1021 mm
Wind speed (annual mean)	19 km hr^{-1}	37 km hr^{-1}
Water stress	−4 to −5 bars	−6 to −8 bars
Permafrost	Universal	Sporadic
Average length of growing period	55 days	90 days
Number of common vascular plants	40	100
Microhabitat diversity	Small	Large
Average areal vascular production	100 g m^{-2} yr^{-1}	200 g m^{-2} yr^{-1}
Average ratio of above- to below-ground biomass	1:8	1:12
Average areal net photosynthetic efficiency	0.5%*	0.5%

* From Tieszen (1972).

to allow biological generalizations, and it seems that at different points in the tundra continuum different environmental components interact to produce equivalent total environments. This point is illustrated in table 8B.8, where altitude, photoperiod, insolation, precipitation and other factors vary between sites but nevertheless interact at each site giving rise to a related and similar vegetation. Table 8B.8 and other collective observations provided by Billings and Mooney (1968) suggest some generalizations which compare high-latitude arctic tundra with mid-latitude alpine tundra. They are presented here as two hypotheses to promote discussion and further research. When equivalent arctic and alpine habitat and vegetation types are compared, for example, polar desert and alpine fellfield, or low arctic tundra and alpine shrub tundra (see Plates 33 and 34), then (1) they have equal photosynthetic efficiencies but alpine tundra is more productive because of greater insolation, and (2) the alpine has a higher below-ground biomass in proportion to that above ground because of its windier, more desiccating environment, and possibly because of the absence of permafrost or the greater depth of the active layer in the alpine (see Ives, Chapter 4A, pp. 185–8).

4. Vegetation as an ecosystem component

The importance of vegetation within the ecosystem is best viewed in terms of *holocoenosis*. The relation of all consumers to the vegetation by grazing, burrowing, trampling, manuring and decomposing, and in other less obvious but equally important ways such as in pollination (Macior, 1968; Kevan, 1970) and in seed dispersal (Samuelson, 1934; Dahl, 1963) should be emphasized.

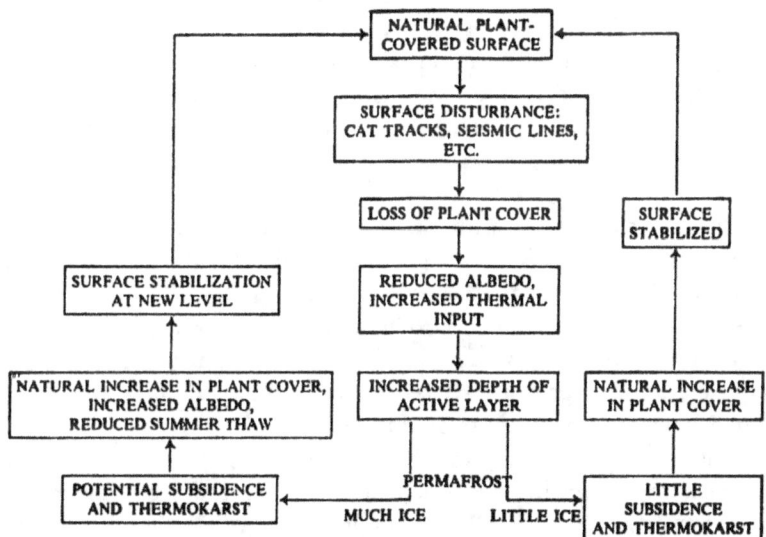

Fig. 8B.9. A word model illustrating the sequence of events in the recovery of damaged arctic tundra vegetation (from Bliss, 1970b).

The most striking holocoenotic effect of vegetation on the ecosystem is seen when vegetation is damaged or removed in areas with permafrost. This has led to the concept of 'fragile' tundra (Bliss, 1970b; Ives, 1970). Fig. 8B.9 illustrates the series of events which follow the damage of tundra vegetation by tracked vehicles. The altered plant cover reduces both surface insulation and albedo, and more heat is absorbed by the soil with the result that there is a greater depth of thaw and both subsidence and erosion may ensue. Such disturbance will ultimately recover due to the natural homeostatic mechanisms in ecosystems which bring about a new equilibrium. Unfortunately little is known about the time factor in these recoveries; it may range from a few to hundreds of years. In fig. 8B.9, well drained sites where the ground contains little ice show slight changes and recovery to a new, little altered equilibrium is rapid. However, on wet sites containing much ground ice greater damage is caused, and recovery will be slow and result in a different vegetation when equilibrium is regained. Some damage to tundra is inevitable if the resources of these regions are to be developed. Oil companies and the military who have been responsible for much of the past damage in the North American tundra (Klein, 1970; Kevan, 1971; Webber and Webber, 1972) are especially active in assessing and reducing this damage. Experiments in reseeding and replacing damaged or removed vegetation with an equivalent insulating and reflecting surface are being made, but the results of these studies are not yet widely published and only time will judge the success of such efforts. It should be pointed out that deleterious effects may not always ensue from vehicle passage. This, of course, involves a value judgement, but some vegetation is resilient and may even be stimulated through increased mineral release as the active layer thaws deeper. Preliminary fieldwork suggests that it may be possible to map resilient vegetation for the routing of tracked vehicles so that susceptible vegetation is avoided or, at least, damage is minimized. Further details on tundra perturbations in relation to permafrost and northern development are given in Fuller and Kevan (1970) and a Royal Society of Canada (1970) publication. These works are particularly timely as they followed the recent proving of oil reserves in the Arctic. (See also Ives, Chapter 4A; Osburn, Chapter 16; Ives et al., Chapter 17A; and Osburn, Chapter 17B.) An appreciation for holocoenosis is the only way that an adequate understanding of the ecosystem and especially of consequences of its modification can be obtained. Research on the possible effects of changing any one part of the system must be viewed in this way.

It is possible to increase tundra productivity by increasing limiting factors such as temperature, nutrients, and water supply. Such programmes must be planned with possible deleterious, holocoentic consequences in mind. Billings and Mooney (1968) pointed out that supplying water to dry alpine tundra in late season may have the effect of preventing the onset of dormancy which is so necessary to winter survival.

Another example of the holocoenotic relations of vegetation and the research to assess man's impact is provided by fig. 8B.10. This figure is the basis for research being conducted in a region of the Colorado alpine where a weather modification pilot project is underway (cf. p. 958). This project is attempting to increase snowfall and ultimately water yield by winter cloud seeding. An increase in winter snowpack may have major consequences for the vegetation since it is a major factor in controlling vegetation type (fig. 8B.8; Billings and Bliss, 1959; Holway and Ward, 1963, 1965). Fig. 8B.10 presents two hypotheses which are being tested by changing the snowpack regimes through the use of snow fences. These hypotheses suggest that each habitat or vegetation type will have a different response to increased snowpack and that Case I, poorly vegetated slopes with greatly increased snowpack, might be severely damaged through the triggered and linked responses of erosion and reduced growth, and that Case II, winter-snow-free slopes with slight increase in snowpack, might become more stable and productive as a result of protection and water supplied by the snow cover. Preliminary results show alpine vegetation to be remarkably resilient. Initial obvious delays in phenology and growth are not apparent at the end of the growing season. This finding fits the known pre-adaptive characteristics of tundra, relating to rapid growth following snowmelt and the rapid development possible in snowbank communities (Billings and Bliss, 1959; Holway and Ward, 1965).

The holocoenotic consequences of disturbing vegetation and other parts of ecosystems have perhaps always been recognized by a few men; but man, who lives in more tem-

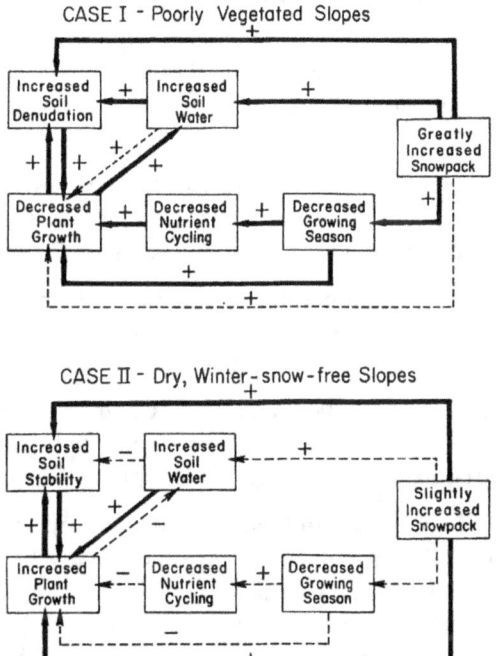

Fig. 8B.10. Hypothetical word models currently being tested by the author (from Webber, 1971b). These models predict the effects of increased snowpack on poorly vegetated slopes and on dry, winter-snow-free slopes in the Colorado alpine. In Case I greatly increased snow creates considerable soil erosion, and in Case II slightly increased snowpack increases both soil stability and primary production.

perate systems than tundra, has generally been unconcerned. However, with increasing technology and population he is becoming concerned, and as he develops the northlands and the high country where systems are more sensitive he must keep uppermost in his mind the concept that the ecosystem is a finite, interacting whole which includes himself, and that a change to any part will to some extent change all other parts. The gloomy picture painted by Klein (1970) and Kevan (1971) of damage from past and current exploitation must not be continued.

5. Summary and conclusions

A number of generalizations have been made about primary productivity patterns of tundra. Any combination of a few of the characteristics used in these generalizations defines tundra on a primary production basis. The generalizations are:

1. Tundra has a low overall net productivity ranging from 0 to an upper limit which may in favourable circumstances reach 1000 g dry weight m^{-2} yr^{-1}; an average value probably lies somewhere between 100 and 200.

2. Temperature is the important factor controlling the level of tundra production. Temperature, triggered by insolation, controls the length of the growing season both on a macroscale (latitudinal and altitudinal gradients) and on a microscale (duration of snow cover and aspect).

3. When temperatures are suitable for growth other factors become limiting. These factors will vary according to local site conditions. The most common factors which limit production are water, light, soil nutrients, surface stability and wind.

4. The average caloric content per unit of tundra plant matter is higher than that of more temperate vegetation. This may have an adaptive value in the severe environment with a short growing season.

5. The proportion of biomass below ground is higher in tundra than most other vegetation types. This is also considered an adaptation to the tundra environment where plants with perennial below-ground storage systems would be favoured.

6. Along a gradient from treeline to polar desert in the Arctic the following interrelated trends can be traced. The environment becomes more severe as interactions of its components produce a shorter growing season, decreasing water supply, and fewer available soil nutrients; the preponderant life forms shift from shrubs and mosses to cushion plants and lichens; primary productivity decreases from a high of near 1000 g m^{-2} yr^{-1} to close to zero; average caloric content of plant matter increases; and the balance of below-ground biomass first increases as shrubs become less common and then decreases as vascular vegetation becomes sparse in the polar desert.

7. A comparison of alpine and arctic tundras shows them to be similar with respect to items 1 to 5, and that the responses in the Arctic to the latitudinal gradient in item 6 are mirrored in the alpine along altitudinal gradients. However, if arctic vegetation types are compared with their alpine equivalents a tentative generalization may be made: alpine tundra is more productive and has a larger below-ground biomass than arctic tundra.

8. All ecological studies require the acceptance and awareness of the importance of the holocoenotic character of the ecosystem. In the tundra biome this becomes especially essential in terms of the fragile nature of the system and the inevitability of environmental modification as these areas are developed to meet the need of man.

The generalizations presented here will in due course be strengthened, modified, or abandoned. The current Tundra Biome effort of the IBP has been evoked many times as the panacea for our knowledge gap; the publications from this research should be studied. Further research is required in all aspects of tundra production and especially in these areas: measurement of gross production or respiration losses; measurement of foliage area, canopy structure and photosynthetic efficiency; measurements of material flow in the vegetation and turnover, decomposition, and nutrient release rates; investigation of the nature of tundra fragility; model building of tundra ecosystems to provide insight into their complexities and function so that sound management practices can be followed.

References

ALEKSANDROVA, V. D. (1958) An attempt to measure the overground and underground productivity of plant communities in the arctic tundra. *Botanicheskii Zhurnal*, **43**, 1748–62 (in Russian).

ALEXANDROVA, V. D. (1970) The vegetation of the tundra zones in the USSR and data about its productivity. *In:* FULLER, W. A. and KEVAN, P. G. (eds.) *Productivity and Conservation in Northern Circumpolar Lands*, 93–114. IUCN Pub. 16. Morges, Switzerland: Int. Conserv. Nature.

ANDERSON, M. C. (1967) Photon flux, chlorophyll content, and photosynthesis under natural conditions. *Ecology*, **48**, 1050–3.

BAZILEVICH, N. I. and RODIN, L. Y. (1971) Geographical regularities in productivity and circulation of chemical elements in the Earth's main vegetation types. *Soviet Geogr.*, **1**, 24–53.

BESCHEL, R. E. (1970) The diversity of tundra vegetation. *In:* FULLER, W. A. and KEVAN, P. G. (eds.) *Productivity and Conservation in Northern Circumpolar Lands*, 85–92. IUCN Pub. 16. Morges, Switzerland: Int. Union Conserv. Nature.

BESCHEL. R. E. and WEBB, D. (1963) Growth ring studies on arctic willows. *In:* MULLER, F. (ed.) *Axel Heiberg Is. Prelim. Rep. 1961–62*, 189–98. McGill Univ.

BILLINGS, W. D. (1970) *Plants, Man and the Ecosystem* (2nd edn). Belmont, Calif.: Wadsworth. 160 pp.

BILLINGS, W. D. and BLISS, L. C. (1959) An alpine snowbank environment and its effects on vegetation, plant development and productivity. *Ecology*, **40**, 388–97.

BILLINGS, W. D. and MOONEY, H. A. (1968) The ecology of arctic and alpine plants. *Biol. Rev.*, **43**, 481–529.

BLISS, L. C. (1962a) Adaptations of arctic and alpine plants to environmental conditions. *Arctic*, **15**, 117–44.

BLISS, L. C. (1962b) Caloric and lipid content in alpine tundra plants. *Ecology*, **43**, 753–7.

BLISS, L. C. (1966) Plant productivity in alpine microenvironments on Mt Washington, New Hampshire. *Ecol. Monogr.*, **36**, 125–55.

BLISS, L. C. (1969) Alpine community patterns in relation to environmental parameters. *In:* GREENRIDGE, K. N. H. (ed.) *Essays in Plant Geography and Ecology*, 167–84. Halifax: Nova Scotia Museum.

BLISS, L. C. (1970a) Primary production within arctic tundra ecosystems. *In:* FULLER, W. A. and KEVAN, P. G. (eds.) *Productivity and Conservation in Northern Circumpolar Lands,* 17–85. IUCN Pub. 16. Morges, Switzerland: Int. Union Conserv. Nature.

BLISS, L. C. (1970b) Oil and ecology of the Arctic. *Trans. Roy. Soc. Can.,* **8**, 361–72.

BLISS, L. C. (1971) Arctic and alpine life cycles. *Ann. Rev. Ecol. System.,* **2**, 405–38.

BRAY, J. R. and CURTIS, J. T. (1957) An ordination of the upland forest communities in southern Wisconsin. *Ecol. Monogr.,* **27**, 325–49.

CLARKE, G. C. S., GREEN, S. W. and GREENE, D. M. (1971) Productivity of bryophytes in polar regions. *Ann Bot.* (n.s.), **35**, 99–108.

COMMONER, B. (1971) *The Closing Circle: Nature, Man and Technology.* New York: Knopf. 326 pp.

COSTIN, A. B. (1967) Alpine ecosystems of the Australasian region: *In:* WRIGHT, H. E. JR and OSBURN, W. H. (eds.) *Arctic and Alpine Environments,* 55–87. Bloomington: Indiana Univ. Press.

CUMMINGS, K. W. and WAYCHECK, J. C. (1971) Caloric equivalents for investigations in ecological energetics. *Mitt. int. Ver. Limnol.,* **18**, 1–158.

DADYKIN, V. P. (1954) Peculiarities of plant behaviour in cold soils. *Voprosy Botaniki,* **2**, 455–72 (in Russian); 473–89 (in French).

DAHL, E. (1963) Plant migrations across the North Atlantic Ocean and their importance for the paleogeography of the region. *In:* LÖVE, Á. and LÖVE, D. (eds.) *North Atlantic biota and their history,* 173–88. New York: Macmillan.

DENNIS, J. G. (1968) *Growth of Tundra Vegetation in Relation to Arctic Microenvironments at Barrow, Alaska.* Unpub. Ph.D. dissertation, Duke Univ. 289 pp.

DENNIS, J. G. and JOHNSON, P. L. (1970) Shoot and rhizome-root standing crops of tundra vegetation at Barrow, Alaska. *Arct. Alp. Res.,* **2**, 253–66.

FULLER, W. A. and KEVAN, P. G. (1970) *Productivity and Conservation in Northern Circumpolar Lands.* IUCN Pub. 16. Morges, Switzerland: Int. Union Conserv. Nature. 344 pp.

GATES, D. M. (1962) *Energy Exchange in the Biosphere.* New York: Harper & Row. 151 pp.

GATES, D. M. (1965) Energy exchange between organisms and environment. *In:* LOWRY, W. P. (ed.) *Biometeorology,* 1–22. Corvallis: Oregon State Univ. Press.

GLEASON, H. W. (1926) The individualistic concept of the plant association. *Bull. Torrey Bot. Club,* **53**, 7–26.

GOLLEY, F. B. (1960) Energy dynamics of a food chain of an old-field. *Ecol. Monogr.,* **30**, 187–206.

GOLLEY, F. B. (1961) Energy values of ecological materials. *Ecology,* **42**, 581–4.

GORDEN, R. W. (1969) A proposed energy budget of a soybean field. *Bull. Georgia Acad. Sci.,* **27**, 41–52.

HADLEY, E. B. and BLISS, L. C. (1964) Energy relationship of alpine plants on Mt Washington, New Hampshire. *Ecol. Monogr.,* **34**, 331–57.

HAMMOND, A. L. (1972) Ecosystem analysis: biome approach to ecosystem research. *Science,* **175**, 46–8.

HEAL, O. W. (1971) *Tundra Biome: Working Meeting on Analysis of Ecosystems* (Kevo, Finland, 1970). London: IBP Tundra Biome Steering Committee. 297 pp.

HEDBERG, O. (1964) Features of afroalpine plant ecology. *Acta Phytogeogr. Suecica,* **49**, 1–144.

HOFFMAN, R. S. and TABER, R. D. (1967) Origin and history of holarctic tundra ecosystems, with special reference to their vertebrate faunas. *In:* WRIGHT, H. E. JR and OSBURN, W. H. (eds.) *Arctic and Alpine Environments,* 142–70. Bloomington: Indiana Univ. Press.

HOLWAY, J. G. and WARD, R. T. (1963) Snow and meltwater effects in an area of Colorado alpine. *Amer. Midl. Nat.*, **69**, 189–97.

HOLWAY, J. G. and WARD, R. T. (1965) Phenology of alpine plants in northern Colorado. *Ecology*, **46**, 73–83.

IVES, J. D. (1970) Arctic tundra: how fragile? A geomorphologist's point of view. *In:* The tundra environment (Symposium of Section III, Roy. Soc. Can.). *Preprint, Trans. Roy. Soc. Can.* (4th ser.), 7, 39–50.

JENKIN, J. F. and ASHTON, D. H. (1970) Productivity studies on Macquarie Island vegetation. *In:* HOLDGATE, M. W. (ed.) *Antarctic Ecology*, Vol. 2, 851–63. London: Academic Press.

JOHNSON, P. L. and KELLEY, J. J. (1970) Dynamics of carbon dioxide and productivity in an arctic biosphere. *J. Ecol.*, **51**, 73–80.

JORDAN, C. F. (1971) A world pattern in plant energetics. *Amer. Sci.*, **59**, 425–33.

KALFF, J. (1970) Arctic Lake Ecosystems. *In:* HOLDGATE, M. W. (ed.) *Antarctic Ecology*, Vol. 2, 651–63. London: Academic Press.

KEVAN, P. G. (1970) *Insect Pollination of High Arctic Flowers.* Unpub. Ph.D. dissertation, Univ. of Alberta. 399 pp.

KEVAN, P. G. (1971) Oil under the tundra in the Mackenzie Delta region. *Can. Field Nat.*, **85**, 79–86.

KLEIN, D. R. (1970) The impact of oil development in Canada. *In:* FULLER, W. A. and KEVAN, P. G. (eds.) *Productivity and Conservation in Northern Circumpolar Lands*, 209–42. IUCN Pub. 16. Morges, Switzerland: Int. Union Conserv. Nature.

KUCERA, C. L., DAHLMAN, R. C. and KOELLING, M. R. (1967) Total net productivity and turnover on an energy basis for tallgrass prairie. *Ecology*, **48**, 536–41.

LAMB, I. M. (1970) Antarctic terrestrial plants and their ecology. *In:* HOLDGATE, M. W. (ed.) *Antarctic Ecology*, Vol. 2, 733–51. London: Academic Press.

LEWIS, M. C. and GREENE, S. W. (1970) A comparison of plant growth at an Arctic and Antarctic station. *In:* HOLDGATE, M. W. (ed.) *Antarctic Ecology*, Vol. 2, 838–50. London: Academic Press.

LIETH, H. (1964) Versuch einer Kartographischen Darstellung der Produktivität der Pflanzendecke auf der Erde. *Geogr. Taschenbuch*, 72–80. Wiesbaden: M. Steiner.

LIETH, H. (in press) The net primary production of the earth with special emphasis on the land areas. *In:* WHITTAKER, R. H. (ed.) *Perspectives on Primary Productivity of the Earth.* Symposium at AIBS 2nd Nat. Congress, Miami, Florida, October 1971.

LONGTON, R. E. (1967) The vegetation of the maritime Antarctic. *Phil. Trans. Roy. Soc. London*, B, 252, 213–35.

LONGTON, R. E. (1970) Growth and productivity of the moss *Polytrichum alpestre* Hoppe in Antarctic regions. *In:* HOLDGATE, M. W. (ed.) *Antarctic Ecology*, Vol. 2, 819–37. London: Academic Press.

LOOMIS, R. S. and WILLIAMS, W. A. (1963) Maximum crop productivities: an estimate. *Crop Sci.*, **3**, 67–72.

LÖVE, D. (1970) Subarctic and subalpine: where and what? *Arct. Alp. Res.*, **2**, 63–72.

MACIOR, L. W. (1968) Pollination adaptation in *Pedicularis groenlandica. Amer. J. Bot.*, **55**, 927–32.

MARK, A. F. and BLISS, L. C. (1970) The high-alpine vegetation of central Otago, New Zealand. *New Zealand J. Bot.*, **8**, 381–451.

MOIR, W. H. (1969) Energy fixation and the role of primary producers in energy flux of grassland ecosystems. *In:* DIX, R. L. and BEIDLEMAN, R. B. (eds.) *The Grassland Ecosystem: a Preliminary Synthesis*, 125–47. Range Sci. Dept, Sci. Ser. 2. Colorado State Univ.

MONTEITH, J. L. (1965) Light distribution and photosynthesis in field crops. *Ann. Bot.* (n.s.) **29**, 17–37.

MOONEY, H. A. and BILLINGS, W. D. (1960) The annual carbohydrate cycle of alpine plants as related to growth. *Amer. J. Bot.*, **47**, 594–8.

ODUM, E. P. (1971) *Fundamentals of Ecology* (3rd edn). London: W. B. Saunders. 574 pp.

ODUM, E. P., CONNELL, C. E. and DAVENPORT, L. B. (1962) Population energy flow of three primary consumer components of old-field ecosystems. *Ecology*, **43**, 88–96.

ODUM, H. T. (1967) Biological circuits and marine systems of Texas. *In:* OLSON, T. A. and BURGESS, F. J. (eds.) *Pollution and marine ecology*, 99–157. New York: Interscience.

ODUM, H. T. and PIGEON, R. F. (1970) *A Tropical Rainforest: a Study of Irradiation and Ecology at El Verde, Puerto Rico.* TID-24270 (PRNC-138). Washington, DC: US Atomic Energy Commission. 1678 pp.

OVINGTON, J. D. (1957) Dry-matter production by *Pinus sylvestris* L. *Ann. Bot.* (n.s.), **21**, 287–314.

OVINGTON, J. D. (1961) Some aspects of energy flow in plantations of *Pinus sylvestris* L. *Ann. Bot.* (n.s.) **25**, 12–20.

OVINGTON, J. D. and LAWRENCE, D. B. (1967) Comparative chlorophyll and energy studies of prairie, savanna, oakwood and maize field ecosystems. *Ecology*, **48**, 515–24.

PHILLIPSON, J. (1966) *Ecological Energetics.* London: Edward Arnold. 57 pp.

POLUNIN, N. (1948) Botany of the Canadian Eastern Arctic. Part III. Vegetation and ecology. *Nat. Mus. Can. Bull.*, **104**, *Biol. Ser.*, 32, 1–304.

RAUNKIAER, C. (1934) *The Life Forms of Plants and Statistical Plant Geography.* Oxford: Clarendon Press. 632 pp.

REIFSNYDER, W. E. and LULL, H. W. (1965) Radiant energy in relation to forests. *US Dept of Agric. Forest Serv., Tech. Bull.*, 1344. 111 pp.

RETZER, J. L. (1956) Alpine soils of the Rocky Mountains. *J. Soil Sci.*, **7**, 22–32.

ROYAL SOCIETY OF CANADA (1970) The tundra environment (Symposium of Section III, Roy. Soc. Can.). *Preprint, Trans. Roy. Soc. Can.* (4th ser.), 7. 50 pp.

SAMUELSON, G. (1934) Die Verbreitung der höheren Wasserpflanzen in Nord-Europa. *Acta Phytogeogr. Suecica*, **6**, 1–211.

SCOTT, D. and BILLINGS, W. D. (1964) Effects of environmental factors on standing crop and productivity of an alpine tundra. *Ecol. Monogr.*, **34**, 243–70.

SCOTTER, G. W. (1963) Growth rates of *Cladonia alpestris, C. mitis, C. rangiferina* in the Talston River region, NWT. *Can. J. Bot.*, **41**, 1199–202.

SIMS, P. L. and SINGH, J. S. (1971) Herbage dynamics and net primary production in certain ungrazed and grazed grasslands in North America. *In:* FRENCH, N. E. (ed.) *Preliminary Analysis of Structure and Function in Grasslands*, 59–124. Range Sci. Dept, Sci. Ser. 10. Colorado State Univ.

THOMAS, M. D. and HILL, G. R. (1949) Photosynthesis under field conditions. *In:* FRANK, I. and LOOMIS, W. E. (eds.) *Photosynthesis in Plants*, 19–52. Ames: Iowa State Coll. Press.

TIESZEN, L. L. (1972) The seasonal course of aboveground production and chlorophyll distribution in a wet arctic tundra at Barrow, Alaska. *Arct. Alp. Res.*, **4**, 307–24.

WALTER, H. and MEDINA, E. (1969) Die Bodentemperatur als ausschlaggekender Factor für die Gliederung der subalpine und alpinen Stufe in den Anden Venezuelas. *Ber. dtsch. bot. Gesell.*, **82**, 275–81.

WARREN WILSON, J. (1957) Arctic plant growth. *Adv. Sci.*, **13**, 383–8.

WARREN WILSON, J. (1967) Ecological data on dry-matter production by plants and plant communities. *In:* BRADLEY, E. F. and DENMEAD, O. T. (eds.) *The Collection and Processing of Field Data*, 77–123. New York: John Wiley.

WEBBER, P. J. (1971a) *Gradient Analysis of Vegetation Around the Lewis Valley, North-Central Baffin Island, Northwest Territories, Canada.* Unpub. Ph.D. dissertation, Queen's University, Kingston. 366 pp.

WEBBER, P. J. (1971b) Monitoring the ecological effects of weather modification. *In:* LILLYWHITE, M. and MARTIN, C. (eds.) Monitoring the ecological effects of weather modification. *Ann. Proc. Inst. Env. Sci.*, **17**, 40–4.

WEBBER, P. J. (1972) Comparative ordination and productivity of tundra vegetation. *In: 1st Annual Symposium of the US Tundra Biome Program* (Lake Wilderness Center, Univ. of Washington), 55–60.

WEBBER, P. J. and WEBBER, M. (1972) Cape Henrietta Maria, Polar Bear Provincial Park. *North*, **19**, 30–7.

WHITTAKER, R. H. (1951) A criticism of the plant association and climatic climax concepts. *Northwest Sci.*, **25**, 17–31.

WHITTAKER, R. H. (1967) Gradient analysis of vegetation. *Biol. Rev.*, **42**, 207–64.

WHITTAKER, R. H. (1970) *Communities and Ecosystems.* New York: Macmillan. 182 pp.

WIEGERT, R. G. and EVANS, F. C. (1964) Primary production and the disappearance of dead vegetation on an old field in southeastern Michigan. *Ecology*, **45**, 49–63.

WIELGOLASKI, F. E. (1972) Vegetation types and plant biomass in tundra. *Arct. Alp. Res.*, **4**, 291–306.

WOODWELL, G. M. and WHITTAKER, R. H. (1968) Primary production in terrestrial communities. *Amer. Zool.*, **8**, 19–30.

WRIGHT, H. E. JR and OSBURN, W. H. (1967) *Arctic and Alpine Environments.* Bloomington: Indiana Univ. Press. 308 pp.

Manuscript received June 1972.

Additional references

BILLINGS, W. D. (1973) Arctic and alpine vegetations: similarities, differences, and susceptibility to disturbance. *BioScience*, **23**, 697–704.

BLISS, L. C., COURTIN, G. M., PATTIE, D. L., RIEWE, R. R., WHITFIELD, D. W. A. and WIDDEN, P. (1973) Arctic tundra ecosystems. *Ann. Rev. Ecol. Syst.*, **4**, 359–99.

DUNBAR, M. J. (1973) Stability and fragility in arctic ecosystems. *Arctic*, **26**, 179–85.

MATVEYEVA, N. W., POLOZOVA, T. G., BLAGODATSKYKH, L. S. and DOROGOSTAISKAYA, E. V. (1973) A brief sketch on the vegetation in the region of Taimyr biogeocenogical station. *In:* TIKHOMIROV, B. A. and MATVEYEVA, N. V. (eds.) *Biogeocenoses of Taimyr Tundra and their Productivity*, Vol. 2, 7–49. Leningrad: Nauka.

CHAPTER 9

Terrestrial vertebrates

Robert S. Hoffmann
Museum of Natural History,
University of Kansas

1. Introduction

It is customary to combine alpine and arctic tundra into a single *tundra biome* (Kendeigh, 1961). With regard to plants, this has some justification, since a number of species are common to both the alpine belt and the arctic tundra (see Billings, Chapter 8A; Webber, Chapter 8B; Löve and Löve, Chapter 10A; Ives, Chapter 10B). Moreover, vertebrate animals inhabiting arctic and alpine regions are faced with many similar environmental problems, such as low temperatures and scarce cover. However, important differences between arctic and alpine environments exist in day-length and solar radiation patterns, topography and moisture, and other physical factors of the environment. Organisms adapted to the environmental conditions common to both arctic and alpine regions, and also capable of surviving under those conditions unique to each, may have 'arctic–alpine' distribution patterns, as with many plants (Major and Bamberg, 1967), although such distributions are less marked among animal groups, particularly vertebrates. A few species of birds occur in both the arctic tundra and the alpine zone, but they are a small part of the two avifaunas. Mammals are even more exclusive; not a single tundra species is common to both the alpine in temperate latitude mountains and the arctic tundra. Only in the ecotone between alpine and arctic tundras that is found on the slopes of high-latitude mountains, such as in northern Alaska, the northern Urals and in north-eastern Siberia, is there an intermingling of the two sorts of mammalian faunas. Alpine mammals and alpine faunas in general are apparently not derived directly from the arctic tundra, but from the adjacent subalpine mountains and the grassland around the base of the mountains (Hoffmann and Taber, 1967). Of course, in addition to *primary* tundra species (those largely restricted to arctic and alpine), there are other more eury-ecious species that may inhabit both environments. These *secondary* tundra species have broader distributional patterns, however, instead of 'arctic–alpine' distributions, and usually live in a wide variety of habitats.

A second general feature of terrestrial vertebrate faunas in arctic and alpine regions is the rarity of amphibians and reptiles. In the North American Arctic, the wood frog (*Rana sylvatica*) is the northernmost amphibian, and is the only poikilothermic vertebrate to live under conditions even approaching the Arctic (Harper, 1956; Darlington, 1957). Similarly, the herpetofauna of the western North American alpine consists of only a few species of toads, and perhaps in a few places the sagebrush lizard (*Sceloporus graciosus*). Arctic herpetofaunas are equally depauperate in the Old World, but reptiles and amphibians are more numerous in the Palaearctic alpine (Swan and Leviton, 1962).

The differences we observe between arctic and alpine vertebrate faunas are due not only to real differences in the habitats available as a result of physical environmental differences, but also to the much patchier 'island' distribution of alpine habitats (Vuilleumir, 1970) in contrast to those of the more continuous circumpolar Arctic. Alpine vertebrates thus face more severe dispersal and colonization problems, and are more likely to be depauperate for historical reasons (Brown, 1971b). In the following discussion, I shall largely confine my remarks to the animals of arctic tundra and of the alpine zone of non-tropical mountains of the northern hemisphere. There are two reasons for this. First, the biology of alpine species in the tropics and in the southern hemisphere is, with few exceptions, very poorly known. Some areas, such as southeast Asia, appear to have a distinctive montane fauna, while others, such as central Africa, seem to lack one (Bourliere, 1955; Coe, 1967). Second, since a terrestrial vertebrate fauna is completely lacking in the Antarctic polar region, in contrast to the Arctic, it is simpler to restrict attention to the ecological theatre of the northern hemisphere, where arctic and alpine regions can be directly compared.

DISTRIBUTION PATTERNS OF ARCTIC AND ALPINE TERRESTRIAL VERTEBRATES IN THE NORTHERN HEMISPHERE

Udvardy (1969) classifies distribution areas of organisms into several different categories. Areas of *seasonal occupation* may be contrasted with areas of *permanent occupation*, *continuous* distributions with *discontinuous* distributions, and so forth. These may be arranged in the form of a matrix (table 9.1).

Table 9.1. The matrix of distributional types, with examples of arctic tundra and alpine birds and mammals that exhibit these sorts of distributions.

Temporal	Continuous distribution	Discontinuous–disjunct	Discontinuous–disperse
Permanent occupation	Peromyscus maniculatus	Ochotona hyperborea	Lagopus leucurus
Seasonal occupation– breeding	Calcarius lapponicus	Anthus spinoletta	Calidris canutus
Seasonal–non- breeding	Cervus elaphus	Buteo lagopus	?
Irregular occupation	Asio flammeus	Cinclus cinclus	Bucephala islandica
Accidental occurrence	Tamiasciurus hudsonicus	Capella gallinago	?

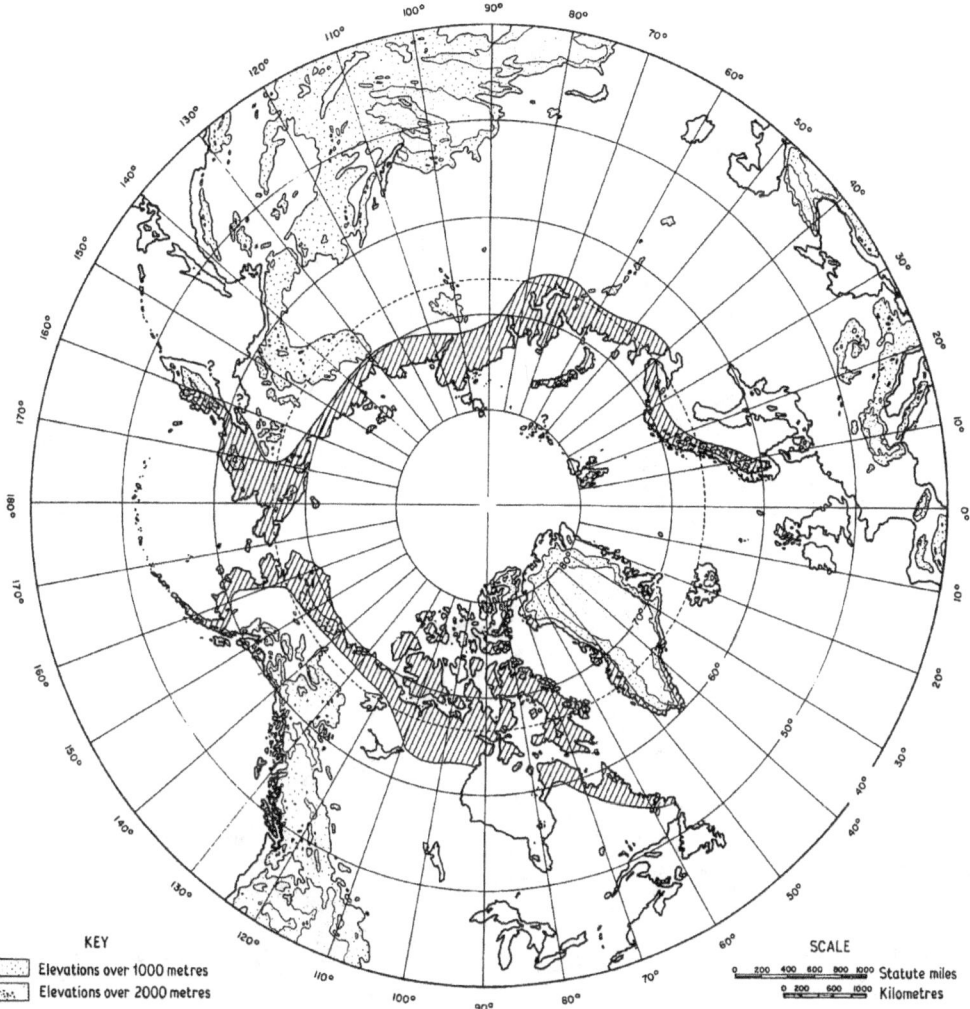

Fig. 9.1a. Nearly continuously circumpolar distribution of the Lapland longspur, or bunting (*Calcarius lapponicus*) (after Voous, 1960).

Polar patterns

The distributions of arctic birds and mammals are frequently continuous, particularly with respect to a single continent. For example, the Lapland longspur (*Calcarius lapponicus*) has an essentially continuous distribution in the wet tundra of North America and Eurasia (Voous, 1960); the two continental populations are disjunct, however, in the region of the North Atlantic and Bering Strait (fig. 9.1a). A few high arctic marine birds and mammals – the arctic skua (*Stercorarius parasiticus*); the black guillemot (*Cepphus grylle*); the ringed seal (*Phoca hispida*) – are apparently continuously circumpolar in distribution, but the majority of these show discontinuities of range associated with unsuitable habitats

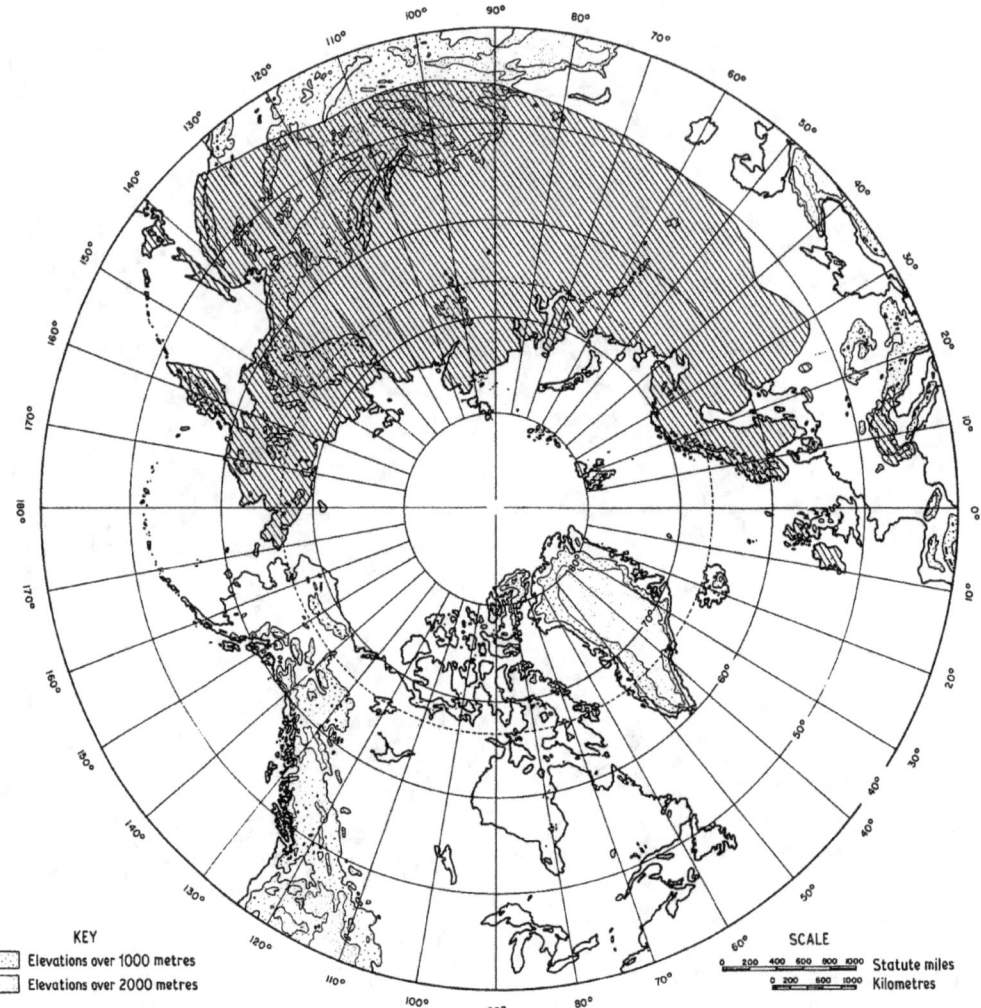

Fig. 9.1b. The boreo-montane, and in part disperse, distribution of the arctic or mountain hare (*Lepus timidus*) in the Palaearctic (after Bannikov, 1954; Flint *et al.*, 1965; and Lyneborg, 1971).

(Davies, 1958; Udvardy, 1963). Narrowly to broadly *disjunct* distributions (defined as a few major areas separated in space) are more common for terrestrial mammals and birds, and grade into *disperse* distributions (defined as more numerous, relatively small but distinctly separated areas; Udvardy, 1969). Certain high arctic shore birds, such as the knot (*Calidris canutus*), are representative of highly disperse distribution areas, 'a consequence not only of the rare occurrence of the breeding habitat, but probably also of the undoubtedly complicated and precarious distribution during the glacial periods which must have brought the species to the verge of extinction' (Voous, 1960).

Montane patterns

The above quotation adds an historical dimension to the spatio-temporal distribution matrix, one that is particularly important in its influence on alpine vertebrates. Alpine birds and mammals inhabit, particularly in more temperate latitudes, permanently or seasonally, areas that are topographic 'islands' (*fide* MacArthur and Wilson, 1967). Thus they have disperse distributions which, having often resulted from climatic change, are *relict* distributions. Such species as the arctic or mountain hare (*Lepus timidus*) are *boreo-montane* in distribution, with broadly continuous distribution areas in the circumboreal tundra and taiga zones, but are also found in disperse areas in the British Isles and the Alps (fig. 9.1b). These more southerly, montane populations are universally interpreted as relicts of a once-continuous Pleistocene population that inhabited ice-free parts of Europe during the most recent glacial period (Würm–Wisconsin). Since much of the area of their present distribution was covered by continental or montane glaciation, mountain hares must have moved northward and upward with the retreat of the last glaciers, coming eventually to occupy those montane areas that are sufficiently high to maintain suitable tundra or taiga habitat, i.e. the patches of habitat, including the vegetation, are also boreo-montane relicts. Sometimes relict species are restricted to tundra, and form a more restricted distributional subset of the boreo-montane type known as *arctic–alpine* distributions. One example is the water pipit (*Anthus spinoletta*), a bird whose breeding distribution is restricted, in North America, to arctic tundra and alpine. In the central and southern Rocky Mountains alpine breeding populations are highly disperse, but as one proceeds northward areas of suitable alpine tundra become larger and more continuous and eventually merge, in northern Alaska, with its arctic tundra breeding habitat. In Eurasia, in contrast, the water pipit occurs only in disperse distribution areas in the Pyrenees, Alps, Carpathians, Caucasus, Tien Shan, Altai, Sayan, Verkhoyansk, and Kamchatka mountains; Eurasian arctic tundra is inhabited by the red-throated pipit (*Anthus cervinus*) while in the alpine region of the Tibetan plateau the water pipit is replaced by Hodgson's pipit (*Anthus roseatus*) (fig. 9.2).

In general, then, arctic vertebrates exhibit circumpolar, continuous or disjunct distribution areas and nearly always belong to Holarctic species or species groups. Alpine vertebrates have disperse distributions and fewer of them are Holarctic; boreo-montane and arctic–alpine forms possess distribution areas that are a mixture of the several types.

HISTORICAL FACTORS INFLUENCING DISTRIBUTION PATTERNS

As noted above, boreo-montane and arctic–alpine distributions are regarded as an historical result of northern hemisphere glaciation. The locations of Pleistocene glaciations, and the frequency and severity with which areas have been influenced (overridden or approached) by continental and montane glaciers, has undoubtedly strongly influenced the present nature of alpine and arctic biotas (see, for instance, Ives, Chapter 10B). However, to assume that all present alpine vertebrates have relict arctic–alpine distri-

KEY

▨ Elevations over 1000 metres
▨ Elevations over 2000 metres

SCALE

0 200 400 600 800 1000 Statute miles
0 200   600   1000   Kilometres

Water or Rock Pipit *(Anthus spinoletta)*
Red-throated Pipit *(Anthus cervinus)*
Hodgson's Pipit *(Anthus roseatus)*

Fig. 9.2. The distributions of three congeneric species of pipits: water, or rock, pipit, *Anthus spinoletta*; red-throated pipit, *Anthus cervinus*; and Hodgson's pipit, *Anthus roseatus* (after Shtegman, 1938, and Voous, 1960). The water pipit has an arctic–alpine distribution pattern in the New World, largely continuous in the arctic but becoming more highly disperse southward in the Rocky Mountains. The isolated alpine breeding populations within the outlined range are not indicated. In the Old World two ecologically different groups of this species are found. The 'rock pipit' inhabits the sea coast of Scandinavia, the British Isles, and the adjacent French coast. The water pipit has a disperse montane distribution,

butions, or that all arctic vertebrates are recent invaders of formerly ice-covered terrain, is a misleading oversimplification.

I have elsewhere (Hoffmann and Taber, 1967) argued that the vertebrate faunas of Holarctic tundra ecosystems evolved fairly recently, since the Late Pliocene, and that 'circumpolar arctic tundra and extensive alpine tundra are a product of the Pleistocene'. Late Pliocene and Early Pleistocene fossil faunas are composed for the most part of species adapted to more temperate conditions, and almost all of these Blancan and Villafranchian mammals belong to species that have since become extinct. 'Cold' elements of the Ice Age faunas, such as the reindeer or caribou (*Rangifer tarandus*), musk-ox (*Ovibos moschatus*), and collared and brown lemmings (*Dicrostonyx torquatus*, *Lemmus sibericus*) first appeared in the early Middle Pleistocene (Gunz II (=Menapian?), Cromerian), and came to dominate the faunas of the Mindel–Kansan, Riss–Illinoian and Würm–Wisconsin in Eurasia (Zeuner, 1959; Kurtén, 1968). In North America, however, primary tundra species did not become established outside Alaska until the last glacial period. This may be due in part to geographic barriers which inhibited colonization of North America southward from Beringia by tundra species (Hoffmann and Taber, 1967; Guthrie and Matthews, 1971) but is also correlated with the different extent to which periglacial tundra developed in the Old and New Worlds.

In Eurasia and Beringia, a very extensive belt of periglacial tundra was formed which extended eastward to Beringia and merged with a cold steppe zone on the south (Moreau, 1955; Vangengeim and Ravskii, 1965; Frenzel, 1968; Zimina and Gerasimov, 1973), while in North America the very existence of periglacial tundra to the south of the continental ice mass has been doubted; at the very least it was much narrower than its Eurasian counterpart.

There is good evidence for a belt of periglacial tundra extending from southern New England to Minnesota (Martin, 1958; Davis, 1967; Watts, 1967) in the Late Wisconsin, but its extent must have been many times smaller than that of the comparable periglacial tundra in Eurasia. Eurasian and Beringian tundra vertebrates able to penetrate the barriers and disperse the long distances to central North America then faced problems of survival in a small and probably relatively unstable biotope. This is reflected in the relatively small list of 'steppe–tundra' species of the North American Pleistocene, as compared with the Eurasian list (table 9.2).

The Eurasian periglacial tundra of the Late Pleistocene differed from its North American counterparts (as well as from recent arctic tundra) not only in extent, but also qualitatively. Much of the Pleistocene periglacial zone of tundra and cold steppe was at much lower latitudes than recent arctic tundra. Thus, total yearly solar energy input

---

occurring in the alpine from Spain to Kamchatka, but also along the coasts of the Kuril and Komandor islands. The red-throated pipit is continuously distributed in Old World tundra, and replaces the water pipit there, but has also recently colonized western Alaska where it and *A. spinoletta* are now geographically sympatric. Hodgson's pipit of the alpine zone of the Tibetan Plateau and the mountains of western China represents a population disjunct from the distributions of the other tundra-dwelling pipits (but cf. Vaurie, 1972).

Table 9.2. Steppe–tundra mammals of the Late Pleistocene (Würm–Wisconsin) period (after Kowalski, 1967, Kurtén, 1968, and others).

Species	Common name	Occurrence in Holarctic			
		Europe	Siberia	Beringia	North America
Sorex tundrensis	Tundra shrew		+	+	
Crocuta (crocuta?) spelaea	Cave hyena	+			
Crocuta crocuta	Spotted hyena		+		
Homotherium (Dinobastis) sp.	Lesser scimitar cat	+	+	?	+
Felis (leo?) spelaea	Cave lion	+			
Felis sp.	Lion		+	+	
Felis pardus	Leopard	+	+		
Gulo gulo	Wolverine	+	+	+	+
Mustela eversmanni (= *nigripes?*)	Steppe polecat (= black footed ferret?)	+	+	?(†)	+
Mustela erminea	Ermine	+	+	+	+
Mustela rixosa	Least weasel	+	+	+	+
Cuon alpinus	Dhole	+	+	+	
Alopex lagopus	Arctic fox	+	+	+	
Canis lupus	Grey wolf	+	+	+	+
Ursus spelaeus	Cave bear	+			
Ursus arctos	Brown bear	+	+	+	+
Mammuthus primigenius	Woolly mammoth	+	+	+	+
Coelodonta antiquitatis	Woolly rhinoceros	+	+		
Equus caballus, E. przwalskii	Wild horse	+	+	+	
Cervus elaphus	Red deer, wapiti	+	+	+	+
Megaloceros giganteus	Giant deer	+	+		
Rangifer tarandus	Reindeer	+	+	+	+
Saiga tatarica	Saiga antelope	+	+	+	
Ovibos moschatus	Musk-ox	+	+	+	+
Bison priscus	Steppe bison	+	+	+	
Bos (Poephagus) sp.	Yak	+	+	+	
Ovis nivicola	Snow sheep		+	+	
Marmota bobac	Steppe marmot	+	+		
Spermophilus major, S. erythrogenys	Large, red-cheeked ground squirrels	+	+		
Spermophilus parryii	Arctic ground squirrel			+	
Spermophilus richardsonii	Richardson's ground squirrel				+
Cricetus cricetus	Common hamster	+	+		
Clethrionomys rufocanus	Red-backed vole		+		
Clethrionomys rutilus	Arctic red-backed vole		+	+	
Lagurus lagurus	Steppe sagebrush vole	+	+		
Microtus oeconomus	Tundra vole	+	+	+	
Microtus gregalis	Narrow-skulled vole	+	+	+	
Lemmus lemmus	Norway lemming	+			
Lemmus sibericus	Brown lemming		+	+	
Dicrostonyx torquatus	Collared lemming	+	+	+	+
Ochotona pusilla	Steppe pika	+	+		
Lepus timidus	Arctic hare	+	+	+	
Total species		33	36	25	13

† Reported from central Alaska (Anderson, 1973).

into the periglacial 'steppe–tundra' ecosystems was greater, growing season longer, and fluctuations in day-length less extreme (i.e. no long polar winter night). Undoubtedly other climatic differences between mid- and high-latitude tundras must have existed. The result was probably a periglacial 'steppe–tundra' ecosystem whose primary (plant) productivity was considerably greater than modern arctic tundra ecosystems. The vegetation also differed in species composition, containing a larger proportion of xerophytic steppe species (*Artemisia, Ephedra, Selaginella, Chenopodiaceae*, etc.) and grasses than do present-day moss-, lichen- and forb-rich arctic tundras.

The great area of the 'steppe–tundra' [connected] central and western Europe with the steppe regions of western and central Asia, with the rich fauna of open country. The composition of the rodent fauna, as well as pollen analysis, shows that the character of Pleistocene vegetation at the time [Würm–Wisconsin] was different from that of the recent arctic tundra in northern Europe and from that of the recent dry steppes in southeastern Europe and southwestern Asia . . . In general, Pleistocene vegetation from the time of the Würm glaciation is most analogous to the vegetation of eastern Asia, which developed under the influence of a cold continental climate in middle latitudes. (Kowalski, 1967)

Thus, this great expanse of 'steppe–tundra' supported a rich fauna of both large and small birds and mammals adapted to cold, open landscape (see pp. 484–94); similar faunas were present in Europe (Butzer, 1964; Kowalski, 1967), Siberia (Vangengeim, 1961), and Alaska (Guthrie, 1968). Guthrie has analysed the palaeosynecology of the large mammal component of the Late Pleistocene community in interior Alaska, at 65°N latitude, which, although in the northern part of the Eurasian–Beringian periglacial 'steppe–tundra', nonetheless 'supported a rather extensive and complex large-mammal "grassland" community'.

Differences between present-day arctic tundra faunas and those of the Pleistocene periglacial 'steppe–tundra' are not due to species replacement, but rather to extinction of many of the large mammals and retraction in ranges of others. The woolly mammoth (*Mammuthus primigenius*), woolly rhinoceros (*Coelodonta antiquitatis*), giant deer (*Megaloceros giganteus*) and cave bear (*Ursus spelaeus*) disappeared completely. The cave lion (*Felis spelaea*), cave hyena (*Crocuta spelaea*) and steppe bison (*Bison priscus*) either became completely extinct, or else survived only as much different forms in temperate and tropical areas (Kurtén, 1968; Kowalski, 1967; Flerov, 1967). Other members of the 'steppe–tundra' faunal complex survived by adapting to the changing environments of either the northern tundra or southern steppe biotopes which evolved in post-Würm–Wisconsin time. Thus, the musk-ox survived in the North American, but not Eurasian, tundra, and reindeer or caribou in circumboreal tundra and taiga. Saiga antelope (*Saiga tatarica*) and wild horse (*Equus przewalskii*), in contrast, survived in the southern Eurasian steppe zone, and the yak (*Bos gruniens*) also retreated southward, first through the steppe, finally becoming restricted to the alpine tundra of the Tibetan plateau (Heptner *et al.*, 1961). Other mammals segregated in the same manner, many of them surviving in both

Old and New Worlds, although in a number of cases the Nearctic range is restricted, centring on the eastern side of the ice-free Beringian refugium (Rausch, 1963).

This model of the dissolution of the periglacial 'steppe–tundra' and its north–south segregation into modern arctic tundra and temperate steppe, separated by the boreal coniferous forest zone, or taiga, provides a reasonable explanation for the extinction of many of these large mammals in postglacial time. To quote Kowalski (1967):

> The Pleistocene vegetation . . . was different from that of the northern tundra as well as from that of the present steppes in Europe and western Asia. The adaptation of mammoth to the cold climate precluded life in the southern steppes, with their warm and dry summers. The present tundra, north of the Polar Circle, is very different from the steppe–tundra of the late Pleistocene; long polar nights and abundant snowfall in winter create quite peculiar conditions of life. Central and western Europe, independent of temperature changes in the Pleistocene, had the same geographical latitude as today and, of course, no polar night; the snowfall was probably slight. The development of forests in the postglacial of Europe and Siberia (Vangengeim, 1961) restricted the habitat of mammoth and finally caused its extinction.

Similarly, for Beringian Alaska Guthrie (1968) states:

> Although there is good circumstantial evidence from other areas that man was involved directly or indirectly in the widespread phenomenon of large-mammal extinctions in the late Pleistocene, the present paucity of the grazing niches in interior Alaska may provide evidence that at least some of these extinctions were due to a shift in climax plant species at the close of the last glaciation. The change in interior Alaska from grassland to coniferous forest and/or shrub tundra appears to have reduced the grazing habitat and increased interspecific competition until some species became extinct.

In summary, in alpine and, particularly, arctic tundra, species of vertebrates have evolved comparatively recently, and these tundra faunas reached a peak of species richness in the 'steppe–tundra' faunas of the Late Pleistocene. Recent arctic tundra faunas are depauperate, due to habitat change and probably human agency, and specialized for the high-latitude arctic tundra environment. Alpine vertebrates were similarly affected, but depauperization of alpine faunas was most intense in peripheral mountain ranges in the Holarctic (Alps, Rockies, etc.). The central and eastern Asian highland, including the Tibetan Plateau, preserves a rich alpine vertebrate fauna (Vaurie, 1972), having been less affected by glacial and postglacial climatic vicissitudes (Hoffmann and Taber, 1967). (For more extensive discussion of man's role in extinction of the megafauna, see Martin, Chapter 11B.)

2. Adaptations of terrestrial vertebrates to arctic and alpine environments

TEMPERATURE ADAPTATIONS

The first association the average person makes with the 'far north' or 'mountain heights' is one of *cold*, and it is true that the alpine and arctic tundras share the characteristic of low mean and extreme temperatures. It is not surprising, then, that warm-blooded vertebrates inhabiting these environments share similar temperature adaptations. Extensive work has been done on these adaptations in arctic birds and mammals, while less has been published for alpine forms. What follows, then, is primarily based on arctic homeotherms, with comparison to alpine species when appropriate. Dunbar (1968) emphasizes that while adaptation to cold has been given much attention, it rarely seems to be limiting for arctic species.

The common problem that alpine and arctic homeotherms face is regulating the balance between body *heat production* and *heat loss* via thermal exchange between the body and a cold environment. This regulation is accomplished in a variety of ways:

A. Control of heat loss (insulation)
   1. Fur or feather thickness
      (a) seasonal changes through moult
      (b) muscular control of position (fluffing)
   2. Control of peripheral circulation
      (a) control of extremities – vascular heat exchangers
      (b) general vasodilation or vasoconstriction
   3. Evaporative cooling
B. Control of heat production
   1. Heat from increased metabolic rate
      (a) increased behavioural activity (running, etc.)
      (b) shivering
      (c) non-shivering thermogenesis
   2. Heat from basal metabolic rate
C. Posture and behaviour

Control of heat loss
The rate of heat loss is directly proportional to the temperature difference across the body surface – ambient air interface (King and Farner, 1961; Gates, 1962). Hence the rate may be controlled by maintaining or changing the temperature gradient across this interface; in turn, this gradient may be modified by changes in body surface temperature and/or adjustments in the covering of fur or feathers. Subcutaneous fat seems to play a less important role in insulating terrestrial animals in the Arctic (Irving, 1972) or alpine, for according to Scholander *et al.* (1950a), it is a 'heavy and poor insulator compared to fur'.

Heat transfer through a layer of fur or feathers may be by: 1) radiation; 2) conduction along individual hairs or feathers, or through the air trapped in the insulative layer; 3) convection of air invading the insulative layer, as by wind-ruffling; and 4) evaporation of water, in case of rain or submergence (Hammel, 1955). Fur and feathers are poor conductors, and if the insulative layer is undisturbed by wind or wetting, the layer of trapped air also conducts heat poorly, so that the effectiveness of the insulation is a function of the amount of air trapped per unit of feathers or fur (fig. 9.3). In mammals, insulative quality depends upon fur depth, and arctic and alpine mammals have, in winter, longer, better insulated fur than temperate and tropical species (Scholander *et al.*, 1950a) (fig. 9.4). In contrast, arctic and tropical birds do not differ in feather depth, probably because of flight requirements, but contour feather structure of arctic birds traps more air per unit mass of feathers than in tropical birds (Irving, 1960).

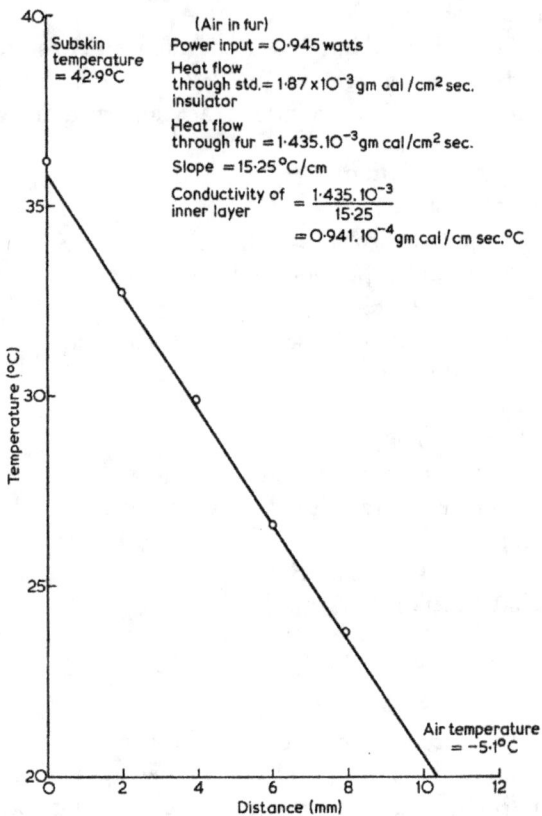

Fig. 9.3. The effectiveness of insulation as a function of fur thickness, measured through a fresh lynx (*Lynx canadensis*) pelt (after Hammel, 1955). The temperature gradient illustrated here was measured by thermocouples in the inner 8 mm of the fur, from a subskin temperature of 42.9°C maintained by a hot plate, in a cold room (−5.1°C).

Maximum insulation is thus achieved when the fur or feathers are held in a position which entraps a dead air layer of maximum depth around the surface; if heat loss must be increased to balance heat production, the insulative layer is reduced by sleeking the fur or feathers closer to the body surface, until a minimum depth is achieved.

Since in both arctic and alpine environments seasonal temperature differences are large, it is advantageous for a resident bird or mammal to modify the thickness of its feathers or fur seasonally. Moult patterns are thus often conspicuous in these animals.

Among birds, the most complex is the moult of the three species of ptarmigan (*Lagopus lagopus*, *L. mutus*, *L. leucurus*), which have three distinctive seasonal plumages – spring, autumn and winter (Plate 35), and in which feather replacement is essentially continuous from early spring until late autumn (Host, 1942; Salomonsen, 1939). Their feather insulation loses heat slowly even in summer; conductance was measured as 0.037–0.047 cm³O₂(gm hr°C)⁻¹ (Johnson, 1968; Herreid and Kessel, 1967).

Increased thickness of the insulation layer in winter is not an unconstrained option open to arctic and alpine birds and mammals of small size, since 'the mechanics of movement

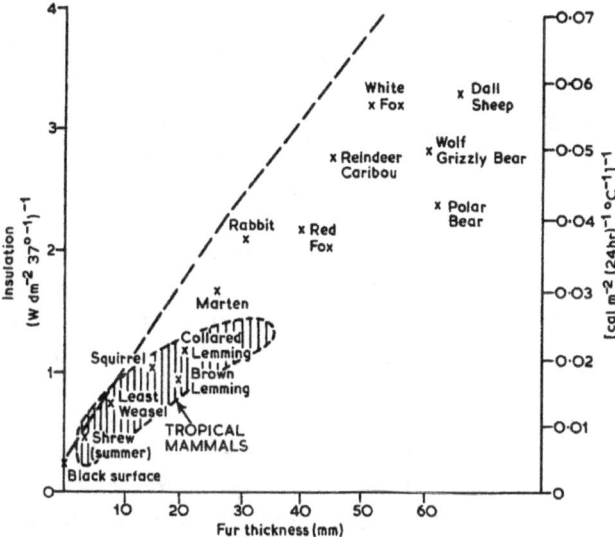

Fig. 9.4. Insulative value of fur in tropical and in arctic mammals (in winter) plotted against fur thickness (after Scholander *et al.*, 1950a). Measurements of insulation were made with a hot plate at 37°C, in a cold room (0°C), as discussed in fig. 9.3.

prevent them from wearing enough fur or feathers for substantial insulation' (Irving, 1960). Small birds may partly circumvent this limitation by fluffing out the feathers, to increase the effective thickness of the insulative layer, and posturally by tucking the head beneath the wing(Steen, 1958; Veghte and Herreid, 1965). Length of the fur is not constrained within such close limits as is feather length, with its close connection to flight adaptations, and shows greater variation as a result of seasonal moults. This is particularly true of larger mammals, where the winter fur may be much longer and less conductive than in summer (Hart, 1956). However, the structure of the fur layer is such that it cannot be fluffed to increase its effective insulation, as can feathers, since increased convective losses of heat cancel increased fur depth (Hammel, 1955).

Birds and mammals both have certain body surfaces poorly to totally uninsulated. These include the bare nostrils, toe pads of arctic foxes (*Alopex lagopus*) and wolves (*Canis lupus*), and palms and soles of brown and polar bears (*Ursus arctos*, *U. maritimus*), the hooves of arctic and alpine deer and bovoids such as caribou and bighorn sheep

(*Ovis canadensis*), and in most mammals the fur covering the nose, lips, face and lower limbs is thin. Some arctic and alpine birds, such as ptarmigan (*Lagopus* sp.) and snowy owls (*Nyctea scandiaca*) possess feathering on the tarsi and upper surface of the foot, but the undersurface of the toes is bare, while the feet and tarsi are bare in many arctic and alpine birds (raven, *Corvus corax*; redpoll, *Carduelis flammea*; rosy finch, *Leucosticte arctoa*; water fowl and gulls). Thus, if body surface temperature in these more exposed areas were as high as the deep-body temperature, heat loss would be rapid in cold weather. At the same time, these surfaces and extremities must remain functional, and protected from freezing. The adaptive strategy adopted by arctic and probably alpine birds and mammals is to control the temperature and volume of blood circulating to these areas. In a cold environment, the amount of blood circulating to the extremities is reduced by constricting the diameter of arterial blood vessels. Moreover, there are many reasons for believing that arterial blood, as it passes from the body core towards the extremities, passes through a vascular heat exchanger. The arterial vessels are closely invested by veins returning from the surface of the extremity. Cooled venous blood is warmed by the arterial blood, which itself is cooled before it reaches the extremity, and 'the heat of arterial blood returns to the veins rather than escaping to the air' (Irving and Krog, 1955). This counter-current heat exchanger produces a typical temperature gradient in the cooled extremity (fig. 9.5); active circulation in the feet of arctic glaucous-winged gulls (*Larus glaucescens*) can be observed at venous temperatures near 0°C, a temperature at which blood circulation would soon be arrested in an unadapted bird or mammal. Moreover, fats from tissues of the extremities exhibit low melting points, a property necessary to tissues which must function normally whilst cold (Irving, 1972). Similarly, the peripheral nerve in the metatarsus of a herring gull (*Larus argentatus*) in a cold environment was found to conduct impulses at temperatures near freezing, and the adaptation was more pronounced when the gull was cold-acclimated (Chatfield *et al.*, 1953).

Insulation by means of fur or feathers, and control of skin temperature, is not only essential in conserving heat in the cold arctic or alpine winter (or summer night!): the insulation of birds and mammals adapted to these regions is remarkably flexible. As noted above, rate of heat loss may be increased by sleeking the fur or feathers close to the body surface. However, this increased heat loss is certainly not sufficient to maintain body temperature equilibrium during a warm summer day in the arctic or alpine, and particularly when the animal's metabolic rate (and resultant heat production) may be increased tenfold during strenuous activity, such as running up a mountainside. Dissipation of this heat load must in part be accomplished by circulatory bypass of the vascular heat exchangers of the extremities, allowing warm arterial blood to pass directly to the surface of the extremities, and by an increase in vessel diameter to increase volume of blood flow to these surfaces. These bare or lightly insulated body surfaces would thus become warmer than the ambient air, and rate of heat loss could be increased in a controlled manner.

Another mode of heat loss is evaporative cooling. Water in the form of sweat or saliva may be evaporated from the skin surface, or from the fur or feathers, as occurs in some temperate or tropical species (including man). However, this form of evaporative

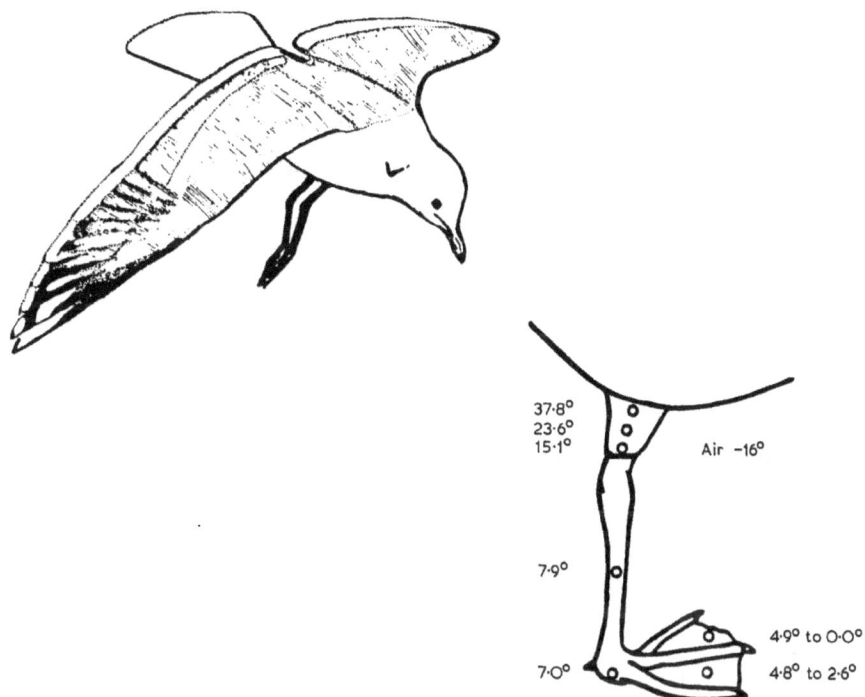

Fig. 9.5. The temperature gradient established in the leg of the glaucous-winged gull (*Larus glaucescens*) as a result of counter-current heat exchange between adjacent arteries and veins in the leg (after Irving and Krog, 1955).

cooling is but poorly regulated and, in an animal faced with balancing rapid change in heat producion (via changing activity) against the rapid changes in air temperature characteristic of some arctic and alpine environments, has another disadvantage: moisture in the fur or feathers greatly reduces its insulative capacity. In contrast, evaporative cooling via moisture contained in exhaled air does not impair the fur/feather layer, and panting may be more closely regulated. Arctic dogs and wolves pant to regulate heat loss in warm weather (Irving and Krog, 1955); other carnivores may as well. Among birds, white-tailed ptarmigan (*Lagopus leucurus*) in the Rocky Mountain alpine have been observed to pant on summer days when the air reaches as moderate a temperature as 21°C (70°F) (Choate, 1963a). However, these birds appear to be poorly adapted to high ambient temperatures, since their evaporative efficiency of 90 per cent is low compared with temperate and tropical birds (Johnson, 1968).

Control of heat production
During the period of rapid and continuous summer moult in the white-tailed ptarmigan, a metabolic rate of 1.30 cm^{3}O$_2$ (g hr)$^{-1}$ (48.8 Kcal/24 hr) was recorded, higher than would

be expected on the basis of the bird's weight (Johnson, 1968). This high rate may reflect the high metabolic cost of the process of feather replacement (Pitelka, 1958a), but it also produces useful additional body heat during the moult period, particularly on cold spring and autumn nights in the alpine (see above). The combined values for rate of heat production and heat loss (see p. 487) result in a lower critical temperature (LCT) of 6.5–11.5°C for white-tailed ptarmigan in summer. This LCT is defined as the point at which heat loss through the animal's maximum insulation just balances heat production by the animal's metabolic rate. At ambient temperatures below the LCT heat

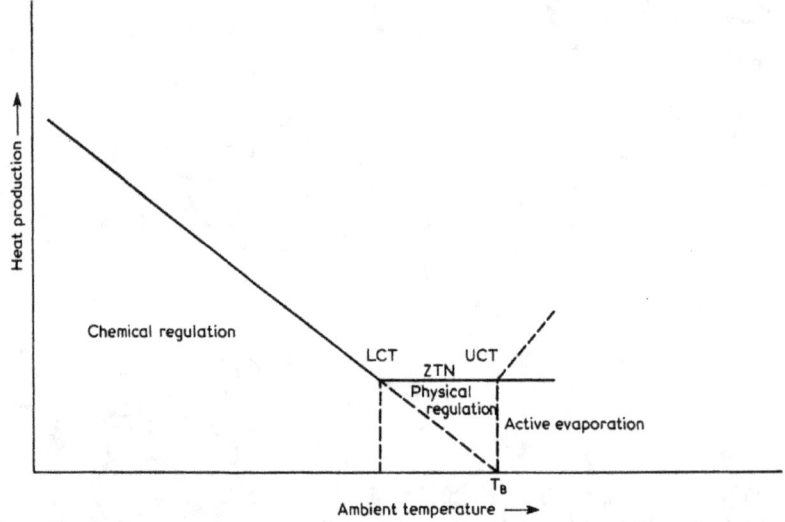

Fig. 9.6. The Newtonian model of metabolic and insulative thermoregulation. The zone of thermal neutrality (ZTN), bounded above by the upper critical temperature (UCT) and below by the lower critical temperature (LCT), is where physical thermoregulation occurs. Above the UCT, active cooling via evaporation water loss occurs. Below the LCT, insulation is maximal, and chemical regulation supplements physical regulation.

will be lost faster than it is produced, and body temperature will decrease unless heat production is increased. This is usually represented by a so-called Newtonian model of metabolic and insulative thermoregulation in which heat production (or metabolic rate) is plotted against ambient temperature (fig. 9.6). The range of temperature over which a constant, minimal metabolic rate is maintained is termed the zone of thermal neutrality (ZTN); its upper end, where the animal must begin to increase heat loss by evaporative cooling, is the upper critical temperature (UCT). In white-tailed ptarmigan this lay between 21° and 38°C, but could not be precisely determined.

The low conductance of the ptarmigan's plumage, and low LCT, would still be insufficient, however, to permit maintenance of body temperature in the winter, where air temperatures as low as −45°C have been recorded in alpine areas (Johnson, 1968) and down to −68°C in the Arctic (Irving, 1960). Compensation is achieved by an increase in thickness, and thus insulative value, of the feathers (Scholander *et al.*, 1950a; Herreid and Kessel, 1967), which reduces the LCT well below that of the ptarmigan in summer plumage (Irving, 1960); increased metabolic activity then permits *L. leucurus* to maintain a normal body temperature (40°C) when the air temperature is −34°C (and probably lower). In contrast, the same species in summer cannot maintain normal body temperature when the air temperature falls below −17°C. The snowy owl, another winter resident of the Arctic, has a similar LCT (Gessaman, 1972).

However, in the arctic or alpine winter, the ptarmigan (350–400 g) is about the smallest bird whose LCT is low enough to withstand the extreme cold. Smaller birds must resort to shivering and other muscular activity to raise metabolic rate and produce enough heat to maintain body temperature (West, 1965), or at least to permit a degree of controlled hypothermia (Irving, 1955; Steen, 1958). Nocturnal reduction in body temperature enables small birds to survive a cold night with a metabolic expenditure 50–75 per cent of normal, an important caloric economy when food is scarce.

Mammals have somewhat different adaptive responses to cold temperatures. The fur layer, in fact, is much less effective in reducing heat loss than a comparable thickness of feathers, so that the smallest mammals capable of withstanding the extreme cold of the arctic winter with only the heat produced from basal metabolism are the arctic fox (*Alopex lagopus*) and arctic hares (*Lepus timidus*, *L. arcticus*). They are about ten times as large, by weight, as the smallest similarly-insulated birds, the ptarmigan (Irving, 1960).

Mammals, though apparently not birds (Hart, 1962; West, 1965), may also produce additional body heat through a process called *non-shivering thermogenesis* (Hart and Jansky, 1963; Jansky, 1967). In response to cold, the metabolic rate of the mammal is elevated biochemically without any associated increase in shivering or other muscular activity. Whether this process is biologically important to wild mammals under natural conditions, as Jansky suggests, is so far unproven.

Another metabolic adaptation by which arctic and alpine animals might increase heat production to cope with the stress of a cold climate is to maintain a higher rate of basal metabolism than comparable temperate and tropical species, or else a higher rate in winter than in summer. However, many data have accumulated to indicate that basal metabolic rates are not subject to selection for temperature adaptation, and that arctic mammals and birds do not differ in this regard from temperate or tropical forms in any manner related to climate (Scholander *et al.*, 1950c; Krog and Monson, 1954; Krog *et al.*, 1955; Irving *et al.*, 1955; Hayward, 1965; Irving, 1972). This permits direct graphic comparison of the zones of thermal neutrality and lower critical temperatures of tropical and arctic birds and mammals (fig. 9.7). From a common baseline of basal metabolism, representing 100 per cent, the differing temperatures at which metabolic rate slopes upward to the left (the LCT), and the steepness of the slope of the line, are functions of 'insulative adaptation'. Tropical species, with less insulation, have higher, less variable LCTs, and their metabolic rates increase more rapidly with decreasing temperature below their LCTs. Some arctic species such as the lemming or snow bunting (*Plectrophenax nivalis*) are not much better insulated, but avoid cold by fleeing to a warmer microclimate beneath the snow or, what amounts to the same thing, by migrating from the arctic or alpine to more temperate latitudes or altitudes (Irving, 1972). The best adapted, most heavily insulated arctic and alpine forms, such as arctic fox, Rocky Mountain goat (*Oreamnos americanus*), and the several species of ptarmigan, have very low LCTs and only slowly rising metabolic rates at even lower temperatures; insulation is their key adaptation.

A cautionary comment should be added in closing this section. The simple Newtonian model of thermoregulation proposed by Scholander assumes only insulative control of

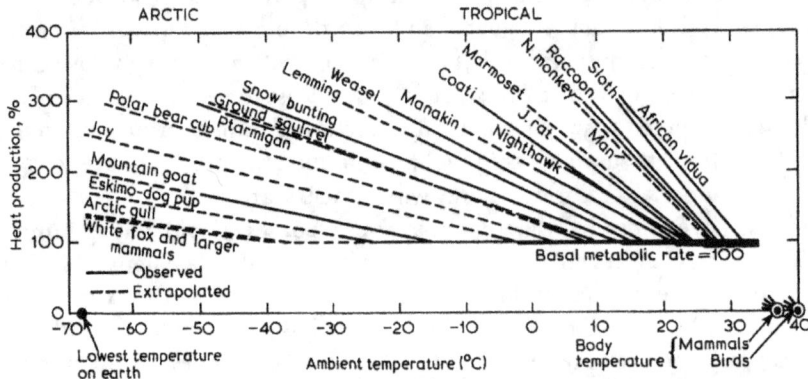

Fig. 9.7. Insulative adaptations of tropical birds and mammals compared with arctic and alpine birds and mammals (after Scholander *et al.*, 1950b; Krog and Monson, 1954; and Johnson, 1968). The horizontal baseline indicates the basal metabolic rate (100 per cent) in the zone of thermal neutrality (ZTN). At each species' lower critical temperature (LCT) the metabolic rate increases, and the rate of increase, and rate of increased heat production, is indicated by the slope of the line. The lower the LCT, and the slope of the rate of increased heat production below it, the better the insulative adaptation of the animal to cold.

heat loss down to the LCT, at which point chemical thermoregulation begins to supplement physical by increasing heat production in a linear fashion. There is evidence (Dawson and Tordoff, 1959; King and Farner, 1961; West, 1962) that the LCT is not a precisely defined point, but that increase in metabolic rate is gradual rather than abrupt, and begins before maximum physical thermoregulation is attained. Nonetheless, the simpler model provides abstractions, the ZTN and LCT, which are useful from a theoretical and comparative standpoint.

Posture and behaviour

Small mammals, lacking the efficient insulation and migratory abilities of small birds, can only survive the arctic or alpine winter by avoiding extremely low ambient temperatures (Irving, 1972). The true hibernators shelter throughout the winter, relying on stored body fat for energy and conserving this limited resource by allowing the body temperature to fall to near that of the ambient temperature in the shelter, which is usually a subterranean burrow. The mammals thus conserve energy by minimizing the temperature differential between their bodies and the environment. These true hibernators include, among alpine mammals, the marmots (*Marmota* sp.) and ground squirrels (*Spermophilus* sp.). In the arctic tundra, there is only one true hibernator, the arctic ground squirrel (*Spermophilus parryii*) found in the tundra of both eastern Siberia and western North America (Nadler *et al.*, 1973). The scarcity of hibernators in the Arctic, and to a lesser extent alpine, as opposed to more temperate environments, is probably related to the problem of locating a frost-free hibernaculum (Dunbar, 1968). In the Arctic, much of the ground in which hibernation burrows might be constructed is occupied by permafrost, and arctic ground squirrels inhabit only well drained, often sandy soils along stream courses and valley slopes (Bee and Hall, 1956) where permafrost lies relatively deep, and where snow accumulations prevent deep freezing of the surface soil layer, thus permitting ground squirrels to hibernate in a relatively warm, stable environment (Mayer, 1953, *in* Pruitt, 1970).

A second kind of response to winter cold is exhibited by mammals that store quantities of food to be used during the winter. Unlike body fat, a stored food source cannot be tapped automatically, and the animal must become active at intervals throughout the winter. Food-storing mammals include pikas (*Ochotona* sp.), deer mice (*Peromyscus* sp.), wood rats (*Neotoma* sp.), several genera of Old World hamsters (*Cricetinae*), and narrow-skulled voles (*Microtus gregalis, M. miurus*). Shelter is ordinarily an insulated nest placed on or above the ground, although sometimes it is subterranean. Apparently some forms in the food-storing group (chipmunks, hamsters) are torpid or inactive most of the time, but intermittently arouse and feed on their stores throughout the winter. On the other hand, deer mice (*Peromyscus maniculatus*), pikas, and others lack the physiological ability to become torpid, and so maintain their normal metabolic rate all winter.

Bears appear to occupy an intermediate position between true hibernators and non-hibernators, since although they store fat (and not food) and exhibit heart rate reduction they do not become truly torpid (Herrero, 1970).

In a third category of small mammals, normal foraging behaviour is maintained during the winter, although some food storage may also occur. A fairly permanent insulating blanket of snow is necessary for the species in this category, since most of them lack warm winter fur (Formozov, 1946). In those places where snow accumulates to sufficient depth, it forms a subnival space, in which the air temperature is not much below 0°C, even though above the snow surface the cold may be intense (fig. 9.8). Thus, the more temperate microclimate in which arctic and alpine shrews and mice live is the key to their survival. Even the moderated temperatures found beneath the snow are below the LCT of many small mammals. Lemmings (*Dicrostonyx torquatus*) and least weasels (*Mustela rixosa*) have LCTs at around 15°C (Scholander *et al.*, 1950b); larger long-tailed weasels (*M. frenata*) also exhibit elevated metabolic rates at relatively high temperatures, partly as a consequence of their long, thin body shape (Brown and Lasiewski, 1972). During periods of activity these species produce sufficient additional body heat, but when inactive, most resort to shivering and/or reduce their heat loss still further by retiring to a well insulated nest lined with dry grasses and old fur (Ognev, 1948; Bee and Hall, 1956; MacLean *et al.*, 1974). Measurement of heat conductivity on one small lemming nest indicated a minimal two to threefold increase in insulation to the animal in its nest

(Scholander *et al.*, 1950c). Nests vary in size and wall thickness (Bee and Hall, 1956), and larger nests, especially if occupied by several animals, may maintain an ambient temperature above the LCT of the small inhabitants.

To sum up, it is significant that both arctic and alpine animals show the same adaptations to their cold environments. This should be regarded not only as a parallel response

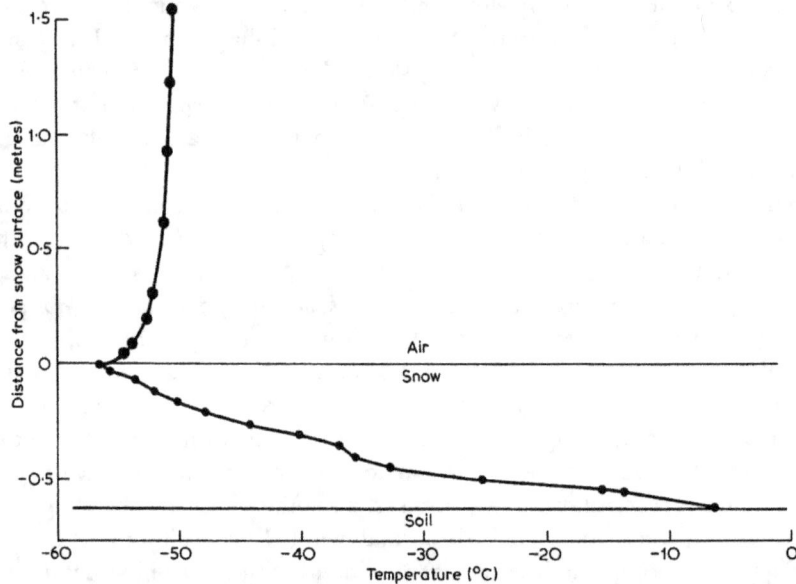

Fig. 9.8. Temperature gradient from above the snow surface on a cold winter day in central Alaska, through the snow blanket to the snow-soil interface (after Johnson, 1953, from Odum, 1959).

to similar environmental challenges, but also a consequence of the evolution of many alpine and most arctic species from continental cold-steppe forms in Siberia and Beringia during the later Tertiary (Hoffmann and Taber, 1967; Guthrie and Matthews, 1971). The northern steppe zone of the interior of continents is also characterized by a severely cold winter climate (Formozov, 1966) and animals adapted to this older biome-type were thus preadapted to the more recently evolved arctic tundra (see pp. 481 ff.).

SNOW ADAPTATIONS

Alpine vertebrates, particularly small mammals, form patterns of dispersion in response to the patterning of winter snowdepth (see Barry and Van Wie, Chapter 2C). As was discussed in the previous section, small mammals, because of their relatively high LCTs, are exposed to a potentially lethal environment if they spend any length of time exposed to very cold air during the alpine winter. Such forms as meadow and water voles (*Microtus* sp., *Arvicola* sp.) are thus restricted to lee slopes and depressions in the alpine, and others such as red-backed and heather voles (*Clethrionomys* sp., *Phenacomys* sp.) to the timberline *krummholz*, where snow regularly accumulates sufficiently to form a moderately warm subnival space. Alternatively, deer mice, pikas, and mountain voles (*Alticola* sp.) live in rocky areas, where accumulations of slide rock, or crevices beneath boulders, provide shelter, but not in snow-free areas, or areas with thin and/or intermittent snow cover, where the ground freezes in winter to considerable depths (see Ives, Chapter 4A).

As in the alpine belt, arctic small mammal populations inhabit the subnival zone beneath snow accumulation areas; the tundra vole (*Microtus oeconomus*) and brown lemming in marshes, wet meadows, and polygon troughs, collared lemmings on better-drained, though still snow-covered, ridges and uplands, and the boreal red-backed vole (*Clethrionomys rutilus*), singing vole (*Microtus miurus*) and narrow-skulled vole in shrubby cover of heath (*Cassiope, Empetrum*), willow (*Salix*) or alder (*Alnus*). Areas with little or no snow are uninhabited during the arctic winter (Pruitt, 1966).

Larger animals do not live beneath the snow, but its spatial distribution and also its quality are of great importance to their ability to move and forage. Both alpine and arctic tundra, because of their open terrain, are characterized by snow which has been moved and worked by wind, and where it accumulates it consolidates into a hard mass (*upsik*). Moreover, much of the winter a portion of the snow cover is moved into the air, or along the surface, by wind; this moving snow (*siqoq*), depending on wind force and direction and topographic relief, becomes stabilized for varying time periods and forms drifts, which may themselves migrate (Pruitt, 1970).

Since larger herbivores must either dig through the snow cover to find food, or consume food plants that project above the snow, they are very sensitive to the hardness, density, and the thickness of the snow. When snow becomes too hard and/or too dense, caribou have difficulty in digging their 'feeding craters' to reach the tundra plants on which they depend. Since over much of the arctic tundra the snow cover is dense and hard due to wind action, 'most caribou leave the tundra during the snow season and migrate to the taiga where the snow is softer and less dense' (Pruitt, 1966). However, the caribou that remain are found in those places on the tundra where, because of local topography, the snow remains relatively soft, or else in those places where the wind is so strong that most or all of the snow is blown away and the tundra is exposed. Alpine populations of caribou perform similar migrations in response to seasonally changing snow conditions, but altitudinally rather than latitudinally (Edwards and Ritcey, 1959). Snow which is too deep for ease of movement, even though soft and fluffy, may also stimulate migration.

Rocky Mountain goats and bighorn sheep among alpine herbivores also respond to snow quality and depth. Goats, capable of browsing in winter, may move to the upper edge of the subalpine forest where conifers on which they feed project above the snow, even though it is deep. Sheep on the other hand are grazers, like caribou, and must seek out areas where they can paw through the snow. Both may, however, utilize ridge tops and other areas blown snow-free by the wind (Brandborg, 1955; Geist, 1971).

Only a few birds are winter residents of the tundra. Ptarmigan are able to find localized small pockets of soft snow on the tundra into which to dive and burrow, thus finding a degree of noctural insulation (Dement'ev and Gladkov, 1952). These arctic and alpine grouse also migrate in winter to areas where their food plants, such as willows, are tall enough to project above the snow surface (Weeden, 1964, 1967).

Tundra predators, too, are affected by snow cover. Many of the birds migrate, either following their avian prey or because their mammalian ones are too effectively protected by the snow cover. Those few that remain, such as the snowy owl, hunt in areas of thin snow cover where prey may be found, or else capture small rodents at the 'ventilator shafts' that they dig from their subnival burrow systems to the surface, so that carbon dioxide which otherwise would accumulate beneath dense thick snow is vented (Formozov, 1946). Mammalian predators, such as wolves, either migrate with their prey or keep to areas of thinner, harder snow over which they can move easily. Others, such as least weasels, are small enough to follow the small mammals right into their subnival havens.

ADAPTATION TO WIND AND INSECT PESTS

These apparently dissimilar topics are considered together here because of the importance of wind to large mammals of the alpine and arctic tundra in reducing insect annoyance. For example, caribou may stand singly on windy hilltops during midday, the wind functioning to suppress insect activity (Pruitt, 1960). Similarly, I have observed Rocky Mountain goats resting during the day at the tops of windswept ridges, often on snowbanks, where personal experience has shown that biting flies and other insect pests are least numerous. The Brooks Range marmot (*Marmota broweri*) of Alaska may remain in its burrows on still, warm cloudless days when mosquitoes are most active, and exhibit greatest above-gound activity themselves on cool, cloudy days with a brisk wind (Bee and Hall, 1956).

Caribou and other ungulates may adopt other strategies to avoid insects. A herd stamping in shallow water creates a spray; caribou have also been described wading into deeper water or running across the tundra to avoid insects. A lone caribou may plunge into the midst of a herd to avoid a nose bot, or warble fly (Pruitt, 1960; Bee and Hall, 1956). Alpine mammals may also move when harassed by insects. When the warm, still, clear weather of early summer encourages biting flies, the red deer (*Cervus elaphus*) of the Scottish moors move upslope to higher ground, where these insects are uncommon (Darling, 1937). In general, by moving upslope to avoid insects, mountain-dwelling mammals are displaying behaviour analagous to the use of water, wallows, or mass-herding by their lowland counterparts.

Wind is, of course, of more general importance to the terrestrial inhabitants of these ecosystems. Since mountain topography causes much local variation in wind (see Flohn, Chapter 2B, and Barry and Van Wie, Chapter 2C), animals need only move short distances to obtain comfortable locations. There are the obvious cooling effects of wind or warming effects of wind avoidance, associated with dissipation of body heat or disruption of the insulative air trapped in the fur or feathers respectively. Thus, wind is a major factor in temperature regulation of alpine and arctic birds and mammals. Birds may also utilize the energy of this windy environment for flight. Winds are also important in shaping the pattern of snow accumulation and snow quality (see Barry and Van Wie, Chapter 2C). Animals may use wind for a combination of reasons; for example, mountain ungulates such as the chamois and Rocky Mountain goat typically make their beds for the day on the edge of a high ridge, where they are cooled by the upslope breezes that at the same time bring a constant supply of scent from below. Finally, in open windy environments most sounds do not carry well. The high-frequency, high-intensity vocalizations of certain diurnal, colonial, or gregarious herbivores in these environments, such as the marmots, ground squirrels, pikas and chamois, are perhaps related to their windiness. Such calls may be more penetrating and hence more easily recognized, and have survival value for the species in reducing probability of predation, or in enhancing social communication (Pattie, 1967b; Nikol'skii, 1971).

SUBSTRATE ADAPTATIONS

Soil type *per se* does not significantly influence terrestrial vertebrates. Of importance instead are characteristics such as rock and soil texture (particle size and per cent coarse fragments), moisture and the presence or absence of ice, either seasonally or as permafrost. However, arctic and alpine soils are highly variable in depth, moisture and annual temperature regime, and bare rock is much more abundant in the alpine than in lowland tundra.

Rock in the alpine

Many alpine animals display definite behaviour patterns with regard to rock type. The true mountain ungulates, such as chamois (*Rupicapra rupicapra*), ibex (*Capra ibex*), Himalayan thar (*Hemitragus jemlaicus*), Rocky Mountain goat, and bighorn sheep typically use cliffs as avenues of escape (Geist, 1971). Morphological adaptations of these mammals to a rocky substrate are obvious: for example, such features as large front feet, hooves with 'elastic' soles, or long toes with a pincer-like grip. Alpine rock-dwelling birds such as the wall creeper (*Tichodroma muraria*) and rock nuthatch (*Sitta neumayeri*) also exhibit feet modified for rock climbing.

Shattered rock provides shelter (see Barry and Van Wie, Chapter 2C) and covered avenues of movement for such characteristic alpine mammals as the pika and the dwarf shrew (*Sorex nanus*) as well as for such widely distributed mammals as the deer mouse. Chipmunks and other mammals normally using trees as escape routes will use rock when it is available. Cracks and fissures are also nesting sites for chipmunks, as well as for bushy-tailed wood rats (*Neotoma cinerea*).

Surface rock protects the burrow-systems of several mammals of the North American alpine from two powerful digging predators – the brown, or grizzly, bear and badger (*Taxidea taxus*). Hoary and yellow-bellied marmots (*Marmota caligata, M. flaviventris*) and the golden-mantled squirrel (*Spermophilus lateralis*) commonly locate burrow openings partly beneath boulders of substantial size. In contrast, the lowland woodchuck (*Marmota monax*), which lives largely outside grizzly bear and badger range, constructs its burrow systems in deep but not particularly rocky soil.

Rock size relative to body size is also important. Large boulders and rock fragments, over 2–3 m in diameter are necessary to shelter marmots (Svendsen, 1973), whereas smaller rocks are suitable for the golden-mantled squirrel and pika. Still smaller fragments shelter deer mice, several species of voles and shrews. The widespread presence of exposed rock and its common use for escape and shelter by mammals are characteristic of alpine environments and on the whole differentiate them from arctic tundra environments.

In the absence of adequate surface rock, as in places on the Tibetan Plateau and other central Asian alpine areas, certain species belonging to groups that are ordinarily rock-dwelling live instead in burrows. For example, Theresa's snow finch (*Montifringilla theresae*) nests in old marmot and pika burrows on high plateaus in the Hindu Kush, whereas the common snowfinch (*M. nivalis*) nests in the same area, only on rocky slopes (Niethamer, 1967).

There is, of course, an ecotone between the alpine and arctic tundra, in those mountains extending to high enough latitudes, such as in the Brooks Range in northern Alaska, in the northern Ural Mountains, and in the mountains of eastern Siberia. In such places, certain species of birds and mammals extend from the lowland tundra through the foothill zone and into true alpine. For example, the boreal red-backed vole and singing vole occur in the lowland tundra only where shelter in the form of birch or willow shrubs (and winter snow cover) exists. However, in the Brooks Range, these species also inhabit rock polygons and boulder fields, whose shelter substitutes for that provided by shrubs in the lowland tundra (Bee and Hall, 1956). Here they live in a habitat which, in the alpine of lower latitudes, is occupied by the red-backed vole (*Clethrionomys gapperi*), montane heather vole (*Phenacomys intermedius*), and long-tailed vole (*Microtus longicaudus*).

Alpine substrate moisture

Soil moisture is also an important habitat factor to alpine species. Valleys, swales and other depressions accumulate winter snow, as noted above. During periods of snowmelt and summer rainfall, these areas also receive the greatest amount of water and are regularly flooded. Soil moisture content is thus high the year round. Certain species of small mammals, such as the montane vole (*Microtus montanus*) and particularly the water voles (*Arvicola*), are adapted to wet alpine tundra, and even swim readily. However, because of the high water table, their burrows are restricted to elevated areas of micro-relief hummocks and the slope margins. Truly fossorial alpine species, such as the northern pocket gopher (*Thomomys talpoides*) and the Promethean vole (*Prometheomys schaposchnikovi*) of the Caucasus are prevented from constructing burrow systems in wet alpine areas by the high water table. At the same time they are limited upslope by the depth of winter snow cover (see above; also Ingles, 1949). Thin or intermittent snow cover results in frozen ground, preventing burrow construction, and gopher winter activity is concentrated at the soil-snow interface, as well as within unfrozen ground beneath deeper snow cover. Soil excavated by tunnelling in winter is packed into subnival tunnels rather than being thrown out on the surface as in summer. Snowmelt later exposes the long, sinuous earth cores on the ground surface, a very characteristic sign of pocket gopher activity in this relatively narrow alpine zone (Plate 36).

Arctic substrate moisture

The distribution of small mammals is strongly associated with soil moisture in the arctic tundra as well. The tundra vole occupies the wettest tundra, marshes, lake margins, and the centres of low polygons, with a high water table and often standing water. These voles also swim readily (Bee and Hall, 1956; Tast, 1966). Brown lemmings inhabit marshes, especially in winter, but occur more abundantly on better-drained soil such as polygon ridges, high-centre polygons, and other places supporting moist meadows. On ridges with better drainage and in areas of frost-scarring and solifluction, as well as rocky uplands, the collared lemming is most common. Finally, the best drained, often sandy soils of steep slopes and bluffs along watercourses overgrown by dwarf willow shrubs harbour narrow-skulled voles (*M. miurus, M. gregalis*). Habitat overlap occurs between species along this gradient of soil moisture at ecotones between the several habitats.

Unlike the alpine belt, no fully fossorial small mammals are found in the arctic tundra. This is probably due to factors of soil moisture and temperature: 1) permafrost closely underlies large areas of the tundra, severely restricting the depth to which fossorial forms could dig even in summer (Holowaychuk *et al.*, 1966; Irving, 1972); 2) winter temperature is sufficiently severe, and snow cover shallow, so that the 'active' surface layer of soil freezes in winter (Lachenbruch *et al.*, 1966).

Arctic and alpine tundra birds also exhibit a pattern of habitat segregation in accordance with degree of soil drainage. Alpine water pipits, for example, nest predominantly in wet but also in dry rocky areas, while horned larks (*Eremophila alpestris*) are more often found in drier, better-drained situations (Verbeek, 1967, 1970; Voous, 1960; see also p. 535). Overlap also occurs in foraging habitats, but pipits in general feed in wetter sites and larks in drier ones; these same species in the Arctic preserve the same habitat relationships (Drury, 1961).

A unique feature of wet habitats in lowland arctic tundra is the large number of water birds, particularly anatids (ducks, geese and swans) and shore birds or charadriiforms (plovers, sandpipers, etc.); alpine regions, with the partial exception of the Tibetan Plateau (Vaurie, 1972), lack this rich avifauna of water birds. For example, on the low tundra around Point Barrow, thirteen species of shore birds are common and widely distributed breeding birds, and an additional five species breed locally or sporadically (Pitelka, 1959). Breeding arctic anatids are similarly abundant in tundra wetlands (Bee, 1958; Ploeger, 1968).

In summary, while there are similarities between alpine and arctic tundra vertebrates in their adaptations to snow conditions and soil moisture gradients, there are also important differences. Alpine species exhibit adaptations to rock in its various forms (talus, cliff, boulder and fellfield, etc.) not seen in lowland tundra species. Also, burrowing habits are better developed in alpine species than in the Arctic, where severe restrictions are placed on this adaptive modification by the presence of permafrost, and completely frozen soil in winter. Conversely, a very important component of the arctic tundra ecosystem is comprised of water birds of several taxonomic groups, which are for the most part unrepresented in alpine ecosystems.

PHENOLOGICAL ADAPTATIONS – SEASON AND DAY LENGTH

The differences between daily and seasonal cycles in high-latitude arctic tundra versus low-latitude but high-altitude alpine environments are striking. At high latitudes in summer, or again in winter, the daily cycle of light and dark is weak (Swade and Pittendrigh, 1967), in comparison with that in spring or autumn, or with the cycle found in low-latitude mountains. On the other hand, the higher the latitude, the stronger the seasonal change in day length. Since duration of solar radiation also strongly influences environmental temperature, it is not surprising to find relatively little mean monthly temperature change in low-latitude alpine regions, but considerable differences between daily maximum and minimum temperatures (Coe, 1967; Mani, 1962). The higher the latitude of an alpine region, the lower its altitude (since both timberline and snowline, or upper limit of vascular plants, are depressed: Swan, 1967) and the more closely its daily and seasonal cycles of radiation and temperature approach that of high-latitude arctic tundra.

Daily patterns

At arctic latitudes diel (twenty-four hour period) cycles of light, and sometimes temperature, are of such low amplitude as to raise serious doubt as to whether or not they can serve to maintain diel activity rhythms. However, extensive observation in both field and laboratory has established that arctic birds (Franz, 1949; K. Hoffmann, 1959) and mammals (M. A. Folk, 1963; G. E. Folk, 1964; G. E. Folk *et al.*, 1966; Swade and Pittendrigh, 1967) do maintain twenty-four hour periodicity of activity, although in some species the periodicity was weak, or phase shift occurred, around the summer solstice (21 June). In the small mammals studied, periodicity was the result of an innate circadian rhythm, entrained by the normal low-amplitude light cycle of the high arctic summer, with its 'midnight sun'. In contrast, studies of the collared lemming under laboratory conditions involving different light and temperature regimes did not reveal any diel periodicity, but instead indicated a pronounced short-term cycle of two to three and a half hours length under all light regimes (Fisher and Needler, 1957; Hansen, 1957), as well as an inverse relationship between total activity and amount of light.

In view of the variability noted in activity rhythms of many small mammals (Calhoun, 1945; R. S. Miller, 1955; Grodzinski, 1962) it is not surprising to find variations in arctic species. Indeed, it is harder to account for persistent twenty-four hour activity patterns in the Arctic, in the absence of any obvious selective advantages accruing to a species which maintains a nocturnal or diurnal pattern in an environment where there is so little difference between 'night' and 'day' for much of the year. It has even been suggested that 'the occurrence of proper phasing of daily periodicity in the Arctic is due not to its functional necessity, but to historical causes . . . they are evolutionary relicts', a legacy to arctic mammals from their temperate zone ancestors (Swade and Pittendrigh, 1967).

Alpine birds and mammals, in contrast to arctic ones, have a marked daily cycle of light and temperature. There is at present no evidence that their activity cycles differ in any respect from those of lowland species.

Seasonal patterns

Despite the differences in patterns of day length, the growing seasons are short in both the arctic tundra and in those alpine areas where snow accumulates to a significant extent. If snowfall is absent due to low-latitude (equatorial mountains) or reduced due to aridity, as in the inner Himalayan ranges, the growing season is lengthened (Swan, 1967).

The short growing season common to arctic tundra and many alpine areas leads to certain common adaptations among the vertebrate inhabitants. Most of the birds are migrant (Irving, 1972); they arrive in late May and early June in 'a floodtide . . . to begin breeding activities in the short summer' (Pitelka, 1959). With only a brief time in which to build their nests, lay and incubate eggs, and rear the young to independence, a selective premium is placed on compressing or eliminating some stages of the breeding cycle. In many species, including most anatids, the birds arrive paired, eliminating the need to spend time in a courtship period; in others, courtship and the pair-bond period are reduced, as in the pectoral sandpiper (*Calidris melanotos*) (Pitelka, 1959). Whether birds arrive already paired or not depends partly on spring phenology. At Barrow, Alaska, in springs when snowmelt was early and undelayed, red-backed sandpipers (*Calidris alpina*) arrived singly, coincident with the appearance of snow-free tundra in late May. Males appeared first, followed by small flocks of mixed sex, and finally female groups: pairing occurred in the first week of June. However, when snowmelt was late or delayed by a spell of cold spring weather, the main influx was delayed, although an advanced guard of a few single birds appeared, and the birds then arrived mostly in pairs (Holmes, 1966a). Migrant alpine birds exhibit similar variations in pattern. Pair formation in rosy finches and white-tailed ptarmigan takes place after the birds reach the alpine belt in mixed flocks (French, 1959a; Choate, 1960). Water pipits may be paired prior to the time territories are occupied (Gibb, 1956; Verbeek, 1970). On the Beartooth Plateau, Wyoming, first arrival coincides with the commencement of snowmelt, and, as with the arctic sandpipers, the first birds occupy snow-free areas as they appear. Severe weather may cause the pipits to leave the alpine temporarily for lower elevations, but they return when the weather improves. Since some single males are seen during this transition period, at least some arrive on the breeding grounds unmated (Verbeek, 1970).

The shortness of the summer season in alpine areas and in the Arctic precludes in most cases more than one nesting attempt, even in those birds which further south may nest several times – horned lark, white-crowned sparrow (*Zonotrichia leucophrys*), savannah sparrow (*Passerculus sandwichensis*): DeWolfe and DeWolfe, 1962; Bent *et al.*, 1968; Verbeek, 1967; DuBois, 1935 – or in which re-nesting occurs if the first attempt should be disrupted (exceptions, however, are to be found: Pitelka, 1959; French, 1959a). Among many alpine and arctic passerines only the female incubates, but both parents feed the young (Drury, 1961; French, 1959a; Verbeek, 1967, 1970), whereas among the three species of ptarmigan, only the *L. lagopus* male shares in the care of the precocial chicks. The young of arctic anatids and shore birds are also precocial, and several patterns of incubation and care of young are found (Van Tyne and Berger, 1959). Either the male or the female may alone incubate, or care for the young, or both may share. For example, in the pectoral sandpiper and white-rumped sandpiper (*Calidris fuscicollis*) the female alone incubates and cares for the young, while males gather in flocks and leave the breeding grounds before the young hatch out (Pitelka, 1959). In the red-backed, Bairds, and semipalmated sandpipers (*Calidris alpina, C. bairdii, C. pusillus*) both parents incubate and initially care for the young, but care devolves increasingly on the males as the breeding cycle proceeds, and it is the females which flock and leave the breeding grounds first (Holmes, 1966a). Moreover, in *C. bairdii* and *C. pusillus*, the males then leave when the young are becoming independent in late July, whereas adult male *C. alpina* remain into late summer. Finally, in the red phalarope (*Phalaropus fulcarius*) it is the male alone who incubates and rears the young, while the females flock and leave the tundra. These patterns of early migratory departure by one or both sexes may be viewed as a device to reduce both intra- and interspecific competition for food, and to increase the probability of survival of offspring by reducing the demands on their food resources (Holmes and Pitelka, 1968).

There is also a high degree of synchrony in the breeding schedules of alpine and arctic birds, which has usually been attributed to the shortness of the summer *per se*, and the adaptive need to begin breeding as soon as possible. Recent work indicates that avian breeding in these environments is instead synchronized to coincide with the seasonal peak in the insect food supply; tundra insect populations in turn emerge synchronously, possibly to reduce the impact of avian predation by 'swamping' the predators (MacLean and Pitelka, 1971).

Aside from producing offspring, a second major energy demand in the bird's annual cycle is the moult. We have already seen how this may affect a larger bird such as the ptarmigan, whose winter plumage must provide sufficient insulation (p. 491). Migratory alpine and arctic species may not require thicker plumage if they winter in temperate or tropical climes, but must still expend much energy to replace feathers. In most birds, moult takes place after the breeding cycle is completed (Pitelka, 1958a; A. H. Miller, 1961; N. Johnson, 1963), so that these two energetically costly events do not overlap. The shortness of the alpine or arctic season requires that either the moult be compressed in time, and commence even before hatching occurs (red-backed sandpiper – Holmes, 1966b; water pipit – Verbeek, 1970), or else that moult be postponed until the birds leave the tundra (pectoral sandpiper and other *Calidris* – Pitelka, 1959; Holmes, 1966a, b; rosy finch – French, 1959a). Those shore birds that postpone the main portion of the moult until they reach winter quarters are all long-distance (transequatorial) migrants, in contrast to those arctic and alpine birds which moult in their summer quarters; the latter fly shorter distances, and commence migration later in the season.

These various patterns (illustrated by a single example in fig. 9.9) may be interpreted as different compromises between competing selective forces, including 1) maximizing energy available for production of offspring and for moult, via time of nesting and post-nesting departure of adults; and 2) attainment of suitable wintering areas, via timing and distance of autumn migration. Which particular pattern has evolved in a given species presumably depends upon the species' basic adaptive type, the energy resources available to the species at different places throughout the year, and interspecific competitive interactions between sympatric forms leading to evolution of 'anti-competition'

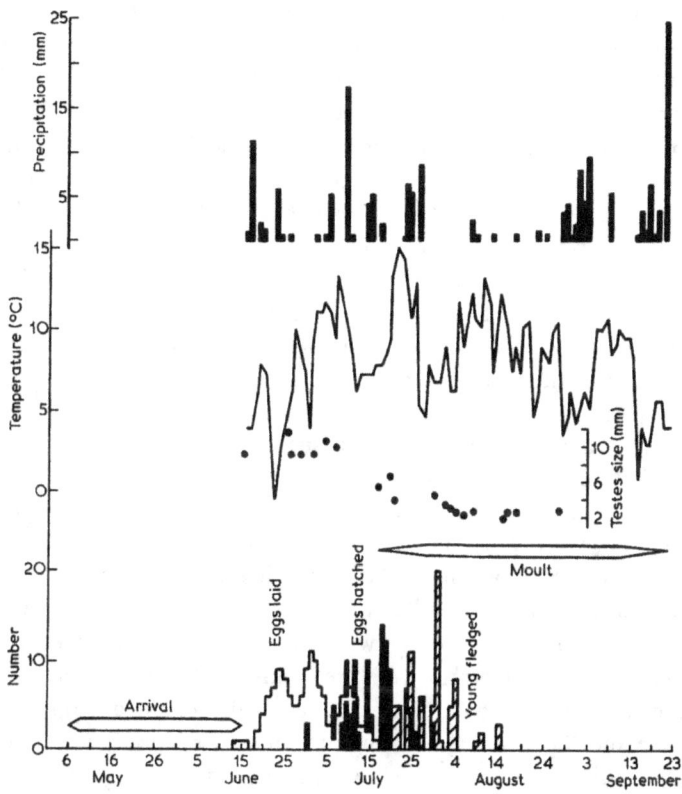

Fig. 9.9. The seasonal cycle of the water pipit (*Anthus spinoletta*) in the alpine zone of the Bear-tooth Plateau (after Verbeek, 1970). The reproductive and adult moult portions of the cycle are plotted against mean precipitation and air temperature on the breeding area. Departure from the breeding area is probably in late October–early November.

mechanisms and patterns of resource allocation (Holmes, 1966a, b; Holmes and Pitelka, 1968; Verbeek, 1970).

The effect of a short growing season may be less apparent on alpine and arctic tundra mammals, most of which are non-migratory. Many of the larger mammals belong to groups which are monoestrous, i.e. have only a single annual breeding effort regardless of the environment they inhabit. Thus, caribou or Rocky Mountain goats breed once a year, as do temperate and many tropical deer and bovoids. The season of birth in arctic and temperate herbivores with long gestation periods is correlated both with plant phenology, and with a pattern of decreasing day length (Marshall, 1937; Bissonette, 1941; Fraser, 1968). The question may be asked: do lowland and alpine populations of a species which live along a single latitude (so that day-length patterns are identical), but with differences in plant phenology due to altitude and exposure, have breeding patterns which are correlated most closely with day length or with plant phenology? A comparison of time of breeding in mule deer (*Odocoileus hemionus*) from 1) the cool, moist seaward side of the Coast Range slopes; 2) the hot, dry interior Coast Range slopes; and 3) alpine and subalpine of the high Sierra, revealed mean fawning dates of about 15 May, 15 April and 15 July, respectively, which were correlated with the stage of development of the vegetation (Leopold *et al.*, 1951; Taber and Dasmann, 1958). This suggests that response to a particular day-length pattern in the timing of the rut is strongly adapted to local plant-growth conditions. This is probably because food high in protein and phosphorus – usually critical diet components – is only available during a brief period of the year. The season of reproduction for alpine and arctic tundra ungulates is directly related to vegetation – a diet high in nutrients is essential for the successful production and nursing of young and for their subsequent growth, prior to the lean times of winter. Thus one finds a high synchrony in breeding behaviour within each population, and a close relation between the season of birth and the season of maximum forage quality (Pruitt, 1960; Kelsall, 1968; Geist, 1971).

Herbivores with short gestation periods, such as microtine rodents, also respond directly to diet quality in breeding phenology (Hoffmann, 1958; Batzli and Pitelka, 1971). Most small rodents of alpine and arctic tundra are, however, polyestrous within the breeding season, with short gestation periods. The number of reproductive efforts per year thus depends upon the length of the breeding season, which in turn is correlated in many instances with latitude and altitude. I first pointed out this relationship (in *Microtus*), and the inverse correlation between litter size, on the one hand, and length of breeding season and annual number of litters, on the other (Hoffmann, 1958). Lord (1960) subsequently showed that there is a statistically significant positive correlation between latitude and the number of young produced per reproductive effort in rabbits (*Sylvilagus*), tree squirrels (*Sciurus*), deer mice, and shrews belonging to four genera, as well as *Microtus*, although litter sizes of ground squirrels (*Spermophilus, Ammospermophilus*) and pocket gophers (*Geomys, Thomomys*) showed no significant correlation with latitude, nor did pikas (*Ochotona*) (Millar, 1973). Lord suggested that increased litter sizes in northern latitudes should be viewed as a compensatory response to higher winter mortality rates in high latitude populations. However, as Spencer and Steinhoff (1968) pointed out, this explanation runs counter to generally accepted evolutionary principles (Lack, 1954; Cody, 1966; G. C. Williams, 1967). Following Williams, they proposed that, since at higher latitudes or altitudes the number of reproductive efforts per season, and lifetime, is reduced, 'it therefore becomes advantageous for an animal to invest its energies in a few, large early litters even though doing so reduces its life expectancy and total reproductive contribution below the maximum achievable by many small litters. This is so because short seasons make it impossible for the animal to realize the returns from the conservative approach within its life span. The most productive strategy is the production of large litters' (Spencer and Steinhoff, 1968).

While this explanation is based on several unverified or oversimplified assumptions (bearing young reduces probability of survival of parent; litter size is independent of age), it is consonant with evolutionary theory. That litter size in many polyestrous small mammals in the alpine belt is larger than at lower elevations or latitudes has been confirmed (Dunmire, 1960; Pattie, 1967a; Spencer and Steinhoff, 1968; Vaughan, 1969; Taber *et al.*, n.d.), thus supplementing Lord's analysis. In addition, it has long been recognized that the number of eggs laid in a nesting effort by birds is positively correlated with latitude (Moreau, 1944; Lack, 1947, 1948, 1954); this in species or groups which raise only one brood per year, as well as multi-brooded species. Lack (1954) accounted for the larger clutch sizes of northern (including arctic) species by postulating that at higher latitudes the increased day length during the breeding season would permit the parent birds to raise more young successfully, because they could gather more food during the longer daylight period. Nevertheless, this explanation could not be applied to altitudinal increase in clutch size at the same latitude (Johnston, 1954) or to small mammals which may be nocturnal, or forage day and night. A general theory to account for geographic and temporal variation in clutch size has been advanced by Cody (1966) which, while it includes Lack's and other earlier theories, is based on the principle of allocation of energy resources available to the population, and thus is similar to the theory of litter size proposed by Spencer and Steinhoff (1968).

There is one important exception to the rule that breeding seasons in alpine and arctic regions are short. This is the adaptation of subnival breeding, found in certain small mammal populations. It is best documented for arctic and alpine populations of the brown lemming, which in certain years may breed from autumn to spring underneath the snow blanket (Curry-Lindahl, 1962; Mullen, 1968). Litters from this winter breeding are small, but the protection offered by the continuous snow cover apparently permits good survival, and subnival breeding may be an important component of the lemming's adaptation to the arctic environment. However, little is known about the factors which lead to subnival breeding in some years but not in others.

A less well-defined form of this phenomenon is found in alpine populations of the montane vole. Breeding may, in some years at least, be initiated beneath the snow in early spring, and continue on into the summer (Hoffmann, 1958; Vaughan, 1969; Pattie, 1967a; Taber *et al.*, n.d.). No instances have yet been recorded, however, of late autumn or winter subnival breeding in alpine or high subalpine populations of montane voles. This may be due to the difficulties of sampling encountered at these seasons, but is more likely to represent a real difference between the vole and lemming populations.

The short growing season influences accumulation of body fat reserves, and food storage activity also. The period of hibernation varies with altitude in some alpine species. Among the yellow-bellied marmots of western Montana, for example, populations in the mountain valleys (about 1000 m above sea level) emerge from hibernation in March, and by the end of August have usually entered hibernation again. In contrast, populations on the alpine Beartooth Plateau of Montana (3000 m above sea level) apparently do not emerge until about mid-May but remain active far into the autumn, long after their lowland relatives have become torpid. Active yellow-bellied marmots have been sighted at 3300 m on 1 October, with the temperature at −1 °C and with light snow falling. The same schedule is followed by the alpine hoary marmot and Columbian ground squirrel (*Spermophilus columbianus*) populations (Manville, 1959).

The short growing season in the arctic tundra and most alpine regions is thus of major importance in shaping the physiological and behavioural adaptations found in the annual cycles of terrestrial vertebrates. Breeding seasons of both birds and mammals are shortened; birds are single-brooded, and small mammals produce a reduced number of litters per season, although breeding may occur under the snow in the case of microtine rodents in certain years. Clutch and litter sizes are often larger in the alpine and arctic than in environments with longer growing seasons. Migration is favoured in species of suitable vagility. Although several proximate factors are probably involved, the ultimate factor producing these adaptations seems to be one of use of limited energy resources in such a temporal pattern that the production and survival of young of the population is maximized. This is an 'r strategy' (r being the intrinsic rate of natural increase of a population) adaptively advantageous to animals living in a rigorously seasonal, climatically unstable, environment (Cody, 1966; MacArthur and Wilson, 1967).

ALPINE ALTITUDINAL ADAPTATIONS

Such alliteration may serve to emphasize the distinctive feature of alpine regions; namely, their high elevation and consequent reduced oxygen tension of the atmosphere, or hypoxia (see Grover, Chapter 14B). While unique in that a hypoxic environment is found only in the alpine, oxygen tension, being an inverse function of altitude above sea level, is not markedly reduced in the relatively low alpine zones of high and intermediate latitudes. However, terrestrial vertebrates such as the yak are known to be permanent inhabitants up to 6000 m in the Himalayas, and birds (alpine chough, *Pyrrocorax graculus*; bar-headed goose, *Anser indicus*) fly even higher for short periods (Swan, 1961). That animals are found living at such elevations is, *ipso facto*, evidence that they have adapted to the hypoxic environment. However, in contrast to some of the adaptations that have already been examined, such as to temperature, 'microclimatic evasion is not possible for low barometric pressure' (Morrison, 1964). The physiological specialization necessary for this adaptation is an increase in the capacity of the blood to transport oxygen at reduced atmospheric pressure. Increased transport capacity may be partly achieved by increasing the number of red blood cells and/or the amount of haemoglobin in the blood cells, depending on species. However, these features appear most frequently in individuals of lowland species which are transported to high altitudes, and can thus be regarded as *acclimatization* rather than adaptation. In alpine rodents studied by Morrison *et al.* (1963a, b) there was no essential difference in range and mean of either cell haemoglobin level or haematocrit from related lowland species. There was, however, an increase in affinity of haemoglobin from high altitude forms for oxygen, as well as increased myo-globin content in various tissues (Morrison, 1964; Reynafarje and Morrison, 1962). Finally, high altitude forms exhibited inherently high heart and respiration rates (Morrison and Elsner, 1962).

These various mechanisms may not, in some alpine species, suffice to completely compensate for reduced oxygen tension, since a reduction in metabolic rate (i.e. oxygen consumption) is generally observed when high altitude populations are compared with lowland ones. This has implications for temperature regulation in alpine small mammals which utilize increased metabolic heat production to balance increased winter heat loss (Hock and Roberts, 1966). In general, however, oxygen tension seems not to be a limiting factor to alpine animals; with increasing elevation food probably runs short before the element used to metabolize it.

3. Ecological niches of arctic and alpine terrestrial vertebrates

In the previous section we have seen the ways in which birds and mammals of the alpine and arctic tundra are adapted to survive over the range of physical conditions they usually encounter in their environments. For each environmental variable (i.e. air temperature, snow depth) there is a range of values over which the species may exist, but also limiting values beyond which it may not survive and/or successfully reproduce. If we imagine all of the environmental variables that may affect a species' ability to exist indefinitely in an area, we will have a large number of variables, each of which can be considered a dimension of the fundamental *ecological niche* of the species (Hutchinson, 1957, 1965). Since more than three dimensions will certainly be involved, the environ-mental 'space' occupied by a species' fundamental niche is a multi-dimensional hyper-volume, or an abstraction by which the ecological niche concept may be formally defined.

So defined, the ecological niche is more broadly conceived than in other recently popular definitions. Grinnell was the first to develop the niche concept, emphasizing initially adaptation to physical environmental factors (1917, 1924), and later essential biotic variables such as food (1928). At this time Elton (1927) also defined the ecological niche as the animal's 'place in the biotic environment, *its relations to food and enemies*', and implicitly suggested a dichotomy between the physical environmental variables of the habitat, and the biotic environmental variables of the niche. This distinction was until recently maintained by most authors, who followed Elton, while generally not apprecia-ting Grinnell's contribution (Udvardy, 1959). For example, Odum (1959) defined the habitat of the organism as 'the place where it lives . . . The ecological niche, on the other hand, is the position or status of an organism . . . the habitat is the organism's "address", and the niche is its "profession" . . .'

Hutchinson's formal definition of the fundamental niche includes both variables of the physical environment, which individuals encounter in their habitat and to which the species must be adapted, later called by Odum (1971) the spatial or habitat niche, and variables of the biotic environment, such as species on which they feed or which feed on them, which individuals encounter in their habitat. This second group of environmental variables will be considered in the present section.

TROPHIC NICHES

Trophic, or food, niches should, as we have seen, be regarded as subsets of the ecological niches of organisms. In most terrestrial ecosystems, a majority of organisms are *producers* – green plants which fix solar energy via photosynthetic processes. Animals that feed on these green plants are classed as *herbivores*; various species possess different anatomical and physiological specializations that influence their use of potential food plants. They may be subdivided into feeders on various plant parts – seeds, fruits, leaves, roots and underground storage organs. Specialized terminology (spermivore, frugivore) is used for the first two subdivisions, but has not been adopted for foliage and root-eaters (foliovore? rhizovore?).

Animals that in turn feed on herbivores are called *carnivores*. Here, too, subdivision is possible, but the only subclasses much used are *piscivore* and *insectivore*, although the *secondary*, or '*top*' *carnivore*, which feeds on other carnivores, is sometimes encountered (Odum, 1959). Finally, *omnivores* feed on both plants and animals.

It should be emphasized that these broad classifications can be only loosely applied. Variations in animal food habits are the rule, and it is especially true of higher vertebrates that monophagy (restriction of diet to a single food source) is rare. Chipmunks, which in spring feed mainly on foliage, are by autumn almost exclusively spermivorous (Beg, 1969). Some small rodents generally thought of as herbivores, such as the deer mouse, are actually omnivorous (Landry, 1970; Whitaker, 1966). Nonetheless, generalizations about trophic niches are useful, if their limitations are recognized.

Arctic trophic webs

The transfer of food energy from one species to another through successive trophic levels is termed a trophic web. Qualitative food webs have been constructed for a number of arctic tundra areas, and differ as a function of the species present in the area. For example, in the Point Barrow area, brown and collared lemmings are the only herbivorous rodents, and caribou no longer occur in the vicinity. Thus, in addition to the two lemmings, the only other herbivores in the food web are insects and about eighteen species of birds, a few of which are partly herbivorous (Pitelka, 1957a). These few herbivores support a carnivore trophic level consisting of pomarine and long-tailed skuas (*Stercorarius pomarinus, S. longicaudus*), glaucous gull (*Larus hyperboreus*), snowy and short-eared owls (*N. scandiaca, Asio flammeus*), masked and arctic shrews (*Sorex cinereus, S. tundrensis*), arctic and red foxes (*A. lagopus, Vulpes vulpes*), least and short-tailed weasels (*M. rixosa, M. erminea*) and insectivorous birds.

A progressively more diverse vertebrate fauna is encountered towards the interior and in the Brooks Range foothills, inland from Point Barrow: at least seventeen species of mammals and thirty-two species of birds are regular components of the food web (Pitelka, 1957a). These include, as herbivores, arctic ground squirrel; the two lemmings;

boreal red-backed, tundra, and singing voles; caribou; and, in addition to the carnivores found at Barrow, a third shrew, the dusky (*Sorex obscurus*), the grey wolf, grizzly or brown bear and wolverine, or glutton (*Gulo gulo*). A comparable degree of food web complexity is also found near the coast at Cape Thompson; thirty-four breeding bird species and nearly the same suite of mammals (Pruitt, 1966). Thus, trophic web complexity of vertebrates is not reduced at coastal localities, as Pitelka (1957a) implied, but the degree of complexity observed in the Brooks Range foothills and at Cape Thompson may approach the maximum diversity and complexity achieved in the arctic tundra ecosystem. It is noteworthy in this regard that the same magnitude of complexity is found in the tundra of the Yamal Peninsula in western Siberia (*cf.* Pitelka, 1957a).

Food webs are conveniently depicted as food energy flow diagrams; several of these may be compared in fig. 9.10. But in such diagrams it is not possible to illustrate the fine structure of an animal's trophic niche. It is widely agreed that two or more co-existing species cannot have the same ecological niches (Gause's principle), and as a corollary to this, that if two species have the same or sufficiently similar niches, one will out-compete and replace the other (competitive exclusion principle) (Hutchinson, 1957; Hardin, 1960). While these ecological principles have been questioned (Cole, 1960; Ayala, 1969), they provide useful working hypotheses by which to examine niche relationships of arctic and alpine birds and mammals.

Consider the case of two or more sympatric species of herbivores, such as occur at Cape Thompson. It is theoretically possible that they eat the same plants, but the plants are so abundant that food is never in short supply, and competition does not occur. (By competition is meant the common requirement by two or more species of limited resources, in excess of the amount available.) But it is known that some of the species at least (brown lemming and caribou) do at times severely graze down the vegetation (Thompson, 1955; Bee and Hall, 1956). One means of reducing or avoiding competition for a limited supply of plant food would be for each species to utilize different plant species, thus partitioning the food resources of the tundra among the several sympatric species.

Qualitative information on food habits of small tundra herbivores is very scarce, and quantitative information practically non-existent (Quay, 1951; Thompson, 1951; Morrison and Teitz, 1953; Pitelka, 1957a; Bee and Hall, 1956; Pruitt, 1966; Stoddart, 1967). It would appear that the microtines living in different habitats eat different foods, but that differences are not yet apparent between sympatric pairs. For example, 'brown lemming and tundra vole occur together; they eat the same kind of food and use the same kinds of nesting places' (Bee and Hall, 1956). It may be that intensive study will reveal that there are in fact significant differences in the plant parts eaten by these two species, and that they do not indiscriminately graze the vegetation in their shared habitat. Studies of sympatric pairs in temperate grasslands (*Microtus pennsylvanicus–M. ochrogaster*; *M. ochrogaster–Sigmodon hispidus*: Zimmerman, 1965; Fleharty and Olson, 1969) show that these rodents are selective and that dietary differences exist, but they also show broad overlap in foods eaten by the two species.

Similar configurations of trophic niches may be seen in arctic tundra carnivores.

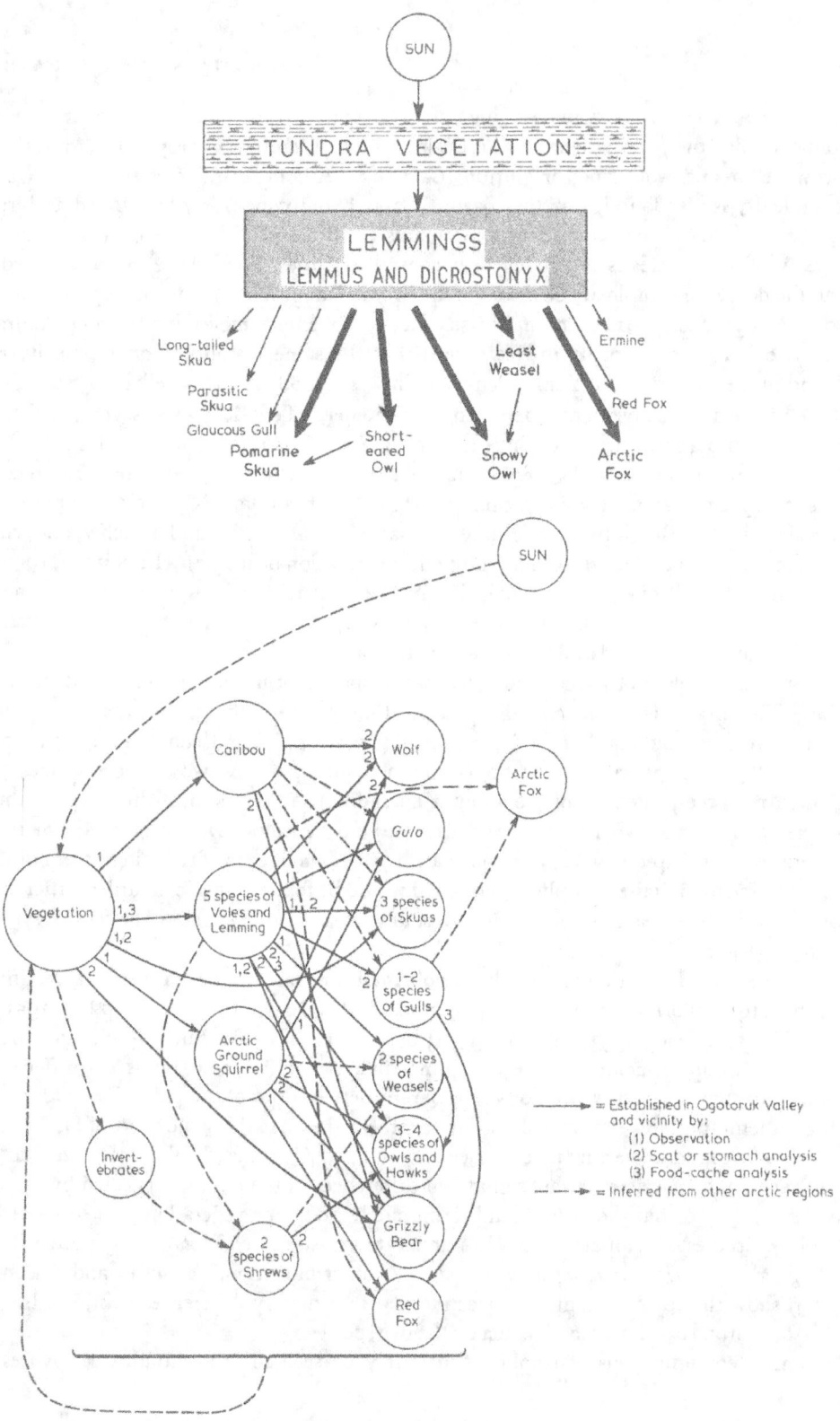

SUN

TUNDRA VEGETATION

LEMMINGS
LEMMUS AND DICROSTONYX

Long-tailed
Skua

Parasitic
Skua

Glaucous Gull

Pomarine
Skua

Short-
eared
Owl

Least
Weasel

Snowy
Owl

Ermine

Red Fox

Arctic
Fox

SUN

Caribou

Wolf

Arctic
Fox

Gulo

Vegetation

5 species of
Voles and
Lemming

3 species
of Skuas

1-2
species
of Gulls

Arctic
Ground
Squirrel

2 species
of
Weasels

3-4
species
of Owls
and
Hawks

Invert-
ebrates

Grizzly
Bear

2
species of
Shrews

Red
Fox

⟶ = Established in Ogotoruk Valley
and vicinity by:
(1) Observation
(2) Scat or stomach analysis
(3) Food-cache analysis
--▸ = Inferred from other arctic regions

The most detailed quantitative study deals with four sympatric species of sandpipers studied at Point Barrow (see p. 502). All four during their brief summer residence on the tundra feed almost exclusively on insects, with a few spiders and small crustaceans added (Holmes and Pitelka, 1968). All show considerable overlap in diets, but with significant differences as well. The semipalmated sandpiper tends to concentrate on small chironomid (midge) larvae; Baird's sandpiper feeds on some chironomids in early June, but switches to upland forms of tipulid (crane fly) larvae, and then adult flies, beetles and spiders. The two larger species – the red-backed and pectoral sandpipers – are more similar in food habits, but even here some differences occur. Differences in prey eaten are reinforced by habitat differences (see below), and by seasonal and yearly differences in dietary overlap.

Seasonally, the greatest degree of differentiation in diets of the four species occurs in late June, when the diversity of available insect prey is greatest. Overlap increases sharply in July, corresponding with a 'sharp peak in surface activity of tundra arthropod populations in mid-July . . .' (MacLean and Pitelka, 1971). Insect prey, for a brief period at least, seems to be superabundant, thus allowing all sandpiper species to eat the same things without interacting competitively. For the prey species, the strategy may be to emerge synchronously and 'swamp' their predators. Finally, by early August many sandpipers have left the tundra, and are on their way southward (see p. 502).

Yearly differences in trophic niche overlap appear to be most important between the two larger species, the red-backed and pectoral sandpipers, since the niches of the semipalmated and Baird are segregated to some extent by diet and habitat. The pectoral sandpiper exhibits pronounced variations in breeding density (×3 or more), both from year to year and place to place, and given favourable conditions, may take advantage of the local opportunity by compressing territory size and forming a high-density breeding population (Pitelka, 1959; Holmes and Pitelka, 1968). In contrast, the red-backed sandpiper shows little variation from year to year in its relatively low breeding densities (Holmes, 1966a, c). The factors to which pectoral sandpiper populations appear to respond positively are early spring snowmelt and availability of food. When insect food is particularly abundant, larger numbers of pectoral sandpipers will be nesting together with the less common red-backs; both utilize the superabundant food sources coextensively, with much dietary overlap. Conversely, in a year when insect food is scarce, the density of pectorals may be similar to, or less than, that of red-backed sandpipers and a much greater partitioning of the relatively scarce food resources occurs (Pitelka, 1959; Holmes and Pitelka, 1968). Fig. 9.11 summarizes some aspects of the trophic niches of these species.

This sort of feeding strategy, based on abundance of prey, is also found in lemming

Fig. 9.10. Partial trophic webs in simple (above) and complex (below) arctic tundra communities (after Pitelka et al., 1955, and Pruitt, 1966). The simple web at Point Barrow, Alaska, has a higher ratio of carnivores to herbivores (5:1) than does the more complex web at Cape Thompson, Alaska (about 2:1). Insectivorous and herbivorous birds are not included in these webs.

predators. However, whereas most of the insectivorous shore birds breed every year, albeit in variable numbers, the carnivorous birds which feed mainly on small mammals of the arctic tundra are adapted to skip a breeding year or two if their prey species are rare. Pitelka *et al.* (1955) reported that on the lowland tundra around Point Barrow,

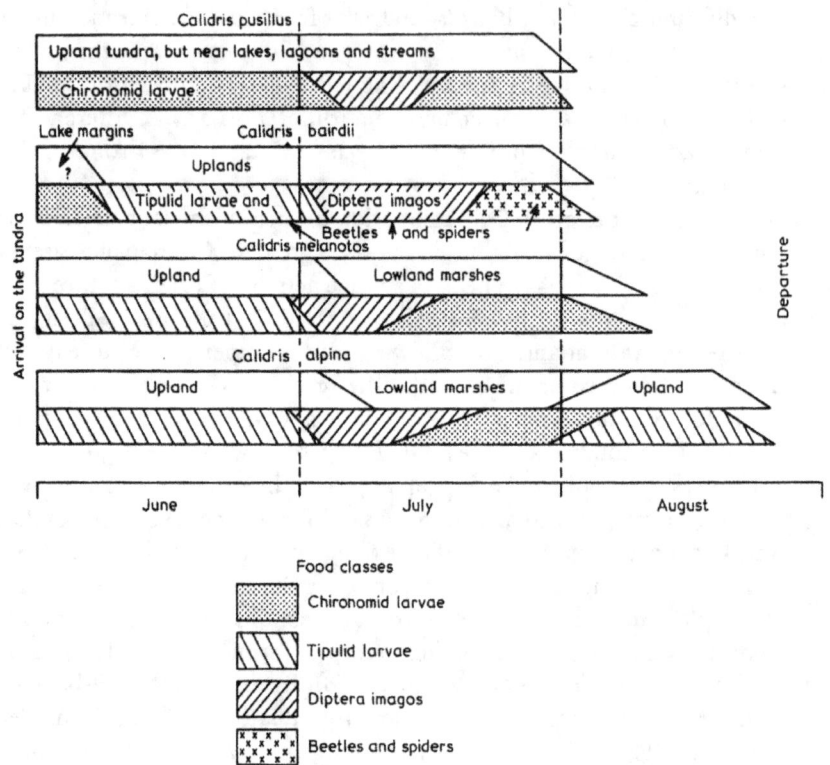

Fig. 9.11. Trophic niches of four sympatric, congeneric species of sandpipers breeding in arctic tundra at Point Barrow, Alaska (after Holmes and Pitelka, 1968). For each species the upper half of the double bar indicates principal habitats utilized, and the lower, main food classes. Late summer food of *C. bairdii* (beetles and spiders) are also utilized earlier in the season, and overlap tipulid larva and diptera imago usage, as indicated.

when brown lemmings were at a low point in their population cycle (see p. 539), pomarine skuas and snowy owls were scarce and they did not nest. The following year lemmings were present in moderate numbers, and both species of avian predators bred; skua density was about 0.015 breeding pairs/ha, while for the much larger snowy owl, density was lower – 0.001–0.002 breeding pairs/ha. In the third year brown lemmings attained a high cyclic peak density; density of pomarine skuas increased to 0.07 breeding pairs/ha, and while the numbers of breeding pairs of snowy owls remained the same as the year before, there was a marked increase in non-breeding individuals present in the area. Finally, short-eared owls appeared on the tundra in large numbers – an average of

0.012–0.015 breeding pairs/ha. Following this peak density, the lemming population 'crashed', declining to a very low density, and in that fourth year of the cycle the skuas and owls again failed to breed at Barrow (Maher, 1970).

The numbers of carnivorous mammals fluctuated in a parallel manner at Barrow during this lemming cycle. In the first year least weasels and red foxes were not seen and arctic foxes were rare; but in the second, when lemmings had increased to moderate numbers, both species of foxes were present, and by the third, peak, year all three mammalian predators were common (Pitelka *et al.*, 1955). When the lemming population declines, the avian predators can move away and seek areas with higher prey density, or in the case of skuas, turn to marine sources of food. However, the foxes and weasels have no alternative but to remain, seeking out the ever-scarcer lemmings. As a result it has been suggested that these carnivores, and especially the least weasel and ermine, may be directly responsible for reducing the declining lemming population to the very low levels characteristic of the year following a cyclic peak (Maher, 1967, 1970).

Among the mammalian predators at least two, the arctic fox and short-tailed weasel or ermine, may exhibit reproductive adaptations similar to avian predators of lemmings. At lower latitudes foxes and ermine regularly breed in their first and second years (Layne and McKeon, 1956), whereas in the Arctic yearling ermines remain reproductively immature (Macpherson and Manning, 1958) and arctic fox vixens rarely produce litters as one- or two-year-olds (Macpherson, 1969). Since the number of surviving young per litter is strongly influenced by food (= lemming) abundance, 'most arctic foxes destined to survive their first winters are born in years of lemming abundance, and it is unlikely that the following year, or even the year after, will bring another such bumper season. Breeding in the first or second years of life may thus be non-advantageous . . . it is possible that early breeding . . . is actively selected against . . . owing to the risks and energy drain of parenthood being counterbalanced so rarely by the production of descendants' in the years of low lemming populations following a cyclic peak (Macpherson, 1969). Thus, whether or not a young female arctic fox breeds may depend upon the stage of the lemming cycle.

Even if arctic foxes do breed when lemmings are scarce, the incidence of prenatal mortality may be increased and entire litters may die, resulting in the absence of young foxes at dens occupied by adult pairs, and the abandonment of dens that had been used at the beginning of the breeding season. Moreover, when the adults are successful in raising some fox whelps, the number weaned per litter is about half that successfully weaned in a peak lemming year (Macpherson, 1969). This is analogous to the variations in fledging success noted in rough-legged hawks (*Buteo lagopus*), and in snowy and short-eared owls – higher when their microtine prey are abundant than when food is scarce (Gross, 1947; Lack, 1954; Lockie, 1955). However, these raptors have been reported to possess another adaption to fluctuating prey density – variable clutch size. When their prey are particularly abundant, short-eared owls may lay a clutch of twice the usual size, though the increase for snowy owls may not be as marked. It has been suggested (Clark, 1940; Lack, 1954) that variation in arctic fox litter size provides a parallel to the variable clutch size in the arctic raptors. There is no statistically sound evidence for variable ovulation rates in foxes leading to varying litter size; observed differences in litters from year to year are more likely a result of changes in prenatal and nestling mortality. However, the net results, production rate of independent young adjusted to the year's food resources, are presumably the same in the raptors and arctic foxes, though achieved by different processes.

In short, we should note that arctic carnivores share certain trophic niche features which may be adaptive to the tundra environment. First of all, there is considerable overlap in diets among different species, the insectivorous birds (and mammals?) concentrating on the abundant arctic dipterans and the larger birds and mammals on the microtine rodents. Given the low taxonomic diversity of prey species in the Arctic, and the relatively simple structure of the tundra community, 'the range of opportunities permitting the variety of food specialization observed at temperate latitudes' is not present (Holmes and Pitelka, 1968). This may also prove to be true for arctic herbivores. Second, the great fluctuations in population density of prey species in the Arctic have led to the evolution of adaptations in both avian and mammalian predators which permit them to exploit a year of food abundance by producing many offspring, but conserve energy in a year of scarcity by not breeding, or breeding on a reduced scale commensurate with the food supply. The factors involved in these fluctuations in herbivore density will be examined later (p. 537).

Alpine trophic webs

It is time now to compare with the arctic tundra the trophic niches of alpine birds and mammals. The complexity of the alpine food web is, as in the Arctic, a function of the number of species present, and there is great geographic variation. The Beartooth Plateau in the central Rocky Mountains provides an example typical of many alpine ecosystems. In addition to insects and other arthropods, the herbivorous mammals include a pocket gopher, three terrestrial squirrels, four microtine rodents, the deer mouse, two lagomorphs, bighorn sheep, and mountain goat (the last introduced). Alpine birds, partly to mainly herbivorous, include the horned lark, rosy finch, white-crowned sparrow and Lincoln's sparrow (*Melospiza lincolnii*) (Pattie and Verbeek, 1966, 1967). Carnivores observed to utilize these prey populations include the ermine, long-tailed weasel, pine marten (*Martes americana*), badger, grizzly bear, coyote (*Canis latrans*), red fox, and bobcat (*Lynx rufus*). Predatory birds include the marsh hawk or hen harrier (*Circus cyaneus*), golden eagle (*Aquila chrysaetos*), three *Buteo* hawks, the red-tailed, Swainson's and ferruginous (*B. jamaicensis, B. swainsoni, B. regalis*), the prairie falcon (*Falco mexicanus*) and sparrow hawk or American kestrel (*F. sparverius*); insectivorous passerines were the water pipit, and rock wren (*Salpinctes obsoletus*). The species components of the trophic web on the Beartooth Plateau thus numbered twenty-two mammals and thirteen birds, or more mammalian but fewer avian species than in a complex arctic tundra web such as exists at Cape Thompson. The comparison is further complicated, however, by occasional sojourns on the Beartooth alpine of an additional group of herbivores: snowshoe hare (*Lepus americanus*), red squirrel (*Tamiasciurus hudsonicus*), bushy-tailed wood rat, long-tailed vole, western jumping mouse (*Zapus princeps*), porcupine (*Erethizon dorsatum*), wapiti, or American elk (*Cervus elaphus*), mule deer, and formerly, bison (*Bison bison*). Vagrant birds which regularly visit the alpine belt in late summer include the rough-legged hawk (down from the Arctic!), American robin (*Turdus migratorius*) (a few may nest occasionally in riparian willows or krummholz), rufous hummingbird (*Selasphorus rufus*), common raven, Clark's nutcracker (*Nucifraga columbiana*), mountain bluebird (*Sialia currucoides*), pine siskin (*Spinus pinus*), savannah sparrow, vesper sparrow (*Pooecetes gramineus*), and chipping sparrow (*Spizella passerina*). In addition, a few species of water birds appear on the alpine ponds and streams; some such as the dipper (*Cinclus*

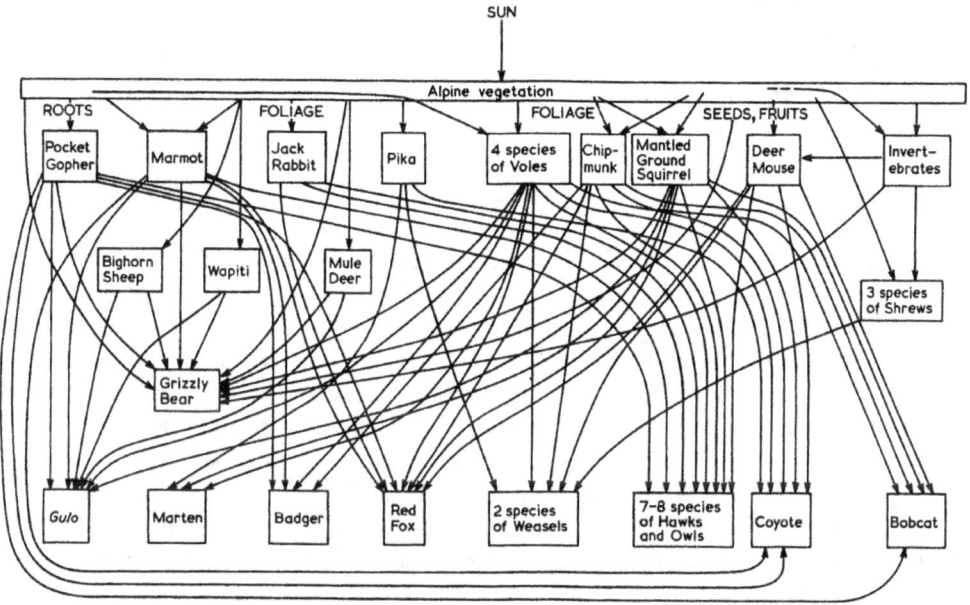

Fig. 9.12. Partial trophic web in the alpine tundra community of the Beartooth Plateau (after Pattie and Verbeek, 1966, 1967, and Taber *et al.*, n.d.). Insectivorous and herbivorous birds are not included.

mexicanus), spotted sandpiper (*Tringa macularia*), Baird's sandpiper, and common snipe (*Capella gallinago*) more or less regularly, and other ducks and shore birds quite erratically. If these late summer visitors and vagrants are considered, the combined number of birds and mammals in the trophic web of the Beartooth Plateau somewhat exceeds that of Cape Thompson or the Brooks Range foothills, but the general level of complexity must be considered as the same in these several ecosystems (fig. 9.12).

Other alpine ecosystems have greater, or less, faunal diversity. The least diverse are small alpine areas isolated by intervening valleys or plains from contact with other alpine mountains (Hoffmann and Jones, 1970). For example, a small area of true alpine tundra caps the Big Snowy Mountains, an isolated range rising from the northern Great Plains in central Montana (Bamberg and Major, 1968). Here only seven mammalian herbivores and four species of birds are known to be alpine residents or breeders, although as in the Beartooth Plateau there are a number of additional vagrants and late-summer visitors. Distance seems to be more important than size in determining faunal richness in these isolated alpine habitats; a much smaller area in the Little Belt Mountains, which is closer to other larger alpine ranges, has a larger number of alpine birds and mammals than the Big Snowies (Hoffmann, 1960; Johnson, 1966).

The largest alpine area in the world is located on the Tibetan Plateau and adjacent central Asian mountains (Meinertzhagen, 1928; Swan, 1963; Vaurie, 1972). This central Asian highland was probably the only alpine area in the northern hemisphere to survive the Pleistocene glacial and interglacial periods without being completely disrupted (Hoffmann and Taber, 1967) and thus has the greatest temporal as well as spatial continuity of any alpine area. Also, while a very large portion of the area is a dry alpine 'desert', in the Himalayan and western Chinese sectors habitat diversity is very great, and the alpine occupies a wider altitudinal belt than anywhere else on the globe (Swan, 1963).

Given these conditions, it is not surprising that the alpine birds and mammals reach a maximum number of species in this region (Hoffmann and Taber, 1967; Vaurie, 1972). Unfortunately, no quantitative studies of the ecology of the Himalayan–Tibetan alpine have been undertaken, and even qualitative descriptions in English are brief, though often well illustrated and illuminating (Bourliere, 1964; Pfeffer, 1968). Swan (1961) has published a schematic food web, in which some of the complexity of this remarkable alpine ecosystem can be perceived, but much is still hidden in links of the web labelled 'pikas' and 'mice', of which there are many species, and many other forms are not included at all. The study of these rich and diverse alpine ecosystems in the central Asian highlands presents a fascinating challenge for the future.

One of the noteworthy features of the trophic niches of arctic tundra vertebrates is their broad overlap in terms of prey species or kinds of foods eaten. Some preliminary data suggest that alpine mammals and birds may have less dietary overlap, and thus may not compete as intensely for food. On the Beartooth Plateau an intensive study of alpine small mammal populations was carried out by Pattie (1967a). The three most common small herbivores on the study quadrat were the deer mouse, montane vole, and water vole (see p. 532). Northern pocket gophers and montane heather voles (*Phenacomys intermedius*) were present in lower numbers. The main food habits of these five species are: gopher – rhizovore; deer mouse – spermivore and insectivore; montane and water voles – foliovore; heather vole – probably foliovore. Greater differentiation of food sources is evident in this crude classification than was possible for a comparable suite of five small rodent herbivores in the arctic tundra of the Brooks Range foothills (see above). Moreover, differences in the kinds of foliage eaten by the montane and water voles may exist; Pattie believed that the former fed on forbs while the latter utilized more grasses and sedges.

Segregation in food habits is also apparent in the four species of alpine birds foraging on this same study area (Verbeek, 1970). Water pipits, the most abundant, are entirely insectivorous, and much of their food in spring and early summer comes from numbed or dead insects on the surface of the extensive snowbanks remaining on the Beartooth

Plateau. Horned larks, too, feed on the snowbanks, but about a quarter of their food is plant material, obtained by foraging on the drier tundra. Rosy finches are even more herbivorous, though their young nestlings are fed insects (French, 1959a); in the spring they feed along the edge of the receding snowbanks, but not on them. White-crowned sparrows, also seed-eaters, avoid the open alpine tundra and snowbanks, and feed in the small willow patches (Verbeek, 1970). No detailed quantitative study of the food of this sympatric group of alpine birds was made, so seasonal shifts, such as occur in the arctic sandpipers, cannot be documented, except in the obvious case of insects of the snowfield surface, and plant food exposed along the margins. This important spring food source literally melted away, and was replaced by foods gleaned from or among the tundra vegetation. At the same time, there were no marked seasonal or yearly changes in occupancy of the alpine belt, as happens in the Arctic. All four species remained until the alpine autumn, and population densities appeared fairly stable from year to year (Verbeek, 1967, 1970; Johnson, 1965; French, 1959a, b).

Carnivorous mammals and raptorial birds occupy what at first seems to be a radically different place in the food web of the Beartooth Plateau compared with their counterparts in the arctic tundra, because of their ability to move down below timberline and then back again on a seasonal or even daily basis. Thus, arctic carnivores such as foxes and weasels are permanent tundra residents, whereas these forms in the alpine are probably summer residents only (Pattie and Verbeek, 1967). The raptorial birds, mostly marsh hawks and golden eagles, are also summer residents, as are skuas and short-eared owls in the Arctic, but the alpine raptors are daily visitors, soaring up above timberline on ascending air currents in the late summer and autumn to hunt the alpine herbivores – principally microtines and marmots, respectively – and returning to the subalpine at night.

Yet these seemingly different temporal patterns have the same adaptive basis: both arctic and alpine raptors are exploiting their mammalian foods at times when they are most abundant, in years of peak density in the Arctic, and the season of peak density in alpine areas. Similarly, the abundance of the principal alpine mammalian predator, the long-tailed weasel, is directly related to the overall abundance of small mammals (Pattie and Verbeek, 1967), just as are arctic weasels, even though the alpine weasels may return to the subalpine belt in winter.

A comparison of arctic and alpine trophic niches could stress the considerable niche overlap in arctic species, and the greater segregation of niches in alpine ecosystems, but this is a tentative conclusion, based on fragmentary data. If it is true, it must stem from the contrasting natures of the two sorts of ecosystems, the arctic tundra having simpler structure and lower taxonomic diversity. The greater complexity of the alpine can be attributed to complex topography, leading to more complex community structure and to an increase in taxonomic diversity due to the numerous species of lower altitudes which are sufficiently adaptable to be able to exploit, at least daily or seasonally, the resources of the alpine environment. Since degree of complexity of community structure is thus seen to be important, we will now examine structural components of the ecological niche.

This term is used to refer to those ecological niche dimensions that are formed from the 'microtopography' of a species' environment. Species must be adapted to a range of values for such things as vegetational density or percentage of coarse fragments in the soil, although the range within which a species can survive and reproduce is usually known only qualitatively, often anecdotally. R. S. Miller (1964) described the range of values for soil depth and texture as niche variables for several species of pocket gophers in Colorado, and Lack (1933) described a high perch for territorial singing as an essential factor for the tree pipit (*Anthus trivialis*); these are examples of structural niche dimensions. This concept is similar to, but broader than, structural habitat, first defined by Rand (1964) for *Anolis* lizards, as the place 'where an individual (and by abstraction, a species) lives', described in terms of 'the distributions of perch heights and diameters of undisturbed individuals under natural conditions' (Rand, 1964; Andrews, 1971). Rand contrasted structural habitat with climatic habitat, i.e. where a species lives described in terms of the range of temperatures and related variables such as altitudes and exposure to solar radiation. Those niche variables encountered within such a climatic habitat have already been discussed (pp. 484–500); structural and climatic niches together comprise the spatial or habitat niche (Odum, 1971).

Arctic structural niches

As remarked earlier, arctic tundra communities have relatively simple structure. Local environmental gradients of soil moisture, texture and slope, with accompanying cryopedologic and geomorphologic processes produce vegetational patterns, giving rise to what structural diversity exists along the relatively featureless arctic coastal plain (Johnson *et al.*, 1966; Brown *et al.*, 1970; Britton, 1957). On the coastal plain the tundra is flat, with poorly drained soils, and seasonally much surface water. Wet meadow and marshes are common, as are ice-wedge polygons, both low and high centre. These habitats are occupied by brown lemmings (Thompson, 1955; Pitelka, 1957b). Areas of microrelief, with better drained soils, are relatively scarce, consisting of an interrupted network of low ridges between polygons and marshes, as well as scattered isolated mounds. These habitats are occupied by collared lemmings, which in this lowland tundra are much less common than *Lemmus*. In such a mosaic of habitats, both species occur along the ridges and troughs of polygonal grounds, and considerable habitat overlap exists, especially when brown lemming populations are high, and they spread into upland habitats (Pitelka, 1957b).

Further inland, toward the Brooks Range foothills, the coastal plain relief gradually increases, and this increase in topographic complexity contributes to greater structural complexity of the tundra community. The first new species to appear in the community is the tundra vole, an inhabitant of marshes on river flood plains, stream and lake borders, where it overlaps broadly with brown lemmings (see above). It also occurs, perhaps less regularly and commonly, in both riparian willow and alder and in drier tussock heath and dry meadow communities which become more common toward the foothills (Pruitt, 1966; Britton, 1957). In the foothills and in the Brooks Range itself, structural complexity of tundra communities reaches a maximum. Marsh and wet meadow communities still occupy limited areas; tussock and moist meadow communities are present, but shrubby upland heath communities and dry meadows grading into even more xeric fellfield communities and rock-dominated communities such as talus, boulder field and cliffs, add to the habitat diversity. Dendritic patterns of riparian willow and alder intersect the greater variety of topography, soils, and vegetation on this inland tundra. The narrow-skulled and red-backed voles both occur in the foothills and mountains, the former occupying riparian areas with overhead cover of dwarf willow and/or alder shrubs on fairly dry, well drained soils, and the latter occupying dry upland areas with shrubby overhead cover of dwarf birch or various species of heath. There is considerable overlap between the two species' habitats, however. They occur together beneath the shelter of willow, alder, birch, and heath shrubs (*Cassiope, Empetrum, Ledum, Arctostaphylos, Vaccinium*), often on tussocky or rough ground and, if there are no shrubs, both may shelter in boulder fields and stabilized overgrown talus slopes. These two upland/riparian microtines also overlap to a lesser degree the previous three: with the tundra vole, and to a lesser extent with the brown lemming along streams and lake margins, and with the collared lemming in drier upland sites on valley slopes and canyon sides, and also on frost-disturbed sites. According to Pitelka (1957a) the collared lemming is found in a wider variety of habitats in the foothills than on the coastal plain. In any case, the greater variety of habitats in the foothill tundra must be causally related to their greater faunal richness, but the complex interdigitation of the habitats may also be responsible for the considerable overlap in structural niches observed between these five sympatric microtines.

As previously noted (p. 508) the number of bird species shows a similar pattern of increase from the lowland tundra of the coastal plain to the foothills tundra. Yet where topographic complexity exists along the coast, as around Cape Thompson, structural complexity of bird and mammal habitats supports faunas as rich as in the Brooks Range foothills. Seven species of shore birds commonly nest on the coastal tundra at Barrow, among them four congeners (Pitelka, 1959), two owls, two skuas, and only two passerine birds, the Lapland longspur and snow bunting. In terms of habitat selection the longspur and bunting correspond roughly to the brown and collared lemmings – moist versus dry tundra.

At Cape Thompson, there are thirteen shore birds, two ptarmigan, two owls, two skuas, the golden eagle, and perhaps twelve passerines (Williamson *et al.*, 1966); the increase is due to those species which utilize habitats not present at Barrow (table 9.3). Thus, golden eagles nest on the inland cliffs; knots, horned larks, wheatears (*Oenanthe oenanthe*) and water pipits on the fellfields and dry meadows, and willow ptarmigan (*L. lagopus*), semipalmated plover (*Charadrius semipalmatus*), wandering tattler (*Tringa incana*), yellow wagtail (*Moticilla flava*), arctic warbler (*Phylloscopus borealis*), red-throated pipit, redpoll, savannah and white-crowned sparrows in the riparian willows and stream-side gravel bars. These new birds are therefore occupying the same habitats that the narrow-skulled and red-backed voles, arctic ground squirrels, Brooks Range marmot, and Dall sheep (*Ovis dalli*) inhabit in the foothill and montane tundra; without this

Table 9.3. Breeding birds of the Cape Thompson area, Alaska, arranged according to breeding habitat preference (after Williamson *et al.*, 1966).

Species	Cliffs, talus	Fellfield, Dry meadow	Heath tussocks	Moist, wet meadow marsh	Riparian
Golden eagle	1				
Willow ptarmigan			2	3	1
Rock ptarmigan			1	2	3
Semipalmated plover				2	1
Golden plover		1	2	2	3
Common snipe				1	2
Wandering tattler					1
Knot		1			
Pectoral sandpiper			2	1	
Bairds sandpiper			3	2	1
Red-backed sandpiper			4	1	2
Semipalmated sandpiper				2	1
Western sandpiper			2	1	3
Long-billed dowitcher				1	
Bar-tailed godwit			1	1	
Northern phalarope				4	
Parasitic jaeger		4	2	1	4
Long-tailed skua		2	1	1	2
Snowy owl		4	1	1	3
Short-eared owl		3	1	1	3
Says phoebe					
Horned lark					
Wheatear		1			2
Arctic warbler					
Yellow wagtail			3		1
Water pipit		1			2
Red-throated pipit				1	1
Common redpoll			2		1
Savannah sparrow			3	2	1
White-crowned sparrow					1
Lapland longspur		4	1	2	3
Snow bunting	1	2			4

1,2,3,4 = Primary, secondary, tertiary and quaternary affinities for these breeding habitats.

increase in structural complexity of the tundra communities, there would not be sufficient 'niche space' to accommodate the larger number of species. Table 9.3 also suggests a considerable degree of overlap in structural niches of many of the species, as was true of the mammals. A degree of niche segregation is indicated, however, by the fact that many of the more closely related species pairs show main affinities for different habitats: for example, the rock ptarmigan (*L. mutus*) and willow ptarmigan, water and red-throated pipits, and Lapland longspur and snow bunting. Competitive interaction may be important in maintaining this niche segregation in areas of sympatry (Weeden, 1965a; Voous, 1960; Drury, 1961) (see below).

Alpine structural niches

Alpine areas present a kind of mirror image to the Arctic. The simplest communities, structurally, are those on steep, well drained mountain tops without surface water. Structural complexity increases with areas of gentler topography, impeded drainage, marshes, ponds and streams. Since there is so much heterogeneity of slope, exposure, rock and soil type among alpine areas (Johnson and Billings, 1962), no major geographic gradients in community complexity exist. However, as in the Arctic, local gradients in those factors, acting directly as well as by controlling snow cover, wind, temperature, and soil moisture, produce a classifiable mosaic of alpine community types (Bamberg, 1961). The following discussion of the flora and fauna of these communities is based on Bamberg and on Taber *et al.*, n.d.).

The simplest communities in which a mature soil and extensive vegetation are dominant features of the habitat are alpine fellfields. Rock-dominated communities, also of simple structure, will be considered separately. Fellfields develop on exposed ridges and summits of mountains, and other local areas which, due to their topographic position, such as saddles through which wind is funnelled, have strong wind and high insolation. This is the driest of the alpine tundra communities because snow is largely blown away in winter, and in summer there is a high evaporation rate; the soil is relatively shallow and there are many scattered surface rocks (Nimlos and McConnell, 1965; Nimlos *et al.*, 1965). On acid soils many of the dominant species are cushion-plants with long taproots and a characteristically horizontal rather than vertical growth. On neutral and basic soils *Dryas* is dominant and the vegetated areas form well-defined patterns, either circular mats on relatively level areas, or terrace faces on steeper slopes. Fellfield plants grow, flower, and set seed quickly on the initial moisture in the soil from spring precipitation. By midsummer they are drying up.

The vertebrate animals of this community are few: the dominant mammal is the deer mouse and the dominant bird is the horned lark. Northern pocket gophers occasionally invade this ecosystem in summer, but apparently cannot winter there because there is not enough snow cover to provide insulation. The only small mammal common in the fellfield is the deer mouse; it is, however, widespread in the alpine and in non-alpine habitats as well. Presumably its adaptability enables it to exist in a habitat where food and cover are both scarce. The deer mouse inhabits fellfield by virtue of its ability to shelter among small rock fragments as well as larger ones; its utilization of burrow systems of other mammals, especially pocket gophers; and its ability to hide within whatever vegetative cover is available. The abundance of deer mice at any given station seems to be directly related to the amount of these kinds of cover present. The horned lark (*Eremophila alpestris*) has neither as great a local population density nor is as widely distributed geographically as the water pipit. It has been found in less than half of the alpine areas studied in the central Rocky Mountains (Taber and Hoffmann, 1961). Horned larks in the alpine are pre-eminently birds of the fellfield and driest, or most disturbed, portions of the dry meadows, but occur in more mesic communities as well.

Dry meadows adjoin the fellfield. They are subjected to moderate rather than severe winds. Thus some but not all winter snow is blown off, and evaporation is moderately but not extremely high. The soil is well drained; there may be midsummer plant dormancy due to drought followed by second flowering in some species in response to summer thunder showers. This community has three aspects: on acid soils, dry sedge (*Carex*) meadows and sedge-mountain avens (*Geum*) meadows occur; on neutral and basic soils, *Dryas*-sedge meadows are characteristic.

The dominant mammal of this community is the pocket gopher, both in point of abundance and of effects upon the vegetation. There is sufficient snow cover to insulate these animals in the winter, during which they remain active, and yet the ground is never saturated for long. When the turf has been broken over gopher feeding-burrows, wind and needle ice deflate the scarp and cause a horseshoe-shaped break which moves uphill. Once the plant cover is removed and mineral soil exposed, fellfield plants invade. Succession is presumed to be very slow because of the long time necessary for the build-up of the humus layer necessary to maintain the dry alpine meadow (Osburn, 1958). Deer mice are moderately abundant in the dry meadows, where pocket gopher burrows are probably important as shelter. A few montane voles winter in the denser dry meadow stands, but none were ever trapped here in summer. The horned lark and the water pipit both occur in this community, but the centre of abundance of the former is the fellfield and of the latter in the moist alpine meadows below.

Moist alpine meadows are found on fair to poorly drained sites which are subjected to only gentle winds. There are a few surface rocks. Soil moisture for plant growth is available throughout the growing season in wet years. This community has two aspects: on acid soils, hairgrass (*Deschampsia*) meadows; and on neutral to basic soils, sedge meadows.

The dominant mammal of this community is the montane vole, whose runways, shallow burrows and winter nests are usually abundant. In some localities the long-tailed vole is found instead. Like the gopher, the montane vole extends into the subalpine and below, where suitable habitats occur. However, its grassland habitat is usually scarce and discontinuous in the subalpine and montane forest belts, and this results in the partial isolation of alpine populations from those of grassland at lower elevations. Pocket gophers are abundant in the well drained aspects of moist meadows but they cannot winter where the soil is saturated. Deer mice are found less commonly than they are in dry meadows; presumably they are heavily dependent on vole and gopher burrows for shelter. The montane heather vole and the dusky shrew are fairly common, and the water and red-backed voles occasional. Water pipits are the dominant birds of moist alpine meadows. Horned larks also occur, and rosy finches forage here early in the breeding season.

Sedge-moss tussock communities are found on flat or gently sloping depressions with impeded drainage, where there is a rather light winter snow cover. Soil moisture is abundant at all times; the gutters or depressions between the tussocks are usually wet.

The dominant mammal is the montane vole, on occasion replaced by the long-tailed vole. The water vole is also found in some years; it spreads out in high years from its centre of abundance in the willow-sedge community below. There is evidence that the water vole and the smaller montane vole do not exhibit competitive exclusion as do the long-tailed and montane voles (Pattie, 1967a). The dusky shrew is occasional; it apparently requires more overhead cover than it finds in this community to be really abundant. Deer mice invade from drier areas in years of high population, but generally speaking it is too wet for them. As in moist meadows, the water pipit is the dominant bird; it is the only one known to nest. Horned larks and white-crowned sparrows occasionally forage in the tussocks, as do rosy finches in the early spring.

Willow-sedge communities were found on sites abundantly watered by melting snowbanks above, generally in valley bottoms which received at least a moderate winter snow cover. The principal features of this community are the presence of permanent water, wet soils and much overhead vegetation. The willows commonly grow up to 0.5–1 m in height and occasionally more, thus forming an important habitat element for several animals.

The dominant mammal of this community is the water vole, whose burrows and runways are made on the banks of the hummocks and along streams. Another abundant mammal is the dusky shrew. Present, though less abundant, are the water shrew (*Sorex palustris*) and the montane vole. The shrubby willows provide habitat for the white-crowned sparrow. The western robin is also common, although less abundant, and Lincoln's sparrow is sometimes present. Water pipits regularly forage among the willow-sedge hummocks and along the stream courses, and also nest.

Rock polygon-stone stripe communities were found on ridges, summits or slopes which were apparently not glaciated, but which were subject to a strong deep freeze-thaw cycle, with attendant sorting, in glacial times. The soil is now well drained in most cases, ordinarily with dry meadow vegetation. This may be modified by the cutting and grazing of pikas (*Ochotona princeps*). The characteristic which distinguishes this from all other communities is the close interspersion of stripes or polygons of vegetated soil and stripes or nets of stone blocks. The rock provides shelter, and the vegetation food, for the characteristic animals.

The dominant mammal of this community is the pika. The deer mouse reaches its greatest abundance here presumably because of the quality of the shelter. The dwarf shrew is found characteristically in the rock gutters. The montane heather vole is not uncommon, while the red-backed vole is occasional. Among birds both the horned lark and the water pipit are characteristic, while the rock wren is also often found.

Talus slopes and cliffs are even more dominated by rock than polygon-stone stripe communities. Adjacent vegetation varies, but is most often fellfield or dry alpine meadow stand-types. There is also a characteristic cliff vegetation. It appears that plants with fibrous root systems are more common in rock outcrop and cliff habitats, while tap-rooted species are more common in talus and rock rubble. Lichen stands are characteristic of all stable rock except that subjected to prolonged snow cover.

Dominant mammals characteristic of these communities are marmots, mountain goats and chipmunks. Golden-mantled squirrels, pikas and deer mice are locally abundant. The two characteristic cliff-dwelling birds are the rosy finch and the rock wren; ravens, golden eagles, and other raptors may also nest on alpine cliffs, but are more likely to be found below timberline.

These last two community types are, to a greater or lesser extent, structurally dominated by rock forms rather than vegetation, and are thus characteristic of the alpine, as
contrasted with arctic tundra. It is worth re-emphasizing that rock is an extremely
important source of shelter in the alpine (except in the krummholz; see below) where
there is no woody litter, where the winter snow cover is irregularly distributed, and where
the ground cover is usually low and often thin as well.

The size of the rock fragments can be related to the occurrence of mammal species.
Marmots are found only where large rocks, usually above 1 m in diameter, occur
together in a well drained location. Because of this need for rock of large size, marmots
are not found in alpine areas where rock is soft and shatters into small, often flat fragments
(plate-rock). The golden-mantled squirrel is a burrowing animal but one which requires
rock beneath which to dig. Like the marmot, it requires rather large rock, but unlike the
marmot, it does not require that this rock be in extensive aggregations. The pika shelters
in assemblages of rocks of medium size, perhaps 15–30 cm in diameter. Rocks of such
size regularly occur in alpine terrain. Polygon-gutters and rock stripes provide pika
habitat, as do talus slopes. The dwarf shrew shares pika habitat, and is found among
blocks of medium size even in alpine areas where the pika is absent. Since this animal
occupies a narrow habitat – beneath the shelter of assemblages of medium-sized rock –
it may easily be overlooked.

The rosy finch is one of the two species of alpine birds specifically adapted to, and
restricted to, alpine habitats for breeding (the white-tailed ptarmigan is the other).
Breeding rosy finches resort to less accessible cliff areas to nest (French, 1959a; Johnson,
1965). Although the rock wren is also present in most alpine areas, it was common in only
a few localities, but it complemented the much more abundant rosy finches in their
occupation of rocky cliff areas.

Krummholz communities include all conifer-dominated associations at the alpine
forest–tundra ecotone, whether the trees are prostrate or upright. They differ from all
other alpine habitats in the presence of fairly tall coniferous vegetation, and in large
amounts of woody surface litter. Here, small conifers and associated shrubs may provide
the necessary overhead protection for small mammals.

Deer mice are common, and maximum population densities of chipmunks in the
alpine belt appear to centre on the krummholz, although they are also found in several
other communities, particularly cliff and talus, and polygon and stripe areas. Red-
backed and montane heather voles are both inhabitants of the montane and subalpine
forest floor at lower elevations. The red-backed vole is often considered a common coniferous forest species, whereas the montane heather vole has long been considered a rare
species, or at least, one difficult to trap. In the alpine belt, both were trapped regularly
in about equal numbers; the centre of abundance for both species was the krummholz,
indicating that the principal alpine populations are directly related to the forest populations of lower altitudes. Dusky shrew was also found in small but significant numbers
in the krummholz, though not as regularly as in willow-sedge hummocks. Snowshoe
hares and mule deer are occasional visitors. The dominant breeding bird in the krummholz is the white-crowned sparrow. Pine siskins, mountain bluebirds, Clark's nutcrackers,

western robins, fox sparrows (*Passerella iliaca*), Oregon juncos (*Junco oregonus*) and blue grouse (*Dendragapus obscurus*) are found in the krummholz to some extent although their main populations are in the subalpine.

Table 9.4 summarizes the occurrence of alpine birds and mammals with respect to communities classified on the basis of structural features. Some birds and mammals are too wide-ranging to be treated conveniently in this manner; their structural niches are

Table 9.4. Alpine birds and mammals of the Beartooth Plateau, arranged according to habitat preferences. Numbers as in table 9.3.

Mammals	Krummholz	Cliff, talus	Rock polygons	Fellfield	Dry meadow	Moist meadow	Sedge tussock	Willow-sedge
S. obscurus	2		4		4	3	3	1
S. nanus		2	1	2				
O. princeps		1	2					
L. townsendii				2	1	2		
S. lateralis		1	2					
M. flaviventris		1	2		3			
E. minimus	1	2	3					
T. talpoides	3				1	2		
P. maniculatus	1	1	1	2	2	3	4	
P. intermedius	1	3	3	3	3	2		
C. gapperi	1	2	3					
A. richardsonii						3	2	1
M. montanus					4	1	3	2

Birds

	Krummholz	Cliff, talus	Rock polygons	Fellfield	Dry meadow	Moist meadow	Sedge tussock	Willow-sedge
E. alpestris				1	2	3		
S. obsoletus		1	2					
A. spinoletta					3	2	1	2
L. arctoa		1	2					
M. lincolni								1
Z. leucophrys	1							2

too large. The white-tailed ptarmigan is a good example; in the central Rocky Mountains it is restricted to, but ranges widely throughout the alpine, from cliff, talus and fellfield communities to sedge-moss tussocks, and even willow hummocks, if the shrubs are not above about 24–45 cm tall. However, in this diversity of communities, rock is always present (Choate, 1963b). Avian and mammalian predators also range widely across the alpine belt, but tend to concentrate their activities in those communities where prey species of appropriate size are most common. Thus, long-tailed weasels and ermines are most frequently seen in moist meadows, tussocks, and willow hummock communities, where microtine rodents were most abundant, while martens and badgers are seen in rock-dominated communities inhabited by marmots and pikas (Pattie, 1967b; Verbeek, 1965).

Patterns of structural complexity

The alpine areas with greatest structural complexity are those in which all of the described community types occur, such as on the Beartooth Plateau. Alpine areas with less topographic diversity have fewer community types, thus less structural complexity. Minimum complexity is found in small alpine areas sloping uniformly upward above timberline to the mountain peak; the alpine may include only krummholz, dry meadow or fellfield, and perhaps cliff and/or talus. In such a simple ecosystem in the central Rocky Mountains, deer mice and water pipits may be the only terrestrial vertebrates living above timberline.

In 'typical' alpine or arctic tundra ecosystems, in the central Rocky Mountains or Alaska's arctic slope, maximal structural niche complexity and species richness is attained in areas of maximal topographic diversity – for example, the Beartooth Plateau and the Brooks Range foothills respectively. In the northern Rocky Mountains, where the alpine belt is found at progressively lower altitudes, it approaches arctic tundra and an ecotone between the two exists. Another factor is operating to enrich bird and mammal faunas in this ecotone; the mixing of 'typical' alpine and arctic species. For example, in the northern Rockies at Chilkat Pass in extreme northwestern British Columbia, about 60 per cent of the birds breeding in timberline communities and above had their main zoogeographic affinity with the Arctic, while about 40 per cent were Rocky Mountain alpine breeders (Weeden, 1960). For example, of the three species of ptarmigan at Chilkat Pass, the willow and rock are arctic birds, while the white-tailed ptarmigan is an alpine species, as has already been noted. Similarly, both the arctic snow bunting and the alpine rosy finch breed among the cliffs and rocky ledges of the area. Mammals also exhibit this phenomenon of ecotonal mixing; in the vicinity of Chilkat Pass, brown lemmings and tundra voles occur sympatrically with long-tailed voles and deer mice. When two species with similar structural and/or trophic niches occur together, ecological theory predicts that competitive interactions will occur. Where the range of the white-tailed ptarmigan overlaps that of the rock and willow ptarmigans, then we may expect some manifestations of this competition for niche space to be apparent; this applies to 'pure' alpine and arctic tundra as well as to their ecotonal communities, and will be considered in the next section.

NICHE SEGREGATION AND INTERSPECIFIC COMPETITION

We have already seen that there is considerable overlap in both trophic and structural components of the ecological niches of arctic birds and mammals, particularly when community structure is simple, but that alpine species may show less niche overlap. Further insight may be gained by studying the niche relationships of a species at several points in its range, where it inhabits communities of differing complexity.

The white-tailed ptarmigan is the only tundra grouse in the central and southern Rocky Mountains. In the northern Rockies, by contrast, it is widely sympatric with the willow and rock ptarmigan (Aldrich, 1963). Where they are the only ptarmigan in the alpine, white-tails occupy nearly the full range of alpine communities present, and feed on a wide variety of plant foods, depending heavily on willow particularly in winter (Choate, 1963a, b; Weeden, 1967). Where the three species are sympatric, white-tailed ptarmigan are restricted in summer to the highest elevations, to cliffs, talus, boulder and

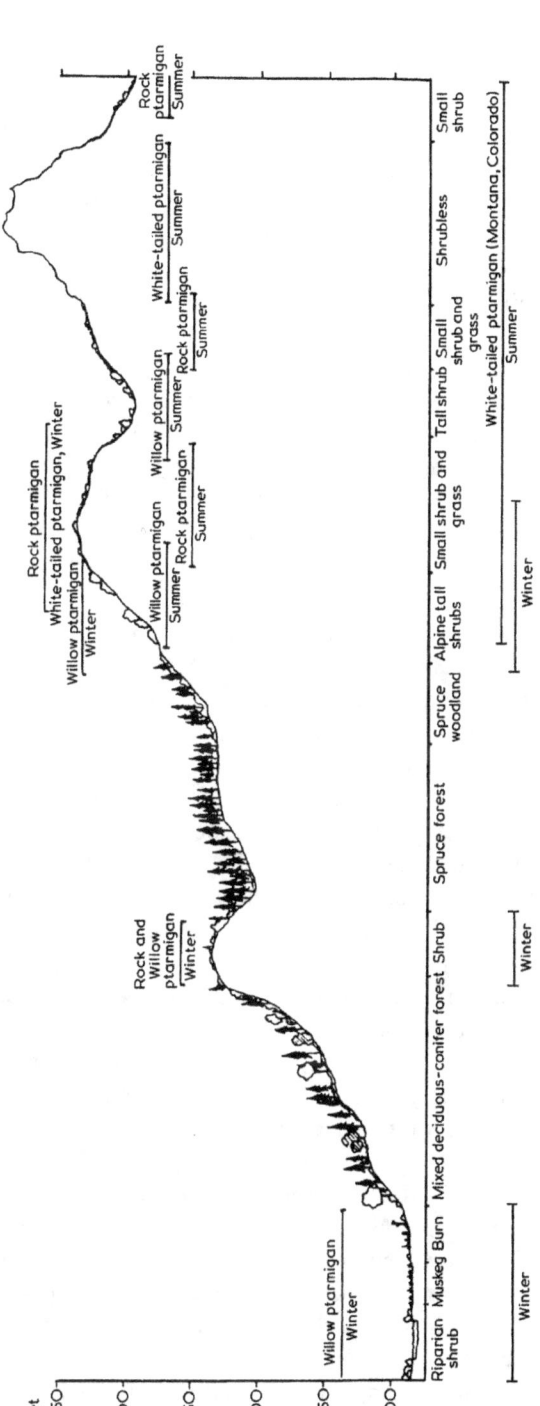

Fig. 9.13. Habitat preferences and structural niches of three sympatric, congeneric species of ptarmigan in central Alaska, showing habitat segregation and overlap, and compared to the much broader structural niche of the white-tailed ptarmigan in the central and southern Rocky Mountains, where other ptarmigan are absent (after Choate, 1963a; Weeden, 1965a; and Braun and Rogers, 1971).

fellfields. The lower alpine valley bottoms, wet meadows and tussocks, and riparian shrubs are occupied instead by willow ptarmigan, and on the middle slopes of the alpine belt, in upland heath tussocks, xeric dwarf shrubs and dry meadows the rock ptarmigan occurs (fig. 9.13) (Weeden, 1965a). The fundamental structural niche of the white-tailed ptarmigan thus is larger (embracing more kinds of alpine communities) than the niche it is able to realize in the presence of two other species of ptarmigan, each of which seems to pre-empt a portion of the white-tails' niche. This distinction between fundamental and realized niches (Hutchinson, 1957, 1965) is basic to discussions of niche segregation and competitive exclusion.

Since the three sympatric ptarmigan are segregated by habitat and structural niche during the summer, there can be no major overlap in their realized food niches during this period either, and they may eat the same kinds of food plants without coming into competition, since they are in different places. However, in the winter all three species may concentrate in upland or riparian communities, where willow, birch and alder shrubs are tall enough to project above the snow surface. In the central and southern Rockies the white-tailed ptarmigan, not having to share this food resource, browse mostly on willows (Choate, 1963a, b; May and Braun, 1972). In areas of sympatry, the willow ptarmigan use this food most heavily, followed by birch, and rarely alder. Rock ptarmigan favour birch, followed by willow, and rarely alder. White-tailed ptarmigan do eat some willow (their main food in the south) as well as birch, but feed most heavily on alder, the food plant rarely used by the other two species (Weeden, 1967). Thus, where structural niche segregation is not operative, trophic niche segregation is.

Examples of this kind could be multiplied. Alpine water pipits inhabit a variety of communities, being most common in moist to wet meadows and riparian shrubs. In the Arctic, in places where the red-throated pipit occupies the wetter habitats, water pipits are excluded and occur principally on fellfields and dry meadows (Williamson *et al.*, 1966; Voous, 1960), but where red-throated pipits are absent, the water pipits' realized niche is again larger (Drury, 1961).

Strictly in alpine environments, competitive exclusion takes somewhat different forms, an example of which will now be considered.

Among certain species of chipmunks of the genus *Eutamias* (= *Tamias*) there is widespread sympatry. About sixteen species of chipmunks are currently recognized as occurring throughout western North America, and one species (*E. sibericus*) is prevalent in Eurasia.

Three species of chipmunks – the least chipmunk (*E. minimus*), red-tailed chipmunk (*E. ruficaudus*), and yellow-pine chipmunk (*E. amoenus*) – are present in the alpine belt in Montana. Thirteen alpine sites are occupied by only one of these three species of chipmunk, although the other species may occur below timberline (table 9.5). Pronounced habitat segregation thus tends to minimize contacts between the different species, even though extensive geographic sympatry occurs.

Here is another case of apparent competitive exclusion of one species from a limited niche by a better adapted congener. Wherever the least chipmunk occurs, it occupies the alpine belt, even though another species may be present in the adjacent subalpine.

For example, in Glacier Park the least chipmunk inhabits the alpine down to the krumm-holz. Just below, the red-tailed chipmunk is found in the subalpine spruce-fir forest, and the yellow-pine chipmunk appears in the Douglas-fir montane forest of the lower elevations. Least chipmunk populations reappear on the plains east of the mountains but are separated from their alpine counterparts by populations of both red-tailed and yellow-pine chipmunks. Evidently it is not simple proximity to the alpine belt that allows a chipmunk species to colonize, since if this were the only factor the red-tailed chipmunk would be the most likely alpine species; the observed distribution suggests that the least chipmunk is best adapted to alpine conditions. Similar patterns of competitive exclusion in chipmunks, both in the alpine and below, have recently been reported by Brown (1971a), Heller (1971), and Sheppard (1971).

Table 9.5. Species of chipmunks in the mountains of Montana.

Area	Alpine species	Species in adjacent areas
Glacier Park	E. minimus	E. amoenus, E. ruficaudus
Beartooth Plateau	E. minimus	E. amoenus, E. umbrinus
Big Snowy Mts	E. minimus	E. amoenus
Flint Creek Mts	E. ruficaudus	E. amoenus
Pioneer Mts	E. ruficaudus	E. amoenus
Highland Mts	E. ruficaudus	E. amoenus
Little Belt Mts	E. amoenus	none
Big Belt Mts	E. amoenus	none
Crazy Mts	E. amoenus	none
Tobacco Root Mts	E. amoenus	none
Bridger Mts	E. amoenus	none
Gallatin Mts	E. amoenus	none

The least chipmunk's superior adaptability is probably due to the physiographic similarity of open alpine landscape with scattered krummholz to open prairie landscape with scattered shrub thickets. None of the forest habitats bear as close a resemblance to the alpine as does the grassland. The least chipmunk's principal niche adaptation includes open shrubby habitat, whereas both the red-tailed and yellow-pine chipmunks are principally adapted to a niche that includes forest. Therefore, in the alpine environment the least chipmunk may have a competitive advantage over its congeners.

If the least chipmunk is absent, the red-tailed chipmunk occupies the alpine belt, and if the red-tailed also is absent, then we find the yellow-pine chipmunk. There is, then, a well-defined hierarchy of occupancy of the 'chipmunk niche' in the alpine areas of Montana.

In contrast to the sympatric species of chipmunks, the ranges of the several species of Holarctic marmots rarely overlap; geographic allopatry is almost complete (fig. 9.14). In North America the two most widespread mountain marmots are the yellow-bellied marmot and hoary marmot. The yellow-bellied marmot inhabits a wide variety of climates and habitats, ranging from the hot, arid foothills of the Great Basin ranges (Hall, 1946) to the cold alpine belt of the Beartooth Plateau. In Montana, its range extends from bunchgrass-prairie habitats on the valley floors right up through the various zonal plant communities to the highest peaks (Hoffmann and Pattie, 1968). In contrast, the hoary marmot is found only in alpine or high subalpine habitats, and in Montana it is largely restricted to the northwestern mountain ranges, where the yellow-bellied marmot is absent.

The two species are geographically sympatric, however, in the Bitterroot and Ana-conda-Pintlar mountains of Montana. In these areas the yellow-bellied marmot does

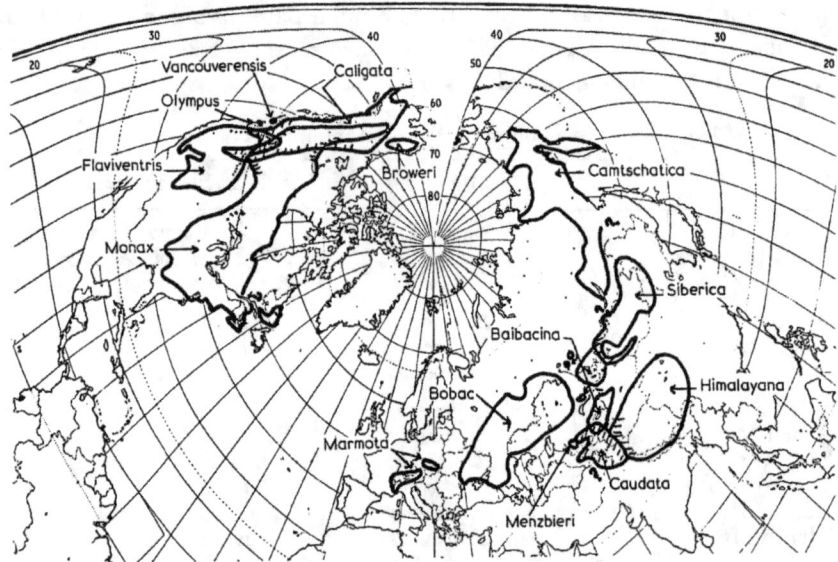

Fig. 9.14. The geographic distribution of fourteen named species of the genus *Marmota*. The species are mostly allopatric, but in areas of geographic sympatry (as between *M. caligata* and *M. monax* or *flaviventris*, or between *M. caudata* and *M. baibacina*) the species are segregated by habitat (based on various sources).

not occur at high elevations, in habitat occupied by *M. caligata*, although it occurs in the alpine habitat throughout the rest of its range, where the hoary marmot is absent. If this is so, then there must be only one 'marmot niche' in the alpine zone. Carruthers (1949) described a similar relationship between the red and grey marmots (*M. caudata, M. baibacina*) in the Pamirs; Jerdon (1874) and Blanford (1888, 1891) describe habitat segregation between *M. caudata* and the Himalayan marmot (*M. himalayana*) in the western Himalaya.

Not all cases of niche segregation are as clear-cut as those discussed above. Montane and water voles both inhabit riparian willows, sedge tussocks, moist meadows, and occasionally dry meadows in the alpine belt. The water vole, judging by its small geographic and altitudinal range, as well as by its restriction to a few wet alpine and high subalpine habitats, has a small niche, although the Palaearctic water vole (*Arvicola terrestris*) is found in many different habitats over a broad range of altitude and area, including alpine and arctic tundra (Ognev, 1950). Montane voles have a much wider altitudinal and geographic distribution, and live in both wet and dry habitats; these facts suggest they have a large structural niche. Findley (1954) regarded the montane vole as well adapted to existence in dry grasslands, but capable of living in wet areas in the absence of the closely related meadow vole (*Microtus pennsylvanicus*). Competitive exclusion between the two *Microtus* seems confirmed (Koplin and Hoffmann, 1968; Murie, 1969, 1971) although the outcome varies regionally (Grant, 1972; Stoecker, 1972). However, not only do montane voles inhabit wet alpine meadows (meadow voles being excluded because of altitude: Negus and Findley, 1959); they are more abundant there and in riparian willows than in dry meadows. It seems that dry intermontane grasslands and dry alpine meadows do not provide interchangeable structural niche parameters for montane voles.

Since both montane and water voles are concentrated in wet habitats, the index of habitat overlap between them (73/200) is twice that of the overlap index between meadow and montane voles at low altitudes (35/200), where strong interspecific competitive interactions have been demonstrated (Koplin and Hoffmann, 1968; Murie, 1969, 1971). This relatively high overlap value may be related to differences in food habits imposed by differences in body size; meadow and montane voles are about the same size, whereas water voles attain more than twice the size of *M. montanus*. As is true for sympatric birds, such as the tundra breeding sandpipers of the genus *Calidris* (Holmes and Pitelka, 1968), or the long-tailed, short-tailed, and least weasels (Rosenzweig, 1966), such size differences might permit montane and water voles 'to partition the available food resources by the selection of different-sized packets of food in a common habitat . . .' (Koplin and Hoffmann, 1968). In addition, habitat overlap between the two species is far from complete; moist meadows support densities of montane voles as large as tussock or willow communities, but very few water voles, while water voles are three times as abundant as montane voles in willow hummocks (see table 9.4). A combination of trophic and structural niche differences probably contributes to overall niche segregation and reduction of competition between these two species.

The model of partial habitat overlap provided by the two species of alpine voles is likely to be applicable to arctic tundra pairs such as the brown lemming and tundra vole. Additional quantitative studies of niche relationships are clearly needed in both arctic and alpine regions.

4. Population characteristics of arctic and alpine terrestrial vertebrates

POPULATION DENSITY

The adaptations discussed in section 2 of this chapter permit individual birds and mammals to survive the severe conditions of the physical environment encountered in the alpine and arctic tundra, and adaptations to specific kinds of foods and habitats (see section 3) permit individuals to obtain an amount of energy sufficient to sustain life and reproduction. Aggregations of such individuals constitute the species populations inhabiting the alpine and arctic ecosystems, and these populations possess characteristics, or 'group attributes' (Odum, 1971), which may be described and compared between species or regions.

Among these characteristics is density or standing crop; the numbers or biomass of individuals per unit of area occupied by the species. Densities of arctic and alpine vertebrates have not often been estimated. It is clear that while at some places and times certain arctic species such as caribou or brown lemmings may be very abundant, the average standing crop in the tundra is low (Haviland, 1926; Dunbar, 1968). Moreover, the peak densities of small mammals such as lemmings are frequently exaggerated: 'swarms' and 'hordes' of lemmings (*cf.* Elton, 1942), when actually sampled quantitatively, yield more conservative estimates. Thus, peak densities of brown lemming (*Lemmus*) populations are on the order of 50–300/ha (Krebs, 1964; Thompson, 1955; Bee and Hall, 1956; Rausch, 1950; Watson, 1956), while collared lemming peaks are lower, ranging up to about 40/ha (Krebs, 1964; Shelford, 1943). The highest estimate to be based (apparently) on quantitative sampling is that of Curry–Lindahl (1962) for arctic-alpine *Lemmus* populations in Sweden – 750/ha.

Quantitative estimates of densities of other species of arctic small mammals have not, to my knowledge, been made. Tundra studies indicate great geographic diversity of species composition (see section 3); for example, on the coastal plain of arctic Alaska the two lemming species alone dominate the total small mammal population (Pitelka, 1957a, b) whereas in the Plateau Province and Brooks Range foothills farther inland, and in the Cape Thompson region of northwestern Alaska, red-backed, tundra, and singing voles are co-dominants, and are more abundant than either species of lemming with which they share the tundra habitat (Bee and Hall, 1956; Pruitt, 1966). It is probable that the maximum combined densities of the small mammal populations are no more than twice that of the values usually attained by lemmings; i.e. of the order of 500 individuals/ha – this is probably a generous estimate. In addition, density of arctic ground squirrels at Cape Thompson can be estimated at about 7 individuals/ha, based on Carl's (1971) work.

The other arctic tundra mammal which locally attains high densities is the caribou, or reindeer (*Rangifer tarandus*). Recent aerial censuses in central Canada revealed densities ranging up to 0.24 caribou/ha, and averaging 0.07/ha (Thomas, 1969). Daily and seasonal changes in caribou density are, of course, of great magnitude owing to the migratory patterns of the species.

Density estimates of alpine small mammal populations are even scarcer than for the arctic tundra. One such study was conducted on the Beartooth Plateau, in the central Rocky Mountains (Montana–Wyoming boundary). Maximum densities on a study quadrat in moist to wet meadow revealed that the three most common small rodents were the montane vole, water vole, and deer mouse, estimated to be 8.5/ha, 2.5/ha, and 17/ha respectively, with average densities considerably less. The combined densities of these three species never exceeded 25 individuals/ha (Pattie, 1967a). Other species occurring in other alpine habitats were estimated (table 9.6) to have similar densities (Taber *et al.*, n.d.). It seems fair to generalize that small mammal populations in the alpine rarely, if ever, attain densities comparable to peak densities found in their arctic counter-parts.

This also seems true of large mammals of the alpine zone. Crude estimates place the density of bighorn sheep (*Ovis canadensis*) at 0.02/ha; snow sheep (*Ovis nivicola*) and mountain goat (*Oreamnos americanus*) at 0.01–0.02/ha (Geist, 1971; Holroyd, 1967; Chernyavskii, 1967). The highest densities reported in alpine ungulates are for Siberian ibex (*Capra siberica*) and tur (*Capra caucasica*): about 0.10–0.15/ha (Chernyavskii, 1967).

Table 9.6. Estimates for population density of small mammals inhabiting the alpine zone of the Beartooth Plateau (after Pattie, 1967, and Taber *et al.*, n.d.).

Species	No./Hectare
Sorex vagrans	5
Sorex nanus	5
Ochotona princeps	25
Lepus townsendii	0.25
Eutamias minimus	2
Spermophilus lateralis	0.5
Marmota flaviventris	5
Thomomys talpoides	25
Peromyscus maniculatus	17
Clethrionomys gapperi	5
Phenacomys intermedius	10
Microtus montanus	8.5
Arvicola richardsonii	2.5

Estimates of breeding bird population densities in the Arctic are nearly as scarce as those of small mammals. The most complete censuses have been made in the Cape Thompson region, and show total avifaunal densities ranging from about 1–10 birds/ha depending upon habitat (Williamson *et al.*, 1966). The lowest densities were in *Dryas* fellfield (<1/ha) and marshy sedge meadow tundra (1.2/ha) and the highest in riparian willows, with census plots in other habitats having intermediate values (table 9.7). The values tabulated are influenced by the area occupied by each vegetation type. For example, riparian willows, though they supported a relatively high bird population density, occupied only a small fraction of the total tundra, and so disproportionality weighted the estimates. Conversely, *Dryas* fellfield and wet sedge meadow, with relatively few birds, together comprised 45 per cent of the tundra area (A. W. Johnson *et al.*, 1966). The densities at Cape Thompson are comparable to those reported by Udvardy (1957) and Baldwin (1955). The most abundant small bird in all habitats except the riparian willows was the Lapland longspur; in the willows, redpolls, savannah and white-crowned sparrows were more common. In many other parts of the Arctic, the Lapland longspur remains the most common bird, but in some places the snow bunting is reported to exceed it in numbers (Drury, 1961). Willow ptarmigan attained higher densities than did rock ptarmigan at Cape Thompson (0.3/ha versus 0.15/ha). Weeden's (1965a, b) data for other areas of Alaska and British Columbia showed the same thing, and the same general order of magnitude for densities. The arctic shore birds are also abundant in suitable wet tundra habitats: Baird's, red-backed, and pectoral sandpiper densities on the order of 0.2–0.6/ha have been reported (Drury, 1961; Pitelka, 1959; Holmes, 1966a; Holmes and Pitelka, 1968). The larger avian predators have, of course, lower

Table 9.7. Estimates for population density of birds in the Cape Thompson area, Alaska (after Williamson et al., 1966).

Species	Plot number*						
	1	2	3	4	5	6	–
	Riparian willows	*Sedge meadow (ridged)*	*Cotton grass tussock*	*Sedge marsh*	*Sedge meadow*	*Low-centre polygon*	*Combined†*
Willow ptarmigan	0.25	—	—	—	—	0.30	0.09
Rock ptarmigan	0.12	0.05	0.15	0.10	—	0.12	0.09
Semipalmated plover	0.12	—	—	—	—	—	0.02
Golden plover	—	0.05	0.07	—	0.50	—	0.10
Wandering tattler	0.12	—	—	—	—	—	0.02
Long-billed dowitcher	—	—	—	0.10	—	—	0.02
Red-backed sandpiper	—	—	—	0.05	—	—	0.01
Semipalmated sandpiper	0.47	—	—	—	—	—	0.08
Western sandpiper	0.12	—	—	0.10	—	0.56	0.13
Pectoral sandpiper	—	—	—	—	—	0.30	0.05
Northern phalarope	—	—	—	0.15	—	0.54	0.12
Long-tailed skua	—	—	—	—	—	—	0.01
Yellow wagtail	0.44	—	—	—	—	—	0.07
Common redpoll	3.92	—	0.07	—	—	—	0.67
Savannah sparrow	1.36	0.15	0.30	0.05	0.50	0.84	0.53
White-crowned sparrow	1.36	—	—	—	—	—	0.23
Lapland longspur	0.69	1.85	2.62	0.69	1.60	2.18	1.61
Total	8.97	2.10	3.21	1.24	2.60	4.84	

* Mean of two years, when possible.
† Unweighted mean of all plots.

densities compared with the smaller birds. Peak densities of both pomarine and long-tailed skuas of about 0.1–0.2/ha have been reported. Short-eared owl densities are lower (0.01–0.02/ha), and for the very large snowy owls, lower still (0.002–0.004/ha) (Pitelka et al., 1955; Williamson et al., 1966).

Dry upland tundra, especially in the High Arctic, apparently supports lower bird (and mammal) densities than lowland tundra, although certain species, such as horned larks and some plovers, have a primary affinity for these 'arctic deserts'.

Alpine bird populations exhibit densities as low as, or lower than, their arctic counterparts. However, this generalization is based on sparse evidence, gathered only in the Rocky Mountains. Quantitative censuses have been made on the alpine tundra of the Beartooth Plateau, where the previously cited studies of small mammals were carried out. On one study quadrat where the habitat mosaic ranged from dry fellfield to marshy hummocks (or tussocks), the breeding densities of water pipits and horned larks were 1.0 and 0.26 birds/ha, respectively, for a total density of 1.26 birds/ha (Verbeek, 1967, 1970). This is very similar to the lowest breeding density found in the Cape Thompson area. Census transects made during the breeding period in areas of the Plateau near the study area provided an overall estimate of 2.70 birds, higher than on the study quadrat (table 9.8). This is in part due to habitat heterogeneity and to sampling error, but partly

Table 9.8. Estimates for population density of birds inhabiting the alpine zone of the Beartooth Plateau (after Verbeek, 1967, 1970, and original data).

Species	Riparian willows	Moist-wet meadow	Dry meadow	Fellfield	Combined*
Horned lark	—	—	0.62	3.2	0.85
Water pipit	2.8	2.3	0.91	0.83	1.35
Robin	0.6	—	—	—	0.05
Lincoln sparrow	0.6	—	—	—	0.05
White-crowned sparrow	4.7	0.2	—	—	0.40
Total	8.7	2.5	1.5	4.0	

* Mean of transects, weighted by distance traversed in each habitat type.

because of the disproportionate contribution made by transect counts conducted in riparian willows. As was true at Cape Thompson, this habitat was scarce, but where it occurred maximum alpine bird population densities were encountered. However, in contrast to the arctic tundra, alpine fellfields appeared to support moderate numbers of horned larks; this species was most abundant in dry upland alpine tundra, while the reverse was true to the water pipit. These complementary habitat selections account for the ratio of 1 horned lark to 4 water pipits found on the Beartooth study quadrat. It should also be noted that neither of the census methods employed on the Beartooth Plateau permitted an estimate of rosy finch numbers. My subjective estimate is that they are less common, on the average, than either water pipits or horned larks; French (1959a) also commented on their widely scattered occurrence on their alpine breeding grounds. Finally, white-tailed ptarmigan attained maximum densities of 0.17/ha and 0.19/ha in Colorado and Montana (Choate, 1963b; Braun and Rogers, 1971), similar to densities of Alaskan rock ptarmigan.

In summary, we may conclude that density or standing crop biomass of alpine and arctic tundra terrestrial vertebrates are lower than those found in most temperate and tropical ecosystems. The exception are temperate grasslands which support the same low order of magnitude of birds and mammals (Wiens, 1969; Hoffmann et al., 1971). The similarity between alpine, arctic tundra, and steppe avi- and theriofaunas in this regard are probably attributable to the basically two-dimensional nature of the structural niches within such ecosystems (see pp. 518–24).

DENSITY FLUCTUATIONS

Densities of species populations are not fixed values and may vary markedly in time and in space in the alpine or arctic tundra in response to environmental variation; the values given in the previous section must be regarded as examples only. Temporal fluctuations may be annual, in response to seasonal change (see p. 501), or they may cover more than one year. Multi-annual density fluctuations may vary in amplitude, in the interval between density peaks, in the regularity with which peak densities occur, and finally, in the degree of synchrony between fluctuations in different areas, or between separate species.

Density fluctuations in the Arctic – 'cycles'

Certain fluctuations in density of animal populations have attracted great attention because of their seeming regularity which has earned them the name 'cycle'. Although the rather loose usage of the term has been criticized (Lack, 1954) it carries the weight of tradition; just as the lay mind associates alpine and arctic with cold, so too the population biologist is conditioned to think of the tundra ecosystems as the place where the 'lemming cycle' occurs. These cycles have fascinated ecologists for many years; so much so that their views of the population dynamics of tundra birds and mammals are often not balanced by the recognition that populations of some alpine and arctic mammals are more stable. However, a brief review of cycles in general, and particularly that of the legendary lemming, are essential to an understanding of arctic population dynamics.

The species which are usually considered cyclic are restricted to relatively few groups of vertebrates, namely, the microtine rodents (lemmings and voles), northern hares and grouse (including ptarmigan), and their predators. Two main types of cycle have generally been recognized, a short cycle with an average period of three to four years, and a long cycle with an interval of about ten years. Arctic tundra lemmings, voles, hares and ptarmigan, as well as microtine rodents (except the muskrat) in more temperate latitudes, have a short cycle, while the longer cycle is restricted to grouse and hares and their predators in middle-high latitudes. Geographically, cyclic fluctuations seem to be restricted to the northern hemisphere. The short cycle is most prominent above 60°N although there is a good deal of variation in this, and the microtines undergo the short cycle at all localities. The generalization that the cycle is most extreme in the north, and diminishes to the south, seems approximately true (Pitelka, 1969).

Following the recognition of lemming cycles in the Arctic (Collett, 1878, 1912; Cabot, 1912), and similar fluctuations in other species elsewhere (Ross, 1861; MacFarlane, 1905; Seton, 1911), a great deal of work has been devoted to attempting to elucidate the causes of cyclic population fluctuations. The diversity of explanatory hypotheses advanced is disconcerting, but a few will be mentioned here. They may best be considered as 'extrinsic' or 'intrinsic' causal mechanisms.

An extrinsic cause for population cycles requires only that the cyclic species be responsive to periodic fluctuations of some sort within its abiotic environment. The possible nature of these fluctuations has long been a matter of speculation, but the factors might include periodic fluctuations in some cosmic factor, such as sunspots (Elton, 1924; Wing, 1935), weather conditions (Elton, 1929; Shelford, 1943), or fluctuations in nutritive value of food (Nordhagen, 1928; Heape, 1931; Rowan, 1950). Sunspots are no longer given serious consideration as the primary cause of cycles (MacLulich, 1936; Elton and Nicholson, 1942). Weather changes alone were doubted by Krebs to be a main cause of lemming cycles, though he conceded that winter weather might be 'a partial cause of the increases and declines'. But Fuller (1967) reanalysed Krebs's data, and constructed a model of weather-lemming interaction, emphasizing critical spring and autumn periods, as well as winter snow cover, which he and Pruitt (1970) feel is adequate to explain the fluctuations Krebs found.

It is thus clear that there is no consensus concerning the role of weather in cyclic fluctuations of lemmings and other alpine and arctic vertebrates. Were it not for the general synchrony between cycles over broad areas, the role of weather would not seem so important, but since sunspots or other cosmic events cannot be appealed to, weather seems to be the only variable which can account for the observed degree of synchrony (D. Chitty, 1952; Krebs, 1964).

Another manner in which climate might influence populations is by affecting in some way the food supply of the population. A number of workers have suggested mechanisms based on the supposed influence of year-to-year variation in weather patterns on quality of vegetation, which in turn affected reproduction and/or mortality of the cyclic herbivores (Elton et al., 1935; Baashuus-Jessen, 1937; Braestrup, 1940; Bodenheimer, 1949; Rowan, 1950). However, none of these extrinsic mechanisms has so far been convincingly documented; while weather is clearly important in affecting some aspects of lemming population dynamics, it may act only as a 'synchronizer', and most recent studies, as well as earlier ones, have focused on several sorts of interactions intrinsic to cyclic populations.

Intrinsic mechanisms involve interaction between the species' ability to increase on the one hand, and those agencies of mortality such as predators or diseases on the other, leading to density oscillations about an equilibrium level. Dymond (1947) stressed the point that the 'biotic potential' of cyclic species is greater than the normal 'environmental resistance', which allows the species to achieve excessive numbers, following which their numbers are reduced by emigration and disease. The 'monotony or relative uniformity of the . . . far northern environment accounts for the regularity with which they increase to the unstable levels of numbers which brings about periodic decline.' Christian (1950) then decided that the 'shock' disease symptoms described by Elton et al. (1931), Findlay and Middleton (1934), Hamilton (1937) and Green and Larson (1938) might be due to a common cause, this being Selye's general adaptation syndrome (Selye, 1946). While these various studies show that diseases or other pathological conditions may be present during a cyclic decline in population density, there is little evidence that disease, 'shock' or other, is anything more than a secondary agency of mortality (D. Chitty, 1954, 1959; Munday, 1961; Krebs, 1964).

Turning to predator-prey interactions, Lack (1954) suggested that the basic cause of the cycles is the 'dominant rodent interact[ing] with its vegetable food to produce a predator-prey oscillation'. Pitelka (1958b) also suggested that the principal cause of decline in brown lemmings in Alaskan lowland tundra was predation, particularly by snowy owls. Pearson (1964, 1966, 1971), working with temperate zone voles, has provided the first quantified data supporting the concept that the microtine cycle results from a predator-prey interaction. Claims of cyclic decline of lemmings in the absence of predation (Krebs, 1964) cast doubt on the general applicability of the concept, but MacLean *et al.* (1974) have concluded from a recent study at Barrow that least weasel predation is at least an essential component of the lemming cycle there.

Pitelka (1964) and Schultz (1964) subsequently modified the predator-prey oscillation by incorporating in it the concept of qualitative food deficiency. This deficiency is not conceived to be the result of extrinsic cosmic or climatic factors (see above), but instead the result of selective grazing of tundra vegetation by lemmings, which ties up essential nutrients and leads to decline in forage quality (particularly phosphorus content). Reduced forage quality reduces lemming survival, as does reduced quantity of tundra vegetation stemming from the grazing of the peak lemming population; increased starvation and losses to avian and mammalian predators result. This cyclic decline permits qualitative and quantitative recovery of the vegetation over a period of several years, the recovery lag being due to the effect of changes in vegetative cover on insulation of the active soil layer which, together with increased deposition of lemming faeces at the peak, may modify the soil microflora (Schultz, 1964). This intrinsic feedback cycle is thus 'a result of interaction between herbivore and vegetation mediated by factors of nutrient recovery and availability in the soil' (Pitelka, 1964), and is termed the 'nutrient-recovery hypothesis'.

Another intrinsic feedback mechanism of an entirely different sort was first proposed by D. Chitty (1955, 1957, 1960). He postulated that the social strife engendered in high density populations resulted in a change in the quality of the population and its subsequent generations which increased their susceptibility to normal kinds of mortality. In this general form, the idea of a change in the quality of a population as a result of high density is very similar to Christian's stress hypothesis. The critical difference is that while Christian initially advocated an adreno-pituitary mechanism, later supplemented by changes in maternal physiology transmissible to offspring via the mother's milk (Christian and Lemunyan, 1958), Chitty (1967) suggested that changes in the direction of natural selection were the reason for the observed changes in quality of the fluctuating population.

A number of data sets support the idea that change in quality of the organism is inherent in the microtine density cycle. Qualitative changes in the spleen; in adrenal and other endocrine glands; in lactation and subsequent stunting of nurslings; in sexual maturation; in behavioural patterns (Christian, 1963; Clarke, 1953; Chitty, 1957; Louch, 1958; Christian and Davis, 1964); and in body weight and fat reserves (Chitty, 1952; Krebs, 1964; Batzli and Pitelka, 1971) have been observed in cyclic populations of voles and lemmings. Schultz (1964) has recorded changes in the quality of the lemming's forage, as have Batzli and Pitelka for voles. These qualitative changes are thought to be due to high population densities, but the manner in which the changes are brought about is hotly disputed. The types of change most recently demonstrated in microtines involve gene frequencies of polymorphic blood plasma proteins (transferrin, leucine aminopeptidase, esterase, and albumin) in several species of voles (Semeonoff and Robertson, 1968; Gaines and Krebs, 1971; Krebs *et al.*, 1973). However, in all of these qualitatively variable systems it was not possible to demonstrate unequivocally that the changes observed were causes of the population density fluctuations; there seems to be no way to distinguish between cause and effect in the available data.

Further field experimentation on wild vole and lemming populations, or other cyclic species (Schultz, 1964; Batzli and Pitelka, 1971; Krebs *et al.*, 1973) will be necessary before the fascinating questions surrounding causes of microtine cycles in the Arctic, and in temperate regions, are resolved.

Even if the question of proximate cause were resolved, another controversy remains, the matter of synchrony in the fluctuations of a species over a wide area, and in the fluctuations of several sympatric species synchronously. If each species population in a given region is considered independent, then fluctuations in these populations would result in a completely aperiodic, unsynchronized pattern of fluctuation on a continent-wide basis. Evidence supporting synchrony, either interspecifically or interregionally, has been presented by Collett (1878, 1912), Elton (1924, 1942), Siivonen (1948), H. Chitty (1950), D. Chitty (1952, 1960), Mackenzie (1952), Moran (1952), Wildhagen (1952), G. R. Williams (1954), Buckley (1954), and Koskimies (1955).

The case for non-synchrony is considered by Pitelka *et al.* (1955) from the point of view of the breeding of the avian predators in the High Arctic. They believed that regional populations of species such as the pomarine skua are highly mobile and exploit the local abundances of microtines that are necessary for their successful breeding. To the authors this suggested that 'on the basis alone of evidence regarding avian predators, their numbers and their demands as species, it may prove possible to maintain that the concept of regional synchrony in the lemming cycle is spurious'. Pitelka (1957a, b) presented evidence that the regular, high-amplitude cycle of brown lemmings on the lowland tundra around Point Barrow did not occur along other parts of the arctic coast or at inland localities, and that populations of sympatric microtines might not fluctuate in phase. Krebs (1964), while admitting that the lemming cycle at Barrow might be unusually strong, countered that 'this does not mean that the same phenomenon may not be occurring [in interior Alaska] to a lesser extent . . . the absolute densities at the "peak" and the "low" may be very different from area to area . . . while the same cyclic process may occur in all these areas . . .'. Krebs, Watson (1956) and Macpherson (1966) all supported the idea of general regional synchrony, as well as interspecific synchrony between brown and collared lemming populations.

Even though the causes of regularity and possible synchrony in cycles of lemmings and other animals remain controversial, the pattern of density change from year to year has now been well documented. In the year after a 'crash' decline from a population peak, numbers remain low through the following winter and summer. However, in the second winter after the decline lemmings typically reproduce beneath the snow, leading to a high population the beginning of the second summer after the decline. However, with snowmelt high mortality ensues, as lemmings are exposed to wetting, a harsher microclimate and the attention of predators. Numbers usually recover by summer's end, and what happens next is variable.

Krebs (1964) found a 'severe decrease' in population density over this third winter, with decline continuing throughout the following summer to a cyclic low in some areas (type G decline, D. Chitty, 1955) whereas in other areas, partial recovery of the populations occurred after melt-off, and decline was resumed in the winter (type H). Pitelka (1957b), on the other hand, found the brown lemming population increasing again at a fast rate throughout the third winter, and peak density reached at the beginning of the

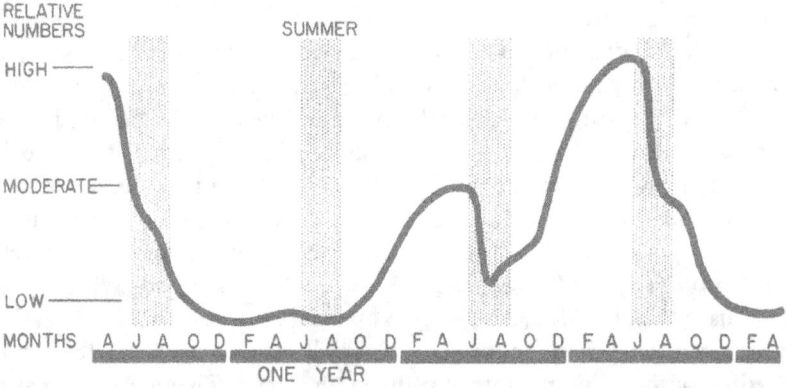

Fig. 9.15. A generalized curve of brown lemming (*Lemmus sibericus*) population density during the course of a single cyclic oscillation at Point Barrow, Alaska (from Pitelka, 1957b).

next summer. Large numbers of lemmings and a badly overgrazed tundra were exposed by snowmelt, and the decline in density was precipitous not only through the summer when many avian predators were active, but continued through the fourth winter as well, aided by persistent resident predatory mammals. Other variations in the cycle have been observed; the 'crash' decline may begin in later winter before snowmelt (Rausch, 1950), or it may be delayed in onset until well after snowmelt (Pitelka, 1957b). Finally, lemmings may remain at very low density for two years after a 'crash', instead of one. The course of a 'typical' lemming cycle is illustrated in fig. 9.15.

My earliest *caveat* that the lemming cycle casts a disproportionately large shadow over the tundra, thus obscuring a balanced view, has probably been forgotten by now. It is worth restating that not all arctic tundra species fluctuate this violently and regularly. Most bird population densities (excluding lemming predators and ptarmigan – Buckley, 1954; Weeden, 1965a, b) show much more moderate changes from year to year (Williamson *et al.*, 1966; Holmes, 1966a; Holmes and Pitelka, 1968), and the arthropods on which most of them depend also do not, as a group, exhibit cyclic fluctuations (MacLean and Pitelka, 1971).

Density fluctuations in the alpine

Simplicity of community structure has been advanced as an explanation for the extreme amplitude of small mammal population density fluctuations in the arctic tundra, but although alpine communities are also quite simple, similar fluctuations have not been identified in the alpine tundra. This suggests that the phenomenon of periodic population fluctuations in these populations is related not to the simplicity of the ecological community, but to the biology of microtine rodents themselves. This hypothesis can be

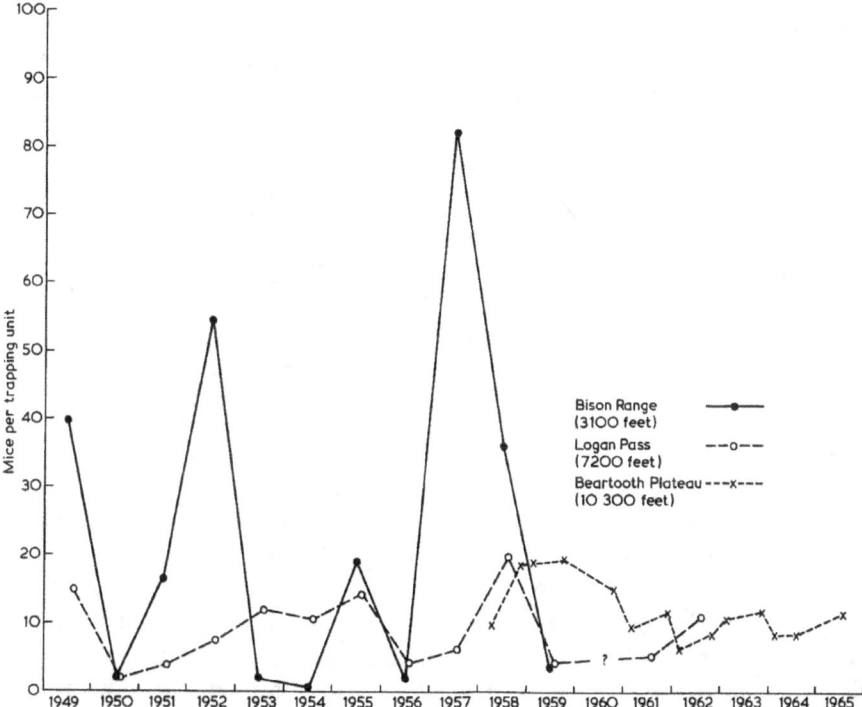

Fig. 9.16. Density fluctuations in alpine small mammal populations at Logan Pass, Glacier National Park, and the Beartooth Plateau, compared with fluctuations in grassland small mammals at the National Bison Range in western Montana (original data). The amplitude of fluctuation at the Bison Range is characteristic of microtine fluctuations in such habitats.

tested by further intensive study of alpine small mammal populations over a number of years. My preliminary data from the Beartooth Plateau indicate that, over a four-year period for which estimates of absolute density were available, and for an additional three years when relative density estimates were made, small mammal populations, including microtines, were never abundant, and showed no evidence of cyclic changes (fig. 9.16). A second data set, spanning eleven years, is available from an alpine area (2100 m) at Logan Pass, Glacier National Park, Montana. During that period, densities of alpine small mammals were low and stable, while during the same time, populations of meadow and montane voles at lower elevations (1100 m) at the National Bison Range were undergoing a typical three–four year cycle (fig. 9.16).

In contrast to this evidence of slight population fluctuations in the alpine, there exist some observations that support the occurrence of a microtine cycle in certain alpine species or at certain times. Racey (1960) suggests that water voles periodically reach great abundance in British Columbia. If the catch records from Logan Pass are analysed for this species alone, some support is given this idea, in that the take fluctuated between zero and eleven, with peaks occurring in 1949, 1953 and 1958. This corresponds roughly to microtine population peaks observed on the Bison Range.

The stability exhibited by alpine montane vole populations is especially puzzling since subalpine populations of the species have a well-marked cycle in many places (Hoffmann, 1958; D. A. Spencer, 1959). Furthermore, Kuns and Rausch (1950) and Negus and Findley (1959) indicated that at times *Microtus montanus* became very abundant in the alpine. However, it seems certain that microtine populations on alpine tundra do not regularly go through the characteristic sequence of 'crash' decline followed by recovery to a cyclic peak in places so far studied. Perhaps the reason lies in the low average population densities of small mammals in the alpine – if the alpine ecosystem could support a greater microtine biomass, perhaps cyclic fluctuations would occur.

5. Secondary productivity in arctic and alpine ecosystems

The *biomass*, or amount of living organisms, in an ecosystem is a measure of the energy that has been 'captured' by the organisms in the system. At the consumer trophic level, the biomass of herbivores and carnivores (and decomposers as well) per unit area or volume of the ecosystem, at a given point in time, is the *standing crop* of consumers. However, the amount of consumer biomass or energy produced by growth and reproduction over a time period such as a summer growing season, or a year, is the *production rate* of the consumers, and is usually referred to as the *secondary productivity* of the ecosystem, often expressed in grams or cal m^{-2} day^{-1} (Evans, 1967; Odum, 1971).

All heterotrophic organisms contribute to the total secondary productivity of an ecosystem, but our concern here is the contribution made by the terrestrial birds and mammals of the alpine and arctic tundra. Their rates of energy storage are dependent in part upon the amount of energy stored in the organisms in the trophic levels they utilize as food (green plants, insects, etc.). Their energy is ultimately yielded to the organisms on the several decomposer levels, whose contribution to the ecosystem's secondary productivity is thus in part determined by the rate at which the birds and mammals store energy.

In the herbivorous and carnivorous vertebrates of terrestrial ecosystems, energy is 'captured' from other organisms by seizing a mouth- (or beak-) full of tissue from some other organism, living or dead, and ingesting the food 'packet'. This packet contains both chemical energy and matter; part of the food may pass through the animal unchanged, but the bulk of it is assimilated; the energy fuels the metabolic machinery of the bird or mammal while the molecular matter is stored in new tissue (including reproductive – eggs and embryos) and retains a portion of the original energy. Sooner or later the energy is lost in the form of heat, and thus flows through and out of the ecosystem, but the matter in the food nutrients may, though it need not, be retained indefinitely within the ecosystem, circulating from soil to green plant to herbivore to carnivore to decomposer and back to the soil. These two aspects of secondary productivity will be considered separately.

ENERGY FLOW

The plant base

The rate of energy flow through terrestrial vertebrate consumers depends first of all on the amount of energy available to them, as noted above. The standing crop and rate of net production, P_n, of green plant biomass in arctic tundra communities is low compared with other terrestrial communities (Bliss, 1962a; Akad. Nauk SSSR, 1971), though in the same range as deserts (Odum, 1971), and close to some temperate grasslands (Sims and Singh, 1971). Alpine communities also share these low rates of above-ground net primary production (Scott and Billings, 1964; Bliss, 1966).

Of course, net above-ground biomass is actually produced only during the growing season, and daily production rates are higher then (Bliss, 1966). Also the above-ground shoots contain only a part of the new plant biomass produced during a growing season; the roots and other below-ground structures of a plant may contribute even more than shoots to total net primary productivity. This component of productivity is more difficult to study, but may be of major importance in mesic alpine areas. Scott and Billings (1964) found that well over half of this large below-ground production was contributed by one plant species, *Geum turbinatum*, a mountain avens of the sort whose roots are an important food of alpine and subalpine pocket gophers (Keith *et al.*, 1959). In contrast, their xeric site produced below-ground biomass only about as rapidly as the shoots at the same site. In arctic tundra, too, below-ground production may exceed above-ground (Bliss, 1962b; Akad. Nauk SSSR, 1971; Webber, Chapter 8B, pp. 460–3).

Root/shoot ratios are somewhat misleading, since the standing crop of above-ground parts in most tundra plants dies and is grown anew each year (shrubs and cushion plants are exceptions) whereas root standing crop biomass may represent the accumulated production of several years (Bliss, 1962b). However, peak standing crop biomass, of both above- and below-ground plant parts, does represent the food available to vertebrate herbivores, and can be measured fairly easily. It is, therefore, useful as an indication of net primary production rate in the community. Peak above-ground standing crops of tundra plants are naturally of rather small magnitude, usually falling between 100–400 g (dry weight) m^{-2} for both arctic tundra and alpine (Rickard, 1962; Scott and Billings, 1964; Bliss, 1966; Brown and West, 1970; Akad. Nauk SSSR, 1971; Wielgolaski, 1972), although shrub tundras may have much higher standing crops because of the accumulation of biomass in woody stems, and some dry high arctic or alpine fellfields have much lower standing crop biomass.

The value of this standing crop of vegetation to herbivores depends not only on the amount available, but also on its energy value. Caloric content of vegetation varies in different parts of the plant, and among species, as well as seasonally, and our fragmentary present knowledge of such variation permits only gross generalizations. Seeds appear to have the highest caloric value, followed by roots, and finally leaves, stems and branches; Golley (1961) gives average figures for these parts, based on pooled samples of fifty-seven species, of 5.07, 4.72, 4.23, and 4.28 kcal g^{-1} (dry weight) respectively. These values were derived from the temperate zone species, and it is significant that average

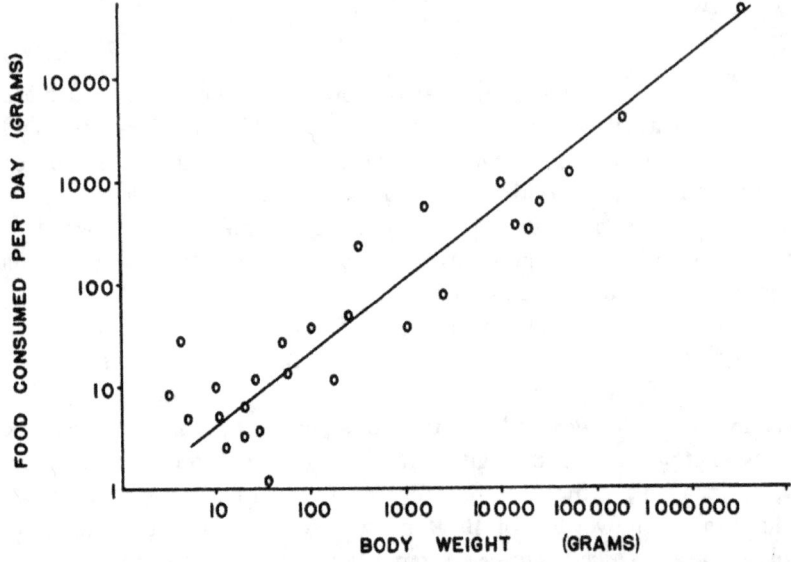

Fig. 9.17. Relationship between food consumption and body weight in mammals (from Davis and Golley, 1963).

energy content of arctic tundra and alpine plant communities are higher than temperate grasslands of comparable standing crop biomass. Thus, three grassland communities cited by Golley have average caloric values of 3.91–4.18 kcal g^{-1}, compared with 4.68–5.27 kcal g^{-1} for alpine communities (Bliss, 1966), or 4.36–4.46 kcal g^{-1} for wet arctic tundra meadow at Point Barrow, Alaska (Brown and West, 1970). Bliss (1962b) commented that 'these high caloric values can be accounted for in part by the relatively high lipid percentages', a matter of no little importance to the herbivores that feed on the tundra plants. Webber in Chapter 8B also discusses the caloric content of tundra plants. The quality of the plant material available to the herbivores (lipid, protein, carbo-hydrate, mineral nutrients, etc.) is, unfortunately, little known.

Food consumption

The rate of energy flow from the plant base into herbivorous vertebrates depends first on the rate of food intake, I. Few measurements are available for tundra herbivores. Morrison and Teitz (1953) found consumption of food (it is not clear in their table 2 if dry or wet is referred to) to range from 0.37–0.81 g g^{-1} live body weight day^{-1} in several species of small mammals (tundra red-backed vole, collared lemming, pika) which weighed from 25–127 g. Larger arctic ground squirrels averaging about 500 g body weight consumed food at a rate of 0.06–0.13 g g^{-1} day^{-1}. These findings agree with the general relationship between food consumption and body weight: that it is linear on a double log plot, and that intake per unit weight is greater in small animals than large ones, because of their greater metabolic requirements (see p. 491). This relationship (fig. 9.17) permits rough estimation of food intake rate of arctic tundra and alpine herbivores, and comparison of these values with net production rate of their plant foods. For example, in the lowland tundra of Point Barrow, Alaska, the overwhelming majority of vertebrate herbivore biomass is concentrated in a single species, the brown lemming. If the average adult brown lemming weighs about 70 g (Bee and Hall, 1956), and consumes about 0.2 g of food (dry weight) g^{-1} (body weight) day^{-1} (fig. 9.17), then one brown lemming will consume 14 g day^{-1}. If the caloric value of this food is about 4.4 kcal g^{-1}, then the lemming's caloric intake will be about 60 kcal day^{-1}, or 0.85 kcal g^{-1} day^{-1}. This estimate is nearly twice the estimated metabolic rate of *Lemmus* and *Dicrostonyx*, (about 0.45 kcal g^{-1} day^{-1}) (Morrison and Teitz, 1953; Brown and West, 1970), but may reflect the excess of calories required for normal maintenance activity, which Davis and Golley (1963) concluded was about twice the resting metabolic rate, as well as excess energy available for production of new lemming tissue.

Assimilation

Not all of the energy ingested is available for metabolic use, however, because not all is assimilated, A. Assimilation efficiencies A/I are poorly known for wild populations. Although some laboratory studies suggest efficiencies of about 90 per cent (Davis and Golley, 1963), the few estimates for natural populations of small herbivores (Golley, 1960; Johnson and Maxell, 1966) were about 70 per cent. This would indicate that about 0.60 kcal g^{-1} day^{-1} are ingested and then assimilated by a brown lemming at Barrow, the remaining 0.25 kcal g^{-1} day^{-1} being lost through faeces.

Grazing pressure

At a cyclic low of brown lemming density, the impact of lemmings grazing the tundra vegetation may be negligible. Thus, a density of one lemming ha^{-1} would result in consumption of about 14 g ha^{-1} day^{-1} of vegetation, or a little over 5000 g ha^{-1} year^{-1}. Net productivity may produce a peak standing crop of 100 g m^{-2}, or 1 million grams dry weight of vegetation per hectare; our lonely lemming might consume 0.5 per cent of the tundra's primary production (cf. Thompson, 1955). On the other hand, a peak density lemming population of 300 ha^{-1} might consume over 4000 g ha^{-1} day^{-1}, or over 1.5 million g ha^{-1}: well in excess of the annual net primary productivity of the tundra vegetation. That this level of grazing exists during peak lemming densities at Barrow is well documented (Thompson, 1955; Pitelka, 1957a, b; Schultz, 1969). Such heavy utilization of the available plant food resources by small herbivorous rodents is unusual; more typically it is very low – in the case of meadow vole and cotton rat populations, studied by Golley (1960, 1961), a utilization efficiency I/P$_n$ of less than 2 per cent.

An energy flow model for arctic tundra

Energy flow from primary producers to herbivores may be summarized by a simple food chain diagram (fig. 9.18). For this simple example the crude estimates made above will suffice, and we may further assume a moderately low lemming density: say 25 ha^{-1}. At such a density the majority of net primary production of the tundra (0.275 g m^{-2} day^{-1}, on an annual basis; 100 g m^{-2} peak standing crop divided by 365 days) is not utilized (NU) by lemmings, but is consumed by other herbivores, or accumulates as standing dead vegetation and litter. About 0.035 g m^{-2} day^{-1} is ingested by the lemming population, but about 0.010 g m^{-2} day^{-1} is not assimilated (NA), and passes out as faeces into the litter. If resting metabolic rate of lemmings is about 0.45 kcal g^{-1} day^{-1}, then a population of this size will suffer respiratory losses no less than 0.08 kcal m^{-2} day^{-1}, the equivalent of 0.018 g m^{-2} day^{-1}, and normal maintenance activity can be expected to increase this respiratory loss (energy losses through urine are ignored here).

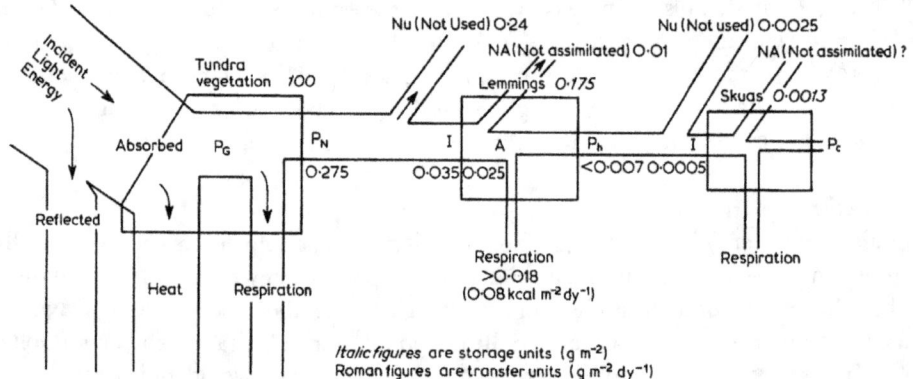

Fig. 9.18. Simplified energy flow diagram for coastal tundra at Point Barrow, Alaska (after Odum, 1971, and sources cited in text). Plant biomass expressed as dry weight, animal biomass as live weight.

Thus, secondary productivity (P$_h$) of this lemming population is less than 0.007 g m^{-2} day^{-1}; perhaps on the order of 0.003 g m^{-2} day^{-1}, or about 100,000 g ha^{-1} yr^{-1}. This represents a live weight biomass equivalent to several hundred young and growing lemmings; the approximately thirteen adult females in our initial hypothetical population, with litter sizes of seven or eight and several litters in the course of the year, could produce such an annual increment to the herbivore biomass. Overall, of the biomass produced by green plants (100 g m^{-2}), about 1 g of new herbivore biomass m^{-2} may result; within the herbivore trophic level the lemmings have an ecological growth efficiency (P/I) of about 10 per cent (see also table 8B.3).

This is only a crude first approximation to one channel of energy flow in the coastal tundra ecosystem of northern Alaska. It is based on incomplete data and unverified assumptions, but work is presently in progress (Brown and West, 1970) that will provide the information on which to construct a more accurate and complete energy flow model (cf. Petrusewicz, 1967). Additionally, this is a static model, in which values obtained at one time (i.e. peak standing crop biomass) are extrapolated on an annual basis. A dynamic model, which employs an appropriate calculus to represent seasonal variation in plant growth and lemming population density, will require much more refined input data.

Higher trophic levels

The new lemming biomass formed (P_h) represents secondary productivity which is available to feed carnivores of higher trophic levels. In northern Alaska the chief lemming predator is the pomarine skua (Pitelka *et al.*, 1955), and in the interests of simplicity only this additional component of the energy flow model will be considered in detail.

At the low brown lemming density postulated for the model, density of the breeding skua population would also be low; 1–2 pairs per square mile (Maher, 1970), or about 0.012 adult skua ha^{-1}. Even at the low lemming density of the model, pomarine skuas consume mostly lemmings, which comprise somewhere between 90–100 per cent of their food items. According to Maher's observations, an adult skua eats about 250 g live weight of lemmings each day. In the approximately 100 days skuas are on their breeding grounds at Barrow, this amounts to 25,000 g per skua, or about 300 g of lemmings per hectare. Added to this are the energy demands of the skua chicks; since a pair lays two eggs, the chick density is the same (0.012 chicks ha^{-1}). Maher calculated that 1 chick would eat about 8500 g live weight of food in growing from hatching to fledging (approx. 40 days), and for the remainder of the period (15 days?), about 220 g day^{-1} (Pitelka *et al.*, 1955; Maher, 1970). A chick's requirements for food would thus be met by about 13,000 g of lemmings, or about half that of the adult requirement. This would mean an additional 150 g ha^{-1} of lemmings added to the adult consumption of 300 g ha^{-1}; allowing for some wastage, perhaps the total would be as much as 500 g ha^{-1}.

This rate of skua predation represents about 5 per cent of the calculated secondary productivity of the lemming herbivores ingested at the next trophic level of the carnivorous skuas, a low value compared to transfer rates into carnivores found in other studies (Golley, 1960; Pearson, 1964). However, although plenty of lemming calories exist to support such a low density skua population, finding the sparsely scattered lemmings is apparently too difficult. Skua breeding success is low at a lemming density of 25 ha^{-1}, apparently because the adults are usually unsuccessful in catching enough lemmings (or other food) to keep the chicks from starving. The same applies to other avian predators, and the effect of low lemming density on breeding cycles of skuas, hawks and owls has been discussed (see p. 512).

As brown lemming density increases, the proportion of net primary productivity consumed by lemmings increases, and their own secondary production also increases. At peak lemming densities, the concentration of lemming predators is such that a much larger proportion of the secondary productivity of lemmings is utilized by the carnivores. Maher (1970) calculated seasonal lemming consumption (25 May–31 August) by pomarine skuas at 77.5 lemmings ha^{-1}; by snowy owls, 7.5 ha^{-1}; by least weasels, 25 ha^{-1}; by glaucous gulls, 2.5 ha^{-1}; and killed but not eaten by predators, 10 lemmings ha^{-1}. The total, 122.5 lemmings ha^{-1}, represents a rate of loss which, if timed properly, may significantly depress a spring lemming population of about 75 ha^{-1}, though it will not depress a spring population having a density of about 90 ha^{-1} or greater (table 9.9).

Table 9.9. Hypothetical effects of predation on different spring densities of peaking brown lemming populations (after Maher, 1970).

	Lemming density, nos./ha		
Spring density/ha	50–75	75–100	100–125
No. of females	31	44	56
Females killed before breeding	26	26	26
Females remaining to breed	5	18	30
Mean litter size	6	6	6
No. of young produced	30	108	180
Total adults remaining	10	36	60
Midsummer density/ha	40	144	240
Predation loss after breeding	70	70	70
Population change after post-breeding predation	−30	+74	+170

Energy flow in more complex communities

The simple structure of the coastal tundra community (see pp. 508 and 518) results in relatively simple energy flow patterns, though these are admittedly much more complex than in our highly oversimplified model. As more data become available it will be possible to integrate important additional energy flow pathways into such a model. For example, another group of carnivores at Barrow are the insectivorous shore birds, which consume during their summer residence 1736 kcal ha^{-1}, whereas the omnivorous Lapland and snow buntings eat a total of 883 kcal ha^{-1}, about half of which is insects, the remainder being seeds and other plant parts (Brown and West, 1970; Bent et al., 1968).

In the structurally more complex communities of the inland and foothill tundra, or at such coastal localities as Cape Thompson, the densities of herbivorous vertebrates may possibly be more stable and have higher mean values than in the simple Point Barrow community, though more data are needed.

Pruitt (1968) found a six- or sevenfold fluctuation in small tundra rodents in several of what I judge to be relatively more complex tundra communities in interior Alaska, and others appeared to show density and biomass fluctuations of lower amplitude. Tundra vegetation in these more complex communities may have a higher net production rate, at least partly because of the contribution of shrubs (Wielgolaski, 1972).

I know of no published analyses of energy flow in alpine communities. As noted above, net primary production rates in alpine vegetation are comparable to those of arctic tundra, but biomass densities of alpine herbivores appear usually to be much lower than in the Arctic, so it may be inferred that the proportion of the available standing crop of vegetation consumed by alpine herbivores is correspondingly low. A major difference is the presence of fossorial, root-eating herbivores in the alpine, in contrast to their absence in the arctic tundra. Exploitation of below-ground plant productivity may significantly modify the pattern of energy flow in alpine communities.

A study of the energy dynamics of pikas does, however, provide interesting insights into the energy flow process in this important member of the alpine herbivore trophic level (Johnson and Maxell, 1966). Pikas store plant material in 'hay piles', for use as both bedding and food (Broadbooks, 1965). Analysis of 'hay pile' samples revealed an energy content of 4.47 kcal g^{-1}, within the range of alpine vegetation generally. Pika stomach contents had a significantly higher caloric content, 4.87 kcal g^{-1}, while 'soft' faeces, produced in the caecum, were higher still (5.00 kcal g^{-1}). Selection of plant foods higher than average in energy may occur when the pika feeds, and the partially digested food is further enriched by the activity of micro-organisms in the caecum. Pikas, as well as other small mammals, may then reingest these special (soft,

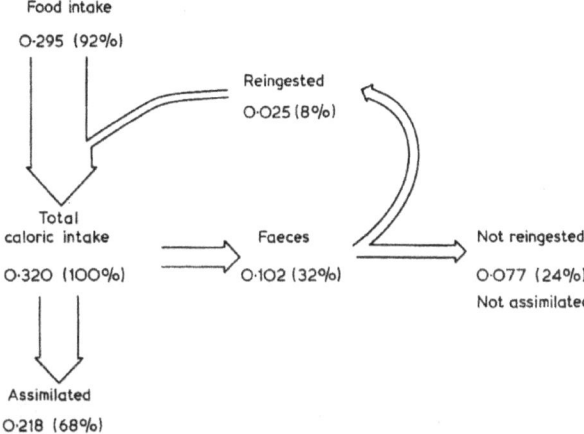

Fig. 9.19. Energy flow in an alpine pika, in k cal g^{-1} day^{-1} (from Johnson and Maxell, 1966).

'night') faeces, a process called coprophagy. The second passage of the material through the digestive tract permits more efficient extraction of nutrients, and is probably aided by the caecal villae found in several arctic and alpine herbivores (pika, collared lemming, heather voles). Coprophagy adds a shunt (fig. 9.19) to the energy flow system, the importance of which requires elucidation in arctic and alpine communities.

MINERAL NUTRIENT CYCLING

In addition to calories to fuel their 'metabolic engines', arctic and alpine vertebrates must derive essential mineral nutrients from their foods. The many macro- and micro-nutrients needed by vertebrates will not be reviewed here, since only a few have been examined in tundra communities. Excluding the atmospheric nutrients (oxygen, carbon from CO_2 for plants), the main reservoir for mineral nutrients in terrestrial ecosystems is the soil. Arctic tundra soils are characteristically low in nitrogen (Bliss, 1962b, 1966; Scott and Billings, 1964), and this appears to limit plant growth. However, in certain alpine soils at least, very high nitrogen levels are encountered in the surface horizon, in the form of organic nitrogen. The low microbial activity found in arctic tundra and alpine soils (Faust and Nimlos, 1968; Brown and West, 1970) probably accounts for a low rate of transfer of organic nitrogen in the soil surface to inorganic nitrate or ammonium deeper in the soil, where it becomes available for plant nutrition. Low microbial activity in turn may be due to cold soils, which may also impede plant growth by reducing assimilation of nitrogen into organic compounds (Dadykin, 1954).

In contrast to nitrogen, 'certain other soil nutrients seem to be present in fairly adequate amounts in some arctic and alpine soils' (Bliss, 1962b). A number of these are currently under intensive investigation at Point Barrow (Brown and West, 1970), and

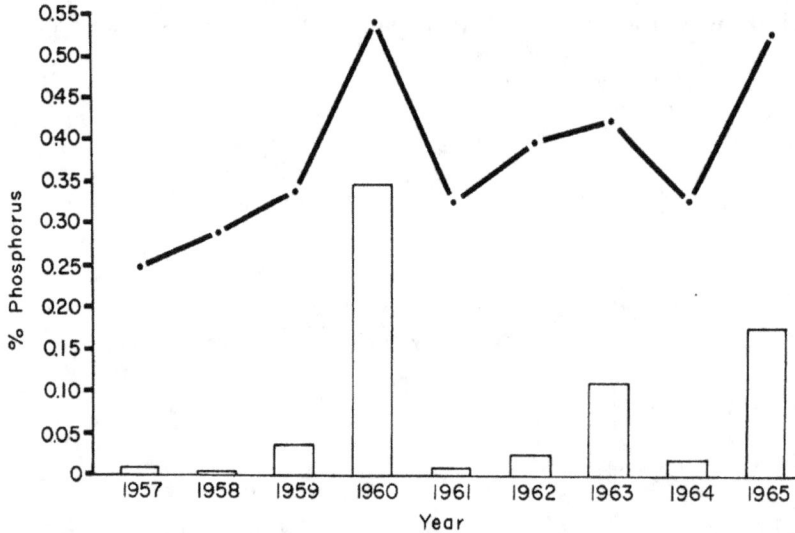

Fig. 9.20. Direct relationship between phosphorus content of vegetation (·—·) and brown lemming population density (bars) at Point Barrow, Alaska (from Schultz, 1969). Data points for phosphorus represent seasonal means.

an interesting pattern is emerging. Nutrient content of the three most important species in the standing crop of tundra vegetation (*Dupontia fischeri, Eriophorum angustifolium, Carex aquatilis*), which also make up the bulk of the brown lemming's diet, were analysed for nitrogen, phosphorus, calcium, magnesium and sodium. Nutrient concentrations in the plants were directly proportional to lemming population density, reaching maximum values in the years when lemmings attained cyclic peaks, and dropping to low values when the lemmings 'crashed' (Schultz, 1969). The trend for phosphorus is shown in fig. 9.20. The rate of microbial decomposition of organic matter on the soil surface showed the same direct relationship.

Schultz has proposed a homeostatically controlled system, discussed briefly on p. 538, in which these essential mineral nutrients are tied up in organic matter (lemming, faeces, dead plants) at the peak of the lemming cycle, and so are unavailable in the next, 'crash', year; net primary production is then low and poor in nutrient quality, and nutrients in the organic litter have not yet been released by decomposition. During the following year or two, the rate of decomposition accelerates, more mineral nutrients are available for plant growth, and both live plant biomass and standing dead plants accumulate, insulating the soil and reducing the depth of the active layer. The shallowness of the thawed soil layer permits freezing action to concentrate mineral solutes in the upper soil layers where they are available to the plants. Finally, this leads again to a

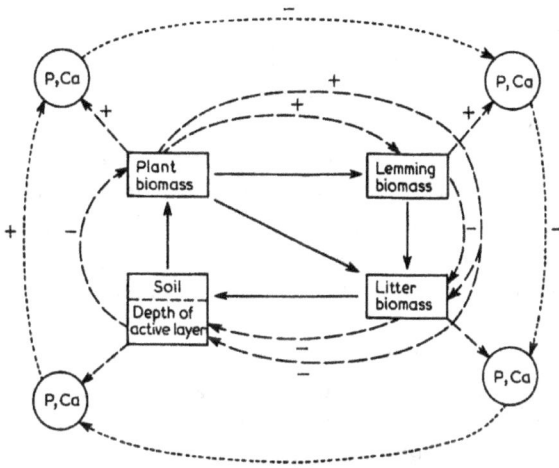

Fig. 9.21. Homeostatic feedback model of mineral nutrient cycling in the arctic tundra (modified from Schultz, 1969). Solid lines ($\longrightarrow$) represent pathways of mineral transfer. Dashed lines ($--\rightarrow$) represent feedback states; *i.e.* increase in plant biomass results in increase (+, amplification) in lemming biomass, but decrease (−, counteraction) in depth of active soil layer. Circles connected by dotted lines ($\cdots\rightarrow$) represent proportions of nutrients (in this case, phosphorus and calcium) in total mineral pool, and their feedback states. Increases in nutrients in the soil compartment leads to increase in plant biomass and nutrient content, which in turn leads to increase in lemming biomass and concentration of the mineral pool in organic matter, instead of soil.

lemming peak, and massive transfer of mineral nutrients from soil and live plant to lemming and litter compartments (fig. 9.21). This hypothesis of mineral cycling on the Alaskan coastal tundra now requires testing; similar approaches to mineral cycling in other arctic and alpine ecosystems may prove rewarding.

6. Some generalizations

I have employed a comparative approach here, in order to reach some general conclusions concerning terrestrial vertebrates of arctic and alpine regions. However, comparisons also emphasize the differences between arctic tundra and alpine vertebrates. Hence, adaptive similarities and differences will be summarized here.

  1. Most arctic tundra vertebrates have circumpolar ranges, often continuous, or

disjunct in relatively few places. Alpine species have highly disperse distributions, usually restricted to one continent, although a few have boreo-montane and arctic–alpine distribution patterns.

2. The arctic tundra vertebrate fauna evolved only during the Pleistocene, probably from a 'cold steppe' adapted fauna in Siberia and Beringia. The centre of evolution of alpine vertebrates has been in the mountains of central Asia, but local adaptation of montane forest and steppe species to alpine conditions of various mountain regions has frequently occurred. Thus, alpine faunas in the Holarctic are much more heterogeneous than the arctic tundra fauna.

3. Arctic and alpine vertebrates exhibit similar adaptations to cold. Large, resident birds and mammals utilize seasonally varying insulative cover, control of peripheral circulation, and evaporative cooling to maintain homeothermy. Smaller resident mammals may hibernate, but most survive in the microhabitat beneath the snow blanket. Smaller birds migrate to lower latitudes or altitudes to avoid cold and food shortage.

4. Adaptations to substrate moisture, snow cover, wind, and insect pests are similar in arctic and alpine regions. Certain species are uniquely adapted to moving on, or sheltering beneath, the various forms of rock found only in the alpine, or to the low oxygen tension of high altitudes.

5. Although day length and seasonal patterns are radically different in the arctic tundra and mid- to low-latitude alpine areas, the patterns each result in a short summer breeding season; arctic and alpine vertebrates show similar adaptations to it, in the form of migration schedules, moult cycles, litter or clutch size, and number of reproductive efforts per season.

6. Both trophic and structural niches are relatively simple in arctic and alpine regions, but overall the arctic tundra seems to have simpler structure and less taxonomic diversity. Niche complexity appears correlated with topographic diversity. The possibility that there is greater niche overlap in the Arctic, in contrast to niche segregation in the alpine, should be investigated. Competitive exclusion phenomena appear to be commoner in alpine areas than in the Arctic.

7. Mean population densities are low in both arctic and alpine regions. However, arctic tundra species periodically reach high densities, but these 'cycles' do not seem to be common or widespread in the alpine. The causes of cyclic fluctuations remain controversial, but may be related in some fashion to complexity of niche structure.

8. Secondary productivity of vertebrate populations is also low in both arctic and alpine regions, as a consequence of low primary production rates, which in turn are largely a function of the short growing season (see 5 above). Patterns of energy flow and nutrient cycling may be implicated in the arctic lemming cycle.

The similarities between vertebrates of arctic and alpine regions are thus seen to be a result of convergent adaptations to cold and snow, to the short summer growing season, and to the relatively simple niche structures of the arctic and alpine ecosystems. These similarities, however, should not obscure the sharp distinctions between the two ecosystems and the vertebrate faunas, in their patterns of evolution and in their adaptations to the unique aspects of arctic and alpine environments.

ACKNOWLEDGEMENTS

My research in the Rocky Mountain alpine was begun during my faculty tenure at the University of Montana, Missoula. I am grateful to my many colleagues there, and in particular Dr Philip L. Wright, then Chairman of the Department of Zoology, for much assistance and encouragement. My principal collaborator in the alpine, and helpful critic of this paper, was Dr Richard D. Taber, now at the University of Washington; our fieldwork was jointly supported by National Science Foundation Grants 5174, G11568, B9486 and B14090. Many others have also contributed in various ways: a partial list includes Sam Bamberg, Clait Braun, Thomas Choate, Richard Johnson, Tom Nimlos, William Osburn, Donald Pattie, Philip Shelton and Nicolaas Verbeek.

References

AKAD. NAUK SSSR (1971) *Biogeocoenoses of the Tundra*. Int. Tundra Biome, Transl. 1. Tundra Biome Center, Univ. of Alaska. 4 pp.

ALDRICH, J. W. (1963) Geographic orientation of American Tetraonidae. *J. Wildl. Mgt.*, **27**, 529–45.

ANDREWS, R. M. (1971) Structural habitat and time budget of a tropical *Anolis* lizard. *Ecology*, **52**, 262–70.

AYALA, F. J. (1969) Experimental invalidation of the principle of competitive exclusion. *Nature*, **224**, 1076–9.

BAASHUUS-JESSEN, J. (1937) Periodiske vekslinger i smaviltbastanden. *Norges Svalbardog Ishavs-Undersøkelser, Medd.*, **36**, 1–15. (English summary, 12–15.)

BALDWIN, P. H. (1955) The breeding ecology and physiological rhythms of some arctic birds at Umiat, Alaska. *Final Rep., Project ONR-121*. Arct. Inst. N. America and Off. Naval Res. (Cited in WILLIAMSON *et al.*, 1966.)

BAMBERG, S. A. (1961) *Plant Ecology of Alpine Tundra Areas in Montana and Adjacent Wyoming*. M.A. thesis, Univ. of Colorado. 163 pp.

BAMBERG, S. A. and MAJOR, J. (1968) Ecology of the vegetation and soils associated with calcareous parent materials in three alpine regions of Montana. *Ecol. Monogr.*, **38**, 127–67.

BANNIKOV, A. G. (1954) *Mlekopitayushchie Mongol'skoi Narodnoi Respubliki*. Moscow: Akad. Nauk SSSR. 669 pp.

BATZLI, G. O. and PITELKA, F. A. (1971) Condition and diet of cycling populations of the California vole, *Microtus californicus*. *J. Mammal.*, **52**, 141–63.

BEE, J. W. (1958) Birds found on the arctic slope of northern Alaska. *Univ. of Kansas Mus. Nat. Hist. Pub.* 10, 163–211.

BEE, J. W. and HALL, E. R. (1956) Mammals of northern Alaska. *Univ. of Kansas Mus. Nat. Hist., Misc. Pub.*, 8. 309 pp.

BEG, M. A. (1969) Habitats, food habits, and population dynamics of the red-tailed chipmunk, *Eutamias ruficaudus*, in western Montana. Ph.D. thesis, Univ. of Montana. 163 pp.

BENT, A. C. *et al.* (1968) Life histories of North American cardinals, grosbeaks, buntings, towhees, finches, sparrows, and allies. *US Nat. Mus. Bull.*, 237. 1889 pp.

BISSONETTE, T. H. (1941) Experimental modification of breeding cycles in goats. *Physiol. Zool.*, **14**, 379–83.

BLANFORD, W. T. (1888, 1891) *The Fauna of British India, including Ceylon and Burma. Mammalia.* London: Taylor & Francis. 617 pp.

BLISS, L. C. (1962a) Net primary production in tundra ecosystems. *In:* LIETH, H. (ed.) *Die Stoffproduktion der Pflanzendecke*, 35–46. Stuttgart: Gustav Fischer.

BLISS, L. C. (1962b) Adaptations of arctic and alpine plants to environmental conditions. *Arctic*, **15**, 117–44.

BLISS, L. C. (1966) Plant productivity in alpine microenvironments on Mt Washington, New Hampshire. *Ecol. Monogr.*, **36**, 125–55.

BODENHEIMER, F. S. (1949) *Problems of Vole Populations in the Middle East; Report on the Population Dynamics of the Levant Vole* (Microtus guentheri). Jerusalem: Azriel Print. Works. 77 pp.

BOURLIERE, F. (1955) *Mammals of the World.* New York: Knopf. 223 pp.

BOURLIERE, F. (1964) *The Land and Wildlife of Eurasia.* New York: Time, Inc. 198 pp.

BRAESTRUP, F. W. (1940) The periodic die-off in certain herbivorous mammals and birds. *Science*, **92**, 354–5.

BRANDBORG, S. M. (1955) Life history and management of the mountain goat in Idaho. *Idaho Dept of Fish and Game. Bull.*, **2**. 142 pp.

BRAUN, C. E. and ROGERS, G. E. (1971) The white-tailed ptarmigan in Colorado. *Colo. Div. Game, Fish and Parks Tech. Pub.*, **27**. 80 pp.

BRITTON, M. E. (1957) Vegetation of the arctic tundra. *In:* HANSEN, H. P. (ed.) *Arctic biology*, 26–61. Corvallis: Oregon State Univ. Press.

BROADBOOKS, H. E. (1965) Ecology and distribution of the pikas of Washington and Alaska. *Amer. Midl. Nat.*, **73**, 299–335.

BROWN, J. and WEST, G. C. (1970) *Tundra Biome Research in Alaska. The Structure and Function of Cold-Dominated Ecosystems.* Hanover, NH: US Tundra Biome. 156 pp.

BROWN, J., PITELKA, F. A. and COULOMBE, H. N. (1970) Structure and function of the tundra ecosystem at Barrow, Alaska. Sect. 1– A word model of the Barrow ecosystem. *In:* FULLER, W. A. and KEVAN, P. G. (eds.) *Productivity and Conservation in Northern Circumpolar Lands*, 41–3. IUCN Pub. 16. Morges, Switzerland: Int. Union Conserv. Nature.

BROWN, J. H. (1971a) Mechanisms of competitive exclusion between two species of chipmunks. *Ecology*, **52**, 305–11.

BROWN, J. H. (1971b) Mammals on mountaintops: nonequilibrium insular biogeography. *Amer. Nat.*, **105**, 467–78.

BROWN, J. H. and LASIEWSKI, R. (1972) Metabolism of weasels: the cost of being long and thin. *Ecology*, **53**, 939–43.

BUCKLEY, J. L. (1954) Animal population fluctuations in Alaska – a history. *Trans. N. Amer. Wildl. Conf.*, **91**, 338–57.

BUTZER, K. W. (1964) *Environment and Archeology.* Chicago: Aldine. 524 pp.

CABOT, W. B. (1912) *In Northern Labrador.* London: John Murray. 292 pp.

CALHOUN, J. B. (1945) Diel activity rhythms of the rodents. *Microtus ochrogaster* and *Sigmodon hispidus hispidus. Ecology*, **26**, 251–73.

CARL, E. A. (1971) Population control in arctic ground squirrels. *Ecology*, **52**, 395–413.

CARRUTHERS, D. (1949) *Beyond the Caspian. A Naturalist in Central Asia.* Edinburgh: Oliver & Boyd. 310 pp.

CHATFIELD, P. I., LYMAN, C. P. and IRVING, L. (1953) Physiological adaptation to cold of peripheral nerve in the leg of the herring gull (*Larus Argentatus*). *Amer. J. Physiol.*, **172**, 639–44.

CHERNYAVSKII, F. B. (1967) Ecology of bighorn sheep and their value as a game animal on the Koryak uplands (in Russian). *Problemi Severa*, **11**, 128–41. (English transl., Nat. Res. Council, Ottawa.)

CHITTY, D. (1952) Mortality among voles (*Microtus agrestis*) at Lake Vyrnwry, Montgomeryshire, in 1936–39. *Phil. Trans. Roy. Soc. London*, B, 236, 505–52.

CHITTY, D. (1954) Tuberculosis among wild voles; with a discussion of other pathological conditions among certain mammals and birds. *Ecology*, **35**, 227–37.

CHITTY, D. (1955) Adverse effects of population density upon the viability of later generations. *In*: CRAGG, J. B. and PIRIE, N. W. (eds.) *The Numbers of Man and Animals*, 57–67. Edinburgh: Oliver & Boyd.

CHITTY, D. (1957) Self-regulation of numbers through changes in viability. *In*: Population studies: Animal ecology and demography, *Cold Spring Harbor Symp.*, **22**, 277–80.

CHITTY, D. (1959) A note on shock disease. *Ecology*, **40**, 728–31.

CHITTY, D. (1960) Population processes in the vole and their relevance to general theory. *Can. J. Zool.*, **38**, 99–113.

CHITTY, D. (1967) The natural selection of self-regulatory behaviour in animal populations. *Proc. Ecol. Soc. Austr.*, **2**, 57–78.

CHITTY, H. (1950) Canadian arctic wildlife enquiry, 1943–49; with a summary of results since 1933. *J. Anim. Ecol.*, **19**, 180–93.

CHOATE, T. S. (1960) *Observations on the reproductive activities of white-tailed ptarmigan* (Lagopus leucurus) *in Glacier Park, Montana*. M.A. thesis, Montana State Univ. 128 pp.

CHOATE, T. S. (1963a) *Ecology and population dynamics of white-tailed ptarmigan* (Lagopus leucurus) *in Glacier National Park, Montana*. Ph.D. thesis, Montana State Univ. 216 pp.

CHOATE, T. S. (1963b) Habitat and population dynamics of white-tailed ptarmigan in Montana. *J. Wildl. Mgt.*, **27**, 684–99.

CHRISTIAN, J. J. (1950) The adreno-pituitary system and population cycles in mammals. *J. Mammal.*, **31**, 247–59.

CHRISTIAN, J. J. (1963) Endocrine adaptive mechanisms and the physiologic regulation of population growth. *In*: MAYER, W. V. and VAN GELDER, R. G. (eds.) *Physiological mammalogy*, Vol. 1, *Mammalian populations*, 189–353. New York: Academic Press.

CHRISTIAN, J. J. and DAVIS, D. E. (1964) Endocrines, behaviour, and population. *Science*, **146**, 1550–60.

CHRISTIAN, J. J. and LEMUNYAN, C. D. (1958) Adverse effects of crowding on lactation and reproduction of mice and two generation of their offspring. *Endocrinology*, **63**, 517–29.

CLARK, C. H. D. (1940) A biological investigation of the Thelon Game Sanctuary. *Nat. Mus. Can. Bull.*, 96, Biol. Ser., 25. 139 pp.

CLARKE, J. R. (1953) The effect of fighting on the adrenals, thymus, and spleen of the vole (*Microtus agrestis*). *J. Endocrin.*, **9**, 114–26.

CODY, M. L. (1966) A general theory of clutch size. *Evolution*, **20**, 174–84.

COE, M. J. (1967) *The Ecology of the Alpine Zone of Mount Kenya*. The Hague: W. Junk. 144 pp.

COLE, L. C. (1960) Competitive exclusion. *Science*, **132**, 348–9.

COLLETT, R. (1878) On *Myodes lemmus* in Norway. *J. Linn. Soc. (Zool.)*, **13**, 327–34.

556 ROBERT S. HOFFMANN

COLLETT, R. (1912) *Norges pattedyr*. Kristiana, Norway: Aschehoug. 744 pp.

CURRY-LINDAHL, K. (1962) The irruption of the Norway lemming in Sweden during 1960. *J. Mammal.*, **43**, 171–84.

DADYKIN, V. P. (1954) Peculiarities of plant behaviour in cold soils. *Voprosy Botaniki*, **2**, 455–72. (Cited in BLISS, 1962b.)

DARLING, F. F. (1937) *A Herd of Red Deer*. London: Oxford Univ. Press. 225 pp.

DARLINGTON, P. J. JR (1957) *Zoogeography: the Geographical Distribution of Animals*. New York: John Wiley. 675 pp.

DAVIES, J. L. (1958) Pleistocene geography and the distribution of northern pinnipeds. *Ecology*, **39**, 97–113.

DAVIS, D. E. and GOLLEY, F. B. (1963) *Principles in Mammalogy*. New York: Reinhold. 348 pp.

DAVIS, M. B. (1967) Late-glacial climate in northern United States: a comparison of New England and the Great Lakes region. *In:* CUSHING, E. J. and WRIGHT, H. E. JR (eds.) *Quaternary paleoecology*, 11–43. New Haven: Yale Univ. Press.

DAWSON, W. R. and TORDOFF, H. B. (1959) Relation of oxygen consumption to temperature in the evening grosbeak. *Condor*, **61**, 388–96.

DEMENT'EV, G. P. and GLADKOV, N. A. (1952) *Ptitsi Sovetskovo Soyuza*, Vol. IV, *Sovetskaya Nauka*. Moscow. 640 pp. (English edn, TT65-50100, US Dept of Comm., 1967.)

DEWOLFE, B. B. and DEWOLFE, R. H. (1962) Mountain white-crowned sparrows in California. *Condor*, **64**, 378–89.

DRURY, W. H. JR (1961) Studies of the breeding biology of horned lark, water pipit, Lapland longspur, and snow bunting on Bylot Island, Northwest Territories, Canada. *Bird-banding*, **32**, 1–46.

DUBOIS, A. D. (1935) Nests of horned larks and longspurs on a Montana prairie. *Condor*, **37**, 56–72.

DUNBAR, M. J. (1968) *Ecological Develoment in Polar Regions. A Study of Evolution*. Englewood Cliffs: Prentice-Hall. 127 pp.

DUNMIRE, W. W. (1960) An altitudinal survey of reproduction in *Peromyscus maniculatus*. *Ecology*, **41**, 174–82.

DYMOND, J. R. (1947) Fluctuations in animal populations with special reference to those of Canada. *Trans. Roy. Soc. Can.*, **41**, Sec. 5, 1–34.

EDWARDS, R. Y. and RITCEY, R. W. (1959) Migrations of caribou in a mountainous area in Wells Gray Park, British Columbia. *Can. Field Nat.*, **73**, 21–5.

ELTON, C. S. (1924) Periodic fluctuations in the numbers of animals, their causes and effects. *Brit. J. Exp. Biol.*, **2**, 119–63.

ELTON, C. S. (1927) *Animal Ecology*. London: Sidgwick & Jackson. 229 pp.

ELTON, C. S. (1929) The relation of animal numbers to climate. *Conf. Empire Meteorologists 1929*, Agric. Sec., 121–9.

ELTON, C. S. (1942) *Voles, Mice, and Lemmings: Problems in Population Dynamics*. New York: Oxford Univ. Press. 496 pp.

ELTON, C. S. and NICHOLSON, M. (1942) The ten-year cycle in numbers of the lynx of Canada. *J. Anim. Ecol.*, **11**, 215–44.

ELTON, C. S., DAVIS, D. H. S. and FINDLAY, G. M. (1935) An epidemic among voles on the Scottish border in the spring of 1934. *J. Anim. Ecol.*, **4**, 277–88.

ELTON, C. S., FORD, E. B., BAKER, J. R. and GARDNER, A. D. (1931). Health and parasites of a wild mouse population. *Proc. Zool. Soc. Lond.*, 657–721.

EVANS, F. C. (1967) The significance of investigations in secondary terrestrial productivity. *In:* PETRUSEWICZ, K. (ed.) *Secondary Productivity of Terrestrial Ecosystems (Principles and Methods)*, Vol. 1, 3–15. Warsaw: Polish Acad. Sci.

FAUST, R. A. and NIMLOS, T. J. (1968) Soil microorganisms and soil nitrogen of the Montana alpine. *Northwest Sci.*, **42**, 101–7.

FINDLAY, G. M. and MIDDLETON, A. D. (1934) Epidemic disease among voles (*Microtus*) with special references to *Toxoplasma*. *J. Anim. Ecol.*, **3**, 150–60.

FINDLEY, J. S. (1954) Competition as a possible limiting factor in the distribution of *Microtus*. *Ecology*, **35**, 418–20.

FISHER, K. C. and NEEDLER, M. E. (1957) Spontaneous activity of the lemming *Dicrostonyx groenlandicus richardsoni Merriam* as indicated in 24-hour records of oxygen consumption. *J. Cell Comp. Physiol.*, **50**, 293–308.

FLEHARTY, E. D. and OLSON, L. E. (1969) Summer food habits of *Microtus ochrogaster* and *Sigmodon hispidus*. *J. Mammal.*, **50**, 475–86.

FLEROV, K. K. (= FLEROW, C. C.) (1967) On the origin of the mammalian fauna of Canada. *In:* HOPKINS, D. M. (ed.) *The Bering Land Bridge*, 271–80. Stanford: Stanford Univ. Press.

FLINT, V. E., CHUGUNOV, YU. D. and SMIRIN, V. M. (1965) *Mlekopitayushchie SSSR*. Moscow: Misl. 437 pp.

FOLK, G. E. JR (1964) Daily physiological rhythms in carnivores exposed to extreme changes in arctic daylight. *Fed. Proc.*, **23**, 1221–8.

FOLK, G. E. JR, FOLK, M. A. and BREWER, M. C. (1966) The day–night (circadian) physiological rhythms of large arctic carnivores in natural continuous light (summer) and continuous darkness (winter). *4th Int. Biometeorol. Cong.* (abstract). (Cited in SWADE and PITTENDRIGH, 1967.)

FOLK, M. A. (1963) The daily distribution of sleep and wakefulness in the arctic ground squirrel. *J. Mammal.*, **44**, 575–7.

FORMOZOV, A. N. (1946) Snow cover as an integral factor of the environment and its importance in the ecology of mammals and birds. (English transl., *Occas. Pap.*, 1, Boreal Inst., Univ. of Alberta, 1964, 200 pp.)

FORMOZOV, A. N. (1966) Adaptive modifications of behavior in mammals of the Eurasian steppes. *J. Mammal.*, **47**, 208–23.

FRANZ, J. (1949) Jahres-und Tagesrhythmus einiger Vögel in Nordfinnland. *Zeits. Tierpsychol.*, **6**, 309–29.

FRASER, A. F. (1968) *Reproductive Behaviour in Ungulates*. London: Academic Press. 202 pp.

FRENCH, N. R. (1959a) Life history of the black rosy finch. *Auk*, **76**, 158–80.

FRENCH, N. R. (1959b) Distribution and migration of the black rosy finch. *Condor*, **61**, 18–29.

FRENZEL, B. (1968) The Pleistocene vegetation of northern Eurasia. *Science*, **161**, 637–48.

FULLER, W. A. (1967) Ecologie hivernale des lemmings et fluctuations de leurs populations. *La Terre et la Vie*, **2**, 97–115. (English summary, 111–12.)

GAINES, M. S. and KREBS, C. J. (1971) Genetic changes in fluctuating vole populations. *Evolution*, **25**, 702–23.

GATES, D. M. (1962) *Energy Exchange in the Biosphere*. New York: Harper & Row. 151 pp.

GEIST, V. (1971) *Mountain Sheep. A Study in Behaviour and Evolution*. Chicago: Univ. of Chicago Press. 398 pp.

GESSAMAN, J. A. (1972) Bioenergetics of the snowy owl (*Nyctea scandiaca*). *Arct. Alp. Res.*, **4**, 223–38.

GIBB, J. (1956) Food, feeding habits and territory of the rock pipit *Anthus spinoletta*. *Ibis*, **98**, 506–30.

GOLLEY, F. B. (1960) Energy dynamics of a food chain of an old-field community. *Ecol. Monogr.*, **30**, 187–206.

GOLLEY, F. B. (1961) Energy values of ecological materials. *Ecology*, **42**, 581–4.

GRANT, P. R. (1972) Interspecific competition among rodents. (Experimental studies of competitive interaction in a two-species system, V. Summary of the evidence for rodent species, and some generalizations.) *Ann. Rev. Ecol. Syst.*, **3**, 79–106.

GREEN, R. G. and LARSON, C. L. (1938) A description of shock disease in the snowshoe hare. *Amer. J. Hygiene*, **28**, 190–212.

GRINNELL, J. (1917) The niche-relationships of the California thrasher. *Auk*, **34**, 427–33.

GRINNELL, J. (1924) Geography and evolution. *Ecology*, **5**, 225–9.

GRINNELL, J. (1928) The presence and absence of animals. *Univ. of Calif. Chronicle*, **30**, 429–50. (Cited in UDVARDY, 1959.)

GRODZINSKI, W. (1962) Influence of food upon the diurnal activity of small rodents. *In:* KRATOCHVIL, J. and PELIKAN, J. (eds.) *Symp. Theriologium, Brno, 1960.* Prague: Czech. Acad. Sci. 383 pp.

GROSS, A. O. (1947) Cyclic invasions of the snowy owl and the migration of 1945–46. *Auk*, **64**, 584–601.

GUTHRIE, R. D. (1966) The extinct wapiti of Alaska and Yukon Territory. *Can. J. Zool.*, **44**, 47–57.

GUTHRIE, R. D. (1968) Paleoecology of the large-mammal community in interior Alaska during the late Pleistocene. *Amer. Midl. Nat.*, **79**, 346–63.

GUTHRIE, R. D. and MATTHEWS, J. V. JR (1971) The Cape Deceit fauna – Early Pleistocene mammalian assemblage from the Alaskan Arctic. *Quat. Res.*, **1**, 474–510.

HALL, E. R. (1946) *Mammals of Nevada.* Berkeley: Univ. of Calif. Press. 710 pp.

HAMILTON, W. J. JR (1937) The biology of microtine cycles. *J. Agric. Res.*, **54**, 779–90.

HAMMEL, H. T. (1955) Thermal properties of fur. *Amer. J. Physiol.*, **182**, 369–76.

HANSEN, R. M. (1957) Influence of daylength on activity of the varying lemming. *J. Mammal.*, **38**, 218–23.

HARDIN, G. (1960) The competitive exclusion principle. *Science*, **131**, 1292–7.

HARPER, F. (1956) Amphibians and reptiles of the Ungava Peninsula. *Proc. Biol. Soc. Wash.*, **69**, 93–103.

HART, J. S. (1956) Seasonal changes in insulation of the fur. *Can. J. Zool.*, **34**, 53–7.

HART, J. S. (1962) Seasonal acclimatization of four species of small wild birds. *Physiol. Zool.*, **35**, 224–36.

HART, J. S. and JANSKY, L. (1963) Thermogenesis due to exercise and cold in warm and cold acclimated rats. *Can. J. Biochem. Physiol.*, **41**, 629–34.

HAVILAND, M. D. (1926) *Forest, Steppe and Tundra.* Cambridge: Cambridge Univ. Press. 226 pp.

HAYWARD, J. S. (1965) Metabolic rate and its temperature-adaptive significance in six geographic races of Peromyscus. *Can. J. Zool.*, **43**, 309–23.

HEAPE, W. (1931) *Emigration, Migration, and Nomadism.* Cambridge: Heffer. 369 pp.

HELLER, H. C. (1971) Altitudinal zonation of chipmunks (*Eutamias*): interspecific aggression. *Ecology*, **52**, 312–19.

HEPTNER, V. G., NASIMOVICH, A. A. and BANNIKOV, A. G. (1961) *Mlekopitayushchie Sovetskova Soyuza. Vol. 1, Parnokopitnie i neparnokopitnie.* Moscow: Vishaya Shkola. 776 pp.

HERREID, C. F. and KESSEL, B. (1967) Thermal conductance in birds and mammals. *Comp. Biochem. Physiol.*, **21**, 405–14.

HERRERO, S. (1970) Bears – their biology and management. *IUCN Pub.* (n.s.) 23. Morges, Switzerland: Int. Union Conserv. Nature. 371 pp.

HOCK, R. J. and ROBERTS, J. C. (1966) Effect of altitude on oxygen consumption of deer mice: relation of temperature and season. *Can. J. Zool.*, **44**, 365–76.

HOFFMANN, K. (1959) Über den Tagesrhythmus der Singvögel im arktischen Sommer. *J. Ornithol.*, **100**, 84–9.

HOFFMANN, R. S. (1958) The role of reproduction and mortality in population fluctuations of voles (*Microtus*). *Ecol. Monogr.*, **28**, 79–109.

HOFFMANN, R. S. (1960) Summer birds of the Little Belt Mountains, Montana. *Montana State Univ. Occas. Pap.*, 1. 18 pp.

HOFFMANN, R. S. and JONES, J. K. JR (1970) Influence of late-glacial and post-glacial events on the distribution of recent mammals on the northern Great Plains. *In:* DORT, W. JR and JONES, J. K. JR (eds.) *Pleistocene and Recent Environments of the Central Great Plains*, 255–94. Lawrence: Univ. of Kansas Press.

HOFFMANN, R. S. and PATTIE, D. L. (1968) *A Guide to Montana Mammals*. Missoula: Univ. of Montana Print. Serv. 143 pp.

HOFFMANN, R. S. and TABER, R. D. (1967) Origin and history of Holarctic tundra eco-systems, with special reference to their vertebrate faunas. *In:* WRIGHT, H. E. JR and OSBURN, W. H. (eds.) *Arctic and Alpine Environments*, 143–70. Bloomington: Indiana Univ. Press.

HOFFMANN, R. S., JONES, J. K. JR and GENOWAYS, H. H. (1971) Small mammal survey on the Bison, Bridger, Cottonwood, Dickinson, and Osage sites. *Grassland Biome, Int. Biol. Progr., Tech. Rep.*, 109. Fort Collins, Colorado. 74 pp.

HOLMES, R. T. (1966a) Breeding ecology and annual cycle adaptations of the red-backed sandpiper (*Calidris alpina*) in northern Alaska. *Condor*, **68**, 3–46.

HOLMES, R. T. (1966b) Molt cycle of the red-backed sandpiper (*Calidris alpina*) in western North America. *Auk*, **83**, 517–33.

HOLMES, R. T. (1966c) Feeding ecology of the red-backed sandpiper (*Calidris alpina*) in arctic Alaska. *Ecology*, **47**, 32–45.

HOLMES, R. T. and PITELKA, F. A. (1968) Food overlap among coexisting sandpipers of northern Alaskan tundra. *Syst. Zool.*, **17**, 305–18.

HOLOWAYCHUK, N., PETRO, J. H., FINNEY, H. R., FARNHAM, R. S. and GERSPER, P. L. (1966) Soils of Ogotoruk watershed. *In:* WILIMOVSKY, N. J. (ed.) *Environment of the Cape Thompson Region, Alaska*, 221–73. Washington, DC: US Atomic Energy Commission.

HOLROYD, J. C. (1967) Observations of Rocky Mountain goats on Mount Wardle, Kootenay National Park, British Columbia. *Can. Field Nat.*, **81**, 1–22.

HOST, P. (1942) Effect of light on the molts and sequences of plumage in the willow ptarmigan. *Auk*, **59**, 388–403.

HUTCHINSON, G. E. (1957) Concluding remarks. *In:* Population studies: animal ecology and demography, *Cold Spring Harbor Symp.*, **22**, 415–27.

HUTCHINSON, G. E. (1965) *The Ecological Theatre and the Evolutionary Play*. New Haven: Yale Univ. Press. 152 pp.

INGLES, L. G. (1949) Ground water and snow as factors affecting the seasonal distribution of pocket gophers, *Thomomys monticola*. *J. Mammal.*, **30**, 343–50.

IRVING, L. (1955) Nocturnal decline in the temperature of birds in cold weather. *Condor*,
57, 362–5.

IRVING, L. (1960) Birds of Anaktuvuk Pass, Kobuk, and Old Crow. *US Nat. Mus. Bull.*,
217. 409 pp.

IRVING, L. (1972) Arctic life of birds and mammals including man. Vol. 2 of *Zoophysiology
and Ecology*. New York: Springer-Verlag. 192 pp.

IRVING, L. and KROG, J. (1955) Temperature of skin in the Arctic as a regulator of heat.
J. Appl. Physiol., 7, 355–64.

IRVING, L., KROG, H. and MONSON, M. (1955) Insulation and metabolism of arctic mammals
in winter and summer. *Physiol. Zool.*, 28, 173–85.

JANSKY, L. (1967) Evolutionary adaptations of temperature regulation in mammals.
Zeits. Säugetierk., 32, 167–72.

JERDON, T. C. (1874) *The Mammals of India; a Natural History of All the Animals Known to
Inhabit Continental India* (2nd edn). London. 335 pp. (Orig. pub. Roorkes, Thomason
Coll. Press, 1867.)

JOHNSON, A. W., VIERECK, L. A., JOHNSON, R. E. and MELCHIOR, H. (1966) Vegetation
and flora. *In:* WILIMOVSKY, N. J. (ed.) *Environment of the Cape Thompson Region, Alaska*,
227–354. Washington, DC: US Atomic Energy Commission.

JOHNSON, D. R. and MAXELL (1966) Energy dynamics of Colorado pikas. *Ecology*, 47,
1059–61.

JOHNSON, H. M. (1953) Preliminary ecological studies of microclimates inhabited by the
smaller arctic and subarctic mammals. *Proc. Alaska Sci. Conf. 1951*, 125–31.

JOHNSON, N. (1963) Biosystematics of sibling species of flycatchers in the *Empidonax
hammondii–oberholseri–wrightii* complex. *Univ. of Calif. Pub. Zool.*, 66, 79–238.

JOHNSON, P. L. and BILLINGS, W. D. (1962) The alpine vegetation of the Beartooth
Plateau in relation to cryopedogenic processes and patterns. *Ecol. Monogr.*, 32,
105–35.

JOHNSON, R. E. (1965) Reproductive activities of rosy finches, with special reference to
Montana. *Auk*, 82, 190–205.

JOHNSON, R. E. (1966) Alpine birds of the Little Belt Mountains, Montana. *Wilson Bull.*,
78, 225–7.

JOHNSON, R. E. (1968) Temperature regulation in the white-tailed ptarmigan, *Lagopus
leucurus*. *Comp. Biochem. Physiol.*, 24, 1003–14.

JOHNSTON, R. F. (1954) Variation in breeding season and clutch size in song sparrows of the
Pacific Coast. *Condor*, 56, 268–73.

KEITH, J. O., HANSEN, R. M. and WARD, A. L. (1959) Effect of 2,4–D on abundance and
foods of pocket gophers. *J. Wildl. Mgt.*, 23, 137–45.

KELSALL, J. P. (1968) Migratory barren-ground caribou of Canada. *Can. Wildl. Ser.*,
Monogr. 3. 340 pp.

KENDEIGH, S. C. (1961) *Animal ecology*. Englewood Cliffs: Prentice-Hall. 478 pp.

KING, J. R. and FARNER, D. S. (1961) Energy metabolism, thermoregulation, and body
temperature. *In: Biology and Comparative Physiology of Birds*, 2, 215–88. New York:
Academic Press.

KOPLIN, J. R. and HOFFMANN, R. S. (1968) Habitat overlap and competitive exclusion in
voles (*Microtus*). *Amer. Midl. Nat.*, 80, 494–507.

KOSKIMIES, J. (1955) Ultimate causes of cyclic fluctuations in numbers in animal
populations. *Papers on Game Research*, 15. Helsinki: Finnish Game Foundation. 29 pp.

KOWALSKI, K. (1967) The Pleistocene extinction of mammals in Europe. *In:* MARTIN, P. S. and WRIGHT, H. E. JR (eds.) *Pleistocene extinctions. The search for a cause*, 349–64. New Haven: Yale Univ. Press.

KREBS, C. J. (1964) The lemming cycle at Baker Lake, Northwest Territories, during 1959–62. *Arct. Inst. N. Amer., Tech. Pap.*, 15. 104 pp.

KREBS, C. J., GAINES, M. S., KELLER, B. L., MYERS, J. H. and TAMARIN, R. H. (1973) Population cycles in small rodents. *Science*, **179**, 35–41.

KROG, H., MONSON, M. and IRVING, L. (1955) Influence of cold upon the metabolism and body temperature of wild rats, albino rats, and albino rats conditioned to cold. *J. Appl. Physiol.*, **7**, 349–54.

KROG, J. and MONSON, M. (1954) Notes on the metabolism of a mountain goat. *Amer. J. Physiol.*, **178**, 515–16.

KUNS, M. L. and RAUSCH, R. L. (1950) An ecological study of helminths of some Wyoming voles (*Microtus* spp.) with a description of a new species of *Nematospiroides* (Heligmosomidae: Nematoda). *Zoologica*, **35**, 181–8.

KURTÉN, B. (1968) *Pleistocene mammals of Europe*. Chicago: Aldine. 325 pp.

LACHENBRUCH, A. H., GREENE, G. W. and MARSHALL, B. V. (1966) Permafrost and the geothermal regimes. *In:* WILIMOVSKY, N. J. (ed.) *Environment of the Cape Thompson Region, Alaska*, 149–63. Washington, DC: US Atomic Energy Commission.

LACK, D. (1933) Habitat selection in birds. *J. Anim. Ecol.*, **2**, 239–62.

LACK, D. (1947) The significance of clutch-size: I, Intraspecific variations; II, Factors involved. *Ibis*, **89**, 302–52.

LACK, D. (1948) The significance of clutch size: III, Some interspecific comparisons. *Ibis*, **90**, 25–45.

LACK, D. (1954) *The Natural Regulation of Animal Numbers*. London: Oxford Univ. Press. 343 pp.

LANDRY, S. O. JR (1970) The Rodentia as omnivores. *Quart. Rev. Biol.*, **45**, 351–72.

LAYNE, J. N. and MCKEON, W. H. (1956) Some aspects of red fox and gray fox reproduction in New York. *New York Fish and Game J.*, **3**, 44–74.

LEOPOLD, A. S., RINEY, T., MCCAIN, R. and TEVIS, L. (1951) The Jawbone deer herd. *Calif. Dept of Fish and Game, Game Bull.*, 4. 159 pp.

LOCKIE, J. D. (1955) The breeding habits and food of short-eared owls after a vole plague. *Bird Study*, **2**, 53–69.

LORD, R. D. (1960) Litter size and latitude in North American mammals. *Amer. Midl. Nat.*, **64**, 488–99.

LOUCH, C. D. (1958) Adrenocortical activity in two meadow vole populations. *J. Mammal.*, **39**, 109–16.

LYNEBORG, L. (1971) *Mammals in colour*. London: Blandford Press. 247 pp.

MACARTHUR, R. H. and WILSON, E. O. (1967) *The Theory of Island Biogeography*. Princeton: Princeton Univ. Press. 214 pp.

MACFARLANE, R. (1905) Notes on mammals collected and observed in the northern Mackenzie River district, Northwest Territories of Canada. *Proc. US Nat. Mus.*, **28**, 673–764.

MACKENZIE, J. M. D. (1952) Fluctuations in the numbers of British tetraonids. *J. Anim. Ecol.*, **21**, 128–53.

MACLEAN, S. F. JR and PITELKA, F. A. (1971) Seasonal patterns of abundance of tundra arthropods near Barrow. *Arctic*, **24**, 19–40.

MACLEAN, S. F. JR, FITZGERALD, B. M. and PITELKA, F. A. (1974) Population cycles in arctic lemmings: winter reproduction and predation by weasels. *Arct. Alp. Res.*, **6**, 1–12.

MACLULICH, D. A. (1936) Sunspots and abundance of animals. *J. R. Astron. Soc. Can.*, **30**, 233–46.

MACPHERSON, A. H. (1966) The abundance of lemmings at Aberdeen Lake, District of Keewatin, 1959–63. *Can. Field Nat.*, **80**, 89–94.

MACPHERSON, A. H. (1969) The dynamics of Canadian arctic fox populations. *Can. Wildl. Ser., Rep. Ser.*, 8. 52 pp.

MACPHERSON, A. H. and MANNING, T. H. (1958) The birds and mammals of Adelaide Peninsula, Northwest Territories. *Nat. Mus. Can. Bull.*, 161. 67 pp.

MAHER, W. J. (1967) Predation by weasels on a winter population of lemmings, Banks Island, Northwest Territories. *Can. Field Nat.*, **81**, 248–50.

MAHER, W. J. (1970) The pomarine jaeger as a brown lemming predator in northern Alaska. *Wils. Bull.*, **82**, 130–57.

MAJOR, J. and BAMBERG, S. A. (1967) A comparison of some North American and Eurasian alpine ecosystems. *In:* WRIGHT, H. E. JR and OSBURN, W. H. (eds.) *Arctic and Alpine Environments*, 89–118. Bloomington: Indiana Univ. Press.

MANI, M. S. (1962) *Introduction to High Altitude Entomology*. London: Methuen. 302 pp.

MANVILLE, R. H. (1959) The Columbian ground squirrel in northwestern Montana. *J. Mammal.*, **40**, 26–45.

MARSHALL, F. H. A. (1937) On the change-over in the oestrous cycle in animals after transference across the equator, with further observations on the incidence of the breeding season and the factor controlling sexual periodicity. *Proc. Roy. Soc. London*, B, 122, 413–28.

MARSHALL, W. H. (1941) *Thomomys* as burrowers in the snow. *J. Mammal.*, **22**, 196–7.

MARTIN, P. S. (1958) Pleistocene ecology and biogeography of North America. *In:* HUBBS, C. L. (ed.) *Zoogeography*, 375–420. Pub. 51. Washington, DC: Amer. Ass. Adv. Sci.

MAY, T. A. and BRAUN, C. E. (1972) Seasonal foods of adult white-tailed ptarmigan in Colorado. *J. Wildl. Mgt.*, **36**, 1180–6.

MAYER, W. V. (1953) Some aspects of the ecology of the Barrow ground squirrel (*Citellus parryi barrowensis*). *Current Biol. Res. in the Alaskan Arctic, Stanford Univ. Pub., Univ. Ser. Biol. Sci.*, **11**, 48–55.

MEINERTZHAGEN, R. (1928) Some biological problems connected with the Himalaya. *Ibis*, 12th ser., **4**, 480–533.

MILLAR, J. S. (1973) Evolution of litter-size in the pika, *Ochotona princeps* (Richardson). *Evolution*, **27**, 134–43.

MILLER, A. H. (1961) Molt cycles in equatorial Andean sparrows. *Condor*, **63**, 143–61.

MILLER, R. S. (1955) Activity rhythms in the wood mouse, *Apodemus sylvaticus*, and the bank vole, *Clethrionomys glareolus*. *Proc. Zool. Soc. Lond*, **125**, 505–19.

MILLER, R. S. (1964) Ecology and distribution of pocket gophers (*Geomyidae*) in Colorado. *Ecology*, **45**, 256–72.

MORAN, P. A. P. (1952) The statistical analysis of game-bird records. *J. Anim. Ecol.*, **21**, 154–8.

MOREAU, R. (1944) Clutch size: a comparative study, with special reference to African birds. *Ibis*, **86**, 286–347.

MOREAU, R. E. (1955) Ecological changes in the Palearctic region since the Pliocene. *Proc. Zool. Soc. Lond.*, **125**, 253–95.

MORRISON, P. (1964) Wild animals at high altitudes. *In*: EDHOLM, O. G. (ed.) *The Biology of Survival* (Zool. Soc. Lond. Symp. 13), 49–55.

MORRISON, P. and ELSNER, R. (1962) Influence of altitude on heart and breathing rates in some Peruvian rodents. *J. Appl. Physiol.*, **17**, 467–70.

MORRISON, P. R. and TEITZ, W. J. (1953) Observations on food consumption and preference in four Alaskan mammals. *Arctic*, **6**, 52–7.

MORRISON, P. R., KERST, K., REYNAFARJE, C. and RAMOS, J. (1963a) Hematocrit and hemoglobin levels in some Peruvian rodents from high and low altitude. *Int. J. Biomet.*, **7**, 52–8.

MORRISON, P. R., KERST, K. and ROSENMANN, M. (1963b) Hematocrit and hemoglobin levels in some Chilean rodents from high and low altitude. *Int. J. Biomet.*, **7**, 45–50.

MULLEN, D. A. (1968) Reproduction in brown lemmings (*Lemmus trimucronatus*) and its relevance to their cycle of abundance. *Univ. of Calif. Pub. Zool.*, **85**, 1–24.

MUNDAY, K. A. (1961) Aspects of stress phenomena. *In*: Mechanisms in biological competition, *Symp. Soc. Exper. Biol.*, 15, 168–89.

MURIE, J. O. (1969) An experimental study of substrate selection by two species of voles (*Microtus*). *Amer. Midl. Nat.*, **82**, 622–5.

MURIE, J. O. (1971) Behavioral relationships between two sympatric voles (*Microtus*): relevance to habitat segretation. *J. Mammal.*, **52**, 181–6.

NADLER, C. F., VORONTSOV, N. N., HOFFMANN, R. S., FORMICHOVA, I. I. and NADLER, C. F. JR (1973) Zoogeography of transferrins in arctic and long-tailed ground squirrel populations. *Comp. Biochem. Physiol.*, 44B, 33–40.

NEGUS, N. C. and FINDLEY, J. S. (1959) Mammals of Jackson Hole, Wyoming. *J. Mammal.*, **40**, 371–81.

NIETHAMER, G. (1967) On the breeding biology of *Montifringilla theresae. Ibis*, **109**, 117–18.

NIKOL'SKII, A. A. (1971) Distantnaya zvukovaya signalizatsiya dnevnikh grizunov otkritikh prostranstv. *Moscow Univ., Zool.*, 907. 20 pp.

NIMLOS, T. J. and MCCONNELL, R. C. (1965) Alpine soils in Montana. *Soil Sci.*, **99**, 310–21.

NIMLOS, T. J., MCCONNELL, R. C. and PATTIE, D. L. (1965) Soil temperature and moisture regimes in Montana alpine soils. *Northwest Sci.*, **39**, 129–38.

NORDHAGEN, R. (1928) Rypear og baerar. Bidrag til diskusjonen om var rypebestands vekslinger. *Bergens Mus. Arb.* (Naturvidenskapelig Rekke), **2**, 1–52.

ODUM, E. P. (1959) *Fundamentals of Ecology* (2nd edn). Philadelphia: W. B. Saunders Co. 564 pp.

ODUM, E. P. (1971) *Fundamentals of Ecology* (3rd edn). Philadelphia: W. B. Saunders Co. 574 pp.

OGNEV, S. I. (1948) *Zveri SSSR i prilezhashchikh stran*, 6. Moscow–Leningrad: Akad. Nauk. 559 pp. (English transl., Off. Tech. Serv. 63–11058, 1963.)

OGNEV, S. I. (1950) *Zveri SSSR i prilezhashchikh stran*, 7. Moscow–Leningrad: Akad. Nauk. 706 pp. (English transl., Off. Tech. Serv. 63–11059, 1963.)

OSBURN, W. S. JR (1958) *Ecology of Winter Snow-Free Areas of the Alpine Tundra of Niwot Ridge, Boulder County, Colorado.* Unpub. Ph.D. thesis, Univ. of Colorado. 77 pp.

PATTIE, D. L. (1967a) *Dynamics of alpine small mammal populations.* Unpub. Ph.D. thesis, Univ. of Montana. 107 pp.

PATTIE, D. L. (1967b) Observations on an alpine population of yellow-bellied marmots (*Marmota flaviventris*). *Northwest Sci.*, **41**, 96–102.

PATTIE, D. L. and VERBEEK, N. A. M. (1966) Alpine birds of the Beartooth Mountains. *Condor*, **67**, 167–76.

PATTIE, D. L. and VERBEEK, N. A. M. (1967) Alpine mammals of the Beartooth Mountains. *Northwest Sci.*, **41**, 110–17.

PEARSON, O. P. (1964) Carnivore-mouse predation: an example of its intensity and bio-energetics. *J. Mammal.*, **45**, 177–88.

PEARSON, O. P. (1966) The prey of carnivores during one cycle of mouse abundance. *J. Anim. Ecol.*, **35**, 217–33.

PEARSON, O. P. (1971) Additional measurements of the impact of carnivores on California voles (*Microtus californicus*). *J. Mammal.*, **52**, 41–9.

PETRUSEWICZ, K. (1967) *Secondary Productivity of Terrestrial Ecosystems (Principles and Methods)*. Warsaw: Polish Acad. Sci. 379 pp.

PFEFFER, P. (1968) *Asia. A Natural History*. New York: Random House. 298 pp.

PITELKA, F. A. (1957a) Some characteristics of microtine cycles in the Arctic. *In:* HANSEN, H. P. (ed.) *Arctic Biology* (18th Biol. Colloq.), 73–88. Corvallis: Oregon State Univ. Press.

PITELKA, F. A. (1957b) Some aspects of population structure in the short-term cycle of the brown lemming in northern Alaska. *In:* Population studies: animal ecology and demography, *Cold Spring Harbor Symp.*, **22**, 237–51.

PITELKA, F. A. (1958a) Timing of molt in Steller jays of the Queen Charlotte Islands, British Columbia. *Condor*, **60**, 38–49.

PITELKA, F. A. (1958b) Population studies of lemmings and lemming predators in northern Alaska. *XVth Int. Congr. Zool.*, Sect. X, Pap. 5. 3 pp.

PITELKA, F. A. (1959) Numbers, breeding schedule, and territoriality in pectoral sandpipers of northern Alaska. *Condor*, **61**, 233–64.

PITELKA, F. A. (1964) The nutrient-recovery hypothesis for arctic microtine cycles: I. Introduction. *In:* CRISP, D. J. (ed.) *Grazing in Terrestrial and Marine Environments*, 55–6. Oxford: Blackwell.

PITELKA, F. A. (1969) Ecological studies on the Alaskan arctic slope. *Arctic*, **22**, 333–40.

PITELKA, F. A., TOMICH, P. Q. and TREICHEL, G. W. (1955) Ecologial relations of jaegers and owls as lemming predators near Barrow, Alaska. *Ecol. Monogr.*, **25**, 85–117.

PLOEGER, P. L. (1968) Geographical differentiation in arctic Anatidae as a result of isolation during the last glacial. *Ardea*, **56**, 1–159.

PRUITT, W. O. JR (1960) Behavior of the barren ground caribou. *Univ. of Alaska, Biol. Pap.*, 3. 48 pp.

PRUITT, W. O. JR (1966) Ecology of terrestrial mammals. *In:* WILIMOVSKY, N. J. (ed.) *Environment of the Cape Thompson Region, Alaska*, 519–64. Washington, DC: US Atomic Energy Commission.

PRUITT, W. O. JR (1968) Synchronous biomass fluctuations of some northern mammals. *Mammalia*, **32**, 172–91.

PRUITT, W. O. JR (1970) Some ecological aspects of snow. *In: Ecology of the Subarctic Regions*, 83–99. Paris: UNESCO.

QUAY, W. B. (1951) Observations on mammals on the Seward peninsula, Alaska. *J. Mammal.*, **32**, 88–99.

RACEY, K. (1960) Notes relative to the fluctuations in numbers of *Microtus richardsoni richardsoni*, about Alta Lake and Pelberton Valley, BC. *Murrelet*, **41**, 13–14.

RAND, A. S. (1964) Ecological distribution in anoline lizards of Puerto Rico. *Ecology*, **45**, 745–52.

RAUSCH, R. L. (1950) Observations on cyclic decline of lemmings (*Lemmus*) on the arctic coast of Alaska during the spring of 1949. *Arctic*, **3**, 166–77.

RAUSCH, R. L. (1963) A review of the distribution of Holarctic Recent mammals. *In:* GRESSIT, J. L. (ed.) *Pacific Basin Biogeography* (10th Pacific Sci. Congr. Symp.), 29–43. Honolulu: Bishop Mus. Press.

REYNAFARJE, B. and MORRISON, P. (1962) Myoglobin level in some tissues from wild Peruvian rodents native to high altitude. *J. Biol. Chem.*, **237**, 2861–4.

RICKARD, W. H. (1962) Comparison of annual harvest yields in an arctic and a semidesert plant community. *Ecology*, **43**, 770–1.

ROSENZWEIG, M. L. (1966) Community structure in sympatric Carnivora. *J. Mammal.*, **47**, 602–12.

ROSS, B. R. (1861) A popular treatise on the fur-bearing animal of the Mackenzie River district. *Can. Nat. Geol.*, **6**, 5–36.

ROWAN, W. (1950) Canada's premier problem of animal conservation: a question of cycles. *New Biol.*, **9**, 38–57.

SALOMONSEN, F. (1939) Moults and sequence of plumages in the rock ptarmigan (*Lagopus mutus* (Montin)). *Videnskab. Medd. fra Dansk Naturalist. Forening* (Haase, Copenhagen), **103**, 95–142.

SCHOLANDER, P. F., WALTERS, V., HOCK, R. and IRVING, L. (1950a) Body insulation of some arctic and tropical mammals and birds. *Biol. Bull.*, **99**, 225–36.

SCHOLANDER, P. F., HOCK, R., WALTERS, V., JOHNSON, F. and IRVING, L. (1950b) Heat regulation in some arctic and tropical mammals and birds. *Biol. Bull.*, **99**, 237–58.

SCHOLANDER, P. F., HOCK, R., WALTERS, V. and IRVING, L. (1950c) Adaptation to cold in arctic and tropical mammals and birds in relation to body temperature, insulation, and basal metabolic rate. *Biol. Bull.*, **99**, 259–71.

SCHULTZ, A. M. (1964) The nutrient-recovery hypothesis for arctic microtine cycles: II. Ecosystem variables in relation to arctic microtine cycles. *In:* CRISP, D. J. (ed.) *Grazing in Terrestrial and Marine Environments*, 57–68. Oxford: Blackwell.

SCHULTZ, A. M. (1969) A study of an ecosystem: the arctic tundra. *In:* VAN DYNE, G. M. (ed.) *The Ecosystem Concept in Natural Resource Management*, 77–93. New York: Academic Press.

SCOTT, D. and BILLINGS, W. D. (1964) Effects of environmental factors on standing crop and productivity of an alpine tundra. *Ecol. Monogr.*, **34**, 243–70.

SELYE, H. (1946) The general adaptation syndrome and the diseases of adaptation. *J. Clin. Endocrin.*, **6**, 117–230.

SEMEONOFF, R. and ROBERTSON, F. W. (1968) A biochemical and ecological study of plasma esterase polymorphism in natural populations of the field vole, *Microtus agretis* L. *Biochem. Gen.*, **1**, 205–27.

SETON, E. T. (1911) *The Arctic Prairies*. New York: Scribner. 308 pp.

SHELFORD, V. E. (1943) The abundance of the collared lemming (*Dicrostonyx groenlandicus* (Tr.) var. *richardsoni* Mer.) in the Churchill area, 1929 to 1940. *Ecology*, **24**, 472–84.

SHEPPARD, D. H. (1971) Competition between two chipmunk species (*Eutamias*). *Ecology*, **52**, 320–9.

SHTEGMAN, B. K. (1938) *Osnovi ornitogeograficheskovo deleniya Palearkitiki* (Fauna USSR, n.s., 19). *Birds*, **1** (2), 1–76 (Russian); 77–156 (German). Moscow–Leningrad: Akad. Nauk.

SIIVONEN, L. (1948) Structure of short-cyclic fluctuations in numbers of mammals and birds in the northern parts of the northern hemisphere. *Papers on Game Research*, 1. Helsinki: Finnish Game Foundation. 166 pp.

SIMS, P. L. and SINGH, J. S. (1971) Herbage dynamics and net primary production in certain ungrazed and grazed grasslands in North America. *In:* FRENCH, N. R. (ed.) *Preliminary Analysis of Structure and Function in Grasslands*, 59–124. Range Sci. Dept, Sci. Ser., 10. Colorado State Univ.

SPENCER, A. W. and STEINHOFF, H. W. (1968) An explanation of geographic variation in litter size. *J. Mammal.*, **49**, 281–6.

SPENCER, D. A. (1959) Biological and control aspects in the Oregon meadow mouse irruption of 1957–58. *Fed. Coop. Exten. Ser.*, 15–25. Oregon State Coll.

STEEN, J. (1958) Climatic adaptations in some small northern birds. *Ecology*, **39**, 625–9.

STODDART, D. M. (1967) A note on the food of the Norway lemming. *J. Zool.*, **131**, 211–13.

STOECKER, R. E. (1972) Competitive relations between sympatric populations of voles (*Microtus montanus* and *M. pennsylvanicus*). *J. Anim. Ecol.*, **41**, 311–29.

SVENDSEN, G. E. (1973) *Behavioral and Environmental Factors in the Spatial Distribution and Population Dynamics of a Yellow-Bellied Marmot Population.* Unpub. Ph.D. thesis, Univ. of Kansas. 93 pp.

SWADE, R. H. and PITTENDRIGH, C. S. (1967) Circadian locomotor rhythms of rodents in the arctic. *Amer. Nat.*, **101**, 431–66.

SWAN, L. W. (1961) The ecology of the high Himalayas. *Sci. Amer.*, **205**, 68–78.

SWAN, L. W. (1963) Ecology of the heights. *Nat. Hist.*, **57**, 22–9.

SWAN, L. W. (1967) Alpine and aeolian regions of the world. *In:* WRIGHT, H. E. JR and OSBURN, W. H. (eds.) *Arctic and Alpine Environments*, 29–54. Bloomington: Indiana Univ. Press.

SWAN, L. W. and LEVITON, A. E. (1962) The herpetology of Nepal: a history, check list, and zoogeographical analysis of the herpetofauna. *Proc. Calif. Acad. Sci.*, **32**, 103–47.

TABER, R. D. and DASMANN, R. F. (1958) The black-tailed deer of the chaparral. *Calif. State Dept of Fish and Game, Game Bull.*, 8. 163 pp.

TABER, R. D., HOFFMANN, R. S., NIMLOS, T. J. and BAMBERG, S. A. (n.d.) *Alpine Ecosystems of the Northern Rocky Mountains.* Unpub. MS.

TAST, J. (1966) The root vole, *Microtus oeconomus* (Pallas), as an inhabitant of seasonally flooded land. *Ann. Zool. Fenn.*, **3**, 127–71.

THOMAS, D. C. (1969) Population estimates and distribution of barren-ground caribou in Mackenzie District, NWT, Saskatchewan, and Alberta – March to May 1967. *Can. Wildl. Ser., Rept. Ser.*, 9. 44 pp.

THOMPSON, D. Q. (1951) Summer food preference of the brown and collared lemmings. *Proc. Alaskan Sci. Conf.*, **2**, 347.

THOMPSON, D. Q. (1955) The role of food and cover in population fluctuations of the brown lemming at Point Barrow, Alaska. *Trans. N. Amer. Wildl. Conf.*, **20**, 166–76.

UDVARDY, M. D. F. (1957) An evaluation of quantitative studies in birds. *In:* Population studies: animal ecology and demography, *Cold Spring Harbor Symp.*, **22**, 301–11.

UDVARDY, M. D. F. (1959) Notes on the ecological concepts of habitat, biotope and niche. *Ecology*, **40**, 725–8.

UDVARDY, M. D. F. (1963) Zoogeographical study of the Pacific Alcidae. *In:* GRESSIT, J. L. (ed.) *Pacific Basin Biogeography* (10th Pacific Sci. Congr. Symp.), 85–111. Honolulu: Bishop Mus. Press.

UDVARDY, M. D. F. (1969) *Dynamic Zoogeography with Special Reference to Land Animals.* New York: Van Nostrand. 445 pp.

VANGENGEIM, E. A. (1961) Paleontologicheskoe obosovanie stratigrafii Antropogenovikh otlozhenii severo vostochnoi Sibiri. *Trudy. Geol. Inst., Akad. Nauk SSSR,* **48,** 1–182.

VANGENGEIM, E. A. and RAVSKII, E. I. (1965) O vnutrikontinental'nom tipc prirodnoi zonal'nosti v Chetvertichnom periode (Antropogene). *In: Problemy Stratigraf. Kainozoia,* 128–41. Moscow: Nedra.

VAN TYNE, J. and BERGER, A. J. (1959) *Fundamentals of Ornithology.* New York: John Wiley. 624 pp.

VAUGHAN, T. A. (1969) Reproduction and population densities in a montane small mammal fauna. *Univ. of Kansas, Misc. Pub.,* 51, 51–74.

VAURIE, C. (1972) *Tibet and Its Birds.* London: H. F. Witherby. 407 pp.

VEGHTE, J. H. and HERREID, C. F. (1965) Radiometric determination of feather insulation and metabolism of arctic birds. *Physiol. Zool.,* **38,** 267–75.

VERBEEK, N. A. M. (1965) Predation by badger on yellow-bellied marmot in Wyoming. *J. Mammal.,* **46,** 506.

VERBEEK, N. A. M. (1967) Breeding biology and ecology of the horned lark in alpine tundra. *Wils. Bull.,* **79,** 208–18.

VERBEEK, N. A. M. (1970) Breeding ecology of the water pipit. *Auk,* **87,** 425–51.

VOOUS, K. H. (1960) *Atlas of European Birds.* London: Nelson. 284 pp.

VUILLEUMIR, F. (1970) Insular biogeography in continental regions. The northern Andes of South America. *Amer. Nat.,* **104,** 373–88.

WATSON, A. (1956) Ecological notes on the lemmings *Lemmus trimucronatus* and *Dicrostonyx groenlandicus* in Baffin Island. *J. Anim. Ecol.,* **25,** 289–302.

WATTS, W. A. (1967) Late-glacial plant macrofossils from Minnesota. *In:* CUSHING, E. J. and WRIGHT, H. E. JR (eds.) *Quaternary Paleoecology,* 89–97. New Haven: Yale Univ. Press.

WEEDEN, R. B. (1960) The birds of Chilkat Pass, British Columbia. *Can. Field Nat.,* **74,** 119–29.

WEEDEN, R. B. (1964) Spatial separation of sexes in rock and willow ptarmigan in winter. *Auk,* **81,** 534–41.

WEEDEN, R. B. (1965a) *Grouse and Ptarmigan in Alaska. Their Ecology and Management.* Federal Aid in Wildlife Restoration Project Rep. V, Proj. W–6–R–5. Juneau: Alaska Dept of Fish and Game. 110 pp.

WEEDEN, R. B. (1965b) Breeding density, reproductive success, and mortality of rock ptarmigan at Eagle Creek, central Alaska, from 1960 to 1964. *Trans. N. Amer. Wildl. Conf.,* **30,** 336–48.

WEEDEN, R. B. (1967) Seasonal and geographic variation in the foods of adult white-tailed ptarmigan. *Condor,* **69,** 303–9.

WEST, G. C. (1962) Responses and adaptations of wild birds to environmental temperature. *In:* HANNON, J. P. and VIERECK, E. G. (eds.) *Comparative Physiology of Temperature Regulation,* 291–333. Fort Wainwright, Alaska: Arctic Aeromedical Lab.

WEST, G. C. (1965) Shivering and heat production in wild birds. *Physiol. Zool.,* **38,** 111–20.

WHITAKER, J. O. JR (1966) Food of *Mus musculus, Peromyscus maniculatus bairdi* and *Peromyscus leucopus* in Vigo County, Indiana. *J. Mammal.,* **47,** 473–86.

WIELGOLASKI, F. E. (1972) Vegetation types and plant biomass in tundra. *Arct. Alp. Res.,* **4,** 291–305.

WIENS, J. A. (1969) An approach to the study of ecological relationships among grassland birds. *Amer. Ornith. Union, Ornith Monogr.*, 8. 93 pp.

WILDHAGEN, A. (1952) *Om vekslingene i bestanden av Smagnagere i Norge, 1871–1949.* Drammen, Norway: Statens Viltundersøkelser. 192 pp.

WILLIAMS, G. C. (1967) Natural selection, the costs of reproduction, and a refinement of Lack's principle. *Amer. Nat.*, **100**, 687–90.

WILLIAMS, G. R. (1954) Population fluctuations in some northern hemisphere game birds (Tetraonidae). *J. Anim. Ecol.*, **23**, 1–34.

WILLIAMSON, F. S. L., THOMPSON, M. C. and HINES, J. Q. (1966) Avifaunal investigations. *In:* WILIMOVSKY, N. J. (ed.) *Environment of the Cape Thompson Region, Alaska*, 437–80. Washington, DC: US Atomic Energy Commission.

WING, L. W. (1935) Wildlife cycles in relation to the sun. *Trans. Amer. Game Conf*, **21**, 345–63.

ZEUNER, F. E. (1959) *The Pleistocene Period.* London: Hutchinson. 447 pp.

ZIMINA, R. P. and GERASIMOV, I. P. (1973) The periglacial expansion of marmots (*Marmota*) in Middle Europe in Late Pleistocene. *J. Mammal*, **54**, 327–39.

ZIMMERMAN, E. G. (1965) A comparison of habitat and food of two species of *Microtus. J. Mammal.*, **46**, 605–12.

Manuscript received March 1972; revised April 1973.

Additional references

ANDERSON, E. (1973) Ferret from the Pleistocene of central Alaska. *J. Mammal.*, **54**, 778–9.

BLISS, L. C. *et al.* (1973) Arctic tundra ecosystems. *Ann. Rev. Ecol. System.*, **4**, 359–99.

CORBETT, P. S. (1972) The microclimate of arctic plants and animals, on land and in fresh water. *Acta Arctica* (Copenhagen), **18**, 7–43.

Development of biota

Origin and evolution of the arctic and alpine floras

Áskell Löve and Doris Löve
Institute of Arctic and Alpine Research,
University of Colorado

1. Introduction

In all subjects connected with biological evolution, the provisional aspect of science is most apparent. So much has to be done, and so much has to be revealed before any firm statement can be made. But there is room both for those who like to have a picture before their minds as to how things might have happened and for the others who prefer to remain sceptical and hold back their opinions until firmer ground is established for one view or the other.

Even when the studies are limited to the field of historical biogeography, considerable insecurity affects every major conclusion. In this field, it would appear that much of the effort has been that of trying to square the beliefs from earlier times with the observations of the past. The controversies on the history of dispersals and climates have also reflected the fact that this is actually a matter of combined biology and geology. Many of the conflicting ideas which still prevail in all books on the subject seem to be caused by an almost complete lack of communication between these basic biogeographical subjects. In addition, biologists observing the present distribution of taxa tend to ignore important observations made by palaeobiologists studying the bygone distribution of these same or related biota, except concerning the immediate past, and vice versa.

Temporary solutions are also characteristic of studies of more limited problems of biogeography as, for example, of the development and history of some of the more restricted plant formations. One of the better known of these are the so-called cryophytes, or the plants covering the cold barren lands or tundras of the high altitudes and latitudes. The origin of the tundra and its plants has long been and will long remain a subject of dispute, but a dispute which can take place between scientists with a similar background knowledge, who will admit, at least provisionally, the superiority of some ideas over others. The task of the makers of the new explanations of the history of these plants, which will be discussed below, is essentially that of bringing together the external evidence of present and past distributions of biota and connecting them with modern geological ideas about the inconstancy of oceans and the drifting of continents, and the internal evidence of the processes of evolution and the conservatism of the species as revealed by experimental cytogenetics. Such an approach, it must be admitted, is likely to require a revolution in biogeographical thinking, a revolution that has been long overdue.

2. Characteristics of tundra and cryophytes

It is customary to use the name tundra for the lands of high altitudes and latitudes which are devoid of trees. Some phytogeographers (Troll, 1948, 1960) prefer to restrict its use to the arctic regions, because other such areas may be only physiognomically similar to it and covered by vegetation of different evolutionary relationships. That is an academic question. The tundra in the wider meaning here accepted occurs only where the summers are short, cold, and usually windy so that growth of trees is prevented. Although winter cold is considerable in some tundra areas, mild winters are characteristic of many oceanic islands where tundra is predominant, as, for example, Iceland and the Falkland Islands, and early or late frosts certainly are more damaging to plant growth than are the most severe winter colds (Levitt, 1966; Mazur, 1969; Weiser, 1970). The treeline is, however, determined not only by the temperature of the summer, but also by wind velocity and duration, amount of ultra-violet light, and several other less well-understood factors.

Throughout the northlands where the tundra is predominant the summer is a single nightless day, or in the southernmost parts of this vast area, the nights are at least short. Even in cloudy situations, which are common in many parts of the Arctic, the daylight is not broken by darkness for the months of the growing season. The alpine tundra is free from these long days, but instead it enjoys a considerable increase of ultra-violet rays, which seem to add their growth restricting effects to those of the cold and wind by turning certain growth hormones into growth retarding chemicals that force trees and shrubs into krummholz and shorten the stems of other plants (D. Löve et al., 1970). A detailed description of the physiological characteristics of cryophytes and the differences between those of arctic and alpine situations is given by Billings and Mooney (1968); these and numerous other attributes of all kinds of plants of high altitudes and latitudes all over the world are thoroughly reviewed by Walter (1968), whereas Bliss (1971) reviews the life cycles of cryophytes, and Savile (1972) their adaptations.

The subsoil of much of the tundra is frequently frozen even in the summer, although permafrost is more common where the winter is continental and cold than where the temperature is affected by proximity to the ocean. In some areas of Canada and Siberia, permanently frozen soil is reported to reach considerable depths, although that is of no consequence to the vegetation, because a thin layer of ice is equally effective in preventing all root growth and interfering with natural drainage. It is more important to notice that in such areas the rapid interchange of frost and thaw produces considerable heaving of the soil resulting in solifluction, which damages roots of plants and leaves them vulnerable to drought before the growing season begins. Thawing also produces a superficial layer of soil which is soggy with cold water, thus restricting plant life only to those species which can endure a cold, waterlogged soil. In well drained areas that are protected from wind by the landscape some low shrubs can survive, but their height is usually limited by a protecting snow cover in the winter and the desiccating winds above it.

The tundra is covered by mosses and lichens interspersed with only a few higher plants, because it is characterized by physiological drought. In wet areas the soil tends to be peaty because of acidity and slow decomposition; in such areas communities of a few species of grasses, sedges, and heath shrubs may cover several square kilometres. Much of the real tundra, however, is covered with only patches of vegetation between which are large tracts of stones, gravel, or tundra ostioles, which are solifluction boils kept open by annual frost action and are invaded only by a few fast-growing annuals, notably *Koenigia islandica* L. (Sörensen, 1942; Rousseau, 1949).

The polar limit of the tundra in the boreal and austral hemisphere, and also its upper limit on mountains, is set by increasing shortness of the growing season and probably also by the absence or thinness of soil. Its limit toward the temperate zones or lowlands is determined by a soil layer thick enough and a summer long enough to support the growth of trees. The term 'tundra' is not applicable to the total barrens of arctic and antarctic lands or mountain peaks which are free of all vegetation. Such completely plant-free areas characterize much of Antarctica south of the 60th parallel, although two higher plants reach Graham Land (Skottsberg, 1954; Greene, 1964, 1970a, b) and three species reach north of the 80th parallel in the extremely cold and dry Canadian Arctic Archipelago (Porsild, 1964). Most high mountains in the temperate zone are totally bald at the summit, and even in the tropics no higher plants reach above 5000–7000 m altitude (Troll, 1959; Hedberg, 1964; Coe, 1967).

Since the arctic and antarctic lowlands actually are characterized by high alpine conditions beginning at sea level, the tundra is in fact a general alpine phenomenon. Therefore, the plants of the tundra wherever it occurs belong to the so-called cryophytes, or cold-loving plants, the largest part of which are the arctic plants. Nowhere outside the arctic regions is the tundra as wide and continuous and rich in plants as in the north-lands of Eurasia and North America.

Alpine plants or cryophytes belong to many families and numerous genera, though the preponderance of monocotyledons, especially grasses and sedges, is significant. These plants are always characterized by dwarf, low creeping or branching types of growth which seem to be conditioned by the extreme exposure or the curtailment of the growing season by early cold or snow cover (Böcher, 1938). It is assumed that such conditions stimulate the plants to adapt to rapid growth and flowering as soon as the short summer arrives, though these phenomena themselves certainly are genetically determined adaptations. Cryophytes are frequently perennial and only rarely annual. In the life form classification by Raunkiaer (1934), they are either so-called hemicryptophytes, whose overwintering buds are protected by the leaf remains of the previous year at or below the soil surface, or chamaephytes, which are mostly dwarf shrubs on whose persistent twigs the overwintering buds are situated above soil level; but geophytes over-wintering buried in the soil are also found. These life forms are frequently characterized also by flower buds that are formed the previous autumn so they can blossom fast on the emergence of the warm season (Sörensen, 1941). Alpine and arctic plants often grow as hemispherical cushions (fig. 10A.1) or tend to have their numerous short stems closely crowded together. The rarity of even low trees which then are crooked or pressed towards the ground is characteristic, and so also is the almost complete absence of annuals. In these characteristics, cryophytes are physiognomically similar to the plants of the hot desert, although in such situations shrubs and annuals are much more common (Went, 1948; Walter, 1968).

It is typical of many cryophytes that their flowers are pollinated by wind rather than by insects or other carriers, although autogamy, or self pollination, frequently occurs even in those species which have showy flowers with apparently relict insect-attracting devices. Allogamy, or cross pollination, is, however, more frequent in austral mountain plants than in those from the arctic tundra, a condition which still remains unexplained. An important group of cryophytes exhibits an apparently normal seed production, but some or all of the seeds are produced without fertilization. This phenomenon is known as agamospermy (Gustafsson, 1946–7; Gustafsson and Nygren, 1958). Several species are obligate agamosperms, so that all their seeds are always formed in this way, as for instance in the genus *Alchemilla* of the Rosaceae (fig. 10A.2) and the genus *Antennaria* of the Asteraceae. The more common condition, however, is faculative agamospermy. Then seeds may be normally formed by the sexual process which is replaced only under extreme conditions, as is known to be the case in the genus *Ranunculus* of Ranunculaceae, the genus *Acetosa* of Polygonaceae, and numerous other groups. Or, more frequently, facultative agamospermy is characterized by predominantly asexual seed production, which is only occasionally replaced by the sexual process; this rare fertilization gives rise to an immense variability in morphology and even in chromosome number, because every one of these actually hybrid combinations can then reproduce agamospermically as clones and give rise to large populations of apparent stability comparable to the sorts of cultivated potatoes. This is typical in the genus *Poa* of the Poaceae and of the genera *Hieracium* and *Taraxacum* of the Cichoriaceae. Such facultative agamospermy has resulted in the thousands of meaningless microspecies distinguished by some taxonomists, who even use them as reason for rejecting an evolutionary species concept without realizing the actual hybrid nature of these agamosperms.

With rare exceptions, cryophytes spread effectively by direct vegetative means so they may be more or less independent of reproduction by seeds over long periods. Most commonly, such reproduction is by means of root-stocks or stolons, which may enable clones of the same genetical individual to cover a considerable area of land, even several square kilometres. Some such plants produce bulbils, usually as modified axillary branches associated with the leaves or leaf axils, but similar structures may replace some or all of the ordinary flowers in the inflorescence and give rise to false vivipary, as, for instance, in the arctic-alpine species *Bistorta vivipara* (L.) S. F. Gray of the Polygonaceae (fig. 10A.3). True vivipary, when seeds germinate while still attached to the parent plant (Gustafsson

Fig. 10A.1. A cushion plant. *Silene acaulis* (L.) Jacq. (drawing by Dagny Tande Lid, from Á. Löve, 1970b).

Fig. 10A.2. An arctic–alpine agamosperm, *Alchemilla vulgaris* L. ssp. *glomerulans* (Bus.) Murb. (drawing by Dagny Tande Lid, from Á. Löve, 1970b).

Fig. 10A.3. A bulbilliferous agamosperm, *Bistorta vivipara*(L.)S.F.Gray(drawing by Dagny Tande Lid, from Á. Löve, 1970b).

and Nygren, 1958; Wycherley, 1953) to secure speedy rooting, is common in some alpine and arctic grasses; it is frequently connected with agamospermy and is due to inherited tendencies which seem to be triggered by light and cold and other environmental stimuli (Schwarzenbach, 1951, 1953, 1956).

One cytological characteristic of cryophytes is worth mentioning at this stage, since it may explain their evolution better than all the other characteristics together and affect each of them. This is their high frequency of polyploids, or of species with high chromosome numbers, a phenomenon which is purely genetical and affects dispersal and survival by increasing considerably the adaptability of the species to all kinds of extreme conditions (Löve and Löve, 1949, 1971).

3. Distribution of cryophytes

Nobody knows how many mountains and mountain chains have risen and eroded away during the history of the earth, and the tundra of most of those that once were has left no traces, either geological or biological. Present mountains, which naturally have been formed at various times, are sometimes so-called island mountains, which have ascended from the plains through the action of volcanism, ranging in age from the Tertiary African giants – Kilimanjaro, Mt Kenya, and Mt Elgon – to the recently formed Mexican Paricutin. However, most high orography is uplift mountains combined into mountain ranges which may be of great length, though they have not necessarily all existed for the same period of time. Some geographers find it possible to classify all high altitude regions into only three great mountain systems, the Eurasian-Australasian, the western American, and the African systems, though from a geobotanical point of view the mountains of these systems carrying tundra seem to require a considerably more complicated classification (Good, 1964).

When relating the distribution of cryophytes to the present distribution of tundras, it becomes evident that although some alpine plants have reached considerable areas that may even span both hemispheres, most such plants are considerably more local or restricted to some few mountain systems. However, the different age of these species is suggested by their areas of distribution, which also seem to have been affected by the varying climates of the past, periodically lowering the snowline, and thus the lower border of the tundras, to unite those cryophyte associations which presently may be effectively separated by forested valleys.

As pointed out by Troll (1959), the plants of the barrens of the tropical highlands are very different from those of the temperate mountains, a fact which he explains on the basis of different ecological conditions and, mainly, as a function of the lacking annual thermal rhythm. Clausen (1963) explained it in terms of the difference in the genetical constitution of the lowland species from which the austral and boreal cryophytes derived. In the classification by Troll (1959), these southern mountains and their cryophytes are distinguished as the Andes Region, which reaches from Colombia, Equador, and Peru to northern Chile and Bolivia; the East African region, which includes the highlands of Ethiopia, the mountains of the volcanic and lake district, to Uganda and into Zambia; the highlands of central Madagascar; the plateau mountains of Indonesia and Indochina; and the Tibetan mountains, ranging from Pamir and Hindu Kush to Chungking. Each of these complexes is characterized by cryophyte groups which have evolved separately from older lowland plants; the vegetation of the Tibetan mountains, for instance, is apparently composed of several old elements which have intermixed, one of these being a remnant of an old flora of the mountain chain north of the Tethys Sea which is also represented in the southern Rocky Mountains of North America. The plants of the mountains of the austral temperate and subantarctic zones seem to be more closely related to the apparently relic and very old lowland cryophyte flora of the antarctic tundra. However, there is an evident relationship between a considerable part of the arctic flora and that of temperate Eurasiatic and North American mountains. A few species involved in this relationship even range to the other side of the Tropics to New Zealand (Du Rietz, 1940) as, for instance, *Carex diandra* Schrank, *Carex echinata* Murr, and *Carex tripartita* All. (for name, see Á. Löve *et al.*, 1971); and Tierra del Fuego (Dusén, 1900; Roivainen, 1954; Raven, 1963). Representative examples of this group of plants are, for instance, *Koenigia islandica* L. from which, in our opinion, *Koenigia fuegensis* Dusén is inseparable (cf. map in fig. 10A.4); *Carex capitata* L. with the North American and amphiatlantic ssp. *arctogena* (H. Sm.) Böcher reaching northern Scandinavia, with which the Tierra del Fuego plant *C. arctogenoides* Roiv. is at least very closely related; *Carex maritima* Gunn.; *Trisetum spicatum* (L.) Richter; and others (for more details, cf. Hultén, 1958, 1959, 1962, 1968; Raven, 1963; Roivainen, 1954; Moore and Chater, 1971).

The general distribution patterns of cryophytes with arctic connections are reasonably well known, thanks especially to the enormous efforts by Hultén (1937, 1958, 1962, 1968 and elsewhere) in mapping the areas of the circumpolar boreal flora. The more detailed and local patterns are, naturally, best known from Europe, where these plants have been observed by hundreds of botanists since Wahlenberg (1814) made his initial studies of the relationships of plants of Scandinavian and central European mountains (Raven and Walters, 1956; Favarger, 1956, 1958). Much is also known about the relationships of the mountain flora of Siberia studied by generations of Russian botanists and most recently by Tolmatchev (1960a, b, 1966a, b), Tolmatchev and Yurtsev (1968, 1970), and Yurtsev (1966). In North America, however, only a handful of geobotanists have studied the cryophytes of the eastern and western mountains, so their floristic composition and relationships are still incompletely understood. The history and relationships of the Appalachian mountain plants are, perhaps, no less thoroughly known than those of other mountain complexes in the world (Löve and Löve, 1965, 1966; D. Löve, unpublished), whereas the knowledge of the details of the distribution of Rocky Mountain plants and their relationships is more patchy (Weber, 1965; Á. Löve *et al.*, 1971).

Almost all discussions of the origin of the cryophytes in arctic and other regions have centred upon concrete studies of their recent distribution. Thus, the relationship between the flora of the central European mountains and that of Scandinavia was observed by Wahlenberg (1814), whereas Forbes (1846) concluded that 'the alpine floras of Europe and Asia, so far as they are identical with the flora of the Arctic and sub-Arctic zones of the Old World, are fragments of a flora which was diffused from the North', and that 'the termination of the Glacial epoch in Europe was marked by a recession of an Arctic fauna and flora northwards'. The correctness of the first part of these observations has been verified by later palaeobotanical and palynological research (Firbas, 1949, 1952; Godwin, 1956; Walker and West, 1970), whereas the hypothesis of the complete northward dispersal of these plants after the glaciation still is open to dispute.

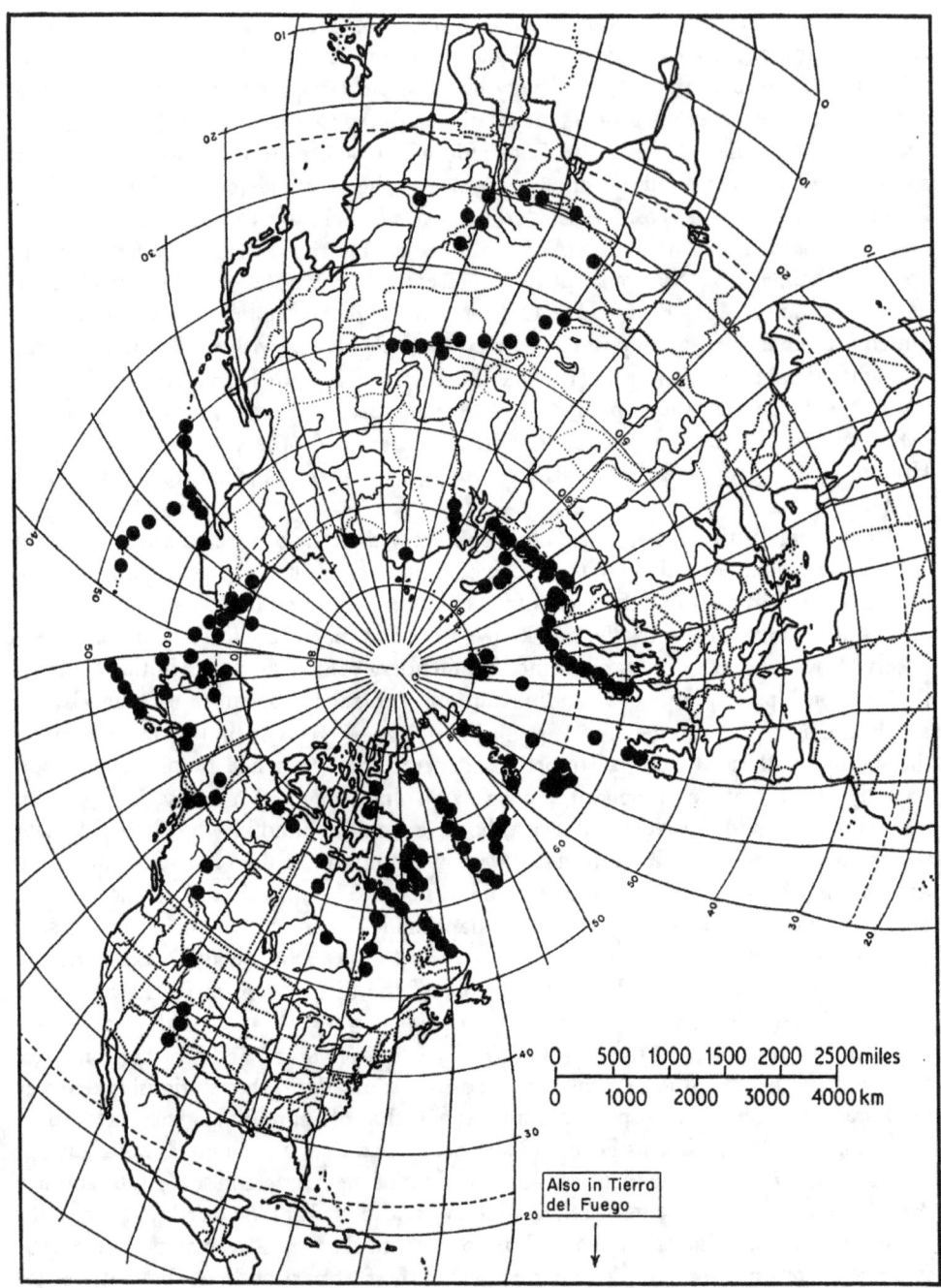

Fig. 10A.4. The distribution of *Koenigia islandica* L.

Darwin (1859) fully agreed with Forbes (1846) as to the dispersal of arctic plants to temperate alpine situations during the glaciations, and so did Hooker and Gray (1880), Wallace (1900), Rydberg (1914), and others. Darwin (1859) also concluded that the present arctic flora must have been more uniformly distributed over the northlands in the Pliocene or earlier, an observation subscribed to by Hooker (1862) and others. Darwin (1859) likewise pointed out, on the authority of Hooker, that several plants of tropical mountains in South America and Africa are distinctly related, if not identical, to arctic species (cf. also Engler, 1904; Hedberg, 1957), and indicated that bipolar plants of boreal origin which presently occur in southernmost South America or in New Zealand had likely arrived at their austral localities through dispersal between tropical mountains during the Pleistocene. This latter opinion was supported by Hooker (1862), and it is still maintained by Raven (1963) as the only plausible explanation available for these bipolar distributions.

Although the Forbes–Darwin–Hooker opinion as to the reasons for the occurrence of arctic species in temperate or even tropical mountains has been almost generally accepted for more than a century, some recent authors have presented deviating ideas. Thus, Hultén (1958) postulates that the northern hemisphere was characterized by a warm climate in the Tertiary during which most of the temperate and subtropical plants evolved and reached circumpolar distributions. During the Pleistocene, many of these species were forced from their northern areas, and the circumpolar areas were broken up into isolated fragments. Hultén is apparently of the opinion that the present alpine flora of temperate regions has not dispersed southwards from the Arctic except in a few cases, but rather that the alpine floras ought to be regarded as being remnants only of the circumpolar Tertiary cryophyte flora which became isolated during the Pleistocene glaciations. Tolmatchev (1960a, b, 1966a, b), who has made very thorough investigations of the arctic–alpine flora of the Asiatic lands of the Soviet Union, agrees with these conclusions, though Tolmatchev and Yurtsev (1968, 1970) emphasize the complex origin and history of both the arctic and alpine floras of Asia and also their relationship with the Tertiary nemoral flora of the mountains. In this we concur. Weber (1965) in a thorough review of the relationships of the Rocky Mountain flora, points out that its complicated nature results from the fact that it includes not only old elements of the Tertiary – which the present writers prefer to regard as remnants of the alpine flora of the north shores of the Tethys Sea – but also circumpolar species which may have occurred in these mountains prior to the Quaternary or dispersed to them during the glacial phases. That such a dispersal has taken place is evident from the occurrence of arctic–alpine plants in Tierra del Fuego, as mentioned above.

4. Genetical considerations

It is appropriate here to introduce a brief discussion of the cytogenetical processes, which are basic for speciation and subspeciation, since they are likely to be fundamental also for the dispersal and adaptation of taxa to new environments. These processes have recently been discussed in considerable detail by Gustafsson (1946–7), Stebbins (1950, 1966), Clausen (1951), Grant (1963, 1964), Mayr (1963, 1970), Á. Löve (1964), Timoféev-Ressovsky *et al.* (1969, 1973), and others, though most of the knowledge in this field is traditionally ignored in biogeographical texts. It has been shown that a species in the biological meaning of the term, as accepted even by classical pre-genetics taxonomists as De Candolle (1813), is a genepool of interfertile populations: it is defined on the basis of its reproductive isolation by biologists approaching it from as different points of view as Turesson (1922), Danser (1929, 1950), Du Rietz (1930), Löve and Löve (1942), Stebbins (1950, 1966), van Steenis (1957), Hultén (1968) and, especially, Dobzhansky (1951), Mayr (1942, 1963, 1970), Grant (1963, 1964), and Á. Löve (1964). Evolutionary differentiation within such a genepool is predominantly and probably entirely confined to the local population, which we prefer to name 'deme' following Huxley (1942), Mayr (1963, 1970) and other recent authors who have redefined a stillborn suffix proposed by Gilmour and Gregor (1939). This differentiation, which has been named subspeciation because it is confined to a single genepool (A. Löve, 1964), is caused by gene mutations, genetic recombination through hybridization, and natural selection. The last is frequently partially replaced by genetic drift, because of the limited number of individuals in most demes, which are likely to be spatially somewhat isolated from other demes and, frequently, have been founded by a single individual (Clausen, 1951). It is the environment that determines the success or failure of every new combination. Because of the decisive influence of the environment on the genetical constitution of the demes, which are forced into harmony with the milieu, the local populations tend to group themselves into a continuous line of demes, which form so-called 'clines' (Huxley, 1939), of which more or less isolated groups of demes in definable environments have been termed 'ecotypes' (Turesson, 1922). This process of constant adaptation of parts of the genepool is the main cause of the fact that plants do not grow where they would but rather where they must (Salisbury, 1929). When this process combines geographical adaptation, isolation and some morphological differentiation, it will result in the development of taxonomically distinguishable varieties and subspecies each well adapted to a special environment. Their genetical combinations within the genepool are, however, not permanently conserved and may disappear into the common genepool of the species by aid of hybridization. It ought to be emphasized that this process is strictly genetical.

This may be the place to point out that although these genetical processes allow certain changes in reaction norms towards the environment, considerable changes in tolerances of a species are so rare as to make it impossible to imagine a rather fast wholesale change in climatical tolerances of plant communities, as recently suggested by Wolfe (1969). Tolerances of all kinds seem to be extremely conservative and extremely complex, making such a change at least a very rare, though not wholly excluded, phenomenon.

Conservation of new and favourable gene combinations is connected with the processes of speciation during which a barrier to reproduction is built up to prevent miscibility. As far as experimental evidence goes, these processes are never genetical but exclusively chromosomal, involving either linear or numerical changes in the chromosomes. The former, or gradual speciation, is based on accumulation of chromosomal rearrangements, probably mainly translocations but also complex inversions; whereas the latter, or abrupt speciation, involves dysploid changes in the basic number of chromosomes or, most frequently, direct polyploidy. Polyploidy, which suddenly produces an effective reproductive barrier between an old genepool and a new one, is either autoploidy derived from the duplication of the haplome (Heilbronn and Kosswig, 1966), or haploid gene set of a single genepool, or alloploidy, derived from a more or less sterile hybrid between two distinct but related genepools. As far as is known, both processes commence sympatrically within one or two demes, and frequently strengthen parapatrically (Lewis, 1966), but their reproductive isolation and changes in adaptive mechanisms soon result in that they continue their further differentiation allopatrically.

Polyploidy is a duplication of the genes of the original genepool or genepools, either all, as in a panautoploid (Löve and Löve, 1949), or at least some, as in a hemiautoploid or alloploid. According to our present knowledge, this phenomenon occurs with low frequency in all populations in all climates, but through natural selection it becomes most frequent in floras of restricted size, as in alpine and arctic regions. Its percentage increases distinctly with an increase in latitude (fig. 10A.5). Although it was long believed to affect hardiness and other tolerances directly under extreme conditions, it has lately been shown to be one of the most effective processes of pre-adaptation because of the increased possibilities of genetical variability with an increased chromosome number (Löve and Löve, 1971). This seems to be the explanation of the high frequency of polyploids in arctic and alpine regions, mentioned above as one of the main characteristics of cryophyte floras.

Polyploidy is pre-adaptation at the species level, making it possible for a taxon in a stable environment suddenly to invade a new area when it becomes available for dispersal, even far outside the tolerances required by the old milieu. It may also be pre-adaptation at the deme level, which is responsible for the development of new demes or clinal ecotypes which invade areas that suddenly become available to dispersal. Such genetically conditioned pre-adaptation may perhaps be caused by a single gene mutation affecting some of the important tolerance factors, although it is more likely to be caused by an accumulation of the necessary adaptational genes.

Pre-adaptation to changes in tolerance factors seems to be much rarer at the diploid than at the polyploid level, probably because considerable genetical changes are required to widen established tolerances so that the plant may invade drastically new environments. Such pre-adaptation mechanisms are not well understood and not easily available to experimental scrutiny. At the polyploid level, however, it seems to be the rule and it is easily explainable, and open to experimental study, on the basis of Melchers's (1946) hypothesis concerning the reasons for the increased adaptability of polyploids to extreme climates. Melchers assumed that hardiness is determined by only two gene pairs, of which F increases the hardiness with $1\,°C$, whereas H decreases it with $1\,°C$ or erases all the effect of F. Both genes are completely dominant, and the recessives, f and h, are without effect. Accordingly, diploids with the constitution $ffhh$ are hardy at $0\,°C$ and so also are individuals with $FFHH$ and $FfHh$. Whereas $Ffhh$ is hardy at $-1\,°C$, $FFhh$

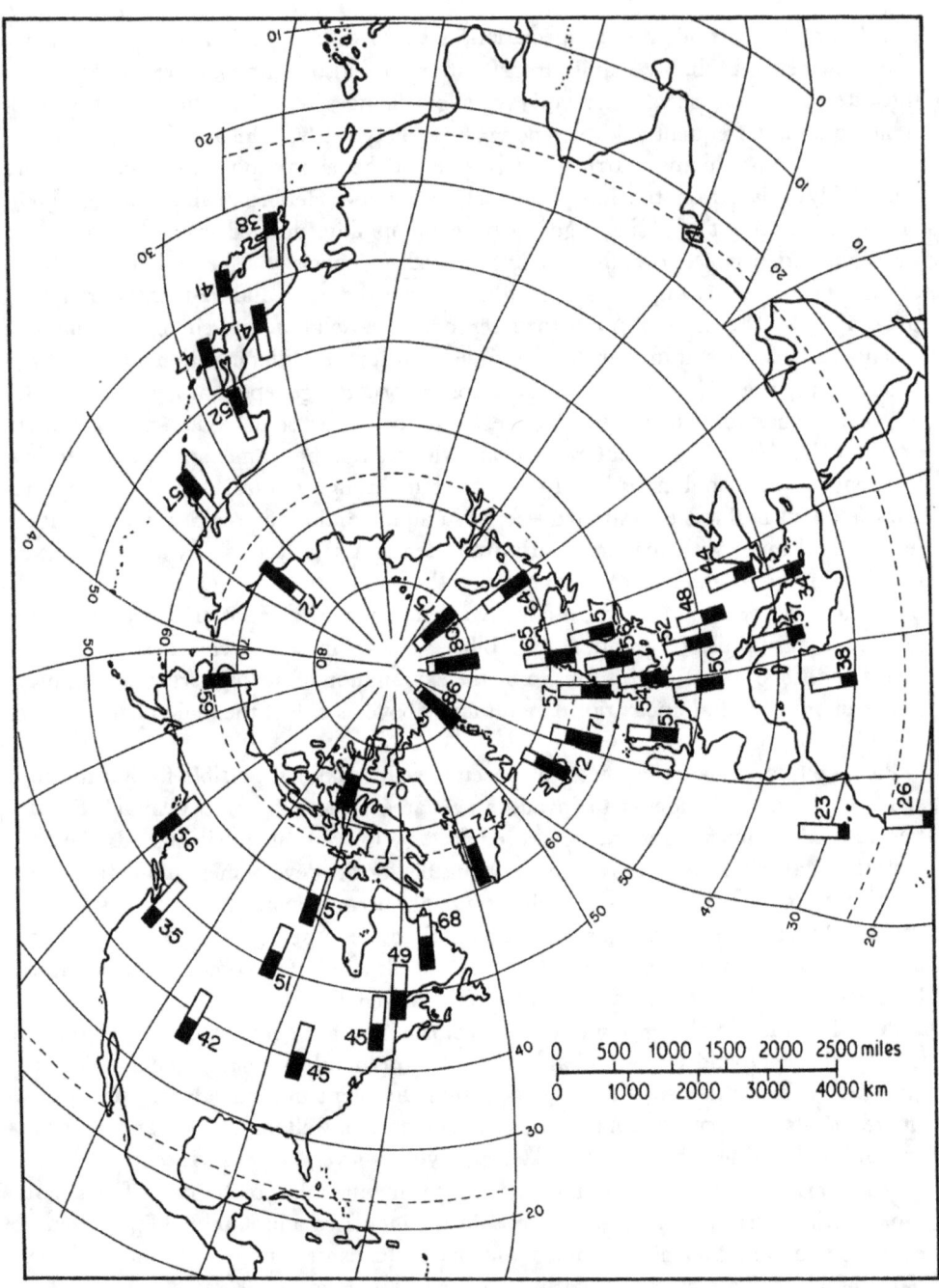

Fig. 10A.5. Frequency of polyploids in the floras of the northern hemisphere. Full column denotes 100 per cent.

at −2°C, *FFHh* at −1°C, *FfHH* at +1°C, and *ffHH* at only +2°C. If the temperature conditions during the critical period of flowering and seed production never drop below 0°C, it is likely that all these combinations will be available in every generation. But if the plant invades a region where temperatures down to −2°C are a rule during this period, then the population will soon become uniformly *FFhh*.

If such a population becomes polyploid, its selective adaptability will be drastically changed. The plants with the constitution *ffffhhhh*, *FFFFHHHH*, and *FFffHHhh* will demonstrate no difference from the original diploid lines and be hardy at 0°C only. Individuals with *FFFFhhhh*, however, will be hardy at −4°C, whereas those with *ffffHHHH* will be able to survive only at +4°C, but intermediate combinations will be most frequent and show intermediate degrees of hardiness. If the polyploid population is formed where the conditions in question are congenial even to the combination characterizing the most tender plants, polyploidy will not effect any change in the reaction of the plants to this environment. However, if the climate, or other environmental conditions connected with the tolerance factors under study, should deteriorate, or the taxon try to invade other areas where these conditions differ, then the pre-adaptive variability of these tolerance factors will become an effective selective property that will allow the population to survive considerably greater cold than the most hardy demes of the original diploid. This seems to be the explanation of the fact that although polyploids are most numerous in the large tropical floras, their frequency is by far highest in the much smaller floras of arctic and alpine regions where only a few species are able to survive (Löve and Löve, 1949, 1971). At the same time, a complete adaptation of such taxa to environmental extremes will result in a loss of genetical variability for such factors and so make the process and dispersal irreversible.

We want to stress our conviction that polyploids have not been formed at high altitudes or latitudes to any higher degree than in the much richer and more amiable tropical floras, and that their high frequency under conditions of cold and short summers is the result of pre-adaptation and selection only.

It ought to be emphasized that adaptation may either be aggressive or passive. In the former case, individuals disperse to a new area and establish a founding population (Mayr, 1942) which adapts to the new conditions on the basis of recombination of their own tolerance genes, or by the aid of recombination through hybridization between the founding individuals and the original genepool, if they are within pollination distance. Aggressive adaptation may be characteristic of the species, but it is most frequent at the deme level. It is the only kind of adaptation that occurs when new volcanoes or newly disturbed areas are invaded by vegetation, and it is especially characteristic of all kinds of weeds. Pre-adaptation is probably rarely involved in the process.

Passive adaptation takes place when the environment of a population changes, either rapidly as in the case of various human activities, or slowly as in climatic variations of different kinds. In such cases, pre-adaptation is probably considerably more effective than new recombinations, and more likely to allow the deme or species to avoid extinction because it has already built up the tolerances needed.

Plastic species are likely to be widely adaptive, both aggresively and passively. Strong

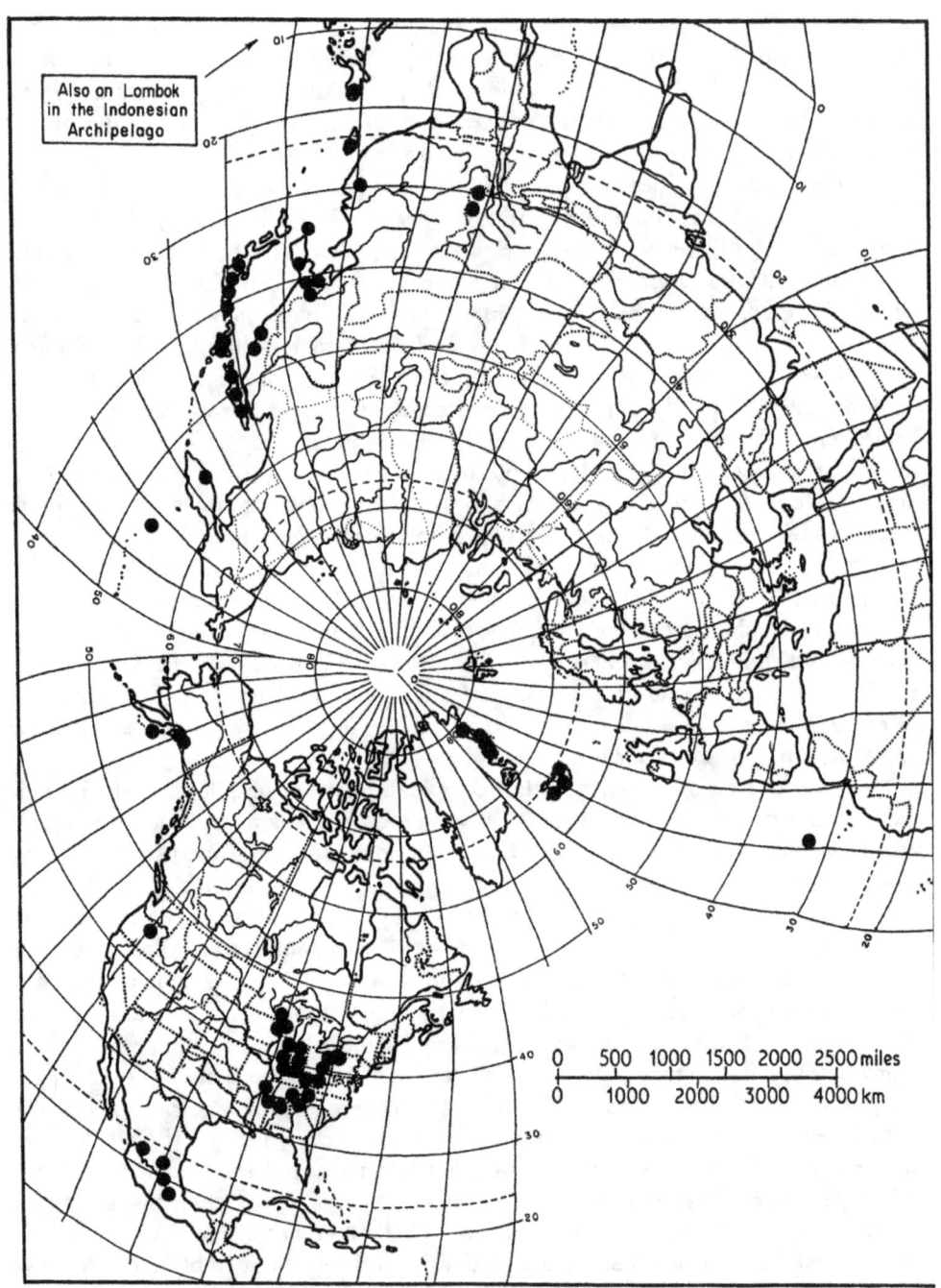

Also on Lombok
in the Indonesian
Archipelago

Fig. 10A.6. The distribution of *Bruoxiphium norvegicum* (Brid.) Mitten

selection through numerous generations will, theoretically at least, diminish their variability and reduce their plasticity, but only old populations that have been reduced to a few demes with limited heterozygosity are likely to become rigid and thus unfit for further adaptation. Nevertheless, most of the major floristic groups of the globe are apparently characterized by rigidity in some tolerance factors that tether their species to certain environments and prevent them from dispersing from a determined setting (Gleason and Cronquist, 1964). Although polyploids seem to be widely adaptive, this characteristic may be reduced through long and strong selection so that only a single one of the critical combinations in Melchers's scheme is available; such rigidity is met with in some high-polyploid ferns, which are at the last step towards extinction which long since reached their diploid ancestors. All loss of plasticity is irreversible, at the level of deme no less than at the level of species.

It might be inferred that plants that have lost or restricted their sexual reproduction would also lose their plasticity and climatic tolerance leading towards extinction. This does not seem to be so, at least not always. A case in point is the primitive moss *Bryoxiphium norvegicum* (Brid.) Mitten (incl. *B. madeirense* Löve & Löve), which is met with from the tropical forests of the Indonesian Archipelago to the high arctic tundra of northeastern Greenland (Löve and Löve, 1953; Lee, 1958; Nakanishi, 1964; Iwatsuki and Sharp, 1966; Pócs, 1966) as is shown on the map in fig. 10A.6. This species probably belonged to the subtropical forests of the northern shores of Laurasia even in pre-nemoral times and was able to both remain in place when the land drifted through the temperate to the cold zone and to disperse southward with its original companion plants. It very rarely reproduces sexually (Löve and Löve, 1953).

Within the limits of the genepool, new combinations may result in new demes almost *ad infinitum* but, although selection may change them considerably during adaptation, they can return to the original genepool if geographical isolation is not added to prevent this. They remain a part of the same biological species, even though they have been geographically isolated for millions of generations, as long as reproductive isolation is not produced, as, for instance, in the case of the two hexaploid taxa of *Platanus* named as two species by Linnaeus (1753), one of which has later been split further by American taxonomists following the morphological species concept. Such demes or races may show considerable variability, as is the case in the human species, but every one of their characteristics may be bred away through hybridization. When taxa have achieved the reproductive barrier of a biological species, however, the selection results in irreversible changes. Although demic variations of such a genepool are considerable, they must stay within the limit of the main taxon which soon becomes extremely stable towards other units at the same level. Speciation is an erratic process, fast in abrupt and slow in gradual species, and many new species will soon succumb to natural selection. But when a species has been firmly established, it is extremely conservative, almost immutable and more durable than the rocks on which it grows, so that most of the species of higher plants presently met with have remained the same, or almost the same, throughout times when mountain chains have risen and eroded away and when continents have drifted considerable distances on the globe.

5. Geological considerations

Most attempts to explain the peculiarities of the cryophyte floras of the world have concentrated upon either the arctic flora and its alpine counterparts, or upon the bipolar distribution of some plants of alpine situations. The oldest idea about the reasons for the present distribution of these plants attributed it entirely to suitable climate and environment, an explanation which is quite inadequate because it fails to account for the great differences in the flora between regions of similar climate and environment. This idea was followed by another, which is still predominant among biogeographers, based on the belief that the past and present geographic relations, the land connections and water barriers that exist at present and may be inferred to have existed in the past, provide an explanation of the floras peculiar to each region or locality not explainable by suitable environment. This was observed by Wallace (1880). Partly before and chiefly since his time a great literature on distribution has accumulated, encompassing a mass of exact knowledge of the present distribution of plants, and the knowledge of their past distribution has also advanced enormously. So also has historical geology, and the data on which our interpretations and theories ought to rest are so vast that no one man can really understand more than a small part of the evidence. Yet in order to interpret it one needs not merely a knowledge of the data, not merely a series of lists and maps of plants and their distribution, but a thorough comprehension of what these data mean, what the floristic differences may or may not signify, what the geological data indicate as probable or reasonable, and what they render in varying degree improbable. Overwhelming floristic evidence might override weak geological observations, as it has done, for example, in the case of the postulate of non-glaciated refugia in Scandinavia and North America (Löve and Löve, 1963), and the other way round. But a floristic list, and especially a list of fossil plants, may mean much or little, according to the accuracy and certainty of the identification of the taxa included, the probable completeness of the record, and the aspects that it represents.

The lack of cryophytes in the recorded floras of northernmost Eurasia and America even in the Late Tertiary is a significant fact in their distribution and origin. Their non-occurrence in the earlier Tertiary of Alaska (Wolfe, 1969) has probably little weight. For the Later Tertiary floras of northern Eurasia observations are extensive and varied, representing different aspects and in general probably most, though by no means all, of the woody plants then living in that region. One might fairly expect that if any real cryophyte formations were met with there in the Late Tertiary, some remains would have turned up among the fossil floras. On the other hand, the Early Tertiary plant record in the present arctic lands appears to be that of a deciduous forest flora of plains and open country of a presumably temperate climate, including the same and similar species now characteristic of the eastern deciduous forest province of North America (Gleason and Cronquist, 1964), the comparable floristic area in eastern Asia, the somewhat reduced floras of western Europe and, in part, the westernmost areas of North America south of the cold regions and north of California. This is the flora which is usually misnamed Arcto-Tertiary but more correctly termed the nemoral flora. It apparently covered the now northern lands in the Early Tertiary, and the region where Iceland now is situated in the Late Miocene, until the onset of the glaciations in the Pliocene.

To infer the entire absence of cryophytes from the whole northern continents from their non-occurrence in the Tertiary record would be going far beyond what the nature of the evidence warrants. The old nemoral flora is not yet sufficiently known to infer that cryophytes, or plants related to them, were not present in the region because they have not been found in the flora collected. One can say only that they were certainly not dominant and probably rare, if present, on the northern plains, and that they could have been met with only in the mountains, and that their progenitors might have grown near the coasts where there were few shadows, or in the forest-free areas or semi-deserts where there was plenty of sunshine. It might indeed be inferred upon other grounds, from the broader relations of the arctic–alpine plants to the ground cover of the present nemoral forests, that the pro-cryophytes gradually replaced the nemoral pediophytes towards the end of the Tertiary and that the nemoral flora still present around the 40th parallel near the eastern North American, eastern Asiatic, western European, and, to a much smaller degree, western American coasts, retreated southward during the entire Tertiary and subsequently became extinct save in the areas of mild climates where they presently survive. This inference would be in accord with the fossil record, which shows that arctic cryophytes had replaced the nemoral flora all over the northlands in the Pliocene and the Early Pleistocene.

The great differences between the distribution of the present nemoral flora and that of the Tertiary (Dorf, 1969) has led to several hypotheses, most of which have postulated tremendous climatic changes during the past 60 million years (Wolfe, 1969). However, all these hypotheses have ignored the necessity of regarding the geological evidence, so far as it goes, as fundamental when dealing with phytogeographical problems; at the same time the genetical stability of the species has been overlooked. The slow progress in biogeographical thought has also been affected by the conservatism among geologists that actually delayed, for almost three scores of years, the general acceptance of the only geological theory which provides a satisfactory basis for explaining all the major biogeographical phenomena. This is the very ingenious and plausible theory of continental drift as set forth by Wegener (1912, 1915), Du Toit (1937), and others. During the past two decades it has at long last been winning general geological acceptance (Carey, 1958; King, 1962; Runcorn, 1962; Blackett et al., 1965; Kay, 1969; Dietz and Holden, 1970a, b; Dewey, 1972), and is slowly fostering a long-needed revolution in biogeographical thinking (Chaloner, 1959; Polunin, 1960; Hawkes and Smith, 1965; Good, 1964; Illies, 1965; Melville, 1966; Löve and Löve, 1967a, b; Adams and Ager, 1967; Sneath, 1967; Tzvelev, 1969; Johnson and Thein, 1970; Smith, 1970; Keast, 1971; Menzel and Martin, 1971; Colbert, 1973; and the critical review by Meyerhoff and Meyerhoff, 1972).

So far as the theory of continental drift pertains to the cryophytes and their evolution, this approach differs from the older prevailing concepts of geological climatic conditions chiefly in assuming a constancy of the Poles and an almost complete immutability of climatic zones through which the continents drift, though the secondary effects of the lands on the climate even during the drift are acknowledged.

It seems to be the opinion of many geologists that all the continents were once joined in a single continent, Pangaea. There is reason to believe that Pangaea remained a single, irregularly-formed landmass for a long period of time, and that it was then balanced and situated around the South Pole with the central part, the present Sahara, at the Pole. This assumption is based on the evidence for continental glaciation on the Tassili Plateau in eastern Algeria in the Late Ordovician about 350 million years ago (Fairbridge, 1970). The long equilibrium was disrupted about 200 million years ago (Dietz and Holden, 1970a, b) when the original landmass divided into the later northern continent Laurasia and the later southern continent Gondwana through the formation of the early Tethys Sea, which at the beginning seems to have gone almost through the geographical South Pole just north of what is now North Africa. The causes of this oldest rift remain unknown, but the spinning movement of the centred continent changed into a spiral movement northward and westward, slowly moving Laurasia and Gondwana towards the other Pole and later splitting them into the present still drifting continents. That is another story of great importance for the understanding of evolution of animals and plants, because it seems to have been the drift that increased the rate of evolution considerably from what had been happening near the coasts of the stable Pangaean continent. During this drift, plants of all kinds arose and evolved, and since mountain chains were doubt-lessly formed on the continents, cryophytes must have developed from time to time. At least most of these early alpine plants later disappeared without a trace when subaerial erosion reduced mountains to produce a temperate climate in which alpine plants could not exist.

For our discussion of the origin of the arctic–alpine cryophytes, it is of interest to visualize the situation of the lands at the end of the Cretaceous, when most of the present genera of boreal plants had been formed (Hawkes and Smith, 1965). At this time the southern continent had been divided into the large Antarctic–Australian plate, South America was separated from Africa by a narrow South Atlantic, Africa had almost reached north to eastern Laurasia where it constricted the Tethys Sea to a narrow and long pre-Mediterranean and caused the uplift of the Alps, and India was on its way to hit the northern continent, push up the Himalayas and jolt the continent northward in such a way that the North Atlantic rift could become complete. However, the North Atlantic was only a small fiord in the far south, as indicated, for example, by the close relationship, at the generic level, of the floras of Nova Scotia and the Iberian peninsula.

It is of fundamental interest for the present discussion to emphasize that at the end of the Cretaceous and in the Early Eocene the northern shores of central Laurasia enjoyed a climate which made it possible for the nemoral flora to dominate the landscape, as shown by studies from Spitsbergen and other northern lands (Manum, 1962). It seems likely that the continent has since rotated somewhat clockwise, because in Alaska there occurred at the same time a rather different and more warmth-loving flora which seems to have been closer related to the Malayan flora of the southern coasts of the continent (Wolfe and Leopold, 1967; Wolfe, 1969). But the geological and plant record from the northlands (Manum, 1962) seems to provide unequivocal evidence that they were then situated at and around the 40th parallel, at least 40 degrees further south than shown by Dietz and Holden (1970a, b) on their Cretaceous map.

The northward drift continued with the speed of a little less than 2 degrees latitude per million years, as documented in the palaeobotanical layers. By the time that the North Atlantic rift had reached Scotland and the Scoresbysund area of Greenland, and volcanism had formed the northwestern part of Iceland in the Late Miocene about 15–20 million years ago (Moorbath *et al.*, 1968; Sigurdsson, 1969), the nemoral flora had dispersed into this area, indicating that these lands were then around the 40th parallel, to which latitude the nemoral flora has always been adapted (Schwarzbach, 1963; Einarsson, 1963; Friedrich, 1966). The rift was completed in the Early Pliocene (Einarsson *et al.*, 1967), at about the same time as South America contacted North America. According to Dietz and Holden (1970a, b), a pre-arctic ocean, or an hypothetical arctic bay, was then met with in the north, without connection southward, whereas Johnson and Heezen (1967) argue that the North Atlantic must have already been completed as a narrow strait between northeastern Greenland and Spitsbergen in the Mid-Mesozoic.

As repeatedly mentioned above, there is no trace of any kind of cryophytes in the flora of the northern lands in the Early Tertiary. Likewise, there is still no evidence of arctic plants from the palaeobotanical layers from the Middle Miocene from Iceland or Alaska. However, remains from western Iceland, of Late Miocene or Early Pliocene age (Lindquist, 1947), include genera of cold-loving shrubs. There is, therefore, reason to believe that in lands north of old Iceland cryophyte vegetation may have developed even in low altitudes when the nemoral flora still occupied this part of the North Atlantic area, and that in Late Miocene the lands north of the conifer zone had become occupied by the species which now characterize the arctic flora. During earlier parts of the Pliocene, this flora had ample time to disperse widely so that most of its species could reach the circumpolar distribution indicated by their present area, which was again split up by the Plio-Pleistocene glaciations.

6. The influence of the glaciations

Phytogeographers agree that the original arctic flora of perhaps 1500 species must have reached a largely circumpolar distribution prior to the onset of the Plio-Pleistocene glaciations. Although there is evidence of cold-temperate climate in Pliocene strata from Iceland (Einarsson *et al.*, 1967), only the northernmost lands may have reached into the real arctic zone at that time.

The northlands had become covered with their special flora before the great climatic deterioration commenced during the Pliocene, and most of the so-called equiformal areas observed by Hultén (1937) had been formed. These areas were, however, greatly disturbed where the glaciers were thrust into the temperate zone, and in the Arctic the areas of most of the formerly circumpolar plants were so radically split that they never recovered, as is evident from the present distribution of numerous species.

When the glaciations began 2 or 3 million years ago, the tundra was pressed southward but was also isolated in pockets surrounded by the ice further north. Some of the plants that had become adapted to the mild arctic conditions of the Early Pliocene may not have kept the plasticity which was required for a relatively fast dispersal away from the increasing cold and the shortening of the growing season, so they succumbed to natural selection without leaving a trace. Other species succeeded in escaping the glaciers either by dispersing south into the lowlands or mountains of the normally temperate zone, which became considerably cooler during the glaciations, or by occupying islands and coast-lands surrounded but not overrun by the ice, while a few very hardy species were able to survive on ice-free nunataks surrounded by the inland ice.

The southern refugia were considerable in size, especially near the coasts of eastern and western North America, in central Alaska, central Europe, and eastern Siberia. These refugia were open southward for dispersal in both directions and allowed the dispersal of some northern species far south into the present lowlands of the temperate zone (Frenzel, 1968; Guillien, 1955; Sjörs, 1956). Refugia were absent, or at least very restricted, south of the ice in the middle of the continents, because the glaciers pre-vented runoff north of the continental divides and caused the formation of enormous lakes that drowned the vegetation before it could succumb to the cold (D. Löve, 1959).

The northern refugia were of various sizes, from small nunataks or islands to areas as large as most of the Canadian Arctic Archipelago extending almost without inter-ruption from the coasts of northern Alaska to the only mildly glaciated northern part of Greenland (Wegmann, 1941; Á. Löve, 1959; Schwarzenbach, 1961; D. Löve, 1962). They were always closed southward and from each other by a more or less considerable ice cover which effectively prevented dispersal, and thus gene flow, between the isolated populations for numerous generations during each glaciation. Small coastal refugia were typical of both the Atlantic coasts south of the permanent ocean ice and also of the larger North Atlantic islands (Löve and Löve, 1963), and of the coasts of the northern Pacific (Karlstrom and Ball, 1969), whereas nunataks are likely to have been predomi-nant near steep coasts, where they seem to be a function of the height of the coastal mountains, degree of slope, and width of the continental shelf (Dahl, 1946; Lindroth, 1965), and also where mountain summits were high enough to reach above the inland ice. Naturally, the number of species of plants and animals and the size of their populations varied considerably, depending not only on edaphic, climatic, and other environmental factors in each refugium, but also on the normal factors of haphazard selection when the refugia closed. Final survival of plants and animals in each refugium certainly was de-termined by the most severe summers of each glaciation rather than by averages of temperatures and hardly by the most severe winter colds.

It should be pointed out that some of the arctic cryophytes which at present are also met with in southern mountains, may have had a very complicated history. In the cases of *Vaccinium gaultherioides* Bigel. (fig. 10A.7, cf. D. Löve and Boscaiu, 1966), *Anthoxanthum nipponicum* Honda (Löve and Löve, 1968), *Beckwithia glacialis* (L.) Löve & Löve, *Cardamine bellidifolia* L. *Saxifraga hyperborea* R. Br., *Dryas octopetala* L., *Diapensia lapponica* L., and

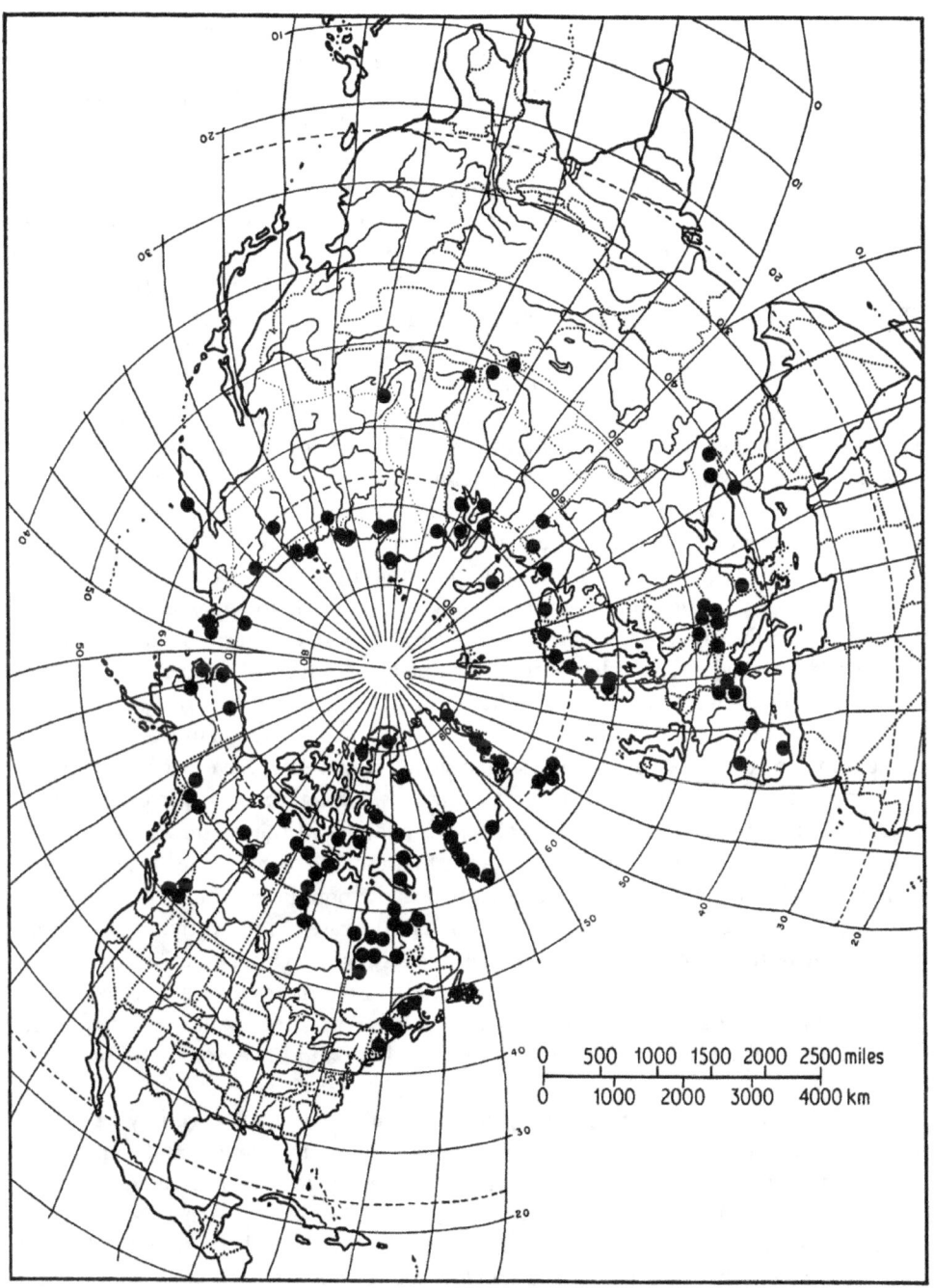

Fig. 10A.7. The distribution of *Vaccinium gaultherioides* Bigel.

some other diploids with mainly arctic distribution and isolated occurrences on mountains in the temperate zone, we suggest that whereas the American alpine populations may be the results of southward dispersal during the glaciations, the populations in central and southern Asiatic mountains are more likely to be relics from Tertiary times, probably the ancestors of the arctic and American populations.

It is also important to remember that although the great majority of arctic plants are species that were formed when these lands were situated at more southern latitudes, either in mountains or on suitable lowlands, a few species certainly have been formed in the Arctic itself. One of the youngest of these is the high-polyploid *Alchemilla faeroeënsis* (Lge.) Bus. (Löve and Löve, 1956), which is restricted to the Faeroes and eastern Iceland in the North Atlantic. Since the separation of these islands cannot have been earlier than the Pliocene or later than during the Early Plio-Pleistocene glaciations, that approximately dates its origin as a relatively recent alloploid. Some other species have a less restricted though limited distribution in the Arctic which may indicate that they originated there even prior to the glaciations and survived not far from their present area; a good example of this group is the dodecaploid American poppy, *Papaver cornwallisense* D. Löve, and probably also the sedge *Eriophorum triste* (Th. Fr.) Hadač & Löve (fig. 10A.8), and some other species with similar distribution areas.

It is assumed that when the climate improved during the interglacials and after the last glaciation the cryophytes which then occupied more southern lowlands retreated northward with the ice, and that they also found some refuge in temperate mountains. This is inferred from studies of the fossil flora from in front of the retreating ice in central and northern Europe and eastern North America. The role of the plants of the northern refugia in the re-vegetation of the northlands may, however, have been considerably greater than that of the flora following the glaciers, a suggestion supported by the fact that some of the temperate alpine species with relatives in the northlands are racially somewhat different, thus indicating that at least some of the plants of the southern refugia never reached the real northlands again, whereas the populations isolated in the north became the main source of the new arctic flora. In the case of the North Atlantic, North Pacific and arctic islands, which have been effectively isolated from other lands by oceans since prior to at least the last glaciation, all the later plant cover must have originated from populations that survived in the countries themselves, with a few exceptions that later may have dispersed by aid of ocean currents (Löve and Löve, 1947; D. Löve, 1963; Dahl, 1963).

The Pleistocene glaciations were repeated several times. Each certainly caused the extinction or repulsion of some elements in parts of the northlands, and some species became extinct. The flora of the oceanic islands was drastically decimated (Löve and Löve, 1956, 1967b). So also was the arctic flora itself, ending up with only about 1000 species of higher plants. The areas of these species, which had been circumpolar prior to the glaciations, became split up into smaller isolates which sometimes became characterized by morphological differences that have warranted their recognition as subspecies and varieties, of which no example is better than that of the arctic–alpine complex that includes the species *Papaver relictum* (Lundstr.) Nordh. (2n = 70), *P. radicatum* Rottb.

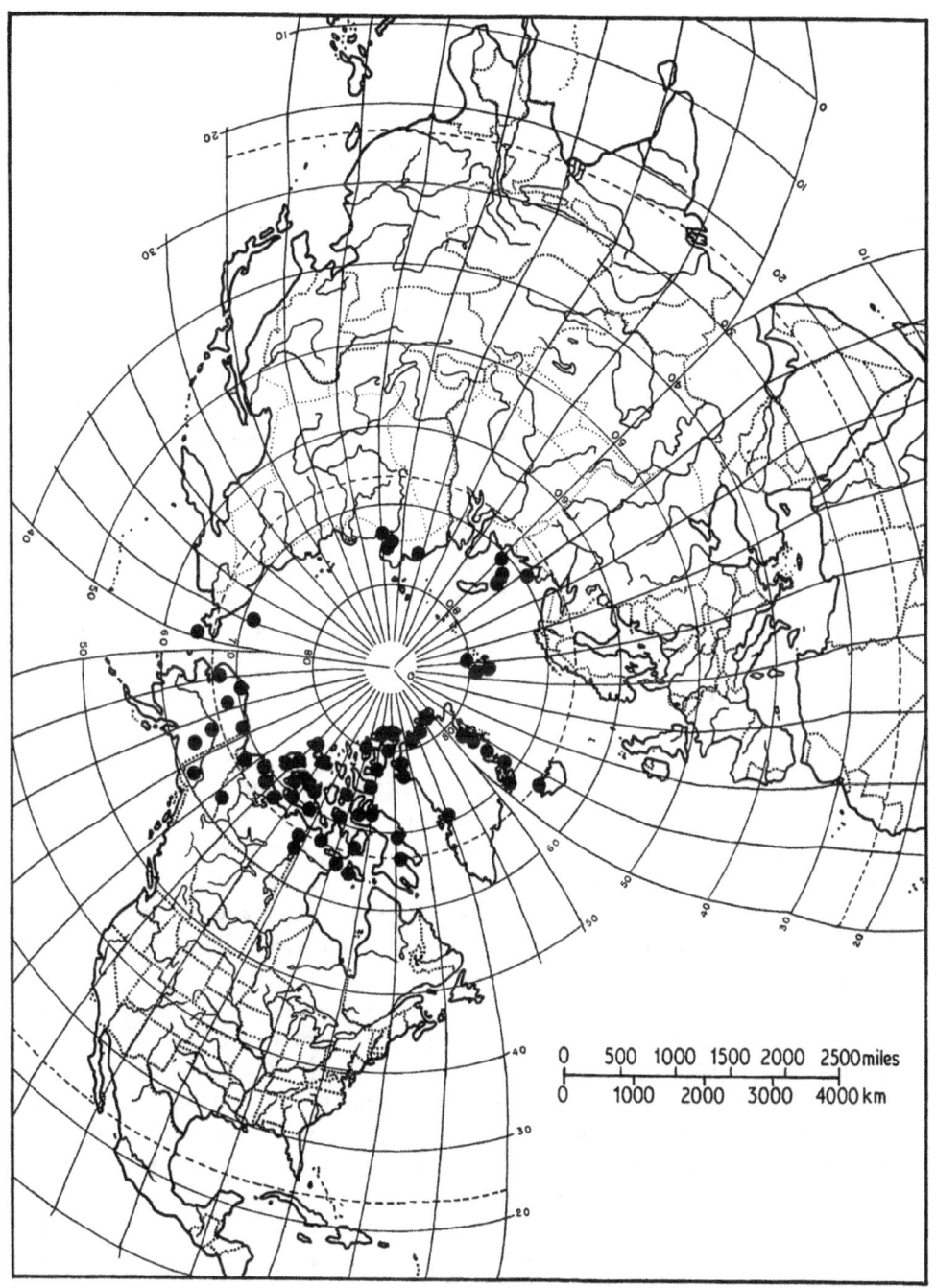

Fig. 10A.8. The distribution of *Eriophorum triste* (Th. Fr.) Hadač & Löve.

s.str. (2n = 56), *P. kluanense* D. Löve (2n = 42) and *P. alpinum* L. s. lat. (2n = 14) (Knaben, 1959a, 6; Á. Löve, 1962a, b, 1970a, b; D. Löve, 1962, 1969).

As inferred above, there is some reason to believe that more polyploids than diploids were able to adapt to the cold climate into which the Laurasian land drifted during the Tertiary. Therefore the frequency of polyploids in the old arctic flora has conceivably been higher than that of the nemoral flora from which it originated. It is, however, unlikely that this frequency reached the 70–80 per cent level presently characteristic of the floras of northern refugia and arctic islands (Löve and Löve, 1971). That high frequency, we believe, is the result of the selective effects of the coldest period of the Pleistocene when the original arctic flora was probably reduced to only two-thirds of its original size.

The plants of the northlands and highlands of the world are listed and described in numerous manuals that cover only limited parts of this remarkable flora. There is no flora manual or even a checklist covering all the alpine plants of the world, and no complete review of their distribution, past or present. The flora of the arctic regions, however, has recently been reviewed by Polunin (1959) utilizing a collective concept of the species, and is being treated in a manual that is only partially published by Tolmatchev (1960–71), who accepts a considerably narrower species concept. Nobody has yet attempted to study all these plants on the basis of the biological species concept, which alone gives a thoroughly safe background for discussions of their evolutionary history.

7. Conclusions

The acceptance of the theory of continental drift as a basis for our understanding of plant dispersals and evolution allows us to imagine the earth and its inhabitants as something reminiscent of the Oriental ivory carving which consists of a series of freely rotating spheres, one inside the other. Such is also our image of the earth. In its centre is the core, outside which float the continents on their plates. Outside the continents is the biosphere with its fauna and flora, which in turn is surrounded by a layer of atmosphere which largely determines the climates of the layer below.

The sphere of the continents may seem more independent than the others, whereas the sphere of vegetation and animal life is distinctly less mobile in relation to the sphere of climates, since plants depend more on the climate than on the rocks on which they grow, and the climate depends on geographical position on the globe. Therefore, the vegetation tends to stay in the same latitudinal position wherever the continents may drift, and the dispersal activities are mainly aimed at retaining this position, with the exception of the rarer cases of adaptation to new environments and new climates. That carries us to the cryophytes and their story.

Cryophytes are produced from ordinary populations of lowland floras wherever conditions arise that are conducive to the selection of plants that are able to withstand the climatic rigours of high altitude or latitude. Such conditions have frequently arisen in the past, when volcanoes were built up or mountain chains formed, but also when the continents drifted into the cold boreal or austral zones. In the case of alpine cryophytes, a small percentage of plants of the lowlands have invaded the new mountain slopes and adapted themselves to the new conditions as races of the species surrounding the mountain; if time is sufficient to extinguish the original lowland races, or to allow the more or less isolated highland races to undergo speciation processes that prevent miscibility with the original populations, the taxa of the mountain may evolve into a distinct group of cryophytes which are unable to return to the original conditions if the mountains erode away. We have records of old cryophytes only when they have been conserved as fossils or when they have been able to survive such changes by invading other alpine regions, whereas nobody will ever know how many alpine plant species have evolved and succumbed when their environmental conditions changed too drastically for them to re-adapt.

It is likely that some plants invading a new mountain from a lowland population have been genetically pre-adapted to such severe conditions prior to their dispersal, since polyploids, which are strongly pre-adapted to fast adaptation to extreme conditions, are predominant in alpine areas all over the world. Such pre-adaptation has also preceded the zonation of vegetation in the boreal hemisphere, when the land of Laurasia drifted northward from the warm to the cold regions with a speed of less than 2 degrees latitude per 1 million years during the Tertiary. The early vegetation of the northern rim of Laurasia was tropical and related to the present Malayan flora, certainly with a background of mountain chains with summits where some cryophytes grew. When these regions reached the warm temperate climate of the 30th–40th parallels, where winter frosts prevent the survival of tender tropical plants, only those species pre-adapted or adaptable to less congenial conditions were left to survive. These were the species of the past and present nemoral flora, which is still frequently misnamed the Arcto-Tertiary flora because its geological remains are now met with in the arctic regions.

When the Laurasian continent reached up to the latitudes where tropical and other tender plants could not survive, these species dispersed southward to retain their latitudinal affiliation, whereas the considerably fewer plants that were able to adapt and continue with the land on which they were located stayed and formed the nemoral floristic group. When the northern coasts of Laurasia had drifted up to the 50th parallel, however, most of their nemoral plants became unable to adapt to a further deterioration of the climate. A new zonation began, when these plants either succumbed or dispersed south to keep at the latitude of their adaptation, whereas the conifers, which had been mixed with and dominated by the deciduous trees, were able to continue and form the circumpolar conifer zone together with a selection of formerly nemoral plants. The cryophytes on the mountains continued to grow where they were, probably also invading lower belts laid bare by the downward dispersal of more tender plants and trees. However, those species of the nemoral forests that had pre-adapted to still more severe conditions, were now protected by the conifer forests. If Laurasia had remained a single continent, the nemoral forests would have been able to disperse southward only on its western and eastern coasts where the so-called Malayan, or north-shore Tethyan, flora was dislocated, because the nemoral plants could not invade the dry and semi-desert or desertlike inlands of the continent. However, at this time the North Atlantic was being formed by a new kind of rift, leaving the humid coasts of what later became eastern North America and western Europe open for invasion by the nemoral flora.

The northward drift continued with the same slow but constant speed, and in the Late Eocene and Early Miocene the northern parts of Laurasia reached into a zone where the climate was hostile also to the hardy conifer forests. Then the cryophytes invaded the deforested lowland, either from the alpine areas where some of them had long prevailed, or they were formed by selection from pre-adapted demes and species of the nemoral flora that had been protected by the conifer forests. There is a reason to believe that this Early Tertiary flora of the northlands was able to disperse all over the lowlands of these early arctic regions, and that it was considerably larger than the present arctic flora. It is evident that it left close relatives in the temperate mountains. It is also open to reason that it must have included considerably more polyploids than did the lowland nemoral flora, though less than at present.

When the periodical Plio-Pleistocene glaciations made the living conditions in the northlands still less favourable, many of the cryophytes of these lands succumbed, and others had their formerly circumpolar area split into smaller and isolated populations. Also, during the glaciations themselves, the cold climate in which the cryophytes can survive again reached temperate lowlands to allow a redispersal southward of these plants. Where the main mountain chains lie north–south, as in eastern Asia and western America, conditions were created that allowed the boreal cryophytes to disperse slowly southward, even all the way to New Zealand and Tierra del Fuego, either from their arctic area or, more likely, from temperate mountains in which they had been formed and where they had stayed when some of their populations originally drifted northward and adapted to the arctic conditions.

Such seems to be the story of the plants we prefer to call arctic–alpine cryophytes. In the northlands of the globe they form the vegetation cover everywhere, and in temperate mountains they also predominate, even as far south as the Himalayas, Caucasus, and the southern Rocky Mountains. They have several representatives in the southern hemisphere, latecomers from the Pleistocene, which mix with cryophytes of austral origin. The strong element of arctic plants in the temperate boreal mountain chains is also mixed with an older element, which apparently was formed in the mountains near the southern shores of the Laurasian continent which were washed by the Tethys Sea. Naturally, all the alpine plants of the Tethys mountains evolved in a similar way to those that originated much later from the nemoral flora of the north shores, though there are cytogenetical indications, from the lower frequency of polyploids, that the climate of these mountains where they originated may have been considerably less severe. However, since these older cryophytes are still much too little known and of less importance in the present connection, we may perhaps be excused if we leave them for a later discussion.

References

ADAMS, C. G. and AGER, D. V. (1967) Aspects of Tethyan biogeography. *Syst. Ass. Pub.*, **7**, 1–336.

BILLINGS, W. D. and MOONEY, H. A. (1968) The ecology of arctic and alpine plants. *Biol. Rev.*, **43**, 481–529.

BLACKETT, P. M. S., BULLARD, E. and RUNCORN, S. K. (1965) A symposium on continental drift. *Phil. Trans. Roy. Soc. London.*, **1088**, I–X, 1–323.

BLISS, L. C. (1971) Arctic and alpine plant life cycles. *Ann. Rev. Ecol. Syst.*, **2**, 405–38.

BÖCHER, T. W. (1938) Biological distributional types in the flora of Greenland. *Medd. om Grønland*, **106** (2), 1–339.

CAREY, S. W. (1958) *Continental Drift. A symposium*. Geol. Dept, Univ. of Tasmania. 363 pp.

CHALONER, W. G. (1959) Continental drift. *New Biol.*, **29**, 7–30.

CLAUSEN, J. (1951) *Stages in the Evolution of Plant Species*. Ithaca: Cornell Univ. Press. 206 pp.

CLAUSEN, J. (1963) Tree lines and germ plasm – a study in evolutionary limitations. *Proc. Nat. Acad. Sci.*, **50**, 860–8.

COE, M. S. (1967) *The Ecology of the Alpine Zone of Mount Kenya*. The Hague: W. Junk. 136 pp.

DAHL, E. (1946) On different types of unglaciated areas during the Ice Ages and their significance to phytogeography. *New Phytol.*, **45**, 225–42.

DAHL, E. (1963) Plant migrations across the North Atlantic ocean and their importance for the paleogeography of the region. *In:* LÖVE, Á. and LÖVE, D. (eds.) *North Atlantic Biota and their History*, 173–88. Oxford: Pergamon Press.

DANSER, B. H. (1929) Über die Begriffe Komparium, Kommiskuum and Konvivium und über die Entstehungsweise der Konvivien. *Genetica*, **11**, 399–450.

DANSER, B. H. (1950) A theory of systematics. *Bibl. Biotheoretica*, **4**, 117–80.

DARWIN, C. (1859) *The Origin of Species by Means of Natural Selection*. London: John Murray. 507 pp.

DE CANDOLLE, A. P. (1813) *Théorie élémentaire de la botanique*. Paris: Déterville. 533 pp.

DEWEY, J. F. (1972) Plate tectonics. *Sci. Amer.*, **226** (5), 56–68.

DIETZ, R. S. and HOLDEN, J. C. (1970a) Reconstruction of Pangaea: Breakup and dispersion of continents Permian to present. *J. Geophys. Res.*, **75**, 4939–56.

DIETZ, R. S. and HOLDEN, J. C. (1970b) The breakup of Pangaea. *Sci. Amer.*, **222** (10), 30–41.

DOBZHANSKY, T. (1951) *Genetics and the Origin of Species* (3rd rev. edn). New York: Columbia Univ. Press. 375 pp.

DORF, E. (1969) Paleobotanical evidence of Mesozoic and Cenozoic climatic changes. *Proc. N. Amer. Paleontol. Convention*, 323–46.

DU RIETZ, G. E. (1930) The fundamental units of biological taxonomy. *Svensk Bot. Tidskr.*, **24**, 333–428.

DU RIETZ, G. E. (1940) Problems of bipolar plant distribution. *Acta Phytogeogr. Suecica*, **13**, 215–82.

DUSÉN, P. (1900) Die Gefässpflanzen der Magellansländer nebst einem Beitrage zur Flora der Ostküste von Patagonia. *Wiss. Ergebn. Schwed. Exp. Magellansl. 1895–1897*, **3** (5), 77–266.

DU TOIT, A. L. (1937) *Our Wandering Continents*. Edinburgh: Oliver & Boyd. 379 pp.

EINARSSON, T. (1963) Some chapters of the Tertiary history of Iceland. *In:* LÖVE, Á and LÖVE, D. (eds.) *North Atlantic Biota and their History*, 1–9. Oxford: Pergamon Press.

EINARSSON, T., HOPKINS, D. M. and DOELL, R. R. (1967) The stratigraphy of Tjörnes, northern Iceland, and the history of the Bering land bridge. *In:* HOPKINS, D. M. (ed.) *The Bering Land Bridge*, 312–25. Stanford: Stanford Univ. Press.

ENGLER, A. (1904) Plants of the northern temperate zone in their transition to the higher mountains of tropical Africa. *Ann. Bot.*, **18**, 523–40.

FAIRBRIDGE, R. W. (1970) An ice age in the Sahara. *Geotimes*, **15** (6), 18–20.

FAVARGER, C. (1956) *Flore et végétation des Alpes, I. Étage alpin.* Neuchâtel: Delachaux & Niestlé. 271 pp.

FAVARGER, C. (1958) *Flore et végétation des Alpes. II. Étage subalpin, avec considérations sur le Jura et les montagnes insubriennes.* Neuchâtel: Delachaux & Niestlé. 274 pp.

FIRBAS, F. (1949) *Spät- und nacheiszeitliche Waldgeschichte Mitteleuropas nördlich der Alpen. Erster Band: Allgemeine Waldgeschichte.* Jena: Gustav Fischer Verlag. 480 pp.

FIRBAS, F. (1952) *Spät- und nacheiszeitliche Waldgeschichte Mitteleuropas nördlich der Alpen. Zweiter Band: Waldgeschichte der einzelnen Landschaften.* Jena: Gustav Fischer Verlag. 256 pp.

FORBES, E. (1846) On the connexion between the distribution of the existing fauna and flora of the British Isles and the geological changes which have affected their area, especially during the epoch of northern drift. *Mem. Geol. Surv. Engl. and Wales*, **1**, 336–432.

FRENZEL, B. (1968) The Pleistocene vegetation of northern Eurasia. *Science*, **161**, 637–49.

FRIEDRICH, W. (1966) Zur Geologie von Brjánslaekur (Nordwest Island) unter besonderer Berücksichtigung der fossilen Flora. *Sonderöff. Geol. Inst. Univ. Köln*, **10**, 1–108.

GILMOUR, J. S. L. and GREGOR, J. W. (1939) Demes: a suggested new terminology. *Nature*, **144**, 333–4.

GLEASON, H. A. and CRONQUIST, A. (1964) *The Natural Geography of Plants.* New York: Columbia Univ. Press. 428 pp.

GODWIN, H. (1956) *The History of the British Flora.* Cambridge: Cambridge Univ. Press. 384 pp.

GOOD, R. (1964) *The Geography of Flowering Plants* (3rd edn). London: Longmans. 534 pp.

GRANT, V. (1963) *The Origin of Adaptations.* New York: Columbia Univ. Press. 616 pp.

GRANT, V. (1964) *The Architecture of the Germ Plasm.* New York: John Wiley. 251 pp.

GREENE, S. W. (1964) Plants of the land. *In:* ADIE, R. Z., PRIESTLEY, R. and ROBIN, G. DE Q. (eds.) *Antarctic Research*, 240–52. London: Butterworths.

GREENE, S. W. (1970a) Studies in *Colobanthus quitensis* (Kunth) Bartl. and *Deschampsia antarctica* Desv. I. Introduction. *Brit. Antarct. Surv. Bull.*, **23**, 19–24.

GREENE, S. W. (1970b) Studies in *Colobanthus quitensis* (Kunth) Bartl. and *Deschampsia antarctica* Desv. II. Taxonomy, distribution and relationships. *Brit. Antarct. Surv. Bull.*, **23**, 63–80.

GUILLIEN, Y. (1955) La couverture végétale de l'Europe pleistocène. *Ann. de Géogr.*, **44**, 241–76.

GUSTAFSSON, Å. (1946–7) Apomixis in higher plants. *Acta Univ. Lund* (n.s.) **42** (3); **43** (2, 12), 1–370.

GUSTAFSSON, Å. and NYGREN, A. (1958) Die Fortpflanzung und Vermehrung der höherem Pflanzen. *Handb. Pflanzenzüchtung*, **1**, 54–85.

HAWKES, J. G. and SMITH, P. (1965) Continental drift and the age of angiosperm genera. *Nature*, **207**, 48–50.

HEDBERG, O. (1957) Afroalpine vascular plants. A taxonomic revision. *Symb. Bot. Upsal.*, **15** (1), 1–411.

HEDBERG, O. (1964) Features of afroalpine plant ecology. *Acta Phytogeogr. Suecica*, **49**, 1–144.

HEILBRONN, A. and KOSSWIG, C. (1966) *Principia Genetica. Grunderkenntnisse und Grundbegriffe der Vererbungswissenschaft*, 2 (2nd rev. edn). Hamburg and Berlin: Verlag Paul Parey. 43 pp.

HOOKER, J. D. (1862) Outlines of the distribution of arctic plants. *Trans. Linn. Soc. London*, **23**, 251–348.

HOOKER, J. D. and GRAY, A. (1880) The vegetation of the Rocky Mountain region and a comparison with that of other parts of the world. *US Geol. Surv. Terr.*, **6**, 1–62.

HULTÉN, E. (1937) *Outline of the History of Arctic and Boreal Biota During the Quaternary Period.* Stockholm: Bokförlags Aktiebolaget Thule. 168 pp.

HULTÉN, E. (1958) The amphi-atlantic plants and their phytogeographical connections. *Svenska Vetensk. Akad. Handl.*, IV, **7** (1), 1–340.

HULTÉN, E. (1959) The *Trisetum spicatum* complex. *Trisetum spicatum* (L.) Richt., an arctic-montane species with world-wide range. *Svensk Bot. Tidskr.*, **53**, 203–28.

HULTÉN, E. (1962) The circumpolar plants, I. Vascular cryptogams, conifers, monocotyledons. *Svenska Vetensk. Akad. Handl.*, IV, **8** (5), 1–275.

HULTÉN, E. (1968) *Flora of Alaska and Neighboring Territories.* Stanford: Stanford Univ. Press. 1031 pp.

HUXLEY, J. (1939) Clines: an auxiliary method in taxonomy. *Bijdr. Dierk.*, **27**, 491–520.

HUXLEY, J. (1942) *Evolution, the modern synthesis.* London: Allen & Unwin. 645 pp.

ILLIES, J. (1965) Die Wegenersche Kontinentalverschiebungstheorie in Lichte der modernen Biogeographie. *Naturwissenschaften*, **52**, 505–11.

IWATSUKI, Z. and SHARP, A. J. (1966) *Bryoxiphium norvegicum* subsp. *japonicum* (Berggr.) Löve et Löve new to the Philippines. *Misc. Bryol. et Lichénol.*, **4**, 29–30.

JOHNSON, B. L. and THEIN, M. M. (1970) Assessment of evolutionary affinities in *Gossypium* by protein electrophoresis. *Amer. J. Bot.*, **57**, 1081–92.

JOHNSON, G. L. and HEEZEN, B. C. (1967) Morphology and evolution of the Norwegian–Greenland sea. *Deep-Sea Res.*, **14**, 755–71.

KARLSTROM, T. N. V. and BALL, G. E. (1969) *The Kodiak Island Refugium: Its Geology, Flora, Fauna and History.* Toronto: Ryerson Press. 277 pp.

KAY, M. (1969) North Atlantic–geology and continental drift. *Amer. Ass. Petrol. Geol. Mem.*, **12**, 1–1082.

KEAST, A. (1971) Continental drift and the evolution of the biota on southern continents. *Quart. Rev. Biol.*, **46**. 335–78.

KING, L. C. (1962) *The Morphology of the Earth. A Study and Synthesis of World Scenery.* Edinburgh: Oliver & Boyd. 711 pp.

KNABEN, G. (1959a) On the evolution of the *radicatum* group of the *Scapiflora* papavers as studied in 70 and 56 chromosome species. Part A. Cytotaxonomical aspects. *Opera Botan.*, **2** (3), 1–74.

KNABEN, G. (1959b) On the evolution of the *radicatum* group of the *Scapiflora* papavers as studied in 70 and 56 chromosome species. Part B. Experimental studies. *Opera Botan.*, **3** (3), 1–96.

LEE, S. C. (1958) The genus *Bryoxiphium* in China. *Acta Phytotax. Sinica*, **7**, 254–62.

LEVITT, J. (1966) Winter hardiness in plants. *In:* MERYMAN, H. T. (ed.) *Cryobiology*, 495–563. London and New York: Academic Press.

LEWIS, H. (1966) Speciation in flowering plants. *Science*, **152**, 167–72.

LINDQVIST, B. (1947) Two species of *Betula* from the Iceland Miocene. *Svensk Bot. Tidskr.*, **41**, 339–53.

LINDROTH, C. H. (1965) Skaftafell, Iceland, a living glacial refugium. *Oikos*, Suppl. 6, 1–142.

LINNAEUS, C. (1753) *Species plantarum, I–II.* Holmiae: Impensis Laurentii Slavii. 1231 pp.

LÖVE, Á. (1959) Origin of the arctic flora. *Pub. McGill Univ. Museums*, **1**, 82–95.

LÖVE, Á. (1962a) Typification of *Papaver radicatum* – a nomenclatural detective story. *Bot. Notiser*, **115**, 113–36.

LÖVE, Á. (1962b) Nomenclature of North Atlantic Papavers. *Taxon*, **11**, 132–8.

LÖVE, Á. (1964) The biological species concept and its evolutionary structure. *Taxon*, **13**, 33–45.

LÖVE, Á. (1970a) Emendations in the Icelandic flora. *Taxon*, **19**, 298–302.

LÖVE, Á. (1970b) *Íslenzk ferdaflóra.* Reykjavík: Almenna Bókafélagid. 428 pp.

LÖVE, Á. and LÖVE, D. (1942) Chromosome numbers of Scandinavian plant species. *Bot. Notiser.* 19–59.

LÖVE, Á. and LÖVE, D. (1947) Studies on the origin of the Icelandic flora, I. Cyto-ecological investigations on Cakile. *Icel. Univ. Inst. Appl. Sci., Dept Agric. Rep.*, B (2), 1–29.

LÖVE, Á. and LÖVE, D. (1949) The geobotanical significance of polyploidy, I. Polyploidy and latitude. *Portug. Acta Biol.*, A (R. B. Goldschmidt Jub. Vol.), 273–352.

LÖVE, Á. and LÖVE, D. (1953) Studies on *Bryoxiphium*. *Bryologist*, **56**, 73–84, 183–203.

LÖVE, Á. and LÖVE, D. (1956) Cytotaxonomical conspectus of the Icelandic flora. *Acta Horti. Gotob.*, **20** (4), 65–291.

LÖVE, Á. and LÖVE, D. (1963) *North Atlantic Biota and their History.* Oxford: Pergamon Press. 442 pp.

LÖVE, Á. and LÖVE, D. (1965) Taxonomic remarks on some American alpine plants. *Univ. of Colorado Stud., Ser. in Biol.*, 17, 1–43.

LÖVE, Á. and LÖVE, D. (1966) Cytotaxonomy of the alpine vascular plants of Mount Washington. *Univ. of Colorado Stud., Ser. in Biol.*, 24, 1–74.

LÖVE, Á. and LÖVE, D. (1967a) Continental drift and the origin of the arctic–alpine flora. *Rev. Roum. Biol., Sér. Bot.*, **12**, 163–9.

LÖVE, Á. and LÖVE, D. (1967b) The origin of the North Atlantic flora. *Aquilo, Ser. Bot.*, **6**, 52–66.

LÖVE, Á. and LÖVE, D. (1968) The diploid perennial *Anthoxanthum*. *Science in Iceland.* 26–30.

LÖVE, Á. and LÖVE, D. (1971) Polyploïdie et géobotanique. *Nat. Can.*, **98**, 469–94.

LÖVE, Á., LÖVE, D. and KAPOOR, B. M. (1971) Cytotaxonomy of a century of Rocky Mountain orophytes. *Arct. Alp. Res.*, **3**, 139–65.

LÖVE, D. (1959) The postglacial development of the flora of Manitoba: a discussion. *Can. J. Bot.*, **37**, 547–85.

LÖVE, D. (1962) Plants and Pleistocene. *Pub. McGill Univ. Museums*, **2**, 17–39.

LÖVE, D. (1963) Dispersal and survival of plants. *In:* LÖVE, Á. and LÖVE, D. (eds.) *North Atlantic Biota and their History*, 189–205. Oxford: Pergamon Press.

LÖVE, D. (1969) *Papaver* at high altitudes in the Rocky Mountains. *Brittonia*, **21**, 1–10.

LÖVE, D. (n.d.) *Mount Washington and its Alpine Flora.* Unpub. MS.

LÖVE, D. and BOSCAIU, N. (1966) *Vaccinium gaultherioides* Bigel., an arctic–alpine species. *Rev. Roumaine de Biol., Sér. Bot.*, 11, 295–305.

LÖVE, D., MCLELLAN, C. and GAMOW, I. (1970) Coumarin and coumarin derivatives in various growth-types of Engelmann spruce. *Svensk Bot. Tidskr.*, **64**, 284–96.

MANUM, S. (1962) Studies in the Tertiary flora of Spitsbergen, with notes on Tertiary floras of Ellesmere Island, Greenland, and Iceland. *Norsk Polarinstitutt, Skrifter*, **125**, 1–127.

MAYR, E. (1942) *Systematics and the Origin of Species*. New York: Columbia Univ. Press. 348 pp.

MAYR, E. (1963) *Animal Species and Evolution*. Cambridge, Mass.: Harvard Univ. Press. 812 pp.

MAYR, E. (1970) *Populations, Species, and Evolution*. Cambridge, Mass.: Harvard Univ. Press. 468 pp.

MAZUR, P. (1969) Freezing injury in plants. *Ann. Rev. Plant Physiol.*, **20**, 419–48.

MELCHERS, G. (1946) Die Ursachen für die bessere Anpassungsfähigkeit der Polyploiden. *Zeit. f. Naturforsch.*, **1**, 160–5.

MELVILLE, (1966) Continental drift, Mesozoic continents and the migrations of the angiosperms. *Nature*, **211**, 116–20.

MENZEL, M. Y. and MARTIN, D. W. (1971) Chromosome homology in some intercontinental hybrids in *Hibiscus* sect. *Furcaria*. *Amer. J. Bot.*, **58**, 191–202.

MEYERHOFF, A. A. and MEYERHOFF, H. A. (1972) The new global tectonics: major inconsistencies. *Amer. Ass. Petrol. Geol. Bull.*, **56**, 269–336.

MOORBATH, S., SIGURDSSON, H. and GOODMAN, R. (1968) K-Ar-ages of the oldest exposed rocks in Iceland. *Earth Planet. Sci. Letters*, **4**, 3.

MOORE, D. M. and CHATER, A. O. (1971) Studies on bipolar disjunct species. I. *Carex*. *Bot. Notiser*, **124**, 317–34.

NAKANISHI, S. (1964) *Bryoxiphium norvegicum* ssp. *japonicum* found in Taiwan (Formosa). *Hikobia*, **4**, 23–7.

PÓCS, T. (1966) *Bryoxiphum norvegicum* subsp. *japonicum* (Berggr.) Löve et Löve in the Indonesian archipelago. *Misc. Bryol. et Lichénol.*, **4**, 35–7.

POLUNIN, N. (1959) *Circumpolar Arctic Flora*. Oxford: Clarendon Press. 514 pp.

POLUNIN, N. (1960) *Introduction to Plant Geography and Some Related Sciences*. London: Longmans. 640 pp.

PORSILD, A. E. (1964) Illustrated flora of the Canadian Arctic Archipelago. *Nat. Mus. Can. Bull.*, **146**, 1–218.

RAUNKIAER, C. (1934) *The Life Forms of Plants and Statistical Plant Geography*. London: Oxford Univ. Press. 648 pp.

RAVEN, J. and WALTERS, M. (1956) *Mountain Flowers*. London: Collins. 255 pp.

RAVEN, P. H. (1963) Amphitropical relationships in the floras of North and South America. *Quart. Rev. Biol.*, **38**, 151–77.

ROIVAINEN, H. (1954) Studienüber die Moore Feuerlands. *Ann. Bot. Soc. Vanamo*, **28** (2), 1–205.

ROUSSEAU, J. (1949) Modification de la surface de la toundra sous l'agents climatiques. *Rev. Can. Géogr.*, **3**, 43–51.

RUNCORN, S. K. (1962) *Continental drift*. New York: Academic Press. 338 pp.

RYDBERG, P. A. (1914) Phytogeography and its relation to taxonomy and other branches of science. *Torreya*, **12**, 73–85.

SALISBURY, E. J. (1929) The biological equipment of species in relation to competition. *J. Ecol.*, **17**, 198–222.

SAVILE, D. B. O. (1972) Arctic adaptations in plants. *Can. Dept Agric. Res. Branch Monogr.*, 6, 1–81.

SCHWARZBACH, M. (1963) The geological knowledge of the North Atlantic climates of the past. *In:* LÖVE, Á. and LÖVE, D. (eds.) *North Atlantic Biota and their History*, 11–19. Oxford: Pergamon Press.

SCHWARZENBACH, F. H. (1951) Ökologische Beiträge zur quartäre Florengeschichte Ostgrönlands. *Ber. Geobot. Forschungsinst. Rübel* (1950), 44–66.

SCHWARZENBACH, F. H. (1953) Die Abhängigkeit der Bulbillenbildung bei *Poa alpina vivipara* vom Photoperiodismus und Frost. *Experientia*, **9**, 96–8.

SCHWARZENBACH, F. H. (1956) Die Beeinflussung der Viviparie bei einer grönländischen Rasse von *Poa alpina* L. durch jahreszeitlichen Licht- und Temperaturwechsel. *Ber. Schweiz. bot. Ges.*, **66**, 204–23.

SCHWARZENBACH, F. H. (1961) Botanische Beobachtungen in der Nunatakkerzone Ostgrönlands zwischen 74° und 75°N Br. *Medd. om Grønland*, **163** (5), 1–172.

SIGURDSSON, H. (1969) Nýjar aldursákvardanir á íslenzku bergi. *Náttúrufr.*, **38**, 187–93.

SJÖRS, H. (1956) *Nordisk växtgeografi.* Oslo, Copenhagen, Stockholm and Helsingfors: Scandinavian Univ. Books. 229 pp.

SKOTTSBERG, C. (1954) Antarctic flowering plants. *Bot. Tidsskr.*, **51**, 330–8.

SMITH, A. C. (1970) The Pacific as a key to flowering plant history. *Harold L. Lyon Arboretum Lecture*, 1, 1–28. Univ. of Hawaii.

SNEATH, P. H. A. (1967) Conifer distribution and continental drift. *Nature*, **215**, 467–70.

SÖRENSEN, T. (1941) Temperature relations and phenology of the northeast Greenland flowering plants. *Medd. om Grønland*, **125** (9), 1–305.

SÖRENSEN, T. (1942) Untersuchungen über die Therophytengesellschaften auf den isländischen Lehmflechen ('Flags'). *Kgl. Danske Vidensk. Selsk. Skr.*, **II** (2), 1–31.

STEBBINS, G. L. (1950) *Variation and Evolution in Plants.* New York: Columbia Univ. Press. 663 pp.

STEBBINS, G. L. (1966) *Processes of Organic Evolution.* Englewood Cliffs, NJ: Prentice-Hall. 203 pp.

STEENIS, C. G. G. J. VAN (1957) Specific and infraspecific delimitations. *Flora Malesians*, ser. I, **5** (3), CLXVII–CCXXXIV.

TIMOFÉEV-RESSOVSKY, N. V., VORONTSOV, N. N. and YABLOKOV, A. V. (1969) *Kratkiy ocherk teorii evolyutsii.* Moscow: Nauka. 407 pp.

TOLMATCHEV, A. (1960a) Der autochtone Grundstock der arktischen Flora und ihre Beziehungen zu den Hochgebirgsfloren Nord- und Zentralasiens. *Bot. Tidsskr.*, **55**, 296–76.

TOLMATCHEV, A. (1960b) O proiskhozhdenii arkticheskoy flory. *Voprosy Botaniki*, **3**, 72–4.

TOLMATCHEV, A. (1960–71) *Arkticheskaya flora SSSR. I–VI.* Moscow-Leningrad: Izdat. Nauk. 102, 273, 175, 96 and 208 pp.

TOLMATCHEV, A. (1966a) Die Evolution der Pflanzen in arktisch-Eurasien während und nach den quartären Vereisung. *Bot. Tidsskr.*, **62**, 27–36.

TOLMATCHEV, A. (1966b) Progressive Erscheinungen und Konservatismus in der Entwicklung der arktischen Flora. *Acta Bot. Hung.*, **12**, 175–98.

TOLMATCHEV, A. and YURTSEV, B. A. (1968) Voznikovenie i razvitie arkticheskoy flory v ich svyazi s istoriey polyarnogo basseina. *In:* BELOV, N. A. *et al.* (eds.) *Kainozoiskaya istoriya polyarnogo basseina*, 133–6. Leningrad: Geograficheskoe Obshchestvo SSR i Vsesoyuznoe Botanicheskoe Obshchestvo SSR.

TOLMATCHEV, A. and YURTSEV, B. A. (1970) Istoriya arkticheskoy flory v ee svyazi s istoriey severnogo ledovitogo okeana. *In:* TOLMATCHEV, A. (ed.) *Severnyi ledovityi okean i ego poberezh'e v kainozoe*, 87–100. Leningrad: Gidromet Izdat.

TROLL, C. (1948) Der asymmetrische Aufbau der Vegetationszonen und Vegetationsstufen auf der Nord- und Südhalbkugel. *Ber. Geobot. Forschungsinst. Rübel* (1947), 46–83.

TROLL, C. (1959) The relationship between climates and plant geography of the southern cold temperate zone and of the tropical high mountains. *Proc. Roy. Soc. London*, B, 152, 529–32.

TURESSON, G. (1922) The genotypical response of the plant species to the habitat. *Hereditas*, **3**, 211–350.

TZVELEV, N. N. (1969) Some problems of the evolution of plant life and the hypothesis of an expanding earth. *Byull. Moskv. Ob. Prir. Otd. Biol.*, **74**, 27–36.

WAHLENBERG, G. (1814) *Flora Carpatorum principalium, etc.* Goettingae: Vandenhoeck. 526 pp.

WALKER, D. and WEST, R. G. (1970) *Studies in the Vegetational History of the British Isles.* Cambridge: Cambridge Univ. Press. 274 pp.

WALLACE, A. R. (1880) *Island Life: or, the Phenomena and Causes of Insular Faunas and Floras, Including a Revision and Attempted Solution of the Problem of Geological Climate.* London: Macmillan. 526 pp.

WALLACE, A. R. (1900) *Studies, Scientific and Social*, Vol. I. London: Macmillan. 526 pp.

WALTER, H. (1968) *Vegetation der Erde in öko-physiologischer Betrachtung, Band II: Die gemässigten und arktischen Zonen.* Jena: VEB Gustav Fischer Verlag. 1001 pp.

WEBER, W. A. (1965) Plant geography in the southern Rocky Mountains. *In:* WRIGHT, H. E. and FREY, D. G. (eds.) *The Quaternary of the United States*, 453–68. Princeton, NJ: Princeton Univ. Press.

WEGENER, A. (1912) Die Entstehung der Kontinente. *Petermanns Geogr. Mitt.*, 185–95, 253–6, 305–9.

WEGENER, A. (1915) *Die Entstehung der Kontinente und Ozeane.* Braunschweig: Sammlung Vieweg. 94 pp.

WEGMANN, C. E. (1941) Geologische Gesichtspunkte zur Frage der Eiszeitüberdauerung von Pflanzen in Grönland. *Mitt. Naturforsch. Ges. Schaffhausen*, **17**, 97–116.

WEISER, C. J. (1970) Cold resistance and injury in woody plants. *Science*, **169**, 1269–78.

WENT, F. W. (1948) Some parallels between desert and alpine flora in California. *Madroño*, **9**, 241–9.

WOLFE, J. A. (1969) Paleogene floras from the Gulf of Alaska region. *US Dept of Interior, Geol. Surv., Open-file Rep.*, 1–111.

WOLFE, J. A. and LEOPOLD, E. B. (1967) Neogene and early Quaternary vegetation of northwestern North America and northeastern Asia. *In:* HOPKINS, D. M. (ed.) *The Bering Land Bridge*, 193–206. Stanford: Stanford Univ. Press.

WYCHERLEY, P. R. (1953) Proliferation of spikelets in British grasses. *Watsonia*, **3**, 41–56.

YURTSEV, B. A. (1966) *Hipoarkticheskiy botaniko-geografccheskiy poyas i proiskhzhdenie ego flory.* Moscow-Leningrad: Izdat. Nauk. 94 pp.

Manuscript received December 1970.

Additional references

COLBERT, E. H. (1973) *Wandering Lands and Animals.* New York: E. P. Dutton. 345 pp.

TIMOFÉEV-RESSOVSKY, N. V., YABLOKOV, A. V. and GLOTOV, N. V. (1973) *Ocherk ucheniya o populyatsii.* Moscow: Nauka. 277 pp.

Biological refugia and the nunatak hypothesis

Jack D. Ives

Institute of Arctic and Alpine Research,
University of Colorado

1. Introduction

In the late nineteenth century Axel Blytt and Rutger Sernander began to search for the basic cause of certain peculiarities in plant species distributions in Scandinavia. Formerly, natural scientists had assumed that with the onset of the single great Ice Age all life had migrated southward or else had been obliterated, and that the present distribution of species reflected the pattern of recolonization during the 10,000 years or so following the melting of the Fenno-Scandinavian ice sheet (tabula rasa theory). The work of Blytt (1876) and Sernander (1908) led to the formulation of the 'nunatak hypothesis'[1] which states that certain plant and animal species survived the Ice Age on nunataks, or in refuges that were never completely inundated by ice. This hypothesis, for instance, best accounted for the anomalous present-day distributions of the so-called unicentric and bicentric members of the amphiatlantic group of vascular plants in the Scandinavian mountains (Hultén, 1958; E. Dahl, 1955) (fig. 10B.1) and for the strong phytogeographic affinities in general between mountain coasts on both sides of the North Atlantic in high latitudes (E. Dahl, 1946).

As more and more detailed botanical work was completed during the twentieth century, the botanical support for the nunatak hypothesis has been progressively strengthened (Nannfeldt, 1935, 1963; Faegri, 1937; Hultén, 1937, 1950; E. Dahl, 1946, 1952; Löve and Löve, 1947, 1951, 1956, 1965, 1967a, b; Á. Löve, 1959; D. Löve, 1962; Nordhagen, 1931, 1936, 1963; Gjaerevoll, 1963). The botanical concepts have been applied to Greenland (Böcher, 1938, 1951; E. Dahl, 1952), to Iceland (Löve and Löve, 1947, 1956, 1967a, b), to Svalbard (Rønning, 1960, 1961), and to northeastern North

---

[1] Nunatak: Eskimo word taken into the technical literature and meaning *sensu stricto* a mountain peak surrounded on all sides by glacial ice.

America (Fernald, 1925; Abbe, 1938). They have also been extended to the study of the distribution of invertebrates (Lindroth, 1931, 1957, 1969) and to evaluation of the overall history of the arctic and North Atlantic flora and fauna (Löve and Löve, 1963a; see also Löve and Löve, Chapter 10A).

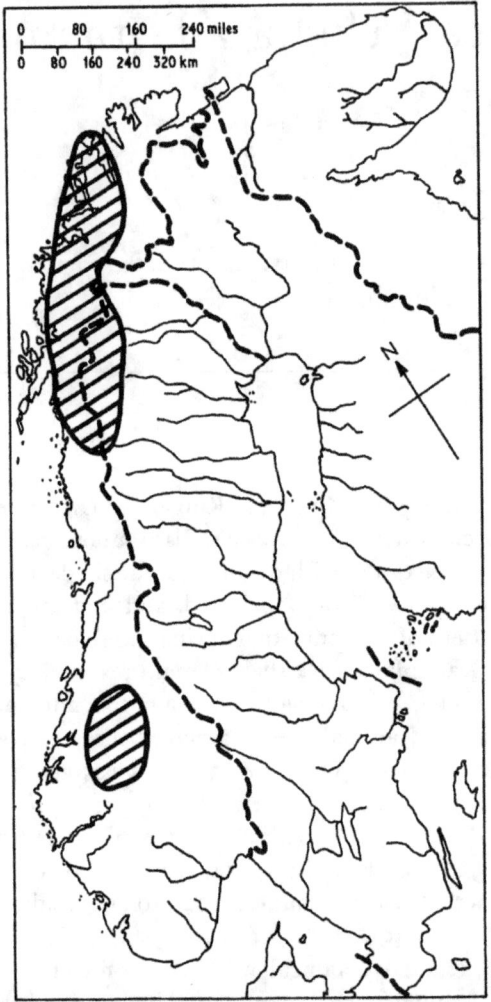

Fig. 10B.1. Distribution of the amphiatlantic bicentric and unicentric species in Scandinavia (from Gjaerevoll, 1963).

Finally, the last decade in North America has witnessed an upsurge of biological interest in the nunatak hypothesis as studies in the High Arctic progress beyond the initial reconnaissance phase. Thus Brassard (1971) has made a major contribution to the study of bryophytes in the Queen Elizabeth Islands, using the nunatak hypothesis to account for many peculiarities of their biogeography; Leech (1966) supports the concept that one arthropod fauna had inhabited Ellesmere Island since before the

Wisconsin glaciation; Macpherson (1965) provided a similar line of reasoning arising from his studies of subspeciation in the tundra mammals.

Savile (1961) has discussed the biogeography of the vascular plants of the Queen Elizabeth Islands and rejects the refugia concept although his stand is strongly contested by Schuster et al. (1959) and Brassard (1971). Of special interest is the collaboration of geologist, botanist and entomologist in the production of a research volume on the Kodiak Island refugium (Karlstrom and Ball, 1969). In this instance an area of Kodiak Island, defined by geological evidence as having been unglacierized throughout at least the last two glaciations, is subjected to biological investigations.

As the biological conviction grew, so it faced increasing opposition from geologists and physical geographers who have argued that all high mountain tops in Scandinavia, and in other localities around the North Atlantic, were ice-covered at least during maximum phases of the Neogene glaciations, and that the biologists must seek an alternate explanation for the anomalous distributions of plant and animal species (Odell, 1933; Flint, 1943, 1953; Hoppe, 1959, 1968, 1971; R. Dahl, 1963, 1966; Holtedahl, 1960, pp. 364–9). It is nevertheless curious that early in this century there were many geologists who on albeit qualitative reconnaissance and even intuitive bases, supported the nunatak hypothesis (Vogt, 1913; Reusch, 1910; Ahlmann, 1919; Coleman, 1920). Their work, however, has been contradicted by more intensive fieldwork and by changes in general geomorphological thought; Hoppe (1968, 1971), in particular, has produced unequivocal evidence of glacial erosion and deposition in an increasing number of the postulated refugia.[1]

That scientists from different disciplines should disagree completely over such a basic question is both intriguing and significant in itself; but the resolution of this issue is fundamental to our proper understanding of the evolution of life forms in the arctic regions and also of the history of glaciation of high and middle-high latitudes. Several other major scientific problems are closely associated: for instance, rates of subspeciation, pattern and speed of plant migration, rates of weathering in cold climates, effectiveness of glacial erosion. Furthermore, progress in any of these associated fields of study will have an impact upon our scientific appreciation of mountain regions in lower latitudes.

This section will outline the protagonist biological arguments in support of the nunatak hypothesis, review the geological counter-attack, discuss some of the ensuing contradictions and differences in interpretation, and synthesize recent geological work in Baffin Island. The final section will constitute an attempt to apply the lessons being learned in Baffin Island to a current assessment of the nunatak controversy.

2. The biological defence

The biological defence of the nunatak hypothesis is principally botanical in scope and can be best introduced by restating the problem that faced Blytt and Sernander almost

---

[1] Emphasis should be drawn to the clear-cut distinction in current Scandinavian thinking between former nunatak refugia (i.e. actual mountain tops) and lowland, or glacial foreland refugia (Hoppe, personal communication, 1972; R. Dahl, personal communication, 1972). In the following text I use 'nunatak' in the broadest sense to include both extremes.

one hundred years ago. The botany of Norway was already sufficiently well known by 1880 for the restricted occurrence of certain vascular plants to west Norwegian mountain massifs, together with their absence from comparable neighbouring environments (arctic–alpine) and from central Europe in general, to force rejection of the tabula rasa theory. Blytt and Sernander pioneered the concept of Ice Age survival of plants within, or close to, their present highly restricted areas. As the discussion was enlarged to embrace all the arctic and subarctic environments around the North Atlantic, the botanical case was vastly strengthened. The following presentation of the argument is largely taken from E. Dahl (1955, 1961).

E. Dahl (1961) refers to the floristic similarity of areas on both sides of the Atlantic Ocean as the most outstanding phytogeographic feature of the North Atlantic region. This floristic similarity is greatest between Svalbard and northeast Greenland, although the affinity between the arctic–alpine floras of Scandinavia, Britain, Iceland, Greenland and Labrador is also pronounced. The affinity is such that it renders difficult the establishment of any phytogeographic boundary between northwest Europe and northeastern North America. The general impression, which is strengthened by more detailed analysis, is that there is a fairly gradual transition from a predominantly European flora in the east to a predominantly American flora in the west. Many individual species occur on both sides of the Atlantic but have large gaps in their distributions across Asia and western North America. Hultén (1958) termed this group 'amphiatlantic'. Some members have a restricted area on the European side of the ocean with wide distribution in North America, termed the West Arctic species in Scandinavia, while others have a reverse pattern and are termed the East Arctic species in North America (Löve and Löve, 1967b). A third group has a more or less symmetrical distribution pattern.

Hultén (1937, 1958) has argued that the source area of the amphiatlantic group was Beringia, from which centre they emigrated east and west, becoming disjunct through extinction in their area of origin. This is heavily contested by E. Dahl and Löve and Löve who maintain that Hultén's hypothesis would require much greater differentiation, at least at the species and subspecies level, than is apparent in the group. Thus the source area of the amphiatlantic plants is regarded as the area where they are most abundant today. To E. Dahl (1955) in the 1950s this raised the question of long-distance dispersal across wide areas of open water, or the possibility of the existence of a land connection until close before the onset of world glaciation. Löve and Löve (Chapter 10A) reconcile this problem with the recent rapid accumulation of evidence and popular support for the concept of continental drift. Thus the botanical argument invokes the need for a major high latitude refuge during the Ice Age with subsequent long-distance dispersal with its attendant problems, or for a large number of small glacial refuges with minimal postglacial dispersal (fig.10B.2). The nunatak hypothesis, *sensu stricto*, embraces the latter concept.

The most impressive non-botanical support for the nunatak hypothesis derives from the work of Lindroth (1931, 1957, 1969) following intensive studies of the biogeography of the carabid beetles, especially flightless forms. Lindroth's work ranges from Scandinavia, Iceland, eastern Canada and Alaska.

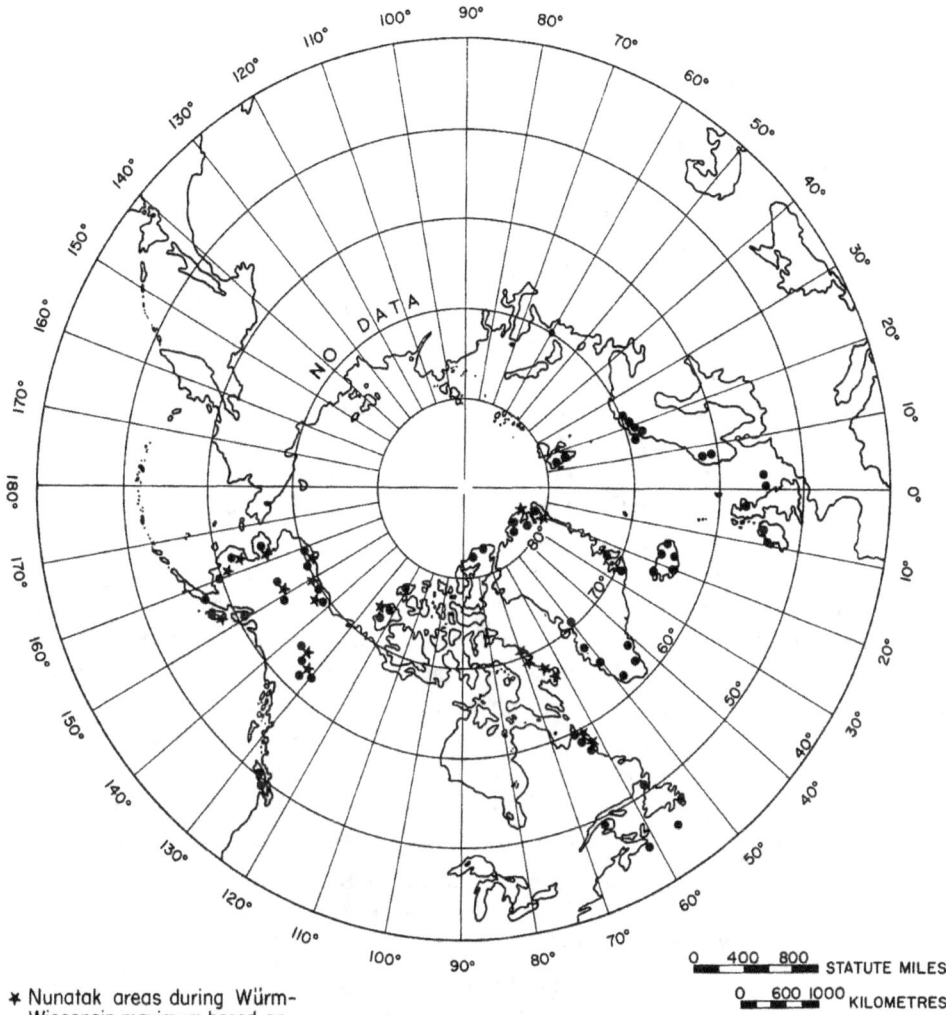

✶ Nunatak areas during Würm–
Wisconsin maximum based on
geological evidence

• Nunatak areas during Würm–
Wisconsin maximum inferred
from biological evidence but
contested by some on geological grounds

Fig. 10B.2. Outline of the presumed distribution of arctic and high latitude refugia in the
northern hemisphere classified according to biological and geological criteria. Most
certain glacial maximum ice-free localities (Würm–Wisconsin) occur where the biological
and geological criteria coincide; e.g. Kodiak Island, interior Alaska, Yukon, Banks
Island, Peary Land. Biological evidence for Labrador is heavily contested (compiled from
various sources).

Before taking the case a step further, it will be profitable to examine and refute the postglacial long-distance dispersal hypothesis. As E. Dahl (1961) and D. Löve (1963) point out, plants may disperse by wind, water, animals and man in a variety of ways. Yet dispersal mechanisms of individual species are remarkably little understood. Many species do exhibit special morphological adaptations to long-distance dispersal, or else have ecological characteristics favouring such dispersal. For instance, anemochorous species have either very small seeds (limit 0.2 mm) or hairs attached to seed or fruit. Halophilous species are assumed to be most adaptable to dispersal by ocean currents. Yet not enough is known to facilitate classification of seeds in terms of dispersal properties, nor is dispersal merely the problem of transport of diaspores across oceans; the diaspores must land on an environment favourable for establishment. In this respect the halophilous species would appear to have an obvious advantage (D. Löve, 1962, has indicated that coconuts may float ashore in Iceland, but unfortunately will not grow there!). Limnic species are considered as a special group because there is evidence for effective dispersal by water birds. Finally, it is assumed that the remaining species have no special adaptations to long-distance dispersal.

If the present distribution of the amphiatlantic species were the result of long-distance dispersal, those species having special dispersal adaptations should show an advantage. Table 10B.1 illustrates the situation for Scandinavia which has an arctic–alpine element comprising some 250 species, of which about 25 are of the West Arctic group. The table gives a two-way grouping according to phytogeographic element and to adaptations to long-distance dispersal. The numbers expected under an assumption of *no* correlation between phytogeographic elements and adaptation to long-distance dispersal are given in brackets. No significant correlation is found; in the West Arctic element there is a slight under-representation of types adapted to long-distance dispersal.

Table 10B.1. West Arctic element in the Scandinavian flora compared with other arctic–alpine elements as to adaptations to long-distance dispersal (after E. Dahl, 1961). See text.

	Adaptation to long-distance dispersal	No adaptation to long-distance dispersal	Total
West Arctic	4 (7.3)	21 (17.7)	25
Other	71 (67.7)	160 (163.3)	231
Total	75	181	256

D. Löve (1963) carried out much more detailed analyses for the total North Atlantic flora with similar results. Thus she was forced to argue that adaptations to long-distance dispersal are of little consequence to the spread of plants over an ocean, and that the question of long-distance dispersal in this area bears little relevance to the present-day distribution pattern of the amphiatlantic plants. The second alternative is the only one that seems botanically acceptable and carries the discussion towards acceptance of

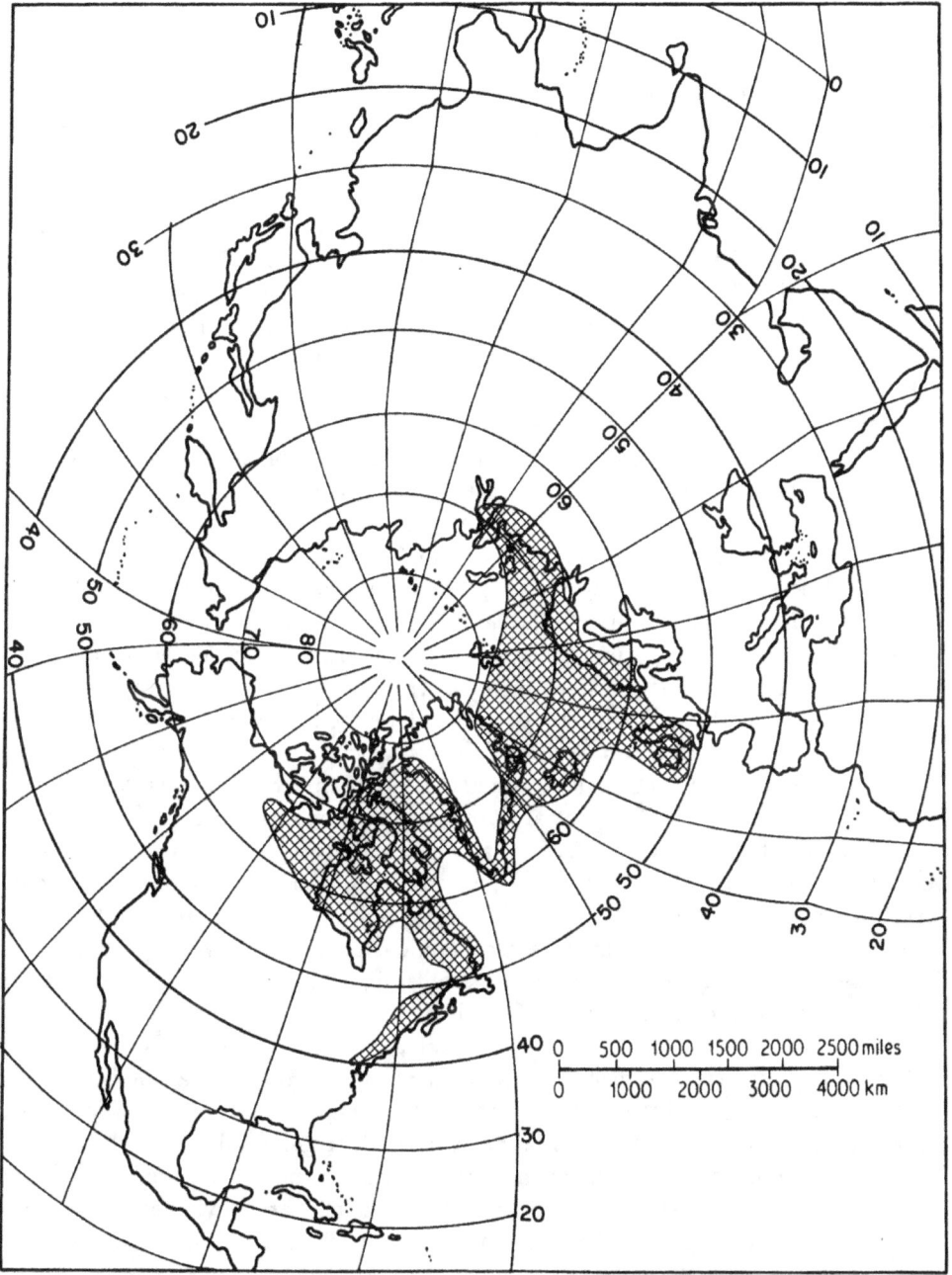

Fig. 10B.3. Example of typical amphiatlantic taxon (*Salix herbacea* L.) that probably survived the Würm–Wisconsin maximum in the North Atlantic area (from D. Löve, unpub., after Hultén, 1958).

glacial survival and towards explanations based on earlier land connections (Löve and Löve, 1967b).

This argument does not necessarily rule out the possibility of occasional chance long-distance dispersal, since it primarily related to entire floral elements, or communities.

The amphiatlantic distribution pattern is essentially a feature of arctic and alpine species (fig. 10B.3); very few southern species exhibit this kind of distribution. By closer analysis, E. Dahl (1955) has shown that a very high proportion of species capable of growing above timberline in south Norway also occur on mountains along the western shores of the Atlantic, the exceptions being mainly polymorphic species which might have differentiated since continental drift occurred, and species with a continuous distribution across the Eurasian lowlands. Of species unable to grow in the birch belt of south Norway, only a few occur in northeastern North America; this element seems to have a high over-representation of types possibly adapted to long-distance dispersal.

The next step in the discussion is to consider where the amphiatlantic plants survived the last glaciation. Emphasis is placed here upon the final glaciation (Wisconsin–Würm, or approximately the last 100,000 years) since it is generally agreed that the Sangamon/Riss–Würm interglacial was of sufficient duration to permit gradual migration of plants into pre-Wisconsin–Würm positions to account for their present-day distribution, assuming last Ice Age survival in refuges. No West Arctic species absent from Scandinavia is present in the Alps, all species of that region having their phytogeographical connections with the north and east. This is a statement of profound significance in any evaluation of the nunatak hypothesis. Plants that occurred along the borders of the Fenno–Scandinavian ice sheet in central Europe during the last glacial age would have had a much easier migration route during deglaciation to the Alps than to Scandinavia, so that if the West Arctic species survived in central Europe it would be surprising if all migrated to Scandinavia and none to the Alps. It is also generally recognized that conditions for northward migration of truly arctic or alpine species into the Scandinavian mountains were very unfavourable owing to high temperatures during the later phases of deglaciation. Furthermore, fossil remnants of typical arctic–alpine plants have never been found in the lowlands of central Scandinavia (E. Dahl, 1955). This further suggests that the West Arctic element now occurring in Scandinavia must have survived the Wisconsin–Würm glaciation close to the Atlantic coast, probably in west and north Norway.

Additional botanical arguments can be used to support the general thesis outlined above. A principal one relates to the occurrence of endemic species that, while rare, are by no means absent from the North Atlantic region under consideration. Segregation into endemic species and subspecies is much more frequent among the Scandinavian alpine flora than among the lowland flora. If the entire biota of Scandinavia had immigrated in late- and post-glacial time, the alpine and lowland elements would be of approximately equal age, but since endemism is regarded as evidence of age, it is difficult to account for the higher proportion of endemics among the alpine element compared with the lowland elements. A further line of evidence may be accredited to Löve and Löve who have shown by prolonged work over many years (Löve and Löve, 1943, 1949, 1953, 1957, 1963a, b, 1965, 1967c, 1971; Á. Löve, 1950, 1953, 1959; D. Löve, 1962) that polyploidy is much more frequent among the arctic–alpine flora than more southerly or 'temperate' floras, indicating progressive adaptation to a severe environment over a long period of time. Another basic characteristic of the arctic–alpine flora is the long-day element, that would have been unable to survive southerly migration with each glacial onset.

So far the presentation of the botanical case has been kept very general. One of many specific examples of anomalous plant distribution patterns will now be introduced to illustrate the type of field evidence that has been accumulated as a basis for the rather general statement given above.

BICENTRIC AND UNICENTRIC DISTRIBUTIONS

E. Dahl (1955) discusses the possibility of using biogeographic techniques in an attempt to indicate the location of last glacial plant refuges. The neighbourhood of refuges, he argues (E. Dahl, 1955, p. 1507), should be characterized by the occurrence of rare plants, and he singles out two Scandinavian areas, the Dovre–Jotunheimen Mountains in the south and west and the Troms–Finnmark area in the north. A great number of vascular plants are found only in these two areas, some in both (bicentric) and some in either one or the other, but not in both (unicentric). This special group includes many West Arctic and endemic species. There have been several attempts to explain this distribution pattern by recourse to ecological arguments (Blytt, 1876; Böcher, 1951; Arwidsson, 1943; E. Dahl, 1952) although they break down whenever the entire group of species and the detail of distribution anomaly are considered. The general consensus among botanists is that the distribution pattern has to be explained in terms of historical factors. Fries (1913) first suggested two refuge areas in Scandinavia during the last glacial age, a northern one from the Lofoten Islands northward, and a southern one on the coast of Möre northwest of Dovre. The bicentric and unicentric species, therefore, were assumed to have survived from the last interglacial through the Wisconsin–Würm maxima in the coast refuges and to have migrated to the adjacent mountains since.

The following species are bicentric (after E. Dahl, 1955): *Poa arctica; Carex parallela; C. arctogena; C. bicolor; C. misandra; Luzula arctica; L. parviflora; Nigritella nigra; Stellaria crassipes; Cerastium arcticum; Sagina caespitosa; Minuartia rubella; Melandrium apetalum; Ranunculus hyperboreus; R. platanifolius; Papaver radicatum; Draba cacuminum; D. fladnisenzis; D. lactea; Braya linearis; Saxifraga aizoon* (aberrant type), *S. hieraciifolia; Rhododendron lapponicum; Euphrasia lapponica; Campanula uniflora.*

The following are northern unicentric: *Woodsia glabella; Trisetum subalpestre; Scirpus pumilus; Carex scirpoidea; C. nardina; C. macloviana; C. holostoma; Platanthera oligantha; Potentilla chamissonis; P. emarginata; P. multifida; P. pulchella; Oxytropis deflexa; Cassiope tetragona; Gentiana detonsa; Polemonium boreale; Arenaria humifusa; A. ciliata* ssp. *pseudofrigida; Ranuculus lyngei; R. sulphureus; Papaver laestadianum; P. dahlianum; Draba crassifolia; Braya purpurascens; Pedicularis flammea; P. hirsuta; Erigeron unalaschkensis; Antennaria carpathica; A. porsildii; Arnica alpina; Crepis multicaulis.*

The remainder are southern unicentric: *Phippsia concinna; Stellaria crassipes* ssp. *dovrense; Papaver relictum; Draba hirta* ssp. *dovrense; Pedicularis oederi; Artemisia norvegica.*

Many of these species are calcicoles although the localities available for calcicole species are not confined to the areas of the centric plants. In many cases summer temperature may be a factor (e.g. *Campanula uniflora*, fig. 10B.4) but the distribution of other centric species cannot be explained in this way. Many bicentric species like *Ranunculus*

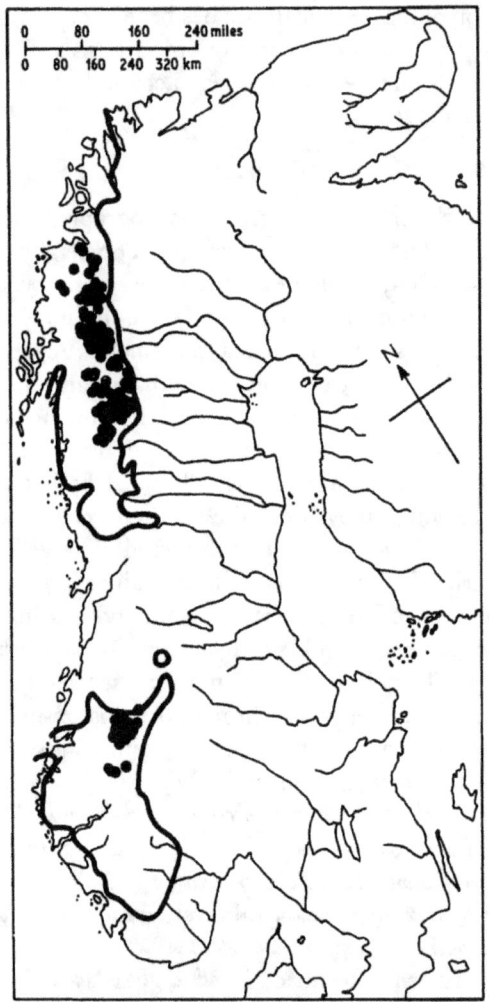

Fig. 10B.4. Distribution of *Campanula uniflora*, a bicentric species, shown with the 22°C mean maximum summer isotherm (after E. Dahl, 1955).

platanifolius, Luzula parviflora, and *Nigritella nigra* (represented in Scandinavia by a taxon differing from that of the Alps in some morphological characteristics and in chromosome number) are subalpine species which in so far as requirements to low summer temperature are concerned might well be able to live along the whole mountain range.

The suggestion that the continentality of centric species is important was more impressive in Blytt's time than today. Closer investigations have revealed that many of these species occur much closer to the ocean than known by Blytt. Most of the species occurring in the southern area are also found in the Trollheimen Mountains northwest of Dovre, with no higher mountains further west and with considerable precipitation. *Saxifraga hieraciifolia, Pedicularis oederi*, and *Euphrasia lapponica* have been found close to the coast. In the north, numerous species grow close to the coast, indicating that continentality is no delimiting factor.

Of course it cannot be proved that the distribution of the centric species is not determined by present-day ecological factors. More information on the ecology of alpine species is needed. But even if it were possible to relate the distribution of all Scandinavian alpine species to ecological factors, several questions would still remain.

If ecological conditions account for the distribution of the centric species, it is evident that they can persist only within a narrow range of conditions. This is in marked contrast to the behaviour of the same species in other areas where they occur widely and in a series of different vegetation types (e.g. *Rhododendron lapponicum, Poa arctica, Carex scirpoidea, Saxifraga paniculata*). This discrepancy might be explained if the populations in question are relicts which have survived in a small refuge. If the population was once small it would have been depleted of genes and hence of variability (see e.g. Dobzhansky, 1947, p. 331). The population would become less adaptable to new conditions, in itself an indication of a relict.

Species able to migrate over considerable distances behind the retreating ice, despite vicissitudes during later phases of deglaciation, must have good spreading ability and be able to persist over a considerable range of ecological conditions. Most centric plants show little adaptation to rapid spreading, and the populations are very stable from year to year.

A closer examination of certain species or groups of species bears out their relict character. Isolation is commonly followed by morphological differentiation. Nordhagen (1931) showed that *Papaver radicatum* s.l. in Scandinavia consists of a number of distinct taxa which he classified as species and subspecies, some endemic, others of a wider range. Nannfeldt (1940) investigated *Poa arctica*, most forms of which are apomictic, and found several distinct taxa which he classified as subspecies, some of a wider range, others apparently endemic. The southern population of the bicentric *Stellaria crassipes* forms its own endemic subspecies *dovrense* according to Hultén (1943). It grows in a small group of the Dovre Mountains and is not known to set seeds. Knaben (1959a, b) has undertaken an extensive study of subspeciation amongst *Papaver radicatum* s.l. which adds further strength to this line of argument.

As more details of the distribution of the Scandinavian alpine plants are learned, the problem becomes more puzzling. Within the centric species of southern Norway, different distribution patterns may be recognized. Some are confined to the Dovre–Trollheimen Mountains (*Artemisia norvegica, Campanula uniflora*), others to the Jotunheimen Mountains (*Rhododendron lapponicum, Braya linearis*). Perhaps valleys have been sufficient obstacles to the spreading of such species.

BIOLOGICAL EVIDENCE FROM NORTH AMERICA

The biological evidence from North America is much more diffuse and ambiguous since the arctic regions are comparatively unknown. There is a marked contrast to north-western Europe, however, in the form of a geological consensus for persistence of nunataks, and even large ice-free regions in northern Alaska and Yukon Territory (fig. 10B.2). Banks Island, in the extreme western Canadian Archipelago, is also believed to have remained ice-free during at least the last two glaciations, while the northwestern Queen Elizabeth Islands show remarkably little evidence of glacierization (Prest, 1970). Biological work has been relatively slight, yet the last decade has seen an upswing of interest in the phytogeography of the High Arctic following the remarkably long lull since Fernald's (1925) work in eastern Canada and northeastern United States was generally discredited by the geological opposition.

Savile (1961) made the first comprehensive effort to discuss the phytogeography of the Canadian Arctic Archipelago. He shows that Ellef Ringnes Island has a remarkably depauperate flora, consisting of only forty-nine vascular plants and five parasitic fungi (bryophytes were not included in his survey), and argues that the adjacent islands with less diversity of habitat probably have even poorer floras. No endemics were noted, and species that do exist are described as dwarf. Although there were to him no convincing indications that Ellef Ringnes Island was overrun by a Wisconsin continental ice sheet, he believes that it cannot have escaped being snow covered. He argues that a light cover of snow and ice would be quickly lost during the postglacial thermal maximum, thus allowing plants to spread along the periphery of the archipelago. The numerous plants that occur southwest and northeast of the islands, but not in them, are taken to indicate that postglacial cold periods, probably accompanied by at least partial snow cover of the outer islands, have driven out many species. Through analysis of plant distribution patterns he argues that no refugia existed in the Canadian Arctic Archipelago and that the region was colonized in postglacial time from the Peary Land refuge, the Yukon–Alaska refugia and from south of the retreating ice sheets.

Ten years later Brassard (1971) has provided a thorough analysis of the mosses of the Queen Elizabeth Islands, with emphasis on northern Ellesmere Island. He comments upon the surprisingly diverse moss flora of northern Ellesmere with 151 species confirmed. Almost half are common or widespread, but the remainder are classed as rare. He notes little correlation between fertility and commonness, and shows that habitat specificity restricts only a few species. He also shows that the moss flora of Ellesmere Island is surprisingly different from that of Peary Land, North Greenland. He proceeds to document the geographical distribution of 233 moss species found in the Canadian High Arctic which he believes indicates, when analysed in conjunction with similar studies of the vascular plants, that ice-free refugia almost certainly occurred in northern Ellesmere Island during at least the Wisconsin glaciation. Some of his detailed evidence includes the documentation of ten high altitude species of bryophytes. Six of these 'are absent from Peary Land, most require a low pH and one, *Grimmia flaccida*, in addition to being highly disjunct, is the plant which reaches the highest altitude in eastern North

America' (Brassard, 1971, p. 256). The high altitude location of *Grimmia flaccida* is Barbeau Peak, 2400 m, northern Ellesmere Island. Brassard (1971, pp. 266–7) provides a more detailed critique and rebuttal of Savile's (1961) conclusions. Certainly, from the biological evidence presented by him, the case for Wisconsin age refugia in northern Ellesmere Island would appear to be very strong.

The work of Kodiak Island (Karlstrom and Ball, 1969) is completely different from that in the Canadian High Arctic. It starts from the position of strength, based upon convincing geological evidence that a considerable area (more than 200 km²) remained ice-free during the Akalura (classical Wisconsin) glacial maximum, that approximately 100 km² remained ice-free during the Karluk (pre-classical Wisconsin) glacial maximum and that small nunatak areas even survived the maximum of the Sturgeon River (pre-Sangamon) glaciation (Karlstrom and Ball, 1969, pp. 41–6). Even considering the highly favourable setting for such nunatak hypothesis protagonists as Hultén and Lindroth, there is a degree of uncertainty amongst the collaborators, in part explainable through their different specialities.

Lindroth (1969, pp. 236–8) undertakes the summing-up and some of his main conclusions are presented here:

(1) The size of the refugium, even during the maximum of each of the last two glaciations, was so large that survival of terrestrial biota could be expected. This also included freshwater lakes.
(2) Given a severe continental climate during glacial maxima only hardy taxa well adjusted to arctic or alpine tundra conditions could have survived.
(3) Despite these assumptions, the present flora and fauna is a mix of survivors and postglacial and late-glacial immigrants, and it is difficult to make a clear-cut 'differentiation between the three groups excepting a small number of individual species. Flightless species of carabid beetles provide the strongest support for the survival hypothesis.

Lindroth makes the general assessment that each line of argument, if taken on its own, is not conclusive, yet their combined testimony 'provides a high degree of probability in favour of the conclusion that survival of hardy animals and plants took place within the limits of the investigated refugium'. The degree of proof, however, is disappointing considering the highly favourable conditions under which some of the most ardent supporters of the nunatak hypothesis were working.

Thus examination of the flora and fauna of a geologically established ice-free area still faces serious problems of identification of those elements that survived. In this sense we should perhaps differentiate between a nunatak or ice-free area established on geological grounds that did not necessarily serve as a plant refuge, and actual refugia – refuges in the sense that absence of ice cover at glacial maxima permitted the survival of plant and animal species. Part of the difficulty faced by Karlstrom and Ball (1969) and their collaborators lies in the fact that plant and animal distributions for Alaska are far from perfectly known. Some botanists will argue, however, that given fifteen collection sites on an island the size of Baffin Island, for instance (approximately the size of Norway and Sweden combined), then it can be presumed that the flora is well known (Hultén, personal communication, July 1963). Yet this will still leave a large element of doubt in the mind of a geologist. It is regrettable that the work on the Kodiak Island refugium did not include studies of the geomorphological evidence and analysis of fine material for clay minerals.

The biological consensus, therefore, is that adherence to the nunatak hypothesis provides the most effective and satisfactory way of accounting for the biogeographic characteristics of the Scandinavian arctic–alpine flora. Yet there remain serious problems. Much of the argument rests upon assumptions of rates of subspeciation which are little understood; there are some fundamental ecological questions typified by the prime ones: could seeds survive refrigeration and germinate subsequent to glacial melting (Porsild et al., 1967; Kjøller and Ødum, 1971)? Even if it could be proven on geological grounds that certain mountain tops remained as nunataks throughout the Wisconsin–Würm maxima, would not the climate be so extremely severe as to preclude all possibility for plant survival? The second question is only partially answered by resorting to south-facing niches, low-lying coastal areas protected from glacial inundation by high mountains, refuges presently below sea level (e.g. Dogger Bank in the North Sea) that may have been dry land during glacial maxima. Yet the location of Möre and Lofoten/Vesterålen adjacent to the narrowest sections of the continental shelf, thus facilitating the rapid calving of accumulating land ice into deep water, seems a striking coincidence.

Finally, in recent years it has become apparent that the biological 'consensus' is by no means monolithic. Hoppe (personal communication, 1972) has pointed out that some of the leading biologists in Sweden are more or less critical of the nunatak hypothesis. Already in 1956 Sjörs took a very cautious position, allowing for the possibility that deglaciation and recolonization *can* explain the present biota distribution patterns; a position which he maintains in his recent book *Ekologisk Botanik* (Sjörs, 1971). Another example is Brinck (1966) who took exception to Lindroth's general thesis and pointed to the 'North Sea Land' as a possible survival area.

3. The geological opposition

Since the early work of Vogt (1913), Reusch (1910), Ahlmann (1919) and Coleman (1920), who believed that they recognized in the landscape 'nunatak forms' (see Plate 37) and absence of evidence of glacial erosion, there has been a heavy and persistent reversal of opinion in favour of total glacial inundation of such high coastal areas as Lofoten, west Norway, Labrador. The reversal has been in part based upon definite field evidence (Hoppe, 1959, 1963, 1971, personal communication, 1971; R. Dahl, 1963, 1966) relating to specific areas (Lofoten, Shetland Islands, Barents Sea) but in part upon mere speculation (Flint, 1943, 1952, 1953) that has penetrated the general literature and has become part of the established geological dogma. The result has been the creation of a 'climate of opinion' which is firmly entrenched and which, aside from the consideration of R. Dahl (1963) leaves E. Dahl's demands for acceptance of firm principles for the identification of glacial erratics like a voice in the wilderness (E. Dahl, 1955).

The geological case against the nunatak hypothesis contains both positive and negative facets. On the positive side there is proof through unequivocal field evidence that a certain mountain top, or offshore island, was glaciated. On the negative side there is a somewhat more tenuous argumentation that lack of geological evidence supporting the existence of an unglaciated enclave indicates, on balance, and by reference to surrounding areas bearing signs of glacial activity, that the area in question was most likely ice-covered also.

Odell (1933) reported the occurrence of 'poorly preserved' glacial striations on summits above 1400 m in the Torngat Mountains of northern Labrador, thus effectively refuting the conclusions of Coleman (1920). Odell's interpretation was accepted by Tanner (1944) and Flint (1943, 1952) and became the basis for widespread belief that all the high mountains along the eastern North American mainland coast were glaciated. Yet, following careful examination of Odell's field area, I have strong doubts concerning the validity of the evidence (Ives, 1957). Discovery of glacial erratics of Labradorean provenance on the high Shickshock Mountains (Flint et al., 1942) led to the contradiction of Fernald's (1925) plea for plant refuges in that area. Even in his latest synthesis Flint (1971, pp. 486–7) accepts the case for total glacial inundation along the West Atlantic coast with the minor qualification that in some instances the 'higher level glaciation may antedate the Late Wisconsin'. Similarly he accepts the prevailing Scandinavian geological view with the statement that 'most, if not all, mountain summits seem to have been overtopped by ice' (Flint, 1971, p. 596).

SPECIFIC GEOLOGICAL EVIDENCE

The prime plant refuge areas of Lofoten and Vesteralen, and of Möre, have become targets for geological study in recent years (Bergström, 1959; R. Dahl, 1963). This investigation has broadened to include the study of distant island groups, such as Shetland, Höpen, Björnöya (Hoppe, 1971) and strategically placed offshore islands, such as Grimsey, north of Iceland (Hoppe, 1968, 1971), that have a critical bearing upon determination of the maximum extent of Würm ice in northwest Europe in general. Prior to these studies the edge of the continental shelf was generally regarded as marking the outer limit of Würm ice, in the case of Vesterålen this being barely 10 km from the present coastline. Work in the hinterland of north Norway (R. Dahl, 1963), in an area (Narvik–Skjömen) where Ahlmann (1919) described 'nunatak topography', has provided important supporting evidence.

The basic approach of these workers has been to first seek out definitive field proof of the passage of glacier ice; striations on bedrock, erratic boulders, roches moutonnées and forms of glacio-fluvial erosion. Bergström (1959) has tackled the alpine forms of the Lofoten and Vesterålen Islands (fig. 10B.5) and has located definitive erratics of mainland provenance up to 720 m above present sea level, some actually upon mountain tops. While acknowledging the difficulty of assigning a date to erratic blocks, he believes

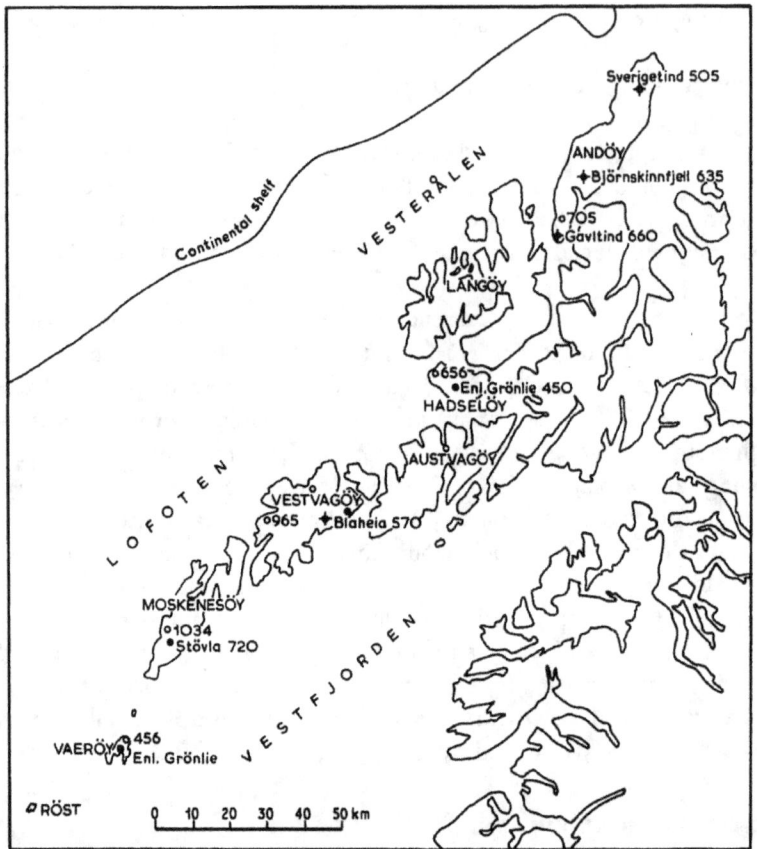

Fig. 10B.5. Map of the Lofoten and Vesterålen, North Norway, showing localities of highest erratics of continental origin (o) and highest neighbouring summit (●) (from Bergström, 1959). The cross indicates that the erratic is lying on the summit.

that their delicately perched positions and minimal weathering rinds indicates a Würm age. Hoppe (1971) has concluded from plotting the distribution of striations on the Shetland Islands that this area was first traversed by ice emanating from the east, presumably Norway, and subsequently supported a local ice cap. This interpretation benefits from the discovery of a block of Tönsbergite at Dalsetter lying well above the marine limit and deriving from the Tönsberg area south of Oslo.[1] Evidence of glacial action in the Svalbard group, including Höpen and Björnöya, and application of accruing knowledge of the regional pattern of glacio-isostatic uplift (Schytt *et al.*, 1968), has led to the conclusion that the shallow Barents Sea supported a Würm ice sheet, thus uniting the main Fenno–Scandinavian inland ice to the northernmost of the Svalbard

---

[1] Hoppe (1970, and personal communication, 1972) qualifies use of the Tönsbergite boulder: 'It thus seems most probable that the boulder was transported by Scandinavian ice during Würm, but it cannot be proved beyond doubt. A more complicated way of transportation, including also movements during earlier glaciations, is not impossible' (Hoppe, 1970, p. 208).

islands. Hoppe (1970) refers to Valentin's (1957) belief that the North Sea floor shows evidence of ice movement from Norway and indicates that the Shetland Islands also were reached by Würm ice from Norway (Hoppe, personal communication, 1971). Two major conclusions derive from this work: first, the glacial map of Würmian north-west Europe has been redrawn; and second, corollary to the first, the continental shelf did not contain the Würm inland ice, but rather seawater of considerable depth was traversed by glacier ice and the ice sheet was considerably more extensive than hitherto thought. Specifically, the prospect of lowland plant refuges situated below present sea level has been infinitely reduced. R. Dahl (1963) has effectively demonstrated that Sandvikfjell (1550 m) and Lapviktind (1475 m) in the Narvik–Skjömen area have been glacierized, definitive erratic blocks having been discovered up to altitudes of 1535 m.

In summary the geological assault on the nunatak hypothesis has, this past decade, received its most effective evidence, the significance of which can be seen from a comparison of fig. 10B.6 and fig. 10B.7. Yet the anomalous distribution of the amphiatlantic plants still demands a rational explanation, and Hoppe's (1963, p. 333) dismissal of the problem remains unacceptable to many botanists:

> One of the strongest arguments for the refugium hypothesis is the existence of plants and animals of the so-called bicentric distribution. There may, however, be other ways of explaining such distributions. Attention may be called to the fact that the supposed refugium areas probably were deglaciated earlier than the rest of northern and western Scandinavia, and thus both flora and fauna have had a longer time to become established.

The impressive extension of the limits of Würm ice in northwestern Europe by Hoppe and his co-workers, however, makes this approach even more difficult considering that the prime characteristics of the centric plants are narrow ecological range, depleted genepool, and inability to migrate freely. It must be stressed, however, that the recent work on the hypothetical 'super ice sheet' covering the very shallow North Sea and Barents Sea (Schytt et al., 1968) is still in progress. While the northern Barents Sea seems undoubtedly to have carried an extensive Würm ice sheet, the precise status of this ice sheet over the southern Barents Sea remains unclear, Grosval'd (1970) arguing for contiguity with northern Scandinavia, and Hoppe (personal communication, 1972) cautioning that absolute proof will require further work.

AMBIGUOUS GEOLOGICAL EVIDENCE

Throughout the long period of contention between opponents and proponents of the nunatak hypothesis a great deal of equivocal evidence has been presented and variously interpreted. Such problematic evidence includes the mountain-top detritus or autochthonous felsenmeer, tors and associated weathering forms, high altitude soil profiles with extensive clay mineral development, and chemical weathering (E. Dahl, 1954, 1955; Hoppe, 1963; R. Dahl, 1963, 1966; Ives, 1958a, b, 1960, 1963, 1966).

Felsenmeer, in particular, has been assumed to represent areas not actively eroded

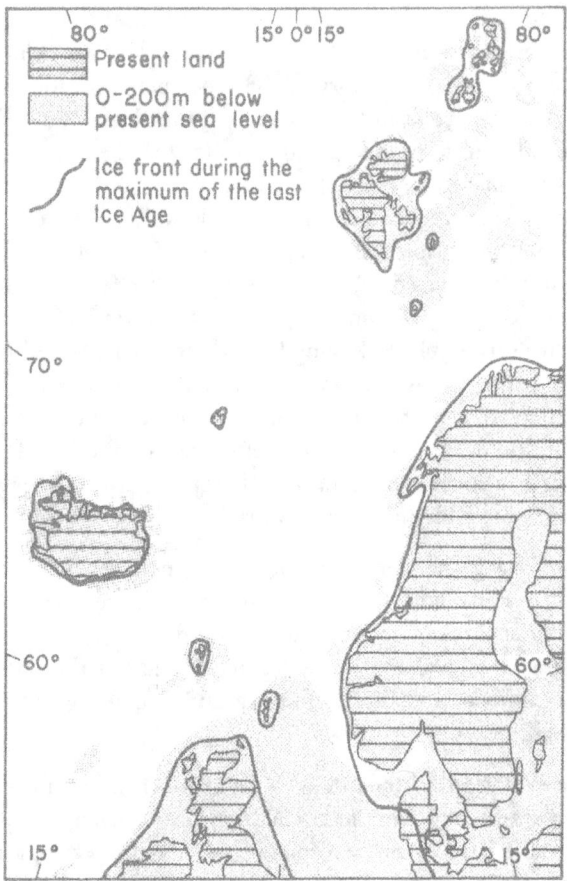

Fig. 10B.6. Map of the outer limits of Würm ice in northwestern Europe (based in part on Hoppe, 1970).

by glacier ice, and the frequently observed seaward inclination of its lower limits has been supposed to parallel the regional slope of the inland ice (E. Dahl, 1955, 1961; Ives, 1963). Felsenmeer is believed to be the product predominantly of mechanical weathering of bedrock in cold climates (see Bird, Chapter 12A). The nunatak hypothesis opponents believe that rapid frost-splitting as the effective agent in the process of its formation has had ample time to produce the deep, mature boulder detritus since the maximum of the last glaciation and arguments such as high frequency of freeze–thaw cycles on mountain tops have been used to support this contention. Alternately, the occurrence of autochthonous boulder fields at altitudes significantly below areas carrying definitive evidence of glacial erosion (Hoppe, 1963; Rapp and Rudberg, 1960) has prompted the conclusion either of rapid formation or of preservation beneath an over-riding ice sheet. This type of argument is weakened by lack of understanding of the actual process, or processes, involved as well as rate of such process, and the possibility that

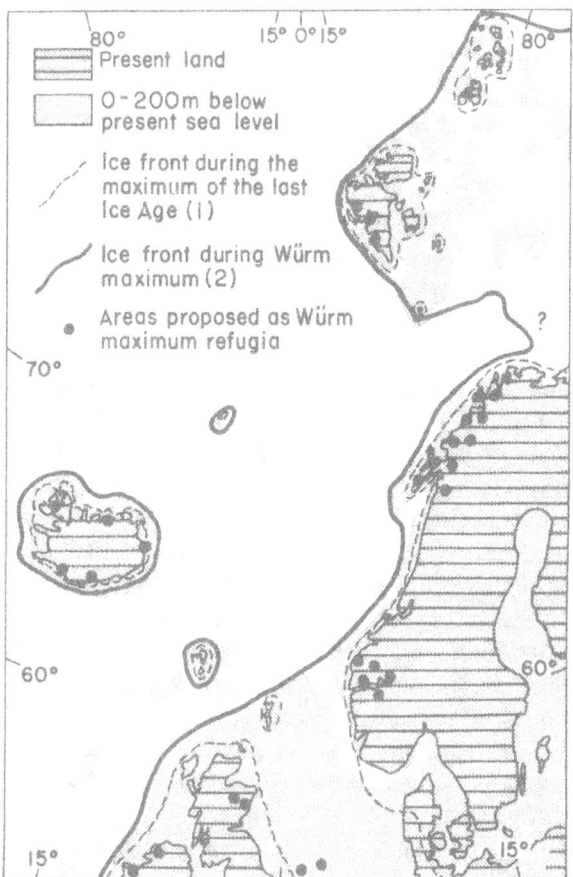

Fig. 10B.7. Improvised map of the outer limits of Würm ice in northwestern Europe (based mainly on Grosval'd, 1970; Hoppe, 1970; and Schytt *et al.*, 1968; proposed refugia are taken from various biological sources). (1) denotes the limit from Fig. 10B.6, (2) revised limit.

certain block fields cannot be compared. This latter point was emphatically brought home to me when, on a field excursion in the Abisko area of northern Sweden in 1960, Drs Anders Rapp and Sven Rudberg demonstrated the presence of boulder fields that occur at a lower elevation than evidence of glacial erosion. The Abisko boulder fields are not nearly so well developed as the mountain-top detritus of the Torngat Mountains, northern Labrador, but rather corresponded with what I have termed 'incipient felsenmeer' in Labrador (Ives, 1958b).

Some progress can be made, however: first, it can be assumed that diurnal fluctuation of the ambient air temperature is no indication of the rate of felsenmeer formation (R. Dahl, 1966; Ives, 1966); and second, that fluctuations of temperature across the freezing point at the air–ground interface are not relevant since below a few centimetres depth there is only a single freeze–thaw cycle each year (Cook and Raiche, 1962; Fahey, 1971). Finally, R. Dahl (1963) has provided convincing proof that in some instances at least erratic boulders do occur within felsenmeer, a viewpoint anticipated by Ives (1957, 1958a, b) in Labrador, although contested by Løken (1962).

The only firm conclusion that can be derived from the form and distribution of felsenmeer, at this point, is that our understanding of its mode and speed of development is scant, and its occurrence on a particular mountain top in no way proves that no glacial cover existed subsequent to its formation. R. Dahl (1963) has indicated that the remarkably consistent regional trend of the lower limits of the felsenmeer can be explained by assuming total cover of ice with vigorous movement in the lower layers conditioned by a strong regional topography with a relatively inert condition in the ice mass above that level. Ives (1963) has indicated that autochthonous boulder fields could have been protected by thin ice carapaces frozen to their bed. What has been said concerning felsenmeer can also be applied, at this point, to weathering pits and tors, although all three forms will be discussed further in the following section.

Some of the above discussion can also be applied to E. Dahl's high altitude soil profiles (E. Dahl, 1954, 1955, 1961; Gjaerevoll, 1963), yet some consideration of the implications of X-ray analysis in the identification of clay minerals is deemed advisable.

E. Dahl (1948, 1961) has argued that the presence of considerable quantities of illite, vermiculite, montmorillonite, and hydrobiotite in soil profiles on high mountain tops (e.g. Gjevilvasskammene, 1640 m, west Norway) indicates chemical weathering of biotite at least an order of magnitude greater than is found in soils of distinctly Late Glacial initiation discovered at lower elevations in southern Norway. He compares the high-level soil development to similar soils of unglacierized Devon and Cornwall, indicating that the Norwegian examples derive from the Tertiary, thereby implying the existence of nunataks throughout the full extent of the Neogene glaciations. This argument has been opposed on two grounds: first, that clay minerals may be derived directly from the bedrock or that they were inherent Tertiary constituents of material moved by glacial action from some other area, and second, that they could have survived in situ in frozen form without suffering extensive erosion despite inundation by glacier ice.

A final bone of contention is that even the supposed unequivocal evidence for glaciation cannot be related to one specific period (E. Dahl, 1955). Hoppe (1959, 1963, 1970) and Bergström (1959) have argued that glacial striations, polish, delicately perched erratic blocks, for instance, must be ascribed to the last glacial period since subaerial erosion through the long Riss–Würm interglacial would have surely obliterated them. Yet striations associated with the Dwyka Tillite (Permo-Carboniferous) indicate preservation of delicate forms through a large proportion of total geological time. Despite claims to the contrary, the actual age of glacial erratics and striations remains unproven in any strict scientific sense.[1]

---

[1] This enigma has been further emphasized by Hoppe (personal communication, 1972) by his reference to the widely known striated surfaces of Precambrian age at Bigganjargga in Varanger Fjord, Norway, in contrast to the very rapid destruction of recently glacially polished surfaces common throughout Scandinavia (e.g. R. Dahl, 1967).

RECENT EVIDENCE FROM NORTHEASTERN NORTH AMERICA

Two major undertakings to construct glacial chronologies for northeastern North America, Labrador–Ungava during the 1950s and Baffin Island/Foxe Basin during the 1960s have had to face the problem of determining absolute maximum vertical and horizontal extent of the ice masses and age of such occurrence. This in turn has driven the discussion into a full consideration of the nunatak hypothesis. Working independently Mercer (1956) argued on geomorphological grounds that the outer part of the high plateau overlooking the southwest coast of Frobisher Bay remained ice-free, at least during the Wisconsin.

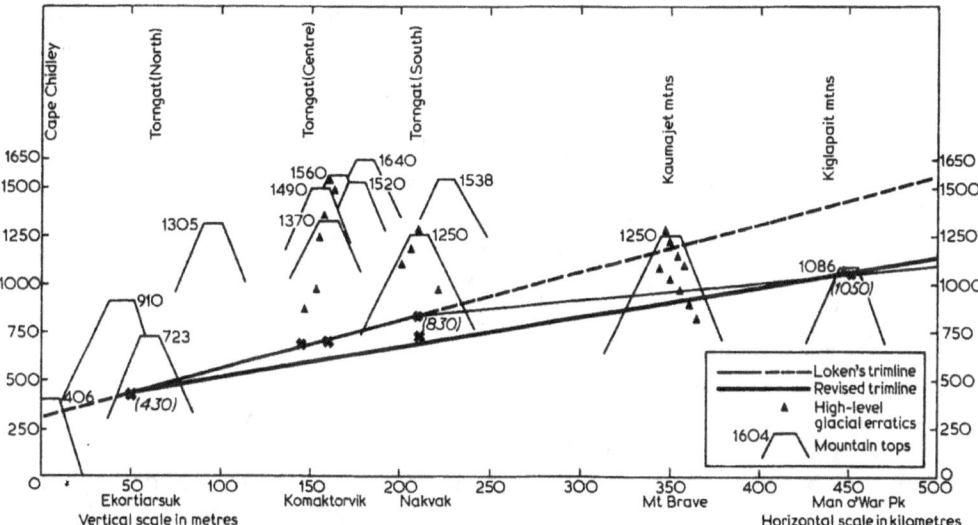

Fig. 10B.8. The north–south slope of the northern Labrador trimlines and high-level erratic according to the interpretations of Løken (1962) and Ives (1962).

In contrast to the work of Hoppe and collaborators in northwestern Europe (Hoppe, 1971) which has provided evidence for greater expanse of glacial ice than was identified previously, the effort in northeastern North America has tended toward a rejection of the Flint–Tanner concept of total glacial inundation, especially during the Wisconsin Stade.

The earliest of this work in the Torngat Mountains of northern Labrador (Ives, 1957, 1958a, b, 1960) led to the recognition of three distinct altitudinal zones of weathering sloping generally toward the Atlantic Ocean. In addition, anomalous blocks, possibly glacial erratics, were encountered on various high summits on top of and within mature mountain-top detritus. Added weight was given to the glacial erratic interpretation by the discovery of gneissic rocks resting on basic volcanics at altitudes in excess of 1000 m near Cape Mugford (Tomlinson, 1958) and of anorthosite resting upon granitic gneiss at 900 m on Mt Thoresby (Johnson, 1969), although both these localities lie considerably further south than the Torngat Mountains. Løken (1962) has heavily challenged the glacial erratic interpretation for the Torngat blocks, proposing that they are erosion residuals from veins and inclusions within a mixed series of granitic gneisses, schists and metasediments (fig. 10B.8). Andrews (1963) identified the lowest two of the weathering zones in the Okak Bay area further south and proposed that the three zones be named Torngat, Koroksoak and Saglek, in order of decreasing altitude and decreasing age from the area where they were first recognized – the Saglek Fiord drainage area of northern Labrador. No dating was possible, although it was recognized that the highest level, whether glacierized or not, had been subjected to subaerial erosion for a very long time, presumably at least during and since the Sangamon interglacial, and Andrews (1963) suggested that the same may be true of the Koroksoak zone. The only definitive radio-carbon date (Løken, 1962) relates to marine mollusc shells close to the head of Ryans Bay proving that some fiords were totally ice-free by 9000 BP while others still carried the tongues of Laurentide outlet glaciers pushing through the great transection troughs from the west.

In the 1960s attention was turned to Baffin Island (Ives, 1964). Preliminary observations and a conceptual framework for glaciation were presented by Ives and Andrews (1963), but the first significant breakthrough was made by Løken (1966) who dated marine mollusc shells in excess of 54,000 BP ($Y-1703 > 54,000$). These shells were recovered from a glacio-fluvial delta deposited into a sea level approximately 80 m above present sea level in the vicinity of Cape Aston. The disposition of the delta was such that Løken concluded that the local area had not been disturbed by glacier ice since delta building had been completed. This conclusion is significant since it indicates that a low-lying area, available as a plant refuge, persisted throughout at least the classical Wisconsin Stade. Since Løken's contribution there has been a succession of radiocarbon dates on marine mollusc shells in association with deltas and lateral and terminal moraine systems, indicating that significant areas of low-lying land, as well as high mountain tops, re-mained ice-free throughout classical Wisconsin, Mid- and Early Wisconsin times (Ives and Buckley, 1969; England, personal communication, 1971; Pheasant, 1971; Ives and Borns, 1971). Clarke and Upton (1971, p. 250) working independently on the petrology and stratigraphy of the Tertiary basalts of Baffin Island, conclude that 'the outcrops show no evidence, either in the form of striations or glacial erratics, of erosion during Pleistocene glaciation'. On an entirely different line of enquiry, Buckley (1969) shows that a consideration of gradients of present and former outlet glaciers and ice sheets makes it highly improbable that the higher coastal mountain summits of northeast Baffin Island were overtopped by Wisconsin, or earlier ice masses. The most relevant work is that of Pheasant (1971) who has not only dated a series of moraine systems as greater than 110,000 years old (amino-acid and uranium series dates) but has identified three altitudinal zones of weathering that he correlates with Torngat, Koroksoak and Saglek zones of northern Labrador. In 1966 Løken had noted that degree of weathering on rock knobs in the vicinity of the Cape Aston delta seemed to compare closely with that within the Korok-soak zone of the Torngat Mountains. In 1966 Ives made an extensive helicopter-supported survey of a wide tract of mountain and fiord land extending from the outer Baffin Bay coast to the interior plateau, some 150–200 km to the southwest (Ives, 1966, unpublished).

Felsenmeer and tors were mapped and, as in Labrador, seaward slope of the lower felsenmeer limit was noted, as was an intermediate zone characterized by degraded tors and an intermediate degree of weathering. Also, large anomalous blocks of gneissic material were located within the felsenmeer that were compared with the controversial blocks resting within the Torngat zone in northern Labrador. In addition, glacially disturbed felsenmeer, showing lineations parallel to the regional iceflow (NE–SW) was located from the air along the southwesternmost tract of the felsenmeer zone, in precisely the same area where the large anomalous blocks were located. This suggests possible division of the felsenmeer zone (Torngat) into two: the older, or northeasternmost, possibly never glacierized, and the younger, or southwesternmost, glacierized, but in the early part of the Neogene. If the two-fold division can be substantiated, the younger zone should presumably extend northeastward below the older zone, thus preserving the altitudinal arrangement. No evidence for this was discovered in the Inugsuin–Clyde fiord area in 1966, although it must be admitted that the arragement was not suspected then, and no attempt was made in the field to distinguish two sections of the Torngat zone. Thus the above discussion is based upon a slender speculation, although this has been given added significance by more recent information deriving from Andrews and his co-workers in the Cumberland Peninsula area (Andrews, personal communication, 1971). It appears that systematic differences in degree of weathering within the felsenmeer zone may be detectable although more data are required. (For a discusion of the glacial chronology of eastern Baffin Island, see Andrews, Chapter 6A.)

Andrews and Miller (1972) have discussed the results of various analyses of fines collected from the three weathering zones both from Cumberland Peninsula and the Inugsuin–Clyde area (Ives, unpublished, 1966). Analyses include mechanical grain-size, percent composition of the major elements, amount of free iron, X-ray diffraction of oriented samples of material less than 2 μm and scanning electron microscope (SEM) examination of sand-size quartz grains.

Andrews and Miller (1972) recognized the problem that local bedrock sources could seriously influence the percent composition of major elements in the surficial samples. Thus local geology was used as a group classification in the study to determine whether or not it had any significant impact upon the results of the analyses. Multiple stepwise discriminant analysis (MSDA) was used to enquire whether the chosen classification of the samples (i.e. classified as belonging to the Torngat, Koroksoak or Saglek weathering zones) was statistically appropriate. The percent composition of 8 major elements was determined (MnO, MgO, CaO, Al_2SO_3, SiO_2, Fe_2O_3, TiO_2, NaO_2). In the analysis 6 of the results were used in straight percent form while Al_2O_3 and SiO_2 were combined to form a ratio. Percent of silt and clay was also used as a variant in the discriminant analysis. Numbers of samples of undisputable classification were Torngat, 13, Koroksoak, 6, and Saglek, 6. An additional 17 samples were used that were derived from either the Torngat or Koroksoak zone but without definite knowledge as to which. Part of Andrews and Miller's objective was to develop a discriminant function that would properly assign these samples to their respective zone.

MSDA indicated that only 2/25 of the samples in the three zones were incorrectly classified and that the molecular ratio of SiO_2/Al_2O_3 was the most useful variable for discrimination between samples deriving from the three zones. On the young moraines of the Saglek zone this ratio averaged 9.34:1, for the Koroksoak zone it was 7.57:1, and for the Torngat zone 7.35:1. These results indicate that silica is moderately mobile in arctic environments, thus agreeing with the results of Church (1970) on the chemical composition of runoff from eastern Baffin Island rivers. The other elements showed a percent increase with age, probably because of the addition of granular disintegration products.

Andrews and Miller (1972) go on to point out that iron oxidation is visually important in Baffin Island, and analysis of free iron in soil horizons provided interesting results. Free iron on young, Neoglacial moraines varied between 0.1 and 0.3 per cent, on till of Late Wisconsin age (8000 BP) 0.4 to 0.6 per cent, and in moraines estimated at 25,000 years old the percentage of free iron was 0.9. Moraines approximately 45,000 years old contained 1.2 per cent free iron. A general feature was that all observed A horizons had slightly greater percentages of free iron than B or C horizons.

X-ray diffraction of samples from the Saglek zone indicated little or no development of clay minerals; the only frequency peaks are small and are limited to illite/muscovite and hydrobiotite. In the Koroksoak and Torngat zones the diffraction patterns were much more complex with larger and more numerous significant peaks. Specifically, material from both zones have significant peaks indicating the presence of hydrobiotite and mixed layer chlorite/vermiculite at 7.5 to 7.7 and 6.31 θ respectively. These peaks are absent from the analysis of Saglek zone soils.

Detailed examination of sand-size quartz grains with the SEM system at 500 to 50,000 magnification was undertaken. Saglek zone samples showed fresh surfaces with typical glaciated patterns (Krinsley and Margolis, 1969). The Koroksoak zone material still showed the glaciated fracture patterns, but surface pitting was evident; the surfaces from Torngat zone material provided no evidence of glacial action but considerable pitting and wasting.

Pheasant (1971) working in the same general area as Andrews and Miller (Cumberland Peninsula) determined various ratios based on weathering criteria between the three major zones. He used such criteria as thickness of weathering rinds, percent of fresh boulders, size and area covered by weathering pits, and showed a 'weathering ratio' of 5:1 between the Koroksoak and Saglek zones. He believes that the significant increase in weathering between the Saglek and Koroksoak zones indicates that the latter has been subjected to subaerial weathering processes for a period of about 500,000 years. This is corroborated by Andrews and Miller (1972); see also Andrews, Chapter 6A.

Finally, work in the Colorado Rocky Mountains may have some bearing on the discussion. Miller (1971) has derived weathering ratios on Pinedale (=Wisconsin) and Bull Lake (=Early or Pre-Wisconsin) deposits indicating that, as a first approximation, weathering ratios are a linear product of age. Madole (personal communication, 1971) has obtained comparable results to Andrews and Miller (1972) from SEM analyses of sand-size quartz grain samples, and Street (personal communication, 1971) has shown a significant relationship between the distribution of felsenmeer and tors and the absence of

Pinedale and Bull Lake till. The Front Range of the Colorado Rocky Mountains exhibits at least four zones of different age classes. The oldest, confined to upland surfaces at the eastern ends of high alpine ridges east of the Continental Divide, that all authorities agree was never glacierized (Richmond, 1965); a younger surface especially typical of the western sections of zone 1 ridges that carries a deep mantle of detritus believed to be very old till (cf. Madole's SEM analysis of quartz grains), and the surfaces occupied by Pinedale and Bull Lake tills. The similarities with Baffin Island and Labrador, at least on first reconnaissance analysis, is striking. Such similarity is important because it adds considerable weight to the contention that the Torngat zone in the northeastern Arctic is very old. However, actual surface weathering differences between the two oldest classes are not well defined, possibly due to former vigorous frost-churning.

In summary, therefore, it would appear that a large amount of new data of various types indicates not only a very complex history of glaciation in the eastern Canadian Arctic, but points with increasing insistence to the conclusion that appreciable areas of high land may have remained as nunataks throughout the entire period of the Neogene glaciations, but even more emphatically, that larger areas were ice-free throughout the Wisconsin Stade. Even so, these postulates must be qualified by the nice distinction between the terms 'glaciated' and 'glacierized'. Ice may never have eroded the highest summits, or low-lying areas characterized by Torngat-type weathering products, but thin, inert, cold ice may have occupied the entire area (Ives, 1962) leaving no trace upon melting. Nevertheless, even this eventually permits a close assessment of the maximum possible thickness of Neogene ice in eastern Baffin Island and northern Labrador, although it leaves in abeyance the question of plant refuges. At least within the context of the Wisconsin Stade (about 100,000 years) extensive areas remained ice-free as potential refugia.

4. Conclusions

It is necessary to re-emphasize that geological investigations over the past ten to fifteen years have produced a series of contrasting conclusions for opposite sides of the North Atlantic Ocean. These contrasting results, of more extensive Wisconsin–Würm ice cover in northwestern Europe, and of less extensive Wisconsin–Würm ice cover in northeastern North America, are not necessarily contradictory. However, several comments seem justified: first, the European results depend upon unproven age assumptions; specifically, striations cannot be dated and stratigraphic data are lacking, even though in a qualitative sense the broad conclusions seem acceptable; second, the North American picture is faced with a similar problem of definitive age determinations on weathering zones, although a major beginning has been made and incomplete Wisconsin–Würm glacial inundation now seems certain.

Much more exacting work on differentiation and age determination of the weathering zones in the Canadian Arctic is required, and comparisons with similar zones in Scandinavia should be made. The controversial blocks on high Torngat and Baffin Island summits cannot be dismissed lightly. They are of less importance to discussion of the nunatak hypothesis than they are to full understanding of the complete glacial-interglacial picture, since it has been agreed that extent of ice during the past 100,000 years is the critical component in the formula for plant and animal phytogeography.

One further contrast between the two sides of the North Atlantic deserves emphasis. Based upon the current literature, it is remarkable that support for the nunatak hypothesis in northwestern Europe lies entirely in biological hands while in eastern North America the proponents based upon actual field investigation are almost entirely the geological workers. An immediate realization is that eastern Baffin Island and northern Labrador deserve a massive biological research input.

Finally, the botanical case itself must be rigorously re-examined. One approach would be the application of detailed and extensive palynological investigations in an attempt to trace the history of plant migration through time (Faegri, 1963). Faegri also provides a stimulating review of the extant plant distribution problem, emphasizing the ambiguous and unproven nature of many parts of the botanical case. Some of his arguments may well be contested by his botanical colleagues, yet statements to the effect that although 'polyploidy may indicate both youth and great age in a taxon – these terms are relative only and cannot be adequately translated into geologic or absolute chronology' (Faegri, 1963, p. 231), and 'as long as we know nothing about the speed of evolution, cytogenetic data cannot contribute very much, nor can the presence or absence of endemic taxa', require careful evaluation. In another section of his presentation Faegri agrees that conditions of plant distribution in the Scandinavian north are much more difficult to explain without resort to the nunatak hypothesis. He then goes on to refer to the separate floral element with *Crepis multicaulis* and *Oxytropis deflexa* as its most significant representatives. Their nearest stations are in the far east of Asia; none are known to occur in the intermediate mountains and especially not in the Urals.

> I would consider it very possible that these species have survived the last glaciation in or near Scandinavia, but can see no reason compelling us to reject the hypothesis that they survived at the edge of the glaciated area in the USSR. The absence of these species there now is easily explained by reference to their ecologic demands and the Postglacial Hypsithermal period forests that would have crowded out these and many other species from their former stations (Faegri, 1963, p. 230).

A second approach is already evolving in Norway and Sweden. With the rapid increase in understanding of the complexities of Würm glaciation in northwestern Europe, alternatives to the nunatak hypothesis are being requested (Sjörs, 1956, 1971; Brinck, 1966; Coope, 1967, 1969) and in turn are encouraging rethinking amongst several protagonists. Lindroth (1969), in particular, states that the rapid growth in knowledge of Würm glaciation opens up three possibilities:

(1) Survival from the last main interglacial (Eemian), that is, through Würm I and Würm II.

(2) Survial from the Würm interstatial period, that is, during Würm II.

(3) Survival only through the Younger Dryas period following an immigration from the south (or possibly northeast) in the preceeding warm Allerød period.

Lindroth indicates that while survival through the Younger Dryas may explain some distributions it cannot explain the basic pattern.

The existing situation is clearly unsatisfactory, therefore, and requires specific research in a wide variety of fields and geographic areas, but especially it demands an overriding interdisciplinary approach. The 1962 symposium on the North Atlantic biota and their history, organized by Drs Áskell and Doris Löve, and held in Iceland, produced a major volume (Löve and Löve, 1963a) giving the status of the nunatak hypothesis at that time. It has also provided impetus to much of the new work reported in this chapter. Despite the unsatisfactory nature of the situation, many considerable advantages have accrued from the continuing controversy, as must be evident from the significant increase in our knowledge of the Neogene environments of high latitude lands around the North Atlantic Ocean.

ACKNOWLEDGEMENTS

I am deeply grateful for fifteen years of encouragment, stimulation and criticism from a small group of the major contributors to the nunatak discussion: to Professors Eilif Dahl, Gunnar Hoppe, Carl H. Lindroth, Áskell Löve and Drs Ragnar Dahl and Doris Löve, protagonist and antagonist alike.

References

ABBE, E. C. (1938) Phytogeographical observations in northernmost Labrador. *In:* FORBES, A. (ed.) Northernmost Labrador mapped from the air, *Amer. Geogr. Soc. Spec. Pub.*, 22, 217–34.

AHLMANN, H. W. (1919) Geomorphological studies in Norway. *Geogr. Ann.*, 1, 1–246.

ANDREWS, J. T. (1963) End moraines and late glacial chronology in the northern Nain–Okak section of the Labrador coast. *Geogr. Ann.*, 45 (2–3), 158–71.

ANDREWS, J. T. and MILLER, G. H. (1972) Chemical weathering of tills and surficial deposits in east Baffin Island, NWT, Canada. *In:* ADAMS, W. P. and HELLEINER, F. M. (eds.) *International Geography, 1972*, Vol. 1, 5–7. Toronto: Univ. of Toronto Press.

ARWIDSSON, T. (1943) Studien über die Gefasspflazen in den Hochgebirgen der Pite Lappmark. *Acta Phytogeogr. Suecica*, 17, 1–274.

BERGSTRÖM, E. (1959) Utagjorde Lofoten och Vesterålen elt refugium under sista istiden? *Svensk. Naturvet.*, 116–22.

BLYTT, A. (1876) *Essay on the Immigration of the Norwegian Flora during Alternative Dry and Rainy Periods.* Christiania: Cammermeyer. 89 pp.

BÖCHER, T. W. (1938) Biological distributional types in the flora of Greenland. *Medd. om Grønland*, 148 (1), 1–326.

BÖCHER, T. W. (1951) Distributions of plants in the circumpolar area in relation to ecological and historical factors. *J. Ecol.*, 39, 376–95.

BRASSARD, G. R. (1971) The mosses of northern Ellesmere Island, Arctic Canada, I. Ecology and Phytogeography, with an analysis for the Queen Elizabeth Islands. *Bryologist*, 74 (3), 233–81.

BRINCK, P. (1966) Animal invasion of Glacial and Late Glacial terrestrial environments in Scandinavia. *Oikos*, **17**, 250–66.

BUCKLEY, J. T. (1969) Gradients of past and present outlet glaciers. *Geol. Surv. Can. Pap.*, 69–29. 13 pp.

CHURCH, M. (1970) Baffin Island Sandar: A study of Arctic fluvial environments. *Geol. Surv. Can. Mimeogr. Rep.*, Proj. 680042. 536 pp.

CLARKE, D. B. and UPTON, B. G. J. (1971) Tertiary basalts of Baffin Island: field relations and tectonic setting. *Can. J. Earth Sci.*, **8**, 248–58.

COLEMAN, A. P. (1920) Extent and thickness of the Labrador Ice Sheet. *Geol. Soc. Amer. Bull.*, **31**, 319–28.

COOK, F. A. and RAICHE, V. (1962) Freeze–thaw cycles at Resolute, NWT. *Geogr. Bull.*, **18**, 64–78.

COOPE, G. R. (1967) The value of Quaternary insect faunas in the interpretation of ancient ecology and climate. *Proc. VII Congr. Int. Ass. Quat. Res.* (New Haven), **7**, 359–80.

COOPE, G. R. (1969) The contribution that the Coleoptera of Glacial Britain could have made to the subsequent colonisation of Scandinavia. *Opusc. Entomol.*, **34**, 95–108.

DAHL, E. (1946) On different types of unglaciated areas during the ice ages and their significance to phytogeography. *The New Phytol.*, **45**, 225–42.

DAHL, E. (1948) Studier over forvitringstyper i strøket Nordfjord-Sunmör og deres relasjon til istidene. *Nor. Geol. Tidsskr.*, **27**, 242–4.

DAHL, E. (1952) On the relation between summer temperature and the distribution of alpine vascular plants in the lowlands of Fennoskandinavia. *Oikos*, **3**, 1–31.

DAHL, E. (1954) Weathered gneisses at the island of Runde, Sunmöre, West Norway, and their geological interpretation. *Nytt. Mag. Bot.*, **3**, 5–23.

DAHL, E. (1955) Biogeographic and geological indications of unglaciated areas in Scandinavia during the glacial ages. *Geol. Soc. Amer. Bull.*, **66**, 1499–519.

DAHL, E. (1961) Pleistocene history of the flora of the North Atlantic region with special reference to Scandinavia. *In: Recent Advances in Botany*, 919–25. Toronto: Univ. of Toronto Press.

DAHL, R. (1963) Shifting ice culmination, alternating ice covering and ambulant refuge organisms? *Geogr. Ann.*, **45** (2–3), 122–38.

DAHL, R. (1966) Block fields, weathering pits and tor-like forms in the Narvik Mountains, Nordland, Norway. *Geogr. Ann.*, **48**, A(2), 55–85.

DAHL, R. (1967) Postglacial micro-weathering of bedrock surfaces in the Narvik district of Norway. *Geogr. Ann.*, **49**, A(2–4), 155–66.

DOBZHANSKY, T. (1947) *Genetics and the Origin of Species* (3rd edn). New York: Columbia Univ. Press. 446 pp.

FAEGRI, K. (1937) Some recent publications on phytogeography in Scandinavia. *Bot. Rev.*, **3**, 425–56.

FAEGRI, K. (1963) Problems of immigration and dispersal of the Scandinavian flora. *In:* LÖVE, Á. and LÖVE, D. (eds.) *North Atlantic Biota and their History*, 221–32. Oxford: Pergamon Press.

FAHEY, B. D. (1971) *A Quantitative Analysis of Freeze–Thaw Cycles, Frost Heave Cycles, and Frost Penetration in the Front Range of the Rocky Mountains, Boulder County, Colorado.* Unpub. Ph.D. thesis, Univ. of Colorado. 329 pp.

FERNALD, M. L. (1925) Persistence of plants in unglaciated areas of boreal North America. *Amer. Acad. Arts Sci. Mem.*, 15, 3, 237–42.

FLINT, R. F. (1943) Growth of the North American ice sheet during the Wisconsin Age. *Geol. Soc. Amer. Bull.*, **54**, 325–62.

FLINT, R. F. (1952) The Ice Age in the North American Arctic. *Arctic*, **5**, 135–53.

FLINT, R. F. (1953) Probable Wisconsin substages and late-Wisconsin events in northeastern United States and southeastern Canada. *Geol. Soc. Amer. Bull.*, **64**, 897–920.

FLINT, R. F. (1971) *Glacial and Quaternary Geology*. New York: John Wiley. 892 pp.

FLINT, R. F., DEMOREST, M. and WASHBURN, A. L. (1942) Glaciation of the Shickshock Mountains, Gaspe Peninsula. *Geol. Soc. Amer. Bull*, **53**, 1211–30.

FRIES, T. C. E. (1913) *Botanische Untersuchungen im nördlichsten Schweden*. Uppsala and Stockholm: Akad. Abh. 361 pp.

GJAEREVOLL, O. (1963) Survival of plants on nunataks in Norway during the Pleistocene Glaciation. *In:* LÖVE, Á. and LÖVE, D. (eds.) *North Atlantic Biota and their History*, 261–83. Oxford: Pergamon Press.

GROSVAL'D, M. G. (1970) Glaciological and geomorphological research in the Eurasian Arctic with special reference to the Barents Sea area. *Musk-Ox*, **7**, 10–31. Saskatoon: Inst. for Northern Studies.

HOLTEDAHL, O. (1960) The Geology of Norway. *Nor. Geol. Unders.*, **208**, 364–639.

HOPPE, G. (1959) Några kritiska kommentarer till diskussionen om isfria refugier. *Sven. Naturvitensk.*, 123–34.

HOPPE, G. (1963) Some comments on the 'ice-free refugia' of northwestern Scandinavia. *In:* LÖVE, Á. and LÖVE, D. (eds.) *North Atlantic Biota and their History*, 321–35. Oxford: Pergamon Press.

HOPPE, G. (1968) Grimsey and the maximum extent of the last glaciation of Iceland. *Geogr. Ann.*, **50**, A, 16–24.

HOPPE, G. (1970) The Würm Ice Sheets of Northern and Arctic Europe. *Acta Geogr. Lodz.*, **24**, 205–15.

HOPPE, G. (1971) Nordvästeuropas inlandsisar under den sista istiden. *Särta Svensk. Naturvitensk.*, 31–40.

HULTÉN, E. (1937) *Outline of the History of Arctic and Boreal Biota During the Quaternary Period.* Stockholm. 168 pp.

HULTÉN, E. (1943) *Stellaria longipes* Goldie and its allies. *Bot. Not.*, 251–70.

HULTÉN, E. (1950) *Atlas of the Distribution of Vascular Plants in NW Europe*. Stockholm. 512 pp.

HULTÉN, E. (1958) The amphi-atlantic plants and their phytogeographical connections. *Kgl. Svensk. Vetensk. Akad. Handl.*, ser. 4, **7** (1), 1–340.

IVES, J. D. (1957) Glaciation of the Torngat Mountains, northern Labrador. *Arctic*, **10** (2), 67–87.

IVES, J. D. (1958a) Glacial geomorphology of the Torngat Mountains, northern Labrador. *Geogr. Bull.*, **12**, 47–75.

IVES, J. D. (1958b) Mountain-top detritus and the extent of the last glaciation in northeastern Labrador–Ungava. *Can. Geogr.*, **12**, 25–31.

IVES, J. D. (1960) The deglaciation of Labrador–Ungava – an outline. *Can. Geogr. Que.*, **IV**, 323–43.

IVES, J. D. (1962) Indications of recent extensive glacierization in north-central Baffin Island, NWT. *J. Glaciol.*, **4** (32), 197–205.

IVES, J. D. (1963) Field problems in determining the maximum extent of Pleistocene glaciation along the eastern Canadian seaboard. *In:* LÖVE, Á. and LÖVE, D. (eds.) *North Atlantic Biota and their History*, 337–54. Oxford: Pergamon Press.

IVES, J. D. (1964) Work of the Geographical Branch, Dept of Mines and Technical Surveys, Canada, in Baffin Island, 1961–1964. *Polar Rec.*, **12** (78), 281–9.

IVES, J. D. (1966) Block fields, associated weathering forms on mountain tops and the nunatak hypothesis. *Geogr. Ann.*, **48**, A(4), 220–3.

IVES, J. D. and ANDREWS, J. T. (1963) Studies in the physical geography of north-central Baffin Island, Northwest Territories. *Geogr. Bull.*, **19**, 5–48.

IVES, J. D. and BORNS, H. W. JR (1971) Thickness of the Wisconsin ice sheet in southeast Baffin Island, Arctic Canada. *Zeits. Gletscherk. Glazialgeol.*, **VII** (1–2), 17–21.

IVES, J. D. and BUCKLEY, J. T. (1969) Glacial geomorphology of Remote Peninsula, Baffin Island, NWT, Canada. *Arct. Alp. Res.*, **1** (2), 83–96.

JOHNSON, J. P. JR (1969) Deglaciation of the central Nain–Okak section of Labrador. *Arctic*, **22** (4), 373–94.

KARLSTROM, T. N. V. and BALL, G. E. (1969) *The Kodiak Island Refugium*. Toronto: Ryerson Press for Boreal Inst., Univ. of Alberta. 262 pp.

KJØLLER, A. and ODUM, S. (1971) Evidence for longevity of seeds and microorganisms in permafrost. *Arctic*, **24** (3), 230–3.

KNABEN, G. (1959a) On the evolution of the *radicatum*-group of *Scapiflora* Papavers as studied in 70 and 56 chromosome species, Part A. Cytotaxonomical Aspects. *Oper. Bot.*, **2** (3), 1–76.

KNABEN, G. (1959b) On the evolution of the *radicatum*-group of *Scapiflora* Papavers as studied in 70 and 56 chromosome species, Part B. Experimental studies. *Oper. Bot.*, **3** (3), 1–96.

KRINSLEY, D. and MARGOLIS, S. (1969) A study of quartz grain surface textures with the scanning electron microscope. *NY Acad. Sci.*, ser. 2, **31**, 457–77.

LEECH, R. E. (1966) The spiders (Araneida) of Hazen Camp 81°49′N, 71°18′W. *Quaest. Entomol.*, **2**, 153–212.

LINDROTH, C. H. (1931) Die Insektenfauna Islands und ihre Probleme. *Zool. Bidr. Upps.*, **13**, 105–589.

LINDROTH, C. H. (1957) *The Faunal Connections between Europe and North America*. New York and Stockholm: John Wiley and Almqvist & Wiksell. 344 pp.

LINDROTH, C. H. (1969) The theory of glacial refugia in Scandinavia. Comments on present opinions. *Not. Entomol.*, **XLIX**, 178–92.

LØKEN, O. H. (1962) On the vertical extent of glaciation in northeastern Labrador–Ungava. *Can. Geogr.*, **VI** (3–4), 106–15.

LØKEN, O. H. (1966) Baffin Island refugia older than 54,000 years. *Science*, **153** (3742), 1378–80.

LÖVE, Á. (1950) Polyploidy in the Arctic. *In: Encyclopedia Arctica*, Vol. 5 (microfilm). Dartmouth College. 9 pp.

LÖVE, Á. (1953) Subarctic polyploidy. *Hereditas*, **39**, 113–24.

LÖVE, Á. (1959) Origin of the arctic flora. *Pub. McGill Univ. Museums*, **1**, 82–95.

LÖVE, Á. and LÖVE, D. (1943) The significance of differences in distribution of diploids and polyploids. *Hereditas*, **29**, 145–63.

LÖVE, Á. and LÖVE, D. (1947) Studies on the origin of the Icelandic flora, I. Cyto-ecological investigations on Cakile. *Icel. Univ. Inst. Appl. Sci., Dept Agric. Rep.*, B (2), 1–29.

LÖVE, Á. and LÖVE, D. (1949) The geobotanical significance of polyploidy, I. Polyploidy and latitude. *Portug. Acta Biol.*, A (R. B. Goldschmidt Jub. Vol.), 273–352.

LÖVE, Á. and LÖVE, D. (1951) Studies on the origin of the Icelandic flora, II. Saxifragaceae. *Svensk. Bot. Tidskr.*, **45**, 368–99.

LÖVE, Á. and LÖVE, D. (1953) The geobotanical significance of polyploidy. *6th Int. Grassland Congress* (State College, Penn., 1952), 240–6.

LÖVE, Á. and LÖVE, D. (1956) Cytotaxonomical conspectus of the Icelandic flora. *Acta Horti. Gotob.*, **20** (4), 65–291.

LÖVE, Á. and LÖVE, D. (1957) Arctic polyploidy. *Proc. Gen. Soc. Can.*, **2**, 23–7.

LÖVE, Á. and LÖVE, D. (1963a) *North Atlantic Biota and their History.* Oxford: Pergamon Press. 442 pp.

LÖVE, Á. and LÖVE, D. (1963b) Utbreidsla og fjöllintni – Distribution and polyploidy. *Flora. J. Icel. Bot.*, **1**, 135–9.

LÖVE, Á. and LÖVE, D. (1965) The North Atlantic flora – its history and late evolution. *10th Int. Bot. Congress* (Edinburgh, 1964), 139–40 (abstract).

LÖVE, Á. and LÖVE, D. (1967a) Continental drift and the origin of the arctic–alpine flora. *Rev. Roum. Biol., Sér. Bot.*, **12**, 163–9.

LÖVE, Á. and LÖVE, D. (1967b) The origin of the North Atlantic flora. *Aquilo, Ser. Bot.*, **6**, 52–66.

LÖVE, Á. and LÖVE, D. (1967c) Polyploidy and altitude: Mt Washington. *Biol. Zentralbl.* **86 Bein.**, 307–12.

LÖVE, Á. and LÖVE, D. (1971) Polyploïdie et géobotanique. *Nat. Can.*, **98**, 469–94.

LÖVE, D. (1962) Plants and Pleistocene. *Pub. McGill Univ. Museums*, **2**, 17–39.

LÖVE, D. (1963) Dispersal and survival of plants. *In:* LÖVE, Á. and LÖVE, D. (eds.) *North Atlantic Biota and their History*, 189–205. Oxford: Pergamon Press.

MACPHERSON, A. H. (1965) The origin of diversity in mammals of the Canadian arctic tundra. *Syst. Zool.*, **14**, 153–73.

MERCER, J. H. (1956) Geomorphology and glacial history of southernmost Baffin Island. *Geol. Soc. Amer. Bull.*, **67**, 553–70.

MILLER, C. D. (1971) *Quaternary Glacial Events in the Northern Sawatch Range, Colorado.* Unpub. Ph.D. thesis, Univ. of Colorado. 86 pp.

NANNFELDT, J. A. (1935) Taxonomical and plant-geographical studies in the *Poa laxa*-group. *Symb. Bot. Ups.*, **15**, 1–113.

NANNFELDT, J. A. (1940) On the polymorphy of *Poa arctica* R. Br. with special reference to its Scandinavian forms. *Symb. Bot. Ups.*, **4**, 1–54.

NANNFELDT, J. A. (1963) Taxonomic differentiation as an indicator of the migratory history of the North Atlantic flora with especial regard to the Scandes. *In:* LÖVE, Á. and LÖVE, D. (eds.) *North Atlantic Biota and their History*, 87–97. Oxford: Pergamon Press.

NORDHAGEN, R. (1931) Studien uber die skandinavischen Rassen des Papaver radicatum Rottb. sovie einige mit denselben verwechselten neue Arten. *Bergens Mus. Arbok. Naturv. Rekke.*, **2**, 1–50.

NORDHAGEN, R. (1936) Skandinavias fjellflora og dens relasjoner til siste istid. *Nord.* (*19 Skand.*) *Naturforskermöt. Helsingfors*, 93–124.

NORDHAGEN, R. (1963) Recent discoveries in the south Norwegian flora and their significance for the understanding of the history of the Scandinavian mountain flora during and after the Last Glaciation. *In:* LÖVE, Á. and LÖVE, D. (eds) *North Atlantic Biota and their History*, 241–60. Oxford: Pergamon Press.

ODELL, N. E. (1933) The mountains of northern Labrador. *Geogr. J.*, **82**, 193–211, 315–26.

PHEASANT, D. A. (1971) *The Glacial Chronology and Glacio-isostasy of the Narpaing–Quajon fiord area, Cumberland Peninsula, Baffin Island.* Unpub. Ph.D. thesis, Univ. of Colorado. 232 pp.

PORSILD, A. E., HARINGTON, C. R. and MULLIGAN, G. A. (1967) *Lupinus arcticus Wats.* grown from seeds of Pleistocene Age. *Science*, **148** (3797), 113–14.

PORSILD, M. P. (1922) The flora of Greenland: its affinities and probable age and origin. *Abstr., Torreya*, **22**, 52–4.

PREST, V. K. (1970) Quaternary geology of Canada. *In:* Geology and Economic Minerals of Canada, *Geol. Surv. Can., Econ. Geol. Rep.,* 1 (5th edn), 675–764.

RAPP, A. and RUDBERG, S. (1960) Recent periglacial phenomena in Sweden. *Biul. Peryglacjal.,* **8,** 143–54.

REUSCH, H. (1910) Norges Geologi. *Norg. Geol. Unders.,* **50,** 1–196.

RICHMOND, G. M. (1965) Glaciation of the Rocky Mountains. *In:* WRIGHT, H. E. JR and FREY, D. G. (eds.) *The Quaternary of the United States,* 217–30. Princeton, NJ: Princeton Univ. Press.

RØNNING, O. I. (1960) The vegetation and flora north of the Arctic Circle. *In: Norway North of 65°,* 50–72. Oslo.

RØNNING, O. I. (1961) Some new contributions to the flora of Svalbard. *Nor. Polarinst. Skr.,* **124,** 1–20.

SAVILE, D. B. O. (1961) The botany of the northwest Queen Elizabeth Islands. *Can. J. Bot.,* **39,** 909–42.

SCHUSTER, R. M., STEERE, W. C. and THOMPSON, J. W. (1959) The terrestrial cryptogams of northern Ellesmere Island. *Nat. Mus. Can. Bull.,* **164,** 1–132.

SCHYTT, V., HOPPE, G., BLAKE, W. and GROSSWALD, M. (1968) The extent of the Würm glaciation in the European Arctic. *Int. Assoc. Sci. Hydrol. Pub.* (General Assembly of Berne), 79. *Manuscript received June 1971.*

SERNANDER, R. (1908) On the evidence of postglacial changes of climate furnished by the peat-mosses of northern Europe. *Geol. Fören. Förh.,* **30,** 465–78.

SJÖRS, H. (1956) *Nordisk växtgeografi.* Oslo, Copenhagen, Stockholm and Helsingfors: Scandinavian Univ. Books. 229 pp.

SJÖRS, H. (1971) *Ekologisk Botanik.* Stockholm: Almqvist & Wiksell. 296 pp.

TANNER, V. (1944) Outlines of the Geography, Life and Customs of Newfoundland– Labrador. *Acta Geogr.,* **8** (1), 1–907.

TOMLINSON, R. F. (1958) Geomorphological investigations in the Kaumajet Mountains and Okak Bay region of Labrador. *Arctic,* **11,** 254–6.

VALENTIN, H. (1957) Glazialmorphologische Untersuchungen in Ostengland. *Abh. Geogr. Inst. freien Univ. Berlin,* **4,** 1–86.

VOGT, T. (1913) Landskabsformene i det ytterste av Lofoten. *Nor. Geogr. Selsk. Arbok.,* **23,** 1–50.

Manuscript received June 1971.

Additional references

BERGSTROM, E. (1973) Den Prerecenta Lokalglaciationens Utbredningshistoria inom Skanderna. *Naturgeografiska Institutionen,* Rep. 16. Stockholm. 216 pp.

STROMQUIST, L. (1973) Geomorfologiska Studier av Blockhav och Blockfält i Norra Skandinavien. *Uppsala Universitet Naturgeografiska Institutionen,* Rep. 22. Uppsala. 159 pp.

Arctic North American palaeoecology: the recent history of vegetation and climate deduced from pollen analysis

Harvey Nichols

Institute of Arctic and Alpine Research,
University of Colorado

1. Introduction

In a paper on Late Quaternary climatic changes in Canada published in 1961, Teras-mae noted that there were preliminary palynological studies underway in the Arctic but that published results were absent. In the last decade some pollen diagrams have appeared but the available picture is remarkably fragmentary and there is an urgent need for more exploration of this very large and significant area. One may bear in mind Livingstone's comment from his Alaskan palynological studies (1955, p. 598) that 'Conditions in tundra regions are not favourable for the registration of even such major climatic phenomena as the post-glacial thermal maximum'.

The present significance of arctic meteorology for the understanding of the atmospheric circulation of the northern hemisphere makes it likely that detailed knowledge of arctic palaeoclimates will elucidate the climatic history of many other parts of the northern hemisphere. Palaeoclimatic data may also be used to test the validity of some of the contemporary climatological hypotheses. Modern arctic and subarctic climatology and its control of environmental processes also provides the closest analogies to the conditions during the Quaternary Full and Late Glacial times in what are now temperate regions.

For the study of human prehistory, the far north may provide particularly clear evidence of the impact of environmental and climatic changes on a technologically primitive society. Similarities between recent Eskimo cultures and the hunting and gathering economies of Upper Palaeolithic or Mesolithic Europe may aid in understanding the relationships of European pre-Neolithic societies to their environment. Prehistoric Eskimo society apparently had little effect on the natural landscape, so that natural ecological changes may be seen more clearly in arctic diagrams than, for example, in diagrams from Europe where landscape modification by man since at least

5000 BP has made the palynological distinction of natural climatic changes very difficult or impossible. If trans-Atlantic correlation of the Late Quaternary palaeoclimate of some regions is firmly established, as appears likely from the preliminary data summarized here, then problematical palaeoclimatic events in both continents may be clarified by comparison with the other sequence.

SPECIAL PROBLEMS OF ARCTIC PALAEOECOLOGICAL RESEARCH

One of the greatest difficulties encountered by anyone undertaking fieldwork of this nature is that of logistics. The absence of roads and railways in the Canadian north, necessitating airplane charter, makes sample collection inconvenient and expensive. The small number of regular commercial flights into the North American Arctic severely limits the number of sites for study, unless very substantial funds are available for aircraft charter, which of course involves at least two separate flights (setting down and picking up) with possibly other flights for supplies and equipment. Investigated sites, therefore, frequently coincide with human settlements and these are often on the Arctic Ocean coast and not necessarily in climatically and vegetationally sensitive areas such as the forest–tundra ecotone. The palaeoecological record may therefore be biased towards complacency. In view of the limited available records, visitors to the Arctic are urged to collect monolithic samples of peat for palynology and macrofossils such as timber which may record different plant distributions in the past. Such materials may prove invaluable to palynologists. Details of field sampling procedures, in addition to those set out below, may be obtained on request from the author.

The difficulties of overland travel in the north also inhibit the collection of modern surface samples of pollen and spores which record the present atmospheric fallout from local, regional and exotic plants. Such contemporary samples indicate the relationship between modern plant cover and numbers of pollen and spores which land on the surface, and sometimes provide surprising results due to differential pollen productivity, efficiency of pollination, and very long-distance aerial transport; they may provide useful analogues with fossil pollen assemblages. Important work of this type has been performed in the Arctic by Bartley (1967), Ritchie and Lichti-Federovich (1967), Terasmae (1967), and Lichti-Federovich and Ritchie (1968).

Davis (1967) compared surface sample pollen assemblages from Canada with Late Glacial palynological counts from New England and the Great Lakes area, and incidentally illustrated another problem which may be more common in arctic areas than elsewhere. Davis's pollen count from Keewatin, which was taken from the published work of Wright et al. (1963), involved analysis of surface peat from Ennadai Lake collected by J. A. Larsen. It is now clear (Nichols, 1967a) that the modern peat surface ceased to grow at this site shortly after 630 ± 70 BP (WIS-133, depth 4 to 6 cm below surface) and that the samples referred to by Davis had a large component of ancient pollen which might invalidate that material as a representative of the modern pollen fallout at Ennadai Lake.

In the cold dry desert climates of the Arctic, exemplified by the Barren Grounds of

Keewatin, the accumulation of incompletely broken-down plant debris as peat is slow. Much of the peat is composed of sedge and grass remains, and *Sphagnum spp.* bog mosses, which frequently make up much of the volume of temperate northern hemisphere peat accumulation do not grow extensively or build up organic deposits quickly in the tundra. The slow accumulation of the peat occurs only in sites with a wet and relatively mild summer climate. Marginal peat growth may occur during relatively warm climatic periods, but ceases during a climatically severe episode of dry cold summers. This loss of the peat record and its contained environmental history has its virtues, however. A period of severe summers may be clearly marked by the slow-down or absence of peat growth, and the presence of windblown sand in the dry humified peat may indicate some local destruction of the plant cover due to climatic deterioration (Nichols, 1967b, 1970; Bartley and Matthews, 1969).

2. Sampling methods

Postglacial peats in the Arctic are generally shallow, and are usually no deeper than 1 or 2 m, only one-fifth or one-tenth the depth of temperate latitude peats representing the same period of time. This abbreviation of the record necessitates closer pollen sampling intervals than are common elsewhere and constrains samples for radiocarbon assay to narrow vertical limits (preferably 2 cm or less). Fortunately, the shallowness of the peat helps in sampling by the monolith method, in which a pit is dug through the organic accumulation to the underlying mineral material, and then peat blocks are cut out in a continuous vertical section. While the organic deposits may have the virtue of being shallow they have the major disadvantage of frequently being permanently frozen (see Ives, Chapter 4A). The wetness and excellent insulating properties of peat allow only shallow thawing, relative to mineral soils such as sand.

After the possibly contaminated outer face (back 15 cm or more) has been removed from any peat exposure, work with hammer and chisel is sufficient to obtain the small samples needed for palynology (*c.*1 or 2 cm³) from an exposed frozen peat face. However, monolith sampling for ¹⁴C dating and macrofossil studies by this method is impractical from even an exposed cliff face of perennially frozen peat. Dynamite can be used to obtain samples from such a situation, but its use is not recommended, except in special circumstances (Nichols, 1967c); blast control is difficult, and at best only a discontinuous series of odd-shaped frozen blocks is likely to be obtained. The samples from Lynn Lake were obtained in this way (Nichols, 1967b). Blasting may be used advantageously in creating a pit in frozen ground which may then be sampled in other ways.

A SIPRE iceborer with attached powerdrive has been used by the Geological Survey of Canada to sample frozen peat; rapid recovery of 7.6 cm (3 in.) diameter cores from frozen ground are reported (Hughes and Terasmae, 1963). The borer weighs 48 kg (105 lb) and the motor 12 or 39 kg (26 or 85 lb) depending on the type used; this equipment is not easily portable by one man over long distances in rough country.

A conventional gasoline-powered chainsaw, without adaptation, is capable of cutting frozen peat easily and accurately, particularly from a cliff or exposed peat face (Nichols, 1967c). The chainsaw is light, 7 kg (15 lb), and relatively small and inexpensive compared to coring devices. Frozen peat blocks large enough for ^{14}C dating and plant macrofossil analysis (c. 20 to 25 cm^2) may be recovered. Peat which forms level ground is more difficult to excavate with a chainsaw without first opening a pit with explosives, but excavation can be achieved with time. The chain is dulled more rapidly when minerals are present in the peat, and presumably severe chain damage would arise from prolonged sawing into frozen sand, etc. A spare sharp chain and/or file should be carried.

The difficulties of terrestrial sampling are avoided by coring lake sediments, though other problems then arise. The equipment commonly used is that preferred for temperate latitude investigations, the Livingstone corer or the Hiller peat sampler. The former device is very suitable for organic lake sediments, though there is difficulty in penetrating minerogenic sediments, and sand layers more than a few decimetres thick are hard to pierce. Modifications of the Livingstone corer and comments on usage are found in Cushing and Wright (1965) and its application in the Arctic is dealt with by Colinvaux (1964a). Ice often acts as a stable drilling platform for lake sediment boring in the north, but the discomfort of working in very cold conditions may make for errors in operation or recording of data. Colinvaux (1964a) notes that metal parts which become frozen together may be freed by burning gasoline. At the height of the summer warmth, when lake ice disappears or becomes unsafe, it is necessary to use a boat, rubber dinghy or raft as a drilling platform. This operation involves difficulties in anchoring and in relocating the borehole (Faegri and Iversen, 1964).

One problem of organic sediment sampling in the Arctic is the slow rate of detrital organic deposition, due presumably to the general depauperation of the plant cover, shortness of the growing season, and lack of nutrients reaching the lake waters. Lakes surrounded by great thicknesses of unconsolidated materials (sands, gravels) may have sediments which are almost totally inorganic, due to mineral inwash, freeze–thaw movement and wind transport. This is exemplified by the small lakes in the outwash delta north of Pelly Lake (Nichols, 1970). Such sediments are very difficult to penetrate by manual sampling devices (Hiller or Livingstone) and produce unsatisfactory samples for ^{14}C dating, since a relatively large volume of sediment is needed to yield the necessary few grams of carbon. These problems become less pronounced outside the cold tundra deserts, such as the Barren Grounds of Keewatin, and indeed the rarity of peat in such regions is a strong inducement to overcome the problems of arctic lake sediment sampling (Nichols, 1967d).

A major source of error for lacustrine work in a few situations may be the 'bottom ice effect' (Nichols, 1967e). This was reported on the basis of only one experience, and it is not known how widespread the phenomenon may be in the Arctic. During several weeks the lacustrine sediments of a small pond remained frozen and attached to the mineral base by 'anchor ice' while the upper ice thawed. The bottom ice with included sediment then floated to the surface and slowly melted, while for several days winds blew the frozen materials from one part of the pond to another and sediment fallout from the melting ice continued. At one time an ice block *c.* 1 m thick rolled over in the water because the base was melting faster than the exposed top, thus reversing the sedimentary sequence in the ice. This situation occurred in a shallow lake, where the depth of winter freezing (probably 3 to 4 m) reached down to include the basal sediments. In deeper lakes the central sediments may be unaffected, though freeze-flotation would still affect shallow, sloping shoreline deposits, which may wholly or partially represent the sedimentary history of the basin and which might be ice-rafted out to the middle of the lake by winds or currents to contaminate the abyssal sediments. This process would presumably lead to some homogenization of lake sediments by mixing of materials of quite different ages, and thus smooth out marked environmental and climatic changes which were originally registered in the sediments. While perhaps now confined to limited areas to the Arctic, this type of disturbance may have been more widespread in the past when quasi-arctic climates affected present-day temperate regions during the Quaternary glaciations. Such effects might perhaps be recognizable from abnormally numerous aberrations of radiocarbon dating, with some reversals of the expected older-to-younger superimposition of ^{14}C dated sediments. One might also expect that in such cases the lacustrine palynological record would be much less sensitive than a neighbouring terrestrial (peat) record. (I know of no illustrations for these hypotheses at present; the well-dated reversals of radiocarbon chronology of shallow Alaskan lake sediments found by Colinvaux, 1964b, are not yet fully understood, and it is possible that the bottom ice effects may have some bearing on the matter.)

SPECIAL CONSIDERATIONS FOR POLLEN ANALYSIS

It is now generally accepted that the low pollen productivity of tundra plants makes possible the relative over-representation of highly productive exotic pollen types such as *Pinus* (pine), *Picea* (spruce) and *Alnus* (alder) on the tundra surface. Tundra pollen appears to be distributed mostly to the immediate vicinity of the producing plant, so that surface sample counts of modern pollen show very variable numbers over small areas of ground (Bartley, 1967; Cole *in* Nichols, 1970). For these reasons it is particularly necessary to analyse 'absolute' numbers of pollen and spores in addition to the conventional relative percentage figures. 'Absolute' counts may be derived from estimates of the numbers of pollen contained in each cubic centimetre of sediment (Davis, 1967) when radiocarbon dates are available throughout the sediment core. I prefer to count pollen numbers per gram (oven-dried weight) because the rate of peat growth is variable (cf. Jörgensen, 1967). During dry periods the accumulation of peat is slow, and therefore more annual accumulations of pollen and spores are contained in each centimetre depth of peat than in a moister period of rapid peat growth. Slow growing oxidized peat is denser than fast growing unoxidized peat, so that a weight basis for 'absolute' pollen counts offers some crude correction factor, compared to the equal volume method. The latter seems to be well suited to analysis of lacustrine sediments which, at least in the temperate lakes studied (Davis, 1967), seem to have accumulated at approximately linear rates during much of postglacial time.

'Absolute' pollen counts are less likely to be influenced by local variations in pollen output productivity than relative percentage counts. It seems that if the local output of, say, arctic birch (*Betula glandulosa*) rises, the relative percentages of regional or exotic pollen will fall, producing the effect of a regional vegetational change due to climatic change. Percentage pollen diagrams from the Arctic are particularly open to challenge on these grounds, but it appears now that there is considerable similarity between the relative and 'absolute' changes in pollen diagrams from some parts of the Arctic (Nichols, 1970), perhaps suggesting that in such circumstances percentage diagrams are more reliable than has been thought, or that the 'absolute' counting method is to some degree invalid.

3. Canadian Arctic and Subarctic Late Quaternary palynology

OGLIVIE MOUNTAINS, YUKON TERRITORY (*c.* 65°N, 139°W, AND 65°N, 140°W):
TERASMAE AND HUGHES (1966)

Pollen diagrams from the northwest of Canada are reported by Terasmae and Hughes (1966), stemming from interest in the Late Pleistocene glaciation of the Ogilvie Mountains in Yukon Territory. The study sites lie in the northern boreal forest (see fig. 11A.1), while to the north of the sites is the alpine forest–tundra section of Rowe (1959). The deepest samples recovered from frozen peat were dated 13,870 ± 180 BP (GSC-296) and 12,550 ± 190 BP (GSC-128), and provide minimum ages for a glacial episode in this locality. From the deposition of basal peat at Chapman Lake (13,870 ± 180 BP) to sometime shortly after 9620 ± 150 BP (GSC-310), there was little *Picea* and *Pinus* pollen, which I suggest denotes absence of boreal forest from the site. Its presence after 9620 ± 150 BP is indicated by considerable numbers of *Picea* and *Alnus* pollen, reaching up to the present-day surface. Terasmae and Hughes note that 'the magnitude of the post-glacial climatic changes appear to have been smaller than in the more southerly regions', but I consider that since the sites apparently lay south of the forest–tundra boundary for much of postglacial time they would be much less sensitive to vegetational and climatic changes than sites nearer the ecotone. One must also note that from 9620 ± 150 BP at 1.5 m (5 ft) depth in the peat to the modern surface there are only six pollen analyses recorded, and there are no more recent ^{14}C dates from Chapman Lake to indicate the time span represented by that 1.5 m of peat; growth may have ceased at any subsequent time. The magnitude of the climatic changes was not stated, but if it was smaller than changes recorded to the south, as suggested, then this is in interesting contrast to the theoretical climatic comments of Lamb (1966) and others (Bryson, personal communication, 1967) that climatic changes are expressed more strongly in the north than in more southerly areas. This latter viewpoint is supported by the very marked warming of the North Atlantic in the late nineteenth/early twentieth century, but the validity of meteorological records of contemporary short-term climatic changes as analogues for long-term postglacial alterations is unknown.

The Gill Lake peat contained *Picea* pollen after 12,550 ± 190 BP (GSC-128) and *Alnus* pollen some time later, suggesting the presence of boreal forest. The variations in these two pollen taxa suggest environmental or climatic changes in the postglacial period but the samples are relatively widely spaced (*c.* 17 cm) and there are no ^{14}C dates after 12,550 BP; the environmental history is, therefore, unclear. Terasmae and Hughes suggest that the palynological sequences in the Yukon can be correlated with Livingstone's (1955, 1957) pollen diagrams from the Brooks Range in Alaska, but on the basis of the few ^{14}C dates from the Yukon the zone boundaries appear to be metachronous. The suggested correlation is that the vegetational sequences recorded by the

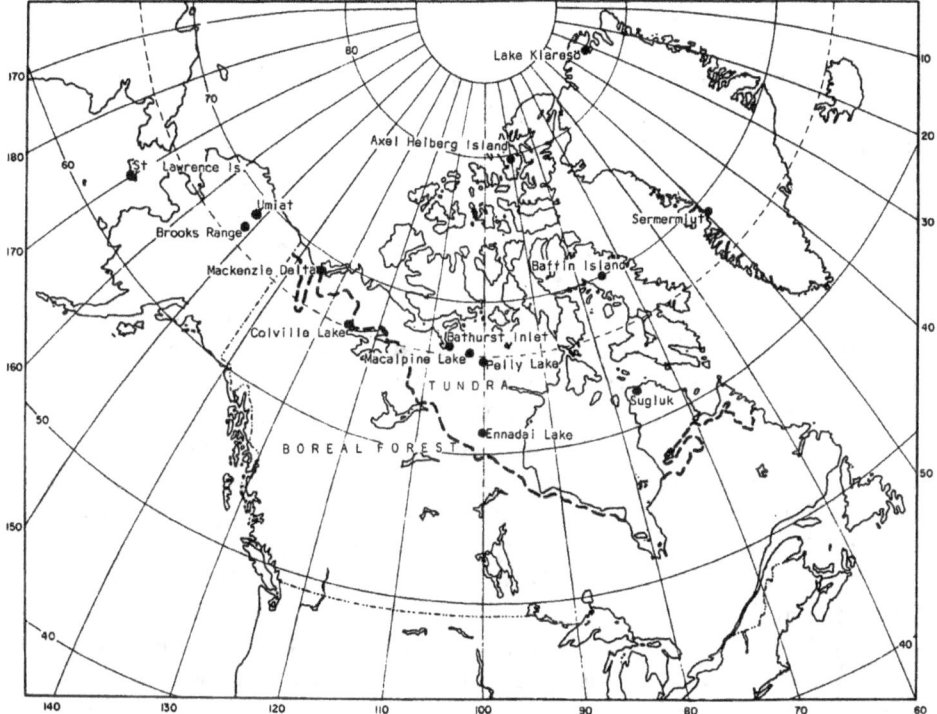

Fig. 11A.1. Location map (from Larsen, this volume, and Rowe, 1959). Northern forest limit marked by broken line.

Yukon and Alaskan diagrams are similar: a basal sedge (*Cyperaceae*) pollen zone, followed by a birch (*Betula*) episode, and then a *Picea* and *Alnus* zone (Zones I, II and III of Livingstone, 1955). Comparison of Livingstone's (1957) ^{14}C chronology with that of Terasmae and Hughes (1966) shows that the transition from Zone I to II in Alaska was dated 7500 ± 250 BP or 8125 ± 250 BP, while in Yukon it was 11,000 to 12,000 years BP. The Alaskan Zone II/III boundary was dated 5900 ± 200 BP, while the beginning of the rise of *Picea* and *Alnus* pollen in the Yukon occurred *c.* 34 cm above (i.e. shortly after) a date of 9620 ± 150 BP (GSC-310), where the rest of the postglacial accumulation to the modern surface was 168 cm. Since these initial datings are not synchronous (though further ^{14}C determinations may clarify matters) it is fair to point out that the similarity in the sedge-birch sequence in the two regions may reflect only some local successional or migratory sequence in the Yukon, though the spruce–alder episode is likely to reflect not only migration but also a regional climatic change. The initial sedge-birch sequence at Chapman Lake is not seen at the Gill Lake site which covers essentially the same period, though the rise of spruce and alder *is* apparent and resembles that at Chapman Lake.

MACKENZIE DELTA (68°N, 133°W, AND 69°N, 132°W): MACKAY AND TERASMAE (1963)

These two pollen diagrams 'are the first for post-glacial deposits in the Canadian Arctic' (Mackay and Terasmae, 1963). The sites for pollen diagrams were at Eskimo Lakes and Twin Lakes which lie just north and south of the forest–tundra boundary. Deglaciation occurred prior to 12,000 ± 300 BP (S-69, Muller, 1962) and 10,800 ± 300 BP (I-483; Geogr. Branch, JRM-60-W2), which are dates on driftwood of unknown origin. The basal peat dates from the two palynological sites are similar – 8000 ± 300 BP (GSC-25) and 7400 ± 200 BP (GSC-16), respectively. These dates indicate that at least 4000 radiocarbon years elapsed after the minimum deglaciation dates before the accumulation of peat allowed the preservation and registration of the environmental history to begin. The sites were apparently above sea level throughout this time, and also 'provide two excellent marine limits for the past 7000 to 8000 years' (Mackay and Terasmae, 1963, p. 232). Since each diagram has only one ^{14}C date there is no guarantee that the deposits have continued to accumulate throughout this period or that the recent past is represented at all; I see no reason, however, to suggest that that was the case in this instance. This comment may be applied to the authors' remark that 'some 7.5 ft (2.26 m) of silty peat accumulated in about 7500 years' (ibid., p. 232), and that 'there is no proof that the rate of accumulation was uniform'. The authors consider that the two pollen diagrams from the Mackenzie Delta show 'rather good correlation' because the basal pollen counts were high in *Betula*, *Salix* (willow), and *Cyperaceae*, followed by higher percentages of *Ericaceae* (heaths) and later by *Alnus*. They note the local presence of *Picea* registered by high percentages (*c.* 90 per cent) of that pollen type at the base of both diagrams *c.* 8000 years BP. A later undated horizon was characterized by the rise of *Alnus* in both diagrams perhaps resulting from a warmer climate. Such a warm episode(s) is suggested from macrofossils of the moss *Ceratophyllum demersum* dated 5400 BP (L-428, Terasmae and Craig, 1958), north of its modern range, and by undated fossil *Picea* stumps buried by peat on Richards Island northeast of the Mackenzie Delta (Porsild, 1938), at Kittigazuit and the Eskimo Lakes area (Mackay and Terasmae, 1963, p. 233).

The summary of the vegetational and climatic history offered by Mackay and Terasmae is that there was a cool-dry climate (compared to the present) from *c.* 8500 to 7500 years ago, when spruce forest occupied the area along with birch. A warmer episode followed, with possibly less available moisture, suggested by slow peat growth and some decay, and poor pollen preservation. Terasmae suggests that a later rise in *Ericacae* pollen indicates an increase in available moisture, that a *Myrica* (bog myrtle) pollen maximum suggests climatic cooling, and that large numbers of *Sphagnum* spores suggest increased available moisture in late postglacial time. Climatic cooling was responsible for pingo formation in the last few thousand years.

These two pollen diagrams are less informative than they might be because they have only one ^{14}C date each, and because the peat stratigraphy is not recorded in sufficient detail to allow the reader to check whether any of the apparent palaeoclimatic changes may be explained by very local vegetational events registered by macrofossils at the site. *Sphagnum* spore numbers were recorded only as present, abundant, or very abundant.

COLVILLE LAKE, MACKENZIE TERRITORY (67°06′N, 125°47′W): NICHOLS (UNPUBLISHED)

A single pollen diagram for Colville Lake was prepared from monolithic blocks of peat (taken by J. A. Larsen for my use) to form a continuous column from the modern surface down to a frozen impenetrable base of calcareous shell marl at 215 cm depth (fig. 11A.2). It is not known whether other organic deposits exist below the calcareous base, dated 6790 ± 75 BP (WIS-275, Bender *et al.*, 1968b); the date of deglaciation for this area is estimated to fall within the period 7000 to 8000 BP (Bryson *et al.*, 1969).

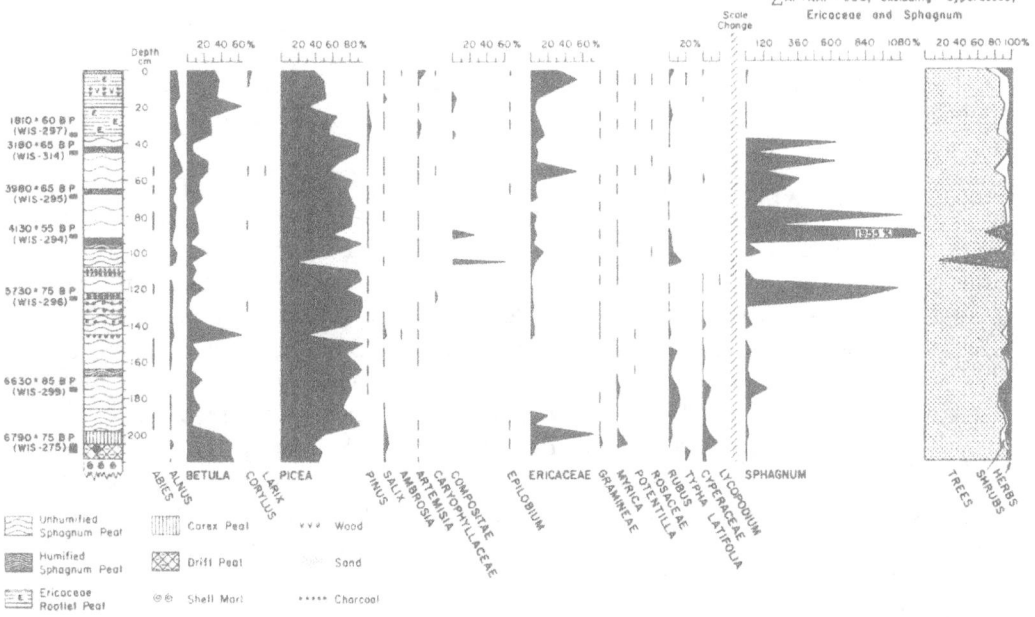

Fig. 11A.2. Pollen diagram from Colville Lake.

The site is in open boreal forest close to the northern limit of trees. *Picea* pollen has over 40 per cent at the base of the diagram at *c.* 6800 BP, when open spruce–birch forest surrounded the site. Between 6790 ± 75 BP (WIS-275) and 6630 ± 85 BP (WIS-299) the spruce forest became denser and the birch component declined, and from then until about 3200 BP the forest limit lay well to the north of the site which was apparently surrounded by closed spruce woodland; there were two clearances of the local forest by fire (marked by sharp reductions in *Picea* pollen percentages and by charcoal in the peat) but the forest quickly recovered.

The greatly increased numbers of *Sphagnum* spores and the less oxidized and fresher state of the *Sphagnum* peat between 4130 ± 55 BP (WIS-294) and 3180 ± 65 BP (WIS-314) is suggestive of more favourable climatic conditions for local peat growth, possibly a shift to cooler wetter summers than previously. This suggests that the climatic boundaries which limit the northern edge of the forest may have shifted southwards (Bryson, 1966; Nichols, 1967b).

Just after 3180 ± 65 BP (WIS-314), *Picea* percentages fell. *Sphagnum* spores almost disappeared from the record, and the peat was very humified and stopped growing. This may be interpreted as a retreat of the forest–tundra limit southwards and across Colville Lake, leaving the site exposed to the cold dry arctic summers characteristic of the tundra, which are inimicable to *Sphagnum* peat growth or sporogenesis (Nichols, 1967b). This severe climatic episode lasted from *c.* 3200 to 1810 ± 60 BP (WIS-297) when slower *Ericaceae* peat growth began again, but the *Picea* percentages did not recover and it seems that since then the forest has not been closed, but has shared the ground with birches and heaths during a period of generally cool dry summer climate.

There is a clear distinction in the rate of peat accumulation in that the growth rate was rapid from 6800 to 3200 BP (160 cm), and slow from 3200 BP to the modern surface (46 cm), which supports this hypothesis of climatic desiccation and cooling. The peat was not dated after 1800 BP, so that the age of the surface peat is unknown.

MACALPINE LAKE (66°35′N, 103°15′W): TERASMAE (1967)

Organic materials were collected by W. Blake Jr and analysed for pollen by Terasmae from sites 320 to 400 km northeast and *c.* 240 km north of the boreal forest–tundra boundary in northeastern Mackenzie Territory. The paper (Terasmae, 1967) is primarily concerned with surface or near-surface pollen samples as indicators of wind-blown pollen fallout from the forest, but it includes a pollen diagram from MacAlpine Lake (fig. 11A.3). The basal peat overlying inorganic sediment was dated 2330 ± 150 BP (GSC-300); Terasmae noted that the discontinuity which separated the organic and inorganic materials was seen elsewhere in the area, and that peat growth was extensive there only in late postglacial time (cf. Nichols, 1969a). The peat was about 55 cm deep, and five pollen counts were recorded between the base and the surface, at intervals of about 10 cm. No ^{14}C date was provided other than that at the base. Terasmae believed that the *Picea* and *Pinus* pollen in the peat were windblown from the boreal forest, which is supported by his surface sample counts and subsequent work (Bartley,

1967; Ritchie and Lichti–Federovich, 1967; Nichols, 1970). The absence of notable change in these two pollen curves led Terasmae to suggest that there had been no change in the meteorological factors responsible for atmospheric pollen dispersal during the last 2000 years (Terasmae, 1967, p. 26). Because ^{14}C dates are lacking above the base at 2300 BP there is no guarantee that continuous peat accumulation has occurred during the subsequent time span, and from other tundra sites (Nichols, 1967b, 1970; Bartley and Matthews, 1969) it is clear that peat growth may have ceased shortly after

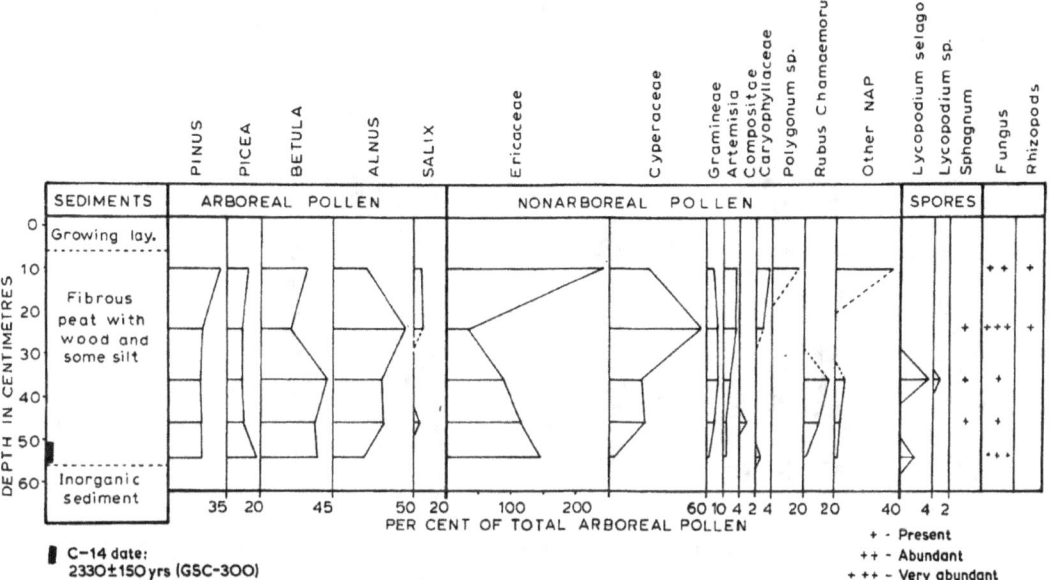

Fig. 11A.3. Pollen diagram from MacAlpine Lake (from Terasmae, 1967).

it began, and quite probably 700 or 800 radiocarbon years ago, due to colder drier summers. More ^{14}C dates and closer-interval pollen sampling (e.g. at 5 or 2.5 cm vertical intervals) would answer this objection, but it should be noted that the Mac-Alpine interpretation is not supported by palaeoclimatic data covering the period 3400 to 900 BP from Pelly Lake, *c.* 100 km to the southeast (Nichols, 1970, p. 56, see below), where clear evidence exists for changes in *Picea* and *Pinus* fallout due to regional vegetational and climatic changes.

PELLY LAKE (66°05′N, 101°04′W): NICHOLS (1970)

Two neighbouring sites named Pelly Lake and Drainage Lake (unofficial name) provided peat monoliths from which two short pollen diagrams were obtained, only fractions of the postglacial history of the area. The region is named 'the Barren Grounds' because of its impoverished flora and fauna, and is characterized by cool, very dry summers and the coldest winters known in North America. This cold dry arctic desert lies about 450 km north and 500 km east of continuous boreal forest, whose northern edge is formed of *Picea mariana* (black spruce) and *Picea glauca* (white spruce), with *Pinus banksiana* (jack pine) to the south of the ecotone.

The fossil *Picea* and *Pinus* pollen represented in the diagrams is considered to have been windblown from the forest, since there is no macrofossil evidence from buried wood, leaves, etc., in the area that conifers ever reached Pelly Lake during the last few thousand years, and because the long distance transport of these pollen types is now well established (Aario, 1940; Faegri and Iversen, 1964). The variations in both relative percentage and 'absolute' numbers of *Picea* and *Pinus* pollen (see fig. 11A.4)

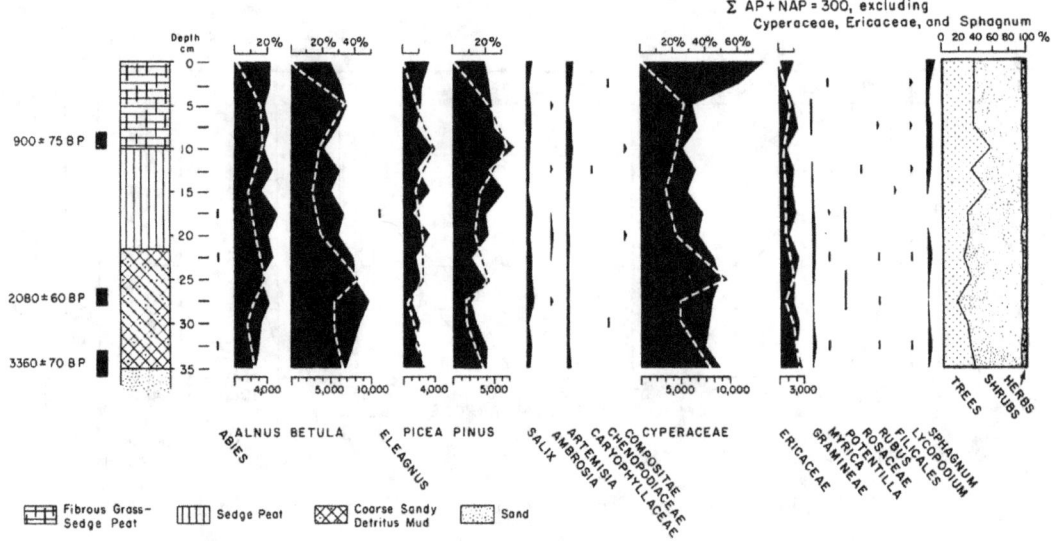

Fig. 11A.4. Pollen diagram from site north of Pelly Lake (from Nichols, 1970), showing relative pollen percentages (solid columns) and absolute pollen numbers per gm oven-dry weight (dashed lines).

are therefore thought to represent changes in the position of the forest boundary to the south and/or changes in the frequency of southerly/westerly winds in the summer which would transport spruce and pine pollen from the boreal forest, as a result of climatic changes. The pine and spruce numbers parallel each other, and their joint maxima and minima are believed to represent the times of maximum climatic change for this area, since the *Picea* and *Pinus* variations are directly due to meteorological expressions of climatic changes, rather than representing the local climatic transgression of some plant growth threshold as in the usual pollen diagram. The north–south movements of the forest–tundra ecotone are thought by analogy with modern conditions to be due to prolonged changes in the mean summer position of the arctic front, which implies an increased summer flow of southerly and/or westerly air of modified tropical or Pacific origin into the Canadian Arctic during warm periods, and a diminution of this effect with increased dominance of arctic air and more northerly winds in the growing season during colder periods (Bryson, 1966; Nichols, 1967b; and see Larsen and Barry, Chapter 5).

At Pelly Lake organic deposition (of detrital mud) began at 3360 ± 70 BP (WIS-216, Bender *et al.*, 1968a) when decreasing numbers of *Picea* and *Pinus* pollen suggest the decreased presence of southerly/westerly winds in summer and the retreat southwards of the northern edge of the forest, indicating cooler summers, as the arctic front achieved a more southerly summer position. Minima of spruce and pine pollen numbers occurred at about 2080 ± 60 BP (WIS-292, Bender *et al.*, 1968a), indicating maximum summer dominance by the cold dry arctic airmass and the most southerly retreat of the forest edge. This severely cold dry episode was followed by increases in *Picea* and *Pinus* to maxima at 900 ± 75 BP (WIS-245, Bender *et al.*, 1968a) when southerly/westerly summer winds increased in frequency and the arctic front and the forest limit moved north.

At this time of warmer summers the second site at nearby Drainage Lake began to accumulate peat at 1060 ± 55 BP (WIS-263, Bender *et al.*, 1968b), due probably to melting of permafrost and snowbanks or possibly to more precipitation associated with the amelioration.

Both sites ceased to accumulate peat shortly after 900 ± 60 BP (WIS-278), at Drainage Lake and 900 ± 75 BP (WIS-245) at Pelly (Bender *et al.*, 1968a); the desiccation of the peat plus minima of spruce and pine pollen suggest very cold dry summers and a southwards retreat of the forest not long after 900 BP (probably from *c.* 700 BP) lasting to the present day. The local tundra taxa provided little comprehensible palynological evidence of climatic change. Increases in *Betula* representation were associated with episodes of climatic cooling, with the possible explanation that increased size and numbers of snowbanks during colder climates following 3500, 2400 and 700 BP may have provided more sheltered habitats for dwarf arctic birches.

This palaeoclimatic sequence is matched by abundant geomorphological signs of palaeoenvironmental conditions which were different from those of the present: dried enclosed lake basins, fossil humus horizons buried by blown sand, lag gravels formed by deflation of fine materials, destruction of plant cover, fossil ice wedges and frost polygons. The implications of the suggested palaeoclimatic sequence for contemporary plant ecology of this area are that the episodes of colder drier summers following 900 BP or prior to 2000 BP may have been responsible for much of the degradation and destruction of the local plant cover.

ENNADAI LAKE (61°10′N, 100°55′W): NICHOLS (1967a, b)

This site lies in the southern part of the tundra region of Keewatin, just north of the sharply defined boreal forest–tundra ecotone (see fig. 11A.1) which coincides with and may be controlled by the mean southern limit of the cold dry arctic airmass in summer (Bryson, 1966; Barry, 1967). The lowest organic material from a vertical series of monolithic peat blocks collected by J. A. Larsen was dated 5780 ± 110 BP (WIS-67, Bender *et al.*, 1966); deeper penetration was prevented by permafrost, but the inorganic base is believed to be not much older (Nichols, 1967b, p. 179). The ^{14}C determination 5780 ± 110 BP provides a minimum date for deglaciation, in an area which was occupied until at least 8000 BP by the Late Wisconsin ice sheet (Falconer *et al.*, 1965). Ennadai Lake is about 480 km from the nearest end moraines mapped by Falconer *et al.*, so that the ice sheet covering that area seems to have disappeared within a maximum of about 2000 radiocarbon years.

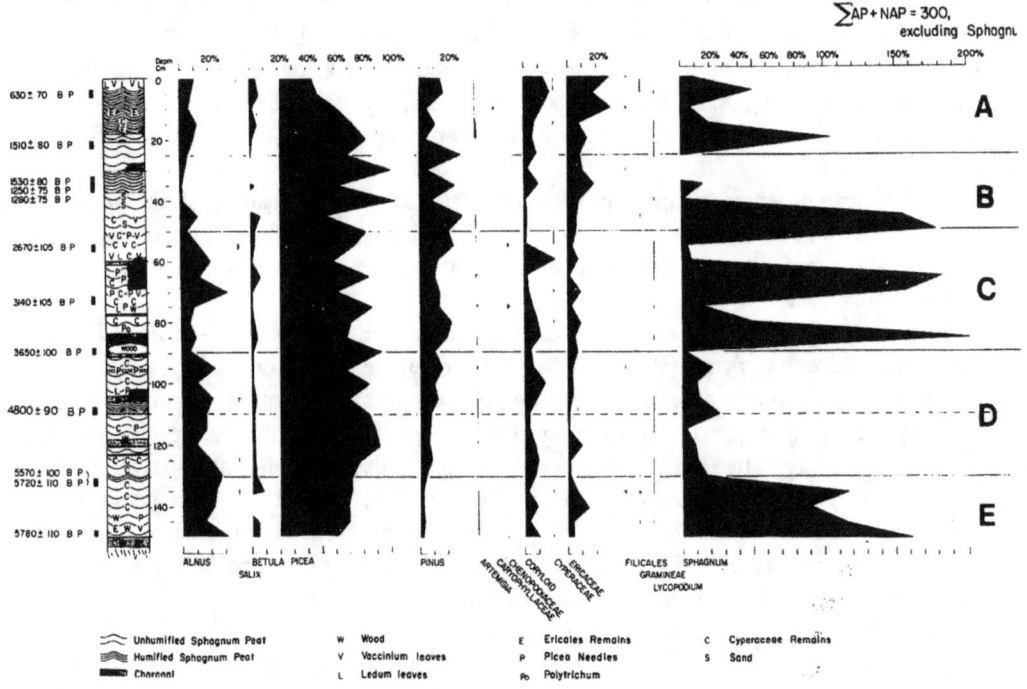

Fig. 11A.5. Pollen diagram from Ennadai Lake (from Nichols, 1967a).

The presence of *Picea* needles in the basal Ennadai materials dated 5780 BP and the high numbers of spruce pollen (fig. 11A.5) demonstrate that the *Picea* forest had migrated from a position probably outside the end moraines before 8000 BP to the site at Ennadai Lake prior to 5780 BP, a migration rate of 480 km in a maximum of 2000 ^{14}C years, giving a minimum mean rate of 24 km per century, or 240 m per year. These *Picea* fossils suggest too that any tundra which may have fronted the ice was of very limited extent and was swiftly replaced by open spruce forest, which may have migrated over the freshly-exposed ground as quickly as the ice retreated. Spruce trees and alders probably grew very close to the ice margin or even on the morainic surface of the decaying glacier itself as they do today at the Malaspina Glacier in Alaska (Sharp,

1958) where the local cooling effect of the ice is extremely limited – apparently to only some scores or a few hundreds of metres from the glacier.

The rapid deglaciation of central Canada and the swift immigration of the spruce forest suggests that the period 8000 to 6000 BP was relatively warm locally. The moderate but increasing percentages of *Picea* pollen from 5800 BP up to about 5000 BP suggests continued migration of spruce trees onto a recently exposed landscape under ameliorating climatic conditions. At 4800 ± 90 BP (WIS-166) a decrease in *Picea* pollen numbers, a stratigraphic change from unhumified to moderately humified peat, and an increase in *Sphagnum* spore numbers, suggested an environmental change which was interpreted as an opening up of the spruce forest locally, and perhaps a southward retreat of the forest limit in Keewatin, as a result of a cooling trend in summer climate. At 5140 ± 100 BP (WIS-112, Bender *et al.*, 1966) there was an environmental change registered in a pollen diagram from Lynn Lake, 450 km to the south of Ennadai, when reduced *Picea* and *Alnus* values and increased *Sphagnum* numbers may have recorded a synchronous climatic cooling (Nichols, 1967b).

The cooling at about 5000 BP was apparently short-lived and, while the recovery from this episode was not radiocarbon dated at these sites, it occurred so little distance above the start of the cooler period that it is estimated to have lasted until only about 4500 BP.

The period from *c.* 4500 to 3500 BP saw a northward movement of the forest, to form charcoal layers of *Picea* and *Larix* (larch or tamarack) wood up to 250 km north of present treeline when the forest burned at about 3500 BP (Bryson *et al.*, 1965). This very substantial macrofossil evidence is interpreted by the authors as a reflection of warmer summers (probably about 3°C above present summer temperatures, see below) when the arctic front was further north than now. The very widespread forest fire was caused by or was followed by a colder climatic episode. One causal explanation for this coincidence is that when colder drier arctic air dominated the northern part of the forest after the climatic cooling of *c.* 3500 BP, the trees were dessicated and perhaps died, so that large areas of woodland were very prone to clearance by lightning fires, and subsequent regeneration under altered climatic conditions would not occur (see Nichols, 1967b, pp. 188–9, for fuller treatment). The burned forest horizon was dated at four sites: 3430 ± 110 (WIS-12), 3540 ± 110 (WIS-52), 3550 ± 120 (WIS-18), and 3650 ± 100 (WIS-80) (Bender *et al.*, 1965, 1966).

This and other macrofossil evidence (Bryson *et al.*, 1965) was added to the palynological and stratigraphic data to produce a tentative reconstruction of the location of the forest–tundra ecotone along the meridian 100°W during Late Quaternary time (fig. 11A.6); this ecotone may also have registered the location of the arctic front in summer, the boundary between the cold dry stable arctic airmass to the north, and the cool moist Pacific and modified tropical air over the boreal forest. On the basis of this diagram a tentative palaeotemperature diagram was drawn (fig. 11A.7). This was based on the untested assumption that when the forest limit moved, for example, 250 km north of Ennadai Lake, the summer temperatures at Ennadai rose to figures similar to those now found 250 km south of Ennadai Lake. A further uncertainty is introduced in that the modern meteorology of this part of Canada is not well known. When the tentative palaeotemperature figures were calculated they were sufficiently close to other estimates for the same climatic episodes (particularly the well established sequence from northern Europe; Godwin, 1956) to be considered worth publication, though they are expected to be corrected by later findings. These postglacial temperature anomalies

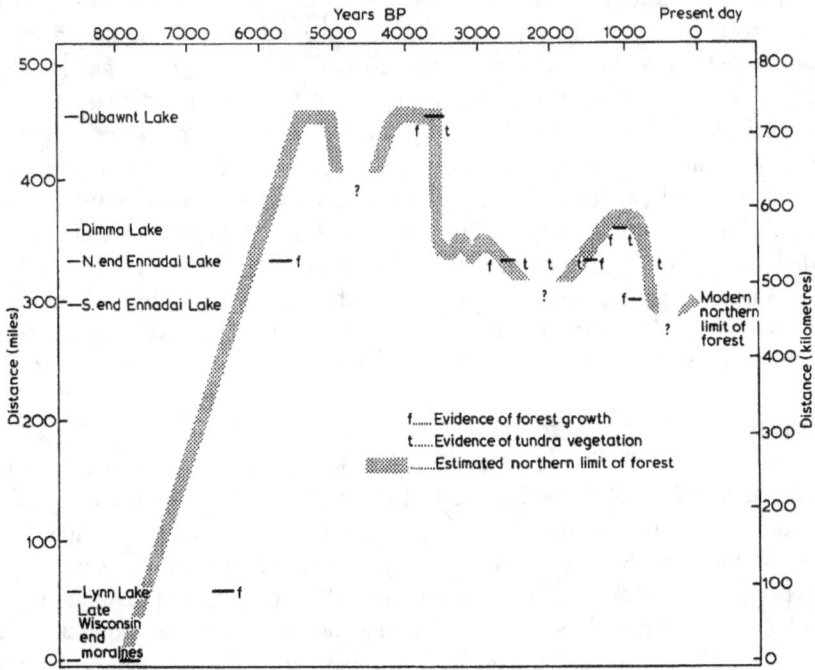

Fig. 11A.6. A tentative reconstruction of the location of the northern limit of continuous forest along the meridian 100°W in central Canada (from Nichols, 1967b). Stratigraphic evidence for forest (*f*) or tundra (*t*) is shown by short horizontal lines which represent the radiocarbon date and the standard deviation. The rest of the diagram is based on the radiocarbon-dated pollen diagrams from the north end of Ennadai Lake and from Lynn Lake, and this palynological and stratigraphic evidence is less unequivocal. For instance, vegetation changes following 5000 BP are derived from radiocarbon-dated palynological evidence, but there is no clear stratigraphic evidence of a change from forest to tundra, so that the magnitude of the ecotonal movements can only be inferred. It is not known how far south of Ennadai the forest retreated between 2500 and 1500 BP and after 600 BP.

are nevertheless of the correct magnitude to explain the forest limit movements (Hare, 1968).

From 3500 to 2500 BP at Ennadai there were varied *Sphagnum* spore percentages, variations in peat humification, and irregularity in *Picea* representation, all pointing to variable environmental conditions. Today, the *Sphagnum* bog mosses grow rapidly and produce abundant spores in the cool moist summers of the boreal forest, but are less fertile and grow more slowly in the cold dry summers of the arctic tundra. Thus at Ennadai the very fresh unhumified pale brown *Sphagnum* peat (which often contains moderate or high percentages of *Sphagnum* spores) seems to represent periods of cool moist Pacific airmass-type summer climate, while dark brown humified (oxidized) peat with low *Sphagnum* spore counts accumulated slowly during the cold dry episodes dominated by arctic airmasses during the growing season at Ennadai. At the present day a southward extension of the average summer position of the arctic front causes the

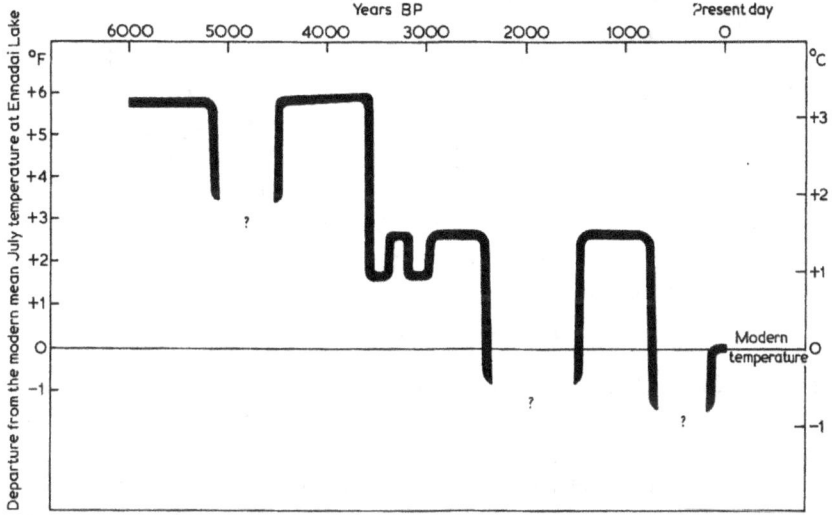

Fig. 11A.7. Estimated changes in the mean July temperatures at Ennadai Lake, Keewatin (from Nichols, 1967b, Fig. 10). The departures from the modern means were calculated from the distance of the forest limit from Ennadai. When, for example, the forest extended 200 km north of Ennadai Lake, it was assumed that the temperature at Ennadai resembled that now found 200 km to the south of that site. The probable error is ±0.5°C. The climatic data were from post-1945 observations covering less than a decade. The magnitude of the cooling after 5000 BP is not known, beyond the fact that temperatures were not as low as modern figures. The deterioration from 2500 to 1500 BP, and following 600 BP, witnessed summer temperatures below modern values, but the amount of cooling is unknown.

associated stormtrack to affect more southerly latitudes than before. Ennadai, lying north of the frontal zone, experiences colder and drier conditions at the same time as sites to the south get colder and wetter. When the frontal zone is far to the north Ennadai enjoys relatively warmer and wetter conditions, while more southern locations are drier since they are missed by the stormtrack (Nichols, 1967b, p. 182).

Between 2600 and 1500 BP at Ennadai there were low *Sphagnum* spore counts, humified peat, irregular *Picea* percentages, increased *Ericaceae* pollen values and the presence of sand grains in the peat, especially at 40 cm. This is interpreted as a retreat of the spruce forest to a position south of Ennadai, while the peat grew slowly and oxidized, and tundra spread around the site. The sand grains probably resulted from a break-down of the local plant cover, which exposed the nearby sandy eskers to attack by wind. It seems likely that a drier, colder climate prevailed during the growing season at Ennadai, suggesting that the arctic airmass extended south in summer.

Some recovery from this severely cold episode was apparent in the pollen diagram by c. 1500 BP, made more clear by macrofossils of burned spruce trees dated 1140 ± 90 BP or 1050 ± 180 BP (WIS-17, Bender et al., 1965) discovered by Bryson et al. (1965) 40 km north of the Ennadai site and 100 km north of modern treeline. After the fire the forest did not regenerate, probably due to climatic deterioration. *Picea* and *Sphagnum* percentages fell after this episode at Ennadai Lake, the number of *Ericaceae* pollen increased, and the peat was humified and contained macrofossils (e.g. leaves) of tundra plants, until peat accumulation ceased very shortly after 630 ± 70 BP (WIS-133, Bender et al., 1967). This evidence suggests the expansion of tundra around Ennadai, and the retreat southward of the forest due to a southerly extension of the dry cold arctic airmass, at about 600 BP at the latest. The absence of peat growth since that time suggests that the area has continued to be dominated by a cold dry tundra climate for the last 600 radiocarbon years.

AXEL HEIBERG ISLAND (79°25′N, 90°30′W): HEGG (1963)

This pollen diagram (fig. 11A.8) from the High Arctic has a ¹⁴C date of 4160 ± 100 BP near the base of the sandy *Carex* (sedge) peat section; Hegg suggests that lacustrine sedimentation began about 4500 BP. The pollen recorded throughout the profile was almost entirely locally derived, e.g. sedge, grass, willow, saxifrage pollen; changes in these taxa may have recorded local environmental changes which are difficult to interpret in terms of regional vegetational and climatic changes.

There was much variation in pollen representation, and the stratigraphy recorded many thin layers of *Carex* and *Drepanocladus* moss peat interspersed with waterlain sand, perhaps resulting from changing water levels in the Expedition River which ran alongside the site. Hegg suggests that between 4000 and 3000 BP the local climate was more favourable to plant growth than it is now. Some unknown time after 3020 ± 120 BP the water level of the 'swamp' was lowered, and the edges dried out, while the drier areas of the local plain became 'heaths', the plant cover became discontinuous, and the wind eroded the 'top layers' (p. 219). Hegg explains this as a result of the river cutting a deeper channel, but it may additionally have been due to a shift to a drier colder climate.

Small amounts of *Pinus* and *Alnus* pollen were recorded discontinuously in the profile, and these are considered exotic since the nearest modern occurrences of the trees are 1900 and 1800 km away to the southwest. More information on regional climatic change might have been gained if these long-distance-transported pollen types had been counted more fully, relative to local taxa, so that a continuous representation of pine and alder pollen might have shown variations due to changing wind trajectories and altered northern limits of the Canadian forest due to climatic change.

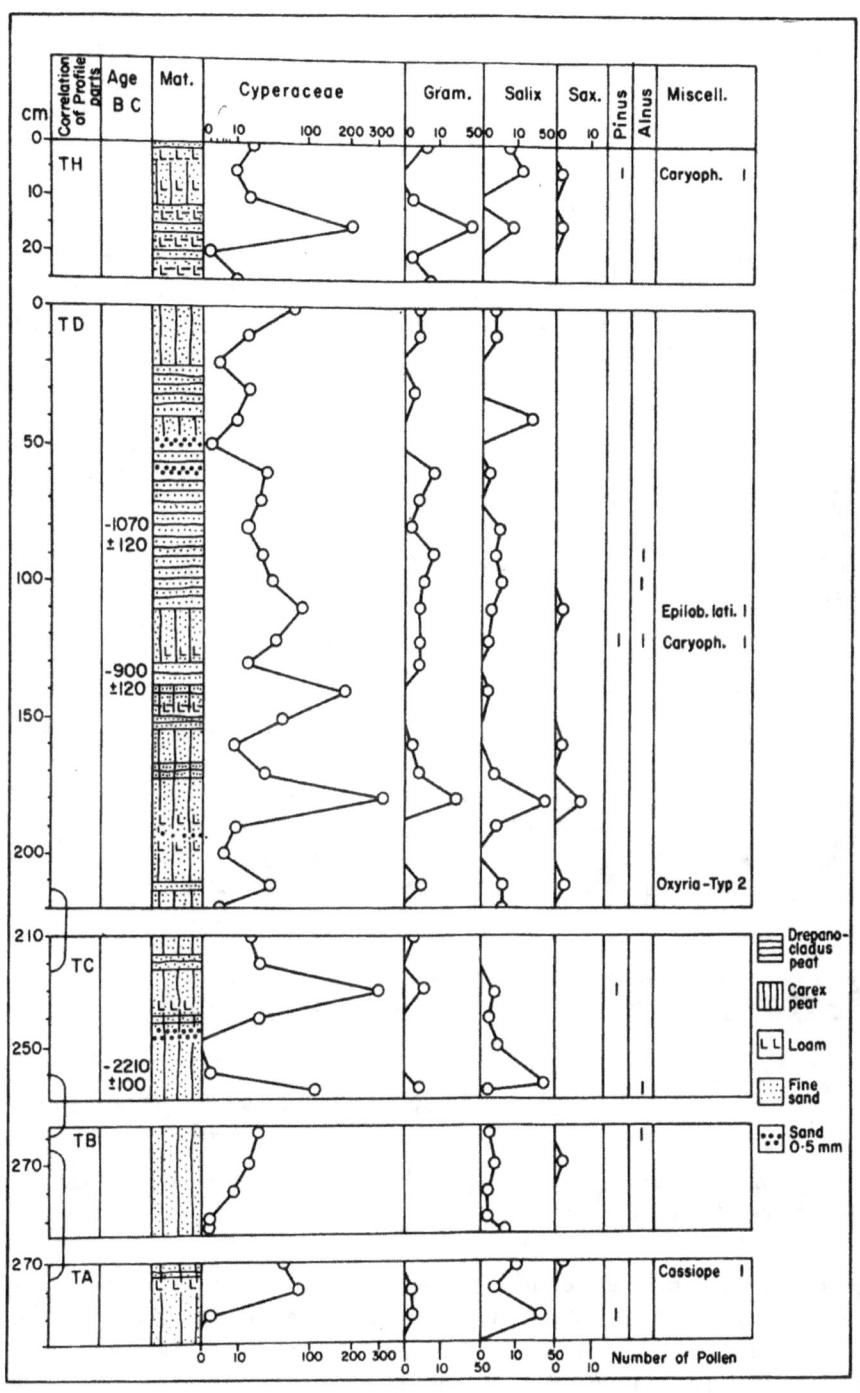

Fig. 11A.8. Pollen diagram from Axel Heiberg Island (from Hegg, 1963).

This diagram illustrates the difficulty of interpreting local vegetational changes in the high arctic tundra at great distances from major ecotones such as the boreal forest-tundra boundary.

SUGLUK (*c.* 62°N, 75°W): BARTLEY AND MATTHEWS (1969)

A number of peat sections from the tundra of northern Ungava provided pollen diagrams which suggested that the area has been covered by treeless tundra throughout the last 4000 radiocarbon years. The coastal sites had polleniferous sediments consisting of thin peaty layers separated by thick horizons of marine sands, deposited by relatively

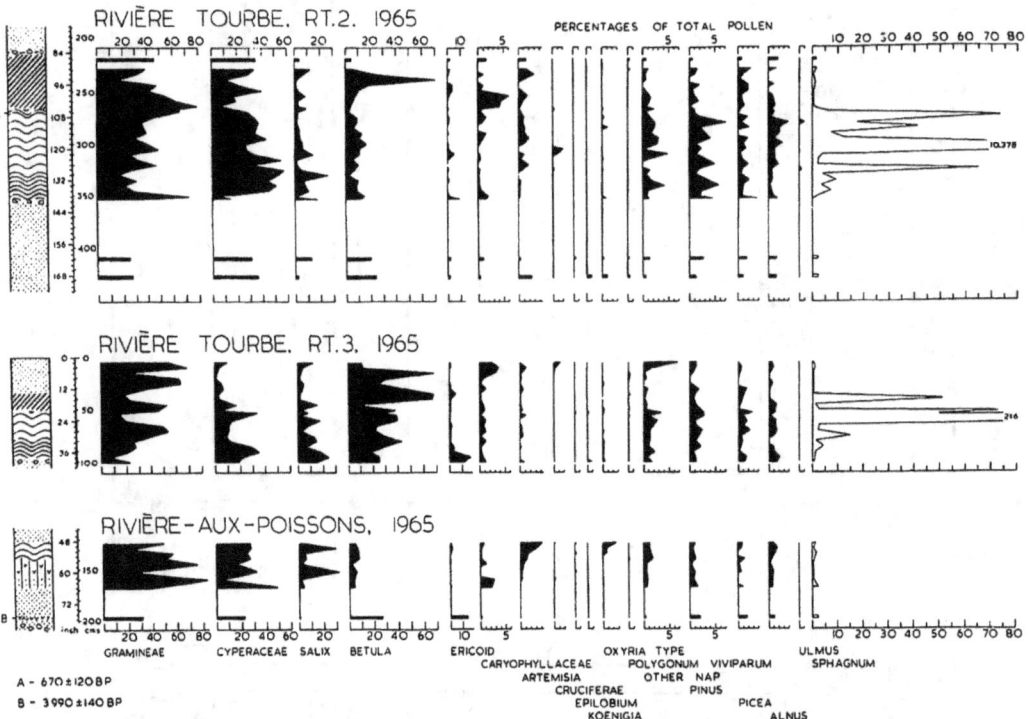

Fig. 11A.9. Pollen diagram from the Sugluk area (from Bartley and Matthews, 1969). Based on percentages of total pollen excluding spores.

higher sea levels, so that, as the authors stress (Bartley and Matthews, 1969), the changes in the pollen diagrams are not continuous, and there was an interval between the lower marine sedimentation and the upper thick peat development. The authors noted the difficulty of separating local effects caused by changing relief and substrate from more general effects of climatic change in their palynological interpretations, and they comment that the stratigraphic evidence is most indicative of climatic change.

A stony layer (probably representing emergence from marine conditions) was overlain by thin peat dated 3990 ± 140 BP (NPL-114); the presence of *Betula* wood and *Empetrum* (crowberry) seeds in this peat may indicate a relatively mild summer climate, with moist maritime conditions and a winter snow cover. Another site at Sugluk had organic peaty layers intercalated in marine sediments; one stratum contained high percentages of *Pinus* and *Picea* pollen (26.0 and 19.5 per cent) indicative of a relatively mild summer climate (and possibly with some over-representation due to marine conditions) which was dated 2840 ± 160 BP (GSC-818). Molluscs from the marine sediments indicate seas warmer than the present.

These mild climatic episodes were followed by colder and drier conditions, with deposition of windblown sand at one site due to a sparser plant cover and the wind erosion of local sandy soils. This severe episode seems to have lasted from after 2800 BP to about 1600 BP, when there was a considerable development of *Sphagnum* peat growth – at 1625 ± 175 BP (1-727), and 1600 ± 140 BP (GSC-537) – indicative of a somewhat warmer moister climate (fig. 11A.9). The authors note that the greater development of peat during this second period need not indicate that this episode was milder than that before 2800 BP, since marine molluscan analyses suggest that the warmest local sea temperatures occurred between about 5200 and 2800 years ago (Matthews, 1967). They note that the land and sea climatic optima may not have coincided, or that local conditions of drainage may have generated the later development of peat.

This milder episode lasted from about 1600 to about 670 ± 120 BP (NPL-125), when *Sphagnum* peat growth ceased in response to cold dry summer climate, which promoted breakdown of the local plant cover and the deposition of windblown sand, probably synchronously over a number of sites. The close correlations which Bartley and Matthews propose may be made with other distant sites are examined below (p. 664).

In the Sugluk area there has been recent growth of *Sphagnum* mosses which the above authors suggest may have reflected the climatic amelioration of the present century.

The peaks of *Betula* pollen seen in some of these pollen diagrams were considered to reflect local changes in relief causing local changes in snow distribution.

BAFFIN ISLAND (70°21′N, 74°59′W), SUGLUK, UNGAVA (62°10′N, 75°57′W) AND BATHURST INLET, NWT (*c*. 67°N, 108°W): TERASMAE, WEBBER AND ANDREWS (1966)

Pollen analyses by Terasmae and plant macrofossil identifications of plant-bearing beds in north-central Baffin Island dated by radiocarbon assay at over 40,000 BP suggest that during this interglacial episode July temperatures were 1° to 4°C higher than now, and precipitation about 50 per cent more than at present.

A pollen diagram (fig. 11A.9) from Sugluk was derived from peat overlying sand, with a basal organic date of 1625 ± 175 (I-727). Low percentages of *Picea* and *Pinus* pollen declined upward towards the undated upper part of the diagram. Another diagram from Bathurst inlet shows minimum values for *Picea* and disappearance of *Pinus* pollen around 2170 ± 140 (GSC-138), and increased values for these genera after that date continuing to maxima considerably after 1850 ± 140 (GSC-137). I suggest that this may have resulted from a colder episode around 2200 BP marked by southward retreat of the boreal forest–tundra ecotone, with a somewhat warmer wetter episode of summer climate following 1600 BP, and a subsequent colder period characterized by reduced spruce and pine percentages perhaps indicative of treeline retreat southward.

SERMERMIUT, JAKOBSHAVN, WEST GREENLAND (69°12′N, 51°08′W):
FREDSKILD (1967a, b)

A coastal peat site in the Greenland tundra provided Fredskild with samples for palynology and plant macrofossil studies reaching back to 3510 ± 120 BP when humus accumulated over a stony sand beach which had recently emerged from the sea. The basal organic samples contained microscopic charcoal from 3510 to 3360 ± 120 BP which he suggested might have blown from forest fires in Canada (see p. 651). The basal plant cover of grasses and chenopods was succeeded by a dwarf shrub-heath episode characterized by *Empetrum* sp. (crowberry) and *Ledum* sp. (Labrador tea), with subsequent dominance by the dwarf arctic birch *Betula nana*. This latter was a period of dry climate, with a maximum of peat humification, following 3360 BP and lasting until 2830 ± 120 BP. Peat then formed amidst a *Salix glauca* (arctic willow) scrub just before 2570 ± 110 BP, apparently marking a climatic change, possibly to a more humid climate or to heavier winter snowfall. Altered moisture conditions are more clearly indicated when the willow scrub was replaced by a *Sphagnum* bog between 2570 and 2350 ± 110 BP, when *Sphagnum* spore counts were very high. Fredskild suggested this might have reflected higher summer or winter precipitation or a raising of the permafrost table, associated with a temperature decrease; his correlation with the north European sequence is discussed below.

By 1910 ± 100 BP a drier dwarf community dominated the site; by 1540 ± 100 BP there was a reversion to a hydrophilous plant cover similar to that following 3570 BP, though this latter change was not as pronounced as the former. This *Sphagnum* peat bog episode recorded high *Sphagnum* spore counts.

Organic accumulation ceased sometime shortly after 1540 BP, and Fredskild believed that Thule people who settled nearby about 950 BP must have peeled turf from the site and removed the uppermost record, though I suggest that it is possible that local conditions were unsuitable for substantial organic accumulation since then. He noted that Meldgaard (1958) provided evidence for a climatic amelioration between the moist period of 1500 BP and the Thule settlements of *c.* 950 BP. Iversen's (1954) pollen diagrams from the Godthab Fjord area also showed the climatic changes of *c.* 2500 and 1500 BP, and he observed that a change to colder conditions about 650 BP can be discerned in Iversen's earlier work (1934) from that area.

Fredskild had noted a small number of pollen grains which were exotic to Greenland (probably windblown from North America) such as *Alnus*, *Pinus*, *Picea*, etc., and these small counts of alien taxa throughout the profile showed small increases coincident with peaks of windblown microscopic charcoal derived probably from North America and perhaps partially from local Eskimo burning. Two of these peaks coincided with *Sphagnum* spore peaks at the times of peat growth due to climatic change at *c.* 2500 and 1500 BP.

LAKE KLARESÖ, PEARY LAND, NORTH GREENLAND (82°19′N, 30°20′W):
FREDSKILD (1969)

In this high arctic desert less than 3 per cent of the land surface was occupied by plants, a result of the low precipitation (*c.* 25 mm per year) and high winds. The lake sediments analysed by Fredskild began to accumulate *c.* 5000 BP, and initially the dominant pollen of grasses and sedges reflected a herbaceous episode, which was replaced by a *Salix arctica* zone beginning *c.* 4800 BP when the climate became more humid than today and the plant cover richer in individuals.

The site lies alongside the 200 km long Independence Fjord which today is frozen all year, but soon after the climatic change at 4800 BP there is the oldest of a series of dates on driftwood which floated over a partially open Arctic Ocean and up the fiord to attest to warmer climatic conditions. The date was 4540 ± 120 BP, but most of the driftwood (which was found in excavated Eskimo hearths) dated between 4000 and 3600 BP. The sediments from 4000 to 3300 BP contained the maximum pollen content, due presumably to the densest plant cover; the episode also saw the most luxuriant growth of the sub-aquatic moss *Calliergon trifarium*, and thus the climate of this period was probably warm and humid with relatively high precipitation, resulting from the open fiords and Arctic Ocean. The local Eskimos (of the Independence-I culture) left remains of the numerous musk-oxen they hunted, another indication of a thriving plant cover (see McGhee, Chapter 15A, p. 845).

The local climatic optimum lasted from 4000 to 3300 BP, when there was a decline in the local output of pollen, and after that there were few pieces of driftwood which had floated up the fiord, all indicative of an impoverished plant cover and freezing seas due to a colder climate. No driftwood was dated younger than 2720 ± 110 BP, while a further reduction in pollen productivity was dated *c.* 2100 BP, both suggesting increasingly cold dry climate with summers reaching modern standards of severity *c.* 2100 BP, marked by the establishment of a high arctic desert which has lasted over 2000 radiocarbon years. Fredskild noted that ^{18}O palaeotemperature data from the Greenland ice sheet also indicated a climatic cooling beginning 3200 BP and lasting just over 1000 years (Dansgaard *et al.*, 1969).

4. Arctic Alaska

A brief summary is presented, since Heusser (1965) has fully reviewed the subject (see Colinvaux, 1967a) and little material has appeared since then.

BROOKS RANGE, ALASKA (*c.* 68°N, 152°W): LIVINGSTONE (1955); AND UMIAT
(69°N, 152°W): LIVINGSTONE (1957)

Four pollen diagrams from the Alaskan tundra allowed Livingstone to recognize a
tripartite pollen zonation: the lowest zone represented tundra conditions with dominant sedges, grasses and composites and some birch and willow; the middle zone was
dominated by *Betula* pollen derived from birch shrubs; and the topmost zone had
much *Alnus* pollen (probably over-represented) and windblown *Picea* pollen from the
spruce forest to the south.

These data were not assigned ^{14}C dates (1955) but the later work at Umiat showed a
somewhat similar sequence, and its dates have been applied to the earlier study. A date
from Zone I, the lower tundra episode, was 8125 ± 250 BP, and from above the transition Zone I to II, the birch zone, was 7500 ± 250 BP, while an assay from below the
Zone II/III boundary was 5900 ± 200 BP. This last date lay below the maximum of
Alnus pollen which Livingstone correlates with a worldwide thermal maximum.

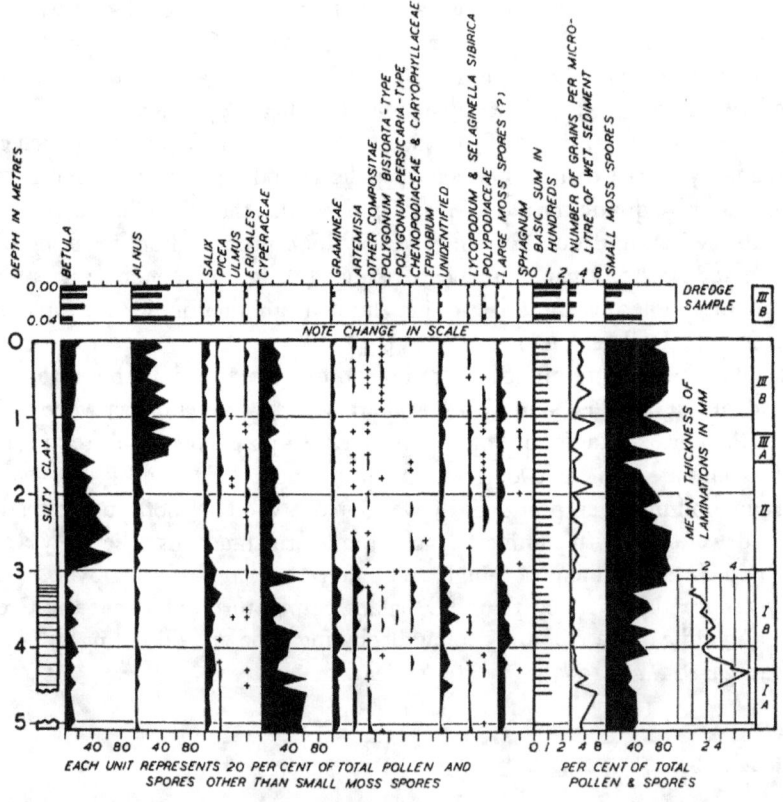

Fig. 11A.10. Pollen diagram from Chandler Lake (from Livingstone, 1955).

ST LAWRENCE ISLAND, ALASKA (63°N, 170°W): COLINVAUX (1967b)

A pollen diagram from Flora Lake has a single ^{14}C date 5650 ± 275 BP near the base of the spruce–alder zone which marked the warmest period of the arctic Alaskan post-glacial episode. Below this the birch and then sedge-grass pollen zones resemble Zones II and I of Livingstone (1955, 1957); see fig. 11A.10. Colinvaux (1964a) noted elsewhere that there was apparently little vegetational change during the last 4000 radiocarbon years in the Alaskan Arctic coastal records, and that the sites chosen for pollen analysis have remained in a tundra state during most of the Holocene or Flandrian (Colinvaux, 1964b, 1967a, b). Other work by him in arctic Alaska is primarily concerned with pre-Holocene events (Colinvaux, 1964a, b).

The palaeoclimatic evidence from arctic Alaskan palynology is particularly unclear at present, though whether this is due to insensitivity in the few sites examined, or whether the Holocene climatic changes were more weakly developed than further south (Livingstone, 1955; McCulloch and Hopkins, 1966) is not known. Macrofossil evidence for forest north of its modern limit is frequent from c. 10,000 to 8300 BP (McCulloch and Hopkins, 1966); this evidence for warm climate is followed by evidence for separate brief warm intervals at 7270 ± 350 BP (cf. Matthews, 1970) and 3600 ± 500 BP, and colder climate setting in before 2700 BP (McCulloch and Hopkins, 1966).

This macrofossil data and the palynological evidence for a weakly developed thermal maximum sometime between 6000 and 3000 BP provides a rather insubstantial framework for a climatic sequence which does not at present appear to be at variance with the schemes for arctic Canada.

5. Summary and correlations

The differences in the sensitivity of the palaeoenvironmental records summarized above presumably reflect distance of the studied sites from vegetational and climatic boundaries; these variations and the usually inadequate degree of radiocarbon dating control does not obscure the fundamental broad similarity of the circumboreal climatic sequences for the Late Quaternary period.

PRE-6000 BP

The evidence for the Early and Middle Holocene or Flandrian indicates relatively warm climates (warmer than present), with spruce forest ranging north of its modern limit in arctic Alaska (10,000 to 8300 BP), the Yukon (Terasmae and Hughes, 1966), in the Mackenzie Delta (Mackay and Terasmae, 1963) and at Colville Lake before 6000 BP, with very rapid peat growth under mild Pacific airmass dominance at that latter site, and with extremely quick melting of the Late Wisconsin ice sheet in the Ennadai Lake area due to climatic warming prior to 6000 BP.

6000–5000 BP

The evidence for a climate warmer than the present continues throughout this period, with the boreal forest limit lying north of its present location at Colville Lake and Ennadai Lake and in Alaska. This and the above episode fall within the 'Climatic Optimum' and particularly within the period of the Atlantic pollen zone of northern Europe (*c.* 8000 or 7500 BP to *c.* 5000 BP), which was warmer than the present (perhaps by 2 or 2.5°C in summer: Godwin, 1956). Terasmae and Anderson (1970) have reported strong pollen and macrofossil evidence for a northward extension of the range of white pine (*Pinus strobus*) by almost 100 km in Quebec (49°01′N, 79°05′W) dated 5030 ± 130 (GSC-585) which was buried by peat. The authors note this as proof of a warmer climate leading up to that time, and I think it is permissible to add that the stratigraphy implies that a wetter cooler peat-forming episode followed 5000 BP and allowed the burial and preservation of the trees.

5000–3500 BP

There was a short episode of summer climatic cooling at Ennadai Lake just before 4800 ± 90 BP (WIS-166) which was estimated to have lasted until maybe 4500 BP (Nichols, 1967b), and probably did not descend to modern arctic temperatures. This event has not been identified in other arctic pollen diagrams, which may mean it was merely a local occurrence, or that it was a minor happening which was registered only in sensitive sites. The peat formation in Quebec after 5030 ± 130 (GSC-585: Terasmae and Anderson, 1970), and the cooler wetter episode at Lynn Lake in the Manitoba forest following 5140 ± 100 BP (WIS-112: Nichols, 1967b) may be expressions of the same short-lived climatic cooling.

At the same time in the Canadian Arctic (Andrews, 1970a, b) and in Europe (Page, 1968) and elsewhere (Mercer, 1967) there were glacial advances and evidence of treeline lowering (Frenzel, 1966) to suggest climatic cooling, and the sub-boreal pollen zone of northern Europe began with the 'Elm decline' as a response to climatic change or human modification of the environment, or both (Frenzel, 1966; Smith, 1961).

Some centuries after this cooling in northern Canada there was a recovery to a climate warmer than now, beginning at Ennadai Lake at *c.* 4500 BP (estimated), which continued to 3500 BP and allowed forest migration 250 km beyond its present limit (Bryson *et al.*, 1965), with estimated summer temperatures *c.* 3°C above modern values (Nichols, 1967b; cf. Mercer, 1967). Treeline was further north at one site in Alaska at 3600 BP (McCulloch and Hopkins, 1966). The period of peat formation at Sugluk (Bartley and Matthews, 1969) dated 3990 ± 140 BP (NPL-114) recorded a warm episode, as did the data from Axel Heiberg Island from *c.* 4000 to 3000 BP (Hegg, 1963). This is broadly synchronous with Fredskild's (1969) evidence for a partially ice-free Arctic Ocean beginning possibly 4540 ± 120 BP, and with the greatest abundance of exotic driftwood dated between 4000 and 3600 BP. Together with palynological evidence this suggests that Peary Land in north Greenland experienced a climatic optimum from 4000 to 3300 BP (Fredskild, 1969).

3500 OR 3300–2000 BP

There is a good deal of agreement on the reality of the cooling of the arctic climates in the last few thousand years. After 3200 BP Colville Lake experienced a forest retreat and dominance of tundra indicative of cold dry arctic airmass summers lasting to 1800 BP (Nichols, unpublished). At Pelly Lake (Nichols, 1970) the cooling began just after 3400 BP, at Ennadai at 3500 BP (Bryson *et al.*, 1965; Nichols, 1967b), at Axel Heiberg (Hegg, 1963) sometime after 3000 BP, at Peary Land (Fredskild, 1969a) at 3300 BP, and in Alaska sometime before 2700 BP (McCulloch and Hopkins, 1966). From then until *c.* 2500 or *c.* 2200 BP the climate continued to cool, in a regular decline (Pelly Lake: Nichols, 1970) or in steps with occasional temporary recoveries (Nichols, 1967b) until an episode of severely cold climate occurred, with its maximum severity *c.* 2100 BP (Nichols, 1970). The effects of this severe spell of climate, which saw the widespread southward extension of tundra at the expense of forest, were seen at almost all the sites summarized above, and the timing was broadly synchronous, though with some sites enduring longer periods of severe climate than others, due no doubt to their location. An exception was MacAlpine Lake (Terasmae, 1967) which registered no climatic changes during an unknown time span following peat genesis at 2300 BP.

The similarities in the Canadian and Greenland sequences are very marked for this episode, and they have been compared (Fredskild, 1967b, 1969; Nichols, 1967b, 1970) with the similar climatic cooling in northern Europe which marked the start of the sub-Atlantic period at *c.* 2500 BP (Godwin, 1956).

2000 BP–PRESENT

Recovery from this severe climatic cooling seems to have begun by 2000 BP (Pelly Lake: Nichols, 1970; cf. Peary Land: Fredskild, 1969) but not to have warmed sufficiently to be registered at most sites until 1800 BP (Colville: Nichols, unpublished), or 1600 BP (Sugluk: Bartley and Matthews, 1969) or 1400 BP (Ennadai: Nichols, 1967b). This warm episode, estimated at Ennadai to be *c.* 1.5°C above present summer temperatures (Nichols, 1967b), culminated in a maximum of summer warmth *c.* 1000 BP at Pelly Lake (Nichols, 1970), when at the same time the forest limit had again moved north (100 km from modern treeline: Bryson *et al.*, 1965).

Cooler summers followed 900 BP at Pelly Lake (Nichols, 1970), and Ennadai Lake (Bryson *et al.*, 1965; Nichols, 1967b) became dominated by tundra, while shortly afterwards there was a very widespread cessation of peat growth due to colder drier (arctic airmass) summers at Pelly (900 BP), at Ennadai (630 BP) and Sugluk (670 BP). Peat growth has not resumed at these sites since, due presumably to continued cold dryness.

This part of the sequence matches the timing and direction of climatic changes in medieval northern Europe and Greenland (Lamb, 1966).

Thus it is possible to argue that the Late Quaternary climatic histories of those Alaskan and Canadian Arctic sites summarized above exhibit a common timing and direction of climatic change, which is comparable to those seen in Greenland and northern Europe, and which in turn have points of similarity with the north Russian sequences which are beyond the scope of this article (see Kind, 1967). It is reassuring that Bartley and Matthews (1969) have used the Keewatin sequence (Nichols, 1967b) as a basis for their interpretation of the Sugluk data. The northern edge of the Canadian boreal coniferous forest in the areas described above exhibits synchrony and parallelism in its movements along its entire east–west length so far examined by palynology. The few studies of the southern limit of that forest unit (Nichols, 1969b; Ritchie, 1969) and an unpublished study of the Peace River by the writer suggest very tentatively that large sections *at least* of the southern treeline experienced movements synchronous and parallel to those of the northern forest–tundra ecotone.

References

AARIO, L. (1940) Waldgrenzen und subrezente Pollen-spektren in Petsamo Lappland. *Ann. Acad. Sci. Fenn.*, A, **54** (8). 120 pp.

ANDREWS, J. T. (1970a) Differential crustal recovery and glacial chronology (6,700 to 0 BP), west Baffin Island, NWT, Canada. *Arct. Alp. Res.*, **2** (2), 115–34.

ANDREWS, J. T. (1970b) A geomorphological study of post-glacial uplift with particular reference to Arctic Canada. *Inst. Brit. Geogr., Spec. Pub.*, 2. 156 pp.

BARRY, R. G. (1967) Seasonal location of the arctic front over North America. *Geogr. Bull.*, **9**, 79–95.

BARTLEY, D. D. (1967) Pollen analysis of surface samples of vegetation from Arctic Quebec. *Pollen et Spores*, **9** (1), 101–5.

BARTLEY, D. D. and MATTHEWS, B. (1969) A palaeobotanical investigation of post-glacial deposits in the Sugluk area of northern Ungava, Quebec. *Rev. Palaeobotan. Palynol.*, **9**, 45–61.

BENDER, M. M., BRYSON, R. A. and BAERREIS, D. A. (1965) Univ. of Wisconsin Radiocarbon dates, I. *Radiocarbon*, **7**, 399–407.

BENDER, M. M., BRYSON, R. A. and BAERREIS, D. A. (1966) Univ. of Wisconsin Radiocarbon dates, II. *Radiocarbon*, **8**, 522–33.

BENDER, M. M., BRYSON, R. A. and BAERREIS, D. A. (1967) Univ. of Wisconsin Radiocarbon dates, III. *Radiocarbon*, **9**, 530–44.

BENDER, M. M., BRYSON, R. A. and BAERREIS, D. A. (1968a) Univ. of Wisconsin Radiocarbon dates, IV. *Radiocarbon*, **10** (1), 161–8.

BENDER, M. M., BRYSON, R. A. and BAERREIS, D. A. (1968b) Univ. of Wisconsin Radiocarbon dates, V. *Radiocarbon*, **10** (2), 473–8.

BENDER, M. M., BRYSON, R. A. and BAERREIS, D. A. (1969) Univ. of Wisconsin Radiocarbon dates, VI. *Radiocarbon*, **11** (1), 228–35.

BENDER, M. M., BRYSON, R. A. and BAERREIS, D. A. (1970) Univ. of Wisconsin Radiocarbon dates, VII. *Radiocarbon*, **12** (1), 335–45.

BRYSON, R. A. (1966) Air masses, streamlines, and the boreal forest. *Geogr. Bull.*, **8** (3), 228–69.

BRYSON, R. A., IRVING, W. N. and LARSEN, J. A. (1965) Radiocarbon and soils evidence of former forest in the southern Canadian tundra. *Science,* **147**, 46–8.

BRYSON, R. A., WENDLAND, W. M., IVES, J. D. and ANDREWS, J. T. (1969) Radiocarbon isochrones on the disintegration of the Laurentide Ice Sheet. *Arct. Alp. Res.*, **1** (1), 1–14.

COLINVAUX, P. A. (1964a) Sampling stiff sediments of an ice-covered lake. *Limnol. and Oceanog.*, **9**, 262–4.

COLINVAUX, P. A. (1964b) The environment of the Bering land bridge. *Ecol. Monogr.*, **34**, 297–329.

COLINVAUX, P. A. (1964c) Origin of ice ages: pollen evidence from Arctic Alaska. *Science*, **145**, 707–8.

COLINVAUX, P. A. (1967a) Quaternary vegetational history of Arctic Alaska. *In:* HOPKINS, D. M. (ed.) *The Bering Land Bridge*, 207–31. Stanford: Stanford Univ. Press.

COLINVAUX, P. A. (1967b) A long pollen record from St Lawrence Island, Bering Sea (Alaska). *Palaeogeogr., Palaeoclim., Palaeoecol.*, **3**, 29–48.

CUSHING, E. J. and WRIGHT, H. E. (1965) Hand-operated piston corers for lake sediments. *Ecol.* **46** (3), 380–4.

DANSGAARD, W., JOHNSEN, S. J., MÖLLER, J. and LANGWAY, C. C. (1969) One thousand centuries of climatic record from Camp Century on the Greenland ice sheet. *Science*, **166**, 377–81.

DAVIS, M. B. (1967) Pollen accumulation rates at Rogers Lake, Connecticut, during late- and postglacial time. *Rev. Palaeobotan. Palynol.*, **2**, 219–30.

FAEGRI, K. and IVERSEN, J. (1964) *Textbook of Pollen Analysis*. New York: Hafner. 237 pp.

FALCONER, G., IVES, J. D., LØKEN, O. H. and ANDREWS, J. T. (1965) Major end moraines in eastern and central Arctic Canada. *Geogr. Bull.*, **7** (2), 137–53.

FREDSKILD, B. (1967a) Palaeobotanical investigations at Sermermiut, Jakobshavn, west Greenland. *Medd. om Grønland*, **178** (4). 54 pp.

FREDSKILD, B. (1967b) Postglacial plant succession and climatic changes in a west Greenland bog. *Rev. Palaeobotan. Palynol.*, **4**, 113–27.

FREDSKILD, B. (1969) A postglacial standard pollen diagram from Peary Land, north Greenland. *Pollen et Spores*, **XI** (3), 573–83.

FRENZEL, B. (1966) Climatic change in the Atlantic/sub-Boreal transition on the Northern Hemisphere. *In: World Climate from 8,000 to 0 BC*, 99–123. London: Roy. Met. Soc.

GODWIN, H. (1956) *History of the British Flora*. Cambridge: Cambridge Univ. Press. 384 pp.

HARE, F. K. (1968) The Arctic. *Quart. J. R. Met. Soc.*, **94** (402), 439–59.

HEGG, O. (1963) Palynological studies of a peat deposit in front of the Thompson Glacier. *Axel Heiberg Island Res. Rep.*, 217–19. McGill Univ.

HEUSSER, C. J. (1965) A Pleistocene phytogeographical sketch of the Pacific Northwest and Alaska. *In:* WRIGHT, H. E. JR and FREY, D. G. (eds.) *The Quaternary of the United States*, 469–83. Princeton: Princeton Univ. Press.

HUGHES, O. L. and TERASMAE, J. (1963) *SIPRE* ice-corer for obtaining samples from permanently frozen bogs. *Arctic*, **16** (4), 271–2.

IVERSEN, J. (1934) Moorgeologische Untersuchungen auf Grønland. *Medd. Dansk Geol. Foren.*, **8**.

IVERSEN, J. (1954) Origin of the flora of Western Greenland in the light of pollen analysis. *Oikos*, **4** (2), 1952–3.

JÖRGENSEN, S. (1967) A method of absolute pollen counting. *New Phytol.*, **66**, 489–93.

KIND, N. V. (1967) Radiocarbon chronology in Siberia. *In:* HOPKINS, D. M. (ed.) *The Bering Land Bridge*, 172–92. Stanford: Stanford Univ. Press.

LAMB, H. H. (1966) *The Changing Climate.* London: Methuen. 236 pp.

LICHTI-FEDEROVICH, S. and RITCHIE, J. C. (1968) Recent pollen assemblages from the Western Interior of Canada. *Rev. Palaeobotan Palynol.*, **7**, 297–344.

LIVINGSTONE, D. A. (1955) Some pollen profiles from Arctic Alaska. *Ecol.*, **36** (4), 587–600.

LIVINGSTONE, D. A. (1957) Pollen analysis of a valley fill near Umiat, Alaska. *Amer. J. Sci.*, **255**, 254–60.

MCCULLOCH, D. and HOPKINS, D. (1966) Evidence for an Early Recent warm interval in Northwestern Alaska. *Bull. Geol. Soc. Amer.*, **77**, 1089–108.

MACKAY, J. R. and TERASMAE, J. (1963) Pollen diagrams in the Mackenzie Delta area, NWT. *Arctic*, **16** (4), 228–38.

MATTHEWS, B. (1967) Late Quaternary land emergence in northern Ungava, Quebec. *Arctic*, **20** (3), 176–202.

MATTHEWS, J. V. JR (1970) Quaternary environmental history of interior Alaska: pollen samples from organic colluvium and peat. *Arct. Alp. Res.*, **2** (4), 241–51.

MELDGAARD, J. (1958) Sermermiut, and its different periods of culture. *In:* Paleo-eskimo cultures in Disko-Bugt, West Greenland, *Medd. om Grønland*, **161**, 2.

MERCER, J. H. (1967) Glacier resurgence at the Atlantic/sub-Boreal transition. *Quart. J. R. Met. Soc.*, **93**, 528–34.

MULLER, F. (1962) Analysis of some stratigraphic observations and radiocarbon dates from two pingos in the Mackenzie Delta area, NWT. *Arctic*, **15**, 278–88.

NICHOLS, H. (1967a) Pollen diagrams from sub-Arctic central Canada. *Science*, **155**, 1665–8.

NICHOLS, H. (1967b) The postglacial history of vegetation and climate at Ennadai Lake, Keewatin and Lynn Lake, Manitoba. *Eiszeitalter und Gegenwart*, **18**, 176–97.

NICHOLS, H. (1967c) Permafrost peat sampling – dynamite and chainsaw. *Arctic*, **20** (1), 54.

NICHOLS, H. (1967d) The suitability of certain categories of lake sediments for pollen analysis. *Pollen et Spores*, **9** (3), 615–30.

NICHOLS, H. (1967e) Disturbance of Arctic lake sediments by 'bottom ice': a hazard for palynology. *Arctic*, **20** (3), 213–44.

NICHOLS, H. (1969a) Chronology of peat growth in Canada. *Palaeogeogr., Palaeoclimat., Palaeoecol.*, **6**, 61–5.

NICHOLS, H. (1969b) The Late Quaternary vegetational and climatic history of Porcupine Mountain and Clearwater Bog, Manitoba. *Arct. Alp. Res.*, **1** (3), 155–67.

NICHOLS, H. (1970) Late Quaternary pollen diagrams from the Canadian Arctic Barren Grounds at Pelly Lake, Northern Keewatin, NWT. *Arct. Alp. Res.*, **2** (1), 43–61.

PAGE, N. R. (1968) Atlantic/early sub-Boreal glaciation in Norway. *Nature*, **219**, 694–7.

PORSILD, A. E. (1938) Earth mounds in unglaciated arctic northwestern America. *Geogr. Rev.*, **28**, 46–58.

RITCHIE, J. C. (1969) Absolute pollen frequencies and carbon-14 age of a section of Holocene lake sediment from the Riding Mountain area of Manitoba. *Can. J. Bot.*, **47** (9), 1345–9.

RITCHIE, J. C. and LICHTI-FEDEROVICH, S. (1967) Pollen dispersal phenomena in Arctic-Subarctic Canada. *Rev. Palaeobot. and Palynol.*, **3**, 255–66.

ROWE, J. S. (1959) Forest regions of Canada. *Can. Dept of Northern Affairs, Forestry Branch Bull.*, **123**, 1–71.

SHARP, R. P. (1958) Malaspina Glacier, Alaska. *Bull. Geol. Soc. Amer.*, **69**, 617–46.

SMITH, A. G. (1961) The Atlantic-Sub-boreal transition. *Linn. Soc. Proc.*, **172**, 38–49.

TERASMAE, J. (1961) Notes on late Quaternary climatic changes in Canada. *Ann. New York Acad. Sci.*, **95** (1), 658–75.

TERASMAE, J. (1967) Recent pollen deposition in the northeastern district of Mackenzie (Northwest Territories, Canada). *Palaeogeogr., Palaeoclimat., Palaeoecol.*, **3**, 17–27.

TERASMAE, J. and ANDERSON, T. W. (1970) Hypsithermal range extension of white pine (*Pinus strobus* L.) in Quebec, Canada. *Can. J. Earth Sci.*, **7** (2), 406–13.

TERASMAE, J. and CRAIG, B. G. (1958) Discovery of fossil *Ceratophyllum demersum* L. in Northwest Territories, Canada. *Can. J. Bot.*, **36**, 567–9.

TERASMAE, J. and HUGHES, O. L. (1966) Late Wisconsin chronology and history of vegetation in the Ogilvie Mountains, Yukon Territory, Canada. *The Palaeobotanist*, **15** (1–2), 235–42.

TERASMAE, J., WEBBER, P. J. and ANDREWS, J. T. (1966) A study of late-Quaternary plant-bearing beds in north-central Baffin Island, Canada. *Arctic*, **19** (4), 296–318.

WRIGHT, H. E., WINTER, T. C. and PATTEN, H. L. (1963) Two pollen diagrams from southeastern Minnesota; problems in the regional late-glacial and post-glacial vegetational history. *Bull. Geol. Soc. Amer.*, **74**, 1371–96.

Manuscript received May 1971.

Additional references

FREDSKILD, B. (1973) Studies in the vegetational history of Greenland. *Medd. om Grønland*, **198** (4). 245 pp.

RITCHIE, J. C. and HARE, F. K. (1971) Late Quaternary vegetation and climate near the arctic tree-line of northwestern North America. *Quat. Res.*, **1**, 331–42.

Palaeolithic players on the American stage: man's impact on the Late Pleistocene megafauna

Paul S. Martin
Department of Geosciences,
University of Arizona

1. Introduction

Four times larger than Australia, the Americas were the largest landmass of favourable habitat unknown to hominids before the evolution of modern men. Despite the attention given by the public to the voyages of European explorers, the discovery of America occurred thousands of years earlier at an unsung moment in the Late Pleistocene. The Palaeolithic pioneers who crossed the Bering Bridge out of Asia (see McGhee, Chapter 15A) indeed took a giant step for mankind. For archaeologist Bordes (1968), 'There can be no repetition of this until man lands on a [habitable] planet belonging to another star.'

Long preoccupied with determining to within 1000 years the date of the discovery of America, archaeologists have theorized less on what would have happened in the first 1000 years afterward. At some time late in the Pleistocene an interbreeding human population in Siberia reached the Arctic Circle, crossed the Bering Bridge and spread south through the Americas. I shall propose that south of the Cordilleran ice sheet this population had to expand very rapidly, that no prior invasion could have taken place, and that the swift extinction of fauna by superpredation would not yield much fossil evidence. My purpose is to accommodate what has been recognized as a major weakness in the hypothesis of Pleistocene overkill, that is, its failure to account for the lack of abundant kill sites.

2. Overkill re-examined

Determining to within 1000 years the time of the extinction of Late Pleistocene large mammals has become a major concern of Pleistocene biologists. At the end of the last Ice Age, it appears that the sudden loss of native American mammoths, mastodons, ground sloths, horses and camels coincided with the first appearance of Stone Age big game hunters (Martin and Wright, 1967).

Nevertheless, not all palaeontologists will concede that Late Pleistocene (Rancholabrean) extinction was unusual in the history of the earth, much less that it was man-caused. For example, Webb (1969) regards it as one of many fluctuations around a basic equilibrium maintained through the Late Cainozoic by evolution and faunal exchanges. Even those willing to regard the matter of human invasion as more than circumstantial have failed to explain precisely how extinctions were brought about. According to Flint (1971, p. 778), 'This argument most frequently advanced against the hypothesis of human agency is that in no territory was man sufficiently numerous to destroy the large numbers of animals that became extinct.' Krantz (1970, p. 165) was more forceful: ' . . . hunting man could not and did not exterminate these animals'. In many parts of Eurasia, where megafaunal losses were minor (Reed, 1970, p. 284), large numbers of artifacts can be found associated with abundant bones of large mammals throughout the Palaeolithic. Despite many more extinctions in the New World, no appreciable numbers of artifacts have been found in kill site association with any of the extinct genera other than the mammoth (Hester, 1967; Jelinek, 1967; Krantz, 1970). The best of the eleven man-mammoth associations in America lack the wealth of cultural material including art objects associated with Old World mammoths in the Ukraine (Haynes, 1971, p. 83). Of the thirty-three genera of large mammals that became extinct at the end of the last Ice Age (Martin, 1967), none are known to have been hunted by drives or killed by jumps as in the case of the American bison. No elaborate traps like those used to capture mammoth in Eurasia have been reported in America.

While individual bones of *Equus* (horse), *Camelops* (camel), *Tapirus* (tapir) or those of other extinct genera may appear in a mammoth kill site, they are not found in a context that clearly demonstrates human predation (Haynes, personal communication, 1972). Despite claims dating back at least to the time of Koch in 1839, no one in the east has succeeded in linking stratigraphically the bones of any extinct genus, even the widespread mastodon, with stone tools of the early hunters. It is not surprising that many authors have ignored and discounted the possibility of human predation as the cause of extinction.

However, removal rates have not been estimated for a predator–prey relationship leading to overkill. Sufficiently rapid rates of killing could terminate a prey population before appreciable evidence could be buried. Poor palaeontological visibility would be inevitable. In these terms the scarcity of known kill sites on a landmass which suffered severe megafaunal losses ceases to be paradoxical and becomes a predictable consequence of the special circumstances which distinguish *invasion* from cultural development within a continent. Perhaps what is remarkable in America is not that so few, but that *any* kill sites of extinct mammals have been found.

Those objecting to the indictment of overkill counterpose that the megafauna was suffering initial losses before the big game hunters arrived. They also claim that the fauna lingered into the postglacial long beyond the time of the Clovis hunters (see fig. 11B.1). Postglacial dates that appear to support this thesis continue to be reported (see nos. 4, 27, 29, 37–43, and 45 in table 11B.1), but on critical inspection many appear discordant with stratigraphic or palynological evidence. Some may be based on uncertain geochemical techniques. Their uncritical incorporation into an extinction chronology is unwarranted (Martin, 1967, pp. 87–8). The sceptic must hold firm in the face of the inevitable proliferation of more dates of similar quality to demand their verification or replication. *No alleged postglacial generic extinction has been documented adequately.* The only extinct large mammals clearly placed in the last 10,000 years by thoroughly convincing radiocarbon, stratigraphic, palynological and archaeological evidence are fossil species of *Bison* (see especially Shay, 1971, and Dibble and Lorrain, 1968).

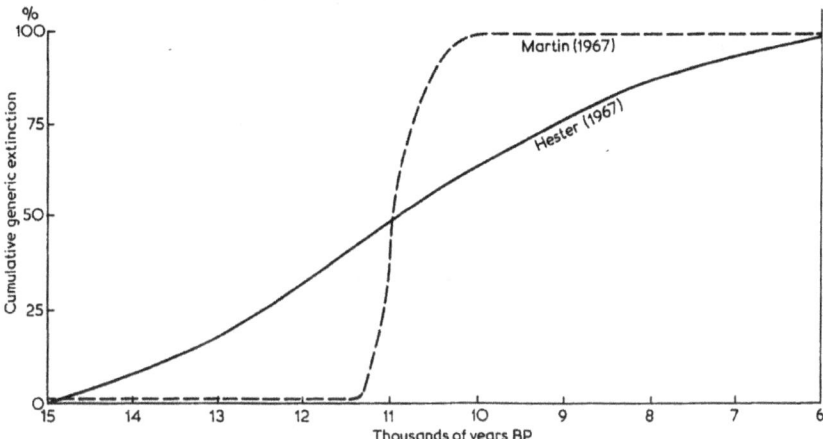

Fig. 11B.1. Generic extinction models for North America.

There remains the question of whether the American megafauna was declining in numbers before 12,000 BP, as has been claimed in the Old World.

SPEEDY, OR SLOW, EXTINCTION?

A comparison of Late Pleistocene and Late Cretaceous events is illuminating. The process of dinosaur extinction at the end of the Cretaceous was hardly instantaneous (Colbert, 1961, p. 258). There was a progressive reduction in numbers of species over an interval of 12 million years, with evidence of transgressive loss at the very end. The last dinosaurs in Montana became extinct half a million years before the last ones in New Mexico (Sloan, 1970). Major changes among small mammals and in land plants are other features which distinguish the ecology of the Late Cretaceous extinction from that of the Late Pleistocene (Sloan, 1970).

In the case of large mammals, Kurtén (1965a) has concluded that: '. . . many of the "sudden" extinctions at the end of the Ice Age are in fact the result of population declines dating back well into the last glaciation or possibly even the last interglacial'. Criteria that might suggest a gradual eclipse include a reduction in range, in abundance, and

Table 11B.1. North American radiocarbon dates associated with extinct genera.

No.	Site	Extinct genus	Sample no.	Years BP	RC* ref.	Comments
1.	Big Bone Lick, Ky.	Bootherium bombifrons	W-1617	17,200 ± 600	9, 507	Plant-bearing silt from cranial cavity.
2.	Fairbanks, Alas.	Bootherium nivicoleus	SI-292	22,540 ± 900	9, 380	Horn sheath. Other samples from similar sites have yielded dates from 12,000 to 28,000 years BP.
3.	Murray Springs, Cochise Co., Ariz.	Camelops, Equus, Mammoth	A-805A A-805B	11,150 ± 450 11,300 ± 500 Ave. 11,230 ± 340	9, 11 9, 11 9, 11	Charcoal (A-805A) and humates (A-805B) from top of unit containing artifacts associated with bones. Bison also identified.
4.	La Mirada, Calif.	Camelops hesternus, Equus, Mammut americanus, etc.	UCLA-1325	8550 ± 100	11, 204	Fossil wood. Extinct Pleistocene mammals have been found immediately above and below this wood-bearing stratum.
5.	Fitchburg Park, Ingham Co., Mich.	Castoroides?	M-1807	10,890 ± 350	10, 64	Wood which may have been gnawed by Castoroides; no bones found.
6.	Middletown, NY	Cervalces scotti	I-4016	10,950 ± 150	12, 94	Rib bone. First New York State record.
7.	El Toro, Calif.	Equus, Paramylodon	UCLA-1324	42,500 ± 4400	11, 204	Fossil wood. Fauna generally similar to that from Rancho La Brea.
8.	Lost Chicken Creek, Alas.	Equus	SI-355	26,760 ± 300	11, 177	Bone from frozen muck deposits. Two species of extinct horse and Panthera atrox occur in vicinity.
9.	Cave 9, Trail Creek, Alas.	Equus	K-1210	15,750 ± 350	10, 318	Organic fraction of scapula of horse which was associated with the heel bone of Bison sp. apparently worked by man.
10.	Cochrane Terrace, Cochrane, Alta.	Equus conversidens	GSC-613 GSC-612	11,370 ± 170 10,760 ± 160	9, 170 9, 169	Bone from middle of three postglacial terraces of Bow River. Bison bison, Ovis canadensis, and Cervus canadensis also identified.

No. / Location	Taxon	Lab no.	Date	Ref.	Comments
11. Cochrane Terrace, Alta.	*Equus conversidens*	GSC-989 GSC-988	11,100 ± 160 5670 ± 150	**12**, 68 **12**, 68	GSC-989 agrees closely with GSC-613 from same pit. GSC-988 is much younger than GSC-612.
12. Lindoe Bluff, Medicine Hat, Alta.	*Equus* *Equus*	GSC-805	11,200 ± 200	**10**, 220	Bone date on both horse and bison. See also *Camelops*, Murray Springs, Ariz.; *Camelops*, La Mirada, Calif.
13. Ester Creek, Alas.	*Felis atrox*	SI-456	22,680 ± 300	**12**, 203	Tendon and organic fraction of left tibia from frozen muck.
14. Seagoville, Tex.	*Geochelone*	W-1719	22,130 ± 600	**9**, 511	Oak and cockleburr from thin lens of organic muck in cross-bedded sand, Trinity River. Minimum terminal date for *Geochelone* in Texas.
15. Grundel Mastodon, Holt Co., Mo.	Mastodon	I-1559	25,100 ± 200	**11**, 77	Charcoal associated with mastodon remains.
16. Boney Spring, Mo.	*Mammut*	I-3922	16,580 ± 220	**12**, 90	Wood (*Picea* sp.) associated with bone bed.
		M-2211	13,700 ± 600		Organic material in tusk; see Mehringer *et al.* (1970, 175).
17. Akron, Ohio	Mastodon	OWU-190	15,315 ± 625	**11**, 137	Spruce wood from beneath partial skeleton of mastodon. Another piece of this log dated 13,300 ± 600 (M-1971). Pollen is sparse and is principally spruce, pine and sedge.
18. Demopolis Dam, Ala.	Mastodon	W-1571	14,650 ± 500	**9**, 507	Log from mastodon-bearing clay.
19. Hallsville, Ross Co., Ohio	Mastodon	OWU-220 OWU-260C OWU-260B OWU-260A	13,180 ± 520 13,695 ± 520 12,685 ± 244 12,835 ± 275	**9**, 321 **11**, 139 **11**, 139 **11**, 139	Spruce wood from extensive marl deposit which has yielded remains of several mastodons. Pollen: 64% spruce, 16% pine.
20. Leap Peat Bog, Monroe, Co., Pa.	Mastodon	I-3929 I-3930	12,160 ± 180 12,020 ± 180	**12**, 94 **12**, 94	Wood associated with Marshalls Creek, mastodon.
21. Thamesville, Ont.	Mastodon	GSC-611	11,380 ± 170	**10**, 216	Bone collagen.

Table 11B.1.—*cont.*

No. Site	Extinct genus	Sample no.	Years BP	RC* ref.	Comments
22. Thaller Mastodon, Gratiot Co., Mich.	Mastodon	M-1743	11,200 ± 400	10, 63	Gyttja from depth of 130 to 140 cm below surface. This date appears too early.
		M-1739	9910 ± 350	10, 62	Bone. Sediments of peat and clay belong to Late Glacial spruce so date is slightly younger than pollen evidence indicates.
		M-1742	7390 ± 280	10, 62	Peat from 80 to 90 cm.
		M-1741	7220 ± 250	10, 62	Wood from depth of 60 cm below surface.
		M-1740	7120 ± 250	10, 62	Wood from depth of 30 cm below surface.
23. Rappuhn Mastodon, Series II, Lapeer Co., Mich.	Mastodon	M-1780	10,750 ± 400	10, 63	Wood stick ⎫ Dates of wood sticks are considered more reliable than those of bone and ivory.
		M-1781	10,750 ± 400	10, 63	Wood stick
		M-1782	10,400 ± 400	10, 64	Wood stick
		M-1778	9900 ± 400	10, 63	Mandibular fragments ⎭
24. Rappuhn Mastodon, Series I, Lapeer Co., Mich.	Mastodon	M-1783	9250 ± 350	10, 64	Left tusk
		M-1746	10,730 ± 400	10, 63	Peat 20–22 in. below surface.
		M-1745	10,450 ± 400	10, 63	Peat 15–16.5 in. below surface.
		M-1744	9640 ± 350	10, 63	Peat 12–13.5 in. below surface. Bones were cut, partially charred, and scattered, interpreted to be results of butchering.
25. Johnstown Masdodon, Ohio	Mastodon	OWU-141	10,190 ± 160	9, 317	Spruce wood, association with mastodon not clear. Pollen spectrum showed predominantly spruce, pine, sedge, and grass.
26. Cole Mastodon, Ohio	Mastodon	OWU-194	9460 ± 305	9, 321	Spruce wood, not clearly associated with death of mastodon. Date may be regarded as minimum since pollen analysis indicates greater age.

	Species	Lab no.	Date	Ref.	Notes
27. Urbana Mastodon, Ill.	*Mammut*	ISGS-17C	9190 ± 200	12, 506	Ivory first dissolved in dilute CH_3COOH; insoluble fraction acidified with H_3PO_4; resulting CO_2 was dated.
		ISGS-17B	8330 ± 200	12, 506	Bone washed with 0.1 N NaOH; then dissolved in 2 N HCl. Insoluble fraction combusted and dated.
		ISGS-17A	7490 ± 200	12, 506	Bone dissolved in 2 N HCl; insoluble fraction combusted and dated.
28. Tupperville, Ont. (Ferguson Farm)	Mastodon / *Mammut*	GSC-614	8910 ± 150	10, 216	Bone collagen. Gyttja from cavities within mastodon skull is 6230 ± 240 (S-16, *Radiocarbon*, 2, 73). See also, *Camelops*, La Mirada, Calif.
29. Whitaker Mammoth, Alliance, Ohio	*Mammut* sp.	OWU-224B / OWU-224A	5560 ± 245 / 5490 ± 235	9, 321 / 9, 321	Peaty gyttja. There is no evidence that animal punched through the peat, and therefore, date clearly postdates burial of the animal. Author also calls it a 'mammoth'.
30. Hurley Mammoth, Cochise Co., Ariz.	*Mammuthus columbi* (?)	A-988	21,210 ± 770	13, 5	CO_2 from bone apatite from *M. columbi* (?). Bones occurred in mudstone of Unit D dated 29,000 ± 2000 BP (A-396A) at Murray Springs. Contamination by exchanged CO_2 is likely.
31. Fairbanks Creek, Alas.	*Mammuthus primigenius*	SI-453	15,380 ± 300	12, 203	Flesh from mummified *M. primigenius*.
32. Lamb Spring, Littleton, Colo.	Mammoth	M-1464	13,140 ± 1000	10, 102	Mammoth bone, associated with worked camel toe bone.
33. Malloy Farm, Lockport, NY	Mammoth	I-838	12,100 ± 400	10, 262	Wood (spruce).
34. Kyle, Sask.	Mammoth	S-246	12,000 ± 200	10, 369	Bone. (See also no. 40.)
35. Bartow Mammoth, Okla.	Mammoth	A-582	11,990 ± 170	13, 3	Rib of mammoth, acid-soluble organic matter.

Table 11B.1.—cont.

No. Site	Extinct genus	Sample no.	Years BP	RC* ref.	Comments
36. Sun River Canyon, Lewis and Clark Co., Mont.	Mammuthus columbi	W-1753	11,500 ± 300	11, 214	Clay enclosing mammoth bone.
	Mammuthus				See also Camelops, Murray Springs, Cochise Co., Ariz.
37. Lehner Mammoth, Cochise Co., Ariz.	Mammoth	A-874C	9980 ± 220	13, 4	CO_2 from carbonyl apatite of bone. The oldest date yet obtained from Lehner mammoth bone, but still younger than associated charcoal.
		A-806D	7930 ± 490	13, 4	Mammoth bone, solution from acid treatment was made basic with NaOH. Organic matter coprecipitated with hydroxides dried and pyrolized to yield CO_2. Sample obviously contaminated.
		A-876C	7780 ± 150	13, 4	Enamel from mammoth tooth treated in same manner as A-874C.
		A-806A:3	5610 ± 350	13, 4	Mammoth bone, greyish-brown residue after gentle treatment in 1 N HCl under vacuum followed by 0.5% NaOH at room temperature. Collagen, if present, is degraded and contaminated by younger organic residue.
		A-806C	1190 ± 90	13, 4	Initial yield of CO_2 from acid treatment of powdered bone.
General comment: These various fractions of carbon from mammoth bone were analysed for comparison to charcoal reliably dated at 11,260 ± 360 (*Radiocarbon*, **8**, 12).

No. & Site	Taxon	Lab No.	Date	Ref.	Remarks
38. Naco Site no. 1, Ariz.	Mammoth	A-195	8980 ± 270	9, 7	Vertebra pyrolized to recover bone organic matter. No collagen present. Contamination by younger humic acids is suspected.
39. Stein Ranch Mammoth, Mont.	Mammoth	A-584	8890 ± 300	13, 3	Mammoth bone, acid-soluble organic matter.
40. Kyle Mammoth, Sask.	Mammoth	A-619	8650 ± 400	13, 3	Vertebra of mammoth, acid-soluble organic matter. Date is significantly younger than that obtained by Can. Geol. Surv. (unpub.) which is not surprising considering that this fraction commonly gives erroneous results. (See also no. 34.) Claimed as a late survivor.
41. Peace River, BC	'N. American elephant' (*Mammuthus?*)	I-2244	7670 ± 150	10, 263	Tusk.
42. Blackwater Draw, Clovis, NM	Mammoth	A-536	6370 ± 160	9, 7	Pyrolized mammoth rib. Date is inconsistent with archaeological, geological, and other radiocarbon data. Sample did not contain collagen.
43. Manhattan Mammoth, Mont.	Mammoth	A-587	6050 ± 750	13, 3	Jaw of mammoth, acid-soluble organic matter.
44. La Grande, Ore.	*Mylodon harlani*	SI-331	11,030 ± 800	11, 178	Skull fragments.
		UCLA-1223	2900 ± 80	9, 480	Distal end of atlatl shaft thought to be of sloth age.
45. Gypsum Cave, Nev.	*Nothrotherium*	UCLA-1069	2400 ± 60	9, 479	Charred greasewood sticks found under 18 in. of apparently undisturbed layer of sloth dung. It appears from above evidence that the two samples of wood thought to be of the same age as sloth dung from that site are not contemporaneous. Redeposition by wood rats is suspected.
	Paramylodon				See *Equus*, El Toro, Calif.; *Mylodon*, La Grande, Oreg.

Table 11B.1.—*cont.*

No. Site	Extinct genus	Sample no.	Years BP	RC* ref.	Comments
46. Rancho La Brea, Los Angeles, Calif.	*Smilodon californicus*	UCLA-1292D	28,000 ± 1400	**10**, 402– 4	All except 1292L are on femurs of *Smilodon*; the latter is on a tibia which Miller (1969) suspects is an artifact. Free amino acids were dated after liquid chromatography of collagen hydrolysate.
		UCLA-1292G	26,700 ± 900		
		UCLA-1292H	23,700 ± 600		
		UCLA-1292A	21,400 ± 560		
		UCLA-1292J	20,500 ± 900		
		UCLA-1292K	19,300 ± 395		
		UCLA-1292I	15,300 ± 200		
		UCLA-1292L	15,200 ± 800		
		UCLA-1292F	14,950 ± 430		
		UCLA-1292C	14,500 ± 190		
		UCLA-1292E	14,200 ± 2100		
		UCLA-1292B	12,650 ± 160		
47. Little Eldorado Creek, Alas.	*Symbos giganteus*	SI-291	>40,000	**11**, 177	Dung from frozen muck near Fairbanks.
48. Saltpeter Cave, Cumberland Co., Tenn.	*Tremarctos floridanus*	SI-459	33,660 ± 3980	**12**, 202	Rib fragments. Northernmost recorded find of extinct Florida spectacled bear.
49. Canyon Creek, Alas.	'Extinct mammals'	SI-356	3750 ± 380	**11**, 177	Wood from same horizon as bones of extinct mammals. All wood dates from Alaskan muck appear to be intrusive (Martin, 1967, p. 88).

*RC = *Radiocarbon*, a publication of the *American Journal of Science*. Dates listed are from vols. **9–13**. For earlier dates and a discussion of each extinct genus, see Martin (1967).

in mean body size. Such a record could suggest a deteriorating environment. Any progressive decline in herbivore populations or productivity should especially be sought in early extinction of large carnivores and scavengers before extinction of their prey.

From caves of the Levant, Kurtén (1965b) identified 356 individual fossil carnivores associated with a 40,000-year cultural succession from the Acheulian to the Natufian. He found a reduction of 20 to 25 per cent in size of large carnivores. Species affected included brown bears (*Ursus arctos*), leopards (*Felis pardus*), and spotted hyenas (*Crocuta crocuta*). Between the Early and Late Würm, these large carnivores decreased in numbers relative to the smaller carnivores. In the Late Würm leopard and spotted hyena were eliminated from most of Europe and the Near East.

In contrast, the distinctive large Pleistocene felid of the New World, the sabre tooth (*Smilodon*), increased rather than decreased in mean body size during the Middle and Late Pleistocene (Kurtén, 1965a). While Kurtén proposed a decrease in sabre-tooth populations of Florida through this interval, his sample size is deficient. It is based on 6 individual *Smilodon* determinations out of 30 Florida felids assigned to early (pre-Wisconsin) faunas and 7 out of 49 assigned to the Wisconsin. No shrinkage in *Smilodon* range prior to 12,000 BP has been detected.

The best known populations of *Smilodon* come from Rancho La Brea. Seven of the tar pits yielded 897 individuals (Marcus, 1960). More than 2400 individual sabre tooths were examined in Miller's study (1968). Their age may be inferred from the radiocarbon dates on the amino acid fractions of 11 sabre tooths between depths of 7 (2.1 m) and 26 feet (7.9 m) in 4 pits. These range from 28,000 to 13,000 BP (Berger and Libby, 1968; Ho *et al.*, 1969), the majority of the determinations being younger than the midpoint, 20,000 BP (see table 11B.1). Evidently, sabre-tooth populations were not visibly depleted thousands of years before the loss of the herbivores.

While the Rancho La Brea environment trapped carnivores selectively, the muck deposits of similar age found near Fairbanks, Alaska (Guthrie, 1968), and in the Yukon (Harington, 1970) did not. Large numbers of bones have been recovered incidental to placer gold mining. Near Fairbanks 4 separate vertebrate faunas of Wisconsin glacial and possibly earlier age contained a minimum of 5965 individual large mammals (Guthrie, 1968). The deposit is believed to represent an especially productive grassy tundra sustained by active deposition of loess deflated from the flood plains of rivers draining icefields. The ratio of carnivores to ungulates, approximately 1:100, is comparable to that found in the Serengeti and other African game parks. Guthrie (1968) concluded that the large herbivore biomass must have been appreciable to support the 4 species of large carnivores (wolf, extinct lion, sabre tooth, and brown bear) represented in the Alaskan fauna.

Had the American ecosystem been under stress before early hunters arrived, one might expect carnivore extinctions to begin sometime before 12,000 BP. At first glimpse, the 13,890 (Y-354b) date Hester (1967, p. 183) and Butzer (1971, p. 506) list for 6 carnivores (*Canis furlongi*, *C. milleri*, *Felis bituminosa*, *F. daggetti*, *F. atrox* and *Smilodon californicus*) from Rancho La Brea suggests such a response. But the first 4 species are poorly known, possibly invalid, and are not dated by radiocarbon elsewhere. The date is on wood not necessarily in direct association with any of the 6 carnivores. From the record at Jaguar Cave, Idaho (Martin, 1967, p. 95), *Felis atrox* presumably survived at least to about 11,580 BP. The youngest date on *Smilodon* is listed above (see also table 11B.1). There is no firm stratigraphic evidence that any of the American carnivores declined in numbers before their prey, much less that they preceded the herbivores to extinction.

Turning to the record of the large herbivores, one also finds that during the last mammalian age, the Rancholabrean, there was no apparent loss of species, or gradual shrinkage of range. On the contrary, during most of the last 200,000 years the mega-fauna (species equalling or exceeding 45 kg in adult body weight) was increasing in diversity, at least in the northern latitudes. The steady Eurasian emigration over the Bering Platform which occurred throughout the Pleistocene intensified. *Sangamona* (extinct deer), *Cervalces* (extinct moose), *Alces* (moose), *Rangifer* (caribou), *Bison* (bison), *Bos* (yak), *Saiga* (Asiatic antelope), *Oreamnos* (mountain goat), *Ovis* (bighorn sheep), *Symbos* (woodland musk-ox), and *Ovibos* (musk-ox) were among the Rancholabrean aliens that preceded the arrival of *Homo* himself (Repenning, 1967). By 15,000 BP North America was as rich in large mammals as Africa today, both continents support-ing a megafauna of some 40 genera.

Butzer (1971) proposed that 11 large species were extinct before 12,000 BP, including the carnivores mentioned above. A major question of negative evidence arises. Despite two decades in which the method has been widely used, many species of the extinct fauna are still associated with only one or two radiocarbon dates. By chance alone, one would expect a portion of these to be older than 12,000 BP. To attribute much weight to such records is highly misleading. Of the 11 species listed by Butzer as extinct before 12,000 BP only *Bison latifrons* may be well enough known to place some confidence in its Late Glacial *absence*.

While a gradual megafaunal decline lasting 10,000 years (Hester, 1967) is not impossible, the evidence for any significant faunal losses in the Americas before 12,000 BP is weak at best. Kurtén's example of long-term population shrinkage of the Old World megafauna throughout the last glaciation stands in sharp contrast with fossil evidence of a flourishing fauna in the New. One must second the opinions of Bordes (1968, p. 218) and Haynes (1971, p. 90) that the Stone Age hunters found abundant game when they found America. Explosive extinction soon followed (fig. 11B.1).

3. Estimating Rancholabrean biomass

The size of the Rancholabrean game herd is not known. While the fossil megafaunal populations reveal no unmistakable symptoms of stress or reduction prior to 12,000 BP, I do not propose that the fossils themselves can be used as a quantitative estimate of population size, secondary productivity or of biomass. On the basis of a painstaking faunal analysis one may conclude that mammoths constituted 10 per cent of a fossil fauna or (perhaps after appropriate weight conversion and other assumptions) 50 per cent of its biomass, but one still lacks an unbiased estimate of the total biomass itself.

Determining rate of deposition of large fossils seems equally unsatisfactory. While radiocarbon dates can suggest sedimentation rates and thereby absolute fallout values for microfossils such as pollen, a similar approach to any hypothetical bone-fallout rate founders on formidable obstacles, notably the non-random distribution of bones within a deposit and their very size, which may encompass 1000 years of lake bottom sedimentation.

Abandoning palaeontology one may attempt two crude but relatively independent methods of estimating past biomass from present range carrying capacity. The first is

Table 11B.2. Large mammal biomass in some African parks and game reserves (after Bourliere and Hadley, 1970, p. 138).

Location	Habitat	No. of species	Biomass (T/km^2)	Biomass (au/mi^2)
Tanangire Game Reserve, Tanzania	Open *Acacia* savanna	14*	1.1	6
Kafue N. Park, Zambia	Tree savanna	19*	1.3	7
Kagera N. Park, E. Rwanda	*Acacia* savanna	12	3.3	26
Nairobi N. Park, Kenya	Open savanna	17*	5.7	32
Serengeti N. Park, Tanzania	Open and *Acacia* savannas	15*	6.3	36
Queen Elizabeth N. Park, W. Uganda	Open savanna and thickets	11	12.0	68
Queen Elizabeth N. Park, W. Uganda	Same habitat, overgrazed	11	27.8–31.5	158–179
Albert N. Park, N. Kivu	Open savanna and thickets, overgrazed	11	23.6–24.8	134–141

* Species actually present are greater than the number given. The densities and standing crops of those not included are very low. *au* = animal unit.

by analogy with animal populations in modern game parks of Africa and elsewhere. The other is by assuming that present livestock plus game populations in the Americas would at least equal the maximum Late Pleistocene carrying capacity.

Estimates of megafauna biomass in various African parks are shown in table 11B.2. The drier parks such as Kruger, Tsavo, and the Henderson Ranch support 10 to 20 animal units per section (1.8 to 3.5 T per km²). In the Americas during the Pleistocene similar values might be expected on drier ranges (400 to 600 mm mean annual precipitation), dominated by mammoth, horse and camel. Carrying capacity would have been much less in the driest regions (under 200 mm annual precipitation).

African game parks supporting the highest biomass, over 100 au per section (over 18 T per km²) occur in savannas along the margin of tropical forest and are dominated by elephant, buffalo and hippo. In the tropical American savannas, one might expect a similar biomass, dominated by mastodons and large edentates. In temperate North America carrying capacity would have been greatest in the warmer, wetter savannas between forest and grassland, along river flood plains, on marshy prairies of the coastal plain and possibly also on the fresh mineral soils forming in the wake of the retreating ice sheet.

Humid tropical and temperate forests mantled by a closed canopy with few natural openings would, like similar habitats today, support very few large herbivorous mammals. If much of eastern North America was extensively forested with conifers and hardwoods, as seems probable 12,000 years ago, the regions of high biomass within the forest would have been limited to the flood plains of the larger rivers and estuaries, the glades of the limestone districts, the coastal salt marshes and similar accessible and highly productive habitats. Williams and Stoltman's map (1965, p. 677) of the distribution of mastodons and fluted points in southeastern United States suggests the Late Glacial distribution of the more productive 'game parks'.

For the 3 million sections in unglaciated North America north of Mexico I propose the following stocking capacities, each of roughly 1 million sections. Savannas, forest and other high to moderately productive habitats: 50 au per section, or 22.7×10^6 metric tons; arctic, boreal, semi-arid temperate ranges, and other low to moderately productive habitats: 10 au per section, or 4.6×10^6 metric tons; closed canopy forest, extreme desert and other habitats unproductive to herbivores: 2 au per section, or 0.9×10^6 metric tons. The total for North America north of Mexico is 62 million animal units of 27.2 million metric tons to be divided among some 100 species of large mammals up to 12,000 years ago.

This is about half the livestock plus game biomass occupying the same area at present. Letting 1 cow, 1 horse or 1 steer = 5 sheep = 5 deer = 4 hogs = 1 animal unit of 1000 lb (450 kg), the twentieth-century figures on biomass range from 120 million (1900) to 148 million (1945). Game populations should add an additional 3 million animal units at most, making an average of 50 per section, or 68×10^6 metric tons. A decline in draft animal numbers since the turn of the century has been more than offset by increases in cattle (World Almanac, 1968, p. 605).

Present cattle populations of America outside the United States are especially high in Brazil (76 million), Argentina (43 million), and Mexico (35 million). They suggest that one may project the 50 au per section average for the United States to the rest of the continent (outside of Canada). Accordingly, the figure for both domestic and wild species in the Americas in this century is 600 million animal units.

Because of major uncertainties about the extent of closed canopy forest which would have been unsuitable for any large herbivores 12,000 years ago, and because of the vast gain in biomass made possible by techniques of livestock management, it seems likely that 600 million animal units must exceed the number stocked naturally in Rancholabrean time. Native animals were better adapted as herbivores than domesticated breeds, an advantage offset by modern methods of pasture management. I shall use the maximum of 600 million in the following model, but 300 million would be perhaps closer to African game park values and a lower limit of 100 million animal units is not impossible. Even 100 million animal units of unsuspecting prey would have constituted an extraordinary resource to the earliest human invaders.

4. Population explosion, 12,000 years ago

Rapid expansion of the population that discovered America could have been prevented by excessive mortality caused by intense predation, by serious parasitism, by disease, or by cultural controls. Spacing of births may be expected in a nomadic population, but extremely slow population growth may be discounted in a favourable new environment.

Predation can be discounted. Lacking any biped in their previous experience, potential predators such as the dire wolf, extinct lion (*Felis atrox*) and sabre tooth would not have been pre-adapted for killing the new competitor. African lions are not serious hominid predators (Quigley, 1971) and the American wolves are not known to have attacked and killed humans, even when the individual was unarmed and defenceless (Mech, 1970, p. 293).

Disease can be discounted. Microbiologists generally regard the Palaeolithic as a healthy episode (Hare, 1967). The reasons are based less on detailed knowledge of skeletal pathologies or the scarcity of major parasites in prehistoric faeces (Fry and Moore, 1967; Heizer and Napton, 1969), than on biological inference.

(1) Lacking closely related hosts, the New World held no major reservoir of hominid diseases. Cholera and African sleeping sickness never became established. The most serious diseases known at present were introduced in historic time, especially smallpox, tuberculosis (Morse, 1961), plague and probably most if not all malaria (Jarcho, 1964). Even yaws and syphilis may not be endemic (Hackett, 1967).

(2) The subarctic filter served as a natural quarantine against importation of flies, fleas, ticks and other human disease vectors.

(3) The Palaeolithic population was presumably below the number necessary to sustain deadly viruses.

(4) When contact with western Europe and Africa was established, the American Indian suffered catastrophic population losses, presumably through shock of initial exposure to Old World diseases (Dobyns, 1966). Before the fifteenth century, life in tropical America must have been reasonably free of disease, as it is now for 'primitive' Indian infants who survive intimate environmental contact in remote villages (Neel, 1970).

The minimum growth rate required to attain the estimated contact (AD 1500) population of the New World in 15,000 years is 1.4 per cent *per 20-year generation*. But slow, imperceptible growth of the sort demographers commonly project into the Palaeolithic

(Lorimer, 1953; Desmond, 1961; Deevey, 1960) would have been impossible in a continent free of endemic hominid diseases, free of experienced hominid predators, and well stocked with unwary game. When they ultimately reached the American heartland the hunters that conquered the Siberian Arctic, crossed Beringia, and threaded the ice-free corridor east of the Canadian Rockies must have multiplied exponentially. Their rate of increase would have been close to the maximum for *Homo sapiens*, a rate which Birdsell (1957) has placed at 3.4 per cent annually, or a doubling in every 20-year generation. Even much lower rates, as the 1.4 per cent per year which Deevey (1960) regarded as maximal for prehistoric people, means a doubling every 50 years. If continued unchecked, a population of millions will be generated from an invading band of 100 hunters within 1000 years.

To avoid the Malthusian difficulty of standing room only, an upper limit for the Palaeolithic population must be considered. Historic and modern hunting–gathering societies in both Australia and Africa attain population densities of from 0.01 to 3.0 people per square mile (0.004 to 1.0 per km²: Lee, 1968, p. 35). This range of values should not be accepted uncritically. The extinction of Pleistocene megafauna in Africa, and especially in Australia, meant reduced game for later hunters. Furthermore, modern hunters do not have a naïve (pre-contact) fauna to prey upon. Finally, most modern 'Stone Age' tribes of the warmer regions are now predominantly herbivorous, familiar with plant phenology and skilled at gathering, roasting, grinding and storing the native plants of their region. The American invaders would not have succeeded as gatherers until they had gained similar knowledge of the unfamiliar species of plants encountered on their southward hegira.

It is not certain how long this would have taken, but for an expanding population, the techniques learned in one region would not serve in the next. Of 379 species of the more common vascular plants found in one part of southern Arizona, only 2 per cent occur in Canada and 17 per cent in central America (from Whittaker and Niering, 1964). Methods of utilizing mesquite beans, agave shoots and saguaro fruits, all important in the economy of historic southwestern Indian tribes, could not be anticipated by knowledge of the plants of the Great Plains or the Colorado Plateau. While the skills and techniques used for preying on megafauna were universal, the skills and techniques needed for harvesting plants were essentially regional. The ecologist will not be surprised to learn that cultural remains postdating the time of extinction, when plant resources were more important, feature distinctive regional traits.

I am arbitrarily adopting an average of 1 person per square mile (0.4 per km²) as a tentative upper limit to the initial population of hunters. As in the case of Pleistocene biomass, modern analogy rather than fossil evidence is the basis for the projection. Had such a density been attained simultaneously, it would represent a total population of 12 million in the 12 million square miles outside of Canada and other glaciated regions, not including parts of the continental shelf subsequently drowned by sea level rise. At Birdsell's 3.4 per cent annually, 340 years (17 generations) would be the minimum time for a band of 100 invaders to saturate the continent. Deevey's 1.4 per cent would require a minimum of 800 years. Presumably a population crash would soon follow the concurrent extinction of megafauna. But one need not propose that the New World population ever attained a total of 12 million at any one time.

Animal invaders expand along an advancing front (Elton, 1958). Populations of the giant African snail, *Achatina fulica*, attain highest values at the time of establishment, declining rapidly after initial introduction (Mead, 1961). Exotic mammals spreading through New Zealand attain peak population densities and maximum reproduction rates along an advancing front (Riney, 1964; Caughley, 1970).

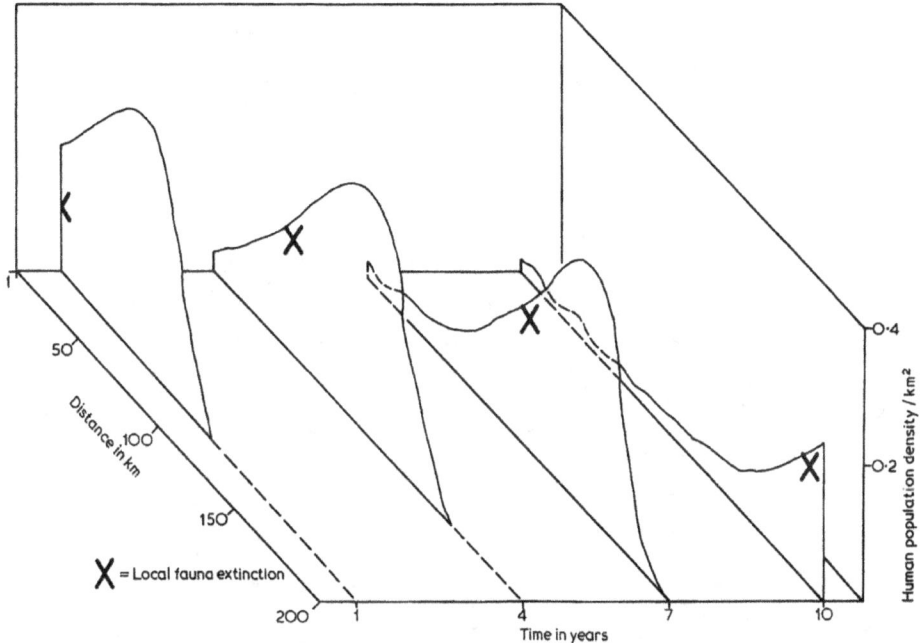

Fig. 11B.2. A model of the southward expansion of the front of high human population density.

I propose that the invasion of the Americas proceeded in a similar fashion (see fig. 11B.2). A population of high density was concentrated along a front whose sweep southward was determined partly by the abundance of big game within and beyond it. In a decade or less, the population of vulnerable large animals would have been greatly reduced or obliterated, the drastically reduced remaining human population would of necessity be developing skills at harvesting local plant and small animal resources, and the front itself would have advanced southward. At high latitudes, it is unlikely that any human populations endured behind the front.

An average southward expansion of the front of no more than 16 km (10 mi) per year would be sufficient to sweep the 20,000 km from the Bering Strait to Tierra del Fuego in 1250 years. At 25 km (15 mi) annually, only 800 years would be needed.

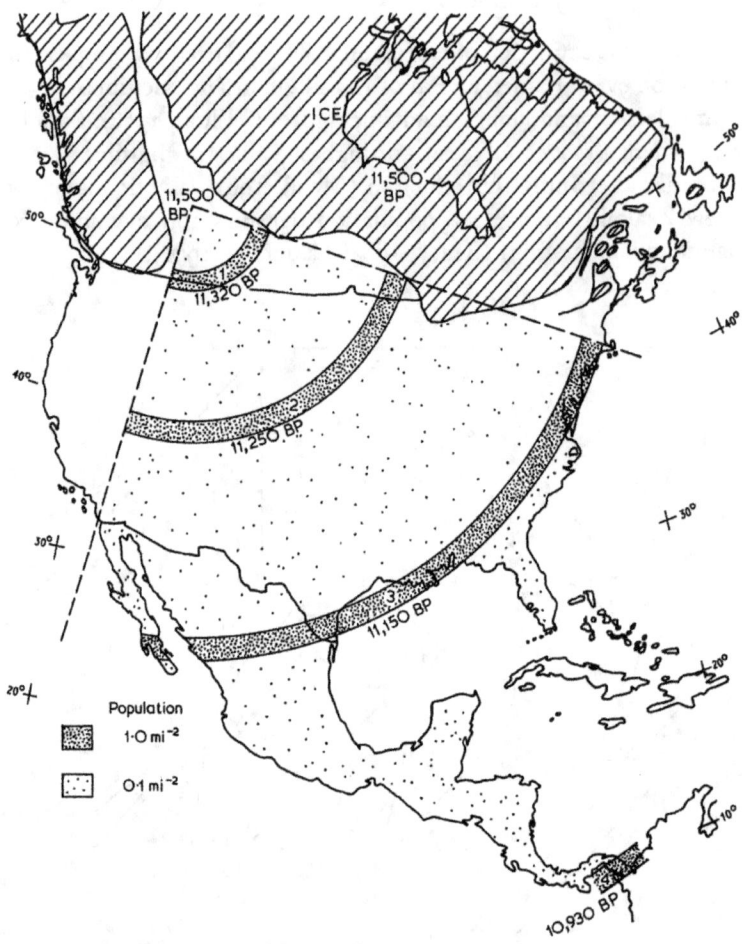

Fig. 11B.3. The advance of the population front from Edmonton to Patagonia in 1000 years assuming a rate of 10 mi yr^{-1} with short delays when the frontal population is less than 50,000. The extinction on the front is 20–100 animal units per mi^2 destroyed by hunters at a density of 1 per mi^2 over a 10-year period (density of 1 mi^{-2} = 0.386 km^{-2}).

	Population ($\times 10^3$) front	Behind front	Total population ($\times 10^3$)
Stage 1	45	6	51
2	150	75	225
3	300	300	600
4	10	490	500

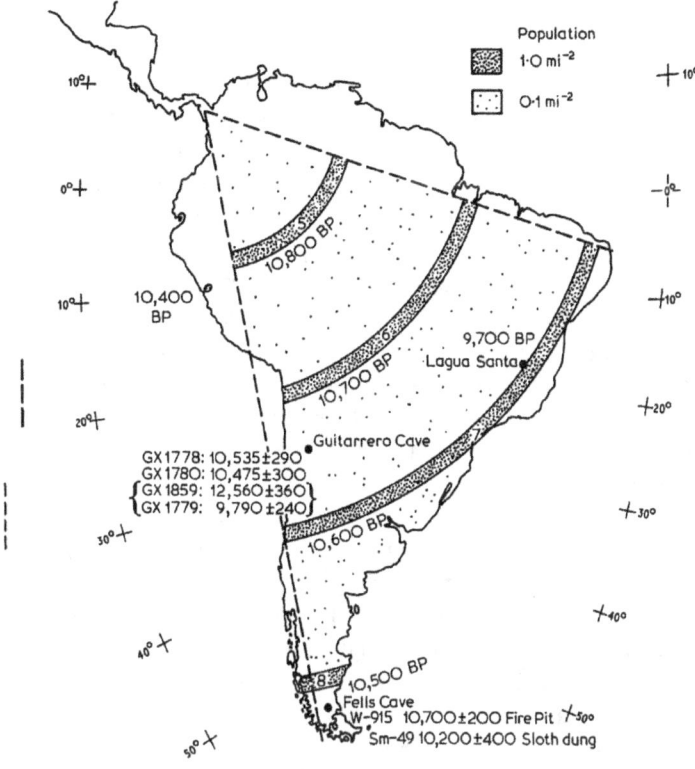

Population
1·0 mi⁻²
0·1 mi⁻²

10,800 BP

10,400 BP

9,700 BP
Lagua Santa

10,700 BP

GX 1778: 10,535±290
GX 1780: 10,475±300
GX 1859: 12,560±360
GX 1779: 9,790±240

Guitarrero Cave

10,600 BP

10,500 BP
Fells Cave
W-915 10,700±200 Fire Pit
Sm-49 10,200±400 Sloth dung

Fig. 11B.3.—*cont.*

At this pace, a maximum population growth rate of 3.4 per cent annually would possibly be briefly limiting only at two nodes. One would be the southern terminus of the Cordilleran ice corridor where up to 150 years might be required to develop a frontal population large enough to expand southward at the required density of 1 per mi² (fig. 11B.3). The other would be at the Isthmus of Panama. The expansion rates proposed are well within the distance covered by groups of Zulus known to have moved from Natal to Lake Victoria (3000 km) and halfway back in half a century (Clark, 1959, p. 168).

OVERKILL AT THE FRONT

Could a human population on the hypothetical front exterminate the megafauna? An average section may have held 50 animal units (22.7 T) divided among some 10 or 15 species ranging in weight from elephant-size (4000 kg) to pronghorn-size (40 kg). By analogy with African game parks, one may assume 2 or 3 of the species constituted at least 75 per cent of the biomass.

The average human population along the front, 1 person per section, had an energy requirement of about 1 million calories per year. Discarding bones, hide and offal, the edible portion of the carcass of a large mammal is roughly 50 per cent, containing 1870 cal kg^{-1}. The annual energy requirement of 1 person can be met by 2 animal units (0.9 T), providing the portion not immediately consumed is stripped, dried, and and successfully stored without spoilage. A population of 100 ranging over 100 sections could eliminate problems of preservation and meet their minimal requirement by killing slightly more than 1 animal unit every other day.

This predation rate, a harvest of roughly 5 per cent of the mean standing crop, may be compared with 18 per cent, the maximum available to wolves (or other predators) from steady-state moose production on Isle Royal (Jordan et al., 1970). Unless the animals killed were reproducing females and unless the herbivores were already supporting a heavy load of natural predation by native carnivores it seems unlikely that human predation at this level would be excessive.

On the other hand vulnerability would increase if the hunters did not prey on all species equally but by preference selecting first one, then another, concentrating on each in turn until its numbers were severely reduced and selectively seeking it even after other prey had become the main energy source.

Finally, vulnerability to extinction would greatly increase if slaughter were not highly efficient, avoiding waste. Such behavioural control seems very unlikely Wasteful slaughter may have been inevitable. In the Olsen–Chubbock site, a bison drive postdating the time of major extinction by 2000 years, 25 per cent of the animals killed were not completely butchered (Wheat, 1967). Mass killing far beyond conceivable immediate needs has been observed repeatedly in historic time (Prufer, 1966). The history of hunting is bloody (Dembeck, 1965; Reed, 1970; Taber, 1969).

Whatever the fuller meaning of these behaviours and their relationship to warfare, it is obvious that men enjoy hunting and killing (Washburn and Lancaster, 1968). A trophy hunter defends the practice by claiming that '. . . the compulsion to hunt is as basic a part of man's nature as the mating urge' (Aitken, 1969, p. 6). African hunting-gathering tribes would seem to share this sentiment. Despite the fact that hunting yielded less than half the number of calories than plant gathering in the same amount of time, Bushmen males, no less than White Hunters, find it a source of prestige and devote appreciable effort to it (Lee, 1968, p. 40). In the half million years in which man has been the enemy of even the largest mammals (Washburn and Lancaster, 1968, p. 299) one may seek the origin of hunting behaviours. Certainly there was sufficient time for selection to bring them under a generalized genetic control. Whether or not this is the case it is apparent that powerful behavioural re-enforcers associated with excitement of the chase and the killing passion have made modern man a super-predator, a species which kills for more than food alone.

I propose that during the invasion of America when the expanding front of big game hunters first overran the fauna, the rate of killing greatly exceeded minimal caloric requirements of the hunters. In a land section initially averaging 50 animal units, the removal of 5 annually (2 T) would probably exceed annual recruitment for certain species, and the removal of 15 (7 T) would certainly eliminate the megafauna within a decade. If the population averaged 1 person per square mile and only 1 person in 4 actually hunted, this individual would have to destroy no more than 60 bison-sized animals annually to ensure multiple extinctions. This does not seem beyond the capability of an aggressive young male presented with a rich fauna that was unaccustomed to human predation.

In brief, unless one insists on believing that the human population explosion of the last 500 years is unique in the history of our species, or that the Palaeolithic invaders lost enthusiasm for the hunt and became vegetarians by choice as they moved south from Beringia, there is no difficulty in predicting rapid extermination of the more vulnerable native American large mammals by overkill. I do not discount the possibility of disruptive side effects, perhaps caused by the introduction of dogs or the destruction of habitat by fire. But even a very large biomass of native large mammals, one averaging 50 animal units per section, or 6×10^8 units (2.7×10^8 T) for the entire continent, is not so vast a resource that it could not be swiftly reduced by an expanding hunting population of modest size. Basic to the model is the assumption that a relatively innocent prey was suddenly exposed to a new superpredator who preferred killing large animals to other outdoor activities. With the extinction of all but the least vulnerable species, such as cervids, a more normal predator–prey relationship would be established accompanied by major cultural changes. Only after extinction were the former hunters obliged to focus their attention on plant resources and to readapt to the distribution of biomass in America in the manner Fitting (1968) has proposed.

The 'swift overkill' model will account for the lack of extinct animals in many obvious kill sites. The big game hunters achieved high density during those few years when their prey was abundant, easily approached and slaughtered. Elaborate drives or traps were unnecessary. Except in the case of certain species of *Bison*, which were hunted over several thousand years, the major extinctions all occurred so soon after the fauna was overrun by hunters that there was little chance for preservation of either the extinct animals, the weapons used to kill them, or the bones of the hunters themselves.

The model is in accord with the relatively homogeneous tool types of the Palaeo-Indian (Mason, 1962). The hunters were wide ranging, subsisting on a resource requiring one basic type of tool kit. As plant resources became the mainstay of populations in the Archaic, regional adaptations assumed greater significance.

The model conforms with the absence of New World Palaeolithic art, either cave paintings, engraved bone or carved ivory. The big game economy came to an end before art forms could be developed. Finally, the model removes any objections that acceptable radiocarbon dates of 11,000 BP on artifacts from the southern tip of South America (cf. Bird, 1970) demand a crossing of the Bering Platform thousands of years earlier (see Bates, 1952, p. 261; Black, 1966; Irwin and Wormington, 1970). At an initial annual growth of 3.4 per cent, limited by a rate of frontal advance not to exceed 25 km annually, a high-density wave of hunters would sweep the continent from the Arctic Circle to Tierra del Fuego in less than 1000 years. At 1.4 per cent annual growth and 16 km per year in average rate of advance, it would take only slightly longer. The earliest record of prehistoric people in North America needs to precede by only a few hundred years the earliest record in South America.

As Birdsell (1957) found in the case of Australia, it appears that prehistorians have overlooked the potential for a population explosion in what ecologists must regard as a uniquely favourable environment, the New World when first discovered. From this perspective, it is possible to make a fresh approach to a major archaeological controversy, the question of 'pre-Palaeo-Indians' or 'Early-Early Man'.

5. The Early-Early Man hunt

Faunal extinction in North America coincided with the rapid continent-wide spread of the oldest known, well defined, cultural complex, the Llano. It includes Clovis spear points, scrapers, knives, occasionally burins and rarely bone tools. The earliest Clovis sites are not significantly older than 11,200 BP (Haynes, 1971). The search for an older occupation that might explain the 'chopper-scraper problem' has long been underway, with little success (Jennings, 1968, pp. 61–70). Nevertheless, new or restated claims of antiquity continue to appear. New World occupation is proposed at least by 15,000 BP (Bordes, 1968; Butzer, 1971; Cruxent, 1968; Lanning and Patterson, 1967) or even earlier (Bryan, 1969; Chard, 1969; Lynch, 1967; MacNeish, 1970, 1971; Müller-Beck, 1966; Orr, 1968; Raemsch, 1968; Stalker, 1969). The growing array of claimants is impressive. The unwary may accept the thesis of human invasion of the Americas by 15,000 years ago as firmly established, overlooking a variety of difficulties ranging from the absence of any 15,000-year or older Palaeolithic record on the Siberian side of the Bering Platform (Klein, 1971) to the question of how the putative invaders could maintain a low archaeological profile for thousands of years prior to 11,200 BP.

After their arctic passage could the discoverers of the New World have remained rare, not preying conspicuously on the megafauna and not reaching the level of archaeological visibility attained in the Old? Few anthropologists claim that the hypothetical pre-Clovis invaders were numerous. Irwin-Williams (1969) specifically states that they were scarce. Just how scarce can best be understood by considering the Stone Age record of the Old World.

A vast quantity of stone tools and cultural debris has been recovered from innumerable Old World sites older than 12,000 BP. Most were sufficiently obvious to be recognized as curiosities by farmers, ditch diggers, peasants and miners who called them to the attention of scientists. Some 17,000 artifactual stones of Palaeolithic origin representing the interval between 25,000 and 12,000 BP have been inventoried in the incompletely studied Kostenki-Borshevo region on the Don where peasants provided most of the clues to site location (Klein 1969). This is only one of a large number of similar Palaeolithic hunting communities in the Ukraine. Associated faunal remains are abundant. Pidoplichko (1969) tabulated 2442 bones from 324 individual mammoths used in construction of 11 dwellings. The famous Gravettian encampment at Predmosti in Moravia is stated to have contained bones of about 1000 mammoths (Augusta, 1963, p. 46).

One might conclude that the discrepancy reflects the advanced state of Old World archaeology which has benefited from a much longer search for antiquities within a smaller area. However, in Africa south of the Sahara where archaeologists are certainly far less numerous than in either Europe or America, some 23 radiocarbon samples, most on charcoal, representing 17 well-defined sites, have been dated at between 12,000 and 30,000 years (table 11B.3). If America were occupied by *Homo sapiens* at the same time, why should so little radiocarbon dated cultural material be evident to the far larger population of archaeologists searching this continent? Surely not because of lack of a food supply – the diversity of large extinct species and the abundance of Pleistocene fossils suggest that until 12,000 years ago the Americas supported a fauna as rich as that of Africa. More detailed comparisons can be made.

In the arid Jordan Valley, Mousterian sites of the Early Würm range from about 0.5 to 1.5 m in depth. They contain remains of a rather generalized fauna of gazelle, fallow deer, wild cattle and pig. To the west on the much wetter western slopes of the coastal mountains above the Mediterranean, caves contain much thicker Mousterian deposits, over 20 m of stone tools, bones and other debris in the case of Ksar'Akil and Tabun (Binford, 1970).

The bioclimatic analogue of the Levant is found at 27° to 36° N in southern California. The contemporaneous fauna would have included extinct bison, mastodon, horses, sloths and camels. Claims of occupation dating 12,000 to 30,000 BP and earlier have been reported from California in the past and very likely more will be forthcoming in the future. But again, palaeocologists encounter the obvious fact that nothing approaching a cultural deposit of 20 m or even a half metre in thickness, comparable to those of the Levant, has been found by the most partisan enthusiast for ancient New World occupation. Despite new excavations conducted by the Los Angeles County Museum under the direction of George Miller, and evidence of man-made cuts on *Smilodon* bone (Miller, 1969), no indisputable pre-12,000 year occupation has yet been found in one obvious place to expect it, the famous tar pits of Rancho La Brea.

Appalachian caves have proved equally obdurate. Guilday (1972) has summarized a decade's active work in search of bone-bearing caves. From Pennsylvania to Georgia he and his colleagues have discovered many interesting southward displacements of Pleistocene mammals, have documented a Late Glacial Bergman effect, and have established new records of extinct peccary and sabre tooth. But in contrast to those of Europe, none of the 35 Appalachian caves studied have yielded human remains in association with the extinct fauna. In North American caves and rockshelters containing rich cultural records such as Frightful Cave in Coahuila, Modoc in Illinois, Rodgers Shelter in Missouri, and Ventana in Arizona, obvious evidence of occupation disappears prior to 11,000 BP. What has been the fate of those early American sites when subjected to critical review?

ON VALIDATING CLAIMS

The nature of their research has made behavioural scientists increasingly aware of interpreter and experimenter effects (Rosenthal, 1968). As Pearson (1902) recognized,

Table 11B.3. Radiocarbon dated sites in Africa south of the Sahara from *Radiocarbon*. Interval of 12,000–30,000 years BP.

Site	Lab. no.	Date	Comments
1. Bushman Rock Series, Transvaal	GrN-4885	47,500	Charcoal and wood. Middle Stone Age, Later Stone Age, Iron Age.
	GrN-5116	53,000	
	GrN-4816	$12,510 \pm 105$	
	GrN-5873	$12,470 \pm 145$	
	GrN-4815	$12,160 \pm 95$	
	GrN-4814	$12,090 \pm 95$	
2. Tshangula Cave, Southern Rhodesia	UCLA-629	$12,200 \pm 250$	Charcoal overlying consolidated ash contemporary with Early Wilton Culture.
3. Luembe River, Angola	UCLA-172	$12,970 \pm 250$	Charcoal, Lower Tahitolian Culture.
4. Kamusongolo Kopje Cave, Zambia	SR-45	$13,300 \pm 250$	Charcoal, Later Stone Age.
5. Magosi Rock Shelter, Uganda	SR-92	$13,870 \pm 130$	Charcoal, Later Wilton occupation.
6. Kisese II, Tanzania	NPL-38	$31,480 {\,\pm\, 1640 \atop -1350}$	Charred ostrich eggshell. Middle part of second intermediate sequence.
	NPL-37	$18,190 \pm 306$	Charred ostrich eggshell. First intermediate/Late Stone Age transition.
	NPL-35	$14,760 \pm 202$	Charred ostrich eggshell. Middle part of Late Stone Age.
7. Leopoldville, Congo	Lv-166	$15,080 \pm 480$	Charcoal, Middle Lupembian inddustry.
8. Pomongwe Cave, Southern Rhodesia	SR-39	$35,530 \pm 780$	Charcoal associated with Middle Stone Age.
	SR-10	$21,700 \pm 400$	Charcoal associated with Middle Stone Age.
	SR-11	$15,800 \pm 200$	Charcoal. First intermediate Stone Age.
9. Leopards Hill Cave, Zambia	UCLA-1291	$16,400 \pm 265$	Charcoal.
10. Holley Shelter, Natal, South Africa	BM-30	$18,200 \pm 500$	Charcoal from hearth with Middle Stone Age implements.
11. Robberg, Cape Province, South Africa	GrN-5889	$18,660 \pm 110$	Shell fragments from layer with sparse Middle Stone Age artifacts.
12. Montagu, Cape Province, South Africa	GrN-5123	$19,100 \pm 110$	Charcoal. Late Middle Stone Age.

Table 11B.3.—*cont.*

Site	Lab. no.	Date	Comments
13. Lion Cavern, Swaziland	GrN-5020	$28,130 \pm 260$	Charcoal, middle stage of Middle Stone Age.
	Y-1827	$22,280 \pm 400$	Charcoal, tools and mining.
14. Twin Rivers, Zambia	UCLA-229	$22,800 \pm 1000$	Travertine. Middle Stone Age industry.
15. Sibebe Shelter, Swaziland	GrN-5314	$22,850 \pm 160$	Charcoal from upper levels of Middle Stone Age.
16. Rose Cottage, Orange Free State	GrN-5300	$25,640 \pm 220$	Charcoal. In sterile layer between Pre-Wilton and Magosian.
17. Revez Duarte, Portuguese East Africa	SR-67	$28,540 \pm 490$	Oyster shell from mound with Middle Stone Age palaeoliths.

validation by independent investigators does not entirely overcome the problem, but it does constitute an essential step toward verification. 'Perhaps the greatest contribution of the skeptic, the disbeliever, in any given scientific observations is the likelihood that his anticipation, psychological climate, and even instrumentation may differ enough so that his observation will be more an independent one' (Rosenthal, 1968, p. 15). Validation by sceptics established the authenticity of the Folsom bison site and also the contemporaneity of ancient man and extinct fauna in the New World.

In the current manhunt for the first invaders of America, the boldest claim of great antiquity is that of Leakey and Simpson near the Calico Hills, California, where flaked cherts have been found in fan deposits of considerable age. An estimate of over 40,000 years and an open invitation to visit the site has been offered (Leakey *et al.*, 1968). A visit to the quarry in October 1970 by sixty leading American geologists and archaeologists sustained the view that the deposits are of great age. But it failed to convince sceptics that the alleged artifacts were definitely man-made and occur in a cultural context (Behrens, 1971).

Another recent case in which a very impressive report long embedded in the record of New World prehistory was not verified under scrutiny is that of Gypsum Cave, Nevada. Two recent radiocarbon dates on wood artifacts, one a portion of a painted dart shaft, failed to establish that prehistoric man and extinct ground sloths coexisted in the cave (Heizer and Berger, 1970), much less that people preceded the last of the sloths as believed by Harrington (1933). It appears that redeposition of cave artifacts by wood rats was more extensive than Harrington recognized at the time of excavation.

The most determined effort at verification was directed toward the Tule Springs site, also outside Las Vegas, Nevada. First discovered by Fenley Hunter in 1933 and reported to contain extinct fauna in association with stone artifacts of Early Man, the area was repeatedly visited by M. R. Harrington, R. D. Simpson, and others from the Southwest Museum. Charcoal believed to have come from ancient hearths was radiocarbon dated at 23,000 BP. Stone tools thought to be of similar age were reported (Harrington and Simpson, 1961).

To replicate this apparently persuasive case of antiquity, a massive effort that combined the talents of scientists in half a dozen disciplines, supported by an estimated quarter of a million dollars in public and private funds, was undertaken by the Nevada State Museum. Earth moving equipment removed 200,000 tons of overburden from two miles of trenches revealing in detail the alluvial and lacustrine beds. Teams of archaeologists and palaeontologists screened outcrops where ancient occupation appeared most likely, including localities previously reported as ancient. Their efforts were guided during site visits by eminent senior archaeologists. Eighty dates provided rapidly during the field seasons by the University of California, Los Angeles, Radiocarbon Laboratory helped determine the strategy of excavation. Faunal, pollen and charcoal analyses revealed the nature of the Late Pleistocene environment (Wormington and Ellis, 1967).

But despite the scope of the efforts and the initial optimism of the investigators, the units that were 13,000 years and older failed to yield sound evidence of human occupation. Even in the younger units, archaeological material numbered only three bone tools, a stone scraper, and five chips or flakes (Shutler, 1965). Alleged charcoal which Simpson and Harrington believed was of cultural origin proved in at least one case to be humified plant material soluble in hydroxide, rather than charcoal from a hearth (Cook, 1964). Similar problems in distinguishing cultural charcoal from natural carbonaceous material have been reported by Riddel (1969).

In brief, efforts at replication by independent investigators failed to sustain either the claim of great age for occupation of Las Vegas Valley or the coexistence of man and sloth in an adjacent cave. Field visitation failed to resolve questions of what constitutes an artifact at the Calico Hills site. The high cost of a full-scale replication, as in the case of Tule Springs, is no less apparent than the fact that such efforts must be intensified. Since there is no way to *prove* that man *was not* in the New World before 12,000 BP, it is incumbent upon claimants that they spare portions of their sites for the most critical inspection by independent investigators and incumbent upon granting agencies that they support such validation efforts.

Meanwhile, the growing collection of new claims, suggestive radiocarbon dates on bone or on samples of uncertain provenance and typological concepts of the proper Old World predecessor of New World artifacts must be examined against the failure of older claims to survive critical attempts at replication. At least four major sources of error can be held responsible for premature claims: 1) taxonomic – the alleged artifacts were not man-made; 2) stratigraphic – rebedded or intruded material was not recognized during the excavation, or associations were secondary; 3) geochemical – samples submitted for dating were unsuitable for accurate age determination; 4) behavioural – there was a clerical error in handling samples, or the principal investigator was misled by an 'excavator effect' akin to an experimenter effect in psychology, or possibly he was victim of a deliberate fraud. Surely no one will insist that there can never be another attempt at a Piltdown hoax.

A parsimonious view regarding spectacular claims in the Early-Early Man hunt will scarcely startle supporters of Hřdlicka. Clewlow (1970) and Graham and Heizer (1967) note that his 50-year-old objections to enthusiastic claims have stood the test of time very well. The enticing question of a pre-12,000-year-old invasion of the Americas resembles other plausible but unproved concepts which pose major theoretical difficulties. These include such perennially newsworthy subjects as the question of extrasensory perception, of flying saucers, and of organic fossils in moon rocks. The major theoretical difficulty in the case of an ancient discovery of America is why the supposed invaders failed to proliferate and to leave a material record at least as rich as that of contemporaneous Old Stone Age cultures in Eurasia and Africa.

If Early Man discovered the New World before 12,000 radiocarbon years ago, he came without Early Woman.

6. Summary

By analogy with other animal invasions, the spread of *Homo sapiens* into the New World can be modelled as a front of high density sweeping the continent, followed by a sharp drop in human population behind the front as prey animals were depleted. Arbitrary values proposed for the model include an average frontal depth of 160 km, an average density within the front of 0.4 person per km^2, an average density behind it of 0.04 per km^2 and an average rate of frontal advance of 16 km annually. Initial population explosion during the formation of the front would not exceed maximum rates of growth, here set at 3.4 per cent annually, or a doubling in each 20-year generation. The total population of North America would not exceed 600,000 people (stage 3 in fig. 11B.3) at any one time.

The model generates a population large enough to overkill a biomass of Pleistocene large animals averaging 9 T per km^2 (50 au per section) or 2.8×10^7 metric tons in the continent. Within a decade in any one region the average annual removal of 2 metric tons per person could, and the removal of 7 metric tons would certainly, decimate the megafauna.

From what is known of the fossil record, extinction occurred suddenly. There was no decline in diversity, biomass, or range of any of the extinct genera of Late Pleistocene large mammals prior to 11,000 years ago. There is no compelling evidence of generic extinction much later. The concept of an advancing front in which the first invading hunters rapidly exterminated most of their prey is conformable with what is presently known of the extinction chronology.

A cultural chronology based on rapid rather than slow population growth differs substantially from current views regarding the prehistory of America. While certain assumptions incorporated into it are untestable, the new chronology itself can be disproved in several ways. It can be rejected if future discoveries establish that the extinction of large mammals was gradual, occurring over a 10,000-year interval, rather than within 1000 years. It can be rejected if extinction of the Rancholabrean megafauna is found to have preceded human invasion. It can be rejected if claims for occupation of the New World prior to 12,000 BP are successfully validated.

In the absence of such findings, I conclude that the New World was free of hominids before 12,000 BP. At that time human immigration precipitated a major and irreversible environmental upset. The chronology of large animal extinctions at the end of the last Ice Age may yet prove to be the best clue to the spread of *Homo sapiens* in the New World.

ACKNOWLEDGEMENTS

My views on the discovery of America benefited from seminar exposure arranged at Yale by John H. Ostrom and R. F. Flint, at the Museum of Comparative Zoology, Harvard University, by P. J. Darlington and H. B. Fell, at the University of California, Santa Barbara, by Joseph H. Connell, at the University of California, Los Angeles, by Malcolm Gordon, at Southern Methodist University by C. Vance Haynes, and at the University of Arizona by Emil W. Haury and T. L. Smiley. I am also grateful for comments, suggestions and counter-arguments received from John E. Guilday, James J. Hester, Arthur J. Jelinek, R. G. Kline, G. S. Krantz, Leslie Sue Lieberman, Everett H. Lindsay, Jr, Austin Long, John H. McAndrews, Robert H. MacArthur, Marian Martinne, Philip Miles, Walter Ogston, Bruce Rippeteau, Jeffrey Saunders, G. G. Simpson, Walter W. Taylor, Norman T. Tessman, S. David Webb, and especially to Robert F. Heizer. Special thanks for editorial aid are due to Betty Fink. This study was supported in part by NSF 7797.

References

AITKEN, R. B. (1969) *Great Game Animals of the World*. New York: Macmillan. 192 pp.

AUGUSTA, J. (1963) *A Book of Mammoths*. London: Hamlyn.

BATES, M. (1952) *Where Winter Never Comes*. New York: Scribners. 310 pp.

BEHRENS, C. (1971) The search for New World man. *Science News*, **99** (6), 98–100.

BERGER, R. and LIBBY, W. F. (1968) UCLA radiocarbon dates VIII. *Radiocarbon*, **10**, 403–4.

BINFORD, S. R. (1970) Late middle Paleolithic adaptations and their possible consequences. *BioScience*, **20**, 280–3.

BIRD, J. B. (1970) Paleo-Indian discoidal stones from southern South America. *Amer. Antiq.*, **35**, 205–9.

BIRDSELL, J. B. (1957) Some population problems involving Pleistocene man. *In:* Population studies: Animal ecology and demography, *Cold Spring Harbor Symposium on Quantitative Biology*, **XXII**, 47–69. Cold Spring Harbor, NY: The Biol. Lab.

BLACK, R. F. (1966) Geologic history, late Pleistocene to recent history of Bering Sea-Alaska coast and man. *Arctic Anthrop.*, **3** (2), 7–22.

BORDES, F. (1968) *The Old Stone Age*. New York and Toronto: McGraw-Hill.

BOULIERE, F. and HADLEY, M. (1970) The ecology of tropical savannas. *In:* JOHNSON, R. F., FRANK, P. W. and MICHENER, C. M. (eds.) *Annual Review of Ecology and Systematics*, Vol. 1, 125–52. Palo Alto, Calif.: Amer. Reviews, Inc.

BRYAN, A. L. (1969) Early man in America and the late Pleistocene chronology of western Canada and Alaska. *Current Anthrop.*, **10** (4), 339–48.

BUTZER, K. W. (1971) *Environment and Archeology: An Ecological Approach to Prehistory* (2nd edn). Chicago and New York: Aldine-Atherton. 703 pp.

CAUGHLEY, G. (1970) Eruptions of ungulate populations with emphasis on Himalayan thar in New Zealand. *Ecology*, **51**, 53–75.

CHARD, C. S. (1969) *Man in Prehistory*. New York and Toronto: McGraw-Hill. 351 pp.

CLARK, J. D. (1959) *The Prehistory of Southern Africa*. Harmondsworth: Penguin Books.

CLEWLOW, C. W. JR (1970) Some thoughts on the background of early man, Hřdlicka and Folsom. *Kroeber Anthrop. Soc. Pap.*, **32**, 26–46.

COLBERT, E. H. (1961) *Dinosaurs: Their Discovery and Their World*. New York: E. P. Dutton. 300 pp.

COOK, S. F. (1964) The nature of charcoal excavated at archaeological sites. *Amer. Antiq.*, **29**, 514–17.

CRUXENT, J. M. (1968) *In: Biomedical Challenges Presented by the American Indian*, 11–16. Scientific Publication 165. Washington, DC: Pan American Health Organization.

DEEVEY, E. S. JR (1960) The human population. *Sci. Amer.*, **203**, 195–204.

DEMBECK, P. (1965) *Animals and Men*. Garden City, NY: Natural History Press. 390 pp.

DESMOND, A. (1961) How many people have ever lived on earth? *In:* NG, K. Y. and MUDD, S. (eds.) *The Population Crisis, Implications and Plans for Action*, 20–38. Bloomington, Ind.: Midland Books.

DIBBLE, D. S. and LORRAIN, D. (1968) Bonfire shelter, a stratified bison kill site, Val Verde County, Texas. *Texas Mem. Mus. Misc. Pap.*, 1. Austin. 139 pp.

DOBYNS, F. (1966) Estimating aboriginal American population: an appraisal of techniques with a new hemispheric estimate. *Current Anthrop.*, **7** (4), 395–416.

ELTON, S. (1958) *The Ecology of Invasions by Animals and Plants*. London: Methuen. 181 pp.

FITTING, J. E. (1968) Environmental potential and the post-glacial readaptation in eastern North America. *Amer. Antiq.*, **33** (4), 441–5.

FLINT, R. F. (1971) *Glacial and Quaternary Geology*. New York: John Wiley. 892 pp.

FRY, G. F. and MOORE, J. G. (1967) *Enterobius vermicularis*: 10,000-year-old human infection. *Science*, **166**, 1620.

GRAHAM, J. A. and HEIZER, R. F. (1967) Man's antiquity in North America: views and facts. *Quaternaria*, **9**, 225–35.

GUILDAY, J. E. (1972) The Pleistocene history of the Appalachian mammal fauna. *In:* The distributional history of the biota of the southern Appalachians, Part III: Vertebrates, *Virginia Polytech. Inst., Res. Div. Monogr.*, 3.

GUTHRIE, R. D. (1968) Paleoecology of a late Pleistocene small mammal community from interior Alaska. *Arctic*, **21** (4), 223–54.

HACKETT, C. J. (1967) The human treponematores. *In:* BROTHWELL, D. and SANDERSON, A. T. (eds.) *Diseases in Antiquity*, 152–69. Springfield, Ill.: Charles C. Thomas.

HARE, R. (1967) The antiquity of diseases caused by bacteria and viruses, a review of the problem from a bacteriologists' point of view. *In:* BROTHWELL, D. and SANDERSON, A. T. (eds.) *Diseases in Antiquity*, 115–31. Springfield, Ill.: Charles C. Thomas.

HARINGTON, C. R. (1970) Ice-age mammal research in the Yukon Territory and Alaska. *In:* SMITH, R. A. and SMITH, J. W. (eds.) *Early Man and Environments in Northwest North America*, 35–51. Calgary: The Student's Press, Univ. of Calgary.

HARRINGTON, M. R. (1933) Gypsum Cave, Nevada. *Southwest Mus. Pap.*, 8. Los Angeles, Calif. 197 pp.

HARRINGTON, M. R. and DE ETTE SIMPSON, R. (1961) Tule Springs, Nevada, with other evidences of Pleistocene man in North America. *Southwest Mus. Pap.*, 18. Los Angeles, Calif. 146 pp.

HAYNES, C. V. JR (1966) Elephant-hunting in North America. *Sci. Amer.*, **214** (6), 104–12.

HAYNES, C. V. JR (1971) Geochronology of man-mammoth sites and their bearing on the origin of the Llano complex. *In:* DORT, W. JR and JONES, J. D. JR (eds.) Pleistocene and recent environments of the central great plains, *Univ. of Kansas, Dept of Geol. Spec. Pub.*, **8**, 77–92. Lawrence: Univ. Press of Kansas.

HEIZER, R. F. and BERGER, R. (1970) Radiocarbon age of the Gypsum Cave culture. *Contr. Univ. of Calif. Arch. Res. Facility*, **7**, 13–18.

HEIZER, R. F. and NAPTON, L. K. (1969) Biological and cultural evidence from prehistoric human coprolites. *Science*, **165**, 563–8.

HESTER, J. J. (1967) The agency of man in animal extinctions. *In :* MARTIN, P. S. and WRIGHT, H. E. JR (eds.) *Pleistocene Extinctions, the Search for a Cause*, 169–92. New Haven and London: Yale Univ. Press.

HO, T. Y., MARCUS, L. F. and BERGER, R. (1969) Radiocarbon dating of petroleum impregnated bone from the tar pits of Rancho La Brea, Calif. *Science*, **164** (3883), 1051–2.

IRWIN, H. T. and WORMINGTON, H. M. (1970) Paleo-Indian tool types in the Great Plains. *Amer. Antiq.*, **34** (1), 24–34.

IRWIN-WILLIAMS, C. (1969). *Summary of Archaeological Evidence from the Valsequillo Region, Puebla, Mexico.* Unpub. MS.

JARCHO, S. (1964) Some observations on disease in prehistoric North America. *Bull. Hist. Med.*, **38**, 1–19.

JELINEK, A. J. (1967) Man's role in the extinction of Pleistocene faunas. *In:* MARTIN, P. S. and WRIGHT, H. E. JR (eds.) *Pleistocene Extinctions, the Search for a Cause*, 193–202. New Haven and London: Yale Univ. Press.

JENNINGS, J. D. (1968) *Prehistory of North America.* New York: McGraw-Hill. 391 pp.

JORDAN, P. A., BOTKIN, D. B. and WOLFE, M. L. (1970) Biomass dynamics in a moose population. *Ecology*, **52**, 147–52.

KLEIN, R. G. (1969) *Man and Culture in the Late Pleistocene: a Case Study.* San Francisco: Chandler Pub. Co.

KLEIN, R. G. (1971) The Pleistocene Prehistory of Siberia. *Quat. Res.*, **1**, 133–61.

KRANTZ, G. S. (1970) Human activities and megafaunal extinctions. *Amer. Sci.*, **58** (2), 164–70.

KURTÉN, B. (1965a) The carnivora of the Palestine caves. *Acta Zool. Fenn.*, **107**, 1–74.

KURTÉN, B. (1965b) The Pleistocene Felidae of Florida. *Bull. Fla. St. Mus.*, **9**, 215–73.

LANNING, E. P. and PATTERSON, T. C. (1967) Early man in South America. *Sci. Amer.*, **217**, 44–50.

LEAKEY, L. S. B., SIMPSON, R. D. and CLEMENTS, T. (1968) Archaeological excavations in the Calico Mountains, Calif.: preliminary report. *Science*, **160**, 1022–3.

LEE, R. B. (1968) What hunters do for a living. *In:* LEE, R. B. and DEVORE, I. (eds.) *Man the Hunter*, 30–48. Chicago: Aldine.

LORIMER, F. (1953) Historical outline of world population growth. *In:* The determinants and consequences of population trends, *Population Studies*, **17**, 5–20. New York: United Nations.

LYNCH, T. F. (1967) The nature of the central Andean preceramic. *Occ. Pap. Idaho St. Univ. Mus.*, 21. 98 pp.

MACNEISH, R. S. (1970) Megafauna and man from Ayachuco, highland Peru. *Science*, **168**, 975-7.

MACNEISH, R. S. (1971) Early man in the Andes. *Sci. Amer.*, **224** (4), 36-46.

MARCUS, L. F. (1960) A census of the abundant large Pleistocene mammals from Rancho La Brea. *Los Angeles Co. Mus. Contr. Sci.*, 38. 11 pp.

MARTIN, P. S. (1967) Prehistoric overkill. *In:* MARTIN, P. S. and WRIGHT, H. E. JR (eds.) *Pleistocene Extinctions, the Search for a Cause*, 75-120. New Haven and London: Yale Univ. Press.

MARTIN, P. S. and WRIGHT, H. E. JR (1967) *Pleistocene Extinctions, the Search for a Cause.* Proc. VII Congr. Inst. Ass. Quat. Res., Nat. Acad. Sci.-Nat. Res. Council. New Haven and London: Yale Univ. Press. 453 pp.

MASON, R. J. (1962) The paleo-Indian tradition in eastern North America. *Curr. Anthrop.*, **3** (3), 227-78.

MEAD, A. R. (1961) *The Giant African Snail.* Chicago: Univ. of Chicago Press. 257 pp.

MECH, L. D. (1970) *The Wolf.* Garden City, NY: Natural History Press. 384 pp.

MEHRINGER, P. J., KING, J. E. and LINDSAY, E. H. (1970) A record of Wisconsin-age vegetation and fauna from the Ozarks of western Missouri. *In:* DORT, W. T. and JONES, T. K. (eds.) *Pleistocene and Recent Environments of the Central Great Plains*, 173-83. Lawrence: Univ. of Kansas Press.

MILLER, G. J. (1968) On the age distribution of *Smilodon californicus* bovard from Rancho La Brea. *Los Angeles Co. Mus. Contr. Sci.*, 131. 17 pp.

MILLER, G. J. (1969) Man and *Smilodon*: a preliminary report on their possible coexistence at Rancho La Brea. *Los Angeles Co. Mus. Contr. Sci.*, 163. 8 pp.

MORSE, D. (1961) Prehistoric tuberculosis in America. *Amer. Rev. Respiratory Diseases*, **83**, 489-504.

MÜLLER-BECK, H. (1966) Paleohunters in America; origins and diffusion. *Science*, **152** (3726), 1191-210.

NEEL, J. V. (1970) Lessons from a 'primitive' people. *Science*, **170** (3960), 815-22.

ORR, P. C. (1968) *Prehistory of Santa Rosa Island.* Santa Barbara, Calif.: Santa Barbara Mus. Nat. Hist. 253 pp.

PEARSON, K. (1902) On the mathematical theory of errors of judgement with special reference to the personal equation. *Phil. Trans. Roy. Soc. London*, **198**, 235-99.

PIDOPLICHKO, I. G. (1969) *Late Paleolithic Dwellings of Mammoth Bones in the Ukraine* (in Russian). Kiev. 162 pp.

PRUFER, O. H. (1966) Some reflections on the extinction of the Pleistocene megafauna. *In:* SEN, D. and GHOSH, A. K. (eds.) *Studies in Prehistory*, 28-40. Calcutta: Firma K. L. Mukhopadhyay.

QUIGLEY, C. (1971) Assumption and inference of human origins. *Current Anthrop.*, **12**, 519-40.

RAEMSCH, B. E. (1968) Artifacts from mid-Wisconsin gravels near Oneonta, New York. *The Yager Mus. Pub. Anthrop. Bull.*, **1**. 8 pp.

REED, C. A. (1970) Extinction of mammalian megafauna in the Old World late Quaternary. *BioScience*, **20** (5), 284-8.

REPENNING, C. A. (1967) Palearctic-nearctic mammalian dispersal in the Late Cenozoic. *In:* HOPKINS, D. M. (ed.) *The Bering Land Bridge*, 288-311. Stanford: Stanford Univ. Press.

RIDDEL, F. A. (1969) Pleistocene faunal remains associated with carbonaceous material. *Amer. Antiq.*, **34** (2), 177-80.

RINEY, T. (1964) *The impact of introduction of large herbivores on the tropical environment.* IUCN Publ. (n.s.) no. 4, 261–73. Morges, Switzerland: Int. Union Conserv. Nature.

ROSENTHAL, R. (1968) *Experimenter Effects in Behavioral Research.* New York: Appleton-Century-Crofts. 464 pp.

SHAY, C. (1971) The Itasca Bison Kill Site, an Ecological Analysis. *Pub. Minn. Hist. Soc.* St Paul.

SHUTLER, R. JR (1965) Tule Springs expedition. *Curr. Anthrop.*, **6** (1), 110–11.

SLOAN, R. E. (1970) *The Ecology of Dinosaur Extinction.* Preprint, Dept of Geol. and Geophys., Univ. of Minnesota.

STALKER, A. MCS. (1969) Geology and age of the Early Man site at Taber, Alberta. *Amer. Antiq.*, **34** (4), 428–33.

TABER, R. D. (1969) *Man as a Hunter: Origins and Cultures.* Univ. of Washington. 30 pp.

WASHBURN, S. L. and LANCASTER, C. S. (1968) The evolution of hunting. *In:* LEE, R. B. and DE VORE, I. (eds.) *Man the Hunter*, 293–303. Chicago: Aldine.

WEBB, S. D. (1969) Extinction-origination equilibria in late Cenozoic land mammals of North America. *Evolution*, **23**, 688–702.

WHEAT, J. B. (1967) A Paleo-Indian bison kill. *Sci. Amer.*, **216** (1), 44–52.

WHITTAKER, R. H. and NIERING, W. A. (1964) Vegetation of the Catalina Mountains, Arizona, I. Ecological classification and distribution of species. *J. Ariz. Acad. Sci.*, **3**, 9–34.

WILLIAMS, S. and STOLTMAN, J. B. (1965) An outline of southeastern United States prehistory with particular emphasis on the Paleo-Indian era. *In:* WRIGHT, H. E. JR and FREY, D. G. (eds.) *The Quaternary of the United States*, 669–83. Princeton, NJ: Princeton Univ. Press.

WORMINGTON, H. M. and ELLIS, D. (1967) Pleistocene studies in southern Nevada. *Nev. St. Mus. Anthrop. Pap.*, 13. 411 pp.

Manuscript received January 1972.

Abiotic processes

Geomorphic processes in the Arctic

J. Brian Bird

Department of Geography,
McGill University

1. Introduction

Recognition that unique geomorphic processes are active in polar terrestrial environments came in the second half of the nineteenth century. Initially attention was focused on the effects of glaciation on arctic and alpine landscapes. However, they only affect a small part of the circumpolar lands and this chapter is restricted to analysis of non-glacierized areas whilst an interpretation of glacial landforms will be found in Hattersley-Smith and Østrem, Chapters 4B and 4C.

By the end of the century the special properties of perenially frozen ground, and the unusual moisture characteristics of soil under these conditions, were being studied. A field excursion to Svalbard in 1910 during the Eleventh International Geological Congress concentrated interest on geomorphic problems of the unglacierized, northern or *periglacial* environments. During the decades that followed research was commonly restricted to national areas, but after 1945 progress was rapid and international in scope. Significant advances included the establishment of a committee on periglacial geomorphology by the International Geographical Union in 1949, the foundation of the International Association for Quaternary Research (INQUA) and the initiation of the scientific journal, *Biuletyn Peryglacjalny* (Periglacial Bulletin) in Poland in 1954.

Geomorphological studies in contemporary periglacial areas have concentrated on describing the unique landforms, related sediments and the responsible processes. Parallel studies have proceeded, particularly in central Europe, of former periglacial regions that existed during the Pleistocene glacial periods. Most recently, significant developments have occurred in applied periglacial studies, particularly with regard to terrain and related engineering investigations and studies of tundra ecosystems.

2. Weathering processes

Physical weathering primarily associated with frost riving is recognized to be of considerable importance in arctic lands and has often been assumed to attain a global maximum in polar lands (Büdel, 1969). When bedding planes, joints and cracks in bedrock are penetrated by surface water that freezes, sufficient pressure is generated by expansion of the ice to disrupt the rock. However, this simple explanation leaves many unanswered problems. Neither the physical system in the rock when freezing occurs nor the relative importance of free, rather than saturated water zones in the rock are fully understood. Since it has been shown that the annual number of temperature cycles about freezing point in rock near the surface is commonly less in the Arctic than further south (Fraser, 1959; Cook and Raiche, 1962) it is clear that other factors must influence the degree of physical weathering.

The significance of the frequency and amplitude of temperature cycles has been examined in controlled laboratory studies (Tricart, 1956; Wiman, 1963; and Potts *et al.*, 1970). When frost cycling occurs in rocks, the magnitude of the temperature change appears to be of minor importance in weathering and there is greater dependence on depth of soil, vegetation cover (if any), ground moisture, the permafrost regime and lithology. It is clear that there is often great variation in rate of weathering over short distances and in apparently homogeneous rocks (Dahl, 1967; Bird, Southampton Island, unpublished).

Frost-induced, physical weathering takes several forms. Macro-weathering in thin-bedded rocks including limestone (Plate 38), sandstone and basalt is rapid in all polar environments and leads quickly to the formation of scree slopes and *felsenmeere* (block fields). Massive crystalline rocks weather less rapidly but are vulnerable to frost action along joints, and blocks several metres across may be forced out of the rock surface by ice accumulating on the underside. Schistose rocks are similarly thrust up along cleavage planes to form 'tombstones'. In both cases further weathering leads to the disintegration of the blocks.

Highly porous rocks, particularly sandstone, are weakened by the growth of ice between mineral grains although the importance of this weathering is disputed. On a smaller scale some rocks, notably granite, exhibit microflaking whilst others show evidence of surface granular disintegration. Other forms of physical weathering, notably unloading, sheeting and spalling, occur in periglacial areas particularly after the withdrawal of an ice sheet and contribute to the total rock disintegration.

Weathering on horizontal rock surfaces leads to the formation of a debris mantle. In the Arctic during the early stages of debris development the permafrost table lies in the rock beneath the debris, but if climatic conditions are stable, a point is reached when the permafrost table and the debris/rock interface coincide, after which rate of frost shattering is much reduced. In the majority of rocks, the regolith in the early phases of weathering is composed of angular blocks, but the blocks themselves experience weathering, and over exceptionally long periods of time the debris particles are reduced in size until a lower limit of probably 3–6 μm, depending on mineral composition, is achieved (Dylik and Klatka, 1952). It has been suggested that the fine grained surficial deposits that cover the sedimentary uplands of the central Canadian Arctic and the lowlands of the Sverdrup Islands may have originated in this way.

A mantle of angular blocks, felsenmeere, forms a significant proportion of polar terrains particularly on uplands, where they are commonly termed *mountain top detritus* (Ives, 1963), and on lowlands underlain by limestone, sandstone and trap rocks. Felsen-meere are not unique to present-day arctic environments although they are often associated with frost riving. Some felsenmeere preserved on granites in the Canadian Shield (Mackenzie District, NWT) are considered to be the product of Tertiary, deep chemical weathering, from which the finer material has been removed and the corestones partly frost shattered;[1] other apparent felsenmeere are underlain by till and are in reality frost-shattered boulders that have been concentrated by wave action or frost churning from the till.[2] Lowland felsenmeere are normally of postglacial age, but upland felsenmeere are commonly much older and may survive glacierization without major changes (but see Ives, Chapter 10B).

Chemical weathering in the polar regions is considered of minor importance due to retarded chemical activity at low temperatures, the aridity of many high arctic areas and the limited release of organic acids from vegetation decay. On Somerset Island, in the central Canadian Arctic, limestone and dolomitic surfaces experience a low rate of solution of about 2 mm/1000 years that is concentrated at the snow/rock interface (Smith, 1969). Observations supporting these figures have been repeated at other arctic sites but solution in maritime polar areas may be at least two magnitudes greater (Corbel, 1959).

Rapid periglacial weathering on steep rock faces leads to scree accumulation at the foot of the slope. Scree development along valley walls in thin-bedded sedimentary and volcanic rocks can be extremely rapid, and it is not unusual in the Arctic to find valley sides over 100 m, and occasionally more than 300 m high, that have been buried in the postglacial period.

3. Mass movement

The mass wasting processes of weathering and gravity-induced transportation in many arctic environments are relatively rapid and the rate of denudation may equal or exceed that of any other environment. Some processes operating in the arctic landscape are identical with those found in other environments; others are restricted to areas where ice is present in the ground for part of the year (annual ground frost) whilst a few processes are confined to areas where permafrost occurs. None of the transfer geomorphic processes is unique to the Arctic although several are best developed under arctic conditions.

Amongst the slow types of mass wasting, *solifluction* (Plate 39) is widely observed in northern environments. The process was named by Anderson (1906) following a visit to Bear Island when he defined solifluction as the slow flow of waste that had been saturated with water from rain or snow. His subsequent use of the term in the Falk-land Islands showed that it might also be applied to certain forms of earth and mud flow. The terminology has subsequently become confused as more than one periglacial

---

[1] Cf. Wilhelmy, 1958, p. 202.
[2] The complex origins of felsenmeere in northern Canada are discussed by Bird (1967, pp. 168–71). The corresponding analysis for north Greenland is made by Nichols (1969, pp. 74–5).

process is involved in the slow movement of debris downslope and because solifluction as originally defined may apply to phenomena in warm climates.

The slow movement of debris in polar environments results from at least too processes. *Soil creep* is induced by fluctuations of moisture and temperature in the surface soil that displace soil particles: on slopes there is a downhill component due to gravity, and soil movement occurs. An important component of soil creep in the Arctic results from heaving by frost action, a phenomenon that has been referred to as frost creep (Kerr, 1881). It attains maximum development in fine grained soils.

A second type of slow movement occurs when a high concentration of water in the upper horizons of the soil reduces the shear strength and leads to viscous flow. The abundant moisture arises from melting of ice in the soil, the presence of a frost table through which surface water cannot pass, and from melting snowbanks. In none of these cases is permafrost a requirement, but soil moisture appears to reach a maximum when there is perennially frozen ground beneath an active layer. This variety of solifluction[1] is also best developed in fine grained soils. Although there are serious difficulties in distinguishing soil creep from other forms of slow movement (Higashi and Corte, 1971), it has been done successfully in east Greenland (Washburn, 1967) and in an alpine environment in Colorado (Benedict, 1970); however, most measurements of solifluction fail to differentiate between creep and moisture-induced flow. As might be expected there is a general relationship between rate of movement and the gradient of the slope. Solifluction may occur on slopes of less than 2°, although at gradients of 3° or less, annual displacement rarely exceeds 3 cm. On steeper slopes the annual rate is highly variable, but 5–15 cm at 15° is not uncommon. The most precise available data (Washburn, 1967) suggest that creep may be of the order of three times more important than viscous flow. Solifluction is concentrated in the spring when flow is rapid for brief periods of several days to a few weeks, coinciding with the period when the moisture content of the upper layers of the ground is at a maximum. Changes during the remainder of the year are usually slight.

In periglacial environments the debris mantle on hillslopes may be displaced more or less uniformly by sheet solifluction. More commonly solifluction is concentrated in discrete lobate streams and tongues which are separated by zones of zero or slow flow. Solifluction tongues have a convex front which may be from a few metres to a hundred metres or more across. When the regolith contains boulders, the front of the lobe at the surface is commonly a boulder wall from 1 to 5 m high. During the summer water often seeps from the base of this bank. In localities where the mantle consists solely of fines the fronts of solifluction streams are generally turf covered, and a series of turf banks or terrace steps cover the slope. Sections in solifluction terraces often show soil horizons and organic layers that have been incorporated into the debris as a more rapid section of the sheet overruns a slower part. The presence of micro landforms usually associated with solifluction does not necessarily imply that waste transfer is currently active everywhere on a slope as short- and long-term climatic changes lead to much local variation.

---

[1] The term gelifluction may be used for this type: originally congelifluction (Dylik, 1951, 1967).

Although individual solifluction sheets may form minor irregularities on slopes, the general effect as rubble is transferred from higher to lower slopes is to smooth the hill profile by exposing some rock knobs to enhanced weathering and burying others under the debris. At the base of the slope solifluction material either accumulates or, if it reaches a stream channel, the fines are removed by running water, leaving characteristic boulder accumulations in the stream bed. *Solifual* deposits are not easily studied in the Arctic because of the difficulty of obtaining sections in frozen ground and because of their similarity on visual inspection to till. Fossil deposits have been described from many Pleistocene periglacial regions; they are known as 'head' in Britain.

Solifluction is essentially a transport process and only occasionally is it an active erosive agent when it moves over rock surfaces. Blocks of weathered rock may be carried forward on top of a solifluction sheet; this is conspicuous at the base of escarpments and cliffs if boulders that break off the rockface roll or slide into solifluction sheets below and are then transported downslope for many hundreds of metres.

Slow mass movement in arctic environments is not confined to solifluction. Valley floors are commonly occupied by 'streams' of angular boulders which are advancing now, or have done so in the past. The process by which they move is not clear although in some cases it is assumed they are really solifluction streams on the surface of which a veneer of boulders is being carried forward; in other cases they may be a variety of rock glacier. The latter are moving, tongue shaped, boulder deposits, which may attain lengths of 3–5 km; some varieties are bulbous at the termini, while others are broader than they are long. Rock glaciers generally rise in cirques or at the foot of talus slopes and move down steep valleys or valley sides at 0.5–2 m yr^{-1}. Movement occurs in the presence of interstitial ice, or mixtures of ice and fines that form when surface moisture penetrates the boulders and freezes at the permafrost table. In some cases rock glaciers are the wasted remnants of ice glaciers that became buried in moraine as they melted. Many hundreds of rock glaciers are known from the mountains of Yukon and Alaska. Others occur in northeast Siberia, eastern Baffin Island and northern Labrador, but they are not commonly found elsewhere in arctic areas (see also Chapter 4C).

When soil above the permafrost table becomes saturated with melt and rainwater, slope failure may result. A semicircular hollow develops upslope and the soil moves downhill carrying with it jumbled tundra blocks. Although arctic earth flows are widely distributed (e.g. Rapp, 1960, in north Sweden; Bird, 1967, p. 187, in Bathurst Inlet, NWT) and old flow scars are common on the mainland and islands of the Canadian Western Arctic, their frequency is not known.

Mudflows occur in arctic mountains particularly where the environment is arid and sedimentary rocks are weathering rapidly as on Axel Heiberg and Ellesmere Islands. They are basically no different from mudflows in other parts of the world. However, a special arctic variety of mudflow may form when surficial sediments containing massive ground ice are exposed in summer. As the exposure thaws, water and soil are released often in the proportion of 2:1 to 3:1; a mudflow develops, thus cleaning the exposure, more ground ice is uncovered, melting continues and the mudflow increases. The process halts when the angle of the rear slopes is no longer steep enough for the flow to uncover ground ice. Susceptible localities for this process are marine cliffs in unconsolidated sediments, in particular around the Beaufort Sea (Mackay, 1971) and on the banks of meandering streams in the Sverdrup Islands (Lamothe and St-Onge, 1961).

Solifluction is commonly assumed to be the dominant form of mass wasting in the Arctic. Whilst this may be so on slopes with a continuous tundra cover over fine grained soils, in areas where plants are widely scattered or absent, the soil is exposed and sheet wash may exceed solifluction as a denuding agent. It can occur directly in rainstorms or beneath and beyond melting snow patches. On fine grained soils, especially where soil stripes are present, surface water may be concentrated into a parallel network of rills. In Vestspitsbergen, Jahn (1960) found that 18 g $m^{-2}yr^{-1}$ was being *washed* downslope below snowbanks, corresponding to denudation of 1 mm in 150 years; and Pissart (1967) has shown in the Queen Elizabeth Islands that even where precipitation is low (<8 cm of water annually) water flowing beneath snow patches may be a significant denudation process.

4. Patterned ground

Polar terrains show a repetition of soil, vegetation and micromorphological features that produce geometrical patterns; following Troll (1944) the phenomenon may be described as *frost patterned ground*. The term is sometimes extended to include virtually all minor periglacial terrain features (e.g. Tricart, 1963) but this may lead to confusion.

Patterned ground is widely distributed in the Arctic. It is frequently highly conspicuous on the surface and early accounts of polar geomorphology contained numerous descriptions of patterned ground and introduced many special terms. Subsequent attempts at rationalization of the terminology and classification of the forms have been only partly successful. The most widely used classification, due to Washburn (1950, 1956), is based on soil sorting and geometry. The processes forming patterned ground are inadequately understood and further detailed, controlled field and laboratory experiments, especially stressing the three-dimensional aspect, are required before an effective genetic classification can be developed. Studies are complicated by the presence of fossil as well as active forms and it is clear that much patterned ground is polygenetic in origin.

Probably the most widespread patterned ground is initiated through contraction of surface sediments associated with either a sudden drop of temperature or dessication. If the resultant crack or fissure is to become a permanent feature in the landscape it must be kept open when the process is reversed and expansion occurs. With temperature changes ice particles migrating into the crack, or dust, sand or organic matter slumping into it, wedge the fissure open. The growth increment of an ice-wedge in one winter is typically 0.5–1.0 mm. The process is repetitive and large wedges which may be several metres across are several thousands of years old. At the surface the top of the wedge appears as a furrow, often with slightly raised sides when developed in silt or peat; in sand and fine gravel, it is generally flush with the surface. Fissures intersect to produce polygons with four, five and six straight or curved sides usually intersecting orthogonally (fig. 12A.1a). The diameter of a polygon is normally 15–30 m although they may exceed

(a)

(b)

(c)

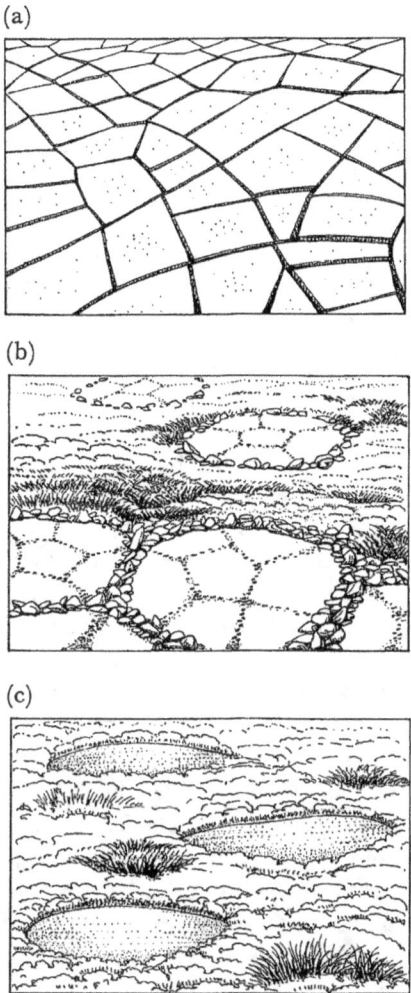

Fig. 12A.1. Three varieties of frost patterned ground: (a) ice-wedge fissure polygons, (b) polygons and circles formed by a net of pebbles with a second type of pattern in the centres, and (c) tundra circles.

this size considerably. The terrain within recently formed silt and peat polygons is initially slightly depressed and may be filled with shallow ponds in summer. Other polygons, generally considered to be older, have domed centres due to peat accumulation. *Ice-wedge fissure polygons*, the type that predominates in silt plains, are characteristic of the lowlands of northern Alaska, arctic Canada around the Beaufort Sea and the coastal plains of arctic Siberia. *Sand-wedge fissure polygons* are generally confined to outwash plains, terraces and elevated deltas and are especially numerous in the eastern Canadian Arctic.

Soil, peat and ice mounds are typical features of arctic terrains (Lundqvist, 1969).

The larger forms, especially pingos (see Ives, Chapter 4A) do not form recognizable patterns; the smaller hummocks and mounds are often distributed in polygonal patterns possibly connected with ice wedging or dessication cracking between the mounds.

Ice-wedge fissures and associated polygonal patterns also develop, but less frequently in exposed sedimentary rocks, notably limestones. The location of the fissures appears to be controlled by jointing and square or rectangular patterns are normal. The sides of the fissures are raised and hence resemble morphologically ice-wedges in silts. Preliminary investigations on Southampton Island, NWT, suggest fallen blocks of rock are mainly responsible for holding rock fissures open. Ice-wedge fissures may also form in organic soils. The arctic peat plains of the northern Hudson Bay and Foxe Basin lowlands commonly show polygonal fissure patterns in peat, that often overlies clear ice, which in turn rests on shattered bedrock.

Fissuring also follows dessication of soils. Cracks developed by this process are commonly 10–50 cm across. They are kept open by the migration of gravel and small pebbles to the crack from an initially random distribution on the surface. The motion, at least in some cases, results from the growth of needle ice beneath the stones on the slightly domed surface between the cracks; in other cases surface fines lift the stones (Chambers, 1967). The end result is a circular or polygon sorted pattern (fig. 12A.1b), in which the fines remain in the centre whilst the coarser particles move to the sides: displacements of 5 cm yr^{-1} have been recorded (Chambers, 1970).

It is clear that several periglacial processes tend to separate coarser and finer particles in exposed soils resulting in other varieties of patterned ground. One of the commonest forms consists of bare areas of soil roughly oval or circular in shape, 0.25–2.0 m in diameter and separated by vegetated areas. Where there are many patches the vegetation between them may form polygonal patterns. The features are known by several names including *tundra circle* and mud circle (fig. 12A.1c). The exposed soil in the centre of the circle acts as a 'wick' carrying moisture to the surface; soil may be forced upwards in the circle resulting in slight doming and accumulation of coarser soil particles on the surface as wind removes the fines. Of the several hypotheses of the origin of tundra circles, one that involves the movement of plugs of fine material towards the surface under hydrostatic pressure generated between freezing surfaces in the ground appears most likely. *Debris islands*, a term introduced by Washburn (1956: following Meinardus, 1912) describes rather similar patches of fines that occur in fields of coarse debris, commonly angular boulders.

Vegetation plays an important part in outlining many patterns in which there is localized soil movement. Under these conditions the bounding nets are stable areas which become vegetated, in contrast to the interior of the patterns which are mobile and unvegetated. Some patterns are largely or entirely developed in vegetation; these include lichen polygons and string bogs in the Subarctic (Allington, 1961) and vegetated clay and turf hummocks (thufur) in the Arctic (Beschel, 1966).

The periglacial geomorphic processes that operate on horizontal surfaces are also active on slopes, with the added features that particles which experience small frost-induced displacement are drawn downhill by gravity and that solifluction and rill wash may be present. The characteristic downslope patterned ground is a series of parallel stripes. They may consist of sorted material in which lines of gravel or coarser debris are separated by fines, or by stripes of vegetation, heath tundra or dwarf shrub separated by gravel microridges. Patterns also develop along contours across the slope. They are formed by steps on the slope in surface debris (*cryoturbation steps*) and are frequently elongated downslope (*stone garlands*). Similar features occurring in a mixture of soil, peat and sod are known as *turf garlands*.

The significance of frost patterned ground lies primarily in the indication it gives of a mobile surface. Individual soil displacements are small but they are repetitive and, for the most part, accumulative; above all they reflect the presence of considerable and varying quantities of ground moisture above the permafrost table. Together the soil and moisture changes have a continuing and significant influence on the distribution and density of the vegetation cover.

5. Periglacial slopes

It has already been shown that the overall effect of periglacial mass wasting is to smooth slopes by reducing exposed bedrock either through concentrated mechanical weathering or burial under solifluction deposits. Local slope irregularities are produced when solifluction lobes overrun stable or slower moving material. Larger irregularities occur on hillslopes when a succession of steep-faced outcrops and extended low-angle solifluction slopes combine to form stepped or terraced profiles (fig. 12A.2). These features are known by several names including altiplanation, cryoplanation and goletz terraces.

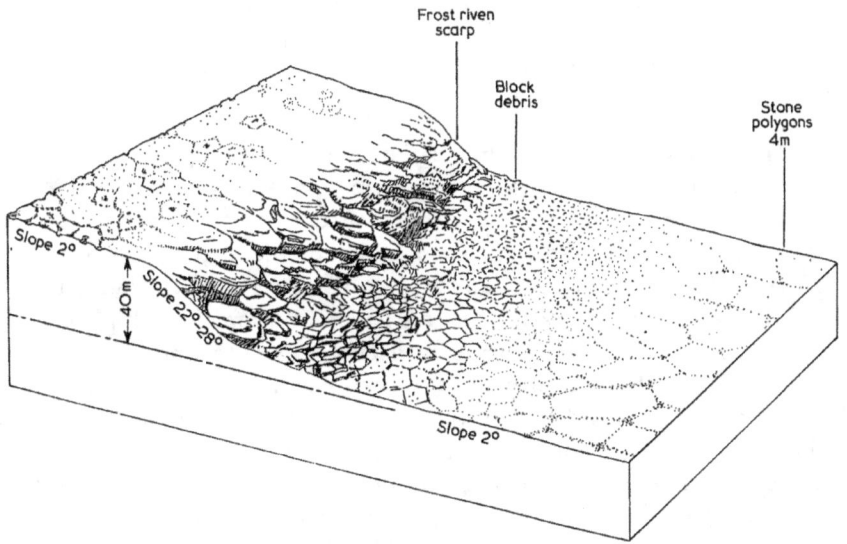

Fig. 12A.2. A characteristic altiplanation terrace retreating under the combined action of frost and nival sapping (after Demek, 1969).

Typically there is an abrupt rock backwall which rises from a few to several tens of metres, whilst the terrace below it is gently sloping, mantled with debris and may attain a width of hundreds of metres. Altiplanation terraces develop when an initial slope irregularity is steepened and retreats under the influence of intense frost and nival weathering; the debris is transported across the terraces by solifluction. They occur in many different rock types when there is adequate joint or bedding planes for frost riving. Where the rock outcrop is on hill summits and crests, it is not unusual for cryoplanation processes to act from both sides of the hill, ultimately leading to the development of tors in the final stages.

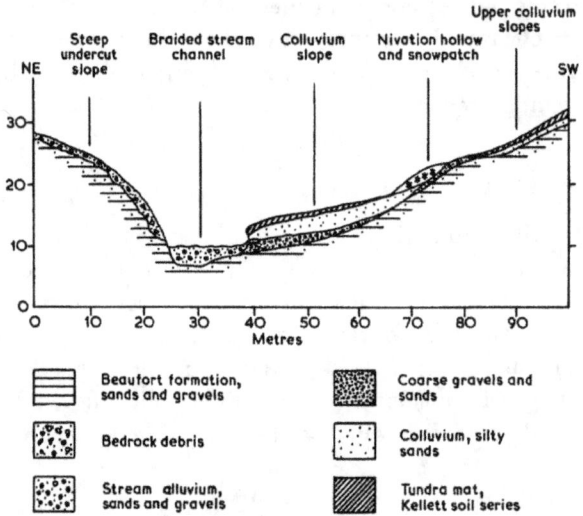

Fig. 12A.3. Generalized relationships between soils, slopes and materials in asymmetrical valleys (from French, 1971).

Demek (1969) has shown that altiplanation terraces are a product of periglacial environments. Although they may form rapidly in postglacial time, most large terraces are of considerably greater age and are found predominantly in areas that escaped glacierization or were only mildly glaciated, including the interior plateaus of Alaska and the Yukon, the Urals and Yakutia in Siberia. Evidence is accumulating that many hillslope irregularities in northern areas are inherited from past altiplanation (e.g. Péwé, 1970), modified to varying degrees by glaciation and other processes.

It has been shown (Pissart, 1966) that the two characteristic slopes that occur on altiplanation terraces are widely distributed in periglacial environments; one set of slopes (from which surface moisture is quickly removed) has typical angles of 30–32°, and the second set, in which solifluction is the dominant transfer agent, is at 5–7°.

Mass-wasting processes and associated vegetation are strongly responsive to local site parameters and it is not unusual to find solifluction widespread on some slopes and nonexistent on the opposite face. This contributes to the development of asymmetrical valleys which are characteristic of many periglacial areas. A general explanatory model of asymmetrical valleys has not been achieved due to the numerous complex processes responsible for their formation, although several have been identified (fig. 12A.3).

6. Aeolian processes

Wind erosion in polar environments results from deflation and abrasion by sand and snow. The latter is effective at low temperatures, as the hardness of snow crystals increases as the temperature drops and a Mohs hardness of 6 is reached between −50° and −80°C, at which level feldspars are attacked. However, evidence for strong wind abrasion is generally slight. Ventifacts with several wind-cut faces have been reported from many arctic regions, and boulders and bedrock faces often reveal smoothing and etching by abrasion: other abrasion forms are not common. The geomorphic influence of wind on boulders is complex; abrasion and a generally dry surface on the windward side is contrasted with a damp environment and lichen growth on the protected face. This may reduce physical weathering but may be associated with enhanced chemical action. Cavernous weathering pits, including tafoni, are fairly common in arid arctic landscapes and are occasionally explained by wind erosion, but chemical weathering associated with deflation is the more usual process.

Wind transport of silts (loess) is found in the Arctic today and was even more widespread in the periglacial Pleistocene regions. Loess covers the surface of considerable areas of arctic and subarctic Alaska (Péwé, 1968); much of it was deposited during the Illinoian and Wisconsin glacial phases but it is also being formed under contemporary conditions. Deltas in the western Canadian Arctic and northern Alaska contain quantities of silt, and brown clouds of dust are whipped up by the wind from vegetation-free areas in the spring and deposited over considerable distances. As well as contributing to the soil, the dark dust has the effect of at first increasing the melting rate of the spring snow; however, thick silts and sand blown onto snow are protective and develop cones as unequal melting proceeds.

Deflation is a selective process removing the finer material and leaving a lag deposit of larger particles on deltas, terraces, outwash plains and eskers. Coarser material up to pebble size is known to be blown by the wind over ice but the phenomenon is rare. The wind may remove dry organic soils, including earth hummocks, and produce fluting in thick peats such as occur in the Canadian Western Arctic. The wind also has a significant, if secondary influence in landscape processes through the control of drifting snow; the effects are more appropriately described under nival processes.

Notwithstanding many examples of wind action, Nichols's (1969, p. 80) statement that 'although most of Inglefield Land is devoid of vegetation the work of the wind is of very minor importance' applies in many other parts of the polar world, and the denudational impact of the wind when compared with other processes is slight.

7. Fluvial processes

The effectiveness of fluvial processes in the polar environment is not fully understood. In virtually all arctic rivers, from the largest to ephemeral streams, a great part of the annual discharge occurs in a period of ten days to three weeks in June or early July. Probably the Colville River in northern Alaska is representative of large streams when it was found in 1962 that 43 per cent of the annual discharge occurred during a three-week period (Arnborg *et al.*, 1967). Cook (1967) and McCann *et al.* (1972), who studied a small stream on Cornwallis Island, NWT, obtained similar figures. The available data suggest that a major part of the annual sediment load is transported in a short period; on the Colville River 75 per cent occurred in three weeks.

Wide fluctuations in discharge are associated with the development of braided sections particularly toward the mouth of the river. In Queen Elizabeth Islands and in northern Greenland the resulting arid appearance of the river channel is heightened by the presence of alluvial fans on the valley sides. Flooding of local rivers in the Arctic does not seem to produce conspicuous geomorphic phenomena, but larger rivers, such as the lower Mackenzie River and many of the northern Siberian rivers that rise in the south where break-up occurs first in the headwaters, are notoriously susceptible to flood and distribute much fine sediment at the time.

In smaller streams the pattern of break-up and initiation of the spring flow differs from the mid-latitudes. Pissart (1967) has shown that early in the spring, water begins to flow beneath snow collected in the stream channel and surface material is washed away. However, even at maximum discharge the bed of the stream may continue to be covered with anchored ice and hence the effective geomorphic role of the running water is diminished.

In permafrost zones, small streams induce thawing in buried ice masses, and *beaded streams*, consisting of a single deep channel joined by pools, are a common feature of arctic coastal plains in which there is often a considerable development of ground ice.

Stream channels in bedrock, particularly when these are bedded sedimentaries, may be quickly entrenched in arctic environments. Frost shattering in the stream bed in the autumn and early winter is followed by the spring flood during which the broken rock is removed, exposing the bedrock to further weathering later in the year (Bird, 1967; Büdel, 1969). This process is highly developed in many limestone and sandstone areas of northern Canada.

8. Marine and lacustrine processes

Ice on rivers, lakes and seas in the Arctic has both protective and erosional functions. The protective role is most obvious on the sea where presence of an ice cover for at least part of the year reduces wave action. In polar seas where there is some drift ice at all times high-energy waves rarely develop as onshore winds automatically bring ice back into the area and dampen the waves. Strong wave action on the shores of enclosed arctic seas, especially in the Siberian and the Canadian archipelagos, is restricted to extremely short periods of time during a few days in the year. Exposed coasts with less *inshore* summer pack ice, as along the north Alaska coast, see correspondingly greater beach activity.

Although direct wave erosion is appreciably less on arctic cliffs than elsewhere, the evidence for this is not easily expressed quantitatively due to active subaerial processes. Cliffs in unconsolidated sediments that contain massive ground ice retreat rapidly by melting, whilst bedrock cliffs are vulnerable to frost riving. Abrasion by floating ice is of limited importance although often conspicuous when enclosed material polishes and striates rock exposures.

Sea ice transports debris that is incorporated into it either by falling from cliffs or being frozen onto the underside of the ice in shallow water. Fine silts and large boulders are both carried in this way. Most of the finer material is deposited ultimately in deep water but the boulders commonly travel short distances and are stranded on tidal flats after the ice melts during the summer. Total quantities of material that are redistributed are not known except from Foxe Basin, where fine materials are incorporated on the considerable scale of 0.05–2.9 g kg^{-1} yr^{-1} (Campbell and Collins, 1958). River ice may play a greater role in sediment transport because it will be moving rapidly as it is carried downstream at break-up and many riverbanks are of unconsolidated debris that are easily weakened by ice abrasion; ice blocks frozen onto the banks also slide into rivers after break-up and carry much sediment with them. Thermal erosion by rivers through melting ground ice in the bank sediments is frequently a significant process in channel development (Czudek and Demek, 1970).

On flat coasts beach material is forced up into mounds, ridges and ramparts when ice is pushed into them. For sea ice this results from onshore winds, but in the case of lake ice, although winds may play some part, expansion and contraction of ice during the winter months is known to produce similar features. In extreme cases the ramparts may be 10 m or more in height and pushed tens of metres and rarely hundreds of metres inland (Owens and McCann, 1970). On coasts which have a moderate to large tidal range, large, permanent beach ridges are uncommon, because of the tide crack along which the energy of ice driven onshore is dissipated, and of the action of high tides in combing down material. In high-range tidal environments boulder barricades, a single or double ridge of boulders, and submerged at high water, are not unusual (Bird, 1967, pp. 217–24).

Blocks of ice are frequently stranded on the shore, where they may be buried by beach deposits; subsequently as the ice melts pits are formed which survive until more beach material is carried into the area. Lakes also experience changes in outline in consequence of the melting of ground ice. In this case the banks of a lake retreat through a combination of wave and thermal erosion. In regions where the wind direction is constant in summer, parallel, elongated (*oriented*) lakes result; when the wind is variable in direction circular lakes are formed.

9. The geomorphic influence of snow in the Arctic

Snow has a complex role in the development of polar landscapes. The duration of the snow cover at an arctic site is highly variable depending on winter temperature, snowfall, and exposure; it may range from a few days to practically the whole year beneath semi-permanent snowbanks. A snow cover, by increasing surface insulation, reduces short-term fluctuations in the ground temperature and consequently inhibits frost riving. It also provides protection for vegetation against desiccation and abrasion by blowing snow, and there is close interaction between snow that is trapped by arctic shrubs and which at the same time protects them. The reverse effect occurs when snow remains on the ground into the late summer; vegetation fails to survive under these conditions and the barren ground, although only exposed for a few weeks in the year, experiences strong geomorphic activity.

Melting snow in the summer is an important source of moisture for solifluction, rill wash and chemical weathering. Late snow patches are concentrated in climatically favourable localities, such as lee slopes and hollows, and the associated geomorphic processes become concentrated in a like manner leading eventually to the formation of asymmetrical slopes. The role of melting snow as a source of moisture is conspicuous along sand and clay terraced riverbanks where snow collects in cracks and hollows. As the snow melts, small streams wash out the fine sediments and, helped by slumping, circular hollows develop on the terrace side. Concentrated chemical weathering on carbonate rock surfaces occurs beneath melting snow, and the etched, hackly limestone surfaces that are widespread in the Arctic are the result.

The ability of snow to contribute to the development of rounded depressions, hollows and ledges on hillslopes has long been recognized (Matthes, 1900). Similar nivation phenomena have been described from many polar environments, but the actual processes and the speed with which they operate are not fully understood (Bird, 1967, pp. 226–9). It is probable that quasi-permanent snowbanks in which the summer temperature reaches 0°C concentrate moisture at the snow/rock interface, and that this leads to chemical weathering beneath the snow and frost riving around the margins. The weathered products are removed in suspension or solution in the water and frequently by solifluction. Nivation ledges develop where horizontal rock structures favour the evolution of benches which trap the snow; nival processes along the backwall lead to its retreat, leaving a residual ledge (St-Onge, 1969).

Snowbanks collected at the foot of steep slopes may be sufficiently compact by early summer that boulders and debris falling off the rear slope slide and roll across the snow to collect in a ridge at the foot. Repetition of the process over many years leads to the development of *protalus ramparts* which may be several metres high and form an arcuate or linear wall in front of the slope.

Snow has a strong capability through avalanches to modify slopes; they are, however, largely confined to mountains and are more suitably discussed under that heading. Slush avalanches, in which snow in stream beds and on glaciers becomes saturated with water and eventually moves downvalley as a mass of snow and water, have been widely described but their significance to periglacial denudation is not known.

10. Sedimentary rocks in the arctic landscape

Rocks of varying lithology react to weathering processes in different ways. In periglacial environments many sedimentary rocks respond in a similar manner and the landscapes developed on thin to moderate bedded limestones, sandstones and trap rocks are often very similar. They are characterized by deep and rapid incision of the rivers into gorges (when the initial surface is sufficiently elevated), the growth of flat, structurally controlled pavements, low plateaus bounded by scarps and extensive felsenmeere.

The high solubility of the carbonate rocks in surface waters when compared with other rocks leads to additional special landforms where they occur. Limestone and dolomite occur over more than one-third million square kilometres in northern Canada and are widely distributed in other polar lands, but their surface morphology has not been widely studied. It is clear that surface solution, probably intensified beneath the wet snow cover in spring, is widespread. It produces an etched, hackly rock surface in which individual pits are rarely more than 2 cm deep. Although lapiés, fissures and enlarged joints have been reported from the Arctic, they are apparently less widespread than in lower altitudes. Low precipitation, many months of below-freezing temperatures and restricted amounts of vegetation limit the rate of denudation, particularly in arctic continental regions.

Continuous deep permafrost inhibits subsurface drainage and underground karst is not developed on any extensive scale. In northern areas karst is limited to the Subarctic and to arctic areas that escaped glaciation or were only weakly glaciated; a distribution that suggests that the present features are relic forms from earlier warmer climates.

11. Conclusion

Although many periglacial features are unique, the totality of landforms in arctic regions of low-moderate relief often differs little from landscapes on similar rocks in mid-latitudes. This is particularly true of the Precambrian Shields of northern Canada, Greenland and northern Scandinavia, where the larger morphological elements in the arctic sector resemble closely those further south. The landscapes of these areas are dominated by inherited landforms from the Pleistocene glaciations and from the Late Tertiary. In contrast, in areas where the rock has weathered rapidly, and which escaped glaciation or were only feebly glaciated, periglacial sediments and slopes assume a major importance. These areas include parts of northern Alaska and Yukon, many of the Canadian northwestern arctic islands and the Siberian arctic islands east of Novaya Zemlya.

When periglacial processes have dominated the terrain for long periods the surface is said to have experienced cryoplanation (Bryan, 1946). Peltier (1950) summarized the views on arctic landscape evolution in a Davisian framework when he described a periglacial cycle. In practice this particular model has limited application because of the overriding significance of rock type and the major climatic changes of the Quaternary. Above all the magnitude of the variation of the physical environments (and the related geomorphic processes) between circumpolar lands must be recognized. The extreme range is between the continental areas of the interior of the northern Canadian islands and Peary Land where climatic, botanical and morphological deserts prevail, and the maritime zones, including the Aleutians and coastal Iceland, where permafrost and the associated processes may be lacking.

The significance of arctic geomorphology lies in the development of minor landforms, the mobility of the soil and above all the close interaction between surficial sediments, soil moisture and temperature, vegetation and the landforms.

References

ALLINGTON, K. R. (1961) The bogs of Central Labrador–Ungava. *Geogr. Ann.*, **43**, 401–17.

ANDERSON, J. G. (1906) Solifluction, a component of subaerial denudation. *J. Geol.*, **14**, 91–112.

ARNBORG, L., WALKER, H. J. and PEIPPO, J. (1967) Suspended load in the Colville River, Alaska, 1962. *Geogr. Ann.*, **49**, A, 131–44.

BENEDICT, J. B. (1970) Downslope soil movement in a Colorado alpine region: rates, processes, and climatic significance. *Arct. Alp. Res.*, **2** (3), 165–226.

BESCHEL, R. E. (1966) Hummocks and their vegetation in the high Arctic. *In:* Proc. Int. Permafrost Conf., *NAS–NRC Pub.*, 1287, 13–20. Washington, DC: Nat. Acad. Sci. and Nat. Res. Council.

BIRD, J. B. (1967) *The Physiography of Arctic Canada.* Baltimore: Johns Hopkins Press. 336 pp.

BIRD, J. B. and BIRD, M. B. (1961) Bathurst Inlet, Northwest Territories, Canada. *Geogr. Branch Mem.*, 7. 66 pp.

BLACK, R. F. (1969) Thaw depressions and thaw lakes. *Biul. Peryglac.*, 19, 131–50.

BRYAN, L. (1946) Cryopedology – the study of frozen ground and intensive frost action with suggestions on nomenclature. *Amer. J. Sci.*, **224**, 622–42.

BÜDEL, J. (1969) Das System der Klimagenetischen Geomorphologie. *Erdkunde*, **23** (3), 1965–83.

CAMPBELL, N. J. and COLLINS, A. E. (1958) The discoloration of Foxe Basin Ice. *J. Fish. Res. Bd. Can.*, **15**, 1175–88.

CHAMBERS, M. J. G. (1967) Investigations of patterned ground at Signy Island, South Orkney Islands, III: Miniature patterns, frost heaving and general conclusions. *Brit. Antarct. Surv. Bull.*, **12**, 1–22.

CHAMBERS, M. J. G. (1970) Investigations of patterned ground at Signy Island, South Orkney Islands, IV: Long-term experiments. *Brit. Antarct. Surv. Bull.*, **23**, 92–100.

COOK, F. A. (1967) Fluvial processes in the high Arctic. *Geogr. Bull.*, **9**, 262–8.

COOK, F. A. and RAICHE, V. G. (1962) Freeze-thaw cycles at Resolute, NWT. *Geogr. Bull*, 18, 64–78.

CORBEL, J. (1969) Vitesse de l'érosion. *Zeit. Geomorphol.*, **3**, 1–28.

CZUDEK, T. and DEMEK, J. (1970) Thermokarst in Siberia and its influence on the development of lowland relief. *Quat. Res.*, **1**, 103–20.

DAHL, R. (1967) Post-glacial micro-weathering of bedrock surfaces in the Narvik district of Norway. *Geogr. Ann.*, **49**, A, 155–66.

DEMEK, J. (1969) *Cryoplanation Terraces, their Geographical Distribution, Genesis and Development.* Prague: Acad. Nakladat. Ceskoslovenske. 80 pp.

DYLIK, J. (1951) Some periglacial structures in Pleistocene deposits of middle Poland. *Bull. Soc. Sci. Lettr. de Lodz*, **3** (2), 6.

DYLIK, J. (1967) Solifluxion, congelifluxion and related slope processes. *Geogr. Ann.*, **49**, A, 167–77.

DYLIK, J. and KLATKA, T. (1952) Recherches microscopiques sur la désintegration périglaciaire. *Bull. Soc. Sci. Lettr. de Lodz*, **6**.

FRASER, J. K. (1959) Freeze–thaw frequencies and mechanical weathering in Canada. *Arctic*, **12**, 40–53.

FRENCH, H. M. (1971) Slope asymmetry of the Beaufort Plain, northwest Banks Island, NWT, Canada. *Can. J. Earth Sci.*, **8**, 717–31.

HAMELIN, L.-E. and COOK, F. A. (1967) *Le périglaciaire par l'image*. Quebec: Les Presses de l'Université Laval. 237 pp.

HIGASHI, A. and CORTE, A. E. (1971) Solifluction: a model experiment. *Science*, **171**, 480–2.

IVES, J. D. (1963) Field problems in determining the maximum extent of Pleistocene glaciation along the eastern Canadian seaboard – a geographer's point of view. *In:* LÖVE, Á. and LÖVE, D. (eds.) *North Atlantic Biota and their History*, 336–54. Oxford: Pergamon Press.

JAHN, A. (1960) Some remarks on the evolution of slopes on Spitsbergen. *Zeit. Geomorphol.*, Suppl. 1, 49–58.

KERR, W. C. (1881) On the action of frost in the arrangement of superficial earthy material. *Amer. J. Sci.* (3rd ser.), **21**, 345–58.

LAMOTHE, C. and ST-ONGE, D. (1961) A note on a periglacial erosional process in the Isachsen area, NWT. *Geogr. Bull.*, **16**, 104–13.

LUNDQVIST, J. (1969) Earth and ice mounds: a terminological discussion. *In:* PÉWÉ, T. L. (ed.) *The Periglacial Environment*, 203–15. Montreal: McGill Univ. Press.

MCCANN, S. B., HOWARTH, P. J. and COGLEY, J. G. (1972) Fluvial processes in a periglacial environment. Queen Elizabeth Islands, NWT, Canada. *Trans. Inst. Brit. Geogr.*, 55, 69–82.

MACKAY, J. R. (1971) The origin of massive icy beds in permafrost, western arctic coast, Canada. *Can. J. Earth Sci.*, **8**, 397–422.

MATTHES, F. E. (1900) Glacial sculpture of the Bighorn Mountains, Wyoming. *US Geol. Surv. Ann. Rep.*, 21, part 2, 167–90.

MEINARDUS, W. (1912) Beobachtungen über Detritussortierung und Struktorboden auf Spitzbergen. *Gesell. Erdkunde, Berlin, Zeit.*, 250–9.

NICHOLS, R. L. (1969) Geomorphology of Inglefield Land, North Greenland. *Medd. om Grønland*, **188** (1). 109 pp.

OWENS, E. H. and MCCANN, S. B. (1970) The role of ice in the arctic beach environment with special reference to Cape Ricketts, southwest Devon Island, Northwest Territories, Canada. *Amer. J. Sci.*, **268**, 397–414.

PELTIER, L. C. (1950) The geographic cycle in periglacial regions as it is related to climatic geomorphology. *Ann. Ass. Amer. Geogr.*, **40**, 214–36.

PÉWÉ, T. L. (1968) Loess deposits of Alaska. *In:* Genesis and classification of sedimentary rocks, *Proc. 23rd Int. Geol. Congress* (Prague, 1968), Sect. 8, 297–309. Prague: Academia.

PÉWÉ, T. L. (1970) Altiplanation terraces of Early Quaternary age near Fairbanks, Alaska. *Acta Geogr. Lodz.*, 24, 357–63.

PISSART, A. (1966) Étude de quelques pentes de l'île Prince Patrick. *Ann. Soc. Géol. Belg.*, **89**, 377–402.

PISSART, A. (1967) Les modalités de l'écoulement de l'eau sur l'île Prince Patrick. *Biul. Peryglac.*, 16, 217–24.

POPOV, A. I. (1962) Periglacial phenomena and the laws of their distribution in the USSR. *Biul. Peryglac.*, 11, 77–84.

POTTS, A. S. (1970) Frost action in rocks: some experimental data. *Trans. Inst. Brit. Geogr.*, 49, 109–24.

RAPP, A. (1960) Recent development of mountain slopes in Kärkevagge and surroundings, northern Scandinavia. *Geogr. Ann.*, **42**, 65–200.

SMITH, D. I. (1969) The solution erosion of limestones in an arctic morphogenetic region. *In: Problems of Karst Denudation*, 99–109. Brno.

ST-ONGE, D. A. (1969) Nivation landforms. *Geol. Surv. Can. Pap.*, 69–30. 12 pp.

TRICART, J. (1956) Étude experimentale du probleme de la gélivation. *Biul. Peryglac.*, 4, 285–318.

TRICART, J. (1963) *Géomorphologie des régions froides*. Paris: Presses Universitaires de France. 289 pp.

TROLL, C. (1944) Strukturböden, Solifluktion und Frostklimate der Erde. *Geol. Rund.*, **34**, 545–694.

WASHBURN, A. L. (1950) Patterned ground. *Rev. Can. Geogr.*, **4**, 5–59.

WASHBURN, A. L. (1956) Classification of patterned ground and review of suggested origins. *Bull. Geol. Soc. Amer.*, **67**, 823–76.

WASHBURN, A. L. (1957) Instrumental observations of mass-wasting in the Mesters Vig district, northeast Greenland. *Medd. om Grønland*, **166** (4). 296 pp.

WILHELMY, H. (1958) *Klimamorphologie der Massengesteine*. Braunschweig: G. Westermann Verlag. 238 pp.

WIMAN, S. (1963) A preliminary study of experimental frost weathering. *Geogr. Ann.*, **45**, 113–121.

Manuscript received November 1971.

Additional references

PRICE, R. J. and SUGDEN, D. E. (1972) Polar geomorphology. *Inst. Brit. Geogr. Spec. Pub.*, 4. London. 215 pp.

WASHBURN, A. L. (1973) *Periglacial processes and environments*. New York: St Martin's Press. 320 pp.

The geomorphic processes of the alpine environment

Nel Caine

Institute of Arctic and Alpine Research,
University of Colorado

1. Introduction

Various lines of evidence suggest that mountain environments should be highly active geomorphically. The steep slopes characteristic of such an environment suggest a high level of gravitational stress and thus rapid sediment movement. This need not, of course be true for the landscape could be adjusted to a high stress level but such an adjustment seems unlikely, given the time required to bring it about (Ahnert, 1970) and the fact of climatic change. A second line of deductive reasoning stems from the morphoclimatic concept which associates geomorphic activity and climatic factors (e.g. Peltier, 1950; Stoddart, 1969). The effects of mountains on temperature and precipitation are well known and should place nonglacial alpine areas in a periglacial system characterized by rapid erosion rates (Budel, 1968, p. 420).

These reasons are supported by direct observation. 'Almost every mountain observer has been struck by the evidence of decay which centres around the larger peaks' (Pearsall, 1950, p. 4) and this is corroborated by more catastrophic events. It is in the mountainous parts of the world that large rock and snow slides occur, like that from Huascaran which buried Yungay, Peru, on 31 May 1970. The amount of geomorphic work accomplished in a single event of this magnitude may be equivalent to many centuries of 'normal' erosion in lowland areas and emphasizes the inherent instability of the alpine landscape. A second line of empirical evidence lies in recent work on the measurement of geomorphic processes themselves, in particular of 'denudation rates' since these can be most easily compared. The initial conclusions of Corbel (1959) about the average rates at which the land surface is being lowered show a consistent difference between mountain and lowland areas. Rivers remove up to five times more sediment per unit area from mountain basins than from lowland ones. Although conclusions about other factors influencing denudation rates tend to be inconsistent (Stoddart, 1969, p. 186), the importance of relative relief is shown by many studies (e.g. Fournier, 1960; Schumm, 1963; Muller and Forstner, 1968; Diaconu, 1969; Litvin, 1969). This work is, however, based on the sediment discharge from river catchments and takes no account of the movement of waste within the basin. Therefore, the confirmation of an erosional difference between mountainous and normal relief by other sets of evidence is important (Young, 1969). Young shows an order of magnitude difference in erosion rates between the two relief types in three sets of evidence: river load estimates (i.e. the denudation rate), reservoir sedimentation and geological data. Only surface process measurements do not clearly establish the contrast.

There are, therefore, good reasons for concurring with the general opinion of geomorphic instability in mountain environments. This chapter reviews the processes responsible for that instability and the sediment movement rates which derive from it.

2. Geomorphically significant characteristics of the alpine environment

Many environmental influences of potential importance to geomorphic processes are examined elsewhere in this volume and need not be treated here. They originate from the physical and biotic milieu of alpine mountains or from their historical development, and tend to produce a geomorphic environment of considerable diversity. This variability in both time and space is perhaps the single most significant geomorphic characteristic of the alpine zone.

PHYSICAL CHARACTERISTICS

Geologic factors
In any environment, the most important geologic factors in the production of the landscape are lithology and structure. These have a passive influence in that they tend to control the response of a landform (through the resistance and strength of its constituent material) to stress-inducing processes of change derived from the subaerial environment. In most alpine landscapes the erosional resistance of the surface material is notable for its variability. A bedrock lithology of high-grade metamorphics and intrusives with a complex structure is usually only partially covered by an assortment of surficial waste of recent age. The resulting high variance in the strength of the surface materials implies a parallel in the response to erosion which could embarrass simple generalizations.

On a general level, structural factors and the relationship between them and mountain form has often been used to classify mountains. The topic has, however, received little attention in recent years with the result that there appears to have been only slight development from the classification of, for example, von Richthofen (1886, quoted in Chorley et al., 1964, p. 601) to that of Fairbridge (1968, p. 752).

Geologic factors also have an active influence, especially on alpine landscapes. The tectonic instability of alpine areas may show as fault-scarps across recent landforms, as in the Southern Alps (Suggate, 1965, pp. 25, 43) and the Andes (Schubert and Sifontes, 1970). A less direct effect of earthquakes is their influence on the timing of major landslides, such as those associated with the Alaska Earthquake of 1964 (Shreve, 1966). Simonett (1967) documents the effect of even a low level of seismicity on the frequency of landslides in an area of high relief. Yet a further active effect lies in the production of mountainous areas. Many of these are the direct result of geophysical processes in the crust, either as orogenic and isostatic responses or as the result of volcanic action.

Physiographic factors

In general terms, the alpine environment is characterized by a high rate of energy transfer and this is particularly evident in the geomorphic part of it, where steep slopes prevail. Steep slopes, especially where associated with high erosion rates, will, however, only survive while the accumulation of waste on them is relatively slight. On the other hand, the disappearance of cliffs from the landscape probably requires so much time that climatic change is likely to intervene before it is accomplished and produce again the conditions initially responsible for cliff development. Steep slopes do not, however, make up the entire alpine area for valley floors and some interfluve areas form low energy enclaves within this high energy environment.

The slope angle control is frequently associated with a lithologic one for very steep slopes ($>35°$) are common only where the material composing them has sufficient strength to stand at high angles. Pippan (1967, p. 203), for example, shows the difference between slopes developed on shales and sandstones (mean $= 21°$), slightly indurated sands (mean $= 18°$) and Pleistocene moraines ($4° <$ mean $< 12°$) in the Alps. Examples of the opposite, of steep slopes related to material of low strength, seem to be restricted to areas of very rapid waste removal, i.e. of great internal negative disequilibrium (Ahnert, 1967, p. 30).

Climatic factors

As an influence upon geomorphic action, climatic factors, especially microclimatic ones, are also highly variable in the alpine. Spatial variations in radiation receipts due to topography and vegetation and temporal diversity due to seasonal influences, especially those affecting snow cover and variations in the albedo of the ground surface, seem particularly important. The effects of this in a geomorphic context have rarely been considered but, given the coupling of radiation, ground surface temperature and evaporation, must be significant in weathering and mass movement.

The effects of altitude on precipitation and temperature are well known. The latter is the factor most likely to be quoted as a reason for including mountain areas in a periglacial zone (e.g. Derruau, 1968, p. 739) but precipitation is at least as influential as air temperature on rates of waste movement. In alpine mountains, the effect of altitude in increasing precipitation from frontal systems is accompanied by a potential for generating local storms which may be even more important in producing high intensity rainfall. Such climatic effects are basic to definitions of altitudinal morphoclimatic zones (e.g. Jennings and Bik, 1962; Bik, 1967; Derruau, 1968). In fact, the climatic variability of mountain areas should make them useful testing grounds of the morphoclimatic concept although a lack of detailed geomorphic and climatic data prevents this at present.

Hydrologic factors

In response to snowmelt, the water budget of an alpine area tends to be markedly seasonal and the river regime fairly predictable, at least in terms of the timing of annual peak flows (Beckinsale, 1969, pp. 468–70). The spring–early summer period, the time of greatest discharge, is therefore likely to be the occasion of greatest fluvial geomorphic activity in alpine river channels. Outside the stream channels, too, the seasonal difference in precipitation type (snow or rain) and the timing of snowmelt is likely to be significant for sediment movement.

BIOTIC CHARACTERISTICS

The accepted view of a periglacial environment in which 'there is little vegetation to bind together the inherently mobile layer in summer' (Sparks, 1960, p. 312) is likely to be an oversimplification, although a discontinuity in geomorphic activity seems to occur, in response to the vegetation change, at timberline. This has been used to account for large-scale benching in the alpine (Thompson, 1962, 1968). The effects of vegetation above timberline remain important, however. Many forms of patterned ground involve patterning of the vegetation cover and so attest the interaction of vegetation, frost action and rates of soil movement. Models of slope wasting which include the effects of vegetation as an influence on the shearing resistance of soil and as an agent favouring greater infiltration and less rainsplash (e.g. Carson, 1969) are applicable to the ground above timberline as well as to that below it.

   Sparks's statement does have some relevance, particularly since large parts of the alpine environment carry little vegetation, other than lichens and mosses. Areas of bedrock and superficial material are, therefore, exposed directly to atmospheric weathering environments and may be expected to show a different response to that of material below a vegetation and soil cover (Keller, 1957, p. 69). Thus, a further source of local variability is introduced into the geomorphic system.

HISTORICAL CHARACTERISTICS

Glacial effects are very evident in the alpine landscape and their bearing on presently acting processes is often obvious. The high relief of the alpine is usually a direct expression of this, even though there appears to have been a great increase in alpine slope stability in the 10,000 years since their deglaciation. The zone of Pleistocene glacial valley deepening has been designated the most geomorphically active one in the world (Büdel, 1968) and, outside polar latitudes, it consists basically of alpine areas.

3. Sediment production and movement in the alpine

Alpine geomorphic processes are considered here as a set of determinants controlling the flow of waste released by bedrock weathering. Transfers of surficial material are ultimately gravity-controlled and so trend along slope profiles and through stream channels. Whether the waste moves as clastic material or as solutes in soil and ground-water, the gravity influence and the mean direction of movement remain the same. Only minor exceptions, such as wind transportation, do not fit this generalization.

This flow model allows an integration of geomorphic processes and is also compatible with other environmental studies, with which it has many conceptual overlaps. However it does entail some disadvantages. In particular, it does not allow the precise definition of alpine periglacial landforms and their mode of development. Knowledge of these has frequently been summarized elsewhere (e.g. Tricart, 1963, 1970; Embleton and King, 1968), so there is no need for a repetition of those treatments here.

MODELS OF ALPINE GEOMORPHIC ACTIVITY

Much recent work in hydrology, ecology and geomorphology emphasizes the measure-ment of material and energy budgets in 'representative' basins (Rapp, 1960a; Slaymaker and Chorley, 1964; IASH, 1965; Bormann and Likens, 1967). Such a basin offers simple, physically meaningful definitions of system boundaries and so should be the fundamental unit of study in dynamic geomorphology (Chorley, 1969). It forms a useful basis for the present review of geomorphic processes, although there is little empirical data available on which to base estimates of the rates at which these processes operate in any single alpine basin. For this reason, I shall attempt to integrate here information available for the Colorado Rockies in general and will use data from other areas only where this is lacking.

For convenience the alpine drainage basin can be considered as two subsystems: the valley wall slopes and the stream channels of the valley floor. This is justified by the

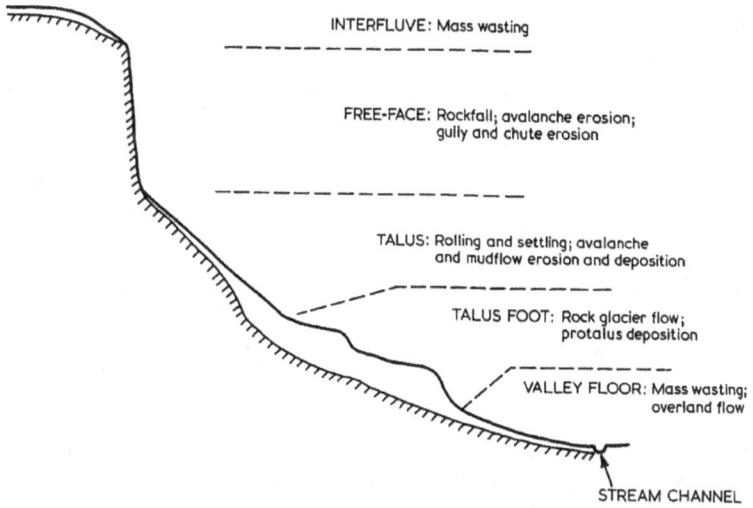

Fig. 12B.1. A hypothetical alpine slope profile.

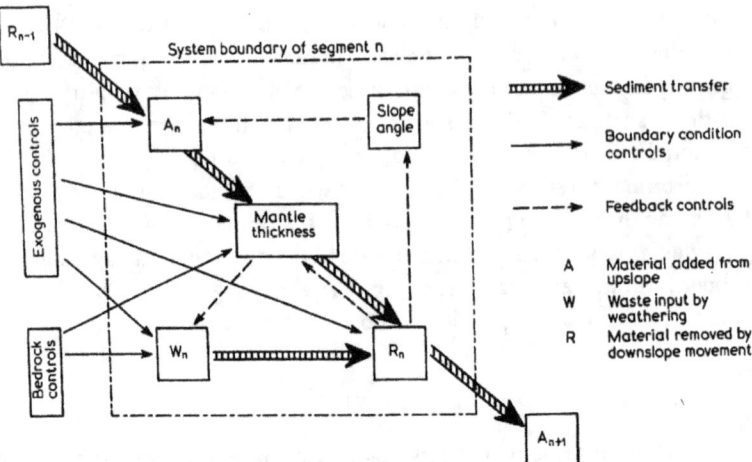

Fig. 12B.2. The hillslope waste budget. This model is intended for only small parts of the slope (on a scale of under 10 m). R is the waste discharged downslope; A is the waste acquired from upslope; W is the waste acquired from the weathering of bedrock *in situ*.

differences in the stress to which the waste is subject (and its resultant behaviour) but it should not be allowed to mask the functional unity of the basin – the main system.

The slope model
Since waste transfer occurs along the slope profile, a diagrammatic alpine slope (fig. 12B.1) may be used in the examination of the flows involved. This slope consists of a

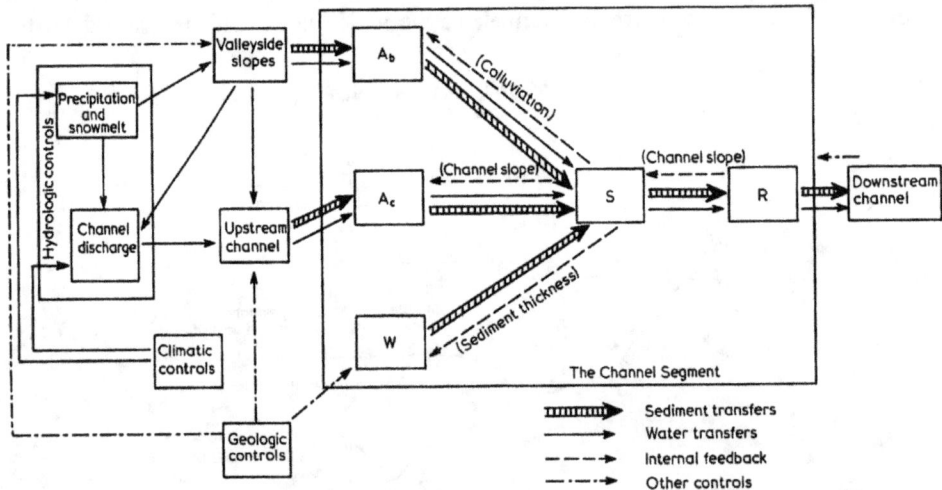

Fig. 12B.3. The stream channel sediment budget. R is the sediment discharged down the channel; Ab is the sediment acquired from the slopes surrounding the channel; Ac is the sediment acquired from upchannel; W is the sediment acquired from erosion and weathering of the channel bed and banks *in situ*; S is the volume of sediment stored within the channel segment.

number of broad segments, akin to those defined by Wood (1942), which often seem to occur in the sequence shown (e.g. Plate 40) but which may also be found as partial or multiple sequences (Young, 1964).

Three interacting sets of processes, involving the input (by weathering, mainly), the transfer (by mass wasting) and the storage (by colluviation) of surficial debris, are of significance here. The manner of their interaction at a point on the slope may be defined in a general way (fig. 12B.2) but the rates at which they operate will vary considerably between different points on the same slope.

The stream channel model

For present purposes, the stream channel may be considered a parallel to the valley wall subsystem for it, too, involves a linear sequence of sediment transfers. Once again, these transfers can be shown diagrammatically (fig. 12B.3) but, because of the relatively simple coupling of stream discharge and sediment movement, some of the external controls and internal feedback links can also be shown here.

4. Alpine slope processes

(a) Input processes

Waste production, the basic input to a geomorphic system, is dominated by weathering processes but includes other sources of material that may be locally significant. The latter are largely atmospheric and provide particulate matter, as well as cyclic salts, which may be important in soil formation.

WEATHERING

It is convenient to treat weathering processes in the classical fashion and distinguish mechanical and chemical weathering, but this should not be allowed to conceal interactions between the two. Mechanical breakdown of rocks, for example, will facilitate chemical alteration by exposing a greater surface area to atmospheric, hydrologic and biologic influences.

The input of waste by bedrock weathering is probably closely controlled by the thickness of the surficial mantle. This has already been invoked as a feedback loop in fig. 12B.2 and has been examined by Keller (1957) and Ahnert (1967) who suggest that the rate of weathering should decrease exponentially as the thickness of the mantle increases. Such a relationship allows the components of an alpine slope model to be ranked in terms of their capacity to provide input to the waste cover. The components with the least capacity for storage are probably the ones most influenced by weathering: on a slope like fig. 12B.1 the cliff face would obviously dominate. Keller's concept of weathering environments is relevant here, too (Keller, 1957, p. 69), and may be applied to frost weathering although it was initially intended for chemical environments alone. It suggests that, because of the importance of water and the frequent aridity of exposed bedrock, the exponential model may not be applicable to the surface. Maximum weathering probably occurs just below the soil surface.

Mechanical weathering

In a high mountain situation, as in other periglacial areas, mechanical weathering in the form of gelifraction (frost riving) is usually thought to be dominant (Derruau, 1968, p. 739; Tricart, 1970, p. 112; Stoddart, 1969, p. 177; Fisher, 1970, p. 112). Despite such frequent mention, however, the process of frost riving is still incompletely known although it seems to involve both the volumetric expansion of water that occurs on freezing and a migration of liquid water toward freezing centres (Scott, 1969). Because of this lack of knowledge, the most usual approach to a classification of frost action is from a consideration of the weathering products, rather than the processes responsible for them.

Tricart (1956) suggests two types of frost action: macro- and microgelivation. The first comprises the production of waste by frost working along structural or textural weaknesses in the rock. It therefore includes the granular disintegration of sandstones and igneous materials into their constituent grains, as well as the riving of large joint-defined blocks. Macrogelivation refers to the fracturing of the bedrock independently of grain boundaries or joint planes. Its product is normally broken, silt-size fragments which are very angular and sharp-edged.

The climatic controls of frost weathering, and particularly the significance of freeze–thaw cycles, continue to be a subject of discussion. Cook and Raiche (1962) demonstrate the inadequacy of air temperature in any estimate of ground freezing and have also shown that the diurnal freeze–thaw cycle may not be as important as once thought. Temperature fluctuations with a period of twenty-four hours are rarely felt below 5 cm in the ground and so will affect only a shallow surface layer. The relative insignificance of the diurnal cycle is also shown, indirectly, by the occurrence of the greatest number of frost cycles (predominantly daily ones) below treeline in alpine areas (Stelzer, 1962). Macrogelivation and coarse block production by frost riving seem to require cycles with a longer period (the annual cycle, perhaps). The distribution of blocky deposits on alpine summits and ridge crests (Rapp and Rudberg, 1960) suggests that macrogelivation is more important at higher altitudes, but this could be the result of a longer exposure to subaerial conditions, especially during past glacial episodes (see Ives, Chapter 10B, pp. 626–31).

General rates of mechanical weathering are difficult to define because little quantitative work has yet been done on them. Average cliff retreat rates in alpine areas vary from practically nothing to about 1.0 mm yr^{-1} which falls within the range suggested by Young (1969) for total denudation in mountainous areas. Talus accumulation in the Front Range of the Colorado Rockies suggests a free-face retreat rate of about 0.76 mm yr^{-1}, with a seasonal peak of activity during the spring and early summer similar to that which Gardner (1970a) reports from the Canadian Rockies.

Mechanical weathering other than that due to frost action is rarely considered in the alpine. Jahn (1964) accounts for 'twin-ridges' by the large-scale failure of the valley sides and Gerber (1969) also suggests the importance of residual stress in the bedrock as a control of the form of alpine valley walls. Such large-scale unloading processes must have been especially significant following the deglaciation of alpine valleys. On a much smaller scale, the products of insolation weathering (such as Ollier, 1969, pp. 14–21, documents) are likely to be indistinguishable from those of frost riving and so would probably not be recognized in the alpine.

Chemical weathering

Chemical weathering in the alpine has also received little attention (Embleton and King, 1968, p. 452) although its significance is demonstrated by work on alpine soil development (Retzer, Chapter 13B), by the small weathering features and rinds used to differentiate glacial deposits (Sharp, 1969) and by the solute content of river waters (Livingstone, 1963). The processes of chemical weathering in the alpine are probably the same as those operating elsewhere (Keller, 1957, Chapter 6; Ollier, 1969, Chapter 3) but their rate of operation may be quite different.

Low temperatures tend to retard chemical disintegration both by the slowing of chemical reactions and by imposing aridity when liquid water is frozen. This may be accentuated by a sparse vegetation cover, by the short period when nutrient extraction and plant metabolism occurs and by the slow rate of humic decay (Pearsall, 1950, p. 72; Wood, 1970). On the other hand, weathering should be accelerated by the increase in CO_2 solubility at low temperatures and by the higher CO_2 content of air in snowbanks (Williams, 1949). These factors may account for the high rates of limestone solution in cold climates (e.g. Corbel, 1964). Further, the amount of exposed bedrock and relatively unweathered material in environments characterized by highly aggressive water and fluctuating temperatures (Keller, 1957, p. 69) must increase general weathering rates in the alpine.

Biotic influences, and particularly microbiological ones, seem to be important, even in the relatively unvegetated environment of the alpine zone. The presence of bacteria, algae and fungi within bedrock weathering rinds is reported by Parfenova and Yarilova (1965) while Silverman and Munoz (1970) go further to estimate the quantitative importance of fungal attack on different igneous rocks. Lichens and mosses, which are known to extract nutrients directly from the substrate by ion exchange (Ollier, 1969, p. 48), form yet another biotic weathering complex of considerable importance in the alpine.

The coincidence of these biotic weathering agents with a zone of aggressive water, thermal stress and frost action, should facilitate weathering and perhaps be reflected in high denudation rates. Chemical weathering rates seem to be similar to those due to frost weathering but, in view of the complexity of controls and rock responses, cannot easily be generalized. For limestone weathering, on which most work has been done, solution rates seem to be about 0.1 mm yr⁻¹. On exposed rock in the Austrian Alps, Bauer (1962) measured a rate of 0.1 mm yr⁻¹ and found that the effect of a vegetation cover (and its associated soil) was to increase this to almost 0.3 mm yr⁻¹. Bögli (1960) and Barsch (1969a) define similar rates from Switzerland and the results of Goodchild (1890) and Sweeting (1966) from the Pennines fall into the same range (0.05 to 0.10 mm yr⁻¹). The weathering of materials other than limestone seems, on the basis of the low solute contents of alpine stream water, to proceed much more slowly than this (e.g. <0.008 mm yr^{-1} in the Wind River Range, Wyoming: Hembree and Rainwater, 1961).

OTHER WASTE SOURCES

Material input to the geomorphic system may also come from the corrosion of bedrock when fragments are impacted against it during transportation in running water, avalanches, rockfalls and during creep of a seasonal snowpack. None of these seem to have more than local importance simply because of their restricted spatial range but neither have they been adequately assessed as sediment sources. The work of Costin *et al.* (1964) suggests that bedrock abrasion by rocks at the base of the snowpack might lead to measurable erosion but its magnitude is not estimated.

The input of solutes in rainwater has been measured on a number of occasions (e.g. Gorham, 1961; Douglas, 1968; Gibbs, 1970) and may be the main source of the solutes in alpine stream water which usually has only a low solute load. Such an atmospheric contribution must be taken into account in any estimate of chemical denudation rates. The contribution of particulate matter from the atmosphere may also be significant. Windom's (1969) study of the dust in glacier ice shows its importance in mid-latitude alpine areas and my own measurements in the Colorado Front Range suggest that it amounts to a layer about 0.005 mm thick each year. While this is less than half the waste input from bedrock weathering, it probably represents an important source of silt and clay for soil formation and for much of the inorganic sedimentation in Colorado alpine lakes.

ZONES OF WASTE INPUT

Waste input to the alpine sediment system is not uniform but, with little information available, only qualitative estimates will be attempted here (table 12B.1).

Table 12B.1. Waste input zones.

Slope segment*	Convex	Free-face	Talus	Concave
Bedrock weathering				
Mechanical	2	4	3	1
Chemical	2	1	4	3
Waste reduction				
Mechanical	2	1	4	4
Chemical	2	1	3	4
Atmospheric sources				
Particulate	{ Slight, but approx. uniform across all segments }			
Solute	{ with modification by rain/wind patterns. }			

*Slope segments, defined in fig. 12B.1, are ranked (from 1 = slight to 4 = high) across the rows and not in absolute terms.

Bedrock weathering, as the main initial source of debris, is likely to be highly concentrated. Temperature and water availability control both mechanical and chemical
weathering and allow most rapid waste production only where the bedrock is close to the
surface. Where this coincides with a situation from which debris can be rapidly removed,
as on a free-face, greatest waste production is to be expected. By comparison with the
free-face, the other parts of the slope profile will provide only slight input although
further reduction in the size of the waste on them may occur rapidly.

The generally slight input of material from the atmosphere may be assumed to act
uniformly across the landscape although such an assumption is probably invalid in
detail.

(b) Transfer processes

The movement of a waste mantle is commonly classified as either 'fast' or 'slow' (Sharpe,
1938; Varnes, 1958). This is useful since it often implies, in addition, the periodicity
of the process concerned: 'fast' processes tend to have a long return period and to be
catastrophic in nature, whereas slow processes act more continuously through both time
and space. It is not, however, entirely satisfactory since the inverse correlation between
rate of operation and length of return period is not perfect. Solution transport, in particular, does not easily fit this classification since it acts rapidly and continuously whenever water is moving in the system. Rockfall from rapidly weathering free-faces may be a
second process which operates both quickly and continuously, although elsewhere it is
more episodic and usually seasonal in nature (Gardner, 1970a).

Nevertheless, the distinction between 'fast' and 'slow' processes will be retained here
as a second level of classification (following the slope segments of fig. 12B.1). Creep
solifluction and other slow flowages of the waste mantle of periglacial environments are
the most important slow processes in the alpine, whilst mudflows, rockslides, rockfalls
and avalanches probably comprise the basic members of a 'fast' set.

THE INTERFLUVE

In many areas the convex slope of the interfluve is not a spatially important part of the
alpine landscape although it is quite common in the Colorado Rockies. Where present,
it is most influenced by slow wasting processes, except at its lower margin where the
increasing slope angle may produce large-scale movements (e.g. Jahn, 1964) or be
partially responsible for mudflow initiation (e.g. Baird and Lewis, 1967). Residual
deposits and soils older than the last glacial are most likely to be found on these interfluves (Retzer, Chapter 13B) which, with the survival of relict patterned ground, suggest
relatively slow change. Waste transfer thus appears to be dominated by soil creep, frost
creep and localized (in both time and space) solifluction, which in turn are largely
controlled by soil moisture. Variations in the soil moisture regimes (and resulting
movement rates) allied with changes in the resistance of the soil to movement and in the
vegetation cover then give rise to the variety of solifluction features and patterned ground

forms described elsewhere (e.g. Tricart, 1963, 1970; Embleton and King, 1968; Davies, 1969; Bird, Chapter 12A).

On Niwot Ridge, Benedict (1970) finds that terraces and lobate features are the most active parts of an alpine interfluve, moving between 4 and 43 mm yr^{-1}, in contrast to the more stable areas which move less than 1 mm yr^{-1}. This suggests a maximum lowering of 0.01 mm yr^{-1} for Niwot Ridge which is comparatively slight for a high relief area (Young, 1969). Similar estimates derive from studies on convex summit slopes elsewhere (e.g. Costin et al., 1960: 0.14 mm yr^{-1} of surface lowering; and Caine, 1963: 40 mm yr^{-1} of lateral movement).

THE FREE-FACE

Steep bedrock cliffs are one of the most characteristic features of the alpine landscape. They appear simple in a generalized form (fig. 12B.1) but, in detail, the cliff profile is usually stepped, in response to structural controls, and its angle may vary from that of the talus to a true vertical. Because of ledges, even the free-face has a storage capacity for waste and it is from this, rather than from bedrock weathering, that material is added to the talus. In three dimensions, the cliff form is yet more variable and its gullies have a particular control on the talus below it (Rapp, 1960b; Rapp and Fairbridge, 1968).

Despite this variability, a two-dimensional profile is adequate for modelling the processes operating on alpine cliffs. These are dominated by the rapid movement of clastic waste in free fall and the bouncing and rolling which follows it. Associated with this is the work performed by snow avalanches whose effects have not yet been examined, except in terms of the deposits below the cliff (Rapp, 1959; Potter, 1969; Caine, 1969a; Luckman, 1971).

Rates of material movement from free-faces are usually based on measurements of talus accumulation which means that not all the measured accumulation need have come from the free-face. Sampling the volume of waste on a winter snow cover avoids this problem, but gives only an underestimate of the transfer rate. Measurements of this kind suggest cliff retreat rates of up to 1.0 mm yr^{-1} in a number of alpine areas (about 0.76 mm yr^{-1} in the Colorado Front Range), with even more rapid rates on very active cliffs. This is corroborated by estimates based on a longer time scale, e.g. 30 m in 30,000 yrs (Caine and Jennings, 1969) and 5 m in 10,000 yrs (Rapp, 1960b, p. 88). The rate of waste removal from the cliff, therefore, seems to occur ten to one hundred times faster than on the interfluves.

THE TALUS SLOPE

Much of the waste from alpine cliffs accumulates below them as a talus, of which three types (rockfall, alluvial and avalanche) have been recognized by White (1968). The form of the talus is controlled by the cliff topography: talus cones or compound talus arise below gullies in the free-face and simple or sheet talus where the cliff is approximately straight (Rapp, 1960b).

Rockfall talus is a simple response to the interaction of gravity and internal friction and so should have a rectilinear form with a slope approximating the angle of internal friction (37° to 40°) (Van Burkalow, 1945; Scheidegger, 1970, p. 84). However, departures from rectilinearity are common (Andrews, 1961; Caine, 1967), perhaps as a response to processes other than simple accumulation or to textural changes in the talus. The concave toe of many talus slopes, reported by early workers (e.g. Marr, 1909, p. 96), may be explained by a crude downslope size sorting of the waste. Alluvial talus is less simple, presumably because it is affected by water and mudflows as well as by direct accumulation. It tends to have a more concave profile form, a lower angle (30° to 38°) a greater volume of fines between the blocks and a more complex microtopography. Avalanche talus is even more concave and of gentler angle (under 30° for large part) than alluvial talus. It develops by the deposition of debris from snow avalanches and by their reworking of material already on the slope. Where large avalanches occur frequently, an avalanche talus tongue may even extend, at low angle, right across the valley floor.

The average movement on many active talus slopes is probably no more than 20 cm yr^{-1} at the surface but is very variable (Rapp, 1960a; Caine, 1963; Wallace, 1968; Gardner, 1969). Some particles may move up to 10 m in a year or even 100 m (Luckman, 1971), whilst others are motionless. This and the isotropic fabric (Klatka, 1961; Caine, 1967, 1969b) of rockfall talus suggests that each particle responds to stress as an individual. In the Colorado Front Range, my own measurements and those of Wallace (1968) suggest that surface movement is less than the average (under 5 cm yr^{-1}). At the head of the talus, accumulation by rockfall amounts to a layer of about 0.02 mm yr^{-1} which balances a downslope movement of 5 cm yr^{-1}, if movement involves only the surficial 50 cm of the talus.

On alluvial talus, the processes associated with low frequency storms predominate, producing an uneven talus surface by the branching and superposition of individual mudflow channels and levées. Mudflows are particularly effective in transporting debris from the cliff and the head of the talus to the lower parts of the slope, or even the valley floor. A single storm caused up to 4 m of gully erosion and involved 17,000 m^3 of talus in mudflow movement in the Ten Mile Range, Colorado (Curry, 1966). This represents a higher level of activity than that reported by Rapp (1962). Estimating return periods for mudflows on the same slope is difficult, but very important. A date of 180 yrs BP for unmodified mudflow deposits in the Front Range (Benedict, 1967) implies that the recurrence interval is to be estimated in centuries and, therefore, that their significance is no more than that of talus creep. Perhaps the distance of downslope movement and the fact that mudflows effectively transfer waste from the talus to the valley floor, and the stream channel, is more significant.

Avalanche talus is akin to alluvial talus in that the reworking of the slope is catastrophic and intermittent. Wet snow avalanches are most effective in reworking debris on the talus to give an unsorted jumble of blocks, many of which may be precariously perched. On such talus slopes are found the small-scale features, such as avalanche debris tails (Potter, 1969; Rapp, 1959), that are diagnostic of avalanche activity.

Again, the variability of the avalanche process makes estimation of the work it performs difficult. In Colorado, accumulation rates on the lower parts of avalanche talus reach 0.3 mm yr^{-1} in the San Juan Mountains, compared with a maximum of 0.6 mm yr^{-1} in New Zealand (Caine, 1969a) and 1.2 mm yr^{-1} in the Canadian Rockies (Luckman, 1971).

THE TALUS FOOT

The foot of the talus is subject to a variety of influences deriving from the discontinuities in sediment type, geomorphic process and waste movement rates which occur at the contact of valley floor and talus. Occasionally the talus may run out onto the valley floor but it is more usual for one of three sets of features specific to this zone to be interposed between the two.

At the foot of rockfall talus, it is common to find a fringing apron of scattered blocks. This develops from rockfalls which give large boulders that roll right across the talus. Below avalanche talus, a similar fringe of blocks seems to be related to rare avalanches of considerable magnitude. Many boulders in such aprons are embedded in the finer sediments of the valley floor and may be of great age. The volume of waste transferred to the valley floor by this process cannot be estimated on the basis of contemporary measurements because of its irregularity. Probably only a small part of all rockfall debris reaches the talus foot in this way.

With a long-lasting winter snow cover, the movement of blocks across the snow to the valley floor leads to the development of protalus ramparts (Bryan, 1934). These are unlikely to form where there is rapid talus accumulation (leading to their burial) or where avalanche processes (leading to their redistribution) occur. Most protalus ramparts are not developing rapidly today and may even be entirely relict.

The remaining features of the talus foot are rock glaciers and similar forms involving a flow of blocky debris away from the talus. These vary considerably in size. A single rock glacier may occupy the entire floor of a glacial cirque and be 2 or 3 km long (Kesseli, 1941) while others (valley wall rock glaciers: Outcalt and Benedict, 1965; and protalus lobes: Richmond, 1962, p. 20) are 100 m or less in size. All of these forms seem to require a deformation of the coarse mantle which is usually approximated by a viscous flow model (Wahrhaftig and Cox, 1959; Caine and Jennings, 1969; White, 1971). Some rock glaciers have been shown to contain a core of glacial ice when it may be assumed that flow in this ice is the cause of movement and the rock glacier itself the final stage in the burial of a true glacier (Lliboutry, 1961). In other cases, rock glacier motion may require only an interstitial matrix of ice and fine sediments (Wahrhaftig and Cox, 1969, pp. 401, 406, 415; Thompson, 1962).

Rock glacier flow involves movement as a single unit (Wahrhaftig and Cox, 1959; Outcalt and Benedict, 1965; Barsch, 1969b) which simplifies the estimation of the magnitude of waste transfer. For three Colorado examples, White (1971) defines average rates of surficial movement at 5.0 cm yr^{-1}, 6.6 cm yr^{-1}, and 9.7 cm yr^{-1}, which is less than that reported by Barsch (1969b) – up to 30 cm yr^{-1} – or Wahrhaftig and Cox (1959) – 76 cm yr^{-1}. From these rates, White also estimates the volumes of debris discharge at 215 m^3 yr^{-1}, 269 m^3 yr^{-1} and 771 m^3 yr^{-1}. These are equivalent to cliff retreat rates of 0.71 mm yr^{-1}, 0.79 mm yr^{-1} and 1.06 mm yr^{-1} which are lower than the estimate of Wahrhaftig and Cox (3.0 mm yr^{-1}), but correspond to the cliff retreat rates quoted earlier.

THE VALLEY FLOOR

In many alpine valleys the concave slope segment of the valley floor is absent, or occupied by the stream channel and standing water. Where it exists as a distinct unit, two types of alpine valley floor are distinguished.

In some areas relatively rapid sediment movement across a single valley floor seems quite common. In Spitsbergen, Jahn (1960) defines two sub-units of such periglacial foot slopes: an area of terrace forms due to solifluction and creep just below the talus; and, lower, a low-angle concave slope influenced by overland flow. This sequence is not clearly developed in the alpine, perhaps because of the slight extent of the valley floors, but wetter areas on which solifluction and overland flow occur are common and obvious around the lower edges of long lasting snowbanks.

In Colorado, most alpine valley floors are relatively inactive. Bedrock outcrops are interspersed with pockets of surficial waste, largely derived from the Late Pleistocene. The transfer of clastic debris onto and across the valley floor in this case is slight, although much of the stream load must be derived from this zone or carried across it.

Rates of waste movement on the valley floor are very variable. On inactive footslopes in Colorado three years of observations have not revealed any movement below the soil surface. In more active situations where vegetation cover is slight, the effects of overland flow may be obvious. However, they are unlikely to match the transfer rates effected by solifluction and frost creep higher on the slope (0.5 cm^3 cm^{-1} yr^{-1} compared with 50.0 cm^3 cm^{-1} yr^{-1} in the case examined by Jahn, 1961). These are probably maximum values approached only locally in the alpine and should not be thought representative of broad areas.

(c) Output processes

Most waste is eventually transferred from the slopes to the fluvial system of the stream channel. While this is an output from the slope, it is also an input of sediment to the stream channel and will be considered as such here.

A second sediment sink is the converse of the atmospheric dust fall. The evidence of wind disturbance of rock fragments is occasionally noted (Caine, 1963) but the quantitative significance of dust removal from the alpine system is unknown. Rapp (1960a, p. 186) considers wind effects to be negligible in the Scandinavian mountains but this is unlikely to be true of drier alpine areas. In situations like those found in Colorado, where little other sediment loss occurs, the atmosphere could be a relatively important sediment sink.

5. Alpine stream channel processes

The physical behaviour of water and sediment in open channels in the Alpine will probably not differ markedly from that at lower altitudes (Tricart, 1970, p. 141) and so need not be treated in detail here. The general topic of sediment transport has been reviewed on a number of occasions in recent years (e.g. Leopold *et al.*, 1964, Chapters 6 and 7; Allen, 1970, Chapter 4) and the empirical evidence of studies on glacial and high mountain streams suggests that they fit the conclusions of these reviews (Fahnestock, 1963; McPherson, 1971a, b). Some characteristics of the mountain environment, however, may have an influence on open channel processes and rates of sediment movement. A simple model for the interaction of the channel, the discharge and the sediment load (fig. 12B.4) may be applied to any open channel and suggests the areas in which such local

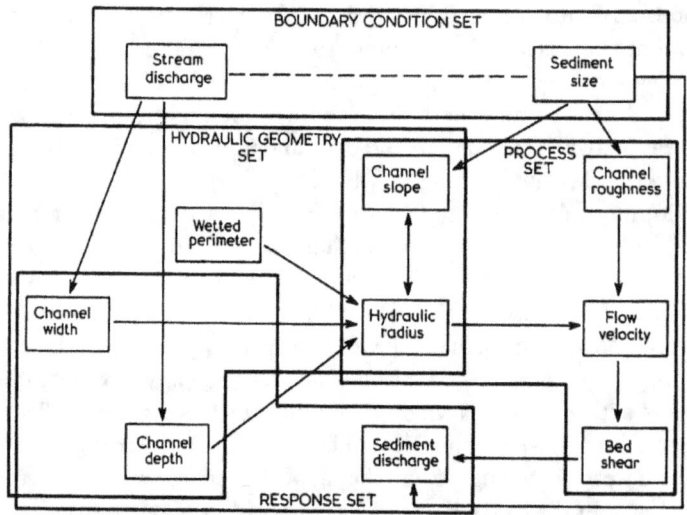

Fig. 12B.4. General relationships between the factors involved in the hydraulic geometry of a channel (after Chorley, 1969). Arrows suggest the direction of influence.

alpine influences might be felt. Water discharge, for example, will be affected by the flashiness of the alpine hydrograph (up to 0.64 m³ sec⁻¹ km⁻¹) (Slaymaker, Chapter 3B) and by the drainage network (particularly by the presence of lakes within it). Sediment size will probably reflect the heterogeneous material supplied to the channel and the Pleistocene history of the landscape. The effects of these two boundary conditions should eventually show in the work of erosion and sediment transportation and deposition. The effect will, however, be made through the same interactions between hydraulic geometry and flow terms that are found elsewhere.

Stream channel processes are considered here in terms of a sediment budget model which represents a continuation of the waste transfers on the slopes around the channel.

SEDIMENT SOURCES

There are three sources for the sediment stored in or in transit through a segment of stream channel (fig. 12B.3). One of these (Ac) will be ignored here since it is a transfer within the channel. The other two sources from the slopes around the channel (Ab) and from the channel itself by erosion of its bed and banks (W) will be considered in turn.

Sediment sources on the hillsides

The sediment derived directly from the valley sides may be subdivided according to its calibre and the tractive power available in the channel. One subset of this material comprises the accumulation of large blocks that is common in alpine channels (but rare at lower altitudes) and may even bury them completely. While this detritus may not be transported by the stream, it will affect the channel shape and roughness and, through that, the work of erosion and sediment transportation (fig. 12B.4). It may have a protective, armouring effect on finer sediments on the bed but, conversely, will also tend to increase the flow resistance of the channel and, through that, the rate of sediment transport. The net effect will, therefore, not be simple and will also vary both spatially and temporally. Waste input to the channel of this calibre will be largely a response to such processes as rockfalls, avalanches and mudflows that are capable of transporting large blocks. Large boulders in the stream channel may remain from former glacial episodes although Stewart and LaMarche (1967) doubt such an explanation, at least as it applies to valley floor deposits below timberline.

Another subset of the stream sediments is the finer particles (predominantly clay, silt and fine sand) and solutes transported to the channel by overland flow and ground and soil water flow. The fine sediments transported by mudflows and other catastrophic slope processes might be included in this subset. These finer sediments will be further transported, at least periodically. Solutes are likely to be removed most rapidly while sand and gravel will move more slowly in a sequence of intermittent steps, each step associated with a peak in stream discharge. The frequency and distance of movement in these steps will be controlled by the periodicity of the hydrologic events responsible for them.

Channel erosion

The second major source of sediment (through bed and bank erosion) is not restricted to the channel itself, for the period of snowmelt, when stream discharges and their associated erosional work are at their highest level, often coincides with the blocking of perennial channels by ice. As the stream breaks out of its channel it erodes the valley floor wherever contact is made with it (elsewhere the flow may be over or through the snow cover). This erosion may be facilitated by the lack of a protective armouring of coarse sediments on the valley floor. It tends to give an uneven cut-and-fill topography and may be a cause of the reticulate drainage patterns of some alpine valleys.

The processes by which sediment is set in motion in the stream channel are those associated with the shear of the fluid moving through the channel (Leopold *et al.*, 1964, pp. 169–84; Allen, 1970, pp. 122–8). They may be aided by local effects, such as that due to frost action on the riverbanks or the variations in roughness associated with channel icing, but even these will only be felt through the links defined in fig. 12B.4.

Input rates

The rate at which sediment is input to any stream channel is particularly difficult to estimate. Bed and bank erosion may be measured directly, and can even be tabulated on a continental scale (Todd, 1970, p. 79), but the amount of waste derived directly from the hillslopes surrounding the channel can only be estimated. A quantitative estimate of Ab (fig. 12B.3) for a Colorado situation is still not possible, although the relative inactivity of the valley floors suggests that the channel gains little clastic sediment from this source. Solute input, on the other hand, is tied to ground and soil water movement and so to the sources of stream runoff. At lower altitudes the study of these is difficult (Carson and Sutton, 1971), and in the alpine it remains almost unconsidered.

Over a period of three years, direct observation in the upper Green Lakes Valley, Colorado Front Range, confirms the low level of waste input from alpine slopes to the stream channel. Even where the channel is steep, little modification occurs at the present time. The most significant erosional input seems to occur when the streams are forced out of their channels by ice blocking in the late spring. Then, an average of about 0.1 m^3 yr^{-1} of debris is acquired and later deposited in the main channel for each kilometre length of the channel (this is equivalent to a denudation rate of 0.00005 mm yr^{-1}).

FLUVIAL TRANSFER PROCESSES

Fluvial transportation in the alpine, like that in other environments, is also a product of the shear exerted on the channel bed and banks by the flowing water (fig. 12B.4). It is very variable, in response to the wide fluctuations in stream discharge found in all small catchments and to the range of available sediment sizes and channel form.

The mechanics of fluvial sediment transfer have been summarized by Leopold *et al.* (1964, Chapter 6), Allen (1970, Chapter 4) and Scheidegger (1970, Chapter 4.5). Transport is usually said to occur in three modes: bed load, suspended load and solute load. Bed load is entrained into motion when the bed shear exceeds a critical drag and is kept moving by the continued application of this force. Particles carried in suspension are entrained by the same force but are supported by turbulence in the moving water and carried downchannel by the flow. The solute load is dispersed more uniformly and carried at negligible cost in terms of energy. For this reason solute transportation is an efficient one (Tricart, 1961; Rapp, 1960a, p. 184; Douglas, 1964) which is unlikely to constitute an internal transfer within the small alpine stream basins.

Although the general problem of sediment movement is common to all environments in which streamflow occurs, the relative significance of its three modes may vary. An increase in the solute content of river water with distance downstream (Livingstone, 1963, p. 5) suggests that the solute load may be insignificant in many alpine zones and this is supported by the survey of Durum *et al.* (1960) and the recent work of McPherson (1971b). The data of other workers (e.g. Viro, 1953, p. 14; Barsch, 1969a, p. 83), however, show that, even in cold climates, solutes may make up more than half the total sediment load. Obviously the geochemical environment and the area of the drainage basin have an influence which precludes any simple generalization.

Similarly, it is often assumed that bed load transportation is relatively more important in alpine areas than elsewhere, but this too appears to be variable. McPherson (1971b) shows the bed load contribution to be less than 0.5 per cent of the total sediment yield of Two O'Clock Creek, and Muller and Forstner (1968) estimate it at only 1 per cent of the sediment load of the Rhine at Lake Constance. Mapes (1969), on the other hand, suggests that bed transportation amounts to between 5 and 12 per cent of the suspended load in the Walla Walla drainage. This variability seems to be due to the flow of the rivers concerned during the period of study and so emphasizes the need to take time into consideration. If low-frequency floods occur during the study period (as in the Walla Walla), the relative significance of bed load transport may seem to increase at the expense of the other modes of sediment transfer.

Transportation rates

Rates of sediment transfer in the alpine environment are reflected in the sediment output from the system and, as such, will be treated later. The processes of sediment movement through the stream channel and of its removal from the basin by the stream are identical.

FLUVIAL OUTPUT PROCESSES

The products of geomorphic activity in an alpine stream basin may be lost either by continued transfer down the channel to subalpine altitudes, or by sedimentation into lakes within the alpine area itself. On a geologic time scale, the latter of these may be no more than a temporary storage but, over a shorter period, it can be considered a separate output. The distinction is made clearer by the need to evaluate the two outputs by different methods.

Sediment sinks

The processes of bed traction and turbulence responsible for the transfer of material in the stream channel above timberline continue its movement out of the alpine system. The relative importance of the three modes of fluvial transportation probably remains the same: suspended load seems to contribute most to the total sediment yield while the significance of bed load and solute load movement varies widely from area to area.

The seasonal variability of the alpine sediment yield is marked. McPherson (1971b) shows the normal coupling of the seasonal hydrograph and the rate of clastic sediment removal from an alpine catchment. The relationship between solute load and stream discharge is less normal, particularly when snowmelt makes a large contribution to streamflow. In the Green Lakes Valley, both the solute content and the stream discharge tend to decline from late May to August, after which the more usual inverse relationship imposes itself.

The second set of waste outputs involves sediment deposition which may occur within the channel itself, when it is likely to be only temporary, or in lake basins. Only the latter is of significance here. In Colorado alpine lakes, the redistribution and mixing of the fine sediments following their initial deposition prevents the formation of a seasonal sedimentary structure although the process of sedimentation presumably reflects the seasonal pattern of fluvial activity outside the lakes.

Stream channel output rates

The correspondence of high sediment yields with mountainous topography need not be a response to the output of waste from the alpine zone. The fluvial sediment load in mountain areas is normally measured at subalpine altitudes and so applies to an area in which the true alpine may be only a small part. Further complications arise from a lack of standardized methods. Some workers use only the solute content of river water (e.g. Hembree and Rainwater, 1961) but it is more common to consider the suspended load component by itself since this is the most important single one (e.g. Litvin, 1969; Muller and Forstner, 1968). The measurement of bed load is notoriously difficult and so its contribution is often ignored. Nevertheless, during major, infrequent floods it may be extremely significant (Stewart and LaMarche, 1967).

In view of the problems and errors associated with stream and sediment discharge measurement, only qualitative statements about sediment loss from the alpine environment in general are possible. Estimated values for the Colorado Rocky Mountains allow comparison with the earlier consideration of other processes but may not be representative of all mountain areas.

The solute content of water flowing from the upper Green Lakes valley in the Front Range is usually less than 12 ppm. Measurements are made at a lake outlet, where the suspended load is negligible (under 3 ppm) for much of the time, so that solute transportation must be the main agent of waste removal through the stream channel sink. Although low, a solute content of 12 ppm is probably representative of small (2 km² area) basins on granitic and metamorphic materials. Hembree and Rainwater (1961) and Douglas (1968, p. 131) record values of 29 ppm and 19 ppm for very much larger alpine drainage basins (up to 240 km²) on similar granitic bedrock. Solute contents as low as 20 ppm are also fairly common in the larger streams of the Front Range at lower altitude (e.g. USGS, 1962, p. 254).

A total solute content of 12 ppm includes the solutes brought into the basin by the precipitation (Rapp, 1960a, p. 167; Hembree and Rainwater, 1961) and so must be reduced to give a chemical denudation rate. Values of 5 ppm of dissolved solids in fresh snow from the Green Lakes correspond closely to the estimates for the Wind River Range made by Hembree and Rainwater (1961) but, after correction by the ratio of precipitation to stream runoff, leave only 1 to 2 ppm as the net solute load for much of the year. This corresponds to a denudation rate of about 0.002 mm yr^{-1} which is not very much lower than that derived by Hembree and Rainwater (1961) for the southwestern part of the Wind River Range (0.0036 mm yr^{-1}). Values about an order of magnitude greater have been estimated for the upper Rhine (Jackli, 1957, quoted in Leopold et al., 1964, p. 93), where the effects of basin size might account for the difference, and for the Two O'Clock Creek drainage (McPherson, 1971b), where the limestone bedrock probably accounts for the difference.

The loss of clastic sediment from the upper Green Lakes Valley amounts to a surface lowering of less than 0.001 mm yr^{-1} which suggests that sedimentation within the lakes of the valley is more significant than the removal of waste by the stream. Values for North Clear Creek in the Front Range (USGS, 1959, p. 224) correspond to a much greater denudation rate (0.015 mm yr^{-1}) and are close to those estimated for Transcaucasia (0.016 mm yr^{-1}: Litvin, 1970) and the upper Rhine (0.014 mm yr^{-1}: Muller and Forstner, 1968). Leaf (1970) quotes higher values for two undisturbed catchments in the Fraser Experimental Forest, Colorado (0.093 mm yr^{-1} and 0.064 mm yr^{-1}). McPherson's (1971b) observations give an even higher denudation rate for a small basin in the Canadian Rocky Mountains (0.48 mm yr^{-1} by the suspended load alone) but even this does not approach the value for the Kosi River in the Himalaya (0.965 mm yr^{-1}: Schumm, 1963).

Alpine lake sedimentation rates

The main sediment sink in many small alpine catchments is probably through sedimentation into glacial lake basins. As with most alpine geomorphic processes, lake infilling should be quite variable and those areas in which fluvial transportation is more active than it is in the Front Range will probably show much more rapid sedimentation rates than those discussed here.

Pennak (1963) gives sedimentation rates for two subalpine lakes on the east slope of the Front Range but, since the sediments contain five times as much organic detritus as those in alpine lakes in the same area and are derived from smaller catchments, the rates (0.19 mm yr^{-1} and 0.28 mm yr^{-1}: Pennak, 1963, p. 6) are not immediately relevant here. The start of recent infilling in Redrock Lake at 3100 m altitude (6900 yrs BP) may, however, be correlated with that in nearby alpine lakes at 3500 m: 75,000 m^3 of recent sediments in the upper Green Lakes Valley would then represent a mean sedimentation rate of 0.14 mm yr^{-1} and a denudation rate of 0.0043 mm yr^{-1} for the basin. This is slightly less than the present annual dust fall. The similarity of the lake sediments and the present dust fall in terms of their mechanical composition also suggests a direct link between the two.

6. Conclusion

It is difficult to define a general model for alpine geomorphic activity. Fig. 12B.5 represents a simple graphical attempt at such but is applicable only to the geomorphic system of the Colorado Front Range. This area is probably not typical, for the mean denudation rate there is about two orders of magnitude less than that found in other alpine regions. In qualitative terms, however, the sediment transfers shown in fig. 12B.5 may retain their relative significance elsewhere and so allow discussion of the alpine geomorphic system in general.

The definition of two subsystems in the alpine drainage basin (the hillslopes and the drainage channels) is useful, but is not the only one that might be made. In terms of sediment transfer rates, the basin could be considered an amalgamation of three subsystems defined by the nature of the sediments involved and their movement: a geochemical subsystem, a fine sediment subsystem, and a coarse detritus subsystem. These

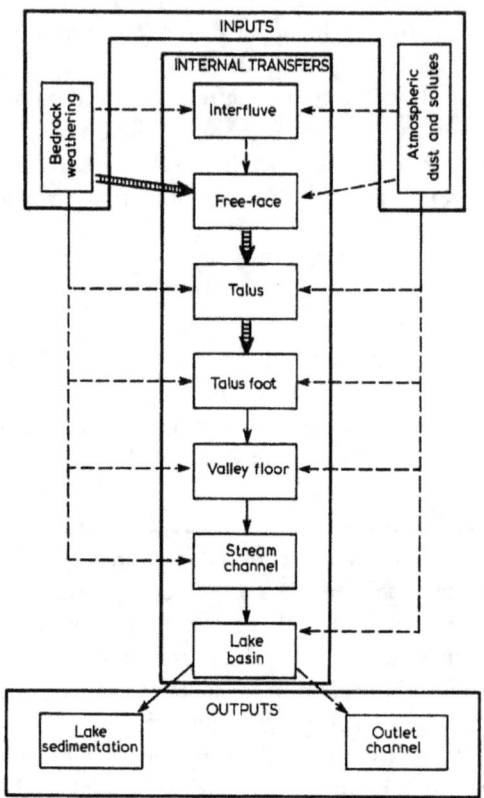

Fig. 12B.5. An alpine sediment transfer model. The arrows show the directions and relative magnitudes of the waste transfers in the alpine system.

overlap in both time and space, and even act together at times (as in the case of mudflow transportation which involves both coarse and fine waste).

Solution weathering and transportation make up the geochemical subsystem and are largely controlled by the water budget. In the Front Range, this subsystem is dominated by a uniform atmospheric input which contributes two or three times as much material as does bedrock and clastic waste weathering. This is unlikely to be generally true of the alpine, however. An increase in catchment size (giving a longer average period of contact between water and the weathering materials) and a different lithology will act to reduce the significance of atmospheric salts. In limestone areas, the solution subsystem will probably be of much greater significance than elsewhere. Waste transfer in solution is particularly efficient although, on the Front Range evidence, involving only a small mass. Responses to fluctuations in the water budget should be made rapidly and, because of this, the geochemical subsystem should achieve a steady state quite quickly.

The second subsystem, involving particles under about 2 mm ($-1\ \phi$) size, contributes to lake infilling and derives largely from an atmospheric dust source in Colorado. This subsystem, too, is controlled by the environmental water budget and responds relatively rapidly to changes in this. About 50 per cent of the present sedimentation in alpine lakes in Colorado seems to derive from an atmospheric source and waste tends to move to this sink quite rapidly.

In terms of volume, the coarse detritus subsystem dominates (fig. 12B.5) and is most characteristic of the alpine environment. It operates on the free-face, talus and talus foot units which are often thought to develop as a closed system in terms of the coarse detritus (e.g. Bakker and Le Heux, 1952; Caine, 1969a). The upper limit of the free-face and the lower edge of the talus or protalus form discontinuities across which only solutes and fine sediments are transported. The analogy to a closed thermodynamic system also suggests that the burial of the free-face by its own talus is akin to the ultimately degraded state of the classical closed system. Such an unchanging ultimate state would probably be approached asymptotically over a long time-scale and so contrasts with the relatively rapid trend to steady state in the other subsystems.

This suggests an alpine geomorphic system consisting of two dynamic subsystems (in the slopes and the stream channels) overlain by three sediment subsystems. All operate in the same geographic area but respond to boundary conditions in different ways and at different rates. It is through this variety of controls and responses (and the effects of time in introducing a 'historical hangover') that the diversity of alpine scenery is produced, rather than through the operation of geomorphic processes peculiar to the alpine.

References

AHNERT, F. (1967) The role of the equilibrium concept in the interpretation of landforms of fluvial erosion and deposition. *In:* MACAR, P. (ed.) *L'évolution des Versants*, 23–41. Belgium: Université de Liège.

AHNERT, F. (1970) Functional relationships between denudation, relief and uplift in large mid-latitude drainage basins. *Amer. J. Sci.*, **268**, 243–63.

ALLEN, J. R. L. (1970) *Physical Processes of Sedimentation.* London: Allen & Unwin. 248 pp.

ANDREWS, J. T. (1961) The development of scree slopes in the English Lake District and central Quebec – Labrador. *Cah. Géogr. Qué*, **10**, 219–30.

BAIRD, P. D. and LEWIS, W. V. (1957) The Cairngorm floods, 1956. *Scot. Geogr. Mag.*, **73**, 91–100.

BAKKER, J. P. and LE HEUX, J. W. N. (1952) A remarkable new geomorphological law. *Kowinkl. Nederlandse Akad. van Wetenschappen, Proc.*, ser. B, **55** (4), 399–410, 544–70.

BARSCH, D. (1969a) Studien zur Geomorphogenese des zentralen Berner Alpen. *Basler Beitr. zur Geogr.*, 9. 221 pp.

BARSCH, D. (1969b) Studien und Messungen an Blockgletschern in Macun, Unterengadin. *In:* Glazialmorphologie, *Zeit. Geomorphol.*, Suppl. 8, 11–30.

BAUER, F. (1962) Karstformen in den Osterreichischen Kalkhochalpen. *Actes 2nd Int. Congr. Speleology* (1958), 299–329.

BECKINSALE, R. P. (1969) River regimes. *In:* CHORLEY, R. J. (ed.) *Water, Earth and Man*, 455–71. London: Methuen.

BENEDICT, J. B. (1967) Recent glacial history of an alpine area in the Colorado Front Range, USA. I: Establishing a lichen-growth curve. *J. Glaciol.*, **6**, 817–32.

BENEDICT, J. B. (1970) Downslope soil movement in a Colorado alpine region: rates, processes and climatic significance. *Arct. Alp. Res.*, **2**, 165–226.

BIK, M. J. J. (1967) Structural geomorphology and morphoclimatic zonation in the Central Highlands, Australian New Guinea. *In:* JENNINGS, J. N. and MABBUTT, J. A. (eds.) *Landform Studies from Australia and New Guinea*, 26–47. Canberra: Australian National Univ. Press.

BÖGLI, A. (1960) Kalklosung und Karrenbildung. *In:* Int. Beitr. zur Karstmorphologie, *Zeit. Geomorphol.*, Suppl. 2, 4–21.

BORMANN, F. H. and LIKENS, G. E. (1967) Nutrient cycling. *Science*, **155**, 424–9.

BRYAN, K. (1934) Geomorphic processes at high altitudes. *Geogr. Rev.*, **24**, 655–6.

BÜDEL, J. (1968) Geomorphology – Principles. *In:* FAIRBRIDGE, R. W. (ed.) *Encyclopedia of Geomorphology*, 416–22. New York: Reinhold.

CAINE, N. (1963) Movement of low angle scree slopes in the Lake District, northern England. *Rev. Géomorph. Dyn.*, **14**, 171–7.

CAINE, N. (1967) The texture of talus in Tasmania. *J. Sedim. Petr.*, **37**, 796–803.

CAINE, N. (1969a) A model for alpine talus slope evolution by slush avalanching. *J. Geol.*, **77**, 92–100.

CAINE, N. (1969b) The analysis of surface fabrics on talus by means of ground photography. *Arct. Alp. Res.*, **1**, 127–34.

CAINE, N. and JENNINGS, J. N. (1969) Some blockstreams of the Toolong Range, Kosciusko State Park, New South Wales. *J. & Proc. Roy. Soc. NSW*, **101**, 93–103.

CARSON, M. A. (1969) Models of hillslope development under mass failure. *Geogr. Ann.*, **1**, 76–100.

CARSON, M. A. and SUTTON, E. A. (1971) The hydrologic response of the Eaton River Basin, Quebec. *Can. J. Earth Sci.*, **8**, 102–15.

CHORLEY, R. J. (1969) The drainage basin as the fundamental geomorphic unit. *In:* CHORLEY, R. J. (ed.) *Water, Earth and Man*, 77–99. London: Methuen.

CHORLEY, R. J., DUNN, A. J. and BECKINSALE, R. P. (1964) *The History of the Study of Landforms*, Vol. 1: *Geomorphology before Davis*. London: Methuen. 678 pp.

COOK, F. A. and RAICHE, V. G. (1962) Freeze–thaw cycles at Resolute, NWT. *Geogr. Bull.*, **18**, 64–78.

CORBEL, J. (1959) Vitesse de l'érosion. *Zeit. Geomorphol.*, N.F., **3**, 1–28.

CORBEL, J. (1964) L'érosion terrestre, étude quantitative. *Ann. Géogr.*, **398**, 412.

COSTIN, A. B., WIMBUSH, D. J. and KERR, D. (1960) Studies in catchment hydrology in the Australian Alps, II: Surface run-off and soil loss. *CSIRO Tech. Pap.*, 14. 23 pp.

COSTIN, A. B., JENNINGS, J. N., BLACK, H. P. and THOM, B. G. (1964) Snow action on Mt Twynam, Snowy Mountains, Australia. *J. Glaciol.*, **5**, 219–28.

CURRY, R. R. (1966) Observation of alpine mudflows in the Ten Mile Range, Central Colorado. *Geol. Soc. Amer. Bull.*, **77**, 771–6.

DAVIES, J. L. (1969) *Landforms of Cold Climates*. Cambridge, Mass.: MIT Press. 200 pp.

DERRUAU, M. (1968) Mountains. *In:* FAIRBRIDGE, R. W. (ed.) *Encyclopedia of Geomorphology*, 737–9. New York: Reinhold.

DIACONU, C. (1969) Résultats de l'étude de l'écoulement des alluvions en suspension des rivières de la Roumanie. *Bull. Int. Ass. Sci. Hydrol.*, **XIV**, 57–89.

DOUGLAS, I. (1964) Intensity and periodicity in denudation processes with special reference to the removal of material in solution by rivers. *Zeit. Geomorphol.*, **8**, 453–73.

DOUGLAS, I. (1968) The effects of precipitation chemistry and catchment area lithology on the quality of river water in selected catchments in eastern Australia. *Earth Sci. J.*, **2**, 126–44.

DURUM, W. H., HEIDEL, S. G. and TISON, L. J. (1960) World-wide runoff of dissolved solids. *Int. Ass. Sci. Hydrol. Pub.* (General Assembly of Helsinki), **51**, 618–28.

EMBLETON, C. and KING, C. A. M. (1968) *Glacial and Periglacial Geomorphology*. London: Edward Arnold. 608 pp.

FAHNESTOCK, R. K. (1963) Morphology and hydrology of a glacial stream – White River, Mt Rainier, Washington. *US Geol. Surv. Prof. Pap.*, 422-A. 70 pp.

FAIRBRIDGE, R. W. (1968) Mountain types. *In:* FAIRBRIDGE, R. W. (ed.) *Encyclopedia of Geomorphology*, 751–61. New York: Reinhold.

FISHER, N. H. (1970) Rock weathering, anatexis, and ore deposits. *Search*, **1**, 111–19.

FOURNIER, F. (1960) *Climat et Erosion*. Paris: Presses Universitaires de France. 201 pp.

GARDNER, J. (1969) Observations of surficial talus movement. *Zeit. Geomorphol.*, N.F., **13**, 317–23.

GARDNER, J. (1970a) A note on the supply of material to debris slopes. *Can. Geogr.*, **14**, 369–72.

GARDNER, J. (1970b) Geomorphic significance of avalanches in the Lake Louise area, Alberta, Canada. *Arct. Alp. Res.*, **2**, 135–41.

GERBER, E. (1969) Bildung und Formen von Gratgipfeln und Felswanden in den Alpen. *Zeit. Geomorphol.*, Suppl. 8, 94–118.

GIBBS, R. J. (1970) Mechanism controlling world water chemistry. *Science*, **170**, 1088–90.

GOODCHILD, J. G. (1890) Note on some observed rates of weathering of limestones. *Geol. Mag.*, **27**, 463–6.

GORHAM, E. (1961) Factors influencing the supply of major ions to inland waters, with special reference to the atmosphere. *Bull. Geol. Soc. Amer.*, **72**, 795–840.

HEMBREE, C. H. and RAINWATER, F. H. (1961) Chemical degradation on opposite flanks of the Wind River Range, Wyoming. *US Geol. Surv. Water Supply Pap.*, 1535-E. 9 pp.

IASH (1965) Symposium of Budapest: Representative and Experimental Areas. *Int. Ass. Sci. Hydrol. Pub.*, 65 and 66. 2 vols, 710 pp.

JACKLI, H. (1957) Gegenwartsgeologie des bundnerischen Rheingebietes – ein Beitrag zur exogen Dynamik Alpiner Gebirgslandschaften. *Beitr. Geol. der Schweiz*, Geotech. Ser., 36. 126 pp.

JAHN, A. (1960) Some remarks on evolution of slopes on Spitsbergen. *Zeit. Geomorphol.*, Suppl. 1, 49–58.

JAHN, A. (1961) Quantitative analysis of some periglacial processes in Spitsbergen. *Zesz. Nauk. Univ. Wroclaw*, Ser. B, **5**. 56 pp.

JAHN, A. (1964) Slopes morphological features resulting from gravitation. *Zeit. Geomorphol.*, Suppl. 5, 59–72.

JENNINGS, J. N. and BIK, M. J. (1962) Karst morphology in Australian New Guinea. *Nature*, **194**, 1036–8.

KELLER, W. D. (1957) *The Principles of Chemical Weathering*. Columbia, Miss.: Lucas Bros. 111 pp.

KESSELI, J. E. (1941) Rock streams in the Sierra Nevada, California. *Geogr. Rev.*, **31**, 203–27.

KLATKA, T. (1961) Indices de structure et de texture des champs de pierres des Lysogory. *Bull. Soc. Sci. Lettr. de Lodz*, **12**, 10. 21 pp.

LEAF, C. F. (1970) Sediment yields from the central Colorado snow zone. *ASCE, J. Hydraul. Div.*, 96, HY1, 87–93.

LEOPOLD, L. B., WOLMAN, M. G. and MILLER, J. P. (1964) *Fluvial Processes in Geomorphology*. San Francisco: W. H. Freeman. 522 pp.

LITVIN, L. F. (1969) Relationships between landscape – geomorphological factors and sediment discharge in rivers of western Transcaucasia. *Soviet Hydrol.*, **1**, 85–8.

LIVINGSTONE, D. A. (1963) Chemical composition of rivers and lakes. *In:* FLEISCHER, M. (ed.) The data of geochemistry, *US Geol. Surv. Prof. Pap.*, 440-G, Chap. G. 64 pp.

LLIBOUTRY, L. (1961) Les glaciers enterrés et leur rôle morphologique. *Int. Ass. Sci. Hydrol. Pub.* (General Assembly of Helsinki), 54, 272–80.

LUCKMAN, B. H. (1971) The role of snow avalanches in the evolution of alpine talus slopes. *Inst. Brit. Geogr., Spec. Pub.*, 3, 93–110.

MCPHERSON, H. J. (1971a) Downstream changes in sediment character in a high energy mountain stream channel. *Arct. Alp. Res.*, **3**, 65–79.

MCPHERSON, H. J. (1971b) Dissolved, suspended and bed load movement patterns in Two O'Clock Creek, Rocky Mountains, Canada, Summer 1969. *J. Hydrol.*, **12**, 221–33.

MAPES, B. E. (1969) Sediment transport by streams in the Walla Walla River basin, Washington and Oregon, July 1962 – June 1965. *US Geol. Surv., Water Supply Pap.*, 1868. 32 pp.

MARR, J. E. (1909) *The Scientific Study of Scenery* (3rd edn). London: Methuen. 372 pp.

MULLER, G. and FORSTNER, U. (1968) General relationship between suspended sediment concentration and water discharge in the Alpenrhein and some other rivers. *Nature*, **217**, 244–5.

OLLIER, C. (1969) *Weathering*. Edinburgh: Oliver & Boyd. 304 pp.

OUTCALT, S. I. and BENEDICT, J. B. (1965) Photo-interpretation of two types of rock glacier in the Colorado Front Range, USA. *J. Glaciol.*, **5**, 849–56.

PARFENOVA, E. I. and YARILOVA, E. A. (1965) *Mineralogical Investigations in Soil Science*. Jerusalem: Israel Progr. Sci. Transl. 178 pp.

PEARSALL, W. H. (1950) *Mountains and Moorlands*. London: Collins. 312 pp.

PELTIER, L. C. (1950) The geographic cycle in periglacial regions as it is related to climatic geomorphology. *Ann. Ass. Amer. Geogr.*, **40**, 214–36.

PENNAK, R. W. (1963) Ecological and radiocarbon correlations in some Colorado mountain lake and bog deposits. *Ecology*, **44**, 1–15.

PIPPAN, T. (1967) Slope studies in the flysch zone and the Tertiary hills of the Pleistocene Salzach Glacier area in Salzburg and upper Austria. *In:* MACAR, P. (ed.) *L'évolution des Versants*, 201–14. Belgium: Université de Liège.

POTTER, N. (1969) Tree-ring dating of snow avalanche tracks and the geomorphic activity of avalanches, northern Absaroka Mountains, Wyoming. *Geol. Soc. Amer. Spec. Pap.*, 123, 141–65.

RAPP, A. (1959) Avalanche boulder tongues in Lappland. *Geogr. Ann.*, **41**, 34–48.

RAPP, A. (1960a) Recent developments of mountain slopes in Kärkevagge and surroundings, northern Scandinavia. *Geogr. Ann.*, **42**, 71–200.

RAPP, A. (1960b) Talus slopes and mountain walls at Tempelfjorden, Spitsbergen. *Norsk Polarinst. Skrifter*, **119**. 96 pp.

RAPP, A. (1962) Kärkevagge. Some recordings of mass movements in the northern Scandinavian mountains. *Biul. Peryglac.*, **11**, 287–309.

RAPP, A. and FAIRBRIDGE, R. W. (1968) Talus fan or cone: scree and cliff debris. *In:* FAIRBRIDGE, R. W. (ed.) *Encyclopedia of Geomorphology*, 1107–9. New York: Reinhold.

RAPP, A. and RUDBERG, S. (1960) Recent periglacial phenomena in Sweden. *Biul. Peryglac.*, **8**, 143–54.

RICHMOND, G. M. (1962) Quaternary stratigraphy of the La Sal Mountains, Utah. *US Geol. Surv. Prof. Pap.*, 324. 135 pp.

SCHEIDEGGER, A. E. (1970) *Theoretical Geomorphology* (2nd edn). Berlin: Springer-Verlag. 435 pp.

SCHUBERT, C. and SIFONTES, R. S. (1970) Bocono fault, Venezuelan Andes: evidence of post glacial movement. *Science*, **170**, 66.

SCHUMM, S. A. (1963) The disparity between present rates of denudation and orogeny. *US Geol. Surv. Prof. Pap.*, 454-H. 13 pp.

SCOTT, R. F. (1969) The freezing process and mechanics of frozen ground. *Cold Regions Science and Engineering Monogr.*, II-D1. Hanover, NH: US Army CRREL. 67 pp.

SHARP, R. P. (1969) Semiquantitative differentiation of glacial moraines near Convict Lake, Sierra Nevada, California. *J. Geol.*, **77**, 68–91.

SHARPE, C. F. S. (1938) *Landslides and Related Phenomena*. New York: Columbia Univ. Press. 137 pp.

SHREVE, R. L. (1966) Sherman Landslide, Alaska. *Science*, **154**, 1639–43.

SILVERMAN, M. P. and MUNOZ, E. F. (1970) Fungal attack on rock: solubilization and altered infrared spectra. *Science*, **169**, 985–7.

SIMONETT, D. S. (1967) Landslide distribution and earthquakes in the Bewani and Torricelli Mountains, New Guinea. *In:* JENNINGS, J. N. and MABBUTT, J. A. (eds.) *Landform Studies from Australia and New Guinea*, 64–84. Canberra: Australian National Univ. Press.

SLAYMAKER, H. O. and CHORLEY, R. J. (1964) The vigil network system. *J. Hydrol.*, **2**, 19–24.

SPARKS, B. W. (1960) *Geomorphology*. London: Longmans. 371 pp.

STELZER, F. (1962) Frostwechsel und Zone maximaler Verwitterung in den Alpen. *Wetter und Leben*, **14**, 210–13.

STEWART, J. H. and LAMARCHE, V. C. JR (1967) Erosion and deposition produced by the flood of December 1964 on Coffee Creek, Trinity County, California. *US Geol. Surv. Prof. Pap.*, 422-K. 22 pp.

STODDART, D. R. (1969) Climatic geomorphology: review and re-assessment. *In:* BOARD, C., CHORLEY, R. J., HAGGETT, P. and STODDART, D. R. (eds.) *Progress in Geography*, Vol. 1, 159–222. London: Edward Arnold.

SUGGATE, R. P. (1965) Late Pleistocene geology of the northern part of the South Island, New Zealand. *New Zealand Geol. Surv. Bull.* (n.s.) **77**. 91 pp.

SWEETING, M. M. (1966) The weathering of limestones with particular reference to the Carboniferous limestones of northern England. *In:* DURY, G. H. (ed.) *Essays in Geomorphology*, 177–210. London: Heinemann.

THOMPSON, W. F. (1962) Preliminary notes on the nature and distribution of rock glaciers relative to true glaciers and other effects of the climate on the ground in North America. *Int. Ass. Sci. Hydrol. Pub.* (Commission of Snow and Ice Symposium, Obergurgl), 58, 212–19.

THOMPSON, W. F. (1968) New observations on alpine accordances in the western United States. *Ann. Ass. Amer. Geogr.*, **58**, 650–69.

TODD, D. K. (1970) *The Water Encyclopedia.* Port Washington, NY: Water Information Center. 559 pp.

TRICART, J. (1956) Étude expérimentale du problème de la gélivation. *Biul. Peryglac.*, 4, 285–317.

TRICART, J. (1961) Observations sur le charriage des matériaux grossiers par les cours d'eau. *Rev. Géomorphol. Dyn.*, **12**, 3–15.

TRICART, J. (1963) *Géomorphologie des Régions Froides.* Paris: Presses Universitaires de France. 289 pp.

TRICART, J. (1970) *Geomorphology of Cold Environments* (transl. E. Watson). London: Macmillan. 320 pp.

UNITED STATES GEOLOGICAL SURVEY (1959) Quality of surface waters of the United States, 1955: Parts 5 and 6. *US Geol. Surv. Water Supply Pap.*, 1401. 305 pp.

UNITED STATES GEOLOGICAL SURVEY (1962) Quality of surface waters of the United States, 1958: Parts 5 and 6. *US Geol. Surv. Water Supply Pap.*, 1572. 365 pp.

VAN BURKALOW, A. (1945) Angle of repose and angle of sliding friction: an experimental study. *Bull. Geol. Soc. Amer.*, **56**, 669–708.

VARNES, D. J. (1958) Landslide types and processes. *In:* ECKEL, E. B. (ed.) *Landslides and Engineering Practice*, 20–47. Spec. Rep. 29. Highway Research Board, USA.

VIRO, P. J. (1953) Loss of nutrients and the natural nutrient balance of the soil in Finland. *Comm. Inst. Forest. Fenn.*, **42** (1). 51 pp.

YOUNG, A. (1964) Slope profile analysis. *In:* Forschritte der internationalen Hangforschung, *Zeit. Geomorphol.*, Suppl. 5, 17–27.

YOUNG, A. (1969) Present rate of land erosion. *Nature*, **224**, 851–2.

WAHRHAFTIG, C. and COX, A. (1959) Rock glaciers in the Alaska Range. *Bull. Geol. Soc. Amer.*, **70**, 383–436.

WALLACE, R. G. (1968) *Types and Rates of Alpine Mass Movement, West Edge of Boulder County, Colorado, Front Ranges.* Unpub. Ph.D. thesis, Ohio State Univ. 199 pp.

WHITE, S. E. (1968) Rockfall, alluvial and avalanche talus in the Colorado Front Range. *In:* Abstracts for 1967, *Geol. Soc. Amer. Spec. Pap.*, 115, 237. 538 pp.

WHITE, S. E. (1971) Rock glacier studies in the Colorado Front Range, 1961 to 1968. *Arct. Alp. Res.*, **3**, 43–64.

WILLIAMS, J. E. (1949) Chemical weathering at low temperatures. *Geogr. Rev.*, **39**, 129–35.

WINDOM, H. L. (1969) Atmospheric dust records in permanent snowfield: implications to marine sedimentation. *Bull. Geol. Soc. Amer.*, **80**, 761–82.

WOOD, A. (1942) The development of hillside slopes. *Proc. Geol. Ass.*, **53**, 128–40.

WOOD, T. G. (1970) Decomposition of plant litter in montane and alpine soils on Mt Kosciusko, Australia. *Nature*, **226**, 561–2.

Manuscript received December 1971.

Arctic soils

Samuel Rieger

Soil Conservation Service,
US Department of Agriculture

Low temperatures and, in most places, the presence of an impermeable perennially frozen substratum are the dominant factors affecting the development of soils of the Arctic. The cold environment modifies soil-forming processes to such an extent that it overshadows, and may even completely obliterate, the effects of relief and time on soil characteristics.

Soils that are wet nearly continuously throughout the thaw period are overwhelmingly dominant in the Arctic. Only in the polar desert of the High Arctic, near the limits of plant growth, are other soils more extensive. Throughout the area dominated by wet soils, however, especially on coarse materials in positions with exceptionally good surface drainage, are patches of well drained soils. The presence of these soils and variations in the properties of the wet soils result, in many places, in a soil pattern of great complexity. Nevertheless, soils of the Arctic can be divided conveniently into two broad groups: 1) poorly drained soils that are usually or always saturated, with thin or moderately thick active layers over solid, ice-rich permafrost, and 2) well drained soils with dry permafrost that usually have moisture contents below field capacity. Soils with intermediate moisture regimes exist, but are of relatively minor extent.

1. Poorly drained soils

OCCURRENCE AND MORPHOLOGY

Soils that are usually wet during the thaw period occupy 85 to 90 per cent of the tundra region, exclusive of mountainous areas with no soil mantle and sparsely vegetated areas in the far north. These soils cover most areas of low relief, ranging from nearly level plains to rolling or hilly uplands and footslopes. They remain saturated or close to the saturation point because the permafrost prevents downward percolation of meltwater, and water added by precipitation and condensation during the summer probably equals or exceeds that lost by evapotranspiration. Lateral movement of water is slow, even on slopes, and appears to take place mainly in the upper portion of the soil (Douglas and Tedrow, 1960).

Poorly drained mineral soils of the tundra all have essentially the same sequence of horizons. They consist, typically, of a mat of organic materials at the surface; a grey, greyish brown, or bluish mineral layer which may be mottled in whole or in part with colours ranging from olive or olive brown to reddish brown and which may contain patches or streaks of organic matter; and a grey perennially frozen layer which may have patches of organic matter in its upper part. The description below, of a soil in north-western Alaska (Holowaychuk *et al.*, 1966), is representative. (Colour and Munsell colour notations are for moist conditions unless otherwise noted. Depths are measured up and down from the surface of the mineral soil.)

Horizon	Depth (cm)	Description
O11	18–8	Black (7.5YR 2/1, wet) to dark brown (10YR 4/3, squeezed) partially disintegrated, finely fibrous peat.
O12	8–0	Very dark grey (10YR 3/1, wet) to dark brown (10YR 3/3, squeezed) partially disintegrated, finely fibrous peat, with inclusions of olive yellow and olive brown silty material; abrupt, slightly wavy lower boundary.
B21g	0–3	Dark grey (N 3/) silt loam, with common large distinct olive brown (2.5Y 4/4), light olive brown (2.5Y 5/4), and grey (N 5/) mottling or staining; non-sticky and non-plastic; few roots; clear, slightly wavy lower boundary.
B22g	3–23	Dark grey (N 4/–3/) silt loam; massive, with tendency toward platiness; non-sticky and non-plastic; some small pebbles; abrupt smooth lower boundary. The lower 3 cm of this horizon has erratic streaks of olive brown (2.4Y 4/6–5/6) fine soft concretionary material intermixed with very dark greyish brown (2.5Y 3/2) partially disintegrated organic material.
C1f	23–28	Frozen very dark grey (N 3/) silt loam with inclusions of partially disintegrated, finely fibrous, very dark greyish brown organic material; some small rock fragments; ice constitutes about 10 per cent of volume; clear lower boundary.
C2f	28–46	Frozen dark grey (N 4/) silt loam with inclusions of dark brown (7.5YR 3/4) partially disintegrated, finely fibrous organic material; ice constitutes about 75 per cent of volume.

Many variations of this pattern exist. The thickness of the organic mat is variable, both locally and regionally; within a radius of 30 m of the described profile it ranged from 8 to 20 cm. In general it is thick in warmer, fully vegetated areas of the tundra and thins in the direction of the polar desert. In some low areas, it may be thick enough so that all of the active layer consists of organic material. Some tundra soils, especially those close to the treeline or those formed in calcareous materials, have a black or dark brown horizon of mixed mineral and organic materials immediately below the organic mat. Mottling may be completely absent, as in the wettest soils, or may extend to the permafrost table. The soils may be of any texture, ranging from gravelly sand to clay, but soils with silty or loamy textures are dominant. There is a wide range in reaction, from strongly acid to strongly alkaline, depending principally on the character of the parent material.

Temperatures in the poorly drained soils are always low. Except for brief periods at the surface, they seldom exceed 10°C even in the upper part of the organic mat (Holowaychuk et al., 1966). Temperatures decrease steadily with depth to 0°C at the permafrost table.

GLEYING

Under saturated anaerobic conditions, iron, which is the principal inorganic colouring agent in soils, is reduced to the ferrous state. Ferrous compounds that form as a result of this reduction are responsible for the grey or bluish cast in poorly drained soils. The reducing process, known as gleying or gleization, is the principal soil-forming process in tundra areas. Except in the very wet positions, such as depressions and troughs between polygons, partial drying occurs in the active layer during prolonged dry periods in the summer and in the zone between freezing fronts during the refreezing period (Brown, 1967). In these periods of less than total saturation, some of the ferrous compounds are converted to hydrated ferric oxides, which remain in the soil as brown or reddish mottles and streaks.

DEPTH OF THAW

The principal factors affecting depth of thaw, or thickness of the active layers, in soils of the Arctic are the climatic regime, the amount of surface insulation, soil texture, and position of the soil in the landscape. These parameters occur in various combinations and may thereby have variable effects. As a result, there is no consistent correspondence between depth of thaw and any one of the factors. (For a discussion of permafrost, see Ives, Chapter 4A.)

Climatic regime
Ordinarily, soils thaw to greater depths under relatively mild maritime climates than under severe continental or far northern climates. In the eastern European tundra, for example, the permafrost table in poorly drained soils with loamy textures is commonly deeper than 1 m (Kreida, 1958), whereas in soils with comparable textures in northern Siberia it is less than 0.5 m deep (Karavayeva and Targul'yan, 1960). The thaw period in the warmer areas may be as long as four months, but is very short at high latitudes. In northern Alaska, for example, thawing in poorly drained soils begins in late June and the

maximum depth of thaw is reached in early August. Refreezing in such areas, both downward from the surface and upward from the permafrost table, begins in late August and is completed in early October (Drew *et al.*, 1958).

Surface insulation

Depth of thaw is inversely related to the thickness of the mat of organic matter that normally accumulates on the surface of poorly drained soils. The insulating effect of the peaty mat may offset the effect of climate. Thus, some poorly drained loamy soils in southwestern Alaska (mean annual air temperature 0.5°C) with peaty mats 40 cm thick have been observed to thaw only to the base of the mat, while in northern Alaska (mean annual air temperature −12°C) soils with mats only 5 cm thick thaw to nearly an equal depth from the surface. Many observations have been made of the increased depth of thaw that follows removal of the insulating organic mat from the surface (Brown, 1963).

The living vegetation on the soil reduces heat absorption by shading and by reflection of radiant heat. In an experiment on a soil covered with *Eriophorum* tussocks in interior Alaska, removal of the living vegetation alone resulted in an increased depth of thaw two-thirds as large as the increase resulting from removal of the vegetation and the 23 cm organic mat (Brown *et al.*, 1969). Burning may have even larger thaw effects than clearing because of charring of the surface. Compression of the living vegetation and the underlying peaty mat, as by the wheels or tracks of a heavy vehicle, also commonly results in deeper thaw.

Soil texture

In general, coarse-grained soils thaw to greater depths than loamy or fine-grained soils. In the tundra of European Russia, permafrost is 1.0 to 1.3 m deep in loamy soils and 1.2 to 1.5 m deep in sandy soils (Kreida, 1958). Similarly, in northern Alaska, poorly drained silty soils with an organic mat 12 to 15 cm thick thaw only to depths of 25 cm from the surface, whereas sandy soils with equally thick organic mats have 40 cm of thaw. With thinner organic mats, the increase in depth of thaw is considerably greater in coarse-grained than in fine-grained soils.

Position of the soils

Soils of low points in the microrelief, which are usually extremely wet, do not thaw as deeply as soils that lose surface water through runoff. The permafrost table is generally shallow, for example, in the troughs between polygons (McMillan, 1960; Brown, 1967) and in the lower parts of solifluction slopes (Smith, 1956). At least some of the reduction in depth of thaw, however, results from the thicker organic mat that commonly develops in wetter situations.

FROST-STIRRING

Poorly drained soils of the Arctic are frequently disturbed by frost processes. Some soils, as on solifluction slopes, are almost continually moving. Others may be stable most of the time, but are subject to occasional disruptions. Physical displacement in the active layer – the result of subsidence, warping, heaving, and lateral flow over permafrost – can be considered to be a normal factor in the development of these soils (Douglas and Tedrow, 1960; Sokolov and Sokolova, 1962; Ugolini, 1966b). In the Arctic, geomorphic processes of this kind play a larger role in the determination of soil properties than in temperate regions (see also Bird, Chapter 12A).

Reasons for soil movement and mixing vary over the tundra landscape. In very cold areas with little snow, fissuring and the development and subsequent degradation of ice-wedge polygons is an important cause. In the early stages of development, the growth of ice wedges displaces the soil mass between the wedges and forces formation of a raised

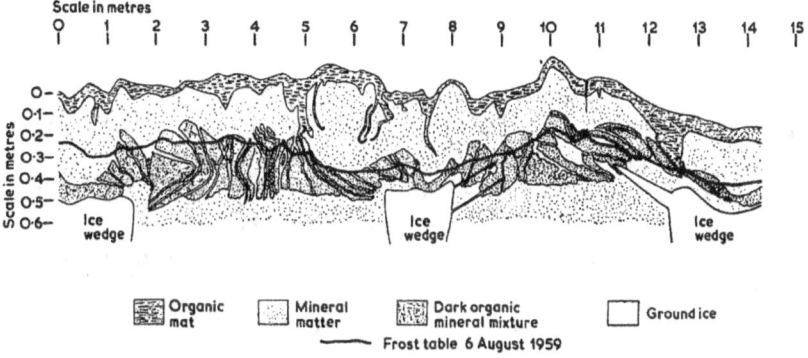

Fig. 13A.1. Section through ice-wedge polygons at Barrow, Alaska, showing irregular soil surface, variable thickness of the surface organic mat, and concentration of organic material at the permafrost table (from Douglas and Tedrow, 1960).

polygon. Continued expansion of the wedge may result in soil flow and, eventually, in deeper burial of the wedge. A new wedge may then form in the overlying material. It is possible that the soil in all parts of a polygon periodically goes through all stages of such a cycle (Karavayeva *et al.*, 1965). Both the mineral and organic portions of the soil are involved (fig. 13A.1).

Heaving and churning processes are not as vigorous in areas of milder climate, where thawing usually is deeper and the winter snow cover is greater (Ignatenko, 1963). Even there, however, frost action commonly results in an irregular microrelief, especially where the heaving potential is increased because of a textural interface within the active layer or because of an exceptionally large water supply in the subsoil (Troll, 1958). Local changes in drainage, or even single-year changes such as a year with unusually high rainfall, may activate soils that have been in a steady state for some time (Ugolini, 1966b).

On steeper slopes, generally those with gradients of 12° or more, soil movement downslope is primarily due to creep, defined as the movement of individual fragments or aggregates by gravity assisted by volume changes associated with wetting, drying and frost heaving. On lower slopes movement is by a combination of viscous flow, in which maximum movement takes place at or near the surface, and slips immediately above the permafrost (Smith, 1956). Maximum flow occurs in fine materials, largely because of the presence of a greater volume of dispersed ice, but the flow pattern in all soils is commonly complex and irregular. It is likely that the greatest amount of movement is in the early summer when the soils are wettest. The cumulative effect of soil creep and soil flow is as much as 6 cm/year on 10 to 14° slopes (Ugolini, 1966b).

Soil flow commonly results in a pattern of lobate sloping terraces, with many evidences of overturning including buried organic layers near the frontal portion of the lobe, abrupt textural boundaries, and tongues of intruding material. In many places, as a result of fissuring of frozen soil and subsequent slips, small pits and patches of bare ground occur in the back portions of the lobes.

FROST CIRCLES

Heaving and thrusting in wet soils over permafrost creates spots of bare ground in the tundra. The spots are of various shapes and range in size up to 2 m, but they generally are roughly circular. They commonly exhibit some sorting of gravel and stones, but many are apparently not sorted. Various names have been applied to them, but in Washburn's (1956) classification they are called sorted and non-sorted circles. Both will be known here as 'frost circles'.

Relatively few frost circles occur in warmer tundra areas with a dense vegetative cover and thick organic mats on the soil surface, but they become increasingly prominent in colder and more continental climates. In the warmer areas such as the European, eastern Siberian, and southwestern Alaska tundras, the circles are completely level, but in the colder areas they are gently domed (Troll, 1958). In general, they are more numerous in fine-textured than in coarse-textured poorly drained soils. On nearly level or gently sloping land they are most common on raised polygons, though they also occur in other positions. On solifluction slopes they occur in the upslope portions of lobes because of fissuring and greater heaving in the wetter part of the lobe.

It is likely that, in the formation of the circles, the soil is forced first upward to the surface, then outward. In gravelly soils, the axis of detritus and rubble beneath the scar is commonly perpendicular (Kreida, 1958); in soils with a contrasting substratum within the active layer, the boundary between strata may be tilted to a nearly vertical position. Buried peat and arms of organic matter extending deep into the mineral material at the edges of the circle are common (Douglas, 1961; Ignatenko, 1967; James, 1970) (see fig. 13A.2).

A characteristic of frost circles is the presence of vesicles or small pores in the upper few centimetres (Svatkov, 1958; Ugolini, 1966b). These probably are caused by needle-ice crystals that form after short periods of freezing. The greater periodic summer freezing in the circles, which are not protected by an insulating organic mat, enhances capillary movement of moisture from the adjacent warmer soil. In places this leads to formation of a saline crust at the surface of the circle (Ignatenko, 1963).

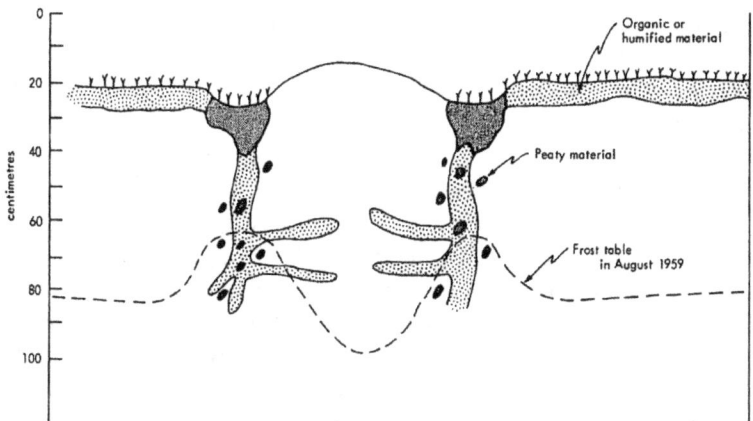

Fig. 13A.2. Frost circle, with arms of organic material at the edges of the circle (after Ignatenko, 1967). Note inverse relationship between thickness of organic mat and depth of thaw.

ORGANIC MATTER DISTRIBUTION

In the absence of any deep root system in the poorly drained soils, organic matter accumulates only at the soil surface. Because the rate of accumulation is closely related to the density of the tundra vegetation, the thickness of the organic mat is greater in areas close to the treeline than in the north, and greater in the troughs between polygons than on the more thinly vegetated elevated centres. Despite low summer temperatures, the deposited organic matter decomposes slowly and exists in soils both as fibrous and as finely divided material.

Frost stirring is undoubtedly responsible for the mechanical breakdown of organic particles, but much of the finely divided material is partially or completely humified. Some of the products of humification are colourless. An upland soil in the polar desert, for example, has 6 per cent organic matter in the upper 12 cm – probably contributed by diatoms and algae – but no apparent darkening (Tedrow, 1966). Colourless humus ranges from 1.5 to 3 per cent in northern Siberian soils (Karavayeva and Targul'yan, 1960) and is fairly uniform in the profile; colouration is noticeable only in zones of maximum accumulation of humus in the upper horizon and at the permafrost table.

In soils with shallow permafrost, patches of organic matter are distributed irregularly below the surface mat, but commonly reach a maximum and may even form a continuous layer in the zone immediately above and below the permafrost table. Both fibrous and finely divided organic materials occur in this subsurface organic layer. This is a normal feature of soils with shallow permafrost in colder tundra areas, but does not occur under maritime climates where the permafrost table is generally more than 1 m deep.

Several explanations have been advanced for this phenomenon. Some Russian investigators (Karavayeva and Targul'yan, 1960; Dimo, 1965) have suggested that the secondary maximum in organic matter results from the downward flow of colloids and dissolved humus toward the frozen permafrost table and that the layer represents a humic-illuvial horizon peculiar to tundra soils. Where the cold front is too deep, according to this belief, dissolved humus is precipitated before reaching the permafrost table and no illuvial horizon develops. This hypothesis, however, fails to account for the presence of fibrous organic material at and below the permafrost table (see fig. 13A.1).

Douglas and Tedrow (1960) note that much of the organic matter at the permafrost table is 8000 to 10,000 years old, and suggest that the organic layer may be the remnants of a surface organic mat formed during a warmer climatic interval. During the subsequent cooling period, they think the growth of ice lenses forced the underlying mineral soil through the organic mat, eventually burying it. They recognize, however, that in some areas burial was the result of solifluction processes.

Other investigators (Mackay *et al.*, 1961; Brown, 1966) believe that the deep organic layer is primarily the result of frost processes that can be observed at the present time. Among these are solifluction and the stirring processes that are involved in the formation of polygons and frost circles. Organic matter from the surface, in various stages of decomposition, is carried along in soil movements and tends to concentrate above the permafrost table, which acts as a floor to soil churning. On solifluction slopes, the highly organic soil may be trapped under the permafrost table as it rises because of a greater thickness of overburden. Where there is no solifluction, organic matter deposited above the permafrost during a warmer period is now below the present shallower permafrost table.

DISTRIBUTION OF SOLUBLE MATERIALS

There is seldom a marked horizon of accumulation of mineral elements anywhere in the profile of the poorly drained soils, except in connection with accumulations of organic matter. Nevertheless, some movement of mineral materials does occur, and some soluble products are lost from the soil. The upper horizons or even the entire active layer of soils formed in calcareous materials may be leached free of carbonates, and base saturation may be higher near the permafrost table than at the surface (Douglas, 1961). Films of iron oxide, presumably originating as soluble ferrous iron in the surrounding soils, may occur in stagnant pools (McMillan, 1960).

In frost circles and other bare areas, an upward migration of water containing soluble salts may result in a salt crust at the surface (Douglas and Tedrow, 1960). Crusts of this type are especially common in far northern areas, where precipitation rates are very low.

ORGANIC SOILS

Organic soils composed of thick accumulations of moss or sedge remains occur in closed depressions and in some nearly level areas and swales on flood plains and low terraces. Generally, in these soils the entire active layer consists of organic materials. Peats in depressions are normally derived from sedges, but *sphagnum* moss peat may occupy the flat tops of raised polygons with sedge peat in the troughs (Brown, 1966). Also, a low-centred polygonal pattern may develop in soils derived from *sphagnum* moss (Holloway-chuk *et al.*, 1966). Isolated ice-core mounds (pingos) are common in areas of peat soils.

2. Well drained soils

OCCURRENCE AND MORPHOLOGY

Well drained soils occur throughout tundra areas, but are of relatively small total extent except in arid regions near the northern limits of vegetation. South of those regions they occur almost exclusively in stony, gravelly, or coarse sandy materials on steep slopes, high narrow ridges, escarpment edges, stabilized dunes and beach ridges, and elevated portions of flood plains (fig. 13A.3). The combination of coarse-grained materials and exceptionally good surface drainage is, with rare exceptions, an essential condition for

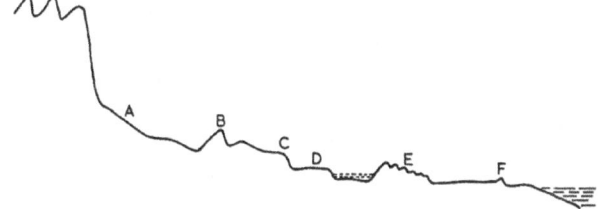

Fig. 13A.3. Diagrammatic tundra landscape, showing some positions in which well drained soils occur: *A* – steep foot slope, *B* – narrow ridge, *C* – escarpment edge, *D* – gravelly terrace, *E* – sand dune, *F* – beach ridge.

their existence. Either bedrock or dry permafrost underlies these soils; where they are developed in thick unconsolidated materials, solid permafrost may occur at greater depths.

The well drained soils differ from most soils of the tundra in that excess moisture is able to escape. As a result, oxidizing rather than reducing processes are dominant, and soluble products of weathering are redistributed or removed from the soil. The rate of weathering, decomposition of organic materials, and movement of materials – that is, the intensity of soil development – depend on temperature and the amount of moisture available. In the cold and relatively arid arctic environment, these processes go on at a much slower rate than in forested areas to the south. As a consequence, the soils that are formed may have the same sequence of horizons as in comparable well drained soils of taiga areas, but have much thinner horizons. In nearly all cases, physical and chemical properties of the soils reflect those of the parent materials.

The well drained soils vary in the kinds of horizons that develop, and in the thickness and degree of development of the horizons. The soil described below, from southwestern Alaska (Rieger, 1966), is representative of the most extensive of these soils. Colours are for moist conditions.

Horizon	Depth (cm)	Description
O1	5–2	Mat of loose litter, roots of shrubs and grasses, and partially decayed organic materials.
O2	2–0	Dark reddish brown (5YR 2/2) well decomposed organic matter; many roots; strongly acid; abrupt smooth lower boundary.
A11	0–8	Dark brown (7.5YR 3/2) very gravelly silt loam; weak very fine granular structure; very friable, smeary when rubbed; many roots; very strongly acid; abrupt wavy boundary.
A12	8–15	Dark brown (7.5YR 3/2) very gravelly silt loam; weak very fine granular structure; friable; roots common; very strongly acid; clear wavy boundary.
B	15–30	Brown (10YR 4/3) very gravelly silt loam; weak very thin platy structure; friable; roots common; many fine tubular pores; dark brown (7.5YR 3/2) krotovinas up to 8 cm in diameter; few roots; very strongly acid; gradual boundary.
C	30–71	Brown (10YR 5/3) very gravelly silt loam; weak thin and medium platy structure, breaking to very fine angular blocks; friable; few roots to 50 cm, none below; strongly acid.

The principal horizons in this soil are the dark upper horizon and the yellower subsoil horizon, which grades into the shaly parent material. The soil is acid and low in exchangeable bases throughout. Chemical analyses indicate the A12 horizon is slightly higher in free iron than both the A11 horizon above it and the B horizon below it. A similar distribution of iron has been noted in other soils of this kind (Hill and Tedrow, 1961; Brown and Tedrow, 1964), indicating that there is a tendency in these soils toward the development of an iron-rich subsurface horizon. The process of podsolization, in which iron, aluminium, and organic carbon are moved in solution from the upper mineral horizon and reprecipitated in a lower horizon, is common in soils of cool, moist, forested regions. In a few places in tundra regions just north of the treeline where large quantities of water move through coarse-grained soils, as in micro-depressions in uplands or protected areas where snow accumulates, the process is also intensive. There the soils have bleached upper horizons over thin horizons of iron accumulation (Brown and Tedrow, 1964). Soils of this kind are more common in tundra areas immediately above treeline in subarctic regions.

Soils developed in materials derived from basic rocks, such as basalt, limestone, and calcareous shales, are similar in appearance to the acid soils described above, but pH values and base saturation are high. In some places, calcium carbonate has accumulated in the lower part of the soil.

With increasing distance from the treeline, the amount of organic matter added to the soil decreases and, because of lower temperatures and lower precipitation, chemical and biological processes such as oxidation and humification proceed at a slower rate. Well drained soils in the colder tundra areas may have only very thin or no dark upper horizons and thin brown B horizons. In the far north, where precipitation rates are very low and the vegetative cover is sparse, well drained soils that have no discernible horizon development are dominant. Other soils with no horizon development are the young sandy soils of active dunes bordering some major rivers (Ricket and Tedrow, 1967) and the very gravelly soils of flood plains, low terraces, and steep colluvial slopes.

SOILS OF THE POLAR DESERT

The polar desert is the region in which vascular plants cover less than 25 per cent of the soil surface. The mean July temperature in this region is less than approximately 5°C, and annual precipitation seldom exceeds 8 cm (Tedrow, 1966). With decreasing temperatures to the north the plant cover becomes even more sparse, and in areas with mean July temperatures of about 2°C or less it ceases to exist. At this point no soil, as it is defined by the US Soil Survey Staff (1960), can be recognized. In the northern hemisphere the polar desert is largely confined to the Canadian and Russian arctic islands and to the northernmost parts of Greenland and Siberia, but areas with similar properties occur in mountains to the south (fig. 13A.4). Areas completely lacking in higher plants probably exist only in Antarctica and in some high mountains.

Soils of the polar desert differ in several aspects from well drained soils of the more densely vegetated tundra. These soils exist in an arid environment, and have many characteristics reminiscent of the soils of warm deserts (Charlier, 1969). Because summer precipitation is usually only 5 cm or less and most of the winter snow is lost by sublimation and wind stripping, the soils are quite dry. Salt efflorescences at the surface are common, especially near sea coasts on marine sediments, and pH values are high. In upland areas vegetation is largely confined to the somewhat wetter troughs between polygons, and to swales, depressions, and slopes sheltered from the wind where blowing snow accumulates. Finer soil particles are removed from the surface by winds, resulting in a concentration of pebbles and stones at the surface (McMillan, 1960; Karavayeva, 1963; Tedrow, 1966).

FROST SCARS

It is obvious that the well drained soils of the Arctic are more stable than the poorly drained soils; the distinct horizons in these soils could not develop in soils subject to constant frost stirring. This stability, however, is seldom continuous over a well drained landscape. The soil surface is generally hummocky, and the vegetative cover is interrupted by many bare patches and, on steep slopes, by stone stripes. The soil profile under the bare patches, or frost scars, has no regular sequence of horizons but generally has only the colour of the parent materials. Textural sorting may occur in these scars. Although the vegetated portions of the soils are normally stable for long periods, they are sometimes subject to frost heaving which results in distortion and convolution of the developed horizons. Surficial cracking, produced by either desiccation or frost, is common in these soils (Ugolini, 1966b). The cracks may be penetrated by roots and may

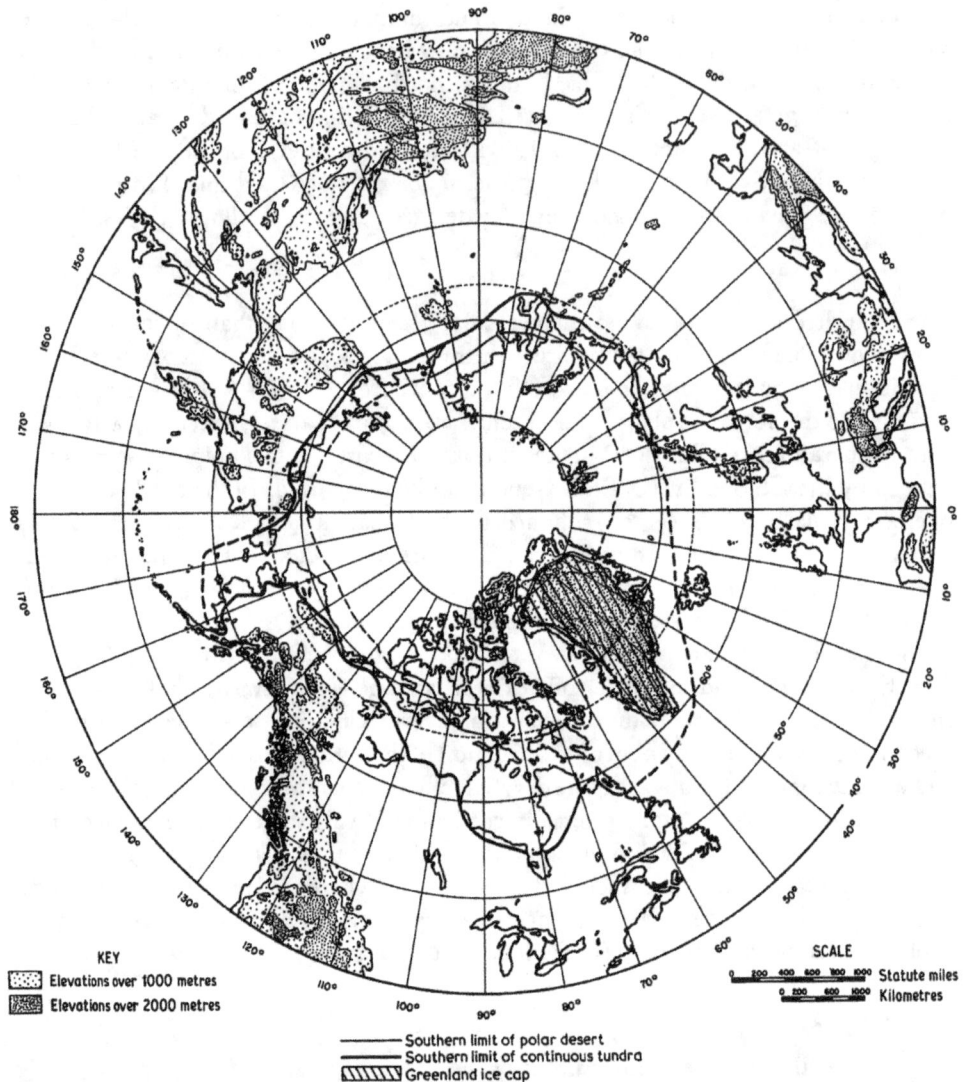

Fig. 13A.4. The extent of the polar desert and the southern limit of continuous tundra (partly after Charlier, 1969). Cf. Webber, fig. 8B.7.

accumulate organic matter. This tends to create an irregular boundary between the A and B horizons, and may even result in obliteration of the B horizon.

Frost scars in well drained soils are probably created originally by soil movements, but they may be enlarged by the formation of needle-ice crystals at the soil surface. These fine vertical crystals form during periods of abrupt freezing where the vegetative cover is thin or absent, and continue to grow as long as the subsurface soil is moist and unfrozen. The crystals, in their growth, exert pressure great enough to push up stones and to tear roots. The uprooted plants are then easily removed by winds. This process of turf exfoliation is most effective in loamy soils that are normally moist in the subsoil and that exist under a climate with frequent diurnal freeze–thaw cycles. It is, therefore, least common in the polar deserts of the High Arctic and most common in alpine areas south of the Arctic (Troll, 1958).

3. Uses of arctic soils

AGRICULTURE AND HUSBANDRY

Conventional farming, as it is practised in forested regions to the south, is probably not feasible in tundra areas except in a few favoured places near the treeline where marginal yields of grasses and some hardy vegetables may be obtained. Vegetables may be grown successfully, however, in greenhouses almost anywhere.

Apart from the wild berries which are gathered by the indigenous peoples of the Arctic, the principal 'agricultural' products of tundra areas are meat and other materials derived from reindeer. It is likely that about 3 million head of domesticated reindeer exist in the Arctic. Management of the herds ranges from highly scientific to a level barely above subsistence hunting of wild reindeer or caribou. Good herding practices and adequate facilities for slaughter and handling of the meat and byproducts result in greatly increased production (Zhigunov, 1961). With scientific management and more efficient utilization of the range, the tundra could be one of the important meat producing regions of the world.

In many areas, the reindeer spend the summer months in the tundra and the winter in the taiga regions to the south. In others, they remain in the tundra throughout the year. In either case, summer grazing on the tundra is most important in preparing the animals for the autumn slaughter, and an essential part of herd management is controlled use of the native vegetation so that adequate forage is available each year. Little information is available on the amount of land required to support an animal during the grazing season on a sustained yield basis. This varies with the kind and quantity of vegetation and the migration pattern of the herd, but it seems likely that the average requirement is between 100 and 200 acres (40 to 80 hectares). A great deal of additional research on growth rates of tundra plants and the nutritional requirements of reindeer is needed to establish carrying capacities of tundra range.

Another animal native to the tundra is the musk-ox. This animal, which had been hunted close to extinction, is useful for its wool as well as its meat. Research on the domestication of the musk-ox is in progress, and may lead to the development of another livestock industry in the Arctic.

ROADS AND STRUCTURES

All roads and structures in the Arctic must be designed to prevent thawing of the underlying permafrost. Failure to observe this rule will inevitably result in failure of the structure, and may cause gullying and other damage to the tundra soils themselves. Under buildings, preservation of the permafrost is usually accomplished by providing an air space under well insulated floors, and, where necessary, by the use of refrigerating coils in the supporting pilings. A thick pad of gravel laid directly on the tundra provides insulation for roads, airfields, and other sites that support heavy equipment. Heated pipelines and utility corridors need to be elevated, except in gravelly areas with dry permafrost.

Removal, disturbance, or even compression of the living vegetation and the mat of organic matter on the soil surface by vehicles results in increased depth of thaw. Roads built without adequate insulation over ice-rich permafrost subside unevenly and soon become impassable. Such a road or even tracks left by ordinary vehicles on slopes can result in channelization of runoff in depressions in the permafrost, and eventually in deep gullies. Even on level land, tracks can remain visible for many years after passage of a heavy vehicle. For these reasons it is essential that only vehicles equipped with low ground pressure treads or wheels, or air-cushion vehicles, be used for offroad movement in the summer.

In planning roads and other structures in tundra areas it is best to make use, to the fullest extent possible, of areas with dry permafrost, such as gravelly terraces along rivers. Structures in such areas are easier to build, are least likely to sustain damage because of thawing permafrost, and have minimum potential for adversely affecting adjacent areas.

4. Classification of arctic soils

There is no universally accepted soil classification system. Instead, different classification schemes have been developed and used by individuals and national soil survey organizations. These vary in quality and complexity. Some are intended to encompass all of the soils of the world, and others are limited to soils of particular regions. Three classification systems that are widely used and that have particular application to soils of the Arctic are discussed in this section.

US SOIL TAXONOMY

The system recently adopted by the cooperative soil survey of the United States (Soil Survey Staff, 1960, 1967) has worldwide application. In common with most modern soil classifications, soils are grouped at the highest levels largely on the basis of characteristics related to their genesis or the soil forming processes involved in their development. Classes are defined, however, in terms of visible or measurable soil characteristics rather than in genetic terms. The degree of development of genetic horizons, along with non-genetic features of importance to the growth of plants, define classes in the lower levels of the taxonomic system.

This soil classification differs from earlier ones in several important respects. Among these are: 1) definitions of all classes entirely by soil characteristics rather than by climatic or vegetative regimes or by soil forming processes presumed to be responsible for these characteristics; 2) completely quantitative rather than qualitative definitions of taxa; 3) use of a coined connotative terminology derived largely from the classical languages; and 4) recognition and definition of a small three-dimensional soil body, the *pedon*, as the ultimate soil unit.

A pedon may have continuous horizons or may consist of intermittent or cyclic horizons as in frost-rived polygons or solifluction lobes with frost scars. Recognition of pedons which include in a single unit the complete assemblage of horizon arrangements in these complex soils is a considerable aid in understanding the genesis of arctic soils and in their classification. It is noteworthy that users of other classification systems (discussed below) have commented approvingly on this concept (Karavayeva *et al.*, 1965; Ugolini, 1966a).

The highest category in the taxonomy is the order. Each order is assigned a connotative syllable which is used as a formative element in developing the names of classes in lower categories. Five orders (of a total of ten in the complete scheme) are represented in the Arctic. These, together with a brief description and the assigned formative element, are listed below:

Entisols – soils that have little or no evidence of development of pedogenic horizons (formative element *-ent*).

Histosols – soils derived mostly from organic materials (formative element *-ist*).

Inceptisols – soils that have been altered by pedogenic processes such as oxidation or gleying, that have lost mineral materials, and that have no horizons of appreciable accumulation of clay or mixtures of iron, aluminium, and organic carbon (formative element *-ept*).

Mollisols – soils with a dark base-rich upper mineral horizon (formative element *-oll*).

Spodosols – soils with a horizon of accumulation of amorphous mixtures of aluminium, iron, and organic carbon (formative element *-od*).

Names of classes in the two categories below the order are formed by connecting formative elements suggestive of properties that are emphasized in the taxonomy. Thus, Inceptisols with characteristics of wetness are in the suborder of *Aquepts*. Other examples of suborders are *Umbrepts*, with the syllable *Umbr* signifying a dark upper horizon that is low in bases, and *Ochrepts*, with *Ochr* indicating the absence of a dark upper horizon of significant thickness. One of the criteria for defining great groups, the third category in the system, is the soil (not air) temperature regime. All soils in the Arctic are in Cryic, or cold, great groups. The formative element, *Cry*, is prefixed to the suborder name to give great group names of *Cryaquepts*, *Cryumbrepts*, etc. Subgroup names are great group names modified by one or more adjectives. The adjective *pergelic* is used to indicate mean annual soil temperatures of 0°C or less, hence permafrost at some depth. Soil properties such as texture, mineralogy, and reaction are used to define classes of families and series, the lowest levels in the taxonomy.

Listed below are the principal subgroups in arctic regions, together with brief descriptions of their properties:

Pergelic Cryaquepts – poorly drained grey or mottled soils with thin (usually less than 20 cm) surface organic mats.

Histic Pergelic Cryaquepts – poorly drained grey or mottled soils with thick surface organic mats.

Ruptic-Histic Pergelic Cryaquepts – poorly drained grey or mottled soils with a thick organic mat in part of each pedon (as in a trough between polygons) and a thin or no mat in other parts; pedons commonly include frost circles.

Pergelic Cryaquolls – poorly drained grey or mottled soils with dark base-rich upper horizons.

Pergelic Sphagnofibrists – fibrous *sphagnum* moss peat, at least 60 cm thick.

Pergelic Cryofibrists – fibrous sedge peat, at least 40 cm thick.

Pergelic Cryohemists – partially decomposed peat, at least 40 cm thick.

Pergelic Cryorthents – well-drained loamy, clayey or gravelly soils with no developed horizons.

Pergelic Cryosamments – well drained sandy soils with no developed horizons.

Pergelic Cryochrepts – well drained brown soils with very thin or no dark upper horizons.

Pergelic Cryumbrepts – well drained soils with dark, acid upper horizons and, usually, brown subsoil horizons.

Pergelic Cryoborolls – well drained soils with dark, base-rich upper horizons.

Pergelic Cryorthods – well drained soils with a thin bleached upper horizon and a brown or reddish brown subsoil horizon in which aluminium, iron, and organic carbon have accumulated.

Ruptic-Entic Pergelic Cryochrepts, Ruptic-Entic Pergelic Cryumbrepts, etc. – well drained soils in which part of each pedon is a frost scar with no recognizable horizons.

Lithic Cryorthents, Lithic Cryochrepts, etc. – well drained soils with bedrock less than 50 cm deep.

It should be noted that, although only general descriptions are given here, rigorously defined criteria must be satisfied for placement in any subgroup. This assures uniformity of classification in all areas. Other advantages of this taxonomic scheme are that the system of nomenclature makes it possible, for those familiar with the meanings of the formative elements, to infer many of the properties of a soil from its name alone, and that the classification clearly indicates relationships in both soil characteristics and genetic processes between soils of the Arctic and soils of other regions (see also Retzer, Chapter 13B).

ZONAL CLASSIFICATION

A classification system that has been used by a number of American arctic pedologists in recent years (Tedrow *et al.*, 1958) is, basically, an extension of the system in use in the United States before adoption of the present taxonomy. That system itself is a modification of a Russian classification of the early twentieth century. In the highest category of this classification soils are arranged in three orders – Zonal, Intrazonal and Azonal.

Zonal soils are those that are believed to reflect fully the influence of climate and living organisms, the 'active' factors of soil genesis, in any geographic zone. Intrazonal soils are those with properties determined largely by some local factor of relief or parent material, such as poor drainage or calcareous rock. Azonal soils have no developed soil characteristics because of their youth, steepness, or some characteristic of the parent rock. It is apparent that judgements may differ as to the proper classification of certain soils; there have been some different opinions, for example, as to whether the areally dominant poorly drained soils of tundra regions are Zonal or Intrazonal soils.

Suborders, the second category of the system, are defined in general terms by the climatic and vegetative characteristics of the zone and, in most cases, by either a highly generalized soil property or the dominant process in soil formation. In the original classification, all soils of the Arctic were grouped in the suborder, 'Soils of the Cold Zone', but Tedrow (1968) has divided this into the Arctic Brown Zone (fully vegetated), the Polar Desert Zone (sparsely vegetated), and the Cold Desert Zone (essentially no higher vegetation). Boundaries between these zones, as with most others, are vague. In this classification, poorly drained soils of the tundra are assigned to the Intrazonal order.

The third and most significant category in the classification is the Great Soil Group. In this category, soils are defined by specific characteristics but, because of the absence of quantitative criteria, boundaries between groups may be hazy. Names of the great soil groups may be in terms of the characteristic vegetation, soil colour or other soil property, the dominant process involved in soil formation, or a combination of these. In their classification of arctic soils, Tedrow and his associates have proposed several new great soil groups, and in some cases subdivisions of these groups, and have carried over some previously recognized great soil groups from temperate regions. The principal great soil groups of the Arctic, together with the closest equivalents in the US cooperative soil survey taxonomy, are:

Lithosols – well drained soils with shallow bedrock (Lithic Cryorthents).
Regosols – deep, well drained soils with no genetic horizons (Pergelic Cryorthents and Cryopsamments).
Polar Desert soils – well drained soils of the Polar Desert Zone with no genetic horizons (Pergelic Cryorthents).
Arctic Brown soils:
  Normal phase – well drained brown soils with or without dark upper horizons (Pergelic Cryochrepts, Cryumbrepts, and Cryoborolls).
  Shallow phase – soils as above with shallow bedrock (Lithic Cryochrepts, Cryumbrepts, and Cryoborolls).
  Moderately well drained phase – soil gradational between normal phase and Tundra soils, with gleying in lower horizons (Pergelic Cryaquepts).
Rendzinas – dark soils over shallow calcareous rock (Lithic Cryoborolls).
Podzols or Podzol-like soils – soils with bleached upper horizons and brown lower horizons (Pergelic Cryorthods).

Tundra soils:

> Upland Tundra – relatively drier poorly drained soils with brown and yellow colouration (mottles) in the upper horizon and with thin organic mats (Pergelic Cryaquepts and Cryaquolls).
>
> Meadow Tundra – wetter soils with dark grey colours and somewhat thicker organic mats (Pergelic Cryaquepts).

Half-Bog soils – saturated soils with organic mats approximately 15 cm to 30 cm thick (Histic Pergelic Cryaquepts).

Bog soils – saturated soils with thick organic accumulations (all Histosols).

The concept of the complex pedon is not reflected in this classification. It is not unusual to have Half-Bog, Meadow Tundra, and Upland Tundra soils all recognized in the same polygon (e.g. Brown, 1967).

USSR SOIL CLASSIFICATION

Russian pedologists have been involved in the study and classification of arctic soils longer than any other national group. Unfortunately, until recently a lack of coordination among investigators has resulted in many poorly defined technical terms and the absence of a standard soil nomenclature. Their systems, however, invariably emphasize the genetic rather than the taxonomic approach to classification. Soils are classified according to climatic and vegetative zones and the processes believed to be involved in their development (Basinski, 1959), although morphological features associated with these processes are used to determine the direction of development and the identification of soils in the field.

Recently efforts have been made to standardize soil systematics and nomenclature in the USSR. A unified classification system summarized by Rozov and Ivanova (1967) may be regarded as a result of these efforts. In this rather complex system, soil groups, the major units, are arranged to reflect the effects of three kinds of forces: 1) the natural environment, including the thermal and hydrologic regimes, the intensity and depth of weathering, and the biological cycle (i.e. the Zone); 2) the soil moisture regime; and 3) the character and distribution of organic matter and the soluble mineral components of the soil. Soil groups in arctic regions are distinguished from each other by difference in the thickness and character of organic matter at the surface and in the profile, degree of gleying, and characteristics of the groundwater. Two zones are recognized in the area north of the treeline – the Tundra Zone (mean July air temperatures 6–10°C) and the Arctic Zone (mean July air temperatures 2–6°C). Names of the soil groups indicate the zone, the water regime, and the kind of organic matter accumulation. The groups are divided into subgroups and lower categories on the basis of forms of humus accumulation and various secondary features, including position in the landscape.

Some soil groups of the Tundra Zone and their rough equivalents in the US cooperative soil survey taxonomy are:

> Sod Tundra – well drained soils of uplands (Pergelic Cryorthents, Cryochrepts, Cryumbrepts, Cryoborolls).

Alluvial Tundra-Sod – well drained soils of flood plains (Pergelic Cryorthents).

Gley Tundra – poorly drained soils (Pergelic Cryaquepts).

Waterlogged (moss-gley) Tundra – poorly drained soils of low positions (Histic Pergelic Cryaquepts).

Bog Tundra – soils with standing free water (Histosols and Histic Pergelic Cryaquepts).

The concept of the unity of complex soil individuals, such as those in polygons, is clearly acceptable to pedologists of the USSR (Karavayeva *et al.*, 1965), but the pedon is not formally identified in classification systems as the ultimate soil unit.

The Arctic has received much less attention from pedologists than other regions. Although the general properties of the soils are known, few detailed studies have been made of the processes involved in their development. Soil patterns in the Arctic are, in many areas, as complex as those in any other part of the world, but only a small number of detailed soil maps are available. Much remains to be learned about the soils themselves and their response to manipulation and treatment for both food production and construction. Recent industrial and other developments in the Arctic, however, have resulted in increased interest in the soils of the region and in their potentials for use. This will undoubtedly lead to more intensive investigation and to greater awareness of suitable methods of utilization of these cold soils.

References

BASINSKI, J. J. (1959) The Russian approach to soil classification and its recent development. *J. Soil Sci.*, **10**, 14–26.

BROWN, J. (1966) *Soils of the Okpilak River Region, Alaska.* Cold Reg. Res. Eng. Lab., Res. Rep., 188. Hanover, NH: US Army CRREL. 49 pp.

BROWN, J. (1967) Tundra soils formed over ice wedges, northern Alaska. *Soil Sci. Soc. Amer. Proc.*, **31**, 686–91.

BROWN, J. and TEDROW, J. C. F. (1964) Soils of the northern Brooks Range, Alaska, 4: Well-drained soils of the glaciated valleys. *Soil Sci.*, **97**, 187–95.

BROWN, J., RICKARD, W. and VIETOR, D. (1969) *The Effect of Disturbance on Permafrost Terrain.* Cold Reg. Res. Eng. Lab., Spec. Rep., 138. Hanover, NH: US Army CRREL. 13 pp.

BROWN, R. J. E. (1963) Influence of vegetation on permafrost. *Proc. Int. Permafrost Conf.*, 20–4. Washington, DC: Nat. Res. Council.

CHARLIER, R. H. (1969) The geographic distribution of polar desert soils in the northern hemisphere. *Geol. Soc. Amer. Bull.*, **80**, 1985–96.

DIMO, V. H. (1965) Formation of a humic-illuvial horizon in soils on permafrost. *Soviet Soil Sci.*, **9**, 1013–21.

DOUGLAS, L. A. (1961) *A Pedologic Study of Tundra Soils from Northern Alaska.* Unpub. Ph.D. thesis, Rutgers Univ. 147 pp.

DOUGLAS, L. A. and TEDROW, J. C. F. (1960) Tundra soils of arctic Alaska. *Trans. 7th Int. Congress Soil Sci.*, **4**, 291–304.

DREW, J. V., TEDROW, J. C. F., SHANKS, R. E. and KORANDA, J. J. (1958) Rate and depth of thaw in arctic soils. *Trans. Amer. Geophys. Union*, **39**, 697–701.

HILL, D. E. and TEDROW, J. C. F. (1961) Weathering and soil formation in the arctic environment. *Amer. J. Sci.*, **259**, 84–101.

HOLOWAYCHUK, N., PETRO, J. H., FINNEY, H. R., FARNHAM, R. S. and GERSPER, P. L. (1966) Soils of Ogotoruk Creek watershed. *In:* WILIMOVSKY, N. J. and WOLFE, J. N. (eds.) *Environment of the Cape Thompson Region, Alaska*, 221–73, and Appendix published separately. Oak Ridge, Tenn.: US Atomic Energy Commission.

IGNATENKO, I. V. (1963) Arctic tundra soils of the Yugor Peninsula. *Soviet Soil Sci.*, **5**, 429–40.

IGNATENKO, I. V. (1967) Soil complexes in Vaygach Island. *Soviet Soil Sci.*, **9**, 1216–29.

JAMES, P. A. (1970) The soils of the Rankin Inlet area, Keewatin, NWT, Canada. *Arct. Alp. Res.*, **2**, 293–302.

KARAVAYEVA, N. A. (1963) Description of arctic-tundra soils on Bolshoi Lyakhovskii I (Novosibirske Islands). *In:* IVANOVA, E. N. (ed.) *Soils of Eastern Siberia*, 123–46. Transl. TT69-55073. Springfield, Va.: US Dept of Comm. Clearinghouse for Fed. Sci. and Tech. Inf.

KARAVAYEVA, N. A. and TARGUL'YAN, V. O. (1960) Humus distribution in the tundra soils of northern Yakutia. *Soviet Soil Sci.*, **12**, 1293–300.

KARAVAYEVA, N. A., SOKOLOV, I. A., SOKOLOVA, T. A. and TARGUL'YAN, V. O. (1965) Peculiarities of soil formation in the tundra taiga frozen regions of eastern Siberia and the Far East. *Soviet Soil Sci.*, **7**, 756–66.

KREIDA, N. A. (1958) Soils of the eastern European tundras. *Soviet Soil Sci.*, **1**, 51–6.

MACKAY, J. R., MATHEWS, W. H. and NACNEISH, R. S. (1961) Geology of the Engigstciak archaeological site, Yukon Territory. *Arctic*, **14**, 25–52.

MCMILLAN, N. J. (1960) Soils of the Queen Elizabeth Islands (Canadian Arctic). *J. Soil Sci.*, **11**, 131–9.

RICKERT, D. A. and TEDROW, J. C. F. (1967) Pedologic investigations on some aeolian deposits of northern Alaska. *Soil Sci.*, **104**, 250–62.

RIEGER, S. (1966) Dark well-drained soils of tundra regions in western Alaska. *J. Soil Sci.*, **17**, 264–73.

ROZOV, N. N. and IVANOVA, YE. N. (1967) Classification of the soils of the USSR. *Soviet Soil Sci.*, **2**, 147–56; **3**, 288–300.

SMITH, J. (1956) Some moving soils in Spitsbergen. *J. Soil Sci.*, **7**, 10–21.

SOIL SURVEY STAFF (1960) *Soil classification: a comprehensive system (7th approximation)*. US Dept of Agric., Soil Conserv. Serv. 265 pp.

SOIL SURVEY STAFF (1967) *Supplement to soil classification system (7th approximation)*. US Dept of Agric., Soil Conserv. Serv. 207 pp.

SOKOLOV, I. A. and SOKOLOVA, T. A. (1962) Zonal soil groups in permafrost regions. *Soviet Soil Sci.*, **10**, 1130–6.

SVATKOV, N. M. (1958) Soils of Wrangel Island. *Soviet Soil Sci.*, **1**, 80–7.

TEDROW, J. C. F. (1966) Polar Desert soils. *Soil Sci. Soc. Amer. Proc.*, **30**, 381–7.

TEDROW, J. C. F. (1968) Pedogenic gradients of the polar regions. *J. Soil Sci.*, **19**, 197–204.

TEDROW, J. C. F., DREW, J. V., HILL, D. E. and DOUGLAS, L. A. (1968) Major genetic soils of the Arctic slope of Alaska. *J. Soil Sci.*, **9**, 33–45.

TROLL, C. (1958) *Structure Soils, Solifluction, and Frost Climates of the Earth*. SIPRE Transl., 43. US Army Snow, Ice and Permafrost Res. Estab. 121 pp.

UGOLINI, F. C. (1966a) Soils of the Mesters Vig District, northeast Greenland, I: The Arctic Brown and related soils. *Medd. om Grønland*, **176** (1). 22 pp.

UGOLINI, F. C. (1966b) Soils of the Mesters Vig District, northeast Greenland, II: Exclusive of Arctic Brown and Podzol-like soils. *Medd om Grønland*, **176** (2). 25 pp.

WASHBURN, A. L. (1956) Classification of patterned ground and review of suggested origins. *Bull. Geol. Soc. Amer.*, **67**, 823–66.

ZHIGUNOV, P. S. (1961) *Reindeer Husbandry*. Transl. TT67-51247. Springfield, Va.: US Dept of Comm. Clearinghouse for Fed. Sci. and Tech. Inf. 348 pp.

Manuscript received May 1971.

Alpine soils

John L. Retzer

Fort Collins,
Colorado

1. Introduction

As the name suggests, alpine soils are those soils found in the high mountain regions above timberline. They occur extensively throughout the major mountain systems of the world such as the Alps, the Andes, the Himalayas, and the Rocky Mountains, and to lesser degrees in other mountainous areas whose crests are above timberline. In the United States their greatest extent is in the Rocky Mountains of Colorado, Wyoming, Montana, and northern New Mexico, as well as the Cascade Range of Washington. Small areas of alpine soils occur in the northeastern states and on higher peaks in other states as well. For the most part this discussion deals with the character of these soils as found in the central Rocky Mountains.

2. Landscape

The term 'alpine landscape' as used here includes that portion of the earth's surface that occurs above the areas of normal tree growth and below those of permanent snow cover. The alpine landscape is an area of steep, rocky topography with abrupt and often large differences in elevation occurring within short distances. In the central Rocky Mountains elevations exceed 4250 m in many locations while the lower boundary is irregular but occurs at about 3350 m to 3475 m. At these elevations normal tree growth ceases (timberline) and only small woody shrubs, such as willows, persist in the sheltered and wetter areas.

The central Rocky Mountains consist of a broad system of mountains and valleys having a general northwest/southeast trend. The highest peak is 4399 m above sea level. There are 53 peaks above 4267 m and 830 exceed 3353 m. The area as measured on an ordinary planimetric map is about 28,340 km², but since the terrain is characteristically very steeply sloping, the total surface area is considerably greater. The Continental Divide assumes prominence near Conejos Pass and steadily rises in elevation into central Montana. Many peaks along the Divide have elevations in excess of 4000 m.

CLIMATE

Extreme variations in temperature and precipitation within short periods of time are characteristic of the alpine areas. Summers are short and cool, while winters are long and cold. Average precipitation is between 65 and 100 cm a year and two-thirds or more of this may fall as snow. Summer and autumn months are relatively dry with most precipitation occurring as short, low-intensity showers. Snowfall during the winter months can be heavy with as much as 75 cm from a single storm not uncommon. Average monthly wind velocities of 13 m s^{-1} are common, and gusts of 45 m s^{-1} occur in the winter months. Strong westerly winds that accompany or closely follow most winter snowstorms remove the snow or blow it into lee-slope positions leaving a large proportion of the alpine soils bare during periods of lowest temperature. This probably accounts for the lower soil temperatures found in the alpine than in adjacent areas below timberline.

A more complete characterization of the climate of the alpine areas is given in Chapters 2 B and C. For the purposes of soil genesis, the low temperatures of the area resulting in long periods of biologic inactivity, the daily variations of temperature affecting physical weathering, and the area's characteristic precipitation pattern with its effect on available soil moisture are of prime significance.

GEOLOGY

Most of the distinctive features of the alpine landscape are directly related to the geology of the region. In the Rocky Mountains granites, pegmatites, and quartzites are common, and massive flows of basalt or other volcanic rock occur in many places. Sedimentary rocks are less common. Since they are more easily weathered most have been removed or deeply buried by erosional processes; however, they do occur at or near the surface of the ground in some places and have contributed to soil parent materials.

During the Laramide revolution the original rocks were exposed, broken, and rearranged. In the present cycle they are again undergoing destruction by weathering, but most are hard and resistant so weathering processes proceed slowly. The more granular rocks have been most susceptible to weathering and in some areas have been reduced to coarse textured saprolite. Glaciation has altered the original rock structure and shape of the landscape in many areas, leaving the customary glacial deposits as soil parent material. Rock structures such as dikes and fault lines act as barriers to free subsurface drainage, and permafrost zones often occur adjacent to them.

The glacial and intraglacial periods of the Pleistocene were as important to landscape formation in the Rocky Mountains as they were to the midwestern states. Since much of the alpine landscape was formed or modified during the glacial epochs, and many of the soil parent materials deposited during this time, the Pleistocene is of special significance to soil genesis. Fig. 13B.1 illustrates this schematically.

The end of the Wisconsin glacial period in the Midwest is considered to be about 11,000 or 12,000 years ago. Presumably this approximate date would be valid for the central Rocky Mountains as well. In consequence at least a part of the region's alpine landscape and the parent materials in which the soils are developing can be considered to be not more than 12,000 years old with many considerably younger. This does not preclude the probability that some landforms and some parts of the regolith may not date from Early to Middle Pleistocene and therefore be of much greater age.

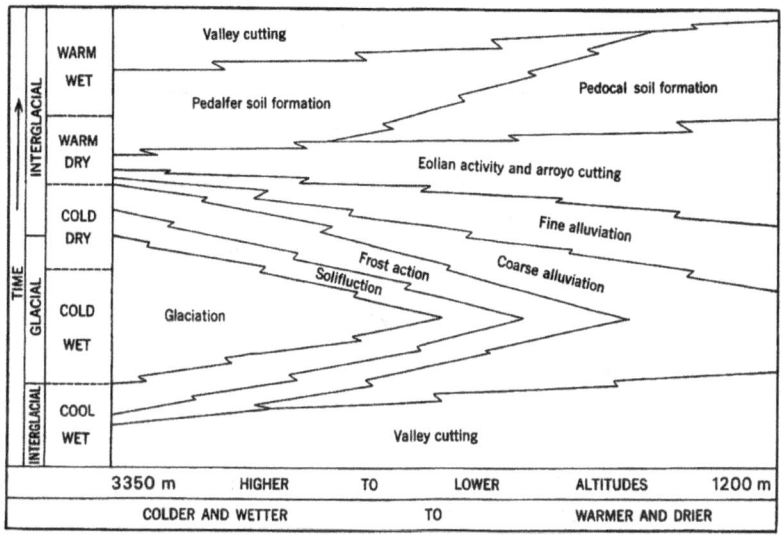

Fig. 13B.1. Schematic illustration of the variation of processes with altitude, time and inferred climate during a typical glacial-interglacial cycle in the La Sal Mountains, Utah (from Richmond, 1962).

Dates marking the beginning of the Pleistocene are much less definite and are open to question. Recent work has identified a mountain glaciation about 3 million years ago in the Sierra Nevada (Curry, 1966). Precise dating is of little importance for our purpose, except to point up the possibility that some parts of the alpine landscape as well as the soils and their parent materials could be of great age. This is particularly true of many of the higher summit areas and broad ridge crests that were not inundated by ice during Pinedale and Bull Lake times, and consequently exhibit landforms reflecting prolonged periods of subaerial, and especially periglacial, erosion.

Detailed soil geomorphologic studies correlate only weak degrees of soil development, or those genetic soil horizons that are known to develop rapidly, with deposits and landscapes of Late Wisconsin (Pinedale) or younger age. Exceptions to this correlation could exist as, for example, very youthful soils occurring at focal points of geologic erosion involving parent materials of considerable age. In such areas geologic erosion has kept pace with soil genesis or nearly so, and well developed soils have never formed.

Both conditions prevail in the alpine areas and in consequence the soil pattern consists mainly of soils with weakly contrasting horizonation, or with horizons that form from the accumulation of humus or by weak translocation of sesquioxides. Illuvial horizons of humus and sesquioxide can develop in relatively short periods of time. The predominance of such soils strongly suggests the youthful nature of both landforms and detrital parent materials. It seems probable that genetic soil development in much of the area is not much older than Late Wisconsin.

Continental glacial periods of the Midwest were followed by interglacial periods when strongly developed soils formed. Subsequent glacial periods frequently did not obliterate, but only buried these soils, and their relic profiles now serve as easily recognizable stratigraphic markers.

Such palaeosols have not as yet been recognized in the alpine areas, although remnants are occasionally found below timberline. If glacial development in the Rocky Mountains paralleled that of the Midwest, such interglacial soils should have formed. It seems probable that due to the scouring action of the ice in the relatively steeply sloping alpine areas, each succeeding glaciation obliterated any interglacial soil development, and strong mass movement under periglacial conditions mixed up and redistributed soil materials on the unglaciated upper surfaces. Thus the present landforms and deposits are most likely those left by the last glacial advance.

LANDFORMS AND GEOMORPHIC PROCESSES

There are several landforms in the alpine landscape that are of importance to soil development because of the characteristic kind and age of detrital deposit associated with them. In addition, some landforms have had an effect on soil climate through their control of soil moisture supplies and their influence on biologic life through their ability to influence soil temperature. Landforms such as relic erosion surfaces (peneplains), cirque basins, ground moraines, outwash fans and terraces, solifluction terraces, and scree and rock glaciers that have considerable acreage are easily recognized and are prominent throughout the alpine.

Landforms involving much smaller acreage and resulting from soil movements due to steep slopes, materials moistened to fluid-like consistency, melting and collapse of ground ice, frost boils and heaves, or wind erosion are also important. They are of particular significance pedogenically since they are the result of forces that tend to destroy or disrupt normal soil horizonation.

The surface configuration and the subsurface drainage in the alpine are largely rock controlled. Surface movement and the rearranging of surface materials by runoff waters or by gravity are common. Subsurface rock depressions that fill with water in the warm season produce frost boils, through the development of thin vertically segregated ice bands during the rapid frost penetration in autumn and winter. Surface heave, resulting from this frost penetration, exceeds 30 cm in moist areas of fine grained material. Fahey (1971), for example, measured a mean seasonal heave of 26 cm across the surface of a non-sorted circle at 3500 m in the Colorado Front Range. A maximum heave of 35 cm was recorded. In addition to the differential seasonal heave which causes frost boils and may leave 'ablation ponds', diurnal events also contribute to the heaving process. Needle ice develops in the soil particularly in autumn and spring when diurnal temperature oscillations about the freezing point are common. This dual effect in mid-latitudes may be compared with the Arctic (see Rieger, Chapter 13A, fig. 1), where diurnal effects are at a minimum, and the equatorial alpine where, for example, Coe (1967) found almost nightly formation of needle ice above 3650 m on Mt Kenya.

Measurements of horizontal and downslope movement of alpine soils on slopes in the Colorado Front Range have been made by Benedict (1970). Maximum rates on sorted stripes, turf-banked lobes and terraces and stone-banked lobes and terraces on Niwot Ridge ranged from 0.4 to 4.3 cm yr^{-1}. Rates of displacement are determined primarily by differences in moisture availability and gradient. Movement is generally confined to the upper 50 cm of the soil. Radiocarbon dating of a buried A horizon in one stone-banked terrace implies an average frontal advance of 0.36 cm yr^{-1} during the past 2470 $\pm$ 110 radiocarbon years with a maximum rate as high as 2.3 cm yr^{-1} between 1100 and 1000 years ago. Beyond the limits of the solifluction lobes and terraces soil movement in the alpine of the Front Range would appear to be very slow indeed.

Wind erosion, water erosion, and dust accumulation are all active in alpine areas. Runoff water moves great quantities of fine earth material downslope during storms or during spring snowmelt. The seasonal melt period usually has a cycle of daily freeze–thaw that on steep slopes results in large volumes of runoff. This rapid runoff undercuts natural vegetation and increases rates of soil removal. Wind action on the crests of exposed slopes may prevent a good vegetative cover from developing and eventually results in soil loss. Conversely a considerable amount of airborne dust from dry regions to the west is known to accumulate on lee slopes and around projecting vegetation in alpine areas.

The total effect of erosive forces on soil development in the alpine has never been completely assessed. It is an ever-present force that for the most part wears away the landscape and contributes to the immaturity of the soil pattern.

VEGETATION

Alpine vegetation consists of shrubs, grasses, sedges and forbs. Most of the alpine, then, is short herbaceous vegetation, but areas of low shrubs and alpine willow occur in the more moist sites. Because soil temperatures are low and soil moisture supplies during the growing season are limited, the plant root systems are shallow with a large proportion of the roots concentrated within the upper 15 cm. The tough dense sod that is formed conserves available moisture supplies, acts as a thermal barrier against harmful daily soil temperature fluctuation, provides a zone favourable to soil microbiologic life, provides plant nutrients, and protects the soil from excessive erosion. Its preservation is essential to a balanced ecological status in alpine areas and is of great significance to soil genesis.

The transition between alpine and subalpine is marked by the scattered growth of scarred, bent timber or krummholz. Its extent is variable, generally occurring at higher elevations on north-facing slopes. The subalpine–alpine boundary is primarily an area of transition, and timberline itself is best described as a zone rather than an abrupt boundary: the forest–tundra ecotone (see Wardle, Chapter 7B). It is an unstable zone in which timber is struggling to grow in a hostile environment, and the balance can be tipped in either direction by relatively minor differences in the character of the soil or its environment.

Character of the vegetation is one of the major factors controlling soil development. The morphology of a soil that has developed under grass is markedly different from that of one developed under trees. The wide differences in vegetative cover between the areas above and below timberline are accompanied by similar differences in soil genesis and morphology.

PERMAFROST

Permanently frozen areas have long been acknowledged as a characteristic phenomenon of arctic and alpine soils (see Ives, Chapter 4A). According to information gathered from miners or road-building crews, permanently frozen soil is found throughout the alpine areas of the central Rocky Mountains (Retzer, 1956; Ives and Fahey, 1971). It is not a uniform condition and its relation to exposure, type of rock, and topographic irregularities is not precisely known.

Moist areas that are insulated by thick organic layers have permanently frozen soils at shallowest depth. Miners report that the alpine soils melt to depths of 1.5–2 m in most summers and refreeze the following winter. Frozen soils extending from 20–120 m in depth have been reported.

The effect of permafrost on soil genesis in the alpine is not clearly understood. It undoubtedly has an effect on soil temperature gradients, on the availability of soil moisture supplies, and the degree of biologic activity in the soil.

3. Soil-forming processes

GENESIS AND MORPHOLOGY

Soils are the product of their environment, just as are plants, animals and people. All the factors of the environment act on all types of rocks or parent materials and eventually produce a kind of soil. The major factors influencing formation of soils are a) climate and b) living organisms, which act on c) parent materials to form a type of soil over d) a long period of time. Topography, or relief, is also a soil forming factor. In reality topography, through aspect and position, modifies the local climate by making more or less water and heat available in a certain area (see Barry and Van Wie, Chapter 2C). Thus, on very steep slopes topography may account for formation of shallow, stony soils. Similarly, high elevations and high latitudes modify local climates and markedly affect soil development.

In areas where there is great variability in climate, topography, elevation, and kinds or numbers of living organisms, there will be a corresponding large variation in the character of the soils that have developed. Since the alpine areas of the central Rocky Mountains have environments that display wide variation in both kind and degree of the soil forming factors, it is only reasonable to expect that the soil pattern will be diverse in terms of both morphology and genesis.

Generally speaking the soils of the alpine areas have profiles that are very irregular in regard to thickness of horizon, depth to bedrock, texture, and the size or amount of coarse fragments. Characteristics such as colour, reaction and structure are usually more consistent in local areas but may vary widely from one part of the region to another. Variations in the regolith caused by irregularities in the bedrock, such as faulting, or by the thickness of the weathered mantle of soil parent material are frequent and may occur abruptly. Variations in the soil profile itself due to disruption and mixing of genetic horizons by frost action and slope creep also occur frequently.

For the most part alpine soils have a low inherent fertility, are excessively drained, medium to coarse textured, contain large amounts of stone, cobble and gravel, and are strongly acid.

PARENT MATERIALS RELATIVE TO SOIL FORMATION

A large percentage of the soils of the alpine areas in the central Rocky Mountains has developed in parent materials weathered from gneiss, schist or quartzite. The degree of metamorphism varies but most of the original parent rocks are gneissic. Most common minerals are quartz, biotite and feldspars, with lesser amounts of muscovite and other minerals.

Other rocks occurring in the alpine and contributing to soil parent material include granite, basalt, rhyolite, tuff and limestone. Sedimentary rocks including sandstones, shales and loamstones occur in local areas, and although their total acreage is small, they have a pronounced effect on soil genesis and morphology in the areas of their occurrence.

Advanced weathering of rocks has not taken place because the rocks are hard and climate is unfavourable. Much of the weathering that has taken place is mechanical rather than chemical.

Chemical weathering is slow, but is nevertheless important. The rate of weathering can be appraised by studying analyses of groundwaters. The data in table 13B.1 show that chemical weathering is considerably less than that generally occurring in warmer, more humid areas.

There are many organic and peaty deposits in the alpine. Generally such deposits are of small acreage, but are distinctive features in the alpine soil pattern. In some bogs mapped in the Fraser alpine area (Retzer, 1962) chemical weathering has produced appreciable quantities of silt and clay. In other places stones found in the 25–50 cm layer above the permafrost have been weathered sufficiently so that they can be crushed in the fingers.

Most of the parent materials for soils in the alpine have been transported and mixed to some degree. In some areas they have been transported only short distances downslope and have retained the characteristics of the bedrock that underlies them. Such material is often said to be residual even though some transportation is known to have taken place.

Most material in the alpine has been and still is subject to slope creep although the rate and amount of such movement has varied from time to time. This movement of material downslope can be demonstrated by the relatively thick belt or apron of accumulated material that has formed at the base of most slopes.

Table 13B.1. Chemical analyses of waters from springs and the main channel of Fool Creek, Fraser Experimental Forest, in 1953.*

Item	Spring No. 23	Spring No. 33	Spring No. 41	Spring No. 3	Lower weir†	Upper weir
Conductivity (EC $\times$ 10^6 at 25°C)	39.20	43.10	36.90	42.20	63.10	37.00
Soluble sodium (per cent)	42.00	33.00	33.00	33.00	27.00	38.00
Boron (B) (ppm)	0.02	0.01	0.01	0.02	0.02	0.02
Sum of cations (meq/l)	0.48	0.46	0.42	0.46	0.71	0.40
Calcium (Ca) (meq/l)	0.14	0.17	0.15	0.15	0.35	0.15
Magnesium (Mg) (meq/l)	0.12	0.12	0.10	0.14	0.15	0.08
Sodium (Na) (meq/l)	0.20	0.15	0.14	0.15	0.19	0.15
Potassium (K) (meq/l)	0.02	0.02	0.03	0.02	0.02	0.02
Sum of anions (meq/l)	0.49	0.49	0.44	0.45	0.71	0.43
Carbonate (Co$_3$) (meq/l)	0.00	0.00	0.00	0.00	0.00	0.00
Bicarbonate (HCo$_3$) (meq/l)	0.30	0.34	0.28	0.28	0.55	0.28
Sulphate (SO$_4$) (meq/l)	0.04	0.05	0.06	0.07	0.06	0.05
Chloride (Cl) (meq/l)	0.15	0.10	0.10	0.10	0.10	0.10
Nitrate (NO$_3$) (meq/l)	—	—	—	—	—	—

*Analysed by J. T. Hatcher, Mary G. Keyes and G. W. Akin, US Salinity Laboratory, Rabidoux Unit, Riverside, Calif.
†Elevation of lower weir in Fool Creek is 2925 m; elevation of upper weir is 3445 m.

Where slopes have contained a variety of parent rock, the phenomenon of slope creep has resulted in the formation of soil parent materials that are mixtures. Such parent materials retain some of the characteristics of all of the parent rocks from which they were derived. From the standpoint of soil genesis they are somewhat different from the materials weathered residually or only locally transported and unmixed.

Other mixed deposits that are important as parent material for soil formation in alpine areas include glacial till, glacial outwash, and alluvial fan deposits. The chemical, physical and mineralogical character of these deposits varies depending upon the mode of origin, the original source of the sediment, and the degree of mixing that has taken place. Although it is possible to make some general statements regarding the character of each of these, such statements would have little relevance to the soil's genesis and morphology at any one place because of the variability of the deposits.

Windlain silt is constantly being deposited in the alpine areas. These materials originate as dust from dry regions and are often carried great distances by the prevailing winds.

They come from a variety of sources, principally from the desert areas to the west and southwest, but consist mainly of silts and clays (20 μm or less) that are alkaline in reaction. They accumulate as thin surficial deposits, averaging about 0.005 mm yr^{-1} in thickness, and are most noticeable during winter and spring months when they are caught on the

Table 13B.2. Some characteristics of six alpine turf soils from various parent rocks.

Parent rock horizons (cm)	Colour (air dry)	pH	Organic carbon (%)	Mechanical separates†			
				Gravel >2 mm (%)	Sand 2–0.05 mm (%)	Silt 0.05–0.002 mm (%)	Clay <0.002 mm (%)
Basalt							
A$_{11}$ 0–5	10YR 2/1 Bl*	6.0	17.02	65.9	34.7‡	47.6	17.7
A$_1$ 5–12.5	10YR 2/2 vd B	5.9	10.72	64.1	35.6‡	44.6	19.8
B 12.5–36	10YR 2/2 vd B	5.8	7.74	61.9	41.5	40.2	18.3
C$_1$ 36–82	10YR 5/4 y B	5.6	1.90	90.0	53.6	34.2	12.2
C$_2$ 82–94	10YR 6/3 p B	5.8	0.33	60.8	46.2	7.9	—
Schist							
A$_{11}$ 0–5	7.5YR 2/0 B	5.2	17.22	13.3	45.2‡	38.1	16.7
A$_1$ 5–12	10YR 2/2 vd B	4.7	11.71	30.5	49.1‡	34.0	16.9
B 12–25	10YR 2/2 vd B	4.5	8.93	17.8	59.9	26.7	13.4
C$_1$ 25–40	10YR 5/4 y B	4.6	1.12	17.6	68.2	26.2	5.6
C$_2$ 40–92	10YR 5/4 y B	4.8	0.40	44.0	75.5	22.3	2.2
Granite							
A$_{11}$ 0–18	10YR 2/2 vd B	5.7	4.32	41.2	64.5	24.2	11.3
B 18–46	10YR 5/4 y B	5.6	0.56	54.4	69.9	29.0	1.1
C 46–60	10YR 5/6 y B	5.3	0.54	36.6	78.7	14.5	6.8
Quartzite							
A$_1$ 0–15	5YR 3/1 vd G	4.5	10.23	15.3	49.9‡	34.3	20.8
B 15–56	7.5YR 5/4 B	4.7	1.40	17.8	74.5	17.9	7.6
C 56–82	7.5YR 5/6 s B	5.0	0.35	9.7	85.6	9.6	4.8
Shale							
A$_1$ 0–12.5	5YR 2/1 Bl	6.0	6.88	23.3	59.6	26.0	14.4
B 12.5–23	10YR 5/4 y B	6.0	2.79	32.8	53.8	30.5	15.7
C$_1$ 23–53	10YR 5/4 y B	6.2	0.85	27.8	61.8	26.5	11.7
C$_2$ 53–76	10YR 5/3 B	8.2	0.32	69.2	57.0	33.9	9.1
Limestone							
A$_1$ 0–15	7.5YR 3/2 d B	7.7	7.96	47.1	64.4	25.1	10.5
B 15–46	7.5YR 4/4 B	8.3	1.24	64.3	65.6	27.7	6.7
C 46–76	7.5YR 5/4 B	8.4	0.62	65.4	66.1	26.7	7.2

*Bl – Black, B – Brown, G – grey, vd – very dark, d – dark, y – yellowish, p – pale, s – strong.
†Gravel computed as percentage of entire sample; sand, silt and clay as percentage of the <2 mm fraction.
‡Mechanical analyses by Retzer. All others, including organic carbon, by V. J. Kilmer.

snow (Windom, 1969). It is not known to what extent this dust influences soil development or to what extent it has contributed to the silt and clay fractions of alpine and subalpine soils.

The abundance of bivalent cations, particularly calcium and magnesium, and the rates at which they are recycled by vegetation, have a pronounced effect on soil genesis. Since a major part of the alpine area is dominated by rocks that do not release large quantities of such cations on weathering, and since chemical weathering in these areas proceeds slowly, the soil parent materials are generally low in basic compounds.

In consequence the most common kinds of soil genesis (most active) in the formation of alpine soils are those that are most active under acid conditions.

Exceptions to this are the soils developing in basic parent materials such as basalt, limestone, volcanic debris, and most sedimentary rocks. Bivalent cations are abundant in such materials or in transported materials derived from them. The processes of soil genesis and the character of soil morphology are radically different in these areas than in other parts of the alpine.

In summary the soil parent materials throughout most of the alpine areas can be said to be medium to coarse textured, relatively thin, heterogeneous, only weakly weathered deposits originating primarily from crystaline rocks. They contain large amounts of stone, cobble and gravel, and are subject to considerable local movement through frost action, free fall, slope creep, and water movement. Table 13B.2 shows the textural characteristics of six alpine turf soils.

BIOLOGICAL LIFE RELATIVE TO SOIL DEVELOPMENT

The characteristic alpine vegetation has been discussed briefly in the section devoted to the characterization of the alpine landscape and is discussed in detail in other chapters. Its significance to soil genesis and morphology should not be overlooked.

Soils developing under short herbaceous vegetation in cool to temperate climates are characterized by the accumulation of organic matter in their surface horizons. A major part of their genesis centres upon the accumulation, decomposition and recycling of organic compounds. It is not surprising therefore that all but the very youthful soils of the alpine have very dark coloured, moderately thick, surface (A_0) horizons that are high in organic matter. In the alpine areas vegetation has only a limited above-ground growth, but root systems are dense and normally near the surface. It is probable that most of the accumulation of organic matter has come from decomposition of the root system.

Very little is known about animal and insect life of the alpine regions relative to their effect on soil genesis. Earthworms and the insects normally found in soils of warmer areas have not been observed in abundance in the alpine, and the soil micro fauna and flora are extremely limited. Presumably the low temperatures are not suitable for their existence. In the Australian Alps, however, Wood (1970) reports evidence of rapid breakdown of litter in 'alpine humus' soils at 1950 m on Mt Kosciusko, and notes that in fact decomposition over a twelve-month period *increased* with altitude, contrary to the general rule. While an increase in soil moisture with altitude may be partly responsible, he attributes this pattern primarily to high biological activity of earthworms and other similar invertebrates and micro-organisms.

Burrowing animals are common in these areas and are of importance in soil development. The highly organic surface horizons of the alpine soils with their supplies of plant roots are sources of winter feed for some burrowing animals, and their activities undoubtedly mix the surface horizons of these soils, as well as improving aeration, promoting crumb and granular structural forms, and adding excrement. Alpine areas that have been uniformly cultivated to depths of 25 to 30 cm by rodents have been observed. The activities of burrowing animals, and especially the pocket gopher, where concentrated, disrupt the protective turf mat and lay areas open to wind and water erosion.

While visible plant and animal life have the most obvious effects on soil genesis in alpine areas, the microbiological life of the soil is very important, and is largely responsible for the decomposition of organic matter. The products of this decomposition exert a strong influence on soil reaction, the mobility of clays, sesquioxides, or humus, and play a major role in the recycling of plant nutrients.

Very little precise information is available regarding kinds of specific activity of microscopic life in the soils of alpine regions. Presumably they would be dominated by those strains that are most adaptable to acid soil conditions, low soil temperatures, and alternately moist and dry soil conditions.

The average period of time each year that such life can function in the upper 20–25 cm of alpine soils is limited to about 45 to 110 days. The incomplete decomposition of organic matter resulting from the short period of microbiologic activity is probably a major factor in the accumulation of the large amounts of organic matter characteristic of the surface horizons of alpine soils (table 13B.2).

CLIMATE RELATIVE TO SOIL GENESIS

The characteristic climate of alpine regions is discussed in Chapters 2B and C. Some distinction must be made, however, between the characteristics of climate as measured above the soil and that which exists within the soil.

Some soil temperature measurements for the alpine areas are available, and while many more are needed to characterize adequately soil climates in these regions, they will at least provide a basis for forming a general picture.

Mean annual soil temperature measured at 51 cm and mean summer soil temperature (June, July, August) are values used in the new taxonomic system of soil classification. In 1961 at the D-1 station on Niwot Ridge (3750 m) operated by the Institute of Arctic and Alpine Research, mean annual soil temperatures at 51 cm averaged $-2.3°C$ and mean summer soil temperature was $4.3°C$. At 5 cm these values were $0.3°C$ mean annual soil temperature and $12.9°C$ mean summer soil temperature. At 15 cm (6 in.), mean annual soil temperature was $-1.8°C$ and mean summer soil temperature was $8.1°C$, while at 30 cm mean annual soil temperature was $-3.1°C$ and mean summer soil temperature was $5.1°C$. Averages for 1953–64 at 15 and 30 cm were respectively $6.8°$ and $6.1°C$ for mean summer temperature, and both $-1.1°C$ for mean annual temperature (Marr et al., 1968).

Although open to some question, $5°C$ is usually accepted as the temperature below which biologic activity is very slow or prohibited. Thus the period during which soil genesis proceeds most rapidly can be approximated by the time soil temperature is $5°C$ or above.

From the data gathered at the Niwot measuring site the soil at 5 cm is above $5°C$ for about 110 days of the year. At 15 cm this period is 93 days; at 30 cm it is 45 days; and at

60 cm it is about 21 days. Obviously, periods of time that are most favourable for rapid soil development are short, and the portion of the regolith favourable to soil forming processes is thin.

Most alpine areas of the central Rocky Mountains have soil temperatures similar to those described above; however, recent temperature measurements made by the National Forest Service in alpine areas near Santa Fe, New Mexico, are somewhat higher with mean annual soil temperatures being slightly greater than 0°C. Presumably this variation will be found to exist in other areas as well when more data are available.

The slightly higher temperatures of some parts of the alpine regions probably have little or no impact on genesis of the soils but may be of some significance in their use.

Table 13B.3. Classification of alpine soils of the central Rocky Mountains.

Order	Suborder	Great group	Subgroup
(1.0) Entisols	(1.1) Aquents	(1.11) Cryaquents	(1.111) Typic Cryaquents
	(1.2) Orthents	(1.21) Cryorthents	(1.211) Typic Cryorthents
			(1.212) Aquic Cryorthents
			(1.213) Lithic Cryorthents
			(1.214) Pergelic Cryorthents
(2.0) Inceptisols	(2.1) Aquepts	(2.11) Cryaquepts	(2.111) Typic Cryaquepts
			(2.112) Aeric Cryaquepts
			(2.113) Aeric Humic Cryaquepts
			(2.114) Histic Cryaquepts
			(2.115) Histic Pergelic Cryaquepts
			(2.116) Humic Cryaquepts
			(2.117) Humic Pergelic Cryaquepts
			(2.118) Pergelic Cryaquepts
	(2.2) Ochrepts	(2.21) Cryochrepts	(2.211) Typic Cryochrepts
			(2.212) Dystic Cryochrepts
			(2.213) Lithic Cryochrepts
			(2.214) Pergelic Cryochrepts
	(2.3) Umbrepts	(2.31) Cryumbrepts	(2.311) Typic Cryumbrepts
			(2.312) Entic Cryumbrepts
			(2.313) Lithic Cryumbrepts
			(2.314) Pergelic Cryumbrepts
(3.0) Mollisols	(3.1) Aquolls	(3.11) Cryaquolls	(3.111) Typic Cryaquolls
			(3.112) Histic Cryaquolls
			(3.113) Pergelic Cryaquolls
	(3.2) Borolls	(3.21) Cryoborolls	(3.211) Typic Cryoborolls
			(3.212) Aquic Cryoborolls
			(3.213) Lithic Cryoborolls
			(3.214) Pergelic Cryoborolls
(4.0) Histosols	(4.1) Fibrists	(4.11) Cryofibrists	(4.111) Typic Cryofibrists

It has an immediate effect on the classification of the soils since the new system provides for a taxonomic change between those soils having mean annual soil temperature of more than 0°C and those that are 0°C or lower. The type of problem which this criterion may present is illustrated by the work of Nimlos *et al.* (1965) in Montana. They suggest that the definition of the 'Cry-' groups (see table 13B.3) with respect to a mean annual soil temperature ≤0°C at 51 cm (20 in.) is too restricting and recommend a value of ≤1.7°C (35°F) in the alpine areas.

Precipitation in alpine areas of the central Rocky Mountains is highly variable but probably averages about 75–100 cm a year, of which about two-thirds falls as snow. Soil moisture supplies measured at the Niwot Ridge site show that for an average of nine years of record the soil is above permanent wilting point in all months of the year, but that it approximates wilting point during the month of September. There are, nevertheless, patches of 'microdesert' on talus slopes and in spots where wind removes the snow cover in winter. With the exception of brief periods in the autumn months, soil moisture supplies should, in general, be adequate for biological needs but, apart

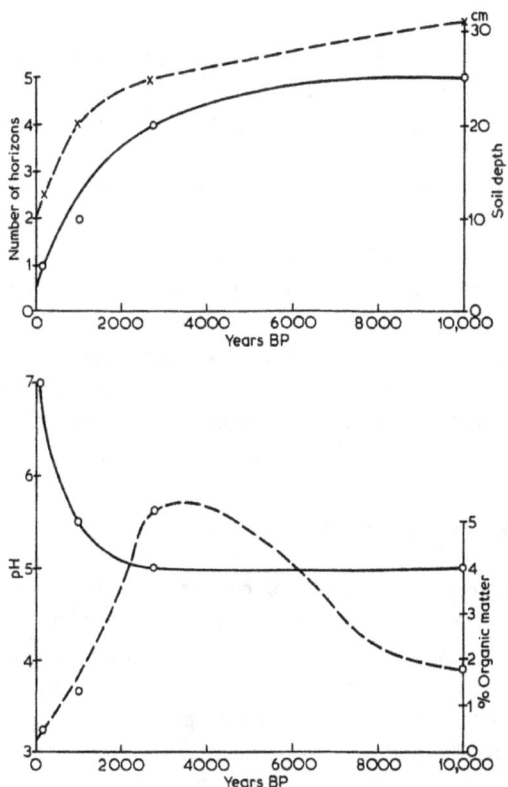

Fig. 13B.2. The change in soil properties with age on Late Pleistocene deposits in the Indian Peaks, Colorado (from Mahaney, 1970). *Above:* soil depth (dashed line), number of horizons (solid line). *Below:* organic matter per cent (dashed line), pH (solid line).

from the spring melt period, there are no prolonged intervals of time when soil temperatures are above freezing with an excess of soil moisture for leaching, except in surface horizons.

TIME RELATIVE TO SOIL GENESIS

In the preceding section dealing the the geology of the alpine regions it was pointed out that most of the soil parent materials are probably not much older than Early or Middle Wisconsin age. Carefully correlated studies of landscape evolution and the resulting soil pattern equate only very weak genetic soil horizonation with these ages. Recent work by Mahaney (1970) on Late Pleistocene and Neoglacial deposits in the Indian Peaks of the Colorado Front Range indicates that soil reaction in subalpine areas reaches a nearly steady state after about 2750 years, but that well developed genetic horizonation requires a much longer time. Fig. 13B.2 demonstrates the time change of horizonation, soil depth and the surface pH and organic content. In this area soil development is apparently limited not only by low temperature but also by the available moisture. Much more rapid rock weathering and soil genesis is to be expected in maritime alpine areas.

Most rock exposed in the high mountains is relatively fresh and unweathered. Consequently much of the soil mass consists of unweathered or only weakly weathered minerals such as quartz, mica, etc. that contribute little to the production of plant life (Retzer, 1956, p. 31).

With respect to the age of soil parent materials it can be said that alpine deposits are young and immature. Therefore soil horizons will generally be weak and immature since there has been insufficient time to develop genetic horizonation. This has been aggravated by the unstable nature of landscapes. Soil horizons that may have developed have been rearranged and mixed by frost disturbance and slope creep, or have been destroyed by processes of erosion. In evaluating time relative to degree of soil development, a clear distinction must be made between chronologic time and degree of genetic development. Particularly in alpine areas degree of pedogenetic development is often less than normally could be equated with the age of the parent soil material found.

RELIEF RELATIVE TO SOIL GENESIS

The generally steep topography of the alpine areas has been discussed relative to characteristics of the alpine landscape. Its effect on soil morphology and genesis due to its influence in regard to physical movement of the regolith has been emphasized.

Topography, or relief, also exerts an influence on soil genesis by its ability to modify climate. Characteristics of aspect, landform and position exert controlling influences on both soil moisture supplies and soil temperature. Excess amounts of soil water in poorly drained areas drastically alter the genetic soil processes, while soils on south-facing slopes may be much warmer than those on north-facing slopes. On very steep slopes topography may account for the formation of shallow stony soils, while in valley areas soils may be much deeper and nearly free of stone.

4. Classification and characteristic morphology of alpine soils

The taxonomic system of soil classification now used by the US Department of Agriculture (Soil Survey Staff, 1960, 1967, 1968) and cooperators in the National Soil Survey Program is based on soil morphology, and the various taxonomic classes are defined in terms of soil morphology. Thus only the physical, chemical and mineralogical properties of the soil that can be determined by the human senses or can be measured by instruments or laboratory methods are used in its classification. Elements of soil genesis are not in themselves used as definitive criteria, but many of the definitive properties selected are representatives of some change in degree or kind of genesis.

There are only a few detailed soil surveys that have been completed in alpine areas.[1] Consequently our knowledge of the details of soil pattern is incomplete, particularly at levels of generalization below the subgroup. Fortunately the higher levels of generalization lend themselves most easily to a study of the major morphological characteristics of the soils of broad regions of the alpine. They also have the greatest relevance to soil genesis.

The following discussion deals primarily with the classification and morphology of alpine soils known to exist in the central Rocky Mountains. Classification is limited to the order, suborder, great group, and subgroup levels. Where possible correlations between the present system of classification and that formerly used (Baldwin et al., 1938; Thorpe and Smith, 1949) are given, and representative profiles have been described for the major subgroups. Table 13B.3 outlines the classification of alpine soils in the central Rocky Mountains in regard to the higher levels of generalization in the classification system. It includes only those kinds of soils that are known to occur, but in many instances the subgroups have not been mapped over a sufficient area to justify the creation of individual soil series.

1.0 ENTISOLS

Entisols (formerly called regosols, lithosols) of alpine areas are cold, well to moderately well drained soils, having no genetic soil horizonation other than light coloured or very thin dark coloured surface horizons (ochric epipedons). They may vary in thickness, colour, texture, reaction, and content of coarse fragments, depending upon the source and character of the parent materials.

Usually they are thin, medium to coarse textured, stony, gravelly, or cobbly soils occupying ridge crests and shoulders where geologic erosion is very strong. Occasionally they may occur on very recent deposits on more level terrain where soil development is in very youthful stages, or they occupy 'snow kill areas' on lee slopes where semi-permanent snowbanks have restricted vegetative cover.

The total area of Entisols recognized in soil surveys of alpine areas is comparatively

[1] In Colorado these are the Fraser Alpine Area covering 282 km² (69,800 acres) above timberline.(Retzer, 1962), the Taylor River area in the Gunnison National Forest, and the Piedra area of the San Juan National Forest where approximately another 600 km² above timberline have recently been surveyed.

small. Their actual area, however, is somewhat larger than most standard soil surveys indicate, since they are often included in land type units and are not recognized as individual soil series.

1.1 *Aquents*

The Aquents of alpine areas are the poorly to somewhat poorly drained Entisols. They have all of the properties of the Entisols and in addition are saturated with water for a significant portion of each year.

In alpine areas these are normally coarse textured or intensely stratified soils where an irregular placement of organic carbon with depth or a coarse textured regolith prevents their classification as Aquepts (described below). They have either fluctuating or permanently high water tables, have colours with hue of 2.5Y or bluer, and usually are strongly mottled. Their extent in the alpine soil pattern is normally very small.

1.11 *Cryaquents*

These are Aquents of cold areas. They have all of the properties of the Aquents, but in addition must have mean annual soil temperatures measured at 51 cm of less than 8.3°C but more than 0°C and have summer soil temperatures of less than 12.8°C if they lack O horizons, or less than 6.1°C if they have an O horizon. All of the Aquents in the alpine areas are in this great group.

1.111 *Typic Cryaquents*

These are the Cryaquents that are developing in parent materials having little or no influence from volcanic ash or other pyroclastic materials. They have all of the other properties of the Aquents.

Following is a profile description of a Typic Cryaquent from an alpine area.

O1   7.5–0 cm – Undecomposed organic material, principally grass remains, roots and leaves.

A1   0–10 cm – Dark greyish-brown (10YR 4/2) dry sandy loam, very dark brown (10YR 2/2) moist moderate fine granular structure; soft dry, very friable moist; 30 per cent gravel; neutral pH 7.0 (BTB); clear smooth boundary (75–125 mm thick).

AC   10–20 cm – Greyish-brown (10YR 5/2) dry sandy loam, dark greyish-brown (10YR 4/2) moist; weak medium sub-angular blocky structure that parts to moderate fine granules; soft dry, very friable moist; 30 per cent gravel; neutral pH 7.0 (BTB); clear wavy boundary (75–125 mm thick).

IIC   20–150 cm – Non-calcareous gravelly loamy sand; common medium sized distinct reddish-brown (5YR 4/4) mottles.

1.2 *Orthents*

These are the well to moderately well drained Entisols of alpine areas. They have all of the properties previously described for the Entisol order, and in addition have organic matter contents that are at a maximum in the surface horizons and decrease uniformly with depth. In alpine regions they are normally medium or moderately coarse textured, often shallow over bedrock, and are stony, gravelly, or cobbly. In places they may be coarse textured providing they contain more than 35 per cent coarse fragments (2 mm or more). They differ from the Aquents in being well or moderately well drained.

1.21 *Cryorthents*

These are the Orthents of cold climates. They have all of the properties of the Orthents and in addition have mean annual soil temperature of less than 8°C and mean summer soil temperature of less than 14°C. This great group includes all of the Orthents in alpine areas.

1.211 *Typic Cryorthents*

In alpine areas these are the Cryorthents that have no mottling accompanied by low chroma within 50 cm, are thicker than 50 cm to any hard continuous bedrock, and have mean annual soil temperatures of 0°C or more or higher. Because their mean annual soil temperature is 0°C or higher they occur most frequently at the lower margins of the alpine or in the southernmost extensions of such regions.

Following is a profile description of a Typic Cryorthent from an area bordering tree-line.

A1 0–15 cm – Pale brown (10YR 6/3) dry loam, brown (10YR 4/3) moist; moderate to strong very fine granular structure; soft dry, very friable moist, non-sticky and slightly plastic wet; 40 per cent coarse fragments by volume; 10 per cent stones and 30 per cent cobble; very strongly acid pH 4.9 (paste); gradual smooth boundary (2.5–20 cm thick).

C1 15–40 cm – Pale brown (10YR 6/3) dry loam, brown (10YR 4/3) moist; weak fine sub-angular blocky structure that parts to moderate very fine granules; soft, dry, very friable moist, non-sticky and slightly plastic wet; 70 per cent coarse fragments by volume; 5 per cent stones and 65 per cent gravel and cobble; very strongly acid pH 4.7 (paste); gradual smooth boundary (10–40 cm thick).

C2 40–100 cm – Pale brown (10YR 6/3) dry loam, brown (10YR 4/3) moist; structure-less; soft dry, very friable moist, non-sticky and slightly plastic wet; 75 per cent coarse fragments by volume; 5 per cent stones, 70 per cent gravel and cobble; very strongly acid pH 4.7 (paste).

1.212 *Aquic Cryorthents*

These are the moderately well drained Cryorthents of alpine areas. They have all of the properties described above for the Typic Cryorthents and differ in having distinct mottling accompanied by low chroma above 50 cm, but not high enough nor sufficiently intense to justify the placement of the soil in the Cryaquept great groups (described below). This subgroup is of minor importance in most alpine soil patterns.

1.213 *Lithic Cryorthents*

These are the Cryorthents of alpine areas that overlie continuous strata of hard bedrock above 50 cm. Unlike the typic subgroup this subgroup is found in areas that are both warmer and colder than 0°C. They are an important subgroup of Cryorthents in alpine areas.

1.214 *Pergelic Cryorthents*

These are the extremely cold Orthents of alpine areas. They have all of the properties of the typic subgroup but have mean annual soil temperature lower than 0°C. They occur in most alpine landscapes above 3650 m.

2.0 INCEPTISOLS

In the alpine areas these are soils that have weak but distinct and continuous genetic horizonation. The distinctive horizonation may take the form of relatively thick dark surface horizons of low exchangeable base status (*umbric epipedons*), or as subhorizons showing definite evidence of genetic weathering and alteration (*cambic horizons*). Normally such subhorizons have redder hue, brighter chroma, or advanced structural forms and grades in well drained areas, or low chroma in the base colour of the matrix accompanied by mottling due to the segregation of iron compounds in the poorly drained areas. They may be distinguished from the Entisols by their recognizable degree of genetic horizonation and from the Mollisols on the low exchangeable base status of their surface horizons.

The Inceptisol order includes most of the soils found in alpine areas in the central Rocky Mountain area and occurs extensively on the gently to steeply sloping parts of the alpine landscape. Typically they have moderately thick, dark coloured, strongly or very strongly acid, organic-rich surface horizons that are underlain by bright coloured sub-angular blocky B2 horizons that have weak evidence of accumulation of sesquioxides and humus. Less extensive areas have the bright coloured B2 horizons but have only light coloured surface horizons. Poorly drained areas with fluctuating water tables have B2 horizons of low chroma accompanied by prominent mottlings of red and yellow. Those with permanent tables have grey, blue and green hues.

These are extensive soils in the alpine and include those commonly called Alpine Turf and Alpine Meadow Soils (Baldwin *et al.*, 1938; Retzer, 1956).

2.1 *Aquepts*

These are the wet Inceptisols of the alpine areas. They occur in areas having high water tables and are saturated with water for significant periods each year. They may have either light or dark coloured surface horizons, but in or immediately below the surface horizons they must have B2 horizons of low chroma with distinct accessory mottling, or they must have chroma of neutral to one if unmottled.

2.11 *Cryaquepts*

The Aquepts of alpine regions occur in this great group. They have all of the characteristics of the Aquepts, and in addition have mean annual soil temperature lower than 8°C, and mean summer soil temperature lower than 13°C if without an O horizon, or less than 6°C if there is an O horizon at their surface. They are only moderately extensive soils in alpine landscapes and are usually associated with small valleys or with swales and depressions.

2.111 Typic Cryaquepts

These are the Cryaquepts of alpine areas that have mean annual soil temperature of 0°C or more or higher, that have chroma of 2 or less in at least 60 per cent of the soil matrix between 15 and 50 cm, have light coloured ochric epipedons, lack organic surface horizons more than 5–10 cm thick, have little or no influence from volcanic ash or pyroclastic material, and lack continuous hard bedrock strata above 50 cm. Since most of the Aquepts of alpine areas have either thick organic surface layers or dark coloured, strongly acid A horizons, the soils of the typic subgroup are relatively inextensive. They occur most frequently at lower elevations or in the southern parts of the alpine regions of the central Rocky Mountains.

2.112 Aeric Cryaquepts

These are the Aquepts of the alpine regions that have all the properties discussed for the typic subgroups except that they have chromas of 2 or less accompanied by mottling in less than 60 per cent of all subhorizons between 15 and 50 cm. At least a part of the soils identified as Alpine Meadow Soils in the 1939 Yearbook would now be classified in this subgroup. Following is a profile description of an Aeric Cryaquept from alpine areas near timberline.

O1   5–0 cm – Undecomposed roots, stems, and leaf remains; abrupt smooth boundary.

A1g   0–15 cm – Pinkish-grey (7.5YR 6/2) dry silt loam, brown (7.5YR 4/2) moist; common numbers of small faint (10YR 3/3) mottles; moderate fine crumb structure; soft dry, very friable moist; non-sticky and plastic wet; 5 per cent coarse fragments by volume; pH 4.7 (paste); clear wavy boundary (12.5–25 cm thick).

B21g   15–30 cm – Light grey (10YR 7/2) dry loam, greyish-brown (10YR 5/2) moist; common numbers of medium sized distinct (10YR 4/1) mottles; moderate very fine granular structure; soft dry, very friable moist, non-sticky and plastic wet; 10 per cent coarse fragments by volume; pH 4.7 (paste); gradual wavy boundary (2–15 cm thick).

B22g   30–48 cm – Pink (7.5YR 7/4) dry silt loam, light brown (7.5YR 6/4) moist, many medium sized prominent (7.5YR 6/6) mottles; weak to moderate fine sub-angular blocky structure; slightly hard dry, very friable moist, non-sticky and non-plastic wet; 10 per cent coarse fragments by volume; pH 4.8 (paste); diffuse wavy boundary (10–23 cm thick).

Cg   48–96 cm – Pine (7.5YR 7/3) dry silt loam, light brown (7.5YR 6/3) moist; many medium sized prominent (7.5YR 6/6) mottles; weak fine sub-angular blocky structure; slightly hard dry, very friable moist, non-sticky and plastic wet; 10 per cent coarse fragments by volume; pH 4.8 (paste) (a metre or so thick).

2.113 Aeric Humic Cryaquepts

These are somewhat poorly drained soils of the lower elevations of the alpine regions bordering timberline that have the properties of the typic subgroup, except that they have chroma of 2 or less in less than 60 per cent of all subhorizons between 15–50 cm, and they have moderately thick, dark coloured, strongly acid A horizons (umbric epipedons). Included in this subgroup is a large part of the Alpine Meadow Soils of the 1938 Yearbook. Following is a profile description of a typical example of these soils from alpine areas.

A11   0–10 cm – Dark grey (10YR 4/1) dry coarse sandy loam, dark brown (10YR 2/2) moist; weak fine and medium sub-angular blocky structure breaking to coarse granules; slightly hard dry, friable moist; 15 per cent stone; very strongly acid pH 4.6; clear smooth boundary.

A12   28–60 cm – Light yellowish-brown (10YR 6/4) dry sandy loam, yellowish-brown (10YR 6/4) moist; 40 per cent of the soil mass is large prominent (2.5YR 5/1) mottles; weak to moderate medium sub-angular blocky structure; slightly hard dry, very friable moist; 20 per cent stone; very strongly acid pH 4.6; diffuse wavy boundary.

Cg   60–150 cm – Light yellowish-brown (2.5YR6/4) dry coarse sandy loam stratified with thin layers of coarse sand, light olive-brown (2.5YR 5/4) moist; 30 per cent of soil mass is medium sized distinct (10YR 5/6 and 5YR 5/1) mottles; massive; slightly hard dry, very friable moist; approximately 40 per cent stone; a free water table at 125 cm; strongly to very strongly acid pH 5.0.

2.114 *Histic Cryaquepts*

These are Cryaquepts of alpine areas that have the properties previously outlined by the typic subgroup, except that they have moderately thick organic surface layers, and may have either light or dark coloured upper mineral horizons. Like the preceding subgroups they are found most frequently near timberline or in the southern extensions of the alpine in the central Rocky Mountains. They are inextensive soils in alpine areas.

2.115 *Histic Pergelic Cryaquepts*

These are the Cryaquepts of the alpine that are like the typic subgroup except that they have moderately thick organic surface horizons, and have mean annual soil temperatures below 0°C. They occur rather more extensively than the Histic Cryaquepts and are more general throughout the central and northern alpine regions in the United States.

2.116 *Humic Cryaquepts*

These are Cryaquepts like the typic subgroup except that they have thick, dark, very strongly acid surface mineral horizons. They are most apt to occur at lower elevations in the alpine near timberline, or in the southern parts of the alpine in the United States.

2.117 *Humic Pergelic Cryaquepts*

These are Cryaquepts of the very cold portions of the alpine. They have the properties of the typic subgroup except that they have moderately thick, dark coloured, strongly acid surface mineral horizons and have mean annual soil temperatures below 0°C.

2.118 *Pergelic Cryaquepts*

These are Cryaquepts of the very cold portions of the alpine. They have the properties of the typic subgroup except that they have mean annual soil temperatures below 0°C.

2.2 *Ochrepts*

In alpine areas these are well to moderately well drained Inceptisols having light coloured surface horizons and subsurface horizons of genetic weathering and alteration (formerly *sols bruns acides*). They are inextensive in most alpine landscapes of the central Rocky Mountains and occur mostly at the lower elevations adjacent to timberline.

2.21 *Cryochrepts*

These are Ochrepts of cold climates and include all of the Ochrepts found in alpine landscapes. They have the properties of Ochrepts and in addition have mean annual soil temperature below 8°C and mean summer soil temperature below 14°C.

2.211 *Typic Cryochrepts*

These are the Cryochrepts of alpine areas that have mean annual soil temperatures of 0°C or more, that lack hard continuous bedrock above 50 cm, that lack mottles in horizons with chroma of 2 or less within 75 cm, that have base saturation (by ammonium acetate) of more than 60 per cent in at least some subhorizon within 75 cm, and that are not dominated by volcanic ash or pyroclastic material.

These soils are not extensive in alpine landscapes but occur occasionally at its lower margins near timberline or in its southern parts of the United States. Following is a description of a typical profile of a soil in this subgroup taken just above timberline.

A1   0–7.5 cm – Brown (10YR 5/3) dry silt loam, brown (10YR 4/3) moist; very weak fine granular structure; soft dry, very friable moist, non-sticky and slightly plastic wet; 55 per cent quartzite boulders by volume; strongly acid pH 5.1 (paste); clear wavy boundary (7.5–15 cm thick).

A3   7.5–20 cm – Light brown (7.5YR 6/4) dry silt loam, brown (10YR 4/3) moist; very weak fine sub-angular blocky structure that parts to very weak fine granules; soft dry, very friable moist, non-sticky and slightly plastic wet; 60 per cent quartzite boulders by volume; very strongly acid pH 4.9 (paste); clear wavy boundary (10–20 cm thick).

B2   20–40 cm – Light brown (7.5YR 6/5) dry silt loam, strong brown (7.5YR 4/5) moist; moderate fine sub-angular blocky structure that parts to very weak fine granules; soft dry, very friable moist, non-sticky and slightly plastic wet; 60 per cent quartzite boulders by volume; very strongly acid pH 4.9 (paste); clear wavy boundary (12–25 cm thick).

C   40–80 cm+ – Very pale brown (10YR 7/4) moist; massive; soft dry, very friable moist, non-sticky and slightly plastic wet; 70 per cent quartzite boulders by volume; strongly acid pH 5.1 (paste).

2.212 *Dystic Cryochrepts*

These are Cryochrepts that have all the properties of the typic subgroup except that base saturation (by ammonium acetate) is less than 60 per cent in all subhorizons of the soil above 75 cm. They are very strongly acid soils. They are not extensive in the alpine, occurring mainly in the lower margins adjacent to timberline, but they are common in the subalpine forested areas below timberline. Following is a description of a soil in this subgroup.

Aoo   5.4 cm – An undecomposed mat of pine needles, bark, and other organic debris (1–5 cm thick).

Ao   4–0 cm – Very dark grey (10YR 3/1) dry, black (10YR 2/1) moist; massive; compact horizon of partially decayed organic matter like the material above but having undergone more complete decomposition. This horizon rests abruptly on that below (2.5–5 cm thick).

A2   0–15 cm – White (10YR 8/1) dry loamy fine sand (10YR 7/1) moist; loose when dry, very friable when moist; moderate medium platy structure breaking to moderate fine and medium crumbs or single grains; this horizon contains a few small faint (7.5YR 4/4) mottles; very strongly acid pH 4.6; lower boundary abrupt and wavy (7.5–18 cm thick).

B2   15–40 cm – Pale brown (10YR 6/3) dry light fine sandy loam, brown (10YR 5/3) moist; soft when dry, very friable when moist; weak fine and medium sub-angular blocky structure breaking to weak medium and coarse granules; very strongly acid pH 4.9; a few iron and manganese pellets; lower boundary gradual and wavy (23–33 cm thick).

C1   50–60 cm – Light grey (10YR 7/2) dry loamy fine sand or fine sand, light brownish-grey (10YR 6/2) moist; massive to single grained; slightly hard when dry, very friable when moist; strongly acid pH 4.9; 70 per cent sandstone rock fragments, lower boundary gradual and smooth.

Dr   60 cm+ – Light grey or nearly white sandstone.

2.213 Lithic Cryochrepts

These are Cryochrepts that have all the properties of the typic subgroup except that hard bedrock occurs above 50 cm. They are not extensive soils and like the typic Cryochrepts occur most frequently in alpine areas that border timberline.

2.214 Pergelic Cryochrepts

These are Cryochrepts of the very cold portions of the alpine. They differ from the typic subgroup in having mean annual soil temperature below 0°C. The soils of this subgroup may or may not be mottled but not to the degree definitive of Aquepts. They are inextensive in most alpine landscapes.

2.3 Umbrepts

In alpine areas these are the well to moderately well drained Inceptisols that have moderately thick, dark coloured, strongly to very strongly acid surface horizons rich in organic matter. They usually have subsurface horizons of genetic weathering and alteration, although such horizons are not mandatory for this suborder. They are extensive soils of alpine areas in the central Rocky Mountains and predominate in most landscapes above timberline.

2.31 *Cryumbrepts*

This great group includes a majority of all the soils of alpine areas. They have all the properties previously described for the Inceptisol order and for the Umbrept suborder. They are the Umbrepts of cold areas and have mean annual soil temperatures below 8°C and mean summer soil temperatures less than 14°C. The group of soils called Alpine Turf in the 1938 Yearbook is now included in this great group (see also Retzer, 1956).

2.311 *Typic Cryumbrepts*

In the alpine areas these are Cryumbrepts that in addition to the dark coloured, acid surface horizons have subsurface horizons of genetic weathering and alteration. They have no hard bedrock above 40 cm and have no mottling above 75 cm in horizons whose chroma is 2 or less. They have mean annual soil temperatures of 0°C or above.

These are extensive soils in alpine areas that meet the temperature requirements such as the southern alpine areas of the central Rockies. In the former system of classification they were part of the group of Alpine Turf soils. A description of a soil from this subgroup follows.

O1   2.5–0 cm – Alpine turf originating from a community of meso and hydrophytic plants; pH 5.4 (CPR); 15 per cent quartzite boulders by volume; abrupt smooth boundary (1–5 cm thick).

A1   0–30 cm – Dark brown (10YR 4/3) dry silt loam, very dark greyish-brown (10YR 2/2) moist; moderate fine granular or crumb structure; soft dry, very friable moist, non-sticky and slightly plastic wet; 30 per cent quartzite boulders by volume; very strongly acid pH 4.4 (paste); numerous particles of mica throughout; abrupt wavy boundary (10–40 cm thick).

B2   30–64 cm – Yellowish-brown (10YR 5/4) dry sandy loam, dark yellowish-brown (10YR 4/4) moist; very weak medium sub-angular blocky structure breaking to weak, very fine granules; soft dry, very friable moist, non-sticky and slightly plastic wet; a few rounded and sub-angular black pellets; very strongly acid pH 4.8 (paste); 40 per cent quartzite boulders by volume; particles of mica throughout; clear wavy boundary (23–40 cm thick).

C   64–86 cm+ – Pale brown (10YR 6/3) dry sandy loam, dark greyish-brown (2.5YR 4/3) moist; structureless; soft dry, very friable moist, non-sticky and non-plastic wet; 40 per cent boulders by volume; particles of mica throughout; very strongly acid pH 5.0 (paste) (0.5–1 m thick).

2.312 *Entic Cryumbrepts*

These are Cryumbrepts that have all the characteristics described above for the typic subgroup except that they lack the subsurface weathered horizons and have only the moderately thick surface horizons of low base status. They occur in the same areas as do the typic Cryumbrepts but usually occupy the younger portions of the landscape.

2.313 *Lithic Cryumbrepts*

In alpine areas these are Cryumbrepts that have all the properties described for the typic subgroup except that hard bedrock occurs above 50 cm and the restrictions regarding mean annual soil temperature and degrees of mottling are waived. These are extensive soils in most alpine landscapes. Following is a profile description of a soil in this subgroup.

A11   0–10 cm – Brown (7.5YR 5/2) dry channery loam, brown (7.5YR 3/2) moist; weak very fine crumb structure; slightly hard dry, very friable moist, non-sticky and non-plastic wet; 20 per cent slate channery and flagstone; strongly acid pH 5.3 (CPR); clear smooth boundary (10–20 cm thick).

A12   10–35 cm – Brown (7.5YR 5/2) dry channery loam, brown (7.5YR 3/2) moist; strong medium sub-angular blocky structure breaking to moderate fine granules; slightly hard dry, very friable moist, non-sticky and non-plastic wet; 20 per cent slate channery and flagstone; strongly acid pH 5.2 (CPR); clear smooth boundary (10–25 cm thick).

R   35–50 cm – Hard, weakly fractured slate.

2.314 *Pergelic Cryumbrepts*

In alpine areas these are Cryumbrepts that have the properties described for the typic subgroup except that their mean annual soil temperature is less than 0°C and the requirement for weathered genetic subsurface horizons is waived.

These are very extensive soils in most alpine landscapes and embrace the central concept of the group formerly called Alpine Turf soils. Nimlos and McConnell (1965), for example, refer to a series of this subgroup as being extensive between 2800 m (treeline) and 3350 m on the Beartooth Plateau in Montana, although they note (Nimlos *et al.*, 1965) that the temperature criterion is not satisfied at an elevation of 3200 m. Following is a description of a soil in this subgroup.

A1   0–20 cm – Very dark grey (10YR 3/1) dry loam, black (10YR 2/1) moist; strong medium and coarse crumb structure; soft dry, very friable moist; very strongly acid pH 4.6; clear wavy boundary.

A3   20–30 cm – Dark greyish-brown (10YR 4/2) dry sandy loam, very dark brown (10YR 2/2) moist; moderate fine and very fine sub-angular blocky structure breaking to moderate to strong fine and medium granules; soft dry, very friable moist; very strongly acid pH 4.6; clear wavy boundary.

B2   30–64 cm – Light yellowish-brown (10YR 6/4) dry coarse sandy loam, yellowish-brown (10YR 5/4) moist; moderate fine sub-angular blocky structure; slightly hard dry, very friable moist; dark coatings on sand grains and dark silt sized pellets; very strongly acid pH 4.6; gradual wavy boundary.

C   64–75 cm – Light yellowish-brown (2.5Y 6/3) dry sandy loam, light olive brown (2.5Y 5/4) moist; massive; slightly hard dry, very friable moist; 70 per cent stone and gravel; very strongly acid pH 4.6; gradual wavy boundary.

R   75 cm+ – Weakly weathered and partially fractured gneiss and schist bedrock; less than 5 per cent fine material in the cracks between the rock.

Table 13B.4 shows the mineralogical characteristics of a soil in this subgroup and table 13B.5 the exchangeable cations.

Table 13B.4. Mineralogy of modal profiles of the Ptarmigan series (Pergelic Cryumbrept) as formed on three alpine areas.

Horizon	Tobacco Root Mountains	Mt Edith	Siyeh Pass
O2	Chlorite, Talc, quartz, feldspar	Vermiculite, quartz, feldspar, amphibole	Illite, chlorite, kaolinite, quartz, feldspar
A1	Vermiculite, biotite, feldspar	Vermiculite, biotite (minor), quartz, feldspar, amphibole	Illite, chlorite, kaolinite, quartz, feldspar
B	Vermiculite, quartz, feldspar	Vermiculite, amphibole, quartz, feldspar	Illite, chlorite, kaolinite, quartz, feldspar
C	Biotite, mixed-layer, feldspar, quartz	Vermiculite, biotite, (minor), quartz, feldspar	Illite, chlorite, kaolinite, quartz, feldspar
Coarse fragments	Quartz, feldspar, biotite (minor), amphibole	Quartz, feldspar, biotite, amphibole	Illite, chlorite, kaolite, quartz, feldspar

Table 13B.5. Exchangeable cations and available phosphorus in two alpine turf soils – a Cryoboroll and a Cryumbrept.

Soil horizons as in table 13B.2	Total base ex. capacity (me)*	Exchangeable cations					Available P (100 kg/km²)
		Ca (me)	Mg (me)	K (me)	Na (me)	H (me)	
Cryoboroll (on basalt)							
A_{11}	58.0	44.16	7.52	0.90	0	5.42	45
A_1	45.2	37.75	5.84	0.60	0	1.01	34
B	35.2	30.34	4.44	0.19	0	0.23	45
C_1	15.7	10.60	2.00	0.10	0.15	2.85	34
C_2	8.4	6.14	1.22	0.18	0.06	0.80	225+
Cryumbrept (on quartzite)							
A_1	29.6	8.08	0.80	0.37	0.02	20.33	23
B	6.6	6.13	0	0.03	0	0.44	34
C	3.0	2.19	0	0.01	0	0.80	68

*Milliequivalents per 100 gm of oven-dry soil.

3.0 MOLLISOLS

In the alpine areas these are soils similar to the Inceptisols in that they have weak but continuous genetic horizonation, but they differ from the Inceptisols in that they must always have moderately thick, friable, dark coloured, organic matter enriched surface horizons that are more than 50 per cent base saturated (*mollic epipedons*). Usually they have subsurface horizons showing evidence of genetic weathering and alteration, but such horizons are not mandatory for the order.

These soils are of limited extent in the alpine regions of the central Rocky Mountains and occur only in those areas where soil parent materials have been derived from bedrocks such as limestone, basalt, or sedimentary rocks in which basic compounds dominate the weathered products. Their genesis centres upon the formation of surface horizons of organic carbon accumulation due to the decomposition of plant remains in an environment dominated by bivalent cations. This order includes chernozem and prairie soils, rendzinas, and some brown forest soils.

3.1 *Aquolls*

These are the somewhat poorly and poorly drained Mollisols. In alpine areas they have the mollic epipedon previously described and in addition have gleyed subsurface B2g horizons that have grey or blue colours or in which horizons with chromas of 2 or less and strongly mottled occur above 100 cm. Iron compounds have been weathered and segregated but not removed from the soil by groundwaters.

3.11 *Cryaquolls*

This great group includes all of the Aquolls in alpine areas. In addition to the properties described for the order and suborder, the Cryaquolls occur only in cold areas where mean annual soil temperature is below 8°C, and mean summer soil temperature is less than 13°C if the soil has no O horizon, or below 6°C if with an O horizon.

They occur in concave or depressional areas in the alpine landscape or where groundwaters moving along relatively impermeable strata come to the surface to form seeps and springs. They are associated with water tolerant types of vegetation and anaerobic bacterial life. Their genesis centres upon their saturation with water for long periods each yeaɪ.

3.111 *Typic Cryaquolls*

In the alpine areas these are the Cryaquolls that lack organic surface horizons more than a few inches thick, that lack strong horizons of calcium carbonate accumulation, and have mean annual soil temperature of 0°C or higher. They are of minor extent and occur at the lower margins of the alpine near timberline, or in the southern extensions of the region in the United States.

3.112 *Histic Cryaquolls*

These are the Cryaquolls that have moderately thick surface horizons of peat overlying the mineral soil. The thickness of the O horizon is not great enough to be classified as a

Histosol but constitutes a histic horizon of the classification system. In other respects they have the properties defined for the typic subgroup.

Like the Typic Cryaquolls they are of limited extent and occur near timberline or in the southern parts of the alpine regions in the central Rocky Mountains.

3.113 *Pergelic Cryaquolls*

These are the Cryaquolls of the colder parts of alpine regions. They differ from the typic subgroups in having mean annual soil temperatures of less than 0°C. The restriction on histic horizons found in the definition of the typic subgroup is waived in the Pergelic subgroup and they may or may not have such horizons.

These are the most extensive subgroup of Cryaquolls in alpine regions of the central Rockies. Following is a profile description of a soil belonging to this subgroup. They were formerly included in the Alpine Meadow soils.

A11 0–20 cm – Dark greyish-brown (10YR 4/2) dry silt loam, very dark greyish-brown (10YR 3/2) moist; strong, fine crumb structure; soft dry, very friable moist; neutral pH 7.0; gradual smooth boundary (12–25 cm thick).

A12g 20–33 cm – Greyish-brown (2.5Y 5/2) dry clay loam, very dark greyish-brown (2.5Y 3/2) moist; common medium distinct light olive-brown (2.5Y 5/6) mottles; weak medium sub-angular blocky structure that parts to medium fine sub-angular blocks; slightly hard dry, very friable moist; 15 per cent flagstone, mostly limestone; neutral pH 7.2; gradual smooth boundary (7.5–25 cm thick).

B2g 33–150 cm – Greyish-brown (2.5Y 5/2) dry clay loam, dark greyish-brown (2.5Y 4/2) moist; common medium distinct olive-brown (2.5Y 4/4) and light olive-brown (2.5Y 56/) mottles; massive; hard dry, very friable moist; approximately 30 per cent limestone, flagstone, and smaller fragments; neutral pH 7.0.

3.2 *Borolls*

These are the well to moderately well drained Mollisols of cold areas. In alpine landscapes they normally have subhorizons showing evidence of genetic alteration in addition to the mollic epipedon, but the presence of such horizons is not mandatory. Mean annual soil temperature is below 8°C.

These are inextensive soils in alpine areas of the central Rocky Mountains and are confined to areas where soil parent materials contain an abundant supply of bases. They occupy well drained sites.

3.21 *Cryoborolls*

All of the Borolls of alpine areas are in this great group. They have all of the properties of Borolls and in addition have mean summer soil temperatures below 3°C.

3.211 *Typic Cryoborolls*

In alpine areas these are the Cryoborolls that have mean annual soil temperature of 0°C or higher, have no hard continuous bedrock above 50 cm, have no distinct or prominent mottles above 100 cm, and are not saturated with water for as long as ninety consecutive days.

This subgroup is inextensive in the central Rocky Mountain alpine regions and occurs mainly at lower elevations near timberline or in southern extensions of the alpine.

3.212 *Aquic Cryoborolls*

In alpine regions these soils have all of the properties of the typic subgroup except that they have mottling above 100 cm and are saturated with water during part of each year.

Like the Typic Cryoborolls they are inextensive and occur only in the warmer parts of alpine regions in the central Rocky Mountains.

3.213 *Lithic Cryoborolls*

In alpine areas the soils of this subgroup have all of the properties defined for the typic subgroups except that bedrock occurs above 50 cm and that they may have mean annual soil temperature either above or below 0°C.

They are inextensive alpine soils in the central Rocky Mountains because the parent materials that are required to develop their mollic epipedons are of limited extent. Since they have less temperature restrictions than the Typic and Aquic subgroups they can occur in any part of alpine regions where bedrock is favourable for their development.

3.214 *Pergelic Cryoborolls*

These are Cryoborolls that have the properties of the Typic Cryoborolls except that mean annual soil temperature is below 0°C. They are the most extensive subgroup of Cryoborolls in alpine regions of the central Rocky Mountains. Following is a description of the profile of a soil from this subgroup.

A1   0–23 cm – Greyish-brown (2.5Y 5/2) dry gravelly heavy loam, very dark greyish-brown (2.5Y 3/2) moist; weak fine sub-angular blocky structure that parts to strong fine crumbs and granules; soft dry, very friable moist; 15 per cent limestone and shale fragments; non-calcareous, mildly alkaline pH 7.6; clear smooth boundary (15–40 cm thick).

C1   23–35 cm – Greyish-brown (2.5Y 5/2) dry gravelly light clay loam, dark greyish-brown (2.5Y 4/2) moist; moderate fine sub-angular blocky structure; slightly hard dry, very friable moist; 25 per cent limestone and shale fragments; non-calcareous except for gravel fragments; neutral pH of matrix material 7.2; clear smooth boundary (10–20 cm thick).

C2   35–75 cm – Olive-grey (5Y 5/2) dry gravelly light clay loam, olive-grey (5Y 4/2) moist; massive; hard dry, very friable moist; 30 per cent limestone and shale fragments; non-calcareous except for gravel fragments; neutral pH of matrix material 7.0; gradual smooth boundary (20–43 cm thick).

R   75–100 cm – Interbedded limestone and slaty shale.

Table 13B.5 illustrates the proportions of available cations in a Pergelic Cryoboroll.

4.0 HISTOSOLS

The classification of organic soils in alpine areas is incomplete. Generally deposits of peat are a minor part of the alpine landscapes and have been included in generalized land type units in most subgroups. This in no way implies that they are unimportant soils. Deposits of peat in the alpine are intimately connected with the control of ground-water and with streamflow in late summer months.

It is anticipated that most alpine peats are fibrous and that decay has been relatively incomplete. They would probably be classed as Cryofibrists in the present system of classification. Following is a description of an organic soil in an alpine area of the central Rocky Mountains developed under rushes, grasses and sedges.

1   0–50 cm – Very dark grey (10YR 3/1) dry fibrous peat, black (10YR 2/1) moist; massive; very strongly acid pH 4.6; the outlines of some leafy plants are still visible. The lower 5–8 cm of this horizon are muck-like in character; the horizon rests abruptly on the horizon below.

2   50–150 cm – Light brownish-grey (10YR 6/2) dry fine gravelly sandy loam, dark greyish-brown (10YR 4/2) moist; massive; slightly hard dry, very friable moist; very strongly acid pH 4.6.

5. Work in other alpine areas

Information on alpine soils is very limited in most parts of the world, making any comprehensive survey virtually impossible at the present time. Nevertheless some discussion of alpine areas outside the central Rocky Mountains is essential.

The Snowy Mountains in Australia have been investigated quite thoroughly (Costin et al., 1952; Costin, 1955). The elevation of the alpine area here is only about 1900 m and annual precipitation amounts to 180–225 cm, falling mainly as winter snow. The major rock types are gneisses and granites and, in contrast to many other alpine areas, slopes are gentle. The most important soil type is the *alpine humus* (Costin et al., 1952) occurring under tall alpine herbfield and sod tussock grassland. Their description evidently corresponds to the Fragilic Cryumbrept subgroup. The soil which is rich in organic content (6–13 per cent in A1) and strongly acid, is of the A–C type, having no illuvial horizon, and is typically 60–75 cm in depth. There is a low content of exchangeable bases yet vigorous herbaceous growth occurs apparently due to rapid decomposition and earthworm activity (Wood, 1970). In spite of the high precipitation totals there is no development of spodic horizons in the alpine or subalpine on freely drained slopes. Fig. 13B.3 demonstrates the relationship between altitude and profile development.

Costin (1955) notes that this soil does not appear to be as well-developed in any other alpine area. In the South Island of New Zealand, alpine humus pedogenesis is limited by steep slopes in the dry eastern alps and by excessive precipitation on the western alps, whereas in North Island there is little profile development on the recent volcanic parent material. In northwest Europe blanket bog or lithosols occur in the oceanic areas, while

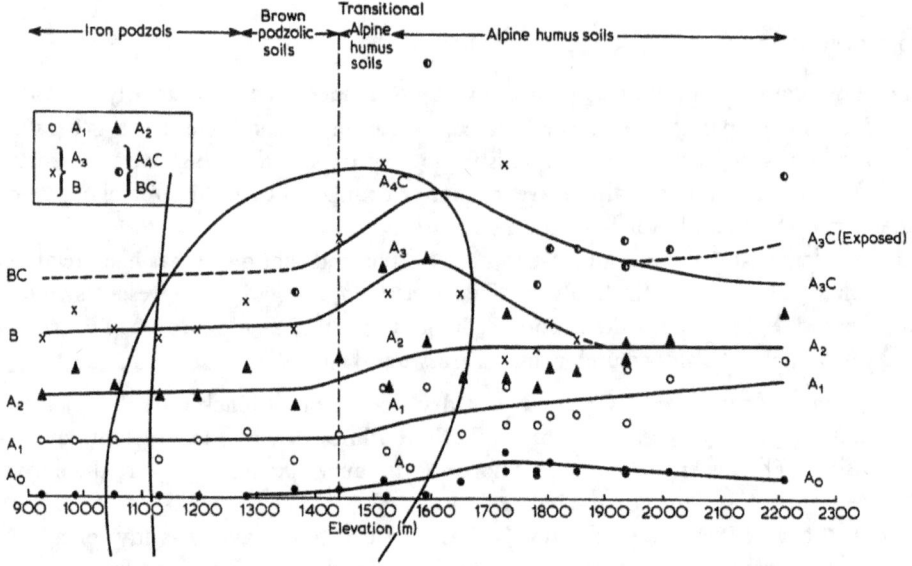

Fig. 13B.3. The variation of profile development with altitude in the subalpine and alpine of New South Wales (from Costin *et al.*, 1952).

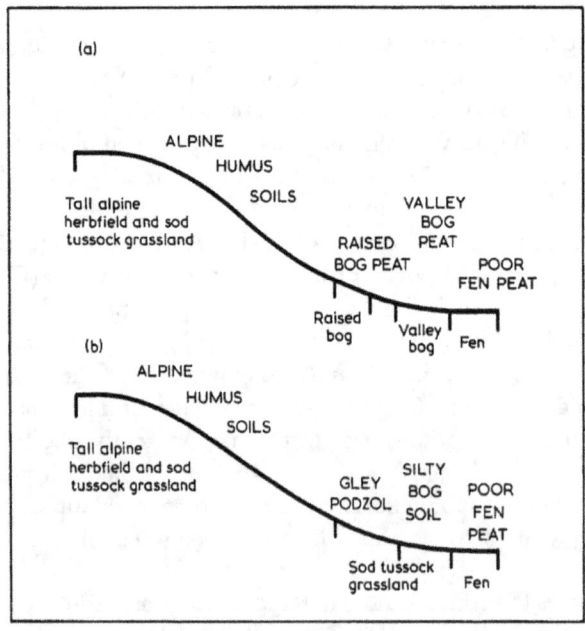

Fig. 13B.4. Topographic soil sequences on freely drained slopes in New South Wales (after Costin *et al.*, 1952). (a) Common sequence on gneissic granite with water table near the surface on the flat all year. (b) Sequence with lower water table conditions.

in Swedish Lapland there is insufficient organic matter due to the short growing season. Alpine humus soils do occur in the European Alps but they are shallow and mainly restricted to acidic parent material. It is only in the wettest areas that alpine humus pedogenesis has taken place on basic rocks. Alpine humus soils are also found in the Pyrenees, where slopes are more gentle and the Pleistocene period was less severe, although they are generally less well developed than in Australia. Costin points out that this occurrence of 'soil mountains' has in fact resulted in special erosion problems due to sheep and cattle grazing and engineering activity.

The catenary relationship of the alpine humus to other soils of the Australian Alps is illustrated in fig. 13B.4. This also shows variations in the basic pattern which arise as result of differences in the level of the water table.

Finally, we should mention the occurrences of Spodosols (formerly podzols) in alpine areas. In the coterminous United States they have so far only been reported on Mt Katahdin, Maine (Bliss and Woodwell, 1965).[1] Here coarse sandy loams with an indurated, iron-cemented B2 horizon occur under sedge heath and krummholz at about 1300 m. They are highly acid (pH about 4.0) and low in nutrients. Similar levels of podzolization are not found even in the Presidential Range of New England. This occurrence is attributed by Bliss and Woodwell to soil stability due to an absence of cryoturbation and the high precipitation, as well as a readily weathered granite bedrock and the presence of heath species in the plant communities. Related soils, with either an iron, or iron-humus, spodic horizon occur widely on acid parent materials in the European Alps where they are associated with grass turf. These soils are in the *Orthod* group; for example, the *Typic Cryorthod* has an illuvial (spodic) humus/sesquioxide horizon and a total thickness of more than 40 cm, no permafrost within 75 cm of the surface, and a mean annual temperature of less than 8.3°C. The groundwater podzols are termed *Cryaquods*.

---

[1] Reference to their occurrence in Colorado (Johnson and Cline, 1965) is now inapplicable as a result of changes in the definition of spodosols (Soil Survey Staff, 1967).

References

BALDWIN, M., KELLOGG, C. E. and THORPE, J. (1938) Soil classification. *In: Soils and Men*, 979–1001. US Dept of Agric. Yearbook.

BENEDICT, J. B. (1970) Downslope soil movement in a Colorado alpine region: rates, processes and climatic significance. *Arct. Alp. Res.*, **2**, 165–227.

BLISS, L. C. and WOODWELL, G. M. (1965) An alpine podzol on Mount Katahdin, Maine. *Soil Sci.*, **100**, 274–9.

COE, M. J. (1967) *The Ecology of the Alpine Zone of Mount Kenya*, 69–79. The Hague: W. Junk.

COSTIN, A. B. (1955) Alpine soils in Australia with reference to conditions in Europe and New Zealand. *J. Soil Sci.*, **6**, 35–50.

COSTIN, A. B., HALLSWORTH, E. G. and WOOF, M. (1952) Studies in pedogenesis in New South Wales, III: The alpine humus soils. *J. Soil Sci.*, **3**, 197–218.

CURRY, R. R. (1966) Glaciation about 3,000,000 years ago in the Sierra Nevada. *Science*, **154**, 770–1.

FAHEY, B. D. (1971) *A Quantitative Analysis of Freeze–Thaw Cycles, Frost Heave Cycles, and Frost Penetration in the Front Range of the Rocky Mountains, Boulder County, Colorado.* Unpub. Ph.D. dissertation, Univ. of Colorado. 305 pp.

IVES, J. D. and FAHEY, B. D. (1971) Permafrost occurrence in the Front Range, Colorado Rocky Mountains, USA. *J. Glaciol.*, **10** (58), 105–11.

JOHNSON, D. D. and CLINE, A. J. (1965) Colorado mountain soils. *Adv. Agron.*, **17**, 233–81.

JOHNSON, P. L. and BILLINGS, W. D. (1962) The alpine vegetation of the Beartooth Plateau in relation to cryopedogenic processes and patterns. *Ecol. Monogr.*, **32**, 105–35.

MAHANEY, W. C. (1970) *Soil Genesis on Deposits of Neoglacial and late Pleistocene Age in the Indian Peaks of the Colorado Front Range.* Unpub. Ph.D. thesis, Univ. of Colorado. 246 pp.

MARR, J. W., CLARK, J. M., OSBURN, W. S. and PADDOCK, M. W. (1968) Data on mountain environments, III. Front Range, Colorado, four climax regions, 1959–64. *Univ. of Colorado Stud., Ser. in Biol.*, 29. 181 pp.

NIMLOS, T. J. and MCCONNELL, R. C. (1965) Alpine soils in Montana. *Soil Sci.*, **99**, 310–21.

NIMLOS, T. J., MCCONNELL, R. C. and PATTIE, D. L. (1965) Soil temperature and moisture regimes in Montana alpine soils. *Northwest Sci.*, **39**, 129–38.

RETZER, J. L. (1950) *Genesis and Morphology of Soils of Alpine Areas in the Rocky Mountains.* Unpub. Ph.D. thesis, Univ. of Wisconsin.

RETZER, J. L. (1956) Alpine soils of the Rocky Mountains. *J. Soil Sci.*, **7**, 22–32.

RETZER, J. L. (1962) Soil survey of Fraser Alpine Area, Colorado. *US Dept of Agric. and Colorado Agric. Expt. Station, Ser. 1956*, 20. 47 pp.

RICHMOND, G. M. (1962) Quaternary stratigraphy of the La Sal Mountains, Utah. *US Geol. Surv. Prof. Pap.*, 324. 135 pp.

SOIL SURVEY STAFF (1960) *Soil classification: a comprehensive system (7th approximation).* US Dept of Agric., Soil Conserv. Serv. 265 pp.

SOIL SURVEY STAFF (1967) *Supplement to soil classification system (7th approximation).* US Dept of Agric., Soil Conserv. Serv. 207 pp.

SOIL SURVEY STAFF (1968) *Supplement to soil classification system (7th approximation), Histosols.* US Dept of Agric., Soil Conserv. Serv.

THORPE, J. and SMITH, G. D. (1949) Higher categories of soil classification: order, suborder and great soil groups. *Soil Sci.*, **67**, 117–26.

WINDOM, H. L. (1969) Atmospheric dust records in permanent snowfields: implications to marine sediments. *Bull. Geol. Soc. Amer.*, **80**, 761–82.

WOOD, T. G. (1970) Decomposition of plant litter in montane and alpine soils on Mt Kosciusko, Australia. *Nature*, **226**, 561–2.

Manuscript received March 1971.

Additional references

KNAPIK, L. J., SCOTTER, G. W. and PETTAPIECE, W. W. (1973) Alpine soil and plant community relationships of the Sunshine area, Banff National Park. *Arct. Alp. Res,* **5** (3), part 2, A161–A170.

SNEDDON, J. I., LAVKULICH, L. M. and FARSTAD, L. (1972) The morphology and genesis of some alpine soils in British Columbia, Canada. *Soil Sci. Amer. Proc.*, **36**, 100–10.

Man in cold environments

Physiological responses to cold environments

Claudia C. Van Wie

Institute of Arctic and Alpine Research,
University of Colorado

1. Introduction

Warm-blooded (or homeothermous) mammals and birds maintain a relatively constant body temperature regardless of ambient conditions or desired metabolic rate. Such temperature regulation permits constant efficiency of the animal's enzymatic processes, and, hence, the activity of the animal is not dependent upon ambient temperatures. This homeostasis is achieved by a series of control mechanisms that regulate heat production and heat loss. In man, the central temperature is generally near 37°C; a fall of only a few degrees greatly reduces enzymatic activity and may result in coma and death, while a rise of only a few degrees may irreversibly damage the cells of the central nervous system.

Homeotherms in extreme temperature environments face many problems in maintaining the desired constant internal temperature. Animals in cold environments have adapted to low temperatures by increasing their insulation, by increasing their heat production, or by lowering their core temperature (hypothermia). Man in a cold climate must rapidly acclimatize through biological change, or modify his environment by clothing and shelter.

In this section the main problems of thermoregulation in man with respect to cold are reviewed. When and how does man produce and lose heat? What are the physiological bases for man's perception of excessive loss of heat, or of cold? What are the physiological mechanisms by which he maintains his body temperature during cold stress? How do these processes interact and how are they controlled?

2. Thermoregulation and heat balance

First, let us consider the mechanisms determining the heat balance in homeotherms. Heat is produced by the basal metabolic processes and by muscular exertion, and is lost by radiation, conduction, convection and evaporation. Aschoff and Wever (1958) analysed the locale of heat production in man. When man is inactive, the brain produces 16 per cent, the chest and abdomen 56 per cent, and the skin and muscles 18 per cent of the body's heat (approximately 72 kcal hr^{-1}); when man is active and working, the brain produces only 3 per cent, the chest and abdomen 22 per cent, and the skin and muscles 73 per cent.

For present purposes, we can consider the body as if it were composed of three con-centric layers: a central core, a middle muscle layer, and an outer skin layer (Crosbie *et al.*, cited in Milsum, 1966, p. 64). Although this is a gross oversimplification, it is a useful conceptual model for discussing heat loss. Basal metabolic heat produced in the organs of the central core is lost directly to the muscle layer. The metabolic heat pro-duced by the muscles can be transferred either to the central core or to the skin. The skin layer, including subcutaneous fat deposits, is primarily an insulative layer across which heat passes.

CORE HEAT LOSSES

Heat is lost from the core to the muscle and skin layers by conduction and convection. Heat loss from the core by conduction occurs whenever the temperature of the muscle layer is lower than that of the core. The rate of conductive heat flow (q) is directly pro-portional to the thermal conductivity (k), the surface area (A), and the difference between the core temperature (t_c) and the temperature of the adjacent layer (t_a), and inversely proportional to the heat flow path length (L):

$$q = \frac{kA}{L}(t_c - t_a)\ cals^{-1}$$

The rate of conductive heat flow between the muscle and skin layers and between the skin and the air is similarly dependent upon (kA/L) Δt.

Heat is also transferred from the central core by the actual motion of materials or by convective heat loss. Convection (q) is dependent upon surface area (A), the convection coefficient (h), and the temperature gradient (Δt):

$$q = h\,A\,\Delta t$$

The convection coefficient is determined by the nature of the surface (flat or curved, horizontal or vertical), by the fluid on contact with the surface (gas or liquid), by the properties of the fluid (density, viscosity, specific heat, and thermal conductivity), and by the rate at which the fluid moves past the surface.

As blood circulates, it carries heat away from the central core to the outer body layers. Part of this heat is then lost by conduction and convection to cooler ambient air; much of the heat, however, is transferred to the cooler venous blood returning from the extremities to the central core (fig. 14A.1). Use of this counter-current exchange mechan-ism enables the body to maintain the extremities at a lower temperature than the central core area; in man, an 8–10°C difference between the hand and rectal temperature is

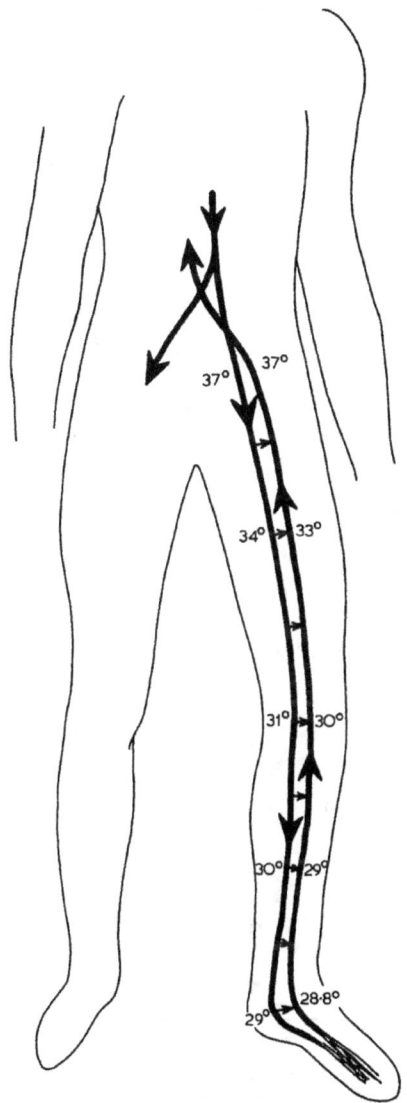

Fig. 14A.1. A hypothetical temperature gradient in an artery and a vein in the leg. When the arterial blood leaves the central core area, it is at approximately 37°C. As it flows into the leg, heat is transferred from the warm arterial blood to the cooler venous blood returning from the foot (arrows). Thus a temperature gradient can be maintained between the extremities and the core of the body.

common. As a result, conductive and convective heat losses to the external air are lessened. These losses are decreased still further when vasoconstriction, a normal response to sudden cold exposure, occurs.

SKIN LAYER LOSSES

As air flow increases, the convective heat loss from the skin also increases. The effects of wind on heat loss are measured either by the windchill index or by the windchill equivalent temperature. The windchill index (kcal m^{-2} hr^{-1}) is a measure of the cooling power of the wind for various combinations of temperature and wind speed. The windchill equivalent temperature is the cooling power of wind such as one would feel on exposed flesh in a light wind (5 mph or 2.23 m s^{-1}). Siple and Passel (1945) determined the relationship between the convection coefficient (h) and the wind velocity (v):

$$h = 10.45 + 10 \sqrt{v} - v$$

Most official windchill charts are based on their work (table 14A.1). Steadman (1971) recently expanded this work by calculating the thickness of clothing needed to insulate 85 per cent of the body's surface at various combinations of temperature and wind speed (fig. 14A.2).

Table 14A.1. Windchill equivalent temperatures (°C) determined from Siple's formula.

Air temperature (°C)	Wind speed				
	Calm	2.5	5	10	20 ms^{-1}
0	19	0	−7	−12	−18
−5	16	−6	−13	−19	−26
−10	13.5	−11	−19	−26	−33
−20	9	−21	−31	−40	−48
−30	4.5	−31	−43	−54	−63
−40	0	−41	−55	−68	−78

In cold surroundings, the predominant mechanism of heat loss is by direct energy transmission to cold objects. For humans, as for other radiating bodies, the net heat exchange, q_{rad}, is described by the Stefan–Boltzmann Law:

$$q_{rad} = \sigma A (t_s^4 - t_e^4) \text{ kcal hr}^{-1}$$

where σ is the Stefan–Boltzmann constant, A the surface or skin area exposed, t_s is the surface or skin temperature (°K), and t_e is the temperature of the radiant enclosure (°K). Thus, a nude with a skin temperature of 31°C will radiate heat at a rate of 100 kcal hr^{-1} to a room with walls at 21°C (Milsum, 1966). Since, at rest, total heat

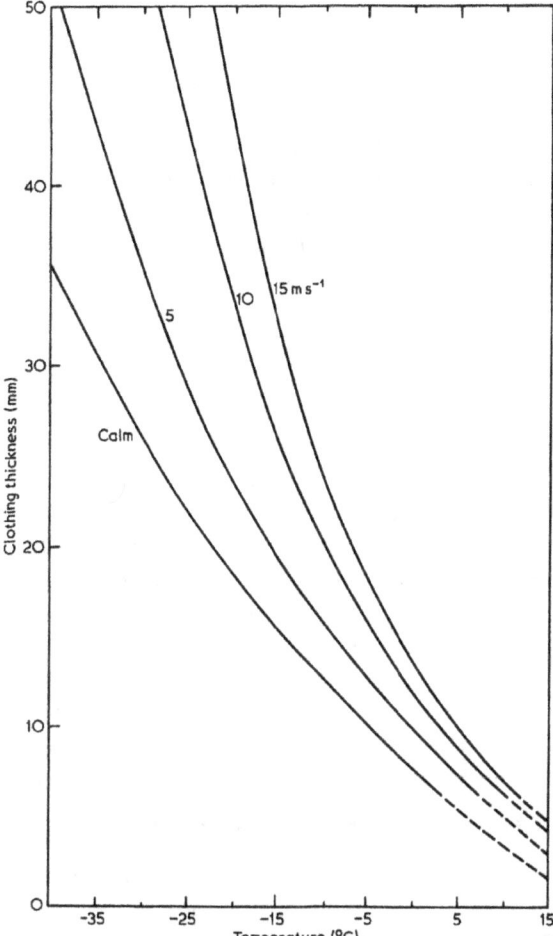

Fig. 14A.2. Thickness of clothing required to insulate 85 per cent of the body's surface at various wind speeds (after Steadman, 1971). The face, 3 per cent of the surface area, is not considered to be insulated. The hands and feet, 12 per cent of the surface area, are considered to be minimally insulated only to prevent tissue damage. Under severe conditions, no amount of clothing is sufficient to keep the model warm.

production is approximately 72 kcal hr^{-1}, this person must either use clothing to modify the radiant heat loss or increase his rate of heat production.

The body also loses heat by evaporation from the skin although this process contributes little to total heat loss at low ambient temperatures. Clothing further decreases evaporative heat losses. However, under cold and extremely dry conditions, heat loss through the lungs by warming the inhaled air and by evaporation can account for more than 20 per cent of the body's heat loss (Steadman, 1971).

PATHOLOGIC EFFECTS OF EXCESSIVE HEAT LOSS

If heat losses are so great that the skin temperature cannot be maintained above freezing, several pathologic conditions result. Chilblains, which are red, swollen, itchy lesions between the joints of the fingers, generally occur when the hands have been exposed intermittently to cold and dampness for long periods of time. A similar condition in the foot, called immersion or trench foot, also results from cold and dampness. The first signs are coldness, numbness, swelling and discolouration, then cramps in the legs and hyperemia accompanied by tingling or pain and motor disturbances. If the condition is sufficiently severe, gangrene may result. When skin tissue is actually frozen or frostbitten, a local burning or stinging is first noticed, and then the area becomes numb. The reddened tissue, killed by freezing, must be replaced (for further discussion, see Kreider, 1964, and Smith, 1970).

Exposure, the medical term for the condition that occurs when the body is not able to maintain a normal temperature, usually results from the victim depleting his available food stores while being severely chilled by subfreezing temperatures or by wind and rain at more moderate temperatures. As the stores of available food nutrients decrease, the basal metabolism needed to keep the core temperature near 37°C cannot be sustained. When the rectal temperature falls to 30°C, consciousness is lost; by 27°C, the heart ceases. The best protection against a lowering of the core temperature or hypothermia is maintenance of body warmth and adequate food and fluid intake. If foods and fluids are available, exercise is an excellent way to elevate heat production; if not, it is best to remain inactive (Wilkerson, 1967).

3. The physiological bases for man's perception of cold

Man and other homeotherms must balance their heat loss with their heat production in order to maintain their body temperature within the observed narrow limits. To do this, homeotherms must have a rapid, accurate feedback control. Thermosensitive cells in the skin, in the hypothalamus, and in the central core are the sensory basis of this feedback system.

Hensel, in his classic studies of the peripheral of skin receptors (1952, 1955; Hensel and Bomen, 1960), describes temperature sensitive cells at the level of the superficial arterial plexus in the dermis. The frequency of discharge or firing of each individual receptor is a function of the ambient temperature and its rate of change. Following a sudden, rapid decrease in temperature, the frequency of firing increases sharply, and a new firing frequency is established at the new temperature. When the temperature rises rapidly, the cold receptors become silent for a period, and then begin firing again at a frequency corresponding to the elevated temperature. The exact response to a change in temperature varies among individual receptors.

The impulses from the skin receptors ascend the lateral spinothalamic tract to central synapses in the posterior hypothalamus, often called the heat conservation centre of the brain. The posterior hypothalamus coordinates incoming impulses from the skin and core receptors and from the anterior hypothalamus, transduces them, and sends out efferent messages to the muscles, blood vessels, and endocrine organs (Benzinger, 1964). Isenschmid and Krehl (1912) first located and described the site where the afferent and efferent centres for shivering are located; destruction of this centre eliminates metabolic responses to cold but not to overheating. Freeman and Davis (1959) have since found that cooling the Krehl-Isenschmid centre in the posterior hypothalamus does not elicit a shivering response, thus implying that the site serves only a synaptic function, and that this area of the brain cannot fulfil a homeostatic function by sensing the internal temperature of the body.

There is a region in the anterior hypothalamus, however, which is sensitive to temperature. Spike potentials in this region (centre A) have been recorded in response to warm stimulation (Nakayama et al., 1961) and a pathway from centre A to the posterior hypothalamus has been observed (Hemingway, 1957). When Hardy (1961) warmed this region of a dog's brain with radio-frequency energy between two implanted electrodes, the dog panted until the artificial warming ceased. After an hour's panting, the dog had lowered its core temperature two degrees below normal. The warming of centre A, which stimulated sweating and vasodilation, inhibited the shivering response and the increase in heat production that normally would have begun when the core temperature fell.

The impulses which cause the conscious sensation of being cold, however, are not inhibited by centre A. When Benzinger (1964) put human subjects in a water bath and slowly lowered the temperature 1° per hour, he noted that each individual had a consistent threshold level at which he felt cold. Varying the cranial temperatures by preexposure in baths of different temperatures had no effect on the threshold level for the conscious sensation of cold. These results, in conjunction with others already discussed, imply that there are at least two afferent pathways for central reception of the impulses from the skin cold receptors; one directly to the hypothalamic centre, and one which bypasses hypothalamic inhibition on its way to the cortical region.

4. Physiological responses to cold stress

When cold, man acts directly or indirectly to change his immediate environment or exercises to increase his heat production. However these behavioural responses are only crude means of regulating the body's temperature and by themselves are insufficient to keep temperature fluctuations within the narrow limits the body can tolerate. It is the autonomic control measures which increase heat production, increase the insulation layer, or permit moderate hypothermia that maintain a nearly constant core temperature.

SHIVERING

Shivering is one of the first and most noticeable reactions to severe cold stress. When man is at rest, the muscles, which form 50 per cent of the total body mass, generate about 18 per cent of the total heat produced. Voluntary exercise of the muscles may increase the heat production tenfold; involuntary exercise or shivering increases the heat production four or five times (Carlson, 1964). However, nearly 90 per cent of the heat produced by shivering is lost by convection because shivering increases the movement of the body and because the active muscles require increased circulation to the periphery (Carlson and Hsieh, 1965). In addition, the heat produced is near the body surface, so that only a thin insulative layer prevents heat loss.

In several experiments when people were exposed to cold for long periods, both the extent and the threshold at which shivering occurs were reduced. In cold bed tests, unacclimated Europeans lived outdoors for six weeks and slept in thin sleeping bags on 0°C nights. At first, constant shivering and cold feet and hands kept the subjects awake. With time, though, as the heat produced by non-shivering thermogenesis increased, they were able to keep their extremities at more comfortable levels and to sleep with lower skin temperatures and with some shivering (Rodahl, *in* Folk, 1966). The diving women of Korea, the Ama, who continue to dive despite winter cold, have a lower shivering threshold than either non-diving Korean women or Korean men (29.2°C, 29.9°C and 31.1°C, respectively) (Hong, 1963).

NON-SHIVERING THERMOGENESIS

Metabolic acclimatization to cold in man must be discussed with caution, for comparisons of people in summer and winter, or before and after prolonged cold exposure, are apt to include changes in diet and physical fitness. In addition, studies of men in cold climates such as the Arctic or Antarctic must take into account each individual's actual environment. For example, Andrew Croft comments: 'skin clothing is very good for a hut to hut or igloo life in a dryish climate like the NW coast of Greenland. We are very much against skins for long journeys, because . . . it takes at least two years getting to know how to wear skin clothing and how not to get over-heated' (Hinks, 1937, p. 93).

Hoffmann (Chapter 9) discusses in detail the ways arctic and alpine fauna adapt to cold environments. The degree to which man can adapt to cold by non-shivering means is not yet known. There is substantial evidence that the means of adaptation varies among groups of people, however. Although Eskimos have the same shivering threshold as Europeans and their overall tissue insulation during maximal vasoconstriction is the same, their non-shivering metabolism is 30–40 per cent greater, and they have a marked ability to withstand hand cooling (Folk, 1966; Meehan, 1955). In the Australian aborigines, who customarily sleep unclothed on the ground even during cold winter nights, metabolic heat production is not increased; instead they tolerate low skin temperatures and moderate hypothermia (Morrison, 1957).

Non-shivering thermogenesis has been studied most extensively in the laboratory rat. During the initial period of cold acclimation, the total daily food consumption of the rat increases although there is a transient decrease in body weight. After forty-five days, growth has resumed and the thyroid, adrenal cortex, heart, kidney, liver, digestive tract, and brown adipose tissue have hypertrophied (Folk, 1966). The basal metabolism of acclimated rats is 20–21 per cent greater than that of the controls. The oxygen uptake of isolated perfused livers from cold-acclimated rats is greater than that of livers from warm-acclimated rats (Chaffee and Roberts, 1971; Hannon, 1963).

The catecholamine, norepinephrine, a hormone of the adrenal medulla and the postganglionic sympathetic nerve fibres, appears to be a major agent in the cold acclimation of rats and may also be important for cold acclimation in man. An intravenous infusion of norepinephrine increases the total vasoresistance, oxygen consumption, blood glucose, and systolic and diastolic blood pressure, and decreases the heart rate and cardiac output; similar physiological effects are listed by Folk (1966) as responses to acute cold exposure. After injecting norepinephrine in the Ama, Kang et al. (1970) reported a slight increase in oxygen consumption for winter experiments, but not for those done in the summer. However, they considered that the magnitude of the response was too small to explain the development of non-shivering thermogenesis in these women. When Wilson et al. (1970) exposed seven subjects to 0°C for three hours, the levels of epinephrine, norepinephrine, and dopamine in the urine increased significantly, indicating that the pituitary-adrenal axis may have been stimulated by this short cold exposure.

The calorigenic response elicited with norepinephrine is closely associated with the level of non-shivering thermogenesis. When this response to norepinephrine is used as the criterion for cold acclimatization in an interspecies survey, it appears that many species of adult mammals, including man, develop some degree of non-shivering thermogenesis after prolonged cold stress (Chaffee and Roberts, 1971). The response of a homeotherm generally falls into one of three groups: 1) thermogenesis increases in response to norepinephrine (rat, guinea pig); 2) the role of catecholamines in non-shivering thermogenesis is not clear because both warm- and cold-acclimated animals respond to norepinephrine (dog, white mouse, hedgehog); and 3) no non-shivering thermogenesis is apparent after cold acclimation although cold resistance may be increased in some cases (birds, miniature pigs, and newborn pigs) (Jansky et al., 1969).

INSULATION

As an initial reaction to cold stress, the vessels in the extremities – the hands, feet, ears, nose, cheeks, and forehead – rapidly constrict (Edwards and Burton, 1960a, b). The constriction of the blood vessels, mediated by norepinephrine, lowers the convective heat loss to the environment by lowering the skin temperature. Occasionally vasodilation (Lewis's hunting reaction) will occur for short time periods and will raise the skin temperature (Lewis, 1930). The skeletal muscle vessels, however, which are less sensitive to norepinephrine than the peripheral vessels, maintain a steady flow of blood to the muscles.

Little mention has been made of the insulating function of adipose tissue. Pugh and Edholm (1955) noted that fat in Channel swimmers appeared to be laid down preferentially in subcutaneous areas rather than in deep fat deposits. Wyndham, Williams and Loots (1968) compared two groups of men, one averaging 59.1 kg and 162.1 cm, and the other 93.8 kg and 172.3 cm. When at rest and scantily clad, the average men increased their basal metabolic rate sharply at temperatures below 20°C; the heavier group did not do so until 10°C. In the temperature range in which the basal metabolism was elevated, the skin temperatures of the average weight group were higher than those of the heavier group. The calculation of heat conductance indicated that the fat men had approximately half the heat conductance of the average group.

5. Concluding remarks

To maintain the internal body within a limited temperature range, an organism must sense all changes which might influence its temperature and then compensate for their expected effect. Continual reassessment of all conditions affecting body temperature is necessary to prevent overcompensation. A system which accomplishes this is called a

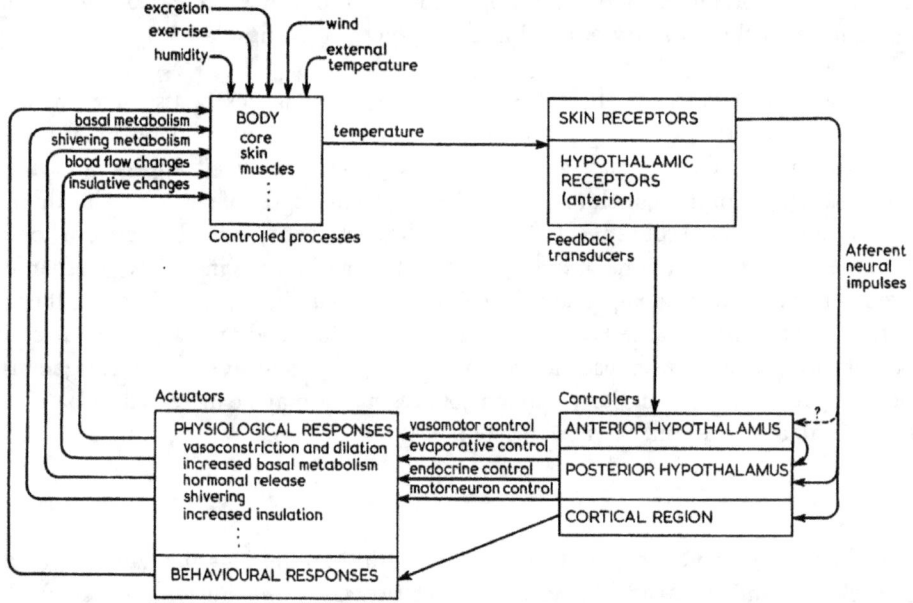

Fig. 14A.3. Schematic diagram of cold thermoregulation in man (after Milsum, 1966).

regulator system for it maintains body temperature at some constant desired reference level. Error in this type of system is measured as the difference between the desired level and the actual level. The main problem faced by a regulator system is the correct programming of all manipulable variables in order to compensate for any disturbance inputs (Milsum, 1966).

A schematic diagram of a regulator system for thermoregulation is shown in fig. 14A.3. The heat losses and heat gains determine the temperature of the body. This body temperature is sensed by the skin receptors and the receptors in the anterior hypothalamus. The hypothalamus and cortical region of the brain then command various physiological and behavioural actions in response to the information from the receptors. If these actions are sufficient to increase the body temperature, the skin and hypothalamic receptors will sense the change, and the controllers will act upon new input.

Many systems are activated in response to cold stress. At first, shivering, vasoconstriction, and the release of epinephrine and norepinephrine from the adrenal medulla and the sympathetic nervous system occur. Basal metabolism and the corresponding consumption of oxygen by the tissues increase. As man habituates to cold, his insulative layers and basal metabolism may change, he may become accustomed to shivering and maintaining his extremities at lower temperatures, or he may tolerate moderate hypothermia.

References

ASCHOFF, J. and WEVER, R. (1958) Kern und Schale im Warmhaushalt des Menschen. *Die Naturwissenschaften*, **45**, 477–85.

BENZINGER, T. H. (1964) The thermal homeostasis of man. *In: Homeostasis and Feedback Mechanisms* (Symposia of the Society for Experimental Biology XVIII), 49–80. New York: Academic Press.

CARLSON, L. D. (1964) Reactions of man to cold. *In:* LIGHT, E. (ed.) *Medical Climatology*, 196–228. Baltimore: Waverly Press.

CARLSON, L. D. and HSIEH, A. C. L. (1965) Cold. *In:* EDHOLM, O. G. and BACHARACH, A. L. (eds.) *The Physiology of Human Survival*, 15–52. New York: Academic Press.

CHAFFEE, R. R. J. and ROBERTS, J. C. (1971) Temperature acclimation in birds and mammals. *Ann. Rev. Physiol.*, **33**, 155–203.

EDWARDS, M. and BURTON, A. C. (1960a) Correlation of heat output and blood flow in the finger, especially in cold-induced vasodilation. *J. Appl. Physiol.*, **15**, 201–8.

EDWARDS, M. and BURTON, A. C. (1960b) Temperature distribution over the human head, especially in the cold. *J. Appl. Physiol.*, **15**, 209–11.

FOLK, G. E. (1966) *Introduction to Environmental Physiology*, 77–136. Philadelphia: Lea & Febiger.

FREEMAN, W. J. and DAVIS, D. D. (1959) Effects on cats of conductive hypothalamic cooling. *Amer. J. Physiol.*, **197**, 145–8.

HANNON, J. P. (1963) Cellular mechanisms in the metabolic acclimatization to cold. *In: Temperature – Its Measurement and Control in Science and Industry*, Vol. 3, Part 3, 469–84. New York: Reinhold.

HARDY, J. D. (1961) Physiology of temperature regulation. *Physiol. Rev.*, **41**, 521–54.

HEMINGWAY, A. (1957) Nervous control of shivering. *Alaskan Air Command Tech. Note*, AAL-TN-57-40.

HENSEL, H. (1952) Physiologie der Thermoreception. *Ergebn. Physiol.*, **47**, 166.

HENSEL, H. (1955) Über die Funktion der Lorenzinischen Ampullen der Selachier. *Experientia*, **11**, 325–7.

HENSEL, H. and BOMAN, K. K. A. (1960) Afferent impulses in cutaneous sensory nerves in human subjects. *J. Neurophysiol.*, **23**, 564–78.

HINKS, A. R. (1937) *Hints to travellers*, Vol. 2 (11th edn). London: Roy. Geogr. Soc. 472 pp.

HONG, S. K. (1963) Comparison of diving and non-diving women in Korea. *Fed. Proc.*, **22**, 831–3.

ISENSCHMID, R. and KREHL, L. (1912) Über den Einfluss des Gehirns auf die Wärmregulation. *Arch. Exp. Path. Pharmak.*, **70**, 109–34.

JANSKY, L. R., BARTUNKOVA, V., KOCKOVA, V., MEJSNAR, Y. and ZEISBERGER, E. (1969) Interspecies differences in cold adaptation and nonshivering thermogenesis. *Fed. Proc.*, **28**, 1052–8.

KANG, B. S., HAN, D. S., PAIK, K. S., PARK, Y. S., KIM, J. K., KIM, C. S., RENNIE, D. W. and HONG, S. K. (1970) Calorigenic action of norepinephrine in the Korean women divers. *J. Appl. Physiol.*, **29**, 6–9.

KRIEDER, M. B. (1964) Pathologic effects of extreme cold. *In:* LICHT, E. (ed.) *Medical Climatology*, 428–68. Baltimore: Waverly Press.

LANGLEY, L. L. (1965) *Homeostasis*. London: Chapman & Hall. 108 pp.

LEWIS, T. (1930) Vasodilation in response to strong cooling. *Heart*, **15**, 177–81.

MEEHAN, J. P. JR (1955) Basal metabolic rate of Eskimos. *J. Appl. Physiol.*, **7**, 537–40.

MILLINGTON, R. A. (1964) Physiological responses to cold. *Weather*, **19**, 334–7.

MILSUM, J. H. (1966) Mathematical modeling of the system: thermoregulation. *In: Biological Control Systems Analysis*, 50–82. New York: McGraw-Hill.

MORRISON, P. R. (1957) Body temperatures in aboriginals. *Fed. Proc.*, **16**, 90.

NAKAYAMA, T., EISENMAN, J. S. and HARDY, J. D. (1961) Single unit activity of the anterior hypothalamus during local heating. *Science*, **134**, 530–1.

PUGH, L. G. C. and EDHOLM, O. G. (1955) Physiology of channel swimmers. *Lancet*, **2**, 761–8.

SIPLE, P. A. and PASSEL, C. F. (1945) Measurements of dry atmospheric cooling in subfreezing temperatures. *Proc. Amer. Phil. Soc.*, **89**, 177–99.

SMITH, A. U. (1970) Frostbite, hypothermia and resuscitation after freezing. *In:* SMITH, A. U. (ed.) *Current Trends in Cryobiology*, 181–208. New York: Plenum Press.

STEADMAN, R. G. (1971) Indices of windchill of clothed persons. *J. Appl. Met.*, **10**, 674–83.

WILKERSON, J. A. (1967) *Medicine for Mountaineering*, 117–26. Seattle: Vail-Ballou Press.

WILSON, O., HEDNER, P., LAURELL, S., NOSSLIN, B., RERUP, C. and ROSENGREN, E. (1970) Thyroid and adrenal response to acute cold exposure in man. *J. Appl. Physiol.*, **28**, 543–8.

WYNDHAM, C. H., WILLIAMS, C. G. and LOOTS, H. (1968) Reactions to cold. *J. Appl. Physiol.*, **24**, 282–7.

Manuscript received March 1972.

Additional reference

MATUSOV, A. L. (1973) *Medical Research on Arctic and Antarctic Expeditions.* Jerusalem: Israel Progr. Sci. Transl. 209 pp. (Original Russian edn, Leningrad, Gidromet. Izdat, 1971.)

Man living at high altitudes

Robert F. Grover

High Altitude Research Laboratory,
University of Colorado School of Medicine

Only a fraction of the earth's surface can support human life, and today man is crowding these limited areas in numbers unprecedented in the world's history. Human populations are not confined to the obviously comfortable environments but are located in seemingly hostile environments as well. This is testimony to man's remarkable adaptability. In view of this, when one finds people living in a rigorous mountain environment, it is justifiable to question whether they live there by choice or by necessity, perhaps economic. In other words, is man a natural and willing component of the alpine environment or merely a reluctant intruder?

1. Prehistoric man at high altitude

It is a reasonable assumption that when early man first came to the western hemisphere some 12,000 years ago, he established residence in areas which provided adequate food and an agreeable climate. If he was responding primarily to these fundamental pressures of survival, it is of considerable interest that he soon became established in the high mountain areas of both North and South America. In the Rocky Mountains, man occupied caves at 1900 m altitude 9000 years ago (Wedel et al., 1968), and hunted game as high as 4000 m 9500 years ago (Benedict, 1970). Ute Indians occupied South Park in Colorado at 3000 m until they were displaced by the arrival of white man (Renaud, 1945). In Mexico, man has occupied the valley of Oaxaca at 1550 m elevation since 8000 BC (Flannery et al., 1967), and the valley of Tehuacan at 2300 m elevation since 9000 BC (MacNeish, 1964). It is known that man has lived in South America for at least 10,000 years. Primitive cultures were established on the Pacific coast of Peru by 8500 BC (Lanning, 1965), while preceramic sites at altitudes above 4000 m date back to 7500 BC (Lynch, 1967). Therefore, there is abundant evidence that man has been willing to reside in high altitude regions since prehistoric times.

2. Alpine climate

TEMPERATURE

Accepting that man has no apparent basic aversion to living at high altitude, then it is the associated cold rather than altitude *per se* which determines the mountainous areas of the world which are actually occupied by man. These are at latitudes within 40° of the equator, and include the Andes of South America, mountainous Ethiopia in Africa, the Caucasus of southern Russia, the Himalayas of Asia, and the Rocky Mountains in the United States. An estimated 10 million people live above 3000 m in these regions. Other high mountain areas, such as the European Alps, are located farther from the equator and consequently cold and snow restrict human residence to comparatively lower altitudes well below 3000 m.

HUMIDITY

At high altitude, the relatively small amount of water vapour in the air provides little filtering of sunlight and consequently the solar radiation can be intense and provide considerable warmth even though the air temperature may be well below freezing. In addition, dry air conducts less heat away from the body, but simultaneously it hastens evaporation; this tendency to dehydration is an added stress at altitude.

ATMOSPHERIC PRESSURE

Physiologically, the dominant feature of high altitude is the hypoxia produced by the reduction in atmospheric pressure. Recently, Dill and Evans (1970) summarized the barometric pressure (P_B) observed at numerous altitude stations; the range was from 760 mmHg (1013 mb) at sea level to 320 mmHg (427 mb) at 6500 m (fig. 14B.1). When air is taken into the lungs, it is rapidly saturated with water vapour at body temperature. For man with a temperature of 37°C, this water vapour pressure (P_{H_2O}) is 47 mmHg (63 mb). Consequently, the combined pressure of the gases in the air taken into the lung will be $P_B - P_{H_2O}$ or $P_B - 47$. Air contains about 21 per cent oxygen (actually 20.93 per cent) at all terrestrial altitudes. Hence, the partial pressure of inspired oxygen (P_IO_2) will be $0.21 (P_B - 47)$. When $P_B = 760$ mmHg (1013 mb) at sea level, then P_IO_2 $= 0.21 (760 - 47) = 150$ mmHg (200 mb). At 5200 m $P_B = 400$ mmHg (533 mb) and P_IO_2 is 74 mmHg (99 mb) or one-half that at sea level (fig. 14B.1). This is the upper limit of altitude which man can tolerate indefinitely. However, very few people choose to live higher than 4500 m where $P_IO_2 = 82$ mmHg (109 mb).

While the above calculations define the severity of hypoxia at altitudes near the equator, an anomalous situation exists in the Antarctic. The pressure of the atmosphere at the earth's surface results from the weight of the air above that point on the earth. Because of the earth's rotation and other factors, the atmosphere is 'less deep' over the Poles than over the equator. As a result, at the South Pole located 2800 m above sea level

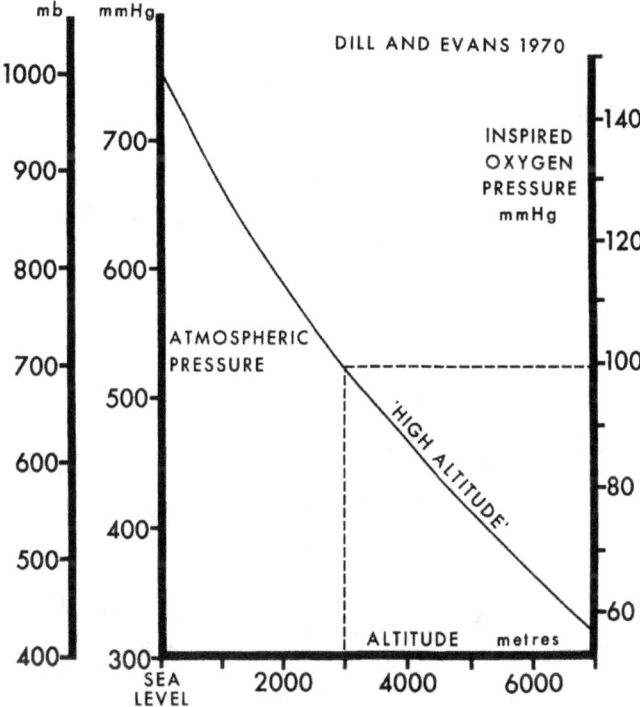

Fig. 14B.1. With increasing altitude above sea level, there is a progressive decrease in total atmospheric pressure (P_B). Values for P_B were measured at many terrestrial altitudes and summarized by Dill and Evans (1970). There is an associated decrease in inspired oxygen pressure $(P_IO_2) = 0.21\ (P_B - 47)$; see text. The level of P_IO_2 is a measure of atmospheric hypoxia which becomes physiologically significant below 100 mmHg. Therefore, 'high altitude' begins at 3000 m.

P_B is only 505–515 mmHg (673–686 mb) (Guenter *et al.*, 1970). These values of P_B would correspond to an altitude of 3400 m near the equator. Consequently, altitude imposes more severe hypoxia in the Antarctic than one might have predicted.

3. Physiology of oxygen transport

NORMAL

For man to live and work at any altitude, his body must have a continuous supply of O_2. Like all mammals, the human body is equipped with a specialized O_2 transport system which takes O_2 from the air we breathe and delivers it to tissues throughout the body. The vehicle for O_2 is the respiratory pigment haemoglobin which chemically binds O_2 reversibly. Haemoglobin is contained in red blood cells which are suspended in blood plasma. In the adult human male at sea level, 100 ml of blood contains 45 ml of red blood cells which in turn contain 15 grams of haemoglobin with the capacity to bind 20 ml of O_2 (fig. 14B.2).

The primary function of the lung is to oxygenate blood. Air containing O_2 enters the lung through the trachea and proceeds through a highly branched system of tubes, the bronchi and bronchioles, to the terminal alveolar air sacs. The alveolar walls contain a meshwork of capillaries, and collectively this alveolar–capillary interface provides a vast area for the diffusion of O_2 from alveolar air to the red blood cells in the capillaries.

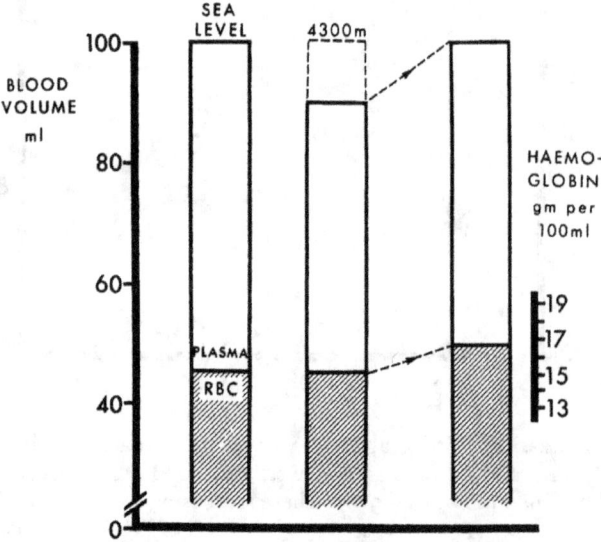

Fig. 14B.2. In normal man at sea level, 100 ml of blood contains 45 ml of red blood cells (RBC) and 55 ml of plasma; the 45 ml of RBC contain 15 grams of haemoglobin (left bar). When this man ascends to 4300 m altitude, approximately 10 ml of water is removed from the plasma. Consequently, the original 45 ml of RBC are now contained in only 90 ml of blood (middle bar). This means that 100 ml of blood will contain 50 ml of RBC and nearly 17 grams of haemoglobin (right bar). Thus, haemoglobin concentration is increased even though there is no change in total haemoglobin.

Once oxygenated, this blood drains via the pulmonary veins into the left side of the heart. The left ventricle, by its pumping action, delivers the blood into the arterial tree whose branches distribute the blood to all tissues throughout the body. Within the tissues, the ultimate ramifications of the arterial tree are the systemic capillaries. These capillaries bring the red blood cells in close proximity to all tissue cells. Because the pressure of O_2 is higher in the blood than in the tissue, the O_2 is released by the haemoglobin and diffuses out of the capillary and into the tissue cells. Thus depleted of much of its O_2, the blood drains into the collecting veins which return it to the right side of the heart. The right ventricle then pumps it into the lungs where the oxygenation process is repeated.

DECREASE AT HIGH ALTITUDE

Since the transfer of O_2 is dependent upon pressure gradients, the entire system will be influenced by the highest pressure in that system, namely, the pressure of O_2 in the inspired air (P_IO_2). At high altitude, P_IO_2 is reduced, the entire O_2 transport system is taxed, and consequently the O_2 supply to the body is compromised. The capacity of the O_2 transport system is the major factor which determines the maximum amount of O_2

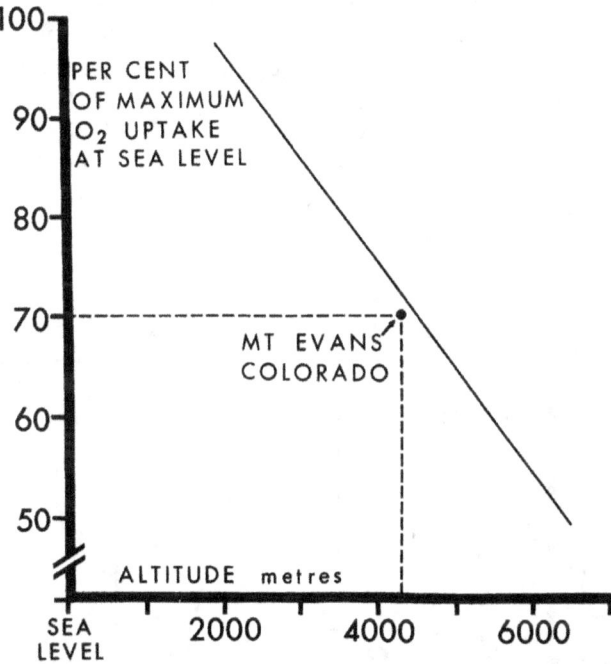

Fig. 14B.3. A man's aerobic working capacity is measured as the maximum amount of O_2 which he can consume per minute (max $\dot{V}O_2$). At high altitude, a decrease in max $\dot{V}O_2$ has been reported by many investigators. Buskirk (1969) summarized these reports which show a 10 per cent decrease in max $\dot{V}O_2$ with each 1000 m increase in altitude above 1500 m. For example, on the summit of Mt Evans, Colorado, at 4300 m, max $\dot{V}O_2$ was only 70 per cent of the value measured at sea level (after Saltin *et al.*, 1968).

which the body can consume per minute (max $\dot{V}O_2$) (Astrand *et al.*, 1964), i.e. the so-called aerobic working capacity. Above 1500 m altitude, max $\dot{V}O_2$ is reduced 10 per cent per 1000 m increase in altitude (fig. 14B.3) (Buskirk, 1969). Thus, at 4500 m altitude, aerobic working capacity is only 70 per cent of what it was at sea level (Saltin *et al.*, 1968). Anyone who has ever been at high altitude has noticed that he tires more quickly, since his reduced capacity for work makes any task relatively more strenuous. Thus, one of the inescapable effects of high altitude on man is a reduction in his daily work output.

4. Physiologic adaptations to hypoxia

HYPERVENTILATION

At high altitude, numerous physiologic mechanisms are activated by the lower P_IO_2 which tend to compensate for and lessen the degree of hypoxia within the body. A reduction in P_IO_2 in the lung is followed by a reduction in the pressure of O_2 in the blood (PaO_2). This lower PaO_2 is sensed by the carotid body chemoreceptors, and this information is transmitted via the glossopharyngeal nerves to the respiratory centre in the medullary portion of the brain stem. The respiratory centre then increases the activity of the respiratory muscles; a greater volume of air is taken into the lungs with each breath, and the number of breaths per minute is also increased somewhat (Weil *et al.*, 1970). This increase in the volume of air inspired per minute (hyperventilation) offsets the decrease in air density associated with the lower atmospheric pressure at high altitude. Consequently, the number of molecules of O_2 taken into the lungs per minute is restored and is virtually the same as at sea level (Christensen, 1937).

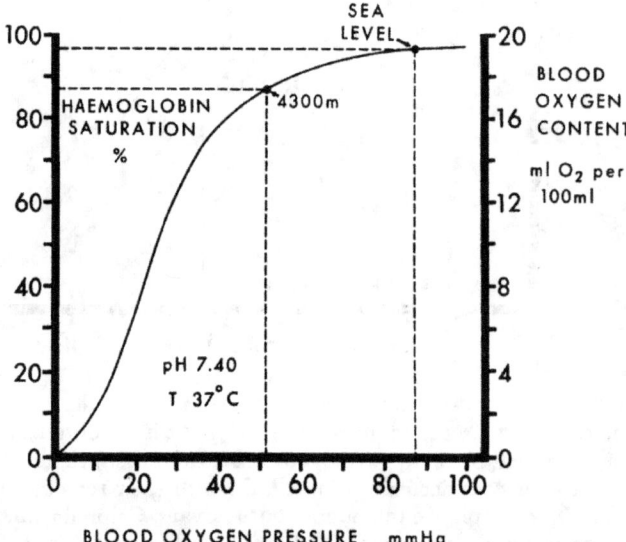

Fig. 14B.4. Oxygen (O_2) is reversibly bound to haemoglobin within the red blood cells. When the pressure of O_2 in the blood is high, then the haemoglobin is 100 per cent saturated with O_2. As the O_2 pressure decreases, the saturation also decreases, but the relationship is sigmoidal. For a man breathing air, the O_2 pressure in arterial blood will be 87 mmHg at sea level and 51 mmHg at 4300 m. However, this will only reduce the haemoglobin saturation from 97 to 87 per cent. Each gram of haemoglobin when 100 per cent saturated will carry 1.34 ml of O_2. Hence, 100 ml of blood containing a normal 15 grams of haemoglobin will have an O_2 content of 19.5 ml O_2 when 97 per cent saturated at sea level. Ascent to 4300 m which reduces saturation to 87 per cent will reduce the O_2 content of 100 ml of blood to 17.5 ml O_2.

While hyperventilation maintains the quantity of O_2 available in the lungs, it has only a minimal affect on PaO_2 which remains lower than at sea level. For example, PaO_2 which is normally about 90 mmHg (120 mb) at sea level is only 60 mmHg (80 mb) at 3100 m altitude in spite of hyperventilation. As a result of the reduced PaO_2, the haemoglobin in the blood binds less O_2 and full saturation is not achieved. (The relationship between PaO_2 and O_2 saturation is not linear but sigmoidal and is defined by the haemoglobin-oxygen dissociation curve which can be modified by changes in temperature and acidity of the blood: see fig. 14B.4.) Thus, during the first day or two at altitude, each 100 ml of blood carries fewer ml of O_2 than at sea level.

INITIAL INCREASE IN CARDIAC OUTPUT

The quantity of O_2 available to the tissues of the body is determined by the quantity (ml) or O_2 in each unit (ml) of blood (CaO_2) and the number of units (ml) of blood pumped by the heart per minute, the so-called cardiac output $(\dot{Q})$. O_2 transport is then $CaO_2 \times \dot{Q}$. Cardiac output, in turn, results from the volume of blood pumped by each contraction (beat, stroke) of the left ventricle, i.e. the stroke volume (SV), and the number of beats per minute, i.e. the heart rate (HR). Using these symbols, $\dot{Q} = SV \times HR$. Under ordinary conditions, SV remains relatively constant, and so an increase in $\dot{Q}$ is achieved by an increase in HR. In normal man, HR at rest is about 60 per minute and increases to about 190 per minute during maximal exercise.

Upon ascent to high altitude, the decrease in PaO_2 caused by the decrease in P_B and P_IO_2 stimulates the medullary portion of the brain directly, and this produces an increase in HR and hence an increase in $\dot{Q}$ (Hultgren and Grover, 1968). This acts to compensate for the decrease in CaO_2 and thereby maintains O_2 transport.

HAEMOCONCENTRATION AND POLYCYTHEMIA

After about one week at high altitude, the amount of water in the blood plasma is decreased (Shields et al., 1969). The mechanisms which bring this about are not clear, but probably involve constriction of peripheral veins (Weil et al., 1971). This decreases the total plasma volume without changing the total volume of red blood cells. Consequently, total blood volume (plasma volume + red cell volume) is also decreased; but as a result, each 100 ml of blood now contains 50 ml of red blood cells and about 17 grams of haemoglobin (instead of the original concentration of 15 grams per 100 ml, see fig. 14B.2). A greater haemoglobin concentration means that each 100 ml of blood can now carry a greater quantity of O_2 so that CaO_2 is no longer reduced (fig. 14B.5). Prolonged exposure to chronic hypoxia at high altitude stimulates the bone marrow to produce more red blood cells and increase the total volume of red blood cells in the body (polycythemia) (Weil et al., 1968). When this occurs, then 100 ml of blood will contain 55 to 60 ml of red blood cells. This tends to increase CaO_2 even further.

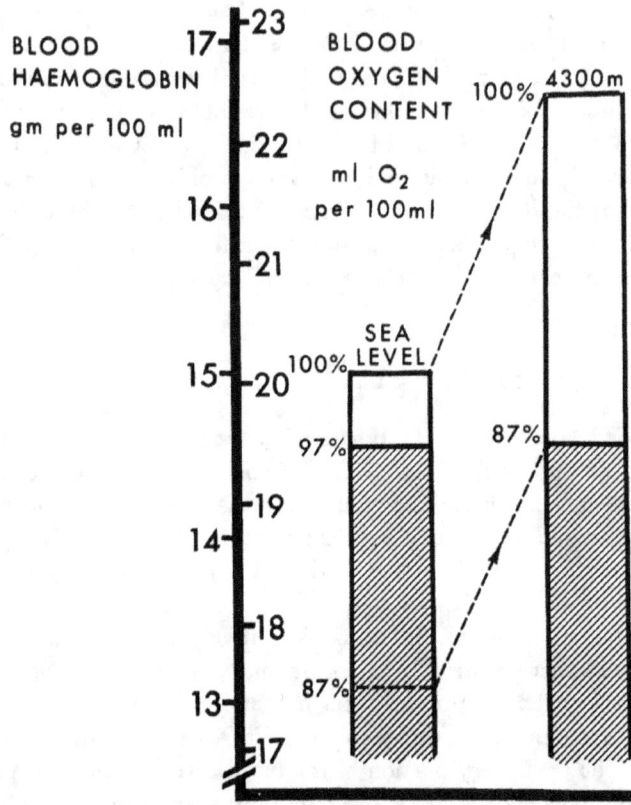

Fig. 14B.5. When 100 ml of blood contains 15 grams of haemoglobin in the red blood cells, it will contain 20.1 ml of O_2 when 100 per cent saturated. At sea level where the normal saturation is 97 per cent, the content will be 19.5 ml O_2 per 100 ml blood (left bar). Ascent to 4300 m altitude lowers the saturation to 87 per cent which would reduce the content to 17.5 ml O_2 per 100 ml blood (left bar). However, haemoconcentration increases the haemoglobin concentration to 16.8 grams per 100 ml blood (fig. 14B.2); this has the potential capacity to carry 22.4 ml O_2 if it were 100 per cent saturated (right bar). However, at a saturation of 87 per cent the content will actually be 19.5 ml O_2 per 100 ml blood. In other words, haemoconcentration at altitude maintains blood O_2 content even though saturation has decreased.

DELAYED DECREASE IN CARDIAC OUTPUT

While compensatory mechanisms are operating to restore CaO_2, there is a concurrent decrease in cardiac output; HR tends to normalize, but SV diminishes with the net result that in the person adapted to high altitude, $\dot{Q}$ is less than at sea level under all conditions (Alexander *et al.*, 1967; Grover, 1969). The mechanisms which reduce SV are not known but may reflect the decrease in total blood volume or a depressant effect of hypoxia on the heart muscle. This means the O_2 transport ($CaO_2 \times \dot{Q}$) is also less than at sea level. Less

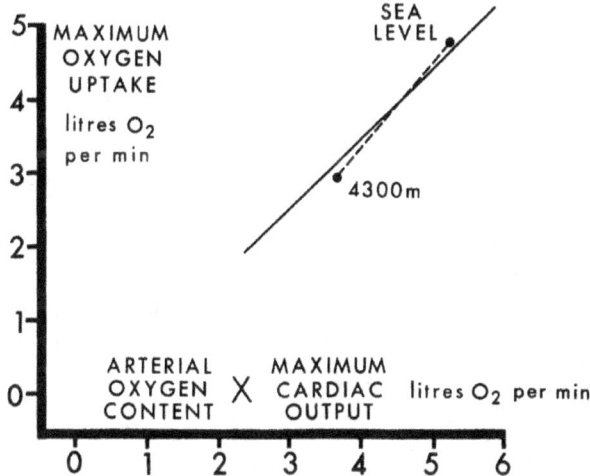

Fig. 14B.6. During exercise, maximum oxygen uptake (max $\dot{V}O_2$) is determined by the maximum quantity of O_2 transported to the tissues. This is the quantity of O_2 in each unit volume of arterial blood times the maximum volume of blood pumped by the heart per minute (maximum cardiac output). The solid line indicates the normal relationship between max $\dot{V}O_2$ and maximum O_2 transport (after Astrand *et al.*, 1964). In an athlete with a max $\dot{V}O_2$ of 4.7 litres O_2 per minute at sea level, ascent to 4300 m reduced his max $\dot{V}O_2$ to 3.0 litres per minute (dashed line). This was primarily the result of a decrease in his maximum cardiac output which reduced his maximum O_2 transport (after Saltin *et al.*, 1968).

O_2 delivered to the body means that less aerobic work can be performed (fig. 14B.6). This situation persists as long as the person remains at high altitude but returns to normal when he descends to sea level.

5. High altitude residents

NATIVE *V.* NEWCOMER

Europeans visiting the high plateaus of the Andes above 4000 m have been impressed repeatedly with the work capacity of the native Indians. Not only do they perform heavy manual labour many hours a day, but for recreation they choose the strenuous sport of soccer. Since the usual visitor cannot match the performance of the native, it has been postulated that the Andean native possesses an unusual physiologic adaptation to high altitude which the newcomer can never acquire. This special degree of acclimatization is said to be inherited and reflects their continuous residence at high altitude since prehistoric times. According to this hypothesis, a man born and raised above 4000 m would still lack complete acclimatization if his parents were not of ancient Indian stock (Monge, 1945, p. 307).

In many ways, the Andean native is physiologically similar to the athlete. Monge (1945, p. 335) recognized this when he wrote that 'athleticism must be the norm for the survival of man in the high altitude'. When Kollias *et al.* (1968) took six well-conditioned athletes from Pennsylvania to the town of Nunoa, Peru, elevation 4000 m, he found that their exercise performance was comparable to that of the local Andean Indians. This suggests that the Andean native's physical superiority is analogous to familiar differences between the athlete and the non-athlete at low altitude. Hence, the apparently unusual performances characteristic of the Andean Indian are minimized when he is compared with a lowlander (athlete) of similar aerobic working capacity (max VO_2). Likewise, Grover *et al.* (1967) found that athletes from sea level were indistinguishable from young men native to 3100 m in the Rocky Mountains, in terms of aerobic working capacity and physiologic responses to submaximal exercise at both high and low altitude. The Himalayan Sherpas have performed admirably as porters on numerous mountaineering expeditions, partly because they too have a high aerobic work capacity. This characteristic accounts for much of the reported difference between Sherpas and non-athletic lowlanders in terms of the physiologic response to exercise (Lahiri *et al.*, 1967).

DEVELOPMENT IN AN HYPOXIC ENVIRONMENT

It appears then, that populations native to high altitude have a higher average aerobic working capacity than most low-altitude populations, although this impression is not based on rigorous population sampling and testing. If true, is this the result of post-natal development modified by chronic hypoxia? Can the capacity of the body's oxygen transport system be increased by the stress of a limited oxygen supply? Recently, Burri and Weibel (1971) showed that in rats born and raised in an hypoxic environment, the structure of the lung was modified to increase the alveolar-capillary gas exchange surface which would facilitate oxygen uptake. Strenuous exercise subjects the working muscles to tissue hypoxia, and herein lies a similarity to high altitude. The aerobic working capacity can be increased significantly in children participating in an extended physical training programme if this is begun well before puberty (Astrand and Rodahl, 1970). Therefore, we cannot exclude the possibility that environmental hypoxia from birth induces the physiologic characteristics observed in high altitude natives.

NATURAL SELECTION

The structure of high altitude populations may also be the result of natural selection. Since the hypoxia of altitude reduces man's working capacity, all tasks are relatively more strenuous than at sea level, and fatigue will appear more frequently. The higher functions of the central nervous system, e.g. judgement and creative thinking, may be impaired, and irritability has been reported frequently at higher altitudes. For these and other subtle reasons, those persons less tolerant of the high altitude environment may choose to leave, whereas those more tolerant will remain. A high aerobic working capacity would increase altitude tolerance by lessening fatigue. Such selection factors could modify the ultimate composition of the population.

6. Maladaptation to high altitude

ACUTE MOUNTAIN SICKNESS

Normal man usually adapts well to altitudes of 3000 m to 4000 m. Once adapted, he remains healthy and productive indefinitely. However, the initial impact of high altitude is often manifested as acute mountain sickness (Shields *et al.*, 1969). This unpleasant and disabling condition is usually precipitated by rapid ascent from near sea level to altitudes above 3500 m. Symptoms appear within hours of ascent. The person spending his first night at high altitude often sleeps poorly and develops a severe headache leading to nausea and vomiting. He is apathetic, somnolent, irritable, and cannot tolerate food. These symptoms are most severe during the first twenty-four hours at altitude, and usually subside within a few days. Occasionally, however, nausea and vomiting persist, leading to serious dehydration and debilitation, in which case the person must descend to a lower altitude.

Obviously persons suffering from acute mountain sickness are unable to perform any sort of task effectively. This proved to be disastrous for troops of the Indian army when they were rushed to altitudes up to 5500 m in the Himalayas to protect their northern border from the Chinese invasion in 1962. It is hypoxia which produces acute mountain sickness, but the pathophysiology is uncertain. Hypoxia causes hyperventilation with an excessive elimination of CO_2; this leads to alkalosis and also a reduction in blood flow to the brain. In addition, water is shifted out of blood and into tissues including the brain (cerebral oedema), and this could produce the observed symptoms. Acute mountain sickness is prevented most effectively by interrupting the ascent at 2800 m for a day or two before continuing on to higher altitude. Other prophylactic measures include a high carbohydrate, low fat diet, drugs which interfere with the elimination of CO_2 (carbonic anhydrase inhibitors), and drugs which reduce the accumulation of fluid in the tissues (diuretics).

HIGH ALTITUDE PULMONARY OEDEMA

On rare occasions, rapid ascent to high altitude combined with strenuous exercise will produce a more serious condition known as high altitude pulmonary oedema (HAPE) (Singh and Roy, 1969). As the term implies, there is an accumulation of fluid in the lungs which interferes greatly with the transfer of oxygen from air to blood. Atmospheric hypoxia is the cause, and the most effective treatment is the administration of supplemental oxygen combined with rest. With this treatment, complete recovery is usual and the person can then remain at high altitude without a recurrence. HAPE occurs in otherwise healthy persons and is not the result of underlying heart disease. Approximately 800 cases have been reported in young men in the army of India. Persons native to high altitude including children are susceptible upon return to high altitude following several weeks at sea level. While the cause is sudden exposure to hypoxia, again the pathophysiology is not known. HAPE may be related to acute mountain sickness, but additional factors must also be involved.

CHRONIC MOUNTAIN SICKNESS

A completely different form of maladaptation to high altitude is chronic mountain sickness (Monge and Monge, 1966). This condition develops in persons who have been at altitude for many years, including fully acclimatized high altitude natives. Essentially the person loses his adaptation to hypoxia. The normal stimulation of breathing by hypoxia is absent in persons with chronic mountain sickness (Sorensen and Severinghaus, 1968). In the absence of normal hyperventilation, the O_2 pressure in the lung alveoli and consequently in the arterial blood (PaO_2) is unusually low. For example, at 4300 m altitude where P_IO_2 is 80 mmHg (107 mb) (fig. 14B.1), the fully adapted newcomer will have a PaO_2 of 51 mmHg (68 mb) (fig. 14B.4), whereas the man with chronic mountain sickness will have a PaO_2 of only 40 mmHg (53 mb). This greater degree of hypoxia in the blood is a greater stimulus for red blood cell production. As a result, at 4300 m in the adapted newcomer, 100 ml of blood contains 52 ml of red blood cells, in contrast to 73 ml of red blood cells in the man with chronic mountain sickness. In addition to this excessively thick and viscous blood (polycythemia), these men have symptoms of headache, dizziness, poor memory, and they tire easily. Temporary relief is provided by removal of some of the excess blood, a pint every two or three weeks, but such persons must leave the hypoxic environment and go to low altitude to remain free of this condition.

As stated earlier, most persons remain in good health at high altitude. In fact, in Azerbaijan in the Caucasus of the USSR, there is a high plateau where people often live to be more than 100 years old (Leaf, 1970). Furthermore, recent evidence suggests that among persons of both European and Spanish American ancestry living at high altitude, fatal heart attacks are rare, and there is less high blood pressure (less systemic hypertension) (Marticorena et al., 1969). Thus, high altitude residents may actually enjoy a degree of protection from these major causes of death in the United States today.

References

ALEXANDER, J. K., HARTLEY, L. H., MODELSKI, M. and GROVER, R. F. (1967) Reduction of stroke volume during exercise in man following ascent to 3100 m altitude. *J. Appl. Physiol.*, **23**, 849–58.

ASTRAND, P.-O., CUDDY, T. E., SALTIN, B. and STENBERG, J. (1964) Cardiac output during submaximal and maximal work. *J. Appl. Physiol.*, **19**, 268–74.

BENEDICT, J. B. (1970) Downslope soil movement in a Colorado alpine region: rates, processes, and climatic significance. *Arct. Alp. Res.*, **2**, 165–226.

BURRI, P. H. and WEIBEL, E. R. (1971) Morphometric estimation of pulmonary diffusion capacity, II: Effect of PO_2 on the growing lung. Adaption of the growing rat lung to hypoxia and hyperoxia. *Respiration Physiol.*, **11**, 247–64.

BUSKIRK, E. R. (1969) Decrease in physical work capacity at high altitude. *In:* HEGNAUER, A. H. (ed.) *Biomedicine Problems of High Terrestrial Elevations*, 204–22. Natick, Mass.: US Army Res. Inst. Env. Med.

CHRISTENSEN, E. H. (1937) Sauerstoffaufnahme und Respiratorische Funktionen in grossen Höhen. *Skand. Arch. Physiol.*, **76**, 88–100.

DILL, D. B. and EVANS, D. S. (1970) Report barometric pressure! *J. Appl. Physiol.*, **29**, 914–16.

FLANNERY, K. V., KIRKBY, A. V. T., KIRKBY, M. J. and WILLIAMS, A. W. JR (1967) Farming systems and political growth in ancient Oaxaca. *Science*, **158**, 445–54.

GROVER, R. F. (1969) Influence of high altitude on cardiac output response to exercise. *In:* HEGNAUER, A. H. (ed.) *Biomedicine Problems of High Terrestrial Elevations*, 223–39. Natick, Mass.: US Army Res. Inst. Env. Med.

GROVER, R. F., REEVES, J. T., GROVER, E. B. and LEATHERS, J. E. (1967) Muscular exercise in young men native to 3100 m altitude. *J. Appl. Physiol.*, **22**, 555–64.

GUENTER, C. A., JOERN, A. T., SHURLEY, J. T. and PIERCE, C. M. (1970) Cardiorespiratory and metabolic effects in men on the south polar plateau. *Arch. Internal Med.*, **125**, 630–7.

HULTGREN, H. N. and GROVER, R. F. (1968) Circulatory adaptation to high altitude. *Ann. Rev. Med.*, **19**, 119–52.

KOLLIAS, J., BURKIRK, E. R., AKERS, R. F., PROKOP, E. K., BAKER, P. T. and PICON-REATEGUI, E. (1968) Work capacity of long-time residents and newcomers to altitude. *J. Appl. Physiol.*, **24**, 792–9.

LAHIRI, S., MILLEDGE, A. S., CHATTOPADHYAY, H. T., BHATTACHARYYA, A. K. and SINHA, A. K. (1967) Respiration and heart rate of Sherpa highlanders during exercise. *J. Appl. Physiol.*, **23**, 545–54.

LANNING, E. P. (1965) Early man in Peru. *Sci. Amer.*, **213** (4), 68–76.

LYNCH, T. F. (1967) Quishqui Puncu: a preceramic site in highland Peru. *Science*, **158**, 780–3.

MACNEISH, R. S. (1964) The origins of New World civilization. *Sci. Amer.*, **211** (5), 29–37.

MARTICORENA, E., RUIZ, L., SEVERINO, J., GALVEZ, J. and PENALOZA, D. (1969) Systematic blood pressure in white men born at sea level: changes after long residence at high altitudes. *Amer. J. Cardiol.*, **23**, 364–8.

MONGE-M., C. (1945) Aclimitacion en los Andes-confirmacions historicas sobre 'agresion climatica' en el desenvolvimiento de las sociedades de America. *Anales de la Facultad de Medicina* (Lima, Peru), **28**, 307–82.

MONGE-M., C. and MONGE-C., C. (1966) Chronic mountain sickness. *In: High Altitude Diseases*, 32–50. Springfield, Ill.: Charles C. Thomas.

RENAUD, E. B. (1945) Archeological survey of South Park, Colorado. *In: Univ. of Denver Archeol. Ser.*, 5th Pap., 1–19. Denver: Univ. of Denver Press.

SALTIN, B., GROVER, R. F., BLOMQVIST, C. G., HARTLEY, L. H. and JOHNSON, R. L. JR (1968) Maximal oxygen uptake and cardiac output after 2 weeks at 4300 m. *J. Appl. Physiol.*, **25**, 400–9.

SHIELDS, J. L., HANNON, J. P., CARSON, R. P., CHINN, K. S. K. and EVANS, W. O. (1969) Pathophysiology of acute mountain sickness. *In:* HEGNAUER, A. H. (ed.) *Biomedicine Problems of High Terrestrial Elevations*, 9–23. Natick, Mass.: US Army Res. Inst. Env. Med.

SINGH, I. and ROY, S. B. (1969) High altitude pulmonary edema: clinical hemodynamic, and pathologic studies. *In:* HEGNAUER, A. H. (ed.) *Biomedicine Problems of High Terrestrial Elevations*, 108–20. Natick, Mass.: US Army Res. Inst. Env. Med.

SORENSEN, S. C. and SEVERINGHAUS, J. W. (1968) Respiratory sensitivity to acute hypoxia in man born at sea level living at high altitude. *J. Appl. Physiol.*, **25**, 211–16.

WEDEL, W. R., HUSTED, W. M. and MOSS, J. H. (1968) Mummy cave: Prehistoric record from Rocky Mountains of Wyoming. *Science*, **160**, 184–6.

WEIL, J. V., JAMIESON, G., BROWN, D. W. and GROVER, R. F. (1968) The red cell mass-arterial oxygen relationship in normal man. *J. Clin. Invest.*, **47**, 1627–39.

WEIL, J. V., BYRNE-QUINN, E., SODAL, I. E., FRIESEN, W. O., UNDERHILL, B., FILLEY, G. F. and GROVER, R. F. (1970) Hypoxic ventilatory drive in normal man. *J. Clin. Invest.*, **49**, 1061–72.

WEIL, J. V., BYRNE-QUINN, E., BATTOCK, D. J., GROVER, R. F. and CHIDSEY, C. A. (1971) Forearm circulation in man at high altitude. *Clin. Sci.*, **40**, 235–46.

Manuscript received April 1971.

Additional references

ASTRAND, P.-O. and RODAHL, K. (1970) *Textbook of Work Physiology.* New York: McGraw-Hill. 307 pp.

LEAF, A. (1973) Every day is a gift when you are over 100. *Nat. Geogr. Mag.*, **103**, 92–119.

The peopling of arctic North America

Robert McGhee
Memorial University of Newfoundland, St. Johns

1. Arctic prehistory

The immediate ancestors of man developed their biological capacities in the warm low-latitude forests of the Late Tertiary era. The subsequent history of human development has seen a series of dispersals into the temperate and cooler regions, as developing cultural capacities supplemented the biological and allowed human populations to adapt to habitats increasingly divergent from their original environmental zone. Aside from insular areas such as Polynesia, the Arctic was the last major region to be occupied by man. In the study of man's development as a cultural animal, the prehistory of the arctic regions is therefore of great interest. In the Arctic, man's biological capacities were clearly insufficient even for simple survival; cultural capacities, especially technological abilities and an accumulated knowledge of the environment, were of supreme importance.

The following sections will trace the technological and economic developments which supported the various prehistoric occupations of the North American Arctic. This is an area where human adaptations occurred relatively recently, where the remains of prehistoric settlements are relatively well preserved and therefore accessible to archaeological study, and where populations with a generally Palaeolithic level of technology and economic adaptation continued to live until the time of European contact. Despite these excellent conditions for the reconstruction of prehistory, it must be admitted that our knowledge is still slight, and much more work must be done before we can present an adequate picture of prehistoric developments in the arctic zone.

2. The arctic environment

Other chapters of this book give intensive descriptions of most aspects of the arctic environment. The present section merely lists the major environmental factors which conditioned and limited the occupation of this zone by prehistoric peoples:

(1) Seasonally cold climate. For at least several months of the year, arctic temperatures are too low for the human body to survive in an unprotected state. A great deal of

energy must therefore be allotted to devising and constructing shelter in the forms of clothing and housing, and in procuring a diet which will maintain body temperature.

(2) Scarcity of plant foods. The Arctic, unlike most other environments, does not provide plant foods sufficient for the maintenance of human life. Arctic peoples must be hunters and fishermen, making a place for themselves near the top of the food chain rather than inserting their economies into several levels as is possible in lower latitudes.

(3) Seasonal scarcity of animal foods. Many species preyed upon by hunters and fishermen are unavailable during the colder seasons, either through migration, hibernation, or because they are protected beneath winter ice. Arctic peoples must therefore organize their lives in terms of a complete annual cycle based on precise knowledge of the expected seasonal availability of various prey species.

(4) Limited number of prey species. Relative to lower latitude zones, the Arctic is poor in the number of species available to human predation. On the other hand, many of these species occur in large numbers and in dense seasonal concentrations, permitting very productive intensive predation techniques which are not often possible in other environmental zones.

(5) Limited supply of fuel and raw materials. The availability of wood was of great importance to most prehistoric populations, who used it as fuel, construction material, and in manufacturing tools and weapons. The scarcity of wood in the tundra and arctic coastal zone necessitated a great deal of technological innovation and an extremely rational use of the materials available. Animal products were extremely important: bones and fat were used as fuel, skins and feathers for clothing, hides for shelter, while bone, antler, ivory and horn were used for most tools and weapons.

3. First occupations of Beringia

Man's first excursions into the colder regions of the earth took place in Eurasia, where we know that by at least 500,000 years ago men of the *Homo erectus* type were living in the cool steppe and forest zones of eastern Europe and northern China. It seems unlikely that any of these populations reached the American continent, since to do so would have required a movement of several hundred kilometres through an arctic environmental zone, across a Bering Platform which was exposed only during glacial periods. We have no evidence that *H. erectus* populations with a Lower Palaeolithic level of culture had learned how to exist in an arctic environment.

Arctic and subarctic adaptations had been developed by the time of the Würm glaciation, however, when populations of the Neanderthal (*H. sapiens neandertalensis*) and *H. sapiens* types inhabited the proglacial areas of Europe and Asia. In the cultures of the European Upper Palaeolithic level, dating between roughly 35,000 and 10,000 BC we can discern *H. sapiens* populations who had learned to cope with most of the arctic environmental factors listed in the previous section. These people were occupying a low-latitude tundra zone using caves and skin tents for shelter, fire and skin-clothing for warmth, and had developed a sophisticated hunting technology using bone, antler and ivory as basic raw materials. The acclaimed cave art of the European Upper Palaeolithic strongly suggests that these people were not merely surviving in an arctic habitat, but were enjoying an economically and intellectually rich existence.

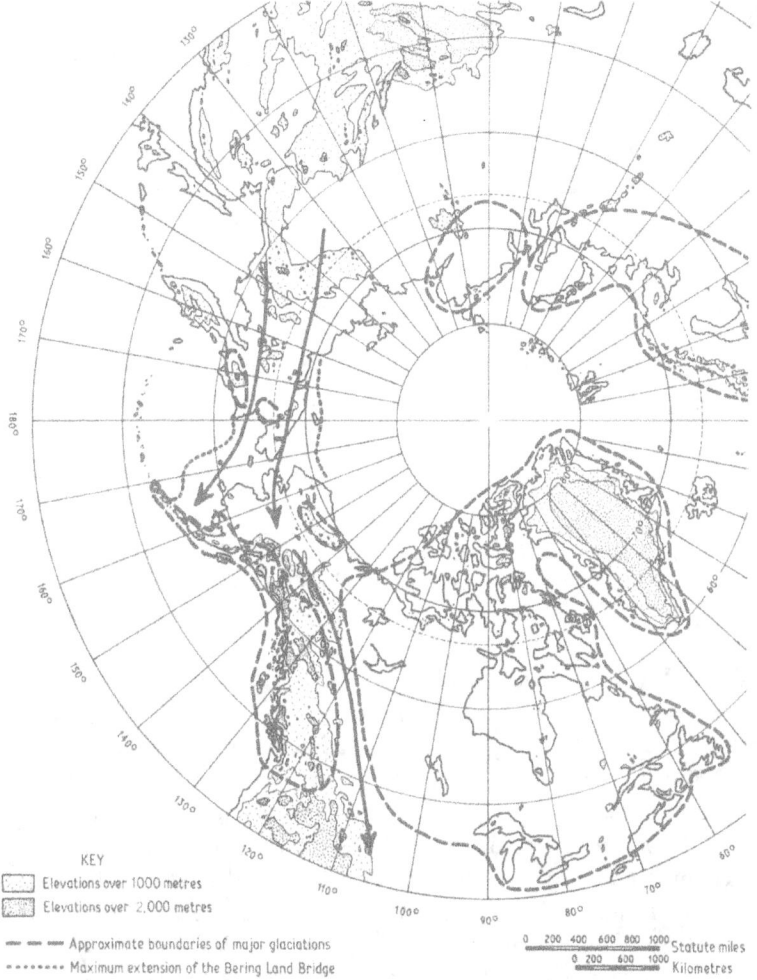

KEY
▭ Elevations over 1000 metres
▭ Elevations over 2,000 metres

— — — Approximate boundaries of major glaciations
••••••• Maximum extension of the Bering Land Bridge

0 200 400 600 800 1000 Statute miles
0 200 600 1000 Kilometres

Fig. 15A.1. Possible human migration routes to North America during the Late Wisconsin.

Our present knowledge of the Upper Palaeolithic period is mainly derived from western Europe, but similar populations inhabited the steppe–tundra zone of eastern Europe and Asia. It is among these populations that we will probably find the ancestors of the American Indians, the people who first passed through the high-latitude tundras of Beringia and then moved southward to readapt their arctic-oriented culture to the

temperate grasslands and forests of central North America (fig. 15A.1). Three disparate
models concerning the migration of ancestral American Indians to America have
recently been proposed by Müller-Beck (1967), Bryan (1969) and Haynes (1969).
Arguments dealing with the number of migrations involved, and the timing of these
migrations in relation to the periodic emergence of a Bering land bridge and of an ice-free
corridor between the Laurentide and Cordilleran ice sheets, are beyond the range of this
paper. We may summarize our present knowledge as follows: 1) Indians were definitely inhabiting America to the south of the Laurentide ice by 12,000 years ago, perhaps
by 25,000 years ago, and possibly as early as 40,000 years ago; 2) these people were the
descendants of Asiatic Upper Palaeolithic populations who were of *H. sapiens* physical
type, and who probably occupied the tundra and tundra–steppe zone which in Wisconsin times stretched from eastern Asia across Beringia and into the unglaciated areas of

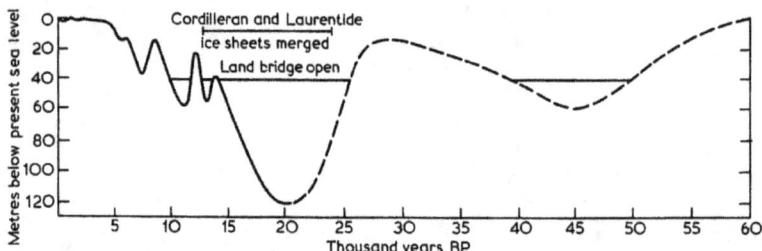

Fig. 15A.2. Speculative reconstruction of sea level changes in the Bering Strait region over the
past 60,000 years (after Hopkins, 1967b, and Müller-Beck, 1967).

Alaska and the Yukon (Hopkins, 1967b, pp. 471–4); 3) these populations occupied
and wandered across Beringia either immediately preceding the Woronzofian marine
transgressions of approximately 35,000 to 25,000 years ago (Hopkins, 1967a, p. 83), or
more likely at some time following this transgression (fig. 15A.2).

Although it is certain that the first human groups to enter America had some type of
arctic adaption, we know virtually nothing of the nature of this adaptation. A few small
surface collections and camp sites found in Alaska yield artifacts, including fluted spear
points, which are similar to those used by early Indians in central North America, but
these may relate to a much more recent occupation of Alaska by people who were unrelated to the early ancestral Indians (Hall, 1969; Schweger, 1969). A single artifact
which may relate to the original entry of man into America has been found recently by
C. R. Harington and W. N. Irving in association with a Late Pleistocene fossil bone
locality in the Old Crow Flats region of the northern Yukon (see fig. 15A.3). This
artifact, a skin scraper made on a caribou bone, has yielded a radiocarbon date of
27,000 $^{+3000}_{-2000}$ years ago (GX-1640; W. N. Irving, personal communication, 1970).

This is the only find from Arctic North America which is dated to the period preceding

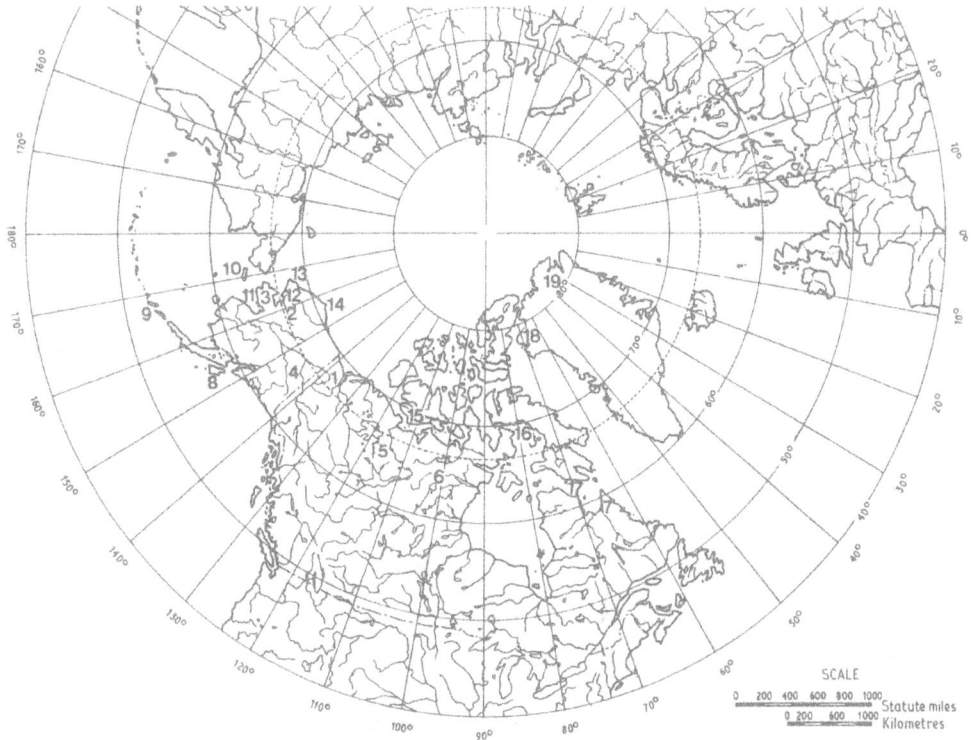

Fig. 15A.3. Locations of sites mentioned in text: 1, *Old Crow Flats*; 2, *Onion Portage*; 3, *Trail Creek*; 4, *Healy Lake*; 5, *Acasta Lake*; 6, *Aberdeen Lake*; 7, *Saglek Bay*; 8, *Kodiak Island*; 9, *Umnak Island (Anangula)*; 10, *St Lawrence Island*; 11, *Norton Sound*; 12, *Cape Krusenstern*; 13, *Point Hope (Ipiutak)*; 14, *Point Barrow (Birnirk)*; 15, *Dolphin and Union Strait*; 16, *Igloolik*; 17, *Southern Baffin Island*; 18, *Thule*; 19, *Independence Fiord*.

the definite arrival of Indians to the south of the Laurentide ice. It suggests that Upper Palaeolithic hunters had crossed Beringia and occupied the unglaciated regions of Alaska and the Yukon by the time of the Mid-Wisconsin interstade, at which time this area was geographically a part of Asia and was isolated from the rest of America by glacial ice. This is the total of our present knowledge concerning Late Pleistocene occupations of the American Arctic, and further speculation on the basis of this meagre evidence does not appear to be useful at the present time.

4. Early adaptations of American Arctic peoples

NATIVE ARCTIC POPULATIONS

During the historic period, three distinct human populations inhabited the North American Arctic. American Indian groups, speaking Athabaskan languages in the area between Alaska and Hudson Bay, and Algonkian languages from Hudson Bay to the

Atlantic coast, inhabited primarily the interior forest zone although some groups also made use of the tundra areas to the north. Eskimos occupied the coasts and adjacent tundra areas from southwest Alaska to Greenland and Labrador, as well as the tip of the Chukchi Peninsula and the islands in Bering Sea. Aleuts inhabited the entire Aleutian Island chain and the western tip of the Alaska Peninsula (fig. 15A.4). Since the archaeological evidence with which we will be dealing relates to the prehistory of these three

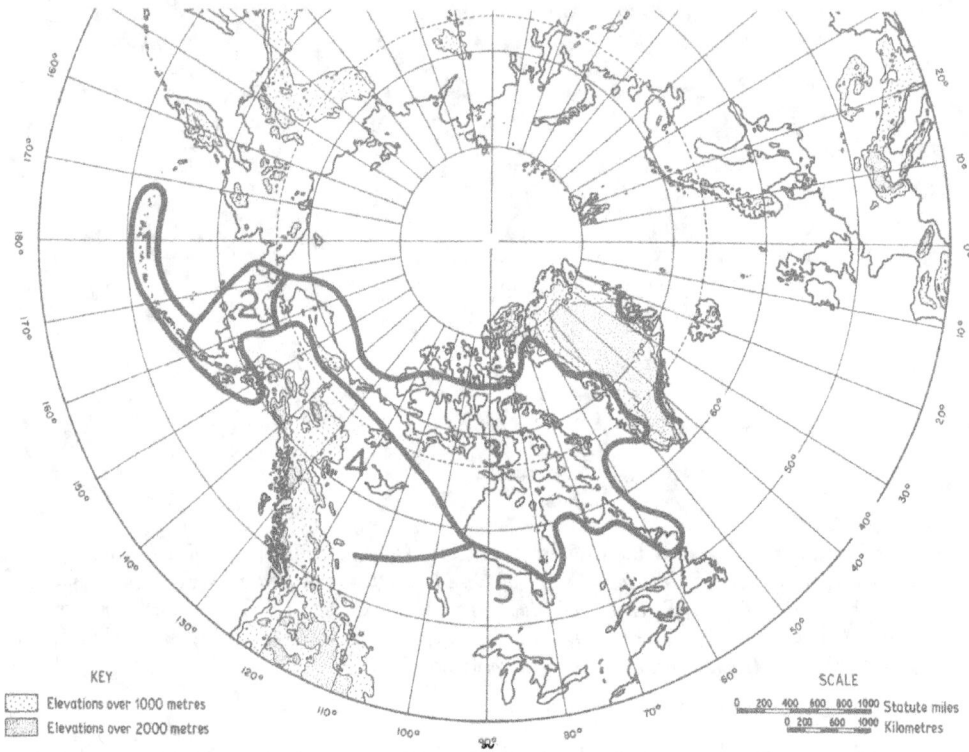

KEY
Elevations over 1000 metres
Elevations over 2000 metres

SCALE
0 200 400 600 800 1000 Statute miles
0 200 600 1000 Kilometres

Fig. 15A.4. Distribution of American Arctic populations in the Historic period: 1, *Aleut*; 2, *Eskimo, Yupik dialects*; 3, *Eskimo, Inupik dialects*; 4, *Athabaskan Indian*; 5, *Algonkian Indian.*

populations, it is useful to summarize our knowledge of the biological and linguistic relationships between northern Indians, Eskimos and Aleuts.

Biologically, the northern forest Indians are very similar to the Indian populations inhabiting the rest of North and South America. Certain physical characteristics indicate the eventual Asiatic ancestry of these populations in the period preceding the original migration across Beringia, but the immediate affinities of the northern Indians are with populations to the south rather than with modern Asiatic populations (Laughlin, 1967, pp. 412–17). The Eskimos and Aleuts, however, belong to a distinct physical type which has been called Bering Sea Mongoloid (Laughlin, 1967, pp. 417–21), and which shows much greater affinities with peoples living on the Asiatic side of Bering Strait than with American Indians. This suggests that the ancestors of the Eskimos and Aleuts reached North America at some time after the original migration of American Indians.

A parallel situation occurs in the distribution of languages spoken by these populations. The Athabaskan and Algonkian languages spoken by northern Indian groups have not yet been demonstrated to show significant relationships to any Asiatic languages, nor to the Eskimo and Aleut languages. The Eskimo and Aleut languages form a related group called Eskaleutian, and are considered to have diverged from a single language which was spoken at some time in the past. It seems likely that Eskaleutian is in turn related to the Chukotan group of languages spoken by peoples living on the Asiatic side of Bering Strait, and that these language groups diverge at an even earlier time (Swadesh, 1962).

EARLY SITES IN THE WESTERN ARCTIC: 10,000 TO 6000 BC

Following the 27,000-year-old date on the Old Crow Flats artifact, there is a 15,000-year gap in our archaeological knowledge of the North American Arctic. Securely dated sites appear only after 10,000 BC at least two millennia after Palaeo-Indian hunting cultures had become established in central North America. These early sites in the western Arctic show cultural affinities with Asiatic sites rather than with the Palaeo-Indians living to the south of the retreating Laurentide ice sheet.

Perhaps the most interesting of the early arctic assemblages is called the Akmak complex, found at the base of the deeply-stratified Onion Portage site. This site is located on the Kobuk River in northwestern Alaska, some 200 km inland from the present Chukchi Sea coast (fig. 15A.3). Over 500 stone artifacts were excavated from an erosion surface beneath stratified cultural deposits dating back to over 6000 BC (Anderson, 1970, p. 62). These artifacts include large core-bifaces which were probably used as heavy-duty cutting and chopping tools, polyhedral blade cores and distinctive narrow Campus-type microblade cores, the blades and microblades removed from these cores, broken microblades apparently designed as lateral insets for weapons, longitudinally-struck burins, various knives and scrapers, and used chert flakes (Anderson, 1970, pp. 17, 57). A small sample of bone associated with these tools yielded a radiocarbon date of 7905 ± 155 BC (Anderson, 1970, p. 70).

The earliest living-surfaces excavated at Onion Portage are deposited above the Akmak material, and dated between 6500 and 6000 BC on the basis of five radiocarbon measurements (Anderson, 1970, p. 62). The artifacts from this layer are termed the Kobuk complex, and suggest continuity with the underlying Akmak material. At the Trail Creek cave site on Seward Peninsula, 200 km southwest of Onion Portage, Larsen (1968, pp. 52-4) found broken microblade fragments of the Akmak type and a few antler weapon points with longitudinal grooves for insertion of microblades as lateral insets. These artifacts, which may be related to the Akmak complex, came from a layer which may also have yielded a bone sample dated at 7120 ± 150 BC. Similar microblades, Campus-type microblade cores, and other Akmak-like artifacts, have been collected from several undated surface sites in the Brooks Range of northwestern Alaska.

The use of microblade insets in antler weapon points, and the presence of burins which were probably used for grooving and cutting bone and antler rather than wood, are distinctive traits of arctic cultures. Anderson (1968, p. 33) suggests that these people were hunters adapted to the interior tundra zone. As we would expect, the cultural-historical relationships of Akmak appear to stretch into Asia rather than south into America. The closest similarities are found in the late Palaeolithic and Mesolithic sites of the Lake Baikal region, dating between roughly 13,000 and 8000 BC. Anderson (1970, p. 70) tentatively suggests that 'cultural crossties between interior Siberia and North-west Alaska may relate to the existence of Beringia, which did not flood until after 10,000 years ago and may reflect the existence during the Siberian Upper Palaeolithic period of a "diffusion sphere" extending from Lake Baikal to Northwest Alaska'. The apparent tundra orientation of the Akmak people suggests to Anderson (1968, p. 31) that these may have been the ancestors of Eskimo peoples who appear in the archaeological record at a much later date.

In the interior of central Alaska, Hadleigh-West (1967) has postulated the existence of a possibly contemporaneous cultural complex. The Denali complex is based on small collections from a few undated sites located on high ridges in the forested zone of the Alaskan interior; during the Denali occupation, this area may have been in a tundra or tundra–forest zone. The stone artifacts include microblades and Campus-type micro-blade cores, large blades, burins made on irregular flakes, bifacial knives, scrapers of various types, and boulder-spall scraping tools. Although there are similarities to Ak-mak, especially in the microblade cores and broken microblades, a comparison of the total assemblages suggests that the two complexes are not identical (Anderson, 1970, p. 65). Hadleigh-West (1967, p. 379) hopes that future research will decide the question of whether or not the Denali complex represents the ancestors of the Athabaskan Indians.

Another collection from the same part of Alaska may be related to the Akmak or Denali complexes. This collection comes from the base of the shallow stratified Village site at Healy Lake, 200 km southeast of Fairbanks (McKennan and Cook, 1968). In association with a series of hearths containing burned bone dated at 7010 ± 150 BC and 9140 ± 170 BC was found an assemblage containing large blades, narrow microblades, heavy endscrapers and a few thin leaf-shaped bifaces. The radiocarbon date of 9140 BC is presently the oldest known from the American Arctic, but the small size of the collection precludes meaningful comparisons.

To complete our survey of early sites in the western Arctic, we now turn to the Aleutian Islands. The site of Anangula is located on a small island off the coast of Umnak, the westernmost large island in the Aleutian chain. Seven radiocarbon dates associated with the occupation range in age between 6500 and 5700 BC (Laughlin, 1967, p. 432). During the Late Wisconsin, Anangula and Umnak Islands formed the tip of an expanded Alaska peninsula, and at the time the site was occupied it may have been separated from the mainland only by a few shallow passes. The site has yielded a large and distinc-tive collection of stone artifacts including polyhedral blade cores, flakes and blades removed from these cores, and cutting, piercing and scraping tools made by unifacially retouching these flakes and blades (Laughlin and Aigner, 1966). The latter category includes burins made by striking a spall transversely across a blade or flake, a technique which was used in Palaeolithic Japanese and Siberian cultures. The remainder of the collection consists of pebble hammerstones, notched pebble line sinkers, grinding and rubbing stones, and a fragment of a small stone lamp or bowl. Recent work at the site has yielded traces of permanent houses, and a collection of refuse bone which suggests that the Anangula people had a littoral economy using the same resources as were used by Aleuts of the historic period (Aigner, 1971).

The Anangula assemblages show no affinities to any known assemblages in either Arctic or temperate North America. The closest similarities, manifested in the blade industry and the transverse burins, are with cultures in Japan and eastern Asia, but these tenuous similarities suggest a rather distant historical relationship. Laughlin (1967) suggests that the ancestors of the Anangula people were coastal hunters who had gradually adapted to the rich marine and inter-tidal resources of the Pacific rim along the southern coast of Beringia. These people may have had little contact with the tundra hunters in the interior of Beringia. With the rising sea levels of 10,000 years ago, some of these coastal hunters were attracted to the southeastern corner of Beringia, where they were able to retain their marine-littoral adaptation in the isolated Umnak Island area. According to this reconstruction, the coastal hunters of Beringia were Bering Sea Mongoloid in physical type, and probably spoke Eskaleutian and Chukotan languages. The retreat of Beringia dispersed this population, and the people who settled at Anangula and in the adjacent area were the ancestors of the historic Aleuts.

In summary, we may state that in the period between 10,000 and 6000 BC, various populations had adapted to at least three major environmental zones in the western Arctic: the Akmak people on the tundra of northwestern Alaska and probably the adjacent areas of interior Beringia, the Denali complex people in the interior tundra or tundra–forest zone of central Alaska, and the Anangula occupants of the Umnak Island coastal zone (fig. 15A.4 provides a convenient summary of the various populations). It has been suggested that these populations may have been ancestral to historic populations who lived in the same geographical areas: respectively Eskimo, Athabaskan Indian and Aleut. Such ethnic identifications are based almost entirely on similarities in general adaptation, and thus are extremely tenuous. At our present state of knowledge, it would appear to be more useful to state only that during this period Alaska was occupied by three or more culturally distinct populations, and that the cultural heritage of these populations was rooted in eastern Asia.

5. Arctic adaptations of American Indians: 6000 BC to European contact

ALASKA AND THE YUKON: THE MOUNTAIN ZONE

After 6000 BC there is a break of over 1000 years in our archaeological knowledge of the western Arctic. The Akmak-Kobuk people no longer used the Onion Portage site, and no other sites indicating the presence of a related culture are known from this period. No occupations of this time are known from Anangula or any other sites in southwestern

Alaska. This was a period of generally warming climate throughout the American Arctic (Anderson, 1968, p. 30; Dansgaard *et al.*, 1969, p. 379). Laurentide ice was retreating rapidly toward the Keewatin Ice Divide (Bryson *et al.*, 1969), and was followed by an advancing treeline which may have reached and surpassed its present position before 4000 BC (Nichols, 1968, p. 387; also Chapter 11A). Proglacial lakes in the central Arctic may have been drained as isostatic rebound and the removal of ice dams established modern drainage systems. By 5000 BC, we may generally consider that geographical and environmental conditions were similar to those of the present day.

When the remains of human occupation again appear in the western Arctic between 5000 and 4000 BC, we can almost certainly ascribe these remains to American Indian populations. The sites known from this period are concentrated within the present forest zone, and the cultural adaptations manifested are those of forest-living people. No close historical connection can be demonstrated between these remains and those of the pre-6000 BC period discussed earlier. The interior sites and assemblages of this period have been grouped into several loose complexes, e.g. the Northern Archaic Tradition (Anderson, 1968) and the Northwest Interior Microblade Tradition (MacNeish, 1959). The present paper will not deal with these complexes, partly because the populations involved did not make extensive use of the tundra and arctic coastal zones, and partly because our present knowledge is very unsatisfactory. We may merely state that the interior forest zone of Alaska and the Yukon seems to have been continuously occupied by American Indians since at least 4000 BC, and that for a presently unknown portion of this period the occupants were Athabaskan-speaking groups.

DISTRICT OF MACKENZIE: THE BARREN GROUNDS

Although some Alaskan Indians of this period did make occasional excursions into the tundra, as indicated by sites at Cape Krusenstern (Giddings, 1960) and Anaktuvuk Pass in the Brooks Range (Campbell, 1962), they do not seem to have developed a pattern of intensive tundra occupation. An intensive use of the tundra zone by Indian populations of this period does appear to have occurred in the Barren Grounds area of the central Canadian Arctic. The earliest dated site in this area is found at Acasta Lake, 130 km southeast of the eastern end of the Great Bear Lake in an area which is presently in the tundra zone. The site consists of a series of hearths concentrated near a large quartzite erratic which was used as a source for stone artifacts; charcoal from one hearth gave a radiocarbon date of 5020 ± 360 BC (Noble, 1971). The stone industry is characterized by a large series of lanceolate weapon points including Agate Basin types and a variety of incipiently stemmed and stemmed forms, plano-convex bifaces, bifacial knives, endscrapers, gravers, twist drill, spokeshaves, transverse burins, wedges, and a variety of cores.

Noble (1971) considers that the Acasta Lake site and a number of related sites may be ascribed to a northern Plano tradition which can be ultimately derived from the Palaeo-Indian hunting cultures of the northern plains in the period preceding 5000 BC (Wormington, 1957). Such a tradition was proposed by MacNeish (1969) in order to account for the lanceolate points, similar to those of early plains cultures, found throughout the western Arctic. More recent work has shown that these lanceolate points can be ascribed to several cultural complexes, culturally diverse as well as temporally distinct. The concept of a northern Plano tradition may still be useful, however, if it is restricted to complexes such as Acasta Lake, representing the remains of caribou hunters in the Barren Grounds area of Mackenzie and Keewatin. In this area, bison hunters of the northern plains may have used the forest zone around Lake Athabaska as a winter retreat, where they could hunt both bison and caribou during the winter months. Some of these groups may have become involved in summer caribou hunting, following the migrating caribou north into the Barren Grounds and utilizing open-country hunting techniques similar to those developed for hunting bison. Eventually, they might have developed a summer utilization of the tundra similar to that of the historic Chipewyan described by Samuel Hearne (in Glover, 1958), involving a winter retreat to the forest where wood was available for fuel and shelter requirements.

Caribou hunters using various styles of lanceolate points seem to have made at least seasonal use of the Barren Grounds for much of the period between 5000 BC and European contact. Lanceolate point sites have been dated on the Great Bear River at 2850 ± 200 BC and 2700 ± 110 BC (MacNeish, 1964, p. 312), on the Coppermine River only 15 km from the arctic coast at 1190 ± 120 and 780 ± 90 BC, AD 160 ± 70, and AD 500 ± 80 (McGhee, 1970a), and around Great Slave Lake at AD 940 ± 160, AD 1070 ± 130, AD 1280 ± 70, and AD 1740 ± 80 (Noble, 1971). Noble feels that the ancestry of the Chipewyan and Yellowknife Indians who roamed the Barren Grounds during the eighteenth century can be traced for at least 2000 years in these lanceolate point hunting cultures Without definitely accepting this ethnic identification, it now appears likely that people with the general Chipewyan type of adaptation have followed the migrating caribou north to the Barren Grounds for most summers of the past 7000 years.

DISTRICT OF KEEWATIN: THE SHIELD AREA

Our knowledge of the interior District of Keewatin is even slighter than that for the western Barren Grounds. Harp's (1961) and Irving's (1968a) surveys in this area disclosed a number of sites representing caribou hunters who used lanceolate projectile points of a type termed 'Keewatin lanceolate', many of which have had burin spalls removed from one or more edges (Harp, 1961, p. 53). No dates have been obtained from any of these 'Complex B' sites, but it has been suggested (Noble, 1971) that they may be of comparable age to the generally similar Acasta Lake complex. This would suggest that tundra hunters followed the caribou eastward into Keewatin shortly after the disappearance of Laurentide ice around the Keewatin Ice Divide. We do not know how long this eastern variant of the northern Plano tradition continued to exist in Keewatin.

Wright (in press) has recently suggested that these tundra hunting people gradually changed their technology as an adaptation to the special requirements of the Canadian Shield area, and to the scarcity on these rocky plains of material suitable for a stone-chipping industry. The result of this transition is a tradition termed the Shield Archaic, characterized by some lanceolate points and side-notched points, heavy scrapers and bifaces, and a few other tool types, all usually made from quartzite or other coarse grained stone. Sites are usually found at caribou hunting localities, suggesting a similar economic adaptation as the ancestral northern Plano hunters. Wright (in press) has excavated one Shield Archaic site at Aberdeen Lake in central Keewatin. This site yielded the characteristic stone industry, as well as the remains of a shallow oval semi-subterranean house with entrance passage and central hearth; material from the hearth gave a radiocarbon date of 1075 ± 90 BC. This find might suggest that some Shield Archaic Indians were actually wintering on the tundra. The evidence is slight, and there is a possibility that gallery forest may have extended down the Thelon River as far as Aberdeen Lake at this time. This suggestion is supported by Nichols's (1968, p. 387) palynological evidence that about 1000 BC the treeline in Keewatin may have been 100 km north of its present position.

Wright considers that the Shield Archaic people abandoned Keewatin around 1000 BC, moving south into the forested Shield areas of northern Manitoba, Ontario, Quebec and Labrador where they may have been ancestral to the historic Algonkian Indians of the area.

LABRADOR

To complete this survey of American Indian adaptations to the arctic zone, we will mention another type of adaptation which occurred in the Atlantic coastal areas of Newfoundland and Labrador. Tuck (1970) has recently proposed the existence on the island of Newfoundland of a Maritime Archaic tradition in the period between roughly 2300 and 1300 BC. The people represented by this tradition were closely related to Archaic stage Indians of the Atlantic Provinces and New England, but had developed an economy based on the seasonal hunting of sea mammals using barbed and toggling harpoons, barbed darts and lances made from bone and ivory. Recent work by Tuck and Fitzhugh (1972) along the coast of Labrador indicates that Indians who may have been related to this Maritime Archaic tradition had penetrated this genuinely arctic coastal zone at least as far north as Saglek Bay (58½°N) in the period around 2000 BC. We do not yet know the nature or extent of this Indian occupation of the eastern Arctic region.

6. Arctic coastal adaptations: 4000 BC to European contact

ESKIMO ORIGINS

The origin and development of human adaptations to the arctic coastal zone is almost as puzzling at the present time as it was in the eighteenth century, when David Crantz suggested that the Eskimos had been driven to the American Arctic by political turmoil in 'Great Tartary between Mongolia and the Arctic Ocean'. The combined evidence of archaeology, linguistics and physical anthropology still points to an Asiatic rather than an American origin for the Eskimos, but the details of the problem remain elusive.

Perhaps the best place to begin an examination of the problem is in southwestern Alaska, the home of the Eskimos' biological and linguistic relatives, the Aleuts.

According to Laughlin's hypothesis discussed earlier, ancestral Aleuts settled in southwestern Alaska as Beringia submerged for the last time, and these people left archaeological remains at Anangula before 6000 BC. If this hypothesis were correct, it would seem likely that a related ancestral Eskimo population would also have been present on the coast and tundra surrounding Bering Sea at the time of its formation. Yet despite a relatively intensive search of this area, no Eskimo or Eskimo-like culture has yet been found until a much later time. The earliest site to appear in coastal southwest Alaska, other than the enigmatic Anangula occupation, is found on Kodiak Island where a small assemblage named Ocean Bay I has been dated to 3553 ± 78 BC (Clark, 1966). The collection is mainly of chipped stone tools including contracting stem points, ovate bifaces, scrapers, a few crudely ground slate items, and a small triangular lamp. Similar material is found on the adjacent coast of the Alaska Peninsula at approximately the same time period (Dumond, 1969, pp. 1110–11). Dumond (1968) has postulated that the people who left these and other remains on the Aleutian Islands were ancestral Aleuts, but this hypothesis has been questioned (Aigner et al., 1971). Since similarities between the Ocean Bay I material and later assemblages are very slight, it would seem to be more useful not to assign ethnic affiliations to Ocean Bay I, as to Anangula, until further evidence accumulates. We now know only that people with a coastal-maritime adaptation were present in southwestern Alaska at 3500 BC as well as at 6000 BC.

This takes us up to the period around 2000 BC, and from this point in time we can trace cultural sequences to the historic Aleut and Eskimo populations. In the Aleutians, the earliest occupation which can almost certainly be ascribed to ancestral Aleuts is found at the base of the deep Chaluka midden, located on Umnak Island only a few km from Anangula. Radiocarbon dating suggests that Chaluka was first occupied around 2000 BC (Laughlin, 1967, p. 439), and the site has probably been occupied almost continuously since that time. A few of the artifact types suggest to Laughlin (1967, p. 441) some continuity with the Anangula assemblage which is 4000 years older. Most of the Chaluka tool types are quite distinctive, including a wide range of chipped stone tools, barbed bone harpoons and dart heads, fish spears, compound bone fishhooks, whalebone mattocks and adze sockets, labrets, stone bowls and lamps (Aigner, 1966; Denniston, 1966). Throughout the 4000 years of occupation at the site the population had a stable maritime adaptation, using most of the extremely rich sea mammal, bird, fish and invertebrate resources of the Aleutian area (Lippold, 1966). The Aleut type of adaptation can therefore be traced back to about 2000 BC, and we can be fairly certain that the population which obtained this adaptation were biologically, linguistically and culturally the ancestors of the historic Aleuts.

At approximately the same time that Chaluka was first settled, or perhaps a few centuries earlier, an entirely different sort of occupation appeared in the tundra areas of western Alaska. Astonished by the minute size of the stone artifacts which are all that is preserved from this occupation, archaeologists (Irving, 1953) named it the Arctic Small Tool Tradition (ASTT). The extremely well made and standardized forms of chipped stone tools, including bifacial points and sideblades, microblades, angle burins, retouched burin spalls, and various small scrapers and knives, have no apparent antecedents on the American continent. Irving (1968b) has recently suggested that the ASTT may have been derived from an Interior Siberian Neolithic culture similar to the Bel'kachinsk culture discovered in the Aldan River region near Lake Baikal (Mochanov, 1970). The Bel'kachinsk culture is dated to the fourth millennium BC, and exhibits the same sort of miniaturization and extremely fine workmanship characteristics of ASTT artifacts, as well as a few artifact types which are similar to ASTT forms. Both culturally and temporally, such a Siberian Neolithic culture would seem to be the likely progenitor of the Arctic Small Tool Tradition.

The environmental region occupied by the early ASTT people (see Plate 41) was the tundra and coastal zone of western Alaska to the north of the Alaska Peninsula. There is no evidence of contact or diffusion of cultural traits between the ASTT people and the contemporaneous occupants of the Aleutians (Chaluka) or of the Pacific coast of southwest Alaska, where a series of ground slate-using cultures can be traced from approximately 2000 BC to the historic period (de Laguna, 1934; Clark, 1966). The ASTT occupation of the area of seasonally frozen coasts and the adjacent tundra suggests an adaptation based on caribou hunting and coastal sea mammal hunting, similar to that characteristic of many historic Arctic Eskimo groups. This adaptation, and the fact that cultural similarities can be vaguely traced from the ASTT through a sequence of prehistoric cultures to the historic Eskimo, suggests to many archaeologists that the ASTT was an Eskimo culture. A few human skeletal remains found in association with an ASTT-derived culture (see below) in the eastern Arctic seem to be of Eskimo physical type (Oschinsky, 1964; Harp and Hughes, 1968), suggesting that the ASTT population was biologically Eskimo. It has also been suggested that the ASTT people spoke an Eskimo language (Dumond, 1965).

The present picture of Eskimo origins might therefore consist of a Siberian Neolithic population, Bering Sea Mongoloid in physical type and speaking an Eskimo language, moving from the interior to the Bering Sea coast where they may have been attracted by rich sea mammal resources, in the period before 2000 BC. In doing so they must have given up the manufacture of pottery and ground stone tools, both of which are present in the Bel'kachinsk assemblages, and learned the techniques of coastal ice hunting which allowed them to stay in the tundra zone throughout the winter without resorting to the shelter of the forest. This movement may have marked the origin of the Eskimo type of adaptation to the arctic region. If this model is correct, we cannot explain the relationship of the ASTT people to the contemporaneous populations of the Aleutians and of the Kodiak Island–Cook Inlet region, whose cultural descendents of the historic period are biologically and linguistically Aleuts and Eskimos respectively. More work must be done in order to clarify this situation.

THE FIRST MAJOR EXPANSION OF ESKIMO CULTURE

The earliest dated ASTT occupation is at the Onion Portage site in northwestern Alaska, where it appears around 2300 BC (Anderson, 1968, p. 31), although a somewhat earlier occupation may be represented by the Denbigh Flint Complex on the Bering Sea coast (Giddings, 1964). In the one or two centuries preceding 2000 BC the ASTT population seems to have undertaken an extreme geographical expansion. ASTT sites earlier than 2000 BC have been dated at Igloolik (Meldgaard, 1962) and southern Baffin Island (Maxwell, in press) in the eastern Canadian Arctic, and in the fiord country of northeastern Greenland (Knuth, 1966). By 1900 BC ASTT material appears in the interior of the Alaska Peninsula (Dumond, 1969), but this is its furthest intrusion into the southwest Alaskan area.

There is some evidence to indicate that the Alaskan ASTT people were primarily tundra caribou hunters and perhaps river fishermen, with only a seasonal interest in coastal hunting of sea mammals (Dumond, 1969, p. 1112). When they expanded eastward across the Canadian Arctic, however, sea mammal hunting was necessarily of prime importance both as a source of food and fuel. At Igloolik, the early ASTT people were wintering on the tundra coast by living in sod-walled houses and burning oil in stone lamps for heat and light (Meldgaard, personal communication, 1969). In northern Greenland an ASTT variant called Independence I seems to have specialized in hunting musk-ox and caribou, and for fuel probably burned the driftwood which is abundant in the area (Knuth, 1966). In general, we can see various local adaptations taking place within the overall tundra-coastal orientation as ASTT people moved into and exploited different regions of the arctic zone.

We do not know the causes of the apparently rapid ASTT expansion across the North American Arctic, but this expansion may have been facilitated or even encouraged by warmer climate at the end of the postglacial climatic optimum. By 1500 BC however, the climate in Arctic Canada and Greenland appears to have cooled significantly (Dansgaard et al., 1969, p. 379; Nichols, 1968, p. 387), and at this time there is no further evidence of contact between Alaskan cultures and the ASTT populations of the central and eastern Arctic. In the period between 1500 and 1000 BC, the Canadian ASTT (Pre-Dorset) people expanded their range to occupy the coastal areas of Nouveau Québec and Labrador as far south as Hamilton Inlet (Taylor, 1968; Fitzhugh, 1970). Some groups appear to have pushed inland through the Barren Grounds of Mackenzie and Keewatin as far as the north shore of Great Slave Lake, probably temporarily displacing the northern Plano or Shield Archaic Indian populations (Harp, 1961; Noble, 1971; Nash, 1969). Southwestern Greenland was now occupied by an ASTT variant known as Sarqaq culture (Mathiassen, 1958).

During the time that the Canadian and Greenland variants of the ASTT remained isolated from Alaskan cultural developments, they gradually became more distinctive as cultural changes accumulated in response to the harsher environmental conditions of the central and eastern Arctic. In the period around 1000 BC, the rate of cultural change rapidly accelerated for no apparent reason, although it may have been related to a brief period of warmer climatic conditions at this time (Dekin, 1969). New forms of chipped stone tools appeared, the fine stone working characteristic of the ASTT declined, ground slate tools became more abundant, ground burins replaced chipped burins, microblades appeared more frequently, and there were stylistic changes in tools and weapons made from antler, bone and ivory (Taylor, 1968; Meldgaard, 1962; Maxwell, in press). Rectangular semi-subterranean houses came into use, and the snow-house may have been invented at this time (Meldgaard, 1962).

The result of these changes was the Dorset culture, a very distinctive eastern Arctic variant of the ASTT. Dorset culture may have originated around Foxe Basin, but within a few centuries it had spread west as far as Dolphin and Union Strait, north to Greenland, down the Labrador coast and around the island of Newfoundland. Our only evidence that the ASTT population was Eskimo in physical type (see above) comes from a few skeletons associated with Dorset culture material in Nouveau Québec and New-foundland. The Dorset occupation of the eastern Arctic can be traced from roughly 1000 BC to AD 1000; although stylistic changes did take place over this span of time, the rate of change decreased greatly after the initial acceleration. For the following 2000 years, Dorset culture seems to have remained a relatively stable adaptation to the unique environmental conditions of the eastern Arctic coastal zone.

FURTHER DEVELOPMENT IN ALASKA

While the ASTT Pre-Dorset and Dorset cultures were slowly developing in the isolation of the eastern Arctic, more extensive changes were occurring among the populations of western Alaska. This difference in the rate of culture change may have been due to the richer resource base of the Bering Sea area, and possibly to the availability of contacts with Asiatic and North Pacific peoples. The ASTT seems to have lasted for a relatively short period in this area, disappearing after 1800 BC at Onion Portage (Anderson, 1968, p. 28), although it may have lasted much longer in peripheral areas such as the Alaska Peninsula (Dumond, 1969, p. 1110). The subsequent sequence of cultures is not yet well known, although we now know that an understanding of this sequence is crucial to our knowledge of Eskimo cultural development. At present, however, we can only give a sketchy outline of the major features of this sequence.

The earliest material to appear in the Bering Strait region after the ASTT occupation has been found at a single site on the beach ridges of Cape Krusenstern. This site, charac-terized by semi-subterranean houses, large triangular and side-notched projectile points, and the bones of large baleen whales, has been dated to 1800 BC and named the Old Whaling culture (Giddings, 1967, pp. 223–45). This is our first evidence of American Arctic peoples who may have hunted the larger sea mammals, perhaps by chasing Bow-head whales from skin boats, harpooning them, and eventually towing them to shore. In any case, the site represents a single isolated occurrence of this type of hunting at this time period, and we have no evidence to indicate that the technique was passed on to later occupants of the area.

The next recognized occupation is named Choris culture, and has been found at several coastal and inland sites in western Alaska where it dates between approximately 1500 and 500 BC (Giddings, 1967, pp. 200–22; Anderson, 1968, pp. 31–2). The Choris people lived in large oval multi-family houses, and seem to have concentrated on hunting cari-bou. Their stone technology appears to be a blend of ASTT forms with other forms which may have been derived from interior Indian cultures. Especially characteristic are the large finely-made projectile points of lanceolate types almost identical to those used by Palaeo-Indian cultures of central North America some 5000 years earlier; the similarity is almost certainly due to chance convergence rather than to historical–cultural relationships. The Choris people were also the first American Arctic people to use pottery, in the form of round-based cooking pots made from organically-tempered clay and decorated on the exterior with cord markings and linear stamps. This pottery is very similar to that used by Siberian Neolithic peoples of the same period, and its appearance in Alaska almost certainly reflects contact across Bering Strait.

The next distinctive culture to be recognized in western Alaska is probably one of the more important in the history of Eskimo development. This is termed Norton culture, dating roughly within the period 500 BC to AD 500, and is found at many coastal sites from the Alaska Peninsula (Dumond, 1969, p. 1110), around Norton Sound and through Bering Strait (Giddings, 1964, 1967), and along the arctic coast to the western Canadian Arctic. Dumond (1965, p. 1246) considers the Norton people to have been at least partial cultural ancestors of all historic Eskimo-speaking peoples, and that the linguistic and cultural diversity of historic Eskimo populations is largely the result of divergence which has occurred since the Norton period. Our present limited knowledge suggests that 'Norton culture' actually represents a number of local or regional populations which shared several elements of material culture. These traits include a chipped stone industry derived from an ASTT ancestry, a ground slate industry which may have been derived from the Pacific area of southwestern Alaska, pottery marked by check-stamped (waffle) decoration and which ultimately must have come from Asiatic peoples. Certain traits which are characteristic of later Eskimo cultures, such as labrets and stone lamps, may also have been introduced from the Pacific area where there is now evidence of contact between Bering Sea and Pacific populations across the Alaska Peninsula (Dumond, 1969, p. 1111).

A very distinctive culture, which may perhaps be thought of as a Norton variant, occurs at a few sites in northwestern Alaska and is well known from the excavation of the large Ipiutak site at Point Hope (Larsen and Rainey, 1948). The site consists of over 600 rectangular semi-subterranean houses, and seems to date from the first few centuries of the Christian era. Although the culture is generally similar to Norton, there is a surprising absence of ground stone tools, pottery and oil lamps. The economy of the Ipiutak people seems to have been based on interior caribou hunting and coastal hunting of seal and walrus; the Ipiutak people, like other Norton populations, do not seem to have hunted the large baleen whales which were the mainstay of later occupants of the area. The most exciting feature of Ipiutak culture is the large collection of carved ivory objects, generally associated with burials and including death masks, carvings of fantastic animals, chains, swivels, and a variety of other artifacts. This material has no counterpart in the American Arctic, and stylistic resemblances have been suggested with the Scytho-Siberian animal style of central Asia (Larsen and Rainey, 1948, p. 149).

Certain features also suggest contact between Ipiutak culture and other Eskimo developments on the Siberian side of Bering Strait, in the area of St Lawrence and

Diomede islands and the coastal area of Chukchi Peninsula. In this area, between roughly 300 BC and AD 500, an elaborate ivory carving industry was being created in two stylistic variants termed Okvik and Old Bering Sea (Collins, 1937; Rudenko, 1961; Giddings, 1967, pp. 151–74). We do not yet know the background of the Okvik–Old Bering Sea people; their culture may have developed from a Norton base, but the use of iron tools and, at a later period, slat armour made from strips of bone, suggests contact with Asiatic peoples. The Okvik–Old Bering Sea development was of great importance for the lifeways of all subsequent Eskimo groups on both sides of Bering Strait. In the deposits of this period we find prototypes of many items of later Eskimo technology including snow goggles, sleds, skin boats, and sea mammal hunting equipment such as toggling harpoon heads and harpoon float gear. Perhaps of greatest importance, these people had learned to hunt large baleen whales in the open sea from skin boats. Intensive open-water hunting of sea mammals provided them with a rich economic base which allowed them to live in large permanent villages. These villages of semi-subterranean houses built of sod over a log framework, underlain by a deep deposit of collapsed houses and accumulated refuse, are the most distinctive mark of the next major expansion of Eskimo culture.

THE SECOND MAJOR EXPANSION OF ESKIMO CULTURE

Returning to the Alaskan side of Bering Strait, the next link in the chain of northern Eskimo development is known as Birnirk culture. Originally identified at a large permanent village site near Point Barrow (Ford, 1959), Birnirk culture has a coastal distribution in northwestern Alaska and seems to date between approximately AD 500 and 900. The Birnirk people hunted caribou and small sea mammals, as well as the large baleen whales which migrate along the North Alaskan coast each spring and autumn. Whaling was probably done from large skin boats (umiat) launched from the edge of ice leads, in the same way that it is still done in several North Alaskan communities. Whaling appears to have been more important to the Birnirk people than it had been to the Okvik–Old Bering Sea populations, and this rich resource allowed the Birnirk people to live in large permanent villages located at good whaling sites.

Although Birnirk culture may have developed from a Norton base, perhaps with some influence from the North Alaskan Ipiutak culture, Ford (1959) sees its formation as largely due to cultural influences from the Old Bering Sea and subsequent Punuk cultures on the Asiatic side of Bering Strait. From this source may be traced the crude black pottery lamps and cooking pots, sometimes decorated with a stamped concentric circle design, the toggling harpoon heads and other gear used for hunting sea mammals, the rectangular semi-subterranean houses with rear sleeping platforms and inclined entrance passages, and many other items of material culture. The technique of hunting Bowhead whales may have been learned from the same source, and the intensification of whaling along the North Alaskan coast served as a springboard for a major territorial expansion.

Shortly after AD 900, there appears to have been a rapid warming trend in the climate of arctic North America. This trend probably resulted in a decrease in arctic pack ice, and may have encouraged an expansion in the population and feeding range of baleen whales and other sea mammals which migrate north through Bering Strait each spring (McGhee, 1970b). There is a noticeable change in the culture of the people inhabiting the northern Alaskan coast at this time, with the development of what is known archaeo-logically as Thule culture (Mathiassen, 1927). Most Birnirk artifact classes continue into the subsequent Thule culture, but there are stylistic changes, a general decrease in the amount of ornamentation applied to functional artifacts, and an increase in the pro-portion of artifacts relating to the hunting of Bowhead whales.

The North Alaskan coast, perhaps as far west as Bering Strait, seems to have been the centre of Thule development. The archaeology of the west Alaskan coast between Bering Strait and Alaska Peninsula is not well known, but Thule influence is recognized as far south as Alaska Peninsula (Dumond, 1969, p. 1113). It is assumed that this influence took the form of the diffusion of a few Thule culture elements among a population developing from a Norton culture base to the historic Eskimos of this region who spoke dialects of the Yupik Eskimo language (Dumond, 1965, pp. 1242–6). Some Thule elements, including coarse black pottery and cairn burials, even penetrated to the Pacific coast area of the Alaska Peninsula and Kodiak Island, and Dumond (1965, p. 1243) suggests that the Yupik Eskimo language also spread to the Pacific area of southwestern Alaska at this time. This hypothesis seems tenuous in view of the strong cultural continuities between Koniag and Pre-Koniag occupations of the past 3000 years recovered by Clark (1966) on Kodiak Island.

The major thrust of Thule expansion took place north of Bering Strait, eastward across the Canadian Arctic as far as Greenland where early Thule sites are dated to around AD 950 (Meldgaard, personal communication, 1969). This area of eastward Thule expansion almost coincides with the area occupied historically by people speaking the Inupik Eskimo language, and we assume that all these people are descended from the early Thule migrants. By tracing early styles of harpoon heads, found in North Alaskan Birnirk deposits as well as at early Thule sites in Greenland, we may suggest that the Thule expansion took the form of a migration eastward along the arctic coast from North Alaska to Amundsen Gulf, and then northward and eastward across Melville Sound, Lancaster Sound and Baffin Bay (McGhee, 1970b, p. 178). This distribution may have followed the summer feeding range of the western Bowhead and eastern Greenland whales, a range which may have been expanded due to decreased pack ice at this time. Thule winter villages of sod houses with wood or whale bone supports (see Plate 42) are scattered along most of the route, and the whale bones and whaling equipment found in the village deposits indicate that the Thule people based a good deal of their subsistence on whaling.

The Thule migrants to the eastern Arctic must have encountered the native inhabitants of the area, the Dorset culture people who had lived here in isolation for the past 3000 years. We know very little about the nature of these contacts. Soapstone pots and lamps, dog-sledding gear, and the knives used to build snow houses first appear at Canadian Thule sites and at slightly later sites in northern Alaska. Were these tools and techniques learned by the Thule people from the Dorset inhabitants of the eastern Arctic? If so, they appear to be isolated examples of Dorset influence, since the material culture of Canadian and Greenland Thule people almost totally reflects their Alaskan ancestry. It seems likely that the Dorset people, at a technological disadvantage which was perhaps emphasized by the changed environmental conditions after AD 900, were overwhelmed by the Thule immigrants and left no cultural or biological descendants.

The Thule whaling economy, however rich and secure during the climatic amelioration of the tenth to twelfth centuries AD, must have become increasingly precarious after AD 1200 when climatic trends began to reverse. In the subsequent period of cooling climate and increasing sea ice, the uniform nature of early Thule culture broke down as various Thule groups began to change their adaptations to suit the changing environmental conditions of specific local areas (McGhee, 1970b, pp. 180–1). In areas such as North Alaska, western Greenland and Labrador, the environmental changes did not preclude a whaling economy, and in these areas a basically Thule way of life continued to the historic period. In the central Arctic however, whaling declined or became impossible as ice increased in extent and in seasonal duration. Previously occupied areas along Lancaster Sound and in northern Greenland were abandoned, and the Thule population moved southward to areas where they could live by hunting caribou, fishing, and hunting the smaller sea mammals. The 'classic' Central Eskimo way of life, involving summer hunting in the interior and winter sealing from snow house villages on the sea ice, seems to have developed during the 'Little Ice Age' of the sixteenth to eighteenth centuries.

During the eighteenth century European influence began to be felt throughout the Eskimo world, first in the form of articles traded into the area, then by epidemic diseases which decimated the populations of some regions, and finally in the constantly accelerating intrusion of European individuals and institutions. By the early twentieth century, purely Eskimo cultures no longer existed in any part of the arctic world.

7. Conclusions

The arctic zone provides conditions which permit intensive predation by human populations with the appropriate hunting knowledge and technology. These conditions have periodically attracted human populations to an area where a rich and rewarding existence could be secured, probably with less labour and hardship than was possible in many subarctic or temperate regions.

The American Arctic has supported such human occupations for at least the greater part of the past 30,000 years (see fig. 15A.5). These populations exploited the arctic zone in terms of three distinct adaptations: a) the early occupants of Beringia between perhaps 30,000 and 8000 years ago, and including the ancestors of the American Indians, appear to have been tundra or tundra–steppe hunters of Late Pleistocene fauna; b) various American Indian groups of the past 7000 years have used the tundra areas of Alaska and central Canada as a summer hunting ground, retreating with the caribou to the forest zone during the winter; c) various Asiatic-derived groups, including ancestral Eskimos and Aleuts of at least the past 4000 years, developed a total tundra-coastal adaptation through intensive hunting of sea mammals.

Although prehistoric population estimates are extremely tenuous, being based merely on an estimation of the number and size of known archaeological sites, it seems likely that at some periods the prehistoric population of the arctic zone was on the same order

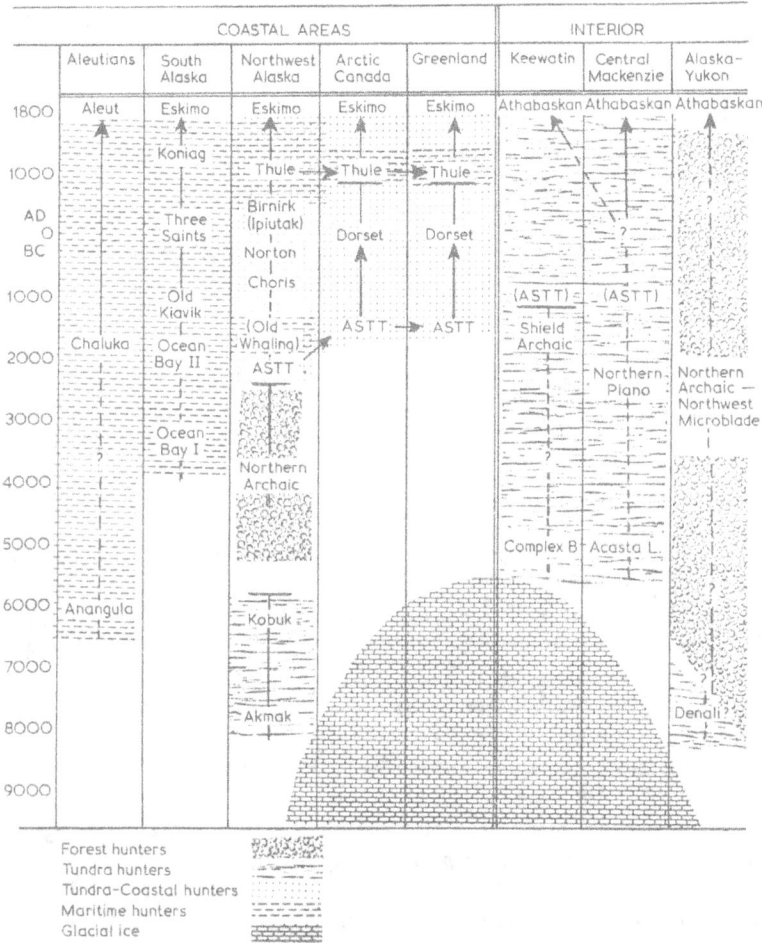

Fig. 15A.5. Summary of North American arctic prehistory.

of size as at present. These prehistoric populations subsisted almost entirely on the resources of the arctic zone without massive imports from temperate regions, and did so in apparent equilibrium with arctic resources. We might expect, however, that these aboriginal adaptations were more fragile and more subject to slight environmental changes than is the present arctic adaptation which is based on a large and stable reservoir of world resources.

Finally, in view of recent concern over the impact of modern technology on the tundra environment and the question of recovery-time related to traces of human activity, it is interesting to note certain archaeological evidence. Archaeology is based on the analysis of traces of human activity; in the American Arctic, such traces survive at least as long as the 27,000-year-old bone scraper from Old Crow Flats. Prehistoric Indian occupations and Eskimo occupations prior to the establishment of permanent villages, however, leave almost no traces on the surface of the present tundra; archaeological sites relating to these occupations are quite difficult to locate. The building of permanent winter houses involving shallow excavation in the earth, on the other hand, has left obvious surface remains over a period of 2000 years around Bering Strait (Old Bering Sea culture), and 1000 years across the Canadian Arctic (Thule culture). Despite the fact that these houses are built entirely of organic materials, they are quite obvious 'scars on the tundra' covered by thick vegetation enriched by the slowly decomposing refuse deposits. Prehistoric villages of this type attract the attention of archaeologist and non-archaeologist throughout the arctic zone. It may be comforting to imagine that our descendants of 1000 years hence will regard the traces of twentieth century arctic technology with similar curiosity and interest.

References

AIGNER, J. S. (1966) Bone tools and decorative motifs from Chaluka, Umnak Island. *Arct. Anthrop.*, **3** (2), 57–83.

AIGNER, J. S. (1971) Report in 'Current Research, Arctic Area'. *Amer. Antiq.*, **36**, 489–90.

AIGNER, J. S., LAUGHLIN, W. S. and BLACK, R. F. (1971) Early racial and cultural identifications in southwestern Alaska. *Science*, **171**, 87–8.

ANDERSON, D. D. (1968) A Stone Age campsite at the gateway to America. *Sci. Amer.*, **218** (6), 24–33.

ANDERSON, D. D. (1970) Akmak, an archaeological assemblage from Onion Portage, northwest Alaska. *Acta Arctica*, Fasc. **16**. 80 pp.

BRYAN, A. L. (1969) Early Man in America and the late Pleistocene chronology of western Canada and Alaska. *Current Anthrop.*, **10** (4), 339–65.

BRYSON, R. A., WENDLAND, W. M., IVES, J. D. and ANDREWS, J. T. (1969) Radiocarbon isochrones on the disintegration of the Laurentide Ice Sheet. *Arct. Alp. Res.*, **1** (1), 1–14.

CAMPBELL, J. M. (1962) Cultural succession at Anaktuvuk Pass, Arctic Alaska. *Arct. Inst. N. Amer. Tech. Pap.*, 11, 39–54.

CLARK, D. W. (1966) Perspectives in the prehistory of Kodiak Island, Alaska. *Amer. Antiq.*, **31** (3), 358–71.

COLLINS, H. B. (1937) The archeology of St Lawrence Island, Alaska. *Smithsonian Inst. Miscell. Publ.*, 96, 1. 431 pp.

DANSGAARD, W., JOHNSEN, S. J., MØLLER, J. and LANGWAY, C. C. JR (1969) One thousand centuries of climatic record from Camp Century on the Greenland Ice Sheet. *Science*, **166**, 377–80.

DEKIN, A. (1969) *Paleo-Climate and Prehistoric Cultural Interaction in the Eastern Arctic.* Paper read at 34th annual meeting of the Society for American Archaeology, Milwaukee, Wisc.

DE LAGUNA, F. (1934) *The Archaeology of Cook Inlet, Alaska.* Philadelphia: Univ. of Pennsylvania Press. 263 pp.

DENNISTON, G. B. (1966) Cultural change at Chaluka, Umnak Island: stone artifacts and features. *Arct. Anthrop.*, **3** (2), 84–124.

DUMOND, D. E. (1965) On Eskaleutian linguistics, archaeology and prehistory. *Amer. Anthrop.*, **65** (5), 1231–57.

DUMOND, D. E. (1968) Eskimos and Aleuts. *Proc. 8th Int. Congress Anthrop. and Ethnol. Sci.*, **3**, 103–7.

DUMOND, D. E. (1969) Prehistoric cultural contacts in southwestern Alaska. *Science*, **166**, 1108–15.

FITZHUGH, W. (1972) Environmental archeology and cultural systems in Hamilton Inlet, Labrador. *Smithsonian Contrib., Archeol.*, **16**. 299 pp.

FORD, J. A. (1959) Eskimo prehistory in the vicinity of Point Barrow, Alaska. *Amer. Mus. Nat. Hist., Anthrop. Pap.*, **47**, Part 1.

GIDDINGS, J. L. (1960) The archaeology of Bering Strait. *Current Anthrop.*, **1** (2), 121–38.

GIDDINGS, J. L. (1964) *The Archaeology of Cape Denbigh.* Providence: Brown Univ. Press. 331 pp.

GIDDINGS, J. L. (1967) *Ancient Men of the Arctic.* New York: Knopf. 367 pp.

HADLEIGH-WEST, F. (1967) The Donnelly Ridge Site and the definition of an early core and blade complex in central Alaska. *Amer. Antiq.*, **32** (3), 360–82.

HALL, E. S. JR (1969) 'Comment' on Bryan (1969). *Current Anthrop.*, **10** (4), 352–3.

HARP, E. (1961) The archaeology of the Lower and Middle Thelon, NWT. *Arct. Inst. N. Amer. Tech. Pap.*, 8. 74 pp.

HARP, E. and HUGHES, D. R. (1968) Five prehistoric burials from Port Aux Choix, Newfoundland. *Polar Notes*, **8**, 1–47.

HAYNES, C. V. (1969) The earliest Americans. *Science*, **166**, 709–15.

HEARNE, S. (1958) *In:* GLOVER, R. (ed.) *A Journey from Prince of Wales' Fort to the Northern Ocean, 1769–1770–1771–1772.* Toronto: Macmillan.

HOPKINS, D. M. (1967a) Quaternary marine transgression in Alaska. *In:* HOPKINS, D. M. (ed.) *The Bering Land Bridge*, 47–90. Stanford: Stanford Univ. Press.

HOPKINS, D. M. (1967b) The Cenozoic history of Beringia: a synthesis. *In:* HOPKINS, D. M. (ed.) *The Bering Land Bridge*, 451–84. Stanford: Stanford Univ. Press.

IRVING, W. N. (1953) Evidence of early tundra cultures in northern Alaska. *Anthrop. Pap. Univ. of Alaska*, **1** (2), 55–85.

IRVING, W. N. (1968a) The Barren Grounds. *In:* BEALS, C. S. and SHENSTONE, A. (eds.) *Science, History and Hudson Bay*, Vol. 1, 26–54. Ottawa: Queen's Printer.

IRVING, W. N. (1968b) The Arctic small tool tradition. *Proc. 8th Int. Congr. Anthrop. and Ethnol. Sci.*, **3**, 340–2.

KNUTH, E. (1966) The ruins of the Musk-Ox Way. *Folk*, **8–9**, 191–219.

LARSEN, H. (1968) Trail Creek. *Acta Arctica*, Fasc. **15**.

LARSEN, H. and RAINEY, F. (1948) Ipiutak and the Arctic whale hunting culture. *Amer. Mus. Nat. Hist., Anthrop. Pap.*, **42**. 276 pp.

LAUGHLIN, W. S. (1967) Human migration and permanent occupation in the Bering Sea area. *In:* HOPKINS, D. M. (ed.) *The Bering Land Bridge*, 409–50. Stanford: Stanford Univ. Press.

LAUGHLIN, W. S. and AIGNER, J. S. (1966) Preliminary analysis of the Anangula unifacial core and blade industry. *Arct. Anthrop.*, **3** (2), 41–56.

LIPPOLD, L. K. (1966) Chaluka: the economic base. *Arct. Anthrop.*, **3** (2), 125–31.

MCGHEE, R. (1970a) Excavations at Bloody Falls, NWT, Canada. *Arctic Anthrop.*, **6** (2), 53–72.

MCGHEE, R. (1970b) Speculations on climatic change and Thule Culture development. *Folk*, 11–12, 172–84.

MCKENNAN, R. A. and COOK, J. P. (1968) Prehistory of Healy Lake, Alaska. *Proc. 8th Int. Congr. Anthrop. and Ethnol. Sci.*, **3**, 182–4.

MACNEISH, R. S. (1959) A speculative framework of northern North American prehistory as of April 1959. *Anthropologica* (n.s.), **1**, 7–24.

MACNEISH, R. S. (1964) Investigations in southwest Yukon. *Papers on the Robert S. Peabody Foundation for Archaeology*, **6** (2). 488 pp.

MATHIASSEN, T. (1927) The archaeology of the Central Eskimo. *Report of the 5th Thule Expedition, 1921–4*, Vol. 4 (1). 327 pp.

MATHIASSEN, T. (1958) The Sermermiut Excavations, 1955. *Medd. om Grønland*, **161** (3).

MAXWELL, M. S. (n.d.) *Archaeology on the South Coast of Baffin Island, Seasons 1960–1963* (in press). Ottawa: Nat. Mus. Man.

MELDGAARD, J. (1962) On the formative period of the Dorset Culture. *Arct. Inst. N. Amer. Tech. Pap.*, 11, 92–5.

MOCHANOV, IU. A. (1970) The Bel'kachinsk Neolithic culture on the Aldan. *Arct. Anthrop.*, **6** (1), 104–14.

MÜLLER-BECK, H. (1967) On migrations of hunters across the Bering Land Bridge in the Upper Pleistocene. *In:* HOPKINS, D. M. (ed.) *The Bering Land Bridge*, 373–408. Stanford: Stanford Univ. Press.

NASH, R. J. (1969) The Arctic small tool tradition in Manitoba. *Dept of Anthrop., Univ. of Manitoba, Occas. Pap.*, **2**. 166 pp.

NICHOLS, H. (1968) Pollen analysis, paleotemperatures, and the summer position of the arctic front in the post-glacial history of Keewatin. *Bull Amer. Met. Soc.*, **49** (4), 387–8.

NOBLE, W. C. (1971) Archaeological surveys and sequences in central district of Mackenzie, NWT. *Arct. Anthrop.*, **8**, 102–35. 122 pp.

OSCHINSKY, L. (1965) *The Most Ancient Eskimos*. Ottawa: Can. Res. Center for Anthrop.

RUDENKO, S. I. (1961) The ancient culture of the Bering Sea and the Eskimo problem. *Anthrop. of the North: Transl. from Russian Sources*, 1. Toronto: Arct. Inst. N. Amer. 240 pp.

SCHWEGER, C. (1969) 'Comment' on Bryan (1969). *Current Anthrop.*, **10** (4), 359.

SWADESH, M. (1962) Linguistic relations across Bering Strait. *Amer. Anthrop.*, **64** (6), 1262–91.

TAYLOR, W. E. JR (1968) The Arnapik and Tyara Sites, an archaeological study of Dorset culture origins. *Soc. Amer. Archaeol. Mem.*, **22**. 129 pp.

TUCK, J. A. (1970) An Archaic Indian cemetery in Newfoundland. *Sci. Amer.*, **222** (6), 112–21.

WORMINGTON, H. M. (1957) *Ancient Man in North America* (4th edn). Denver: Denver Mus. Nat. Hist. 322 pp.

WRIGHT, J. V. (n.d.) *The Shield Archaic* (in press). Ottawa: Nat. Mus. Man.

Manuscript received April 1971.

Prehistoric occupation of the alpine zone in the Rocky Mountains

Wilfred M. Husted

Mid-Atlantic Region,
National Park Service, Philadelphia

The study of prehistoric occupation of alpine zones in the United States is an area of inquiry that has not received the attention due such a surprisingly rich archaeological resource. Admittedly most sites are not spectacular, and a majority produce small numbers of artifacts. However, the apparent large numbers of sites that exist above timberline, the great time depth represented by the artifacts, and the interpretive possibilities present in this material make investigation of alpine zone archaeology a rewarding endeavour. Nevertheless the mountain environment above timberline has remained unknown archaeologically until fairly recently.

Past investigations consisted of surface surveys of more or less limited duration and extent. Ives's (1941, 1942) investigations centred in the Front Range south of Rocky Mountain National Park in northern Colorado. Moomaw (1954, 1957) studied stone alignments and collected surface materials from sites above timberline in the park. Some information on archaeological sites was gathered incidental to investigations of other alpine phenomena in the Sangre de Cristo range of New Mexico (Wendorf and Miller, 1959). Husted (1962) studied collections from alpine zone sites and conducted surveys above timberline in Rocky Mountain National Park. The locations of these and more recent investigations mentioned below are given in fig. 15B.1.

Within the past few years there has been an increase of interest in mountain archaeology. Preliminary surveys of the Beartooth Mountains in Montana (Berry, 1970), the Sapphire and Bitterroot Mountains of western Montana (Fredlund, 1970), and mountain ranges in Arizona and New Mexico (Morris, 1970) have been accomplished. Archaeological excavations have been initiated at several sites in the alpine zone of the Front Range in northern Colorado (Benedict, 1969, 1970; Olson and Benedict, 1970).

Although intensive investigation of the Front Range alpine zone has only begun, initial findings and data accumulated during the various surveys in the past indicate that the prehistory of the Rocky Mountains of northern Colorado and adjacent Wyoming is intimately tied to that of the northwestern Plains. The archaeological sequence determined for the Plains is closely paralleled by that from the mountains. The similarity between the two is such that homogeneity of culture between the mountains and the Plains probably obtained during much of the prehistoric period.

A recurring theme in many investigations to date has been the question of utilization of high mountain regions as refuges during the Altithermal. Interest in the archaeology

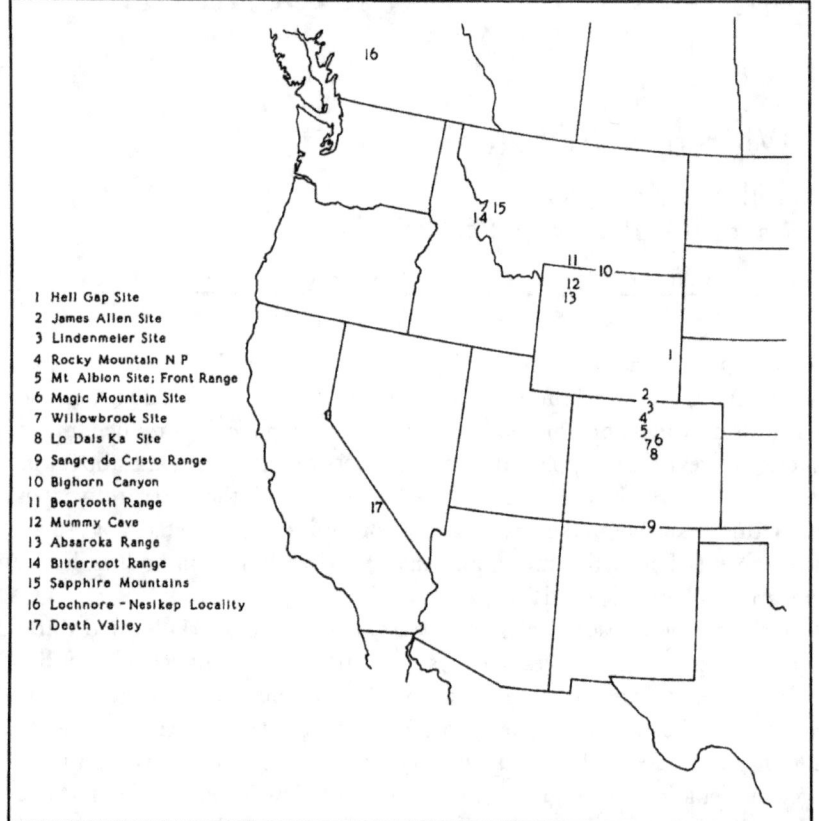

1 Hell Gap Site
2 James Allen Site
3 Lindenmeier Site
4 Rocky Mountain N P
5 Mt Albion Site; Front Range
6 Magic Mountain Site
7 Willowbrook Site
8 Lo Dals Ka Site
9 Sangre de Cristo Range
10 Bighorn Canyon
11 Beartooth Range
12 Mummy Cave
13 Absaroka Range
14 Bitterroot Range
15 Sapphire Mountains
16 Lochnore - Nesikep Locality
17 Death Valley

Fig. 15B.1. Locations of sites and areas mentioned in the text.

of the middle and northern Rocky Mountains seems to have been stimulated by a substantial increase in the knowledge of the prehistory of the high Plains immediately to the east, and the realization that there was indeed a human reaction to an environmental deterioration of such proportions that large interior areas were left virtually unoccupied. The occurrence of an Altithermal of a severity great enough to force man from the Plains has not been universally accepted (Conner, 1968, pp. 13–15). However, a conspicuous and consistent gap in the western Plains radiocarbon chronology, the appearance in the central and northern Rockies of artifact complexes not found to the east or west at coincident time levels, and the sudden appearance of a distinctive artifact complex over a large area of the Plains at about 3000 BC, make it readily apparent that the hiatus in human occupation of the Plains was real and not a product of sampling error or interpretation.

Any study of the archaeology of the Front Range limited to the alpine zone would be unrealistic. Such an approach would lend the impression that the prehistory of the region above timberline was in some manner unique or at least very different from the lower lands around the mountains. This is not entirely untrue; however, the alpine tundra now appears to have formed only a part of the total environmental range exploited by the prehistoric inhabitants of northern Colorado and adjacent regions to the north, south and west after 3000 BC. The similarity between archaeological materials from the mountains and the Plains suggests that the alpine zone was one stop in a cycle of seasonal movement between the mountains, valleys and plains. Under this assumption, this section will discuss the archaeological evidence from the Front Range alpine zone in terms of changes in use through time and the causes of these changes. In order to do this it will be necessary to dwell at some length on the archaeology of the northwestern Plains.

Earlier it was concluded that occupation of the alpine region was transitory and limited for the most part to travel back and forth across the Front Range between the Plains and the mountain parks to the west (Husted, 1962). Some sites in high valleys with no access to passes suggested hunting camps, but these were few in number and very small, yielding few artifacts. No evidence of intensive occupation was noted although a considerable amount of stone artifacts, chipping debris, and pottery had been accumulated over the years by personnel and visitors in Rocky Mountain National Park. This material came from widely scattered locations with large concentrations being collected from only a few sites. The situations of these sites were such that the greater number of artifacts found was attributed to favourable camping locations along trans-mountain routes.

It now appears that there was an indigenous population in the Front Range and surrounding area during a portion of the Altithermal and later. Recent investigations have revealed the presence of game drives and campsites on high ridges and passes (Benedict, 1969). Diagnostic artifacts from sites on the tundra compare favourably with those from excavated sites at lower elevations in the mountains and Plains where cultural continuity from about 3000 BC onward is strongly suggested. The alpine archaeological sequence appears to be practically identical to that documented on the northwestern Plains of Colorado, Montana and Wyoming, with few exceptions. This is not to say that patterns of use of these two environments were identical. The alpine tundra zone differs greatly from the grassed Plains in climate and resources. Differences in exploitation of each would be expected.

There are few radiocarbon-dated sites in the Front Range. In order to arrive at a chronology for this region, diagnostic artifacts from the tundra sites must be crossdated with identical artifacts from dated sites elsewhere. In the northwestern Plains the projectile point has been the only implement demonstrated to have diagnostic value for crossdating. These tools were used throughout the prehistoric period, and changes in form occurred within relatively short periods of time. With one or two exceptions a particular projectile point form or type persisted no longer than about 500 years. Each type had a circumscribed geographic distribution. Some types were found to extend beyond the northwestern Plains, but wherever found in datable contexts, the age of specimens of the same type from widely separated loci within the geographic range of distribution had similar ages. That is, the oldest generally were not more than 400 to 500 years older than the youngest. As noted above, there are some exceptions.

Most of the archaeological complexes comprising the sequence for the northwestern Plains carry the name of the projectile point type common to a particular complex. This sequence appears well established although some yet to be discovered manifestations probably exist. A succession of complexes with a reasonably adequate ladder of radiocarbon dates has been identified, and a picture of prehistoric cultural developments is beginning to emerge. For the present purpose this sequence can be divided into three segments or time periods: an early period of big game hunting dating from 9500 to 9000 BC until about 5500 BC, an interval of non-occupancy from 5500 BC to 3000 BC corresponding to the Altithermal, and the period since 3000 BC during which a hunting and gathering economy predominated.

The earliest period is best illustrated at the Hell Gap Site near Guernsey in eastern Wyoming. Fig. 15B.2 illustrates the general sequence of archaeological complexes in the northwestern Plains. The lower six complexes in this column were represented at Hell Gap (Irwin and Brew, 1968). Whether or not cultural continuity or relationship is represented by this progression is debatable. However, it is generally agreed that each complex represented people whose principal means of obtaining food was hunting. Probably a wide range of animal and vegetable foods were exploited, but the evidence for hunting outweighs that for other economic pursuits. There is some evidence that toward the end of this period a trend toward greater utilization of plant resources was underway (Husted, 1969).

Excavations in Bighorn Canyon in north-central Wyoming and adjacent Montana revealed a series of five complexes not known on the Plains proper (Husted, 1969). The exact nature of their relationship to earlier Plains culture was not determined; however, a general relationship was indicated. All five complexes fell within a very limited time span as indicated by radiocarbon dates. The oldest were dated 8690 ± 100 BP (SI-98) and 8600 ± 100 BP (SI-101). The most recent had carbon-14 ages of 7160 ± 180 BP (SI-240) and 8040 ± 220 BP (SI-241). The latter two dates are suspect because the second oldest of the five complexes was dated at 7960 ± 150 BP (SI-308). Radiocarbon ages for some of these complexes found elsewhere in the Bighorn and Absaroka ranges suggest that the three more recent manifestations most likely date between 8000 and 7500 years ago or slightly later.

Grinding stones and hand stones were associated with all but one of the Bighorn Canyon complexes (Husted, 1969). These implements indicated the use of vegetable foods, probably plant seeds, that were prepared by grinding. Grinding stones were rare in earlier complexes on the Plains although a few examples were found (Agogino, 1962). The use of grinding implements was well established in the Bighorn Canyon complexes; however, hunting played a large role in the economy as was indicated by the projectile points and other stone tools.

It was suggested that the Bighorn Canyon complexes represented local adaptations to an increasingly arid environment. It was hypothesized that there was a general northward migration up the Plains toward Canada beginning sometime after 7000 BC and that the Bighorn Canyon complexes represented offshoots from the main body on the Plains (Husted, 1969). Investigations in the province of Alberta, Canada, strongly

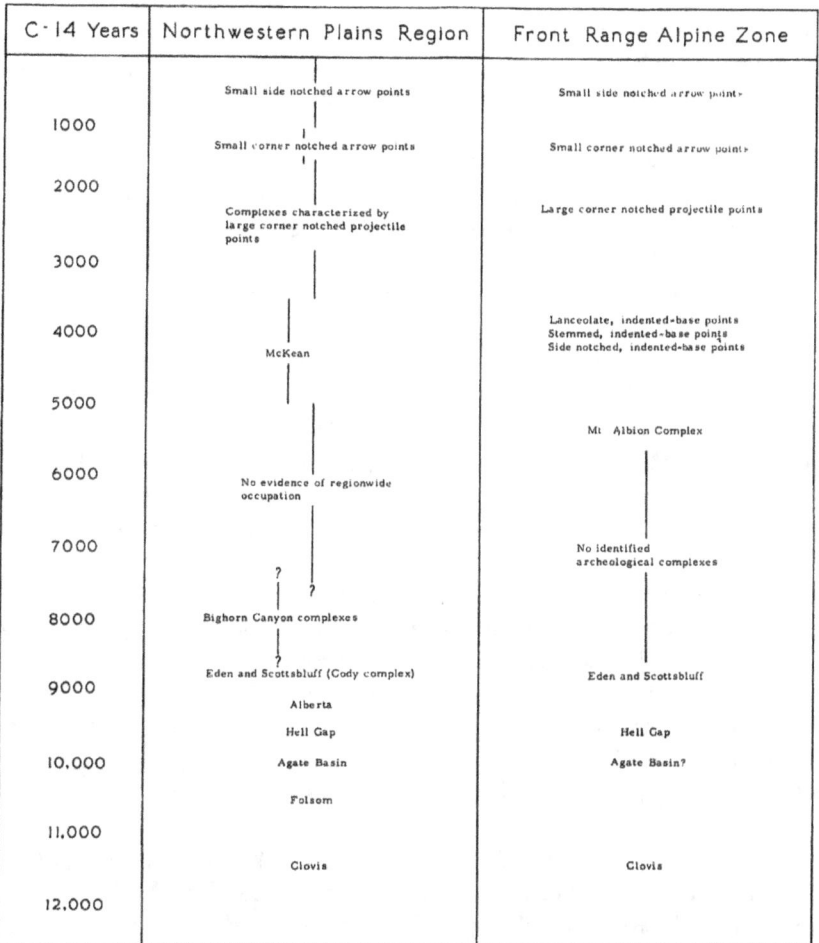

Fig. 15B.2. Schematic comparison of archaeological complexes in the northwestern Plains and the Front Range alpine area.

suggest that even the northern Plains were eventually abandoned. No archaeological complexes have been discovered to fill the gap between that which appears to represent a late phase of the Cody complex (Husted, 1969) and the Oxbow complex (Wormington and Forbis, 1965).

It was concluded that the Plains hunters ultimately were forced to abandon the Plains and moved westward into the Rocky Mountains of Alberta, British Columbia, and the interior plateau to the west. A series of complexes from sites along the Frazer River in the Lochnore–Nesikep locality of British Columbia (Sanger, 1963, 1966, 1967) contained projectile points bearing many similarities to earlier Plains types and others found on the Plains after 3000 BC. It was suggested that these complexes were intermediate to some of the earlier Plains manifestations and what will be termed the McKean complex found on the Plains after 3000 BC (Husted, 1969). The dating of the Lochnore–Nesikep assemblages was not in complete agreement with this hypothesis. However, Sanger (personal communication, 1969) agreed that his material probably did represent such an intermediate phase but that the mechanisms were unclear. Additional evidence was excavated in the Absaroka range west of Cody, Wyoming (Husted and Edgar, 1968; Wedel et al., 1968). A comprehensive report of these excavations is not yet available and a detailed discussion of the evidence is beyond the scope of this paper. Suffice it to say that cultural continuity between early Plains hunters and the later hunter-gatherers represented by the McKean complex was indicated. Adaptations to a changing environment and dwindling resources initiated on the Plains before abandonment were accomplished in the northern Rocky Mountains during the Altithermal. Once conditions on the Plains had improved to a point permitting reoccupation, the former hunters, now hunter-gatherers, returned to the Plains and mountain environments to the south in Colorado, Wyoming, and elsewhere in the west (Husted, 1968, 1969).

Reoccupation of the Plains was evidenced by what has been termed the McKean complex. Evidence of this complex is found as far east as southeastern Manitoba in Canada (MacNeish, 1958), southward down the Plains to at least northern Colorado, and in the mountains of Montana, Wyoming, Colorado, and Idaho (Husted, 1969). Radiocarbon ages for the complex range from slightly more than 5000 years to about 2500 years. However, a majority cluster between 5000 and 3500 years. Those younger than about 3500 years are here considered to be too recent.

Mummy Cave, in the Absaroka range west of Cody, Wyoming, produced the largest assemblage of McKean complex material culture yet recovered east of the Rocky Mountains. This cultural stratum was dated by radiocarbon at 4420 ± 150 BP and produced numerous projectile points, coiled basketry, plant fibre cordage, netting, leather trimmings, grinding stones, and other perishable and imperishable materials. Rock-filled roasting pits were associated with the level (Wedel et al., 1968, p. 184. fig. 1). Three general forms of projectile points were recovered: lanceolate, indented-base; stemmed, indented-base; and sidenotched, indented-base. These three forms occur in considerable variety, and one, two or all three may be found at a particular location.

Grinding stones may or may not be found at a particular McKean complex station. One site may represent a hunting camp where few or no grinding stones were used (Mulloy, 1954). Milling stones may be numerous at other locations (Gant and Hurt, 1965). Some sites represent bison traps where numbers of animals were killed in a single operation (Bentzen, 1961, 1962; Frison, 1968b, 1970). The presence or absence of grinding stones at sites producing the same complex indicates a hunting-gathering economy in which resources were exploited at different times of the year. The bison traps evidence communal efforts in the food quest.

By about 3500 radiocarbon years ago large cornernotched projectile points had replaced the varied forms of the McKean complex. However, the remainder of the material culture inventory remained much the same as it was in the earlier period. Coiled basketry, fibre cordage, grinding stones, hand stones, leather, and wood items continued in use. The rock-filled roasting pit remained in use (Frison, 1962, 1965, 1968a; Wedel et al., 1968). The similarity of the basic artifact complex to that of the preceding period indicated cultural continuity. The continuation of bison trapping or jumping (Frison, 1970) also points to a continuation of the same basic culture.

Radiocarbon dates for these complexes associated with various stypes of large corner-notched projectile points range from about 3500 years ago to between 2000 and 1500 years ago (Husted, 1969). By about AD 650 the small, and often serrated, cornernotched arrowpoint had replaced the larger cornernotched projectile point. This smaller style was widespread with only slight variation in form over great distances. It was found from the mountains of Wyoming westward and southward to Colorado, southern Idaho, Utah and across Nevada to Death Valley, California (Husted and Mallory, 1967). Radiocarbon dates for this point form range from 1290 ± 100 BP (GX-0526) at the Willowbrook Site west of Denver, Colorado (Leach, 1966, p. 46) to 970 ± 150 BP (M-1003) at the nearby LoDaisKa Site (Irwin and Irwin, 1961).

Small, triangular arrowpoints with side notches eventually replaced the corner-notched style. The transition seems to have occurred earlier in the northern Wyoming–southern Montana region. Here one form of sidenotched arrowpoint appeared by AD 900 (Husted, 1969). Somewhat later, probably around AD 1000–1100, a slightly differing form appeared in layer 37 of Mummy Cave (Wedel et al., 1968, fig. 1). Pro- . ably by AD 1300–1400 the typical late Plains sidenotched point was in use over the area in which the small cornernotched arrowpoint had been found previously.

At Mummy Cave Shoshoni pottery accompanied the small sidenotched arrowpoints. Elsewhere arguments have been presented in support of the conclusion that a Shoshoni or at least a Utaztekan continuum was present in western Wyoming for the past 5000 years (Husted, 1969). The McKean complex was considered to be the base of this continuum which culminated in the historic Shoshoni of Wyoming and the Utaztekan-speaking peoples elsewhere in the west such as the Ute of Colorado.

Other peoples inhabited the Plains and Rocky Mountains of Colorado. The Arapaho and Cheyenne were in the region in historic times. Pottery from the Front Range in Rocky Mountain National Park indicates that the Plains Apache travelled the mountains (Husted, 1964). However, all of these peoples were more properly associated with the Plains. The Ute of Colorado and Shoshoni of Wyoming were the indigenous peoples inhabiting the Plains and mountains. Both were ultimately forced from the Plains by intruding peoples and were more commonly considered to inhabit the mountains and country to the west of the Continental Divide by the historic period.

As fig. 15B.2 illustrates, alpine zone sites in the Front Range produced artifacts representing nearly the full range of archaeological complexes found on the north-western Plains. Artifacts in the museum at Rocky Mountain National Park and others collected by the author in the park and the area to the south demonstrated that the high mountain passes, ridges, and valleys above timberline had been travelled by man beginning some 11,000 to 11,500 years ago.

Clovis fluted points (fig. 15B.3l) have been found in all states west of the Rocky Mountains and are plentiful to the east. One specimen of this type is known from the Front

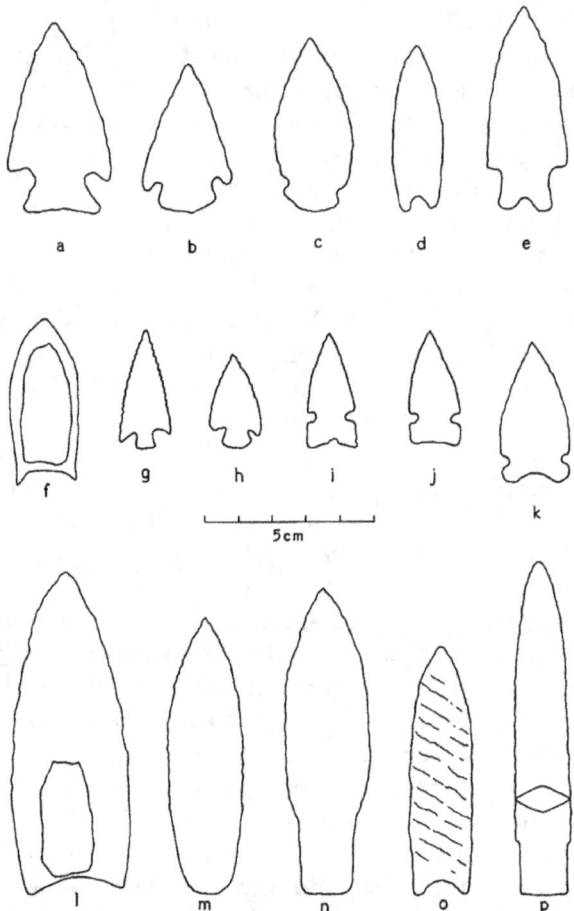

Fig. 15B.3. Outlines of diagnostic projectile point styles from the northwestern Plains and Rocky Mountains (see text for details): a, *Northwestern Plains* (1500 BC *to* AD 0–500); b, *Northwestern Plains* (1500 BC *to* AD 0–500); c, *Mt Albion*; d, *McKean*; e, *McKean*; f, *Folsom*; g, *Plains* (AD 650); h, *Plains* (AD 650); i, *Front Range*; j, *Front Range*; k, *McKean*; l, *Clovis*; m, *Agate Basin*; n, *Hell Gap*; o, *Allen*; p, *Eden*.

Range. It was recovered from the surface above timberline at the eastern end of Trail Ridge in Rocky Mountain National Park (Husted, 1962). The Clovis point is commonly associated with the bones of the mammoth. Since the mammoth was not a mountain animal, the scarcity of Clovis points and accompanying artifacts would not be surprising.

The Folsom point (fig. 15B.3f) appeared after the Clovis point on the northwestern

Plains. A radiocarbon age of 10,780 ± 375 BP (I-141) for the Lindenmeier Site in northern Colorado is the oldest Folsom complex date obtained so far (Haynes and Agogino, 1960, p. 5). By this time the mammoth had become extinct in the area and the large *Bison antiquus* became the principle prey of the Folsom hunters. A Folsom point was found on Trail Ridge in the early 1930s by park personnel (Russell Langford, personal communication, 1961). Benedict (1970) reported the finding of a Folsom point in a valley south of the park at an elevation of 3300 m.

Evidence of other early complexes in the Front Range is scarce. One questionable Agate Basin point base (fig. 15B.3m) was collected from a high slope above Chapin Pass in the park. Several fragments of Hell Gap points (fig. 15B.3n) are known from alpine zone sites here, and the stems of two Eden points (fig. 15B.3p) were found in Forest Canyon Pass near the western end of Trail Ridge.

The scant evidence of early cultures high in the Front Range indicates that this environment held little interest for peoples adapted to the hunting of mammoth and bison. The relatively greater number of finds of early points on Trail Ridge strongly suggests that this high east–west trending ridge was a desirable route between the Plains and the parks to the west.

Benedict (1970) found no concrete evidence that Palaeo-Indian populations persisted longer in the Front Range than on the Plains. The latest Palaeo-Indian projectile point type from the region appears to have been the Allen point (fig. 15B.3o) named after the type site south of Laramie, Wyoming. The site contained bones of *Bison occidentalis* or *Bison antiquus* and was radiocarbon dated at 7900 ± 400 BP (M-304, Mulloy, 1959). A few points morphologically similar to Palaeo-Indian types were found in Rocky Mountain National Park, but no ages were available for them (Husted, 1962, Plate V). The available evidence indicates that the Front Range region was abandoned by man sometime after 6000 BC. Had this region served as an Altithermal refuge, surely some indications would have appeared during Benedict's (1969, 1970) investigations. It is suggested that withdrawal initially was northward up the Plains. The data from Bighorn Canyon suggests that the people began moving toward and into the mountains in the northern Wyoming area but with some continuing northward to the plains of Alberta, Canada (Husted, 1969).

Radiocarbon dates indicate that the Front Range was essentially unpopulated until sometime between 4000 BC and 3500 BC (Benedict, 1969). This interval of abandonment would thus extend from about 6000 BC or somewhat later to at least 4000 BC, a period of 2000 to 1500 years. Reoccupation of the Front Range alpine zone was indicated by the Mt Albion complex distinguished by medium to large projectile points with side notches and convex bases (fig. 15B.3c; Benedict, 1969). Artifacts of the Mt Albion complex were recovered from game drives and campsites on ridge crests above timberline. Excavations were conducted at two sites on Mt Albion in the Front Range south of Rocky Mountain National Park. Site 5 BL 67, a game drive and campsite on Albion Ridge at about 3500 m elevation, was dated by radiocarbon at 5800 ± 125 BP (I-3267) and 5730 ± 130 BP (I-3817, Benedict, 1969, p. 22). A buried charcoal unit at site 5 BL 70, a campsite, yielded Mt Albion complex artifacts. Carbon-14 ages of 5650 ± 145 BP

(I-3023) and 5350 ± 130 BP (I-4419, Olson, *in* Benedict, 1969, p. 24) were obtained for the unit. Projectile points of the Mt Albion complex were collected from the surface of seven sites in the alpine region in Rocky Mountain National Park (Husted, 1962).

Significantly the Mt Albion complex is not found east of the foothills, and apparently it is absent west of the Front Range. The north–south extent of the complex has yet to be determined. The absence of earlier occupations, that is between about 7500 and 6000 years ago, indicates that the complex arrived in the Front Range from some yet to be determined source. Benedict suggests refuge areas to the north or west (1969, p. 30). The apparent absence of anything like the Mt Albion complex east or west of the Continental Divide north of the Front Range suggests that the source might be the mountainous areas to the north. There is no conclusive evidence to support this suggestion, but excavations at the Magic Mountain Site in the foothills west of Denver, Colorado, allow some speculation.

Projectile points of the Mt Albion complex appear to be identical to type MM3 points from the lower zones of the Magic Mountain Site (Irwin-Williams and Irwin, 1966, fig. 20). The Magic Mountain complex containing these points is thought to date prior to 3000 BC. A majority of the MM3 specimens came from Zone E although significant numbers were recovered from Zones D and F, the latter being the lowest. Although the authors believe the next higher complex to be unrelated to the Magic Mountain complex, certain projectile point forms from the higher unit, especially types MM5, MM6, MM7, MM17 and MM18 (Irwin-Williams and Irwin, 1966, figs. 22, 24, 25), bear certain resemblances to type MM3 and a direct relationship certainly is not out of the question. The overlap in specimens of the Magic Mountain and Apex complexes in Zones E and F make assignment of a particular type to one or the other complex very difficult. The Magic Mountain complex was seen in Zones E and F, and the succeeding Apex complex was assigned to Zones E-C. Under the circumstances, it is here felt that the earlier complex was ancestral to the Apex complex.

The Apex complex has definite relationships to the San Jose complex of New Mexico (Bryan and Toulouse, 1943), the Moab complex of northeastern Utah (Hunt and Tanner, 1960), and the Concho complex of Arizona (Wendorf and Thomas, 1951). These three are based on surface collections but serve to show the areal distribution of some varieties of stemmed, indented-base, and sidenotched, indented-base projectile points.

Higher in the Front Range McKean complex materials have been found in association with game drives and campsites (Benedict, 1969, p. 7). As discussed above, lanceolate, stemmed, and sidenotched, indented-base projectile points (fig. 15B.3d, e, k) are included in the McKean complex. They differ from those of the Apex complex in certain respects, but the similarities are obvious, and a relationship between the two, whether close or distant, is proposed here. It is further suggested that the appearance of the McKean complex in the Front Range marks the beginning of an indigenous cultural tradition that was to endure to the historic period. The source of this tradition is believed to lie northward in the Rocky Mountains of Canada and to have begun a southerly movement sometime after 4000 BC. The Mt Albion and Magic Mountain complexes may have been the vanguard of what appears to have been a major population shift as evidenced by the great number and vast spread of the McKean and closely similar complexes at about 3000 BC.

The McKean complex appears to have endured until about 3500 radiocarbon years ago. Sites on the northwestern Plains dating from about 1500 BC to between AD 0–500 produced a similar artifact content except that projectile points were now of moderate to large size with corner notches (fig. 15B.3a, b). Several varieties of large cornernotched points were collected in Rocky Mountain National Park and the area to the south (Husted, 1962, Plate III). However, it would be unrealistic to attempt to date these projectile points by comparison with sites of known age. During the 1500 to 2000 years that this style of point was in use, changes in form were subtle. The close similarity between forms of differing age make them unreliable as diagnostics. All that can be said is that the higher mountains, including the alpine zone, probably were seasonally occupied during the 1500 years or so that large cornernotched projectile points were in use.

The small, often serrated, cornernotched arrowpoint (fig. 15B.3g, h) that appeared on the Plains by about AD 650 was found in considerable numbers on alpine zone sites in Rocky Mountain National Park (Husted, 1962, Plate IV). This point form has been found in foothills sites along with cord-roughened pottery. The occupations have been attributed to Woodland peoples, the implication being that a population from the east moved into the foothills region (Irwin and Irwin, 1959; Irwin-Williams and Irwin, 1966). The cord-roughened pottery is not native to the area, but the projectile points are indigenous. The type can be traced from the Rocky Mountains of Colorado and Wyoming westward and southward across southern Idaho, Utah, and Nevada to Death Valley. It is considered that the complex containing these projectile points represents Utaztekan peoples (Husted and Mallory, 1967). In the Front Range region these probably were Ute. Plain undecorated pottery found with the small arrowpoints and cord-roughened pottery has been identified as Fremont (Irwin and Irwin, 1959, p. 129; Irwin-Williams and Irwin, 1966, p. 213). Others identify it as Shoshoni (Nelson, 1967; Breternitz, 1970, p. 164). Some of the ceramics may well be Shoshoni, but some could just as well be Ute since the Front Range and foothills were Ute homeland. Be that as it may, the occupations identified as Woodland by some actually represent an indigenous population, probably Ute. The cord-roughened ceramics indicate contact with an unrelated people yet to be identified.

The tradition of game driving at locations above timberline first encountered in the Mt Albion complex may still have been active during the period from AD 650 to AD 1000. Projectile points described as small, cornernotched, and often serrated were excavated at the Mt Albion site. Low rock walls and circular, rock walled pits occurred at the site. Radiocarbon ages of 970 ± 100 (M-1542, Crane and Griffin, 1968, p. 102) and 1230 ± 360 (SI-302, Long and Mielke, 1967, p. 375) were obtained for two pits at the site. A third date of 670 ± 150 (SI-301) is too recent for the small, cornernotched, serrated points. The third pit may have been more recent than the others or the carbon sample may have been contaminated.

The small, triangular, sidenotched arrowpoint (fig. 15B.3i, j) has also been found at game drives and campsites in the Front Range (Benedict, 1969, p. 9). The sites have not been identified as to cultural affiliation because no identifiable pottery was recovered. However, the fact that two of the sites were game drives seems sound evidence that at least these represent a continuation of the tradition begun some 4000 or more years ago. The occurrence of flat-slab milling stones and hand-stones at some of these sites (Benedict, 1969, p. 11) is in accord with this interpretation. These implements were associated with the Mt Albion and later complexes in the Front Range and were typical of northwestern Plains complexes from McKean complex times onward. Other unrelated peoples of the Plains used the small sidenotched arrowpoint. Some of these, such as the Arapaho, Plains Apache, and Cheyenne, probably visited the high mountain country. Nevertheless the Front Range region was within the territory dominated by the Ute Indians. It appears reasonable to assume that a majority of the recent sites in the region would be relatable to the Ute, especially the game drive sites.

Benedict (1969, p. 12) describes the game drive as the outstanding feature of the alpine region. This certainly appears to be true, and from the evidence accumulated so far, it seems safe to conclude that the appearance of the game drive sometime between 6000 and 5000 years ago marked the beginning of systematic use of the mountain environment above timberline. Prior to the appearance of game drives, from 9000 BC until about 6000 BC, man travelled the high country but appears to have spent little time above timberline. The interval from 6000 BC to the appearance of the game drive cannot be accounted for archaeologically, and it must be concluded that the Front Range, like the Plains to the east, was abandoned by man during the Altithermal.

Man returned to the Front Range some 500 or more years before he reoccupied the Plains. The arrival of the Mt Albion complex in the high mountains and adjacent foothills indicates that conditions in the montane region had improved while the Plains environment remained hostile to man. The Mt Albion complex was at the base of a tradition of alpine zone hunting that is here considered to have continued uninterrupted to the historic period. Benedict (1969, pp. 12–13) implies that the peoples involved were restricted to the mountains and foothills, moving from game drive to game drive along the Continental Divide in the summer, and wintering in rockshelters and campsites at lower elevations. This seems too narrow an interpretation considering that there was no evidence of such a cultural pattern in the area in historic times. The Ute did camp and hunt in the higher elevations, but they also hunted the Plains and the lower country and other mountain ranges over much of western Colorado. It appears more likely that the alpine region game drives were operated seasonally and for only a short period of time. Like the neighbouring Shoshoni, the Ute probably made use of a number of resources in an annual cycle that would take them to the Plains to hunt bison and back to the mountains and mountain valleys for other resources. Total exploitation of the environment was practised.

The alpine zone is similar to the arctic regions in some respects, a major difference being that the Arctic comprises a vast area of many thousands of square miles while alpine regions might be viewed as islands of tundra environment surrounded by a different climatic zone. The Eskimo was adapted to the Arctic and spent his entire life within that environment. The economy was geared to the seasonal changes and consequent variations in available food resources. On the other hand, the Indian of northern Colorado was not restricted to a single environmental zone. His movements to and from the Plains and valleys to the crests of the mountain ranges took him through several climatic zones. The Indian would move to different zones at various times of the year as particular resources became available. The alpine zone probably was visited only during the summer.

If comparisons are restricted to the arctic and alpine zones, some degree of similarity in method of exploitation obtains. The Eskimo depended almost entirely on animal resources. In the arctic interior of Alaska the caribou was the main source of food. These animals were secured in communal hunts involving impound drives and lake drives. The drives necessitated the construction of a corral and wings of posts, rocks or blocks of ice to direct the animals into the corral or a body of water where they were dispatched with arrows or spears (Spencer, 1959).

The Front Range game drives would have required communal efforts in construction and operation. The target of the drives probably was the mountain sheep. It has been determined that these animals cannot be corralled successfully (George Frison, personal communication, 1968), and the drives probably were designed to direct the sheep toward concealed hunters.

Similarities in the manner of use of the alpine and arctic zones end, for all intents and purposes, with the game drives. When a caribou hunt was completed, the interior Eskimo took up the hunting of other animals, trapping, and fishing. The Indians of the Front Range probably moved to lower elevations to hunt and collect, thus leaving the tundra region for another climatic zone.

In the light of presently available evidence it must be concluded that, prior to 6000 to 5500 years ago, human occupation of the Front Range alpine zone was limited to brief stays during transits of the Continental Divide between the Plains and the mountain parks. There is no evidence of a mountain-adapted culture in northern Colorado during this time. Also there is a lack of any evidence of human occupation from about 6000 BC until 4000 to 3500 BC. Like the high Plains to the east, the mountains must have been abandoned during much of the Altithermal.

Systematic exploitation of the alpine environment began only after 4000 BC. A tradition of game driving began at this time and is here considered to have continued unbroken to the historic period. The population involved in this hunting most likely was the Ute Indians and their ancestors.

References

AGOGINO, G. A. (1962) Comments. *In:* MASON, R. J. The Paleo-Indian tradition in eastern North America, *Curr. Anthrop.*, **3** (3), 227–78.

BENEDICT, J. B. (1969) *Prehistoric Man and Environment in the High Colorado Mountains.* Progress Rep. NSF Grant GS-2606. 32 pp. Mimeographed.

BENEDICT, J. B. (1970) Altithermal occupation of the Front Range alpine region. *1st Meeting, Amer. Quat. Ass., Bozeman* (abstracts), 8.

BENTZEN, R. C. (1961) *The Powers–Yonkee Bison Trap 24 PR5*. A report of an investigation by members of the Sheridan Chapter, Wyoming Archaeol. Soc., Sheridan. 14 pp.

BENTZEN, R. C. (1962) *The Mavrakis-Bentzen-Roberts Bison Trap 48SH311*. A report of an investigation by members of the Sheridan Chapter, Wyoming Archaeol. Soc., Sheridan. 15 pp.

BERRY, L. C. (1970) Report on a preliminary survey of the Absaroka Range in Custer National Forest, southern Montana. *1st Meeting, Amer. Quat. Ass., Bozeman* (abstracts), 9.

BRETERNITZ, D. A. (1970) Archaeological excavations in Dinosaur National Monument, Colorado–Utah, 1964–1965. *Univ. of Colorado Stud., Ser. in Anthrop.*, 17. 167 pp.

BRYAN, K. and TOULOUSE, J. H. JR (1943) The San Jose non-ceramic culture and its relation to a Puebloan culture in New Mexico. *Amer. Antiq.*, **8** (3), 269–80.

CONNER, S. W. (1968) The Northwestern Plains: An Introduction. *In:* The Northwestern Plains, A Symposium, *Center for Indian Stud., Rocky Mountain College, Occas. Pap.*, 1, 13–20. Billings.

CRANE, H. R. and GRIFFIN, J. B. (1968) University of Michigan Radiocarbon Dates XII. *Radiocarbon*, **10** (1), 61–114.

FREDLUND, D. E. (1970) Archaeology in the Sapphire and Bitterroot Mountains of western Montana. *1st Meeting, Amer. Quat. Ass., Bozeman* (abstracts), 48.

FRISON, G. C. (1962) Wedding of the Waters Cave, 48H0301. A stratified site in the Big Horn Basin of northern Wyoming. *Plains Anthrop.*, **7** (18), 246–65.

FRISON, G. C. (1965) Spring Creek Cave, Wyoming. *Amer. Antiq.*, **31** (1), 81–94.

FRISON, G. C. (1968a) Daugherty Cave, Wyoming. *Plains Anthrop.*, **13** (42), 253–95.

FRISON, G. C. (1968b) Site 48SH312: An early Middle Period bison kill in the Powder River Basin of Wyoming. *Plains Anthrop.*, **13** (39), 31–9.

FRISON, G. C. (1970) The Kobold Site, 24BH406: a post-altithermal record of buffalo jumping for the Northwestern Plains. *Plains Anthrop.*, **14** (47), 1–35.

GANT, R. D. and HURT, W. R. (1965) The Sturgis Archaeological Project: An Archaeological Survey of the Northern Black Hills. *South Dakota Mus. News*, 1–54. Univ. of South Dakota.

HAYNES, V. and AGOGINO, G. (1960) Geological significance of a new radiocarbon date from the Lindenmeier Site. *Proc. Denver Mus. Nat. Hist.*, 9, 1–23.

HUNT, A. P. and TANNER, D. (1960) Early man sites near Moab, Utah. *Amer. Antiq.*, **26** (1), 110–17.

HUSTED, W. M. (1962) *A Proposed Archeological Chronology for Rocky Mountain National Park Based on Projectile Points and Pottery*. Unpub. M.A. thesis, Univ. of Colorado. 109 pp.

HUSTED, W. M. (1964) Pueblo pottery from northern Colorado. *Southwestern Lore*, **30** (2), 21–5.

HUSTED, W. M. (1968) Wyoming. *In:* CALDWELL, W. W. (ed.) The Northwestern Plains: A Symposium. *Center for Indian Stud., Rocky Mountain College, Occas. Pap.*, 1, 63–8. Billings.

HUSTED, W. M. (1969) Bighorn Canyon Archeology. *Smithsonian Inst., River Basin Surveys, Pub. in Salvage Archeol.*, 12. 138 pp.

HUSTED, W. M. and EDGAR, R. (1968) *The Archeology of Mummy Cave, Wyoming: An Introduction to Shoshonean Prehistory*. MS on file at the Whitney Gallery of Western Art, Cody, Wyoming. 447 pp.

HUSTED, W. M. and MALLORY, O. L. (1967) The Fremont Culture; its derivation and ultimate fate. *Plains Anthrop.*, **12** (36), 222–32.

IRWIN, H. T. and BREW, J. O. (1968) Archeological investigations at the Hell Gap Site near Guernsey, Wyoming. Washington, 1963 projects, *Nat. Geogr. Soc. Res. Rep.*, 151–6.

IRWIN, H. J. and IRWIN, C. C. (1959) Excavations at the LoDaisKa Site in the Denver, Colorado, area. *Proc. Denver Mus. Nat. Hist.*, 8, 1–156.

IRWIN, H. J. and IRWIN, C. C. (1961) Radiocarbon Dates from the LoDaisKa Site, Colorado. *Amer. Antiq.*, **27** (1), 114–15.

IRWIN-WILLIAMS, C. and IRWIN, H. J. (1966) Excavations at Magic Mountain: A diachronic study of Plains-Southwest relations. *Proc. Denver Mus. Nat. Hist.*, 12, 1–241.

IVES, R. L. (1941) A cultural hiatus in the Rocky Mountains Region. *Southwestern Lore*, **7** (3), 42–6.

IVES, R. L. (1942) Early occupation of the Colorado Headwaters region. *Geogr. Rev.*, **32** (3), 448–62.

LEACH, L. L. (1966) Excavations at Willowbrook: A stratified site near Morrison. *Southwestern Lore*, **32** (2), 25–46.

LONG, A. and MIELKE, J. E. (1967) Smithsonian Institution Radiocarbon Measurements IV. *Radiocarbon*, **9**, 368–81.

LOWIE, R. H. (1924) Notes on Shoshonean ethnography. *Amer. Mus. Nat. Hist. Anthrop. Pap.*, **20** (3), 185–314.

MACNEISH, R. S. (1958) An introduction to the archaeology of southeast Manitoba. *Nat. Mus. Can. Bull.*, *157*, Anthrop. Ser., 44, 1–184. Ottawa.

MOOMAW, J. C. (1954) Ancient stone 'walls' in the Colorado Rockies. *Southwestern Lore*, **20** (1), 5–6.

MOOMAW, J. C. (1957) Aborigines of the Colorado Highlands. *Southwestern Lore*, **23** (3), 35–7.

MORRIS, E. A. (1970) High altitude archaeological sites in Arizona and New Mexico. *1st Meeting, Amer. Quat. Ass., Bozeman* (abstracts), 96.

MULLOY, W. (1954) The McKean Site in Northeastern Wyoming. *Southwestern J. Anthrop.*, **10** (4), 432–60.

MULLOY, W. (1959) The Janes Allen site near Laramie, Wyoming. *Amer. Antiq.*, **25** (1), 112–16.

NELSON, C. E. (1967) The archaeology of Hall-Woodland Cave. *Southwestern Lore*, **33** (1), 1–13.

OLSON, B. L. and BENEDICT, J. A. (1970) *Field Work at the Rollins Pass Sites 1969.* Rep. to US Forest Serv. and Smithsonian Inst., Washington, DC. 10 pp. Mimeographed.

SANGER, D. (1963) Excavations at Nesikep Creek (EdRk: 4). A stratified site near Lillooet, British Columbia: Preliminary Report. *Nat. Mus. Can. Bull.*, *193*, Contrib. to Anthrop. *1961-2*, Part 1, 130–61. Ottawa.

SANGER, D. (1966) Excavations in the Lochnore-Nesikep Creek locality, British Columbia: Interim report. *Nat. Mus. Can. Anthrop. Pap.*, 12, 1–22. Ottawa.

SANGER, D. (1967) Prehistory of the Pacific Northwest Plateau as seen from the interior of British Columbia. *Amer. Antiq.*, **32** (2), 186–97.

SPENCER, R. F. (1959) The North Alaska Eskimo: A study in ecology and society. *Smithsonian Inst., Bur. Amer. Ethnol. Bull.*, 171. 490 pp.

WEDEL, W. R., HUSTED, W. M. and MOSS, J. H. (1968) Mummy Cave: Prehistoric record from Rocky Mountains of Wyoming. *Science*, **160** (3824), 184–6.

WENDORF, F. and MILLER, J. P. (1959) Artifacts from high mountain sites in the Sangre de Cristo Range, New Mexico. *El Palacio*, **66** (2), 37–52.

WENDORF, F. and THOMAS, T. H. (1951) Early Man sites near Concho, Arizona. *Amer. Antiq.*, **17** (2), 107–14.

WORMINGTON, H. M. and FORBIS, R. G. (1965) An introduction to the archaeology of Alberta, Canada. *Proc. Denver Mus. Nat. Hist.*, 11, 1–248.

Manuscript received May 1971.

Man's impact on the environment

CHAPTER 16

Radioecology

William S. Osburn Jr

Division of Biology and Medicine,
US Atomic Energy Commission, Washington

1. Introduction

In order to carry out its mission to develop uses of nuclear energy with utmost safety to man and his environment the United States Atomic Energy Commission has supported extensive physical and biological research pertaining to major ecosystems throughout much of the world. The results of the largest bioenvironmental project are contained in a 1250 page volume titled *Cape Thompson* (Wilimvosky and Wolfe, 1966) and represent an excellent example of providing detailed baseline ecological information prior to a nuclear experiment in an arctic area.

Above-ground nuclear testing in the late 1950s introduced various radionuclides into the atmosphere of the northern hemisphere. Eventually this man-generated radioactivity could be detected within some components of every major ecosystem on earth. Thus portions of the planet were tagged with 'tracer' levels of specific radionuclides – too low to harm even sensitive organisms (in fact usually detectable only by very sophisticated instrumentation[1] and/or by laborious reduction of large amounts of water, sediment, plant or animal tissue to concentrate the radioactive materials), but high enough to detect and trace within or between various landscape systems. Thus investigators have been able to study pathways (and the bioenvironmental forces responsible) or 'tracer' movement and learn the devious nuclide routes, sinks and cycles. Conversely, while following these tracers – initially because of potential hazard to man and his environment – a number of ecological unknowns (erosion mechanisms and rates, mineral cycling, food chain relationships, etc.) have been resolved.

The ultimate objective of being able to predict quantitatively the fate and precise effect of radioactive materials under any reasonable set of environmental circumstances has not been realized. However, radioecological research has been sufficient to allow a

---

[1] Methods of measuring quantities of radioactive materials permit counting of mere hundreds of atoms per sample. For example, radon in an air sample can be detected down to about 350 atoms per litre of air, or approximately 10^{-16} part per million.

number of rules, principles, or generalizations to be made concerning the distribution and potential effects of radionuclides in arctic and alpine regions.

This chapter will concern the gross patterns of radioactivity presently existing or being formed within arctic and alpine regions, environmental and biological factors associated with radionuclide deposition and movement, several terrestrial and aquatic food chains, effects of ionizing radiation on organisms, and the actual or potential radiation dose to man and other organisms will be considered.

2. Patterns and levels of radioactivity

LANDSCAPE PATTERNS OF RADIOACTIVITY

Interrelations of physical-chemical characteristics of fallout particles, mechanisms responsible for bringing fallout to ground level, efficiency of ground level conditions to intercept (or remove) and retain fallout, and factors which influence ground level redistribution (blowing dust, water erosion, type and number of plants or animals) are so numerous, and so variously important on a time basis (daily or seasonal), that rules to predict patterns of fallout quantitatively have not been formulated.

A global network of fallout pots has been maintained since the early days of atmospheric nuclear testing. These pot data suggest that fallout deposited throughout the arctic region was reasonably uniform and much less in amount than in lower latitude regions (List and Telegadas, 1961). However, gaps in our knowledge of fallout distribution patterns do exist in arctic and especially mountainous regions because data collection points were too far apart in the Arctic – and none existed in a US alpine site – to establish the uniformity of fallout.

The most important mechanisms and processes responsible for fallout deposition and movement in arctic and alpine areas are now reasonably well known. Deposition of fallout was initially correlated with meteorological conditions (United Nations, 1962, 1964). As atmospheric testing occurred in the mid-latitude regions it seemed reasonable to expect the greatest amount of fallout to come down in these regions. When data indicated that fallout deposition could be grossly related to the amount of precipitation a region received, it followed that the sites which would need the closest observation would be mid-latitude regions with relatively greater amounts of precipitation. However, it was not long after the atmosphere was sufficiently 'loaded' with fallout from atmospheric testing that radioanalysis of samples of arctic and alpine biota indicated surprisingly high levels, and that these high latitude and high altitude areas too would need to be thoroughly researched in order to establish the mechanisms and degree of contamination that various organisms and man might be accumulating.

Several methods are available to predict gross patterns (landscape scale) of radioactive fallout in arctic or alpine landscapes. In addition a number of techniques exist which can be used to anticipate levels of radionuclide concentration within organisms (or their organs) living within these landscapes. Prediction techniques are developed for terrestrial and freshwater ecosystems and, in the case of the Arctic, for marine and estuarine situations. Each technique – snowfield patterns (terrestrial ecosystems), vegetational indicators (terrestrial ecosystems), general ecological knowledge (all ecosystems), the critical pathway approach (largely marine and freshwater ecosystems), the specific nuclide concept (marine and freshwater ecosystems), food chain dynamics, modelling included (all ecosystems) – will be briefly discussed. The first two techniques are on a gross or macro scale while the latter concern organisms and are on a more refined or micro scale.

Snow patterns and levels of radioactivity
In nearly all arctic and alpine regions most of the precipitation is in the form of snow. Snow has proved to be an effective mechanism to scrub fallout from the atmosphere. Further, the heaviest snowstorms typically occur in the spring months (at least in the northern hemisphere). During this period radioactive fallout injected by above-ground nuclear testing escapes from the stratosphere and enters the troposphere or 'weather region' via the tropopause 'break' in middle latitudes. For a general survey see Reiter (1971). Thus it is readily available to be washed out by spring snowstorms.

In both the arctic and the alpine regions strong winds sweep some areas clear of snow and deposit this snow in drifts throughout most of the cold season. Such drifts may concentrate fallout nuclides to relatively high levels. In addition, for the same reasons that some sites collect snow, these same sites tend to accumulate other windblown material (dust, pollen, etc.). Thus in some arctic and alpine localities, due to disruption of soil by needle ice and digging by mammals, annual dropping of leaves, and abundant production of pollen, there is much debris available for fallout attachment, repeated resuspension of debris and finally deposition in 'snow accumulation sites'. By simply delimiting snow accumulation sites, therefore, the general pattern of fallout radioactivity imprinted upon the landscape can be revealed. A series of air photographs of snowfields will provide a standardized procedure for determining actual or potential patterns of all types of fallout on the landscape.

Although snow drift patterns may vary according to wind direction, characteristics of the falling snow, wind velocity during and after snowfall, condition of the soil, time of year (vegetation may be variously dense as some leaves are 'evergreen' while others may drop at the end of a growing season or be in a particular stage of leaf fall), digging animals, needle ice and other factors, the general topographically controlled pattern is typically very similar year after year. Hence in any region which has snow and winds to drift the snow, snowfield patterns serve as a stable, obvious, and sharply delimited system on which to superimpose a sampling grid for locating sites of radionuclide concentration. The number of soil samples required to locate regions of possible contamination and to determine levels can be materially reduced by this approach.

Plant communities and levels of radioactivity
It is generally considered that discrete assemblages of plants (plant communities) develop in response to various sets of environmental conditions. If one further assumes that the particular set of environmental conditions responsible for the development of a

particular plant community is also responsible for depositing or retaining fallout, then deposition patterns should be related to vegetation patterns. Osburn (1963), working in an alpine area, and Hanson (1968), working in an arctic area, have reported data which support the above commentary. Thus plant communities offer functional units to study for predicting fallout distribution. The environmental factors which seem to control the development of plant communities and are responsible also for the gross distribution of fallout are snow or snow cover, wind, exposure, needle ice activity and soil moisture. These factors operating in various combinations result in particular plant communities with particular radionuclide burdens. Two of these factors, wind and needle ice, will be discussed further.

Wind is directly and indirectly an extremely important factor in the development and maintenance of many ecosystems in both arctic and alpine environments. Although the desiccating effect of wind on plants (increasing evapotranspiration) influences the structure of plant communities, of more importance is the effect of wind in removing snow from ridges, crests and peaks to form drifts of various depths. Snow drifts produce a specific set of environmental characteristics. For instance, deeply drifted snow melts late in the spring and reduces the growing season to very short lengths (in fact, absent in some years), thus limiting the plant community to species which can tolerate short growing periods. Conversely, snowfields can protect plants and animals from intense winter cold and windchill. Also, snowfields may protect animals from many predators. Wind, too, is extremely important in the initial deposition, resuspension and migration of particulates (radioactive nuclides included) within tundra regions. Areas directly beneath and downslope from snowfields are usually supplied with an abundance of moisture from meltwater. In fact, boggy areas with poor oxygenation often exist.

Needle ice is known by a number of different names: Haarfrost, Haareis, Nadeleis, Stengeleis, Eisfilamente, Effloreszenzeis, Barfrost, Mushfrost, Pipkrake, Rouste, and Shimobashira (Troll, 1958). At least two types of needle ice have been observed and described by researchers (Hole, 1970). The type of needle ice which plays a major role in disturbance of surface soil consists of a thin layer of frozen surface soil underlain by coarse ice needles and columns. The condition favourable to the development of this formation is a sharp sudden drop in temperature from about 10°C to −10°C. A thin layer of surface soil is frozen *in situ* and is pushed up by the growing ice needles as segregation occurs by continued freezing. The base of the ice needles forms a smooth moist, barely frozen surface. Needle ice originates at the contact of a superficially dried soil layer with moist undersoil. Water is transferred from the moist soil to form long ice needles which vary from one to several centimetres in length. In some regions the needle lengths may exceed 40 cm. All in all needle ice is a very significant mass wasting mechanism. Not only does it destroy young seedlings (as some species are more prone to frost heave than others a particular type of plant community can result), but the surface soil crust is pushed up so that wind pick-up of soil surface debris is more effective. Also when the ice needles melt, the loose crumbly soil surface is readily subject to wind or water erosion. Thus needle ice not only can keep radionuclides from leaching into the soil but actually keep them on or near the surface, hence readily erodable.

The intensity of needle ice lift varies with soils of different characteristics, exposure, moisture, sky temperature, radiation, humidity, wind, water temperatures, soil heat, water flow, and vegetation cover – factors which may determine the formation of plant communities and patterns of radionuclide distribution.

3. Regional fallout

In ecology an 'endpoint' of knowledge is the ability to predict accurately. To predict whether specific ecosystems are 'concentrators' or 'dispersers' of radioactive fallout one needs to establish the regional fallout level – the amount that would be deposited at a particular spot if it was neither concentrated nor diluted. Regional fallout may be defined as the average amount of fallout per unit area directly deposited (wet and dry) from the atmosphere without significant ground level redistribution (Osburn, 1969).

In order to predict the various fates of fallout within the various components of an ecosystem, the amount of fallout which has entered the ecosystem (or unit being researched) needs to be determined. With this knowledge one can then total the fallout amounts in various components of the system and compare this sum with the entry amount. This comparison provides an index of the accuracy with which one can account for radionuclide distribution – a materials balancing. It is also an assessment of all the measuring techniques used to establish the amount of fallout within ecosystem components – soil, plants, animal populations, etc.

An ecological accounting technique used to verify the amount of fallout measured within a *Sphagnum* bog or regional deposition system will now be described. Bergh *et al.* (1961) and Osburn (1967), working in arctic and mountainous regions respectively, have developed criteria to determine regional fallout, i.e. to delimit fallout standard or reference ecosystems. As site selection entails several additional considerations in mountain rather than arctic landscapes, only alpine site selection will be described.

Basically, sites should be sufficiently permeable to absorb all precipitation which falls upon them, subject to little leaching, and occupy reasonably level terrain so that water from higher ground does not overflow them. In both arctic and alpine areas strong winter winds create major snow and therefore fallout redistribution. Thus, sites to be selected should be little influenced by wind. Sites which best fulfil these criteria are sections of a *Sphagnum* bog which have shrubs (*Salix* or *Alnus*) of adequate density and height to anchor all of the snow which falls into it. In addition to successfully anchoring the snow, these deciduous shrubs lose their leaves before the major snowfall season and so do not intercept snow before it reaches ground level. Also, these particular sites should be surrounded by baffle areas, i.e. areas of sufficient vegetation to cause resuspended snow to be dropped before it reaches the site selected for sampling.

Another problem in locating standard reference sites concerns contamination by water flowing into a sampling site. This problem is overcome by restricting sampling to mounds 40 cm or more in height – well above the level of overland flow. However, as more time elapses, relatively mobile nuclides such as ^{90}Sr will require sampling at progressively greater depths. If, however, the initial nuclide percentages are known one may be able to sample and analyse only for ^{137}Cs, thus not necessitating deep sampling.

To establish a quantitative characterization of these key sites a 'fish eye' (or canopy) camera may be used to photograph the overhead and surrounding vegetation. In the Front Range of the Colorado Rockies, sites which neither lost nor gained snow from wind deflation or deposition had the following characteristics (Osburn, 1969): at an angle of 30° from vertical the canopy above the sample site was (in the winter and spring) nearly open (0–15 per cent closed), while coverage between 30° and 60° varied considerably more (10–50 per cent) and the canopy coverage between the angles of 60° and 90° usually was 30–90 per cent closed. Canopy directly overhead was considered the most critical. If coverage is greater than 25 per cent the site should be rejected. Canopy coverage intercepted by lines extending outward at 60° and 90° from the site centre are of importance as an index of snow anchoring efficiency of the surrounding shrubs. If

Table 16.1. Fallout accountability: Long Bog, 1962–3

		Amount	$Ci\ dec^{-2}$	
Contribution source		Gross beta	^{137}Cs	^{90}Sr
a.	Residual *Sphagnum* cores (October 1962)	2840 ± 147	541 ± 24	304 ± 12
b.	Winter snow deposition (mid-September 1962–April 1963)	5781 ± 449	72	46
c.	Spring dust storm (April 1963)	8041 ± 410	64 ± 4	11 ± 2
d.	Summer rainfall (May–September 1963)	10,876		
e.	Summer dry fallout (May–September 1963)	4229	66	39
f.	*Sphagnum* cores (October 1963)	$30,493 \pm 1204$	745 ± 31	412 ± 19
	Radioactivity increments (September 1962–September 1963):			
	a. as measured (a subtracted from f)	27,653	204	108
	b. summation of b, c, d and e	28,927	202	96

canopy coverage is less than 30 per cent, shrub density is insufficient to stabilize the snow and during periods of high wind velocity much snow is lost.

In order to validate the adequacy of the above technique to locate true regional fallout depositional sites an ecological accounting was involved. Briefly, a series of *Sphagnum* cores were collected and radiocounted in 1962, and again in 1963 (see table 16.1). Thus the increase in radioactivity level was determined and compared with the amounts of radioactivity deposited during the 1962–3 time interval by four separate processes (winter snow, spring dust storms, summer wet and summer dry fallout). Agreement was surprisingly close – within the errors of estimate. This close agreement indicates that *Sphagnum* moss cores can be used to indicate regional fallout levels reliably. Furthermore it was demonstrated that attention must be given to more than precipitation when making fallout estimates. The level of fallout attributable to at least one process heretofore given little attention in fallout estimates – spring dust (particularly the included pollen) – was surprisingly high. The level of dry fallout – 30 per cent of the total – was another minor 'ecological surprise'.

Thus the regional amounts of radioactive fallout can be ascertained in arctic and alpine areas.

4. General ecological knowledge as indicators of radionuclide levels in animals

Although data are meagre and no specific report has attempted to demonstrate how general ecological knowledge can be used to anticipate fallout levels in animals, the following (Osburn, 1967) gives an indication that general knowledge would be of considerable value.

Typically, the amounts of radioactivity contained in the various body components of a number of alpine animals reliably reflect the food source, behaviour or habitat of the animals, regardless of their taxonomic positions. For example, the large differences noted in levels of radioactivity of the skin and feathers of various species of birds may be explained by the differences in behaviour patterns, especially behaviour exhibited in the search for food. The radioactivity (picocuries g^{-1}) of the skin and feathers of the following list of birds (and one mammal) was determined and graded according to the relative amounts of time each species tends to spend flying: bat, *Myotis*, spl. (846); magpie, *Pica pica* (118); goshawk, *Astur atricapillus* (66); pipit, *Anthus spinoletta* (37); sparrow hawk, *Falco sparverius* (31); and the great horned owl, *Bubo virginianus* (21).

Birds which spend a larger amount of time searching among pine needles or plants on the ground tend to encounter and retain larger amounts of fallout radioactivity. Examples are: red crossbills, *Loxia curvirostra* (250); chickadee, *Penthestes gambeli* (110); Clark's nutcracker, *Nucifraga columbiana* (92); Junco, *Junco canicaps* (67): and the rosy finch, *Leucosticte australis* (59).

When all vertebrate animals utilizing the Boulder (Colorado) city watershed were categorized by feeding habits, flesh of predators and scavengers contained relatively low levels of gross beta radioactivity, with seed-eating animals next higher, followed by grazers, and finally insectivorous animals. The extended range was well over a factor of 10.

It is of interest to note that in the case of two species, pika (*Ochotoma princeps*) and pocket gophers (*Thomomys talpoides*), juvenile individuals, relegated to less optimum sites than were the adults, contained much higher radiation burdens.

Research by French (1967) strongly suggests that granivores ingest less radioactive material than herbivores living in the same contaminated environment because herbivore food has a much larger surface area per amount needed to sustain life.

General ecological knowledge not only allows one to anticipate relative nuclide concentrations and environmental or biota 'hot spots', but it allows general rules to be formulated concerning conditions which might modify radiation exposure. The following rules, developed by an alpine study, will probably apply to many arctic regions. The intensity of radiation from fallout exposure is usually influenced by one of the following conditions which typically prevails:

(1) Only a few species, usually in low numbers, occupy the most radioactive sites.
(2) Organisms which occupy these sites of relatively high fallout concentration are extremely radioresistant.

(3) Environmental factors tend to serve as an ameliorating mechanism when organisms are in their most radiosensitive stages.

In general, sites which tend to concentrate high levels of fallout materials are locales which support very few kinds of organisms. For instance, deep snow accumulation sites, typically, are relatively radioactive; however, these sites do not become snow-free until late summer and since few plant or animal species can tolerate such extensive periods of snow cover little vegetation other than highly radioresistant mosses exist in these areas.

In addition to snowfield 'hot spots', fermentation sites where litter is being converted to soil may accumulate relatively large amounts of fallout materials. The most prevalent plants in these layers of decaying litter are radiation 'insensitive' species of fungi and bacteria; likewise the most abundant animals are various species of mites which too are radioresistant.

Organisms tend, fortuitously, to be in their least radiosensitive stage during periods of higher radiation levels. Plants exist as seeds, are dormant and have reduced physiological activity, and are generally radioresistant in the winter. During spring, when concentration of fallout from melting snowfields is high, plants enter the most radiosensitive stages of their life history (rate of cell division increases sharply): seeds germinate, plants break dormancy and new growth is resumed. However, plant exposure to radiation is lessened by the abundant spring soil moisture. The top 10 cm of soil may easily contain over 2–3 cm of water, hence is capable of nullifying the penetration of beta and alpha particles and reducing the penetration power of gamma rays.

Even the small mammals – perhaps one of the most radiosensitive groups – receive less radiation exposure in the spring. Although fallout debris certainly could greatly contaminate plant surfaces brushed against and eaten by small mammals, two factors preclude greater nuclide contamination. The natural process of spring shedding would lessen the amount of radioactive particles adhering to the fur. Again, the spring diet of animals tends to shift from the old contaminated plant foods to younger, newly developing, and less contaminated plant foods.

Many exceptions of course do exist. However, the more thoroughly the animal behaviour is known the easier it is to predict points and times of fallout contamination. Simply by outlining the typical 'year around' animal activities and plant responses in conjunction with the annual climatic sequence, one can anticipate 'critical' periods, i.e. periods of organism radiosensitivity. For this reason a chapter on arctic and alpine radioecology is really incomplete without a section which might be titled 'an ecological alpine or arctic year'. In this section the annual environmental regime should be outlined along with the accompanying 'behaviour' of the biota. From this the relationship of organisms to their environment (the ecology) is revealed.

THE CRITICAL PATHWAY APPROACH AND THE SPECIFIC NUCLIDE CONCEPT

In marine environments two procedures – the critical pathway approach and the specific activity approach – to estimate radiation exposure to man have been developed. These two procedures, described below, with modifications will also be useful procedures in freshwater and perhaps terrestrial environments.

The critical pathway approach is a procedure whereby a sequence of events is evaluated from introduction of radioactive materials into the marine environment until it reaches man. The sequence of events which must be considered includes: the kind and amounts of radioactive materials introduced; the physical and chemical form of the introduced radionuclides; the initial mechanical dilution; subsequent dilution by ocean-mixing processes; availability of the introduced material to the marine biota; concentration factors; consumption of marine products; 'standard man'; external exposure; and applications. Eventually, the radionuclide that contributes the most significant fraction of the dose limit to a critical group of individuals can be identified and used to design a minimum but effective environmental surveillance programme.

The specific activity approach states that if the maximum permissible specific activity (ratio of the amount of radionuclide to total nuclide of the chemical element considered) is not exceeded in the environment where foods are produced, then there is no mechanism by which the specific activity in the foods (or people who eat the foods) could exceed the International Commission Radiological Protection (ICRP) permissible limit. Briefly, the distribution patterns of artificially introduced radionuclides can be predicted by obtaining qualitative and quantitative information on the distribution of stable elements. In order to apply the specific activity approach, information is required concerning: type and amounts of radioactive materials introduced; physical–chemical form of the radionuclides introduced; dilution of nuclides; relative availability to the biota; concentrations of radionuclides in marine organisms and of the seawater; and quantities of marine products used for food. The application of this approach is based on two main assumptions: 1) that a radioisotope introduced into the environment readily equilibrates with the stable isotope(s) of the same element, such that biological concentrating mechanisms will be unable to discriminate between the radioactive and stable forms; and 2) that the quantity of each stable element in each body organ is 'fixed' and does not fluctuate with the intake of that element. Thus the distribution of radionuclides will correspond to the distribution of stable elements in the environment.

By analysing organisms for stable element amounts one can predict the amount of radioactive isotope they (or a particular organ) are likely to concentrate. An example follows:

$$\frac{^{90}\text{Sr pCi/ml}}{\text{Sr }\mu\text{g/ml}} \text{ water} = \frac{^{90}\text{Sr pCi/g}}{\text{Sr }\mu\text{g/g}} \text{ fish bone}$$

5. Food chains or food webs

Interactions between groups of organisms involving the effects of eating can be arranged in a trophic (nourishment) hierarchy. Green plants (primary producers) which convert solar energy into chemical energy are at the bottom. Those animals which eat plants (primary consumers) are next, the carnivores or meat eaters (secondary, tertiary, etc.

consumers) on top. Dead organic matter produced at these levels is consumed by yet other organisms which release nutrients back to the soil and the cycle continues. By plotting the nourishment relationships among all the species of an ecosystem, a complex network or food web appears. While the transfer routes linking living forms within an ecosystem may form a web, it is often possible to follow the passage of a nutrient through a linear succession, or food chain. Typically, food chains in arctic and alpine areas are short and consequently relatively easy to follow. Arctic marine ecosystems are excepted as marine food chains tend to be rather long.

It is a common misconception that all radioactive materials are concentrated as they move upward in food chains. In 'lay' literature one frequently encounters diagrams in which levels of radioactivity found in the water are increased many thousands of times with each step in the food chain. As man is atop most long food chains one can assume that man is in a critical position. Not only is he atop food chains, he is among the most radiosensitive organisms. It is this very fact that led the various groups which regulate the uses of nuclear materials to formulate guidelines to protect man, i.e. the assumption being that if a particular release level or dose is safe for man it will be safe for other organisms.

A more realistic view regarding reconcentration, or biomagnification, of radionuclides may be gained from a general appraisal of the results of tracing 16 'common' nuclides in an arctic environment.

Of the nearly 200 radionuclides of some 35 different elements produced in nuclear explosions (President's Science Advisory Committee, 1965), a high percentage have half-lives of only a few seconds, so time alone precludes these from biological accumulation. Another group of radionuclides (physiologically inert), though they may be ingested, simply pass through the digestive system of many higher organisms. In other cases some nuclides may be taken in and become physiologically involved but often they are discriminated against. Concentration factors of most elements for which we have information decrease from plants to mammalian herbivores (Reichle *et al.*, 1970). However, in specific situations some radionuclides are concentrated to relatively high levels. Typically these nuclides have a relatively long half-life, are physiologically important and their stable counterparts are scarce in the local environment. ^{137}Cs, ^{90}Sr, and ^{55}Fe are three such radionuclides and consequently they have received a proportionate amount of research attention.

Some sixteen radionuclides were traced through a simple food web of an Alaskan arctic ecosystem by W. C. Hanson to identify pathways and retention on three successive trophic levels, lichen–caribou–wolf or man (US Atomic Energy Commission, 1971). Each successive trophic food level discriminated against radium-226 (1620 year half-life), lead-210 (22 years), cobalt-60 (5.25 years), antimony-125 (2.7 years), europium-155 (1.8 years), ruthenium-106 (1.0 year), manganese-54 (312 days), cerium-144 (285 days), yttrium-88 (107 days), zirconium-95 (65 days), ruthenium-103 (40 days), and niobium-95 (35 days). However, cesium-137 (30 years), sodium-22 (2.60 years), silver-110 m (260 days), and zinc-65 (243 days) were concentrated on each level. Radium-226 was highly concentrated in the bone tissue by caribou but not by wolf. Many of the discriminated radionuclides, as well as radium, localize in bone and are effectively removed from the third trophic level because bones are not usually eaten by wolf or man.

Sodium-22 is concentrated at all trophic levels by approximately tenfold for lichen–caribou and twofold for caribou–wolf, whereas cesium-137 varies from one to two times for lichen–caribou and two to three times for caribou–wolf. Radionuclide accumulations in lichen vary by species and location; caribou and wolf radionuclide accumulations are age, sex and season dependent.

Each individual situation must be appraised separately. In anticipating accumulation of radionuclides especially in aquatic environments the following should be considered for each particular element involved: 1) its physiological importance to the organism; 2) the physical–chemical state of the element and its acceptability to the organism; 3) the amount of the element in the environment and the presence of elements which may inhibit or enhance its uptake; 4) the morphology of the organism, its life history and its role in the food web; and 5) the physical and chemical characteristics of the environment.

Investigations of environmental radioactivity in the arctic regions of Finland, Norway, Sweden, United States, Canada and the Soviet Union, carried out since the late 1950s, indicate conclusively that special food chains exist in these high latitude regions which lead to relatively high radioactive contamination of plants, animals and man. The artificial radionuclides of this contamination originate from nuclear weapon tests while the natural radionuclides are decay products (largely ^{210}Po and ^{210}Pb) of the noble gas radon – always present in the atmosphere.

Both terrestrial and aquatic food chains will be examined, and artificial nuclides (^{90}Sr, ^{137}Cs, ^{55}Fe) and natural radionuclides (^{210}Po and ^{210}Pb) will be considered.

CHARACTERISTICS OF SELECTED RADIONUCLIDES

Since the environmental behaviour of radioactive isotopes depends to a large degree on their particular physical and chemical properties, we must take special account of such characteristics in seeking ecological understanding of radiation impact. The subsequent discussion treats the major natural and man-made nuclides in this context, apart from tritium or plutonium. These merit consideration, however, in view of the public attention which they have received, and a brief summary of their characteristics follows.

Tritium is a naturally occurring radionuclide of hydrogen with a physical half-life of about 12 years. It is produced naturally by various interactions of cosmic rays with atoms of the upper atmosphere, is then oxidized to water and falls to the earth as rain. It is also produced artificially by numerous reactions taking place within thermonuclear weapons and in power reactors.

Although tritium is produced in relatively large amounts in nuclear reactors and nuclear explosions, it is considered to be one of the least hazardous of the radioactive isotopes. This is largely due to the fact that the low energy beta-ray has low penetrability and hence entails essentially no danger from external exposure. Tritium, acquired internally as water, is excreted relatively rapidly. Rather large doses of tritiated water have been tolerated by experimental animals without obvious injury, and as yet no evidence exists to indicate that tritium is concentrated in moving from soil or water to plants, plants to animals, and on up food chains.

Plutonium is produced in nuclear tests, at AEC production facilities, and in power reactors. The physical half-life of plutonium-239 is about 24,400 years. It deposits preferentially in the skeleton and liver, but absorption of plutonium from the gastro-intestinal tract is low. Absorption through unbroken skin is essentially zero. Absorption rate via wounds is also slow and the plutonium tends to encapsulate and remain at the site where it can produce a tumorogenic response. Hence inhalation presently appears to be the major route of entry of plutonium into the circulatory system. Absorption from the lung may vary from less than 1 per cent of the amount deposited to as much as 10–15 per cent depending upon such variables as chemical form and particle size. A large fraction of the plutonium deposited in the lung is eliminated at a very slow rate – half-life 1 to 4 years. Following inhalation exposure the relative plutonium concentration which can be expected in tissue in descending order is lymph nodes, lung, liver and skeleton.

Ecological uptake of plutonium appears to be a second-order risk. The *discrimination factor* from soils to plants (concentration per gram of plutonium in plant material to that per gram of soil) appears to be 10^{-4} to 10^{-5} depending on plant species, soil type and the chemical form of the plutonium. However some aquatic biota may concentrate plutonium by a factor of 1000 to 3000. In addition, recent reports by Romney *et al.* (1970) and Mork (1970) suggest plutonium may be more biologically available than formerly thought.

A TERRESTRIAL FOOD CHAIN

The movement of ^{137}Cs[1] in an Alaskan lichen–caribou–man or wolf food chain, equivalent to the Lapland lichen–reindeer–man chain, is discussed.

The short arctic food chain lichen–caribou–man effectively concentrates ^{137}Cs. This food chain efficiency is due to several coinciding factors: lichens are grown in great abundance in these cold, nutrient-deficient regions; they have an extremely slow growth rate and a huge capacity to absorb nutrients from the air and rain; caribou eat lichens almost exclusively for about eight months of the year and Eskimos eat large amounts of caribou flesh and organs.

The first link (producer or first trophic level) in this chain constitutes lichens for eight months of the year and vascular plants for the remaining four months. Lichens (symbionts of algae and fungi) are primitive plants which have no real roots and must obtain nutrients directly from the air and rainwater. Since nutrients are obtained slowly and

[1] 137-Cesium is a rare alkali metal with about a 30-year physical half-life and with chemical characteristics very similar to those of potassium, hence its behaviour in physiological processes is analogous to that of potassium. It is for the most part readily soluble in water but strongly absorbed and retained in soil. Thus aquatic plants readily take up cesium while uptake by terrestrial plants is very limited.

since growing conditions are only favourable for short periods of time in the Arctic, these lichens grow only a few millimetres per year (Miettinen, 1969). As their morphology favours effective interception and retention of all kinds of atmospheric debris they also efficiently retain radioactive contamination. They absorb all long half-life nuclides: natural ones such as ^{210}Pb and ^{210}Po, as well as artificial ones such as ^{90}Sr, ^{137}Cs and ^{55}Fe, are equally absorbed. Residence times for these nuclides within the lichens have been variously estimated. Although ^{137}Cs and ^{90}Sr have physical half-lives of about 28 to 30 years it appears that their lichen resident time is much less. This is because ^{137}Cs can be removed by 'wash-out' and because of physiological decay of the lichens. Estimates of the biological half-life of ^{137}Cs range from a high figure of about 17 years (Liden and Gustafson, 1967) to a low of about 4 to 5 years (Miettinen and Häsänen, 1967). Grazing can shorten this half-life considerably; not only are nuclides removed by grazing animals, but the reduced height of the lichens offers less total surface to retain materials. A biological half-life of about 10 years presently seems realistic (Eberhardt and Hanson, 1969). Radiolead (^{210}Pb) decays via radiobismuth to polonium-210, which is the most important natural alpha emitter.

During the summer months caribou diet shifts from lichens to vascular plants. As most vascular plants renew growth each year (the old plant surfaces dying and being replaced with 'fresh' leaf surface) their radionuclide load is reduced and reflects the seasonal deposition of fallout. The new radiation load is a summation of three processes: atmospheric fallout after new growth is resumed (resuspension included); uptake of radionuclides from the soil; and/or uptake from nuclides held within overwintering plant buds. With less fallout in their food the radionuclide burden of the caribou drops continuously all summer long. The biological half-life of ^{137}Cs has two components. The fast component of excretion in reindeer is 50 per cent in 1.5 days and for the slow component 50 per cent in 20 days (Nevstrueva et al., 1967). Other investigators report 27 ± 5 days (Liden and Gustafsson, 1967), 33 days (Ekman and Greitz, 1967) and 27 days (Hanson, 1967) for the slow component. Temperature and potassium level in the diet may affect ^{137}Cs elimination rate. During the autumn and winter the caribou diet reverts almost wholly to lichens; consequently, the radionuclide burden builds again.

The concentration of ^{137}Cs in various lichen communities may vary appreciably depending largely upon the topographic situation, although other circumstances may also be active. Lichen communities such as those where *Cornicularia divergens* predominates are usually found on ridge tops where sweeping winds blow away the snow cover and expose the plants to atmospheric fallout deposition (Hanson, 1968). In sheltered situations, where the lichens may become warmed by the sun, one may expect greater concentration of ^{137}Cs as the snow (bearing ^{137}Cs) is trapped on the lichens, melts, and evaporates leaving the ^{137}Cs behind. Also lichens such as *Cetraria richardsonii*, which have the property of rolling up into a ball during dry periods and unrolling during periods of high humidity, are conducive to retention of nutrients and fallout such as ^{137}Cs.

In general lichen species which grow best in sites lightly snow-covered all winter and are shielded from direct atmospheric fallout will usually contain proportionally lesser amounts of fallout. Lichen communities which occupy deep depressions and which become deeply covered with snow may or may not concentrate ^{137}Cs. Data by Lund et al. (1962) indicate that ^{137}Cs moves rapidly down through the snow. If the snow is in intimate contact with the lichens, one might expect the plants to increase in ^{137}Cs concentration during the winter. However, it is possible, although no definitive experiments have yet been conducted, that the cesium will move through the snow and a disproportionate amount of ^{137}Cs may be deposited onto the ground surface. The Norwegian experiments were conducted in very clean snow. Much wind-driven snow, especially in alpine areas, contains debris, particularly that snow which accumulates in the early portion of the snow season; at the end of summer the ground surface often contains large amounts of dead plant litter and particles of loose soil. If these snowbed communities are dominated by lichens, much of the debris may be entrapped during the melting of the snowfields, and if the ^{137}Cs is not tightly bound to the debris, much of it may be released to the lichens. However in some snowbeds in arctic and alpine areas the ground cover will contain few lichens. The plants which are found here typically are dormant, and remain so until after the snow has melted. Thus these plants will have a very minimum surface area (the perennating organ) to entrap fallout cesium, so that cesium will become incorporated in the soil or in the litter. Thus the amount of fallout cesium which a plant may receive is dependent upon a number of variables; topographic exposure, physiological state, morphology, etc.

The efficiency of ^{137}Cs transfer from lichens to caribou flesh (the producer to consumer link in the food chain) is not universally agreed upon – especially as to whether or not it is a useful figure. However, it appears that the ratio of ^{137}Cs in meat to ^{137}Cs in lichen (60 per cent moisture) varies from 2 to 6 (Hanson et al., 1967; Miettinen, 1969). Much of this variance may be due to different grazing habits in different regions.

As caribou cannot paw deeply into snow to obtain lichens, they graze on communities of lichens which are exposed to the winter winds or have a shallow snow cover. As mentioned earlier, each lichen community may contain quite different amounts of fallout. Thus the snow cover and the preference of the caribou can influence the level of the radionuclides found in the stomach contents and later the muscle of the caribou.

Data collected from 1961 through 1965 by Hanson et al. (1967) consistently demonstrated that the ^{137}Cs content of the lichens in the caribou rumen were much higher than in the collected lichen samples. Several explanations exist. The caribou may not have grazed upon the lichen communities the investigators assumed they were utilizing. The saliva or other juices which were mixed with the lichens contained relatively high amounts of ^{137}Cs and increased the cesium content of the lichens in the rumen. The lichens which were collected by the investigators did not duplicate the grazing of the caribou – either in selection of particular plants or portions of plants eaten. It seems most reasonable that the latter was true, i.e. caribou probably were grazing only on the tops of the lichens. This assumption was borne out by analysing portions of the lichens – the top 4 cm proved to contain much more cesium (during this interval of time).

Although it would appear to be a safe assumption that the lichen content of the rumen reliably reflects the ^{137}Cs contents of caribou food, data by Wilson (1968) indicate, at least in cattle, that much of the soluble fallout cesium is quickly removed from plants while in the rumen. Hence some degree of uncertainty does exist. However when calculating the increase in ^{137}Cs concentration from the lichens to caribou flesh these assumptions are conservative. Extensive data indicate that there is about a factor of 2 increase of ^{137}Cs per g dry weight in the lichens of the rumen to the amount of ^{137}Cs found in the muscle.

The third link or trophic level (primary to secondary consumer), caribou to man (or caribou to wolf) is of great interest for a number of reasons. For example, the values of body burdens in groups of large meat consumers are amazingly similar – at least during the years of heavy fallout 1962–4. This similarity included figures from Sweden, Finland, the Soviet Union and Alaska (Miettinen, 1969). The ratio of ^{137}Cs (kg reindeer meat)$^{-1}$ ^{137}Cs (kg body weight of man)$^{-1}$ in various regions, particularly in the Soviet Union, varies from 2.4 (in Murmansk) to 0.5 (in Yakutsk).

Seasonal variations in reindeer and caribou and in human body burdens of ^{137}Cs are considerable. In Sweden the maximum in man is found in June, the minimum in December. In Finland the maxima and minima are earlier and in Alaska the maximum is found in late autumn as the caribou killed in the spring are high in ^{137}Cs (after eating radio-enriched lichens all winter) and spring-killed animals are eaten throughout the summer.

In addition, greater variability may be contemplated as lifestyles are changing and with these changes body burden changes may be expected. Home freezing units in some of the Lapp and Alaska regions may cause significant shifts in body burdens. With the use of a freezer people need not harvest large amounts of animals in the spring (and dry the meat) when the animals have higher concentrations of ^{137}Cs.

^{137}CS CYCLING IN AN AQUATIC ECOSYSTEM

It was observed in 1961–2 that the ^{137}Cs content of fish in Finnish lakes was high (1–5 $\times$ 10^{-9} Ci per kg fresh weight) compared with values reported from northern Germany (10–100 $\times$ 10^{-12} Ci per kg fresh weight). This is remarkable in that cumulative fallout in Finland was only one-half to two-thirds of that in Germany. This led to a systematic study by Kolehmainen, Häsänen and Miettinen (1966) of factors governing the ^{137}Cs uptake in fish. They studied water, plankton, plants and fish from water sources representing widely different limnological types from eutrophic to oligotrophic lakes in Finland during 1964 and 1965. Their research shows that the four main factors determining the ^{137}Cs body burdens in fish (the last link in the food chain before man) are as follows:

(1) ^{137}Cs content of water – a minor factor as observed differences in lakes varied only about two- to three-fold.
(2) The limnological type of lake (largely potassium content). This was the major factor influencing ^{137}Cs concentration and it effected ten- to one-hundred-fold differences in the same fish species in different lakes.

(3) The quality of food eaten by the fish – a minor factor effecting two- to three-fold differences.

(4) The biological half-life of ^{137}Cs in fish – an important factor varying from 20 to 200 days at 15°C in different species and effecting up to ten-fold differences in various species in the same water system.

Each factor will be discussed briefly.

(1) A survey of Finnish surface waters was carried out by the Institute of Radiation Physics in 1964. ^{137}Cs values were generally between 0.2 and 2×10^{-12} Ci l^{-1}. Thus, variations in the amount of fallout were small compared with up to one-hundred-fold variations observed in the body burdens of fish.

(2) Analysis of samples of aquatic biota for ^{137}Cs and for potassium often resulted in wide variation even from the same lake at different times. However, the ^{137}Cs content per gram of dry weight and especially per gram of potassium changed regularly, being lowest in samples from the eutrophic lakes and highest in the oligo-trophic lakes. The difference between extreme values was ten- to five-hundred-fold. Another clear-cut (inverse) relationship was of the ^{137}Cs content of fish and the potassium content of the water. Down to a potassium content of 1 mg l^{-1} the ^{137}Cs content of pike remained low ($<3 \times 10^{-9}$ Ci kg^{-1}), but with the potassium content slightly less (0.6 mg l^{-1}) the ^{137}Cs reached a high value (21.3 $\times$ 10^{-9} Ci per kg fresh weight). This 'threshold response' was general among fish species.

(3) ^{137}Cs content of food. Small perch contained only about one-half the level of larger perch, the difference likely due to variations in the diet. The small perch consumes mainly plankton and bottom animals while the larger perch consume mainly small fish. However, roach, bream and whitefish also have diets similar to small perch yet have lower ^{137}Cs content. In order to understand this it is necessary to study the fourth factor affecting body burdens of fish.

(4) The biological half-life of ^{137}Cs in different fish species. The biological half-life of ^{137}Cs of fish (the 'slow' component) varies from 20 to 200 days at 15° centigrade. At 5° centigrade the excretion of ^{137}Cs is slowed to about one-half of the value at 15°C. As the annual mean temperature in Finnish lakes is low (around 4–6°C) the average half-life of ^{137}Cs is approximately a year.

In perch the half-life of ^{137}Cs does not change noticeably in fish of different age groups, but in roach (*Leuciscus rutilus*) the old fish have a ^{137}Cs half-life figure twice as long as the younger fish. In rainbow trout (*Salmo iridaeus*) the change is even greater. Pike (*Esox lucius*), too, has a relatively long half-life burden of ^{137}Cs.

The results of this research project demonstrated rather clearly:

(1) ^{137}Cs contamination in biota is accentuated in two ways in oligotrophic waters: the enrichment of ^{137}Cs is increased because of the low potassium level, and it remains longer.

(2) The biological half-life of ^{137}Cs in fish (about 1 year) was much longer than that found in lower animal groups such as the mayfly nymph (about 8 days) and the

midge larvae (about 4 days). Thus the activity of lower animals closely follows that of the ^{137}Cs levels of the water.

(3) In fish the bulk of the cesium body burden is obtained almost entirely through food, the contribution via osmotic mechanisms in the gills being minimal. A ^{137}Cs content of about 1×10^{-12} Ci l^{-1} in oligotrophic lakes effects predatory fish body burdens of 5 to 20×10^{-9} Ci kg^{-1} and in other fish about 1 to 5×10^{-9} Ci kg^{-1}, in eutrophic lakes correspondingly 0.2 to 2 and 0.1 to 1×10^{-9} Ci per kg fresh weight, respectively.

^{55}Fe IN MARINE FOOD CHAINS

The first report of ^{55}Fe in biological material was published by Palmer and Beasley (1965). This report described ^{55}Fe in the lichen–caribou–Eskimo food chain. However, iron contributes greater amounts of radioactivity in marine food chains.

Concentrations of ^{55}Fe in marine fish are about 1000 times higher than values reported for ^{137}Cs, ^{60}Co and ^{110}Ag. The high levels of ^{55}Fe are due to the low concentrations of stable iron in seawater (10 μg l^{-1}), i.e. iron uptake is a result of physiological demand. It appears that a clear relationship exists – the lower the iron content of the water, the higher the specific activity of the fish. This seems to be true for both fresh water and salt water.

6. Natural radionuclides in the arctic environment

Uranium-238 and thorium-232 (and their decay daughters) plus potassium-40 generate the bulk of natural ionizing radiation in our environment (Adams and Lowder, 1964; Fowler, 1965; and Eisenbud, 1963). See Osburn (1965) for a discussion of the kinds and amounts of naturally occurring radionuclides in bedrock and soils and general pathways of movement.

Despite the relatively large number of references concerning natural radionuclides (Klement, 1965), essentially no information exists relating the levels of naturally occurring nuclides in the environment (rock, soil, water and air) with levels concentrated within plants and on up a food chain to man. Hence, recently published data by Professor Miettinen of Finland, Wayne Hanson of Los Alamos Laboratories and Thomas Beasley of the International Laboratory of Marine Radioactivity, Monaco, concerning biological uptake of various naturally occurring nuclides are very welcome.

Kunasheva (1959) has reported the results of a number of interesting studies regarding radium in the biosphere. From his data it would appear that as one proceeds from soils, to plants, to animals a stepwise reduction of one order of magnitude exists, although large variations within categories do exist. Further, Kunasheva noted that plants collected from acid soils tended to have higher concentrations of radium, monocotyledonous plants averaged less radium than dicotyledonous plants; if radium content is expressed on the basis of percent plant ash, the radium content of plants is in close proximity with that of the soils; and as plants increased in size, so did their absolute concentration of radium.

Radiopolonium and radiolead, two naturally occurring nuclides, are somewhat analogous to artificial radioactive fallout nuclides in their pathways of movement. Much of this similarity is because a forerunner of ^{210}Po and ^{210}Pb is a gas (radon) which escapes from the ground and enters the atmosphere. In the atmosphere radioactive decay of the radon may occur and the daughter products (^{210}Po and ^{210}Pb) may become attached to particles and return to the earth.

As radon and daughter products contribute a substantial part of the total radiation in the biosphere these nuclides will be briefly reviewed. The radioecology of the two most important daughters, ^{210}Po and ^{210}Pb, will be discussed in greater detail.

Each of the three natural radioactive decay series (^{238}U, ^{235}U, and ^{232}Th) contains a gaseous intermediate member (Radon, abbreviated Rn). Without this gaseous emanation the entire disintegration series would take place within the bedrock or overlying mantle of soil with radioactivation of the air being minimal. Other than a limited amount of cosmogentically produced nuclides the intermediate radon gases, ^{222}Rn, ^{219}Rn, ^{220}Rn and their radioactive daughter products, comprise the natural radioactivity in the atmosphere.

RADON IN THE SOIL

The half-lives of ^{219}Rn (2.9 sec.) and ^{220}Rn (52 sec.) are very short compared with ^{222}Rn (3.8 days). The brief lives of ^{220}Rn and ^{219}Rn so severely limit their travel that practically none escapes the ground surface. Consequently the atmosphere content is almost entirely of ^{222}Rn and descendants. Once radon gases escape from mineral grains further migration is influenced by the porosity of the rock or soil, the type of fluid filling the pores, and climatic conditions. Factors which aid in keeping radon in the ground (heavy precipitation, temperature, freezing, snow cover, soil compaction, alluvial deposits, or any process tending to seal the ground) will cause a build-up of radon within the ground and diminish it in the air.

Once in the air radon distribution is subjected to a host of variables. However, the gas decays relatively quickly. All ^{222}Rn descendants are solids that are initially electrically charged and are rapidly and irreversibly absorbed onto airborne debris. As solids, the products are subjected to removal and deposition processes quite different from the gaseous parent. The atmosphere contains many kinds, amounts and sizes of aerosols of various origins, the proportions of which change hourly, daily, seasonally, regularly and irregularly. Not only are many aerosols man-made, but many are naturally generated from dust storms, fires, hydrocarbons from trees, ocean spray, volcanic activity and pollen.

The average removal rate of radon from the atmosphere is 15 days. Atmospheric removal processes responsible for the return to earth of various radon daughter nuclides may be categorized as dry or wet return. In general there is reasonably good correlation between total precipitation and total deposition – for both natural and man-generated fallout. However, the type of precipitation, its duration, size of raindrops or snow crystals, distance of transport, and other factors affect the efficiency of precipitation to return particles to the ground. In addition, features such as topography and vegetation type influence the pattern whereby the particulates are deposited and held. Much of the atmospheric debris, carrying its load of radionuclides, is eventually deposited upon plant surfaces.

Comparatively few studies have concerned the efficiency of natural fallout and interception retention by plants (Hill, 1960). However, from a number of studies regarding fallout from atmospheric nuclear weapons testing (Rickard, 1965; Osburn, 1967; and Romney *et al.*, 1963) and from fallout simulation studies (Dahlman *et al.*, 1971; Murphy and McCormick, 1970) it is evident that plant density is almost certainly the single most important feature to use in predicting radionuclide interception and retention. Of course, the pattern of plant distribution and the density of individual plants is important. The season of the year plays a significant role especially in regard to deciduous plants. Studies by Wilson (1968) and by Osburn (1967) indicate that with increasing density of plants (measured as dry weight production) the amount of fallout intercepted increases. However, this increase reaches a plateau. It is likely that at this point the density of the plant is such that the wind does not move readily through it but over and around, hence the plant 'combs' less fallout from the passing winds. Quantitative data on this feature are needed.

Other factors which need to be considered besides the chemical–physical characteristics of the fallout are exposure of the plant, texture, growth rates, seasonal cycles, persistence or longevity, physiological functions, and environment. For example, plants with glaborous, glaucous, pubescent or glanduliferious surfaces, or combinations of these, should vary in particulate interception and retention ability. Climatic factors such as the types, duration, frequency and intensity of precipitation certainly will alter the amount of fallout intercepted and retained by various species of plants. Lichens and mosses can be expected to intercept and retain fallout almost completely.

RADIOPOLONIUM AND RADIOLEAD IN AN AQUATIC ECOSYSTEM

The major source of radiopolonium and radiolead in the hydrosphere is from precipitation. The natural radionuclide ^{210}Pb and its daughter product ^{210}Po are descendants of gaseous ^{222}Rn, and are continuously produced in the atmosphere. The average concentration of ^{210}Pb in precipitation is about 2.5×10^{-12} Ci l^{-1} while ^{210}Po averages about one-tenth of this amount of 0.25×10^{-12} Ci l^{-1}. However this ratio may vary considerably if the period between rains is extensive.

Upon deposition in freshwater lakes or in the oceans the content of both radiolead and radiopolonium decreases greatly. In freshwater the average concentration of ^{210}Pb is about 0.11×10^{-12} Ci l^{-1} – a reduction in lead concentration from atmosphere to water of approximately twenty-three times. ^{210}Po concentration is reduced, from the air to water, by about a factor of 4 to about 0.06×10^{-12} Ci l^{-1}. However these concentrations and ratios of radiolead and radiopolonium vary widely, even in the same lake at different times. Presumably this variance is related to the rainfall regime, mixing conditions and biological activity. Concentrations of both ^{210}Pb and ^{210}Po in fresh waters are reduced rapidly as one moves from river mouths into the open sea. Here the levels of ^{210}Pb and ^{210}Po become similar – about 0.01 to 0.02 × 10^{-12} Ci l^{-1} (Beasley, 1969).

The research of Miettinen (1969) and colleagues offers information which seems to indicate relationships between natural radionuclides and aquatic ecosystems. A number of fresh and brackish water lakes were sampled. From twelve lakes sampled concentrations of ^{210}Pb were between 0.02 and 0.36 × 10^{-12} Ci l^{-1}, with an average of 0.11 × 10^{-12} Ci l^{-1} and the average polonium-lead ratio was 0.53. There appears to be no correlation between the limnological type of lake and the concentrations of polonium and radiolead. Even concentrations in the same lake may vary considerably, presumably because of the rainfall regime, mixing conditions and biological activity – though no seasonal pattern emerged.

A number of fish caught from the same freshwater lakes from which water samples were collected were analysed along with a number of fish caught from the Gulf of Finland. The highest polonium and radiolead concentrations were found in plankton-eating fish such as Cisco (*Coregonus albula*) and Baltic herring (*Clupla harengus*). Next highest in polonium and radiolead were bottom-feeding fish such as the roach (*Leuciscus rutilis*). In general predator types of fish such as turbot (*Lota vulgaris*) show lowest concentrations of polonium and radiolead. However the predator fish – pike (*Esox lucin*) was an exception.

When one considers the amount of radiolead and polonium which may be transferred to man by eating fish we would expect a sharp decrease because about half to three-quarters of the polonium and radiolead will be located in the fish entrails and bone, hence not passed on to man.

Likens and Johnson (1969) measured the background radiation in a number of Alaskan arctic and subarctic aquatic habitats. They found the distribution of background radiation to correspond closely to the pattern observed in Antarctic, New England and Wisconsin lakes. That is, relatively low count rates if the detector was more than a metre from any boundary; in lakes deeper than 4 m mid-depth values consistently approached 150 counts per minute.

7. Radiation effects in arctic and alpine regions

Effects of ionizing radiation given in chronic or acute dosages, from internal or external emissions and of both natural and artificial radiation sources, have been investigated at all levels of biological organization from intra-cellular, cellular, tissue, organ, organism, population and assemblages of populations or ecosystem (Osburn, 1970).

Effects at the organism level of organization have included death, life-shortening, impairment of physiological processes, metabolism imbalance, interruption of biological rhythm, dormancy, home range size change, behavioural changes, aestivation, hibernation, productivity, trophic transfers, tropism, sense impairment (including navigational ability) of a wide array of plants and/or animals from the most primitive single-celled to those of high complexity and evolutionary advancement. Many stages within the life history of many groups of organisms have been investigated, e.g. in the case of insects, egg, larva, pupae, immature, and adult.

At the ecosystem level sensitive indicators of response to ionizing radiation have been changes in: composition of organisms (diversity), nutrient cycling, biomass, sensitivity to disease and insect attack, ecesis (ability of plants or animals to invade and colonize an area), onset of plant dormancy, spring dormancy release, population numbers, and many others.

In an effort to alter the radiation sensitivity of organisms, field and laboratory experiments have included, besides irradiation, subjecting the various organisms to severe environmental stresses (high and low temperatures, rapid temperature fluctuations, intense light, shade, oxygen deficiencies and surpluses, predation, disease, nutrient deficiencies, toxins, crowded situations, pesticides, alteration of material transfer between links within food webs: environmental stresses applied singly and in various combinations). Experiments have been conducted during all phases of an organism's development and life span including their natural daily and seasonal activity patterns.

No effects have been detected at levels of radiation exposure which organisms are presently receiving in arctic and alpine regions. Experimental 'effect levels' of radiation were usually several orders of magnitude greater.

Few arctic or alpine plants and animals have been directly tested to determine their radiosensitivities. Though one would not expect arctic–alpine fauna or flora to be affected from present levels of radiation, several sensitive prediction techniques which are, or may be, used to predict or detect radiation effects in arctic or alpine species will be discussed.

PREDICTION OF RADIATION EFFECTS IN PLANTS

The use of interphase chromosome volume and the tolerance to harsh environments are used to predict radiosensitivity of plants. The former is the single most promising technique so far developed to predict effects of ionizing radiation on plants; it was developed and perfected by Sparrow and associates (Sparrow, 1962). A positive correlation between the volume of cell nuclei (or chromosome volume) and radiosensitivity of plant meristem cells has been demonstrated. The volume is reasonably constant within plant species. Those with larger chromosome volumes at interphase are proportionally more radiosensitive. Plant-radiosensitivity-regression curves, relating interphase chromosome volume to radiation exposure, have been prepared using various endpoints such as slight effect, severe growth inhibition, and lethality. The regressions can then be employed in estimating the radiation tolerance of species for which directly determined data are unavailable.

Environmental tolerance and radioresistance
Woodwell's supposition contends that plants adapted to tolerate harsh environmental conditions are also more resistant to radiation damage: '... it would seem that characteristics that confer resistance to certain types of environmental extremes also confer

resistance to damage by radiation' (Woodwell, 1965). Additional information to sub-stantiate the above is offered by Brayton and Woodwell (1966), wherein it was noted that in a continuously irradiated forest community a sorting by size or life form (trees, shrubs, tall herbs, prostrate plants) occurred along the radiation gradient, smaller forms of life being generally more radiation resistant than larger ones.

Primitive life forms, such as lichens as contrasted to vascular plants, are many times more resistant to given doses of ionizing radiation.

PREDICTION OF RADIATION EFFECTS IN ANIMALS

Two techniques used to predict radiosensitivity in animals are: 1) modification of inter-phase chromosome volume (ICV) technique; and 2) chromosome analysis.

Modification of ICV (interphase chromosome volume) technique
Estimates of LD_{50} (30–60 days) – lethal dose for 50 per cent of a test group in 30 to 60 days – ranging from 150 to 1500 rads (radiation absorbed dose) have been reported for mammals, but no satisfactory explanation has been advanced for these differing radiosensitivities. As a consequence, it has not been possible to predict such responses as LD_{50} for an untested species. However Dunaway *et al.* (1969) and associates at the Oak Ridge National Laboratory, Oak Ridge, Tennessee, are actively pursuing this prediction objective. Their research is briefly reviewed.

Comparative data from studies of radiation effects and haematology in closely related mammal species were incorporated with information from other species in an effort to supply a predictive index for LD_{50}. Because of a reasonably reliable predictive capability for plants, a correlation between interphase chromosome volume and LD_{50} was tested for mammals but could not be confirmed. However an index obtained by utilizing ICV and the elimination constant (b) of ^{59}Fe gave a close prediction of LD_{50} at the subfamily level. A more complicated formula utilizing mean corpuscular haemoglobin, oxygen consumption, red blood count, and mean survival time at 1200 rads provides fairly good estimates for all taxa tested, but the predictive advantage is lessened by the need to know the mean survival time at one dose.

Chromosome analysis
Excepting some small groups of Pacific Island people, arctic man – Eskimos and Laplanders – has almost certainly been exposed to higher levels of radiation fallout than any other groups in the world. Hence it is well to compare this man-generated burden with the natural radionuclide burden. Further, one should consider the possi-bility of detecting any effects from exposure to these levels of ionizing radiation.

The annual radiation dose from all sources to the Lapp population has been around 260 millirem (rem = roentgen equivalent man). The dose contribution for artificial fallout nuclides, largely ^{137}Cs, ^{55}Fe, and ^{14}C is about 30 per cent of the mean annual dose. Thus, unless the fallout situation changes significantly (additional testing) the radiation dose due to the fallout will be less than 20 per cent of the total radiation dose to one generation. Hence, the contribution of the natural radiation sources to the annual dose was greater than that of the artificial sources even in the period of highest exposure for man-generated fallout.

With the development of cytogenetics, a very sensitive tool known as karyotyping is emerging whereby we can assess possible genetic changes due to stresses such as ionizing radiation. Analysis of chromosomal aberrations is presently one of the most sensitive indicators of radiation dose, and use of this technique holds much promise as it might lead to a more direct answer to the question of what is the lowest radiation level wherein one can detect an effect.

In 1969 (O'Farrell et al., 1970) a cooperative study was organized between scientists from the Terrestrial Ecology Section, Battelle–Northwest, and the Human Cytogenetics Section, Oak Ridge National Laboratory, to document chromosomal aberrations in Eskimos chronically ingesting radionuclides via the now classic lichen–caribou–man food chain.

Chromosome analysis was carried out on cultured peripheral leukocytes obtained from 21 male Alaskan Eskimos who were known to carry an excessive body burden of ^{137}Cs. Samples were taken twice during the year in accord with the highest and lowest body burdens observed. The mean high body burden was 937×10^{-9} Ci and the mean low body burden was 228×10^{-9} Ci. A total of 22 ring and dicentric chromosome configurations were observed in a total of 10,150 metaphase spreads scored. In addition, karyotype analysis of selected cells showed the occurrence of several symmetrical translocations and inversions.

Similar analysis of peripheral leukocytes was also carried out on a control population of Alaskan Eskimos that were known to carry a very low body burden of ^{137}Cs, due to differences in dietary habits. In this latter instance only 1 dicentric chromosome was observed in a total of 3300 metaphase spreads. Although the chromosome aberration yield in the population with the high body burden is low, it appears to be significant.

The interpretation of these preliminary data requires extreme caution owing to the multiplicity of factors which could be responsible for the appearance and maintenance of these aberrations in a small, closed society such as mountain Eskimos at Anaktuvuk Pass. But the aberrations do exist and effort must be expended to evaluate their significance. As the research programme expands other ethnic groups will be examined. In addition biological materials will be obtained from caribou and canines such as wolf and fox. These animals, serving as ecological equivalents of man, can be obtained in relatively large numbers and by destructive techniques their burdens of radioactivity precisely determined; hence better correlations between chromosomal anomalies and radioactivity can be made.

The major study objective will be to obtain suitable information on the cytogenetics of northern people, simultaneously obtaining necessary correlative information on body burdens, medical therapy history, food habits, and genealogy, to form a frame of reference to judge the importance of such genetic anomalies.

If successful it is possible that karyotyping of key plants and animals will afford an extremely sensitive baseline whereby various environmental (pollution) genetic stresses can be evaluated.

SOME RADIATION EFFECTS IN AN ARCTIC MARINE ENVIRONMENT

The amounts of radioactive material allowed to be discharged to the marine environment has, in the past, emphasized safety to man. The implicit assumption is simply if it is safe for man it is also safe for all other fauna and flora. This is not because man is egocentric but simply because he appears to be the most radiosensitive organism. However it is possible to conjecture that certain marine organisms eventually might receive significant radiation doses even when radionuclides are released according to recommended International Commission Radiological Protection guides. These 'unexpected' concentrations could arise because of nuclide reconcentration efficiencies of the animals themselves or some part of their habitat. Though the available information relevant to marine organisms is often quite general and sometimes conflicting and quite limited for arctic marine organisms, it is possible to make some general conclusions.

Effects of radiation on marine species – acute exposure

In agreement with the greater amount of available data relating to terrestrial animals, it is apparent that radiosensitivity tends to increase with the increasing biological complexity of organisms. There is a relationship between radiosensitivity and phylogeny and ontogeny of the organisms (Nat. Acad. Sci./Nat. Res. Council, 1971a, b). Further, radiosensitivity varies according to stage of life history development. In general radioresistance increases regularly with increase in age. This may well vary by a factor of 10 or more. Immediately after conception, during early stages of cell division, most organisms are relatively radioresistant. As organ formation is initiated radiosensitivity increases.

Ionizing radiation effects may also interact with other 'effect-producing' factors. Some radiation effects may be intensified by particular stress while some effects may be reduced (ecological compensation): sometimes the effects may be synergistic. Though rules by which interactions may be predicted have yet to be formulated and tested, experiments by Angelovic, White and Davis (1969), summarized below, indicate some general complexities.

Thirty days after acute irradiation fish (*Fundulus heteroclitus*) were more tolerant of radiation at low water salinities and lower temperatures. At 60 days after irradiation the apparent protective effect of low salinity had disappeared at 27°C though it was still present at 22° and 17°C. At 12°C it was, surprisingly, the high salinity environment which appeared to provide protection against radiation effects. Thus the water temperature and salinity interact to modify the effects of radiation.

From a more realistic viewpoint it is more pertinent to consider the effects of chronic irradiation. Of course any environmental factor which might reduce the rate of organism development could be expected to intensify the radiation effect simply because added time would mean greater radiation dose. Under particular sets of circumstances radiation 'stimulation' has been reported – usually as to growth rate. Again, in another set of experiments concerning temperature, salinity and irradiation, of 9 effect parameters measured all were significantly altered by temperature, 5 by salinity, and only 2 by irradiation. The interactions between irradiation and salinity and irradiation and temperature affected 4 and 8 of the parameters respectively. The interaction between temperature and salinity did not contribute significantly to the variation of any parameter during the experiment. The three-way interaction – irradiation, temperature, and salinity – affected 7 of 9 parameters. These experiments demonstrate in a convincing manner that as many environmental parameters as possible should be studied simultaneously if the ecological effects of environmental variations are to be properly understood.

As stated earlier, the highest dose rates will be received by animals which either accumulate large quantities of radioactivity directly from the water or via a food chain, or which live in a habitat that has a high efficiency for accumulating radioactive materials; sediments for example.

It is generally assumed that there is no safe radiation dose especially from a genetic standpoint, i.e. any dose of radiation can induce mutations and mutations are usually harmful. Nevertheless, at the population level genetic damage does not long remain in the population. This is because these new aberrations tend to make the organism less fit to compete and they are quickly eliminated through natural selection (Blaylock, 1966).

PREDICTING ECOSYSTEM RECOVERY

Early predictions (twenty-five years ago) of 'ecosystem recovery' after exposures to high levels of ionizing radiation varied from 'near zero' recovery to 'complete' recovery. Considerable data and a wealth of experience in a wide array of regions has since been obtained by AEC staff biologists and by those from universities who have participated in research programmes. Among those areas which have been studied for recovery evidence are:

(1) Ecosystems directly damaged by nuclear device testing: a) Trinity Site, Alamagordo, New Mexico, first nuclear test; b) Bikini and Eniwetok Atolls (and, indirectly, Rongelap Atoll) in the Marshall Islands; and c) Nevada Test Site, Nevada, and a number of cratering experimental sites.

(2) Ecosystems subjected to relatively massive exposures of ionizing radiation from gamma ray sources (cobalt-60 or cesium-137) given in both chronic (continuous, or long-term) and acute (prompt, or short-term) doses: a) Brookhaven Ecology Forest, Long Island, New York, oak-pine forest; b) Yucca Flat, Nevada desert plant and animal enclosures; c) Terrestrial Ecology Project, Puerto Rico Nuclear Center, a tropical rainforest; d) Savannah River Ecology Laboratory experiments, old fields, pine and deciduous forests; and e) Colorado State University Project Prairie Grassland.

(3) Ecosystems subject to relatively high radiation from radioactive wastes: a) Oak Ridge, White Oak Lake bed – a meadowland community established after draining of a lake which has been used as a settling pond for radioactive waste from ORNL facilities between 1945 and 1955.

(4) Ecosystems directly damaged by neutron irradiation from unshielded reactors: a) Marietta, Georgia southeastern pine forest adjacent to Lockheed Materials Testing Reactor; and b) Oak Ridge National Laboratory, southeastern pine forest community in vicinity of Oak Ridge Health Physics Research Reactor.

The one result common to the above studies is that recovery tends to follow 'ecologically normal' and predictable successional trends; i.e. patterns following damage from radiation are similar to those following fire, logging, or stripping of topsoil and other natural catastrophic events. Though it is dangerous to extrapolate ecological responses from region to region it would seem that in this case – recovery follows natural successional patterns in ecosystems ranging from a tropical rainforest to a prairie grassland – it would be true in arctic and alpine.

Additional research is needed to develop sensitive indicators of radiation damage to provide an early warning system in all major ecosystems. Investigations need to be continued to develop radioecological rules and principles particularly in regard to interactions (especially synergistic effects) between radiation and the many pollutants and variations of environmental conditions found not only in the marine environment, but also freshwater and terrestrial environments.

References

ADAMS, J. A. S. and LOWDER, W. M. (1964) *The Natural Radiation Environment*. Chicago: Univ. of Chicago Press. 1069 pp.

ANGELOVIC, J. W., WHITE, J. C. JR and DAVIS, E. M. (1969) Interactions of ionizing radiation, salinity, and temperature on the estuarine fish, *Fundulus heteroclitus*. *In:* NELSON, D. J. and EVANS, F. C. (eds.) Symposium on radioecology, *US AEC Rep.*, CONF-670503, 131–41. New York: US Atomic Energy Commission Operations Office.

BEASLEY, T. M. (1969) *Lead-210 in selected marine organisms*. Unpub. thesis, Oregon State Univ.

BERGH, H., FINSTAND, G., LUND, L., MICHELSON, O. B. and OTTAR, B. (1961) *Radiochemical Analysis of Precipitation, Tap Water, and Milk in Norway 1957–1958. Methods, Results and Conclusions*. Lillestrom: Norwegian Defense Res. Estab. 115 pp.

BLAYLOCK, B. G. (1966) Chromosomal polymorphism in irradiated natural populations of *chironomus*. *Genetics*, **53** (1), 131–6.

BLAYLOCK, B. G. (1969) The fecundity of a *Gambusia affinis affinis* population exposed to chronic environmental radiation. *Radiation Res.*, **37** (1), 108–17.

BRAYTON, R. D. and WOODWELL, G. M. (1966) Effects of ionizing radiation and fire on *Gaylussacia baccata* and *Vaccinium vacillans*. *Amer. J. Bot.*, **53** (8), 816–20.

DAHLMAN, R. C., DINGER, B. E., HARRIS, W. F., HENDERSON, G. S., MANN, L., PETERS, L. N., SOLLINS, P., TAYLOR, F. G. and WITHERSPOON, J. P. (1971) Plant-atmosphere relations. *In: Ecol. Sci. Div. Ann. Progr. Rep.*, 2–15. ORNL-4759. Oak Ridge, Tenn.: Oak Ridge Nat. Lab.

DUNAWAY, P. B., LEWIS, L. L., STORY, J. D., PAYNE, J. A. and INGLIS, J. M. (1969) Radiation effects in the Soricidae, Cricetidae and Muridae. *In:* NELSON, D. G. and EVANS, F. C. (eds.) Symposium on radioecology, *US AEC Rep.*, CONF-670503, 173–84. New York: US Atomic Energy Commission Operations Office.

EBERHARDT, L. L. and HANSON, W. C. (1959) A simulation model for an arctic food chain. *Health Phys.*, **17**, 793–806.

EISENBUD, M. (1963) *Environmental Radioactivity.* New York: McGraw-Hill. 430 pp.

EKMAN, L. and GREITZ, U. (1967) Distribution of radiocesium in reindeer. *In:* ABERG, B. and HUNGATE, F. P. (eds.) *Radioecological Concentration Processes*, 655–62. Oxford: Pergamon Press.

FOWLER, E. G. (1965) *Radioactive Fallout, Soils, Plants, Food, Man.* Amsterdam: Elsevier. 317 pp.

FRENCH, N. R. (1967) Comparison of radioisotope assimilation by granivorous and herbivorous mammals. *In:* ABERG, B. and HUNGATE, F. (eds.) *Radioecological Concentration Processes*, 665–73. Oxford: Pergamon Press.

HANSON, W. C. (1967) Radioecological concentration processes characterizing arctic ecosystems. *In:* ABERG, B. and HUNGATE, F. P. (eds.) *Radioecological Concentration Processes*, 183–91. Oxford: Pergamon Press.

HANSON, W. C. (1968) Fallout radionuclides in northern Alaskan ecosystems. *Arch. Environ. Health*, **17** (4), 639–48.

HANSON, W. C., WATSON, D. G. and PERKINS, R. W. (1967) Concentration and retention of fallout radionuclides in Alaskan arctic ecosystems. *In:* ABERG, B. and HUNGATE, F. P. (eds.) *Radioecological Concentration Processes*, 233–45. Oxford: Pergamon Press.

HILL, C. R. (1960) Lead-210 and polonium-210 in grass. *Nature*, **187** (4733), 211–12.

HOLE, J. F. (1970) *The Effects of Needle Ice Damage on Vegetation and Protection of New Seedlings on Sloping Areas from Erosion and Frost Heave, A Bibliography.* Hanover, NH: US Army CRREL. 29 pp.

HOLTZMAN, R. B. (1966) Natural levels of lead-210, polonium-210, and radium-226 in humans and biota of the Arctic. *Nature*, **210**, 1094–7.

KLEMENT, A. W. JR (1965) Natural environmental radioactivity: an annotated bibliography. *US AEC Rep.*, WASH-1061, iii. New York: US Atomic Energy Commission Operations Office. 125 pp.

KLEMENT, A. W. JR (1970) Natural environmental radioactivity: an annotated bibliography. *US AEC Rep.*, WASH-1061 (Suppl.). New York: US Atomic Energy Commission Operations Office. 72 pp.

KOLEHMAINEN, S., HÄSÄNEN, E. and MIETTINEN, J. K. (1966) ^{137}Cesium levels in fish of different types of lakes in Finland during 1963. *Health Phys.*, **12**, 917–22.

KUNASHEVA, K. G. (1959) The amount of radium in plants and animals. *Trav. Lab. Biogeochim. USSR*, **16**, 30.

LIDEN, K. and GUSTAFSON, M. (1967) Relationships and seasonal variation of ^{137}Cs in lichen, reindeer and man in northern Sweden 1961–1965. *In:* ABERG, B. and HUNGATE, F. P. (eds.) *Radioecological Concentration Processes*, 193–208. Oxford: Pergamon Press.

LIKENS, G. E. and JOHNSON, P. L. (1969) Measurements of background radiation in aquatic habitats in Alaska. *In:* NELSON, D. J. and EVANS, F. C. (eds.) Symposium on radioecology, *US AEC Rep.*, CONF-670503, 319–29. New York: US Atomic Energy Commission Operations Office.

LIST, R. J. and TELEGADAS, K. (1961) Global atmospheric radioactivity; May-June 1960. *In:* Fallout program quarterly summary rep., *US AEC Rep.*, HASL-111, 150–5. New York: US Atomic Energy Commission Operations Office.

LUND, L., MICHELSON, O. B., OTTAR, B. and WIK, T. (1962) *A Study of Sr-90 and Cs-137 in Norway 1957–1959.* Kjeller: Norwegian Defense Res. Estab. 129 pp.

MIETTINEN, J. K. (1969) Enrichment of radioactivity by Arctic ecosystems in Finnish Lapland. *In:* NELSON, D. J. and EVANS, F. C. (eds.) Symposium on radioecology, *US AEC Rep.*, CONF-670503, 23–31. New York: US Atomic Energy Commission Operations Office.

MIETTINEN, J. K. and HÄSÄNEN, E. (1967) ^{137}Cs in Finnish Lapps and other Finns in 1962–1965. *In:* ABERG, B. and HUNGATE, F. P. (eds.) *Radioecological Concentration Processes*, 221–31. Oxford: Pergamon Press.

MORK, H. M. (1970) *Redistribution of Plutonium in the Environs of the Nevada Test Site.* Lab. Nuclear Med. and Rad. Biol., Univ. of California. 29 pp.

MURPHY, P. G. and MCCORMICK, J. F. (1970) Ecological effects of acute beta irradiation from simulated fallout particles upon a natural plant community. *In:* BENSEN, D. W. and SPARROW, A. H. (eds.) *Survival of Food Crops and Livestock in the Event of Nuclear War*, 454–81. Washington, DC: US Atomic Energy Commission Office of Information.

NAT. ACAD. SCI./NAT. RES. COUNCIL (1971a) *Radioactivity in the Marine Environment* (a summary report). WASH 1185. Washington, DC. 28 pp.

NAT. ACAD. SCI./NAT. RES. COUNCIL (1971b) *Radioactivity in the Marine Environment.* Washington, DC. 418 pp.

NEVSTRUEVA, M. A., RAMZAEV, P. V., MOISEEV, A. A., IBATULLIN, M. S. and TEPLYKH, L. A. (1967) The nature of ^{137}Cs and ^{90}Sr transport over the lichen–reindeer–man food chain. *In:* ABERG, B. and HUNGATE, F. P. (eds.) *Radioecological Concentration Processes*, 209–15. Oxford: Pergamon Press.

O'FARRELL, T. P., BREWEN, J. G., SOLDAT, J. K., EBERHARDT, L. L. and WOGMAN, N. A. (1970) *Chromosomal Aberrations in Alaskan Eskimos.* Rep. Washington, DC: US Atomic Energy Commission. 5 pp.

OSBURN, W. S. JR (1963) The dynamics of fallout distribution in a Colorado alpine snow accumulation ecosystem. *In:* SCHULTZ, V. and KLEMENT, A. W. JR (eds.) *Radioecology*, 51–71. New York: Reinhold.

OSBURN, W. S. JR (1965) Primordial radionuclides: their distribution, movements, and possible effect within terrestrial ecosystems. *Health Phys.*, **11** (12), 1275–95.

OSBURN, W. S. JR (1967) Ecological concentration of nuclear fallout in a Colorado mountain watershed. *In:* ABERG, B. and HUNGATE, F. P. (eds.) *Radioecological Concentration Processes*, 675–709. Oxford: Pergamon Press.

OSBURN, W. S. JR (1968) Forecasting long-range recovery from nuclear attack. *In:* NAT. ACAD. SCI./NAT. ACAD. ENG./NAT. RES. COUNCIL, *Proceedings of the Symposium on Postattack Recovery from Nuclear War*, 107–35. Washington, DC.

OSBURN, W. S. JR (1969) Accountability of nuclear fallout deposited in a Colorado high mountain bog. *In:* NELSON, D. G. and EVANS, F. C. (eds.) Symposium on radioecology, *US AEC Rep.*, CONF-670503, 578–86. New York: US Atomic Energy Commission Operations Office.

OSBURN, W. S. JR (1970) Nuclear Power and Environmental Effects Studies. *US AEC Congressional Lecture Ser.*, 4. Washington, DC: US Atomic Energy Commission. 45 pp. Mimeographed.

PALMER, H. E. and BEASLEY, T. M. (1965) Iron-55 in humans and their food. *Science*, **149**, 431–2.

PRESIDENT'S SCIENCE ADVISORY COMMITTEE, ENVIRONMENTAL POLLUTION PANEL (1965) *Restoring the Quality of Our Environment*. Washington, DC: US Govt Printing Office.

REICHLE, D. E., DUNAWAY, P. B. and NELSON, D. G. (1970) Turnover and concentration of radionuclides in food chains. *Nuclear Safety*, **11** (1), 43–55.

REITER, E. R. (1971) Atmospheric transport processes, Part 2: Chemical tracers. *US AEC Crit. Rev. Ser.* US Atomic Energy Commission. 382 pp. (Available from NTIS, Springfield, Va., as TID-25314.)

RICKARD, W. H. (1965) Field observations on fallout accumulation by plants in natural habitats. *J. Range Mgt.*, **18** (3), 112–14.

ROMNEY, E. M., LINGBERG, R. G., HAWTHORNE, H. A., BYSTROM, B. G. and LARSON, K. H. (1963) Contamination of plant foliage with radioactive fallout. *Ecology*, **44**, 343–9.

ROMNEY, E. M., MORK, H. M. and LARSON, K. H. (1970) Persistence of plutonium in soil, plants, and small mammals. *Health Phys.*, **19**, 487–91.

SPARROW, A. H. (1962) The role of the cell nucleus in determining radiosensitivity. *Brookhaven Lecture Ser.*, 17, BN: 766. Upton, NY: Brookhaven Nat. Lab. 88 pp.

TROLL, C. (1958) Structure soils, solifluction, and frost climates of the earth. *SIPRE Transl.*, 43. US Army Snow, Ice and Permafrost Res. Estab. 121 pp.

UNITED NATIONS (1962) Report of the United Nations Scientific Committee on the effects of atomic radiation. *General Assembly Official Records*, 17th Session, suppl. 16 (A/5216). New York. 441 pp.

UNITED NATIONS (1964) Report of the United Nations Scientific Committee on the effects of atomic radiation. *General Assembly Official Records*, 19th Session, suppl. 14 (A/5814). New York. 120 pp.

UNITED STATES ATOMIC ENERGY COMMISSION (1971) Environmental aspects. *In: Fundamental Nuclear Energy Research*, 107–39. Washington, DC.

WILIMVOSKY, N. J. and WOLFE, J. N. (1966) *Environment of the Cape Thompson Region, Alaska*. Div. of Tech. Inf., US Atomic Energy Commission. 1250 pp.

WILSON, W. D. (1968) *A Mathematical Model of the Transport of Cesium-137 from Fallout to Milk*. Unpub. Ph.D. dissertation, Colorado State Univ. 139 pp.

WITHERSPOON, J. P. and TAYLOR, F. G. JR (1971) Retention of 1-44u simulated fallout particles by soybean and sorghum plants. *Health Phys.*, **21**, 673–8.

WOODWELL, G. M. (1965) Radiation and the patterns of nature. *Brookhaven Lecture Ser.*, 45, BNL 924 (T-381). Upton, NY: Brookhaven Nat. Lab. 55 pp.

Manuscript received June 1972.

CHAPTER 17

The impact of twentieth-century technology

For little more than a decade there has been an upsurge of concern over the irreparable damage that is occurring, and may continue to occur at an accelerating rate, to the arctic ecosystems. This concern, as indicated in the Introduction, has been brought into sharper focus by planned oil and mineral developments in the Alaskan and Canadian arctic sectors. The potential for damage is obviously great and varied, ranging from physical disruption of the land surface and its flora and fauna to significant changes in the economies and ways of life of native peoples. Because of the broad geographic integrity of the circumpolar arctic lands the discussion is easily focused, whereas the identification of problems in diverse and disjunct alpine areas has remained essentially regional or local. Nevertheless, the similarities between the two systems seem to justify retention of the term 'arctic–alpine' in the present context.

A complete coverage of these considerations is impossible given our present limited information on them. Accordingly, the topic is presented as a series of case studies in order at least to illustrate the variety of problems now emerging. Before discussing these environmental questions it is necessary to enter certain caveats. There is always a risk of one-sided value judgements in interpreting environmental modifications. Man and his attendant technology will continue to consume non-renewable natural resources at an accelerating rate at least until world population and consumption patterns have stabilized. Necessarily, therefore, the 'natural' biomes form a diminishing part of the biosphere. One cannot argue, for instance that a group of indigenous peoples must be preserved untouched by the twentieth-century world regardless of their location astride extensive mineral resources. Two quotations from a recent report of a conference at Ditchley Park compiled by Dr Brian Roberts (1971)[1] illustrate the enigma of the native peoples of the northlands:

You cannot take the Eskimos up a high mountain and show them the false glitter of

[1] References for Section A are collected on pp. 923–4 and for Section B on pp. 950–1.

this world and say 'This is not for you'. You have to go the whole way in either direction – no culture contacts and no education, or alternatively bring the Eskimos as gently as possible into the mainstream of the world . . .

and,

It is doubtful whether there is now any possibility of preserving their old culture, so that complete assimilation with the white population must apparently be the ultimate objective. It is neither possible nor desirable, in the view of the administration of Alaska, to try to put the clock back.

Next we need to consider to what extent an environment can support disturbance. The concept of the fragility of the tundra has probably been too uncritically accepted as a convenient shibboleth by conservationists, in order to exclude totally resource development from the Arctic. Undoubtedly, wet tundra, which may contain frozen water up to five or six times in excess of dry soil weight within the upper 40 metres, is fragile by any definition, and the proposed trans-Alaska pipeline would pass through, or over, such terrain in a large proportion of its span. This problem is rare in the alpine. Conversely, great tracts of the Arctic are underlain by coarse sands and gravels or bedrock that are not nearly so susceptible to heat-induced disturbance. Degree of disturbance potential could be mapped on a regional scale according to water content, grain size and vegetation physiognomy. However, there are other reasons behind the concept of the fragile tundra that relate specifically to plant adaptation and community structure (cf. Billings, Chapter 8A) and primary productivity considered both as daily rates and total season's production (cf. Webber, Chapter 8B). Bliss (1970) has argued that the very limited number of plant species in the Arctic is a real indication of fragility of the ecosystem since adverse impact on a very few species may damage the whole: the potential for recovery may be proportional to plant community variability. Dunbar (1973) emphasizes that a proper consideration of 'fragility' hinges on one's definition of stability. A steady-state situation, which may pertain to arctic lakes and perhaps also to tundra vegetation, is vulnerable to excessive disturbance since the system cannot adapt. Another type of stability is represented by a system which can absorb disturbance and return to its previous state. Tundra animal populations undergo drastic fluctuations but recover, perhaps as a result of their ability to draw on outside areas. Dunbar concludes that simplicity of an ecosystem, coupled with slow growth rates, leads to vulnerability to disturbance which can be tolerated only if the spatial scale is large (as in the Arctic) and the time scale is sufficiently long.

These general concepts should be borne in mind as the following case studies are examined. One obvious conclusion can be anticipated here: much more intensive applied research is necessary before rational resource development alternatives can be posed.

Small-scale examples (1)
The impact of motor vehicles
on the tundra environments

Jack D. Ives

Institute of Arctic and Alpine Research,
University of Colorado

Hitherto, the mountain regions of the world have formed a refuge where a relatively small and privileged group (privileged in the sense of possessing that combination of good health, motivation and aesthetic appreciation) could enjoy a wilderness experience. In comparison with the vastness of the area available, their numbers were few and their environmental impact insignificant. There were a small number of exceptions, such as the more popular mountain resort areas of Europe, yet generations of hikers and climbers failed to make any impact comparable to that which may occur in a single year during the 1970s.

This change has been brought about largely by two mid-twentieth-century developments: the increasing affluence of western society and the rapid application of World War II and Korean War technology to the production of an increasing range of 'off-the-road' vehicles. The ATV (all-terrain-vehicle), whatever its form, whether snowmobile, four-wheel drive automobile, or tracked and floatable personnel carrier, has facilitated mountain access by ever-increasing numbers of people. In many instances the new-comers are not trained in wilderness lore, so that mechanical breakdown and human fatigue sometimes place them woefully unprepared in potentially dangerous situations. Search and rescue operations in turn have additional mechanical impact.

Much of the damage caused by trail bikes and four-wheel drive vehicles (Plate 43) results from illegal operation, but in the absence of regulations, a clash between hiker and trail-biker is inevitable. Physical damage to vegetation cover and soil is already visible in areas within ready reach of urban centres and will continue to grow without enforceable legislation (see Webber, fig. 8B.9).

Perhaps the greatest potential for environmental damage by the oversnow vehicle lies in the fact that it provides man with his greatest possible mobility at a time when nature is least mobile. Little is known about the psychological need of wildlife for the silence and imperturbability (from man) of the hitherto snowbound winter; yet the harsh shattering of the winter sanctuary by the devastating roar of the snowmobile leaves naturalists with due cause for concern.

This section will concentrate on the snowmobile since it represents perhaps the most dangerous potential for impact, a potential rendered all the more difficult to assess because of the many indirect aspects and the current critical absence of research.

Some of the earliest attempts to apply mechanized transport to oversnow logistics followed close on the heels of Henry Ford himself. Captain Robert Falcon Scott tested motor sledges during his quest for the South Pole in 1911. Much later (1947) Colonel P. D. Baird led a massive and largely successful mechanized trek through the Canadian Subarctic and Arctic using 'weasels' during Operation Muskox. Paul Emile Victor's Expéditions Polaires Françaises yielded notable results in Greenland in the early 1950s, shortly thereafter being overshadowed by the US Army Corps of Engineers. Sir Ernest Shackleton's dream of a trans-Antarctic expedition was realized, again with oversnow vehicles, by the British/New Zealand combination of Sir Vivian Fuchs and Sir Edmund Hilary. These are but a few of the arctic and antarctic developments that rapidly grew in number from 1950 onwards. But they were all typified by large, heavy, extremely expensive tracked vehicles most easily adapted to the unlimited expanses of the world's great ice sheets. They were scientific, logistic and military ventures backed by large sums of government money; there were paralleled somewhat by big business mineral exploration such as the search for oil in the North American Arctic. Yet two fairly immediate developments ensued: 1) the application of tracked vehicles to commercial ski resort maintenance, and 2) the development of much smaller, cheaper and lower-powered machines, initially to facilitate more efficient one-man or two-man oversnow manoeuvrability in support of polar expeditionary work, but soon to provide the threshold for a breakthrough into the civilian recreational field. Eliassen, in Canada, was producing a motor toboggan by 1960; it was used extensively in Baffin Island, amongst other arctic locations, in support of scientific reconnaissance in 1962 and 1963. But by 1962–3 Dr Gunnar Østrem was installing wooden A-frame huts on Norwegian glaciers to act as shelters for the much smaller, more efficient (and less expensive) Ski-doo. The Ski-doo replaced the Eliassen toboggan in the Baffin Island research expeditions in 1964 and has been used each year since. At last a small, relatively high-speed, low-priced ($600), easily transportable machine reached the commercial market. Since 1962 there has been a 2500 per cent growth in snowmobile sales in North America and nearly 400 models have been produced by over 50 manufacturing companies; the growth continues at an accelerating pace.

Despite the lack of systematic research, mentioned above, and due largely to the very novelty of the potential environmental hazard, a hypothetical catalogue of prospective and current damage must be compiled, for comparison with a similar list of benefits to be derived from thoughtful application of the new oversnow manoeuvrability. It is possible that some types of disturbance may not be detrimental. Hogan (1972) found that snowmobiles in the Mohawk Valley delayed snowmelt by up to one week on north-facing slopes, which may serve to protect steep roadways from the peak runoff and subsequently increase infiltration into the soil through the increased water storage.

Many aspects of table 17A.1 are thus difficult to evaluate. Sound assessment of the nature and degree of the hazard will require a good many years of comparative and experimental scientific effort. Also, many of the hazards depend for their assessment upon value judgements. What is a 'predator' and what degree of 'control' is necessary? How much noise is acceptable to what species of wildlife? Another characteristic of the negative side of the balance sheet is the heavy implication of improper or irresponsible use. Many snowmobile enthusiasts object to legislation and other forms of restriction on the ground that only a maverick minority causes the damage, and why should the multitude be regulated because of the misdeeds of the few? Such a defence is hardly tenable. A large

Table 17A.1. Hypothetical list of disturbances and benefits associated with the use of over-snow vehicles.

Environmental disturbance	Positive application
1 Direct damage to trees, seedlings and other plants.	1 Ecological and other scientific research.
2a Direct harassment of wildlife by hunting and other forms of irresponsible use.	2a Range patrol.
2b Reduction of 'predator' numbers.	2b Legal predator control, including domesticated dogs.
3 Indirect and unwitting disturbance of large mammal mating areas.	3 Wildlife rescue.
4 Damage to property (oversnow theft from empty summer houses); harassment of ski-tourists.	4 Human succour, search and rescue.
5 Change in snow density characteristics, indirect effects on spring runoff, soil moisture, plants and small mammals.	5 Forestry and harvesting. Water storage may be increased by packing.
6 Noise pollution and its effects on wildlife and humans.	6 Recreation.
7 Air pollution and attendant spread of litter.	7 Exploration and prospecting.

city needs a police force to protect life and property from the malevolence of a criminal minority. Mankind has been traditionally environmentally destructive in the application of developing technology and the malevolent minority may be large in number and impact.

But perhaps the major area of contention is man's interference with man through his use of the snowmobile. Absolute incompatibility between the cross-country skier, or snowshoer, and the mechanized, crash-helmeted noise distributor can scarcely be challenged. The United States mountain wilderness areas, from which all motor vehicles are excluded, may be touted as the answer, yet the 3 per cent of the Colorado mountains, for instance, set aside as wilderness, is hardly enough.

As with many aspects of the conservationist – environmentalist issue, the real problems are compounded by emotional reactions on the part of opposing groups. Undoubtedly conflict will grow as more and more snowmobile enthusiasts penetrate the mountain areas. The growing record of death and injury accruing from New England, eastern Canada, and the north-central United States, due largely to irresponsible and unregulated use, should be a serious warning. Regulatory legislation is slowly creeping ahead: Colorado now requires registration of oversnow vehicles and limits their use. France, through its Environment Minister, Robert Poujade, has eliminated snowmobile use except for emergency or maintenance services in alpine resorts (*Christian Science Monitor*, Vol. 64, No. 43, 17 January 1972, p. 8); similar action has been taken in Switzerland. It would appear that growth of snowmobile use in areas such as the Rocky Mountains will force serious consideration of zoning to exclude snowmobilers from areas suited to other activities or necessary wildlife sanctuaries.

Manuscript received May 1972.

Small-scale examples (2)
The snowmobile in Eskimo culture

William S. Osburn Jr

Division of Biology and Medicine,
US Atomic Energy Commission, Washington

The impact of a recent technological development, the mechanical snow vehicle, or snowmobile, is only beginning to be documented. This summary is taken largely from a study by Dr Wayne Hanson (1973) of Colorado State University. During a five-year interval Hanson watched the arrival of the snowmobile into Alaska and its rapid sweep through the Eskimo cultures, displacing the dogteam and creating many changes in the lifestyle of these arctic people. The centre of study was Anaktuvuk Pass, Alaska (60°01′N, 151°46′W), home of the last remaining band of Nunamiut Eskimos, formerly semi-nomadic caribou hunters of the Brooks Range. Like the relationship of the buffalo to the North American Plains Indian, the relationship of the caribou to the Eskimo's subsistence took many forms – food, clothing, income from selling caribou skin facemasks and from the carnivores which followed the herds, fur and later bounties.

During the first years of Hanson's visits to the village at Anaktuvuk Pass the Eskimos were apparently enjoying a stable period of gradual acculturation with varying amounts of subsistence hunting and trapping, yet living very closely tied and in harmony with their environment. Dogs were prestigious property and generally received good care, though occasionally abused when functioning as a 'psychological escape valve' for venting of Eskimo frustration or aggression. Regardless, the dog played an essential role. The maintenance of good dogteams required considerable effort. The better hunters – those who ranged further from the village and pursued more game – would take their dogs some distance away in early autumn simply to fatten them on Dall sheep. Dogs in good condition were strong and worked well and could withstand the long gradual physio-logical deterioration brought on by lack of high energy foods and hard work. During the winter lean, hard, frozen caribou haunches would scarcely repay the energy expended by the dogs in eating and digesting. Not only did the dogs require considerable time and effort to maintain – daily feeding – but the sledges did also. During the coldest part of the winter spruce wood runners replaced the steel ones, and before each morning's trip several layers of melted snow water would need to be applied to the runners. Women too devoted much time to the dogs. They used them to haul fuel and ice and took care of the chores associated with pup raising.

The people were fully occupied by the requirements of their culture which allowed relatively little time above the requirements placed on them simply to subsist in the arctic environment.

It is especially interesting to note that during this period, 1962–5, there were few conflicts within the Nunamiut community. Men were preoccupied with hunting and trapping, the women engaged in fuel hauling and managing the homes.

Though a dwindling fuel supply – the Eskimos were using willows at a more rapid rate than they were being regenerated – indicated that a pressing problem was approaching, all was relatively calm. Then in the winter of 1964 the first snowmobile appeared. Not only did this 'iron dog' appeal to the natural curiosity of the Eskimos, it performed with remarkable success in quickly running trap lines and hunting down caribou. Of course prestige played a role in the desire for the machines. The low level of maintenance for the new machine also made it highly desirable.

During the winter of 1966 there were five snowmobiles in the community but dogteams were still the major means of transportation. Friction broke out in the village between the 'have-nots' and the 'haves'. The 'have-nots' insisted the caribou were being driven up into the mountains by the machine hunters, hence the game was inaccessible to dog-teamers. The mechanically-aided hunters also had the advantage of being able to get into the field faster when a passing band of caribou were sighted, and not only could these machines reach the caribou sooner, it seemed as if the caribou had not yet associated the snowmobile with danger and allowed hunters mounted on snowmobiles closer approach than they did hunters approaching with dogteams. In addition to these advantages the snowmobiles could range further in a day's time from the village, cover more territory, and more easily obtain the now critically scarce fuel, thus compounding resentment between the groups.

The winter of 1967 and 1968 saw a significant increase in snowmobiles from five to sixteen. To some the snowmobile seemed the salvation of the Eskimo group. However, the change in the term for the snowmobile from 'iron dogs' to 'gasoline dogs' should have foretold trouble. Gasoline prices were eight times those in the continental United States, and most of the snowmobile owners still owed money from the original purchase of their machines and/or for repairs.

By the summer of 1969 the transition from dogteam to snowmobile was complete, as the last three families had purchased machines. Despite extensive wear and tear of running the machines over bare ground, nobody walked anywhere – it was too easy to jump on a machine and go after willows, water, to the permafrost cellar, or simply just go. The dependence of the hunting Eskimo upon his iron dog led to the neglect and disappearance of the living dogs. When the machines started to fail, the Eskimo found himself trapped by an alien item from another culture. Within a brief period of six years the snowmobile entered and totally disrupted the Nunamiut culture. The Anaktuvuk Eskimos are no longer in harmony with their environment. Regardless of whether or not they decide to go back to their dogs they almost surely have gone too far towards acceptance of white man's culture – plywood houses, fuel oil, radio, education, etc. – to ever regain their trademark as *the people of the land*.

This specific example can probably be repeated throughout that sector of the Arctic occupied by Eskimos. At the same time, it must be recognized that native peoples have long since ceased to maintain traditional lifestyles *in toto*. The impact throughout the nineteenth and early twentieth centuries of Christian missionaries, the traders, the whaling fleets and the policeman, had already set in motion irreversible changes in the traditional ways of life. Even though federal welfare will at least eliminate starvation, it is vital to recognize how the introduction of a single artifact, without thought or precaution, can cause serious disruption in the lives of native peoples.

Manuscript received January 1972.

Small-scale examples (3)
The snowmobile, Lapps and reindeer herding in Finnish Lapland

Ludger Müller-Wille
University of Münster

Reindeer herding, besides hunting, trapping and fishing, has been the general means of subsistence for the native peoples from northern Scandinavia to the eastern edge of Siberia. These peoples have developed special working methods, technology and economic skills to suit their particular natural and cultural environments. During the last century, modern western technological elements have been brought to these peoples. Their influence did not affect the existing situation immediately, but rather very gradually. The established pattern of reindeer herding was one in which nuclear families kept close watch over their herds almost all of the year. Reindeer were used for transportation (draft or riding), for clothing and for food. This system remained until after 1945 when western technology penetrated very strongly into the isolated subarctic and arctic regions of Europe and Siberia.

Since World War II, the traditional cultures of the circumpolar regions have been experiencing a rapid change caused by the introduction of various elements of western civilization, especially gas-powered vehicles (Pelto et al., 1968; Pelto and Müller-Wille, 1972; Pelto, 1973). To give a more accurate picture of the process of change, two communities, Utsjoki and Sevetti in northern Finland, have been chosen as examples. Their situations show differences in the adaptation and application of the same technological element – the snowmobile – in ecologically different settings (Müller-Wille and Pelto, 1971; Pelto and Müller-Wille, 1973). Today, these two communities are bi-ethnic, consisting of Lapps and Finns. The reindeer herders, all Lapps, still represent more than 30 per cent of the population. Altogether there are about 36,000 Lapps in northern Scandinavia of which 3000 live in Finland (Vorren and Manker, 1962).

Diffusion of the snowmobile

The process of adopting any innovation is a learning process. The stages in this process are 'awareness, interest, evaluation, trial, final adoption, and application' (Rogers,

1962). In Utsjoki, these stages in adopting the snowmobile were passed through especially rapidly, because the advocates or innovators were leaders in reindeer herding and had strong influence in the reindeer association and in the community.

The small, versatile, one-person snowmobile, developed in Canada in the late 1950s, was first imported to northern Finland in early 1962 (Pelto *et al.*, 1968). Information and enthusiasm about this vehicle was quickly communicated throughout Lapland. The usefulness and applicability of the snowmobile for transportation and deer herding was readily acknowledged especially among Lappish reindeer herders. As early as late 1962, four reindeer herders from Utsjoki employed their snowmobiles tentatively on trips in the mountains and for gathering and driving reindeer into corrals for roundups. It took the Utsjoki herders only a few months to reach the 'trial' stage, and already by the winter of 1963–4 the snowmobile had been adopted into back-country transportation and reindeer herding methods.

The former means of winter transportation in this roadless country – the reindeer sled – was suddenly abandoned. Draft reindeer were slaughtered or sold to people outside the area for use in racing (cf. the situation in Alaska where former sled dogs are now used in Christmas races: Hall, 1971, p. 247). The use of skis has also decreased. In Utsjoki, the fast adoption of the snowmobile and later of the cross-country motorcycle (since 1968) in reindeer hunting was made possible by the favourable ecological setting. Open mountain tundra offers ideal conditions for the manoeuvrability and application of these off-road vehicles. This is not the case in Sevetti to the south, where dense pine woods cover the area. In addition, the mechanical aptitude and pragmatic attitude of innovative-minded Lappish reindeer herders facilitated the change from the traditional to more or less mechanized reindeer herding (Müller-Wille and Aikio, 1971).

Today, the snowmobile is an accepted element of the daily routine of subarctic and arctic communities during the winter. In northern Finland it can be used from October to May, and is not only applied in reindeer herding but also in fishing, hunting and trapping, general transportation, and even in recreation by those who have secure economic positions (e.g. government employees) and are not dependent for their livelihood on the traditional economy. The use of snowmobiles for recreation is often the origin of conflicts between native Lapps, who still rely on their traditional economy, and immigrant Finns, who almost always take over wage labour positions. Reindeer herders fear that uncontrolled snowmobile recreation and tourism might harm their herds and interfere with their work.

Changes in reindeer herding

The introduction of the snowmobile (and the motorcycle for summer work) caused considerable changes in the management of the reindeer herds and in the working methods. In the years before 1962, Lappish reindeer herders in Utsjoki and Sevetti watched their family herds all year round very closely and kept them in more or less defined areas. In winter men in reindeer sleds and on skis stayed for long periods in the mountains. With trained dogs they tried to keep their herds under control to avoid interference with other herds. Thus the reindeer owners achieved full knowledge of the condition of their herds. The knowledge enabled them to decide how many and which animals could be slaughtered during the roundup, which was held once a year before Christmas. This roundup was the main event of the year for the reindeer herders. In the mountains communication between herders was difficult and took valuable time. The single roundup gave them the opportunity to exchange useful information (Paine, 1970).

The drive of the herds to the corral was a slow process which was accomplished by leading reindeer sleds, with skiers and dogs either in front, at the side, or in the rear of the herd. But at that time there was no need to speed up the process because connections via road etc. with other areas did not yet exist. After World War II the expansion of the road network in northern Finland caused a technological and economic lag due to the incompatibility between the traditional herding methods and the broadening of the market for reindeer meat which required modern methods.

Today, when a roundup is to be held, several herders on snowmobiles leave from different starting points to search the grazing area of the herds, which are now allowed to graze unattended. The deer are then driven to a point already agreed upon (a ridge top). This procedure may be repeated until a large enough herd for a roundup (c. 1000 to 1500 head) has been gathered, and the herd is then driven to the roundup place by snowmobilers at its rear and sides. In this way a herd can cover up to 25 km a day. A drive may take several days, depending on the distance to the corral. The corral is built so that the herd can be guided through a funnel-shaped fence into the inner round corral. Usually the deer are kept for a while in the funnel to calm down because the final drive into the main corral is very strenuous for both deer and herders. In the funnel the snow-mobiles' manoeuvrability is limited and the vehicles frighten the animals when they come too close. Therefore the Lapps decided to combine traditional with modern methods in this situation and developed two techniques. With the first, snowmobiles drive very slowly at the rear of the herd, and skiers and men on foot form a chain around the circling animals, urging them carefully toward the opening. With the second, trained reindeer dogs, often riding along on their master's snowmobile, are released in the funnel to encourage the herd to enter the main corral. The use of herd dogs at this stage was abandoned at the beginning of the snowmobile age, but after a re-evaluation it was again taken up by the Lapps in Utsjoki (Müller-Wille and Pelto, 1971).

Response to different ecological settings

As mentioned above, the ecological setting of Sevetti is quite different from the open country of Utsjoki to the north and required different approaches and responses. In the Sevetti area, coniferous forests, steep hills, and a complex drainage system called for a combination of traditional and modern working methods. The above description of the application of snowmobiles in reindeer herding was first introduced by Utsjoki Lapps. Reindeer herders from Sevetti were eager to learn from their experiences. They invited Utsjoki reindeer herders to demonstrate their new herding methods (Pelto et al., 1968). The methods developed in Utsjoki were suited to that particular landscape and could not easily be transferred to different ecological conditions without modification. In the densely forested area, snowmobiles are hard to manoeuvre. Constant turning and circling causes higher fuel consumption and considerable risk of breakdowns and repairs which means higher maintenance costs. In the forests the Skolt reindeer herders discovered it was more practical to employ a combination of snowmobiles plus herdsmen on skis. Herd dogs continue to be indispensable in all stages of roundup work. Although more work is involved, men and dogs can drive a reindeer herd into the corral more easily than snowmobiles can alone.

The natural differences of these two communities partly clarify why the snowmobile was adopted earlier in Utsjoki and later in the southern districts, as well as explaining the difference in working methods (Müller-Wille and Pelto, 1971).

Problems and prospects

ECONOMIC ASPECTS

The adoption of the snowmobile in reindeer herding has speeded up the working process by making more roundups possible during one season. New roads connecting the corrals with central settlements have attracted more buyers, resulting in higher sales and income for the Lappish herders. But the sudden changes have demanded more cash to buy, apply and repair snowmobiles. This required considerable resources in reindeer numbers. To be a successful reindeer herder today means to be constantly involved in activities around the herds and being in the middle of decision-making. The development since 1962 indicates that some reindeer owners with smaller herds (up to 10 head) have left this

Table 17A.2. Annual average expenses for motorized reindeer herder, Utsjoki, 1968–70 (from Müller-Wille and Aikio, 1971).

| | Snowmobile | | Motorcycle | |
	fmk*	Equivalent number of reindeer†	fmk*	Equivalent number of reindeer†
First year				
Acquisition	3600	20	1800	10.0
Upkeep	700	4	100	0.5
Fuel	1800	10	100	0.5
Total	6100	34	2000	11.0
Second year				
Upkeep	1000	6	100	0.6
Fuel	1800	10	100	0.5
Total	2800	16	200	1.0

* fmk = Finnmark: $1 = c. 4 fmk (1970).
† 1 reindeer = c. 180 fmk (1968–9).

field because they could not meet the financial burden. Table 17A.2 gives a picture of the expenses a reindeer herder has to face. He has to own some 150 to 250 adult deer to cover the costs of acquisition and upkeep of snowmobile and motorcycle (both vehicles are considered indispensable in today's reindeer herding) and to provide his family with a decent living. The Lapp with fewer reindeer relies either on a side income (mostly fishing, tourism, or seasonal wage labour) or on his spouse's wage, perhaps as a government employee or bank clerk.

This situation indicates that there is a concentration of herd ownership. The losses for the 'poorest', but also for some 'richer' reindeer herders, have been considerable since

Table 17A.3. Patterns of reindeer sales and herd growth in Utsjoki before and after the application of the snowmobile (from Müller-Wille and Aikio, 1971).

Period	Reindeer slaughtered and sold	Net increase in herds
	(%)	(%)
1956–61		
Owner A*	25.5	1.7
B	15.5	34.3
C	22.7	13.5
Total association (104 members)	23.7	7.8
1963–9		
Owner A	43.7	−16.8
B	32.6	−17.4
C	43.0	−20.2
Total association (111 members)	32.7	−10.0

*A, B and C have the biggest herds in Utsjoki.

the introduction of the snowmobile. A comparison of pre- and post-snowmobile periods also shows a change in the management of the herds, concerning reproduction balance and sales policy (table 17A.3). In contrast to the period 1956–61 reindeer herders sold more deer during the years 1963–9 in order to mechanize. At the same time, the balance of reproduction dropped from a surplus of 7.8 per cent to −10 per cent (in some years due to natural circumstances). However, this trend will probably not continue in the near future after the first large investments for modernization have been made. The concentration of herds will provide a secure economic base for a small group of herders. Today, less manpower is needed in herding work and those who have had to leave this occupation are forced to look for other work, which usually means emigrating to industrial cities in the south.

SOCIAL ASPECTS

Technological change is often more directly observable than social change. How much the introduction of the snowmobile has influenced the social situation in northern Finland is difficult to say. In combination with the whole set of western technological innovations the impact is felt in a variety of social fields. One factor of social change is the broadening of communication between the Lapps and their neighbours. Longer distances which were formerly difficult and rarely travelled are now covered in a short time. Nowadays neighbours and relatives visit each other more often – the flow of information is widened. This situation provides the Lapps, who are a minority group in Scandinavian and Finnish societies, with more intra-ethnic contacts, thereby strengthening their unity and identity as a distinct cultural group (Eidheim, 1971). The adaptation to western technology and economy by the Lapps leads to their partial integration into the national economic systems of Scandinavia and Finland. It is also the first time that they have been able to compete on the same economic level with their technologically more advanced neighbours. Today, Lapps feel that they are on the same level with Finns and Scandinavians.

The situation discussed here shows that traditional Lappish reindeer herders who have recently become acquainted with western culture have taken over those elements which fit into their existing style of life and which could be adjusted to it in a 'natural' way. That means that an innovation is accepted and acknowledged from inside the group. Lappish reindeer herders in northern Finland, for example, have understood the advantages of the snowmobile and have applied them on their own in their particular situation. Some Lapps, however, by being forced to leave their traditional role as reindeer herders, have suffered a considerable loss which in some cases has resulted in social and cultural alienation from their own ethnic background.

Manuscript received January 1972.

Small-scale examples (4)
The impact of man as a biped

Jack D. Ives

Institute of Arctic and Alpine Research,
University of Colorado

It is evident that man does not need a machine before he can cause damage. Trampling, whether in tennis shoes or heavy climbing boots, especially where concentrated, can obliterate the natural alpine environment. The United States Wilderness Act will not serve to preserve wilderness unless strenuous efforts are made to limit overcrowding by foot-sloggers.

Such an approach must be coordinated with alpine ecosystem research designed to determine the carrying capacity of the various plant communities that make up the alpine environment, as well as the overall carrying capacity of discrete recreational areas subjected to varying forms of activity.

The following brief review derives largely from the pioneer studies of Dr Beatrice Willard in Rocky Mountain National Park between 1958 and 1962 (Willard, 1960, 1963; Willard and Marr, 1970; Marr and Willard, 1970).

Willard and Marr (1970), in referring to the increased tourist access to alpine areas in Rocky Mountain National Park made possible by the opening in 1932 of Trail Ridge Road, a modern paved highway which for 20 km lies above an elevation of 3200 m, make the obvious point that significant disruption of the environment consequent upon this influence is to be expected since the alpine landscape evolved without man over millions of years. Tourists number some 20,000 to 28,000 per day (in 6000 to 8000 cars) throughout July and August.

Willard initiated research on visitor impact around four main parking areas above treeline where about 95 per cent of the visitors congregate. Extensive vegetation sampling on the range of affected alpine plant community types was undertaken and tourist activity patterns were observed. She concluded that trampling was the most important form of human activity that resulted in the most serious damage to alpine tundra communities. A few individuals walking across the tundra without design over several years, or even following a specific route at regular intervals throughout several seasons, did only minor damage. However, large numbers of visitors leaving their parked cars to stroll several hundred paces to a mountain viewpoint can cause extensive damage in as short a time as ten days. Another conclusion is that some alpine ecosystems are damaged much more easily than others, the extent of the damage being directly proportional to the moisture content of the associated soil systems. Thus sedges of wet sites are most easily damaged and the sedges of drier alpine turf stands are the most resistant, with cushion plants in alpine fellfield communities intermediate. Willard's scale of visitor effects is reproduced as table 17A. 4.

The observed tendency of tourists to follow trails has led to hard-top path construction by the National Park Service. This approach seems eminently successful in controlling damage.

Table 17A.4. Scale of visitor impact on alpine communities (from Willard and Marr, 1970).

Degree

0 – no known impact; total vegetation cover = 100% of natural.

1 – receiving visitor impact; but no alteration visible; total vegetation cover = 100% of natural.

2 – ecosystem obviously affected; vegetation cover 85–90% of natural.

3 – ecosystem definitely altered; plants show reduced vitality; normal growth persists in protected sites; soil exposed and eroding; total vegetation cover 25–85% of natural.

4 – ecosystem radically altered; vegetation absent except in protected sites; 'A' horizon exposed over most of the area, and eroding; vegetation cover only 4–25% of natural.

5 – ecosystem virtually destroyed; plants existing only in very sheltered sites, if at all, and not growing normally; 'B' and 'C' horizons exposed to erosion; total vegetation cover 0–5% of natural.

The remarks so far relate to trampling effects on highly concentrated areas adjacent to car parks above treeline. Nevertheless, it can still be claimed that Rocky Mountain National Park is largely untouched, despite the more than 50 million visitors since 1916. The current high impact relates to 2 per cent of the total park alpine area and involves 95 per cent of all visitors.

The alpine ecological research programme of INSTAAR is providing some additional measures of disturbance potential. The work undertaken some 60 km south of Trail Ridge on Niwot Ridge in 1971 to 1973 as part of the United States Tundra Biome research results in its own impact through trampling by scientists and assistants, and the installation of instruments. To ensure control of the extent of this type of damage, a well demarcated set of stakes was installed in the most intensively used area in July 1971 to protect both the experimental and control vegetation study plots and to delimit walkways. Wearing of tennis shoes or bare feet was enforced. By mid-August distinct new footpaths could be detected on air photographs flown in support of the research effort. It was nevertheless considered essential to channel the impact (an average of eight persons moving between study plots each day for seventy days) so that its effects could be contained and assessed, rather than distributed across a wider area doing much less but unmeasured and unknown disturbance to potential future study sites.

Antithetically, the rate of recovery of damaged vegetation must be examined. Willard (personal communication, 1970) found that where natural vegetation was obliterated to expose the dark 'A' horizon, seed set and recolonization was especially hampered, owing to the virtual baking of seedlings under conditions of very high surface temperatures induced by dark soil and intense incoming solar radiation in a relatively transparent alpine atmosphere. Estimated rates of recovery following total damage (degree 5) range from 100 to 500 years or more varying with plant community type. This again introduces the concept of the fragile tundra, and it must be stressed that precise long-term studies over a wide range of plant communities are few. Removal of turf cover on Niwot Ridge by Dr W. S. Osburn (personal communication, 1971), for instance, indicates a much faster rate of vegetation recovery than extreme conservationists would have us believe. Systematic research in this general area is urgently needed to establish the applicability of such findings, particularly as pressure on alpine areas increases.

Manuscript received May 1972

References

BLISS, L. C. (1970) Oil and the ecology of the Arctic. *In:* The tundra environment, a symposium of sect. III, Royal Soc. Can. *Preprint, Trans. Roy. Soc. Can.* (4th ser.), 7, 1–12.

EIDHEIM, H. (1971) *Aspects of the Lappish Minority Situation.* Oslo and Bergen: Tromsø (Universitetsforlag). 86 pp.

HALL, E. S. (1971) The 'Iron Dog' in northern Alaska. *Anthropologica,* **13** (1–2), 237–54. Ottawa.

HANSON, W. S. (1973) *Fallout strontium-90 and cesium-137 in northern Alaskan ecosystems during 1959–70.* Unpub. Ph.D. dissertation, Colorado State Univ.

HOGAN, A. W. (1972) Snowmelt delay by oversnow travel. *Water Resources Res.,* **8**, 174–5.

LACHENBRUCH, A. H. (1970) Some estimates of the thermal effects of a heated pipeline in permafrost. *US Geol. Surv. Circ.,* 632.

MARR, J. W. and WILLARD, B. E. (1970) Persisting vegetation in an alpine recreation area in the southern Rocky Mountains, Colorado. *Biol. Conserv.,* **2** (2), 97–104.

MÜLLER-WILLE, L. (1971) Snowmobiles among Lapps: An Ethnographical Report on a Technological Innovation (Utsjoki, Finland). *Nord-Nytt* (Lyngby), **4**, 271–87.

MÜLLER-WILLE, L. and AIKIO, O. (1971) Die Auswirkungen der Mechanisierung der Rentierwirtschaft in der lappischen Gemeinde Utsjoki (Finnish-Lappland). *Terra,* **83** (3), 179–85. Helsinki.

MÜLLER-WILLE, L. and PELTO, P. J. (1971) Technological change and its impact in Arctic regions: Lapps introduce snowmobiles into reindeer herding (Utsjoki and Inari, Northern Finland). *Polarforschung,* **41** (1–2), 142–8.

PAINE, R. (1960) Lappish decisions, partnerships, information management, and sanctions – a nomadic pastoral adaptation. *Ethnology*, **9** (2), 52–67.

PELTO, P. J. (1973) *Snowmobile Revolution. Technology and Social Change in the Arctic.* Menlo Park, Calif.: Cummings. 143 pp.

PELTO, P. J. and MÜLLER-WILLE, L. (1972) Snowmobiles: technological revolution in the Arctic. *In:* BERNARD, R. H. and PELTO, P. J. (eds.) *Technology and Social Change*, 166–99. New York: Macmillan.

PELTO, P. J. and MÜLLER-WILLE, L. (1973) Reindeer herding and snowmobiles: aspects of a technological revolution (Utsjoki and Sevettijärvi, Finnish Lapland). *Folk*, **14–15**, 119–44.

PELTO, P. J., LINKOLA, M. and SAMMALLAHTI, P. (1968) The snowmobile revolution in Lapland. *Suomalais-ugrilaisen seuran aikakauskirja*, **69** (3). 42 pp. Helsinki.

ROBERTS, B. (1971) *The Arctic Ocean.* Ditchley Pap., 37. Oxford: The Ditchley Foundation. 49 pp.

ROGERS, E. M. (1962) *Diffusion of Innovations.* New York: Free Press. 367 pp.

VORREN, Ø. and MANKER, E. (1962) *Lapp Life and Customs.* London: Oxford Univ. Press. 183 pp.

WILLARD, B. E. (1960) *Ecology and Phytosociology of the Tundra Curves Area, Trail Ridge, Colorado.* Unpub. Master's thesis, Univ. of Colorado. 144 pp.

WILLARD, B. E. (1963) *Phytosociology of the Alpine Tundra of Trail Ridge, Rocky Mountain National Park, Colorado.* Unpub. Ph.D. dissertation, Univ. of Colorado. 243 pp.

WILLARD, B. E. and MARR, J. W. (1970) Effects of human activities on alpine tundra ecosystems in Rocky Mountain National Park, Colorado. *Biol. Conserv.*, **2** (4), 258–65.

Large-scale examples

William S. Osburn Jr

Division of Biology and Medicine, US Atomic Energy Commission,
Washington

Regardless of the gloomy thoughts on the inevitability of small- and large-scale disruptions of tundra environments, there are indications that actual and potential misuse
of this 'last frontier', at least in areas under United States jurisdiction, will not be as
extensive as many have feared. An Interagency Arctic Coordinating Committee, with
the National Science Foundation designated as the lead group, is developing a five-year
plan to ensure that man's impact upon the arctic landscape will be minimal. This five-
year plan will receive annual evaluation and be extended for at least another five years.

Many national and Alaskan environmentally concerned groups are giving much
attention to arctic Alaska. The Alaska Conservation Society, Alaska Chapter of the
Sierra Club, Alaska members of the Wilderness Society, the Alaska Sportsmen's Council,
and representatives of resource management agencies in Alaska have formed an Alaska
Wilderness Council.

Perhaps the single most encouraging sign of arctic salvation concerns the passage in
the United States of the National Environmental Policy Act of 1969. Section 102(2)C
of this bill requires all federal agencies to prepare an Environmental Impact Statement
discussing the expected impact of their activities on the environment – adverse impacts
that cannot be avoided, alternatives to the proposed action, the relationship between
short- and long-term uses, and any irreversible commitment of resources involved.
The detailed statements will include comments of local, state, and federal agencies, and
will be made available to the President's Council on Environmental Quality and to the
public. This Act has been variously argued – either it will result in nothing but a monu-
mental paper shuffle, or it is the most important environmental bill ever enacted in the
United States. Regardless of weaknesses in the bill it is having a profound effect. Two
examples follow. The first concerns the construction of a 1270 km pipeline in Alaska.
The second concerns testing of a 5 megaton nuclear device on Amchitka Island. Finally,
the effects of a nuclear accident in the Arctic are discussed.

1. The Trans-Alaska Pipeline System

On 13 March 1968, Atlantic Richfield Company announced an oil strike on the North Slope of Alaska near Prudhoe Bay. More strikes followed and the present estimate is that this area may rival the recoverable reserves of Kuwait's Bergan Field, the world's largest oilfield. The Trans-Alaska Pipeline System (TAPS) is a project designed to transport this oil from northern Alaska to the port of Valdez for shipment to the United States. The potential impact of the 1270 km pipeline is presently being hotly contested. Proponents of the pipeline point to rapidly increasing oil consumption, coupled with correspondingly decreasing reserves available in this country. The United States presently imports about 20 per cent of the oil it uses. Thus it is becoming more and more dependent upon oil imports from politically unstable areas of the world – a situation which may be regarded as dangerous to national security. Many have commented emotionally upon the need for the pipeline, predicting dire consequences for US citizens – the price of fuel will rise precipitously, many will be without heat, and/or native Alaskans will be on relief – if the pipeline is not constructed. Opponents challenge this stand in a number of ways: the United States will need to increase oil imports with or without the North Slope oil; would it not be better to keep the oil in the ground for a true reserve? Again, the oil tankers and the pipeline itself would be vulnerable in case of a war. Dark forebodings have been predicted for the Alaskan environment by conservationists – migration patterns of caribou will be disrupted and caribou might even become extinct; tundra 'ecology' will be irreparably damaged; and oil-line breaks, variously caused, could wipe out various types of fishing and duck nesting sites. Additional basic questions are asked. As oil reserves are finite, should we not spend more dollars and time on researching more efficient, less polluting, and longer lasting energy sources? Are there alternatives to the trans-Alaska project, perhaps a trans-Canada pipeline? Pondering further, other questions are asked. Is a large fraction of North Slope oil earmarked for Japan? Perhaps we should alter our lifestyles and reduce our dependence upon 'useless gadgets' and ecologically damaging conveniences.

A preliminary Environmental Impact Statement was drafted by the Department of the Interior in 1971 and distributed to federal agencies and conservation groups for comment. Much time has since been devoted to hearings in Alaska and in Washington DC. Even the former Secretary of the Interior was distressed by the quality of the draft statement, prepared by his former colleagues. Relatively few facts have been presented but this paucity has been replaced by a plethora of emotion. 'What if' or 'ecocatastrophic' games, and what some environmental activists call ecopornography (so misrepresented as to be ecologically obscene) or the Madison Avenue approach, appear in major newspapers on an almost daily basis. In some instances cartooning has replaced scientific evaluation. In this section, I shall attempt an evaluation of the available factual information.

GETTING THE OIL OUT OF ALASKA

The problem of capitalizing on the potential wealth of Alaska has two separate aspects: getting the oil out of the ground and transporting it out of the Arctic. Difficulties of both

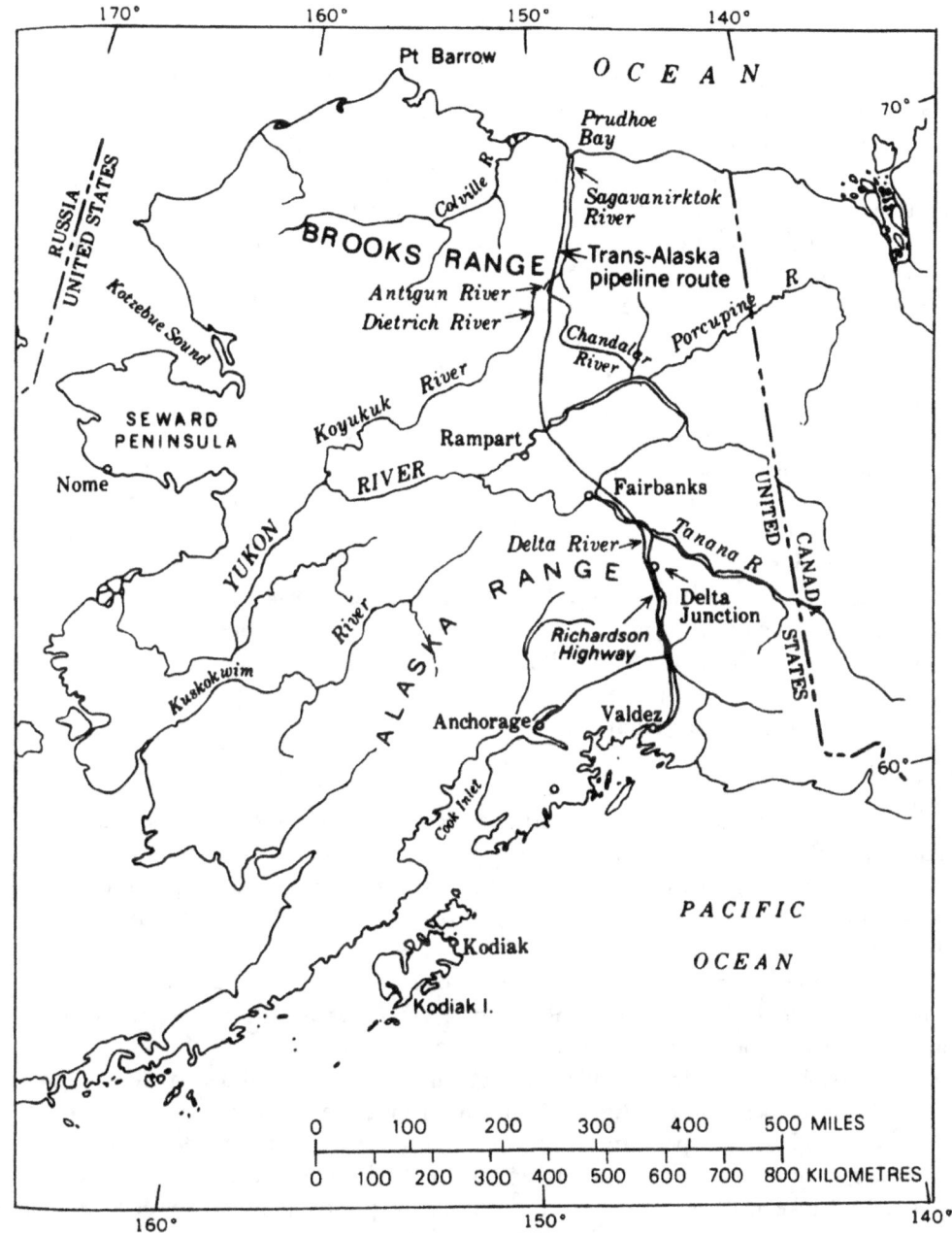

Fig. 17.1. Route of the trans-Alaska pipeline (from Kachadoorian and Ferrins, 1973).

aspects are intensified by the topographic, climatic, and ecological characteristics of the region.

Various methods of getting the oil out of Alaska have received discussion: 1) a trans-Alaska or trans-Canada railroad; (2) a trans-Canada pipeline; 3) ice-breaking super-tankers and/or nuclear-powered submarine tankers; and 4) a trans-Alaska pipeline. Apparently the first three considerations have been abandoned but will need to be discussed now that the Department of the Interior has made available the second draft of the pipeline Environmental Impact Statement. The rationale for eliminating the first three considerations follows. Planners insist that any railroad would be far more environmentally destructive than a pipeline because of greater need for cutting, filling, and grading to provide the maximum 2 per cent grade, as compared with the 30 per cent grade that a pipeline can achieve. Further, a railroad is less efficient in its capacity to carry oil.

The Canadian pipeline route would be considerably longer than the trans-Alaska route and would traverse greater stretches of permafrost and sections of the Arctic Wildlife Refuge. However, it could connect with already existing lines near Edmonton, Alberta, and has received the most serious discussion by planners as an alternative to the trans-Alaska pipeline.

The use of ice-breaking supertankers and/or nuclear-powered submarine tankers have their own set of problems. Oil spills have been difficult to cope with in temperate regions, and in the chill of the Arctic oil spills might present far worse clean-up difficulties.

The Alaska pipeline route would traverse wet and dry tundra, pass through areas of permafrost, glacier bearing mountain ranges, and swampy taiga (fig. 17.1).

HISTORY

In October of 1968 three companies, Atlantic Richfield, Humble Oil and Refining, and the British Petroleum Oil Corporation, formed a group for planning and design of the pipeline. Later (August 1970), with four other companies, the Alyeska Pipeline Company was formed to act as the prime contractor to plan, construct, operate and maintain TAPS.

The pipeline would be about 1.2 m in diameter and extend for 1270 km (789 miles). The project would include pumping stations, campsites for use during construction, airfields during construction and operation, a communication system, and material source sites. At the terminus, Valdez, storage tanks and dock facilities would need to be built. Prior to pipeline installation in the northern portion a supply road would have to be constructed. Thus the permanent pipeline right-of-way including the terminal and pumping stations would occupy approximately 20 km² of the total land area of Alaska.

It would take at least three years to complete the project and employ between 5000 and 10,000 men at a cost of more than $1000 million.

At capacity, approximately 2 million barrels (159 litres per barrel) a day would move through the pipeline. The oil would be hot, perhaps as high as 82°C, because it would be hot from the wells and friction from its flow through the pipe and pressure from pumping would tend to maintain its initial temperature.

ENVIRONMENTAL IMPACTS

Environment impact is a very controversial subject and interpretations range from horrendous to insignificant – according to the viewpoint of 'conservationists' or 'oilmen', respectively. A number of items and viewpoints can be discussed.

The first issue is that if all environmental variables are accounted for and even if nothing goes wrong, the road and pipeline will represent a major intrusion into one of the last big wilderness areas under United States jurisdiction.

Permafrost hazards

Any disturbance of the ground surface may be magnified because the dark soil effectively absorbs the sun's rays and a large 'scar' may develop, as these wounds tend to heal quite slowly (cf. p. 465). In fact, the speed of ecosystem recovery and associated ecosystem manipulations are key issues. Lachenbruch (1970) states 'to bury a hot pipeline in permafrost ground composed of unconsolidated sediments with relatively high ice content could give us grave concern, not because of the melting by itself, but because of the consequences of the melting if safeguards were not designed into the engineering system.' As a result of this and much additional discussion the pipeline design now includes insulating the pipe so that this effect will be minimal. This is one prime example conservationists give to demonstrate that a general industrial slowdown results in better environmental protection design.

Earthquake hazards

Pipeline proponents say the complete system will remain safely operational under the most severe earthquakes ever recorded in Alaska. Opponents contend that Valdez was destroyed by the 1964 Alaska quake (8.0 on Richter scale), and that no accurate earthquake warning system has ever been devised, and that if safeguards fail damage from broken pipelines could be extensive.

Effects on wildlife

Conservationists submit that sections of pipeline above ground could wreak havoc upon caribou migration patterns. Habitats of Dall sheep, grizzly bears, black bears, moose, foxes, wolves, swans, falcons, golden eagles, etc., could be variously disturbed. Spawning beds of salmon, char and grayling could be endangered by construction activities and potentially from oil spills.

Proponents cite their numerous ecological surveys and studies and insist the route and construction schedule is such that effects upon wildlife will be minimal. For example, construction activities are coordinated with various calving, spawning, and nesting seasons. Opponents warn that intensive research over short intervals cannot allow for long-term effects and that it may take many years to predict environmental effects properly.

Effects on native Alaskans

Much concern has been expressed that new oil money could 'ruin' the ancient native cultures of Alaska. Increased social services would become available and natives could make up their own minds if they wished to remain or take up new ways of life. Opponents cite the situation that $900 million in oil lease money sits in banks, gathering $200,000 in interest a day, and the legislature cannot decide what to do with it, yet 57,000 natives live in squalid circumstances.

Oil spills

Major potential environmental impact would result from oil spills, from pipeline ruptures, and from routine tanker operations or accidents. Conservationists argue that because of the volume and temperature of the oil and fragility of the terrestrial, freshwater and saltwater ecosystems involved, any pipeline breakage or tanker mishap could cause an ecological disaster of catastrophic dimensions. Offshore drilling spills could even 'cause the ice cap to melt'. The record of previous oil exploration activities in Alaska has not been particularly good.

Engineers seemingly have great confidence in techniques to ensure leak prevention, detection and suppression. An impressive computerized surveillance system, backed up with round-the-clock air and field manned inspection, strategically placed remote control cut-off valves and drain-off surge tanks has been developed.

The present situation (January 1972)

Alyeska viewpoint. Over a million dollars have been spent on ecological research along the pipeline route.

> We've conducted experiments in the north with both warm and cold pipe. We've studied wildlife – its breeding and migratory habits. We've carried out far-reaching botanical experiments to determine the best methods for reseeding and restoring the tundra. We've even examined our entire proposed route to make sure that we will not disturb areas of archaeological importance. In short, we've done our homework. (Alyeska Pipeline Service Company advertisement: *Time*, 11 January 1971.)

Opposition viewpoint. It is difficult to characterize this group, but certainly a number of conservation organizations must be considered – Wilderness Society, Friends of the Earth, Environmental Defense Fund, Alaska Action Committee, Environmental Action, Izaac Walton League of America, Sierra Club, Sport Fishing Institute, Trout Unlimited. Though impossible to state concisely a view for 'conservationists', one can perhaps summarize by saying their views range from a pessimistic 'the project should be abandoned' to an optimistic 'let's go slowly and carefully'.

LEGAL STATUS

As of February 1971, some 2 million manhours and more than $35 million had already been expended by oil companies on research, special studies, field investigations and route surveys, project development and engineering. Also $900 million was paid to the state of Alaska in 1969 for oil leases on the North Slope drilling grounds.

The project is now stalled by two injunctions. The first concerns a controversy over native land claims which are under a federal 'land freeze'. The second is the result of a suit filed by several conservation groups (The Sierra Club, Wilderness Society, Friends of the Earth and Environmental Defense Fund).

The Department of the Interior demanded a detailed environmental safeguard report from Alyeska from which Interior personnel developed an Environmental Impact Statement to conform with Section 102(2)C of the National Environmental Policy Act. This report – sixteen boxes of data standing 5 feet high – arrived in Interior Secretary Morton's office in July 1971 (*Time*, 26 July 1971). The second draft was released to the public in April 1972. In May 1972 federal government approval was given for pipeline construction, but the case was still being argued in the courts in May 1973 since the width of the proposed right-of-way violated a 1923 law (see Postscript, p. 951).

Many articles are appearing which 'review' the pipeline situation. Polarization between 'conservationists' and 'oilmen' is becoming more pronounced as may be seen by quotations taken from a variety of articles: 'Most people are fed up with ecologists'; 'That ecology (tundra) is not delicate, it'll eat you up'; 'Outsiders want to turn the whole place into a national park'; 'The US of America – it runs on oil. And boy, does it take a lot to keep it running. Three gallons a day for every man, woman and child in this country'; 'An oil spill may start the polar ice cap melting and if it does much of the east coast of the US will be inundated.' In short, for each pro argument one can find an equally emotional con argument. Cartoons reproduced by Klein (1971) graphically depict the opposing views of oil and environmental interests.

This extreme polarization will certainly test the workability of the National Environmental Policy Act and the President's Council on Environmental Quality. Regardless of the outcome of this project, the NEPA of 1969 will almost certainly help to develop a more rational exploitation of the Alaskan, and hopefully also other, arctic regions.

2. Project 'Cannikin' and the Amchitka Bioenvironmental Program

Amchitka Island, located at the western end of the Aleutian chain some 2250 km southwest of Anchorage, Alaska, is approximately 5–6 km wide and 60 km long. It supports a maritime tundra vegetation (Amundsen, 1972) and is a wildlife refuge. In 1967, Columbus Laboratories of Batelle Memorial Institute conducted ecological studies on Amchitka under contract to the US Atomic Energy Commission. This was related to the Commission's underground testing programme which began with 'Long Shot' (an 80 kiloton device) in 1965, followed by 'Milrow' (1.2 megatons) in October 1969 and 'Cannikin' (5 megatons) in November 1971. The Amchitka Bioenvironmental Program, which was conducted in collaboration with scientists from several universities, the Smithsonian Institution and the Fish and Wildlife Service of the US Department of the Interior, was designed to: characterize the ecological features of the island and surrounding waters; predict, document, and evaluate the effects of nuclear testing activities on the biota and the environment; recommend measures for minimizing adverse effects of nuclear testing activities on the biota and the environment; and identify needs and recommend methods for environmental restoration of areas disturbed by AEC activities.

Although the National Environmental Policy Act of 1969 was not in force when this programme was initiated, the bioenvironmental research programme has served as the basis for preparation of the Environmental Impact Statement required under this Act.

BIOENVIRONMENTAL STUDIES

The bioenvironmental programme included studies of the soils, terrestrial vegetation, avian population, freshwater streams and lakes, and of the marine environment surrounding the island. Also included were archaeological surveys and radioecological studies of the terrestrial, freshwater and marine environment.

The objectives of the soil studies were to characterize the major soil types of Amchitka, and to identify and define the physical characteristics of shock sensitive soils. From these studies and from expected ground shock data, mass movement and other soil alterations were predicted. Following detonation, the effects were documented and compared to predictions. Soil characteristics were also evaluated relative to assisting in reclamation of disturbed sites.

The plant studies developed baseline information concerning the vegetation of the island. Plots were established for long-term investigations of the effects of man's past and present activities upon terrestrial plant communities. The feasibility of and best methods for revegetating areas that were disturbed during site preparation and testing were also investigated.

Avian ecology studies were designed to determine the kinds and numbers of birds on the island. Their habitats were identified, breeding periods and nesting sites established, and food habits determined. Estimates of avian 'cost' of testing at various seasons of the year were made. Special emphasis was devoted to endangered species such as the bald eagle and peregrine falcon.

The freshwater studies were designed to determine the limnological characteristics of the lakes, ponds and streams. The food habits and breeding cycles were determined for all species of freshwater fish, including detailed studies of salmon spawning activities on the island. Streams and lakes near the planned test site were selected for intensive study. Here the variety, abundance, distribution, movement, growth, survival and mortality of fish populations and primary productivity were documented.

A team of professional archaeologists surveyed the entire island for evidence of early human occupation. Seventy-eight sites were identified along the sea coast and six of these – nearest the expected test site – were carefully excavated. This work was done in compliance with the Antiquities Act of 1906.

In support of the marine programme, existing information on the most important commercial species of marine fishes and mammals that are harvested around Amchitka was compiled and summarized. These data include information on the seasonal patterns of fishing effort, fishing sites, catch statistics, life histories, and migratory habits of the species of major commercial importance. Some exploratory fishing was conducted in search of possible unexploited fishery resources and nursery grounds of commercially important species near Amchitka, and to develop additional information on the migratory patterns of valuable species. Baseline physical oceanographic information was determined along with current patterns and water mass transport rates relative to the Amchitka region. Detailed ecological studies of the Bering Sea and Pacific Ocean waters nearest to the test area included chemical analysis of water, primary productivity measurements, plankton studies, and investigations of the attached algae and inshore benthic fauna. Particular emphasis was devoted to evaluating the distribution and abundance of the sea otters' food organisms. Predictions were made of test effects and, subsequent to the test, studies were conducted to document and evaluate test effects. This effort included observations for kills due to shock and for any submarine fault movements. Also, specimens were collected for radiochemical analysis.

The objectives of the marine mammal investigations included surveys of the marine mammal population of the general area, and special studies of the sea otters. The sea otter studies include evaluations of population dynamics, experiments to determine the physiological effects of shock waves, and investigations of behaviour which may be related to the test activities. Predictions were made of anticipated test effects and subsequent studies were conducted to quantify these estimates.

The radioecology effort included documentation of the background levels of specific radionuclides present in the soils, water, plants and animals, resulting from natural and worldwide fallout sources. Food chains, whereby certain radioactive materials could be transferred to man or other important species in the remote case of a venting, were identified.

MARINE FOOD CHAINS

Although the Cannikin test was designed so that the venting of underground radio-activity was extremely unlikely, as a precautionary measure attention was given to the possible consequences if the maximum credible prompt release (venting) of a few per cent of the total radioactivity were to occur after the detonation. Since there are no permanent human inhabitants, either on Amchitka or the nearest islands, the principal question about the fate of vented radionuclides concerned the possibility of contaminating marine life, especially fish that are used as food by man. The Battelle Columbus Laboratories have developed mathematical models and computer programmes to predict the dispersion of radionuclides in the sea, uptake by marine organisms, and ultimately the internal radiation dose to man consuming the sea food (Bloom and Raines, 1971).

These modelling calculations were based on extremely cautious assumptions: that all vented radionuclides would reach one particular segment of the sea by a predicted fallout pattern and by runoff of fallout from the land to the sea; that all of the radionuclides would remain within the volume of water in which they were deposited; that all of man's food fish would remain in this volume of seawater until caught and eaten so that the radionuclides in their bodies would have reached an equilibrium level; and that man would consume this food at a rate of 30 kg per year commencing thirty days after the detonation and continuing throughout his lifetime.

PREDICTIONS OF EFFECTS FROM THE NUCLEAR TEST

As a part of their Environmental Impact Statement, complying with Section 102(2)C of the National Environmental Policy Act of 1969, the AEC compiled a rather extensive group of bioenvironmental predictions (United States Atomic Energy Commission, 1971). These were examined and commented upon by a number of federal agencies and other groups. A summary of the predictions follows. In several cases, where data were insufficient for clear-cut predictions, various qualifications as appeared in the Impact Statement are included. The Battelle Columbus Laboratories also made a series of predictions (Kirkwood and Fuller, 1971), and as these did not have to be compiled and circulated to other agencies months before the test, they were made from data procured almost up to shot time.

In a letter to the Honorable John N. Irwin II, Mr Train, Chairman of the President's Council on Environmental Quality, discussed the potential environmental hazards associated with Cannikin, though these considerations became a major point of controversy in court hearings and were ultimately inserted in the Congressional Record. They will not be summarized here.

Though numerous conservation and congressional groups voiced disapproval of the test shot and emphasized the potential bioenvironmental costs, few supplied specific estimates. However, an attempt will be made to discuss the range of predicted effects and compare them with those made by the AEC. At the end of this section an attempt will be made to compare predictions with the costs which actually resulted.

The following predictions were made on the assumption that the Cannikin test would be in the autumn, which was indeed the case.

Radioactivity

The Cannikin test has been designed to retain the radioactivity underground, so that the chance of prompt release of radioactivity is remote. Groundwater will gradually transport radionuclides away from the explosion source and may eventually discharge them to the ocean. Three different times – depending on various assumptions – are discussed. The most likely time to elapse before discharge to the sea exceeds 1000 years and the least likely is 2 to 3 years.

Upon reaching the sea floor interface, dilution would further reduce the tritium concentration. Other isotopes are retarded by absorption and will be below maximum recommended permissible concentration levels before reaching the sea. Radioisotopes of noble gases will seep to the earth's surface and be rapidly dissipated to levels well below applicable Radiation Concentration Guide levels.

Terrestrial ecosystems

The main impact on the terrestrial ecosystem will be the occurrence of rock- and earth-slides along coastal areas and the possible resultant effects on bird nesting habitats. About 12,000 m³ of rock and peat materials were dislocated by the previous nuclear test, Milrow. Because the yield of Cannikin will be larger and because the instability of the soils in the Cannikin area will be greater, the disturbances of rock and peat will probably be more extensive. Numerous massive slumps of fossil sand dune material may occur along the Bering Sea shoreline, near the Cannikin site. If there is severe damage to sea stack and sea cliff nesting areas of bald eagles and peregrine falcons, it could affect their reproductive success in subsequent breeding seasons. Ornithologists estimated that fifty-five pairs of bald eagles and nineteen pairs of peregrine falcons nested on Amchitka during 1970.

Extrapolation from Milrow experience indicates that Cannikin could damage one or two eagle and one or two falcon nesting sites so severely that they would be unsuitable as future nesting locations. This amount of damage to nesting sites would not affect the reproduction potential or populations of either avian species, nor should there be any measurable effects on population density or reproduction potential of the other avian species as a result of the Cannikin test. Rockfalls may also cause damage to one archaeological site situated on a cliff edge near the Cannikin site.

It is probable that a saucer-shaped surface depression will form above the Cannikin emplacement location as it did in the case of Milrow. It could be as large as about 1200 m in diameter from crest to crest, and anywhere from a few metres to 30 m deep at the centre. As with Milrow, this will cause numerous circumferential tears and cracks in the turf.

Topographical changes resulting from Cannikin will alter soil drainage in some locations and this will cause changes in some plant community structures, but this is not expected to be extensive. As during Milrow, the most spectacular effects on vegetation should be the disturbance of moss mounds by ground shock effects. However, because of differences in the composition and distribution of plant communities, Cannikin will affect fewer moss mounds than Milrow.

Freshwater ecosystems

Amchitka Island contributes very little to salmon fisheries. Pink salmon are the most abundant species of salmon on Amchitka and silver salmon are encountered occasionally; red salmon are rare. Freshwater and intertidal spawning is well distributed around the island, but only a small number of fish are found. This is probably because the streams are small and their drainages short.

Experience gained during Milrow and Long Shot, coupled with the water overpressure pulse predictions, indicates that the Cannikin test should not affect Dolly Varden char, or salmon populations on the island. Some small non-commercial fish (three-spined sticklebacks) will be killed in some of the lakes adjacent to the Cannikin surface zero. Since the lakes that will be affected are few and are not nursery areas for salmon smolts, this effect is considered of minor importance. Changes of plankton levels in nearby ponds may occur, but this would be transient and of short duration.

As far as the overall freshwater ecosystems are concerned the principal ecological effects of Cannikin are likely to be those produced during site construction and operations where sediments, and possibly escaped waste materials, may be carried by natural drainage pathways into nearby streams and ponds. While some waste materials have reached streams and ponds, extensive efforts are being made to minimize such mishaps.

Marine ecosystems

Field studies which included observations during the Milrow test have shown that for the Cannikin test the effects of overpressure in the marine ecosystem should be minor.

Estimates of the total population of sea otters at Amchitka vary depending upon when and under what conditions the counts were made. During the period of 1969 to 1970, otter counts ranged from 2500 to 4000.

Controlled tests indicated that overpressures greater than 7 kg cm^{-2} (100 lb in^2) may rupture the eardrums of sea otters and ultimately lead to death by starvation in otters so disabled. Overpressures of this magnitude at the sea floor may extend out to radii of about 8 km from Cannikin; however, at any given location, the overpressure depends on water depth and decreases to zero at the surface. The number of otters within an 8 km radius, of course, differs from time to time, but estimates based on counts have ranged from 200 to 700, including the pertinent sections of both the Bering and Pacific coasts.

Typically during midday, otters intersperse dives for food-gathering with periods of grooming and rest on the surface. Again, estimates vary about the fraction of animals that might be down on the sea floor during midday; on a typical autumn day, the latest estimates for the area of interest range from 10 per cent to 15 per cent.

Combining these latter two estimates yields a prediction that 20 to 100 sea otters might be exposed to overpressures from Cannikin that could be severe enough to rupture their eardrums and ultimately result in their death. To place this prediction in perspective, it should be noted that during the period 1962 to July 1970, the Alaska Department of Fish and Game killed 901 Amchitka sea otters for their pelts. This is in accord with conventional game management practices as a means of keeping animal populations in balance with their food supply. In so far as can be determined, the removal of about 1250 sea otters from Amchitka since 1967, in transplants, harvests, and for experimental use, has not had a prejudicial effect on their population. Therefore, even if shock effect of Cannikin should cause some mortality as discussed above, the net effect on the Amchitka population should not be of long duration.

The Amchitka sea lion and harbour seal populations are not expected to be affected by Cannikin. During the autumn, most of the sea lions are at hauling-out areas too far from surface zero to suffer any adverse effects from ground motion or overpressure (e.g. at Column Rocks, some 21 km distant in the Pacific, or at Ivakin Point about 23 km away, on the Bering coast). Small groups of harbour seals may be somewhat closer to surface zero. If a few such individuals should be killed by ground motion or overpressure from Cannikin, early restoration of the island seal population would be anticipated by normal population growth of groups beyond the range of damage.

Bioenvironmental programme scientists are confident that populations of commercial fish and shellfish species will not be endangered from Cannikin. There is uncertainty, however, as to the extent to which overpressure and underpressure pulses in the sea could cause injury to fish. Those species such as cod and rockfish having air bladders that are not connected to the alimentary canal are considered to be the most vulnerable to abrupt pressure changes. In any event, the possible losses would be localized close to Amchitka Island.

Construction activities

The area of Amchitka is about 300 km². The Department of the Interior's resident biologist estimates that 4 km² of land have been disturbed by various operational activities associated with preparation for Milrow and Cannikin. These activities include road building and improving, drill-site development, and the development of material supply areas. The disturbance includes 0.3 km² of lake and 12 km of stream pollution, mostly from spills or leaks of drilling mud. During the early phases of construction, there was some illegal disturbance of middens despite warnings and some off-road disturbance by tracked vehicles.

Approximately half of the disturbed 4 km² is on land used previously for roads, aircraft taxiways, or gravel pits by the military during World War II. About 2 km² – 0.7 per cent of the island area – are new disturbances.

Field representatives of the AEC and the Department of the Interior at Amchitka have developed plans for continued monitoring and measures to be carried out so that Amchitka Island will be restored to a satisfactory condition before the AEC leaves the site following the test.

After the successful completion of Cannikin, the AEC intends to vacate Amchitka, leaving that island as nearly as possible in the condition in which it was found to be in 1967. All debris created on Amchitka as a result of its activities will be removed from the island or buried in a location that has been designated for that purpose. All structures erected on Amchitka by the AEC will be removed except for specific buildings which the Department of the Interior wishes left in place. Existing paved areas such as the airport runways, taxiways and roads will be left in place. Some buildings and debris left at Amchitka as a result of the World War II activities will have been removed by the AEC incidental to its general clean-up activities. To the extent feasible, areas that have been disturbed through AEC activities will be restored by levelling or contouring, and re-seeding. Restoration methods and plant species to be used for revegetation will be selected with the approval of the United States Fish and Wildlife Service representatives.

The test day storm

The day before the test Amchitka Island was lashed by winds of hurricane force, well above 50 m s⁻¹, and on test day winds were still strong. Whether or not these high winds directly caused the death of various forms of wildlife is problematical. However, indirectly it almost surely caused additional deaths to some birds, sea otters and perhaps seals.

In response to the winds birds would take shelter on the ground, especially along cliff faces leeward to the wind. Thus more birds were likely to be on the ground near 'surface zero' than one might normally expect. Hence when the ground surface directly above the nuclear device was momentarily thrown into the air there was increased probability of injury to birds.

The violent winds caused the sea to become very rough. Under these conditions sea otters, and perhaps some seals, may tend to concentrate in greater numbers in protected coves. Also, the otter might well haul out onto the rocks along the shore. Here, where the cliffs were close to the sea, the otters were exposed to rockfall from cliffs shaken by the nuclear explosion. However, it is equally possible that the sea otters would leave the proximity of the island (where the waves were choppy) to ride out the storm on the swells. No one seems to have observed them during these high wind periods.

These same high winds and waves would also result in decreased accuracy of counting the number of fish, birds, or otters that might be killed. It is difficult enough to spot floating fish or otters even when seas are moderately calm. The high seas would also prevent investigators from placing pens of fish in the ocean at various depths and distances from surface zero. Hence they would be denied this indirect method of calculating the number and kinds of fish that might have been killed.

In delaying the shot into November it is quite possible autumn storms brought addit-ional rains and that the soils contained more water than would have been the case earlier. This increased weight would augment the likelihood of soils reposing on steep slopes to creep, flow, or slide in response to vibrations from the nuclear test.

POST-CANNIKIN

In the final hours before the nuclear test not only was the Supreme Court called upon to rule whether or not the test should proceed, but newspapers, radio and television carried extensive information on it. During the final week this writer noted only two reports in the news media in which an ecological holocaust was not implied. One instance involved an interview with the Chairman of the Atomic Energy Commission, Dr Schlesinger, who predictably enough backed the AEC statements which had been discussed in the Environmental Impact Statement. The other non-calamity statement was at the end of a long article titled 'Hawaii braced for A-test'. The sentence said simply that 'President Nixon had Mr Train's report and the other documents when he made his decision to let the test proceed as scheduled' (*The Washington Daily News*, 4 November 1971).

An article in the *Washington Post*, 5 November 1971, originating from a reporter actually on Amchitka the night before the test, was typical of most which appeared on news media on test day, in that it carried no information to reassure the public that the AEC had done their bioenvironmental homework and that perhaps the shot would not be catastrophic. In fact the article recounted the 'worst estimate' AEC had ever made concerning whether or not a particular test, ten years earlier, might vent.

The first reports back from Amchitka were rather forthright statements such as that issued by the AEC Chairman: 'We have conducted the test with complete safety' (*Atlanta Constitution*, 8 November 1971). However, a number of 'backlash' articles appeared, ranging from cartoons to insertions into the Congressional Record. Their general theme was that the public had listened too long to 'scare' ecologists and/or environmentalists and that perhaps not only was the Amchitka test safe, but possibly a host of other so-called environmentally deleterious projects needed re-evaluation. In that ecologists were frequently indicated as a scare group, these vindictive reviews could have serious repercussions: the fright generated largely by 'instant', 'short-term' or 'situation ecologists' could seriously damage the reputation of all ecologists.

Reports confirming the early statements of the bioenvironmental costs of Cannikin have been periodically released (United States Atomic Energy Commission, 1972), and it appears that the AEC Environmental Impact Statement predictions were reasonably accurate. Disregarding the general predictions of no earthquakes nor tsunamis being triggered, specific predictions of biological impact will be discussed.

Since Cannikin, numerous surveys have been made, and are continuing, in an attempt to observe and record any effects on biota of the island and its surrounding marine environment. Based upon the effects observed at this time, there will be no permanent damage to the island's population of any mammals, birds, fish, invertebrates, or plant life.

As of 28 November 1971, a total of 19 dead sea otters had been recovered. Two additional injured otters were observed in the sea but were not recovered. Also, 2 abandoned otter pups were seen but not recovered. It is assumed that all of these died, making a total of 23, of which 16 were on the Pacific side and 7 on the Bering side. Of the 19 recovered sea otters, 12 underwent autopsies. Findings indicated that 7 were killed by pressure effects in water, 2 died from rockfalls, and 3 were fatally injured by vertical acceleration (upthrust of the ground). It should be assumed that these 23 casualties represent only a fraction of the otters killed by the detonation. Adverse wind conditions and other factors made 100 per cent recovery of dead or fatally injured animals impossible. Preliminary visual censuses have indicated a considerably reduced number of sea otters in the Bering Sea off the Cannikin site, and estimates of those killed have been as high as 1000 (Laycock, 1972). However, the post-shot counts made to date are not adequate to support any final judgements as to the precise degree of population reduction. Weather and sea conditions at this time of the year make reliable population estimates impossible. Further judgement as to the effect of Cannikin on the island otter population must wait until future counts are made during the same period of the year and under comparable conditions to the 1971 counting.

Four dead seals were found on the beach, and all were autopsied. All apparently died from pressure effects. A survey of Seal Beach, which is about 2.5 miles from Cannikin Surface Ground Zero (SGZ) on the Pacific side of the island, showed a normal number of seals (about 75) on the beach several days after the detonation.

Concerning marine fish, nearly 300 dead rock greenlings and several specimens of other species were found on the Bering Sea beach adjacent to Cannikin Surface Ground

Zero. Several cod and one rockfish were found during Pacific Ocean beach surveys. Based upon the beach distributions of dead fishes and the lower catch per unit effort in post-shot bottom trawls in shallow Bering Sea waters, it is believed that only a small percentage of the total number of fish killed was actually collected.

A scuba survey of ocean-bottom transects near Cannikin SGZ on both the Pacific and Bering sides of the island showed no significant difference in numbers of sea urchins post-shot, as compared to those counted pre-shot. (This survey was done because urchins are a part of the diet of sea otters.)

Autopsies of 16 dead birds found on the island (15 on the beach) indicated that all but 1 died of the effects of the explosion, 7 from pressure effects and 8 from vertical acceleration. The other bird apparently died of natural causes. No eagles or peregrine falcons were among the dead birds.

Three bald eagle nesting sites along the Bering coast and 2 on the Pacific coast (out of about 100) were lost in cliff falls. In addition, another eagle nesting site on each coast appears unstable and subject to weather damage. Eagles often change nesting sites, and it is not believed that the losses will affect the population. No peregrine falcon nesting sites used in 1971 were lost, but one used in 1969 and another used in 1970 were destroyed.

Large numbers of freshwater fish in lakes near Cannikin SGZ were killed from detonation-related causes. Current estimates are that approximately 10,000 three-spined sticklebacks and some 700 Dolly Varden char were killed from all causes. The stickleback is a small (about 5–8 cm in length) freshwater fish which is one of the common food items of the Dolly Varden, a game fish. Some fish were stranded on shore when water was thrown from the lakes by ground shock; others were stranded on the exposed bottoms of lakes that drained after the detonation. Still others found dead in the lake waters were presumably killed by shock effects. The kills are not expected to result in long-term damage to these fish populations.

All in all, it appears as if the AEC predictions made in the Environmental Impact Statement were reasonably accurate. If later evaluations indicate the predictions were inaccurate it would seem that the Atomic Energy Commission is obligated to conduct research to determine why, i.e. to evaluate what happened that was not expected, to predict how long it will be for the unexpected bioenvironmental cost to recover, if ever, and to determine how this increased cost can be minimized.

The Cannikin Environmental Impact Statement was challenged repeatedly: 'It is inadequate', 'It does not discuss adverse comments made by various individuals'. Seldom were these 'concerned' individuals more than just that, 'concerned', however. It is regrettable that the AEC chose not to use the names of their experts, also the names of their challengers. How else can we establish reliable environmental forecasters – so urgently needed to guide us during this period when environmental procedures are being established? The subject matter – cost to wildlife, chances of radiation escaping, possible pathways of nuclide movement, potential effect of particular levels of radiation, chance of triggering a tsunami, or an earthquake, etc. – is difficult to evaluate even by those most qualified. Volumes of literature are available containing principles of general rules to predict ecological concentration of radionuclides in terrestrial, freshwater and marine environments. This large amount of predictive data was almost totally ignored by 'adversaries'. The only consistent item was 'we all know radiation may be reconcentrated many thousands or millions of time'. Thus it was made to appear that if *any* radiation escaped it would become reconcentrated to lethal levels – or hazard levels in all sorts of fauna and flora.

Though the Amchitka bioenvironmental studies cost approximately $5 million, one aspect of the Environmental Impact Statement was deficient – it failed to discuss the future natural ecological succession. The tacit assumption was that the island would remain about as it is presently as to the kinds and numbers of plants, birds, fish, etc. – a wildlife refuge.

No studies were conducted concerning the Norway rat, the only terrestrial mammal on the island, which was apparently introduced inadvertently by man during World War II. The US Department of the Interior, in their poisoning and shooting programme to destroy the last remnants of fox (which had been purposely introduced in 1923 with the thought of fur harvest), almost surely reduced the rat population. Thus ornithologists were probably measuring an abundance of birds at a time when rat populations were low. Simple observation indicates the Norway rats are now rapidly extending their territories (almost surely populations are rising rapidly). Based on relatively little knowledge of rat behaviour, population controls, kinds and types of preferred or available food, etc., the rats may materially reduce and perhaps eliminate a number of species of birds. One species I would expect to be reduced is the ptarmigan – the only game bird on the island. The Norway rat may well reduce the wildlife carrying capacity of the island more than all other man-nuclear activities combined.

In retrospect, the AEC has used Amchitka Island – a wildlife refuge – to test a nuclear warhead. The environmental cost, i.e. the impact of the test upon the Amchitka environment, was relatively small and in general within the predicted limits.

It is hoped that the AEC upon departing the island follows the motto of a good picnicker and leaves the area in better appearance than it found it. It is possible this will be done. Research projects are being conducted by University of Alaska and University of Tennessee scientists which will determine methods whereby plant communities will be recolonized. Not only will sites disturbed by the AEC – bulldozed sites, road construction, dumping of garbage, etc. – be recolonized, but sites of digging, construction, etc., during World War II will be investigated, and at least an idea will be gained as to how long it may take the disturbed ecosystems to recover.

Another project involves Norway rats. As discussed above, the impact of these rats will probably be the most significant cause of eventual ecosystem change on the island. Rat populations are rapidly spreading and several species of 'ground nesting birds' are likely to disappear from the island. Perhaps even the peregrine falcon and eagle numbers will be determined by the number of sites which the rats are unable to reach – the steepest cliffs and sea stacks. Two techniques to eliminate the rats, one using a genetic strain of sterile male Norway rat, the other a specific poison, have been suggested in preliminary research proposals to the AEC. If the AEC approves these proposals they might possibly play the role of a modern pied piper and lead the way to eliminating rats from Amchitka;

more important than simply complying with the legal requirements of the National Environmental Policy Act of 1969 but fulfilling the very spirit of the Act.

3. Clean-up from a nuclear accident in the Arctic: the Greenland 'Broken Arrow' incident

The Nuclear Age was ushered in under the spectre of a mushroom cloud. As nuclear destructive potential is so awesome, research and regulation of nuclear activities have received extremely critical attention. In fact this extreme caution, depth and persistence of radiation research led a committee report of the National Academy of Science, National Research Council (1960) to state: 'Despite the existing gaps in our knowledge, it is abundantly clear that radiation is by far the best understood environmental hazard . . . Only with regard to radiation has there been determination to minimize the risk at almost any cost.'

Despite this caution, and though the United States Atomic Energy Commission has compiled an impressive safety record, 'nuclear' accidents have happened. At least one major accident involving nuclear materials has occurred in the Arctic. In order to see the implications let us examine the clean-up of nuclear contamination in a land of severe environmental conditions and short food chains.

On 21 January 1968, a giant eight-engined bomber of the United States Strategic Air Command crashed on the ice of Bylot Sound 11 km west of Thule airfield, Greenland. The impact of the crash detonated the chemical explosives of the four armed nuclear weapons the aircraft was carrying. Some several thousand kilograms of plutonium were released. The contamination was blown in all directions and impinged upon materials of the weapon and aircraft and was also caught up in the splashing, burning fuel. As the burning fuel fell to the snow surface it was quickly extinguished, and the plutonium became fixed in the crust formed by the refreezing of the melted snow. The ice was completely shattered and disoriented at the point of impact, thus raising the possibility that some of the plutonium might have penetrated into the water of the bay.

Within five days a technical advisory group of American, Danish and Greenland scientists assembled at Thule. Simultaneously, joint United States and Danish policy-setting groups were meeting in Copenhagen and Washington DC. In order to set limits of acceptable decontamination, these authoritative committees needed information rapidly, and heavy demand was placed upon the field assemblage. The initial effort was concerned with partitioning the contamination according to the various vectors and disseminating modes and delimiting zones of disposition.

PLUTONIUM DISTRIBUTION AND AMOUNT

Simple autoradiographic studies and portable instrument measurements established unequivocally that the depth-distribution of plutonium in the snowpack was strictly a function of the depth of blackening and melting of the surface. More plutonium contamination was found and its distribution was to a greater depth in areas where more fuel collected and burned with the snowpack melted to the ice surface in the most highly contaminated area. Contours of plutonium distribution were established using the Lawrence Radiation Laboratory Field Instrument for Detection of Low Energy Radiation (LRL Fidler instrument). However, absolute contamination levels were established by taking representative samples in each contour and returning them to Los Alamos for plutonium and americium analysis. Total amounts of plutonium were obtained by integrating the surface concentration as a function of area, the total area being about 700×150 m. This information indicated about 3150 ± 639 g of plutonium on the surface (excluding that on the aircraft debris) of which 99 per cent was in the blackened snow and would be removed by removing the snowpack of the area. This snowpack reduced to water would be about 2.3 million litres of water.

It was felt that the ultimate distribution of the plutonium in the event that large sections of the blackened ice were allowed to break up and enter the bay would be influenced by the form, particle size and fixation. Hence detailed studies were undertaken to obtain the information. The studies showed the plutonium to be in the form of oxide particles with a wide size distribution (count median diameter was 2 ± 1.7 nm). The particles were associated with or were adhering to particles and pieces of inert debris of many kinds (metal, glass, nylon fibres, plastic, rubber, paint, etc.) and sizes. The mass median diameter of the inert particles with which plutonium was frequently associated was about four to five times larger than the plutonium particles.

CONTAMINATION IN THE ICE AT IMPACT POINT

The ice at impact point was about 1 m thick. When the ice was fractured upon impact it was displaced downward into the water, randomly oriented and returned to the surface where it froze into position. Thus a large number of core samples carefully scanned cm by cm were required to locate and map the plutonium distribution. The plutonium distribution pattern in terms of contamination per m² of surface area was highly erratic. It was calculated that the plutonium in the ice would have to be dispersed in about 50,000 m³ of water to be at the maximum permissible concentration. Again, as in the snow it was felt the form and fixation of the plutonium in the fractured area would have a bearing on questions regarding its ultimate ecological distribution. The microscopic and autoradiographic observations of residues filtered from ice core samples showed fine particles of plutonium oxide impinged into or onto particles similar to those found upon examination of the blackened snow.

CONTAMINATION BENEATH THE SURFACE

One very difficult question involved the possibility that contamination might have penetrated the ice to be concentrated later by some biological process. Several mechanisms whereby the contamination could have been dispersed beneath the ice were postulated and exploratory work was done to substantiate these postulations. Due to the numerous unknown quantities and inherent inaccuracies, no attempt was made to determine the amount of plutonium which might have entered the water. Later during the summer season it was established that some contaminated aircraft debris actually penetrated the ice. However, at this time tests were conducted to see how rapidly plutonium might be removed from contaminated debris. It was found that likelihood of rapid removal was quite small. This would make the possibility of high concentrations at any given time very small.

ATMOSPHERIC AND GENERAL AREA CONTAMINATIONS

Attempts to calculate meteorological transport and deposition of long-range contamination suggested contamination levels on landmasses south and east of the crash site would be radiologically insignificant but probably measurable by chemical analysis of surface samples. Hence a large number of surface samples were secured from twenty-five locations over much of Greenland.

REMOVAL OF DEBRIS FROM THULE

Upon completion of on-site evaluation and recovery operations by the Strategic Air Command (SAC), effort was directed towards the problems of clean-up.[1]

The clean-up operations had to be relatively rapid as warmer weather would cause the ice to break up, melt, and the crash site would disappear into the bay. Some of the environmental conditions under which operations would take place were: complete polar darkness, the sun would not be seen until mid-February; periodic storms sweeping the area, with temperatures dropping below −60°C, and with winds gusting to 20 m s^{-1} the wind chill climbed to an unbearable degree; because of the position of the north magnetic pole, compasses were nearly useless, hence reference points to obtain accurate headings to develop grid lines was impossible; and few people had experience in arctic operations. Equipment designed for milder climes would break down with relentless regularity. Batteries on trucks, radiation detection units and flashlights had very short lives. Ironically, though the accident was of a sophisticated, complex, nuclear age endeavour, its recovery would depend on the most primitive methods. Initial surveys of the area would depend entirely upon dogsleds and Greenland drivers.

It would have been desirable to construct roads, then the camp, then engage in search activities, delimit the contamination area, and launch the follow-up programme. However, because of the press of time, all actions had to be undertaken concurrently. In order to establish rapid access to the site a heliport was established. Rotor blades created dense snow clouds reducing visibility to zero. To solve this problem sheets of plywood were flown in. However, rotor blades would raise these sheets and a method had to be devised to keep them down. This was solved by using water to freeze them to the ice surface. All other activities were to encounter a similar series of problems.

---

[1] The Chief Scientific Consultant for the clean-up operation, Wright Langham, was killed in an air crash in May 1972. This account stands as a tribute to his endeavours (W. S. Osburn).

After delimiting a 'zero line' of contamination, the point out from the contamination area at which the radiation reading became zero, ropes were strung between reflectorized stakes and the area was physically enclosed. Initial search teams entered the area. They were instructed not to move any of the wreckage components. Reconstructing the dynamics of the occurrence would be possible if the parts were identified in relation to each other. Thus the four nuclear weapons were located and a public announcement to this effect was made on 29 January. Then the search began for the contaminated wreckage. This search finally resorted to mass manpower rather than sensitive technical equipment. Men, shoulder to shoulder, systematically swept through the area many times. All parts were placed in barrels, cans and drums to be sent back to the United States. This phase was completed by 20 February.

Simultaneous with clean-up operations Danish and American scientists were organizing advisory and working groups. In addition to determining the extent and pattern of nuclear contamination discussion of long-range marine and terrestrial ecological studies were considered.

Huge road graders supplemented by hand-raking were used to scrape the blackened contaminated snow into windrows. Continuous belt loaders then filled large plywood boxes. The contents of the boxes were moved by flatbed trucks to the airbase and transferred to large 96,000 litre tanks. The strain on facilities for vehicle repair and maintenance was increased by having to add an additional vehicle decontamination procedure. This was caused by contaminated snow particles stuck on warm oily engine surfaces. In all, sixty-seven tanks were filled with contaminated snow and aircraft debris, and an additional four tanks with general contaminated 'operational' material such as tyres and lumber. Evaluations had to be made to ensure that no problem of achieving critical mass would be involved by placing too much plutonium in any one container.

During this clean-up time all personnel were carefully monitored. No reading higher than zero was considered satisfactory for man, except American personnel, excluding their clothing. It was found that these extra-strict decontamination proceedings generated numerous questions. One of the most significant was the fear of sterility and impotence. When it was explained that these fears were groundless, no further questions were asked!

By 15 March the total area had been cleared. A radiation survey indicated that the radiation hazard had been reduced to negligible proportions, perhaps 25 curies $\pm$ 50% remaining (Aarkrog, 1970).

On 20 March the fencing and posting of the crash site was completed. In a final effort to ensure that all aircraft and weapons had been collected, the area in and around the site was systematically ploughed and reinspected by long lines of airmen. Areas unsafe for such procedure were surveyed by helicopter using radiation detection equipment suspended from a cable. Approximately 78 km^2 were so surveyed. Decontamination of roads, vehicles and loading areas and sealing of tanks continued until 10 April when the last member of the Disaster Control Team departed. Thus this phase of the project was completed with an approximately three-week margin of safety, i.e. 1 May was estimated as about the end of safe heavy equipment operations upon the ice.

CONCENTRATION OF THE WASTE AND ITS RETURN TO THE UNITED STATES
FOR TERMINAL STORAGE

Transportation requirements for a removal operation of this size created a problem of great magnitude. The job was turned over to the Air Force Logistics Command. It was finally decided to transfer radioactive waste into smaller transportable tanks for return to the United States. Frozen waste in the tanks was to be melted and the liquid pumped into smaller tanks made from modified engine containers of about 6800 litres capacity. These tanks would be shipped to the United States and carefully transferred by rail to the disposal grounds at the Savannah River Laboratory in South Carolina.

Subsequent to the approval of the disposal plan, a large number of airmen were trained to melt, transfer, radio-monitor, package and transport the debris from Greenland. On Friday 13 September 1968 the ship *Marine Fiddler* left Greenland and entered international waters, and the gigantic clean-up was 'completed'.

The operation had been handled with meticulous care. An indication that confidence had been established was that no group asked to be reassured that all was well. Regardless of the care exercised in any clean-up operation man is realizing that ecological surprises can occur. Hence several radioecological programmes to locate and trace plutonium in marine and terrestrial ecosystems were designed and carried out. These programmes were largely conducted by one zoologist, one marine biologist, one hydrographer, two physicists, two assistants for sampling (all Danish) and one American arctic ecologist (Herman and Vibe, 1970; Vibe, 1970; Aarkrog, 1970, 1971; and Hanson, 1973).

Ecological sampling was carried out during two major periods. The first was immediately after the accident and the second during the summer after the ice left the bay. Sampling during the first period consisted of examinations at six stations on the sea ice. Three stations were placed from the southern side of the fiord and outward toward the area of impact, while the remaining three stations were placed some 100 m west, north and east of the crash site.

At each station 6 snow samples were taken on the ice, 2 water samples close to the bottom, 1 plankton sample from bottom to surface, and 1 sample of 0.1 m² sea bottom including bivalves, snails, worms, etc. In addition, samples of the bottomside of the sea ice were taken at stations 3 and 6. Each bottom and plankton sample was split. One sample was analysed in Denmark and the other in the United States. Samples of ringed seal, walrus, arctic fox, and dog were procured.

The summer study was the part of the programme in which a very intensive and detailed sampling effort was carried out. The tasks of the team were: to collect samples of seawater, bottom material, bottom animals (food elements for walrus and bearded seal), *crustacea* (food elements for little auks, narwhals and seals), fish (food elements for several sea birds), mussels and other invertebrates from the littoral zone (food elements for arctic fox and eider), sea birds, marine mammals, seaweed, lichen, eiderdown and dust; to investigate the shores of Saunders Island, Wolstenholme Island, the Eiderduck

Islands, the Manson Islands, the southern shore of Bylot Sound, and the northern shore of Wolstenholme Fiord.

As the natives obtain most of their food via a marine food chain, emphasis was given to the marine radioecological portion of the study. Samples of water, sea bottom, and animals were collected. These comprised: 250 water samples from bottom to surface; 10 samples of bottom material from zone 1; 10 samples of bottom material with animals, including bivalves, from zone 1; 14 samples of bottom material from zone 2; 25 dredge hauls for bottom animals at 14 stations in zone 2; 6 trawl hauls for *crustacea* and fish in zone 2; and 16 plankton hauls in zone 2. In addition a number of seals, walrus, eiders, guillemots, eiderdown, excrements from eider, and dust from several islands was collected.

From analysis of the above samples it was concluded that the plutonium levels in no instances were such that they could be considered harmful to man or to higher animals in the Thule district or any other part of Greenland. However, the accident in Bylot Sound measurably raised the plutonium level in the marine environment as far out as 20 km from the point of impact.

Regardless of whether or not hazardous levels existed or could possibly be concentrated, it is important to examine the data for a number of reasons: to determine the major pathways of plutonium movement in a marine environment, especially which organisms are most sensitive indicators of a contaminated environment; to attempt an ecological accounting, for instance, if 25 curies of plutonium (Ci Pu-239) was estimated to escape – where are these quantities, and can one add up the various contributions and substantiate the estimate?

Levels of plutonium radioactivity will be discussed in the following ecosystem components: seawater, bottom sediments, seaweed, plankton, *crustacea*, bivalves, bottom animals, sea birds, seals, walrus, and human urine.

For seawater, the analysis gave no indication of significant amounts of particulate activity, i.e. the plutonium was in solution. The median fallout background level of plutonium in seawater (from five locations far from Thule) was 4×10^{-15} Ci Pu-239 litre^{-1} as compared with a median level of 5×10^{-15} Ci Pu-239 litre^{-1} at Thule. Hence the accident caused only a slight increase in the plutonium concentration of the seawater of Bylot Sound. Assuming the total water of the Sound to be 3×10^{13} l and the mean concentration of 239-plutonium to be 2×10^{-15} Ci litre^{-1} (Aarkrog, 1971, uses this figure rather than the above), the total 239-Pu content from the accident dissolved in the sea would be about 0.06 Ci. This would be only 0.2–0.3 per cent of the amount estimated to have come into the bay – 25 curies. Sedimentation experiments on melted ice cores showed 85–95 per cent of the debris and associated plutonium oxide sank immediately in water, and that only 1 per cent was suspended as fine particles in the water phase. However, plutonium content in seawater samples failed to reveal more than about one-third of the 'expected' amount.

It was assumed that most of the plutonium not removed by the decontamination of the accident area would be found on the sea bottom. However it was not certain that the main part of the activity would be concentrated immediately below the impact point or that appreciable amounts might have drifted away on the ice before sinking to the bottom.

The median deposition of plutonium in the area, immediately below and adjacent to the impact point, was about 0.13 Ci km^{-2} and the area was about 30 km^2. Hence about 4 Ci of plutonium could be accounted for. The highest level of plutonium from any of the samples indicated a concentration of 1.3 Ci km^{-2} and was situated about 1 km northwest of the point of impact. Hence it is possible that another area might account for significantly more plutonium than the impact region.

Although levels of plutonium could be assigned to various regions of the bay, the number of sampling points was so limited in regard to the extremely wide variability that it is impossible to add reliably the various levels per area to estimate total plutonium on the sea bottom. Regardless, a crude and likely conservative estimate would be 3 to 6 curies.

It was concluded that the plutonium levels in both seaweed and plankton were hardly significantly different from the fallout background. The plutonium activity in the bivalves was very unevenly distributed. The standard error of double determinations was far greater than the analytical error. Hence the bivalves could not be considered in equilibrium with the seawater and a calculation of a concentration factor (from seawater to bivalves) would be meaningless. This very inhomogeneous distribution is undoubtedly due to the plutonium being in particulate form. Additionally bivalves are always contaminated by sediments. The highest levels of contamination were in samples obtained near the crash point. Some samples were 1000 times the fallout background (fallout background was considered to be 5×10^{-12} Ci Pu kg^{-1}). Regardless of variability the bivalves seemed to be very sensitive organisms for the detection of plutonium in the marine environment. If one accepts the estimate of bottom animal biomass as 450 g m^{-2} and assuming all animals to be bivalves one can account for another 0.15 Ci, or about 5 per cent of the estimated deposition in the sea bottom sediments.

Trawl hauls demonstrated that *crustacea* (mainly shrimps) occur in appreciable quantities. The median level of plutonium in the *crustacea* samples was three orders of magnitude above the fallout background (levels due to fallout were taken from Danish water samples and = 2×10^{-12} Ci kg^{-1}).

Though great uncertainty existed because of analytical error and the possibility that much of the radioactivity was to particulate contamination, plutonium concentration factors between seawater and soft tissue may be made for bivalves and shrimp, and it would appear that the bivalve concentration is much greater, namely 3500 against 750.

In fish the highest level (470 pCi kg^{-1}) was, as expected, in a bottom-dwelling fish, the Greenland halibut. The median level in fish was 36 pCi kg^{-1}. However the low number of fish sampled (ten) and the lack of fish number or weight precludes an estimate of the total amount of plutonium to be found in fish.

The median level of plutonium in the intestinal content of sea birds was 3.2 pCi kg^{-1}, very close to the mean level in zooplankton, an important constituent of their diet. Samples of excrement from eiders usually contained relatively high levels of plutonium, but as the samples were scraped from stone surfaces some contamination from dust might have occurred. However, as eiderducks consume relatively large amounts of bivalves in their diet one can reasonably expect the excrement to contain relatively large amounts of plutonium. In general, contamination of sea birds was hardly above fallout background.

One exception to the very low plutonium levels in birds existed. This concerned eiderducks. This is of special interest because Eskimos collect the eiderdown and process it by a rather dusty method that might conceivably constitute a small risk from inhalation of contaminants adhering to the down. The two samples contained 84 and 184 pCi Pu kg^{-1}. Again the sample size is small and in the absence of an estimate of the number of eiderducks which might be contaminated, no estimate can be made of the total plutonium contamination found in this ecosystem compartment.

From the International Commission on Radiological Protection recommendations, the daily permissible intake of insoluble 239-Pu is 200 pCi, or 73,000 pCi annually. Thus 'it is extremely unlikely that any Greenlander occupied with the cleaning of down might reach the maximum permissible intake of 239-Pu into the lungs.' Samples of human urine were collected three times – just after the accident, in September of 1968, and in February of 1969. It was concluded that it was unlikely that any Greenlander in the Thule district had been exposed to significant internal levels of plutonium as a result of the accident.

HAZARD EVALUATION

The International Commission on Radiological Protection (ICRP) has not given maximum permissible concentrations (MPC) for marine samples. However, if food habits and concentration factors in food chains are known it is possible to estimate 'equivalent' permissible levels. From the ICRP recommendations for drinking water it is calculated that the maximum permissible daily intake of Pu-239 is 0.1 microcurie. Now if a Greenlander eats 100 g of bivalves daily – which is almost surely an upper limit to consumption – the MPC in bivalves becomes 1 microcurie Pu kg^{-1}. Even the hottest sample of bivalves contained only 10 per cent of this pessimistically estimated MPC value.

Most emphasis was placed upon marine aspects, as contamination was associated with the sea ice and the marine eocsystems are of overwhelming importance to the Eskimo populations of that region. Terrestrial ecosystems of the Thule region supply relatively little food for the Eskimos. However a cloud of smoke and debris, estimated to contain 1 to 5 curies of plutonium, drifted east-northeastward from the crash site. Hence a radioecological surveillance study was conducted concerning adjacent terrestrial systems (Hanson, 1972).

During the period of 26 July–26 August 1968 samples of foliose and fruticose lichens were collected from twenty-seven sampling sites in the Thule region. Lichen communities were selected because of their proven ability to retain appreciable amounts of airborne materials. Most lichens were found in rills, stone stripes or other irregular surface features that provided protection from the strong winds, and contained organic material and moisture. The term community included all lichen species within a sample that was designated by the dominant lichen species at that location. *Cetraria nivalis* proved to be the most common fruticose community.

It was concluded that plutonium concentrations slightly above background and attributable to the Greenland 'Broken Arrow' incident occurred in about one-third of the lichen samples collected at locations adjacent to and downwind from the crash site. In Thule lichen communities which had been subjected to worldwide fallout deposition only, when weighed for the differing precipitation regimes at Thule and Fairbanks, Alaska, indicated that fallout deposition was nearly the same. No attempt was made to estimate the total lichen biomass per unit area nor the areal extent of a particular contaminated lichen community. Hence it was not possible to estimate what fraction of the plutonium which escaped the crash site via a cloud could be accounted for in the landscape.

References

AARKROG, A. (1970) Project Crested Ice, radioecological investigations. *USAF Nuclear Safety* (spec. edn), **65** (2), 74–9. New Mexico: Kirtland Air Force Base.

AARKROG, A. (1971) Radioecological investigations of plutonium in an arctic marine environment. *Health Phys.*, **20**, 31–47.

AMUNDSEN, C. C. (1972) Plant ecology of Amchitka Island. *Amchitka Bioenvironmental Program, Final Rep.*, BMI-171-139. Columbus, Ohio: Battelle Memorial Inst. 27 pp.

BLOOM, S. G. and RAINES, G. E. (1971) Simulation studies as related to the ecological effects of underground testing of nuclear devices on Amchitka Island. *Amchitka Bioenvironmental Program, Ann. Progress Rep.*, BMI-171-138, 2–23. Columbus, Ohio: Battelle Memorial Inst.

DUKE, J. A. (1971) Symposium on Amchitka Island bioenvironmental studies. *BioScience*, **21**, 593–711.

FULLER, R. G. (1971) Amchitka Biological Information Summary. *Amchitka Bioenvironmental Program, Rep.*, BMI-171-132, 1–16. Columbus, Ohio: Battelle Memorial Inst.

HANSON, W. S. (1973) Plutonium in lichen communities of the Thule, Greenland, region during the summer of 1968. *Health Phys.*, **22**, 39–42.

HERMAN, F. and VIBE, C. (1970) Project Crested Ice. Ecological survey. *USAF Nuclear Safety* (spec. edn), **65** (2), 70–3. New Mexico: Kirtland Air Force Base.

KIRKWOOD, J. B. and FULLER, R. G. (1971) Bioenvironmental effects predictions for the proposed Cannikin nuclear test on Amchitka Island. *Amchitka Bioenvironmental Program, Rep.*, BMI-171-141, 1–28. Columbus, Ohio: Battelle Memorial Inst.

KLEIN, D. R. (1971) Reaction of reindeer to obstructions and disturbances. *Science*, **173** (3995), 393–8.

LACHENBRUCH, A. H. (1970) Some estimates of the thermal effects of a heated pipeline in permafrost. *US Geol. Surv. Circ.*, 632.

LAYCOCK, G. (1972) Amchitka revisited. *Audubon*, **74** (1), 113–15.

NATIONAL ACADEMY OF SCIENCE COMMITTEE (1960) The biological effects of atomic radiation, summary report, *NAS/NRC Committee Rep.*, 28. Washington, DC: Nat. Acad. Sci. and Nat. Res. Council.

UNITED STATES ATOMIC ENERGY COMMISSION (1971) *Environmental Impact Statement, Cannikin*, 1–70 (plus 30 pages of other agency comments). Washington, DC: US Atomic Energy Commission.

UNITED STATES ATOMIC ENERGY COMMISSION (1972) *Project Cannikin, D + 30 Day Report, Preliminary operational and test results summary*, 1–28. Las Vegas, Nevada: US Atomic Energy Commission Operations Office.

VIBE, C. (1970) Project Crested Ice. Ecological background. *USAF Nuclear Safety* (spec. edn), **65** (2), 64–9. New Mexico: Kirtland Air Force Base.

Manuscript received January 1972; revised 1973.

Additional references

GLACIOLOGY DIVISION (1973) *Hydrologic Aspects of Northern Pipeline Development.* Task Force on Northern Oil Development, Rep. 73–3. Ottawa: Dept of Environment, Water Res. Branch. 664 pp.

JACKMAN, A. H. (1973) The impact of new highways upon wilderness areas. *Arctic*, **26**, 69–73.

KACHADOORIAN, R. and FERRIANS, O. J. JR (1973) Permafrost-related engineering geology problems posed by the Trans-Alaska pipeline. *In: Permafrost (North American Contributor, 2nd Int. Conference)*, 684–7. Washington, DC: Nat. Acad. Sci.

POSTSCRIPT

The Alaska pipeline controversy was still being hotly contested in the late summer and early autumn of 1973 when, with dramatic suddenness, decisions were reached. On 17 October 1973, the *Washington Post* reported that the Arabs were to cut oil flow to the United States. The next day Congressional conferees reached an accord on a pipeline bill and the essence of accord was 'to brush aside environmental lawsuits and let a group of oil firms start work immediately'. Work began in near total darkness in winter 1973. Thus, after five long years of energy-environmental confrontation, we find that the demand for energy has overridden the environmental protestors. Hopefully this bill will not be a damaging chink in the future shield of NEPA to protect the US arctic environment, but it does demonstrate how willing we are to trade possible long-term environmental costs for short-term comfort. The resolution of the pipeline controversy is perhaps timely.

CHAPTER 18

Postscript

Jack D. Ives and Roger G. Barry
Institute of Arctic and Alpine Research,
University of Colorado

The contributors to this book have shown that our knowledge of arctic and alpine environments is extensive, yet far from complete and characterized by a high degree of unevenness in both between-disciplinary and areal senses. The pressures facing these environments in the coming decades will be increasingly severe and in order to minimize the impacts of resource development, tourism and pollution on them, their biota and native peoples, it is vital that our present understanding of arctic and alpine ecosystems be as fully documented as possible. This will then provide a base for both newly developing long-range ecosystem research and for rational resource development programmes. Such information is a prerequisite for decision making and intervention in both physical and social process-response systems (Chorley, 1971). In the case of the North American Arctic, concern over these problems has been reflected by the Royal Society of Canada (1970) Symposium on the Tundra Environment and by the National Workshop on People, Resources and the Environment North of '60 (Canadian Arctic Resources Committee, 1972). The latter particularly has attempted to document needs and priorities in research, legislation and management. Government response in Canada to growing public concern for the preservation of environmental quality has resulted, for instance, in the establishment in 1972 of National Parks in the Cumberland Peninsula of Baffin Island and in the Yukon. The Canadian Department of the Environment is also rapidly implementing a mapping programme (Arctic Land Use Research) for the Mackenzie River Valley and adjacent areas. The ensuing Land Use Information Map Series on a scale of 1 : 250,000 depicts available information on resources (particularly wildlife) and human activities (including hunting, trapping, fishing and recreation). Maps will be available, for instance, for the entire Yukon Territory by May 1974.

In the United States the request for permission to begin construction of the trans-Alaska oil pipeline has been pushed as far as the Supreme Court and withheld over the legal implications of the need for a right-of-way in excess of 54 feet wide across federally owned land (March 1973). It appears unlikely that pipeline construction will begin in 1973, and the present restriction necessitates enactment of a new law by Congress. The Alaska oil pipeline has proved by far the largest single element in the current awakening of interest in what has been described as the threat to the fragile arctic tundra. It is unavoidable that the greatest known single reserve of oil in North America will be developed. Too much emphasis must not be placed, however, on the simple picture of a single pipeline reaching from Prudhoe Bay, on the north coast, through the Brookes Range to Valdez. Nor is it a question of alternate routes or methods of transport. Such is the scale of the proven and potential resources that would it seem inevitable that the North Slope–Mackenzie Valley–trans-Canadian route and the Aleyeska route will both be needed and that large sea-going tankers will also be used, if somewhat later. Gas and oil in the Canadian Arctic Islands must also assume great importance and consideration of the geological structure of northern Siberia and its vast continental shelf suggests even greater gas and oil potential there. Overall indications are that the next one or two decades will witness a large-scale transformation of the entire circumpolar arctic zone.

It is at least comforting that the potential for catastrophic environmental damage has been recognized, in addition to private industry's primary concern with resource exploitation for profit. This combination has undoubtedly influenced the large role of Tundra Biome research within the United States International Biological Program commitments. Four seasons of intensive fieldwork have already been completed, primarily at Point Barrow, Alaska, but also at Prudhoe Bay and at the secondary alpine tundra sites of Eagle Summit and Niwot Ridge, Colorado Rocky Mountains. This has already consumed nearly $6 million with the National Science Foundation providing the major financial support, but with substantial contributions from the oil industry and the State of Alaska. On the international scale twelve countries are participating in the tundra programme, the major goal of which is to compare biological function and structure in different tundras of the circumpolar lands. Unlike previous interdisciplinary programmes, the Tundra Biome studies are concentrating research efforts on *processes* that affect the rates of energy and matter flow among the major ecosystem components. Fig. 18.1 shows the locations of the major circumpolar sites in the northern hemisphere at which IBP and related field programmes are being conducted.

Within the United States Tundra Biome Program three major objectives guide an interdisciplinary effort which includes over 50 separate projects and more than 200 personnel:

(1) To understand how the tundra ecosystem functions, particularly the wet coastal tundra of northern Alaska.

(2) To analyse the behaviour of the cold-dominated ecosystem types represented in the United States so that the results can be compared with similar studies underway in other countries.

(3) To use basic environmental knowledge to study problems of degradation, maintenance and restoration of the tundra ecosystems.

The administrative and research structure of the Tundra Biome Program has been centred around an attempt to achieve a digital computer simulation of the whole

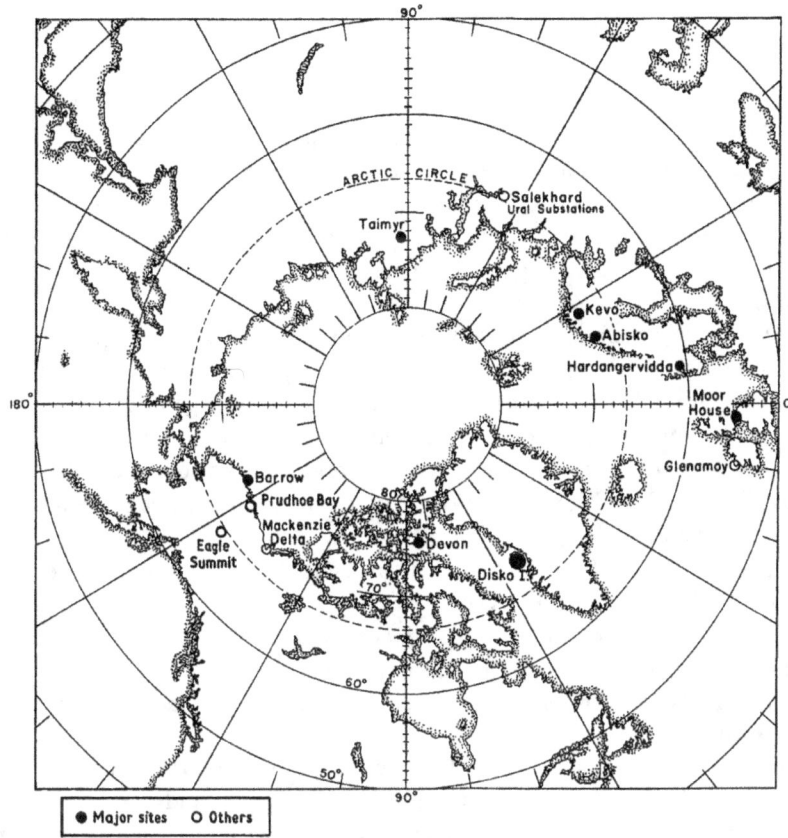

Fig. 18.1. Circumpolar intensive tundra sites at which IBP and related projects have been carried out (sites in Austria and Colorado, USA, not shown).

Vegetation types (see Wielgolaski, 1972)	IBP sites
Shrub-moss-lichen snowbeds	Taimyr Peninsula, USSR
	Disko Island, Greenland
	Hardangervidda, Norway
Dwarf shrub bogs	Taimyr Peninsula, USSR
	Abisko, Sweden
	Moor House, UK
Scrub heaths	Taimyr Peninsula, USSR
	Kevo, Finland
	Hardangervidda, Norway
	Devon Island, Canada
	Patscherköfel, Austria
Monocotyledonous bogs and mires	Glenamoy, Ireland
	Moor House, UK
	Taimyr, USSR
	Hardangervidda, Norway
	Devon Island, Canada
	Barrow, Alaska
Tundra meadows	Disko Island, Greenland
	Hardangervidda, Norway
	Eagle Summit, Alaska
	Niwot Ridge, Colorado
Fellfields	Disko Island, Greenland
	Eagle Summit, Alaska
Shrub meadows	Disko Island, Greenland
	Hardangervidda, Norway
	Taimyr, USSR
Subalpine forest	Kevo, Finland

ecosystem (fig. 18.2) featuring modular association of independent submodels of different levels of mechanistic understanding (Miller and Tieszen, 1972; Nakano and Brown, 1972). Distinct habitats are represented by separate system models, connected at the key points of exchange of biomass and nutrients. All terrestrial habitats connect with similar representations of the major aquatic habitats. Thus, both within- and between-habitat flows of energy and nutrients are being measured and modelled to generate total system understanding.

The 1973 summer will be the final field season after which will begin the major data analysis and writing-up phase. Thus it can be anticipated that the next six to ten years will witness a tremendous increase in the scientific knowledge of at least some aspects of the arctic environments. Several recent comprehensive reports contain details of plans and initial results of both the United States and International Tundra Biome programmes. These include Brown and West (1970), Heal (1971), Wielgolaski and Rosswall (1972) and Wielgolaski (1972).

The Arctic Institute of North America, by establishing its Arctic Development and Environment (ADE) Program (AINA, 1971, 1972) is also beginning to play an increasingly important role in the drive to focus scientific understanding into the area of resource development; a particularly important concept since the Institute is a private and international organization with extensive contacts in the scientific, government and business worlds.

In contrast to the concentration of arctic ecosystem research, studies of the alpine ecosystems match the geographic distribution of alpine lands and are thus widely fragmented and totally lacking in coordination. Even here, however, international concern is beginning to result in at least an initial focusing of attention. UNESCO is in the process of establishing its Man and the Biosphere Programme (MAB) as a logical follow-up to the International Biological Programme whereby the basic ecosystem research results deriving from the latter will hopefully be used in the context of man's use of his environments. Awareness of the growing pressure on mountain lands led the UNESCO directorate to establish, as one of fifteen international panels of experts, a group to prepare a major statement on the impact of human activities on mountain ecosystems. The inaugural meeting of this group was held in Salzburg and the Austrian Tirol from 30 January to 6 February 1973, and a report should be published shortly. One of its major recommendations was that the forest–tundra ecotone, the world over, be recognized as an ecotonal belt, or environmental gradient, that is especially vulnerable to human impact.

Another important international development for mountain research was the creation, through the efforts of Dr Carl Troll in 1968, of the International Geographical Union Commission on High-Altitude Geoecology. Dr Troll's initial task was a study of the tree-lines of the world. Eurasia was considered at a special meeting of the Commission held at Mainz in November 1969, the proceedings of which have been published under Dr Troll's direction (Troll, 1972). The proceedings of a second Commission meeting held in Calgary in August 1972 were published in *Arctic and Alpine Research* (Ives and Salzberg, 1973).

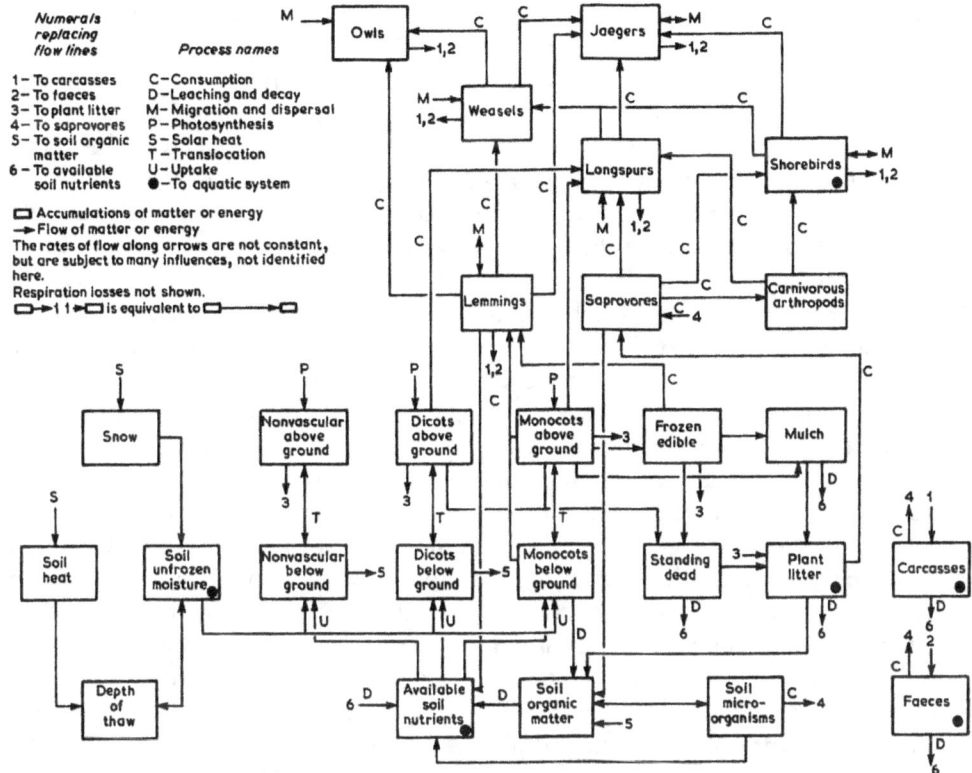

Fig. 18.2. 'Word model' and flow diagram for the wet coastal tundra terrestrial ecosystem (from US Tundra Biome Program, 1971).

It is perhaps especially unfortunate that extreme geographical and political fragmentation characterize the world's alpine areas. Not only does this present a major difficulty in the way of a balanced scientific treatment, as will be evident from the body of this book, but it also renders improbable the realization of coordinated land-use policy developments. One specific research development is perhaps worthy of comment within the context of this postscript, since it relates to a series of major experiments to use mountain ranges to augment the water supply of surrounding areas by artificial means – namely by winter cloud seeding.

Primarily because of rapidly growing water shortages within the Colorado Basin, the Bureau of Reclamation of the United States Department of the Interior was charged by Congress in 1965 to examine the feasibility of developing weather modification techniques as a means of augmenting water supply. Initial meteorological research under the leadership of Professor Lewis Grant of Colorado State University (Mielke *et al.*, 1970; Chappell *et al.*, 1971) led to the conclusion that winter snow accumulation could be increased by as much as 30 per cent through 'seeding' suitable weather systems with silver iodide. The San Juan Mountains of southwestern Colorado were selected for major technological experimentation over four winter seasons beginning in autumn 1969. Extensive meteorological instrumentation was installed and the seeding activity was focused upon a network of ground-based generators. To ensure adequate means of statistical testing of the resulting data a research design was developed such that only half of the suitable weather systems would be seeded while additional instrumentation was established in areas outside of the immediate target area of some 3370 km.

As with the trans-Alaska pipeline, the cloud-seeding experiment emerged at the time that environmental concern was becoming one of the major preoccupations of the North American public. Thus, in addition to the weather modification project itself, the Bureau of Reclamation undertook a commitment to mountain ecological research through a four-year contract that has resulted in the formation of an interdisciplinary and inter-institutional team involving Colorado State University, INSTAAR, and Fort Lewis College, Durango. Some fifteen specific research projects, including studies of alpine tundra primary productivity, tree biomass analysis, climatic change, geomorphic processes and an examination of relationships between elk activity and snowfall, in addition to an assessment of silver toxicity, were also initiated in 1969. A second contract, signed in 1970, charges INSTAAR with the development of an avalanche forecast model applicable to the northern San Juan Mountains. The problem of assessing the ecological impact of winter cloud seeding on such a complex range of environments making up the San Juan Mountains is immense. Not least is the great spatial and temporal variability of snowfall in mountainous terrain when compared with a planned 15 per cent artificial augmentation. Nevertheless, one outcome will be a rapid growth in the available knowledge of mountain and alpine ecosystems. Initial reports that indicate the magnitude of the problem and progress to date include Cooper and Jolly (1969), Colorado State University (1971, 1973), Teller (1972) and Ives et al. (1972).

It will be possible during the next decade to document a more complete synthesis of the characteristics of arctic and alpine environments, building on our systematic knowledge and on the results of holistic studies of the International Biological Programme (see, for example, Bliss et al., 1973) and of the Man and Biosphere Programme. One of the four major programme areas of the latter has been devoted to a study of inter-relationships between human activities and the mountain and tundra environments. Hopefully this increased activity in applied scientific research will be closely related to resource development and land-use policy considerations.

In so far as the more complex mountain landscapes are concerned, some pertinent guidelines for a synthesis are already emerging from the work conducted under the auspices of the Commission on High-Altitude Geoecology. Although appearing too recently to be incorporated here, their results contribute greatly to many of the topics discussed in the preceding chapters. The success of these symposia is a fitting tribute to the scholarship of Carl Troll to whom we dedicate this volume.

References

ARCTIC INSTITUTE OF NORTH AMERICA (1971) *The Arctic Dilemma: Man and His Environment vs. Resources Development* (E. Gourdeau, M. E. Britton and J. C. Reed). Washington and Montreal. 26 pp.

ARCTIC INSTITUTE OF NORTH AMERICA (1972) *Position Paper: United Nations Conference on the Human Environment* (Stockholm, June 1972). Washington and Montreal. 18 pp. Mimeographed.

BROWN, J. and WEST, G. (1970) *Tundra Biome Research in Alaska: The Structure and Function of Cold-Dominated Ecosystems*. US Int. Biol. Progr.-Tundra Biome Steering Committee, London. 297 pp.

CANADIAN ARCTIC RESOURCES COMMITTEE (1972) *Arctic Alternatives* (ed. P. H. Pimlott, K. M. Vincent and C. E. McKnight). Ottawa: Mail-O-Matic Printing. 391 pp.

CHAPPELL, C. F., GRANT, L. O. and MIELKE, P. W. JR (1971) Cloud seeding effects on precipitation intensity and duration of wintertime orographic clouds. *J. Appl. Met.*, **10**, 1006–10.

CHORLEY, R. J. (1971) Forecasting in the earth sciences. *In:* CHISHOLM, M., FREY, A. E. and HAGGETT, P. (eds.) *Regional Forecasting*, 121–37. London: Butterworths.

COLORADO STATE UNIVERSITY (1971) *The San Juan Ecology Project, Interim Report* (H. L. Teller, J. D. Ives, H. W. Steinhoff, C. P. Reid and P. J. Webber). Contract 14-06-D-7052, Bureau of Reclamation, US Dept of Interior. Fort Collins, Colo. 401 pp. Mimeographed.

COLORADO STATE UNIVERSITY (1973) *The San Juan Ecology Project, Interim Report* (H. L. Teller, J. D. Ives, H. W. Steinhoff, H. E. Owen, C. P. Reid and P. J. Webber). Contract 14-06-D-7052, Bureau of Reclamation, US Dept of Interior, Fort Collins, Colo. 173 pp. Mimeographed.

COOPER, F. and JOLLY, C. (1969) *Ecological Effects of Weather Modification: A Problem Analysis*. School of Natural Resources, Univ. of Michigan. 160 pp. Mimeographed.

HEAL, O. W. (1971) *Proceedings of the Tundra Biome Working Meeting on Analysis of Ecosystems* (Kevo, Finland, September 1970). Tundra Biome Steering Committee, Int. Biol. Progr., London. 297 pp.

IVES, J. D. and SALZBERG, K. A. (1973) Geoecology of the high-mountain regions of the Americas and related problems. *Proc. Symposium Int. Geogr. Union Comm. on High-Altitude Geoecology* (Calgary, August 1972). *Arct. Alp. Res.*, **5** (3), part 2. 199 pp.

IVES, J. D., HARRISON, J. C. and ALFORD, D. L. (1972) *Development of Methodology for Evaluation and Prediction of Avalanche Hazard in the San Juan Mountain Area of Southwestern Colorado. Interim Report*. Contract 14-06-D-7155, Bureau of Reclamation, US Dept of Interior. Boulder, Colo: INSTAAR. 88 pp. Mimeographed.

MIELKE, P. W. JR, GRANT, L. O. and CHAPPELL, C. F. (1970) Elevation and spatial variation in effects from wintertime orographic cloud seeding. *J. Appl. Met.*, **9**, 476–88; *Corrigendum*, **10**, 842.

MILLER, P. C. and TIESZEN, L. L. (1972) A preliminary model of processes affecting primary production in the arctic tundra. *Arct. Alp. Res.*, **4**, 1–18.

NAKANO, Y. and BROWN, J. (1972) Mathematical modelling and validation of the thermal regimes in tundra soils, Barrow, Alaska. *Arct. Alp. Res.*, **4**, 19–38.

ROYAL SOCIETY OF CANADA (1970) The tundra environment. Papers presented to a symposium of Section III, *Trans. Roy. Soc. Can.*(4th ser.), 7. 50 pp.

TELLER, H. L. (1972) Current studies in the ecological effects of weather modification in Colorado. *3rd Conf. on Weather Modification* (Rapid City, South Dakota, 26–29 June 1972). Amer. Met. Soc.

TROLL, C. (1972) Geoecology of the high-mountain regions of Eurasia. *Proc. Symposium Int. Geogr. Union Comm. on High-Altitude Geoecology* (Mainz, November 1969), Wiesbaden: Franz Steiner Verlag. 299 pp.

UNITED STATES TUNDRA BIOME PROGRAM (1971) *The Structure and Function of the Tundra Ecosystem*. Vol. 1, *Progress Report and Proposal Abstracts*. US Int. Biol. Progr. 282 pp.

WIELGOLASKI, F. E. (1972) Vegetation types and plant biomass in tundra. *Arct. Alp. Res.*, **4**, 291–306.

WIELGOLASKI, F. E. and ROSSWALL, TH. (1972) *Proceedings of IVth International Tundra Meeting* (Leningrad, October 1971). Tundra Biome Steering Committee, Int. Biol. Progr., Stockholm. 320 pp.

INSTAAR Mountain Research Station
Front Range, Colorado
18 April 1973

Additional references

BLISS, L. C., COURTIN, G. M., PATTIE, D. L., RIEWE, R. R., WHITFIELD, D. W. A. and WIDDEN, P. (1973) Arctic tundra ecosystems. *Ann. Rev. Ecol. Syst.*, **4**, 359–99.

BRITTON, M. E. (1973) Alaskan Arctic Tundra. *Arct. Inst. North Amer.*, *Tech. Pap.*, 25. Ottawa. 224 pp.

DUNBAR, M. J. (1973) Stability and fragility in arctic ecosystems. *Arctic*, **26**, 179–85.

REED, J. C. (1973) Oil and gas development in Arctic North America through 2000. *Arct. Inst. North Amer.*, *Tech. Pap.*, 62. Ottawa. 82 pp.

Glossary

ABRUPT SPECIATION   The sudden development of reproductive isolation through numerical chromosome changes, either by *dysploidy* or, more frequently, *polyploidy*.

ACHEULIAN   A 'culture' or 'industrial tradition' of Mid-Pleistocene time.

ADAPTATION   The hereditary adjustment of a plant or animal to the *environment*.

ADRET (French)   A mountain slope which faces approximately equatorwards, therefore receiving considerable light and warmth during the day. See UBAC.

ADVECTION   Transport of an atmospheric property such as heat or moisture by the mass motion (velocity field) of the atmosphere. Describes predominantly horizontal, large-scale motions of the atmosphere.

AEOLIAN   The uppermost part of the alpine belt above the limit of vascular plants. Organic matter is transferred into this belt by the wind.

AGAMOSPERMY   The formation of seeds without a *sexual process*. It may be either facultative, when at least some populations or individuals produce seeds through normal *fertilization*, or obligate, when normal fertilization is totally absent. Agamospermy is more common in *cryophytes* than in other plants.

ALBEDO   The ratio of the amount of electromagnetic radiation reflected by a surface to the amount incident upon it; often expressed as a percentage. The wavelength interval referred to is commonly the solar spectrum.

ALLELE   Any of the alternative states of a *gene*.

ALLOGAMY   The cross fertilization, or fusion of male and female *gametes* or sex cells from different individuals. The process of transferring *pollen grains* between such individuals is called *cross pollination*. A cross-fertilizing species is said to be allogamous.

ALLOPATRY   The mutually exclusive distribution of population or taxa, which usually inhabit adjacent geographical regions.

ALLOPLOIDS   *Polyploid* taxa formed by the duplication of the chromosome sets of a more or less sterile hybrid between two well defined species characterized by two unlike chromosome sets. Panalloploids, or alloploids in the strict sense, derive from the rare and very sterile hybrid between two species, the chromosomes of which have differentiated so that they cannot pair. Hemialloploids derive from the more frequent hybridization between species the chromosomes of which still are able to pair to some extent.

ALPINE   A high-altitude belt, above the treeline on mountains.

ALTITHERMAL   A postglacial period of greater warmth and aridity than today; more or less equivalent to the so-called 'climatic optimum' in northern Europe.

ALVEOLUS   The terminal air sacs in the lung where gas exchange occurs between air and blood.

AMPHIATLANTIC   Taxa which are met with on both sides of the Atlantic Ocean.

ANCHOR ICE   Submerged ice found attached or ground to bottom, irrespective of the nature of its formation.

ANEMOCHOROUS   Describes plants whose seed dispersal is by wind, and especially plants which save their seeds during the winter and liberate them for dispersal during the spring.

ANEUPLOID   A chromosome number that differs from the exact multiple of the haploid or monoploid number by more or less than a single set of chromosomes.

ANTHOCYANIN   Water-soluble pigments in the cell sap of plants, usually red (acid) or blue (alkaline).

ANTICYCLONIC   With a direction of rotation about the local vertical opposite to that of the earth's rotation; clockwise in the northern hemisphere and counter-clockwise in the southern hemisphere.

APOMIXIS   Reproduction in the floral structures without the usual union of egg and sperm.

ARCTIC   The area of land and sea extending southward from the Pole as far as the northern limits of the coniferous forest and the southern limits of pack ice cover respectively, for several months of the year.

ARCTIC–ALPINE   Refers primarily to taxa which are common to these environments.

ARCTIC BARRENS   The arctic lands without vascular plants. The term is also used for land where some few spots only are covered by plants, preferably in sheltered localities. It is used incorrectly in the case of the so-called 'Tundra Barren Grounds' of Keewatin, NWT

ARCTO-TERTIARY FLORA   The fossil Tertiary flora which is now met with in strata in the arctic regions. These plants never grew under arctic conditions, but the strata drifted into high latitudes after the flora had been fossilized. The species and genera involved are identical or closely related to the present *nemoral flora*.

AREA   The total range of a taxon, such as the area of a species or genus. A limited geographical region.

AUFEIS   See ICING.

AUTHOCHTHONOUS (FELSENMEER)   Refers to rock units of which the dominant constituents were formed *in situ*.

AUTOGAMY   Self-fertilization, or the fusion of male and female *gametes* or sex cells from the same individual. The process of transferring *pollen grains* between male and female organs of the same flower or individual is called self-pollination. Autogamy results in *homozygosity* and inbreeding and is typical of obligately self-fertilizing species.

AUTOPLOIDS   Fertile polyploid taxa formed by the duplication of the same haploid chromosome set or *haplome*. *Panautoploids*, or autoploids in the strict sense, originate from the duplication of the chromosome number of a genetically *homozygous* individual, preferably obligately *autogamous* populations, whereas *hemiautoploids* originate from fertile hybrids between somewhat distinct taxa within the same *genepool*.

AUTORADIOGRAPHY   A photographic technique for detecting radioactive substances.

BAROCLINIC ZONE   A three-dimensional zone in the atmosphere where surfaces of
constant pressure intersect surfaces of constant density. A front is a pronounced
baroclinic zone.

BARREN GROUNDS   See ARCTIC BARRENS.

BASE SATURATION   The (percentage) extent to which a certain soil is occupied by
exchangeable bases, e.g. if base saturation is 80, 4/5 of the exchange capacity is satisfied
by bases and 1/5 by hydrogen.

BERGMANN'S RULE   Homeothermal animals, for example, birds and mammals, exhibit
a tendency to greater size and a lower ratio of body surface to body weight in the
colder parts of their range.

BIFACE   A chipped stone tool, shaped by removing flakes from both sides of the artifact.

BIOLOGICAL EVOLUTION   The transformation of the form and mode of existence of an
organism in such a way that the descendants differ from their predecessors, resulting
in the multitudiny of life on the earth.

BIOLOGICAL SPECIES CONCEPT   A concept of the category of species based on the
reproductive isolation of the constituent populations from other species.

BIOLOGICAL SPECTRUM   List of percentages of the different *life form* classes represented
in the total flora of any defined region. A measure of climate.

BIOLOGICALLY EFFECTIVE IRRADIANCE   An absolute measure of photon irradiance
weighted for biological effectiveness (in terms of photochemical reactions when protein
or nucleic acid chromatophores are involved).

BIOMASS   Weight of living organisms in an *ecosystem*, expressed either as fresh weight or dry
weight.

BIOME   A principal and widespread *ecosystem* type, such as grassland or tundra.

BIOSPHERE   The part of the earth's crust and surrounding layer of air and water which is
inhabited by living beings.

BIOTA   The plants and animals of an area together. Flora and fauna.

BIVALENT CATIONS   Positively charged ions with a valency of two, e.g. Ca^{++}, Mg^{++}.

BOIL, SOLIFLUCTION   See TUNDRA OSTIOLE.

BOULDER-SPALL TOOL   A scraping tool, often used in the preparation of moosehide,
made by striking a disc-shaped spall from the end of a rounded cobble or boulder.

BOWEN RATIO   The ratio of the energy expended in the upward flux of sensible heat
from a surface to that expended in evaporation.

BP   See RADIOCARBON DATING.

BULBIL   A small bulb or modified bud that accomplishes vegetative reproduction. Plants
carrying bulbils are said to be bulbilliferous.

BULL LAKE GLACIATION   Defined from the Bull Lake Till at Bull Lake on the east
flank of the Wind River Mountains, Wyoming, USA. Correlated with Early Wisconsin
glaciation by Richmond (1965) *in:* WRIGHT, H. E. and FREY, D. G. (eds.) *The Quaternary
of the United States* (Princeton: Princeton Univ. Press), p. 220.

BURIN   A chipped stone tool used by many prehistoric arctic peoples for slicing and cutting
hard organic substances such as bone or ivory. The burin is made by driving a
characteristically long and narrow spall from the edge of a stone flake, leaving a sharp
working corner.

CAECAL VILLAE Small absorptive projections in the caecum, a branch of the gut, which are important in the digestive process.

CALCICOLES Plants which grow best on calcareous soil.

CANDLING Ice with crystals which are long and vertical. It is a common form during the melt of lake ice.

CAROTID CHEMORECEPTOR Specialized tissue which senses the oxygen pressure of blood in the carotid artery.

CATENA An association of soils in the field, developed from the same parent materials under the same climatic conditions, on the basis of differences in drainage or relief.

CHAMAEPHYTE A plant of which the perennating buds are situated less than 30 cm above the surface of the soil.

CHANNERY A thin flat fragment in soil between about 2 and 150 mm in thickness.

CHROMA A term used in designating the colour of a soil. The notation for chroma is a number (0 for grey and increasing to a maximum of about 20). See HUE.

CHROMOSOME A structure of the cell containing most of the hereditary material, or *genes*, which are linearly arranged on several chromosomes in each species. A certain chromosome number is typical of each species, diploid (2n) in the body cells and haploid (n) in the sex cells. A basic chromosome number (x) is the lowest haploid number in a *polyploid series* within a genus.

CIRQUE A steep-walled basin high on a mountain, often described as 'armchair shaped', formed by glacial erosion; also 'corrie', 'cwm', 'botn' etc.; the characteristic landform of glaciated mountains.

CLIMAX The type of community maintained under the prevailing climatic and edaphic conditions. The climatic climax is the ultimate stage of ecological development of communities permitted by the climate of a region. The edaphic climax is controlled mainly by the prevailing soil conditions.

CLINE A gradient within a continuous population in the frequencies of different characteristics.

CLONE All the individuals derived by asexual reproduction from a single sexually produced individual.

COLD LOW A low pressure area with, for a given level, colder air near its centre than in the surrounding air. The cyclonic intensity increases with height in the troposphere.

COLLOID (SOIL) Mineral or organic particles that are small enough so that surface forces are important in determining their properties; in soils, these are commonly considered to be particles smaller than 0.002 mm.

COLLUVIAL SLOPE Sloping land at the foot of steep hills or mountains made up of deposits of unconsolidated material that has been moved over short distances by gravity or water, or both; includes talus material and local alluvium.

CONTINENTALITY INDEX An index describing how much the climate of a place is influenced by the small heat capacity of the land in comparison with that of the ocean. It is generally expressed in terms of the annual temperature range.

COPROLITE Faecal pellets or castings.

CORIOLIS FORCE A force per unit mass due to the earth's rotation. It acts as a 'deflecting force', normal to the velocity of a moving particle, to the right of motion in the northern hemisphere and to the left in the southern hemisphere. It does not alter the speed of the particle.

CROSS POLLINATION   See ALLOGAMY.

CRYOPHYTES   The plants which are adapted to life in the cold alpine or arctic regions of the world.

CRYOPLANATION PROCESSES   Processes of intensive frost action (e.g. solifluction) which lower the land surface. Also called equiplanation and altiplanation.

CURIE   The quantity of a radioactive substance that undergoes 3.70×10^{10} radioactive transformations per second.

CYCLOGENESIS   Developing or strengthening in the atmosphere of cyclonic circulation.

CYCLONIC   With direction of rotation about the local vertical the same as that of the earth's rotation; counter-clockwise in the northern hemisphere, clockwise in the southern hemisphere.

CYTOGENETICS   The comparative study of chromosomal mechanisms and behaviour in populations and taxa, and their effect on inheritance and evolution.

DEAD ICE   Part of a glacier or ice sheet which is no longer flowing.

DEGREE-DAYS   A measure of the difference between the daily mean temperature and a given standard: one degree-day for each degree (°C or °F) difference for one day, e.g. freezing/melting degree-days refer to a standard of 0°C.

DEME   A local population, or the community of potentially interbreeding individuals at a given locality.

DEPTH HOAR   Ice crystals which form by sublimation beneath the snow surface (within the snow).

DIASPORE   Any spore, seed or fruit, etc., which is able to produce a new plant when dispersed.

DIFFUSE RADIATION   Radiation which reaches the earth's surface after it has been scattered from the direct solar beam by air molecules or particulate matter.

DIOECIOUSNESS   The condition of having unisexual flowers borne on separate individuals, as in willows.

DIPLOID   Having a double set of chromosomes, as is typical of most individuals derived from a fertilized egg cell. The chromosome number (2n) in the body cells of an individual, representing two homologous chromosome sets, one paternal, the other maternal, as opposed to the *haploid* (n) number of the sex cells. In a *polyploid series*, the chromosome sets are multiplications of the lowest haploid number met with, the *monoploid number* (x); in such a series the diploid (x) represents the lowest diploid number (2n).

DIRECT RADIATION   That part of global radiation which is received at the earth's surface 'direct' from the sun, i.e. without being scattered in the atmosphere.

DISCHARGE HYDROGRAPH   A graph used in hydrology to illustrate the change of discharge of a river with time.

DISCLIMAX   Occurs when a stable community, not the climatic or edaphic *climax*, is maintained by man or domestic animals.

DISJUNCT   Refers to a separation, for example in the geographic distribution of a *taxon* or a community.

DISPERSAL   The spreading of plant *propagules* from one place or an *area* to another. It is a one-way movement each generation, as opposed to *migration*, or the seasonal back-and-forth movement of certain animals.

DOMINANT   Of an allele determining the phenotypic appearance of the *heterozygote*.

DYSPLOID   Refers to different basic chromosome numbers within related genera or species. These numbers are stable, in contrast to populations with *aneuploid* chromosome numbers.

ECOCLIMATOLOGY   The study of climate near the ground, especially with respect to the climatic environment of fauna and flora.

ECOSYSTEM   A complex of living organisms and their environment, which is linked by energy flow and material cycling. It may be used in a collective sense, for example pine forests, or for a concrete habitat such as a pond. It is a broad conceptual unit ranging from small microcosms to the entire world, the biosphere. An ecosystem seldom has distinct boundaries and is never a perfectly closed system in nature.

ECOTONE   A transition zone of vegetation between two communities, which has characteristics of its own and of both types of adjacent vegetation.

ECOTYPE   A local, ecological race adapted to a particular restricted habitat as a result of the genetical response to natural selection.

EDENTATES   Placental mammals with teeth reduced or absent.

EKMAN LAYER   The transition layer (approximately 10–1500 m above the surface) between the surface boundary layer where the shearing stress is constant, and the free atmosphere, where the *geostrophic wind* approximation is applicable.

ENDEMIC   Restricted to a certain area.

ENVIRONMENT   The sum total of effective factors to which an organism responds.

EPACRID   Plants of the *Epacridaceae*; shrubs or small trees of the southern hemisphere which are ecological analogues of the *Ericaceae* (heaths).

EQUIFORMAL AREAS   More or less concentrical areas of taxa which have spread from the same place of origin.

EQUILIBRIUM LINE   The line above which a glacier has a net gain of mass over the year and below which there is a net loss.

ERRATIC   A term referring to anomalous rocks and boulders not belonging to the immediate surroundings which have been transported there by some natural means, e.g. glacial erratic. E. Dahl (1955) expresses the desirability of restricting the term 'glacial erratic' to surface boulders that are unequivocally 'foreign' to the county rock (bedrock) on which they lie, thus excluding the possibility of misinterpretation of weathered-out inclusions.

ESKER   A low ridge of stratified fluvio-glacial deposits, usually with much sand and gravel. These features vary in height and length and several hypotheses exist regarding their origin.

ESPELETIA   A genus of the *Compositae*, analogous to the large *Senecios* of the African alpine, which occurs in the alpine (páramo) of the Andes.

EUSTATIC   Relating to or characterized by worldwide change of sea level (independent of movements of the land).

EVAPOTRANSPIRATION   The process of evaporation of water from a soil surface together with transpiration by plants. *Actual* evapotranspiration depends mostly on the availability of water. *Potential* evapotranspiration is the maximum amount of water that would be removed, in relation to the available energy, if the moisture supply were unlimited.

EXCHANGEABLE BASES (EXCHANGEABLE IONS, EXCHANGEABLE CATIONS)   Cations or bases adsorbed onto colloids in the soil.

FELLFIELD   Arctic–alpine plant community consisting of a large proportion of boulders with lichens, mosses, cushion-form plants and some vascular plants.

FELSENMEER (BLOCKFIELD, MOUNTAIN-TOP DETRITUS, BLOCKMEER)   Extensive accumulations of frost-shattered rocks on horizontal and gently sloping terrain. The debris is commonly angular and boulder-sized although smaller, subrounded detritus also occurs.

FERTILIZATION   The fusion of two *gametes* of opposite sex to form a zygote, which will grow into a seed and a new individual.

FIELD CAPACITY   The amount of water retained in the soil after excess water has drained away and the rate of downward movement has decreased substantially after irrigation or rain.

FIRN LINE (ANNUAL SNOWLINE)   The uppermost line on the glacier to which the snow-fall of the winter season melts during the following summer ablation season.

FÖHN (FOEHN)   A wind occurring on the lee side of a mountain range. The air is warm and dry because of adiabatic compression as it descends the mountain slopes. Known as chinook in North America and by a variety of other local or regional names.

FOREST–TUNDRA ECOTONE   The transition belt between dense conifer forest and tundra, at high latitudes or high altitudes.

FOUNDING POPULATION   A small *population* invading a new *area*, including only a limited fraction of the total *genepool* of the parental population of species.

FRAZIL ICE   Fine spicules or plates of ice, suspended in water which is often turbulent.

FREQUENCY OF POLYPLOIDS   The percentage of polyploid species within a flora. It increases with an increase in altitude and latitude or in severity of climate.

FRONT (FRONTAL ZONE)   A three-dimensional zone in the lower atmosphere separating airmasses of different density. See also BAROCLINIC ZONE.

FROST SCARS   Patches of soil over permafrost that are either barren or partially covered by lichens, and that result from frost stirring or ice crystallization processes.

GAMETE   A mature reproductive cell capable of fusing with a cell of similar origin but of opposite sex to form a *zygote*. A gamete is haploid (n).

GELIFRACTION (FROST CRACKING, FROST SPLITTING, FROST RIVING)   Occurs when moisture soaks into hard rocks via joint and cleavage cracks during the thaw season. The water freezes as the temperature is lowered to the critical level and the accompanying expansion of the ice causes the rocks to split.

GENE   A unit of genetic material localized in the chromosome.

GENEPOOL   The totality of the *genes* of a given species.

GENETIC RECOMBINATION   A process whereby two or more parents with different genetic characters give rise to a progeny in which *genus* in which the parents differed are associated in new combinations.

GENETICS   The science studying the mode of inheritance and the behaviour of *genes*. Frequently used in a collective sense including all kinds of studies of heredity and inheritance.

GENOME   A sum of all genes of the haploid phase of a taxon at any level of ploidy; the haploid *genetype*.

GENOTYPE   The totality of genetic factors that make up the genetic constitution of an individual.

GEODETIC (DATUM)   A datum which consists of the latitude and longitude of an initial point, the azimuth of a line from the same point and two constants which are needed for defraction of the terrestrial spheroid. This datum is the basis for computations of horizontal control surveys in which curvature of the earth is considered.

GEOGRAPHIC BARRIER Any terrain or water that prevents geneflow between populations.

GEOGRAPHIC ISOLATION The separation of genepools or parts of genepools by *geographic barriers*; the prevention of gene exchange between a population and others by *geographic barriers*.

GEOPHYTE A plant of which the perennating buds are buried deeply in the soil and situated on a rootstock, rhizome, corm, bulb or tuber.

GEOSTATIONARY (ATS) SATELLITES Satellites with an orbit around the earth such that they are always located above the same point on the earth's surface.

GEOSTROPHIC WIND The horizontal wind velocity for the state where the *Coriolis* acceleration and the horizontal pressure force are exactly in balance. The wind becomes more or less geostrophic above the friction layer (about 1500 m).

GEOTRIPTIC WIND Unaccelerated flow in the atmospheric boundary layer, in which there is a balance between pressure force, *Coriolis* acceleration and friction force.

GLACIATION LIMIT The (average) lower limit of glacierization in an area. 'Glacierization limit' or 'glaciation level' have been suggested as preferable terms since confusion may arise with the use of the term 'glaciation limit' to refer to the outer limit of glaciation during a glacial phase.

GLACIER SURGE A sudden rapid advance of a glacier, in which the ice in the lower reaches moves several kilometres in a few months or, at most, two or three years.

GLACIO-ISOSTASY When ice on a land area achieves a certain thickness isostatic depression of the crust begins and may eventually become 1/4 to 1/3 of the thickness of the ice. The reverse process, 'isostatic recovery', occurs following melt of the ice sheet.

GLEYING (GLEISATION) A soil-forming process under conditions of poor drainage resulting in reduction of iron and other elements and in grey colours and mottles.

GLOBAL RADIATION The sum of *direct* solar radiation and *diffuse* sky radiation which is received on a unit horizontal surface.

GRADUAL SPECIATION The slow development of reproductive isolation through successive and cumulative linear changes of the individual chromosomes, frequently through *inversions* and *translocations*.

GRAVER A chipped stone tool with a sharp corner or point which is designed for piercing or engraving organic substances.

HADLEY CIRCULATION A direct thermally driven and zonally symmetric circulation composed in each hemisphere of equatorward movement of the trade winds between about latitude 30° and the equator, with rising wind components near the equator, poleward flow aloft, and descending components around latitude 30°.

HALOPHILOUS Tolerant of salt.

HAPLOID Having only a single set of chromosomes. The chromosome number (n) of the *gametes* represents a single, paternal or maternal, chromosome set, as opposed to *diploid* (2n). The lowest haploid number in a *polyploid series* is said to be *monoploid* (x).

HAPLOME A single *monoploid* chromosome set of a *diploid* species or a *polyploid series*, as contrasted to *genome*, which may include several such sets.

HARDINESS   The ability to withstand severe climates, especially frost during the growing
season.

HEMIALLOPLOID   See ALLOPLOIDS.

HEMIAUTOPLOID   See AUTOPLOIDS.

HEMICRYPTOPHYTE   A plant of which the perennating buds are situated in the surface
of the soil.

HETEROSTYLY   A trait in which a species has two types of flowers: one type with long
styles, one type with short styles.

HETEROZYGOTE   An individual with different genetic factors (*alleles*) at the corresponding
loci of the two parental chromosomes; an individual with two different segments at
corresponding places of homologous chromosomes, or with two not completely homologous chromsomes; hybrids. This phenomenon is called heterozygosity, and the
individuals are said to be heterozygous.

HILL REACTION RATE   The amount of oxygen evolved per unit time by the splitting of
water molecules in isolated chloroplasts in light.

HOLOCENE (FLANDRIAN, RECENT)   The most recent stage of the Pleistocene, extending
from the close of the last Ice Age up to the present day.

HOMOZYGOTE   An individual with identical genetic factors (*alleles*) at the corresponding
loci of the two parental chromosomes; or with the same linear arrangement within
these chromosomes, or with two completely homologous chromosomes. This
phenomenon is called homozygosity, and the individuals are said to be homozygous.

HORIZON, SOIL   A layer of soil approximately parallel to the soil surface, distinguishable
from adjacent layers by a distinctive set of characteristics.

HUE   One of the terms designating the colour of a soil. The symbol in the Munsell system
consists of a number from 0 to 10 followed by a letter abbreviation of a colour
(R = red, YR = yellow-red or orange, Y = yellow). Within each letter range the
colour becomes less yellow and more red as the number decreases. See CHROMA.

HYBRIDIZATION   The crossing of individuals belonging to two unlike natural populations,
or of two genetically different individuals of the same population, or of taxa at any
level which leads to hybrid progeny.

HYDROLACCOLITH   A genetic term referring to all ice intrusions.

HYPERBOREAL   Belonging to the extreme north.

HYPOXIA   A deficiency of oxygen resulting from an inadequate quantity or pressure of
oxygen.

ICE FOG   A fog due to minute ice particles; it usually forms in urban areas in calm
conditions when the temperature is below about $-30°C$.

ICING (AUFEIS, NALED, TARYN ICE)   Extrusive ice, formed subaerially, such as ice
formed on river flood plains or below springs that flow all winter.

IDIOMORPHIC   Applied to minerals of igneous rocks which are bounded by their own
crystal faces.

ILLUVIAL HORIZON   A soil horizon in which material carried in suspension or solution
from an overlying horizon has been deposited or precipitated; a horizon of accumulation.

INTERTROPICAL CONVERGENCE   The dividing axis between the southeast and the
northeast trade winds. Actual convergence occurs only intermittently along this axis.

INVERSION (1) Reversal of the linear order of the *genes* in a *segment* of a *chromosome*.

INVERSION (2) A meteorological term indicating a variation from the normal increase or decrease with altitude of the magnitude of a property of the atmosphere. It is particularly applied to a reversal of the normal decrease of temperature with altitude.

ISOBASE   This line joins points of equal postglacial emergence on coastal sites or of equal postglacial uplift on glacial lake shorelines, where these elevations result from postglacial emergence or uplift occurring during the same time span (glacio-isostatic recovery).

ISOCHRONE   A line connecting points on the surface of the earth at which a characteristic time or interval is the same.

ISOSTATIC BALANCE (ISOSTASY)   Theoretical balance of all large parts of the earth as if they were floating on an underlying layer which is denser; areas of less dense crust rise topographically above areas of denser material.

ISOSTATIC COMPENSATION (ADJUSTMENT)   The process in which lateral transport at the surface of the earth by erosion, deposition and similar processes is compensated by lateral transport in a layer beneath the crust.

ISOTOPE   An atom with an atomic weight which differs from that of other atoms of the same element.

ISOZYME   One of a series of different chemical forms of the same enzyme any of which can catalyse the same reaction within a plant cell.

JÖKULHLAUP (Icelandic – GLACIER BURST)   Violent, short-lived variation in the discharge rate of a glacial river caused by rapid drainage of lakes dammed up by the ice. Also, special cases where volcanic processes induce abnormal glacier melt discharges.

KARST   A type of terrain with a distinct, unique assemblage of landforms resulting mainly from solution. Most often found in areas of pure and massive limestone.

KATABATIC WIND   A downslope wind caused by the drainage of cold air under gravity.

KETTLE LAKES   See MORAINE LAKES.

KRUMMHOLZ (German)   The crooked and creeping form of trees at *timberline*. Strict European usage confines the term to genetically deformed species.

LABRET   An ornament worn in the lip or cheek, generally used among the Alaskan Eskimos of the historic period.

LANCEOLATE   With a long shaft and a spear head.

LAPIES   A rill-like erosion feature formed by solution of limestone in a *karst* area.

LAPSE RATE   The role of decrease with altitude of the magnitude of an atmospheric property (which is temperature unless otherwise specified). Moist adiabatic lapse rate (or saturated adiabatic lapse rate) is the rate of decrease of temperature with altitude of an air parcel lifted in an adiabatic process, while the air is kept saturated by evaporation or condensation of water with the latent heat released by or to the air.

LAURASIA   Theoretical continent in the northern hemisphere which broke up in Permo-Carboniferous time to form the present *northern* continents. See PANGAEA.

LAYERING   Vegetative reproduction, especially of conifers near timberline, where the lowest boughs on the central stem take root and produce a new central stem. Candelabra form results from layering in an annulus around the central stem which eventually dies out leaving a circle of daughter trees derived directly from the same parent.

LEACHING   Washing or draining by percolation.

LICHENOMETRY   The method of dating rock surfaces by measuring maximum thalli of suitable lichen species; based upon the assumption that after initial growth phase, growth rate remains approximately uniform for a significant time interval until senescence.

LIFE FORM   The vegetative form of a plant, as determined by the mode of survival of its penennating buds. Used in connection with the determination of the *biological spectrum*.

LIPID   A fat or fatlike compound not soluble in water.

LOCAL POPULATION   See DEME.

LONG-DAY ELEMENT   Long-day plants that need a daily time span of illumination over twelve hours for flowering to occur.

LONG-WAVE RADIATION (INFRA-RED RADIATION)   Electromagnetic radiation with a wavelength from 0.8 μm to an indefinite upper boundary, sometimes arbitrarily defined as 100 μm.

MASS BALANCE (GLACIER)   The mass balance at any time on a glacier is the algebraic sum of the accumulation and ablation. It is the change in mass per unit area relative to the previous summer surface.

MASSENERHEBUNG EFFECT   The continentality (mainly temperature) effect of a mountain massif which displaces climatic and biologic limits to higher altitudes than is typical of the same latitude.

MEDULLA   The brain stem, which contains the respiratory control centre.

MESOPHYTE   A plant which will grow where moisture conditions are medium.

MICROBLADE   A thin, parallel-sided flake of stone with extremely sharp edges, manufactured by a specialized stoneworking technique involving series of microblades driven from a prepared core.

MIGRATION   The seasonal or annual movement of animals, especially birds, from one area to the other and back.

MISCIBILITY   The ability of two populations to mix and exchange genes through hybridization.

MITOCHONDRIA   Small structures or organelles in the cytoplasm of a cell which are involved in respiratory processes.

MOH HARDNESS   The Moh hardness scale is an empirical scale, used for determining the hardness of a mineral as compared with a standard. The scale is:

1. talc; 2. gypsum; 3. calcite; 4. fluorite; 5. apatite; 6. orthoclase; 7. quartz; 8. topaz; 9. corundum; 10. diamond.

MOIST ADIABATIC LAPSE RATE   See LAPSE RATE.

MONOPLOID   The lowest *haploid* number in a *polyploid* series (x).

MORAINE LAKES (KETTLE LAKES)   Both types result from ice marginal deposition. Moraine lakes – lakes dammed by glacier moraines; kettle lakes – formed by blocks of of ice embedded in glacial drift that subsequently melt.

MOTTLING   Differences in soil colour found in imperfectly drained soils, which usually show alternate streaks of oxidized and reduced materials.

MOUNTAIN-TOP DETRITUS   See FELSENMEER.

MUTATION   Any detectable and heritable change in the genetic material not caused by *segregation, genetic recombination,* or linear or numerical alterations of the *chromosomes,* which is transmitted to succeeding generations.

NATUFIAN   A culture in the Near East, dating from approximately 9500–7500 BC.

NEMORAL FLORA   A flora adapted to a mild-temperate, moist climate, with a mull-type humus layer and a brown forest soil profile; so-called deciduous forest of eastern North America, eastern Asia, and western Europe, and the fossil so-called *Arcto-Tertiary flora.*

NEOGENE   The later of the two periods into which the Cainozoic era is divided, comprising the Miocene, Pliocene, Pleistocene and Holocene (Recent) epochs.

NET RADIATION   The difference between the fluxes of incoming and outgoing all-wave-length radiant energy at a surface.

NOBLE GAS   A gas which does not react chemically; for example helium, argon, neon, krypton and xenon.

NUNATAK (Eskimo)   A mountain surrounded on all sides by glacier ice. Sometimes used for any area of land surrounded by ice today, or during past glacial maxima; nunatak hypothesis.

OEDEMA   An abnormal accumulation of fluid in tissue.

OLIGOTROPHIC   Providing inadequate nutrition.

OSTIOLE   See TUNDRA OSTIOLE.

PALAEOMAGNETISM   Faint magnetic polarization of rocks which could have been maintained since sediment accumulated or magma solidified, in which the magnetic particles were oriented with respect to the earth's magnetic field as it was at that place and time.

PALAEOSOL   A buried soil, e.g. a soil formed during an interglacial period and covered by deposits of subsequent glacial advances.

PALYNOLOGY   The science identifying *pollen grains*; the study of the historical and climatical significance of pollen grains accumulated in a sedimentary deposit.

PANALLOPLOID   See ALLOPLOIDS.

PANAUTOPLOID   See AUTOPLOIDS.

PANGAEA   The theoretical large continent in the northern hemisphere which broke up to form the present continents in Permo–Carboniferous time.   See LAURASIA.

PANMICTIC   Of *populations* that breed randomly so that the whole population or species forms a single *deme.*

PARAPATRIC   Mainly *allopatric populations* whose *areas* are in contact so that hybridization is geographically possible.

PEDIOPHYTE   A plant of the temperate lowlands.

PEDON   A three-dimensional body large enough to permit complete description of a soil, including those with cyclic or intermittent horizons; surface area ranges from 1 to 10 m² depending on the variability in the horizons.

PERIGLACIAL   Referring to the belt of land of indeterminate width adjacent to the margin of an ice sheet, or a former ice sheet. Expanded to refer to landscape subjected predominantly to cold-climate processes. Also, periglacial processes – geomorphic processes characteristic of cold climates.

PHENOTYPE   The totality of characteristics of an individual (its appearance) as a result of the interaction between its *genotype* and the *environment.*

PHOTOSYNTHESIS   The food-making process by which green plants and other autotrophs fix sunlight into chemical energy.

PINEDALE GLACIATION   Defined from the Pinedale Till, at the southwest base of the
   Wind River Mountains, Wyoming. Correlated with Late Wisconsin glaciation. See
   BULL LAKE GLACIATION.

PINGO   A large frost mound up to 50 m in elevation caused by bulging-up of frozen ground
   under hydrostatic pressure from below; characteristic of fine sediments in the continuous
   permafrost zone.

PIPKRAKE (NEEDLE, STALACTITE ICE)   Needle-like ice crystals 2–6 cm long which
   form singly or in bundles just below the ground surface and perpendicular to it. They
   form in a few hours during freezing when abundant soil moisture is present. The
   crystals lift soil particles, vegetation, gravel and pebbles. It is usually a diurnal
   formation.

PLANETARY WAVE NUMBER   The number of major waves in the belt of tropospheric
   westerlies (in the atmospheric circulation). There are usually between one and five of
   these waves around the globe.

PLASTIC SPECIES   A species which easily adapts to various conditions.

PLEISTOCENE REFUGIA   Favourable areas outside the area covered by ice, where
   populations of plants and animals survived periods of glaciation.

PLOIDY   An abbreviated term referring to the number of chromosome sets. See POLY-
   PLOIDY.

POINT   General term for a stone tool designed for mounting on the end of a projectile or
   stabbing weapon. The terms 'lanceolate', 'notched', 'stemmed', etc., refer to the
   design for hafting the point to its shaft.

POLAR DESERT   The area at high latitudes with very low temperatures and precipitation
   rates, and in which less than 25 per cent of the surface is covered by vegetation; in
   general, the area in which mean temperatures of the warmest month are between
   5°C and 2°C.

POLLEN GRAIN   *Haploid* (male) microsporangium of a plant, collectively, simply called
   pollen.

POLLINATION   The carrying of pollen to the female organ of a plant, usually by wind or
   flying animals, especially insects.

POLLINATION DISTANCE   The distance within which pollen can be carried between two
   stands or local populations of a species.

POLYCYTHEMIA   An excessive number of red cells in the blood.

POLYMORPHISM   Existence together in the same place of two or more distinct forms of a
   species in proportion such that the rarest of them cannot be maintained by recurrent
   mutation.

POLYNYA   An extensive area of open water within otherwise close pack ice, within a fast
   ice sheet, or between the fast ice edge and the pack ice.

POLYPHENISM   The occurrence of several *Phenotypes* in a population.

POLYPLOID SERIES   Chromosome numbers of a group of related species that form a
   series based on the multiplication of the basic number which is the lowest *haploid*,
   or *monoploid* number, with multiples of 2. The series will include diploids, tetraploids,
   hexaploids, etc.

POLYPLOIDY   A condition in which the number of chromosome sets in the nucleus is a
   multiple (higher than by 2) of the lowest *haploid*, or *monoploid* (x) number. The individual
   is said to be polyploid.

POPULATION   The total number of individuals of a taxon; the total number of individuals of a taxon in a particular habitat. See DEME.

POTENTIAL EVAPOTRANSPIRATION   See EVAPOTRANSPIRATION.

POTENTIAL TEMPERATURE   The temperature which an air parcel would reach if brought to a pressure of 1000 mb by a dry adiabatic process.

POTENTIAL VORTICITY   Potential vorticity $= \dfrac{\zeta + f}{\Delta p}$ where ζ is relative vorticity (cyclonic positive), f is the Coriolis parameter, and Δp is the depth of a column of air between surfaces of constant potential temperature. Potential vorticity is conserved in dry adiabatic motion.

PREADAPTATION   The possession of the necessary properties to permit a shift into a new niche or habitat. A structure is preadapted if it can assume a new function without interference with the original function.

PRESSURE MELTING POINT   The temperature at which a solid melts at a given pressure.

PRIMARY PRODUCTIVITY   The rate of storage of radiant energy by *photosynthetic* and chemosynthetic activity of producer organisms (mainly green plants). The energy is stored as organic substances which can be used as food materials.

PROPAGULE   A part of a plant that implants a new individual.

Q_{10} Increase in respiratory rate of each 10°C increase in temperature.

RACE   A group of infertile taxa connected by common descent or origin and belonging to the same genepool; such a group with a defined *area*.

RADIANT TEMPERATURE   The absolute temperature of the surface of a radiating body.

RADIOCARBON DATING   The method is based on measuring the radiocarbon activity of biogenic material such as wood, peat and shells. In calculation of the age of a sample, the conventional half-life is taken to be 5567 years. The determined date is stated in terms of measured probabilities because of the random nature of radioactive decay. A date is described as the time interval within which the true date will lie with a certain probability. The error given is calculated on the counting statistics and refers to one standard deviation, within which there is a 68 per cent probability that the true date lies. The results are given as radiocarbon age BP (before present), taken as 1950. Because of errors inherent in the technique radiocarbon ages are not identical to actual ages. The practical limit of radiocarbon dating is up to about 30,000 to 40,000 years.

RADIOECOLOGY   Study of various aspects of ionizing radiation in the environment, with particular reference to the responses of organisms and ecosystems to radiation.

RADIONUCLIDE   Any radioactive element. It occurs in the form of various *isotopes*. Radioactive isotopes are unstable atoms due to an excess or deficiency of neutrons in the nucleus (i.e. of different atomic weight).

RAFFINOSE   A non-reducing trisaccharide sugar which is common in plants occurring in cold climates.

RAWINSONDE   A balloon sounding of upper air pressure, temperature, relative humidity and wind velocity involving radar tracking.

RECESSIVE   A genetical character or *gene* that fails to express its presence in the phenotype of a *heterozygote*.

REFUGIUM   A centre just outside or within a glaciated region, where *biota* could survive. See PLEISTOCENE REFUGIA.

REPRODUCTIVE ISOLATION   The internally conditioned prevention of gene exchange or *miscibility* of taxa at the level of or above species.

RESPIRATION   The process by which organisms derive energy from a food source for vital activities.

RICHTER SCALE   A scale used in seismology to describe the amount of energy released by an earthquake. A magnitude of 2 corresponds to a shallow shock; a magnitude of 7 is the lower limit of a major destructive earthquake.

RIGID SPECIES   A species which is adapted to a narrow zone of tolerances; a taxon which has lost its ability to adapt to new conditions.

RIME   Opaque, granular ice formed by the rapid freezing of supercooled water drops as they strike an object.

SCARIFICATION   A process to abrade a seed coat in order to promote germination.

SECONDARY PRODUCTIVITY   The rate of storage of radiant energy at consumer levels. See PRIMARY PRODUCTIVITY.

SEGREGATION   The separation of *alleles* in the formation of gametes.

SELF POLLINATION   See AUTOGAMY.

SENSIBLE HEAT   A common term in meteorology for enthalpy. For a perfect gas, enthalpy is defined as $c_p T$, where c_p is the specific heat at constant pressure and T is the absolute temperature.

SESQUIOXIDES   Often used in soil science to refer to Fe_2O_3 (iron or ferrous oxide) and Al_2O_3 (aluminium oxide).

SEXUAL PROCESS   Production of offspring by means of sexual congress or *fertilization*.

SHANNON–WEINER FUNCTION   In ecology, a diversity index (D) based on information theory. It describes the average uncertainty of predicting the species of an individual selected at random in a community. $D = 3.322 \ (\log_{10} N - \frac{1}{N} \Sigma n_i \log_{10} n_i)$ where N is the number of individuals of all species and n_i is the number of individuals of *i*th species.

SHORT-WAVE RADIATION   An imprecise term usually used to refer to the visible and near visible radiation spectrum with wavelength about 0.3 to 1.0 μm, as distinct from infra-red (long-wave) radiation.

SNOWLINE   The orographic snowline refers to the lowest elevation of large snowbanks at the end of the summer season. The climatic snowline refers to the lower limit of perennial snow on fully exposed flat surfaces. On a regional scale, local irregularities of the orographic snowline occur in a zone which is 100 to 200 m or more wide. The spatially averaged elevation is referred to as the regional snowline.

SOLAR CONSTANT   The rate at which solar radiation is received on a surface exposed perpendicularly to the sun's rays outside the atmosphere, when the earth is at its mean distance from the sun. Recent data give a value of 1.95 cal cm^{-2} min^{-1} (1.36 kw m^{-2}).

SOLIFLUAL DEPOSITS   The rubble transferred downhill by *solifluction*. Debris moved by this and other periglacial processes is also known as congeliturbates, cryoturbates and Head (Great Britain).

SOLIFLUCTION (CONGELIFLUCTION)   Flow, or movement, of saturated or supersaturated soil under gravity. In cold climates frequently occurs in conjunction with frost creep.

SOLIFLUCTION BOIL   See TUNDRA OSTIOLE.

SOLS BRUNS ACIDES   Acid brown soil found on base deficient materials and on well drained slopes on slates or shaley tills.

SPALLING   Breaking off in layers which are parallel to a surface.

SPECIATION   The acquisition of *reproductive isolation* by a *population* or group of populations (*demes*) resulting in the multiplication of species. This isolation is acquired through linear or numerical chromosome changes, the former being *gradual*, the latter *abrupt*.

SPECIFIC HUMIDITY   The ratio of the mass of water vapour to the total mass in a system of moist air.

SPOROGENESIS   The production of spores.

STANDING CROP   The amount of standing (erect) biological material, living and dead, present (contrast fallen dead).

STRING BOG   Narrow parallel ridges of marsh vegetation, mainly sphagnum, separated by pools of standing water.

SUBALPINE   The natural belt below the treeless alpine belt from the upper altitudinal treeline to the closed montane forest at lower elevations.

SUBARCTIC   The natural zone between the polar tree limit and the economical forest line to the south. The zone between true *arctic* and true *taiga* each with its typical vegetation and fauna.

SUBSPECIATION   Pertaining to the process of forming subspecific races of a species or genepool, without *reproductive isolation*.

SUN-SYNCHRONOUS SATELLITE   A satellite whose orbit is such that it passes over a certain point on the earth's surface at the same local time each day.

SYMPATRY   The occurrence of two or more populations or taxa in the same area; more precisely, the existence of a population in breeding condition within the cruising range of individuals of another population.

SYNUSIA   A group of plants having the same life form and occupying the same habitat; e.g. a layer of shrubs in a forest.

TAFONI   Pits and circular (cavernous) hollows 10 cm to several metres deep on exposed rock faces variously described as due to chemical and physical weathering. Generally restricted to arid environments and to medium-coarse grained, acidic crystalline rocks.

TAIGA   The open northern portion of the boreal forest which is composed of open woodland of coniferous trees with lichen growing on the ground. It is sometimes used in a regional sense, corresponding to the subarctic zone.

TARYN ICE   An *icing* which survives the summer.

TAXON   A group of organisms recognized as a formal unit at any level of a hierarchic classification.

TETHYS SEA   A seaway in pre-Tertiary time which separated Europe and Africa and extended across southern Asia.

THERIOFAUNA   Mammals which bear live progeny.

THERMAL ROSSBY NUMBER   The non-dimensional ratio of the inertial force due to the *thermal wind* and the *Coriolis force* in the flow of a fluid heated at the centre and cooled at the rim.

THERMAL WIND   Theoretical wind vector which is in *geostrophic* balance with the gradient of mean temperature in an atmospheric layer between two pressure surfaces. It is directed parallel to the *thickness lines* with cold air to left in the northern hemisphere.

THERMOKARST   A *karst*-like topography found in permafrost areas and caused by melting of the ground ice.

THEROPHYTE   An annual plant which survives the unfavourable season in the form of a seed or a similar *propagule*.

THICKNESS   The thickness of the atmospheric layer between two isobaric (constant pressure) surfaces. It is proportional to the mean temperature of the layer.

TIMBERLINE   The latitudinal or altitudinal limit of tall tree growth. This differs from 'species limit', 'tree limit', and 'limit of economically useful trees'.

TOLERANCE   The ranges of intensity of factors of the environment within which an organism can function.

TOPOCLIMATE   The climate of a place.

TOR   An exposure of rock *in situ*, upstanding on all sides from the surrounding slopes: it is formed by differential weathering of a rock bed, and the removal of the debris by mass movement.

TRANSLOCATION   The shift of a segment of a *chromosome* to another chromosome. This is always reciprocal and frequently called segmental interchange.

TRANSMISSIVITY (OF A SURFACE)   A measure of the quantity of radiation which is transmitted through a certain medium.

TREELINE   See TIMBERLINE.

TRIMLINE   A line which marks the margins of a former ice or snow cover.

TUNDRA   The vegetation of arctic, antarctic and alpine regions beyond the *timberline*.

TUNDRA OSTIOLE   A small area of wet and clayey soil in arctic and alpine regions, surrounded by vegetation but kept moving through frost action so that only a few annuals are able to invade it for a part of the summer.

TURF HUMMOCKS (THUFUR)   Low dome-shaped mounds commonly 30 cm high, of living mosses, grasses and sedges underlain by peat and with core of mineral soil.

UBAC (French)   A mountain slope which faces approximately polewards and is therefore shaded from the sun's rays. See ADRET.

UNIFACE   A chipped stone tool shaped by removing flakes from only one side of the artifact.

VAPOUR PRESSURE   In meteorology this invariably refers to the partial pressure of water vapour in the atmosphere.

VENTIFACTS   Pebbles that have been bevelled, frequently on several sides, by windblown sand.

VIVIPARY   The production of small plantlets in the inflorescence of a plant. These may drop to the ground and become established.

WILTING POINT (WILTING COEFFICIENT, CRITICAL MOISTURE)   The soil moisture content at the stage when plants are in a permanently wilted condition.

ZONAL (WALKER) CIRCULATION   Atmospheric circulation along latitude circles in contrast to north–south or meridional circulation.

ZYGOTE   A fertilized egg; the diploid cell that results from the fertilization of an egg cell.

Index

Plates

Plate 1. The forest–tundra ecotone showing the transition from open lichen-woodland of the Subarctic to full tundra, central Labrador–Ungava. Principal trees and shrubs include black and white spruce (*Picea mariana* and *P. glauca*), tamarack (*Larix laricina*) and shrub birch (*Betula glandulosa*). Photograph by Jack D. Ives.

Plate 2. Loveland Pass, Colorado, showing the alpine transition from upper montane forest to alpine tundra. Principal tree species are limber pine (*Pinus contorta*), subalpine fir (*Abies lasiocarpa*), and Engelmann spruce (*Picea engelmannii*). Summit elevations exceed 4000 m. Photograph by Jack D. Ives.

Plate 3. Krummholz, or dwarfed Engelmann spruce, at the upper reaches
of the forest–tundra ecotone on the east slope of the Front Range,
Colorado Rocky Mountains, September 1969. Note the tree-island snow
drift ecosystem. The foreground lies at an elevation of 3500 m; the highest
summit, Navajo Peak on the centre skyline, reaches a height of over 4000 m.
Photograph by Jack D. Ives.

Plate 4. Avalanche released by artillery fire, December 1971, near Silverton in the San Juan Mountains of southwestern Colorado. The tremendous forces involved in avalanche activity have a major impact on the form of the treeline in high mountain areas; they also pose a serious threat to human occupation. Photograph by Richard Armstrong.

Plate 5. The redistribution of snow by the wind is a significant ecological parameter, especially in areas of steep slopes. Variation in solid precipitation from point to point is extremely large in mountainous terrain; wind redistribution ensures even greater variability and exerts a major control on plant and animal patterns. The small avalanches resulting from the build-up of snow cornices constitute a serious hazard to the cross-country skier. Photograph by Jack D. Ives.

(a)

(b)

Plate 6. The glacier burst, or jökulhlaup, from Skeidararjökull, southeast Iceland, in August 1954, as seen (a) before and (b) at the climax of the event. The white spots on the outwash plain seen in (b) represent icebergs over 30 m high torn from the front of the glacier 4 km distant at the point where the river Skeidara emerges. The jökulhlaup has a major impact on outwash plain development. Photograph by Jack D. Ives.

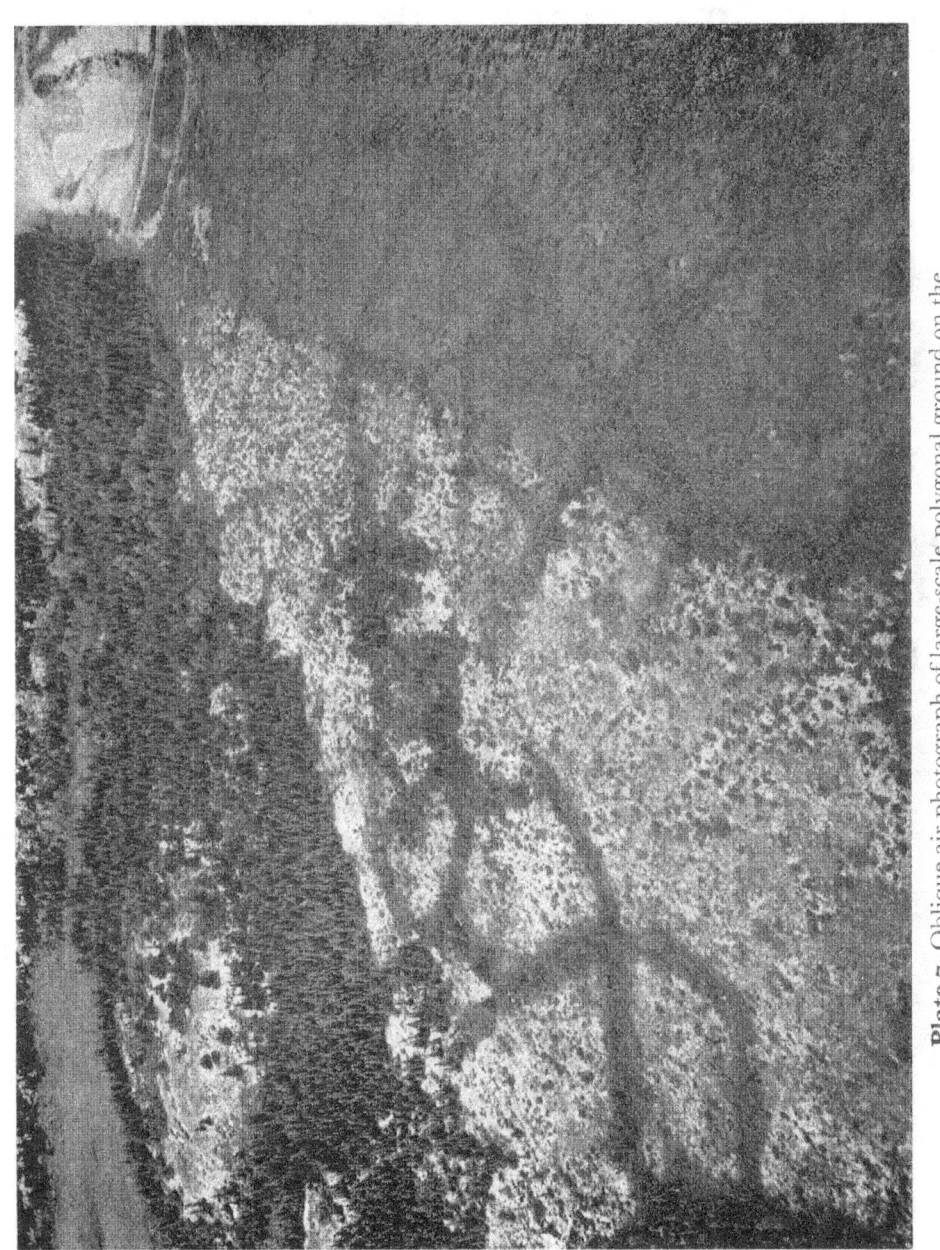

Plate 7. Oblique air photograph of large-scale polygonal ground on the Donnelly Dome area, Alaska. Fossil ice-wedge polygon vegetation cover consists of mixed evergreen-deciduous scrub and shrub. Photograph by T. L. Péwé.

Plate 8. Massive ground ice of the Mackenzie Delta area overlain by zone containing intrusive ice. Note especially the large ice-wedge to the left of upper centre. Photograph by J. R. Mackay.

Plate 10. (*below, opposite*) East Greenland pingo in process of destruction. This process is caused by three factors: a) rupturing of the top during growth, b) melting of the ice body by penetrations of summer warmth through the weakened mineral cover, and c) the sub- or intrapermafrost water rising through and thus melting the ice body. Finally, overflow from the lake will break down the containing walls, destroying the pingo's original symmetry. Photograph by Fritz Müller.

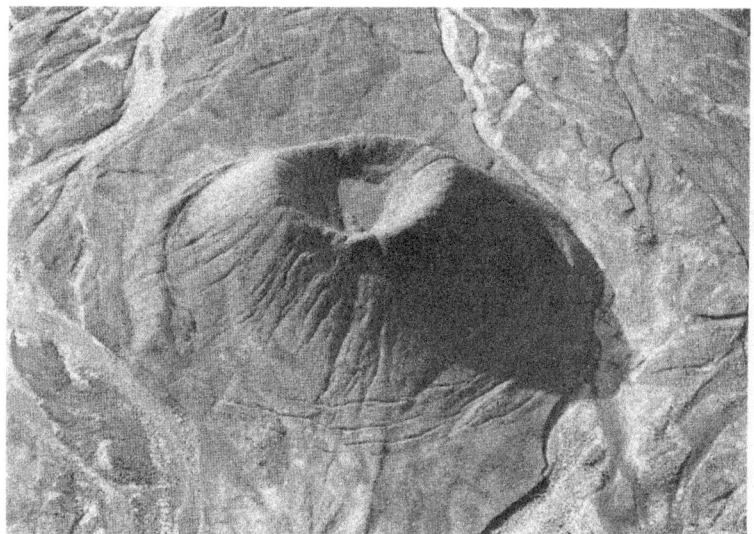

Plate 9. Open-system, or East Greenland type pingo. These occur in the gravel bottom of wide glacial valleys, but also on slopes with an angle of up to 8°. Heights range up to 38 m and base circumference up to 1.6 km. They develop where sub- or intrapermafrost waters penetrate into the permafrost zone, forced up by hydrostatic pressure. Photograph by Fritz Müller.

Plate 11. In the permafrost zone sewerage disposal and piped water facilities pose a problem to human occupation. This Inuvik 'utilidor' combines water and sewerage lines in an electrically heated system, insulated and raised above the ground surface to prevent localized degradation of the permafrost. Photograph by J. R. Mackay.

Plate 12. Rannoch Arm and the middle reaches of Cambridge Fiord with Baffin Bay and the Bruce Mountains in the distance (70°N). The light-toned areas surrounding the snowfields and ice patches represent rock lichen-free zones, or zones with minimum rock lichen development indicating ice recession over the past 300–400 years. Plotting of the distribution of the light-toned areas indicates that as much as 70 per cent of the interior plateau of Baffin Island was mantled by permanent ice and snow during the most recent phase of the Neoglacial. Photograph by Dept of Energy, Mines and Resources, Ottawa (Royal Canadian Air Force Photo T241L-180, 30 July 1948).

Plate 13. Cirque glaciers and ice sheet with major outlet glacier north of Borup Fiord, Ellesmere Island, Northwest Territories, Canada. Photograph by Dept of Energy, Mines and Resources, Ottawa (Royal Canadian Air Force Photo T407R-67).

Plate 14. The Arikaree Glacier, Colorado Front Range, as seen from the summit of Navajo Peak (40°N), October 1972. Alimentation is primarily a combination of wind drift across the Continental Divide to the immediate right of the glacier and avalanching from the higher slopes of Arikaree Mountain, the triangular peak in the middle distance. A rotor effect, causing reversal of local wind direction from westerly to easterly in the lee of the Divide, is considered a significant factor. This is one of a small group of near-inactive glaciers that make up the most southerly examples in the Rocky Mountains. The regional snowline lies above the summit level (4000 m+). Photograph by Jack D. Ives.

Plate 15. Vertical air photograph (part of A19430/91) showing the transient snowline on two glaciers in the Canadian Rockies (117°55′W, 57°40′N). Based upon the topographic map 82°N, 12°W over the same area, it is simple to determine which contour line is closest to the transient snowline. For both glaciers it was the 7250 ft (2210 m) contour (the upper exposed part of the right-hand glacier was not considered in this determination). Photograph by Dept of Energy, Mines and Resources, Ottawa.

Plate 16. End moraines in the Jostedalen Valley, western Norway. The glacier tongue has retreated since the 1750s when it covered most of the valley bottom. Since then the glacier retreated but small re-advances produced archlike moraine ridges. These are emphasized on this vertical photograph (Nor-fly S.65–260. 1257) by the drainage system. The glacier tongue covered the lower end of the lake in about 1930, but is now situated approximately 2 km further upvalley.

Plate 17. An ice-cored moraine in front of the Vaktposten Glacier in Swedish Lapland (68°0′N, 18°25′E). The left-hand part of the ridge is damming a small lake in front of the glacier. At least three distinct ridges can be recognized here; excavations proved the existence of buried ice under 2–3 m of till. Photograph by W. Karlén.

Plate 18. Rock glacier on Mt Marmot, 2608 m (8557 ft), 9 km south of Jasper, Alberta, Canada. The material is originally talus debris that is obviously mixed with snow (note small snowbanks) that eventually forms interstitial ice. This ice–rock mixture is thought to be plastic and makes movement possible. Photograph by Dept of Energy, Mines and Resources, Ottawa (part of air photo A/19668/50).

Plate 19. A rock glacier in the San Juan Mountains of southwestern Colorado. In this instance the rock debris has covered the entire floor of a glaciated valley. The material is stable, as indicated by a heavy development of rock lichens on the individual boulders, implying that movement is very slight. Photograph by Jack D. Ives.

Plate 20. Air photograph from the south of the Numbur-Karyolung Range, south-southwest of Mt Everest, eastern Nepal. The view shows the two main original valleys of the Solu (Shorong) Valley in which it is possible to recognize the end moraines of the last glaciation (to the right at about 3300 m, to the left at about 3500 m). Photograph by Erwin Schneider, Lech-am-Arlberg, Austria.

Plate 21. *Nothofagus solandri* var. *cliffortioides* trees up to 7.5 m tall with *Chionochloa* tussocks in the foreground, timberline at 1350 m, Craigieburn Range, New Zealand (from Wardle, 1965a). Photograph by A. P. Underhill.

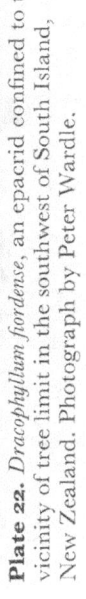

Plate 23. *Cyathea* tree fern at 3350 m on Mt Giluwe, Southern Highlands, Territory of Papua and New Guinea. The section shows the thick layer of old leaf bases and adventitious roots covering the stem, and the *paleae* or scales densely packed among the developing fronds. Photograph by Peter Wardle.

Plate 22. *Dracophyllum fiordense*, an epacrid confined to the vicinity of tree limit in the southwest of South Island, New Zealand. Photograph by Peter Wardle.

Plate 24. Shrubs of *Olearia colensoi* up to 1.5 m tall, in subalpine scrub at 1150 m in the Tararua Range, New Zealand (from Wardle, 1965a).

Plate 25. Arctic tundra vegetation on the Truelove lowland at sea level, Devon Island, Northwest Territories, Canada, at 76°N. On the left is a relatively dry beach ridge with *Dryas integrifolia*, *Saxifraga oppositifolia*, and light coloured areas of lichens. The wet meadow on the right is characterized mainly by *Carex stans* var. *aquatilis*. Photograph by James A. Teeri.

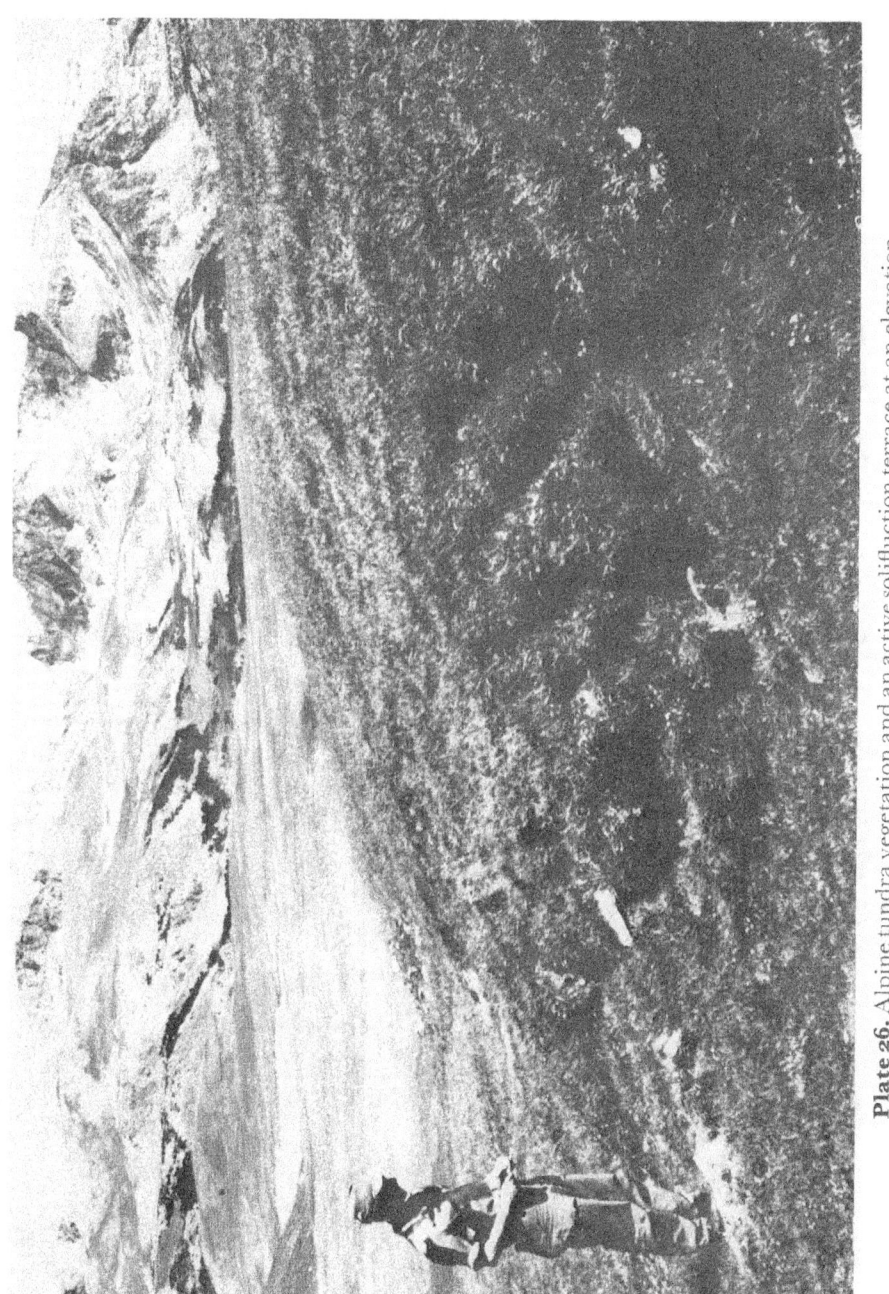

Plate 26. Alpine tundra vegetation and an active solifluction terrace at an elevation of about 3200 m in the Beartooth Mountains, Wyoming, at 45° N (from Johnson and Billings, 1962).

Plate 27. Plant of *Oxyria digyna* in the alpine scree of the Beartooth Mountains, Wyoming, 45°N. Photograph by W. D. Billings.

Plate 28. Plant of the purple saxifrage, *Saxifraga oppositifolia*, in full bloom on the polar desert of Cornwallis Island (75°N), Northwest Territories, Canada, in early June 1971. Photograph by W. D. Billings.

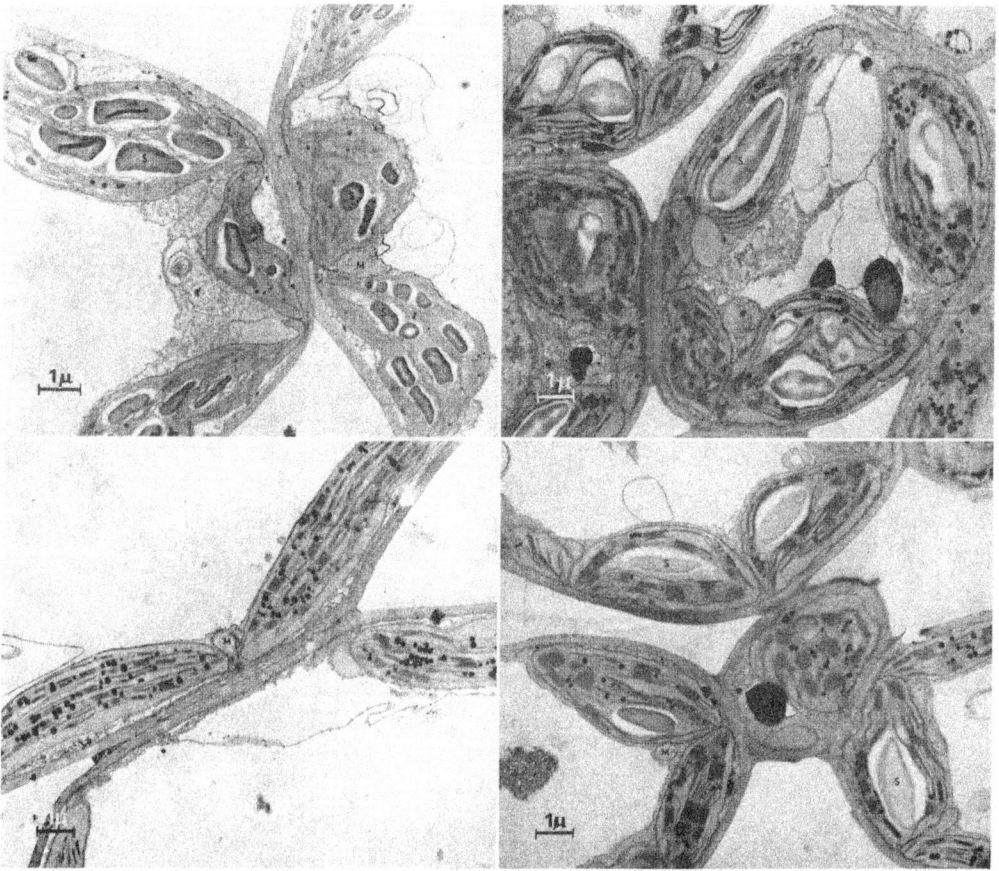

Plate 29. Sections of chloroplasts in leaves of *Oxyria digyna*, an alpine plant from the Sierra Nevada of California (left), and of chloroplasts in leaves of *Encelia virginiensis*, a desert plant from the east side of the Sierra (right). All plants grown at low temperatures ($11°/8°C$) for six months. Top photographs were from the end of the light period at 7.45 am. In *Oxyria* chloroplasts, starch grains (S) accumulate during the light period and are degraded and the food translocated in the dark. In *Encelia* chloroplasts, starch (S) accumulates during the day but is not degraded during the dark period and the products translocated; the starch grains still remain because of the inability to degrade starch and translocate food at low temperatures while *Oxyria* can do this. Magnification: 10,000×. From Chabot and Billings (1971).

Plate 30. Infra-red gas analyser system measuring net photosynthesis in alpine tundra plants at 3300 m in the Medicine Bow Mountains, Wyoming. Photograph by W. D. Billings.

Plate 31. Polar desert, Devon Island, Northwest Territories, Canada. This landscape of sorted ground consisting of small stone nets with fine clay and mud centres has very scant vegetation cover. Plants are mostly restricted to the perimeter of the stone nets where cryoturbation is least. Photograph by P. J. Webber.

Plate 32. High arctic tundra, Point Barrow, Alaska. Plant cover in this region is generally continuous and largely herbaceous prostrate shrubs. An ice-wedge polygon net can be seen. Photograph by P. J. Webber.

Plate 33. Low arctic tundra, Tuktoyaktuk, Mackenzie Delta, Northwest Territories, Canada. This shrub-tundra with dwarf willows and birches can cover much of the landscape on mesic surfaces, but dry ridges and poorly drained sites will be different. Photograph by P. J. Webber.

Plate 34. Alpine shrub tundra, San Juan Mountains, southwestern Colorado. This is the low latitude equivalent of the low arctic tundra of Plate 33. Photograph by P. J. Webber.

(a)

Plate 35. Seasonal changes in plumage of the ptarmigan (*Lagopus leucurus*): (a) winter; (b) early summer (male bird) showing the change beginning from the head and neck downward; (c) mid-summer (female). Photographs by C. E. Braun.

(b)

(c)

Plate 36. Earth cores left by subnival burrowing activities of the northern
pocket gopher (*Thomomya talpoides*). Subnival burrows are filled with
dirt from subsurface tunnelling leaving the characteristic core. Similar
structures are also formed by other fossorial alpine rodents. Photograph by
R. E. Stoecker.

Plate 37 (*above, opposite*). The classical nunatak form showing a series of
rock spires north of Ayr Lake, northeastern Baffin Island. The prominent
peak in the centre-right background is Eglington Tower. Summit altitudes
exceed 1800 m. The diagnostic value of the unglaciated appearance of the
summits has been heavily challenged in recent years although the relief
shown here has most likely never been completely submerged by moving
ice. Photograph by Jack D. Ives.

Plate 38 (*below, opposite*). Strong postglacial frost shattering of limestone
in an arctic environment. The low ridge results from ice accumulating in a
major joint that was widened by thermal contraction. Photograph by
J. Brian Bird.

Plate 39. Aerial view of a series of solifluction lobes advancing over bedrock close to the Foxe Basin coast, Longstaff Bluff, Baffin Island. Where the till has a high clay-silt content lobes may exceed 1 km in length. Photograph by Jack D. Ives.

Plate 40. A single slope illustrating four of the segments which make up a common sequence: a) a convex interfluve (on the skyline); b) an almost vertical free-face, here cut through Tertiary volcanic materials; c) a rockfall talus, modified by a single mudflow from a shallow gully in the free-face; d) a litter of angular blocks and ramparts at the talus foot (from the slope to the right of the photograph). Photograph by Nel Caine.

Plate 41. Remains of an ASTT camp in arctic Canada. Stone artifacts are found on the surface of an elevated gravel beach in association with the remnants of a stone tent ring. Photograph by National Museum of Canada, Ottawa.

Plate 42. Remains of a Thule semi-subterranean house after excavation. The flagstone floor and platform are revealed beneath a mound of earth and sod which represent the collapsed wall and roof covering of the house.' Photograph by National Museum of Canada, Ottawa.

Plate 43. Damage to wet alpine tundra by illegal use of four-wheel drive vehicle incurred early in the summer season while snowbanks are still melting out. James Peak, Colorado Front Range. Photograph by Jack D. Ives.

Plate 44. Wet maritime tundra of the Alaska North Slope, Point Barrow, Alaska. Note the abundant surface water and outline of the ice-wedge polygons in the foreground. The tracks of motor vehicles are plainly visible in terrain comparable to much of that to be traversed by oil pipelines. The United States Tundra Biome intensive study site is situated in the centre. Photograph by Jerry Brown.

Plate 45. Niwot Ridge, alpine tundra study site, United States Tundra
Biome, International Biological Programme. X marks the site of the
'Saddle' intensive study plots and microclimatological instrumentation.
Y locates the Niwot Ridge 3750 m permanent climatological
station established in 1951. View toward the west showing the Indian
Peaks section of the Continental Divide. Note the well developed stone-
banked lobes and terraces on the far side of the Saddle. Permafrost occurs
in scattered patches along the ridge, especially under north-facing slopes
and areas blown free of snow in winter. Photograph by Jack D. Ives.

Plate 46. San Juan Ecology Project, Williams Lake alpine study basin. The lake lies just above treeline at an altitude of 3565 m; the Continental Divide crosses the broad col immediately beyond the lake. The numbers locate the intensive research sites: (1) talus slope, 3617 m, (2) *Geum* slope, 3617 m, (3) willow thicket, 3579 m, (4) dry meadow, 3658 m, (5) exposed ridge, 3746 m, and (6) wet meadow, 3502 m. Photograph by Jack D. Ives.

Plate 47. San Juan Ecology Project, Eldorado Lake alpine study basin.
The Continental Divide runs along the transverse ridge in the middle
distance. The numbers locate the intensive research sites: (1) dwarf willow,
3830 m, (2) wet meadow, 3828 m, (3) willow slope, 3846 m, (4) west-
facing *Geum* slope, 3849 m, (5) talus slope, 3901 m, (6) exposed ridge, 3901 m,
(7) south-facing *Geum* slope, 3849 m, and (8) patterned ground,
3870 m. Photograph by Jack D. Ives.

For Product Safety Concerns and Information please contact our EU representative GPSR@taylorandfrancis.com
Taylor & Francis Verlag GmbH, Kaufingerstraße 24, 80331 München, Germany

www.ingramcontent.com/pod-product-compliance
Lightning Source LLC
Chambersburg PA
CBHW080006210526
45170CB00015B/1839